Practical Text Mining and Statistical Analysis for Non-structured Text Data Applications

Practical Text Mining and Statistical Analysis for Non-structured Text Data Applications

Gary Miner
Tulsa, OK, USA

Dursun Delen
Tulsa, OK, USA

John Elder
Charlottesville, VA, USA

Andrew Fast
Charlottesville, VA, USA

Thomas Hill
Tulsa, OK, USA

Robert A. Nisbet
Santa Barbara, CA, USA

Major Guest Authors:

Jennifer Thompson
Woodward, OK, USA

Richard Foley
Raleigh, NC, USA

Angela Waner
Tulsa, OK, USA

Linda Winters-Miner
Tulsa, OK, USA

Karthik Balakrishnan
San Francisco, CA, USA

ELSEVIER

AMSTERDAM • BOSTON • HEIDELBERG • LONDON • NEW YORK • OXFORD
PARIS • SAN DIEGO • SAN FRANCISCO • SINGAPORE • SYDNEY • TOKYO
Academic Press is an imprint of Elsevier

Academic Press is an imprint of Elsevier
225 Wyman Street, Waltham, MA 02451, USA
The Boulevard, Langford lane, Kidlington, Oxford, OX5 1GB, UK

Library of Congress Cataloging-in-Publication Data

Practical text mining and statistical analysis for non-structured text data applications / Gary Miner ... [et al.]. – 1st ed.
p. cm.
Includes bibliographical references and index.
ISBN 978-0-12-386979-1
1. Data mining. I. Miner, Gary.
QA76.9.D343P73 2012
006.3′12–dc23
2011036169

British Library Cataloguing-in-Publication Data
A catalogue record for this book is available from the British Library.

ISBN: 978-0-12-386979-1

For information on all Academic Press publications visit our Web site at www.books.elsevier.com

Printed and bound in The USA

12 13 14 10 9 8 7 6 5 4 3 2 1

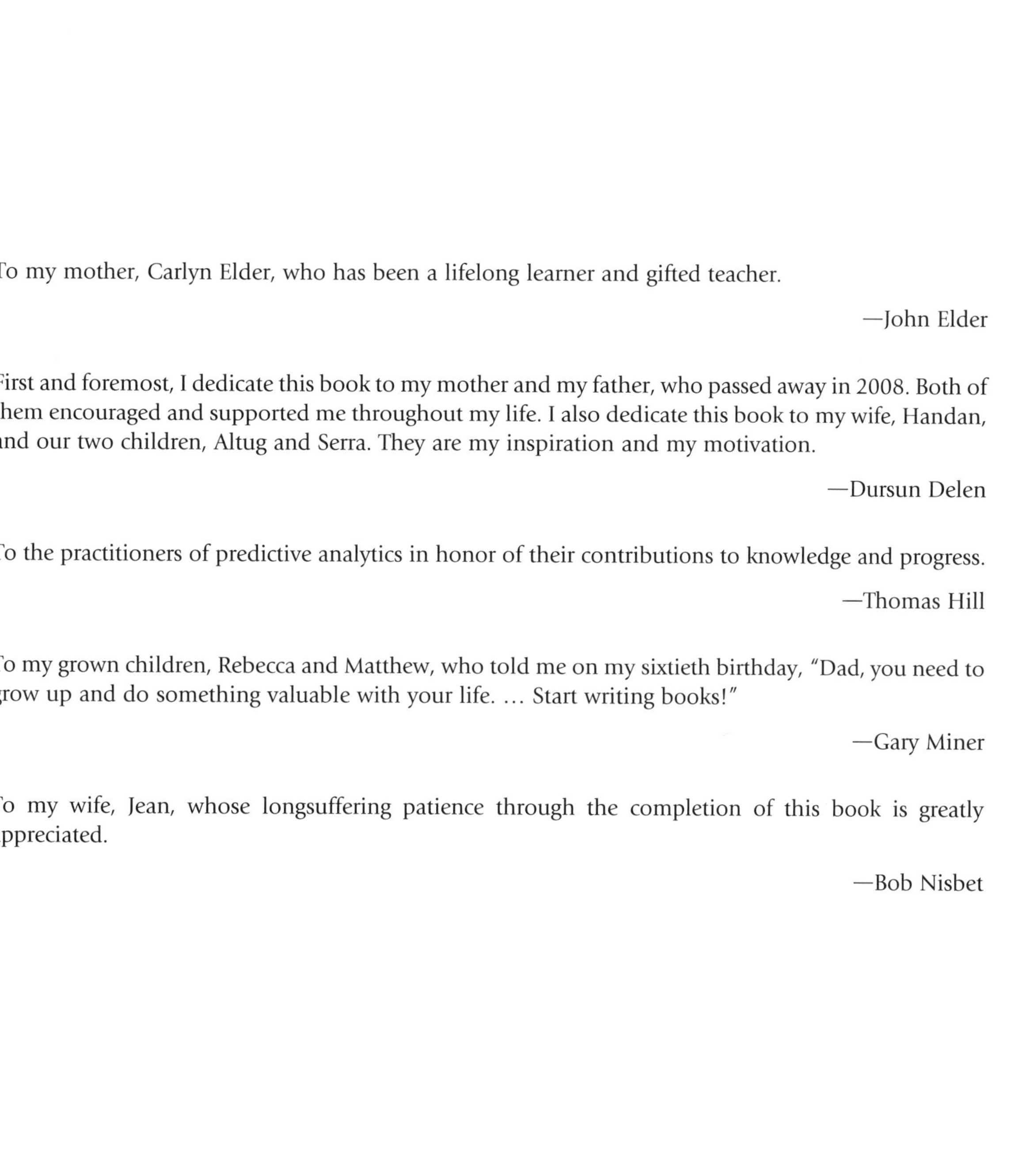

To my mother, Carlyn Elder, who has been a lifelong learner and gifted teacher.

—John Elder

First and foremost, I dedicate this book to my mother and my father, who passed away in 2008. Both of them encouraged and supported me throughout my life. I also dedicate this book to my wife, Handan, and our two children, Altug and Serra. They are my inspiration and my motivation.

—Dursun Delen

To the practitioners of predictive analytics in honor of their contributions to knowledge and progress.

—Thomas Hill

To my grown children, Rebecca and Matthew, who told me on my sixtieth birthday, "Dad, you need to grow up and do something valuable with your life. ... Start writing books!"

—Gary Miner

To my wife, Jean, whose longsuffering patience through the completion of this book is greatly appreciated.

—Bob Nisbet

Contents

Endorsements for *Practical Text Mining & Statistical Analysis for Non-structured Text Data Applications*

Karl Rexer, Ph.D.

"Practical"—that's it! The first word in the title really sums it up. If you want big theoretical academic discussions, go read another book. When you want real help extracting insight from the mountains of text that you're facing, this is the book to turn to for immediate practical advice. It will give you new ideas and frameworks to address your text mining problems. This book will help you get the job done!

The tutorials (case studies) that Gary and his coauthors have assembled really bring the book alive. They show real-world applications of text mining, including the complexity of text analysis, how real-world issues shape the analytic plans and implementations, and the great positive impact that text mining can have.

—**Karl Rexer, Ph.D.**
President, Rexer Analytics
October, 2011

Karl Rexer is president of Rexer Analytics (www.RexerAnalytics.com), a Boston-based consulting firm specializing in data mining and analytic CRM consulting. Karl and his teams have delivered analytic solutions to dozens of companies, including customer attrition measurement and prediction, multicountry fraud detection programs, customer segmentations, targeting and measurement analytics for million-piece direct marketing initiatives, multiyear customer satisfaction tracking, sales forecasting, product allocation optimization, market basket analyses, product cross-sell modeling, and text mining. Every year, Rexer Analytics conducts and freely distributes the widely read *Annual Data Miner Survey*. In 2011, over 1,300 data miners participated in the 5th Annual Survey.

Karl is a leader in the field of applied data mining. He is frequently an invited speaker at conferences and universities. Karl has served on the organizing committees of numerous international Data Mining and Business Intelligence conferences (e.g., KDD, BIWA Summit, Oracle Collaborate). He also reviews papers, is a conference moderator, and serves on awards committees to select the best applied data mining papers. Several software companies have sought Karl's input, and in 2008 and 2010, he served on SPSS's (IBM) Customer Advisory Board. Since his election in 2007, Karl has also served on the board of directors of the Oracle Business Intelligence, Warehousing, & Analytics (BIWA) Special Interest Group. In 2011, LinkedIn named Karl number 1 in their list of "Top Predictive Analytics Professionals." Prior to founding Rexer Analytics, Karl held leadership and consulting positions at several consulting firms and two multinational banks.

Richard De Veaux, Ph.D.

The authors of *Practical Text Mining and Statistical Analysis for Nonstructured Text Data Applications* have managed to produce three books in one. First, in 17 chapters they give a friendly yet comprehensive introduction to the huge field of text mining, a field comprising techniques from several different disciplines and a variety of different tasks. Miner and his colleagues have produced a readable overview of the area that is sure to help the practitioner navigate this large and unruly ocean of techniques. Second, the authors provide a comprehensive list and review of both the commercial and free software available to perform most text data mining tasks. Finally, and most importantly, the authors have also provided an amazing collection of tutorials and case studies. The tutorials illustrate various text mining scenarios and paths actually taken by researchers, while the case studies go into even more depth, showing both the methodology used and the business decisions taken based on the analysis. These practical step-by-step guides are impressive not only in the breadth of their applications but in the depth and detail that each case study delivers. The studies are authored by several guest authors in addition to the book authors and are built on real problems with real solutions. These case studies and tutorials alone make the book worth having. I have never seen such a collection of real business problems published in any field, much less in such a new field as text mining. These, together with the explanations in the chapters, should provide the practitioner wishing to get a broad view of the text mining field an invaluable resource for both learning and practice.

—Richard De Veaux
Professor of Statistics; Dept. of Mathematics and Statistics;
Williams College; Williamstown MA 01267
October, 2011

Richard D. De Veaux is an internationally known educator and lecturer in data mining and statistics who serves as a consultant for many Fortune 500 companies and federal agencies. He is a professor of statistics at Williams College, where he has taught since 1994. He also taught in the Engineering School at Princeton University and the Wharton School of Business at the University of Pennsylvania. Dr. De Veaux is an elected Fellow of the American Statistical Association (ASA), and in 2008, he was named Mosteller Statistician of the Year by the Boston Chapter of the ASA. He was a founding officer of the ASA Section on Statistical Learning and Data Mining (SLDM) and is currently the representative of the SLDM to the council on Sections. In 2006 and 2007, Dr. De Veaux was the William R. Kenan Jr. Visiting Professor for Distinguished Teaching at Princeton University. In addition to numerous journal articles, he is the coauthor of several critically acclaimed and best-selling textbooks, including *Intro Stats* (3rd ed. with Paul Velleman and David Bock), *Stats: Data and Models* (3rd ed. with Velleman and Bock), *Stats: Modeling the World* (3rd ed. with Velleman and Bock), *Business Statistics* (2nd ed. with Norean Sharpe and Velleman), and *Business Statistics: A First Course* (1st ed. with Norean Sharpe and Velleman), all published by Pearson.

Gerard Britton, J.D.

In writing *Practical Text Mining and Statistical Analysis for Nonstructured Text Data Applications*, the six authors (Miner, Delen, Elder, Fast, Hill, and Nisbet) accepted the daunting task of creating a cohesive operational framework from the disparate aspects and activities of text mining, an emerging field that they appropriately describe as the "Wild West" of data mining. Tapping into their unique expertise and applying a wide cross-application lens, they have succeeded in their mission.

Rather than listing the facets of text mining simply as independent academic topics of discussion, the book leans much more to the practical, presenting a conceptual road map to assist users in correlating articulated text mining techniques to categories of actual commonly observed business needs. To finish out the job, summaries for some of the most prevalent commercial text mining solutions are included, along with examples. In this way, the authors have uniquely presented a text mining resource with value to readers across that breadth of business applications.

—Gerard Britton, J.D.
V.P., GRC Analytics, Opera Solutions LLC
October, 2011

Gerard J. Britton is vice president of Text Analytics in the Governance Group at Opera Solutions LLC. Mr. Britton, an experienced attorney, prosecutor, and public and private sector executive, has over 15 years of experience bringing technical and analytics strategies to the mission-critical challenges of governance, compliance, fraud, investigations, and electronic discovery. He has managed the design and development of text analytics solutions in the electronic discovery field, including advanced predictive models to enable early case assessments and machine-assisted issue coding of documents. He maintains the "Critical Thought in Legal, Compliance and Investigative Analytics" blog at http://postmodern-ediscovery.blogspot.com/ and is a frequent speaker at leading industry events on topics related to analytics applications in investigations, compliance, and litigation.

James Taylor

Text Mining is one of those phrases people throw around as though it describes something singular. As the authors of *Practical Text Mining and Statistical Analysis for Non-structured Text Data Applications* show us, nothing could be further from the truth. There is a rich, diverse ecosystem of text mining approaches and technologies available. Readers of this book will discover a myriad of ways to use these text mining approaches to understand and improve their business. Because the authors are a practical bunch the book is full of examples and tutorials that use every approach, multiple commercial and open source tools, and that show the power and trade-offs each involves. The case studies are worked through in detail by the authors so you can see exactly how things would be done and learn how to apply it to your own problems.

If you are interested in text mining, and you should be, this book will give you a perspective that is broad, deep and approachable.

—James Taylor
CEO Decision Management Solutions,
October, 2011

James is the CEO of Decision Management Solutions, and is the leading expert in how to use business rules and predictive analytics to build Decision Management Systems. James is passionate about helping companies improve decision making and develop an agile, analytic and adaptive business. He provides strategic consulting, working with clients to adopt decision making technology. James is also a faculty member of the International Institute for Analytics and has led Decision Management efforts for leading companies in insurance, banking, health management and telecommunications. James is the author of "Decision Management Systems: A practical guide to using business rules and predictive analytics" (IBM Press, 2011) and previously wrote Smart (Enough) Systems: How to Deliver Competitive Advantage by Automating Hidden Decisions (Prentice Hall) with Neil Raden.

Foreword 1

FOREWORD No. 1: for "Practical Text Mining and Statisticall Analysis for Non-structured Text Data Applicationsns" from the viewpoint of a person trained and having a life-long internationally noted career in traditional statistics, especially Discriminant Analysis and the Jackknife procedure, and Co-Author Gary Miner's mentor in statistical analysis:

Text mining and analysis is a new area of research that is less than 15 years old. So we can expect new editions of this book as the field matures. This book concerns the topics of extracting information from text and finding interesting patterns in the results. The book covers working with text documents. The introduction explains that the Library of Congress has many "documents" that include text (books, reports, emails, etc.), recordings (sounds), and images (including stills and motion pictures). Apparently text mining does not include mining sounds and images, although I can visualize it expanding to include scripts and lyrics. Of course, songs and musical contributions involve much more than text. This may be a direction for the future.

Much of text mining is observational, and as such, it may pay to be cautious about drawing conclusions. For example, in data mining one may wish to make inferences about consumer behavior (e.g., purchases in a market chain). However, since the market may not have all brands of all products, no information can be inferred about nonstocked brands. Also, words and phrases are inherently unordered, so many statistical methods that rely on continuous data will not apply. Methods based on tables and proportions will be the order of the day.

There are always concerns about new areas of science. As scientists become familiar with the pitfalls and benefits, practices will change and new methods will be added. For example, the standard of evidence for an association in a text mining context might be quite different from that accepted as legal evidence. The adequacy of accuracy in websites is an area in which concern has been noted. This is not an area that text mining has been concerned with, but some "facts" distributed on the web can be badly distorted. One need only examine political candidates' websites. Additionally, one need only look at the letters to the editor of local newspapers to observe misinformation and disinformation. I hope that text mining can include some figures of merit on documents (based on accuracy of information, completeness of information, and currency of information, among other things). In meta-analysis, studies that don't measure up are minimized or discarded. A key concern is the set of texts that aren't included in the database.

The main steps are as follows:

- *Define the purpose of the study:* This is all too easy to omit. The researcher may be so familiar with the field that the purpose is very clear to him or her, but someone else in the field may have a different idea.

- *Determine the availability of the data:* This implies a full literature review and cataloguing of data sets, and knowing the definitions of the terms in each set is imperative as well. As the data sets are typically documents, each author will likely have his or her own definition.
- *Prepare the data:* Proper encoding of the documents makes life much easier. Methods discussed in this book make it much simpler to use. This will include feature extraction, entity extraction, and reduction of dimension. The last may not mean any fewer variables, but they will be combined in ways that reduce the number of items in the models—definitely a nontrivial task.
- *Develop and assess the models:* One must consider how the models apply in the context of the problem. Assessing model fit, for example, will evaluate how well the models predict the data.
- *Evaluate the models:* This is close to the previous bullet.
- *Deploy the results:* In this step, the researcher shares his or her work with the public. For example, with insurance fraud, one would demonstrate that fraud could be reduced using these models. The models would identify key words in fraudulent claims that could be used in future work—in some instances, keeping the results confidential to avoid alerting the bad guys.

All of these steps are detailed in the chapters.

English (and any language) has a great deal of redundancy, so it is not necessary to mine each word. Many words carry great meaning in context, so there may not be a need to have all words encoded. For example, in a medical context, the terms *tumor*, *cancer*, and *lesion* may be used more or less interchangeably.

The book has 28 tutorials, which should be read as the chapters are read. I also strongly recommend that the reader have access to one of the programs mentioned in the book. Find a general purpose text miner and work through the exercises.

Be alert for items of interest. Get some examples of your own; this will likely be the best way to learn text mining.

Peter A. Lachenbruch, Ph.D.
Oregon State University, Professor of Public Health (2006–present),
Past president of the American Statistical Association

Dr. Peter Lachenbruch received his Ph.D. in biostatistics from UCLA. He has held positions on the faculties of the University of North Carolina (1965–1976), the University of Iowa (1976–1985), and UCLA (1985–1994). He was employed by the FDA/CBER from 1994 to 2005 and retired from there as the director of the Division of Biostatistics. He is currently Professor of Public Health at Oregon State University (2006–present). He is a Fellow of the American Statistical Association and a former elected member of the International Statistical Institute. He has held many professional offices and was the president of the American Statistical Association for 2008.

He has statistical interests in discriminant analysis, two-part models, model-independent inference, statistical computing, and data analysis. He is known for making the jackknife re-sampling method an accepted procedure, and in recent years he has published on validation of neural networks using hybrid resampling methods. He has application interests in rheumatology, psychiatry, pediatrics, gerontology and accident epidemiology. He has more than 180 publications in these fields. Dr. Lachenbruch serves on the Editorial Boards of Statistics in Medicine, Methods of Information in Medicine, Journal of Biopharmaceutical Statistics, and Statistical Methods in Medical Research.

Foreword 2

Practical Text Mining and Statistical Analysis for Nonstructured Text Data Applications follows in the tradition of the popular *Handbook of Statistical Analysis & Data Mining Applications* (Elsevier, 2009), which was authored by three of this new text's six authors: Robert Nisbet, John Elder, and Gary Miner. Three additional authors—Thomas Hill, Dursun Delen, and Andrew Fast—contributed to this book, each bringing substantial experience in text mining.

Practical Text Mining was written to be used as an application-oriented text mining handbook. At some thousand pages in length, the authors have created a text that will to my mind serve as the standard resource in the field for many years. The authors have managed to provide readers with a comprehensive, yet thoroughly understandable, overview of the theory and scope of text mining. In addition, they provide 28 tutorials that assist readers in actually applying theory to a wide range of practical applications using up-to-date statistical techniques.

Text mining is one of those disciplines that are now emerging as cutting-edge technologies. Only a few years ago, standard desktop computers did not have the physical capability to handle huge amount of textual material or to engage in the complex analysis of page-length material. Storage capacity and in particular MHz and RAM were insufficient to appropriately analyze large or complex textual situations. Recently, however, PC-type computers have become powerful and fast enough to engage in sophisticated text mining analysis, which itself has the capability of assisting researchers to better understand a host of social and biological issues.

Several text mining applications now exist to enable researchers to engage in text mining activities. The book does not limit itself to using only one text mining package but rather provides step-by-step instructions on how to work with several of the foremost applications, including StatSoft's *STATISTICA* Data Miner and Text Miner, SAS Enterprise Miner & Text Miner, and IBM-SPSS Modeler Premium. The authors also give readers using RapidMiner, Topsy, Weka, and Salford System's STM™, CART®, and TreeNet® tutorials employing these applications. For all of the tutorials, readers are guided through such text mining processes as classification, prediction, named entity extraction, feature selection, dimensionality reduction, singular value decomposition, clustering, focused web crawling, web mining, and other associated techniques.

Part I takes the reader through the historical background of text mining, provides general text mining theory, and presents common applications and their associated statistical and data management tools. Part II is called the text mining laboratory in which the 28 tutorials are provided. Each tutorial is authored by the book's authors and/or by guest authors having special expertise in the subject of the tutorial and the software used for it. Code is provided for the tutorials where applicable, and annotated screenshots are displayed throughout in order to make it easier for readers to replicate the specific tutorial example, as well as to enable them to work through their own project analysis.

Text mining has an extremely wide range of applications. From the original applications of evaluating text to determine if it was authored by the individual claimed as author or if more than one author was involved in writing the textual material or to determining when it was written, text mining is now reaching into nearly every academic field. The tutorials presented in this book give the reader a thorough flavor of both what text mining can do now and solid hints as to its future capabilities. And importantly, they give clear instructions on how to actually engage in meaningful text mining and related statistical analysis.

Of the number of statistics books that are published each year by the major publishers, only a few books stand out as really being important, meaning that they positively influence how future research is done in the subject area of the text. I believe that *Practical Text Mining* is just such a book. The text offers the reader everything I believe is important for this type of book: the historical background of the subject, the basic principles, the methodological techniques, interpretations, caveats, and methods of comparative evaluation. It also provides the code and annotated screenshots needed to enable readers to replicate examples and to utilize the book as a reference for years to come. Miner and his colleagues have put together just the sort of text needed for understanding and learning the complex subject of text mining. Well done!

Joseph M. Hilbe, J.D., Ph.D.
Arizona State University and Jet Propulsion Laboratory, 16 October, 2011

Joseph M. Hilbe, Ph.D., is emeritus professor at the University of Hawaii, adjunct professor of statistics at Arizona State University, Solar System Ambassador with NASA/Jet Propulsion Laboratory at California Institute of Technology, and a faculty member with the Institute for Statistics Education (Statistics.com), for which he is also a member of the three-person advisory council.

Dr. Hilbe is an elected Fellow of the American Statistical Association (ASA) and was a founding member of the ASA Health Policy Statistics Section executive committee. He is also an elected member (Fellow) of the International Statistical Institute (ISI), the world association of statisticians, for which he chairs both the ISI Sports Statistics and the ISI Astrostatistics committees. In addition, he heads the International Astrostatistics Network, the global association of astrostatisticians. Dr. Hilbe has authored well over one hundred journal articles and is currently on the editorial boards of six statistics journals. He is also editor-in-chief of the Springer Series in Astrostatistics and has written many best-sellers, including *Negative Binomial Regression* (Cambridge University Press, 2007, 2011) and *Logistic Regression Models* (Chapman & Hall/CRC, 2009). With James Hardin, Hilbe is author of *Generalized Estimating Equations* (Chapman & Hall/CRC, 2002) and three editions of *Generalized Linear Models and Extensions* (Stata Press, 2001, 2007, 2011). He is also author of *R for Stata Users* (Springer, with R. Muenchen, 2010) and of the soon to be published *Astrostatistical Challenges for the New Astronomy* (Springer) and *Methods of Statistical Model Estimation* (Chapman & Hall/CRC, with Andrew Robinson).

Foreword 3

The field of text mining—the process of finding patterns in unstructured data—has been an active area of research for more than two decades. Of course, even *defining* text mining is not easy, and as this book so systematically lays out, text mining is defined broadly enough to include everything from basic search to natural language processing (NLP). Unfortunately, the predictive analytics algorithms most analysts use are designed to build models from structured data, not unstructured data. Because it is difficult to convert unstructured data into the more useful structured format, unstructured data are often left unexplored.

Today, with a growing collection of unstructured data, increasing computer power, and enhanced software capabilities, text mining now tops most lists of future trends in analytics. However, merely topping a list still leaves undone the actual building and deploying of text mining solutions. Even in the most straightforward of applications, text mining is a very complex process. At the core of these complexities lies the fact that interpreting the ambiguities and subtleties of language is difficult. Moreover, most analysts embarking on text mining projects do not have an academic background in linguistics or text analysis and therefore must learn "on the job."

This book is a practitioner's book and is organized similarly to the acclaimed *Handbook of Statistical Analysis and Data Mining Applications* by Nesbit, Elder, and Miner. It begins with an overview of text mining and NLP, a daunting task in its own right. The history chapter provides an excellent background of text mining. Most users will find the "The Seven Practice Areas of Text Analytics" chapter insightful and useful for explaining text mining to peers and decision makers. But this book goes beyond providing only a survey of the field: It also provides practical steps for building models using textual data, including the important steps of data preparation and modeling.

Many of us learned data mining and text mining by reading textbooks, going to conferences, and taking courses. But we learn more deeply by "doing"—that is, by incorporating unstructured data in models for projects where we *must find a solution*. Practical experience provides invaluable insight into the difficult areas of data mining and text mining when there is no theory to draw from. Valuable principles and "rules-of-thumb" are learned by trial and error, particularly in feature creation, feature selection, and model assessment. We all learn that we must do much more than merely load data into a tool and turn the crank. It is here, in learning the so-called "art" of text mining, that watching a respected colleague solve a problem step-by-step is so valuable—precisely because it reveals the principles by which he or she arrives at each decision needed for the solution.

This book's 28 tutorials fill this great practical need by demonstrating how text mining can be used to solve real-world problems, providing "rules of thumb" at each stage of the text mining process, and showing how to accomplish these steps using state-of-the-art commercial software. While each tutorial uses a single software tool, most can be translated readily for use with other tools. Additionally, the tutorials included in the book use both commercial tools (IBM SPSS Modeler, Salford System STM/CART/Treenet, SAS Text Miner, and *STATISTICA*) and open source tools (GATE/Weka, R, and RapidMiner), thus appealing to a wider range of practitioners.

From notebook markings in aircraft maintenance records to survey free-form text fields, I vividly remember working on multiple projects in the early 2000s, where text was a key part of the data that had been collected. We knew the text would be valuable, but we didn't have the expertise to know how to extract the concepts or the time to learn text mining through trial and error. At that time, the most valuable jump-start to these projects would have been informal conversations with colleagues who had worked these kinds of problems before and who could provide the benefit of their experience to us. This book has that same feel: It's the well-read and clear-speaking colleague describing how to jump-start the text mining process over several dinners. Bon appétit!

Dean Abbott
Abbott Analytics, Inc.

Dean Abbott is president of Abbott Analytics, Inc. in San Diego, California. Mr. Abbott is an internationally recognized data mining and predictive analytics expert with over two decades of experience applying advanced data mining algorithms, data preparation techniques, and data visualization methods to real-world problems, including fraud detection, risk modeling, text mining, personality assessment, response modeling, survey analysis, planned giving, and predictive toxicology. Mr. Abbott is also chief scientist of SmarterRemarketer, a start-up company focusing on behaviorally and data-driven marketing attribution and web analytics.

Mr. Abbott is a highly regarded and popular speaker and course instructor at dozens of conferences. He often serves on the program committee for the KDD Industrial Track and Data Mining Case Studies workshop, and serves on both the Data Mining Certificate Program Advisory Board and the Data and Analytics Advisory Board at UCSD. Mr. Abbott has taught applied data mining and text mining courses using IBM SPSS Modeler, SAS Enterprise Miner, Statsoft *STATISTICA*, Tibco Spotfire Miner, IBM Affinium Model, Megaputer Polyanalyst, Salford Systems CART, and RapidMiner.

Acknowledgments

The authors are grateful for the contributions of many whose time and expertise greatly improved the quality and reach of this book. We would like to thank Jennifer Thompson, Angela Waner, Linda Winters-Miner, Karthik Balakrishnan, Susan Trumbetta, and Irving Gottesman as guest authors who went beyond the call of writing just one tutorial, but also provided figures, paragraphs, or editorial assistance for individual chapters; Raymond Mitten for the front cover design; Greg Sergeant for his design advice; John Dimeo of Elder Research, Inc. (ERI) for significant contributions to the chapter on focused crawling; Sarah Will of ERI for graphical assistance; Julie McAlpine Platt, editor-in-chief at SAS Press Series, for making all the arrangements for SAS's endorsement of this book; Richard Foley of SAS for not only a tutorial, but writing the SAS-text miner portion of Chapter 6; Oliver Jouve, director of business analytics for IBM for IBM-SPSS Modeler case studies; Kurt Peckman, SPSS senior market manager for IBM Software Group of IBM-SPSS Chicago, who provided software with which Bob and Gary made screenshots of the latest version of IBM-SPSS Modeler Premium; Eric Siegel for an early challenging, yet encouraging, review of the manuscript; several researchers at ERI for patient and helpful editing suggestions; Paul Lewicki, Win Noren, Jim Landrum, and Rob Eames of StatSoft for supporting this effort; Gavin Becker, Jennifer Radda, Mark Rogers, and Paul Prasad Chandramohan of Elsevier, spanning the globe from Boston to the United Kingdom to India, who worked with the authors tirelessly in the final stages to get this book to the printer.

All of these people contributed incrementally to the final copy, but we would like to acknowledge and honor particularly Lauren Schultz, our Elsevier publisher, who not only recognized but embraced the mission of this book and shepherded the manuscript through the submission stage of the Elsevier publishing process. She worked with us to identify and embody into practical principles the most salient points of text mining from among the vast diversity of concepts and techniques that currently characterize this rather disorganized discipline.

Lauren was vital also to our previous book, the *Handbook of Statistical Analysis & Data Mining Applications*, and we expect that this book will also prove popular among professionals who want to apply text mining to solve practical problems in the real world. To you Lauren, we especially dedicate this book. You will be our friend for life.

Gary D. Miner, Ph.D.

Dursun Delen, Ph.D.

John F. Elder IV, Ph.D.

Andrew Fast, Ph.D.

Thomas Hill, Ph.D.

Robert A. Nisbet, Ph.D.

October 25, 2011

Preface

Many people with photographic (or eidetic) memories are slightly nuts. At least, that is the way they are portrayed often in our modern culture in films and books, and on TV. The character of Raymond Babbit (in the film 'Rain Man') was based on a man who was autistic and almost non-functional in his youth apart from his treatment regimen. But, he could memorize prodigious amounts of text information, memorizing sometimes as many as 8 books a day. Lisbeth Salander, the main character in Stieg Larsson's Millenium Trilogy, was characterized as a girl with a photographic memory who thinks she is a freak. Other characters with photographic memories in television shows, like Dr. Lexie Grey in *Grey's Anatomy* and Shawn Spencer of *Psych*, appear rather normal in most respects, but may have severe disabilities in other areas (Shawn Spencer forgets birthdays).

The initial point of this extended reference to instances of people with photographic memories is that 'normal' people process text very differently. The brains of normal humans are pattern recognition factories. We read a book like *'The Girl With The Dragon Tattoo*, and remember the impression (formed from mental patterns) that Lisbeth Salander is strange, but interesting, while we forget the prose that Stieg Larsson used to portray her. We treat all text information in our society in a similar fashion.

Data in our data warehouses can be mined in many clever ways to uncover faint patterns of relationship 'hidden' in mountains of information.[i] Data in these databases are really highly summarized bits of information describing factoids about a person or an entity that were drawn from a much broader context; we think of that context usually in the form of free-form text. It has been estimated that as much as 80% of the information available in the world today is stored in the form of free-form text, and most of that has been written since 1950. This text is unstructured (thus, free-form) and must be 'read' in its entirety in order to understand all of its content. The other 20% of the information that occurs in structured databases is usually summarized from unstructured textual input sources. To that extent, we can view all stored information in the world as existing actually or potentially in the form of unstructured text.

The ability to access large amounts of text information in our memories occupies a tiny corner of our attention, although we decorate our books, films, and TV shows with many examples of it, pointing to significant benefits of its use. Even so, few people recognize the importance of text as unstructured information in our culture, and few people recognize the dangers of its possible consequences. Will the analysis of huge amounts of unstructured data with sophisticated text mining tools available today

[i] Nisbet, R., J. Elder, and G. Miner. "Handbook of Statistical Analysis & Data Mining Applications", Academic Press, 2009.

permit society as a whole to actualize its fascination with photographic memories? Will the challenges of such a massive input of information into our society be accompanied by analogies of the dysfunctional characteristics of our many of our examples and models of photographic memory? The "Terminator" series of films explored an extreme proposition that a vast complex computer system could learn so much (largely from text) that it could become self-aware, and turn on humanity. Is any shadow of that danger awaiting us in the future?

Another issue to keep in mind while reading this book is that a large number of text documents are copied many times and distributed (e. g. news stories). Considering that fact that many text mining tools generate automated reports consisting of numbers of entities (e. g. number of documents with a given word), the question is raised: How much text is actually being *written* and how much text is simply generated by copying. Adobe PDF format is a standard storage format in many libraries; shared PDF resources between libraries greatly multiplies the number of text documents, but how many are unique? Google Books is helping to answer this question.

A final issue to consider is the fact that much of our modern communications occur by cell phone and by computer blogs and tweets, in which long conversations occur over many disjointed instances. How will historians study "what happened today"? Will text mining tools evolve to incorporate speech recognition and develop abilities to track related elements of a series of disjointed instances of cell phone calls or blogs and tweets?

This book will help to expose you to the many ways text mining applications can open up the voluminous recesses of information that exist in online text (e. g. Google Books) and in corporate records (e. g. emails and reports), and to prepare you for the future of text mining. We hope that access to this new frontier of information management will be accompanied by checks, balances, and controls to prevent dysfunctional responses to text mining from dominating our culture in the future. In this book, we will tell you why to use text mining, and how to perform text mining operations, but it is up to you to use it wisely and leverage it to good ends.

About the Authors

Gary D. Miner, Ph.D.

Miner.Gary@gmail.com

Dr. Gary Miner received a B.S. from Hamline University, St. Paul, Minnesota, with biology, chemistry, and education majors; an M.S. in zoology and population genetics from the University of Wyoming; and a Ph.D. in biochemical genetics from the University of Kansas as the recipient of a NASA predoctoral fellowship. During the doctoral study years, he also studied mammalian genetics at the Jackson Laboratory, Bar Harbor, Maine, under a College Training Program on an NIH award; another College Training Program at the Bermuda Biological Station, St. George's West, Bermuda, in a Marine Developmental Embryology course, on an NSF award; and a third College Training Program held at the University of California, San Diego, at the Molecular Techniques in Developmental Biology Institute, again on an NSF award. Following that he studied as a postdoctoral student at the University of Minnesota in behavioral genetics, where, along with research in schizophrenia and Alzheimer's disease, he learned what was involved in writing books from assisting in editing two book manuscripts of his mentor Irving Gottesman, Ph.D. (Dr. Gottesman returned the favor 41 years later by writing two tutorials for this *Practical Text Mining* book.)

After academic research and teaching positions, Dr. Miner did another two-year NIH postdoctoral in psychiatric epidemiology and biostatistics at the University of Iowa, where he became thoroughly immersed in studying affective disorders and Alzheimer's disease. In 1985, he and his wife, Dr. Linda Winters-Miner (author of several tutorials in this book), founded the Familial Alzheimer's Disease Research Foundation, which became a leading force in organizing both local and international scientific meetings and thus bringing together all the leaders in the field of genetics of Alzheimer's from several countries. Altogether he spent over 30 years researching and writing, along with colleagues, papers and books on the genetics of Alzheimer's disease, including *Familial Alzheimer's Disease: Molecular Genetics and Clinical Perspectives* and *Caring for Alzheimer's Patients: A Guide for Family & Healthcare Providers*. During the 1990s, he was a visiting clinical professor of psychology for geriatrics at the Fuller Graduate School of Psychology & Fuller Theological Seminary in Pasadena, California also appointed to the Oklahoma Governor's Task Force on Alzheimer's Disease, and was also appointed associate editor for Alzheimer's disease for The Journal of Geriatric Psychiatry & Neurology, which he still serves on to this day.

By 1995, most of these dominantly inherited genes for Alzheimer's had been discovered. The one that he had been working on since the mid-1980s with the University of Washington in Seattle was the last of these first five, and the first to be identified on Chromosome 1. At that time, Dr. Miner decided to pursue interests in the business world, and since he had been analyzing data for over 30 years, going to work for StatSoft, Inc. as a statistician and data mining consultant seemed like the perfect "semiretirement" career. Interestingly (as his wife had predicted), he discovered that the "business world" was much more fun than the "academic world," and at a KDD—Data Mining meeting in 1999 in San Francisco, he decided that he would specialize in data mining. Incidentally, that is where he met Bob Nisbet, who told him, "You just have to meet this bright young rising star, John Elder!" Bob quickly found John and they were introduced. As Gary delved into this new "data mining" field and looked at statistics textbooks in general, he saw the need for "practical statistical books" and started writing chapters and organizing various outlines for different book possibilities. Gary, Bob, and John kept running into one another at KDD meetings, and eventually they decided to write a book on data mining. At the 2004 Seattle KDD meeting they got together for an early breakfast meeting with the purpose of reorganizing Gary's book outline, which eventually became the *Handbook of Statistical Analysis and Data Mining Applications*, published by Elsevier in 2009. The success of the Data Mining Handbook has led to this second book, *Practical Text Mining*. In the final analysis, thanks go to Dr. Irving Gottesman, Gary's "mentor in book writing," who planted the seed back in 1970 while Gary was doing a postdoctoral with him at the University of Minnesota.

Dursun Delen, Ph.D.

Dr. Dursun Delen is the William S. Spears Chair in Business Administration and Associate Professor of Management Science and Information Systems in the Spears School of Business at Oklahoma State University (OSU). He received his Ph.D. in industrial engineering and management from OSU in 1997. Prior to his appointment as an assistant professor at OSU in 2001, he worked for a privately owned research and consultancy company, Knowledge Based Systems Inc., in College Station, Texas, as a research scientist for five years, during which he led a number of decision support and other information systems–related research projects funded by federal agencies, including DoD, NASA, NIST and DOE.

His research has appeared in major journals, including *Decision Support Systems*, *Communications of the ACM*, *Computers and Operations Research*, *Computers in Industry*, *Journal of Production Operations Management*, *Artificial Intelligence in Medicine*, and *Expert Systems with*

Applications, among others. He recently published three books, *Advanced Data Mining Techniques*, with Springer, 2008; *Decision Support and Business Intelligence Systems*, with Prentice Hall, 2010; and *Business Intelligence: A Managerial Approach*, with Prentice Hall, 2010. He is often invited to national and international conferences for keynote addresses on topics related to data/text mining, business intelligence, decision support systems, and knowledge management. He served as the general cochair for the 4th International Conference on Network Computing and Advanced Information Management (September 2–4, 2008, in Seoul, South Korea), and he regularly chairs tracks and minitracks at various information systems conferences. He is the associate editor-in-chief for the *International Journal of Experimental Algorithms*, the associate editor for *International Journal of RF Technologies*, and is on the editorial boards of five other technical journals. His research and teaching interests are in data and text mining, decision support systems, knowledge management, business intelligence, and enterprise modeling.

John Elder, Ph.D.

Dr. John Elder heads the United States' leading data mining consulting team, with offices in Charlottesville, Virginia; Washington, D.C.; Baltimore, Maryland; and Manhasset, New York (www.datamininglab.com). Founded in 1995, Elder Research, Inc. focuses on investment, commercial, and security applications of advanced analytics, including text mining, image recognition, process optimization, cross-selling, biometrics, drug efficacy, credit scoring, market sector timing, and fraud detection.

John obtained a B.S. and an M.E.E. in electrical engineering from Rice University and a Ph.D. in systems engineering from the University of Virginia, where he's an adjunct professor teaching Optimization or Data Mining. Prior to 16 years at ERI, he spent five years in aerospace defense consulting, four years heading research at an investment management firm, and two years in Rice's Computational & Applied Mathematics Department.

Dr. Elder has authored innovative data mining tools, is a frequent keynote speaker, and was cochair of the 2009 Knowledge Discovery and Data Mining Conference, in Paris. John was honored to serve for five years on a panel appointed by President Bush to guide technology for National Security. His book with Bob Nisbet and Gary Miner, *Handbook of Statistical Analysis & Data Mining Applications*, won the PROSE award for Mathematics in 2009. His book with Giovanni Seni, *Ensemble Methods in Data Mining: Improving Accuracy through Combining Predictions*, was published in February 2010. John is grateful to be a follower of Christ and the father of five children.

Andrew Fast, Ph.D.

Dr. Andrew Fast leads research in text mining and social network analysis at Elder Research. Dr. Fast graduated magna cum laude from Bethel University and earned an M.S. and a Ph.D. in computer science from the University of Massachusetts–Amherst. There, his research focused on causal data mining and mining complex relational data such as social networks. At ERI, Andrew leads the development of new tools and algorithms for data and text mining for applications of capabilities assessment, fraud detection, and national security.

Dr. Fast has published on an array of applications, including detecting securities fraud using the social network among brokers and understanding the structure of criminal and violent groups. Other publications cover modeling peer-to-peer music file sharing networks, understanding how collective classification works, and predicting playoff success of NFL head coaches (work featured on ESPN.com).

Thomas Hill, Ph.D.

VP Analytic Solutions, StatSoft Inc.

Thomas Hill received his Vordiplom in psychology from Kiel University in Germany and earned an M.S. in industrial psychology and a Ph.D. in psychology and quantitative methods from the University of Kansas. He was associate professor (and then research professor) at the University of Tulsa from 1984 to 2009, where he taught data analysis and data mining courses. In a parallel—and more application-oriented—path of his carrier, he also has been vice president for Research and Development and then Analytic Solutions at StatSoft Inc., where he has been involved for over 20 years in the development of data analysis, data and text mining algorithms, and the delivery of analytic solutions.

Dr. Hill has received numerous academic grants and awards from the National Science Foundation, the National Institute of Health, the Center for Innovation Management, the Electric Power Research Institute, and other institutions. He has completed diverse consulting projects with companies from practically all industries and has worked with the leading financial services, insurance, manufacturing, pharmaceutical, retailing, and other companies in the United States and internationally on identifying and refining effective data mining and predictive modeling solutions for diverse applications.

He is the author (with Paul Lewicki) of *Statistics: Methods and Applications*, the *Electronic Statistics Textbook* (the number 1 online resource on statistics and data mining, published in hard copy in 2005), has published widely on innovative applications of data mining and predictive analytics, and has contributed to the *Handbook of Statistical Analysis and Data Mining Applications* (published by Academic Press in 2009).

Robert A. Nisbet, Ph.D.

Consulting Data Miner
Bob2@rnisbet.com

Dr. Nisbet was trained initially in ecosystems analysis. He has over 30 years of experience in complex systems analysis and modeling as a researcher (University of California, Santa Barbara). He entered business in 1994 to lead the team that developed the first data mining models of customer response for AT&T and NCR Corporation. While at NCR Corporation and Torrent Systems, he pioneered the design and development of configurable data mining applications for retail sales forecasting and Churn, Propensity-to-buy, and Customer Acquisition in Telecommunications and Insurance. In addition to data mining, he has expertise in data warehousing technology for Extract, Transform, and Load (ETL) operations; business intelligence reporting; and data quality analyses. He is lead author of the *Handbook of Statistical Analysis & Data Mining Applications* (Academic Press, 2009). Currently, he functions as a data scientist and independent data mining consultant.

Introduction

BUILDING THE WORKSHOP MANUAL

Computer scientists, artificial intelligence practitioners, and even physicians are chasing the goal of the intelligent machine in order to understand how humans work and how to fix the problems that vex our lives. The development of text mining technology in computer science is analogous to the discovery of DNA, how it works, and the mapping of the human genome. Gene mapping is aimed at understanding the bewildering complexity of the structure and function of the human body for the (primary) purpose of building the "workshop manual" that will permit us to fix its physical problems and prolong its life. The practice of text mining is aimed at understanding and applying insights from the most complex analytical processing system in the universe—the human brain—to the analysis of written language. This book is one step in the direction of building the text mining "workshop manual" that will enable us to "fix" many problems of miscommunication and serve as the basis for contributing clarity and meaning in all mediums of human communication involving written text.

COMMUNICATION

The case could be made that text mining in principle is just an exercise in communication. All text documents are objects that exist for the purpose of directed communication between a "source" and a "receiver." Viewed in this way, text mining flows out of the developments of information theory. We view it this way in Chapter 1.

The vast body of information expressed by written text stored in our libraries and databases is very analogous to the "deep" memory to which only very few people "blessed" with eidetic (or photographic) memory have much direct access. The rest of us must depend on "mechanisms" like text mining to reveal some of this information, make sense of it, and enable us to act upon it.

Each document in a body of written texts was created for some purpose of communication. We need to be able to "listen" to and understand those communications. Similarly, we use "mechanisms" (e.g., ears) to understand speech. Speech provides a window into human thought that can be particularly helpful to our understanding of thought disorders such as schizophrenia. Typically, schizophrenia symptoms affect both receptive and expressive language, and speech samples have long informed scientific investigations of the disorder. Tutorial-I analyzes spoken communications converted to text to distinguish schizophrenics from nonschizophrenics. This tutorial points to a fascinating new frontier in text mining, in which spoken communications are combined with speech recognition technology to analyze information in text form with text mining methodologies. In the past, speech analytics applications have depended upon the concept of key word spotting (KWS) to analyze phone calls by suspected terrorists and for call center analysis (Alon, G. 2005). We might add KWS technology in voice communications to the list of applications of named entity extraction technology.

The dependence on the analysis of mechanisms of communications is like the analysis of spoken communication. Spoken communication differs from written communication primarily in the transfer medium. Written communication composes a group of symbols written on some medium to refer to words; spoken communication composes a group of sounds spoken by the medium (the person) to refer to words. We might consider spoken communication that can be combined with speech recognition technology and analyzed in text form as a new frontier of text mining analysis.

Analysis of communications (written and spoken) can offer very important guidance and insights to *significantly* improve our lives. How can we do this? *It's all in the structure.* All communication requires some sort of structure. Random gestures or words (spoken or written) can convey no communication other than the presence of randomness. Every language possesses a certain structure designed to order and connect words (or pictographs) to convey some meaning. In English, for example, the structure of a sentence reflects the way people think. The ordered words in a sentence convey meanings of the thoughts represented by the writers of the text.

THE STRUCTURE OF THIS BOOK

This book is a structure of structures. There is obvious structure in a body of text that consists of sentences, paragraphs, chapters, and sections, all of which compose this book. But there is a less obvious and far more important structure to this book: It is a learn-by-doing excursion into the art and science of text mining. Following that design motif, we divided the book into three sections to facilitate this learning approach. Part I contains a short history of text mining and basic principles of text mining. Part II contains 28 tutorials, where you can begin to "stretch" your text mining "legs," using free text mining software. Part III consists of seven "deep dives" into important text mining methods and an exquisite excursion into applications of text mining in the insurance industry, followed by a chapter on the future of text mining, and finally ending with a summary chapter. That description appears to "wrap up" the book, except it doesn't contain the single most important feature of this book: the "main thing."

Keeping the main thing as the main thing. In many books, tutorials (if they exist) appear to be tacked onto the end of a book. But in this book (and in our companion book, Nisbet et al., 2009), the tutorials are the central focus—the main "thing" in the book—and they belong in the middle of the book, where you can start applying the text mining principles on the basis of the basic principles presented initially. This structure serves an even deeper goal: that of providing a complete learning environment, in which you use the free software provided, create initial text mining models, and feed back initial results (or lack of them) to cause your learning experience to grow dynamically.

PART I: BASIC TEXT MINING PRINCIPLES

Chapter 1 presents a history of text mining, beginning with the development of early systems for *information retrieval* in the form of the library card catalog, through technologies of *information extraction,* to methods of *information discovery*. You will learn the relationships between goals of library science and precepts of information theory and the grafting of powerful accessory technologies of artificial intelligence and data mining to develop modern methods of text mining.

Chapters 2 to 5 present the structure, methodology, and processes of text mining to prepare you to begin to learn by doing in the tutorials. Chapter 2 builds a hierarchical structure (a *taxonomy)* of text mining activities in many areas of text analysis. You will learn where text mining fits into a hierarchical structure and the 'Seven basic practice areas' of information processing. Chapters 3 and 4 cover basic text mining theory and applications, and Chapter 5 presents the text mining process, composed of the common sequence of data processing operations followed in a text mining project. Chapter 6 introduces you to the three most common data mining tool sets used in business, all of which have text mining capabilities:

- IBM-SPSS Modeler Premium
- SAS-Enterprise Miner and Text Miner
- *STATISTICA* Data Miner and Text Miner.

Some if the vendors of these tool packages have set up special programs to provide you with free copies of these packages for a specified period of time.

The discussion of these tools is not intended to present a full complement of all features and functions of the tools, but rather to show you how to use the tool interfaces to do text mining. There are many other data mining capabilities of these tools, which are discussed in Nisbet et al. (2009).

PART II: TUTORIALS

We placed the tutorials after a discussion of basic text mining principles because learning a new skill (text mining and tool use) is best done in a learn-by-doing framework.

PART III: ADVANCED TOPICS

This part includes detailed discussions of the analytical techniques of classification, clustering, and prediction operations used in text mining to help you understand what is going on behind the scenes in each of the tools used in the tutorials. Several advanced topics are explored (including entity extraction and singular value decomposition) to add the final pieces to your understanding of how text mining is performed. The last element presented in Part III is a contributed chapter by Karthik Balakrishnan that outlines a comprehensive list of the applications of text mining in a specific industry: property and casualty insurance.

TUTORIALS

People with hearing impairments can't use their ears to communicate, and those with sight impairments can't use their eyes. But many of these sensory-impaired people still find ways to communicate. One such person was Helen Keller. Natively, most of us are like Helen Keller: We can't see this vast body of written text, and we can't hear its meaning. Text mining can help us to perceive what some of this text can communicate to us, but we need the right tools to "fix" or work around our miscommunication problems (as Helen Keller worked around her deafness and blindness). Text mining tools and methodologies provide some of those tools. The tutorials in this book will teach you how to use these tools to perform many text mining tasks and compose them in the form of graphics to demonstrate how to relate text mining results to help grow your operations. Figure I.1 shows how text mining is the central theme of this book.

As shown in Figure I.1, text mining draws upon many techniques in the broader field of text analytics. Text mining is the practical application of many techniques of analytical processing in text analytics. Within the central theme of text mining stands the central issue of this book: learning how to apply text mining models to solve practical problems in an organization. The tutorials in this book evolved out of the need to serve that issue. Within that central issue lays the very heart of our intent: to lead you up the learning curve in text mining as quickly as possible by using a learn-by-doing approach. This is the primary goal of this book.

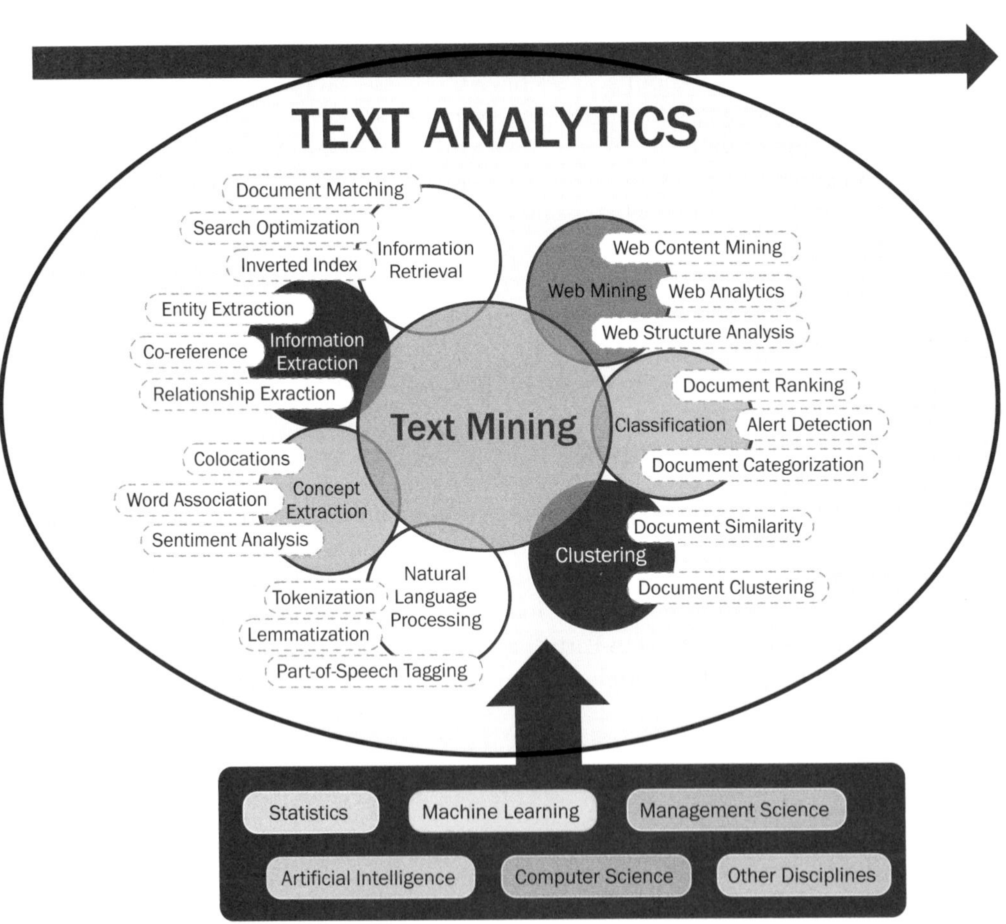

FIGURE I.1

Text mining draws upon contributions of many text analytical components and knowledge input from many external disciplines (shown in the 'bar with upward arrow' at the bottom), which result in directional decisions affecting external results (shown by the arrow point to the right arrow at the top).

WHY DID WE WRITE THIS BOOK?

The previous book in this series (Nisbet et al., 2009) focused on data mining, with a token chapter and two token tutorials on text mining. This was done because text mining is a part of data mining, but a separate volume was needed to do it justice. Consequently, this book focuses on text mining, with a token chapter on web mining and focused web crawling. Both of the web mining and other web analytical procedures are broad enough in content to justify separate books to cover them properly.

WHAT ARE THE BENEFITS OF TEXT MINING?

Web mining, web analytics, and text analytics are all terms that are interrelated with text mining, which can provide enhanced information about analyzed data, showing how much you can expect the model to "lift" productivity (e.g., sales), compared to alternative actions like doing nothing. Analytical modeling is an iterative process, just like sculpture. When we are satisfied that we have the best model (among alternatives), we can use the model (or *deploy* it) to make decisions as individuals or in organizations, corporations, and government agencies to improve our lives by making more efficient use of our resources. This extra lift is called "gains" when we view it as a cumulative number, which provides us with great practical value, as illustrated in Figure I.2.

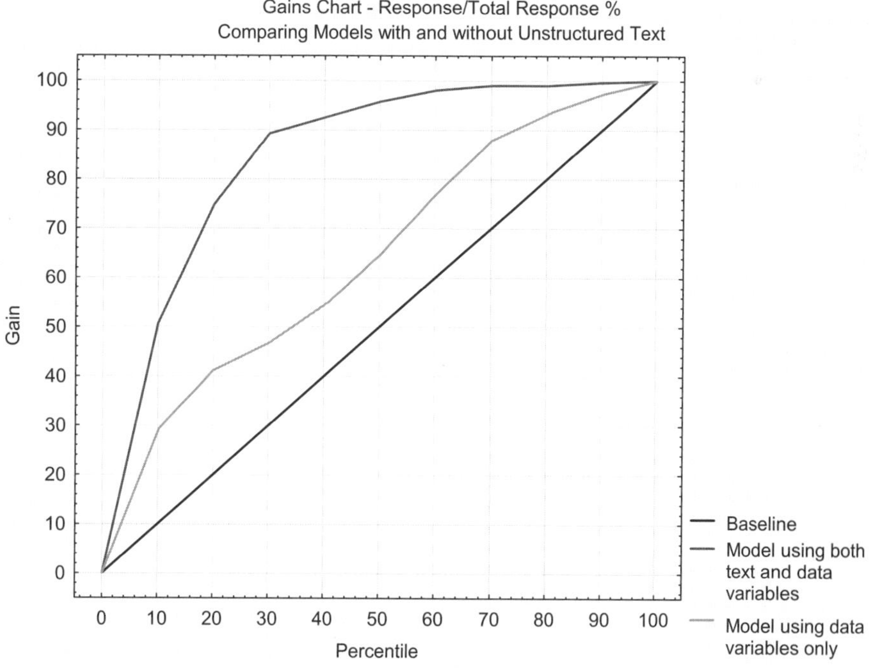

FIGURE I.2

This cumulative gains chart shows how much better the model is using indexed keywords from text mining, compared to the model using only structured text values (from databases). The total cumulative gain is the integral (area) between the middle gains curve (model with database values only) and the top gains curve (model with indexed keywords).

BLAST OFF!

The text mining technology and tools described in this book will help you launch yourself into textual "hyperspace" almost immediately by using the tutorials. There, you can analyze the bewildering array of "text features," interpret the set of concepts they represent as specific examples, and "map" their meanings back to the real world around us. Our brains can group the meanings represented by the text documents and increase our understanding of those concepts and their environments (contexts). This understanding can be orchestrated by us and others to provide new information, generate new insights, and create new applications of them to build better lives. So, dear readers, fire your rockets, prepare for liftoff, and away we go!

References

Alon, G. (2005). Key word spotting—the base technology for speech analytics. National Speech Communication white paper. Rishon Lezion, Israel. http://www.nsc.co.il/pdf/kws-whitepaper.pdf.

List of Tutorials by Guest Authors

E. Analysis of Survey Data for Establishing "Best Medical Survey Instrument" Using Text Mining: Central Asian (Russian Language) Study Tutorial 2: Potential for Constructing Instruments That Have Increased Validity
Jeremy La Motte, M.D.
Ruth Slagg-Moore, DO
Sanjay Thomas, M.D.
Chris Jenkins, M.D.
Linda A. Miner, Ph.D.

F. Using eBay Text for Predicting ATLAS Instrumental Learning
Anne Ashby Ghost Bear, EdD
Linda A. Miner, Ph.D.

G. Text Mining for Patterns in Children's Sleep Disorders Using *STATISTICA* Text Miner
Jennifer Thompson
Karen Spruyt, Ph.D.

H. Extracting Knowledge from Published Literature Using RapidMiner
Dursun Delen, Ph.D.

I. Text Mining Speech Samples: Can the Speech of Individuals Diagnosed with Schizophrenia Differentiate Them from Unaffected Controls?
Jennifer Thompson
Susan L. Trumbetta, Ph.D.
Gary D. Miner, Ph.D.
Irving I. Gottesman, Ph.D.

J. Text Mining Using STM™, CART®, and TreeNet® from Salford Systems: Analysis of 16,000 iPod Auctions on eBay
Dan Steinberg
Mikhail Golovnya
Ilya Polosukhin

K. Predicting Micro Lending Loan Defaults Using SAS® Text Miner
Richard Foley

L. Opera Lyrics: Text Analytics Compared by the Composer and the Century of Composition—Wagner versus Puccini
Gary Miner

M. CASE STUDY: Sentiment-Based Text Analytics to Better Predict Customer Satisfaction and Net Promoter® Score Using IBM®SPSS® Modeler
Marie-Claude Guerin
Eric Martin, Ph.D.
Olivier Jouve

Part III: Chapters

1

PART

Basic Text Mining Principles

CHAPTER 1

The History of Text Mining

CONTENTS

PREAMBLE

Why write a chapter on the history of text mining? There are at least three reasons: to provide the context in which text mining developed; to show the development paths followed in text mining techniques, which can point to how to expand and improve text mining techniques in the future; and to avoid making the mistakes of the past. Text mining developments were initiated by the need to catalog text documents (e.g., books in a library). But soon, development shifted focus to text data extraction using natural language processing (NLP) techniques. Practical needs to extract textual information continue to drive the development outwardly today, but the quest for the "Intelligent Machine" drives it inwardly. The ability to understand and extract knowledge from text is a key requirement in the goal of artificial intelligence researchers to create machines that can at least simulate closely and relatively accurately the most complex thinking machine in the universe: the human brain.

This chapter introduces you to the origins of text mining and the breadth of the text mining methodological landscape. And because of that breadth, the field of text mining development is rather

Practical Text Mining and Statistical Analysis for Non-structured Text Data Applications. DOI: 10.1016/B978-0-12-386979-1.00001-3

fragmented and disorganized. We try to bring some order out of this "chaos" to both introduce the multiple threads of development and orient you in your search for knowledge that resides in the huge mountain of textual data available in the world today.

In 2001, the Internet claimed to provide access to about 10 million web pages (over 100 terabytes). As of 2009, that volume had grown to over 150 billion web pages (or about 1,500 terabytes). That represents a growth rate of about 40 percent per year. This corpus of text includes home pages, corporate public pages, personal pages, and Usenet newsgroups, as well as electronic books and Ph.D. theses. It does not, however, include the vast number of corporate web pages behind firewalls. Within a business, both of these volumes of text must be considered in the process of information retrieval. It is no wonder that one of the hottest high-tech growth sectors is Internet search engines.

Of course, this growth is not linear, and the present rate may not continue into the future, but the present accumulation of text in the form of web pages is an example of how fast we are accumulating textual data in the world. A compounding factor is the burgeoning e-book industry and the conversion of huge numbers of paper text copies to electronic forms (e.g., Google Books).

The Library of Congress is the largest library in the world, with more than 120 million items on approximately 530 miles of bookshelves. The collections include more than 18 million books, 2.5 million recordings, 12 million photographs, 4.5 million maps, and 54 million manuscripts. This body of text also is growing at a prodigious rate.

As recently as 1999, it was observed that text mining had the peculiar distinction of having a name and a lot of hype but almost no practitioners (Hearst, 1999). Perhaps one reason for this stark admission is that the common definition of *text mining* used by Hearst excluded information retrieval. Text mining is often defined in the relatively narrow context of discovering previously unknown information that is implicit in the text but not immediately obvious. This exclusion is convenient for the classroom, but it ignores the long history of automated text processing and its enabling effect on text mining.

Even a cursory analysis of the literature on text mining will show that beginning in 1999, text mining research and applications developed rapidly. In this chapter, we consider the development of several cognate disciplines, which formed the "roots" from which modern data mining arose.

THE ROOTS OF TEXT MINING: INFORMATION RETRIEVAL, EXTRACTION, AND SUMMARIZATION

Modern computer technology is based firmly on the pioneering work of Charles Babbage on his difference engine, which was first proposed in 1822. We cannot fully understand the historical development of computers without recognizing his seminal contribution to the theory of mechanical calculating machines. Likewise, the practice of text mining did not spring out of "whole cloth" either. That is, many of the developments in technology and applications contributed to the modern text analysis systems. In order to fully understand the significance of modern text mining technology, we must view it in the context of the technological history from which it sprang. These developments were spurred by the significant challenges posed by the burgeoning volume of textual information being generated in the world. Not the least of these challenges was the problem of accessing required information buried in the accumulation of vast quantities of text stored in various forms.

Two similar processes are used to access information from a series of documents. The first is *information retrieval*, which distinguishes a set of documents as an answer to a logical query using a set of key words, possibly augmented by a thesaurus. The second process is *information extraction*, which aims to extract from documents specific information that is then analyzed for trends or other data.

This problem consisted of two basic tasks:

- Access to information contained in various text documents required some sort of summarization processing to reduce the body of text to a tractable size.
- Once the documents were summarized acceptably, the job of classifying the myriad of document summaries into some sort of logical arrangement was daunting.

Approaches to access textual information developed in three venues:

- Library science for text summarization and classification
- Information science
- Natural language processing

Library Book Summarization and Classification

One of the earliest examples of text summarization and classification was the library catalog. The earliest library catalog is attributed to Thomas Hyde (1674) for the Bodelian Library at the University of Oxford. The index card was introduced by Melvil Dewey in 1876 to form a library card catalog (Figure 1.1).

Another step in the development of text processing was the summarization of text to generate abstracts. One of the earliest attempts to summarize a large body of text was science abstracts, which began in 1898 as a joint collaboration between the Institution of Electrical Engineers and the Physical Society of London. Luhn (1958) adapted a computer to generate document abstracts. This method performed a word frequency analysis on an early IBM 701 computer (the first commercial computer, built with vacuum tubes), from which a relative measure of significance was derived. The number of relatively significant words was counted for each sentence and combined with the linear distances between the words to produce a metric of sentence significance. The most significant sentences were

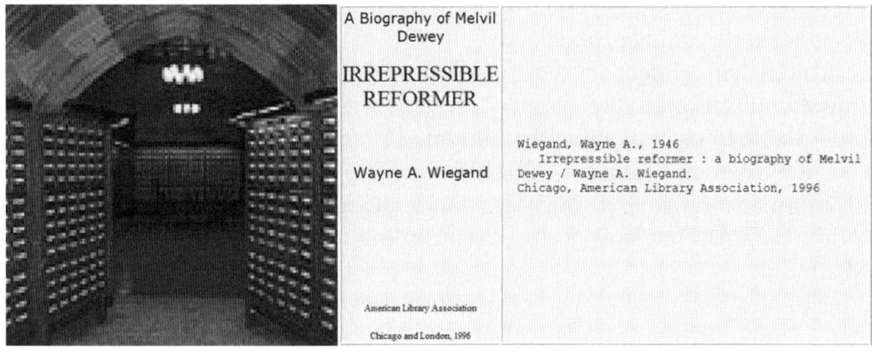

FIGURE 1.1
The library card catalog at Yale University and an index card. *Source: http://commons.wikimedia.org/wiki/File:Yale_card_catalog.jpg*

then extracted to form an abstract for the document. This approach was novel in that it applied some structure to information extracted from unstructured text, and it was performed in an automated fashion on a computer. From that beginning, text mining for information retrieval and manipulation gravitated toward library science. What better milieu could have been found for automated text handling? Science abstracts became available in electronic form in 1967.

During the next 40 years or so, library scientists greatly expanded on the work of Luhn. Doyle (1961) likened a large group of documents in a library to a group of wares in a supermarket. A supermarket does not have a card catalog, but a shopper could usually find the required food item by simply scanning the aisle signs and shelf labels. This convenience was not feasible for a large group of documents in 1961. Doyle picked up on the work of Luhn to suggest a new way to classify information in a library in the form of word frequencies and associations. He claimed that this system could become a highly systematic and automated method for rapid browsing of information in a library. He was right!

Despite the recognized benefits of online library catalogs, the conversion of the card catalog to computers was slow due to the computer costs and the limitations of the technology serving them. At UC Santa Barbara (for example), the many oak wood card catalog cabinets in the campus library were not completely replaced until the mid 1990s.

Information Science

From the time of Thomas Dewey until the dawn of the computer age in the 1940s, the producer of information had to publish it in some form. Subsequently, the published text was then abstracted and indexed manually to form the card catalogs in libraries. The prospective user of this text information was faced with the daunting task of laborious manual searches to find the required information to support research and applications. The advent of the computer permitted electronic storage of text and automated searches. The advent of personal computers (PCs) led to the development of ingenious user interfaces with "hyperlinks" that provided almost instant access to the original text, not just the abstracts. Thus, information science developers revolutionized the interface between producers and the users of text information.

Claude Shannon's (1948) inauguration in the 1940s of the new discipline of information theory may stand as one of the most significant developments of the twentieth century. Shannon showed that the digital interface between the "source" and the "channel" of text information was optimal because the creation of the text (the "coding") and the use of the text (in the "channel") could be described and optimized by two separate coding problems with no loss of generality. This insight led to the invention of his source-encoder-channel-decoder-destination model in cryptography during World War II, but it also had profound implications in the design of modern digital storage and communication of all types of information. It is safe to say that without Shannon's information theory and the research it spawned, there would be no Internet (Gallager, 2001). Information flow in the Internet is enabled by an application of information theory (cyclic redundancy checking and other error-correction protocols). Moreover, modern data compression protocols (including the .gif and .jpg image files) owe their very existence to Shannon.

Applications of information theory to printed text developed along several lines in the 1950s. The science of bibliometrics arose to provide a numerical means to study and measure texts and information. One of the most common bibliometric applications was the formation of the citation index, which analyzes the references to one text document contained in other text documents. The citation

Results: 4 | Page 1 of 1 Go | Sort by: Times Cited

Use the checkboxes to remove individual items from this Citation Report or restrict to items processed between 1900-1914 and 2008 Go	2004	2005	2006	2007	2008	Total	Average Citations per Year
	0	8	10	14	7	39	6.50
1. Title: Automated water analyser computer supported system (AWACSS) Part I: Project objectives, basic technology, immunoassay development, software design and networking Author(s): Tschmelak J, Proll G, Riedt J, et al. Source: BIOSENSORS & BIOELECTRONICS Volume: 20 Issue: 8 Pages: 1499-1508 Published: FEB 15 2005	0	4	5	6	3	18	4.50
2. Title: Automated water analyser computer supported system (AWACSS) Part II: Intelligent, remote-controlled, cost-effective, on-line, water-monitoring measurement system Author(s): Tschmelak J, Proll G, Riedt J, et al. Source: BIOSENSORS & BIOELECTRONICS Volume: 20 Issue: 8 Pages: 1509-1519 Published: FEB 15 2005	0	3	3	7	2	15	3.75
3. Title: Monitoring freshwater sediments Author(s): Alcock S, Barcelo D, Hansen PD Source: BIOSENSORS & BIOELECTRONICS Volume: 18 Issue: 8 Pages: 1077-1083 Published: AUG 1 2003	0	1	2	1	0	4	0.67
4. Title: Bioreporters: gfp versus lux revisited and single-cell response Author(s): Kohlmeier S, Mancuso M, Tecon R, et al. Source: BIOSENSORS & BIOELECTRONICS Volume: 22 Issue: 8 Pages: 1578-1585 Published: MAR 15 2007	0	0	0	0	2	2	1.00

FIGURE 1.2

An example of a *Science Citation Index* page. *Source: http://images.isiknowledge.com/WOK45/help/WOS/h_citationrpt.html#h_index_example*

index could be used to gauge the importance of a given document in terms of its reference by other documents. Figure 1.2 shows a page from the *Science Citation Index*, which contains a list of science articles and the total number of citations of each article for the years 2004 to 2008. The average number of citations per year can be used as a metric of importance for an article. Groups of important articles can be used to track the pathways of development in a given discipline. This application of text processing is analogous to the calculation of word frequencies in text mining.

Other applications of information theory produced various indexing systems, document storage and manipulating applications, and search systems. The future development of text mining was rooted firmly in the successes of these early text manipulation and search systems in information science.

A hybrid discipline, natural language processing (NLP), developed from elements of linguistics and information science in attempts to understand how natural human language is learned and how it can be modeled. NLP began as an attempt to translate language on a computer. These early efforts failed, and the focus turned to the processing of answers to questions (Bates, 1995). The availability of computers in the 1960s gave birth to NLP applications performed on computers (computational linguistics).

Natural Language Processing

The birth of computational linguistics in natural language processing (NLP) occurred amid a background of intense philosophical debate in linguistics, which continues to this day. Ancient ideas of language learning were based on the notion of Plato that the word mappings of a young child were innate, not learned (Tomasello, 2008). In contrast, empiricists like Locke and Hobbes believed that language was learned from abstracted sense impressions and assumed that it could be based on a deductive system, like physics. Their ideas were strongly affected by the philosophy of Aristotle. The work of Luhn (1958) on abstract generation at IBM was an example of this approach. B.F. Skinner took this idea one step further and promoted the behaviorist idea that language is learned through operant conditioning involving reinforcement and consequences (Skinner, 1957).

Platonic philosophy, however, was not dead in NLP. Noah Chomsky (1959) strongly objected to the behaviorist's purely deductive system by calling it "largely mythology" and a "serious delusion." He maintained that we must return to Plato to understand how children learn. Chomsky championed the idea of "generative grammar": rule-based descriptions of syntactic structures. Although many have disagreed with Chomsky's ideas, producing alternative linguistic formalisms or taking issue with his methods of discovering linguistic data, much of the NLP work since 1957 appears to be profoundly marked by his influence.

Between 1980 and 1990, Chomsky's generative linguistics approach ruled as the dominant philosophy in NLP. Between 1990 and 2000, the previous empiricist approach reestablished itself as a viable alternative to the rationalist approach of Chomsky (Dale et al., 2000). The major technical innovation was the application of machine learning technology to create parsing algorithms (which split sentences into *tokens* that can be words, numbers, or symbols) and stemming algorithms (that reduce words to their stems, bases, or roots) that are domain-independent.

Clustering in Natural Language Processing

Clustering is an automated process that groups all input documents into clusters, based on similarities. It is an *unsupervised* process, where no prior information is available about the documents. Early cluster analysis was focused on providing an adaptive method of browsing a document collection when a query could not be created (Cutting et al., 1992). Subsequently, clustering was applied to query-based clustering on document collections using a hierarchical clustering method (Tombros et al., 2002). The basic idea in clustering of document collections is to form some sort of similarity or distance measure and then group documents together so the similarities or distances meet some objective function.

The next phase of natural language processing was concerned primarily with understanding the meaning and the context of the information, rather than focusing just on the words themselves. Subsequent developments in natural language processing moved into bibliometrics to consider the context of documents (Salton et al., 1994). The development of modern text mining technology followed the same path, and it is from these developments in natural language processing during the 1990s that modern contextual text mining arose.

By the year 2000, NLP practitioners could choose to employ either the domain-independent parsing and stemming methods to produce features or newer text categorization techniques that relied on the process of *tagging* documents with a small number of concepts or key words. The tagging approach reflected more of Chomsky's philosophy and depended more on the domain of application (Sanger and Feldman, 2007). The challenge before NLP in 2000 was to devise a way to integrate these two approaches at the process level; the same challenge is still before us today. One approach to meeting this challenge is described by Dale and colleagues (2000) and shown in Figure 1.3.

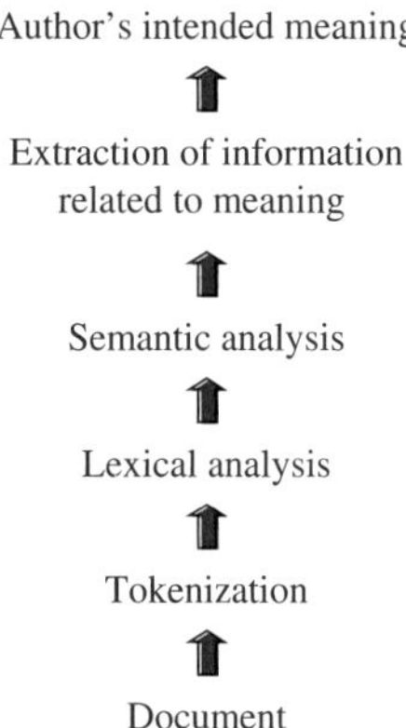

FIGURE 1.3
Stages of analysis in natural language processing. *Source: Based on Dale et al., 2000.*

The necessary information extraction for the meaning of the text can be performed using techniques based on Chomsky's approach as described by Lochbaum and colleagues (2000). To this extent, the Platonic approach based on Chomsky's ideas and the Aristotelian approach based on behaviorists' ideas may be merged following techniques suggested by Lockbaum. This appears to be the best path for development of modern text mining applications. At this point, it is interesting to note that we can recognize the same struggle between Aristotelian and Platonic approaches in the history of NLP contributions to text mining, as we can see also in the broader history of data mining (Nisbet et al., 2009).

INFORMATION EXTRACTION AND MODERN TEXT MINING

Early text mining activities in library science, information science, and NLP were concerned primarily with various forms of *information retrieval* and *information summarization*, such as indexes, abstracts, and grouping of documents. Later developments in text mining focused on *information extraction*. The content and relationship information is extracted from a corpus of documents. These information extraction operations are performed by *tagging* each document for the presence of certain content facts or relationships between them. Information extraction consists of an ordered series of steps designed to extract terms, attributes of the terms, facts, and events (Sanger and Feldman, 2007). We might define modern text mining partly in terms of powerful methods of information extraction developed since 2000. We must keep in mind, however, that there is a considerable body of information extraction activities that extend back as far as 1987 in the Message Understanding Conferences (MUCs).

The MUCs were initiated by the Naval Ocean Systems Center (NOSC), with assistance from the Defense Advanced Research Projects Agency (DARPA). These conferences reported the results of previous text analyses by various groups to fill in templates provided in the form of an exam. The submissions were evaluated against a test key. The first two conferences in 1987 and 1988 were focused on research projects funded to analyze military messages, particularly those of naval fleet operations (Grishman and Sundheim, 1996). Subsequent conferences held from 1989 to 1997 focused on terrorism in Latin America, joint ventures in microelectronics production, negotiation of labor disputes and corporate management succession, and airplane crashes and rocket/missile launches.

The enormous influence of the MUCs is reflected in the many concepts that became part of information extraction engines in text mining, including the following:

1. Defining the processes of named entity recognition (often referred to in text mining as proper name identification). This process became included in the text mining phase *syntactic analysis*.
2. Formalizing the test metrics *recall* and *precision*. Recall was defined as the ratio of correct slots filled to the total slots in the template. Precision was defined as the ratio of correct slots filled to the sum of the correct and incorrect slots filled (with some slots left unfilled). Figure 1.4 shows the relationship between recall and precision.

The concepts embedded in the metrics of recall and precision were adopted also in data mining and applied to classification problems, but the definitions are slightly different (Nisbet et al., 2009).

3. The importance of *robustness* in ML models. MUCs stressed the need for portability of solutions. The robustness (or generalizability) of a model is a measure of its successful application to data sets other than the one used for training and testing. This concept had profound influence on the development of subsequent data mining theory in general.
4. The importance of "deep understanding" to minimize the danger of local pattern matching is reflected in subsequent data mining developments to minimize overtraining of ML models.

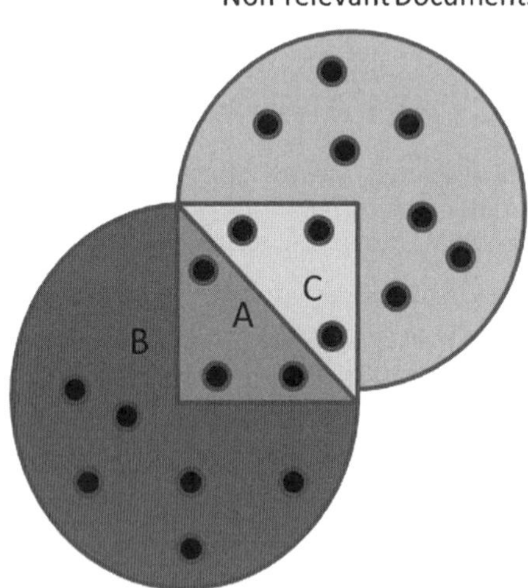

FIGURE 1.4

The relationship between recall and precision criteria for model evaluation. The blue triangles represent the sample of documents in the training set; the dark blue documents are relevant documents correctly classified; and the light blue triangle includes nonappropriate documents that were correctly classified. Recall is the ratio of the number of documents in A to those in B. Precision is the ratio of the number of documents in A to those in C.

5. The importance of making distinctions of coreference among noun phrases in the MUCs led to extensive research in text mining for resolving anaphoras, the process of matching pairs of NLP expressions that refer to the same entity in the real world (Sanger and Feldman, 2007).
6. The importance of word-sense disambiguation led to the development of stochastic content-free grammars in probabilistic text mining models (Collins, 1997).

Web Mining

The Internet passes and connects to one of the largest bodies of unstructured text on earth. At a presentation of the IEEE International Conference on Tools with Artificial Intelligence (ICTAI) in 1997, the question was asked, "Is there any benefit in applying data mining and AI to the World Wide Web?" (Cooley et al., 1997). Rob Cooley formalized his presentation in the form of his dissertation, which focused on finding patterns of information on the World Wide Web (Cooley, 2000). That was the beginning of the flood of research of practical applications of web mining.

One of the earliest practitioners of web mining was Jesus Mena (1999), who cited many possible applications of data mining technology to mining your website. Mena's book is remarkable because it covered such a wide range of possible applications of web mining and because it contained absolutely no references! Therefore, it is difficult to map Mena's ideas into the broader history of data mining and text mining. Nevertheless, the book stands as an excellent overview of web mining in its infancy, viewed through the perceptions of one man.

Focused Web Crawling

One of the most interesting application of web mining is the hunt for relevant web pages in the vast collection of text on the Internet. Web crawling is the process of accessing only specific web pages on the Internet that are relevant to a predefined topic (or set of topics). Web crawling was first proposed by Menczer (1997). His initial strategy was to permit the web crawling program to adapt to web pages as it proceeded and enabled it to learn to define relevance in terms of the web pages it found. Another early strategy was proposed by Pinkerton (1994).

Strategies

One of the early strategies use in focused web crawling was predicting the probability that candidate links are relevant to the chosen topic and avoid having to download all of the pages below a preselected probability level (Pinkerton, 1994). Menczer and colleagues (2004) showed that simple strategies such as those were very effective for short crawls, but more complex methods, such as reinforced learning and evolutionary adaption, were necessary to serve longer crawls. Diligenti and colleagues (2000) suggest using the complete content of the pages already visited to infer the similarity between the driving query and the pages that have not been visited yet.

We can see that elements of text summarization clearly belong to the first phase of text mining development. The second phase can be viewed as information extraction, but the development of major elements of this phase extended back to the mid-1980s. We are tempted to define the third phase of text mining as the development of new techniques to support information discovery of useful patterns in bodies of text information. The problem with using this sequence of phases as a means to classify text mining advances is that there is a large degree of overlap among the developments of all three of them. Will Phase III be the last one? We asked that question in our previous

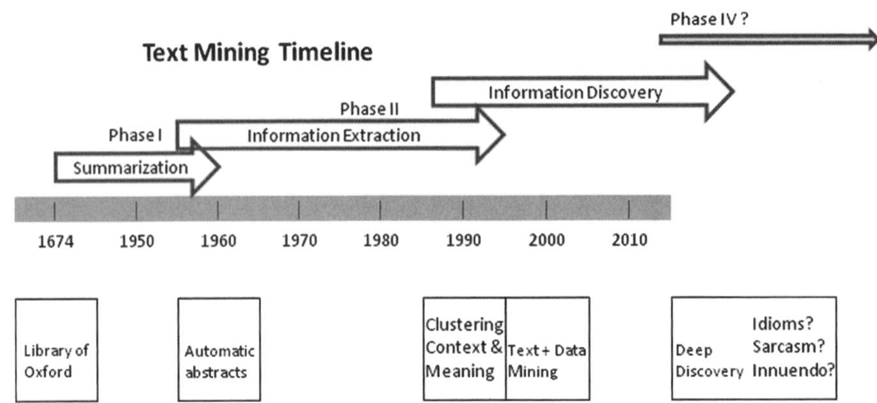

FIGURE 1.5
Timeline of the phases of development in text mining.

book on data mining phases. The answer is the same: Probably not (Nisbet et al., 2009). The next phase of text mining will focus on the "deep" discovery of meaning "between the lines," such as idioms, sarcasm, and innuendo (Figure 1.5).

MAJOR INNOVATIONS IN TEXT MINING SINCE 2000

Text mining has changed drastically since the 1990s. Many new techniques were developed that merge cognate areas of data mining, statistical analysis, and statistical learning theory with text mining. In addition, previous developments in information summarization and information extraction were leveraged as a foundation on which to build new information about pattern discovery techniques.

Modern Information Extraction Engines

The result of these developments was to redefine the tasks in many modern information extraction (IE) engines as shown in Figure 1.5. The *tokenization* module shown in Figure 1.6 accomplishes the task of splitting the input document into named entities (or *terms*) such as words, phrases, and sentences (and sometimes paragraphs). The task in this component is sometimes called *zoning*. All IE engines have a tokenization module.

The *morphological and lexical analysis* module assigns tags to the terms and disambiguates words and phrases. The *semantic analysis* module establishes relationships between terms by shallow and (sometimes) deep parsing. The *domain analysis* module is becoming increasingly popular (Feldman and Sanger, 2007), combining the terms and their relationships and (in advanced systems) providing anaphora resolution.

Discovery in Text Mining

Despite the importance of the evolutionary development of IE engines in the history of modern text mining systems, we can't fully characterize modern text mining solely in terms of their IE functionality. If we can't characterize modern text mining systems as purely IE engines, perhaps we can characterize modern text mining analyses as advanced IE activities that result in the discovery of new information

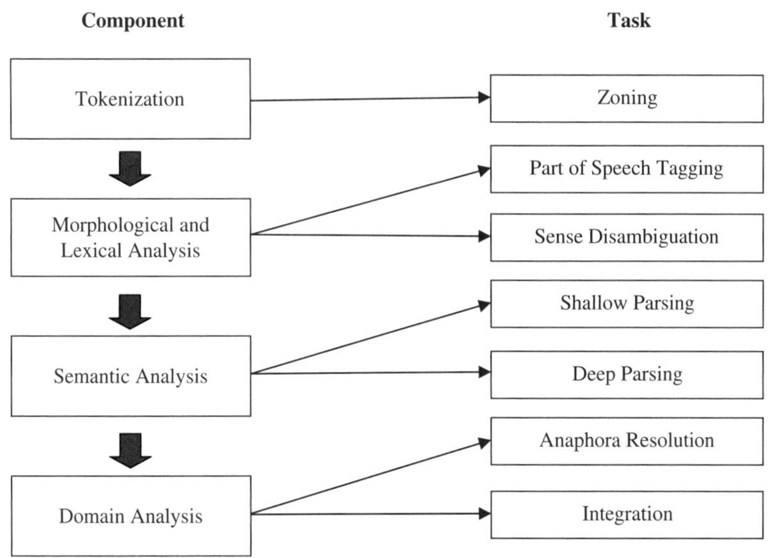

FIGURE 1.6
Major components and tasks in many modern information extraction engines in text mining. *Source: Feldman and Sanger, 2007.*

that is often predictive in nature, rather than just descriptive of the text. Text mining developments of this kind also extend back nearly as far as the MUCs. One of the earliest predictive text mining applications was the adaptation of market basket analysis introduced by Agarwal and colleagues (1993) to define the concept of *frequent sets* (see Figure 1.5). Rajman and Besancon (1998) describe the adaptation of Agarwal's *a priori* market basket algorithm to the analysis of text data. Feldman and Sanger (2007) define a frequent set as a collection of concepts extracted from a collection of documents that have a co-occurrence above a minimum level of *support*. In a natural language application, support is defined as the number or percent of documents associated with a certain rule. Associated with the support value is the *confidence,* which is defined as the percentage of time that the rule is true.

Because of the evolutionary development of IE engines and discovery text mining, it is difficult to pinpoint the exact date for the beginning of the new era of text mining. This difficulty is augmented by the rapid development of IR activities in the direction of text categorization in various areas. An early work by Hayes et al. (1988) presented a system for classifying text in news stories. Another early application of text classification was used to classify text in images using word-length distributions (Hull and Li, 1993). A system for classifying web pages was proposed by Attardi et al. (1999).

In classical data mining circles, one of the seminal papers on discovery text mining was presented by Dörre and colleagues (1999) in the Industrial Track at the KDD-99 conference on Knowledge Discovery and Data Mining. This paper highlighted the use of the new IBM Intelligent Miner for Text toolkit to illustrate the distinction between text mining and traditional data mining. The primary focus of their discussion was on summarization, feature extraction, and the classification of documents (all information retrieval and processing activities in early text mining efforts), but they also introduced the

Topic Categorization tool, which assigned documents to preexisting categories. The function of the tool was to build a contextual model of documents using the preexisting categories as "targets." In this way, the tool functioned very much like a traditional statistical or data mining algorithm. This tool enabled the process of context mining, compared to the purely content mining character of previous textual analyses. Consequently, the year 2000 appears to be a pivotal time in the development of new and powerful text mining capabilities.

The Impact of Domain Knowledge on Text Mining

In addition to the sharp increase in text categorization studies and the evolutionary development of IE engines and discovery text mining, an old issue rose once again to the forefront of attention. In studies of natural language processing in the latter decades of the twentieth century, the notion of including domain (or background) knowledge in processing was hotly contested. In the development of text mining, however, it was warmly embraced. The text mining system FACT (Finding Associations in Collections of Text) developed by Feldman and Hirsh (1997) was one of the first systems to incorporate domain knowledge into discovery operations on a collection of documents. FACT was based on the query language KDTL (knowledge discovery in text language), which influenced later document search systems. Domain knowledge was used by Hotho and colleagues (2003) in the form of *domain ontologies* and *domain lexicons*. An ontology can be viewed as the set of all concepts of interest and the relationships between them in a given domain knowledge base. One example is the gene ontology (GO) knowledge base assembled by Princeton University beginning in 1998. The GO project developed three structured controlled vocabularies that describe various gene products across all of their functions and processes independent of species. The GO ontology enables biologists to query the database with GO tools to discover similar protein-related processes (for example) across many different species and studies. Hotho and colleagues (2003) also described the use of a lexicon to expand the scope of a defined concept in the ontology. These developments were reflected in the addition of a final process in information extraction engines, *domain analysis*, which included anaphora resolution (Feldman and Sanger, 2007).

Machine Learning Applications in Text Categorization

The final element in the development of modern text mining systems was contributed (gradually) by machine learning (ML) technology. Similar to the application of ML techniques to data mining in the 1980s and early 1990s, ML began to be applied to text categorization studies in the early 1990s. Before ML algorithms can be applied successfully to text documents, a set of input features (e.g., terms) must be defined, and all of the documents must be classified into one of a set of categories. Then the classified training set can be submitted to the ML algorithm. After the model is trained, unclassified documents can be submitted to the model for classification, using the patterns of the categories learned in the training operation. The most common ML algorithms used for text categorization are neural networks, decision trees (or decision rules), and support vector machines (SVM). The earliest use of ML algorithms for text categorization was reported by Li and colleagues (1991) for neural networks, Apte and colleagues (1994) for decision trees (actually rules induced from a tree), and Cortes and Vapnik (1995) for SVM.

Bag-of-Words versus High-Dimensional Vector Spaces

One of the most important advances in the early 2000s was the extension of the bag-of-words concept to a higher-dimensional space of "features" defined by nonlinear functions. Cortes and Vapnik (1995) had introduced the concept of applying Statistical Learning Theory to text mining, but this advancement was not developed fully until after the year 2000. Since then, kernel-based learning methods have been applied to various information extraction tasks. For simple data representations (e.g., "bag-of-words") in which features can be easily extracted, some basic kernel functions such as linear kernel, polynomial kernel, and Gaussian kernel are often used.

Issues of High-Dimensional Spaces in Text Categorization with ML Techniques

The total number of terms (words or phrases) used to characterize a document can become huge, creating very high-dimensional spaces for predictive analysis. This is not so much a problem for IR methods, which learn to recognize patterns in text in a manner very similar to that of parametric statistical techniques. That is, they reduce the characteristic inputs for each document to one or more values that represent the entire document (e.g., distance or similarity) analogous to the calculation of the mean and standard deviation. But ML techniques like neural networks and decision trees must process high-dimensional spaces very differently; they learn to recognize patterns case by case, rather than by analysis of the entire database of inputs.

Two challenges must be faced in analyzing data in high-dimensional text spaces. The first challenge is the so-called curse of dimensionality (Bellman, 1957). As the number of variables increases, the complexity of algorithms to analyze them increases exponentially until they become intractable in many real-world applications. The second challenge is that the absolute distance (and, therefore, the ability to discriminate) between the nearest neighbor and the farthest point diminishes as the dimensionality increases (Beyer et al. 1999). The challenge is to assign some metric of difference (or similarity) between points projected into the space.

Kernel-Based Learning Methods

A kernel uses a nonlinear function to "map" text terms (words or phrases) to a higher-dimensional "feature space" (Renders, 2004). Initially, the dot-product was used as a linear function to map terms into feature space. Alternatively, different kernels can be defined by using different nonlinear functions in the mapping process. Tree kernels are defined by dependency or parsing functions. For example, Collins and Duffy (2002) demonstrated how to incorporate decision tree kernels into text mining. A good description of the use of kernel methods in NLP is provided by Renders (2004).

Support Vector Machines

An early ML approach to meet both of these challenges in high-dimensional space was to convert the effects of all variables in a record (or case) into a dot-product (or inner product) equal to the sum of the cross-products of each variable in two adjacent records in a list. This inner product reflected the joint similarity between the two records. In advanced data mining algorithms that function as kernel learning machines (e.g., support vector machines—SVMs), the inner product is replaced by a kernel function that provides an expanded solution space. The dot-product or kernel function also functions as a means

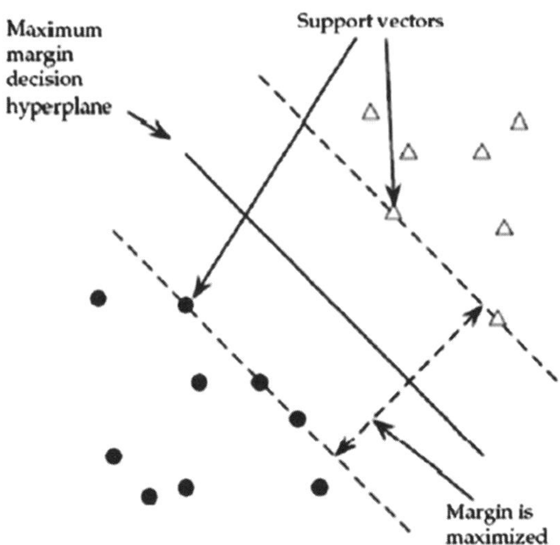

FIGURE 1.7
The maximal separable hyperplane in an SVM. *Source: Vidhya and Aghila, 2010.*

to reduce the dimensionality from many variables to one (for a dot-product), or a very few for each record in an SVM, called *support vectors*. This property renders SVMs relatively immune to the effect of the curse of dimensionality and commends SVMs for high-dimensional problems in text categorization.

SVM operates by finding a hypersurface in the space of possible inputs. The hypersurface attempts to split the positive examples from the negative examples by maximizing the distance between the nearest of the positive and negative examples to the hypersurface. Intuitively, this makes the classification correct for testing data that are near to, but not identical to, the training data.

The dotted lines shown in Figure 1.7 are those vectors of data plotted in the defined hyperspace (support vectors) that separate the two groups of data and represent the greatest "margin" between them. The plane in the middle between the support vectors (the solid line) represents the decision boundary plane (maximum margin plane or maximal separable hyperplane) between the two data clusters. This decision plane is used as a basis for separating the two groups of data into two clusters.

Neural Networks

An artificial neural network (ANN) is a crude simulation of the way human neurons can be "trained" to pass neural impulses in a manner that facilitates some bodily action. The general model of an ANN is based on the basic model of a human neuron proposed by McColloch and Pitts in 1943. This concept was refined through several stages of development until the modern concept of the multilayer perceptron (MLP) was developed in the late 1980s. The McColloch-Pitts model consisted of only two layers, an input layer and an output layer, with some link process in between that enabled some linear classification problems to be solved. The MLP has at least one "hidden" layer between the input and output layers, with link processes between all layers. This multistage linkage permitted nonlinear classification to be performed. Both kinds of ANNs are used in text mining, characterized by their method of learning as *supervised* or *unsupervised.*

Supervised Learning

Supervised neural networks attempt to approximate known values in the data set. The inputs to a neural network for text categorization are expressed as feature values (e.g., word frequencies), and the outputs are the categories or predicted value (e.g., document similarity). The inputs and outputs are connected with link weights assigned initially at random by the processing algorithm. The processing is iterative, where the outputs (e.g., categories) are compared with known values. The error in categorization or prediction is used to adjust the link weights between inputs and outputs for the next iteration. Processing continues until some minimum error value is reached.

Unsupervised Learning

Unsupervised neural networks are used for classification of documents where the categories of a test set are not known. The earliest unsupervised neural networks were proposed by Kohonen (1982) and described comprehensively by Kohonen (1995) as a technique for iteratively partitioning classification space. This process was described as a *self-organizing map* (SOM). As processing progresses, word (or document) distances are evaluated with respect to each other and then assorted into defined regions in the classification space. SOM applications included many industrial and business classification problems. One of the early applications of SOM to text mining was WEBSOM for Internet document exploration in Finland in 1995. Modern implementations of SOM are available in many text mining systems today.

THE DEVELOPMENT OF ENABLING TECHNOLOGY IN TEXT MINING

In addition to algorithms for text mining, other technologies were developed to enable the visualization of text mining results (link analysis), selecting appropriate features (feature selection and reduction of dimensionality), and methods for automatic processing of text information.

Link Analysis

Link analysis is a form of association analysis, in which two or more entities (persons or activities, or things) are associated graphically with the presence of one or more "links" between two entities, and the strength of that link is depicted as lines of different thickness connecting the entities arranges along the circumference of a circle. Figure 1.8 shows the combined association of four people with six organizations that function as channels of communication. On inspection, Figure 1.8 shows that Suspects 2 and 3 are on the same email distribution lists (both are connected with the activity of Email) and are connected to (do business at) the same bank. Suspects 3 and 4 work together at the same company (they are both connected with the activity, Work). Suspects 1 and 2 frequently view the same Internet blog on current affairs in the world (indicated by the very wide line between them and the Blog activity). Suspect 4 also visits the same Internet blog, but rarely (the line connecting him with the blog activity is smaller in width that of Suspects 1 and 2). Suspects 1 and 2 also call the same phone number relatively frequently (indicated by the relatively wide line between them and the Phone activity). Finally, Suspects 2 and 3 are both connected by U.S. citizenship). On interpretation, this link

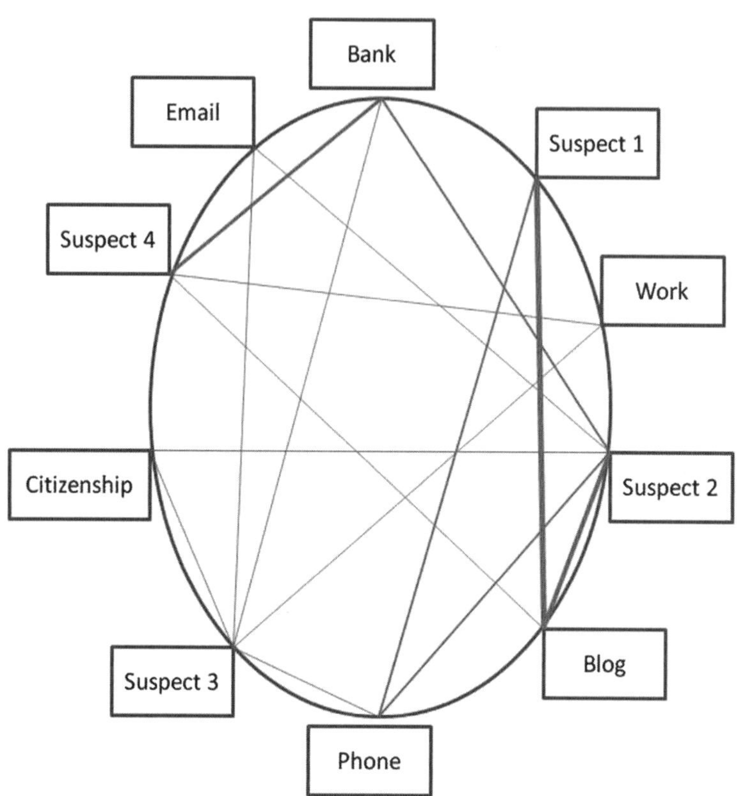

FIGURE 1.8
Link analysis among four suspected terrorists.

analysis suggests that Suspects 1 and 2 are associated with the same channels of communication more frequently than expected and may use the same blog for coded communication. Suspect 2 is a U.S. citizen, but Suspect 1 is not. There is some association of Suspects 1 and 2 with organizations and channels of communications in common with the other two suspects, but these associations appear to be coincidental, particularly when analysis of other information shows that the common phone number linking Suspects 1, 2, and 3 is the Avis rental car number. This link analysis provides investigators with presumptive evidence that Suspect 1 could be operating with Suspect 2 as a handler in a terrorist organization, which communicates often with him in the form of coded blog messages, prompting Suspect 2 to engage in certain banking activities for the sake of the organization.

Thus, link analysis permits us to zero in on a few important associations between members of a group. This operation reduces the "noise" in the body of associations and identifies the best ones to consider further. A similar approach (in principle) is followed by operations that reduce the number of possible predictors of an outcome to those predictors that have the highest likelihood of being important factors of a model. The result of the selection of the important features is a reduction in the number of dimensions in the high-dimensional space that they compose. This reduction in dimensionality increases the likelihood that the modeling algorithm will find the global minimum in the error surface on which it searches for a solution.

Feature Selection and Reduction of Dimensionality

In about 2005, many data mining and text mining tools began to offer special tools that functioned to identify those variables that had a relatively high explanatory value (or high predictive value) in models. Some of these methods are discussed in Chapter 10. Singular value decomposition (SVD) is also used by many systems to reduce the "noise" among several groups of data. SVD is part of the method of principal components analysis, which is used to reduce the number of factors to a smaller number of factor groups (principal components) by specific operations in linear algebra, analogous to finding the least common denominator among a series of divisors in a group of numbers. An excellent intuitive geometric explanation of SVD is found at http://www.ams.org/samplings/feature-column/fcarc-svd.

Another useful technique for linear feature extraction is multidimensional scaling (MDS), which is a form of *ordination*. Ordination is similar to data clustering, in which various objects are ordered on the basis of multiple variables such that similar objects are placed near each other in the order and dissimilar objects are separated in the order. This process maps high-dimensional vectors of variables to a lower dimensional space while attempting to preserve pairwise distances between points. MDS is particularly appropriate for analysis of a collection documents (a *corpus*). When class naming information is available for the document categories, supervised linear feature extraction methods may be used to reduce dimensionality, as well as increase the separation between classes. One classic example is linear discriminant analysis (LDA), which uses a linear transformation of data to minimize within-class scatter and maximize between-class scatter (Radovanovic and Ivanovic, 2008).

Automation

One of the most useful innovations in data mining was the development of automated and semi-automated approaches to search for optimum models. Automatic neural networks and decision trees made data mining much easier and more efficient in the 1990s, and they are featured in packages from SPSS, SAS, Statsoft, and KXEN. This interest in automation was carried over into text mining and has been implemented in many ways, such as the automatic discovery of synonyms (Senellart and Blondel, 2008, in Berry and Castellanos, 2008).

Statistical Approaches

Statistical approaches treat entity extraction as a sequence-labeling problem, with the goal of finding the most likely sequence of entity labels given an input sequence of terms. Statistical approaches have recently been shown to be the more accurate approach to doing entity extraction (Ratinov and Roth, 2009). At the heart of these models is a set of probabilities capturing whether a particular feature of a word corresponds to a particular class label. The most popular statistical approach uses *supervised classification*. Hidden Markov models (HMMs) were the first statistical model of sequences to be applied to entity extraction (Bikel et al., 1997). The assumption behind sequence models like HMMs is that text is originally generated in label-word pairs but that the labels have been lost (or "hidden"). HMMs are a *generative* model—that is, they attempt to recreate the original generating process responsible for creating the label-word pairs.

Sophisticated Methods of Classification

Various machine learning approaches to classification were applied to text mining, such as Naïve Bayes methods, Rocchio Classification, and K-Nearest Neighbor.

Naïve Bayes Classifiers

These are simple probabilistic classifiers that are based on Bayes' theorem, which states that the probability of an event's occurrence is equal to the intrinsic probability (calculated from present available data) times the probability that it will happen again in the future (based on knowledge of its occurrence in the past). This idea is used in Naïve Bayes classifiers only to the extent that the proportion of each class in a data set reflects the "prior" probability that any new object will belong in one class or another. These classifiers make a very simplistic (naïve) assumption that all of the objects to be classified are completely independent of one another in the terms used to characterize them. In spite of its simplistic design and its naïve assumption (which almost never occurs in the real world), these classifiers can be remarkably efficient and accurate, particularly when the number of variables is high.

Rocchio Classification

The Rocchio classification is based on a method of "relevance feedback" that stemmed from the development of the SMART Information Retrieval System in 1970. This system was based on the principle that most people have a general idea about which documents in a corpus are relevant and which are not. Queries for relevant documents are structured to include an arbitrary percentage of relevant and nonrelevant documents as a means of increasing the recall and precision of the search.

The K-Nearest Neighbor Method

This is a relatively simple way to group (cluster) documents together based on their distance to the *K* nearest "neighbor" documents, as measured in terms of the text variables applied to it. K-nearest neighbor is appropriate for a corpus that contains multiple groups of relatively similar documents. The major problems with this algorithm are it is relatively slow to complete the clustering operation, and it weights all of the features (variables) equally, even when some features are much more important than others.

EMERGING APPLICATIONS IN TEXT MINING

Some new and exciting applications in text mining are developing, such as Semantic Mapping and Web search enhancement, which promise have increasingly important effects of the depth of analysis and quality of text mining results.

Semantic Mapping

The concept of using semantic relationships in text mining is not new. Latent semantic indexing (LSI) has been used since 1988 (Deerwester et al., 1988) to correlate semantically related terms that are latent in a collection of text documents. Landauer and colleagues (1998) developed latent semantic analysis (LSA), which exploits the co-occurrence of words in text segments. LSA uses single-value decomposition to reduce the dimensionality of the word through a text-segment matrix. Similarity measurements between word representations allow the inference (mapping) of indirect relationships between words that appear in similar contexts.

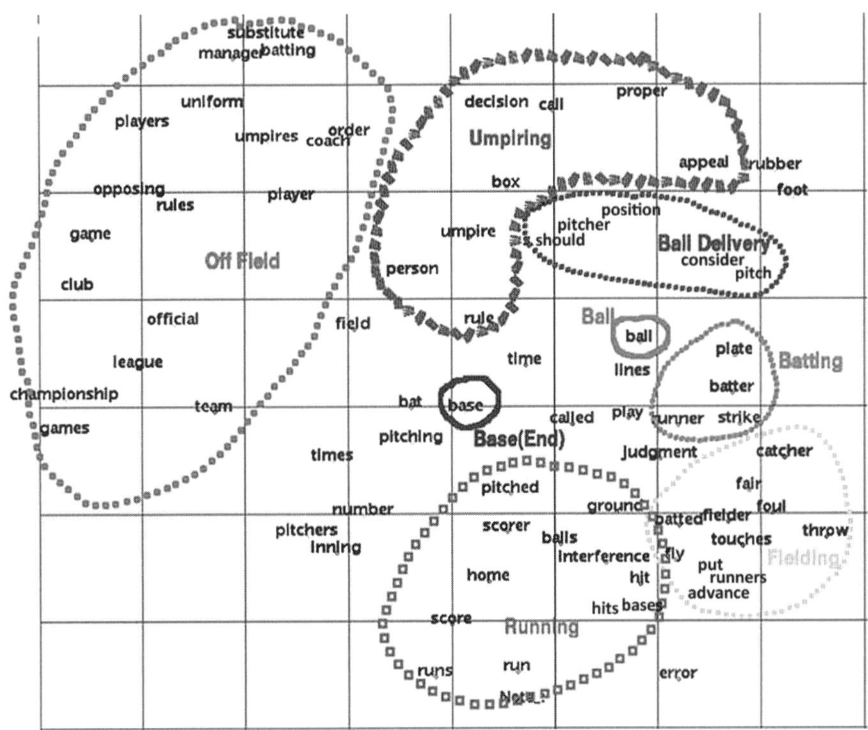

FIGURE 1.9

A semantic map of the rules of baseball. *Source: Smith and Humphreys, 2006.*

These indirect semantic relationships can be shown in visual maps. One such comparison, between the rules of baseball and those of cricket provided by Smith and Humphreys (2006) is shown in Figures 1.9 and 1.10. From this comparison of semantic maps between the two sports, we see that inferences were made to relate the cricket terms *batsman, wicket,* and *hat* in the concept group of batting to the terms associated with batting in baseball: *plate, batter, strike,* and *runner.*

Enhancing Web Searches

A recent development in web searches is the *meta-search engine.* A meta-search engine sends out search requests to other search engines and displays all of the returns. One such meta-search engine, CatS, was used to refine the results of a browsing session by offering an opportunity to browse through a category tree built by the search engine.

SENTIMENT ANALYSIS AND OPINION MINING

Interest in sentiment analysis and opinion mining can be traced back to an early paper by Yorick and Bien (1984), but there has been a relative explosion of interest in this topic since about 2001, including

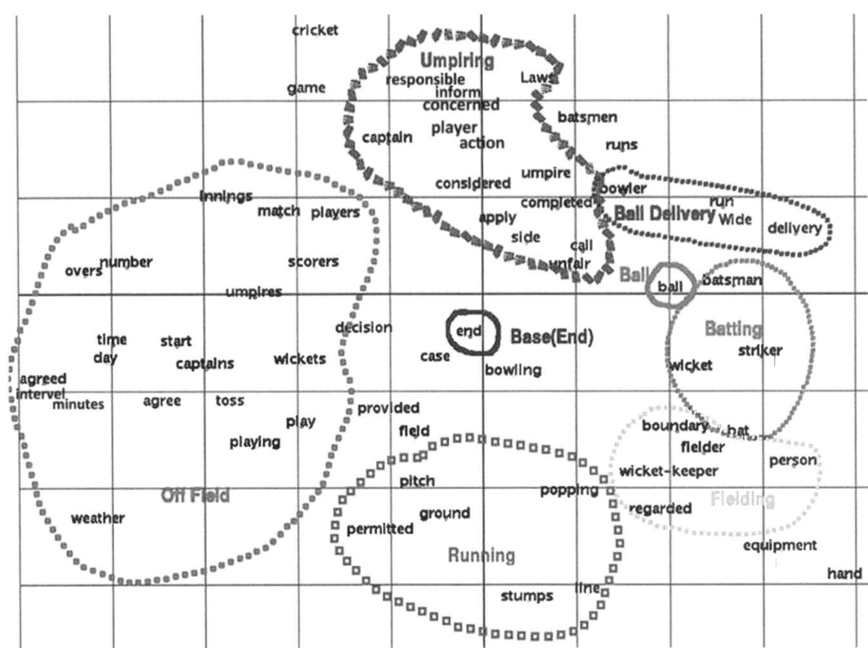

FIGURE 1.10
A semantic map of the rules of cricket. *Source: Smith and Humphreys, 2006.*

Pang and Lee (2008), who cite some factors behind this "land rush" of developments, including the following:

- The rise of machine learning methods in NLP and information retrieval
- The availability of data sets for machine learning algorithms to be trained on, due to the blossoming of the World Wide Web, and specifically the development of review-aggregation websites
- The realization of the fascinating intellectual challenges and commercial intelligence applications that the area offers

Identifying "what other people think" has always been an important piece of information for most of us during our decision-making process. *Consumer Reports* is a favorite source of this information; many people would not dream of making a major purchase without consulting it. Following this motive, many people make their opinions known on one of thousands of blogs on the Internet. Table 1.1 shows the results from two surveys of more than 2000 American adults.

Twitter is one of the most popular social networking services on the Internet, permitting conversations consisting of individual "tweets" among groups of people. Twitter recognized the power of its historical data to permit the study of the timeline of change in inferred characteristics of their data, such as "mood." Figure 1.11 shows the output graph of the online Twitter Sentiment tool.

Table 1.1 Percentage of Americans Who Have Looked for Reviews Online

Online Activity Related to Sentiment or Opinion	%
Have done online research on a product at least once	60
Americans who do online research on a typical day	15
Online reviews that had a significant influence on purchase	73 to 87
Additional money willing to be spent for a 5-star vs. a 4-star establishment	20 to 99
Have provided an online rating at least once	32
Have provided a comment or review of a product or service	30

Source: Adamic and Glance, 2005.

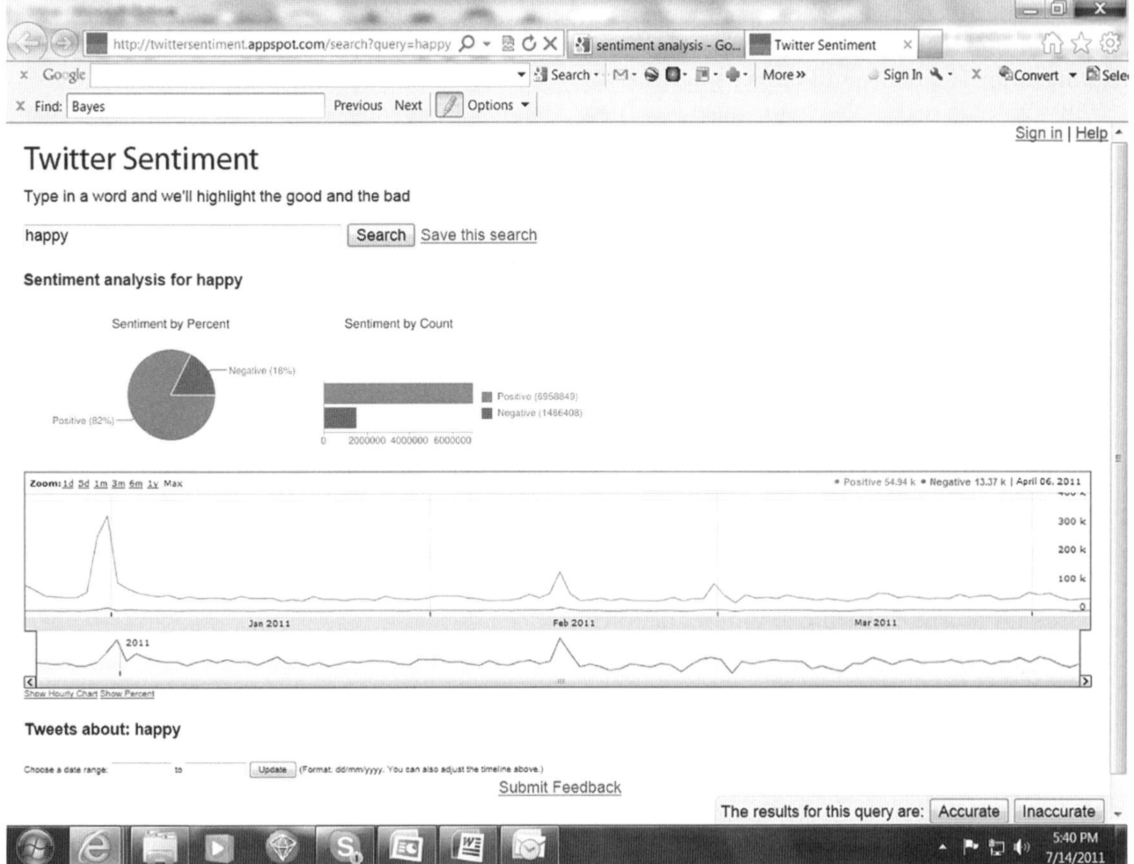

FIGURE 1.11
Time-series report of the mood "happy" on Twitter Sentiment. *Source: http://twittersentiment.appspot.com/search?query=happy*

IBM'S WATSON: AN "INTELLIGENT" TEXT MINING MACHINE?

Beginning in 2007, IBM developed the world's most advanced question answering machine that was able to understand a question posed in everyday human language expression (natural language) and respond with a precise, factual answer. Watson is a form of a text mining machine, dedicated to answering spoken or written questions. This machine can process the language of the questioner, understand the meaning of the questions, and produce an answer in terms of words and concepts stored in its memory. Watson is the near embodiment of the computer "Hal" in Arthur C. Clarke's film *2001—A Space Odyssey*. The name "Hal" was formed by selecting one letter beyond IBM in each position. Watson is a major step in that direction. Some feared that this advance would lead to the development of the "Skynet System" of the *Terminator* film series that would become self-aware and decide to eliminate humanity.

The Test

To test Watson, the producers of the game show *Jeopardy!* agreed to pit Watson against some of the show's best former players. During a three-day game that pitted all-stars Ken Jennings and Brad Rutter against an IBM supercomputer, the machine beat the men every time. Watson finished with $77,147, Jennings with $24,000, and Rutter with $21,600.

IBM is not in the game-playing business. They did build "Deep Blue" to compete in chess with Gary Kasparov (Deep Blue won), and Watson does carry "deep" thinking forward to "deep" response to questions. But there are many business reasons to justify why IBM created both of these machines. They believe that Watson (or its successors) will have a major impact on how they do business in health care, finance, and customer service. The ability to respond to customers' questions with "intelligent" answers can be leveraged to create significant profits for the company.

WHAT'S NEXT?

We expect this rather rapidly developing trend of interest in the application of sentiment and opinion to continue at least into the near future. Many other new trends in text mining are discussed in detail by Berry and Kogan (2010):

- Social network analysis
- Multilingual text mining
- Spam classification
- Use of K-means clustering to group documents
- Anomaly detection
- Trend detection
- Analysis of streaming text data

POSTSCRIPT

Now that you have been introduced to the depth and breadth of text mining history and applications, we can expose you to enough of the theory and practice of it to prepare you to do some tutorials. As with our previous book (*Handbook of Statistical Analysis & Data Mining Applications*), the underlying theme of this book is to conduct you up the learning curve as quickly as possible to enable you to do it yourself. With that goal in mind, let's get to it!

References

Adamic, L. A., and Glance, N. (2005). The political blogosphere and the 2004 U.S. election: Divided they blog. In Proceedings of LinkKDD, pp. 36–43.

Agarwal, R., Bayardo, R. J., and Srikant, R. (1993). Mining association rules between sets of items in large databases. In Proc. ACM SIGMOD Conf. on Mgt. of Data, Washingdon, D.C.: 207–216. ACM Press, New York.

Apte, C., Damerau, F. J., and Weiss, S. M. (1994). Automated Learning of Decision Rules for Text Categorization. *ACM Transactions on Information Systems*, **12**(3): 233–251.

Attardi, G., Gulli, A., and Sebastiani, F. Automatic Web Page Categorization by Link and context Analysis, *European Symposium on Telematics, Hypermedia and Artificial intelligence*, Varese, 1999.

Bates, M. (1995). Models of Natural Language Understanding. *Proc. Natl. Acad. Sci.* USA 92:9977–9982.

Bellman, R. E. (1957). Dynamic Programming. Princeton University Press, Princeton, NJ. Republished 2003: Dover, ISBN 0486428095.

Berry, M. W., and Castellanos, M., eds (2008). Survey of Text Mining II: Clustering, Classification, and Retrieval.

Berry, M. W., and Kogan, J., eds. (2010). Text Mining: Applications and Theory edited by Michael W. Berry and Jacob Kogan © 2010, John Wiley & Sons.

Beyer, K. S., Goldstein, J., Ramakrishnan, R., & Shaft, U. (1999). When is "nearest neighbor" meaningful? Proc. Int. Conf. on Database Theory (pp. 217–235).

Bickel, P., Götze, F., and van Zwet, W. (1997). Resampling fewer than n observations: gains, losses and remedies for losses. *Statist. Sinica*, 7, 1–31.

Chomsky, N. (1959). Bad News: Noam Chomsky—A Review of B. F. Skinner's Verbal Behavior. *Language*, 35(1):26–58.

Collins, M., and Duffy, N. (2002). New Ranking Algorithms for parsing and Tagging over Discrete Structure, and the voted perceptron. In *Proceedings of ACL-02*, pp. 263–270. Philadelphia.

Collins, M. (1997). Three generative, lexicalized models for statistical parsing. In Proc. 35th Ann. Mtg. Assoc. Comp. Ling., Madrid: 16–23, ACM Press, NY.

Cooley, R. (2000). Web Usage Mining: Discovery and Application of Interesting Patterns from Web Data. PhD thesis, University of Minnesota.

Cooley, R., Srivastava, J., and Mobasher, B. (1997). Web mining: Information and pattern discovery on the world wide web. In Proceedings of the 9th IEEE International Conference on Tools with Artificial Intelligence (ICTAI'97).

Cortes, C., and Vapnik V. M. (1995). "Support Vector Networks," *Machine Learning*, Vol. 20, pp. 273–297.

Cutting, D. R., Karger, D. R., Pederson, J. O., and Tukey, J. W. (1992). Scatter/Gather: a cluster-based approach in browsing large document collections. In Proc. 15th Ann. Intern. ACM-SIGIR Conf. Res. And Dev. in Inf. Retrieval (Copenhagen): 318–329. ACM Press, NY.

Dale, R., Moisi, H., and Somers, H. (2000). Handbook of Natural Language Processing. CRC Press, New York, NY.

Deerwester, S., Dumais, S. T., Landauer, T. K., Fumas, G. W. and Beck, L. (1988) Improving Information Retrieval with Latent Semantic Indexing, Proceedings of the 51st Annual Meeting of the American Society for Information Science, 25:36–40.

Diligenti, M., Coetzee, F., Lawrence, S., Giles, C. L., and Gori, M. (2000). Focused crawling using context graphs. In Proceedings of the 26th International Conference on Very Large Databases (VLDB), pp. 527–534, Cairo, Egypt.

Dörre, J., Gerstl, P., and Seiffert, R. (1999). Text Mining: Finding Nuggets in Mountains of Textual Data. In Proceedings of KDD-99, 5th ACM International Conference on Knowledge Discovery and Data Mining. San Diego, ACM Press, New York: 398–401.

Doyle, L. B. (1961). Semantic road maps for literature searchers, *Journal of ACM*, 8:553–578.

Feldman, R., and Hirsh, H. (1997). Finding associations in collections of text. In Machine learning and data mining: method and applications. Michalski, R. S., Bratko, I., and Kubat, M. eds. John Wiley & Sons, New York, NY. pp. 223–240.

Gallager, R. G. (2001). Claude E. Shannon: A Retrospective on his life, work and impact. *IEEE Trans. on Info. Theory*, 47(7):2681–2695.

Grishman, R., and Sundheim, B. (1996). Message Understanding Conference – 6: A brief history. In Proceedings of the 16th International Conference on Computational Linguistics: 466–471. Center for Sprogteknologi, Copenhagen, Denmark. Available at http://acl.ldc.upenn.edu/C/C96/C96-1079.pdf

Hayes, P. J., Knecht, L. E., and Cellio, M. J. (1988). A News Story Categorization System. In Proceedings of ANLP-88, 2nd Conference on Applied Natural Language Processing. Austin, TX, Association for Computational Linguistics, Morristown, NJ: 9–17.

Hearst, M. A. (1999). Untangling text mining. Proc. Annual Meeting of the Association for Computational Linguistics ACL99. University of Maryland, June.

Hotho, A., Staab, S., and Stumme, G. (2003). Text Clustering Based on Background Knowledge. Institute of Applied Informatics and Formal Descriptive Methods, University of Karlsruhe, Germany: 1–35.

Hull, J. J., and Li, Y. (1993). Word Recognition and Result Interpretation Using the Vector Space Model for Information Retrieval. 2nd Annual Symposium of the IEEEComputer Society on Document Analysis and Information Retrieval. Las Vegas, NV. pp. 147–155.

Kohonen, T. (1982). Self-organized formation of topologically correct feature maps. *Biological Cybernetics*, 43:59–69.

Kohonen, T. (1995). Self-Organizing Maps. Springer, Berlin.

Landauer, T. K., Foltz P. W., and Laham, D. (1998). An introduction to latent semantic analysis. *Discourse Processes* 25 (1998), pp. 259–284.

Li, W., Lee, B., Krausz, F., and Sahin, K. (1991). Text Classification by a Neural Network. In Proceedings of the 23rd Annual Summer Computer Simulation Conference. D. Pace, ed. Baltimore, Society for Computer Simulation, San Diego, CA: 313–318.

Lochbaum, K. E., Grosz, B. J., & Sidner, C. L. (2000). Discourse structure and intention recognition. In Dale, R., Moisl, H., & Somers, H. (Eds.), Handbook of Natural Language Processing, Marcel Dekker, Inc., New York, pp. 123–146.

Luhn, H. (1958). The automatic creation of literature abstracts. *IBM Journal of Research and Development*, 2(2), 159–165.

McCulloch, W. S., and Pitts, W. (1943). A logical calculus of the ideas immanent in nervous activity. *Bulletin of Mathematical Biophysics*, 5:115–133.

Mena, J. (1999). *Data Mining Your Website*. Digital Press, Boston, MA.

Menczer, F. (1997). ARACHNID: Adaptive Retrieval Agents Choosing Heuristic Neighborhoods for Information Discovery. In D. Fisher, ed., Proceedings of the 14th International Conference on Machine Learning (ICML97). Morgan Kaufmann.

Menczer, F., and Belew, R.K. (1998). Adaptive Information Agents in Distributed Textual Environments. In K. Sycara and M. Wooldridge (eds.) Proceedings of the 2nd International Conference on Autonomous Agents (Agents '98). ACM Press.

Menczer, F., Pant, G., and Srinivasan, P. (2004). Topical Web Crawlers: Evaluating Adaptive Algorithms. *ACM Trans. on Internet Technology* 4(4): 378–419.

Nisbet, R., Elder, J., and Miner, G. *Handbook of Statistical Analysis & Data Mining Applications*. 2009. Academic Press, Chicago, IL. p. 824.

Pang, B., and Lee, L. (2008). Opinion mining and sentiment analysis. Now Publ. Inc. ISBN 1601981503.

Pinkerton, B. (1994). Finding what people want: Experiences with the WebCrawler. *In* Proceedings of the First World Wide Web Conference, Geneva, Switzerland.

Radovanovic, M., and Ivanovic, M. (2008). Text Mining: Approaches and Applications. *Novi. Sad. J. Math.* 38:227–234.

Rajman, M., and Besancon, R. (1998). Text Mining – Knowledge Extraction from Unstructured Textual Data. In Proceedings of the 6th Conference of the International Federation of Classification Societies. Rome: 473–480.

Ratinov, L., and Roth, D. (2009). Design Challenges and Misconceptions in Named Entity Recognition. In Proceedings of the Annual Conference on Computational Natural Language Learning (CoNLL).

Renders, J. (2004). Kernel Methods for Natural Language ProcessingI Learning Methods for Text Understanding and Mining, 26–29 January 2004, Grenoble, France.

Salton, G., Buckley, A. C., and Singhai, A. (1994). Automatic analysis, theme generation and summarization of machine-readable texts. *Science, 264*, 3(June 1994), 1421–1426.

Sanger, J., and Feldman, R. (2007). The Text Mining Handbook. Cambridge Univ. Press, New York, NY. p. 410.

Shannon, C. E. (1948). A mathematical theory of communication. *Bell Systems Tech. J.* 27:379–423.

Skinner, B. F. (1957). Verbal Behavior, Acton, Massachusetts: Copley Publishing Group.

Senellart, P., and Blondel, V. D. (2008). Automatic discovery of similar words. In Berry, M. W., and Castellanos, M., eds. Survey of Text Mining II. Springer-Verlag, London. p. 238.

Smith, A., & Humphreys, M. (2006). Evaluation of unsupervised semantic mapping of natural language with Leximancer concept mapping. *Behavior Research Methods, 38*(2), 262.

Srivastava, J., and Mobasher, B. (1997). Web Mining: Hype or Reality? 9th IEEE International Conference on ToolsWith Artificial Intelligence (ICTAI '97).

Tomasello, M. (2008). Origins of human communication. MIT Press.

Tombros Villa, R., and Rijsbergen, C. J. (2002). The effectiveness of query-specific hierarchic clustering in information retrieval. *Information Processing & Mgt.* 38(4):559–582.

Vidhya, K. A., and Aghila, G. (2010). Text Mining Process, Techniques and Tools: an Overview International Journal of Information Technology and Knowledge Management 2(2):613–622.

Yorick Wilks and Bien, Janusz. (1984). Beliefs, points of view and multiple environments. In Proceedings of the international NATO symposium on artificial and human intelligence, pp. 147–171, New York, NY, Elsevier North-Holland, Inc.

CHAPTER 2

The Seven Practice Areas of Text Analytics

CONTENTS

PREAMBLE

Presently, text mining is in a loosely organized set of competing technologies that function as analytical "city-states" with no clear dominance among them. To further complicate matters, different areas of text mining are in different stages of maturity. Some technology is easily accessible by practitioners today via commercial software (some of which is included with this book), while other areas are only now emerging from academia into practical use.

We can relate these technologies to seven different *practice areas* in text mining that are covered in the chapters in this book. In summary, this book is strongest in the practice area of document classification, solid in concept extraction and document clustering, reasonably useful on web mining, light on information extraction and natural language processing, and almost silent on the (most popular) practice area of search and information retrieval.

The unifying theme behind each of these technologies is the need to "turn text into numbers" so that powerful analytical algorithms can be applied to large document databases. Converting text into

Practical Text Mining and Statistical Analysis for Non-structured Text Data Applications. DOI: 10.1016/B978-0-12-386979-1.00002-5

a structured, numerical format and applying analytical algorithms both require knowing how to use and combine techniques for handling text, ranging from individual words to documents to entire document databases.

Next, we provide a decision tree to help you determine which practice area is appropriate to satisfy your needs. Finally, we provide tables to relate the practice areas to appropriate technologies and show which chapter in this book deals with that subject area. That is the most organization that we can impose on the current disordered state of text mining technology. Our goal in this book is to provide an introduction to each of the seven practice areas and cover in depth only those areas that are *accessible for nonexperts*. We will follow that theme in Part I of the book to provide you with the basics you need to perform the tutorials. Very quickly, you will be learning by doing.

WHAT IS TEXT MINING?

Text mining and *text analytics* are broad umbrella terms describing a range of technologies for analyzing and processing semistructured and unstructured text data. The unifying theme behind each of these technologies is the need to "turn text into numbers" so powerful algorithms can be applied to large document databases. Converting text into a structured, numerical format and applying analytical algorithms require knowing how to both use and combine techniques for handling text, ranging from individual words to documents to entire document databases.

To date, text mining has resisted a more comprehensive definition because the field is emerging out of a group of related but distinct disciplines, as described in Chapter 1. Figure 2.1 shows the six other major fields that intersect with text mining. Due to the breadth and disparity of the contributing disciplines, it can be difficult even for text mining experts to concisely characterize. Text mining is something of the "Wild West" of analytics, since there are a number of competing technologies with no clear dominance among them (but much braggadocio). To further complicate matters, different areas of text mining are in different stages of maturity.

Our goal in this chapter is to bring clarity to the field by providing a framework and vocabulary for discussing the seven different *practice areas* within text mining. Due to the breadth of text mining, no single book can hope to fully cover the field. Our target audience is nonspecialist text-mining practitioners—analysts who have the technical expertise to handle challenges involving text but have limited experience or background with text processing. Consequently, this book provides an introduction to each of the seven practice areas, but it covers in depth only those areas that are *accessible for nonexperts*, yet not ubiquitous. We reference other resources in areas that are less mature or require additional expertise or, in the case of search technology, are already very useable in their current incarnations.

There are seven different text mining practice areas—that is, seven very different things that a client, speaker, boss, or colleague could have in mind when talking about text mining. The seven practice areas are defined in Figure 2.1. This book is strongest in the practice area of *document classification*, solid in *concept extraction* and *document clustering*, reasonably useful on *web mining*, light on *information extraction* and *natural language processing*, and almost silent on the (most popular) practice area of *search and information retrieval*.

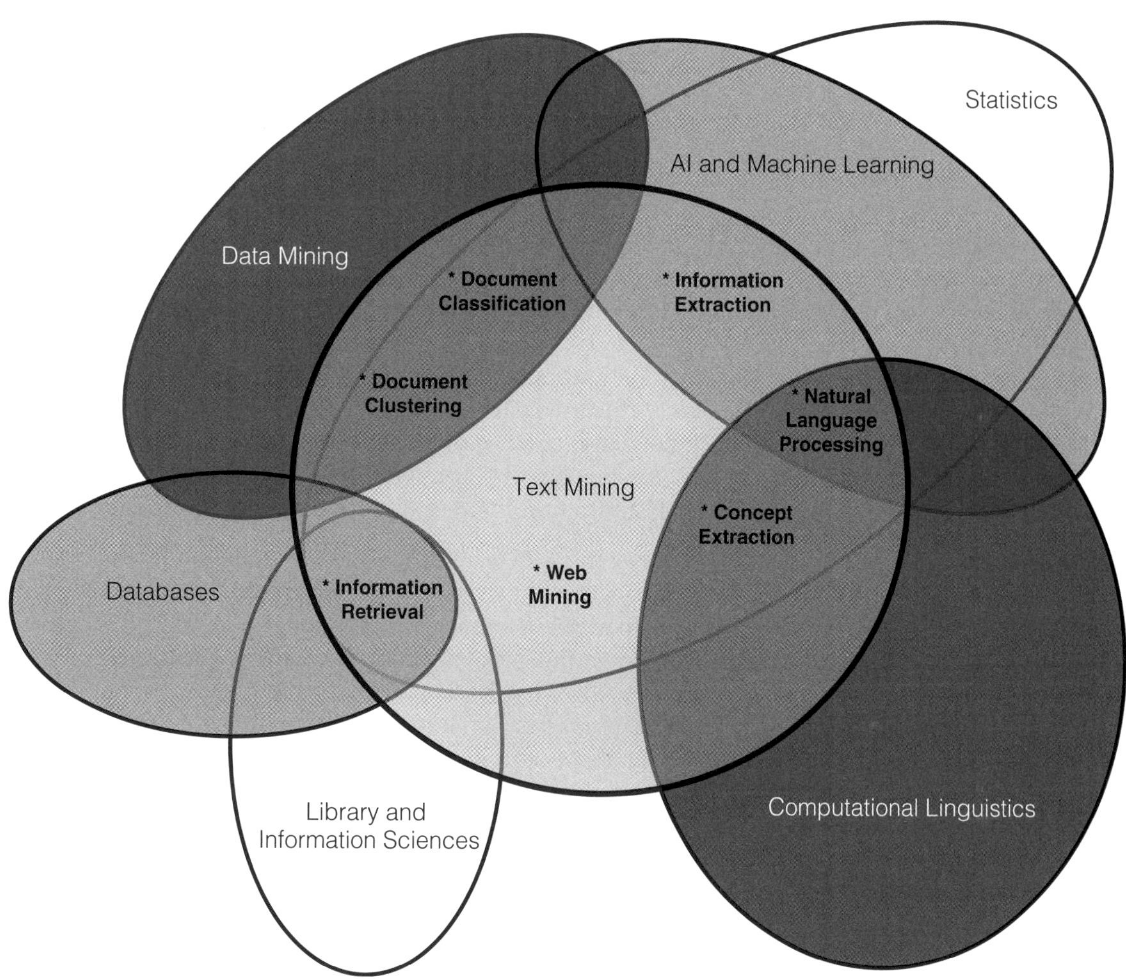

FIGURE 2.1
A Venn diagram of the intersection of text mining and six related fields (shown as ovals), such as data mining, statistics, and computational linguistics. The seven text mining practice areas exist at the major intersections of text mining with its six related fields.

THE SEVEN PRACTICE AREAS OF TEXT ANALYTICS

Text mining can be divided into seven practice areas, based on the unique characteristics of each area. Though distinct, these areas are highly interrelated; a typical text mining project will require techniques from multiple areas. This book views text mining through the eyes of practitioners. Instead of emphasizing the academic or technical differentiators between the practice areas, our focus is on guiding readers toward answers to the problem they are facing. We have inductively identified seven practice areas based on five resource and goal questions that text mining practitioners must answer when facing a new problem. The five questions will be defined soon; meanwhile, the seven practice areas are as follows:

1. **Search and information retrieval (IR):** Storage and retrieval of text documents, including search engines and keyword search.
2. **Document clustering:** Grouping and categorizing terms, snippets, paragraphs, or documents, using data mining clustering methods.
3. **Document classification:** Grouping and categorizing snippets, paragraphs, or documents, using data mining classification methods, based on models trained on labeled examples.
4. **Web mining:** Data and text mining on the Internet, with a specific focus on the scale and interconnectedness of the web.
5. **Information extraction (IE):** Identification and extraction of relevant facts and relationships from unstructured text; the process of making structured data from unstructured and semistructured text.
6. **Natural language processing (NLP):** Low-level language processing and understanding tasks (e.g., tagging part of speech); often used synonymously with computational linguistics.
7. **Concept extraction:** Grouping of words and phrases into semantically similar groups.

These seven practice areas exist at the key intersections of text mining and the six major other fields that contribute to it. Figure 2.1 depicts, as a Venn diagram, the overlap of the seven fields of text mining, data mining, statistics, artificial intelligence and machine learning, computational linguistics, library and information sciences, and databases; it also locates the seven practice areas at their key intersections. For example, the practice area of text classification (the most thoroughly covered in this book) draws from the field of data mining, and the practice area of information retrieval (most popular, but least covered in this book) draws from the two fields of databases and library and information sciences. Tables 2.2 and 2.3 provide alternative methods for identifying the practice areas based on algorithms and desired products.

FIVE QUESTIONS FOR FINDING THE RIGHT PRACTICE AREA

Figure 2.2 is a decision tree depicting how answering a few straightforward questions can direct you to the appropriate text mining solution. Five questions—only two to four of which need to be answered, depending on your problem—best split the major branches of text mining. They identify the seven practice areas, which are depicted as the leaf nodes of the tree highlighted in blue in Figure 2.2. Rarely will a single pass through the tree solve any text mining problem. A text mining solution usually consists of multiple passes through the data at different levels of processing—starting with raw input documents and moving toward fully encoded text. At each step, a group of questions must be answered to determine the appropriate processing task. These questions are detailed in the following sections. In addition, Table 2.1 lists typical desired outcomes for text mining algorithms and their corresponding practice areas.

Question 1: Granularity

This question finds the desired granularity (level of detail of focus) of the text mining task. While documents and words are both integral to successful text mining, an algorithm virtually always emphasizes one or the other. Note that in this book we use the term *document* to describe the unit of text

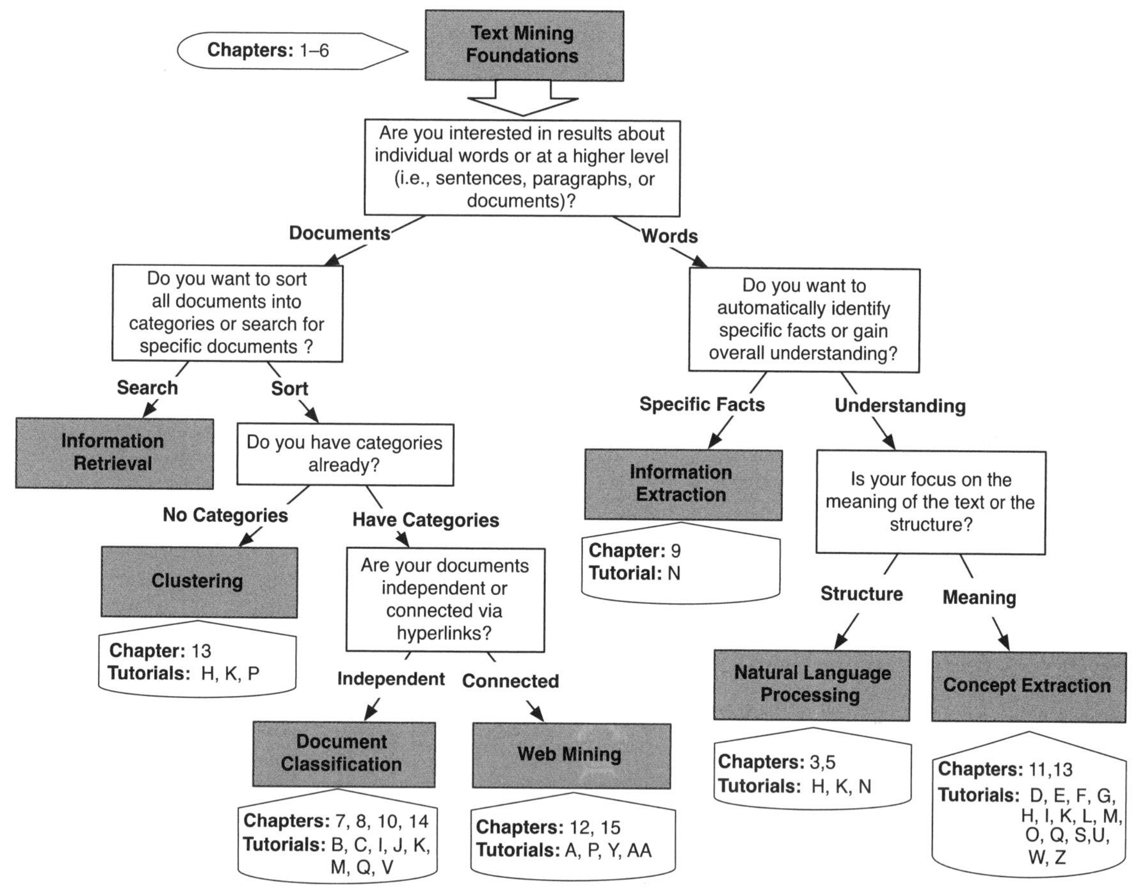

FIGURE 2.2
A decision tree for finding the right text mining practice area by answering 2 to 4 questions about your text resources and project goals.

under analysis. This is a broader definition than is usually employed. In practice, this could be mean typical documents, paragraphs, sentences, "tweets" on social media, or other defined sections of text.

To determine the granularity of your text mining problem, ask yourself about the desired outcome: Is it about characterizing or grouping together words or documents? This is the biggest division between classes of text mining algorithms.

Question 2: Focus

Whether you are interested in document or words, the next question in the decision tree of Figure 2.2 regards the focus of the algorithm: Are you interested in finding specific words and documents or characterizing the entire set? The two practice areas separated by this question—search and information extraction—both concentrate on identifying specific pieces of information within a document database, whereas the other solutions attempt to cluster or partition the space.

Table 2.1 Text Mining Topics and Related Practice Areas

Topic	Practice Area (Number)
Keyword search	Search and information retrieval (1)
Inverted index	Search and information retrieval (1)
Document clustering	Document Clustering (2)
Document similarity	Document Clustering (2)
Feature selection	Document classification (3)
Sentiment analysis	Document classification (3); Web mining (4)
Dimensionality reduction	Document classification (3)
eDiscovery	Document classification (3)
Web crawling	Web mining (4)
Link analytics	Web mining (4)
Entity extraction	Information extraction (5)
Link extraction	Information extraction (5)
Part of speech tagging	Natural language processing (6)
Tokenization	Natural language processing (6)
Question answering	Natural language processing (6), Search and information retrieval (1)
Topic modeling	Concept extraction (7)
Synonym identification	Concept extraction (7)

Table 2.2 Common Text Mining Algorithms and the Corresponding Practice Area

Algorithm	Area	Chapters	Tutorials
Naïve Bayes	Document classification	7, 15	F, X, Z
Conditional random fields	Information extraction	9	
Hidden Markov models	Information extraction	9	
k-means	Clustering	8, 13	F, H, L, O
Singular value decomposition (SVD)	Document classification, clustering	8, 10	K, L, O, Y
Logistic regression	Document classification	7, 8	Q
Decision trees	Document classification	7, 8	B, J, K
Neural network	Document classification	8	I
Support vector machines	Document classification	7	R, Z
MARSplines	Document classification		X, Y
Link analysis	Concept extraction	8	See*
k-nearest neighbors	Document classification	8	X, Z
Word clustering	Concept extraction	8, 13	D, E, G, M, P, Q, U
Regression	Classification		A

**See Tutorial Y in* Handbook of Statistical Analysis and Data Mining Applications, *by Nisbet, Elder, and Miner.*

Table 2.3 Finding a Practice Area Based on the Desired Product of Text Mining

Desired Product	Practice Area
Linguistic structure	Natural language processing
Topic/category assignment	Document classification
Documents that match keywords	Information retrieval
A structured database	Information extraction
"Needles in a haystack"	Document classification
List of synonyms	Concept extraction
Marked sentences	Natural language processing
Understanding of microblogs	Web mining
Similar documents	Clustering

Question 3: Available Information

If you are interested in documents, the next question regards the available information at the time of analysis. This is equivalent to the supervised/unsupervised question from data mining. A supervised algorithm requires training data with an answer (outcome label) for positive and negative examples of the classes you're trying to model (such as distinguishing "interesting versus not interesting" articles for an analyst studying a specialized topic). An unsupervised algorithm does not require any labeled data, and it can be applied to any data set without any available information at analysis time. Supervised learning is much more powerful when possible to use—that is, when enough example cases with target outcomes are known.

Question 4: Syntax or Semantics

If you are interested in words, the major question is about syntax or semantics. Syntax is about what the words "say," while semantics is about what the words "mean." Because natural language is so fluid and complex, semantics is the harder problem. However, there are text mining algorithms to address both areas.

Question 5: Web or Traditional Text

The rise of the Internet (including blogs, Twitter, and Facebook) is largely responsible for the prominence that text mining holds today by making available a vast number of previously unreachable text documents. The structure and style of web documents provide both unique opportunities and challenges when compared to nonweb documents. Though many of the algorithms are theoretically the same for web and traditional text, the scale of the web and its unique structural characteristics justify defining two different categories.

THE SEVEN PRACTICE AREAS IN DEPTH

We have categorized text mining into seven subdisciplines, based on the answers to the preceding questions:

1. Search and information retrieval
2. Document clustering

3. Document classification
4. Web mining
5. Information extraction
6. Natural language processing
7. Concept extraction

The following are brief descriptions of the problems faced in each practice area, a guide to the resources available in this book, and references to other resources if you wish to delve deeper into any of the areas.

Search and Information Retrieval

Search and information retrieval covers indexing, searching, and retrieving documents from large text databases with keyword queries. With the rise of powerful Internet search engines, including Google, Yahoo!, and Bing, search and information retrieval has become familiar to most people. Nearly every computer application from email to word processing includes a search function. Because search is so familiar and available to the practitioner, we have not covered it in this book. Instead, Table 2.4 lists three resources that you might find helpful in the area of search and information retrieval.

Document Clustering

Document clustering uses algorithms from data mining to group similar documents into clusters. Data mining has been a very active field for nearly two decades, and clustering algorithms preceded that, so clustering algorithms are widely available in many commercial data and text mining software packages. We explore document clustering in Chapter 13 and in tutorials G, H, K, P, and X.

For more background information on clustering, see our handbook on data mining: see *Handbook of Statistical Analysis and Data Mining Applications* by R. Nisbet, J. Elder, and G. Miner.

Document Classification

Document classification assigns a known set of labels to untagged documents, using a model of text learned from documents with known labels. Like document clustering, document classification draws from an enormous field of work in data mining, statistics, and machine learning. It is one of the most

Table 2.4 Additional Resources on Search and Information Retrieval

Resource	Emphasis
Search Engines: Information Retrieval in Practice, by Bruce Croft, Donald Metzler, and Trevor Strohman	Emphasis on the practical aspects of building a search engine, including an example search engine. Also includes an overview of the theory and technology behind search engines.
Introduction to Information Retrieval, by Christopher D. Manning, Prabhakar Raghavan, and Hinrich Schütze	Comprehensive coverage of information retrieval, with more of an emphasis on the theory and mathematical origins of the field.
Solr 1.4: Enterprise Search Server, by David Smiley and Eric Pugh	Solr is a widely used open source search engine package from the Apache Software Foundation. This book thoroughly covers implementing Solr.

prominent techniques used in text mining and is a major emphasis of this book. Document classification and related techniques are discussed in Chapters 7, 8, 10, and 14 and in tutorials B, C, G, H, I, J, K, M, P, Q, and X.

For more background information on the theory and practice of classification, see *Handbook of Statistical Analysis and Data Mining Applications*, by R. Nisbet, J. Elder, and G. Miner.

Web Mining

Web mining is its own practice area due to the unique structure and enormous volume of data appearing on the web. Web documents are typically presented in a structured text format with hyperlinks between pages. These differences from standard text present a few challenges and many opportunities. As the Internet becomes even more ingrained in our popular culture with the rise of Facebook, Twitter, and other social media channels, web mining will continue to increase in value. Though it is still an emerging area, web mining draws on mature technology in document classification and natural language understanding. Web mining is covered in Chapters 12 and 15 and in tutorials A, P, Y, and AA.

For more details about web mining, see *Mining the Web: Analysis of Hypertext and Semi Structured Data*, by Soumen Chakrabarti.

Information Extraction

The goal of information extraction is to construct (or *extract*) structured data from unstructured text. Information extraction is one of the more mature fields within text mining, but it is difficult for beginners to work in without considerable effort, since it requires specialized algorithms and software. Furthermore, the training and tuning of an information extraction system require a large amount of effort. There are a number of commercial products available for information extraction, but all of them require some customization to achieve high performance for a given document database. Information extraction is covered in Chapter 9 and in tutorial N.

For more information, see the proceedings of the Message Understanding Conferences (MUC).[1] The MUC were sponsored by the Defense Advanced Research Projects Administration (DARPA) for the express purpose of evaluating different systems on an information extraction task. They provide the earliest summary of the field. More recently, the Conference on Natural Language Learning (CoNLL)[2] has included a shared task for evaluating information extraction approaches in many languages.

Natural Language Processing

Natural language processing (NLP) has a relatively long history in both linguistics and computer science. Recently, the focus of NLP has moved further into the text mining realm by considering statistical approaches. NLP is a powerful tool for providing useful input variables for text mining such as part of speech tags and phrase boundaries. A few areas of NLP are discussed in Chapters 3 and 5 and in tutorials H, K, and N.

---

[1] http://www-nlpir.nist.gov/related_projects/muc/proceedings/muc_7_toc.html

[2] http://ifarm.nl/signll/conll/

For more thorough coverage, we heartily recommend *Foundations of Statistical Natural Language Processing*, by Chris Manning and Hinrich Schütze. This superb book is for both novice and expert readers.

Concept Extraction

Extracting concepts is, in some ways, both the easiest and the hardest of the practice areas to do. The meaning of text is notoriously hard for automated systems to "understand." However, some initial automated work combined with human understanding can lead to significant improvements over the performance of either a machine or a human alone. These techniques are discussed in Chapters 11 and 13 (on clustering) and tutorials D, E, F, G, H, I, K, L, M, O, Q, S, U, W, and Z.

INTERACTIONS BETWEEN THE PRACTICE AREAS

The seven practice areas overlap considerably, since many practical text mining tasks sit at the intersection of multiple practice areas. A visualization of this overlap between practice areas is shown as a Venn diagram in Figure 2.3. For example, entity extraction draws from the practice areas of

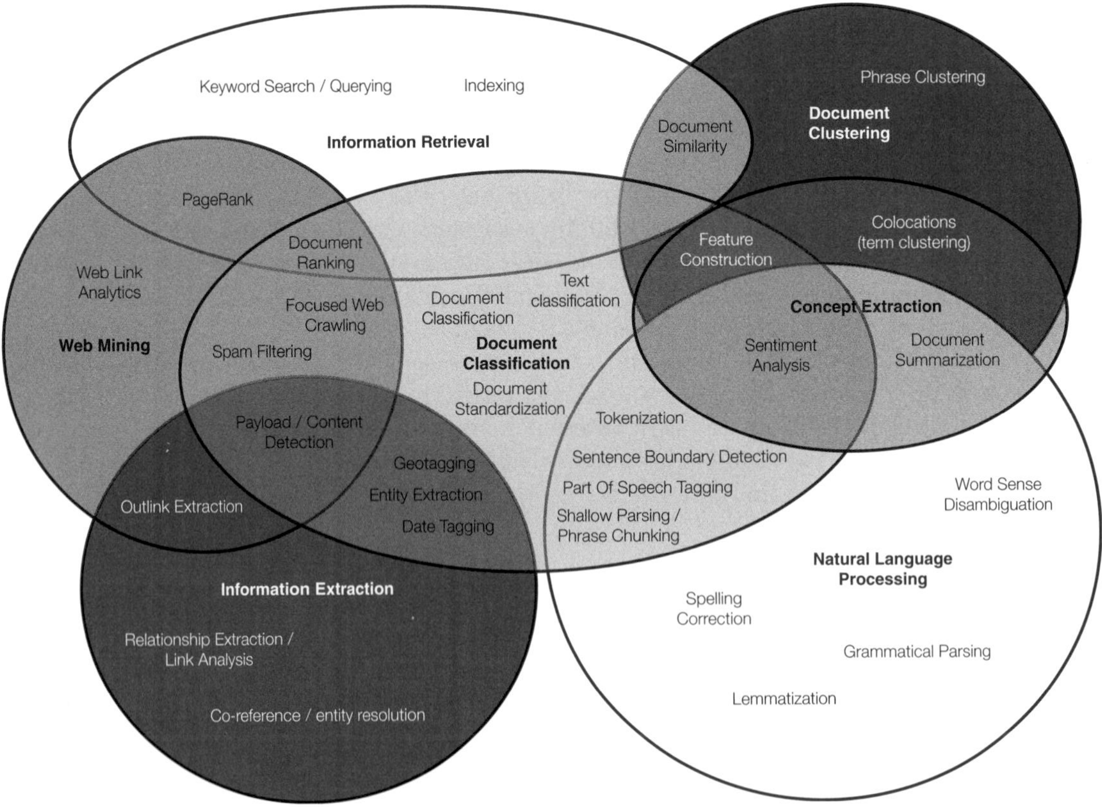

FIGURE 2.3
Visualizing the seven text mining practice areas (ovals) and how specific text mining tasks (labels within ovals) exist at their intersections.

information extraction and text classification, and document similarity measurement draws from the practice areas of document clustering and information retrieval.

SCOPE OF THIS BOOK

As we have just seen, text mining covers a diverse set of applications and algorithms. We have chosen to focus this book on techniques that are readily available for nonspecialists to apply immediately given the proper tools. Consequently, the areas of document classification, clustering, and concept extraction have the strongest representation in the book and the largest number of chapters and tutorials using these methods. This can be seen in Figure 2.2 with the list of the related chapters and tutorials. These three areas use techniques and algorithms that are drawn directly from data mining and are well represented in software for data mining and statistical analysis.

Web mining is a new and exciting application area for text mining practitioners. The Internet is rapidly changing with new information sources such as Facebook and Twitter. Because of this constant change, it has taken longer for a consensus to form over which methods perform best. We provide an introduction in the area of web mining with a limited number of chapters and tutorials. Interested readers are encouraged to explore the area on their own, and because of its high demand and rapid change, it may be possible to quickly become a leader in the field. Also, web mining borrows heavily from the areas of document classification, clustering, and concept extraction, allowing us to focus on those topics more.

Information extraction and natural language processing are becoming more accessible but still require significant amounts of domain expertise in linguistics to be successful. Of the seven areas, information extraction and natural language processing also are the most distinct technically, often requiring specialized software to achieve strong performance. Because of these challenges, we have chosen not to focus heavily on these two areas and instead provide an introduction and an avenue for exploration of these areas.

Finally, we provide minimal coverage of search and information retrieval. Since Google and other search engines have become such an integral part of our lives, search has become a key part of nearly every major software package. Consequently, search has become commoditized and is familiar enough to most users to skip the coverage of it here. If you are interested in building your own search engine, we have listed some excellent technical resources in Table 2.4.

An overall diagrammatic model that summarizes the scope of this book is presented in Figure 2.4. As shown in Figure 2.4, text mining draws upon many techniques in the broader field of text analytics. The central theme of this book is learning how to apply the diversity of powerful text mining models to solve practical problems in an organization. The tutorials in this book evolved out of the goal of driving you up the learning curve in text mining as efficiently as possible, using a learn-by-doing approach. That is the primary goal of this book.

SUMMARY

The term *text mining* can mean many different things to different authors, vendors, speakers, and clients. This chapter creates a rational taxonomy for the field, based on the perspective of a practitioner—a

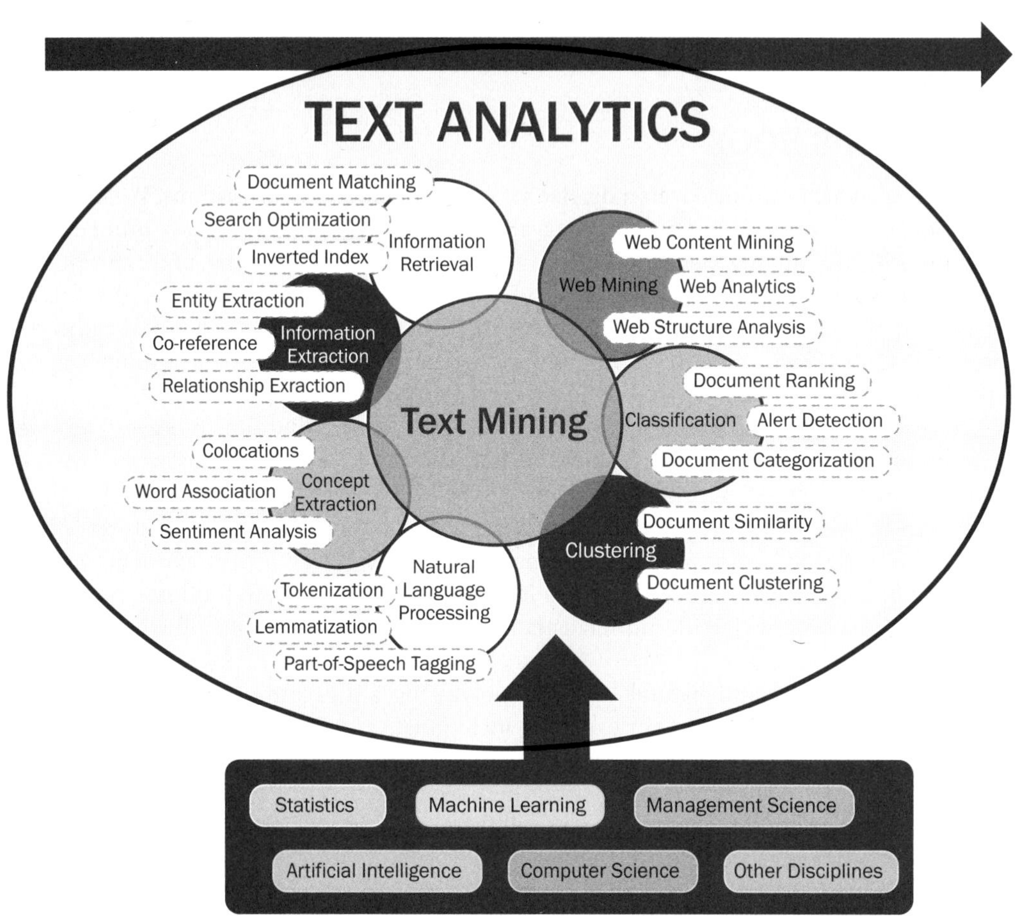

FIGURE 2.4

Text mining is the thematic center of this book, drawing upon contributions of many text analytical components and knowledge from many external disciplines (shown in blue at the bottom), which result in directional decisions affecting external results (shown by the blue arrow at the top).

person with some text data and an application goal. We define seven "practice areas" for text mining, based only on the practical distinctions in data and goal for an analyst trying to solve a given problem.

Chapter 1 described the history of text mining and how it is related to (borrows from and influences) six other fields. Figure 2.1 displays the overlap of those six fields with text mining and reveals the seven practice areas of text mining that are at the key intersections of the fields. An inductive model, in the form of a decision tree (Figure 2.2), asks the five key questions a practitioner needs to answer to be guided to the appropriate practice area for his or her text-based problem. The tree reveals not only the practice area most appropriate for a given text challenge but also the chapters and tutorials of this book that address that type of application. This allows the reader to jump right to the areas in the book that are most useful for his or her work. It further reveals where the book's coverage is strong and where it is light. The areas most covered in this book are those that have arrived just past the cutting edge of research into development—that is, those that are within reach of a technical nonspecialist who is willing to learn and yet are not ubiquitous (like search is).

Finally, when the practice areas are themselves generalized to oval regions in a Venn diagram (Figure 2.3), individual text mining tasks, such as lemmatization, can be located at the intersection of the seven practice areas, further helping to focus a user on the appropriate resources to use for a task. Where the book's coverage is incomplete, recommended high-quality external resources are listed.

POSTSCRIPT

Text mining is proving to be extremely useful, and this taxonomy of the wide-ranging field is designed to help analysts hone in on the practice area and resources for that area that are most helpful to achieving high productivity on the particular text application challenge they are facing.

A common claim among data miners is that 80 to 90 percent of the project time is consumed by data preparation steps. The same is true for text mining. In contrast to data mining, where some of the data are in text format, *all* of the data for text mining are in text format. The initial challenge is to transform these text data into a numerical format for subsequent analysis. In the next chapter, you will be introduced to the steps necessary to preprocess text data to create data structures that can be analyzed numerically.

References

Chakrabarti, Soumen. Mining the Web: Analysis of Hypertext and Semi-Structured Data, Morgan Kaufmann, San Francisco, 2002.

Croft, Bruce, Donald Metzler, and Trevor Strohman. Search Engines: Information Retrieval in Practice, Addison-Wesley, Boston, MA, 2009.

Manning, Christopher D., Prabhakar Raghavan, and Hinrich Schütze. Introduction to Information Retrieval, Cambridge University Press, New York, 2008.

Manning, Chris, and Hinrich Schütze. Foundations of Statistical Natural Language Processing, Cambridge, MA: MIT Press, Cambridge, MA, 1999.

Nisbet, R., J. Elder, and G. Miner. (2009). Handbook of Statistical Analysis and Data Mining Applications, Elsevier, Burlington, MA.

Smiley, David, and Eric Pugh. Solr 1.4: Enterprise Search Server. Packt Publishing, Birmingham, England, UK, 2009.

CHAPTER 3

Conceptual Foundations of Text Mining and Preprocessing Steps

CONTENTS

PREAMBLE

After you determine what you want to do with text mining, you must compose the set of steps that you can follow to perform it. In the next chapter, you will learn about one of the most important innovations in text processing: the generalized vector-space model. Documents are represented by a string of values (a *vector*), in which each element in the string represents a unique word or word group contained among the list documents. Each element of the vector contains either a 1 or a 0 (for the presence or absence of the word in the document), or a count of the number of times the word is present in a document. When you have a data structure that can be analyzed numerically, you are ready to hit the data mining "trail."

INTRODUCTION

Before starting a text mining project, it is first necessary to understand the conceptual foundations of text mining and then to understand how to get started leveraging the power of your data to drive decision making in your organization. This chapter highlights many of the theoretical foundations of text mining algorithms and describes the basic preprocessing steps used to prepare data for text mining algorithms.

Practical Text Mining and Statistical Analysis for Non-structured Text Data Applications. DOI: 10.1016/B978-0-12-386979-1.00003-7

The majority of text data encountered on a daily basis is *unstructured*, or free, text. This is the most familiar type of text, appearing in everyday sources such as books, newspapers, emails, and web pages. By unstructured, we mean that this text exists "in the wild" and has not been processed into a structured format, such as a spreadsheet or relational database. Often, text may occur as a field in a spreadsheet or database alongside more traditional structured data. This type of text is called *semistructured*. Unstructured and semistructured text composes the majority of the world's data; recent estimates place text data at 75 to 80 percent of the data in the world. This is easy to believe considering the amount of information stored as text on the Internet alone.

Given the sheer volume of text available, it is necessary to turn to automated means to assist humans in understanding and exploiting available text documents. This chapter provides a brief introduction to the statistical and linguistic foundations of text mining and walks through the process of preparing unstructured and semistructured text for use with text mining algorithms and software.

SYNTAX VERSUS SEMANTICS

The purpose of text mining algorithms is to provide some understanding of how the text is processed without having a human read it. However, a computer can only examine directly the individual characters in each word and how those words are arranged. A computer cannot know what information is being communicated by the text, only the structure or syntax of the text. *Syntax* pertains to the structure of language and how individual words are composed to make well-formed sentences and paragraphs. This is an ordered process. Specific grammar rules and language conventions govern how language is used, leading to statistical patterns appearing frequently in large amounts of text. This structure is relatively easy for a computer to process. On its own, however, syntax is insufficient for fully understanding meaning.

Semantics refers to the meaning of the individual words within the surrounding context. Common idioms are useful to illustrate the differences between syntax and semantics. Idioms are words or phrases with a figurative meaning that is different from the literal meaning of the words. For example, the sentence "Mary had butterflies in her stomach before the show" is syntactically correct and has two potential semantic meanings: a literal interpretation where Mary's preshow ritual includes eating butterflies and an idiomatic interpretation that Mary was feeling nervous before the show. Clearly, the complete semantic meaning of the text can be difficult to determine automatically without extensive understanding of the language being used.

Fortunately, syntax alone can be used to extract practical value from the text without a full semantic understanding, due to the close tie between syntax and semantics. Many text mining tasks, such as *document classification* and *information retrieval*, are concerned with ranking or finding specific types of documents in a large document database. The main assumption behind these algorithms is that syntactic similarity (similar words) implies semantic similarity (similar meaning). Though relying on syntactic information alone, these approaches work because documents that share many keywords are often on the same topic. In other instances, the goal is really about semantics. For example, *concept extraction* is about automatically identifying words and phrases that have the same meaning. Again, the text mining approaches must rely on the syntax to infer a semantic relationship. The generalized vector-space model, described next, enables this.

THE GENERALIZED VECTOR-SPACE MODEL

The most popular structured representation of text is the vector-space model, which represents text as a vector where the elements of the vector indicate the occurrence of words within the text. This results in an extremely high-dimensional space; typically, every distinct string of characters occurring in the collection of text documents has a dimension. This includes dimensions for common English words and other strings such as email addresses and URLs. For a collection of text documents of reasonable size, the vectors can easily contain hundreds of thousands of elements. For those readers who are familiar with data mining or machine learning, the vector-space model can be viewed as a traditional feature vector where words and strings substitute for more traditional numerical features. Therefore, it is not surprising that many text mining solutions consist of applying data mining or machine learning algorithms to text stored in a vector-space representation, provided these algorithms can be adapted or extended to deal efficiently with the large dimensional space encountered in text situations.

The vector-space model makes an implicit assumption (called the *bag-of-words* assumption) that the order of the words in the document does not matter. This may seem like a big assumption, since text must be read in a specific order to be understood. For many text mining tasks, such as *document classification* or *clustering*, however, this assumption is usually not a problem. The collection of words appearing in the document (in any order) is usually sufficient to differentiate between semantic concepts. The main strength of text mining algorithms is their ability to use *all* of the words in the document—primary keywords and the remaining general text. Often, keywords alone do not differentiate a document, but instead the usage patterns of the secondary words provide the differentiating characteristics.

Though the bag-of-words assumption works well for many tasks, it is not a universal solution. For some tasks, such as *information extraction* and *natural language processing*, the order of words is critical for solving the task successfully. Prominent features in both entity extraction (see Chapter 13) and natural language processing include both preceding and following words and the decision (e.g., the part of speech) for those words. Specialized algorithms and models for handling sequences such as finite state machines or *conditional random fields* are used in these cases.

Another challenge for using the vector-space model is the presence of *homographs*. These are words that are spelled the same but have different meanings. One example is the word *saw*, which can be a noun describing a tool for cutting wood (e.g., "I used a saw to cut down the tree") or the past tense of the verb *to see* (e.g., "I saw the tree fall"). Fortunately, homographs do not typically have a large effect on the results of text mining algorithms for three reasons:

1. Using multiple senses of a single homograph in a document is rare.
2. If used, homographs are typically not the most meaningful feature for the task.
3. Other words appearing in the document are usually sufficient to distinguish between different use cases (for example, the word *cut*).

Though the situation is rare, it is possible for homographs to result in text with completely different meanings to be grouped together in vector space. One extreme example is the following pair of sentences:

> "She can refuse to overlook our row," he moped, "unless I entrance her with the right present: a hit."
>
> Her moped is presently right at the entrance to the building; she had hit a row of refuse cans!

> "She *can* refuse **to** overlook our row," he moped, "unless I entrance her **with the** right *present*: **a** hit."
>
> Her moped **is** *presently* right **at the** entrance **to the** building; she had hit **a** row **of** refuse *cans*!

FIGURE 3.1
Example sentences with the nine homographs underlined, the stopwords in **bold**, and two stemmed words in *italics*. If you ignore the stopwords, there are 11 shared words, and the sets of distinctive words are only {he, I, our, unless} versus {had}.

These two sentences are nearly the same according to the vector-space model. That is, they are nearly identically located in high-dimensional word space. The two sentences are even somewhat more similar than they at first appear, as shown in Figure 3.1. First, most text preprocessing uses a normalization process called *stemming* to remove pluralization and other suffixes, which would make "can" and "cans" identical in the sentences, as well as "present" and "presently." The two sentences thus share homographs. They also share a few other words (e.g., "she," "her"), increasing their similarity in a normal way, and some of their distinct words—which would otherwise increase diversity—are *stopwords* (such as "at" and "with"), which are removed during processing and not used in the vector space model. So they are distinguished only by the words {he, I, our, unless} vs. {had}, despite depicting very different scenes. The preprocessing steps just mentioned are described in more detail next.

PREPROCESSING TEXT

One of the challenges of text mining is converting unstructured and semistructured text into the structured vector-space model. This must be done prior to doing any advanced text mining or analytics. The possible steps of text preprocessing are the same for all text mining tasks, though which processing steps are chosen depends on the task. The basic steps are as follows:

1. Choose the scope of the text to be processed (documents, paragraphs, etc.).
2. Tokenize: Break text into discrete words called *tokens*.
3. Remove stopwords ("stopping"): Remove common words such as *the*.
4. Stem: Remove prefixes and suffixes to normalize words—for example, *run, running,* and *runs* would all be stemmed to *run*.
5. Normalize spelling: Unify misspellings and other spelling variations into a single token.
6. Detect sentence boundaries: Mark the ends of sentences.
7. Normalize case: Convert the text to either all lower or all upper case.

Choosing the Scope of Documents

For many text mining tasks, the scope of the text is easy to determine. For example, emails or call log records naturally translate into a single vector for each message. However, for longer documents one needs to decide whether to use the entire document or to break the document up into sections, paragraphs, or sentences. Choosing the proper scope depends on the goals of the text mining task: for *classification* or *clustering* tasks, often the entire document is the proper scope; for *sentiment analysis*, *document summarization*, or *information retrieval*, smaller units of text such as paragraphs or sections might be more appropriate.

Tokenization

The next preprocessing step is breaking up the units of text into individual words or tokens. This process can take many forms, depending on the language being analyzed. For English, a straightforward and effective tokenization strategy is to use white space and punctuation as token delimiters. This strategy is simple to implement, but there are some instances where it may not match the desired behavior. In the case of acronyms and abbreviations, the combination of using an unordered vector space and separating tokens on punctuation would put different components of the acronym into different tokens. For example U.N. (abbreviation for the United Nations) would be tokenized into separate tokens U and N losing the meaning of the acronym. There are two approaches to address this problem: smart tokenization, which attempts to detect acronyms and abbreviations and avoid tokenizing those words, and adjusting the downstream processing to handle acronyms (for example) in separate tokens prior to converting to a vector space.

Stopping

For many text mining tasks, it is useful to remove words such as *the* that appear in nearly every document in order to save storage space and speed up processing. These common words are called "stopwords," and the process of removing these words is called "stopping." An example stopword list from a popular open-source text mining library is shown in Figure 3.2. Stopping is a commonly included feature in nearly every text mining software package. The removal of stopwords is possible without loss of information because for the large majority of text mining tasks and algorithms, these words have little impact on the final results of the algorithm. Text mining tasks involving phrases, such as *information retrieval*, are the exception because phrases lose their meaning if some of the words are removed.

A special case of stopword removal is the detection of header and "boilerplate" information. For example, many corporate emails also include a lengthy legal disclaimer at the bottom of the message. For short messages, a long disclaimer can overwhelm the actual text when performing any sort of text mining. In the same way, email header information such as "Subject," "To," and "From," if not removed, can lead to skewed results. Removal of this type of text depends on the situation where it appears.

a, an, and, are, as, at, be, but, by, for, if, in, into, is, it, no, not, of, on, or, such, that, the, their, then, there, these, they, this, to, was, will, with

FIGURE 3.2
English stopwords used by the open-source Lucene search index project.

Stemming

Stemming is the process of normalizing related word tokens into a single form. Typically the stemming process includes the identification and removal of prefixes, suffixes, and inappropriate pluralizations.

For example, a typical stemming algorithm would normalize *walking, walks, walked, walker,* and so on into *walk*. For many text mining tasks, including classification, clustering, or search indexing, stemming leads to accuracy improvements by shrinking the number of dimensions used by the algorithms and grouping words by concept. This decrease in dimensionality improves the operation of the algorithms.

One of the most popular stemming algorithms, the Porter stemmer, used a series of heuristic replacement rules to transform word tokens into their stemmed form (Porter, 1980).[1] These rules have been refined in later versions into what is now known as the Snowball Stemmer. The Snowball Stemmer is available in nearly 20 different languages and is a major component of the popular Lucene search engine index released by the Apache Software Foundation.[2] As with most text applications, the general case works well but requires some special cases that the Porter Stemmer calls "dictionary lookups." For example, words ending in *y* are often stemmed as *ability* → *abilit*, but *sky* loses its meaning when the *y* is removed and requires a special case. Similarly, the suffix *-ly* (as in *quickly*) is usually removed, but for words such as *reply*, removing *-ly* is not appropriate.

Lemmatization is a more advanced form of stemming that attempts to group words based on their core concept or *lemma*. Lemmatization uses both the context surrounding the word and additional grammatical information such as part of speech to determine the lemma. Consequently, lemmatization requires more information to perform accurately. For words such as *walk*, stemming and lemmatization produce the same results. However, for words like *meeting*, which could serve as either a noun or a verb, stemming produces the same root *meet*, but lemmatization produces *meet* for the verb and maintains *meeting* in the noun case.

Spelling Normalization

Misspelled words can lead to an unnecessary expansion in the size of the vector space needed to represent a document. In a recent project involving hand-entered text, we found over 50 different spellings of the phrase "learning disability" (Figure 3.3). There are a number of different approaches for correcting spelling automatically. First, dictionary-based approaches can be used to fix common spelling variations such as differences between American and British English (e.g., color and colour). Second, fuzzy matching algorithms such soundex, metaphone, or string-edit distance can be used to cluster together words with similar spellings. String hashing functions such as soundex or metaphone are found in popular database software and text mining software. Finally, if none of those approaches is sufficient, word clustering and concept expansion techniques (see Chapters 8 and 13) can be used to cluster misspellings together based on usage. These approaches can also be used in combination; for example, if two tokens are used in similar contexts (word clustering) and have a small string-edit distance or similar soundex, then they are likely to be the same word despite variations in spelling. Spelling normalization may not be required if the text is mostly clean and misspellings are rare. For messy text, such as text gathered from the Internet, spelling normalization is invaluable.

Sentence Boundary Detection

Sentence boundary detection is the process of breaking down entire documents into individual grammatical sentences. For English text, it is almost as easy as finding every occurrence of punctuation

---

1 Also available at http://tartarus.org/~martin/PorterStemmer/def.txt

2 http://lucene.apache.org

learning disablitiy, learning deisability,
learning disablity, learining disabilities,
learning disabiality, learning disabilty,
learning disabiilty, learning disabilitty,
learning disabilty, learning disablety,
learning disabilitiy, learnoing disability,
learning disabilties, learning disabiltiy,
learning disabilty, learning disibility,
learnings disabilty, learningdisability,
larning disabilities, learning disabilitiies,
learning disabilitties, learning disibilities,
learning diasability, learning dasability,
learnning disability, learning disabilities,
lerning disability, learning disabilites,
learneing disability, learnining disability,
learning disaibilities, leraning disability,
learning disaiblity, learnings disability,
learning disabilitys, learning disabillity,
learnings disabilities, learning diasbility,
learning disabiliites, learning dsiability,
learning disabliity, learning disibilty,
learning disbilities, learning disbality,
learning disbility, learning disabilities,
learning disabilities, learningi disability,
lerniung disabilities, learning disabiliities,
learning disaability, learning disablities

FIGURE 3.3
Evidence that spelling normalization is needed. This table shows 52 misspellings of the phrase "learning disability" found during a recent text mining project.

like "."; "?"; or "!" in the text. However, some periods occur as part of abbreviations or acronyms (as noted before on tokenization). Also, quotations will often contain sentence boundary punctuation, but the actual sentence will end later. These conditions suggest a few simple heuristic rules that can correctly identify the majority of sentence boundaries. To achieve near perfect (~99 percent) accuracy, statistical classification techniques are used. More on text classification techniques can be found in Chapter 7.

Case Normalization

Most English texts (and other Romance languages) are written in mixed case—that is, text contains both upper- and lowercase letters. Capitalization helps human readers differentiate, for example, between nouns and proper nouns and can be useful for automated algorithms as well (see Chapter 9 on named entity extraction). In many circumstances, however, an uppercase word at the beginning of the sentence should be treated no differently than the same word in lower case appearing elsewhere in a document. Case normalization converts the entire document to either completely lower case or completely upper case characters.

CREATING VECTORS FROM PROCESSED TEXT

After text preprocessing has been completed, the individual word tokens must be transformed into a vector representation suitable for input into text mining algorithms. This vector representation can take one of three different forms: a binary representation, an integer count, or a float-valued weighted vector. Following is a (very) simple example that highlights the difference among the three approaches. Assume a collection of text with the following three documents:

Document 1: My dog ate my homework.
Document 2: My cat ate the sandwich.
Document 3: A dolphin ate the homework.

The vector space for these documents contains 15 tokens, 9 of which are distinct. The terms are sorted alphabetically with total counts in parentheses:

a (1), ate (3), cat (1), dolphin (1), dog (1), homework (2), my (3), sandwich (1), the (2).

The binary and integer count vectors are straightforward to compute from a token stream. A binary vector stores a 1 for each term that appears in a document, whereas a count vector stores the count of that word in the document.

Doc 1: 0,1,0,0,1,1,1,0,0 (notice that though "my" appears twice in the sentence, the binary vector still only contains a "1")
Doc 2: 0,1,1,0,0,0,1,1,1
Doc 3: 1,1,0,1,0,1,0,0,1

The integer count vectors for the three documents would look as follows:

Doc 1: 0,1,0,0,1,1,2,0,0 (notice that "my" appearing twice in the sentence is reflected)
Doc 2: 0,1,1,0,0,0,1,1,1
Doc 3: 1,1,0,1,0,1,0,0,1

Storing text as weighted vectors first requires choosing a weighting scheme. The most popular scheme is the TF-IDF weighting approach. TF-IDF stands for *term frequency—inverse document frequency*. The term frequency for a term is the number of times the term appears in a document. In the preceding example, the term frequency in Document 1 for "my" is 2, since it appears twice in the document. Document frequency for a term is the number of documents that contain a given term; it would also be 2 for "my" in the collection of the three preceding documents. Equations for these values are shown in Figure 3.4.

The assumption behind TF-IDF is that words with high term frequency should receive high weight unless they also have high document frequency. The word *the* is one of the most commonly occurring words in the English language. *The* often occurs many times within a single document, but it also occurs in nearly every document. These two competing effects cancel out to give *the* a low weight.

$$tf \bullet idf(t,d) = tf(t,d) \bullet idf(t)$$

$$tf(t,d) = \sum_{i \in d}^{|d|} 1\{d_i = t\}$$

$$idf(t) = \log\left(\frac{|D|}{\sum_{d \in D}^{|D|} |t \in d}\right)$$

FIGURE 3.4
Equations for TF-IDF: term frequency (TF)—inverse document frequency (IDF).

SUMMARY

Once the input text has been processed and converted into text vectors, you are ready to start getting results from applying a text mining algorithm, as described in the remainder of the book. Even though you may seek to discover the semantics or meaning of the text through text mining, algorithms can only directly use the syntax or structure of the document. Fortunately, syntax and semantics are usually closely tied together, and text mining can work well in practice.

POSTSCRIPT

Similar to the early days of data mining, the term *text mining* can refer to many very different methods that can be applied to many use cases. Before you can start down the text mining trail, it is best to understand how the various use cases are similar and how they are different. It is a waste of time and resources to think that your application will involve specific methods of data preparation when your application really belongs in a different category of use cases. The purpose of the next chapter is to pair the appropriate use cases with the appropriate technologies to accomplish them.

Reference

Porter, M. F. 1980. An algorithm for suffix stripping. *Program* **14**(3): 130–137.

CHAPTER 4

Applications and Use Cases for Text Mining

CONTENTS

PREAMBLE

Text mining appears finally to have hit the "mainstream"—that is, it has become a methodology and technology of general interest to solve real-world business or research problems. As was the case in the early days of data mining, however, the term is overused today, referring to very different methodologies to address various diverse use cases. The purpose of this chapter is to provide a survey and

Practical Text Mining and Statistical Analysis for Non-structured Text Data Applications. DOI: 10.1016/B978-0-12-386979-1.00004-9

overview of 5 common use cases and to pair them to the "right" approaches and technologies from the broad area generally referred to as text mining.

WHY IS TEXT MINING USEFUL?

In general, most information available in the "real world" exists as written or recorded words. This is probably the reason why text mining is capturing significant attention in the business and research communities concerned with practical applications.

The Explosion of Text Stored in Electronic Format

While this statement may seem obvious, considering the dominance of word processing during the last 30 years, the explosion of available information, opinions, narratives, and so forth in written text form was not necessarily obvious, given the widespread availability of telephones and the emerging adoption of cell phones at the time. During the 1970s and into the early 1980s, most people communicated by telephone, much information was disseminated through speeches or lectures, and written text was available only in printed form and could only be accessed through tedious library searches.

Much has changed since then. The Internet, texting applications for cell phones, and very inexpensive storage technology to maintain truly vast repositories of text at very low cost have all changed the nature of how information is disseminated and exchanged, how opinions are expressed, how research is published, and how interpersonal communications are channeled.

Text Data and Information

Fundamentally, all of the forms of text mining discussed in this book address the issue of how to process the exponentially growing amount of text data to extract meaningful and useful information. While the amount of text generated in electronic form explodes, the amount of information that can be processed by one person, entity, or group of individuals remains constant. If, for example, one were to receive ten one-page letters a day, it is possible to read them all, think about their contents, and plot a strategy for how to respond. However, an informal survey of professionals from various domains and industries usually reveals that most individuals in positions with some decision-making responsibilities easily receive well over 100 and sometimes 200 (nonspam) emails per day. Obviously, a way to deal with that flood of text data must be developed.

As more and more text data become available, our challenge is to design new strategies to extract relevant and important information from that data efficiently without the need for anyone to read the text. All applications for text mining deal with solutions to meet that challenge in one form or another.

For example, for a manufacturer of commercial analytics software solutions, the determination of popular trends and sentiments with respect to competing analytical solutions requires the analysis of available financial data (numbers) and reports. But it is of much greater interest to identify trends, growing markets, competitors' product strategies and roadmaps, and so on. The problem is, that information is not available in any numerical table but must be gleaned from the large corpus of written words that make up competitors' marketing information, themes discussed in blogs, articles in trade journals, and related materials. Other examples and typical use cases exist in every domain where success depends on processing effectively and drawing correct conclusions from a corpus of text. In short, text mining is and will become increasingly relevant to practically *all* activities and domains where formal analytical approaches add value.

Use Cases

Five basic types of analytical text mining applications and use cases can be identified that address most important text processing issues:

1. *Extracting "meaning" from unstructured text.* This application involves the understanding of core themes and relevant messages in a corpus of text, without actually reading the documents.
2. *Automatic text categorization.* Automatically classifying text is an efficient way to organize text for downstream processing.
3. *Improving predictive accuracy in predictive modeling or unsupervised learning.* Combining unstructured text with structured numeric information in predictive modeling or unsupervised learning (clustering) is a powerful method to achieve better accuracy.
4. *Identifying specific or similar/relevant documents.* Efficiently extracting from a large corpus of text those documents that are relevant to a particular topic of interest or are similar to a target document (or documents) is a vitally necessary operation in information retrieval.
5. *Extracting specific information from the text ("entity extraction").* Automatically extracting specific information from the text (such as names, geographical locations, and dates) is an efficient method for presenting highly focused information for downstream analytical processing or for direct use by decision makers.

EXTRACTING "MEANING" FROM UNSTRUCTURED TEXT

An intuitive notion of this use case in text mining implies that various methodologies can extract specific contents or contextual meaning from a large corpus of small text documents or from a small corpus of large text documents that cannot be read and summarized in a practical manner.

Sentiment Analysis

Sentiment analysis seeks to determine the general sentiments, opinions, and affective states of people reflected in a corpus of text. Analysis of these sentiments can address the following concerns:

What are my customers saying about me? Customer feedback is a very useful source of information on customer satisfaction. For example, it is useful for organizations to be able to extract the body of main "themes" and affective responses associated with their products from customer feedback and reviews or from public blogs that are relevant to the respective products or services (Li and Wu, 2010). Many companies follow very carefully how their brands and products are portrayed in news stories and public forums and blogs in order to identify any strengths, problems, or issues pertaining to their products reflected in these public forums.

In short, text mining can be an extremely useful tool to measure the sentiments or "pulse" of what customers or stakeholders are thinking or feeling that is reflected in a large corpus of textual documents. This pulse represents their emotional (affective) responses to respective products, and the leveraging of that knowledge is *integrally related* to the success or failure of a company. Perhaps nowhere is this better illustrated than in customer reviews available at websites for reviews about cars or the customer reviews sections now commonly found at most merchants' websites.

What are the emerging areas of concern or interest in a specific target group? Like the preceding ones, some applications make a corpus of text available that is relevant to special concerns or sentiments of a specific group of individuals of interest. For example, websites where teenagers can discuss their concerns and interests can provide a wealth of near-real-time information about what topics, themes, and fashions are popular and which trends are fading and which ones are emerging. The value of this information to inform marketing efforts to sell products to teenagers can be enormous (Fukuhara, 2005; Piccalo, 2005).

Analyzing open-ended responses to survey questions. While the preceding concerns center around the processing of textual information available via the web, applications to address these concerns occur frequently in traditional questionnaire surveys with open-ended questions (e.g., consumer research). Such questions are included in order to capture themes, opinions, and sentiments that were not anticipated by the designers of the survey and not encoded into respective questions (with numeric response formats). The task for those applications is to identify in the auxiliary open-ended responses the themes that were missed by the structured portion of the survey.

Trending Themes in a Stream of Text

Similar approaches for "extracting meaning from the text" are used in areas that are not so much concerned about "sentiments" but are more interested in detecting changes, trends, and unusual events expressed in a corpus of text. For example, Hotz (2011) reports that when a 5.8 earthquake hit Viriginia on August 23, 2011, the first Twitter messages reached New York before the first shock waves did (as determined by the social media company SocialFlow). Unexpected or unusual events often are first reflected in text before they are measured numerically.

Warranty Claim Trends

For example, warranty claims for cars, computers, and so on are usually summarized in brief narratives describing what went wrong and what repairs were made to fix the problem (Sureka, De and Varma, 2008). Typically, the complexity of modern cars or electronics makes it impossible to anticipate all possible failure modes that can occur in the everyday use of the product. Yet, it is in the best interests of manufacturers to learn about emerging patterns of potentially dangerous failures as quickly as possible. Thus, the stream of warranty claim narratives, along with the narratives describing the final repairs, can serve as a useful source of information that can be monitored for emerging "new" or "unusual" themes.

Insurance Claims, Fraud Detection, and So Forth

Likewise, insurance claims are usually associated with narratives explaining what happened, describing details of the circumstances that led to the initial claim, and subsequent (e.g., claims adjusters') notes annotating the processing of the claim (Sullivan and Ellingsworth, 2003). These narratives can be analyzed in a similar manner in order to identify emerging themes or anomalies (e.g., potential fraud, areas of unexpected exposure). In general, the goal here is to extract meaning and themes from the corpus of documents in order to create alert systems that are sensitive to changes in information about how a product performs or how it is used. Chapter 14 focuses on some of these applications of text mining in the property and casualty insurance industry.

SUMMARIZING TEXT

Another use case of text mining is the extraction of meaning when the goal is to quickly summarize one or a few very large documents. There are two types of text summarizations. One type summarizes themes across the chapters or paragraphs of the text, in which case the individual paragraphs or chapters can be considered different documents of a larger corpus (the entire text). The goal of this type is to identify the different themes across the various documents (e.g., as just described) or to identify common dimensions or relationships among individuals, events, and so on.

The second type summarizes the contents of a large text document into a meaningful narrative which *cannot* be accomplished effectively (yet) using automatic text mining methods and algorithms. In short, it is not realistic to expect that present computer algorithms are capable of summarizing the "essence" of a very large book into a single paragraph. At present, this can be done only in other highly subjective ways.

COMMON APPROACHES TO EXTRACTING MEANING

Among the preceding applications, the common problem is that an *a priori* list of terms or themes is not available. Instead, a heterogeneous corpus of text must be analyzed to extract the themes and meanings with respect to the dimensions of interest. A typical approach to accomplishing this goal is performing a statistical analysis of words and word frequencies across documents. Examples of these statistical analyses include: statistical natural language processing, statistical summaries of word or word counts, and statistical analysis of dimensions of meaning.

Statistical Natural Language Processing

This analysis begins with simply counting words or phrases and the frequency with which they occur in each document. Usually, various preprocessing steps are applied to the text to do the following:

- Correct misspellings.
- Reduce different grammatical forms (i.e., declinations and conjugations) of the same words to common roots (the process called *stemming*).
- Translate foreign language text as required (perhaps using machine translation algorithms).
- Build lists of abbreviations and synonyms so that common abbreviations and their nonabbreviated counterparts are counted as instances of the same word.
- Eliminate words that are very general in meaning and occur with high frequency across all documents (e.g., *the, to, and*). These words are sometimes called *stopwords* (or a *stoplist*) because they are not counted. This approach is described in greater detail in Chapter 6 and in Manning and Schütze (2002) and Weiss and colleagues (2010).

Example: What Is *Bosseln* (or *Boßeln*)?

To illustrate the basic mechanism of how useful meaning can be extracted very quickly from a corpus of text, even if the language is unknown to the reader and the subject area is entirely unfamiliar, let us consider the activity known as *bosseln* (or *boßeln*, using the older letter "ß" instead of the modern spelling that replaces "ß" with "ss").

Table 4.1 Forty-Nine Paragraphs from the Wikipedia Article on *Bosseln* (or *Boßeln*)

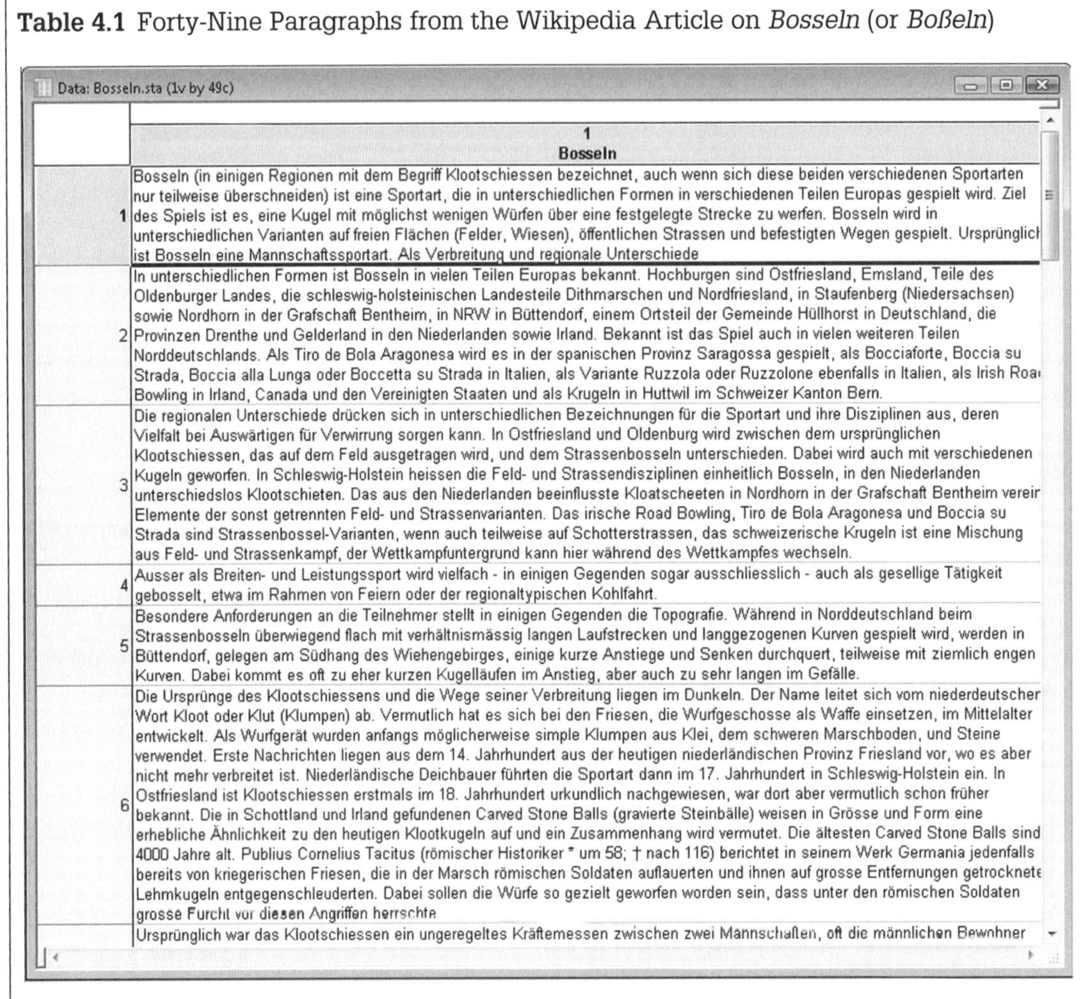

Data: Bosseln.sta (1v by 49c)

	1 Bosseln
1	Bosseln (in einigen Regionen mit dem Begriff Klootschiessen bezeichnet, auch wenn sich diese beiden verschiedenen Sportarten nur teilweise überschneiden) ist eine Sportart, die in unterschiedlichen Formen in verschiedenen Teilen Europas gespielt wird. Ziel des Spiels ist es, eine Kugel mit möglichst wenigen Würfen über eine festgelegte Strecke zu werfen. Bosseln wird in unterschiedlichen Varianten auf freien Flächen (Felder, Wiesen), öffentlichen Strassen und befestigten Wegen gespielt. Ursprünglich ist Bosseln eine Mannschaftssportart. Als Verbreitung und regionale Unterschiede
2	In unterschiedlichen Formen ist Bosseln in vielen Teilen Europas bekannt. Hochburgen sind Ostfriesland, Emsland, Teile des Oldenburger Landes, die schleswig-holsteinischen Landesteile Dithmarschen und Nordfriesland, in Staufenberg (Niedersachsen) sowie Nordhorn in der Grafschaft Bentheim, in NRW in Büttendorf, einem Ortsteil der Gemeinde Hüllhorst in Deutschland, die Provinzen Drenthe und Gelderland in den Niederlanden sowie Irland. Bekannt ist das Spiel auch in vielen weiteren Teilen Norddeutschlands. Als Tiro de Bola Aragonesa wird es in der spanischen Provinz Saragossa gespielt, als Bocciaforte, Boccia su Strada, Boccia alla Lunga oder Boccetta su Strada in Italien, als Variante Ruzzola oder Ruzzolone ebenfalls in Italien, als Irish Roa Bowling in Irland, Canada und den Vereinigten Staaten und als Krugeln in Huttwil im Schweizer Kanton Bern.
3	Die regionalen Unterschiede drücken sich in unterschiedlichen Bezeichnungen für die Sportart und ihre Disziplinen aus, deren Vielfalt bei Auswärtigen für Verwirrung sorgen kann. In Ostfriesland und Oldenburg wird zwischen dem ursprünglichen Klootschiessen, das auf dem Feld ausgetragen wird, und dem Strassenbosseln unterschieden. Dabei wird auch mit verschiedenen Kugeln geworfen. In Schleswig-Holstein heissen die Feld- und Strassendisziplinen einheitlich Bosseln, in den Niederlanden unterschiedslos Klootschieten. Das aus den Niederlanden beeinflusste Kloatscheeten in Nordhorn in der Grafschaft Bentheim verei Elemente der sonst getrennten Feld- und Strassenvarianten. Das irische Road Bowling, Tiro de Bola Aragonesa und Boccia su Strada sind Strassenbossel-Varianten, wenn auch teilweise auf Schotterstrassen, das schweizerische Krugeln ist eine Mischung aus Feld- und Strassenkampf, der Wettkampfuntergrund kann hier während des Wettkampfes wechseln.
4	Ausser als Breiten- und Leistungssport wird vielfach - in einigen Gegenden sogar ausschliesslich - auch als gesellige Tätigkeit gebosselt, etwa im Rahmen von Feiern oder der regionaltypischen Kohlfahrt.
5	Besondere Anforderungen an die Teilnehmer stellt in einigen Gegenden die Topografie. Während in Norddeutschland beim Strassenbosseln überwiegend flach mit verhältnismässig langen Laufstrecken und langgezogenen Kurven gespielt wird, werden in Büttendorf, gelegen am Südhang des Wiehengebirges, einige kurze Anstiege und Senken durchquert, teilweise mit ziemlich engen Kurven. Dabei kommt es oft zu eher kurzen Kugelläufen im Anstieg, aber auch zu sehr langen im Gefälle.
6	Die Ursprünge des Klootschiessens und die Wege seiner Verbreitung liegen im Dunkeln. Der Name leitet sich vom niederdeutsche Wort Kloot oder Klut (Klumpen) ab. Vermutlich hat es sich bei den Friesen, die Wurfgeschosse als Waffe einsetzen, im Mittelalter entwickelt. Als Wurfgerät wurden anfangs möglicherweise simple Klumpen aus Klei, dem schweren Marschboden, und Steine verwendet. Erste Nachrichten liegen aus dem 14. Jahrhundert aus der heutigen niederländischen Provinz Friesland vor, wo es aber nicht mehr verbreitet ist. Niederländische Deichbauer führten die Sportart dann im 17. Jahrhundert in Schleswig-Holstein ein. In Ostfriesland ist Klootschiessen erstmals im 18. Jahrhundert urkundlich nachgewiesen, war dort aber vermutlich schon früher bekannt. Die in Schottland und Irland gefundenen Carved Stone Balls (gravierte Steinbälle) weisen in Grösse und Form eine erhebliche Ähnlichkeit zu den heutigen Klootkugeln auf und ein Zusammenhang wird vermutet. Die ältesten Carved Stone Balls sind 4000 Jahre alt. Publius Cornelius Tacitus (römischer Historiker * um 58; † nach 116) berichtet in seinem Werk Germania jedenfalls bereits von kriegerischen Friesen, die in der Marsch römischen Soldaten auflauerten und ihnen auf grosse Entfernungen getrocknete Lehmkugeln entgegenschleuderten. Dabei sollen die Würfe so gezielt geworfen worden sein, dass unter den römischen Soldaten grosse Furcht vor diesen Angriffen herrschte.
	Ursprünglich war das Klootschiessen ein ungeregeltes Kräftemessen zwischen zwei Mannschaften, oft die männlichen Bewohner

The document for this example is available at http://de.wikipedia.org/wiki/Bo%C3%9Feln. For this example, each paragraph from this article was transferred to an input text corpus, with 49 paragraphs in all. Table 4.1 shows the text.

Instead of translating this article in full and then reading it, we can apply a simple statistical approach to determine what *bosseln* (or *boßeln*) is all about. After ignoring some common and "non-diagnostic" words as previously described (e.g., the equivalents of the English *to, is*, etc.) and applying a German stemming algorithm, the words or terms occurring with the top frequencies and their English translations are shown in Table 4.2.

In short, there are a few "untranslatable terms" such as *klootschiess(en)* which occurs with the highest relative (to the number of paragraphs) frequency, or *bosseln* (or "street-bosseln"), which seems to

Table 4.2 Top Words/Terms Extracted from Wikipedia Article on *Bosseln (Boßeln)*

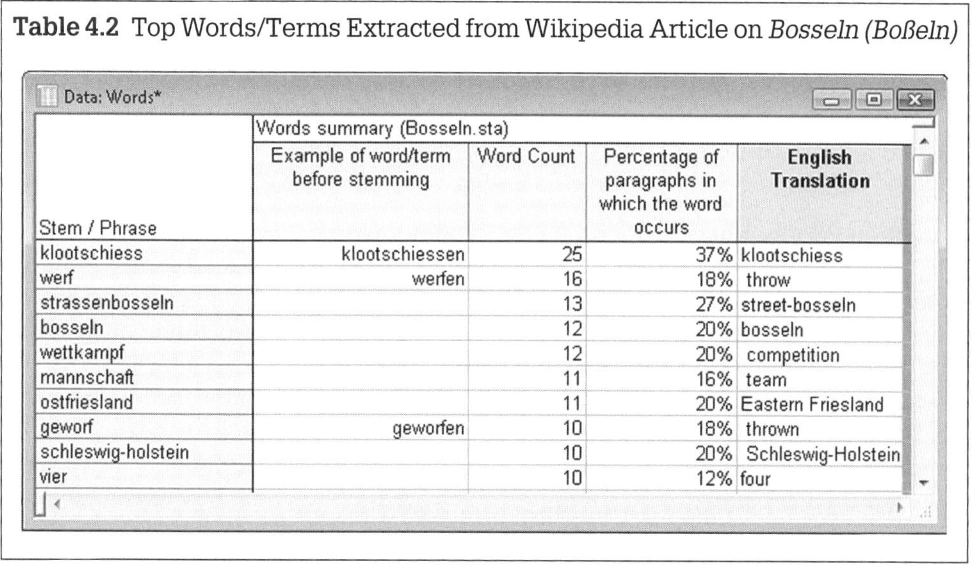

Data: Words*

	Words summary (Bosseln.sta)			
Stem / Phrase	Example of word/term before stemming	Word Count	Percentage of paragraphs in which the word occurs	**English Translation**
klootschiess	klootschiessen	25	37%	klootschiess
werf	werfen	16	18%	throw
strassenbosseln		13	27%	street-bosseln
bosseln		12	20%	bosseln
wettkampf		12	20%	competition
mannschaft		11	16%	team
ostfriesland		11	20%	Eastern Friesland
geworf	geworfen	10	18%	thrown
schleswig-holstein		10	20%	Schleswig-Holstein
vier		10	12%	four

indicate that this activity can happen on a street. The other terms include *throw* and *thrown*, *competition*, *team*, names of geographical locations (*Schleswig-Holstein* and *Eastern Friesland*), and the number *four*.

What Have We Learned about Bosseln?

It appears that *bosseln* is a competitive team (sport) activity that involves throwing something and is played mostly in the most northern part of Germany (Schleswig Holstein); also, the number 4 is somehow significant. As it turns out, if you *were* to translate the entire article to English and read it, all of these pieces of information correctly describe the sport.

EXTRACTING INFORMATION THROUGH STATISTICAL NATURAL LANGUAGE PROCESSING

To summarize, this simple example illustrates how meaningful information or "meaning" can be extracted from an unfamiliar text (and language) by indexing the unique terms and phrases in the corpus and computing statistical summaries (relative frequencies with respect to the number of paragraphs or documents). In short, by simply translating ten words or terms from German into English, one can learn the following:

- What the text that describes an activity called *bosseln* is all about (it is a team sport that involves throwing something)
- Where it is played (northern Germany, in the German state of Schleswig-Holstein, and the region of Eastern Friesland)
- What it apparently relates to or perhaps (and actually) is similar to (*klootschiessen*)

Thus, without reading the article itself, this procedure extracted meaningful information or meaning about *bosseln*.

Computing Statistical Summaries for a Data Matrix of Raw Word/Term Counts

In more general terms, any text can be *numericized*—that is, turned into a matrix of word (terms, phrases) counts. At that point, the corpus of text can be represented by a data matrix, where each column represents a word or term and where each row represents a document. Much of the active research around text mining focuses on the following:

- What to count (words, phrases, nouns, names of locations)
- How to identify automatically (algorithmically) the items to be counted (the algorithms and methods to parse text) and extract "entities," such as simple words or phrases, terms describing locations, names, references to specific companies, or grammatical structures and syntax elements—for example, to determine automatically "who did what to whom"
- How best to process and summarize the counts to extract efficiently information that is relevant to the specific goals of the respective project (e.g., compute word frequencies, relative word frequencies, or indices that summarize the document complexities, relevance to a search term, etc.)

In fact, all of the techniques discussed in this book deal with one or more of these aspects of text mining, and the software available for text mining implements some or all of the methods to perform these tasks.

Example: Analyzing Advertisements of Small Aircraft

Let us consider another example "corpus" and another one where the factors and dimensions that define the meaning of the corpus are not commonly known. Specifically, the following word frequency matrix was extracted from a corpus of "for sale" listings describing small single-engine Cessna aircraft that are at least ten years old, which was advertised on a popular website for buying and selling airplanes. Table 4.3 shows a subset of the terms extracted from the corpus. Each row represents one advertisement, and the numbers in each column are counts of the simple word frequencies—that is, how often the respective term is used in the advertisement.

Table 4.3 Excerpt of a Data Matrix of Word Frequencies

	5	6	7	8	9	10	11	12	13	14	15	16	17	18	19	20
	accents	accessories	activated	additional	adf	adjust	aero	air	airframe	airplane	airport	airspeed	allover	alt	alternate	altimet
4		1	1			4		4							3	
5		1	1			4		4							3	
6			1		1	1										
7				1					1							
8					1			1								
9											1					
10		1			1	3		5				1			4	
11		1			1	3		4				1			4	
12	1				1						1	1				
13							2									
14																
15					1				1		1					
16		1			1							1			1	
17				1												
18	1				1				1				1			
19					1			1	2			1				
20					1											

As in the previous simple example, a straightforward approach to extract meaning from a data matrix of this type relies on the subsequent processing of this data matrix.

Word Frequencies across Documents

The specificity of various words is a useful property for making distinctions between different documents in a corpus, such as aircraft for sale listings. If a particular term exists in all documents with approximately the same frequency, then it is not useful for distinguishing topics, themes, and dimensions that vary among the individual documents (i.e., for sale listings). For example, if this corpus describes only propeller-driven airplanes (and all advertisements contain the word *propeller*), then *propeller* is not useful to distinguish among different aircraft advertisements.

If, however, approximately half of the individual advertisements include *propeller*, while the other half use *jet*, then these two terms are useful to distinguish among the individual documents in the corpus. Hence, these terms are useful for extracting "meaning" with respect to the variety and types of aircraft advertised in this corpus of advertisements.

Correlations of Word Frequencies across Documents

Another way to analyze this matrix is to explore the relationships between the patterns of frequencies across documents. For example, one may find that the terms *ballistic, recovery, system, BRS*, and *Cirrus* all appear with greater than expected frequency across the corpus of documents. In other words, the advertisement documents tend to have either all of these terms in them or none of them; thus, one can infer that it is very likely that Cirrus aircraft are equipped with a BRS or ballistic recovery system. This analysis provides another meaningful characteristic (dimension of meaning) to distinguish among the individual advertisements.

Relating Word/Term Frequencies to Known Dimensions of Interest: What Makes an Airplane Expensive?

In addition to analyzing word co-occurrence across documents, you can analyze the correlation or "covariation" among the dimensions of meaning and other words, terms, or phrases. For example, a simple "model" of the asking price for the single-engine Cessna airplanes in the corpus of advertisements described earlier may be derived as shown in Figure 4.1.

By reviewing the tree structure of word/term occurrences in Figure 4.1, you can learn very quickly what types of equipment are expensive or what makes a Cessna single-engine airplane expensive. Following this approach, we learn the following:

- The average asking price for all Cessna single-engine aircraft is $163,000.
- The most expensive airplanes are associated with the presence of the term *Stationair*, where the mean asking price is $335,000, or nearly twice the overall average asking price. As it turns out, the *Stationair* is indeed the most expensive of the "classic" single-engine Cessna airplanes.
- The next most expensive "class" of aircraft, with an average asking price of $248,000, is associated with *Stationair=No, HIS=Yes*, and *DME=No*. With a bit of research, you can learn that *HSI* is a *horizontal situation indicator*—a complex and somewhat complicated "high-end" navigational instrument, and a *DME* is *distance measurement equipment*, a technology mostly superseded a long time ago (and thus less valuable when present in an aircraft) with the wide acceptance of GPS (global positioning system) technology.

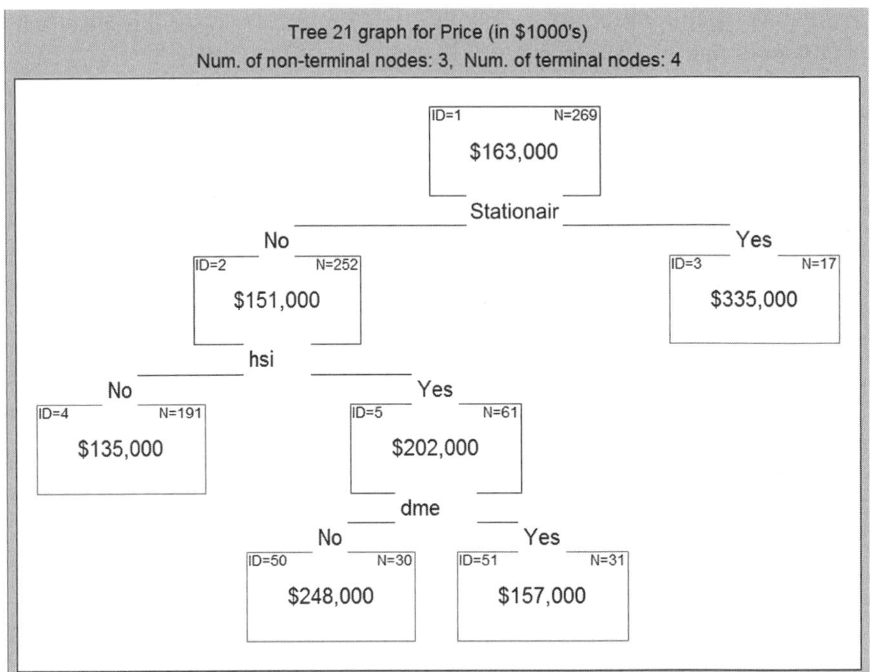

FIGURE 4.1
A simple regression tree model relating the presence of words in an airplane advertisement to the asking price.

To summarize, this simple example illustrates again how meaningful information can be extracted from a complex and unfamiliar corpus of text without actually reading the advertisements. There are many applications where a similar approach and analyses could yield very valuable information. Other examples include pricing real estate (based on descriptions of the respective properties), identifying underpriced properties, and correlating between various dimensions of meaning and other real estate features that organize and stratify a corpus of real estate listings in a meaningful way. The statistical analysis of these dimensions of meaning is described in greater detail below.

STATISTICAL ANALYSIS OF DIMENSIONS OF MEANING

In practice, the raw data matrix of word counts is often transformed to make it more suitable for subsequent analyses, including computing the inverse document frequencies, reducing dimensionality, or applying clustering methods to the data matrix (see Chapters 7 and 13). In all cases, however, the final outcome will be a series of "numericized" terms or combinations of terms or dimensions, required by subsequent numeric processing algorithms.

For example, a common application is to apply singular value decomposition (SVD) to the data matrix (of transformed frequencies) in order to extract relevant dimensions (see also Chapter 11). The logic of these analyses is to identify the terms that "go together" based on the correlations of word frequencies across documents.

Suppose in the data matrix shown earlier (but derived from the reviews of cars by owners published on the Internet), you find a dimension marked by the co-occurrences or lack of co-occurrences of the words *sporty*, *acceleration*, *horsepower*, and so on. Thus, documents in the corpus of reviews appear to be distinguishable along the dimension that could be characterized best as *performance*. Over time, and in the context of other cars, dimensions marked by terms like *efficiency*, *gas mileage*, *cost*, and so on may become more prevalent and relevant. Thus, a change or difference in "meaning" or sentiment could be detected across reviews for different cars or across reviews over time, indicating a shifting sentiment and value system in consumers of cars.

Clustering

Several other methods and use cases may be apparent to users experienced in the application of data analysis methods. For example, it may be of greater interest to identify atypical documents—that is, those that appear to express a different sentiment or reflect on a theme that is unusual, with respect to the corpus of text. When monitoring warranty claims in order to detect new types of problems, it is often useful to derive models of typical claims using clustering or other methods. Following this approach, novelty detection (the detection of new types of problems occurring among the warranty claims) could be accomplished by flagging and counting the specific warranty narratives that are dissimilar to the majority of claims.

Using Dictionaries and Known Dimensions of Meaning (Affect and Sentiment)

A good example is represented by product reviews contributed by owners/users to various Internet sites, which are usually based on one's prior knowledge and expectations of the product. Car reviews will be based often on price, performance, reliability, practicality, and overall expressions of likes versus dislikes. Similar product reviews for computers are likely to focus on technical features (i.e., speed, memory) or suitability for specific applications, such as playing games. Reviews by users of health-related products are likely to be focused on their effectiveness and potential effects.

Dictionaries of Terms That Express Likes and Dislikes

Given a specific domain, such as cars and car reviews, it is possible with relatively little effort to develop dictionaries of terms with obvious meaning or implication with respect to the sentiment expressed about a car. For example, phrases like "fast acceleration," "responsive handling," and "good gas mileage" or terms like "roomy," "efficient," and "reliable" all have obvious implications and meaning with respect to the sentiment that is expressed (*like the car and would recommend it* vs. *dislike the car and would not recommend it*).

It is worth noting at this point that many of the written reviews and blogs published at various sites are typically not very complex with respect to the syntax and size of the vocabulary used in the corpus. Put another way, the language used in such reviews (and in most domains where casual consumers or reviewers compose a few paragraphs to express their opinion) is very simple. This means that it is practical and often quite easy to create a dictionary of known terms and words (the "verbiage") associated with positive reviews and negative reviews.

Counting Expressions of Positive and Negative Sentiments

Once such a dictionary of known terms, and the specific sentiments they describe, has been compiled, one can now add to or replace the simple count of words with a count of sentiments. So one could count all of the terms across all of the documents that invoke positive sentiments (e.g., statements of "like") versus negative sentiments (statements of "dislike") and construct from those counts a simple index to measure the overall affective positive versus negative responses expressed in different documents in the corpus—for example, for different cars, car manufacturers, and so on. Fundamentally, one can think of those terms as synonyms of "likes" and "dislikes" and count those synonyms as an index that likely is related to the degree and intensity with which the respective affect is expressed.

Using Statistical or Analytical Methods to Derive an Index of Positive and Negative Sentiments

In an *a priori* dictionary of terms, for example, it may not be obvious which specific terms express a "strong" versus a "lukewarm" endorsement of a product. In this case, any numeric index or overall rating may be used to build models that relate the word/term frequencies to that overall rating (by the author of each document) of like versus dislike. For example, reviews of cars or other consumer items published on the web are usually accompanied by a single rating (e.g., 1 to 5 stars). One could use methods such as multiple regression to build a linear model with weights to relate the dictionary terms that express liking and disliking to the overall rating. In effect, this would allow one to create weights for each term or phrase and how strongly it is related to (a predictor of) an overall positive rating.

Classifying Words and Phrases Based on Relevant Dimensions of Meaning

With some experience, and usually relatively little effort, it is often possible to create a dictionary of terms and phrases that are known—a priori—to have a certain meaning with respect to the sentiment of interest, such as the overall expressed liking and recommendation of a car in a car review. Given such a dictionary, one can derive easy to summarize single number indices of the sentiment of interest (e.g., like vs. dislike). Such indices then provide efficient numeric summaries that can be tracked over time or used to compare different consumer items, brands, and so on.

Of course, one drawback of such dictionaries is that they may become "stale" at an unknown rate. So while terms like *gnarly*, *groovy*, or *cool* may at some time have been used to describe a good thing, other terms and new words have since replaced them, since commonly used vernacular and the affective meaning of terms are constantly changing. Put another way, the *connotation* associated with words or terms is a reflection of the ever changing subjective cultural or emotional meaning and implication of a word. Often, even the *denotative* or descriptive meaning of terms can change or become entirely obsolete due to technological changes. For example, the term *letter-quality* was once a highly relevant quality characteristic applicable to computer printers, but as laser printers have become omnipresent, this term is now rarely if ever used.

As a practical matter, the creation of and continuous reliance on a relevant dictionary to support sentiment analysis is more of an ongoing *process* than a one-time task.

BEYOND STATISTICAL ANALYSIS OF WORD FREQUENCIES: PARSING AND ANALYZING SYNTAX

A common criticism of the statistical natural language processing approach based on word frequencies alone is that it may miss important meaning reflected in individual documents because it ignores specific syntaxes of sentences and the order in which terms appear in them. This problem is actually a good example and presents a good use case for text mining to determine initially what items to count and how to count them. After these questions are answered, the text miner can concentrate on determining which methods are most effective at extracting information from the text and for converting it to useful numeric indices. For example, the sentences "Dog bites man" and "Man bites dog" are identical if we only counted words. While we might be able to link documents discussing man and dog to other dimensions or insights, or be able to distinguish them from documents discussing "man" and "horse," obviously meaning is lost by not being able to distinguish the former less interesting sentence from the one that seems more interesting (a man biting a dog).

In order to gain a deeper insight into the topics contained in an individual document, it is necessary to analyze the specific syntax of each sentence. This is an active area of research that is obviously very important for the improvement of document searches to identify those that are of particular relevance to an area of interest (discussed further later). For the purposes of automatic text categorization, it is often not necessary to increase the complexity and dimensionality of the task through automatic syntax analysis. The choice of whether or not to use it depends on the complexity of the text that is being processed. As also just noted, most responses to open-ended survey questions are not very elaborate syntactically, nor are narratives that accompany warranty claims, insurance claims, or discussions found on public forums. Furthermore, it might be argued that the sophistication of written communications is typically becoming poorer rather than greater, so detailed analyses of a corpus of text identifying specific syntactical commonalities seldom add much to the effectiveness of a solution.

In other applications discussed below, it is very important to consider the syntax as well as the words and the order in which they are used in a text when the task is to accurately identify documents that are similar or relevant for a highly specific topic (e.g., searching patents). In short, if the interest is to identify stories where a man bit a dog, then we must filter out those cases explicitly where the reverse occurred. To do so, the syntax elements must first be identified that differentiate the two events, and then the methodology must include counting the separate occurrences of the two types of "man–dog interactions."

REVIEW

In summary, common use cases in text mining for extracting meaning from a body of unstructured text include the following:

- Sentiment analysis of unsolicited writings relevant to a particular topic of interest
- Warranty claims analysis
- Extraction of key dimensions of meaning from consumer reports
- Advertisements, customer complaints, warranty claims, or similar analyses to detect common and shifting expressions of values, opinions, and affective responses regarding an area of interest

Statistical natural language processing in these use cases evolved initially from the simple process of identification and counting of the words or phrases used in each individual document to the analysis of patterns of word frequencies across documents, developing methodologies to compare documents using common dimensions or clusters. Further evolution in complexity enabled the derivation of meanings of words and phrases by inspecting the specific words and documents showing the greatest variability in frequencies across documents, combined with finding patterns of co-occurrence with other words and terms across documents.

Automatic Categorization of Text

A separate use case involves the classification of unstructured text to relate to specific predefined categories for subsequent processing.

Automatic Email Routing, Spam Filtering

Automatic text classification is the essential problem that must be solved in order to distinguish between unwanted email spam and welcome email. Further classifications are often useful in larger organizations to route emails to the appropriate departments or recipients through some form of automated email preprocessing.

Fraud Detection

Fraud detection based on a corpus of text (e.g., narratives attached to insurance claims) can be considered a type of automatic text categorization problem, where the task is to automatically route narratives (e.g., claims) for further investigation.

Automatic Text Translation

A common application of text processing relies upon the accurate interpretation of actionable content in an input text or corpus of text (or spoken command) to create a facility of automatic translation of text documents from one language to another. Rather than include automatic text translation as a use case of text mining, it properly belongs to the discipline of computational linguistics, and a discussion of that is beyond the scope of this book.

IMPROVING ACCURACY IN PREDICTIVE MODELING

A common application of text mining is found in an increasing number of domains, to improve accuracy in predictive modeling, thereby enhancing the effectiveness of traditional predictive modeling and data mining activities. Data mining and predictive modeling techniques have been refined to help understand or predict the key performance indicators that make each business successful. These techniques have been embraced in many industries during the past decade, including banks, insurance companies, manufacturing companies, retailers, and service providers.

For example, in a fraud detection project at an insurance company, sophisticated algorithms might be deployed to identify unusual claims or "unusually average" claims that can be flagged for further investigation and verification. The obvious next step in such applications is to leverage all available relevant textual information in order to build better predictive models of an outcome of fraudulent claims and to build more comprehensive and meaningful models of market segmentation (Sullivan and Ellingsworth, 2003).

More recently, some success has been reported (Hotz, 2011) analyzing the themes and evolving sentiments expressed in Twitter text messages in order to predict the outcomes of specific political races or stock market trends.

USING STATISTICAL NATURAL LANGUAGE PROCESSING TO IMPROVE LIFT

The term *lift* is often used in predictive modeling to express numerically how much better a prediction model performs compared to other models or a random expectation (see, for example, Nisbet et al., 2009). It is easy to see how narratives attached often to insurance claims, credit applications, or customer feedback may contain significant information that is not available in any of the numerical indices that are routinely recorded into (structured) fields of a database. If that information is extracted with text mining methods and included in the modeling effort, higher predictive accuracy may be achieved.

Numericizing Text

The goal of a text mining project to achieve a better predictive model or to identify more useful customer segments is to develop a deployable model that will turn text into numerical indices that can be used for automatic scoring (predictions). Thus, a common approach is to numericize the text using the methods just described. These methods accomplish various tasks, including the following:

1. Building a data matrix based on word or phrase counts or transformed counts
2. Computing various numeric indices based on those counts (relative frequencies, or document indices that capture, for example, grammatical complexity, or relevance determined a priori with respect to some target dimensions or categories
3. Merging of those indices (or raw counts) with other numeric indicators and information available for use in predictive modeling

The numericizing activities should be limited to only those features or dimensions of meaning that result from feature selection methods (see Chapter 9). The dimensions of meaning can be related to combinations of features, and both can be transformed (with both linear or nonlinear functions) to match their distributions. Often, this transformation operation improves the lift of the model (see Chapter 8).

Predictive Modeling

The predictive modeling process considers the numeric indices or (transformed) counts derived from the text as being just another set of predictor variables. For example, in a predictive modeling analysis to forecast the likely cost of an auto insurance claim, numeric indices that reflect whether or not the respective claim narrative includes the terms "neck injury," "back injury," and "rehabilitation" may all have diagnostic value for the prediction of the final cost of settling the respective claim. Therefore, the input data file used for building the model may include simple word or phrase counts or binary (0/1) indices to denote the presence or absence of those terms.

In general, one may think of the variables derived from the text by numericizing as just another type of coded or transformed set of predictors that were computed during the data preparation step. It is a common data preprocessing task in predictive modeling to recode variables to derive meaningful indicators for modeling (e.g., to convert the address where an accident happened and the claimant's home address to a driving distance between the two addresses), and from that perspective, deriving numerical indices from text is fundamentally the same process: Text is recoded into numeric indices.

Model Deployment

Deploying a model and computing predictions from a model that involves predictor variables derived (recoded) from text require two separate steps: First, the identical (to the modeling step) data preprocessing has to be repeated in order to recode text to numeric indices. For example, if the predictive modeling found that the mention (in a claim narrative) of "neck injury" is related to higher medical costs and a coded binary predictor called *"Neck Injury Mentioned"* is an important predictor in the model. Second, the deployment of the trained model will generate a prediction based on the variable created during the data preprocessing step.

Again, this is really not different from what needs to be accomplished in order to deploy any predictive model: Certain predictor variables have to be preprocessed and/or recoded before they can be used by the model to compute a prediction.

Example: Predicting Insurance Fraud from Claims Processing Notes

Fraud is a significant and expensive problem for insurance companies. Practically all insurance providers have rules and procedures in place to automatically refer some insurance claims to a special investigative unit (SIU). For example, Progressive insurance employs 235 special investigations professionals in the United States (according to their website, http://www.progressiveagent.com/claims/special-investigations-unit.aspx). The Insurance Research Council and the National Insurance Crime Bureau document with regularity the staggering amounts of money—more than $30 billion every year—that are lost by insurers (and of course the insured, who ultimately pay for it) due to insurance fraud (Holm, 2011; Insurance Information Institute, 2011).

Predictive Modeling

An analytical approach to automatically estimating the potential for fraud in insurance claims would start by identifying all of the available information associated with a claim. Typically, detailed information regarding the nature of the claim, type of damage and injuries (for example), and other details are recorded into the claims database. Also, historical data of closed claims are available along with whether or not the respective claim had been referred to the special investigate unit (SIU) and, if so, if any fraud was identified.

Using the historical data, prediction models can be built using any of the predictive modeling algorithms (such as logistic regression, neural nets, or tree-based models; see Nisbet et al., 2009) based on the structured data recorded into the claims database.

Text Mining

In order to improve the accuracy of the prediction model, the available unstructured text information such as claim agents' notes, narratives, accident descriptions, descriptions of injuries, and so on can be aligned with the other available data so for each claim, both structured information and text information are available. Next, the text fields can be numericized as described earlier—for example, by counting specific word frequencies in each text source. In practice, it is usually advisable not only to include the words and terms identified by the text mining software tool as "diagnostic" but also terms and phrases identified as important and diagnostic of fraud potential by experienced SIU personnel.

Red Flag Variables

For example, often certain "red flag" variables and terms are present that experience has shown to be indicative of claims deserving further review (see, for example, Francis, 2003). In claims involving injuries, if there is immediate mention of and involvement of a lawyer, or certain types of injuries and treatment modalities are reported (given certain injuries), then the probability of fraud may be increased. Typically, a list of phrases and terms such as "back injury," "neck injury," or "MUA" (*manipulation under anesthesia*—a procedure sometimes applied by chiropractors for pain control that involves manipulation and soft tissue movement performed on patients under anesthesia) is identified a priori by experienced SIU investigators as relevant and important and so are included in the indexing (numericizing of terms).

Rebuilding and Deploying the Model

Finally, after text mining has extracted key phrases, terms, and indices derived from the key phrases and terms, these statistics can be added as input predictor variables to the predictive modeling for fraud. If the new models can be shown to improve the accuracy of the prediction of fraud in a test sample, then the automatic indexing of text or notes associated with new claims can be added as a standard (text) data preprocessing step to the deployment of the predictive model.

USING DICTIONARIES TO IMPROVE PREDICTION

In many applications, the domain of interest restricts the nature and complexity of the narratives and the specific words, abbreviations, and phrases that are most likely used. For example, in a warranty claims application using narratives written by technicians describing the problem and the nature of the remedy for the problem, the number of distinct words and terms used in the corpus of text is likely limited to relevant technical terms, parts (that failed), and so on. With experience, it is possible to build a detailed dictionary of words, synonyms, phrases, and terms specifically tailored to those narratives and then restrict the numericizing of the narratives to those terms.

Custom Dictionaries as Portable Prediction Models

Another common application of dictionaries describing a particular domain or set of dimensions relevant for a predictive modeling domain is to create dictionaries of terms, phrases, and synonyms that can be applied directly to similar projects, without the need to build new dictionaries. For example, a complete dictionary of parts, failure modes, and functional groupings used in cars and commonly referenced in warranty claim narratives for one specific automobile or class of automobiles could be "ported" to warranty applications (e.g., to predict the failure times of components) involving other types of cars. In short, the sometimes time-consuming step of identifying the diagnostic words and phrases, common abbreviations, and synonyms in a new corpus of text can be bypassed by developing (or purchasing) a carefully maintained dictionary that can be used to enumerate directly the most important terms and phrases.

IDENTIFYING SIMILARITY AND RELEVANCE BY SEARCHING

Document searching is an application of text mining that is very different from those discussed previously. The general issue here is how to effectively search a very large corpus of text documents to identify those that are relevant to a specific set of search terms or a particular target document (i.e., given a specific document, determining which documents in the corpus are relevant and similar).

Applications of Document Searching

Typical applications of these methods occur in the familiar web searches. Obviously, this is a very competitive area of inquiry with tremendous commercial value, so many of the most advanced methods for performing efficient searches through documents are likely not published or patented. The diversity and complexity of these applications render them beyond the scope of this book. Other more tractable applications of document searching include identifying similar content in diverse documents and querying text.

Identifying Similar Content in Diverse Documents

A common application of document searching requires identifying specific documents in a large corpus of text with content that is relevant to a specific input document or set of topics. Searching patents is a good example. The task there is to locate in the vast list of existing patents those that are relevant to a new patent application in order to verify that the inventions described in the new patent are indeed "new." Patent searches are also very useful for competitive intelligence to identify patents filed by competitors and thus to determine where the research and development occur.

Similar applications exist in most areas of science, where the volume of new publications has exceeded the human capacity to process it. For example, a researcher concerned with a specific issue in plant genetics will be interested in relevant publications in the area of human or animal genetics that discuss similar research. Thus, the task is to identify across a very diverse corpus of scientific publications those that address a specific research topic from different "points of view."

Querying Text

A similar application of document search is to identify and link common names, events, or places across the documents in a corpus of text. For example, one might query a large collection of emails for a specific name and its common abbreviations, bank account numbers, airport names, and so forth in order to identify the relevant emails that describe a particular person's travels and financial dealings. Obviously, such applications are invaluable for intelligence applications to "make sense" of the movements and relations among individuals and groups of interest (who may pose a threat).

PART OF SPEECH TAGGING AND ENTITY EXTRACTION

Part of speech (POS) *tagging* and *entity extraction* describe the two-step process of first annotating or "tagging" text for specific parts of the narrative that describe places, dates, names, and so on to prepare the text for efficient search or query operations and to identify and extract *specific* places, dates, or names.

Part of Speech Tagging

A large amount of active research is ongoing at various institutions to identify efficient algorithms and general standards for how to *tag* words, terms, and phrases in documents so searches through those documents are more relevant and efficient (see, for example, Chapter 13; Feldman and Sanger, 2007; Weiss et al., 2010). For example, suppose a document contains the following sentence:

Jim bought 300 shares of IBM in 2006.

After processing this sentence through an entity extraction and tagging algorithm, the sentence might be "annotated" as follows:

<ENAMEX TYPE="PERSON">**Jim**</ENAMEX> **bought** <NUMEX TYPE="QUANTITY">**300**</NUMEX> **shares of** <ENAMEX TYPE="ORGANIZATION">**IBM**</ENAMEX> **in** <TIMEX TYPE="DATE">**2006**</TIMEX>.

So the words or terms in the sentence are now preceded with tags that identify the type of *entity* that it describes; for example, *IBM* describes an organization, *2006* describes a date, and so on.

If all of the sentences in the entire corpus of text are tagged in this manner, it becomes much easier to perform efficient searches (or queries) of the corpus of text to extract, for example, all of the documents that mention the *organization* by the name of *IBM* and the *person* by the name of *Jim*, and so on. Thus, the corpus of text has been turned into a structured database that can be queried using the values for the entities, making it much easier to identify relevant documents, compute indices of relevance, and display (to the user) the specific places in the document where the entities of interest are found.

Entity Extraction and Indexing

The task of automatically indexing a corpus of text is not simple or unambiguous. For example, in the sentence shown earlier, the number *300* is a *quantity*, while *2006* is a date. This is not immediately apparent unless the algorithms that process this sentence also take into consideration the syntax of the sentence. Also, the specific entities that are useful for indexing must be determined beforehand or built over time to be suitable for the specific purposes of the application of interest. Chapter 13 discusses in greater detail the different approaches and available resources to accomplish this task.

SUMMARY

This chapter provided an overview of the types of applications where (and how) text mining algorithms and analytical strategies can be useful and add value. In general, text mining techniques were developed in order to extract useful information from a large number of documents (a large document corpus) without requiring humans to actually read and summarize the text. Typical applications for text mining discussed in this chapter include the automatic analysis of consumer sentiments expressed in unsolicited product or other reviews ("sentiment analysis"), automatic categorization of text to "route" documents consistent with some assigned category (e.g., email spam filters), leveraging the information contained in the text to build better predictive models for important business outcomes (e.g., fraud detection, credit scoring), identifying and grouping similar documents, and preparing documents to allow for more effective searches in order to extract specific information.

A common approach to text mining to support predictive modeling or automatic text categorization is based on the simple counting of words or phrases in the documents in a document corpus and using statistical or predictive modeling techniques to convert each document into a vector of numbers that represent word/term frequencies or indices reflecting the presence of specific words or combinations of words (phrases) in each document in the corpus. Once each document in the document corpus has been "numericized" in this manner, various statistical, data mining, and/or predictive modeling methods can be applied to leverage that information in data analyses or predictive modeling—for example, to predict fraud, identify spam emails, or find the most relevant dimensions and clusters of words and terms that distinguish and organize the corpus of text (e.g., to identify dimensions of meaning).

Other methods typically identified as text mining techniques and activities focus on effective parsing of text to make it easier to extract information about specific parts of the text (parts of speech). Here, a common approach is to "tag" or annotate the text to make it more accessible for searches; extraction of specific information such as dates, places, and so on; or just to identify accurately which documents describe topics similar to some relevant search term.

The overview provided in this chapter is not comprehensive, and new applications or specific issues for text mining arise all the time. However, the use cases described here are common and represent a significant proportion of the applications and use cases for text mining, since these techniques are increasingly applied to address real-world issues.

POSTSCRIPT

When you select the right use case to fit your needs, you can proceed to the next question: What series of steps do I follow to analyze my data? We will provide a suggested text mining process model in Chapter 5, which conducts you step-by-step through the text mining analysis. Chapter 6 introduces you to three industry standard text mining tools for performing text mining operations.

References

Feldman, R., and Sanger, J. (2007). The text mining handbook. Cambridge University Press, NY (US).

Francis, L. (2003). FCAS, MAAA. Martian Chronicles: Is MARS better than neural networks? March. Available at www.data-mines.com.

Fukuhara, T. (2005). Analyzing concerns of people using Weblog articles and real world temporal data. In Proceedings of the 2nd Annual Workshop on the Weblogging Ecosystem: Aggregation, Analysis and Dynamics. Chiba, Japan, May 10th 2005.

Holm, Eric (2011). Insurance study sees widespread fraud in NYC. Wall Street Journal. January 4, 2011.

Insurance Information Institute (2011). *Insurance Fraud: The Topic*. Insurance Information Institute, April 2011, http://blogs.wsj.com/metropolis/2011/01/04/insurance-study-sees-widespread-fraud-in-nyc/?blog_id=147&post_id=11448.

Hotz, R. L. (2011). *Decoding our chatter*. Wall Street Journal, October 1, 2011.

Li, N., and Wu, D. D. (2010). Using text mining and sentiment analysis for online forums hotspot detection and forecast. *Decision Support Systems*, 48, 354–368.

Manning, C. D., and Schütze, H. (2002). *Foundations of statistical natural language processing*, 5th ed. Cambridge, MA: MIT Press.

Nisbet, R., Elder, J., Miner, G. (2009). Handbook of statistical analysis and data mining applications. Academic Press, Boston, MA.

Piccalo, G. (2005). Fads are so yesterday. Retrieved December 6, 2010, from http://articles.latimes.com/2005/oct/09/entertainment/ca-trend9. Los Angeles Times, October 9, 2005.

Sullivan, D., and Ellingsworth, M. (2003). Text mining improves business intelligence and predictive modeling in insurance. Information Management Magazine, July, http://www.information-management.com/issues/20030701/6995-1.html.

Sureka, A., De, S., and Varma, K. (2008). Mining automotive warranty claims data for effective root cause analysis. Lecture Notes in Computer Science, Volume 4947/2008, 621–626.

Verbeke, Charles A. (1973) Caterpillar Fundamental English. Training and Development Journal, 27, 2, 36–40.

Weiss, S. M., Indurkhya, N., and Zhang, T. (2010). Fundamentals of predictive text mining. Springer.

CHAPTER 5

Text Mining Methodology

CONTENTS

PREAMBLE

The process that you follow in an endeavor is at least as important as what methodology you follow to accomplish it. There are several data mining process models used commonly by data miners, but there is no accepted process model for text mining. This chapter presents a proposed process for text mining, which will guide you in the performance of any of the five text mining application areas described in the previous chapter.

TEXT MINING APPLICATIONS

Text mining applications are so broad in their scope and so varied in their goals that it is difficult to express the accomplishment of it in general terms. Compared to other well-established statistical methods, text mining is a relatively new and unstandardized analytical technique for knowledge discovery. Therefore, it is challenging to create a road map of operations to perform its methodology. A *methodology* is a documented and somewhat standardized process for executing and managing complex projects that include many interrelated tasks (i.e., extracting knowledge from textual data sources) by the use of a variety of methods, tools, and techniques. A well-designed and properly followed/ implemented methodology can help to ensure consistent and successful results. In essence, a methodology is the manifestation of many experiences (both good and not so good) in a given discipline.

Applications of text mining are driven primarily by trial-and-error experiments based on personal experiences and preferences. While data mining methodologies are relatively mature (e.g., CRISP-DM,

Practical Text Mining and Statistical Analysis for Non-structured Text Data Applications. DOI: 10.1016/B978-0-12-386979-1.00005-0

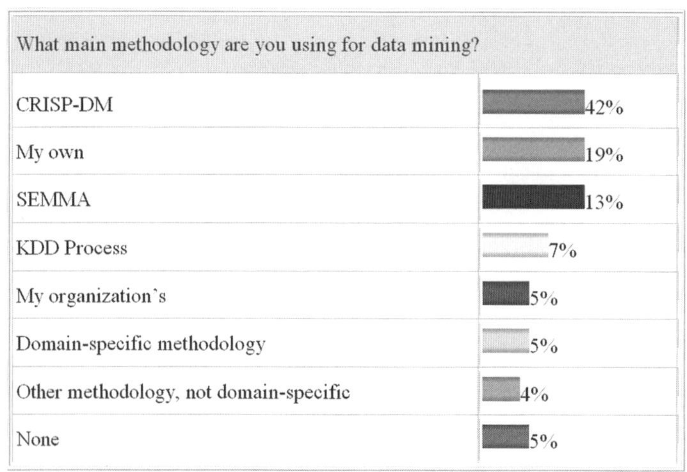

FIGURE 5.1
A user poll on the popularity of data mining methodologies. *Source: KDNuggets.com*

SEMMA, KDD), no commonly accepted methodologies have reflected the essence of best practices in text mining in any domain. The most significant reasons for this void include the following:

- Text mining means different things to different people; even the definition of it and what it encompasses are very unsettled and debatable subjects.
- The unstructured nature of the data opens up a wide range of exploratory avenues.
- There are many different types of unstructured data, some of which can be classified as semistructured (e.g., HTML pages, XML documents, etc.).
- The sheer size of the available data encourages premature sampling and simplification activities.

As the older brother of text mining, data mining went through a similar process of self-definition during the early 1990s that developed several well-known methodologies, including CRISP-DM, SEMMA, and KDD. Even though these three popular methodologies overlap significantly in the way they express and relate the data mining process, they differ in the way they define and scope data mining activities. Among the three, CRISP-DM is the most popular (Figure 5.1).

CROSS-INDUSTRY STANDARD PROCESS FOR DATA MINING (CRISP-DM)

In CRISP-DM, the complete life cycle of a data mining project is represented with six phases: business understanding (determining the purpose of the study), data understanding (data exploration and understanding), data preparation, modeling, evaluation, and deployment. Figure 5.2a shows a slightly modified rendition of the six phases of the CRISP-DM process flow (Chapman et al., 2000), indicating the iterative nature of the underlying methodology.

Within the six phases, CRISP-DM methodology provides a comprehensive coverage of all of the activities involved in carrying out data mining projects. Because the primary distinction between data mining and text mining is simply the type of data involved in the knowledge discovery process, we adopt CRISP-DM

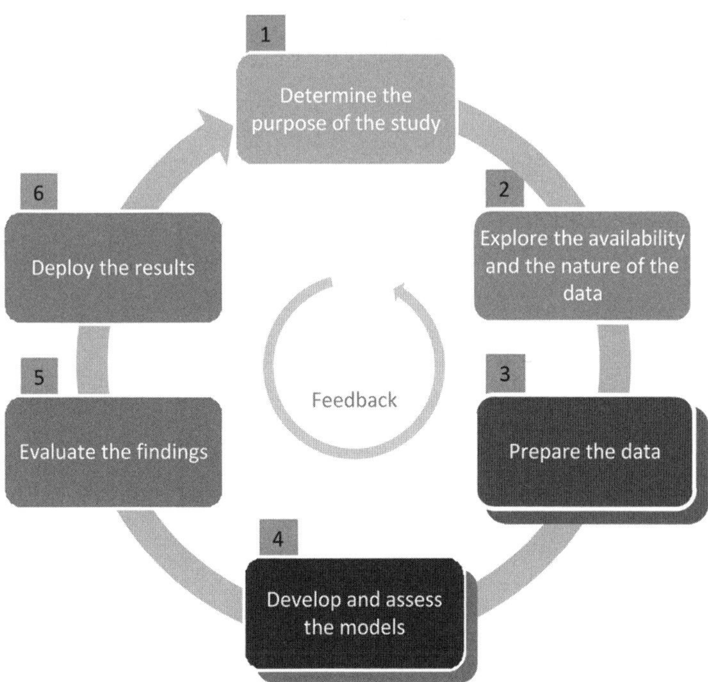

FIGURE 5.2a
A cyclic form of a proposed process flow for text mining, based on CRISP-DM. The feedback loop indicates that the findings and lessons learned at any phase in the process can trigger a backward movement for corrections and refinements, and the completion of a process may lead to new and more focused discovery processes.

as a foundation upon which to derive the text mining methodology followed in this book. Figure 5.2b shows the linear arrangement of phases of a proposed text mining process flow, based on CRISP-DM.

Phase 1: Determine the Purpose of the Study

Like any other project activity, text mining study starts with the determination of the purpose of the study. This requires a thorough understanding of the business case and what the study aims to accomplish. In order to achieve this understanding and define the aims precisely, we must assess the nature of the problem (or opportunity) that initiated the study. Often, we must interact closely with the domain experts in order to develop an in-depth appreciation of the underlying system, its structure, its system constraints and the available resources. Only then can we develop a set of realistic goals and objectives to govern the direction of the study.

Phase 2: Explore the Availability and the Nature of the Data

Once the purpose of the study is determined, we are ready to assess the availability, obtainability, and applicability of the necessary data in the context of the specific study. Some of the tasks in this phase include the following:

- Identification of the textual data sources (digitized or paper-based; internal or external to the organization)

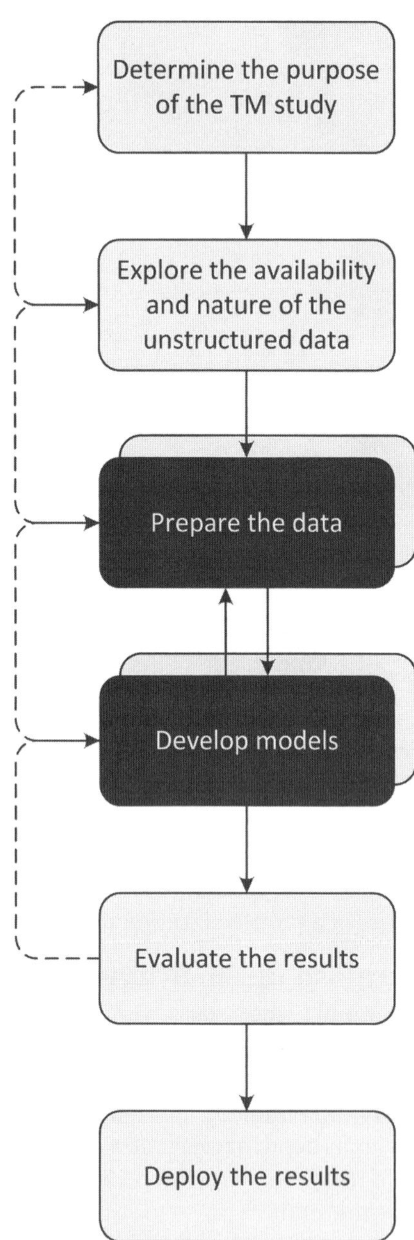

FIGURE 5.2b
A linear expression of the proposed text mining process flow, based on CRISP-DM. The two shaded boxes represent the most significant differences (and require more detailed explanations) for the CRISP-DM methodology as it applies to text mining.

- Assessment of the accessibility and usability of the data
- Collection of an initial set of data
- Exploration of the richness of the data (e.g., does it have the information content needed for the text mining study?)
- Assessment of the quantity and quality of the data. Once the exploration is concluded with positive outcomes, the next phase is to collect and integrate large quantities of data from various sources, which will be used in the study.

Phase 3: Prepare the Data and Phase 4: Develop and Assess the Models

Phases 3 and 4 present the most significant differences between data mining and text mining. In fact, many believe that text mining is nothing but data mining with a more laborious data collection and processing phase. In Figure 5.2b, Phases 2 and 3 are illustrated with shadowed boxes, indicating a more granular, in-depth, text mining–specific delineation. This is discussed after Phases 5 and 6.

Phase 5: Evaluate the Results

Once the models are developed and assessed for accuracy and quality from a data analysis perspective, we must verify and validate the proper execution of all of the activities. For example, we must verify that sampling was done properly and then repeat the steps to validate. Then (and only then) can we move forward to deployment. Taking on such a comprehensive assessment of the process helps to mitigate the possibility of error propagating into the decision-making process, potentially causing irreversible damage to the business. Often, as the analyst goes through these phases, he or she may forget the main business problem that started the study in the first place. This assessment step is meant to make that connection one more time to ensure that the models developed and verified are actually addressing the business problem and satisfying the objectives they were built to satisfy. If this assessment leads to the conclusion that one or more of the business objectives are not satisfied, or there still is some important business issue that has not been sufficiently considered, we should go back and correct these issues before moving into the deployment phase.

Phase 6: Deploy the Results

Once the models and the modeling process successfully pass the assessment process, they can be deployed (i.e., put into use). Deployment of these models can be as simple as writing a report that explains the findings of the study in a way that appeals to the decision makers, or it can be as complex as building a new business intelligence system around these models (or integrating them into an existing business intelligence system) so they can be used repetitively for better decision making. Some of the models will lose their accuracy and relevancy over time. They should be updated (or refined) periodically with new data. This can be accomplished by executing a new analysis process every so often to re-create the models, or, more preferably, the business intelligence system itself can be designed in a way that it refines its models automatically as new and relevant data become available. Even though developing such a sophisticated system that is capable of self-assessing and self-adjusting is a challenging undertaking, once accomplished, the results would be very satisfying.

Phases 3 and 4: Text Mining Process Specifications—A Functional Perspective

Figures 5.2a and 5.2b present the proposed text mining process in terms of process flow through Phases 3 and 4. Figure 5.3 represents a high-level context diagram of the text mining methodology from a functional architecture perspective. This context diagram is meant to present the scope of the process, specifically emphasizing its interfaces with its environment. In essence, it draws the boundaries around the process to explicitly show what is to be included (and/or excluded) from the representation of the text mining process.

Within the context of knowledge discovery, the primary purpose of text mining is to process unstructured (textual) data and structured and semistructured data (if relevant to the problem being addressed) to extract novel, meaningful, and actionable knowledge/information for better decision making. The inputs arrow in Figure 5.3 (on the left edge of the box) to the text-based knowledge discovery process box are the unstructured, semistructured, or structured data that are collected, stored, and made available to the process. The outputs arrow (from the right edge of the box) represents the context-specific knowledge products that can be used for decision making. The constraints (or *controls*) arrow entering at the top edge of the box represents software and hardware limitations, privacy issues, and the difficulties related to processing of the text that is presented in the form of natural language. The enablers entering the bottom of the box represents software tools, fast computers, domain expertise, and natural language processing (NLP) methods.

Figure 5.4 shows that Figure 5.3 can be decomposed into three linked subprocesses that we call "activities." Each has inputs, accomplishes some transformative process, and generates various outputs.

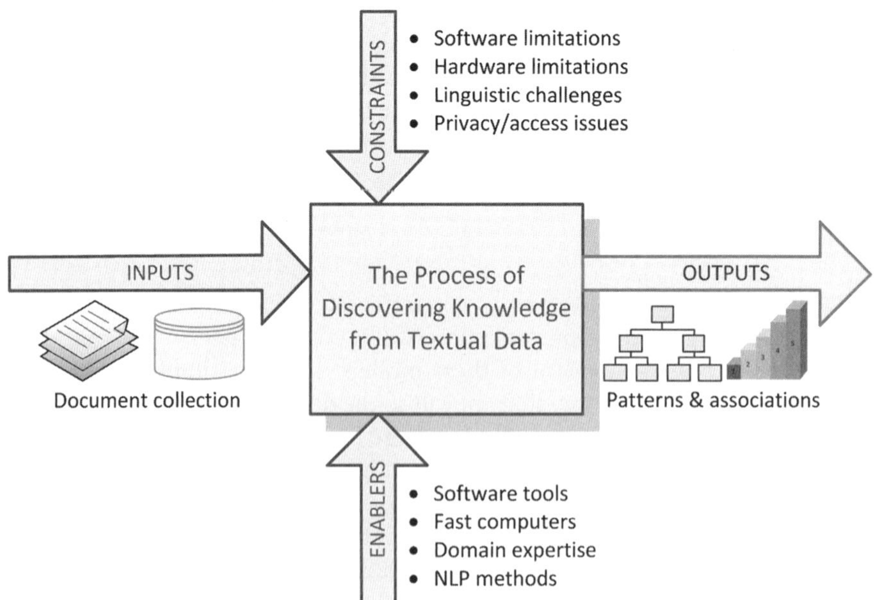

FIGURE 5.3

A high-level context diagram for the text mining process.

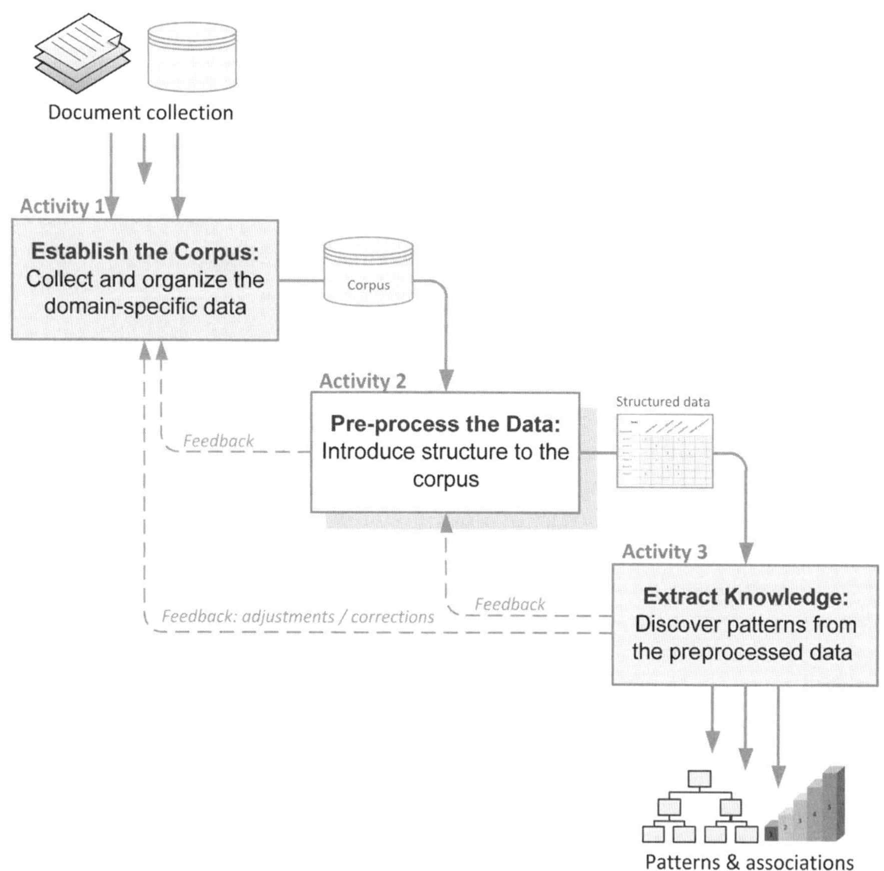

FIGURE 5.4
A detailed view of the context diagram for text mining.

If, for some reason, the output of a subprocess is not what was expected or emerges at an unsatisfactory level, feedback loops redirect information flow to a previous task to permit adjustments and corrections.

Phase 3, Activity 1: Establish the Corpus

The purpose of Activity 1 in Phase 3 is to collect all of the documents that are relevant to the problem being addressed (see Figure 5.4). The quality and quantity of the data are the most important elements of both data mining or text mining projects. Sometimes in a text mining project, the document collection is readily available and is accompanied by the project description (e.g., conducting sentiment analysis on customer reviews of a specific product or service). But usually the text miner is required to identify and collect the problem-specific document collection using either manual or automated techniques (e.g., a web crawler that periodically collects relevant news excerpts from several websites). Data collection may include textual documents, HTML files, emails, web posts, and short notes. In addition to normal textual data, voice recordings may be included by transcribing using speech-recognition algorithms.

Once collected, the text documents are transformed and organized in a manner such that they are all represented in the same form (e.g., ASCII text files) for computer processing. The organization of these documents can be as simple as a collection of digitized text excerpts stored in a file folder, or it can be a list of links to a collection of web pages in a specific domain. Many commercially available text mining software tools could accept these web pages as input and convert them into a flat file for processing. Alternatively, flat files can be prepared outside the text mining software and then presented as the input to the text mining application.

Phase 3, Activity 2: Preprocess the Data

In this activity, the digitized and organized documents (the corpus) are used to create a structured representation of the data, often referred to as the *term–document matrix* (TDM). Commonly, the TDM consists of rows represented by documents and columns representing terms. The relationships between the terms and the documents are characterized by indices, which are relational measures, such as how frequently a given term occurs in a document. Figure 5.5 illustrates a simplified example of a TDM.

The goal of Activity 2 is to convert the list of organized documents (the corpus) into a TDM where the cells are filled with the most appropriate indices. The assumption we make here is that the "meaning" of a document can be represented with a list and frequency of the terms used in that document. But are all terms equally important when characterizing documents? Obviously, the answer is "no." Some terms, such as articles, auxiliary verbs, and terms used in almost all of the documents in the corpus, have no distinguishing power and therefore should be excluded

Documents \ Terms	market share	resource utilization	project schedule	acquisition	material requirement	...
Article 1	1			1		
Article 2		1				
Article 3			3		1	
Article 4		1				
Article 5			2	1		
Article 6	1			1		
...						

FIGURE 5.5
A term-by-document matrix

from the indexing process. This list of terms, commonly called *stopterms,* is often specific to the domain of study and should be identified by the domain experts. On the other hand, one might choose a set of predetermined terms under which the documents are to be indexed; this list of terms is conveniently called *include terms* or *dictionary*. Additionally, synonyms (pairs of terms that are to be treated the same) and specific phrases (e.g., "Supreme Court") can also be provided so the index entries are more accurate. Figure 5.6 shows a more detailed view of the TDM with its four tasks.

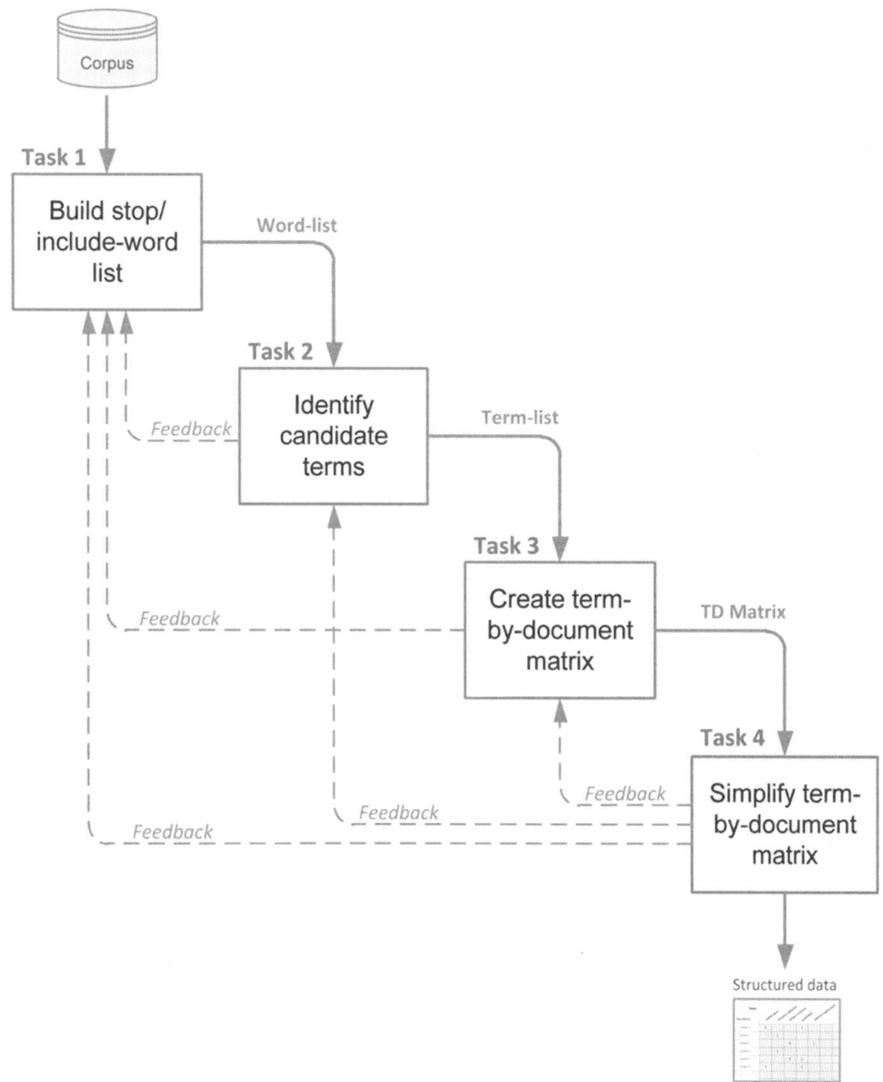

FIGURE 5.6
Decomposition of Activity 2 (preprocess the data) to the task level.

Task 1

The first task generates stopterms (or include terms) along with synonyms and specific phrases.

Task 2

The term list is created by *stemming* or *lemmatization*, which refers to the reduction of terms to their simplest forms (i.e., roots). An example of stemming is to identify and index different grammatical forms or declinations of a verb as the same term. For example, stemming will ensure that *model*, *modeling*, and *modeled* will be recognized as the term *model*. In this way, stemming will reduce the number of distinct terms and increase the frequency of some terms. Stemming has two common types:

1. *Inflectional stemming:* This aims to regularize grammatical variants such as present/past and singular/plural (this is called *morphological analysis* in computational linguistic). The degree of difficulty varies significantly from language to language.
2. *Stemming to the root:* This aims to reach a root form with no inflectional or derivational prefixes and suffixes, which may lead to the least number of terms.

Task 3

Create the TDM. In task 3, a numeric two-dimensional matrix representation of the corpus is created. Generation of the first form of the TDM includes three steps:

1. Specifying all the documents as rows in the matrix
2. Identifying all of the unique terms in the corpus (as its columns), excluding the ones in the stop term list
3. Calculating the occurrence count of each term for each document (as its cell values)

If the corpus includes a rather large number of documents (as is commonly the case), then it is common for the TDM to have a very large number of terms. Processing such a large matrix might be time consuming, and, more importantly, it might lead to extraction of inaccurate patterns. These dangers of large matrices and time-consuming operations pose the following two questions:

- What is the best representation of the indices for optimal processing by text mining programs?
- How can the dimensionality of this matrix be reduced to a more manageable size to facilitate more efficient and effective processing?

To answer question #1, we must evaluate various forms of representation of the indices. One approach is to transform the term frequencies. Once the input documents are indexed and the initial term frequencies (by document) have been computed, a number of additional transformations can be performed to summarize and aggregate the extracted information. Raw term frequencies reflect the relative prominence of a term in each document. Specifically, terms that occur with greater frequency in a document may be the best descriptors of the contents of that document. However, it is not reasonable to assume that the term counts themselves are proportional to their importance as descriptors of the documents. For example, even though a term occurs three times more often in document A than in document B, it is not necessarily reasonable to conclude that this term is three times as important a descriptor of document *B* as it is for document *A*.

In order to have a more consistent TDM for further analysis, these raw indices should be *normalized*. In statistical analysis, normalization consists of dividing multiple sets of data by a common value in order to eliminate different effects of different scales among data elements to be compared. Raw frequency values can be normalized using a number of alternative methods. The following are few of the most commonly used normalization methods (StatSoft, 2010):

- **Log frequencies.** The raw frequencies can be transformed using the log function. This transformation would "dampen" the raw frequencies and how they affect the results of subsequent analysis.
- $$f(wf) = 1 + \log(wf) \quad \text{for } wf > 0$$

  In the formula, *wf* is the raw term frequency and *f*(*wf*) is the result of the log transformation. This transformation is applied to all of the raw frequencies in the TDM where the frequency is greater than zero.
- **Binary frequencies.** Likewise, an even simpler transformation can be used to enumerate whether a term is used in a document.

  $$f(wf) = 1 \quad \text{for } wf > 0$$

  The resulting TDM matrix will contain only 1s and 0s to indicate the presence or absence of the respective terms. Again, this transformation will dampen the effect of the raw frequency counts on subsequent computations and analyses.
- **Inverse document frequencies.** In addition to normalized frequency of terms, the importance of a given term in each document (relative document frequency or *df*) is also an important aspect to include in the analysis. For example, a term such as *guess* may occur frequently in all documents, whereas another term, such as *software*, may appear only a few times. The reason is that one might make *guesses* in various contexts, regardless of the specific topic, whereas *software* is a more semantically focused term that is likely to occur only in documents that deal with computer software. A common and very useful transformation that reflects both the specificity of terms (relative document frequencies) as well as the overall frequencies of their occurrences (transformed term frequencies) is the so-called *inverse document frequency* (Manning and Schutze, 1999). This transformation for the *i*th term and *j*th document can be written as:

$$idf(i,j) = \begin{cases} 0 & \text{if } wf_{ij} = 0 \\ (1 + \log(wf_{ij}))\log\dfrac{N}{df_i} & \text{if } wf_{ij} \geq 1 \end{cases}$$

where wf_{ij} is the normalized frequency of the *i*th term in the *j*th document, df_i is the document frequency for the *i*th term (the number of documents that include this term), and *N* is the total number of documents. You can see that this formula includes both the dampening of the simple-term frequencies via the log function (described previously) and a weighting factor that evaluates to 0 if the term occurs in all of the documents [i.e., $\log(N/N = 1) = 0$], and to the maximum value when a term only occurs in a single document [i.e., $\log(N/1) = \log(N)$]. It can be seen easily how this transformation will create indices that reflect both the relative frequencies of occurrences of terms, as well as their document frequencies representing semantic specificities for a given document. This is the most commonly used transformation in the field.

This brings us to how to reduce the dimensionality of the TDM (question #2). Because, the TDM is often very large and rather sparse (most of the cells filled with zeros), this answer is more tractable to handle. Several options are available for reducing such matrices to a manageable size:

- A domain expert goes through the list of terms and eliminates those that do not make much sense for the context of the study (this is a manual, labor-intensive process).
- Eliminate terms with very few occurrences in very few documents.
- Transform the matrix using singular value decomposition.

Singular Value Decomposition

Singular value decomposition (SVD) is a method of representing a matrix as a series of linear approximations that expose the underlying meaning-structure of the matrix. The goal of SVD is to find the optimal set of factors that best predict the outcome. During data preprocessing prior to text mining operations, SVD is used in latent semantic analysis (LSA) to find the underlying meaning of terms in various documents.

In more technical terms, SVD is closely related to principal components analysis in that it reduces the overall dimensionality of the input matrix (number of input documents by number of extracted terms) to a lower dimensional space (a matrix of much smaller size with fewer variables), where each consecutive dimension represents the largest degree of variability (between terms and documents) possible (Manning and Schutze, 1999). Ideally, the analyst might identify the two or three most salient dimensions that account for most of the variability (differences) between the terms and documents, thus identifying the latent semantic space (is this term the same as *lower dimensional space*?) that organizes the terms and documents in the analysis. When these dimensions are identified, they represent the underlying "meaning" of what is contained (discussed or described) in the documents. For example, assume that matrix A represents an $m \times n$ term occurrence matrix, where m is the number of input documents and n is the number of terms selected for analysis. The SVD computes the $m \times r$ orthogonal matrix U, $n \times r$ orthogonal matrix V, and $r \times r$ matrix D, so $A = UDV'$ and r is the number of eigenvalues of $A'A$.

Phase 3, Activity 3: Extract the Knowledge

Novel patterns are extracted in the context of the specific problem being addressed, using the well-structured TDM, and possibly augmented with other structured data elements (such as numerical and/or nominal variables, potentially including the time and place specifications of the documents). These are the main categories of knowledge extraction methods in text mining studies:

- Prediction (e.g., classification, regression, and time-series analysis)
- Clustering (e.g., segmentation and outlier analysis)
- Association (e.g., affinity analysis, link analysis, and sequence analysis)
- Trend analysis

Classification

Arguably the most common knowledge discovery topic in analyzing complex data sources is the *classification* of certain objects or events into predetermined classes (or categories). The goal of classification is to assign the data instance into a predetermined set of classes (or categories). As it applies to the domain of text mining, the task is known as *text categorization,* where for a given set of categories and a collection of text documents, the challenge is to find the correct topic (subject or concept) for each document. This challenge is met with building models developed with a training data set that include both the documents and actual document categories. Today, automated text categorization is applied in a variety of contexts, including iterative (automatic or semiautomatic) indexing of text, spam filtering, web page categorization under hierarchical catalogs, automatic generation of metadata, detection of genre, and many others.

The two main approaches to text classification are expert systems (via the use of knowledge engineering techniques) and classification modeling (via the use of statistical and/or machine-learning techniques). With the expert system approach, an expert's knowledge about the categories is encoded into the classification system using a declarative representation in the form of production rules. With the machine-learning approach, a generalized inductive process is employed to build a classifier by "learning" from a set of preclassified examples. As the number of documents increases at an exponential rate and as the availability of knowledge experts becomes scarcer, the popularity trend is shifting toward the machine-learning–based automated classification techniques.

Clustering

Clustering is an unsupervised process whereby objects or events are placed into "natural" groupings called *clusters*. An unsupervised process is one that uses no pattern or prior knowledge to guide the clustering process. Text categorization is a supervised process, where a collection of preclassified training examples is used to develop a model based on the descriptive features of the classes in order to classify a new unlabeled example. In the unsupervised clustering process, the problem is to group an unlabeled collection of objects (e.g., documents, customer comments, web pages) into meaningful clusters without any prior knowledge.

Clustering is useful in a wide range of applications, from document retrieval to enabling better web content searches. In fact, one of the prominent applications of clustering is the analysis and navigation of very large text collections, such as web pages. The basic underlying assumption is that relevant documents tend to be more similar to one another than to irrelevant ones. If this assumption holds, the clustering of documents based on the similarity of their content improves search effectiveness (Feldman and Sanger, 2007).

The two most popular clustering methods are scatter/gather clustering and query-specific clustering. *Scatter/gather* method uses clustering to enhance the efficiency of human browsing of documents when a specific search query cannot be formulated. In a sense, the method dynamically generates a table of contents for the collection and adapts and modifies it in response to the user selection. On the other hand, *query-specific clustering* method employs a hierarchical clustering approach where the most relevant documents to the posed query appear in small tight clusters that are nested larger clusters containing less similar documents, creating a spectrum of relevance levels among the documents. This method performs consistently well for document collections of relatively large sizes.

Association

Association is the process of finding affinities/correlations among different data elements (objects or events). In the retail industry, association analysis is often called market basket analysis. The primary idea in generating association rules is to identify the frequent sets of "things" that go together in a specific context. A famous example in retail is the association of beer and diapers in the same shopping cart (related to Monday night football games on TV).

In text mining, associations refer specifically to the direct relationships between concepts (or terms) or sets of concepts. An association rule, $X \Rightarrow Y$, relating two frequent concept sets X and Y, can be quantified (or substantiated) by the two basic measures, support and confidence. *Confidence* is the percent of documents that include all the concepts in Y within the same subset of those documents that include all the concepts in X. *Support* is the percentage (or number) of documents that include all the concepts in X and Y. For instance, in a document collection the concept "project failure" may appear most often in association with "enterprise resource planning" and "customer relationship management" with support of 4% and confidence of 55%, meaning that 4% of the documents in the corpus had all three concepts presented together in the same document, and 55% of the documents that included "project failure" also included "enterprise resource planning" and "customer relationship management."

In an interesting text mining study, association analysis was used to study published literature (news articles, academic publication and web postings) to map out the outbreak and progress of bird flu (Mahgoub et al., 2008). The principal goal of this study was to automatically identify the associations among the geographical areas, spreading across species, and countermeasures (i.e., treatments).

A special case of association analysis is where the concepts are associated with one another in an orderly means (e.g., a sequence in which the concepts tend to appear) or over a specific time period. This type of association analysis is called *trend analysis*, which is briefly explained in the following section.

Trend Analysis

The main goal in trend analysis is to find the time-dependent changes for an object or event. Often, trend analysis in text mining is based on the notion that various types of concept distributions over time are functions of the specific document collections; that is, different collections of the same topic representing different time intervals may lead to different concept distributions. It is therefore possible to compare the time-varying changes in two concept distributions that are otherwise identical except that they are from different subcollections. One notable direction of this type of analysis is having two collections from the same source (such as from the same set of academic journals) but from different points in time. Delen and Crossland (2008) applied trend analysis to a large number of academic articles (published in three highly rated academic journals) to identify the evolution of key concepts in the field of information systems.

EXAMPLE 1: AN EXPLORATORY LITERATURE SURVEY USING TEXT MINING

The explosion of text in various literature domains has rendered literature search and review as a very complex and voluminous operation. When creating new text in the process of extending the body of knowledge, it has always been crucial to gather, organize, analyze, and assimilate existing information from the literature in a particular discipline. The thoroughness of a literature search is increasingly difficult to attain in the face of the increasing abundance of potentially significant research reported in related fields—and even in previously unrelated fields. What was unrelated previously might indeed be relevant now.

In new streams of research, the researcher's task may be even more tedious and complex. Trying to ferret out relevant work that others have reported may be difficult, at best, and perhaps nearly impossible when traditional, largely manual reviews of published literature are required. Even with a legion of dedicated graduate students or helpful colleagues, adequate coverage of all potentially relevant published work is problematic.

Example 2: Semiautomated Analysis

In a study, Delen and Crossland (2008) proposed a method to greatly assist and enhance the efforts of the researchers by enabling a semiautomated analysis of large volumes of published literature with text mining. Using standard digital libraries and online publication search engines, the authors downloaded and collected all of the available articles for the three major journals in the field of information systems: *MIS Quarterly* (MISQ), *Information Systems Research* (ISR), and *Journal of Management Information Systems* (JMIS). In order to work with the same time span for all three journals (to enable future comparative studies), the journal with the most recent starting date for the availability of a digital version was used as the start time for this study. For each article, extracted data included the title, abstract, author list, published keywords, volume, issue number, and year of publication. The extracted data, including all of the articles and their above mentioned features, was loaded into a simple database file. Also included in the combined data set was a field that designated the journal type of each article to serve future segmentation operations and discriminatory analyses. Editorial notes, research notes, and executive overviews were omitted from the collection. At the end, over 900 articles were included in the corpus of their study. Table 5.1 shows a snapshot of how their data were organized in a tabular format.

Table 5.1 A Snapshot of the Data Represented in a Tabular Format

Journal	Year	Author(s)	Title	Vol/No	Pages	Keywords	Abstract
MISQ	2005	A. Malhotra, S. Gosain and O. A. El Sawy	Absorptive capacity configurations in supply chains: Gearing for partner-enabled market knowledge creation	29/1	145-187	knowledge management supply chain absorptive capacity interorganizational information systems configuration approaches	The need for continual value innovation is driving supply chains to evolve from a pure transactional focus to leveraging interorganizational partner ships for sharing
ISR	1999	D. Robey and M. C. Boudreau	Accounting for the contradictory organizational consequences of information technology: Theoretical directions and methodological implications	2-Oct	167-185	organizational transformation impacts of technology organization theory research methodology intraorganizational power electronic communication mis implementation culture systems	Although much contemporary thought considers advanced information technologies as either determinants or enablers of radical organizational change, empirical studies have revealed inconsistent findings to support the deterministic logic implicit in such arguments. This paper reviews the contradictory
JMIS	2001	R. Aron and E. K. Clemons	Achieving the optimal balance between investment in quality and investment in self-promotion for information products	18/2	65-88	information products internet advertising product positioning signaling signaling games	When producers of goods (or services) are confronted by a situation in which their offerings no longer perfectly match consumer preferences, they must determine the extent to which the advertised features of
...	...	...	...	...	...	...	...

Using these data, the authors conducted two studies. The first exploratory study was to look at the time-varying perspective of the three journals (i.e., evolution of research topics over time). In order to conduct a time-based associative study, they divided the 12-year period (from 1994 to 2005) into four 3-year periods for each of the three journals. This framework led to 12 text mining experiments with 12 mutually exclusive data sets consisting of abstracts of articles. For each of the 12 data sets, text mining was used to extract the most descriptive terms from these collections abstracts. Results were tabulated and examined for time-varying changes in the terms published in these three journals.

A second exploration used the complete data set (including all three journals and all four periods) and employed clustering to extract knowledge from the abstracts. Clustering is arguably the most commonly used text mining technique employed today. It was used in this study to identify the natural groupings of the articles and then list the most descriptive terms that characterized those clusters. Singular value decomposition was used to reduce the dimensionality of the TDM, and then an expectation-maximization algorithm was used to create the clusters. Several experiments were performed to identify the *optimal* number of clusters, which proved to be nine. After the construction of the nine clusters, the content of those clusters was analyzed from two perspectives: representation of the journal type (Figure 5.7) and representation of time (Figure 5.8). The idea was to explore the potential similarities and differences among the three journals and potential changes in the importance associated with those clusters. The importance was evaluated according to two questions: 1) are there clusters that represent different research themes specific to a single journal? and 2) is there a time-varying characterization of those clusters? Several interesting patterns discovered in this study are illustrated in Figures 5.7 and 5.8. Figure 5.7 shows the representation of the three journals (in terms of

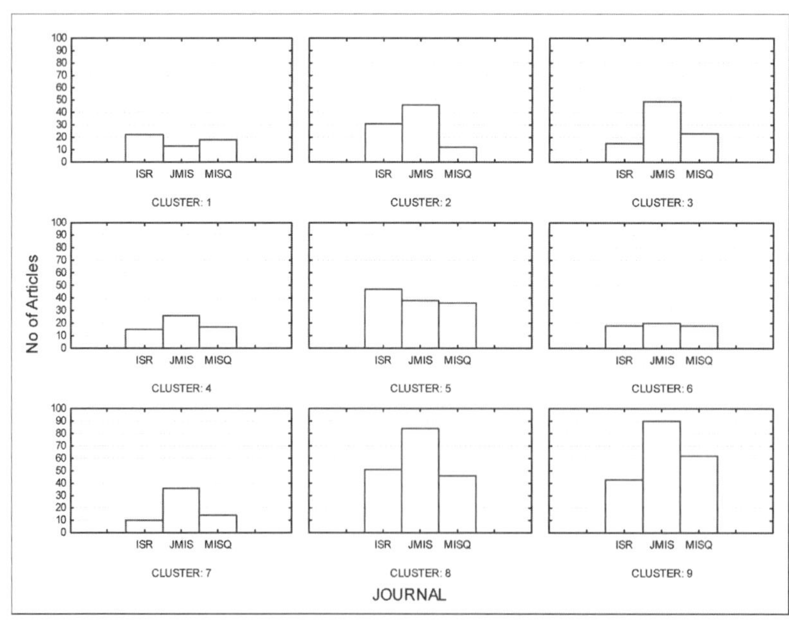

FIGURE 5.7

Distribution articles for the three journals over the nine clusters.

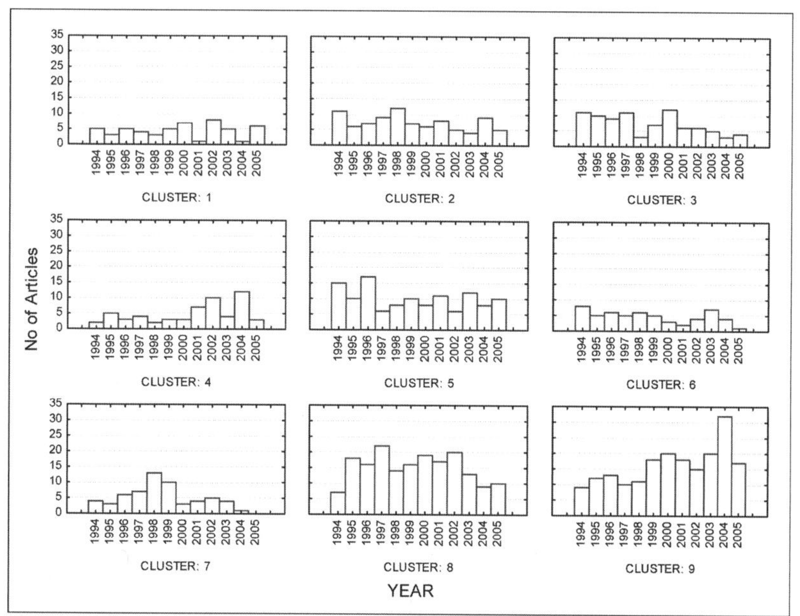

FIGURE 5.8

Distribution of articles over time for each of the nine clusters. *Source: Delen and Crossland, 2008.*

number of articles included) for each of the nine clusters, indicating the overlap (or lack thereof) among the three journals in terms of publishing topics that are represented in nine clusters. Figure 5.8 illustrates the distribution of the articles during the 12-year period for each of the nine clusters.

POSTSCRIPT

Chapters 1–5 present the history, theory, application areas, and a proposed process model to accomplish text mining projects. Chapter 6 introduces you to three common text mining tools that you might use for text mining projects. These text mining software packages are used also in one or more tutorials, although other tools are used also.

References

Chapman, P., J. Clinton, R. Kerber, T. Khabanza, T. Reinartz, C. Shearer, and R. Wirth. (2000). "CRISP-DM—Step-by-step data mining guide." SPSS, Chicago, IL.

Delen, D., and M. Crossland, "Seeding the Survey and Analysis of Research Literature with Text Mining," *Expert Systems with Applications* Vol. 34, No. 3, 2008, pp. 1707–1720.

Feldman, R., and J. Sanger. (2007). *The Text Mining Handbook: Advanced Approaches in Analyzing Unstructured Data.* Boston: ABS Ventures.

Mahgoub, H., D. Rösner, N. Ismail, and F. Torkey. (2008). "A Text Mining Technique Using Association Rules Extraction." *International Journal of Computational Intelligence,* Vol. 4, No. 1, pp. 21–28.

Manning, C. D., and H. Schutze. (1999). *Foundations of Statistical Natural Language Processing.* Cambridge, MA: MIT Press.

StatSoft. (2010). *Statistica Data and Text Miner User Manual.* Tulsa, OK: StatSoft, Inc.

CHAPTER 6

Three Common Text Mining Software Tools

CONTENTS

PREAMBLE

In Chapters 1–5, we present the history and fundamentals of theory and process of text analytics. In this chapter, we list many of the very large number of commercial and open-source/freeware text analytics tools that are available today for text mining, web analytics, and web mining. Our challenge in the tutorial provided by this book is to focus not only on common text analytics problems but also on three of the most popular commercial tools available to solve them: SAS-Enterprise Miner, IBM Modeler, and *STATISTICA* Data Miner.

INTRODUCTION

Text analytics capabilities have been available from the three major commercial data analytics companies for many years, and they are relatively complete and mature in their execution. In the last few years, many commercial and open-source (freeware) tools have become available, but most of them provide only rudimentary text analytics capabilities, or they are dedicated to specific application niches. Few of these tools will perform all of the common text analytics tasks required for a text analytics project.

Practical Text Mining and Statistical Analysis for Non-structured Text Data Applications. DOI: 10.1016/B978-0-12-386979-1.00006-2

RapidMiner, an open-source tool, is used in tutorials H and O. Tutorial J uses the Salford commercial data mining tool, combined with an open-source text mining tool to provide a very comprehensive example of text mining. You can see an extensive PowerPoint presentation of this solution on the enclosed DVD.

IBM SPSS MODELER PREMIUM

IBM SPSS pioneered predictive text analytics by integrating text analytics in its predictive workbench modeler in 2002. IBM® SPSS® Text Analytics (IBM-TA) offers powerful text analytical capabilities that use advanced linguistic technologies and NLP to rapidly process a large variety of unstructured text data and extract and organize the key concepts and sentiments. IBM-TA offers three facilities to help process text quickly, efficiently, and accurately within specific language and application domains.

Prebuilt Dictionaries

In contrast to some NLP systems for text mining, IBM-TA doesn't require any specific external dictionaries to start discovering new or unknown concepts. IBM-TA is delivered with a set of vertical templates for specific domains, such as customer/employee/bank/product satisfaction, market intelligence, security intelligence, and genomics. Each of these templates contains specific linguistic resources, such as dictionaries for terms, synonyms, and text link analysis rules that are specific to each domain. This product permits you to develop and refine these linguistic resources to your context. Fine-tuning of the linguistic resources is often an iterative process and is necessary for accurate concept retrieval and categorization.

Text Analysis Package Organization Framework

IBM-TA provides a framework for storing language resources in a text analysis package (TAP). A TAP serves as a template for text response categorization. Using a TAP is an easy way for you to categorize your text data with minimal intervention, since it contains the prebuilt category sets and the linguistic resources needed to code a vast number of records quickly and automatically. Using the linguistic resources, text data are analyzed and mined to extract key concepts. Based on key concepts and patterns found in the text, the records can be categorized into the category set you selected in the TAP. You can make your own TAP, or you can update an existing one.

Language Weaver

In addition to TAPs and prebuilt dictionaries, IBM-TA provides Language Weaver, a facility that enables you to translate Arabic, Chinese, Persian, and several other languages into English. You can perform your text analysis on translated text and deploy these results to people who could not have understood the contents of the source languages. Because the text mining results are automatically linked back to the corresponding foreign language text, your organization can focus the much needed native speaker resources on only the most significant results of the analysis.

The IBM Modeler Interface

Figure 6.1 shows the main window of the graphical user interface (GUI) of the IBM Modeling canvas. The string of icons along the top composes the main toolbar. The section at the upper right of the window is the Manager pane, where you can select the Streams pane (where available streams are stored), the Objects pane (where tables and graphs are stored), and the Models pane, where trained

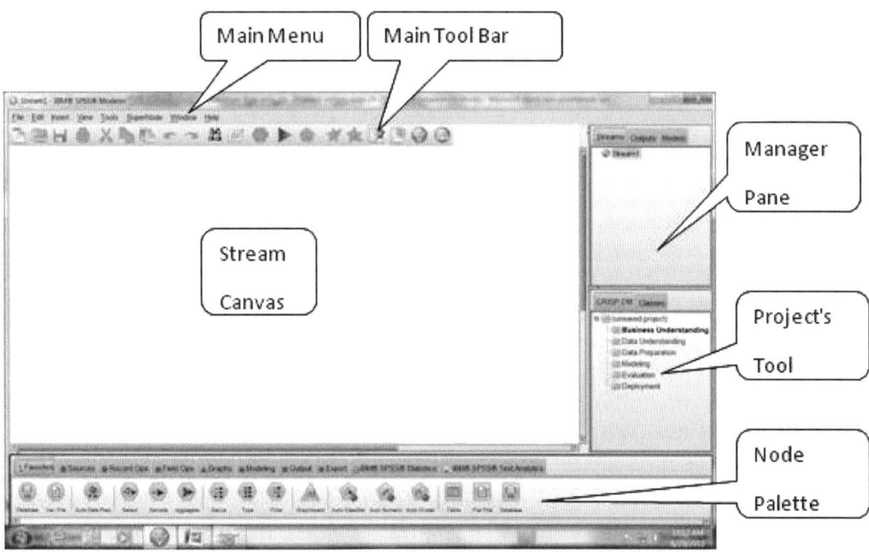

FIGURE 6.1
The main window showing the six information sections.

models are stored. Just below the Manager pane is the Project's Tool pane, which contains two tabs: the CRISP-DM view tab, which permits you to display streams and outputs in the order of the CRISP-DM process model, and the Classes view tab, which permits you to organize your work in terms of project objects, including streams, nodes, models, reports, and other non-Modeler files. Along the bottom of the main window is the Node Palette, which is composed of up to ten tabs for different types of nodes (depending on licensing) that each contain a series of processing nodes. The large pane in the middle of the main window is the stream canvas, which processes data streams that consist of a series of nodes connected by arrows. Nodes from the Node Palette are moved to the stream canvas by drag-and-drop or by right-clicking the node and selecting Add to the canvas.

Adding Nodes to Modeling Streams

There are two ways to add nodes to a stream:

1. Double-click on the node in the Node palette.
2. Right-click on the node in the Node Palette.

For both ways, you can combine adding and linking of nodes by highlighting the node to be connected to and then performing the Add Node operation.

Connecting Nodes in Modeling Streams

There are five ways to connect nodes on the stream canvas:

1. Highlight the source node in the stream to connect from, and double-click a new node on the Node Palette, and a new node will be connected with an arrow from the source node.

2. Click on the source node to highlight it, right-click on the source node, and select Connect; then click on the destination node.
3. Click on the source node to highlight it, press F2, and click on the destination node; then manually connect the nodes.
4. On a three-button mouse, click the middle button, and drag an arrow from the source node to the destination node.
5. On a two-button mouse, press ALT, and drag the arrow from the source node to the destination node.

Accessing Nodes

There are ten tabs in the Node Palette displayed along the bottom of the screen:

Favorites—lists only the commonly used nodes
Sources—lists 10 types of source data
Record Ops—nodes for 9 operations you can perform at the records level
Field Ops—nodes for 17 operations you can perform at the field level
Graphs—nodes for 9 graph types for displaying data
Modeling—nodes for 29 modeling algorithms you can use to train models

Adding Data Sources

There are ten ways to add a new data source. The following are the most common ways to access data sources:

Var(iable) File Node—for loading data in delimeted files (user specified)
Database—for accessing database tables via ODBC connection
Enterprise View—for accessing database tables directly via the call level interface of the database management system
SAS File—reads data files in SAS format

Running Streams (or Parts of Them)

You can run the entire stream on the stream canvas by clicking on the right-facing triangle (will be colored "red" on the computer screen) on the main toolbar (▶). Or you can run the stream (or any part of it) by right-clicking on a node and selecting the "Run from Here" option at the bottom of the pop-up menu.

Using the Help File

On the main menu (along the top of the screen), click on the Help menu option.
Select Help Topics.
Enter the help topic in the Search box at the upper left corner of the screen.

You can learn about the other nodes in the Node Palette by entering the node name in the Search box.

Modeler Text Analytics Node Tab

You can incorporate the power of text analysis into your streams by using the nodes in the Text Analytics tab on the Node Browser. For Figures 6.2 through 6.6 show only the relevant parts of the whole user interface screen; this was done in order to zoom in far enough on the nodes to read the text associated with the objects displayed.

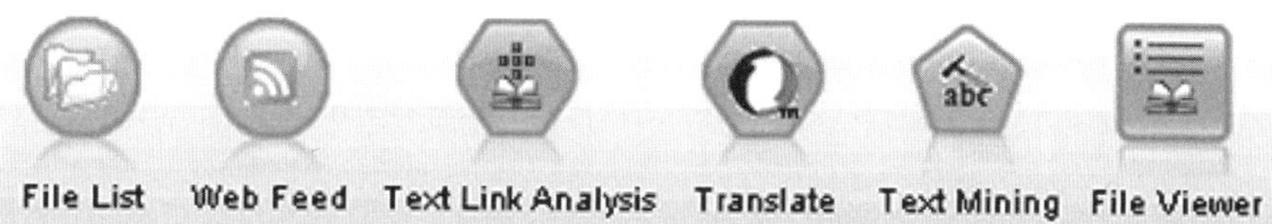

FIGURE 6.2
IBM-TA tab on the Node Palette.

Figure 6.2 shows the six text processing operations that you can perform with IBM-TM. These nodes are contained in the IBM-TA tab on the Node i. The **File List source node** generates a list of document names as input to the text mining process. This node is useful when the text resides in external documents rather than in a database or other structured file. The node outputs a single field with one record for each document or folder listed, which can be selected as input in a subsequent Text Mining node.

The **Web Feed source node** makes it possible to read in text from web feeds, such as blogs or news feeds in RSS or HTML formats, and use data from them in the text mining process. The node outputs one or more fields for each record found in the feeds, which can be selected as input in a subsequent Text Mining node.

The **Text Mining node** uses linguistic methods to extract key concepts in eight different languages, allows you to create categories with these concepts and other data, and offers the ability to identify relationships and associations between concepts based on known patterns (called text link analysis). The node can be used to explore the text data contents or to produce either a concept model or category model. The concepts and categories can be combined with existing structured data, such as demographics, and applied to modeling.

The **Text Link Analysis node** extracts concepts and also identifies relationships between concepts based on known patterns within the text. Pattern extraction can be used to discover relationships between your concepts, as well as any opinions or qualifiers attached to these concepts. The Text Link Analysis is used to extract events, facts, or sentiments.

The **Translate node** can be used to translate text from supported languages, such as Arabic, Chinese, and Persian, into English or other languages for purposes of modeling. This makes it possible to mine documents in double-byte languages that would not otherwise be supported and allows analysts to extract concepts from these documents even if they are unable to speak the language in question. The same functionality can be invoked from any of the text modeling nodes, but using a separate Translate node makes it possible to cache and reuse a translation in multiple nodes.

When mining text from external documents, the **File Viewer node** can be used to generate an HTML page that contains links to the documents from which concepts were extracted.

Integrating Modeling Operations with Cross Industry Standard Process for Data Mining

Located in the bottom right corner of the interface, the Project Tools panel will display where in the CRoss Industry Standard Process for Data Mining (CRISP-DM) outline each stream in the project belongs. In a rental car project for modeling customer behavior, the project streams are displayed in the Streams tab of the Manager pane, and they are distributed to their appropriate place in the CRISP-DM process in the Classes tab of the Project Tools pane (Figure 6.3).

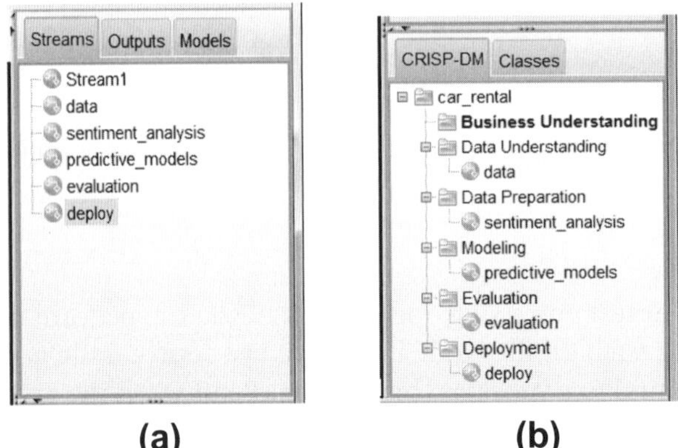

FIGURE 6.3
Streams in the Car_Rental project displayed in the Streams tab of the Manager pane (a) and in the Classes tab of the Project Tools pane (b).

How to Interpret Streams

Figure 6.4 shows the data stream of the Car_Rental project, which merges two data sources and creates display tables of the input sources and the merged output. This stream can be interpreted according to the sequence of its operations as follows:

Customer data (customer_dta.csv) and the customer satisfaction survey returns (satisfaction_survey.csv) are input to the Merge node, shown by the arrows.
Data records from both input sources are merged together in the Merge node by combining records into a single output.
The output of the merge operation and each of the data inputs are directed to a Table node. Double-clicking on the Table node will display the contents of the table.

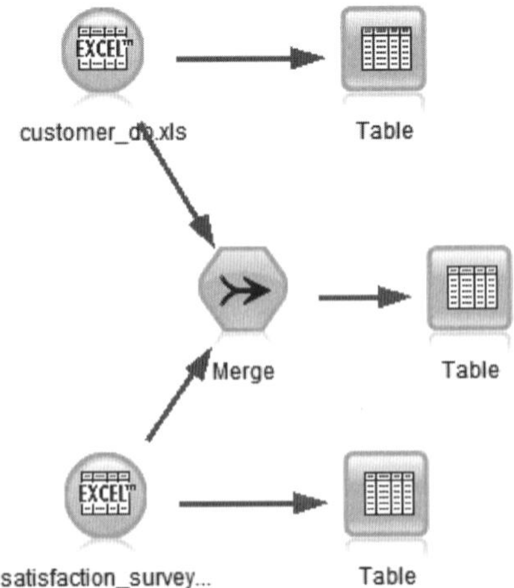

FIGURE 6.4
The data.str stream in the Car_Rental project.

The output tables are included in the data.str for data display purposes only. They will be removed when the data merge operation is included in subsequent streams, as in Figures 6.5 and 6.6.

Other Properties of the Interface

SuperNodes serve as a "shorthand" representation for a string of nodes in a stream. You can collapse a string of nodes into a SuperNode by holding down the CTRL key, selecting all of the nodes in sequence, and then clicking the SuperNode icon on the main toolbar (★). To expand a SuperNode, click the other star icon next it. Figure 6.5 shows the predicted_models stream in full display, and Figure 6.6 shows the same stream using a SuperNode to contain the data merge operation. Note that the SuperNode was renamed to Data Merge by right-clicking on the SuperNode icon and selecting Rename and Annotate. The Data Merge SuperNode in Figure 6.6 represents the operations in the customer_dta.str stream, which was included in the predicted_models.str stream (without the output tables).

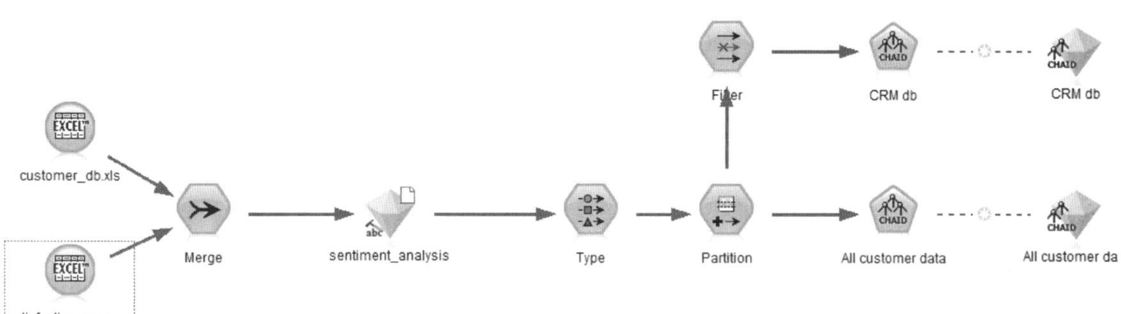

FIGURE 6.5
The predicted_models.str stream in full display mode.

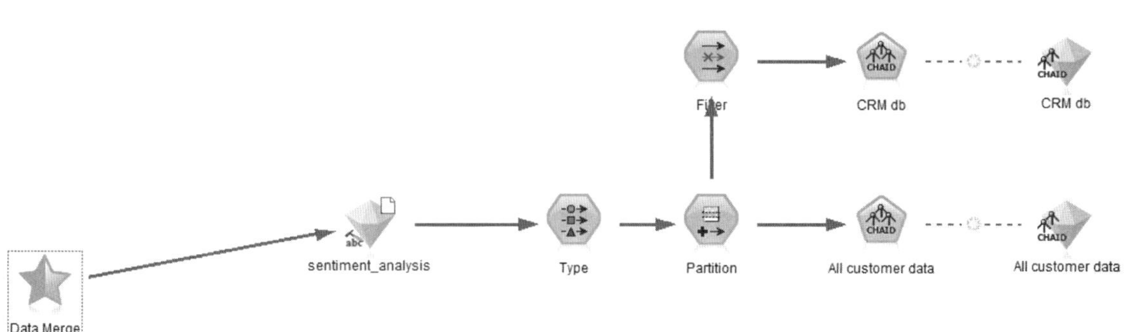

FIGURE 6.6
The predicted_models stream in SuperNode display mode.

Outputs

Various ways to output data from modeling streams are listed in the Output tab on the Node Palette:

Table—for viewing output data in tabular format
Matrix—creates a table showing relationships between fields
Analysis—creates a report showing accuracies of predictive models
Data Audit—creates a summary table showing statistics and handling of outliers, missing values, and other extreme values
Transform—performs interactive data transforms
Statistics—calculates univariate statistics
Means—compares means between pairs of fields or between groups within a field
Report—creates a report from your data
Set Globals—creates global variables for your streams

SAS TEXT MINER

With SAS Text Miner software, you can uncover underlying themes or concepts contained in large document collections; automatically group documents into topical clusters; classify documents into predefined categories; and integrate text data with structured data to enrich predictive modeling endeavors. This chapter assumes the user has basic knowledge and terminology of SAS® and SAS-EM (SAS-EM), but extensive knowledge is not necessary.

SAS-TM is a plug-in for the SAS-EM environment. SAS-EM provides a rich set of data mining tools that facilitate the prediction aspect of text mining (Figure 6.7). The integration of SAS-TM within SAS-EM combines textual data with traditional data mining variables. The integration provides the ability to add text mining nodes into SAS-EM process flow diagrams. SAS-TM encompasses the parsing and

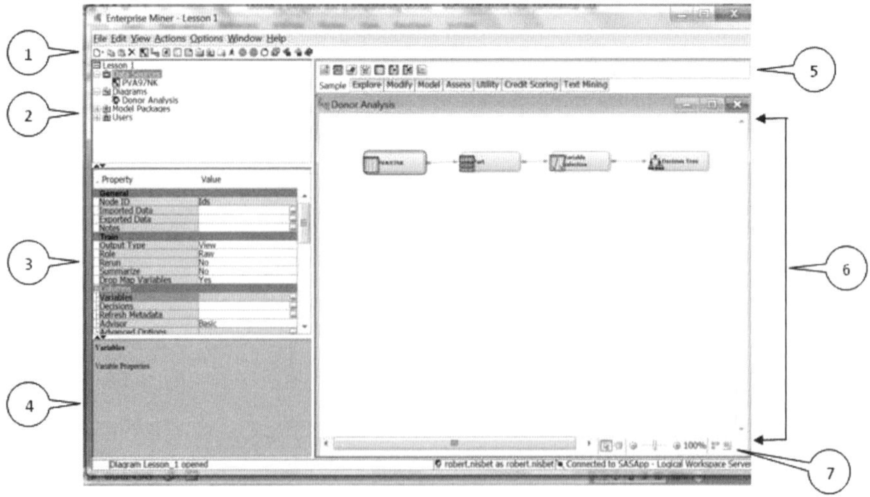

FIGURE 6.7
The SAS-EM main interface, showing the seven sections of operations.

exploration aspects of text mining and prepares data for predictive mining and further exploration using other SAS-EM nodes. SAS-TM supports various sources of textual data: local text files, text as observations in SAS® data sets or external databases, and files on the web. Figure 6.8 shows the SAS-EM interface with the Text Mining tab.

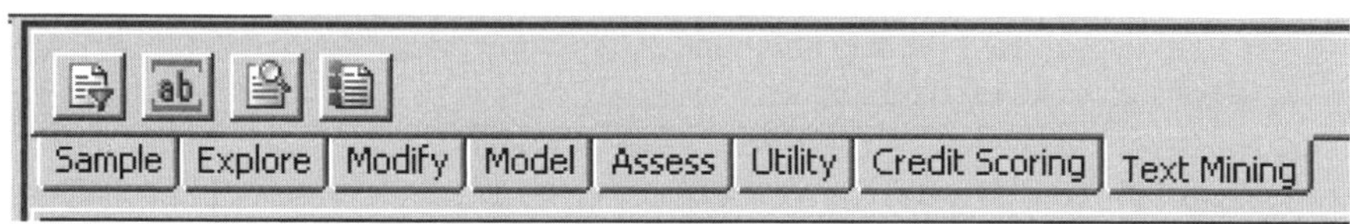

FIGURE 6.8
SAS-TM function bar illustrating the Text Mining tab selected.

1. Toolbar shortcut buttons: Use the toolbar shortcut buttons to perform common computer functions and frequently used SAS Enterprise Miner operations. Move the mouse pointer over any shortcut button to see the text name. Click on a shortcut button to use it.
2. Project panel: Use the Project panel to manage and view data sources, diagrams, results, and project users.
3. Properties panel: Use the Properties panel to view and edit the settings of data sources, diagrams, nodes, and users.
4. Property Help panel: The Property Help panel displays a short description of any property that you select in the Properties panel. Extended help can be found from the Help main menu.
5. Toolbar: The Toolbar is a graphic set of node icons that you use to build process flow diagrams in the Diagram Workspace. Drag a node icon into the Diagram Workspace to use it. The icon remains in place in the toolbar, and the node in the Diagram Workspace is ready to be connected and configured for use in the process flow diagram.
6. Diagram Workspace: Use the Diagram Workspace to build, edit, run, and save process flow diagrams. In this workspace, you graphically build, order, sequence, and connect the nodes that you use to mine your data and generate reports.
7. Diagram Navigation toolbar: Use the Diagram Navigation toolbar to organize and navigate the process flow diagram.

When you select the Text Mining tab on the SAS-EM interface, you can access four text mining nodes. As part of SAS-EM, SAS-TM follows the concept of nodes, where each node provides a customizable task. SAS-TM consists of four of these customizable nodes: the Text Filter node, the Text Miner node, the Text Parsing node, and the Text Topic node.

Text Filter Node

The Text Filter node (Figure 6.9) enables you to focus on the terms and documents that are most likely to enhance your model. For most of your text mining projects, you will want to follow the Text Parsing node with the Text Filter node. The Text Filter node allows for exploratory analysis through concept linking, drop/keep ability, and synonym analysis. Furthermore, this is where spell checking is done, so you can also add your own customized dictionary, as well as select term weightings and frequency

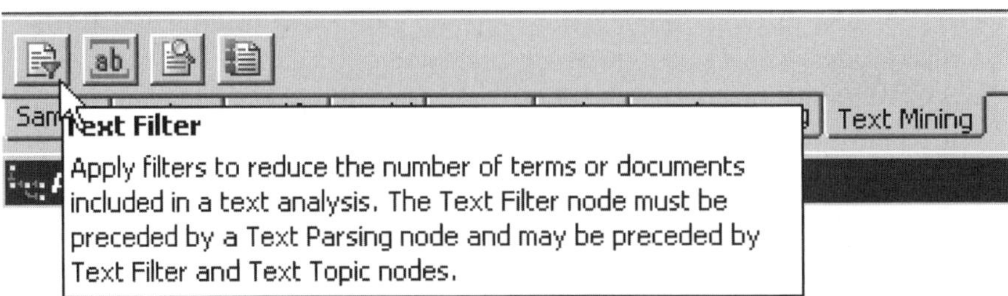

FIGURE 6.9
The Text Filter node, which is used to apply "filters."

weightings. This way, you can eliminate extraneous information caused by the presence of noise terms and other terms that are not pertinent to your analysis.

If your model would be improved by focusing on a subset of the collection, then the Text Filter node can remove documents that do not fit your criteria. The end result of the Text Filter node is a compact, yet information-rich, representation of your collection. For example, you can put a Text Filter node after a Text Topic node and filter for documents that contain specific topics.

Text Mining Node

The Text Mining node (Figure 6.10) is usually the first node in the processing stream. This node incorporates advanced natural language processing, term analysis and filtering, dimension reduction, and clustering in a single interactive node. The document clustering consists of expectation maximization and hierarchical options. Dimension reduction is accomplished through the use of single value decomposition (SVD).

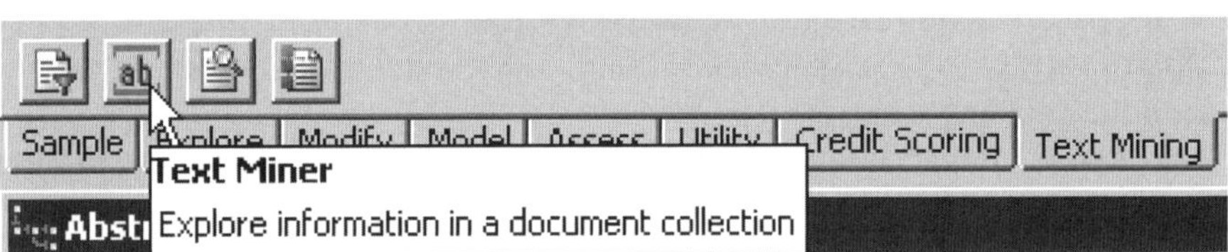

FIGURE 6.10
SAS-TM button used to explore information in documents.

Text Parsing Node

The Text Parsing node (Figure 6.11) enables you to use advanced natural language processing software to represent each document as a collection of terms. A term can represent a single word, either with or without its part of speech; multiword phrases; and custom or built-in entities. Additionally, you can choose to have certain terms represented by their synonyms or stems. The Text Parsing node gives you control over exactly what types of terms to include in your analysis. This node is the first step of any text mining analysis.

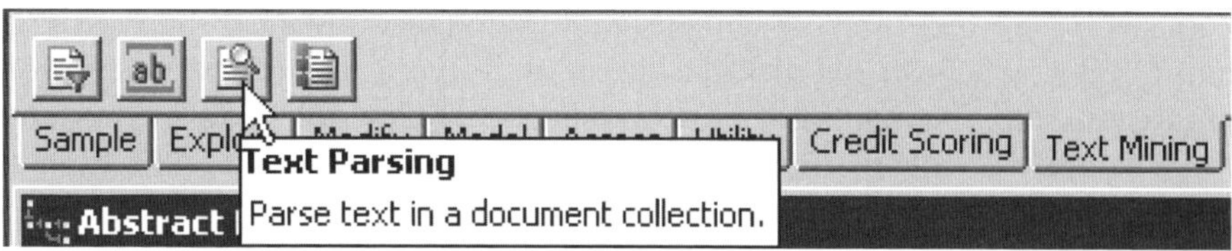

FIGURE 6.11
SAS-TM button used to do "text parsing" in a document.

Text Topic Node

The Text Topic node (Figure 6.12) enables you to define, discover, and modify sets of topics contained in your collection. A *topic* is a collection of terms that are strongly associated with a subset of your collection. Each document can contain zero, one, or many topics. Also, terms that are used to describe and define one topic can be used in other topics. There are no rules to determine how many topics you should create. The number of topics that you create is your preference and will vary from problem to problem.

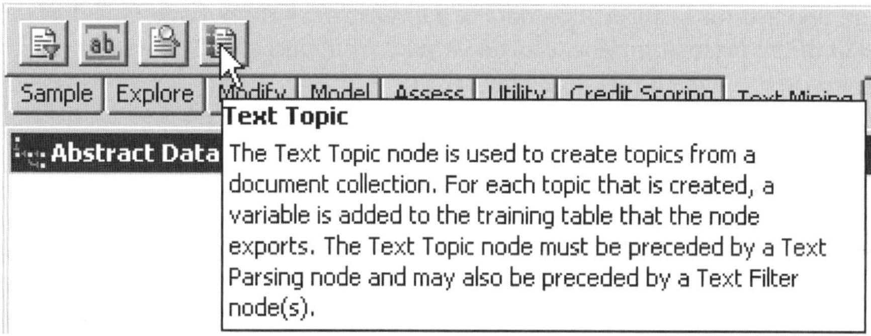

FIGURE 6.12
SAS-TM button used to "create topics" from a document collection.

The Text Topic node must be preceded by a Text Parsing node. If it is also preceded by a Text Filter node, then it will use the term weights set in that node. Otherwise, the Text Topic node will create term weights with the Log frequency weighting and Entropy term weighting settings.

Topic discovery is done through the use of a rotated single value decomposition. The topic node allows for user-defined topics as well as adjusting term weights, term cutoffs, and document cutoffs for better term definitions.

ABOUT THE SCENARIOS IN THIS SAS SECTION

You will use the Text Parsing, Text Filter, and Text Topic nodes to analyze the abstracts from a collection of SAS® Users Group International (now called the SAS® Global Forum) papers. The goal of this example is to determine whether any themes are present in the papers. You will use the Text Topic node to create a set of topics that will describe the document collection. Additionally, you will create a user-defined topic that finds all of the abstracts related to dynamic web pages.

All data come with SAS® Text Analytics, and the libraries are already defined when you install SAS® Text Miner. For further information on this demonstration and other SAS-TM and Enterprise Miner demonstrations, please visit http://support.sas.com/documentation/onlinedoc/txtminer/getstarted42.pdf.

The Abstract data table contains title and abstract data from a series of SAS® Users Group International meetings from 1998 through 2001. In this example you will parse the data with particular settings and view the reports that are generated for the terms that are identified. This chapter will explain how to perform the following steps:

- Create a new project where you will store all your work
- Define the ABSTRACT data as a SAS-TM data source
- Create a new process flow diagram
- Modify and run the Text Parsing node
- Explore, modify, and run the Text Filter node
- Create a user-defined topic, as well as discover new topics with the Text Topic node

Prerequisites for this Scenario

Before you can perform the tasks in this chapter, administrators at your site must have installed and configured all of the necessary components of SAS-TM4.2. You must also do the following (Figure 6.13):

To create a project, do the following (Figure 6.14):

1. Open SAS® Enterprise Miner.
2. Click **New Project** in the SAS-EM window. The Select SAS® Server window opens.

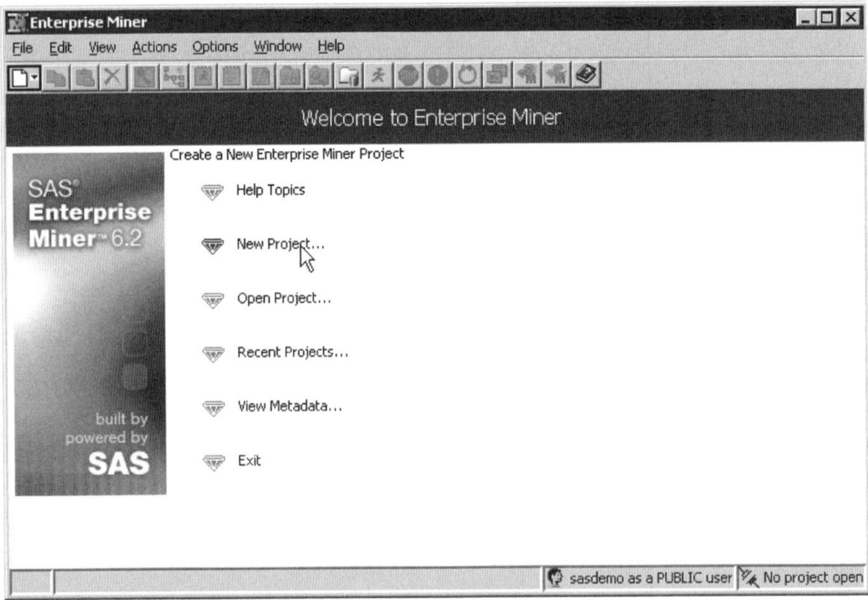

FIGURE 6.13
SAS-EM selection for starting a new project.

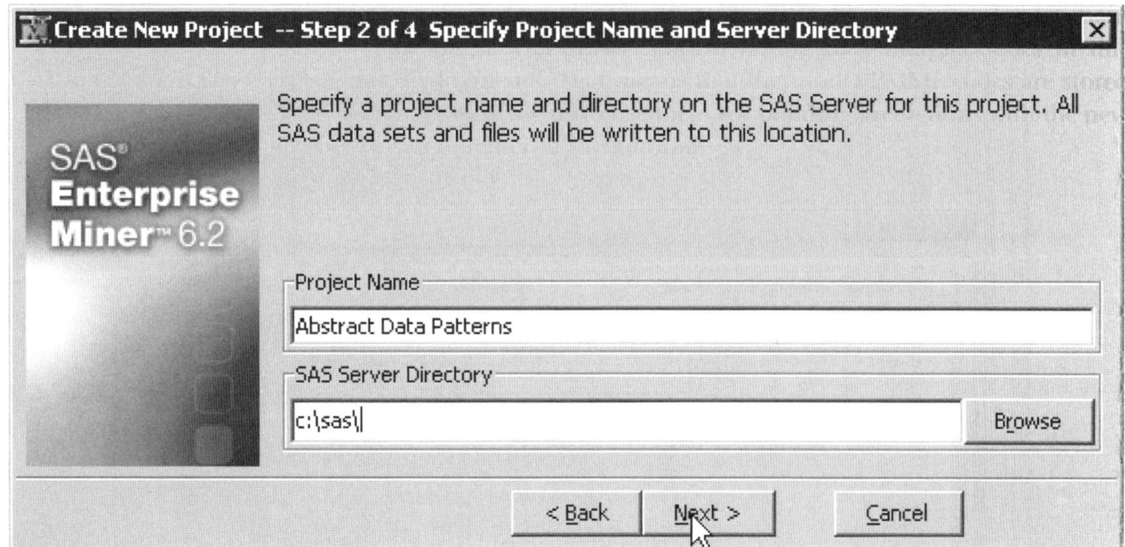

FIGURE 6.14
SAS-EM/TM "Create New Project" window is used for giving the project a unique name and selecting the SAS Server Directory pathway.

3. Click **New Project**. The specify Project Name and Server Directory page opens.
4. Type a name for the project, such as **Abstract Data Patterns** in the Project Name dialog box.
5. In the SAS® Server Directory dialog box, type the path to the location on the server where you want to store data for your project. Alternatively, browse to a folder to use for your project.
6. Click **Next**. The Register the Project page opens.
7. Click **Next**. The New Project Information page opens.
8. Click **Finish** to create your project

To create a data source, do the following (Figure 6.15):

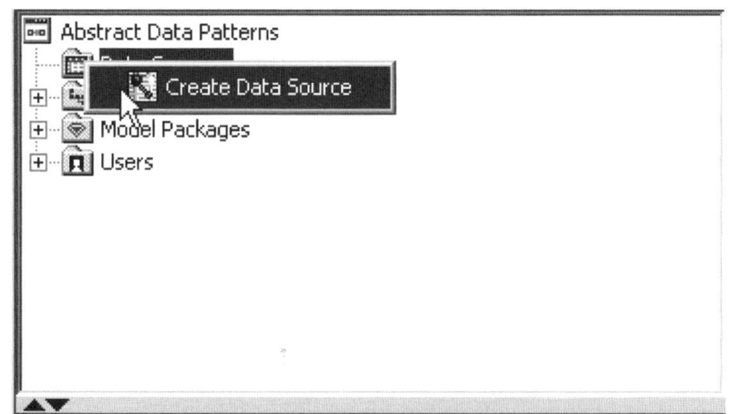

FIGURE 6.15
SAS-EM/TM window for selecting or creating a "data source."

1. Right-click the Data Sources folder in the Project panel and select **Create Data Source** to open the Data Source Wizard window (Figure 6.16).

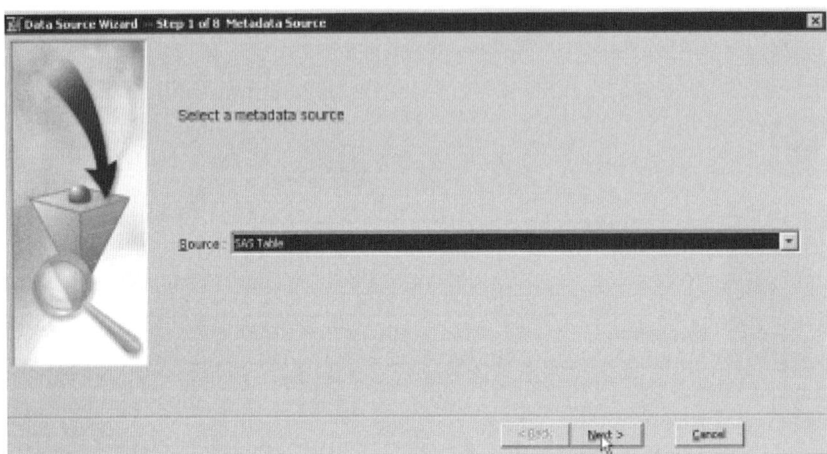

FIGURE 6.16
SAS-EM/TM Data Source Wizard window for selecting a metadata source.

2. Select **SAS® Table** in the Source drop-down menu of the Metadata Source window.
3. Click **Next**. The Select a SAS® Table window opens.
4. Click **Browse**.
5. Click the SAS® library named **Sampsio**. The Sampsio library folder contents are displayed on the Select a SAS® Table dialog box.
6. Select the **Abstract** table, and then click **OK**. The two-level name SAMPSIO.ABSTRACT is displayed in the Table box of the Select a SAS® Table page (Figure 6.17). Click **Next**.

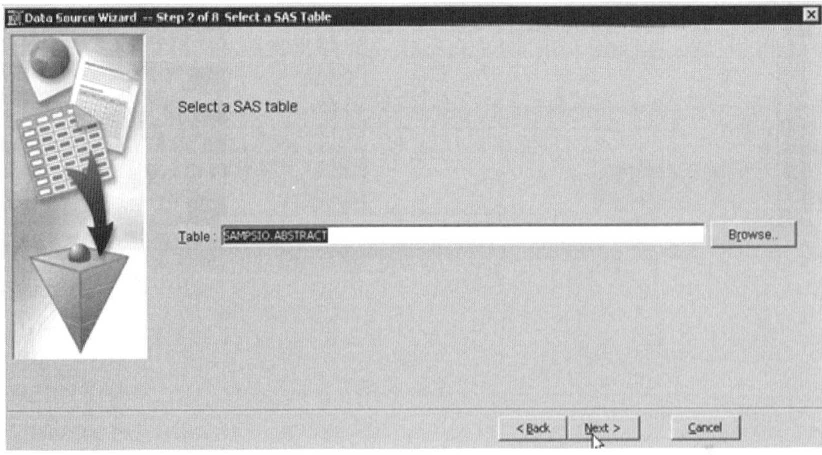

FIGURE 6.17
SAS-EM/TM window for selecting a SAS Table.

7. The Table Information page opens. The Table Properties panel displays metadata for you to review. Click **Next**.
8. The Metadata Advisor Options page opens. Click **Next**.
9. The Column Metadata page opens.
10. The default variables for both **TEXT** and **TITLE** should be set to **Text** by default. If this is not the case, then you need to change their roles to **Text** using the drop-down list. Click **Next**.
11. The Create Sample page opens. Click **Next**.
12. The Data Source Attributes page opens. Click **Next**.
13. The Summary page opens. Click **Finish,** and the ABSTRACT table is added to the Data Sources folder in the Project panel.

To create a diagram, complete the following steps:

1. Right-click the Diagram folder in the Project Panel and select **Create Diagram** as illustrated in Figure 6.18 with the SAS-EM/TM Project Panel where the 'Diagram Folder' is highlighted to bring up the 'Create Diagram' which is selected.

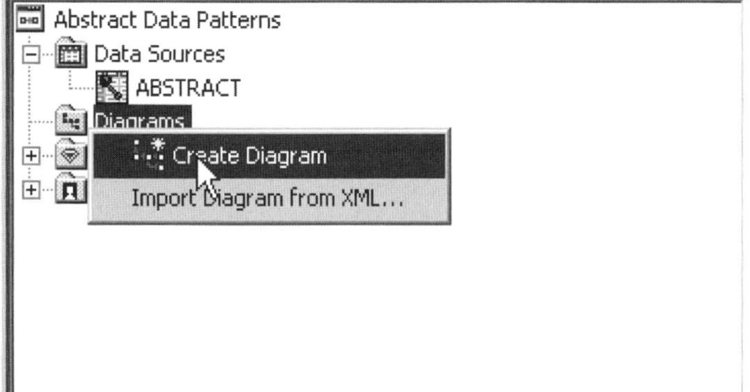

FIGURE 6.18
The Create Diagram dialog selection.

2. Type **Abstract Data** in the Diagram Name box as illustrated in Figure 6.19.

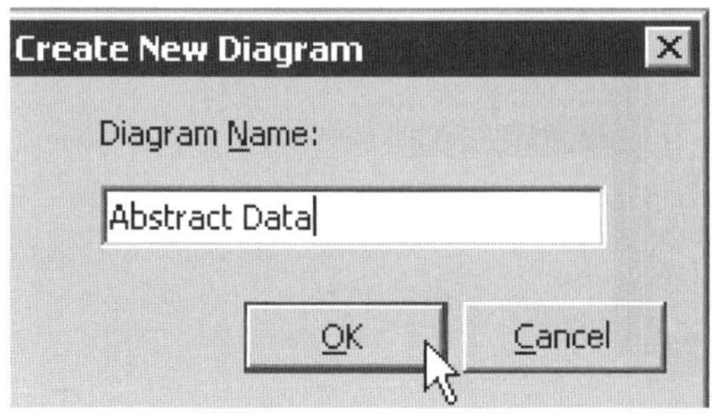

FIGURE 6.19
Create New Diagram dialog box.

3. Click **OK**. The empty Abstract Data diagram opens in the diagram workspace.
4. Choose all defaults by clicking Next in the Create Diagram Wizard.

Once finished, the Abstract Data Patterns diagram opens and is ready for your Text Miner Diagram to be created. First we need to add the data source. Drag and drop the **ABSTRACT** data source from the Data Sources list into the diagram workspace.

Using the Text Parsing Node

In this section you will use the Text Parsing node to parse the document into meaningful terms. You will create multiterm words that represent a single word, create a synonym list, and add words to a stoplist so you can parse the documents into meaningful terms.

Select the **Text Mining** tab in the Enterprise Miner Toolbar, and drag a Text Parsing node into the diagram workspace. Connect the Abstract data source to the Text Parsing node (Figure 6.20).

FIGURE 6.20
Connecting the ABSTRACT data source to the Text Parsing node.

Select the Text Parsing node to highlight it and then select the ellipsis for the **Ignore Parts of Speech** property. Select all parts of speech except for **Noun**, and click **OK**. To do this step, hold the Control key and click every entry but **Noun** so they become highlighted.

Change the **Noun Groups** to **No**. This will mean Text Miner will not try to detect noun groups. Select the ellipsis in the Stop List Property (Figure 6.21). You will see a set of predefined stoplist words, and you can delete any of the predefined stoplist words you want to. For this example we will add four terms. Press the add button six times; this will add six new rows. Add the following words to the newly inserted rows:

term
system
software
paper
application
user
information

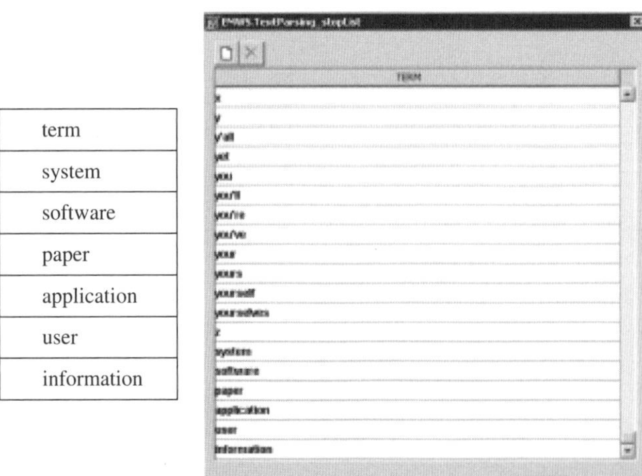

FIGURE 6.21
SAS-EM/TM Stop List Property that is used to modify the stoplist words.

Press **OK** to save, and close the stoplist window.

Select the ellipsis in the Multiwords Term List Property (Figure 6.22). You will see that it is used as an example. Click the add button two times and enter the following multiword terms:

Term	Role	# Documents
web page	Noun	
website	Noun	

FIGURE 6.22
SAS-EM/TM Multiwords Term List Property dialogs.

Press **OK** to save and close the Multiterm window. Select the ellipsis in the Synonym List Property (Figure 6.23). You will see that it is used as an example. Click the add button two times and enter the following synonyms:

Term	_Parent_	_Category_
web page	web page	Noun
website	website	Noun

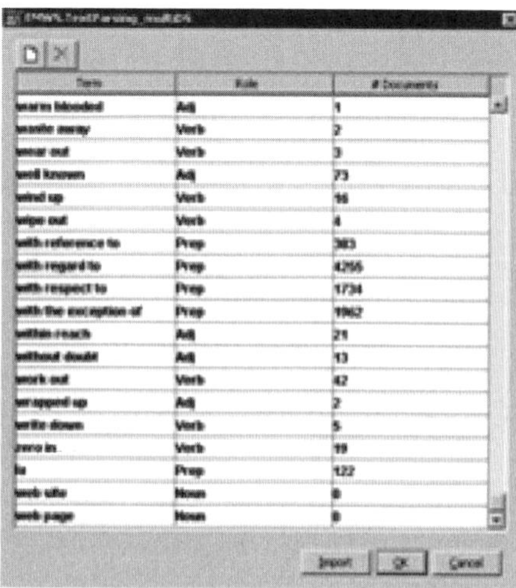

FIGURE 6.23
SAS-EM/TM Synonym List Property dialog.

Press **OK** to save and close the Synonym window. Right-click the Text Parsing node and select **Run**. Click **Yes** in the Confirmation dialog box.

When the node has finished running, select **Results** in the Run Status dialog box. In the Results window, find the Terms table and click anywhere inside it to make it active. The Terms table presents terms that have been parsed by the Text Parsing node, the term's role, the number of documents in which it appears, whether the term was kept or rejected, and other attributes. A term might be dropped from analysis if it appears on a stoplist or was ignored for another reason. Scroll through the list of terms, and note that the majority of terms occur in fewer than ten documents. Close the Results window.

Using the Text Filter Node

In this section you will filter the parsed data to eliminate low-frequency words and other terms that contain little information value. To accomplish this task, you will omit terms that appear in fewer than ten documents. Also, you will interactively remove frequently occurring words that contain little information value.

Select the **Text Mining** tab in the Enterprise Miner toolbar, and drag a Text Filter node into the diagram workspace. Connect the Text Parsing node to the Text Filter node (Figure 6.24).

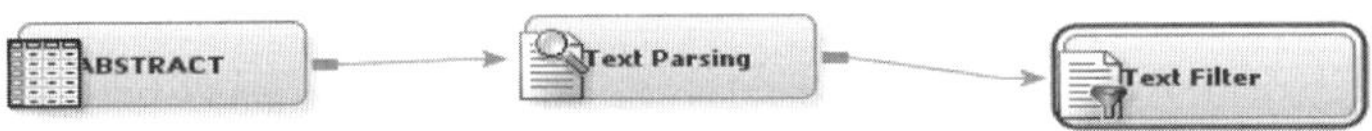

FIGURE 6.24
Connecting the Text Parsing node to the Text Filter node.

Click the Text Filter node in the diagram workspace, and notice the **Minimum Number of Documents** property in the **Term Filters** section of the Train properties. Since we discovered in the results from the Text Parsing node the number of terms that appear in fewer than ten documents, we are going to use this fact to help reduce noise in the data. Change this property from the default value to 10. This ensures that only terms that occur in at least ten documents are included in your analysis.

To enable spell checking, set the **Check Spelling** option in the **Spelling** section of the Train properties to **Yes**. Because you are dealing with professionally written documents, you might think there is no reason to perform this operation. However, this spell checking might suggest terms that should be treated as synonyms. The algorithm used to find misspellings frequently identifies terms that are slight variants of one another in spelling and meaning. Checking for spelling should not remove a large number of terms from this data set but might help when you deal with a different collection of documents. If you are dealing with informal document sets such as emails or customer comments, then spell checking will prove even more beneficial.

To run the Text Filter node, right-click on the node and select **Run**. When you have successfully run the Text Filter node, click the **Results** button in the Run Status window. Find the Terms window within the Results window, and click anywhere inside the Terms window to make it the active window. Select the **#Docs** column to sort terms by document frequency. Scroll down and notice that terms that appear in fewer than ten documents are dropped by the Text Filter node. Terms in ten or more documents might have been dropped for other reasons.

Now find and maximize the Number of Documents by Weight window. This window shows a scatterplot of the terms, with the number of documents on the *x*-axis and the term weight on the *y*-axis. In the window, notice that there is a row of points that all have a weight of zero. These points represent the terms that were dropped from analysis. When you hold the mouse over a point on the graph, a tooltip appears that notes the term represented, the number of documents it appears in, and the assigned weight value. Notice that the point representing the term **data** appears separated from the rest of the points. Because the term **data** appears significantly more often than any other term, you will drop this term from analysis in the following steps. Close the Results window.

Using the Text Filter Interactive Viewer

We now want to work with the interactive viewer to review and possibly change terms and term relationships before our topics are created. For this example, you are going to exclude the single term **data** because it occurs more than twice as often as the next most frequent word. This will give you more relevant topics when you run the Text Topic node in the next section.

Select the Text Filter node, and then click the ellipsis for the **Filter Viewer** property to open the Interactive Filter Viewer. In the Interactive Filter Viewer, you can refine the parsed and filtered data that exist after the Text Filter node has run. The refinement is achieved by filtering documents based on the results of a search expression or modifying the keep or synonym status of a term. The following image shows the Terms table of the Interactive Filter Viewer.

To exclude the term **data** from the analysis, click the check box in the **KEEP** column of the Terms window. The check box should be unchecked. Close the Interactive Filter Viewer. When you do this, the Save window will ask you to save your changes. Make sure that you click **Yes** or you will have to redo step 3.

Using the Text Topic Node

In this section you will use the Text Topic node to create a set of topics. These topics will include a user-created topic and topics that are created by the Text Topic node. Then, you will view all of these topics together with the documents in which they are contained.

Select the **Text Mining** tab in the Enterprise Miner toolbar, and drag a Text Topic node into the diagram workspace. Connect the Text Filter node to the Text Topic node (Figure 6.25).

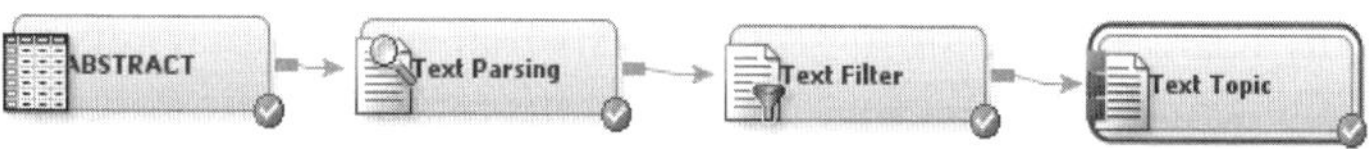

FIGURE 6.25
A Text Topic node being added to the flow pathway.

Before running the Text Topic node, you are going to create a user-defined topic that identifies abstracts that discuss dynamic websites (Figure 6.26). To do this, click on the ellipsis button next to the User Topics property of the Text Topic node. This opens the User Topics window, where you can create your own topics. In the User Topics window, there are four columns, labeled **_Role_**, **_Term_**, **_Weight_**, and **_Topic_**. For this example, you can leave the **_Role_** column empty. To add a row, click the **Add** button at the top of the window. To delete a term, select that row and click the **Delete** button at the top of the window. To create the topic about dynamic websites, enter the three terms with their corresponding weights and topics, as shown in the following chart. (Leave the _Role_ column blank.)

Role	_Term_	_Weight_	_Topic_
	dynamic	1.0	Internet
	web page	0.9	Internet
	website	0.9	Internet

FIGURE 6.26
Mechanics of creating a user-defined topic.

The weight of a term is relative and can be any value between 0 and 1. More important terms should have a higher weighting than less important terms. After you have entered the three terms shown in the chart, click **OK** to save your changes. You are now ready to run the Text Topic node.

Right-click the Text Topic node and select **Run** to run the Text Topic node with all of the other settings at their default values. Click **Yes** in the Confirmation dialog box. When the node finishes running, select **Results** in the Run Status dialog box.

From the Results window, expand the Number of Terms by Topic chart. The default settings created 25 multiterm topics in addition to the one you just defined. Close the Results window.

Select the Text Topic node, and then click the ellipsis button for the **Topic Viewer** property to open the Interactive Topic Viewer window. The topic **Internet** that you created should be at the top of the **Topics** list. Notice that the **Category** of this topic is given as **User,** while the rest of the topics are given as **Multiple**. You can view all of the documents in your topic by right-clicking anywhere in the first row and selecting **Select Current Topic**, if it is not already selected. The middle pane shows you that only the terms *dynamic, web page,* and *website* are in this topic. It also shows how many documents each term is in and how frequently each term appeared.

The bottom pane contains the observations from the Abstract data set that belong to your topic. You can read each of these by right-clicking anywhere in the bottom pane and selecting **Toggle Show Full Text** from the menu. Close the Interactive Topic Viewer.

One problem that you might have noticed is that many of the topics appear to be closely related because they have many terms in common. This suggests that some topics can be merged together. In the **Topics** list of the Interactive Filter Viewer, there are four topics (topics 2, 3, 5, and 10) that contain both **user** and **application** as descriptive terms. These topics can be merged to form one user-created topic. To merge the topics, double-click the first topic you want to rename to make it the active topic. You can rename the topic by deleting all of the text in the **Topic** field and replacing it with **User Applications**. Repeat this process for the other three topics that contain both **user** and **application**. When you rename a topic, the **Category** of the topic changes to **User**. Close the Interactive Filter Viewer and save the changes you made.

Right-click the Text Topic node and select **Run** to rerun the Text Topic node. Click **Yes** in the Confirmation dialog box, and click **OK** in the Run Status dialog box when the node is finished running. Click the ellipsis next to the **Topic Viewer** property to see the new topics that have been created. As you can see, there are now two user-defined topics: **Internet** and **User Applications**. However, the Text Topic node still created 25 topics. If you are not satisfied with these topics, you can continue to merge multiple topics together until the topics are distinct enough for your needs.

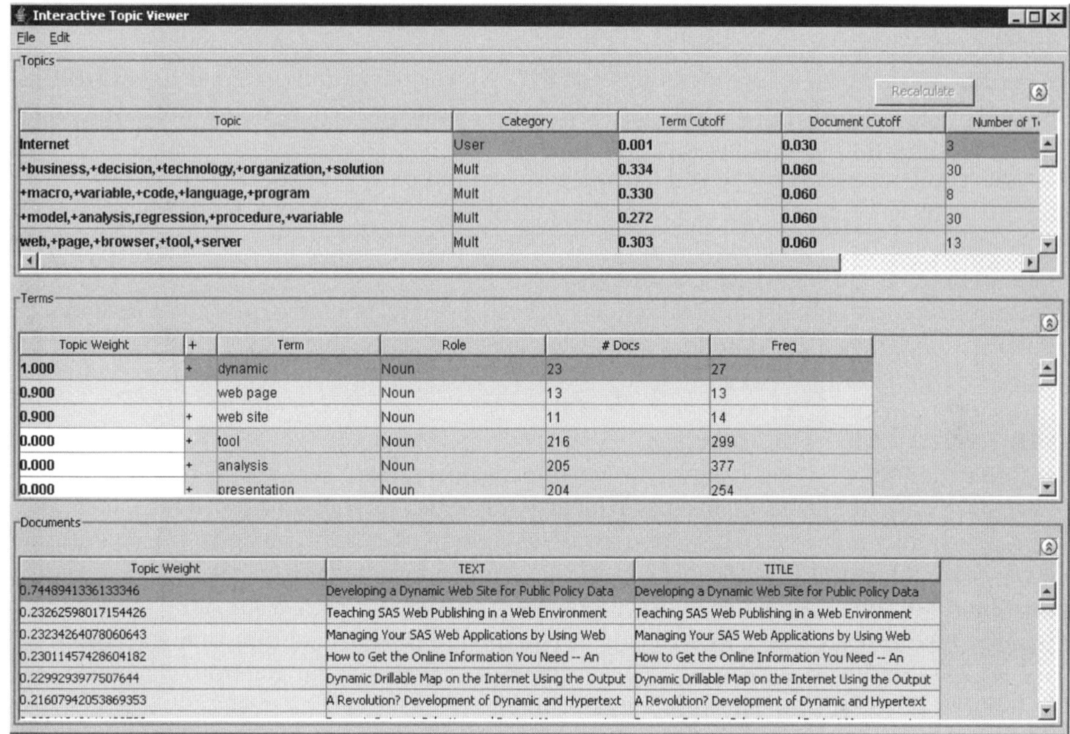

FIGURE 6.27
Interactive Topic Viewer dialog.

The Interactive Topic Viewer (Figure 6.27) allows you to investigate the topics and make changes to Topic Weights, Term Cutoff, and Document Cutoff so you can refine your categories. As you can see in this example, the topics discussed at this SAS® User Group (SAS® Global Forum)—through the abstract, not the title—included such topics as programming, regression analysis, making business decisions, and many more.

TIPS FOR TEXT MINING

The following paragraphs provide some tips for dealing with either 1) large collections of documents and / or 2) very long documents.

Processing a Large Collection of Documents

Using the Text Mining nodes to process a large collection of documents requires a lot of computing time and resources. If you have limited resources, it might be necessary to take one or more of the following actions:

- Use a sample of the document collection.
- When using the Text Miner node, set some of the Parse properties to **No**, such as **Find Entities**, **Noun Groups**, and **Terms in Single Document**.

- When using the Text Parsing node, set some of the Detect properties to **No**, such as **Find Entities** and **Noun Groups**.
- In the Text Miner node, reduce the number of SVD dimensions or roll-up terms. If you are running into memory problems with the SVD approach, you can roll up a certain number of terms, and then the remaining terms are automatically dropped.
- Use the Ignore properties of the Text Parsing node to limit parsing to high information words. You can do this by ignoring all parts of speech other than nouns, proper nouns, noun groups, and verbs.
- You can also use the Parse properties in the Text Miner node to ignore all parts of speech other than nouns, proper nouns, noun groups, and verbs.
- Structure sentences properly for best results, including correct grammar, punctuation, and capitalization. Entity extraction does not always generate reasonable results.

Dealing with Long Documents

SAS-TM uses the "bag-of-words" approach to represent documents. That means that documents are represented with a vector that contains the frequency with which each term occurs in each document. In addition, word order is ignored. This approach is very effective for short, paragraph-sized documents, but it can cause a harmful loss of information with longer documents. You might want to consider preprocessing your long documents in order to isolate the content that is really of use in your model. For example, if you are analyzing journal papers, you might find that analyzing only the abstract gives the best results. Consider using the SAS® DATA step or an alternative programming language such as Perl to extract the relevant content from long documents.

STATISTICA TEXT MINER

STATISTICA Text Miner is an integrated analytics tool of the STATISTICA Data Miner and STATISTICA Enterprise solutions suite. As with all other functionality of the STATISTICA platform, the features and functions of the Text Miner module can be accessed via an interactive graphical (point-and-click) user interface, as well as programmatically via the API (application programming interface).

STATISTICA Text Miner was designed to support the development and deployment of text "models." The tools allow users to develop indexing rules for processing corpora; to enumerate words and text phrases, using stemming, stopwords (rules), and synonym lists, and so on; and to deploy those models for scoring efficiently new (streaming) text. In this way, the information extracted from new documents can be incorporated into (automated) predictive or other models.

Design Philosophy

As discussed in Chapter 4, the term *text mining* can describe a diverse range of activities and goals, as well as approaches and best practices. In the STATISTICA Text Miner, users can move quickly from accessing available text sources, aligning them with available structured/numerical data sources, and then to deploy models in automated solutions. Here are some examples:

1. Users can point the application to documents stored in a data base (text fields, e.g., insurance claims adjuster notes, warranty claims, etc.) or organized as separate documents residing in a particular file

directory or location. The application can also crawl hierarchical file structures or websites to retrieve documents of specific types.

2. The program will index the respective document corpus, taking into account custom synonym lists, exclusion ("stop") word lists, or specific phrases (inclusion lists); indexing takes advantage of multicore computing hardware.
3. After initial indexing, numerous options are provided for reviewing and combining words, terms, and phrases and to apply various transformations to the initial word/term counts; users can also apply latent semantic indexing methods (see Chapter 15) and other techniques for concept extraction.
4. A list of final words, terms, and phrases, as well as document scores (see Chapter 15) can be deployed as a deployment "object" or model for "scoring" (e.g., for computing inverse document frequencies, component scores) new document corpora. In this manner, text can automatically be converted into numbers ("numericized") to support batch or real-time scoring solutions involving text.

Figure 6.28 shows a screenshot of the STATISTICA Text Miner application integrated into the STATISTICA Data Miner client tools.

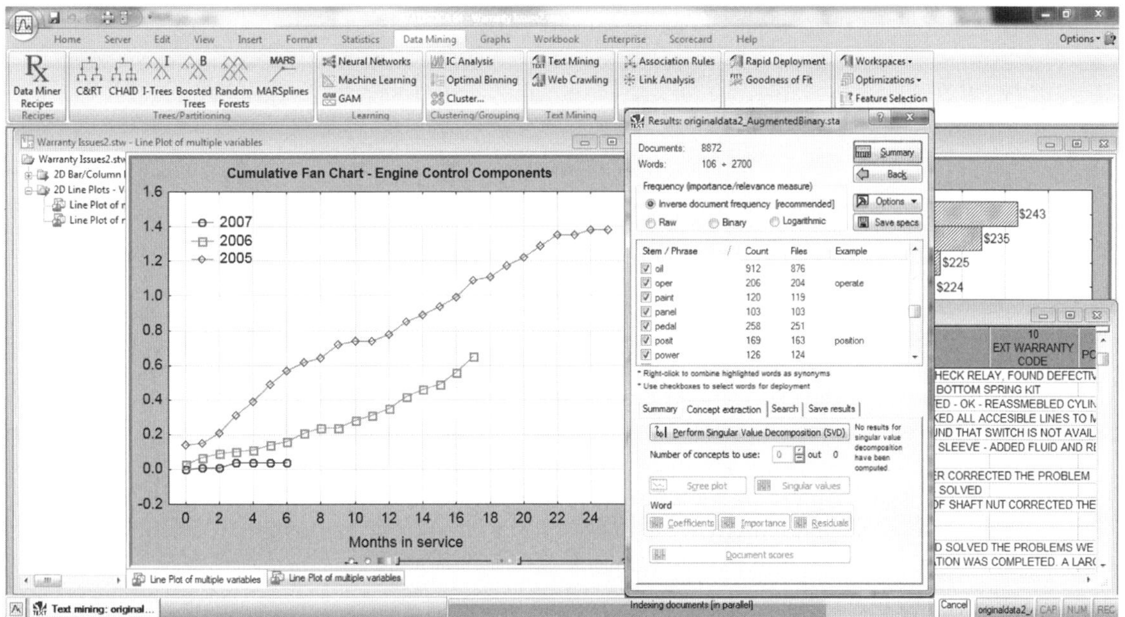

FIGURE 6.28
STATISTICA Text Mines interactive user interface.

Figure 6.29 shows a diagram of how STATISTICA Text Miner tools integrate into the overall STATISTICA platform and solution in order to create and deploy predictive models involving unstructured text to a "production system"—that is, for automated and routine batch or real-time scoring of (customer) data.

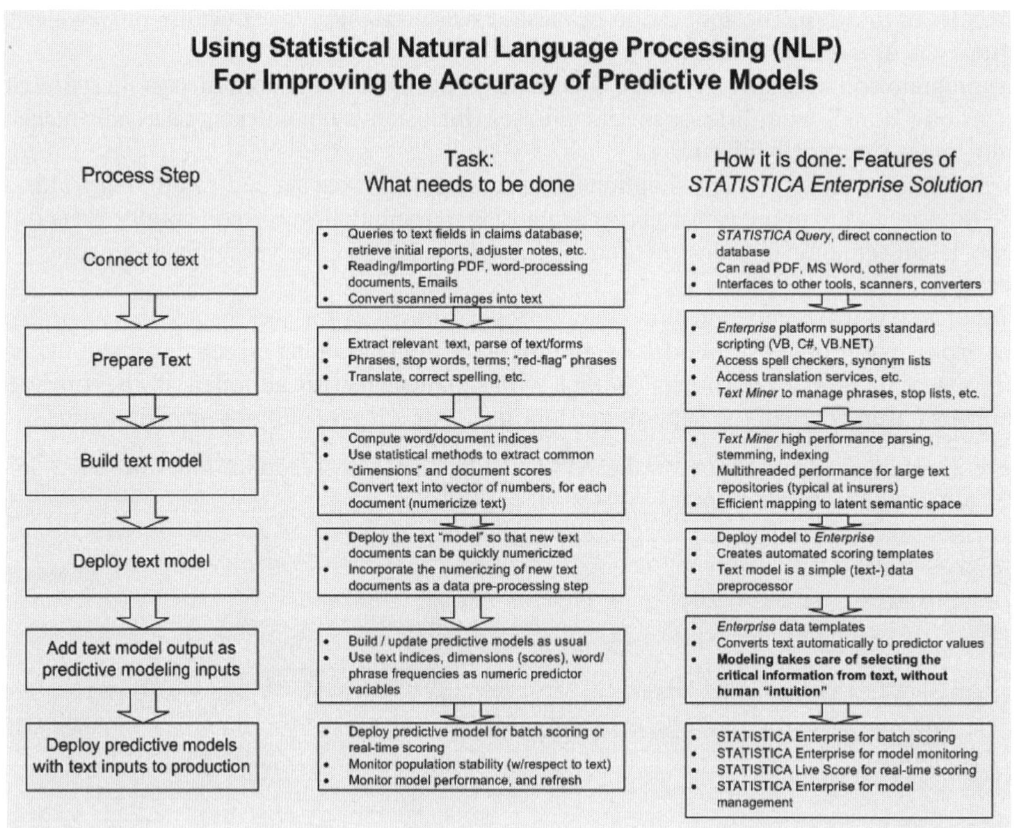

FIGURE 6.29

Integration of STATISTICA Text Miner tools with STATISTICA solutions.

Figure 6.29 shows the steps for incorporating unstructured text sources into predictive models for batch scoring or real-time scoring and how these steps are accomplished using the STATISTICA platform. For example, such a system could be used to automatically score insurance claims narratives (claims adjuster notes) for fraud or subrogation probability; to automatically classify and/or monitor warranty claims reports; or to predict the success of a marketing campaign based on sentiments expressed in blogs, tweets, or other social network sites.

An Example of an Integrated Solution

Figure 6.30 shows a (prototype) custom screen that insurance claims adjusters and managers can use to track and assign open claims. On the left side of each screen is a list of five key performance outcomes: *Fraud*, *Reserve Change*, *Right Track*, *Subrogation*, and *Complexity*. Each of these outcomes is "connected" to a prediction model based on the available structured and unstructured information shown on the right side of each screen.

The predictions (e.g., *Fraud—No; Reserve Change—$3,500*, and so on) on this screen are computed from prediction models, built via STATISTICA Data Miner, and managed via the STATISTICA Enterprise

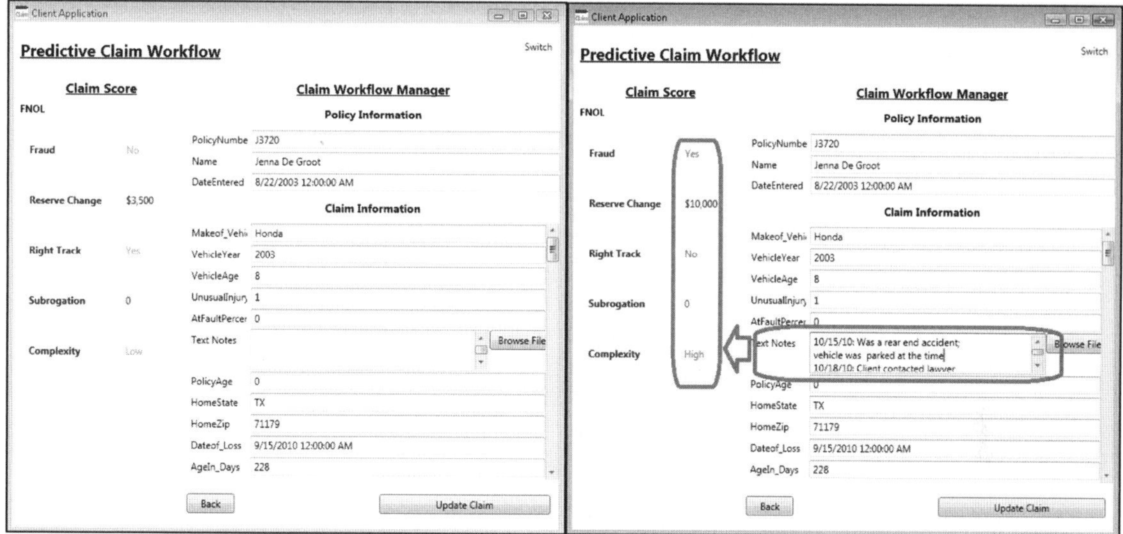

FIGURE 6.30

Five predictive models using unstructured text input.

platform. Users (claims adjusters, managers) can enter new information about claims as it becomes available and score in real time all of the information to derive new estimates (predictions) with respect to the five key outcomes.

In the screen shown to the right, text was entered describing details for an automotive claim. As a result of entering the respective text information (*"rear-end accident, client contacted lawyer"*), all predictions changed as highlighted in Figure 6.30.

Text Models as Data Preprocessing Steps

In this system, text mining models are deployed as data preprocessing steps. These steps include (1) scoring of specific terms and phrases according to the selected text mining model (configuration); (2) alignment of the respective "numericized text" (vector of numbers) with the unstructured information; and (3) predictive modeling to derive a more accurate prediction. Following is a brief overview of the key features and options of the STATISTICA Text Miner to help you start using the STATISTICA Data Miner to create text scoring "models" suitable for deployment.

Connecting to Text Data

The STATISTICA Text Miner can:

- access data in text fields in STATISTICA data files
- access data in in-place database connections to text fields in databases
- point the text miner application to retrieve the documents from the respective location(s)

Figure 6.31 provides a screenshot of the application, pointing to multiple text fields in an input data spreadsheet.

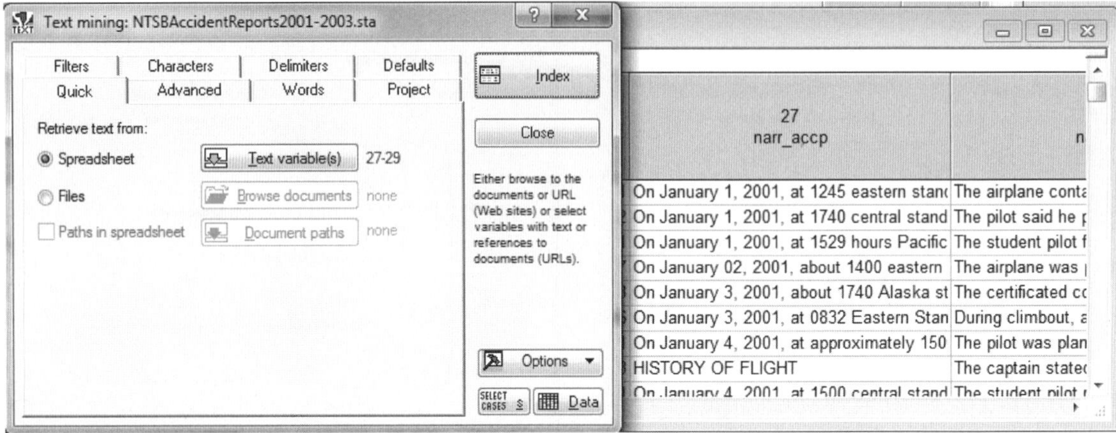

FIGURE 6.31
Connecting to text data sources.

Parameters for Directing the Indexing of Terms

There are a large number of options for configuring the indexing of terms and phrases in the corpus of text documents to be analyzed.

Stemming Language

Stemming is the process of reducing words to their stems or roots (e.g., *walking, walked, walks* to the same root form *walk*). The STATISTICA Text Miner supports various stemming languages. Figure 6.32 shows a list of stemming languages available in the off-the-shelf software.

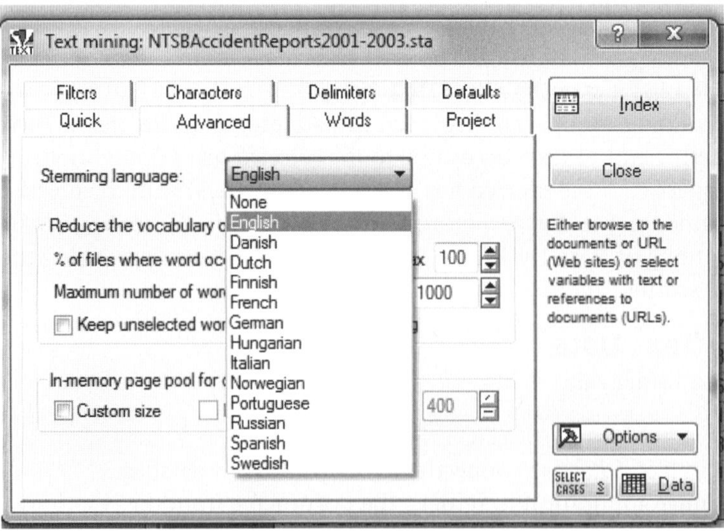

FIGURE 6.32
Advanced text miner configuration options.

Document Frequencies

There are also options on the screen shown in Figure 6.33 to limit the indexing of words and terms only to those that appear in a certain percentage of the input documents. Note that words that occur in every document or in only one of the documents will (usually) not be useful in subsequent predictive modeling or other analyses.

Custom Terms, Phrases, Synonyms, Stoplists

Figure 6.33 shows the screen where files with custom dictionaries can be defined. Phrases, custom terms, synonym lists, and so on can be maintained in text files. In other cases, when scripting automated applications, those files can also be sourced from external databases (e.g., of terms indicating certain sentiments, etc.).

During indexing, STATISTICA Text Miner will enumerate terms and phrases that are consistent with the lists specified in these files. The help documentation of STATISTICA (always accessible via F1 or by clicking on the question mark in the upper right corner of each dialog) provides additional technical details.

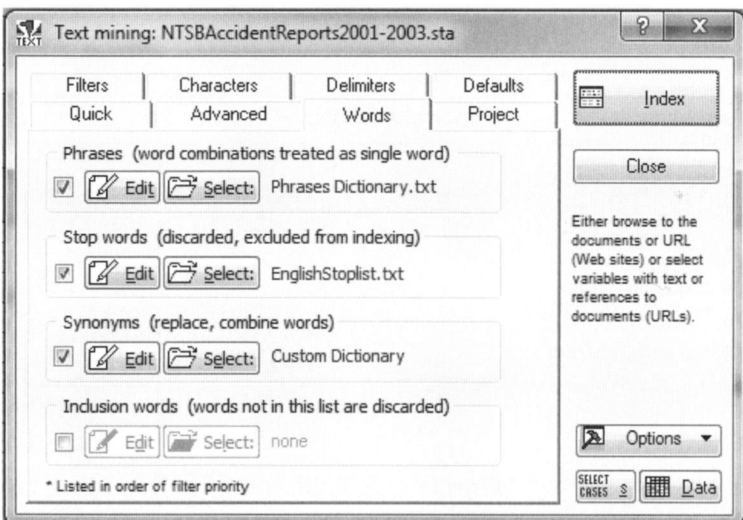

FIGURE 6.33
Custom dictionaries in STATISTICA Text Miner.

Filters, Characters

In addition, there are numerous options for specifying minimum and maximum word lengths, stem lengths, the minimum number of vowels, the maximum number of consonants, and so on. The list of valid characters in words (that are to be indexed, or else they are ignored) can also be configured.

Specifying Which Part of a Document to Scan

The *Delimiters* tab in the screenshots allows users to identify specific phrases in a document that bracket the text that is to be scanned (or indexed). This is useful when the input consists of electronic "forms" where the relevant text is bracketed between specific paragraph headers (e.g., after a header *"Place a brief description of the auto accident below"*; the application can then start indexing only the text *past* that phrase and before the next header). Of course, because there is a comprehensive API and scripting interface to *all* functionality in STATISTICA (which also includes a Visual Basic development environment), you can write scripts and templates for preprocessing any text prior to indexing it (or to fix spelling errors using any of the available automatic spell checkers, etc.).

Creating New Projects and Scoring Existing Projects

The *Project* tab of the STATISTICA Text Miner application allows users to determine where the text "model" of indexed words and terms is to be stored or whether to index the text corpus using a previously saved project. When you click the tab, Figure 6.34 displays. The choices available on this tab provide the options for the following:

- Indexing new documents to create a new project (*Create a new project*)
- Applying a previous project file with specific terms, phrases, and so on
- Using singular-value-decomposition–based (see Chapter 11) latent dimensions (models) to index or "score" a new document corpus

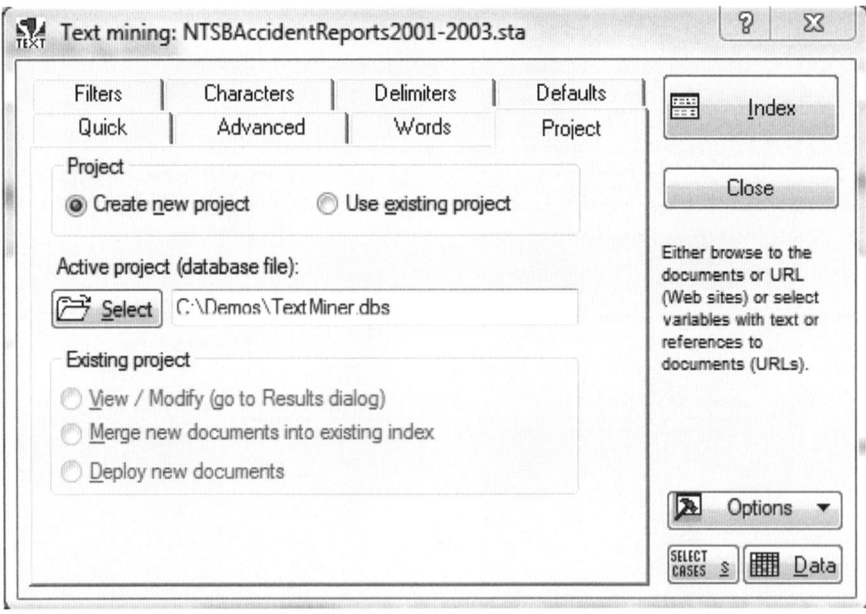

FIGURE 6.34
The STATISTICA Text Miner project location.

Deploying Text Models

For example, in order to create a deployment of a text model (for batch scoring a new document corpus), a user can simply point the application to the new text corpus and select the previously created project that defines exactly how the new corpus is to be scored (what to "look for" and how to compute document scores, etc.). This process can all be recorded automatically into a macro and then turned into a reusable scoring template to be deployed via the STATISTICA server solution. For example, the process can be scheduled to run every night or in a real-time scoring scenario as new text becomes available.

Note that the term *project* as used in this chapter refers to a specific text mining project (i.e., work flow); such text mining projects can also be saved, retrieved, and restarted at a later time. The interface allows the users to save workspaces with any number of linked documents, data sources, and so on and in a sense "freeze" them to be retrieved/reconstructed later—including all originally open documents, to continue the work where it was left off.

Managing Index Results and Writing the Results Back to a Database

After indexing, the results dialog contains numerous options for reviewing the results of the raw index frequencies and various derived indices, including *Inverse document frequencies* and document scores based on singular value decomposition (see Chapter 11). Figure 6.35 shows several results dialog options.

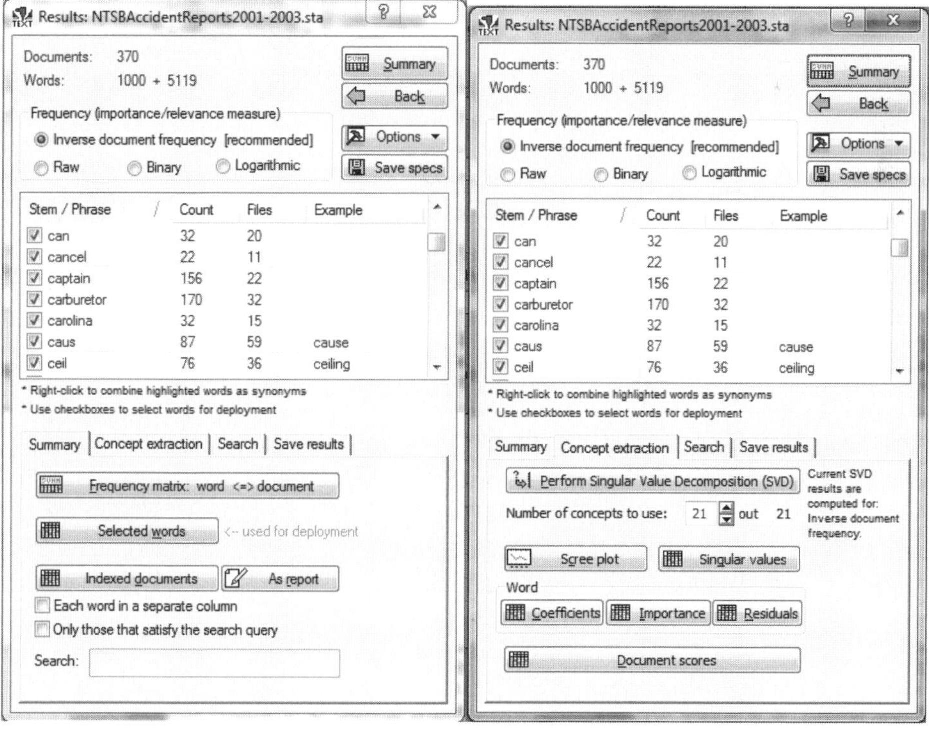

FIGURE 6.35
STATISTICA Text Miner results dialog options.

Users can sort the list of stems/phrases in ascending or descending order, based on any of the columns shown in the dialog by clicking on the respective column header (e.g., to sort the terms in alphabetical order, by their raw frequencies across documents, by the numbers of files in which they occur, etc.). Also, the user can create the final list of what is to be deployed and/or to be used in singular value decomposition to derive latent semantic dimension. This is done by un/checking the words and phrases that are to be retained, the s (*"Concepts"*). Multiple words or terms can be combined into phrases by highlighting the respective words or phrases and then right-clicking to *Combine Words*; this is illustrated in Figure 6.36.

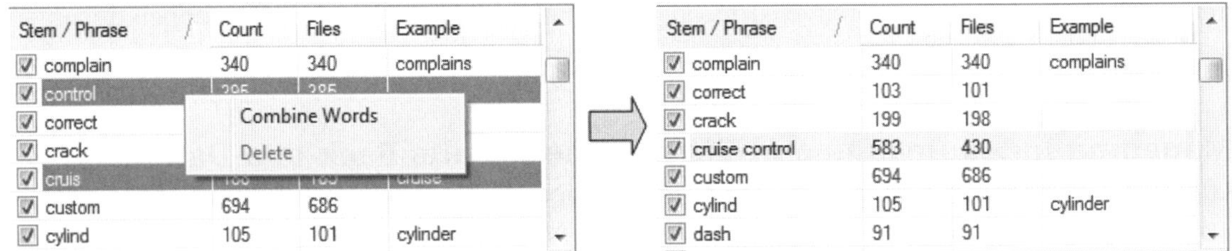

FIGURE 6.36
Combining words/terms into phrases.

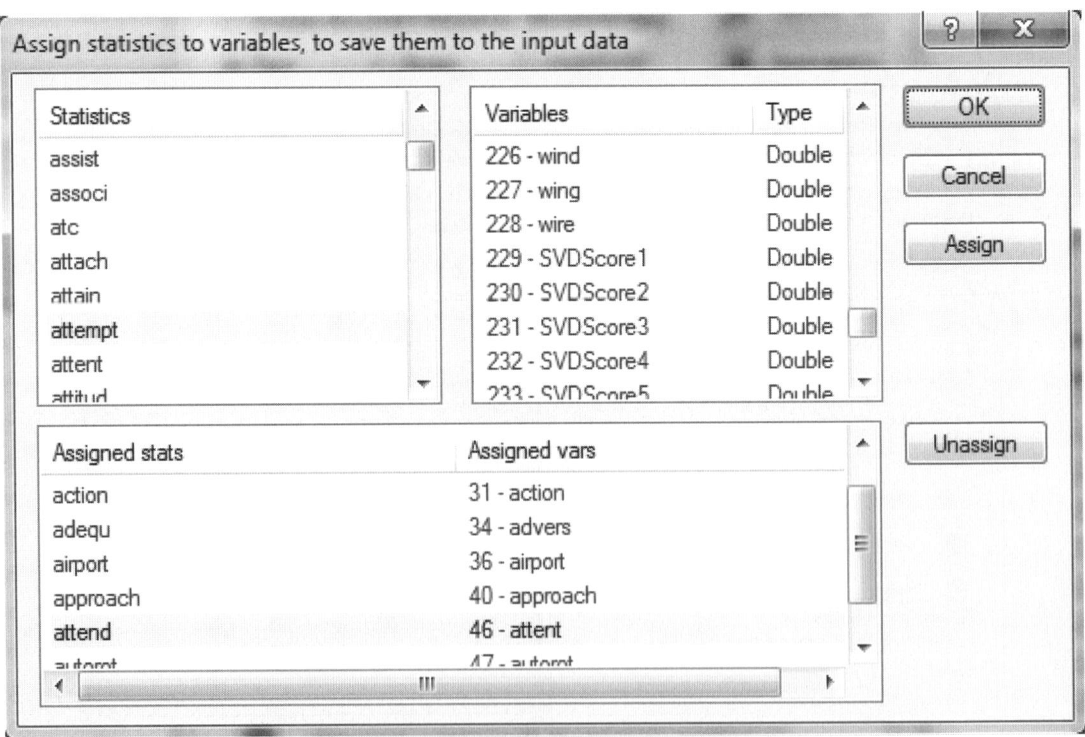

FIGURE 6.37
Writing back results to the input data or database tables.

Writing Selected Terms and Statistics Back to the Project File, Input Data

Selected terms or document (SVD) scores can be written back to data tables. Macros can be created to store and apply identical steps and selections, including write-backs, which can be used, for example, for batch scoring of new text.

Saving Results (Indices) for Subsequent Analyses, Modeling

The options on the *Save results* tab allow users to create data spreadsheets with word counts, inverse document frequencies, and/or document scores (based on singular value decomposition of selected terms). These numerical indices can be written into stand-alone data files or back into the original data source or data base tables. Figure 6.37 shows the screen for assigning specific terms or words to variables or fields in the input table to which to write back the respective statistics (counts, inverse document frequencies, SVD document scores, etc.). This dialog will come up when selecting *Write back current results (to selected variables)* on the *Save results* tab of the *Results* dialog.

SUMMARY: STATISTICA TEXT MINER

STATISTICA Text Miner is a general text indexing "engine" designed to support efficient automated decision support solutions that incorporate unstructured text. The program is fully integrated into the STATISTICA Enterprise and Data Miner platforms and optimized for efficient scoring of large or streaming document corpora on multicore server platforms.

POSTSCRIPT

This book is designed to prepare you with enough background information and exposure to common tools in order to enable you to do several tutorials. This preparation is part of the layered structure of instruction designed to help you learn text mining by doing it. We suggest that you pick several tutorials that are most similar to potential uses for text mining in your organization or in areas that interest you. While you are working through the tutorials, you will be exposed to some advanced techniques, which will be explained in further detail in the final learning layer presented later in the book. So dive into the practice of text mining; don't be afraid of launching into techniques that you don't understand yet. The central theme of this book is to enable you to learn as you go.

PART 2 Introduction to the Tutorial and Case Study Section of This Book

You may be interested in only a certain domain area and thus would prefer to go directly to tutorials or case studies that discuss your area, instead of having to flip through all the tutorials in this section of the book. If you cannot find a tutorial topic that is exactly in your area, you should be able to find one that is in an "enlarged arena" that is close to your interests. The following list will assist you in finding a tutorial to start your study. We should point out, however, that some of the latter tutorials in the list, namely V, W, X, and Z are more advanced, and will in some cases suggest, after initially working through the data set, that you go back to a previous tutorial and try the method illustrated there to this new data set. Thus, if these latter ones are your prime interest, we might suggest that you work through some of the earlier ones to get a feel for the text analytics process before tackling the last 4 or 5 in the list.

CLASSIFYING DOCUMENTS & FINDING PATTERNS in LITERATURE:
Tutorial – X
Tutorial – H
Tutorial – G

CRM – CUSTOMER RELATIONS:
Tutorial – Q
Case Study – M

INSURANCE INDUSTRY:
Tutorial – C

EDUCATION:
Tutorial – F
Tutorial – R

ENTERTAINMENT INDUSTRY:
Tutorial – O
Tutorial – L

FINANCE – BANKING:
Tutorial – K

MARKETING & SALES – Including Web:
Tutorial – J
Tutorial – A

MEDICAL & MEDICAL DELIVERY:
Tutorial – D
Tutorial – E
Tutorial – G
Tutorial – I
Tutorial – U
Tutorial – V
Tutorial – W
Tutorial – Z

MUSIC – OPERA:
Tutorial – L

PRODUCT RECALL – QUALITY CONTROL:
Tutorial – S

PSYCHOLOGICAL TESTING & DECEPTION DETECTION:
Case Study – N
Tutorial – F

SOCIAL MEDIA:
Case Study – AA – Twitter
Tutorial – BB – Twitter
Tutorial – A – various social media
Tutorial – P – Twitter
Tutorial – Y – Twitter

SOCIETAL ISSUES:
Tutorial – U
Case Study – AA
Case Study – Y

TRANSPORTATION – AIRLINE INDUSTRY:
Tutorial – B
Tutorial – T
Tutorial – BB

One word of caution as you read and work through these tutorials: it might be a good idea to always keep in mind the problem of "false positives." What are false positives (or false negatives)? In traditional statistics (called "Frequentists" and involving p-values, t-tests, etc.) the phenomenon of "multiple testing" like doing a t-test between all possible pairs of variables in your data set to find "one" or "some" that are significant, can easily lead to false positives. This is because one is using, generally, a significance cut off of 5%, thus if you run 100 t-tests, 5 of these would be significant just "by chance" of the statistical method; these 5 could all be "false positives." In traditional statistics one needs to ask just "one hypothesis" of the data set, and then only test for that hypothesis. Any further analyses would be considered exploratory; if anything of interest is found from this exploration then it can be made the hypothesis of the next experiment where a new data set is gathered. Similar things can happen in our modern "Statistical Learning Theory" methods which comprise most of our Predictive Data Analysis whether we call it data mining or text mining. By subjecting the data in a predictive analysis to multiple algorithms or models we can keep adding more and more models until we find one that "seems to be best." However, if parts of the data set have not been with held for "test" and "validation" data sets, it is easy to choose a model that has the highest accuracy, but yet have chosen a "false positive" in the same way that happens with multiple t-tests.

In many of the scenarios presented in some of the following tutorials, the author has taken us through various algorithms with the goal of allowing you, the reader, to learn "how to do it." But if the data set is small, and no or not enough of the data have been available to make both a good training data set and also a good testing, and ideally also a 3rd validation set, then one has to be cautious in proclaiming a model with good accuracy to be valid. It may be valid, or it may not. In those cases, only additional data gathering and further analysis will answer this question.

It is clear that during the past century where most statistical analyses were done with traditional p-value statistics, not enough attention was paid to the problem of false positives; we see this in contradictory medical studies, always in the news, where the results of the study of today may be completely opposite of the studies of last year. Meaning that many days the consumer has a difficult time deciding if "they should eat asparagus or carrots"!!! Well, the same thing can happen in the use of our modern predictive analytical methods; thus the need to pay proper attention to really understanding the data, and the train, test, and validation sub sets of data. This problem has just been addressed in an eloquent way by Young and Karr (2011) where they propose ways to improve how studies and data analysis are conducted.

Reference

Young, S. D., & Karr, A. (2011). Deming, data and observational studies. *SIGNIFICANCE: Statistics Making Sense.* Volume 8 / Issue 3 (September 2011).

CASE STUDY: Using the Social Share of Voice to Predict Events That Are about to Happen

Tom Emerson
Senior Research Scientist, Topsy Labs

Rishab Ghosh
Cofounder and Vice President of Research, Topsy Labs

Eddie Smith
CRO, Topsy Labs

CONTENTS

Historical activity levels of what people are talking about within the social web provide excellent proxies for predicting what events are about to occur. Using Mideast uprisings as examples, this tutorial will show how monitoring activity levels for specific keywords related to Mideast uprisings provides accurate predictors of events that are about to happen. We discuss the statistics, the methods used to access the data, the results, and the utility from a business perspective.

It is now possible to access deep historical counts of tweets by keywords (and also references to URLs) so the data can be used to back-test a variety of hypotheses around how social media comments impact the things that occur in real life. Everyone is aware of conceptually how Twitter was used to communicate on the ground activities during the Mideast uprisings, but was it really the volume of activity that caused events to flare up or did the events themselves cause the volume of social mentions to skyrocket?

To test the correlation between social volume and real-world events during the Mideast uprisings, we employed share of voice (SOV) analyses to measure the relative change in activity for a given group of related keywords mentioned in Twitter over time. SOV analysis sums up the total number of mentions for a keyword and divides the number of mentions for a keyword by the summed amount of the group of related keywords being analyzed so the relative percentage of that keyword's mention over time can be analyzed. This is a useful technique for measuring the relative importance of something in the social web over time in a given category of related keywords or phrases.

ANALYSIS

The analysis covers three different dimensions of Mideast activities during the January to March 2011 period. The first analysis examines overall activity by the Mideast country, followed by an analysis of

Practical Text Mining and Statistical Analysis for Non-structured Text Data Applications. DOI: 10.1016/B978-0-12-386979-1.00007-4

Tunisian activity mapped to a chronology of real-world events. A quick analysis of Iranian terms is provided in an SOV analysis, and then specific hashtag dates representing events across the Mideast are examined. Overall Mideast SOV analysis shows interesting trends by country hashtag over time (Figure AA.1).

A few general observations about this SOV analysis tells us the following:

- Tweets related to Iran had historically dominated Mideast social communication but lowered in their importance when the Tunisian uprising started.
- Tweets related to Tunisia began appearing on December 17, 2010, which is when Mohammed Bouazizi, a 26-year-old man trying to support his family by selling fruit and vegetables in the central town of Sidi Bouzid, doused himself in paint thinner and set himself on fire in front of a local municipal office. This act provided the seed for viral communication.
- Tweets related to Egypt actually spiked on January 1, 2011, well before major Egyptian protests began rearing their head in the general media. But Tunisian tweet activity washed out the Egyptian activity. Could the Egyptian spike have been an early warning of things to come in Egypt?
- Tweets relating to Bahrain begin to increase on February 24, and they persisted through the beginning of March.
- Tweets relating to Saudi Arabia initially flare up on February 20, 2011; are overtaken by Bahrain; and then expand in relevant importance during the first week of March.

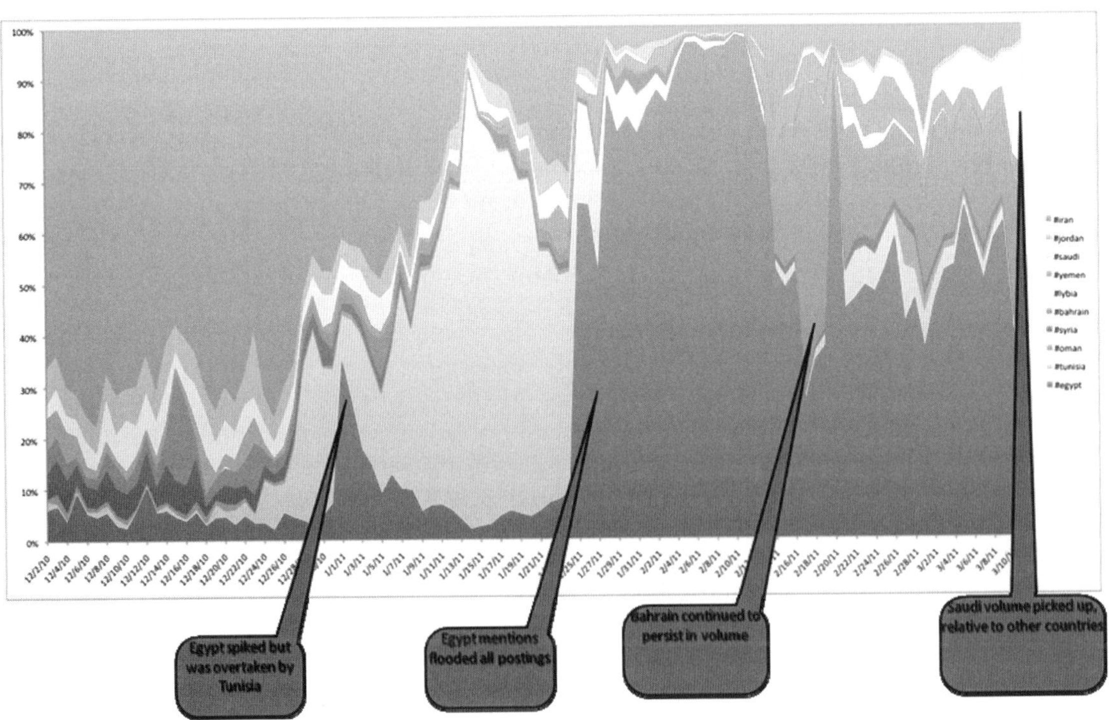

FIGURE AA.1

Mideast share of voice.

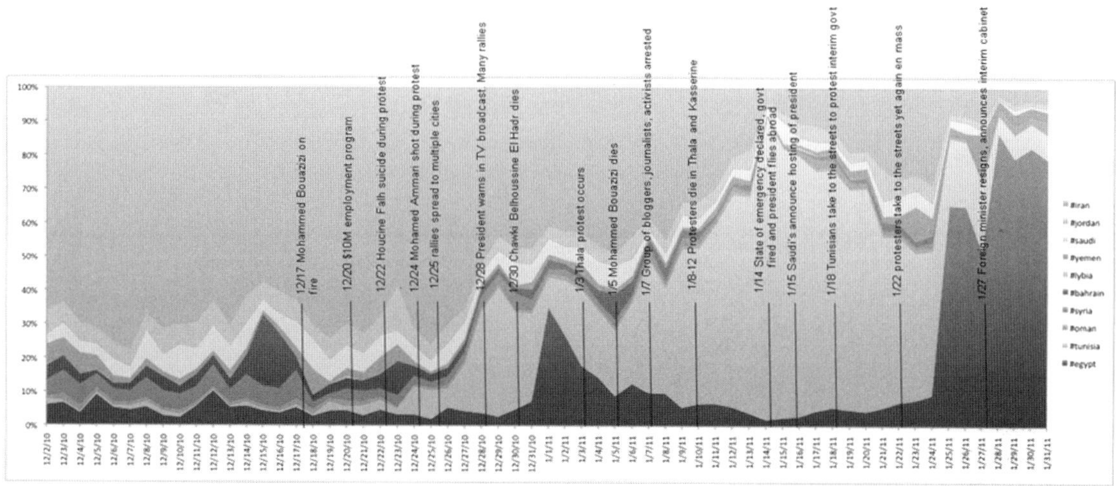

FIGURE AA.2

Tunisian share of voice mapped to real-world events.

Let's take a look specifically at the mapping of Tunisian SOV as it related to a real-world chronological timeline of Tunisian uprising events. A complete history of Tunisian events can be seen at http://english.aljazeera.net/indepth/spotlight/tunisia/2011/01/201114142223827361.html. Figure AA.2 superimposes a sampling of real-world events so they can be mapped against relative SOV.

As can be seen from Figure AA.2, there is definitely a correlation between events occurring and the communication coming out of these events. It's not clear that the social communication made the events occur, but it's certainly clear that the social communication was amplifying the communication of each real-world event and could have been used as a leading indicator of events unfolding on the ground.

Since Iranian terms seemed to be the historically dominant set of Mideast terms, an analysis was done to determine what terms contributed to the Iranian tweet volume (Figure AA.3). What's clear from the Iranian analysis is that the Bahman detention had a very significant impact in what people were talking about, leading to mass protests in Iran that were repressed by the Iranian government. Unlike Tunisia and Egypt, the social activity for Bahman protests was significant but very short lived. Perhaps this is due to the government's tight control over the population in general and Internet censoring.

When people begin rallying around common causes in Twitter, they typically come up with a hashtag to describe the area of interest. This allows postings to be grouped around a common "key" so everyone's communication can be easily referenced. In the Mideast, the syntax "#(date)" evolved as a way to organize people around specific protest dates. Figure AA.4 shows SOV analysis around some of the hashtags used to represent rallies across different countries.

It is interesting to note that some hashtag dates became significant rally cries, while others peaked, faded, and eventually disappeared so the rally cry for protest wasn't heard—at least not in a way that the protest made news so the authors were aware of the protest occurring.

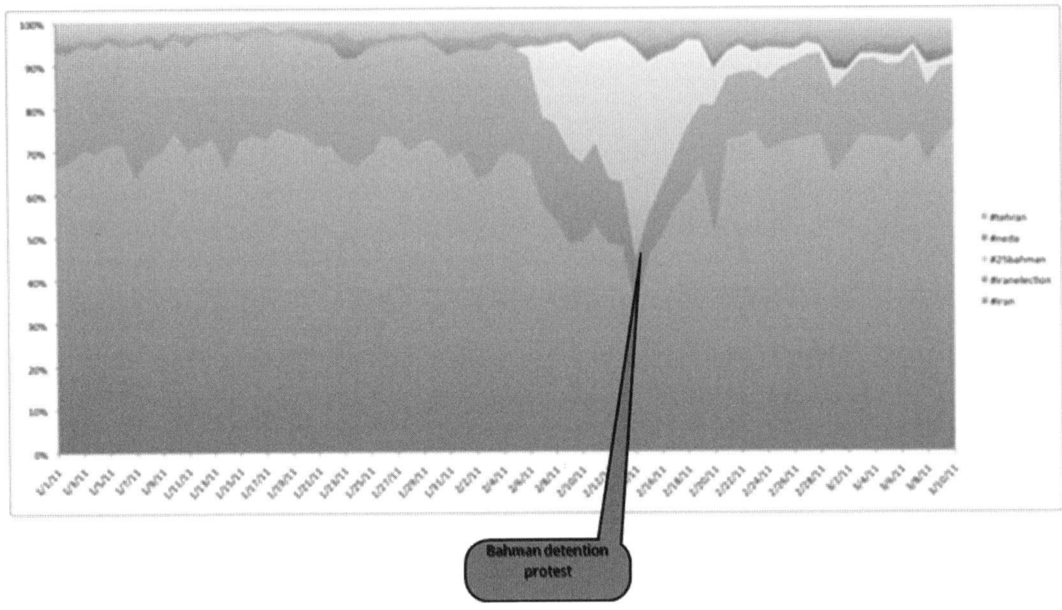

FIGURE AA.3
Iranian share of voice.

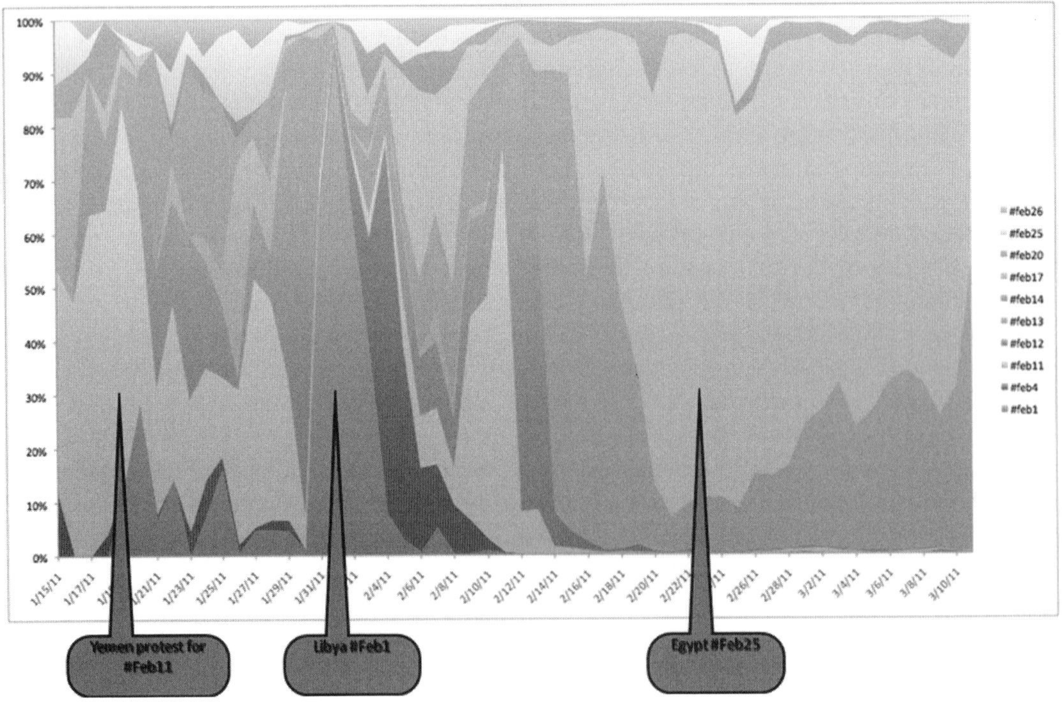

FIGURE AA.4
Significant Mideast protest dates: February 2011.

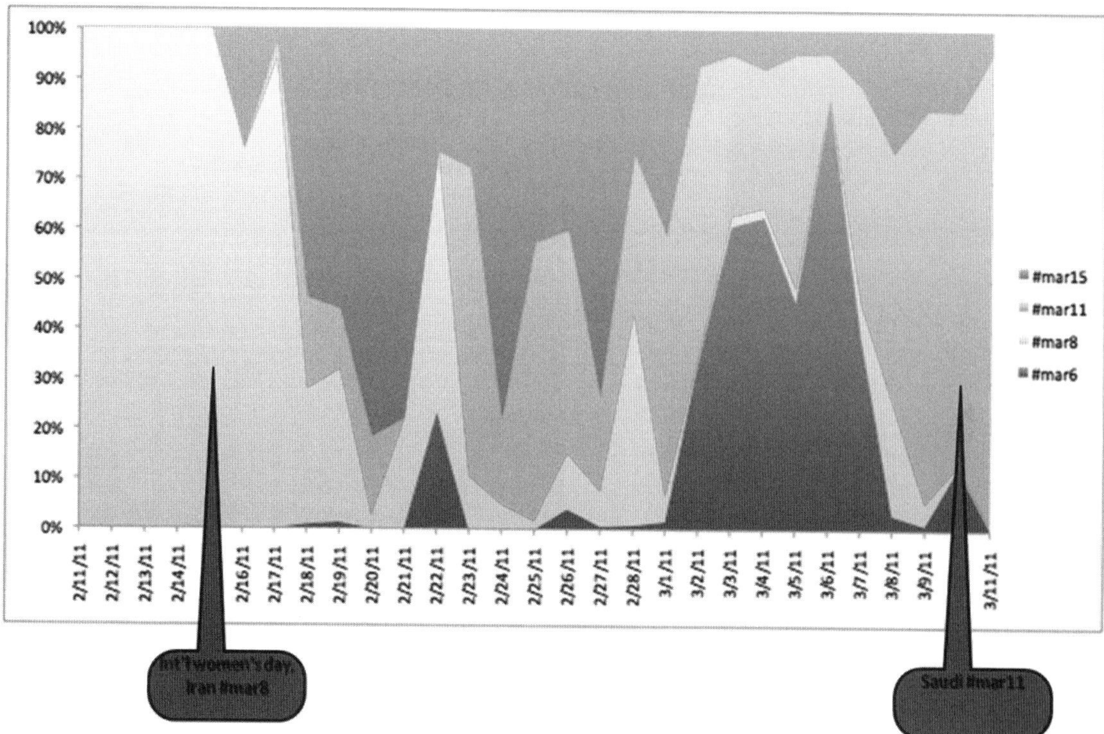

FIGURE AA.5

Significant Mideast protest dates: March 2011.

It is clear that the protest dates SOV analysis can be used to identify efforts to organize protest dates, providing an excellent leading indicator of the interest for each potential protest. Figure AA.5 presents additional data points for March 2011, and, as with February, the data support the predictive identification of events that are about to occur based upon the organizational communication occurring within Twitter.

SUMMARY

SOV analysis is a powerful way to measure the relative impact that communication about specific topics has over time. As was seen with the Mideast analysis, when the topics are related, the SOV analysis provides leading indicators with the information that something is about to occur in the related category of terms. There are implications for using SOV analysis as leading indicators of things about to happen, which can be used by marketers and financial firms to predict outcomes.

For marketers, monitoring terms related to a product will provide insights into significant communication about that product; terms related to competitors provide insights into significant occurrences impacting key competition. For financial firms, monitoring terms related to a portfolio of companies or stock tickers will provide insights into significant events related to each company, impacting trading strategies for that firm.

TUTORIAL BB

Mining Twitter for Airline Consumer Sentiment

Jeffrey Oliver Breen
President and CEO, Cambridge Aviation Research, Cambridge, MA

CONTENTS

Practical Text Mining and Statistical Analysis for Non-structured Text Data Applications. DOI: 10.1016/B978-0-12-386979-1.00008-6

NOTE: All of the code to collect and analyze data and produce the output plots seen in this tutorial are not presented here, only the parts of the code for the specific steps illustrated in the tutorial. The complete code for this tutorial is available from the author's website (http://www.jeffreybreen.com/code/).

INTRODUCTION

This tutorial was originally presented as a first-time introduction to R for the savvy audience of the Boston Predictive Analytics Meetup Group. As such, its primary focus is to highlight R as a tool for getting data easily and synthesizing results quickly. Even though we ignore most of the complexities of text mining, implement a very naïve algorithm for scoring sentiment, and rely on a very small sample of tweets, we are nonetheless able to find an interesting result that compares well to a widely respected study of customer satisfaction.

WHAT IS R?

The R Project for Statistical Computing is a GNU project centered around an open source implementation of the S language that was developed in the 1980s by John Chambers at Bell Labs. Suitable for both scripted and interactive use, the core language is designed to handle real data and modeling work. It contains data types for scalars, vectors, arrays, lists, matrices, tables, complex numbers, and more.

But R is much more than a programming language. It has been adopted by practitioners across a broad spectrum of fields. This rapidly growing community is largely responsible for its ever-increasing utility. Through R's package mechanism, users are able to add and share new functionality and data sets. Many of the contributed packages come from leaders in their respective fields.

Commercial versions and support are available from Revolution Analytics, and RStudio has recently launched a cross-platform, open source IDE. Commercial vendors are increasingly providing R support for their products, from data warehouse vendor IBM/Netezza to Oracle, IBM/SPSS, SAS, Tibco, StatSoft, and high-performance computing vendor Platform Symphony.

LOADING DATA INTO R

One of R's strengths is reading data from other packages and sources. The base distribution includes functions to read a variety of file formats, including comma-separated and fixed-width text files, along with files from other traditional statistical software such as SAS, SPSS, and Stata. Add-on packages are available to access many other file formats, relational databases, Hadoop, and specialized data sources on the web and elsewhere.

THE TWITTER PACKAGE

One such package is the twitteR package by Jeff Gentry. Designed to access Twitter's JSON API and supporting OAuth authentication through its companion ROAuth package, twitteR makes searching Twitter as simple as can be. The library is available from the Comprehensive R Archive Network. Most

IDEs have a menu command to find and install new packages, but even from the command line, it's a one-line affair:

```
> install.packages('twitteR', dependencies=T)
```

R will download and install the twitteR package and any package it depends on, such as the library to parse JSON. (The ">" above is R's command prompt, which is included throughout this tutorial to distinguish our input from R's output.) Once installed, just load twitteR with the `library()` command:

```
> library(twitteR)
Loading required package: RCurl
Loading required package: bitops
Loading required package: RJSONIO
```

Again, notice that R will automatically resolve dependencies and load any required packages. Once loaded, one line is all you need to search Twitter and fetch up to 1,500 results:

```
> delta.tweets = searchTwitter('@delta', n=1500)
```

The `searchTwitter()` function returns an R `list`—in this case 1,500 elements long:

```
> length(delta.tweets)
[1] 1500

> class(delta.tweets)
[1] "list"
```

A `list` in R is a collection of objects. Its elements may be referenced by name or by number (position). Double square brackets are used to refer to individual elements. Let's take a closer look at the first element in the list:

```
> tweet = delta.tweets[[1]]

> class(tweet)
[1] "status"
attr(,"package")
[1] "twitteR"
```

So `tweet` is an object of type `status` that is provided by the twitteR package. The documentation for the `status` class (accessed by typing `?status`) describes accessor methods like `getScreenName()` and `getText()`, which do what you would expect:

```
> tweet$getScreenName()
[1] "Alaqawari"

> tweet$getText()
[1] "I am ready to head home. Inshallah will try to get on the earlier flight to Fresno. @Delta
@DeltaAssist"
```

EXTRACTING TEXT FROM TWEETS

So now that we know how to extract the text from one tweet, how should we process all 1,500? In most traditional programming languages, we would write a loop. We can do that in R, too, but it's not recommended. R is an interpreted language, so any code you write to perform a loop is likely to run much more slowly than equivalent code in compiled language like C. (Most of R's built-in functions are compiled and are therefore very fast.) Fortunately, R provides a facility to iteratively apply functions to each object in a collection through the "apply" family of functions: `apply()`, `lapply()`, `mapply()`, `sapply()`, `tapply()`, `vapply()`, and so on. Unfortunately this family has grown organically over time and lacks consistent naming and calling conventions.

THE PLYR PACKAGE

Rice University's Hadley Wickham has written the plyr package to overcome the limitations of the apply family of functions. While the package provides some nice bells and whistles that we will see as we work with it, its primary advantage is that it provides a simple and consistent naming convention to its `apply()`-like functions.

The first letter specifies the data type you're starting with ("`d`" for `data.frame`, "`l`" for `list`, "`a`" for array, etc.); the second letter specifies the data type you want as output; and the rest of the name is always "`ply`". So if you have an array ("`a`") and you want a list ("`l`") as output, use "`a`" + "`l`" + "`ply`" = `alply()`.

In our case, we have a list ("`l`") but only need a simple array ("`a`") as output, so we use `laply()`. But we don't have a simple function like `sum()` or `median()` to run on each element. Instead, we need to call a function—the `getText()` accessor method—on each `status` object. Such a function is so easy to write we can even write it as an "anonymous" function, in place:

```
> delta.text = laply(delta.tweets, function(t) t$getText())

> length(delta.text)
[1] 1500

> head(delta.text, 5)
[1] "I am ready to head home. Inshallah will try to get on the earlier flight to Fresno. @Delta @DeltaAssist"
[2] "@Delta Releases 2010 Corporate Responsibility Report - @PRNewswire (press release) : http://tinyurl.com/64mz3oh"
[3] "Another week, another upgrade! Thanks @Delta!"
[4] "I'm not able to check in or select a seat for flight DL223/KL6023 to Seattle tomorrow. Help? @KLM @delta"
[5] "In my boredom of waiting realized @deltaairlines is now @delta seriously..... Stil waiting and your not even unloading status yet"
```

By using `laply()`, we requested an array as output, but since our result had only one dimension (column), plyr automatically simplified it to a vector, which you can think of as a one-dimensional array:

```
> class(delta.text)
[1] "character"
```

If we really need an array for some reason, this default behavior can be overridden with a `.drop=F` parameter. With the text of our tweets extracted into a simple vector, let's turn our attention to estimating its emotional sentiment.

ESTIMATING SENTIMENT

Sentiment analysis is an active area of research involving complicated algorithms and subtleties. For the purposes of this tutorial, we err on the side of simplicity and estimate a tweet's sentiment by counting the number of occurrences of "positive" and "negative" words.

To assign a numeric score to each tweet, we'll simply subtract the number of occurrences of negative words from the number of positive occurrences. Larger negative scores will correspond to more negative expressions of sentiment, neutral (or balanced) tweets should net to zero, and very positive tweets should score larger, positive numbers.

LOADING THE OPINION LEXICON

First, we need to find a source that categorizes words by sentiment. A Google search for "sentiment analysis" or "opinion mining" yields a number of sources of such word lists. Hu and Liu's "opinion lexicon" categorizes nearly 6,800 words as positive or negative and can be downloaded from Bing Liu's website (http://www.cs.uic.edu/~liub/FBS/opinion-lexicon-English.rar).

The lexicon consists of two text files, one containing a list of positive words and one containing negative words. Each file begins with some documentation, which we need to skip and is denoted by initial semicolon (";") characters. R's built-in `scan()` function makes short work of reading these files:

```
> hu.liu.pos = scan('data/opinion-lexicon-English/positive-words.txt',
         what='character', comment.char=';')

> hu.liu.neg = scan('data/opinion-lexicon-English/negative-words.txt',
         what='character', comment.char=';')
```

These objects are simple character vectors, just like our `delta.text`:

```
> class(hu.liu.neg)
[1] "character"
> class(hu.liu.pos)
[1] "character
> length(hu.liu.neg)
[1] 4783
> length(hu.liu.pos)
[1] 2006
```

R's `c()` function (for "combine") allows us to add a few industry- and Twitter-specific terms to form our final `pos.words` and `neg.words` vectors:

```
> pos.words = c(hu.liu.pos, 'upgrade')

> neg.words = c(hu.liu.neg, 'wtf', 'wait', 'waiting',
                'epicfail', 'mechanical')
```

IMPLEMENTING OUR SENTIMENT SCORING ALGORITHM

To score each tweet, our `score.sentiment()` function uses `laply()` to iterate through the input text. It strips punctuation and control characters from each line using R's regular expression-powered substitution function, `gsub()` and uses `match()` against each word list to find matches:

```
score.sentiment = function(sentences, pos.words, neg.words, .progress='none')
{
   require(plyr)
   require(stringr)

   # we got a vector of sentences. plyr will handle a list
   # or a vector as an "l" for us
   # we want a simple array of scores back, so we use
   # "l" + "a" + "ply" = "laply":
   scores = laply(sentences, function(sentence, pos.words, neg.words) {
   # clean up sentences with R's regex-driven global substitute, gsub():
      sentence = gsub('[[:punct:]]', '', sentence)
      sentence = gsub('[[:cntrl:]]', '', sentence)
      sentence = gsub('\\d+', '', sentence)
      # and convert to lower case:
      sentence = tolower(sentence)

      # split into words. str_split is in the stringr package
      word.list = str_split(sentence, '\\s+')
      # sometimes a list() is one level of hierarchy too much
      words = unlist(word.list)

      # compare our words to the dictionaries of positive & negative terms
      pos.matches = match(words, pos.words)
      neg.matches = match(words, neg.words)

      # match() returns the position of the matched term or NA
      # we just want a TRUE/FALSE:
      pos.matches = !is.na(pos.matches)
      neg.matches = !is.na(neg.matches)

      # and conveniently enough, TRUE/FALSE will be treated as 1/0 by sum():
      score = sum(pos.matches) - sum(neg.matches)

      return(score)
   }, pos.words, neg.words, .progress=.progress)

   scores.df = data.frame(score=scores, text=sentences)
   return(scores.df)
}
```

ALGORITHM SANITY CHECK

Let's quickly test our `score.sentiment()` function and word lists with some sample sentences:

```
> sample = c("You're awesome and I love you",
             "I hate and hate and hate. So angry. Die!",
             "Impressed and amazed: you are peerless in your
                  achievement of unparalleled mediocrity.")

> result = score.sentiment(sample, pos.words, neg.words)

> result
```

```
score text
1  2 You're awesome and I love you
2 -5 I hate and hate and hate. So angry. Die!
3  4 Impressed and amazed: you are peerless in your
                         achievement of unparalleled mediocrity.
```

Not surprisingly, our simple algorithm completely misses the sarcasm of the third sentence, but our code seems to be working as intended.

Let's try it with a couple of real tweets:

```
score text
1 -4 @Delta I'm going to need you to get it together. Delay on
                         tarmac, delayed connection, crazy gate changes... #annoyed
2  5 Surprised and happy that @Delta helped me avoid the 3.5 hr
                         layover I was scheduled for. Patient and helpful agents.
                         #remarkable
```

Our algorithm again misses the tinge of sarcasm in the second tweet, but at least this one is, on balance, mostly positive.

DATA.FRAMES HOLD TABULAR DATA

Our `score.sentiment()` function returns tabular data with multiple columns and multiple rows. In R, the `data.frame` is the workhorse for such spreadsheet-like data:

```
> result
score text
1  2 You're awesome and I love you
2 -5 I hate and hate and hate. So angry. Die!
3  4 Impressed and amazed: you are peerless in your
                         achievement of unparalleled mediocrity.
> class(result)
[1] "data.frame"
```

A `data.frame` is technically composed of lists, so each column, row, and element can be accessed by a position (a number) or a name (a string). While we normally only use column names, rows are named, too, defaulting to their sequential position within the `data.frame` (as a string):

```
> colnames(result)
[1] "score" "text"

> rownames(result)
[1] "1" "2" "3"
```

We can extract the `score` column by name using array notation and the row, column convention:

```
> result[,'score']
[1] 2 -5 4
```

The dollar sign ($) convention is more common:

```
> result$score
[1] 2 -5 4
```

Or we can reference its position (again using the row, column convention):

```
> result[,1]
[1] 2 -5 4
```

Note that a missing specification returns all elements, so `result[,1]` yields values from each row in the first column.

Similarly, individual elements can be accessed by name, position, or any combination:

```
> result[1,1]
[1] 2

> result[1,'score']
[1] 2

> result['1','score']
[1] 2
```

Finally, positions can also be specified as ranges or vectors of offsets:

```
> result[1:2, 'score']
[1] 2 -5

> result[c(1,3), 'score']
[1] 2 4
```

SCORING THE TWEETS

To score the text of Delta's tweets, just feed it into our `score.sentiment()` function:

```
> delta.scores = score.sentiment(delta.text, pos.words,
                                 neg.words, .progress='text')
|===========================================================| 100%
```

The `.progress='text'` parameter is passed to `laply()` to provide a text progress bar as feedback—a handy feature for long-running processes and provided by all of plyr's functions.

Now that we have our results in a `data.frame`, we should add columns to identify the airline, since we will later combine scores from other airlines (Figure BB.1). To create a new column in a `data.frame`, simply refer to it while assigning a value:

```
> delta.scores$airline = 'Delta'
> delta.scores$code = 'DL'
```

score	text	airline	code
1	I am ready to head home. Inshallah will try to get on the earlier flight to Fresno. @Delta @DeltaAssist	Delta	DL
0	@Delta Releases 2010 Corporate Responsibility Report - @PRNewswire (press release) : http://tinyurl.com/64mz3oh	Delta	DL
1	Another week, another upgrade! Thanks @Delta!	Delta	DL
0	I'm not able to check in or select a seat for flight DL223/KL6023 to Seattle tomorrow. Help? @KLM @delta	Delta	DL
-3	In my boredom of waiting realized @deltaairlines is now @delta seriously..... Stil waiting and your not even unloading status yet	Delta	DL
1	Hmmm... I just got 'upgraded' from my reserved exit row seat to a knee banger. ATL-PVD just got longer. What gives @Delta?	Delta	DL
0	@Delta 7 days I'm trying to book a flight with you. Still w/o success. Starting to get really upset. What can I do?	Delta	DL
-2	its amazing how horrible @Delta there service is horrendous and their staff unprofessional! #angry	Delta	DL
0	With @DeltaAssist, do you believe @Delta has seen greater returns through #SM efforts? http://bit.ly/kXn9qZ	Delta	DL
-3	@AmericanAir app froze at TSA Checkpoint. Had to leave line to get paper boarding pass; never have this problem with @Delta app #fail	Delta	DL
1	How the hell did I end up Sky Priority? I'll take it. Thanks @Delta. The only thing that could make it better would be upgrade	Delta	DL

FIGURE BB.1

Spreadsheet created showing text, score, the airline, and the airline's code letters.

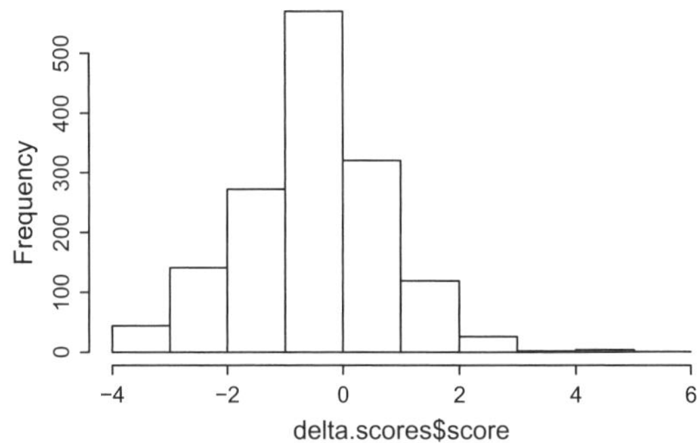

FIGURE BB.2
Histogram of Delta scores$score.

R's built-in `hist()` function will create and plot a histogram of sentiment scores (Figure BB.2) for our tweets about Delta:

```
hist(delta.scores$score)
```

ggplot2 is an alternative graphics package by Hadley Wickham (the author of plyr) that generates much more refined output (Figure BB.3). It is based on Wilkerson's "grammar of graphics," building visualizations from individual layers. While ggplot2 normally requires data in a `data.frame`, its `qplot()` function provides a simplified interface, accepting vectors just like base R's plotting functions:

```
> q = qplot(delta.scores$score)
> q = q + theme_bw()
```

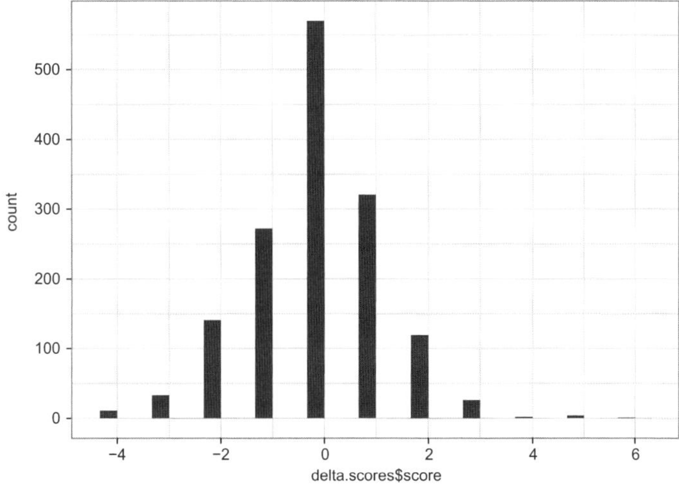

FIGURE BB.3
ggplot2 is an alternative graphics package by Hadley Wickham (the author or plyr) that generates a more refined output.

Thanks to R's object orientation, the normal "+" operator is used to combine plot options and layers.

REPEAT FOR EACH AIRLINE

We can similarly capture, extract, and score tweets for American Airlines, JetBlue, Southwest, United, and U.S. Airways. R's `rbind()` function can then combine all the rows into a single `data.frame`:

```
> all.scores = rbind(american.scores, delta.scores, jetblue.scores,
                     southwest.scores, united.scores, us.scores)
```

COMPARE THE SCORE DISTRIBUTIONS

Let's use ggplot2 to take a look at the score distributions for all the airlines. Since our results are already in a single `data.frame`, we will eschew `qplot()` and build our visualization in the normal ggplot2 way: layer by layer.

First, create the graph with a call to the `ggplot()` function, at which time we can specify the `data.frame` to use and how to map its data columns to plot aesthetics (e.g., x, y, line color, fill color, symbol size, etc.). In this case, we want to plot the sentiment score along the x-axis (`x=score`) and will differentiate each airline's bars with a different fill color (`fill=airline`):

```
> g = ggplot(data=all.scores, mapping=aes(x=score, fill=airline))
```

The `geom_bar()` function creates the bar graph layer itself. Since `geom_bar()` is often used to display histograms, by default it will automatically bin and compute frequencies for you. We haven't computed anything, so we're happy to let the graph layer do it for us, but we specify `binwidth=1` so it doesn't try to pick bins smaller or bigger than our simple integer scores warrant:

```
> g = g + geom_bar(binwidth=1)
```

For any given score, we will have several bars to display, one for each airline. By default, `geom_bar()` will stack the bars on top of one another, but that would make it very difficult to compare them. We can easily move them next to one another by specifying `position="dodge"`, but ggplot2's faceting feature is better yet. It will move each airline into its own separate subplot, and it couldn't be easier to invoke:

```
> g = g + facet_grid(airline~.)
```

Finally, we ask for a cleaner display (with a plain white background) and a nice color palette from Cindy Brewer's ColorBrewer.org (Figure BB.4):

```
g = g + theme_bw() + scale_fill_brewer()
```

Looking at the score distributions in Figure BB.4, some asymmetry is evident. For example, the bars at +1 are much larger than the −1 bars for Southwest and JetBlue. But given the simplicity of our sentiment scoring algorithm, let's focus on the extreme tails, since bigger differences should be more likely to capture real differences in sentiment.

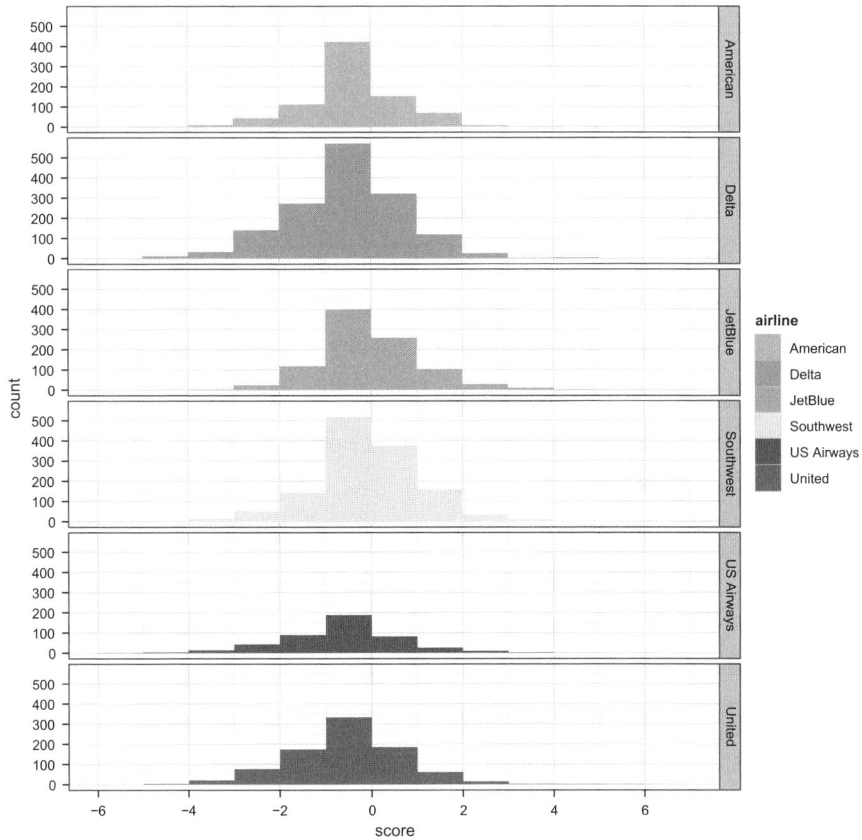

FIGURE BB.4

Graph with cleaner display (with a plain white background) using a color palette from Cindy Brewer's ColorBrewer.org.

IGNORE THE MIDDLE

Let's create two new boolean columns to focus on tweets with very negative (`score <= -2`) and very positive (`score >= 2`) sentiment scores:

```
> all.scores$very.pos.bool = all.scores$score >= 2
> all.scores$very.neg.bool = all.scores$score <= -2

> all.scores[c(1,6,47,99), c(1, 3:6)]
   score airline  code very.pos.bool very.neg.bool
1     -1 American   AA         FALSE         FALSE
6     -3 American   AA         FALSE          TRUE
47     2 American   AA          TRUE         FALSE
99    -5 American   AA         FALSE          TRUE
```

We want to count the occurrence of these strong sentiments for each airline. We can easily cast these `TRUE/FALSE` values to numeric `1/0`, so we can then use `sum()` to count them:

```
> all.scores$very.pos = as.numeric( all.scores$very.pos.bool )
> all.scores$very.neg = as.numeric( all.scores$very.neg.bool )

> all.scores[c(1,6,47,99), c(1, 3:8)]
   score airline  code very.pos.bool very.neg.bool very.pos very.neg
1     -1 American  AA          FALSE         FALSE        0        0
6     -3 American  AA          FALSE          TRUE        0        1
47     2 American  AA           TRUE         FALSE        1        0
99    -5 American  AA          FALSE          TRUE        0        1
```

We can use plyr's `ddply()` function to aggregate the rows for each airline, calling the `summarise()` function to create new columns containing the counts:

```
> twitter.df = ddply(all.scores, c('airline', 'code'), summarise,
             very.pos.count=sum(very.pos),
             very.neg.count=sum(very.neg))
```

As a single, final score for each airline, let's calculate the percentage of these "extreme" tweets that are positive:

```
> twitter.df$very.tot = twitter.df$very.pos.count +
              twitter.df$very.neg.count

> twitter.df$score = round(100 * twitter.df$very.pos.count /
                   twitter.df$very.tot)
```

The `orderBy()` function from the doBy package makes it easy to sort the results. Note that it preserves the original row names:

```
> orderBy(~-score, twitter.df)
   airline    code very.pos.count very.neg.count very.tot score
3  JetBlue     B6             146             28      174    84
4  Southwest   WN             207             72      279    74
1  American    AA              80             57      137    58
2  Delta       DL             152            185      337    45
6  United      UA              82            102      184    45
5  US Airways  US              38             62      100    38
```

COMPARE WITH ACSI'S CUSTOMER SATISFACTION INDEX

Now these results are from a very small sample of consumers who chose to tweet at a particular point in time. It would be neither fair nor valid to draw any firm conclusions from such a small sample. But it is tantalizing to think that such a simple measure might capture something real—especially with JetBlue and Southwest leading the legacy airlines by such a clear margin.

But rather than relying on our personal brand experience or other anecdotal evidence, let's compare our results with the American Customer Satisfaction Index (Figure BB.5). Each year the ACSI conducts tens of thousands of interviews to measure consumer satisfaction with hundreds of companies and organizations. They kindly publish their top-level results on their website (http://www.theacsi.org/).

Airlines

	Base-line	95	96	97	98	99	00	01	02	03	04	05	06	07	08	09	10	11	Previous Year % Change	First Year % Change
Southwest	78	76	76	76	74	72	70	70	74	75	73	74	74	76	79	81	79	81	2.5	3.8
All Others	**NM**	70	74	70	62	67	63	64	72	74	73	74	74	75	75	77	75	76	1.3	8.6
Airlines	**72**	**69**	**69**	**67**	**65**	**63**	**63**	**61**	**66**	**67**	**66**	**66**	**65**	**63**	**62**	**64**	**66**	**65**	**-1.5**	**-9.7**
Continental	67	64	66	64	66	64	62	67	68	68	67	70	67	69	62	68	71	64	-9.9	-4.5
American	70	71	71	62	67	64	63	62	63	67	66	64	62	60	62	60	63	63	0.0	-10.0
United	71	67	70	68	65	62	62	59	64	63	64	61	63	56	56	56	60	61	1.7	-14.1
US Airways	72	67	66	68	65	61	62	60	63	64	62	57	62	61	54	59	62	61	-1.6	-15.3
Delta	77	72	67	69	65	68	66	61	66	67	67	65	64	59	60	64	62	56	-9.7	-27.3
Northwest Airlines	69	71	67	64	63	53	62	56	65	64	64	64	61	61	57	57	61	#	**N/A**	**N/A**

FIGURE BB.5

Contents of the acsi.df data.frame containing airline customer satisfaction scores scraped from the American Customer Satisfaction Index (ACSI) website (http://www.theacsi.org/).

SCRAPE THE ACSI WEBSITE

Duncan Temple Lang's XML package provides many useful functions for scraping and parsing data from the web. But one really shines: A single call to `readHTMLTable()` will download a web page from a URL, parse the HTML, extract any tables, and return a list of populated `data.frames`, complete with headers.

We specify `which=1` to retrieve only the first table on the page, and `header=T` to indicate that the table headings should be used as column names:

```
> acsi.url = 'http://www.theacsi.org/index.php?
option=com_content&view=article&id=147&catid=&Itemid=212&i=Airlines'


> acsi.df = readHTMLTable(acsi.url, header=T, which=1, stringsAsFactors=F)


> acsi.df
                     Base-line 95 96 97 98 99 00 01 02 03 04 05 06 07 08 09 10 11 Previous      First
                                                                                  Year%Change   Year%Change
1 Southwest                 78 76 76 76 74 72 70 70 74 75 73 74 74 76 79 81 79 81         2.5          3.8
2 All Others                NM 70 74 70 62 67 63 64 72 74 73 74 74 75 75 77 75 76         1.3          8.6
3 Airlines                  72 69 69 67 65 63 63 61 66 67 66 66 65 63 62 64 66 65        -1.5         -9.7
4 Continental               67 64 66 64 66 64 62 67 68 68 67 70 67 69 62 68 71 64        -9.9         -4.5
5 American                  70 71 71 62 67 64 63 62 63 67 66 64 62 60 62 60 63 63         0.0         10.0
6 United                    71 67 70 68 65 62 62 59 64 63 64 61 63 56 56 56 60 61         1.7        -14.1
7 US Airways                72 67 66 68 65 61 62 60 63 64 62 57 62 61 54 59 62 61        -1.6        -15.3
8 Delta                     77 72 67 69 65 68 66 61 66 67 67 65 64 59 60 64 62 56        -9.7        -27.3
9 Northwest Airlines        69 71 67 64 63 53 62 56 65 64 64 64 61 61 57 57 61  #         N/A          N/A
```

The preceding tabulation is a "plain screen dump" of the airlines and yearly data. Figure BB.6 shows how it looks from ACSI (http://www.theacsi.org/).

	Base-line	95	96	97	98	99	00	01	02	03	04	05	06	07	08	09	10	11	PreviousYear%Change	FirstYear%Change
Southwest	78	76	76	76	74	72	70	70	74	75	73	74	74	76	79	81	79	81	2.5	3.8
All Others	NM	70	74	70	62	67	63	64	72	74	73	74	74	75	75	77	75	76	1.3	8.6
Airlines	72	69	69	67	65	63	63	61	66	67	66	66	65	63	62	64	66	65	-1.5	-9.7
Continental	67	64	66	64	66	64	62	67	68	68	67	70	67	69	62	68	71	64	-9.9	-4.5
American	70	71	71	62	67	64	63	62	63	67	66	64	62	60	62	60	63	63	0.0	-10.0
United	71	67	70	68	65	62	62	59	64	63	64	61	63	56	56	56	60	61	1.7	-14.1
US Airways	72	67	66	68	65	61	62	60	63	64	62	57	62	61	54	59	62	61	-1.6	-15.3
Delta	77	72	67	69	65	68	66	61	66	67	67	65	64	59	60	64	62	56	-9.7	-27.3
Northwest Airlines	69	71	67	64	63	53	62	56	65	64	64	64	61	61	57	57	61	#	N/A	N/A

FIGURE BB.6

The airline data as they appear on the ACSI web page.

Since we are only interested in the most recent results, we only need to keep the first column (containing the airline names) and the nineteenth (containing 2011's scores):

```
> acsi.df = acsi.df[,c(1,19)]
```

Unfortunately, the headings in the original HTML table do not make very good column names, but they are easy to change:

```
> colnames(acsi.df)
[1] "" "11"

> colnames(acsi.df) = c('airline', 'score')

> colnames(acsi.df)
[1] "airline" "score"
```

As some final cleanup, add two-letter airline codes and ensure that the scores are treated as numbers:

```
> acsi.df$code = c('WN', NA, NA, 'CO', 'AA', 'UA', 'US', 'DL', 'NW')

> acsi.df$score = as.numeric(acsi.df$score)
Warning message:
NAs introduced by coercion
```

The "`NAs introduced by coercion`" warning message indicates that the now-defunct Northwest's score of "#" couldn't be translated into a number, so R changed it to `NA` (as in "not applicable"). R was built with real data in mind, so its support of `NA` values is robust and (nearly) universal.

```
> acsi.df
  airline score code
1  Southwest 81 WN
2  All Others 76 <NA>
3  Airlines 65 <NA>
4 Continental 64 CO
5  American 63 AA
6   United 61 UA
7  US Airways 61 US
8   Delta 56 DL
9 Northwest Airlines NA NW
```

COMPARE TWITTER RESULTS WITH ACSI SCORES

In order to compare our Twitter results with the ACSI scores, let's construct a new `data.frame` that contains both (Figure BB.7). The `merge()` function joins together two `data.frames` using common fields (as specified with the `by` parameter). Columns with different data but conflicting names (like our two "`scores`" columns) can be renamed according to the `suffixes` parameter:

code	airline	very.pos.count	very.neg.count	very.tot	score.twitter	score.acsi
AA	American	80	57	137	58	63
DL	Delta	152	185	337	45	56
UA	United	82	102	184	45	61
US	US Airways	38	62	100	38	61
WN	Southwest	207	72	279	74	81

FIGURE BB.7

Table comparing the Twitter results with the ACSI score.

```
> compare.df = merge(twitter.df, acsi.df, by=c('code', 'airline'),
        suffixes=c('.twitter', '.acsi'))
```

Unless you specify `all=T`, nonmatching rows will be dropped by `merge()`—like SQL's INNER JOIN—and that's what happened to top-scoring JetBlue.

GRAPH THE RESULTS

We will again use ggplot2 to display our results, this time on a simple scatterplot (Figure BB.8). We will plot our Twitter score along the x-axis (`x=score.twitter`), plot the ACSI customer satisfaction index along the y-axis (`y=score.acsi`), and use color to distinguish the airlines (`color=airline`):

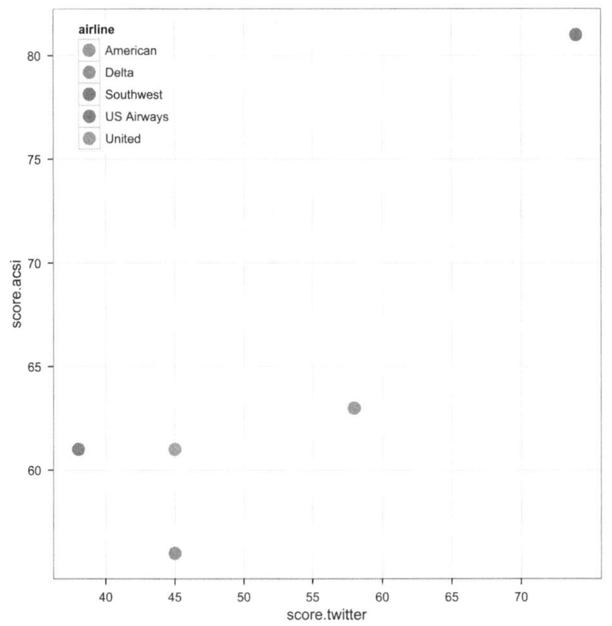

FIGURE BB.8

Using ggplot2 to display the Twitter scores versus the ASCI scores on a simple scatterplot.

```
> g = ggplot(compare.df, aes(x=score.twitter, y=score.acsi)) +
geom_point(aes(color=airline), size=5) +
theme_bw() + opts(legend.position=c(0.2, 0.85))
```

Like R itself, ggplot2 was built for performing analyses, so it can do a lot more than just display data. Adding a `geom_smooth()` layer will compute and overlay a running average of your data. But specify `method="lm"`, and it will automatically run a linear regression and plot the best-fitting model (`lm()` is R's linear modeling function) (Figure BB.9):

```
g = g + geom_smooth(aes(group=1), se=F, method="lm")
```

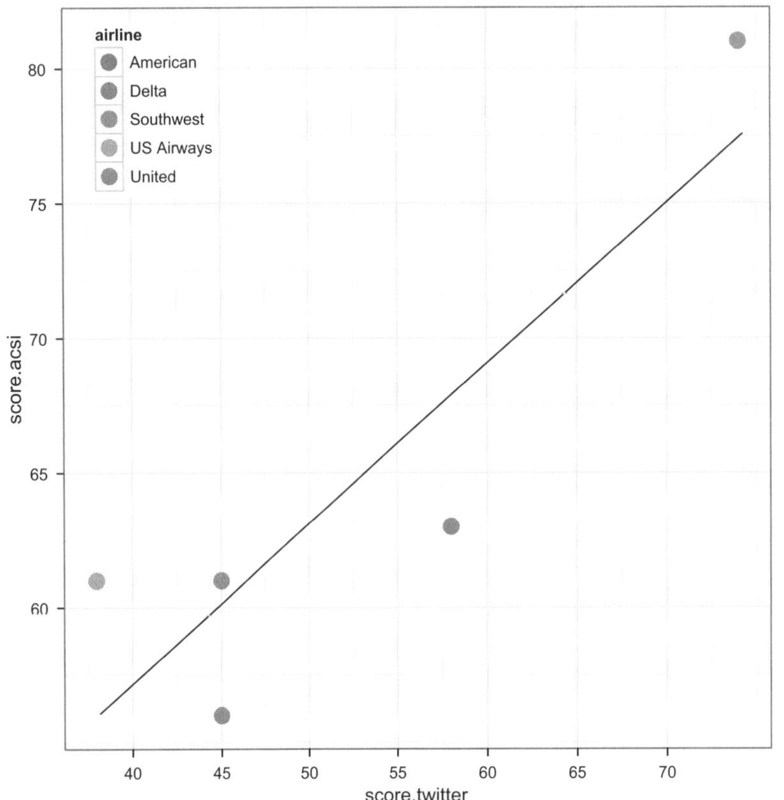

FIGURE BB.9
Linear regression plotted with the ggplot2 tool.

Considering how few data we collected over such a short time and the crudeness of our sentiment scoring algorithm, this correspondence to the highly regarded ACSI survey is remarkable indeed!

NOTES AND ACKNOWLEDGMENTS

R's community is the source of much of its strength. This tutorial would not have been possible without the packages contributed by Hadley Wickham, Duncan Temple Lang, and especially Jeff Gentry.

Thanks to John Verostek for organizing the very interesting Boston Predictive Analytics Meetup Group and for providing valuable input to this presentation. Thanks also to Gary Miner for the invitation to include this tutorial in this book and for valuable feedback on the manuscript.

References

M. A. Harrower and C. A. Brewer, 2003, "ColorBrewer.org: An Online Tool for Selecting Color Schemes for Maps," *The Cartographic Journal*, 40(1): 27–37.

Minqing Hu and Bing Liu, 2004, "Mining and Summarizing Customer Reviews." *Proceedings of the ACM SIGKDD International Conference on Knowledge Discovery and Data Mining (KDD-2004)*, August 22–25, 2004, Seattle, Washington.

Bing Liu, Minqing Hu, and Junsheng Cheng, 2005, "Opinion Observer: Analyzing and Comparing Opinions on the Web." *Proceedings of the 14th International World Wide Web conference (WWW-2005)*, May 10–14, 2005, Chiba, Japan.

Hadley Wickham. ggplot2: *Elegant Graphics for Data Analysis (Use R)*. Dordrecht: Springer, 2009.

Hadley Wickham, 2011, "The Split-Apply-Combine Strategy for Data Analysis." *Journal of Statistical Software*, 40(1), 1–29. http://www.jstatsoft.org/v40/i01/.

Leland Wilkerson. *The Grammar of Graphics*. Dordrecht: Springer, 1999.

Slocum et al. *Thematic Cartography and Visualization*. Upper Saddle River: Prentice Hall, 2008.

TUTORIAL A

Using *STATISTICA* Text Miner to Monitor and Predict Success of Marketing Campaigns Based on Social Media Data

Vladimir Rastunkov PhD, and Mark Rusch
Vice President of StatSoft, Inc.

CONTENTS

INTRODUCTION

Companies in the current business environment have to spend money on advertising their products and services through various types of social media—Twitter, Facebook, YouTube, various sites with product reviews (e.g., amazon.com)—and maintain their own websites. Most chief marketing officers can ask the question "How can we distribute available funds in order to maximize sales by leveraging user-generated content captured in social media to increase sales or production adoption?"

THE KEY ISSUE

- There are many sources of data inside a social network that can provide different kinds of information that in turn can be used for different purposes.
- At the core of a social network are people and their relationships or connections.
- Data associated with people profiles and the associated user-generated content (UGC) such as group associations, "friend" connections, "follower" or subscriber information, and the frequency at which an interaction takes place can be a leading indicator.
- This information can be crucial when planning external viral social media campaigns or when trying to ascertain the effectiveness of a current marketing campaign.

Practical Text Mining and Statistical Analysis for Non-structured Text Data Applications. DOI: 10.1016/B978-0-12-386979-1.00010-4

- The participants in a social network can also be analyzed to identify patterns, clusters, and sentiment in order to determine what the network as a whole "cares" about.
- The premise of this software solution is to (1) analyze what they care about and translate this information into sales lift and (2) to determine where marketing investments should be adjusted in order to maximize lift.

When asked about their main barriers in measuring ROI on social media, CMOs typically have difficulty tying social to conversion and sales metrics, identifying the right metrics and how to track them, and getting CEOs to buy in on metrics

A recent study indicated that over 90 percent of CMOs plan on using some form of user-generated content to inform product and service decisions. Since such a large number of CMOs plan on using this form of content to make more informed decisions, we developed a solution to fulfill this unmet need.

This tutorial discusses various data that can be collected from social media sources and how it can be analyzed with various data and text mining tools. Specifically, we consider the use of univariate and multivariate quality control charts, predictive modeling, performing "what-if" analysis, and performing sentiment analysis.

STEP 1: COLLECTING DATA

The first step is to identify the sources of data that will be monitored, both structured and unstructured (text and other user-generated content). It may seem from the start that the data from social media come in a format of text narratives, reviews, and opinions from countless forums, review sections, and blogs. This is true to some extent, but before starting with processing this bulk of unstructured data, let us try to find some numerical information that is available as well.

The website is the most common way to communicate with customers in order to promote products and services. It is also a source that provides valuable information on the number of visits per unit of time, pages visited, time on each page, and other social and demographic information provided by the IP address. If the website is promoted, then there will be information on the visits achieved through those instruments. Other indicators like conversion rates, add to cart, and visit to purchase could be available if the website works as an online store.

Facebook is a network service that keeps users' profiles, maintains messages exchanges, and so on. Such profiles could also be created by brands and companies. This could provide valuable data on the number of people who like the product or company (Like button). Also, it is a way to promote special coupons for Facebook users (that can be tracked through the sales process).

YouTube allows you to post the product's videos and count the number of subscribers, channel views, and total views. Very similar information is available on the separate videos layer, such as the number of views and the number of people who like it.

Twitter is a blogging service where a company can create a separate thread. This makes it possible to obtain data like number of tweets, number of followers, and number listed. Of course, the tweets themselves are coming in an unstructured text format. Customer reviews and comments on blogs also provide text data. Figure A.1 summarizes sources of structured data available from social media.

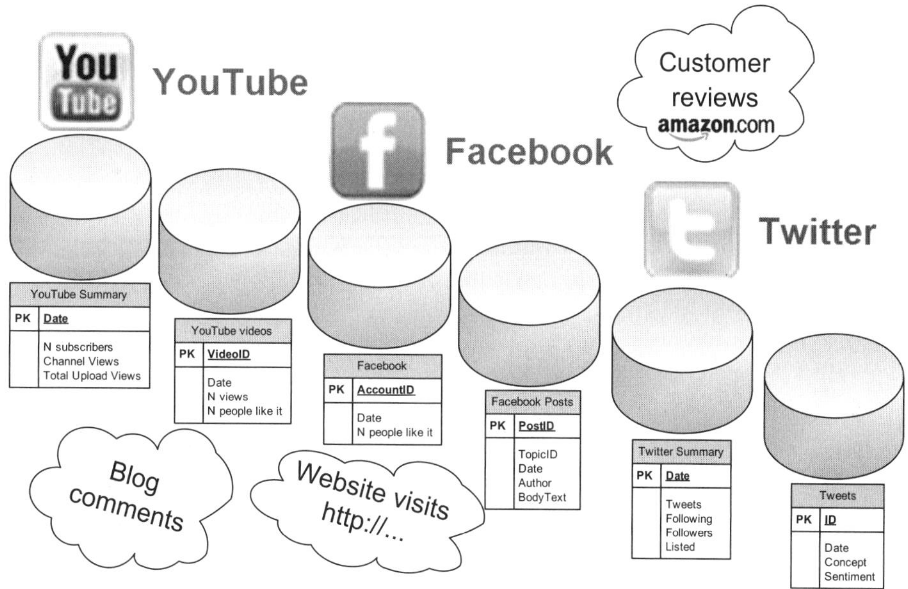

FIGURE A.1
Structured (numerical) and unstructured data from social media.

The data in Figure A.1 could serve as input or as target variables, depending on the business problem. For example, it will be a target whenever we want to optimize the response from social media during a marketing campaign. On the other hand, it could be input whenever we model the number of sales based on social media data. So the whole data set would also include fields such as marketing expenses and numbers of sales. For this tutorial, these data in the following format were simulated (Figure A.2):

1. Date
2. Product A—total sales of product A on that date

Data: Explore_Summary Table.sta* (12v by 1096c)

	1 Date	2 Product A	3 Product B	4 YouTube	5 Website	6 Facebook	7 Twitter	8 Website visits count	9 Facebook N people like Product A	10 Facebook N people like Product B	11 YouTube N subscribers	12 Month
1	6/15/08 12:00 AM	89	108	50	850	50	50	72	17	0	7	6
2	6/16/08 12:00 AM	86	120	50	850	50	50	193	15	10	14	6
3	6/17/08 12:00 AM	85	115	50	850	50	50	133	17	0	0	6
4	6/18/08 12:00 AM	88	107	50	850	50	50	90	0	7	4	6
5	6/19/08 12:00 AM	86	105	50	850	50	50	85	0	4	2	6
6	6/20/08 12:00 AM	87	103	50	850	50	50	44	5	0	0	6
7	6/21/08 12:00 AM	87	111	50	850	50	50	88	41	7	4	6
8	6/22/08 12:00 AM	87	103	50	850	50	50	48	11	0	3	6
9	6/23/08 12:00 AM	88	119	50	850	50	50	182	15	6	16	6
10	6/24/08 12:00 AM	84	106	50	850	50	50	92	0	0	0	6

FIGURE A.2
Preview of the data.

3. Product B—total sales of product B on that date
4. YouTube—expenses on marketing via YouTube instrument
5. Website—expenses associated with website maintenance and promotion
6. Facebook—expenses on marketing via Facebook instrument
7. Twitter—expenses on marketing via Twitter
8. Website visits count—visits during 1 day
9. Facebook N people like product A—number of people who clicked "Like it" button per day
10. Facebook N people like product B—number of people who clicked "Like it" button per day
11. YouTube N subscribers—number of new subscribers
12. Month—the number of months in a year

Unstructured or text information is usually presented in the format in Table A.1.

Table A.1 Format in Which Unstructured Data Are Usually Collected

ID	Date/Time	Author ID/Name	Text Body	Other Fields
Unique Number	mm/dd/yyyy hh:mm:ss.ss	Unique Number/Text	Text	...

The following notes could be made related to the text data:

1. New text messages from blogs and customer reviews will be appearing during relatively short periods of time (several days, or weeks at best).
2. Text messages could appear within minutes and even seconds.
3. Although social media is an evolving field, not all of the products or services could become a subject of user discussions on different forums, blogs, and so on.

How can structured and unstructured data sets be linked together? Obviously, text fields couldn't be simply concatenated with one another. So in order to link the unstructured data with numerical data, the first should be numericized. Various techniques could be implemented, depending on the main task to be solved with the use of unstructured data. In some cases the frequencies of key words are calculated and used, and in others the singular value decomposition method is implemented. For the problem at hand, we are trying to measure (at least primarily) the level of customer support and interest for the products or services. So one of the major inputs would be defined via sentiment ratings calculated from customers' feedback. Once the text narratives are numericized, the unstructured data looks like Table A.2. The format in Table A.2 could be easily transformed (agrregated) to the format in Table A.3.

Table A.2 Format of Unstructured Data after Sentiment Analysis

ID	Date/Time	Author ID/Name	Text Body	Sentiment Rating
Unique Number	mm/dd/yyyy hh:mm:ss.ss	Unique Number/Text	Text	Number

Table A.3 Data Format after Aggregating Sentiment Ratings	
Date/Time	**Sentiment**
mm/dd/yyyy	Number

So in the end of this data transformation process, unstructured data form a separate column (multiple columns) that is available for analysis. For the purposes of this tutorial, the workflow will consist of two parts: analysis of structured data and sentiment analysis of text data. Since it is nearly impossible to simulate the text stream associated with two years of marketing campaigns and sales history, only user-generated content from a real business environment could provide a consistent historical database.

STEP 2: MONITORING THE SITUATION

The next step is to monitor the situation using univariate and multivariate tools. One of the key processes in the business environment is monitoring: detecting nonrandom patterns before they significantly affect the core business. Figure A.2 shows the historical data of sales, marketing expenses, and the response from social media. Those fields could also be viewed as separate processes.

In order to implement a univariate monitoring of those, we will create Shewhart quality control charts for individual observations and moving ranges. Select the following analysis (Figure A.3): *Statistics* → *Industrial Statistics* → *QC Charts* → *QC Charts for Var Lists.*

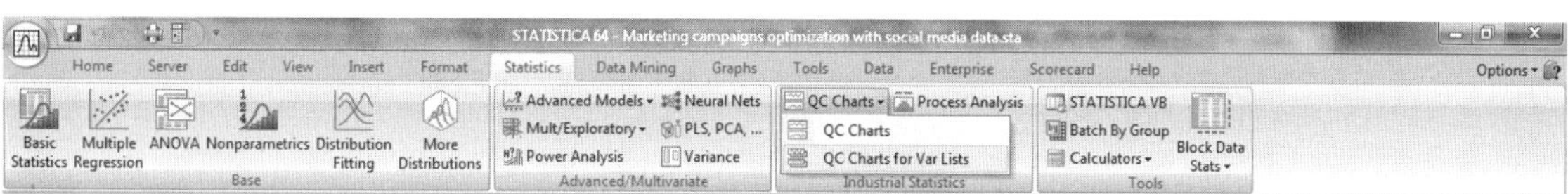

FIGURE A.3

Selecting QC charts analysis on the ribbon bar.

In the next step, the analysis should be selected. Different types of charts serve various purposes, such as maintaining specification limits and detecting trends. For initial exploration we will use "Individuals & moving range" as the simplest and yet very illustrative example of such monitors (Figure A.4).

On the variable selection dialog (Figure A.5), let us specify the key processes we want to monitor: website visits count, Facebook N people like Product A, Facebook N people like Product B, and YouTube N subscribers (Figure A.6).

The next dialog allows the user to specify the method of control limits definition, warning lines parameters, and other options related to X and R/S charts (Figure A.7). We will continue with the default set of parameters and click "Summary."

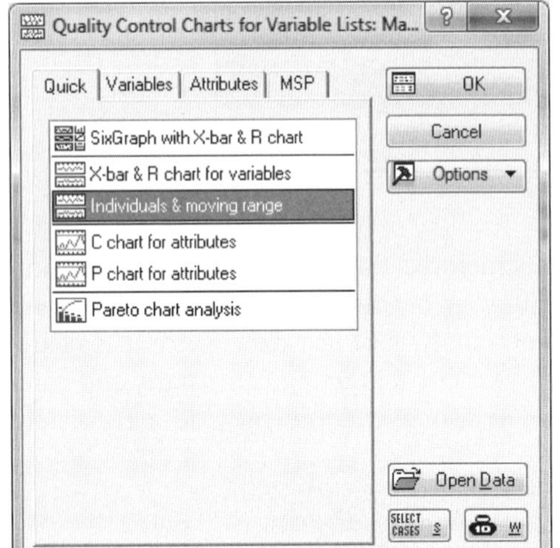

FIGURE A.4
Chart type selection dialog.

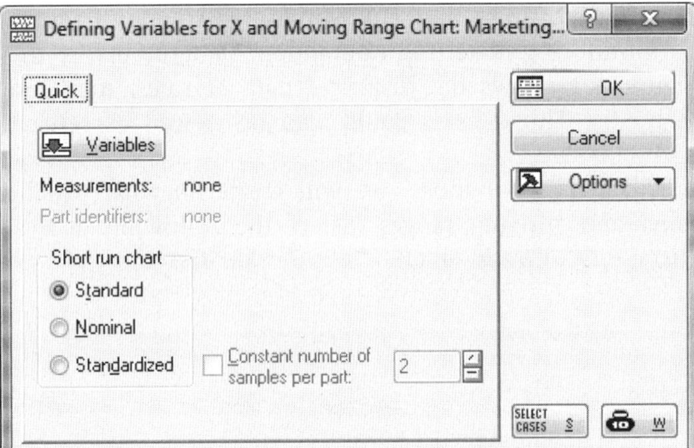

FIGURE A.5
Variable selection dialog.

FIGURE A.6
Variable selection.

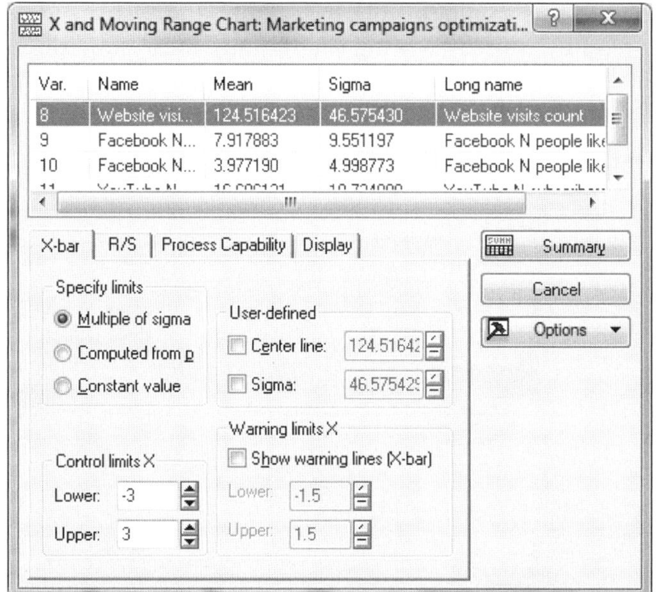

FIGURE A.7
Parameters specification dialog for QC analysis.

Quality control charts are stored into the workbook (this is the default setting for the output of results that can be modified through *Home* → *Options* → *Analysis/Graphs* → *Output Manager* (Figure A.8).

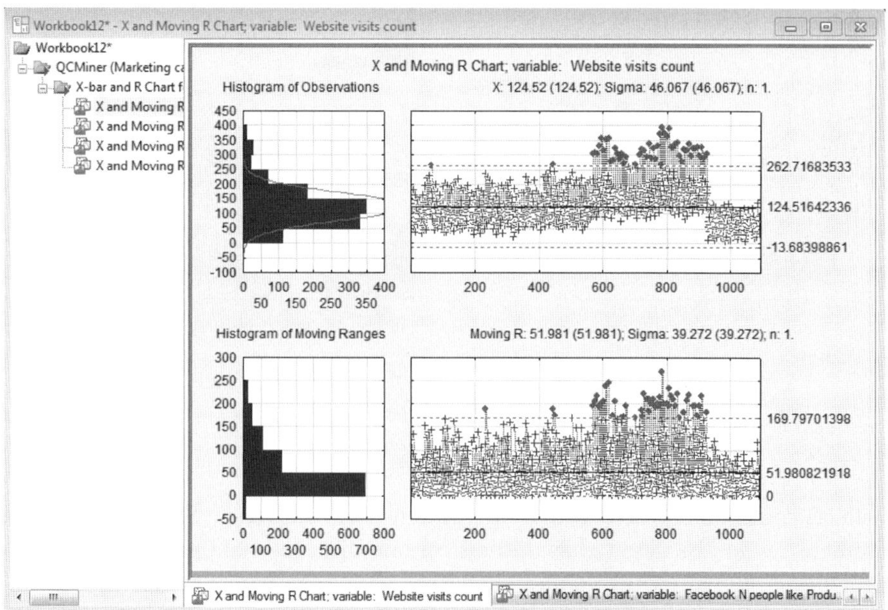

FIGURE A.8
Workbook with QC analysis results.

Significant drops in the number of daily visits could be observed from the website data (Figure A.9). YouTube, on the other hand, recently started to bring significantly more subscribers daily. Those facts should be thoroughly analyzed before making further decisions on marketing investments.

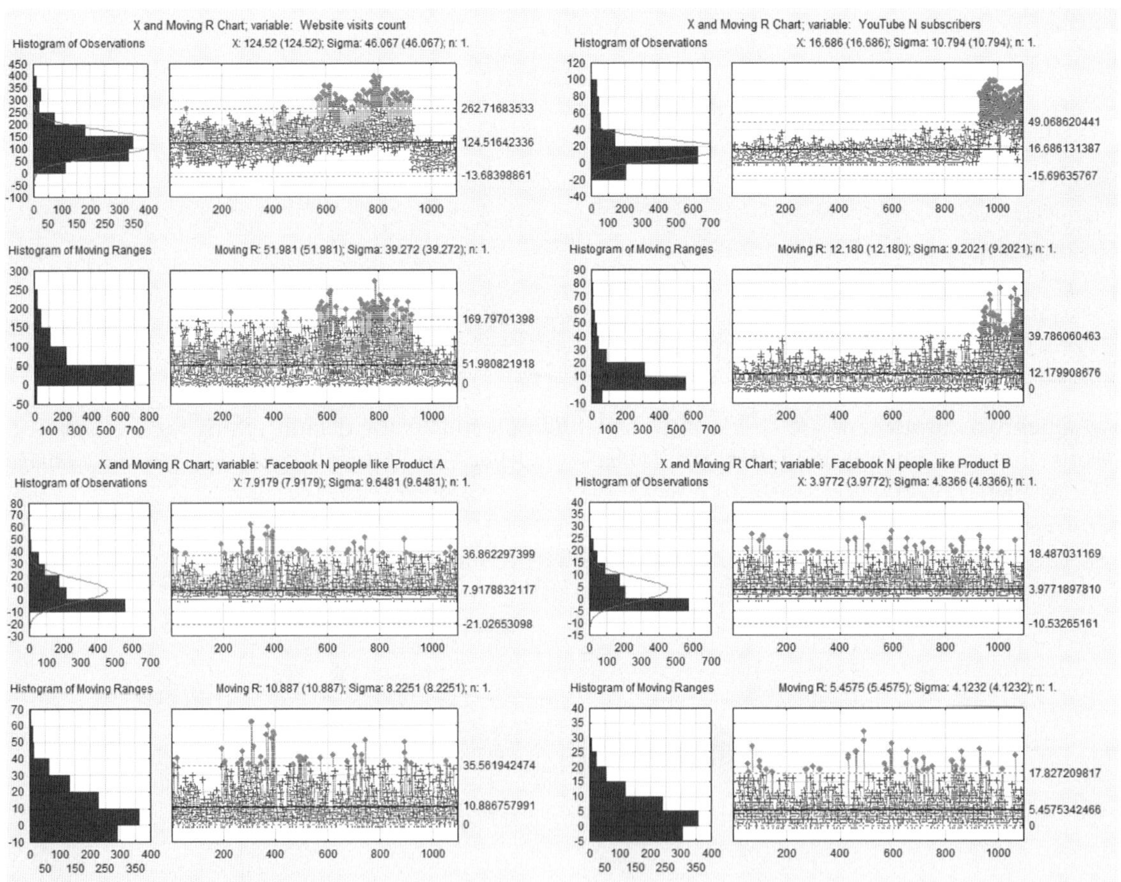

FIGURE A.9
QC charts that identify meaningful shifts.

Not all effects could be captured with a one-dimensional approach. Simultaneous changes in the group of parameters won't be detected as outliers on one-dimensional QC charts if those changes are within control limits. On the other hand, such simultaneous change is not a random effect. The multivariate QC charts are available under Industrial Statistics. On the models selection dialog, we will select "Hotelling T^2 Chart for Individuals" (Figures A.10 and A.11).

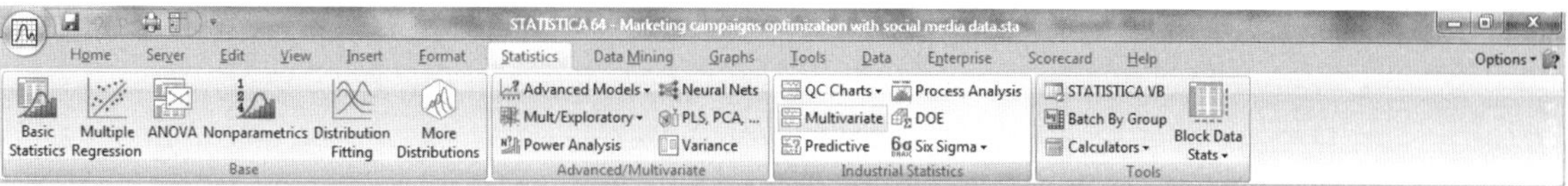

FIGURE A.10

Selecting multivariate QC charts analysis on the ribbon bar.

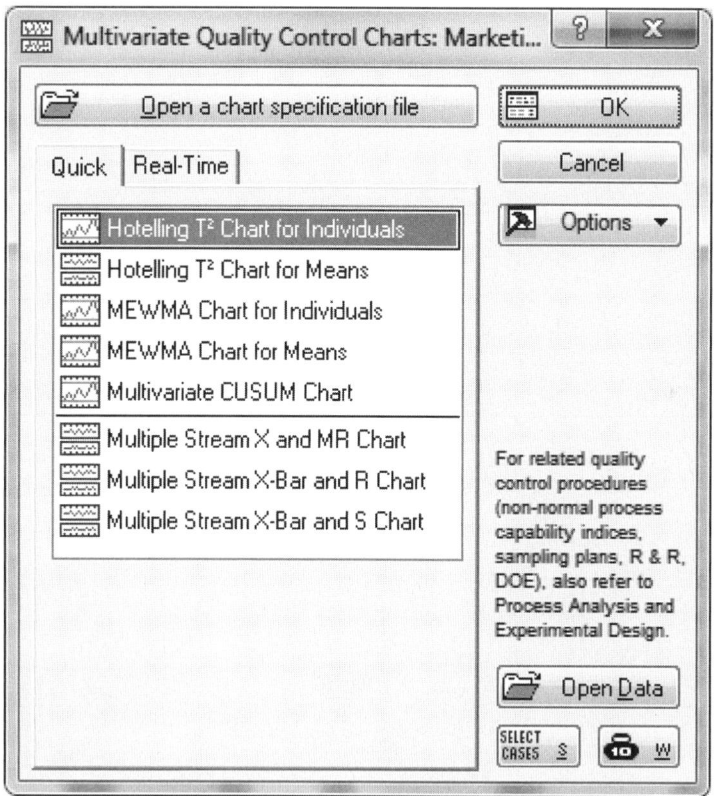

FIGURE A.11

Multivariate quality control charts. Models selection dialog.

The variable list for analysis is the same as in the univariate case (Figure A.12). By clicking the "OK" button, we go directly to the results analysis (Figure A.13). The Hotelling T^2 Chart allows us to not only see the outliers in the multivariate space, but it also provides tools to "drill into" those outliers and identify the driving variables responsible for the outlier under consideration.

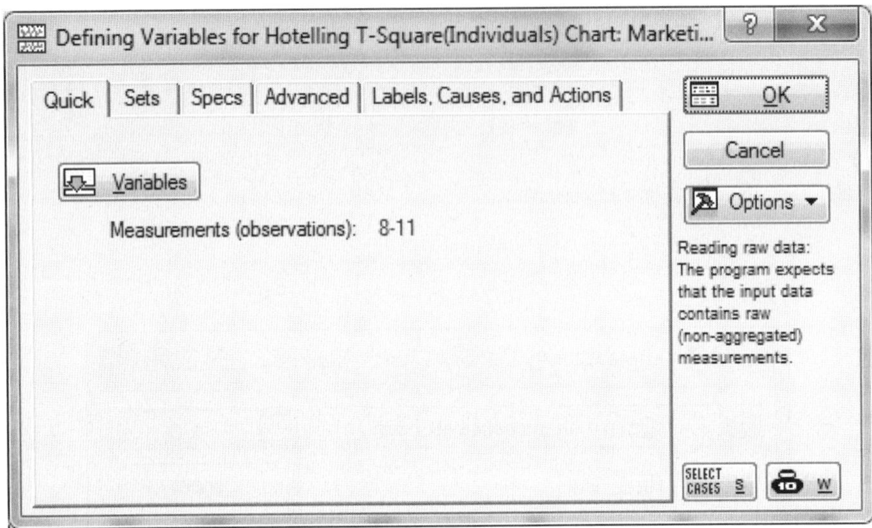

FIGURE A.12

Variable selection dialog.

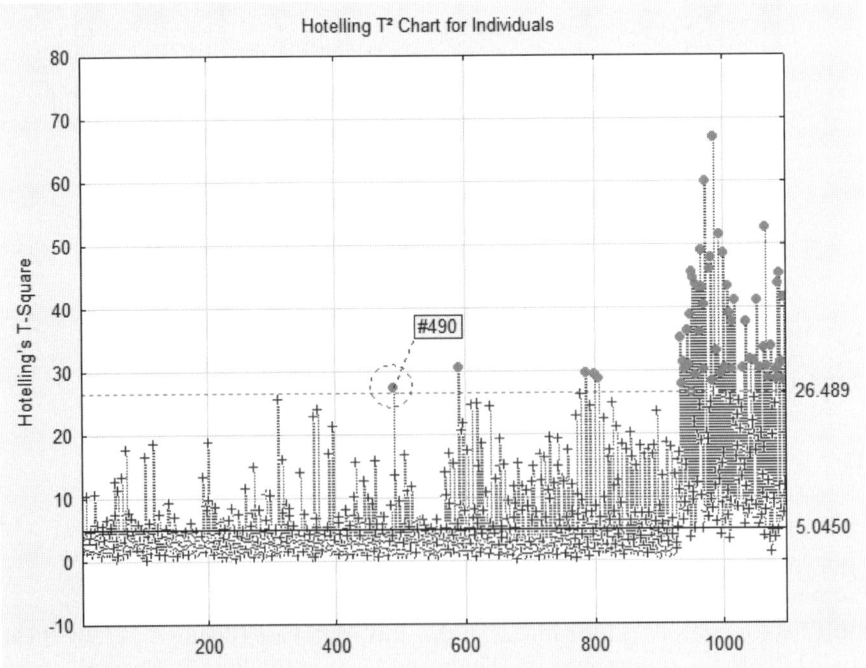

FIGURE A.13

Multivariate Hotelling T^2 chart for individuals.

An example we are going to consider—Case #490—is highlighted in Figure A.13. To perform the drill-down analysis, we would need to get back to the analysis (by default it is minimized on the bottom control panel), switch to the *Brushing* tab, and select the correct case number in the "Include/exclude samples" list (Figure A.14).

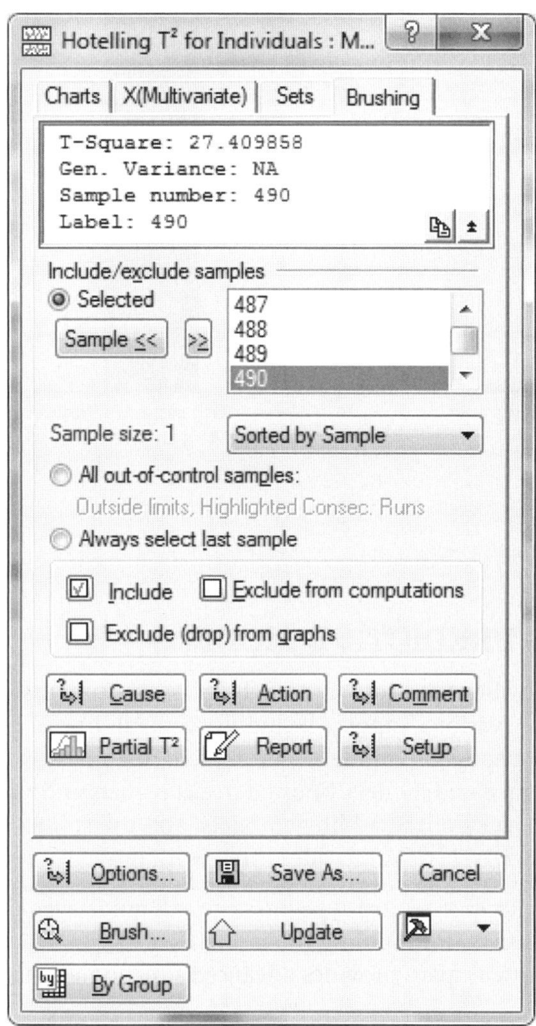

FIGURE A.14
Hotelling T^2 chart for individuals. Results dialog.

The *Partial* T^2 button allows you to review T^2 statistics connected to a particular outlier broken down by inputs from different variables (Figure A.15). This result tells us that the main reason for this outlier is connected to the variable "Facebook N people like Product A." This analysis approach could be repeated for any other outliers.

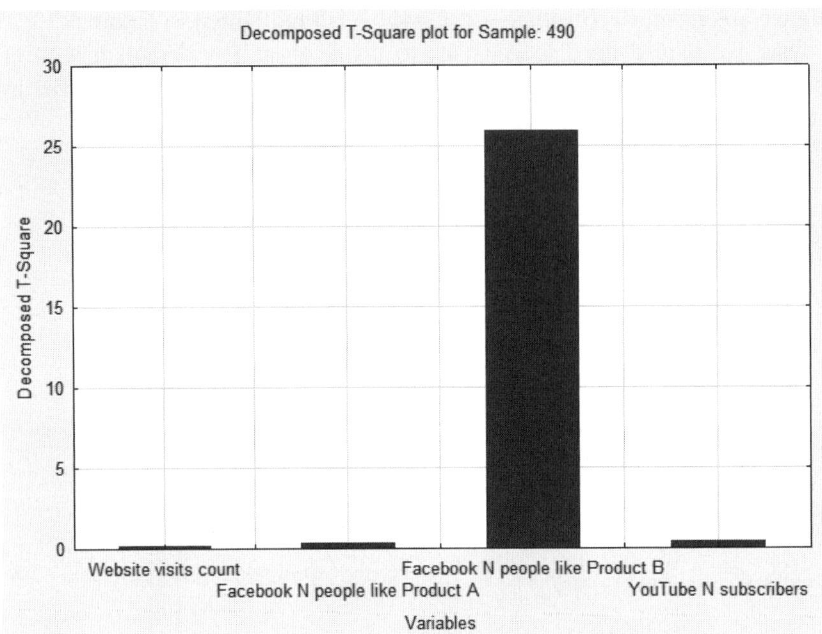

FIGURE A.15
Decomposition of T^2 statistics for the case #490.

STEP 3: CREATING PREDICTIVE MODELS

STATISTICA Data Miner provides the user with a rich set of instruments for modeling dependencies with continuous target and various inputs, including general linear models, generalized linear/ nonlinear models, general regression models, automated neural networks, general classification/ regression tree models, general chaid models, boosted tree classifiers and regression, random forests for regression and classification, generalized additive models, and marsplines.

For building the model, STATISTICA provides three workflows:

1. *STATISTICA* Data Miner Recipes: wizard-type interface will guide even the inexperienced user easily through all the principal steps for building the model.
2. *STATISTICA* Data Miner workspace provides advanced functionality to construct your own work and dataflow.
3. *STATISTICA* Data Miner graphical user interface provides instruments for manual model building with the ultimate set of options.

In this tutorial we'll restrict ourselves with the Data Miner workspace interface.

First of all, a new workspace should be created. Several different data mining project templates are available. For this example we'll use regression models project: *Data Mining → Workspaces → General Modeler and Multivariate Explorer → Advanced Comprehensive Regression Models Project*. As you can see in Figure A.16, all of the principal nodes are already inserted into the workflow.

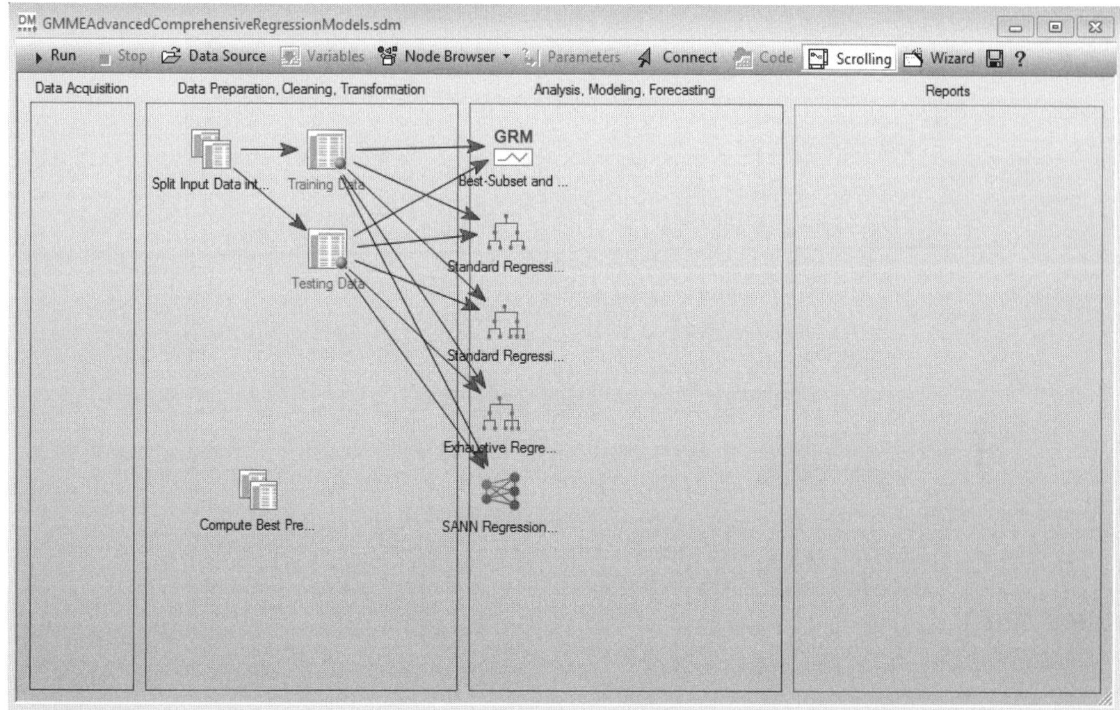

FIGURE A.16
Advanced comprehensive regression models project template.

In the next step, the data file needs to be attached to the workspace of the data acquisition filed. To attach the data, click the *Data Source* button on the data miner workspace main menu and select the data source (the in-place database connection could be used as the data source as well) (Figure A.17).

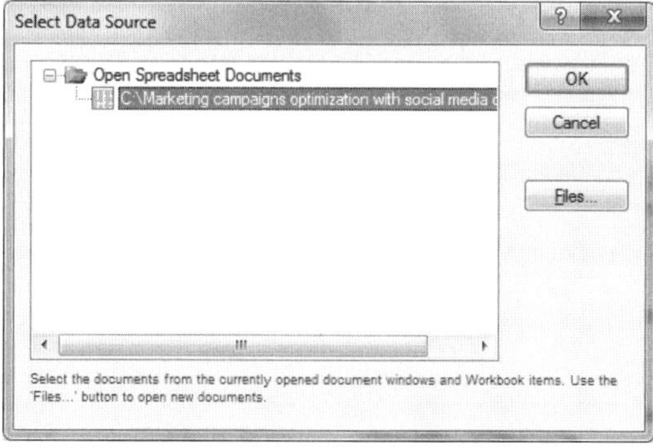

FIGURE A.17
Select data source dialog in the Data Miner workspace.

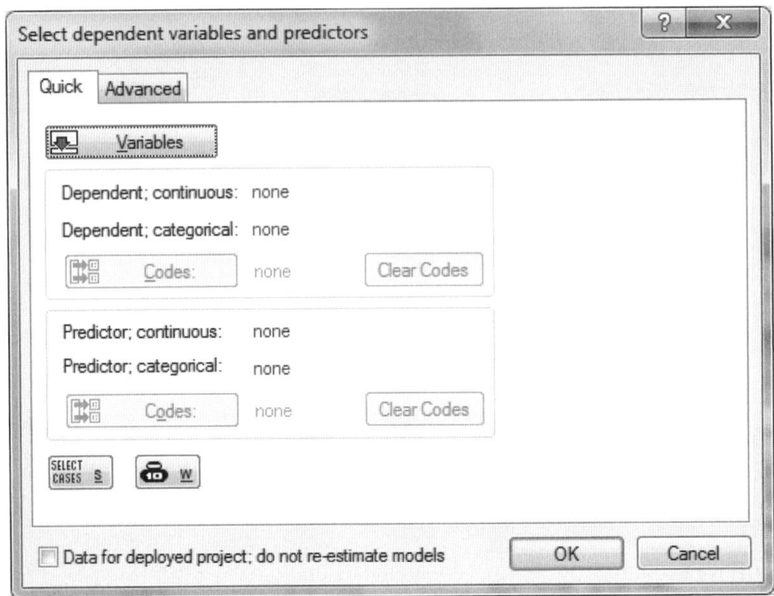

FIGURE A.18
Variable selection dialog.

The next step is the variable selection dialog (Figure A.18). Select the variables as shown in the screenshot (Figure A.19).

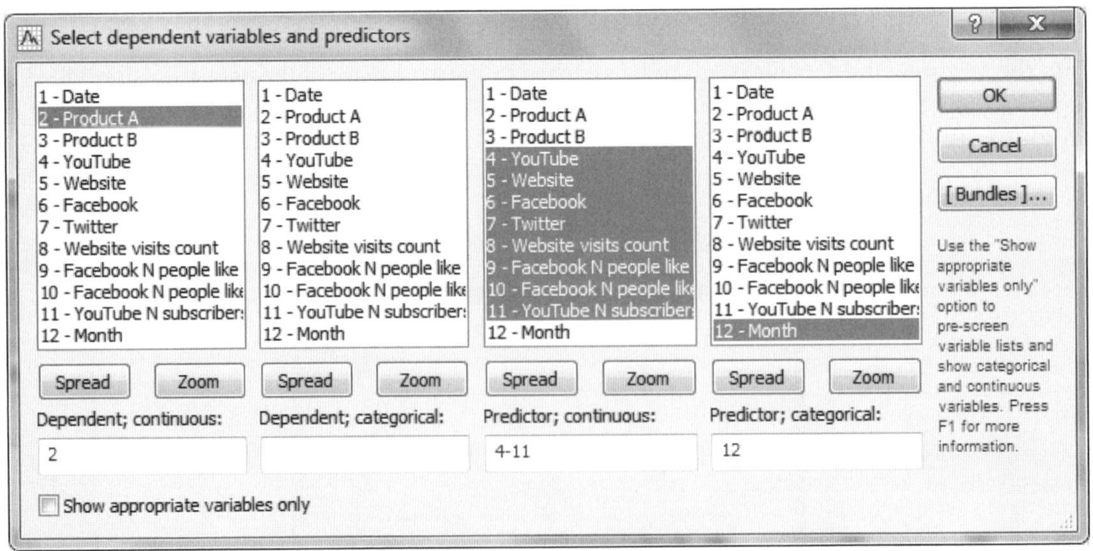

FIGURE A.19
Selecting the variables.

After the data source is attached to the Data Miner workspace, click the Connect button on the Data Miner workspace main menu, click on the data source (the arrow will be created starting on the data source node), and then point the end of the arrow on the Split Input Data node (Figure A.20).

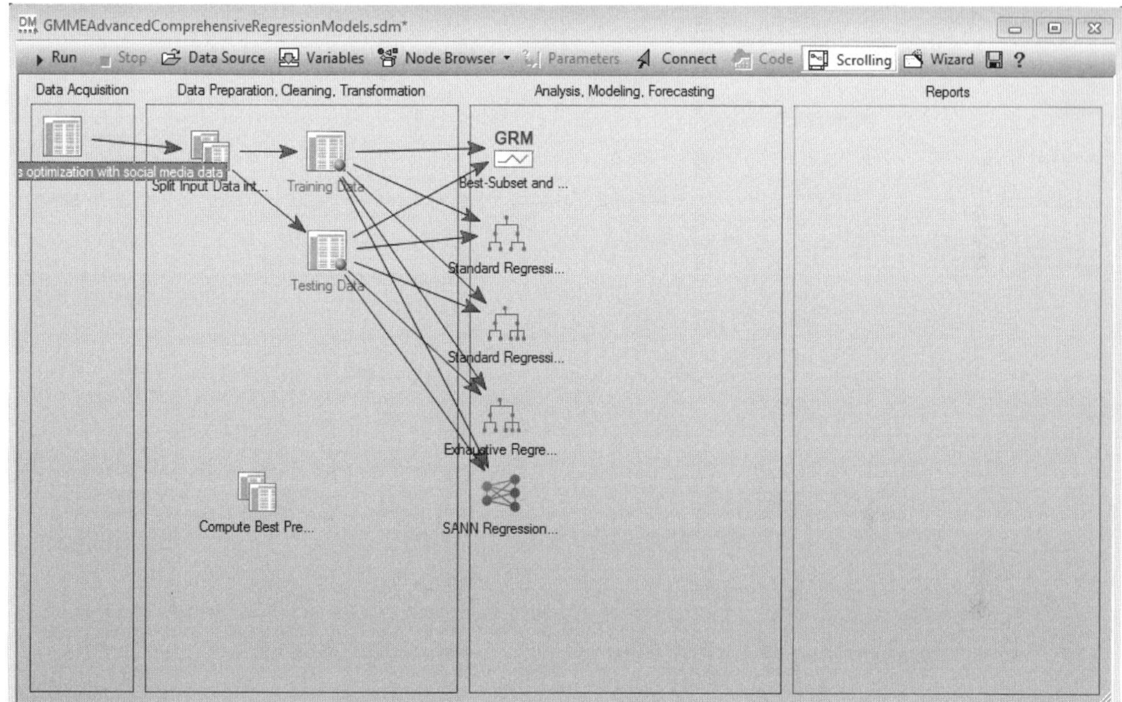

FIGURE A.20
Data Miner workspace with data source in place.

Before running the models, let us change the level of detail to include all of the results. To do this we click on every node in the *Analysis, Modeling, Forecasting* field, and select the *Detail of computed results reported* to be *All results* (Figure A.21).

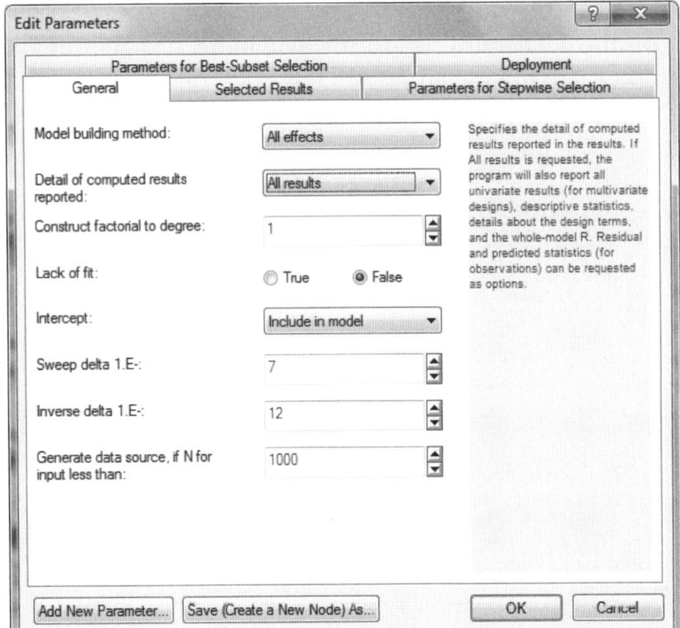

FIGURE A.21

Analysis node—parameters edit dialog.

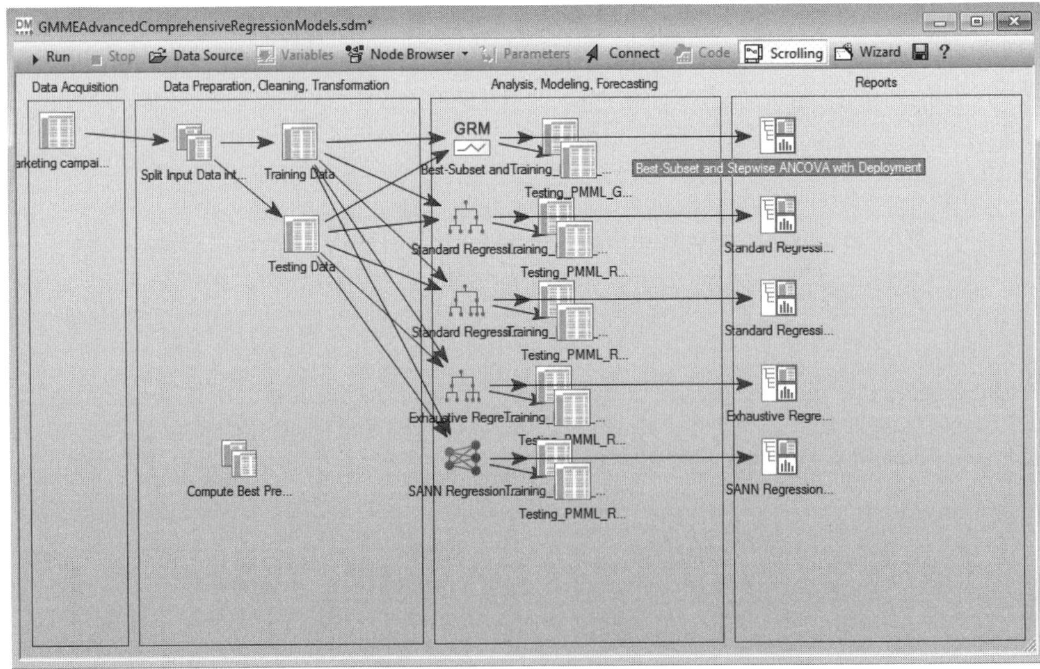

FIGURE A.22

Data Miner workspace with analysis and modeling results.

Now that everything is ready, we can click on the Run button on the data miner workspace main menu and go to results exploration (Figure A.22). It is worth noting that analysis nodes in this workspace are marked as "... with deployment." That means that the model PMML codes are stored with each of those nodes (Figure A.23), allowing us to use them for building predictions with the new data. This functionality will be considered later in this tutorial.

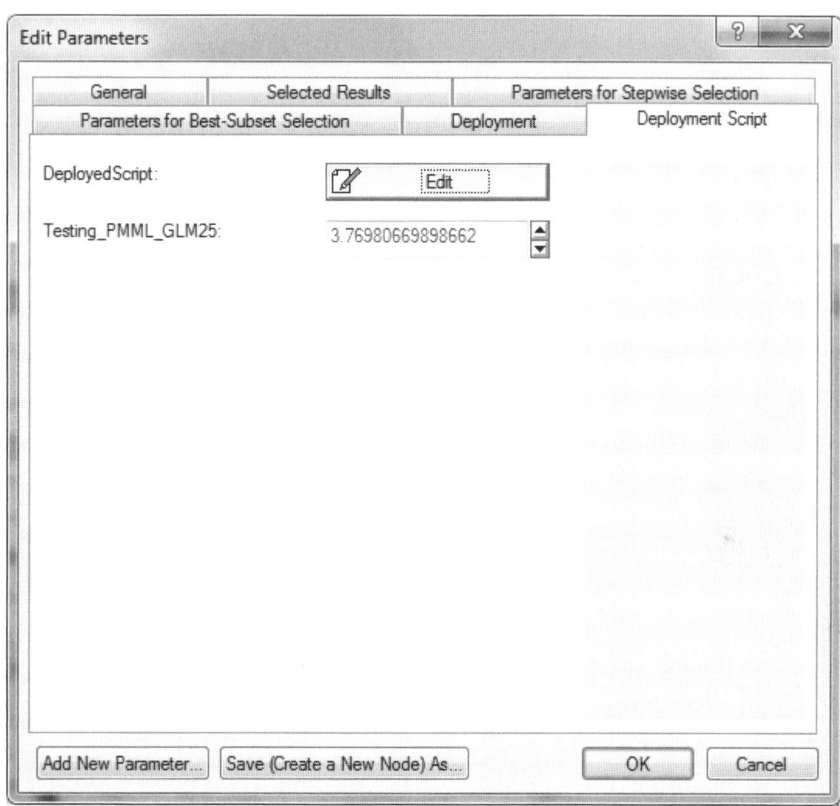

FIGURE A.23
Analysis node—parameters edit dialog. Deployment script tab.

STEP 4: PERFORMING A "WHAT-IF" ANALYSIS OF THE MARKETING CAMPAIGNS

Models built on the previous step represent primarily dependence of sales on expenses on marketing campaigns and social media data from the past. In order to make those models actionable, we need to provide new (planned) inputs and estimate the sales amounts. In some cases those inputs could be strictly determined, but more often they allow some controllable variation. In the latter case, the "what-if" analysis should be performed in order to obtain the sales distribution in response to inputs variation (Figure A.24).

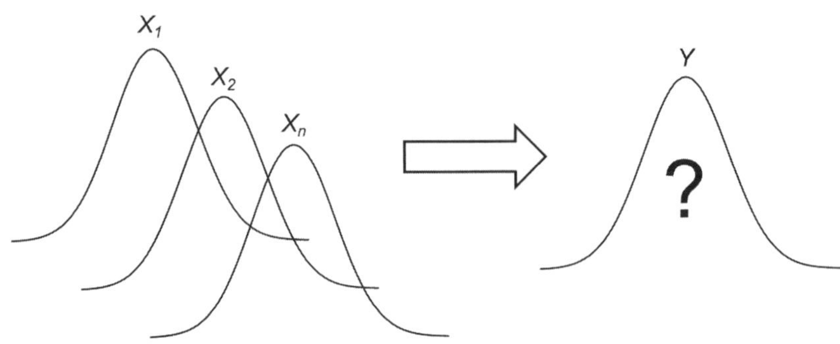

FIGURE A.24
The "what-if" analysis.

To prepare the inputs for the "what-if" analysis, we will use the *STATISTICA* Design Simulation module. The variables that can be controlled in our example are expenses on various marketing instruments: YouTube, websites, Facebook, and Twitter.

The *STATISTICA* Design Simulation module could accept correlation matrices on the input. So before running it, we will create a correlation matrix for those four variables. With the same historical data we can run *Basic Statistics* → *Correlation Matrices*. Variable selection is performed with the *One variable list* button. After that we can click the *Matrix 1* button on the *Advanced* tab. As a result, the matrix in Figure A.25 is created.

Data: Correlation matrix (4v by 8c)

	Marketing campaigns optimization with social medi			
	1 YouTube	2 Website	3 Facebook	4 Twitter
YouTube	1.00000	-0.55984	0.84878	1.00000
Website	-0.55984	1.00000	-0.04046	-0.55984
Facebook	0.84878	-0.04046	1.00000	0.84878
Twitter	1.00000	-0.55984	0.84878	1.00000
Means	138.13869	1050.09124	224.72628	138.13869
Std.Dev.	154.23934	351.23265	150.36005	154.23934
No.Cases	1096.00000			
Matrix	1.00000			

FIGURE A.25
Correlation matrix.

Now we run STATISTICA Design Simulation module (*Statistics* → *More Distributions* → *Design Simulation*). By default, this module normal distributions with locations at 0's and scales equal to 1 for simulation. Before running the analysis, we need to create a list of distributions with custom parameters (Figure A.26).

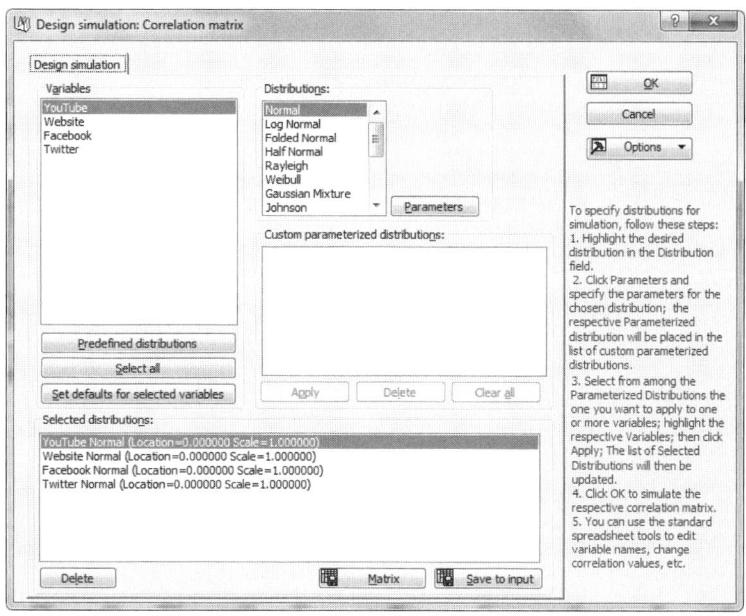

FIGURE A.26

Design simulation. Distribution selection dialog with default settings.

With the *Delete* button on the bottom we can delete default distributions and the parameters assignment. The *Parameters* button allows us to specify manual parameters for distributions. For the purpose of this example we've defined the set of parameters in Figure A.27.

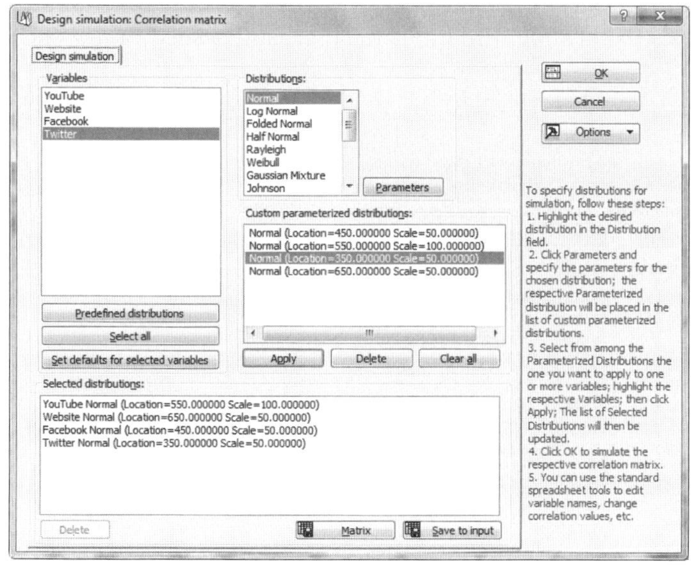

FIGURE A.27

Design simulation. Distribution selection dialog with manual settings.

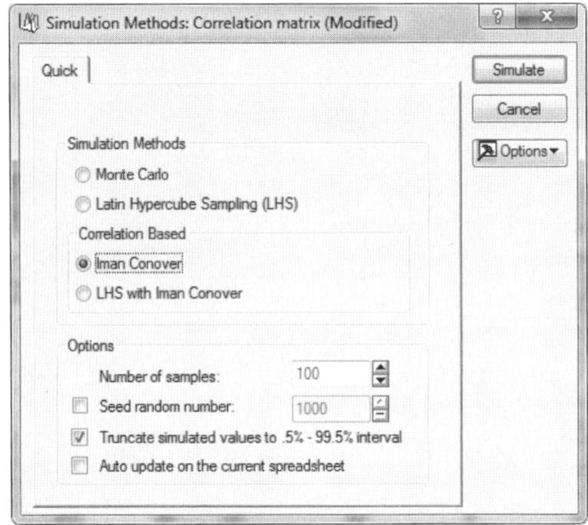

FIGURE A.28
Design simulation. Simulation method selection dialog.

Next, we click *OK*. This dialog allows you to select the method for simulation (Figure A.28). It would be natural to assume that expenses on various marketing instruments are correlated with each other (suffice it to say that the overall budget on marketing is always fixed). In order to preserve this correlation, we select the *Iman Conover* method. Depending on your particular simulation needs, the number of samples could be increased from its default value equal to 100.

Next, after we click *Simulate*, a new spreadsheet appears on the screen (Figure A.29). At this point the spreadsheet is not ready for deployment yet. Fixed factors should be added. The simplest way is to use the batch trasformation tool available from *Data* → *Transformations* → *Transforms*. In the formulas field we should enter precise variable names (coinciding with initial data set) and desired values, such as in Figure A.30.

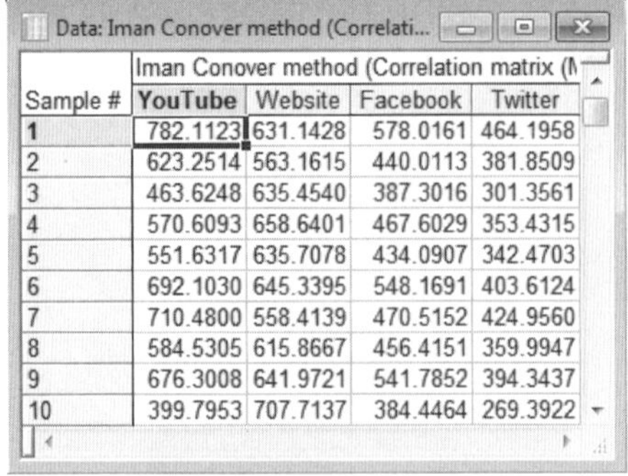
Data: Iman Conover method (Correlati...

	Iman Conover method (Correlation matrix (M			
Sample #	YouTube	Website	Facebook	Twitter
1	782.1123	631.1428	578.0161	464.1958
2	623.2514	563.1615	440.0113	381.8509
3	463.6248	635.4540	387.3016	301.3561
4	570.6093	658.6401	467.6029	353.4315
5	551.6317	635.7078	434.0907	342.4703
6	692.1030	645.3395	548.1691	403.6124
7	710.4800	558.4139	470.5152	424.9560
8	584.5305	615.8667	456.4151	359.9947
9	676.3008	641.9721	541.7852	394.3437
10	399.7953	707.7137	384.4464	269.3922

FIGURE A.29
Simulation result.

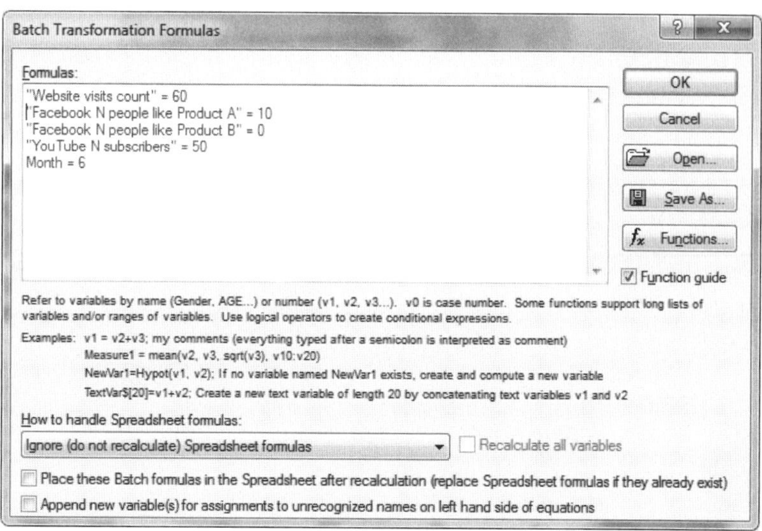

FIGURE A.30
Batch transformation dialog.

Since new (to this data set) variable names were specified, *STATISTICA* will prompt the user to confirm the addition of new variables to the spreadsheet (Figure A.31). Figure A.32 shows that new variables were successfully added to the initial spreadsheet.

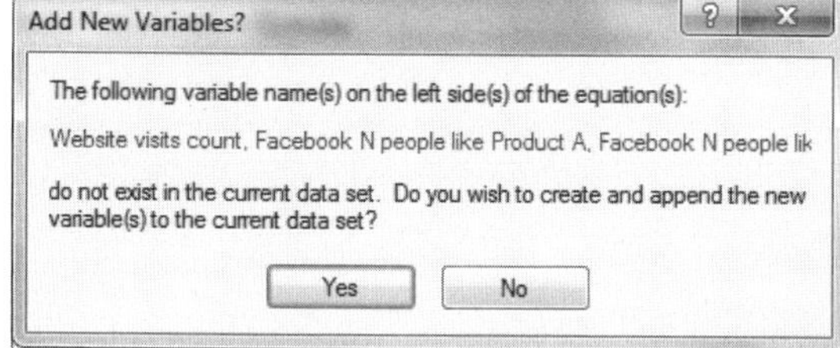

FIGURE A.31
Confirmation of adding new variables to the spreadsheet.

Data: Iman Conover method (Correlation matrix (Modified))*

	Iman Conover method (Correlation matrix (Modified))								
Sample #	YouTube	Website	Facebook	Twitter	Website	Faceboo	Faceboo	YouTube	Month
1	782.1123	631.1428	578.0161	464.1958	60	10	0	50	6
2	623.2514	563.1615	440.0113	381.8509	60	10	0	50	6
3	463.6248	635.4540	387.3016	301.3561	60	10	0	50	6
4	570.6093	658.6401	467.6029	353.4315	60	10	0	50	6
5	551.6317	635.7078	434.0907	342.4703	60	10	0	50	6
6	692.1030	645.3395	548.1691	403.6124	60	10	0	50	6
7	710.4800	558.4139	470.5152	424.9560	60	10	0	50	6
8	584.5305	615.8667	456.4151	359.9947	60	10	0	50	6
9	676.3008	641.9721	541.7852	394.3437	60	10	0	50	6
10	399.7953	707.7137	384.4464	269.3922	60	10	0	50	6

FIGURE A.32
Spreadsheet with new variables added.

To finalize the spreadsheet before deployment, the target variable with missing data should be added (Figure A.33). Now, following the same procedure of adding data sources to the Data Miner project, we attach the created spreadsheet and select the variables (Figure A.34).

Data: Iman.sta (10v by 100c)

Iman Conover method (Correlation matrix (Modified))

Sample #	Product A	YouTube	Website	Facebook	Twitter	Website visits count	Facebook N people like Product A	Facebook N people like Product B	YouTube N subscribers	Month
1		782.1123	631.1428	578.0161	464.1958	60	10	0	50	6
2		623.2514	563.1615	440.0113	381.8509	60	10	0	50	6
3		463.6248	635.4540	387.3016	301.3561	60	10	0	50	6
4		570.6093	658.6401	467.6029	353.4315	60	10	0	50	6
5		551.6317	635.7078	434.0907	342.4703	60	10	0	50	6
6		692.1030	645.3395	548.1691	403.6124	60	10	0	50	6
7		710.4800	558.4139	470.5152	424.9560	60	10	0	50	6
8		584.5305	615.8667	456.4151	359.9947	60	10	0	50	6
9		676.3008	641.9721	541.7852	394.3437	60	10	0	50	6
10		399.7953	707.7137	384.4464	269.3922	60	10	0	50	6

FIGURE A.33

Finalized spreadsheet ready for deployment.

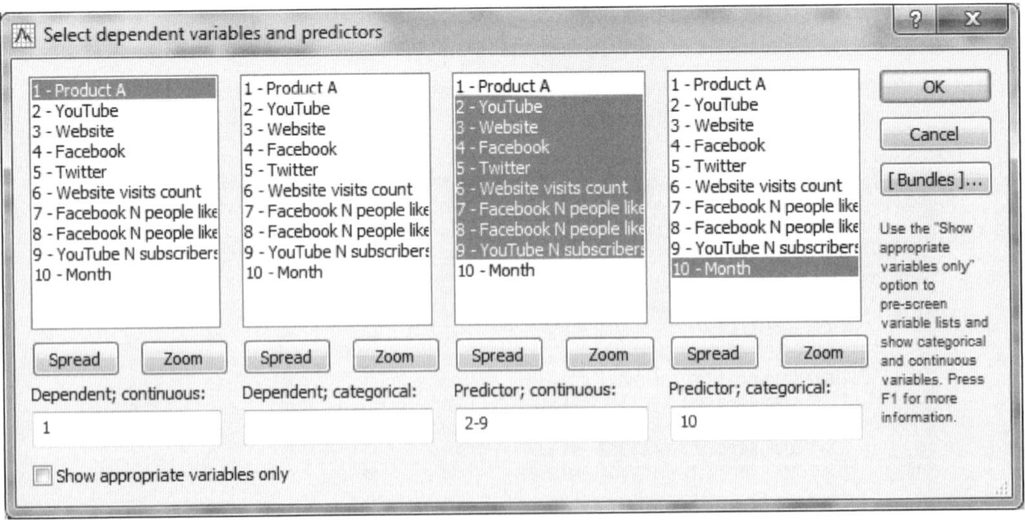

FIGURE A.34

Variable selection for deployment.

It is important to check the *"Data for deployment; do not reestimate models"* checkbox when attaching data for deployment (Figure A.35). Because the data for deployment is now attached to the workspace, a couple of additional operations should be performed in order to link it to analysis nodes: previous links from training/testing data should be disabled (right click on each link and select *Disable*) and the attached datasource should be connected to the analysis nodes (click *Connect* button, select data source, and point to analysis node). The resulting workspace is shown in Figure A.36.

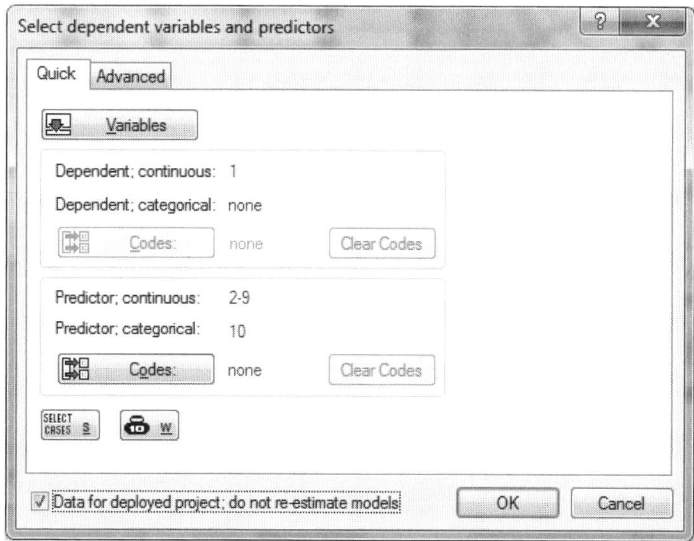

FIGURE A.35
Checking "Data for deployed project; do not re-estimate models" option.

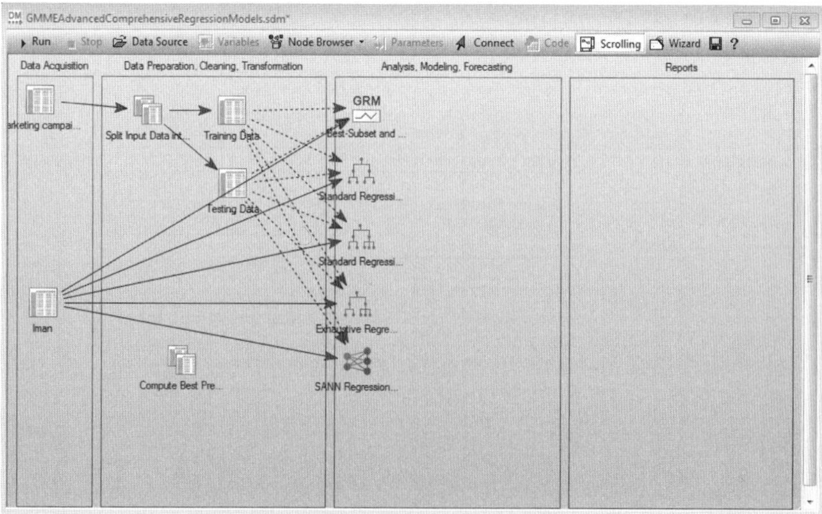

FIGURE A.36
Workspace with data for deployment linked to respective models.

Next, as we click the *Run* button, new spreadsheets are added to the workspace (Figure A.37). Each of the new nodes represents a spreadsheet with predictions (Figure A.38). Those spreadsheets contain three variables (residuals and observed are empty, since we provided the spreadsheet with target variable filled with missing data).

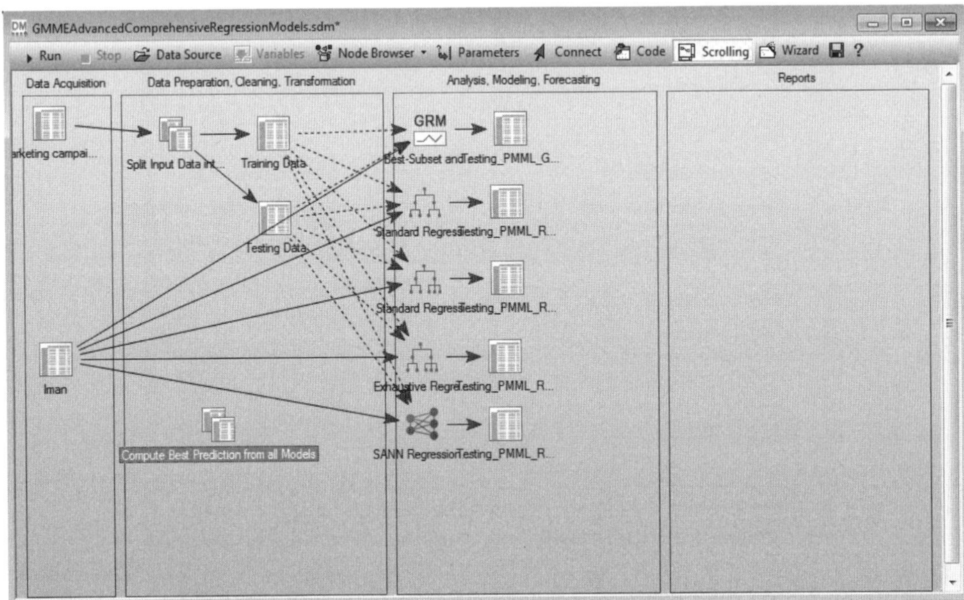

FIGURE A.37

Workspace after running models on new data.

Data: Best-Subset and Stepwise ANCOVA with Deploym...

Input spreadsheet generated from deployment

	1 GeneralRegressionModelPred	2 GeneralRegression ModelRes	3 Product A
1	145.87447		
2	130.70545		
3	123.91350		
4	132.88600		
5	129.40499		
6	142.02261		
7	134.59590		
8	131.92272		
9	141.23245		
10	122.99386		

FIGURE A.38

Prediction spreadsheet.

Now predictions can be collected into one spreadsheet. The box plot in Figure A.39 demonstrates the distributions achieved through "what-if" analysis. The tree methods resulted in single value predictions. The general regression and neural network models achieved the distributions of outputs.

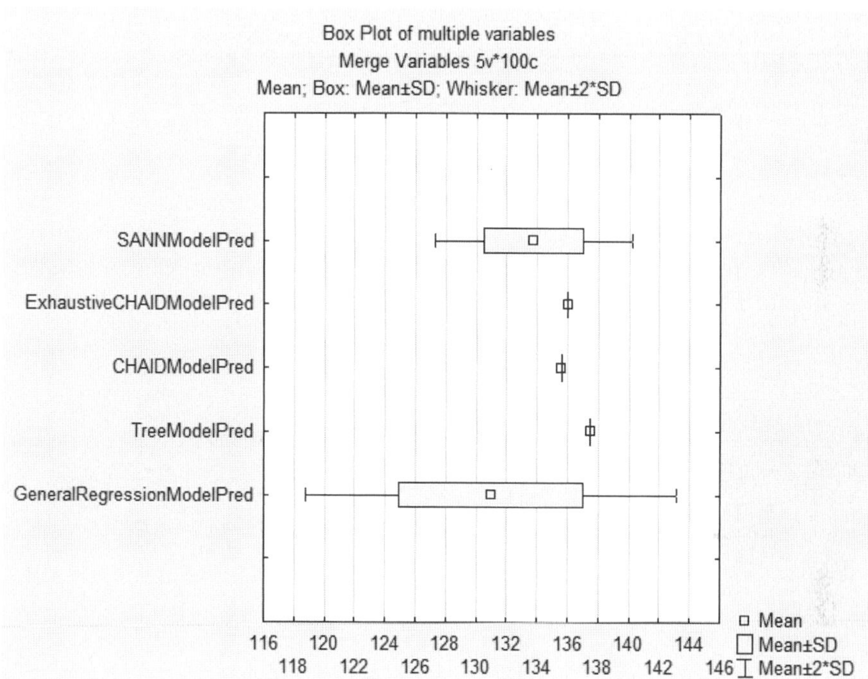

FIGURE A.39
Box plot of predictions.

STEP 5: PERFORMING SENTIMENT ANALYSIS

Sentiment analysis is a way to measure the level of agreement or support of some particular topic. In other words, it is a way to measure positive and negative sentiments in the text narrative. As an object for analysis, we'll consider the narratives extracted from one of the live blogs (comments on: Apple: iPad 2 is "dramatically faster" [live blog], Link: http://news.cnet.com/8601-13579_3-20037801-2.html?communityId=2070&targetCommunityId=2070&blogId=37&tag=mncol#ixzz1ShDCkoL1). The data are collected in the file (Figure A.40).

Data: Live blog data for sentiment analysis.sta (8v by 102c)

	1 ID	2 UserID	3 UserName	4 Comments	5 DatePosted	6 TimeZone	7 Body	8 People like this
1	1	10467348	vamsi_ar	13	3/1/11 2:08 PM	PST	Woooo Hooooo, Come on Apple. Bring us more exciting new	12
2	2	10470839	travisp11	67	3/2/11 9:40 AM	PST	Yeah, tell us you will improve conditions at your manufacturin	23
3	3	10471121	dansterpower	1622	3/2/11 10:22 AM	PST	@ travisp11 -- Sorry to spoil them apples, but your compliant	10
4	4	10471138	puterhead	58	3/2/11 10:25 AM	PST	I would buy it is you were to say, "tell us how you are going	9
5	5	10471522	Renegade Knight	7723	3/2/11 11:23 AM	PST		2
6	6	10471623	chatins	378	3/2/11 11:39 AM	PST	Jobs pulled off some shock and awe! March 11 availability for	1
7	7	10484366	battmail	40	3/5/11 7:18 AM	PST	travisp11 You are a coward! Only a coward has the audacity t	1
8	8	10469747	MichaelNguyen	1	3/2/11 5:43 AM	PST	if i wasnt in school right now i'd be at home sitting in my chaiir waiting for this. [	
9	9	10469775	boyapple35	3	3/2/11 5:53 AM	PST	I can only access twitter at that time :(Are you going to tweet, what's happenin	
10	10	10469776	chinalife	10	3/2/11 5:53 AM	PST		9

FIGURE A.40

Data extracted from live blog.

To perform the sentiment analysis, we will use *STATISTICA* Text Miner, together with lists of positive and negative terms (http://www.cs.uic.edu/~liub/FBS/opinion-lexicon-English.rar). Those lists of words should be concatenated into one file with positive-negative indicator, as shown in Figure A.41.

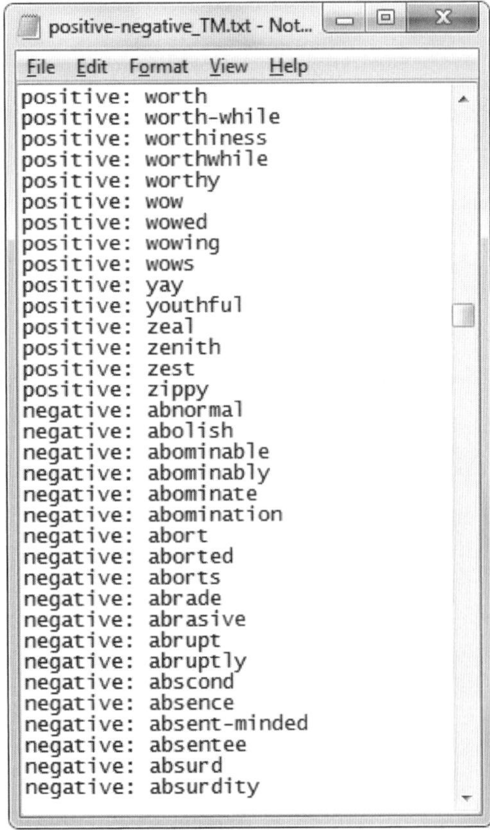

positive-negative_TM.txt - Not...

File Edit Format View Help

```
positive: worth
positive: worth-while
positive: worthiness
positive: worthwhile
positive: worthy
positive: wow
positive: wowed
positive: wowing
positive: wows
positive: yay
positive: youthful
positive: zeal
positive: zenith
positive: zest
positive: zippy
negative: abnormal
negative: abolish
negative: abominable
negative: abominably
negative: abominate
negative: abomination
negative: abort
negative: aborted
negative: aborts
negative: abrade
negative: abrasive
negative: abrupt
negative: abruptly
negative: abscond
negative: absence
negative: absent-minded
negative: absentee
negative: absurd
negative: absurdity
```

FIGURE A.41

List of positive and negative words.

Now everything is ready for text mining/sentiment analysis. Run *Data Mining* → *Text Mining* → *Text & Document Mining*. Select Body as the *Text variable* (Figure A.42). Switch to the *Words* tab (Figure A.43). Next, check the *Synonyms* checkbox, and click the *Edit* button (Figure A.44).

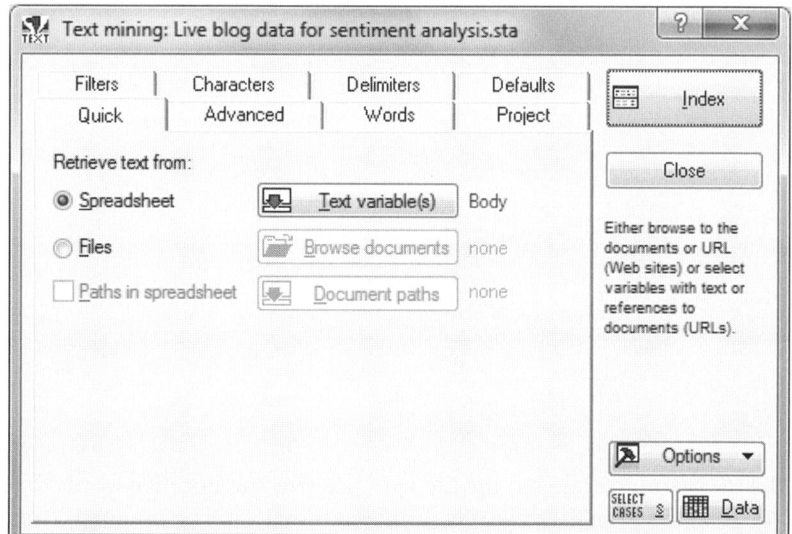

FIGURE A.42
STATISTICA Text miner. Start dialog.

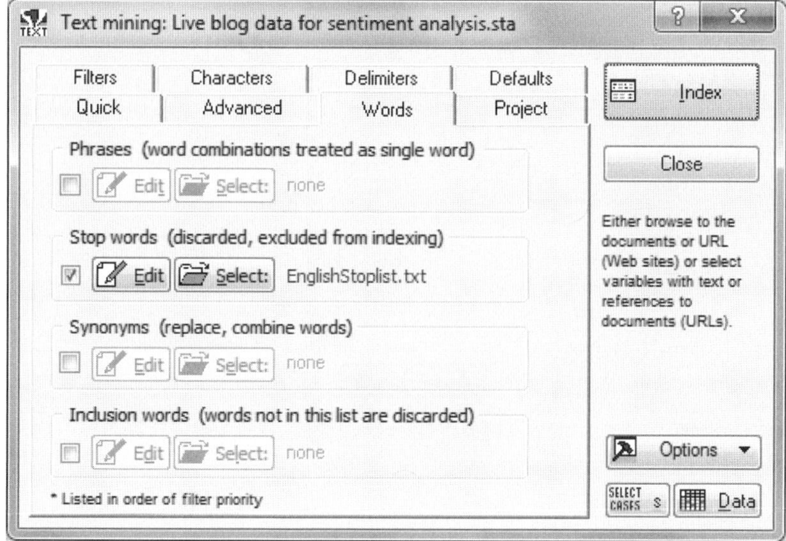

FIGURE A.43
STATISTICA Text Miner. Start dialog. Words tab.

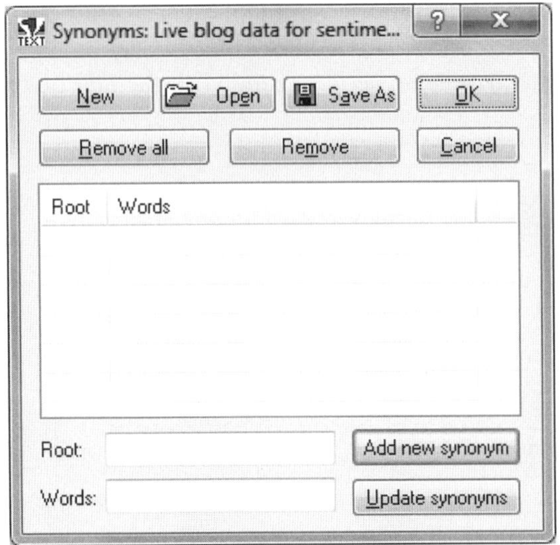

FIGURE A.44
Synonyms selection dialog.

Now click Open, and point the file opening dialog to the file with positive and negative words that has just been created (Figure A.45). After that, the list of words should automatically be updated. Then click *OK*. Next check the *Inclusion words* checkbox, and click Edit. Here we'll specify the list of two words (Figure A.46): positive and negative.

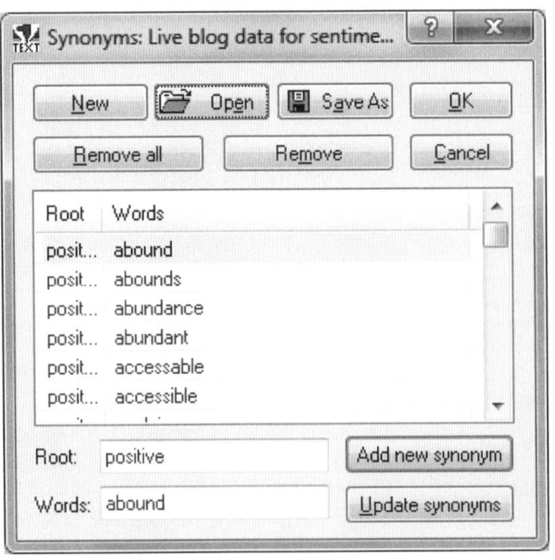

FIGURE A.45
Synonyms selection dialog with data inserted from a text file.

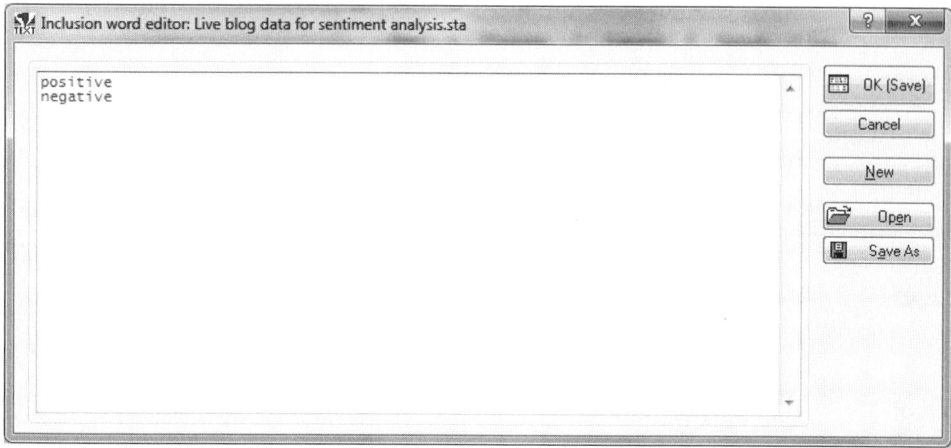

FIGURE A.46
Inclusion word list.

Click *OK*. Now, since everything is ready for analysis, we need to select Index and go to results analysis (Figure A.47). Go to the *Save Results* tab and select *Write back current results*. After adding the appropriate number of variables for results, the final spreadsheet will look like Figure A.48. Usually in the sentiment analysis a single number is used. So the difference *Sentiment* = *posit-negat* could be calculated (Figure A.49).

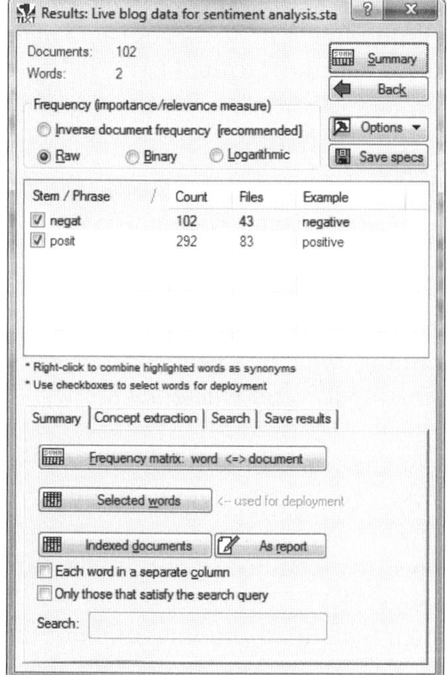

FIGURE A.47
Results dialog.

Data: Live blog data for sentiment analysis.sta (10v by 102c)

	1 ID	2 UserID	3 UserName	4 Comments	5 DatePosted	6 TimeZone	7 Body	8 People like this	9 negat	10 posit
1	1	10467348	vamsi_ar	13	3/1/11 2:08 PM	PST	Woooo Hooooo, Come on Apple. Bring us more exciting new	12	0	2
2	2	10470839	travisp11	67	3/2/11 9:40 AM	PST	Yeah, tell us you will improve conditions at your manufacturin	23	0	2
3	3	10471121	dansterpower	1622	3/2/11 10:22 AM	PST	@ travisp11 -- Sorry to spoil them apples, but your compliant	10	2	1
4	4	10471138	puterhead	58	3/2/11 10:25 AM	PST	I would buy it is you were to say, "tell us how you are going	9	5	6
5	5	10471522	Renegade Knight	7723	3/2/11 11:23 AM	PST		2	0	1
6	6	10471623	chatins	378	3/2/11 11:39 AM	PST	Jobs pulled off some shock and awe! March 11 availability for	1	1	4
7	7	10484366	battmail	40	3/5/11 7:18 AM	PST	travisp11 You are a coward! Only a coward has the audacity t	1	10	4
8	8	10469747	MichaelNguyen	1	3/2/11 5:43 AM	PST	if i wasnt in school right now i'd be at home sitting in my chaiir waiting for this. [		0	1
9	9	10469775	boyapple35	3	3/2/11 5:53 AM	PST	I can only access twitter at that time :(Are you going to tweet, what's happenin		0	3
10	10	10469776	chinalife	10	3/2/11 5:53 AM	PST	[illegible]	9	5	15

FIGURE A.48

Sentiment analysis results.

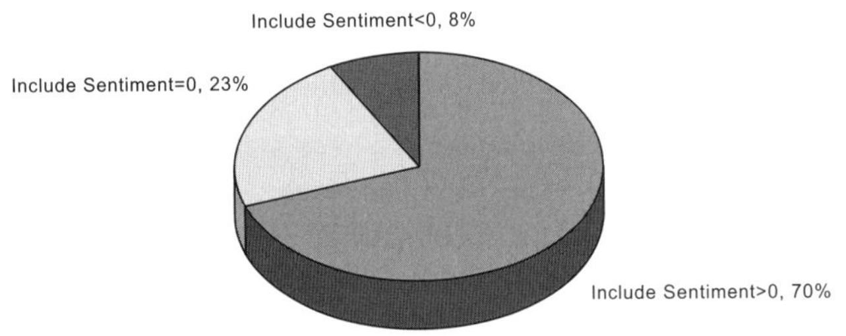

FIGURE A.49

Positive—neutral—negative sentiments pie chart.

SUMMARY

This tutorial walks the user through principal modeling and optimization steps in marketing campaigns optimization using data from social media. Whenever the structured and unstructured (text) data are accumulated or updated across multiple databases or data streams, they can be modeled and scored in order to predict business outcomes.

Consider this approach if:

- The business is challenged with determining the effectiveness of its marketing investment by leveraging user-generated content.
- The business is challenged with leveraging user-generated content produced in the social media environment in order to optimize a current or future marketing investment mix.
- One wants to optimize future marketing investments by performing "what if" scenario modeling to determine if increased spending will drive increased customer adoption and where the money should be spent for maximum lift.
- Your campaign is having the desired effect (factor in sentiment analysis via text mining).
- You would like to monitor campaign effectiveness in real time to detect sentiment shifts by monitoring charts or dashboards.
- Performing "what if" scenario modeling across multiple channels to determine what impact future investment levels will have on sales/sentiment.
- You want to "see" what will happen when marketing investment is adjusted and then monitor the impact of the change in real time.

TUTORIAL B

Text Mining Improves Model Performance in Predicting Airplane Flight Accident Outcome

Jennifer Thompson
Woodward, OK, USA

Thomas Hill
Tulsa, OK, USA

CONTENTS

INTRODUCTION

Keywords from unstructured text can be powerful predictors in data mining. When text mining is used to extract these key terms, we can see substantial gains in model performance. Unstructured text is often more descriptive than the results from a one-size-fits-all questionnaire. This is particularly true of the examples used in this tutorial.

The National Transportation Safety Board is a federal agency dedicated to investigating civil aviation accidents, as well as major accidents in other modes of transportation. These investigations are used to make safety recommendations to help prevent future accidents. The data they collect include some standard items such as date and time of the incident, the mode of transportation involved, and the severity of the crash. Additionally, a free-form text account of the incident is included. These unstructured data may be the most telling information available. Without text mining, it is quite difficult and time-consuming to try to find patterns in the unstructured text.

Practical Text Mining and Statistical Analysis for Non-structured Text Data Applications. DOI: 10.1016/B978-0-12-386979-1.00010-4

THE DATA

The NTSB had data available for 3,235 aviation incidents from 2001 to 2003. The data included variables such as the date and time of the event, weather conditions, geographical location, information about the aircraft, type of damage sustained in the incident, injuries sustained, and free-form text descriptions of the event.

Data: NTSBAccidentReports2001-2003.sta* (26v by 3235c)

	1 regis_no	2 event_id	3 Aircraft_Key	4 event_date	5 event_time	6 event_dow	7 event_month	8 event_year	9 light_cond	10 air_temp
1	N2184N	20010105X00043	1	1/1/2001	1245	Mo	January	2,001	DAYL - Day	-6
2	N737WQ	20010108X00064	1	1/1/2001	1740	Mo	January	2,001	NDRK - Night/Dark	-13
3	N94LW	20010113X00297	1	1/1/2001	1529	Mo	January	2,001	DAYL - Day	14
4	N45CF	20010110X00082	1	1/2/2001	1400	Tu	January	2,001	DAYL - Day	18
5	N19771	20010221X00479	1	1/3/2001	1740	We	January	2,001	DUSK - Dusk	-19
6	N933CA	20010405X00701	1	1/3/2001	832	We	January	2,001	DAYL - Day	-13
7	N26HV	20010110X00098	1	1/4/2001	1500	Th	January	2,001	DAYL - Day	2
8	N435JL	20010126X00361	1	1/4/2001	1547	Th	January	2,001	DAYL - Day	-2
9	N68472	20010108X00062	1	1/4/2001	1500	Th	January	2,001	DAYL - Day	-12
10	N727SP	20010111X00288	1	1/4/2001	1046	Th	January	2,001	DAYL - Day	0

The project has a few goals that can be achieved with text and data mining. These goals are all components of a central theme: gaining understanding of aviation accidents to aid in prevention. The smaller goals are as follows:

1. Finding the structured variables that best explain the type of damage and injuries sustained
2. Indexing the unstructured text and finding key terms
3. Finding terms from the unstructured text that are most related to the type of damage and injuries sustained
4. Determining if model performance gains are possible in a predictive model by using key terms from unstructured text

With *STATISTICA* Data Miner and Text Miner, the goals can be met to give better understanding of aviation accidents. This understanding can help to shape the industry and make it safer.

TEXT MINING THE DATA

Open the example data set, *NTSBAccidentReports2001-2003.sta* in *STATISTICA*. This example has three free-form text variables: *narr_accp, narr_accf,* and *narr_cause*. The cause narrative likely has some useful key words. We will use *STATISTICA* Text Miner to index the words in this narrative. From the *Data Mining* menu, select *Text Mining* to open the *Text Mining* dialog.

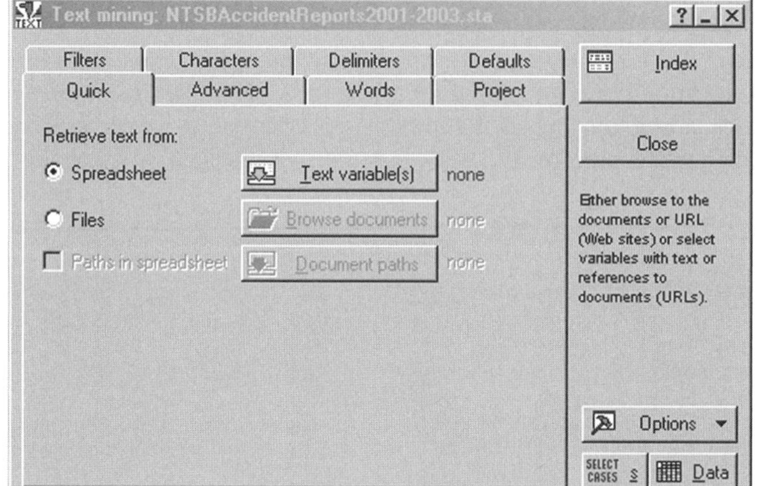

FIGURE B.1
Text Mining quick tab window.

On the "Quick" tab, as illustrated in Figure B.1, select "Text variable(s)" to open the "Select variables" containing text dialog.

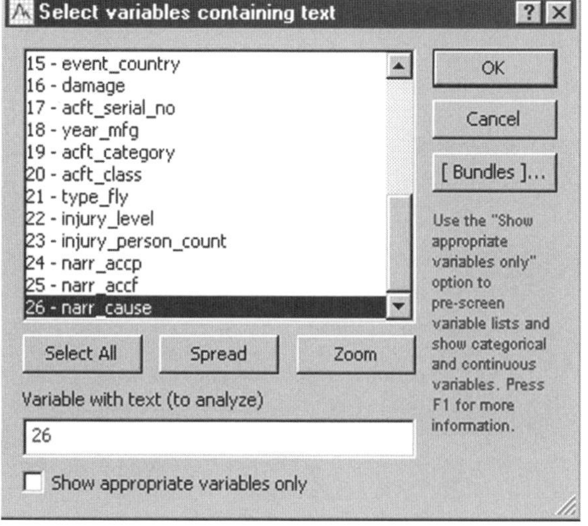

FIGURE B.2
Select variables window.

Select the last variable, "narr_cause" as the *Variable with text (to analyze)*, as illustrated above in Figure B.2. All other default settings can be used. Click *Index* on the *Text Mining* dialog to begin indexing the text.

TEXT MINING RESULTS

When indexing completes, the *Results* dialog is displayed, as shown in the next diagram, Figure B.3. Two hundred two words have been indexed from this unstructured text variable. Some of the words seen in the dialog include *abort, accident, action,* and so on. The results give the total count of each word and the number of files in which it appears. Note that common and often not useful words like I, you, this, and so forth are excluded from indexing by default.

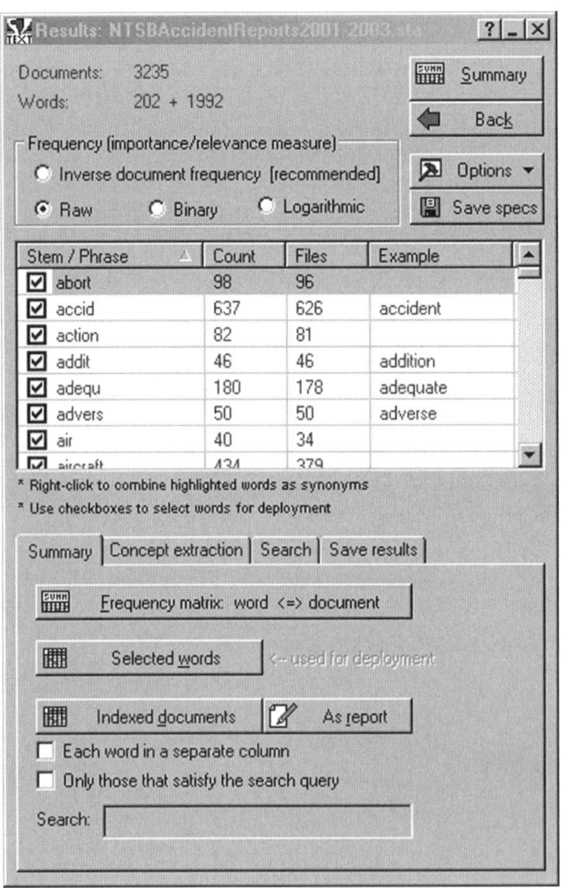

FIGURE B.3
Results of text mining window.

Change the Frequency (importance/relevance measure) option to Inverse document frequency. On the Concept extraction tab, select Perform Singular Value Decomposition (SVD). The buttons on this page will then become active. First select the Scree plot to see the concepts extracted from the unstructured text.

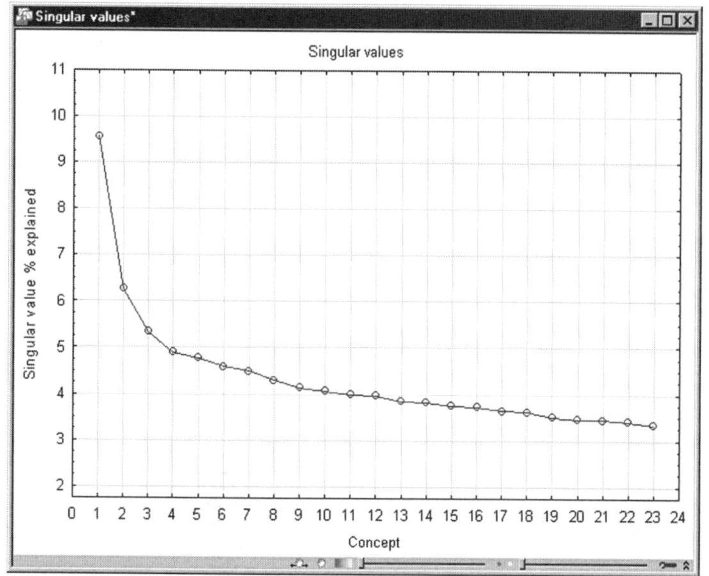

FIGURE B.4

Scree type plot for singular values decomposition.

The first four concepts, as illustrated in Figure B.4, give the highest percent explanation of the singular values. These first four concepts will be the most interesting and beneficial to explore. Change the *Number of concepts to use* to 4, since the Scree plot showed these to extract the most information from the text (see Figure B.5).

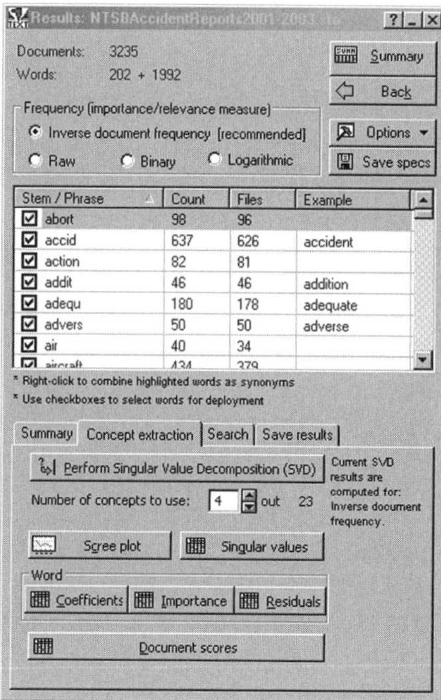

FIGURE B.5

Results dialog after "inverse document frequency" has been selected and the singular values decomposition has been performed.

Select *Coefficients* (see Figure B.6) to create the spreadsheets of singular value coefficients. On the *Data* menu, check the *Input* option in the *Mode* section. Now the output spreadsheet can be used as input in a graphical analysis.

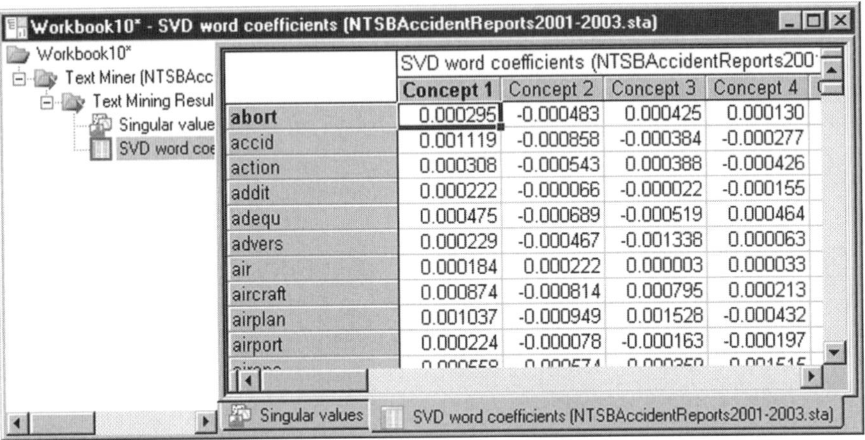

FIGURE B.6

Workbook of results spreadsheets and graphs from the singular values decomposition word coefficients analysis.

From the *Graphs* menu, select *Scatterplot* to open the *2D Scatterplots* dialog (see Figure B.7).

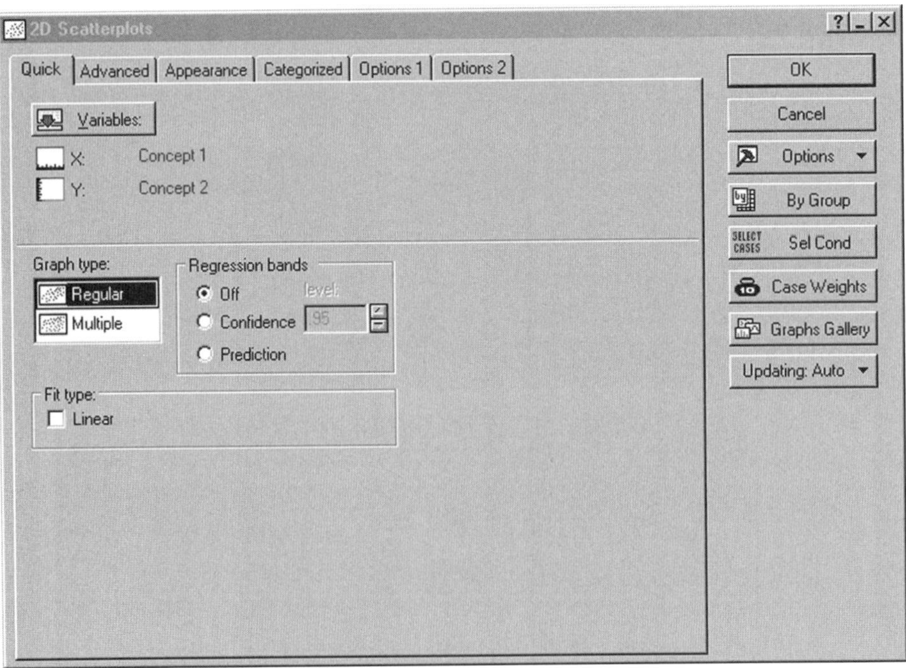

FIGURE B.7

2D Scatterplots dialog (obtained from Graphs pull-down menu).

Select variables to plot *Concept 1* by *Concept 2*. When the plot is made (see Figure B.8), use the *Brushing* tools from the *Edit* menu to label the scatterplot. The *Brushing* tool icon, , is found in the *Customize Graph* section of the *Edit* menu on the top left.

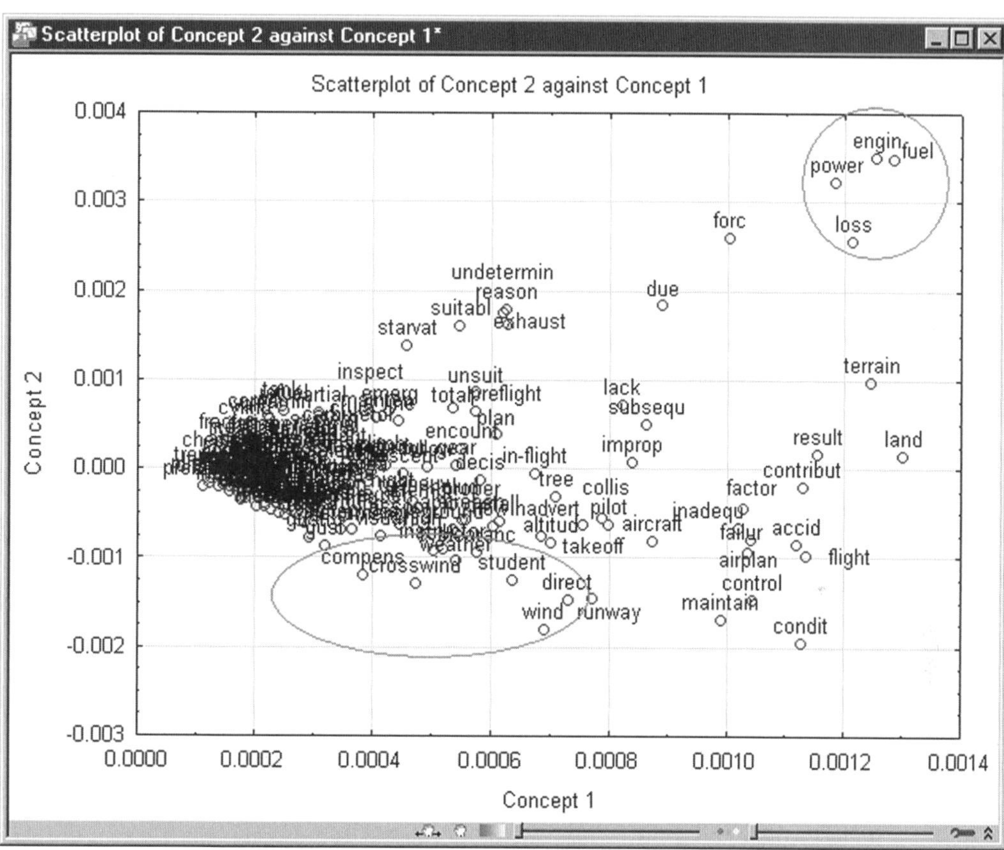

FIGURE B.8
Scatterplot of Concept 2 versus Concept 1.

The labeled scatterplot shows some interesting terms grouped together. The terms *power, loss, engine*, and *fuel* are grouped together, as are *crosswind, wind, student*, and *compensate*. The second group may indicate that student pilots have a tendency to overcompensate when crosswinds are present. This clustering of terms is certainly interesting. It is worth further study. Possibly additional training of student pilots would help in safety, such as flight simulation practice with crosswinds and more classroom time devoted to the subject.

Now on the *Save results* tab, select *Generate a new spreadsheet with current results* to open the *Add selected input variables to output* dialog. Select variables that we will later use in analysis, including *event_id, event_month, light_cond, air_temp, wind_dir_deg, weather_cond_basic, damage, acft_category, type_fly, injury_level, and injury_person_count* (see Figure B.9).

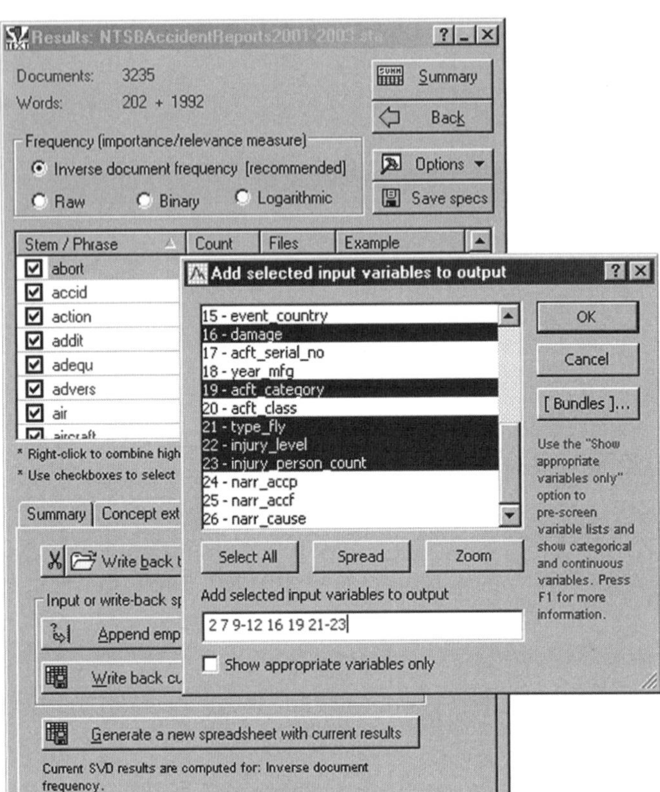

FIGURE B.9

Add selected input variables to output window.

On *OK* a new spreadsheet will be generated with the selected variables, the indexed words, and the four concepts extracted from SVD (see Figure B.10). Note that the indexed words only give values where the word did show up in the text. The remaining fields are missing data. This makes it easy to visualize where the terms are used in the text narratives.

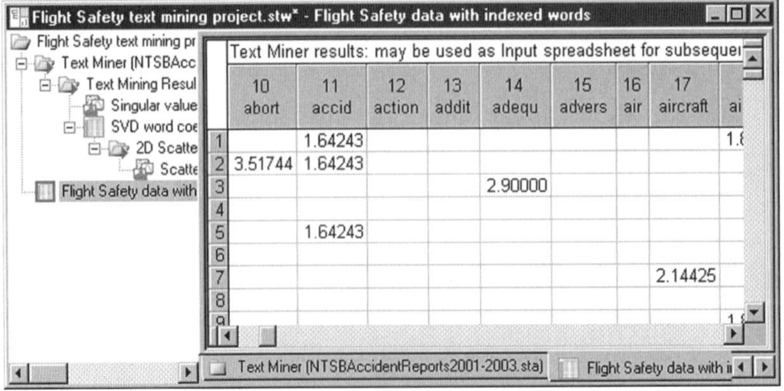

FIGURE B.10

Workbook of results for this text mining project.

DATA PREPARATION

For analysis purposes, the data should not have missing data. So the data will need some preparations before the analysis. Some of the original variables include missing data. These missing cells also need to be dealt with before the analysis begins. The majority of analysis tools will exclude cases where missing data are found.

From the *Data* menu, select *Process Missing Data...* from the *Filter/Recode* drop-down menu to open the *Process Missing Data* dialog (see Figure B.11).

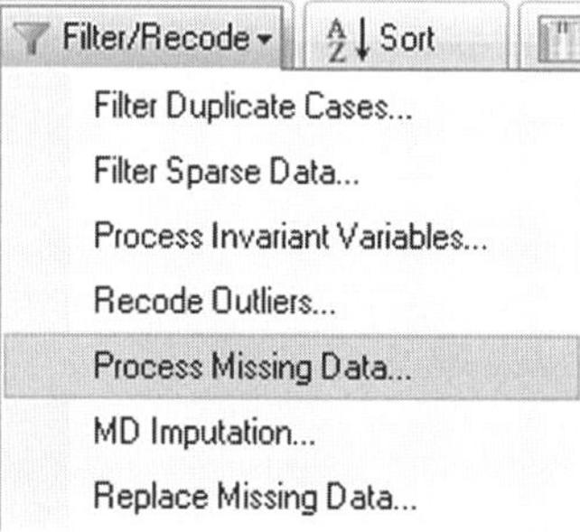

FIGURE B.11
Filter/Recode drop-down menu.

Select variables: *light_cond, air_temp, wind_dir_deg, weather_cond_basic, acft_category,* and *type_fly*. In the *Missing Data Parameters* field, change the *Recode Action* and *Recode Value* columns as in the screen shot below (see Figure B.12). For text variables, the missing values will form a new category, *Other,* and for continuous variables missing values are filled in with the mean.

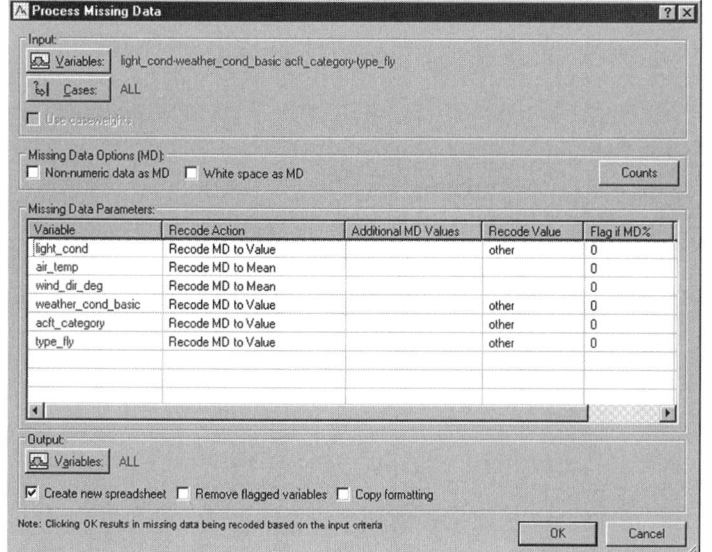

FIGURE B.12
Process Missing Data window.

Click *OK* to fill in the missing data and create a new spreadsheet.

USING TEXT MINING RESULTS TO BUILD PREDICTIVE MODELS

These data have two variables of interest: *damage* ranging from *None* to *Destroyed* and *injury_level* ranging from *None* to *Fatal*. We will focus on the variable *damage*. With the indexed terms as variables, the data set has a large number of possible predictor variables. Not all of these terms are good predictors of the variable of interest, *damage*. Let's narrow the list of possible predictor variables for Data Mining.

From the *Data Mining* menu, select *Feature Selection* from the *Tools* section to open the *Feature Selection and Variable Screening* dialog (see Figure B.13). Click *Variables* to open the *Select dependent variables and predictors* dialog. Select *Damage* as the *Dependent; categorical* variable. Select *air_temp, wind_dir_deg,* and all the indexed terms as *Predictors; continuous*. Select *light_cond, weather_cond_basic, acft_category,* and *type_fly* and *Predictors; categorical.*

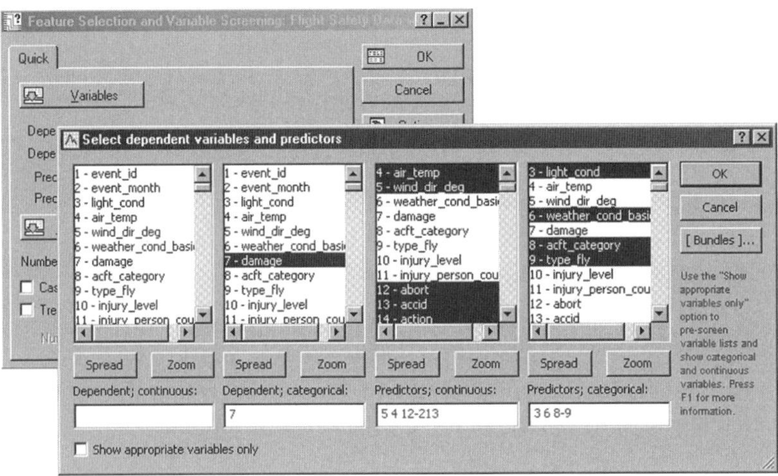

FIGURE B.13

Variable selection window obtained by clicking the *Variables* button in the Quick tab of the Feature selection dialog.

Advance to the *FSL Results* dialog (see Figure B.14). Change the *Criterion for selecting predictors* option to *Display best predictors with p* <, and change the value to .05.

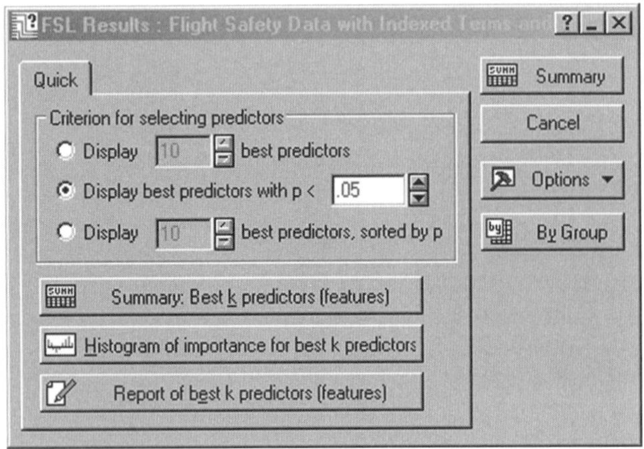

FIGURE B.14

Feature Selection Results window—quick tab.

Create the *Summary: Best k predictors (features)* output. On the *Data* menu, select *Sort,* and sort the data descending by *Chi-square.*

Data: Best predictors for categorical ...

	Best predictors for cate	
	Chi-square	p-value
type_fly	618.9659	0.000000
turbul	276.7159	0.000000
acft_category	260.2357	0.000000
land	225.1597	0.000000
weather_cond_basic	151.2459	0.000000
instrument	141.9037	0.000000
pilot	126.7129	0.000000
visual	121.5360	0.000000
spin	110.8065	0.000000
crew	104.7762	0.000000
low	102.0079	0.000000
stall	100.7928	0.000000
lookout	99.6703	0.000000
altitud	98.7263	0.000000
light_cond	93.3458	0.000000
taxi	81.5488	0.000000
fog	79.1682	0.000000

FIGURE B.15
Best predictors results spreadsheet from the feature selection process.

The spreadsheet lists (see Figure B.15) the predictors in order of their relationship to the variable, *damage*. The variable, *type_fly,* with levels, *Personal, Instructional, and so on* is found to be the most strongly related variable. This was one of the structured variables from the original data. The keyword from the unstructured text with the strongest relationship found in feature selection is *turbul,* the stem word for "turbulence."

Keywords strongly related to the level of damage reported in aviation incidences are *turbulence, land, instrument, pilot, visual, spin, crew, low, stall, lookout, altitude, taxi, fog,* and so on. It may be interesting to further review the cause narratives that contained one or more of these key terms.

Back on the *Text Mining Results* dialog, *Search* tab, you can output the cases where key terms were found. I want to further investigate the term *stall.* Enter this term next to *Words* (see Figure B.16):

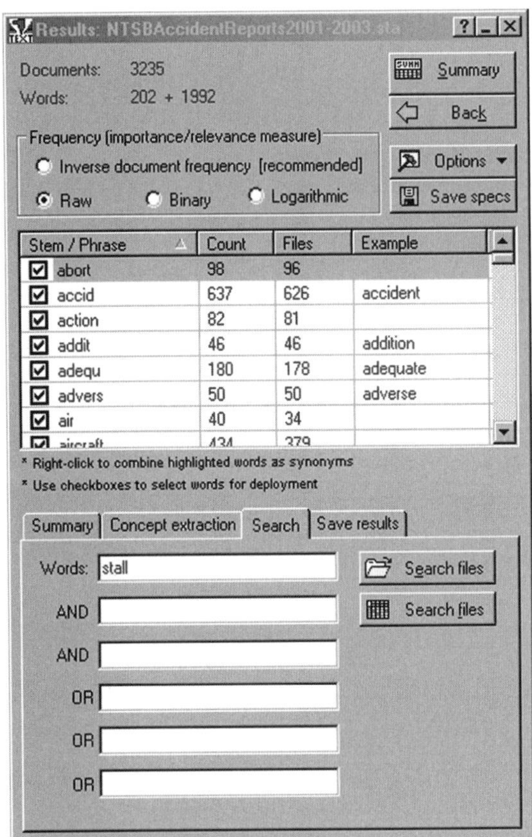

FIGURE B.16

Results of text mining dialog; search tab selected.

Select *Search files* to output a spreadsheet listing the files that included the term *stall* (see Figure B.17).

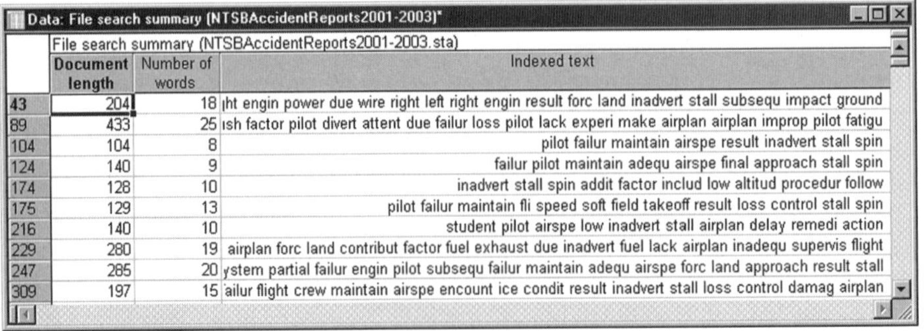

FIGURE B.17

Data file search summary window.

This output allows you to further explore specific terms. This output may reveal patterns that are particularly interesting.

The terms listed in the *Best Predictors* output are all significantly related to the variable of interest: *damage*. Select *Histogram of Importance for best k predictors* to get a graphical representation (see Figure B.18) of this information.

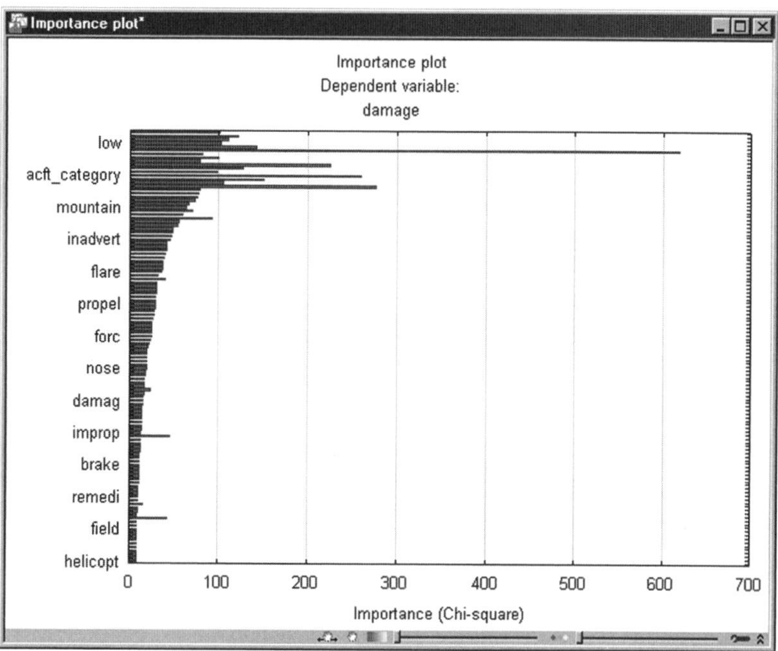

FIGURE B.18
Importance plot, one of the results of the feature selection process.

And, finally, select *Report of best K predictors (features)* to create a list of important predictors (see Figure B.19). This list of important predictors will be used in Data Mining tools to build predictive models.

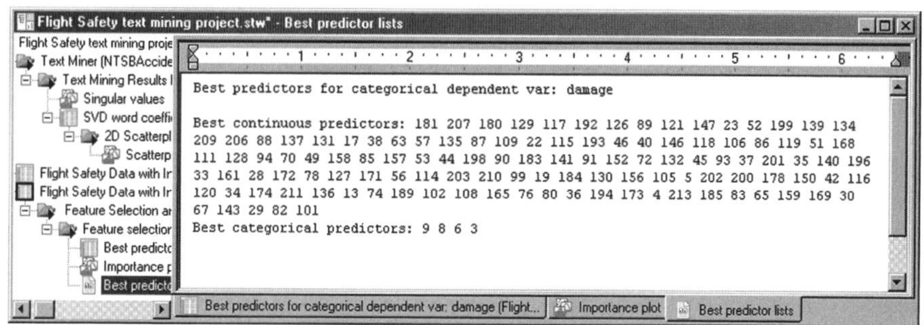

FIGURE B.19
Workbook of results showing the best predictor variables selected by the feature selection process.

From the *Data Mining* menu, select *Boosted Trees* to open the *Boosted Trees Specifications* dialog. Select *Variables* to open the *Select dependent vars, categorical, and continuous predictors* dialog (see Figure B.20).

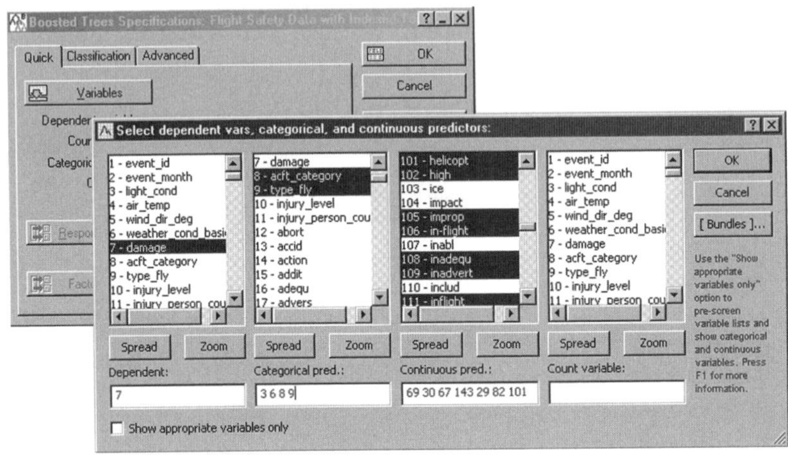

FIGURE B.20
Variable selection dialog.

Select *damage* as the Dependent variable. Copy and paste the variable lists from the Best predictors list report output for both Categorical pred. and Continuous pred. Click OK to begin building trees. When the process completes, the Boosted Trees Results dialog is displayed (see Figure B.21).

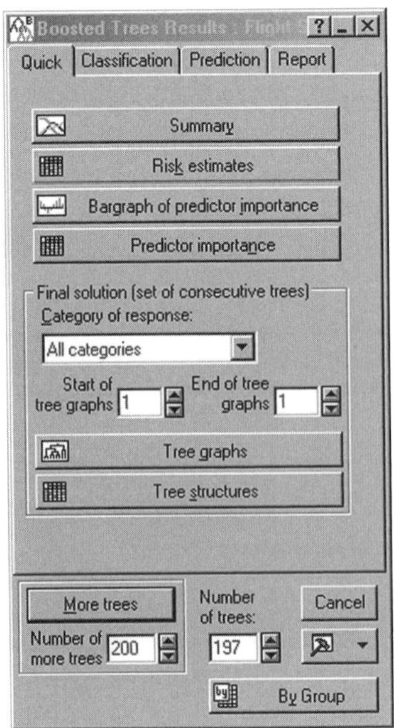

FIGURE B.21
Boosted Trees results dialog.

Select *Summary* to output the *Summary of Boosted Trees* plot (see Figure B.22).

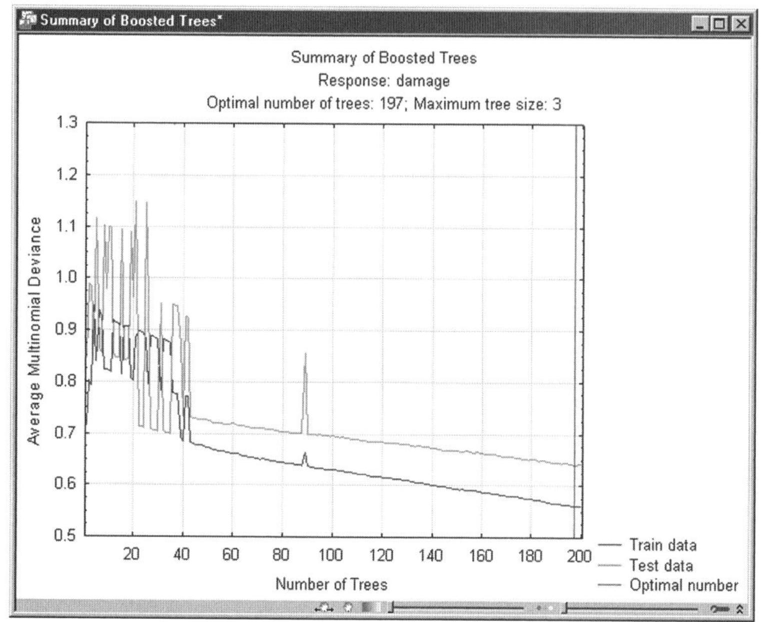

FIGURE B.22
Summary graph of Boosted Trees results.

This plot shows how the predictive error is decreasing in both the train and test samples of the data as each additional tree is added to the model. You can see that when the model stopped building trees, the test data error was still decreasing. Likely the model can be improved by adding additional trees. Select *More trees* to continue building trees for this model. Then create the *Summary* plot again (See Figure B.23).

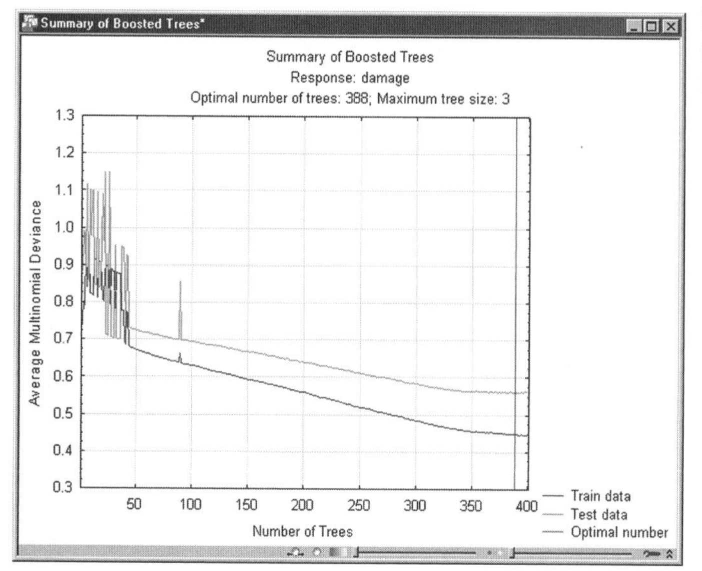

FIGURE B.23
Summary of Boosted Trees for "damage" with a higher number of trees.

The trend is beginning to level off with 400 trees. The optimal number of trees is automatically determined to be 388 trees. We will continue with this model. Select *Bargraph of predictor importance* to create the importance plot.

Figure B.24 is the importance plot for the boosted trees model, zoomed in on the top 20 predictors. The importance plot shows how much individual variables contribute to the boosted trees model. Many of the structured variables are important to the boosted trees model. Some of the unstructured key terms that are also important include *turbulence, control, taxi, intent, pilot, visual, in-flight, direct, terrain, engine, maintain, associate, night,* and *weather*. *Turbulence* and *weather* are logical reasons for accidents. *Visual* and *night* are key terms related to the pilot's ability to see. Any of these terms may be interesting to explore further.

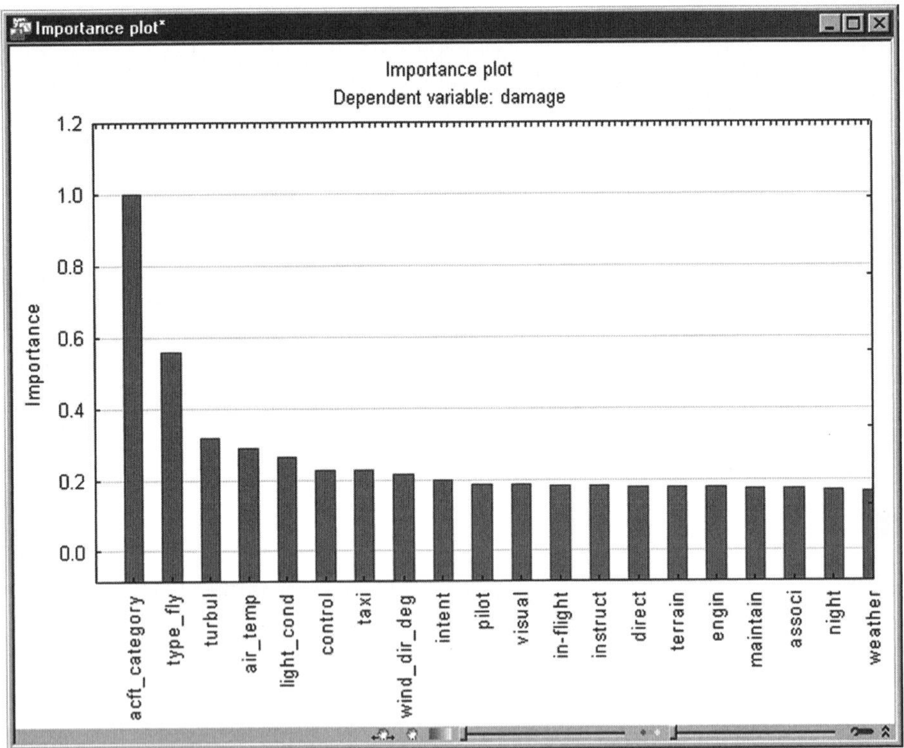

FIGURE B.24
Importance plot results for dependent variable "damage."

From the *Graphs* menu, select *Means* to open the *Means with Error Plots* dialog. Select *Variables* to open the *Select Variables for Means with Error Plots* dialog (see Figure B.25).

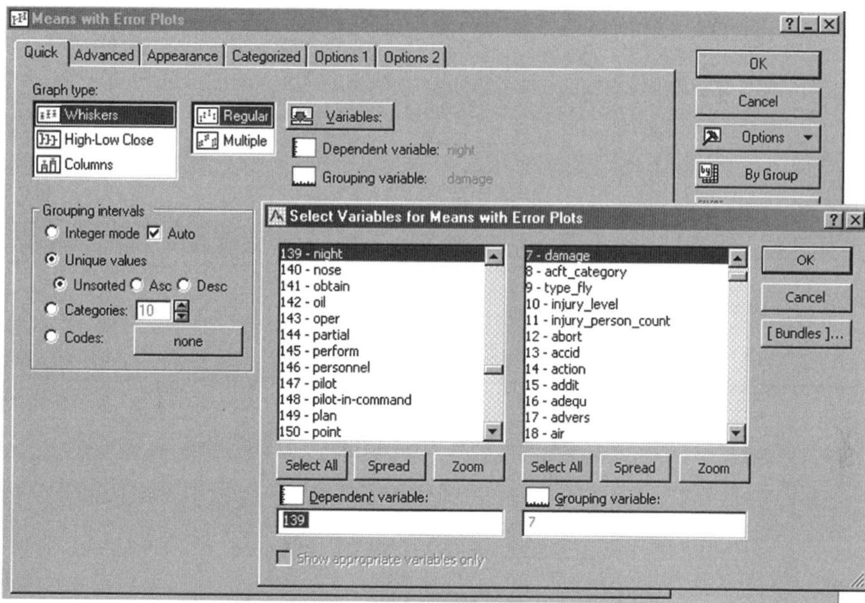

FIGURE B.25

Select variables window from the graph window of "Means with error plots."

Select *night* as the *Dependent variable* and *damage* as the *Grouping variable* (see Figure B.25).

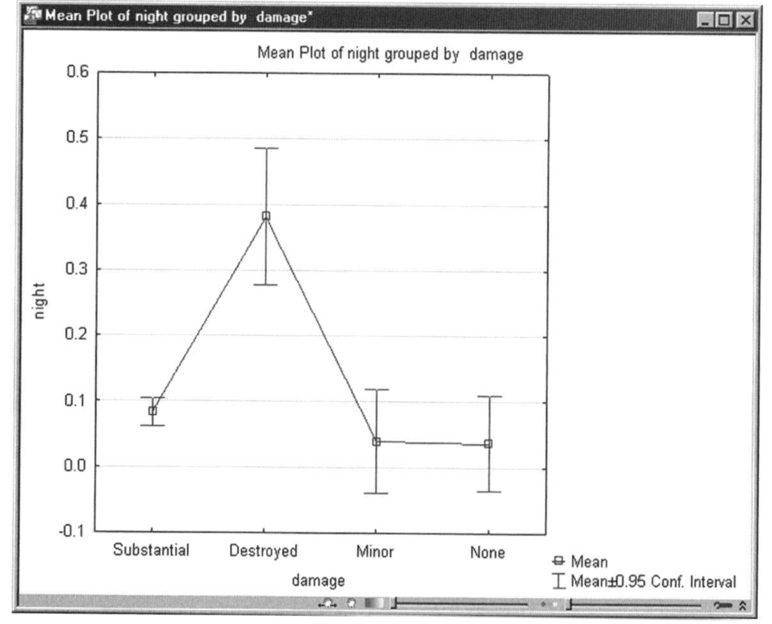

FIGURE B.26

Mean plot.

The means plot (see Figure B.26) shows that the term, *night*, was used most on average when the damage was rated as *Destroyed*.

On the *Boosted Trees Results* dialog, *Classification* tab, change the *Sample* to *All samples*. Select *Predicted vs. observed by classes* to make the classification matrix along with other pieces of output (see Figure B.27).

Data: Classification matrix (Flight Safety Data with Indexed Terms and missing data removed in Flight ...

Classification matrix (Flight Safety Data with Indexed Terms and missing data removed i
Response: damage
All samples;Number of trees: 388

	Observed	Predicted Substantial	Predicted Destroyed	Predicted Minor	Predicted None	Row Total
Number	Substantial	1809	432	205	130	2576
Column Percentage		92.91%	54.89%	68.11%	65.33%	
Row Percentage		70.23%	16.77%	7.96%	5.05%	
Total Percentage		55.94%	13.36%	6.34%	4.02%	79.65%
Number	Destroyed	98	335	29	20	482
Column Percentage		5.03%	42.57%	9.63%	10.05%	
Row Percentage		20.33%	69.50%	6.02%	4.15%	
Total Percentage		3.03%	10.36%	0.90%	0.62%	14.90%
Number	Minor	12	10	48	14	84
Column Percentage		0.62%	1.27%	15.95%	7.04%	
Row Percentage		14.29%	11.90%	57.14%	16.67%	
Total Percentage		0.37%	0.31%	1.48%	0.43%	2.60%
Number	None	28	10	19	35	92
Column Percentage		1.44%	1.27%	6.31%	17.59%	
Row Percentage		30.43%	10.87%	20.65%	38.04%	
Total Percentage		0.87%	0.31%	0.59%	1.08%	2.84%
Count	All Groups	1947	787	301	199	3234
Total Percent		60.20%	24.34%	9.31%	6.15%	

FIGURE B.27
Classification matrix.

This output shows where correct and incorrect model classifications were made. For example, when the *damage* was observed *Substantial*, it was predicted *Substantial* 1,809 times, or 70.23%. The output gives column, row, and total percentages along with the raw counts, allowing you to review the model performance.

For each class, the model did make more correct classifications than incorrect ones. The overall model accuracy rate can be computed by adding the total percentage entry for each correct classification: $55.94\% + 10.36\% + 1.48\% + 1.08\% = 68.86\%$.

For comparison, let's look at the boosted trees model, where only the structured text is used to build the model. Click Cancel on the *Boosted Trees Results* dialog to return to the *Boosted Trees Specifications* dialog. Change the variable selection to no longer include the indexed terms from text mining (see Figure B.28). Then build the model again.

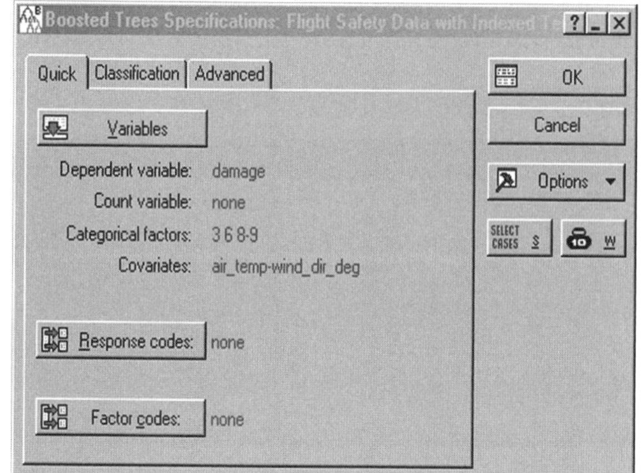

FIGURE B.28
Boosted Trees dialog with dependent variable, categorical factors, and covariates selected.

On the *Boosted Tree Results* dialog, again, make the *Summary* graph output (see Figure B.29).

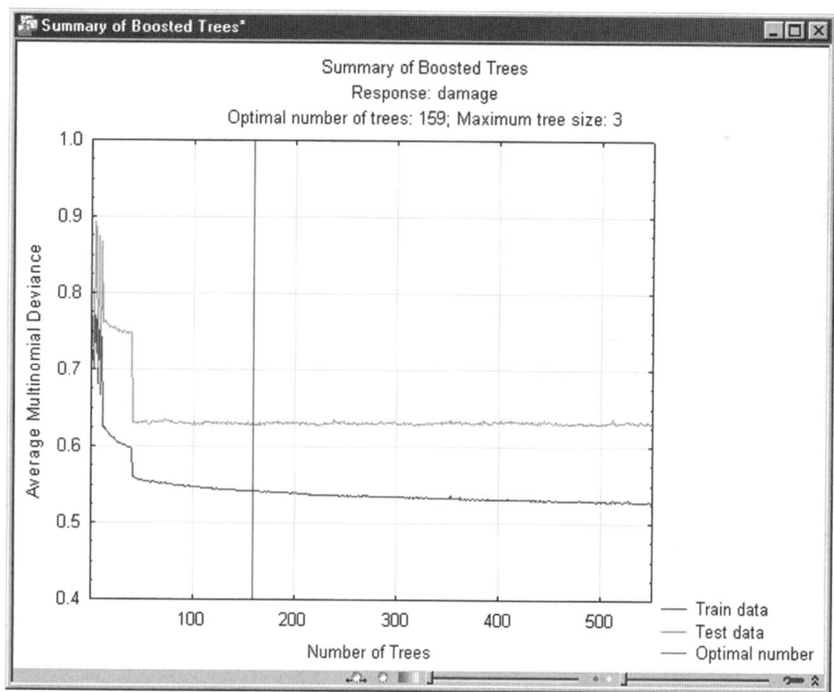

FIGURE B.29
Summary graph of the Boosted Trees computation process.

Additional trees are not needed for this model. The train and test error has leveled off.

On the *Classification* tab, select *Predicted vs. observed by class* to create the classification matrix (see Figure B.30).

Data: Classification matrix (Flight Safety Data with Indexed Terms and missing data removed in Flight ...

Classification matrix (Flight Safety Data with Indexed Terms and missing data removed
Response: damage
All samples;Number of trees: 159

	Observed	Predicted Substantial	Predicted Destroyed	Predicted Minor	Predicted None	Row Total
Number	Substantial	1700	510	171	195	2576
Column Percentage		87.18%	70.44%	75.33%	58.56%	
Row Percentage		65.99%	19.80%	6.64%	7.57%	
Total Percentage		52.57%	15.77%	5.29%	6.03%	79.65%
Number	Destroyed	213	206	25	38	482
Column Percentage		10.92%	28.45%	11.01%	11.41%	
Row Percentage		44.19%	42.74%	5.19%	7.88%	
Total Percentage		6.59%	6.37%	0.77%	1.18%	14.90%
Number	Minor	23	4	25	32	84
Column Percentage		1.18%	0.55%	11.01%	9.61%	
Row Percentage		27.38%	4.76%	29.76%	38.10%	
Total Percentage		0.71%	0.12%	0.77%	0.99%	2.60%
Number	None	14	4	6	68	92
Column Percentage		0.72%	0.55%	2.64%	20.42%	
Row Percentage		15.22%	4.35%	6.52%	73.91%	
Total Percentage		0.43%	0.12%	0.19%	2.10%	2.84%
Count	All Groups	1950	724	227	333	3234
Total Percent		60.30%	22.39%	7.02%	10.30%	

FIGURE B.30

Classification matrix when the unstructured data, e.g. the text variables, were removed from the analysis.

In comparison, when *Substantial damage* was observed, the model without unstructured text accurately predicted this 1,700 times or 65.99%. When terms from unstructured text were used, the accuracy was 70.23%. The overall accuracy for the new model is 52.57% + 6.37% + 0.77% + 2.1% = 65.81%, compared to 68.86% for the model with keywords from unstructured text added.

Lift and gains charts can be made from the *Boosted Tree Results* dialog as well. Figure B.31 is an overlaid gains chart showing the two models for *damage*.

Higher gains are seen when the unstructured text is used to build the model. Keywords in the accident narrative are therefore significantly related to the variable of interest: *damage*.

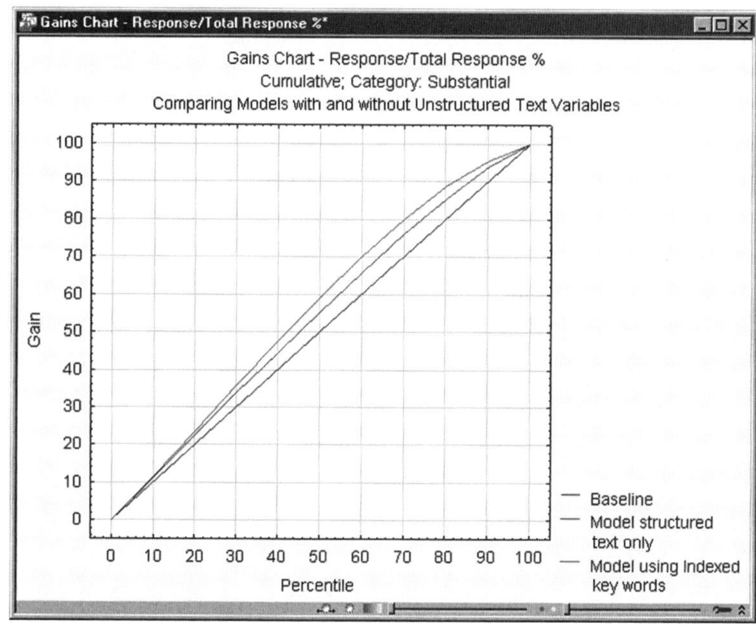

FIGURE B.31

Gains chart using "structured text only."

The same steps can be used to build predictive models for the other variable of interest: *injury_level*. The gains are far more pronounced when modeling the type of injury sustained: *none, minor,* or *fatal*. The gains plot in Figure B.32 shows this graphically.

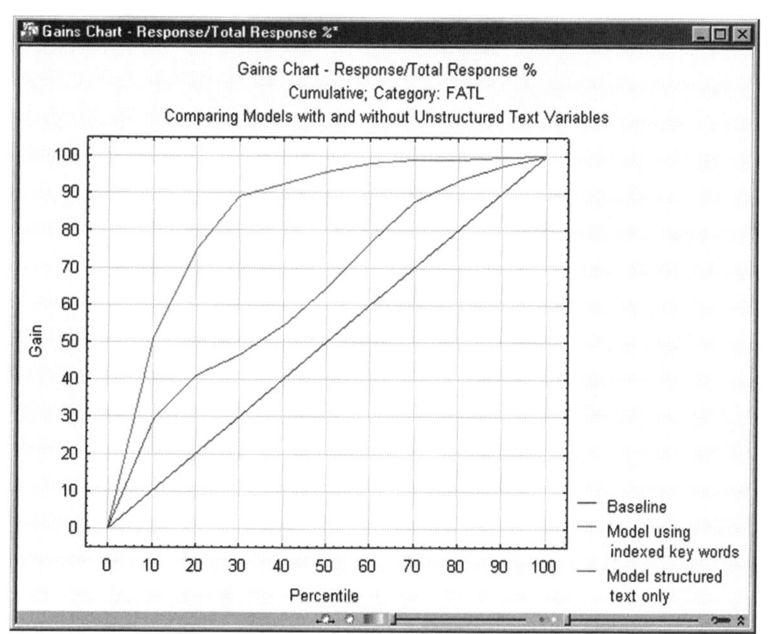

FIGURE B.32

Gains chart showing comparison of using both "structured text only" versus "Indexed keywords."

Insurance Industry: Text Analytics Adds "Lift" to Predictive Models with *STATISTICA* Text and Data Miner

Dev Kannabiran
StatSoft Data Mining Consultant

Thomas Hill
StatSoft Vice President for Development

Gary Miner
StatSoft Senior Statistician and Data Mining Consultant

CONTENTS

INTRODUCTION

The insurance industry, like most others, is severely affected by fraud and is working hard to equip itself with all of the possible tools necessary to identify and control fraud. With fraud being an ever growing business and the complexity of fraud evolving with time, insurance industries find it hard to identify fraud using regular predictive analytics tools. As discussed in Chapter 14, although there are some telltale signs of certain fraud, such as the swoop and squat, these nuggets of information are unfortunately hidden in claims notes as textual/unstructured data. This tutorial will focus on how we can assign numerical indices to unstructured data using text mining and using it in conjunction with structured data to build powerful and systematic fraud detection models and also emphasize the lift that text mining adds to the predictive models.

Practical Text Mining and Statistical Analysis for Non-structured Text Data Applications. DOI: 10.1016/B978-0-12-386979-1.00011-6

DATA DESCRIPTION

The data sets used in this tutorial are structured to resemble auto insurance data that include policy data, claims data, and some unstructured data in the form of claims notes. Though these data sets are not "real," they contain fragments of real problems that are masked to ensure confidentiality. However, the workflow shown in this tutorial is real and is a good approach to building useful fraud detection models.

Data files are needed for this tutorial. (Please find these on the DVD found inside the back cover of this book are illustrated in the folder diagram below.)

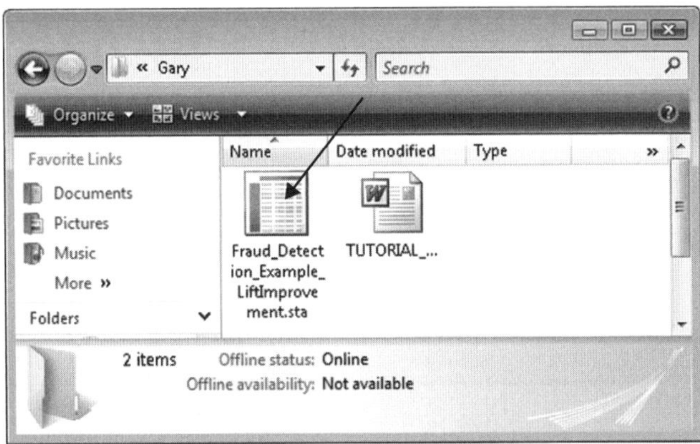

PART A: COMPARING THE *LIFT* OF PREDICTIVE MODELS WITH AND WITHOUT TEXT MINING

To show the lift added by text mining, we will use the dataset called the "Fraud_detection_ Example_LiftImprovements.sta." Select this file by going to the FILE pull down menu on the STATISTICA screen, and then releasing on "open", as shown in Figure C.1.

FIGURE C.1

Opening a data file. File -> Open. The next screen shot, shown in Figure C.2, illustrates where the data file name pathway needs to be entered.

Figure C.2 shows the dialog where the data file is selected, and Figure C.3 shows the spreadsheet of data for this tutorial.

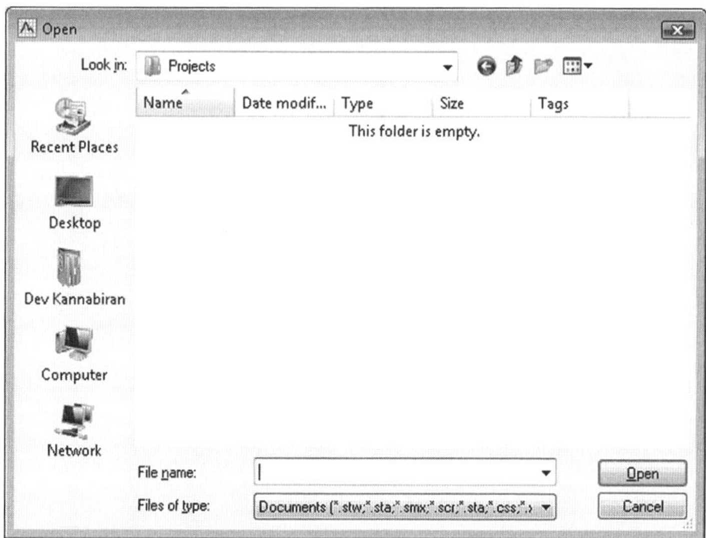

FIGURE C.2
Select the path of the data file and select "Fraud_detection_Example_LiftImprovements.sta."

Data: Fraud_Detection_Example_LiftImprovement.sta (89v by 2732c)

	77 Minor_Impact_Collision	78 Rental_Vehicle_Involved	79 Unusual_Injury	80 CaseNo	81 Cause	82 Fraud	83 Complexity	84 Right Track	S
42956	0	0	0	1548	Striking other car from rear	No	Low	Yes	
42957	0	0	0	1549	Striking other car from rear	Yes	High	No	
43061	0	0	0	1550	Striking other car from rear	No	Low	Yes	
43146	0	0	0	1551	Striking other car from rear	Yes	High	No	
43316	0	0	0	1552	Striking other car from rear	No	Low	Yes	
43375	0	0	0	1553	Striking other car from rear	No	Low	Yes	
43409	1	0	1	1554	Leaving garage, driveway or alley	Yes	High	No	
43449	0	0	0	1555	Leaving garage, driveway or alley	No	Low	Yes	
43468	1	0	0	1556	Striking other car from rear	No	Low	Yes	
43521	0	0	0	1557	Striking other car from rear	No	Low	Yes	
43524	0	0	0	1558	Leaving garage, driveway or alley	Yes	High	No	
43541	0	0	0	1559	Striking other car from rear	No	Low	Yes	
43546	0	0	0	1560	Backing	No	Low	Yes	
43584	0	0	0	1561	Mechanical failure - real or alleged	Yes	High	No	
43604	0	0	0	1562	Leaving garage, driveway or alley	Yes	High	No	
43611	1	0	0	1563	Skidding on ice or snow	Yes	High	No	
43621	1	1	0	1564	Leaving garage, driveway or alley	No	Low	Yes	
43622	1	1	0	1565	Leaving garage, driveway or alley	No	Low	Yes	
43623	1	1	0	1566	Leaving garage, driveway or alley	No	Low	Yes	
43624	1	1	0	1567	Leaving garage, driveway or alley	Yes	High	No	
43665	0	0	0	1568	Opening door in traffic	No	Low	Yes	
43724	1	1	0	1569	Failure to yield at traffic control device	No	Low	Yes	
43750	0	0	0	1570	Failure to yield at traffic control device	Yes	High	No	
43753	1	0	0	1571	Leaving garage, driveway or alley	No	Low	Yes	
43794	0	0	0	1572	Failure to yield at traffic control device	No	Low	Yes	
43798	0	0	0	1573	Failure to yield at traffic control device	Yes	High	No	
43799	0	0	0	1574	Failure to yield at traffic control device	Yes	High	No	
43816	0	0	0	1575	Failure to yield at traffic control device	No	Low	Yes	

FIGURE C.3
Data file.

Once we have the file open, we can see that variable 81 ("Cause") has claims notes in the form of unstructured data. We will now assign numerical indices to the unstructured data using text mining so it can be used in conjunction with other structured data. We can do this from either the "Ribbon Bar Format" shown in Figure C.4, or the "Classic Menu Format" shown in Figure C.5.

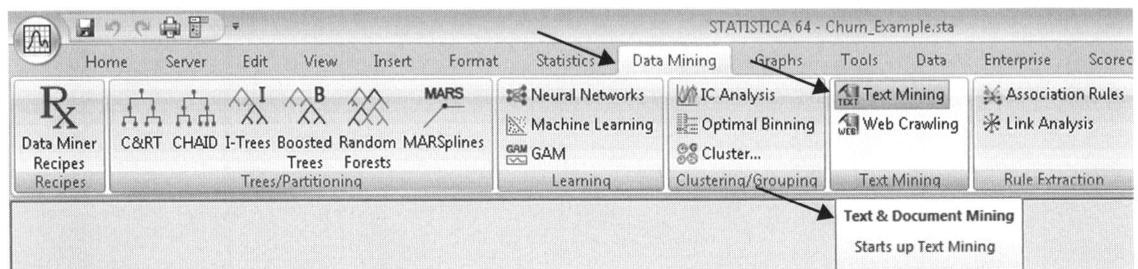

FIGURE C.4
Select "Text Mining" from the Ribbon Bar format.

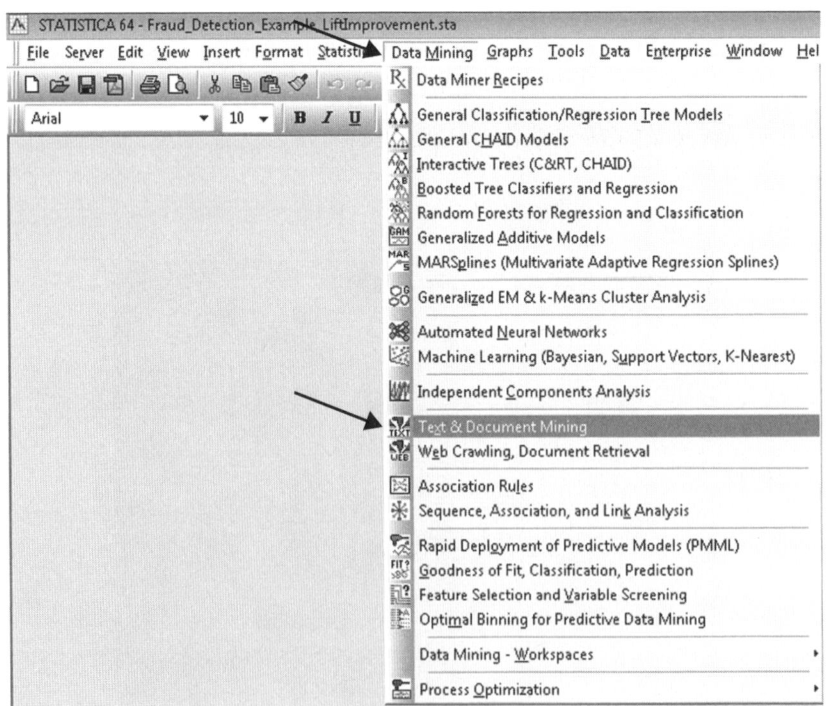

FIGURE C.5
Select "Text Mining" from the Classic Menu format.

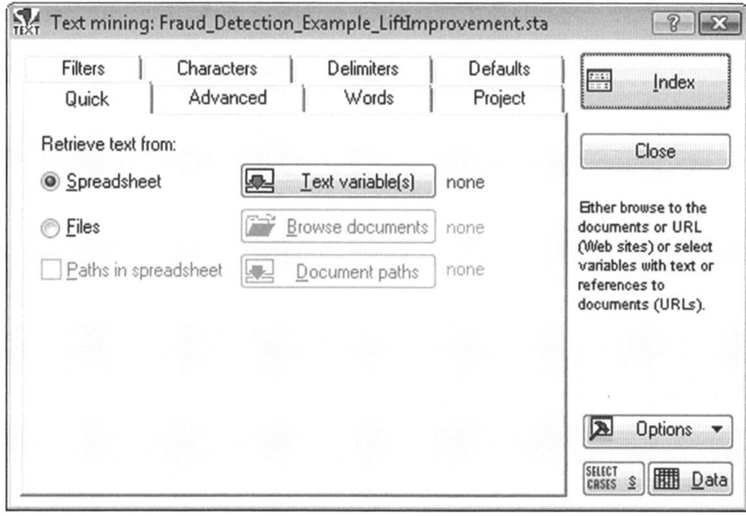

FIGURE C.6
The "Text Mining" opening dialog window.

Select the spreadsheet option under "Retrieve text from." (see Figure C.6) Click on the "Text Variable(s)" button to open the "Select variables containing text" dialog (see Figure C.7).

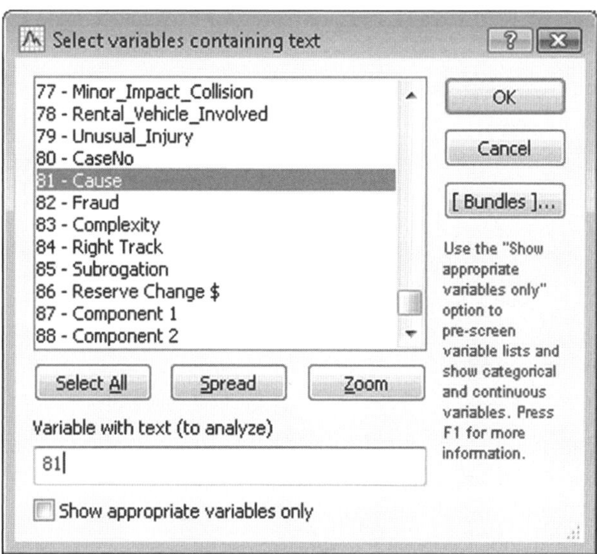

FIGURE C.7
Select variables containing text window.

Select Variable 81 (the text variable). Click on the OK button on the "Variable Selection" window. Click on "Index." Figure C.8 illustrates a warning dialog that will appear; click "yes" on this to continue.

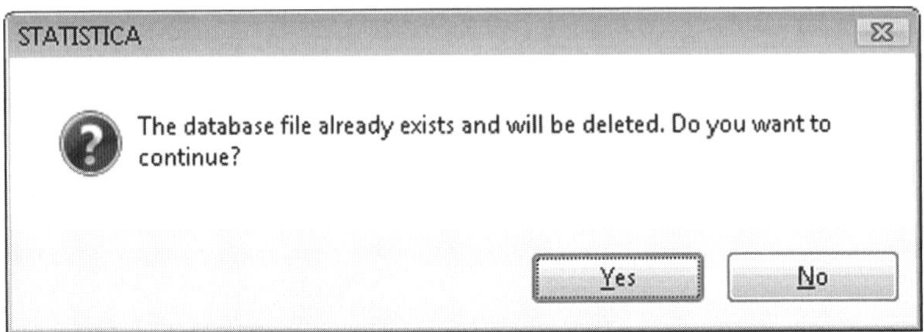

FIGURE C.8
A warning dialog will or might appear (based on whether you have created a text miner database before). Click "yes" to continue on through to the next step.

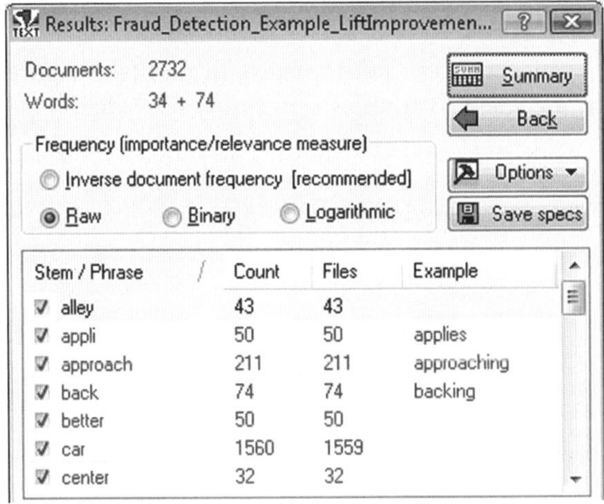

FIGURE C.9
The "Results" window for the Text Mining analysis.

The results window (see Figure C.9) displays the keywords and their counts/frequencies, as well as the number of files they occurred in.

Once the input documents have been indexed and the initial word frequencies (by document) computed, a number of additional transformations can be performed to summarize and aggregate the information that was extracted. In this tutorial we use the inverse document frequency transformation (see Figure C.10) that reflects both the specificity of words (document frequencies) as well as the overall frequencies of their occurrences (word frequencies).

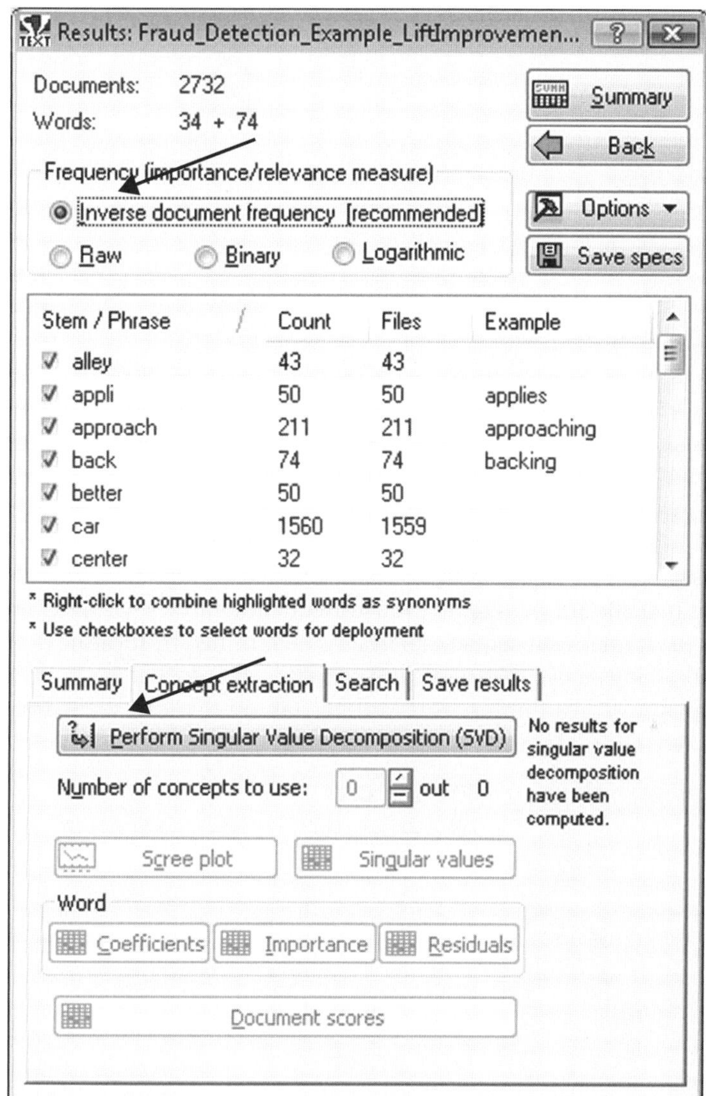

FIGURE C.10

Run singular value decomposition.

The singular value decomposition, which is similar to principal components analysis, is used to reduce the dimensionality of the problem, thus allowing us to work with a few concepts instead of all the actual words. (For more information, see Chapter 15.)

Under the Concept extraction tab, click the "Perform Singular Value Decomposition" button.

After running the SVD, click on the "Scree plot" button.

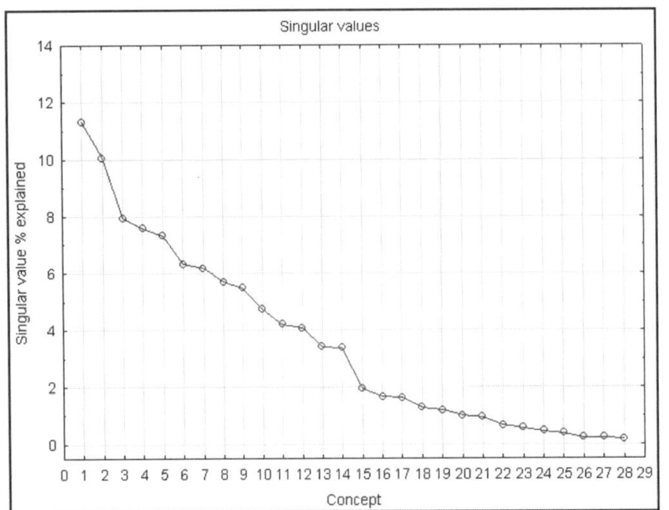

FIGURE C.11
Scree plot from the singular value decomposition analysis.

The Scree plot (see Figure C.11) is used to decide on the number of singular values that are useful and informative and that should be retained for subsequent analyses. Usually, the number of "informative" dimensions to retain for subsequent analysis is determined by locating the "elbow" in this plot. In this tutorial we are going to use the top three concepts.

To resume the analysis, click on the "Text Mining" tab on the lower left corner or hold CTRL+R. Then click on the "save results" tab, as shown in Figure C.12.

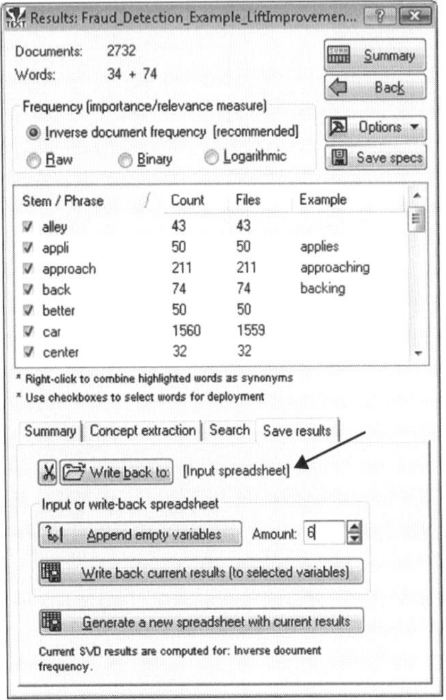

FIGURE C.12
Save tab dialog of the Text Mining results window.

Since the first six concepts capture about 60 percent of all the available information, we will select six concepts and put "6" in the "amount" window (see Figure C.12 above) so we can "append 6 new empty variables" into our data sheet. When we do this, a "warning window" will appear, as shown in Figure C.13; click "ok" on this to continue. After this we will need to click the "Write back current results" button, as shown in Figure C.14, in order to get these results into our spreadsheet.

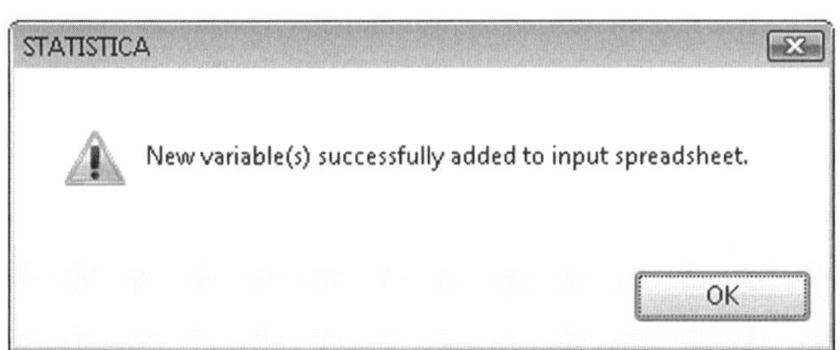

FIGURE C.13
Warning window that informs us that the "new variables" have been succcessfully added to the input data sheet.

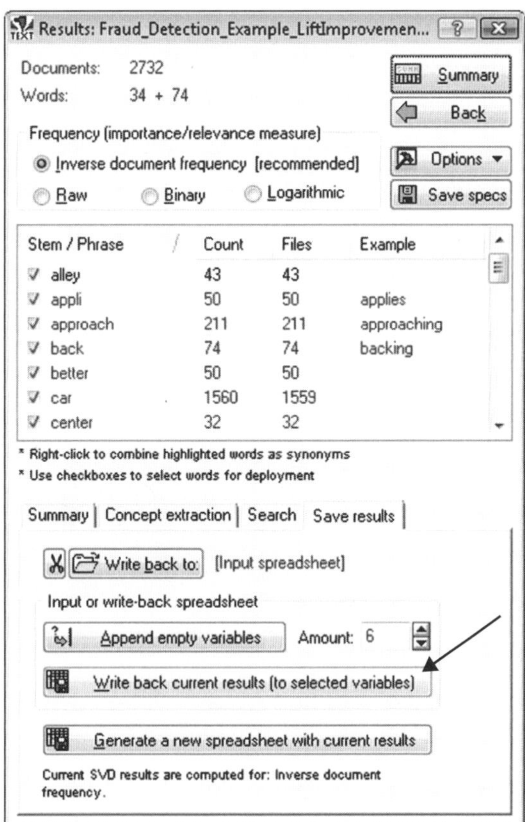

FIGURE C.14
Click on the "Write back current results" button on the Text Mining Results window to allow the six new variables, which we call "concepts," to be written to the six new variables/columns, now empty, in our data sheet.

The next four figures, Figures C.15 through C.18, illustrate the process of selecting what will be written back to the data spreadsheet.

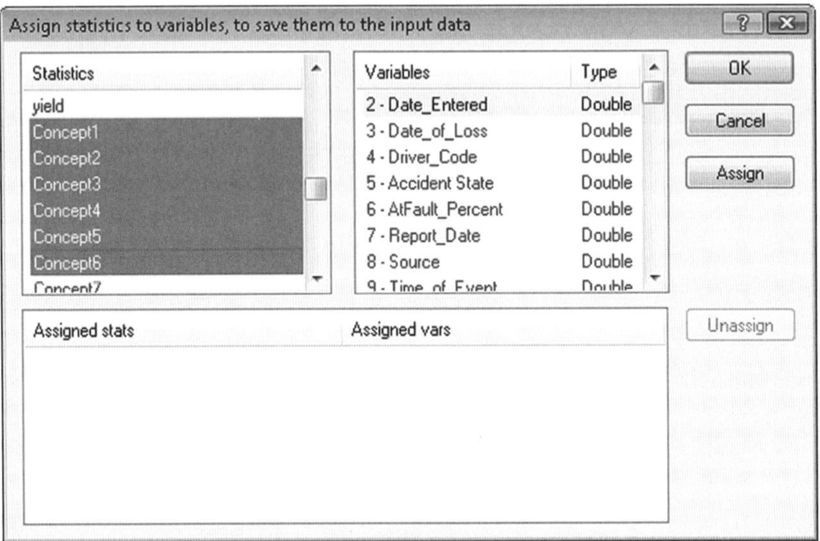

FIGURE C.15
Select Concept1 to Concept6 (by holding Ctrl) under statistics (left pane).

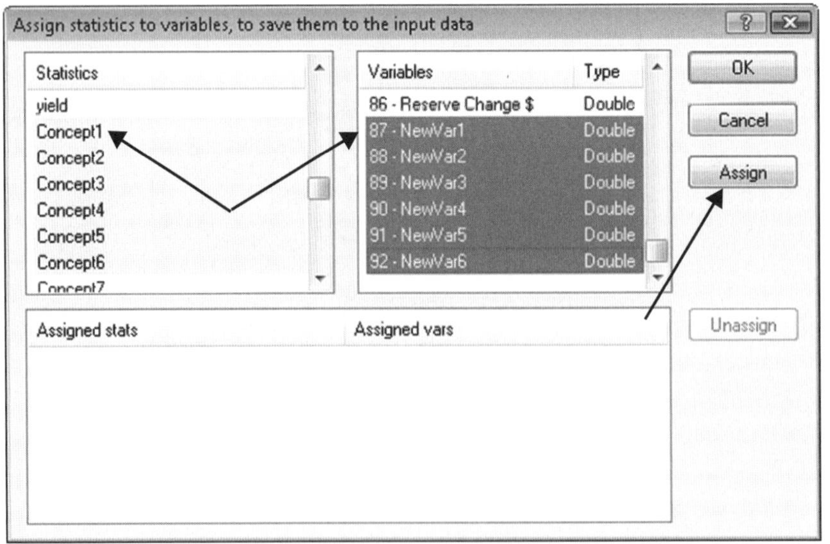

FIGURE C.16
Select NewVar1 to NewVar6 under variables (right pane). Then click on the Assign button to assign statistics to variables.

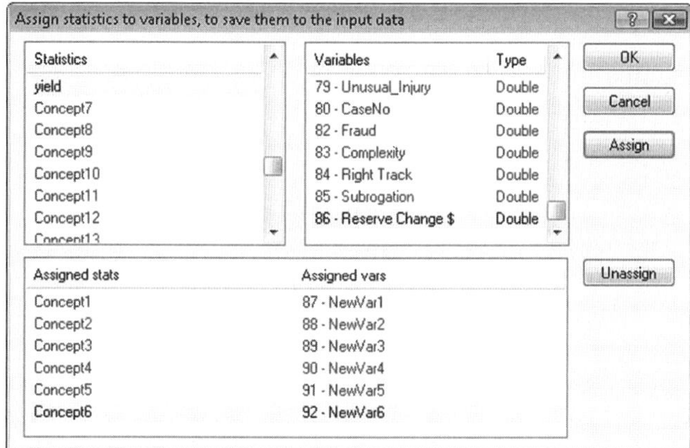

FIGURE C.17

Once the concepts are assigned to the new variables, click OK, which now enters the values for these concepts into the new variables in the data sheet.

Data: Fraud_Detection_Example_LiftImprovement.sta (92v by 2732c)

	84 Right Track	85 Subrogation	86 Reserve Change $	87 Concept1	88 Concept2	89 Concept3	90 Concept4	91 Concept5	92 Concept6
2017	Yes	0	$500	0.00008	0.000068	0.001555	-0.00169	0.000053	-0.00579
2025	Yes	1	$500	0.00008	0.000068	0.001555	-0.00169	0.000053	-0.00579
2039	Yes	0	$500	0.000111	0.000089	0.002188	-0.01305	5.066E-6	-0.00335
2108	No	1	$10,000	0.00008	0.000068	0.001555	-0.00169	0.000053	-0.00579
2112	Yes	0	$500	0.00008	0.000068	0.001555	-0.00169	0.000053	-0.00579
2113	No	1	$10,000	0.00008	0.000068	0.001555	-0.00169	0.000053	-0.00579
2139	No	0	$10,000	0.00008	0.000068	0.001555	-0.00169	0.000053	-0.00579
2164	Yes	0	$1,500	0.00008	0.000068	0.001555	-0.00169	0.000053	-0.00579
2173	Yes	1	$500	0	0	0	0	0	0
2181	No	0	$5,000	0.00008	0.000068	0.001555	-0.00169	0.000053	-0.00579
2245	Yes	0	$500	0	0	0	0	0	0
2302	No	0	$10,000	0.00008	0.000068	0.001555	-0.00169	0.000053	-0.00579
2305	Yes	0	$500	0.00008	0.000068	0.001555	-0.00169	0.000053	-0.00579
2368	No	0	$10,000	0.00008	0.000068	0.001555	-0.00169	0.000053	-0.00579
2436	No	1	$5,000	0.00008	0.000068	0.001555	-0.00169	0.000053	-0.00579
2499	Yes	1	$1,500	0.00008	0.000068	0.001555	-0.00169	0.000053	-0.00579
2572	No	0	$7,500	-2.7E-19	-1.1E-18	-5.6E-18	1.06E-17	2.27E-17	1.14E-17
2590	Yes	0	$1,500	0.00008	0.000068	0.001555	-0.00169	0.000053	-0.00579
2643	No	0	$5,000	-2.7E-19	-1.1E-18	-5.6E-18	1.06E-17	2.27E-17	1.14E-17
2653	No	0	$7,500	0.00008	0.000068	0.001555	-0.00169	0.000053	-0.00579
2668	Yes	0	$500	0.00008	0.000068	0.001555	-0.00169	0.000053	-0.00579
2674	Yes	0	$1,500	0.00008	0.000068	0.001555	-0.00169	0.000053	-0.00579
2759	Yes	1	$1,500	0.00008	0.000068	0.001555	-0.00169	0.000053	-0.00579
2807	No	1	$5,000	0.00008	0.000068	0.001555	-0.00169	0.000053	-0.00579
2835	No	0	$5,000	0.00008	0.000068	0.001555	-0.00169	0.000053	-0.00579
2876	Yes	1	$500	0.00008	0.000068	0.001555	-0.00169	0.000053	-0.00579
2881	No	0	$5,000	0.00008	0.000068	0.001555	-0.00169	0.000053	-0.00579

FIGURE C.18

"Concept1," "Concept2," "Concept3," "Concept4," "Concept5," and "Concept6" data columns have now been filled with data in the input spreadsheet (see columns 87–92).

Now we will take this data spreadsheet and perform some "predictive analytics" with the data. Our first attempt will be to use the Boosted Trees algorithm, but the "text material" will not be used in this model.

BOOSTED TREES (WITHOUT TEXT MATERIAL)

First, select "Boosted Trees" algorithm, as illustrated in Figure C.19.

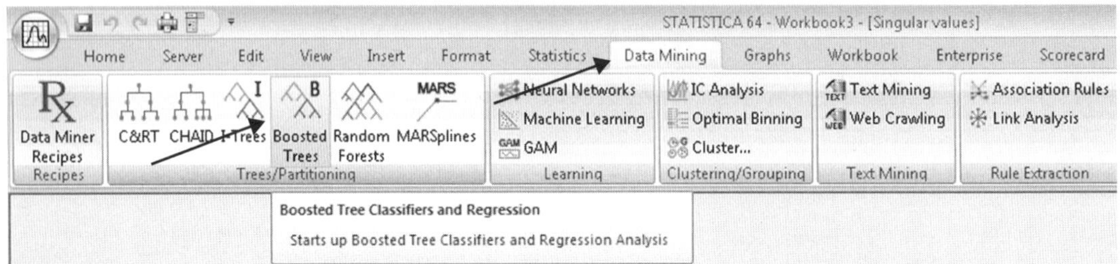

FIGURE C.19
Selecting Boosted Trees from the Data Mining menu.

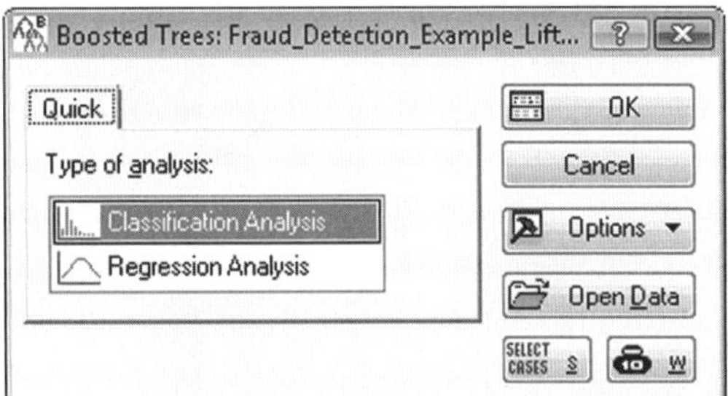

FIGURE C.20
The "Boosted Trees" dialog window.

Select "Classification Analysis" (see Figure C.20) under "Type of Analysis," and click the OK button to continue the analysis. This will bring up the "variable selection window" in the "Quick tab" of the "Boosted Trees Specification" window (see Figure C.21).

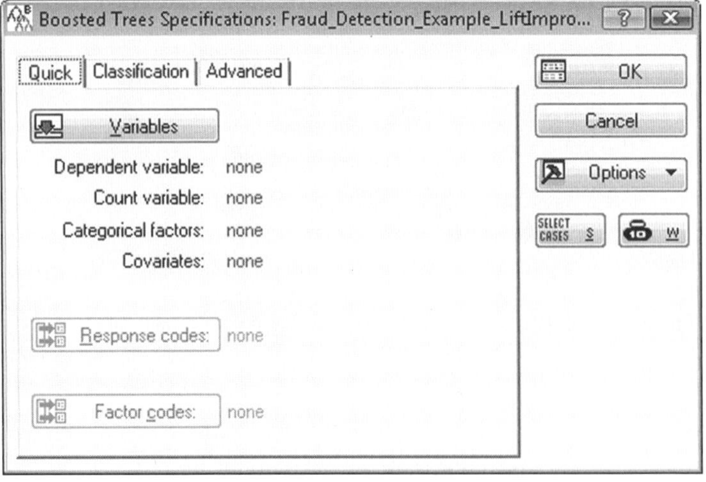

FIGURE C.21
"Boosted Trees Specification" window.

The following series of figures, Figures C.22 through C.25, show how to set up "bundles" of variables; bundles make it easier to select large groupings of variables.

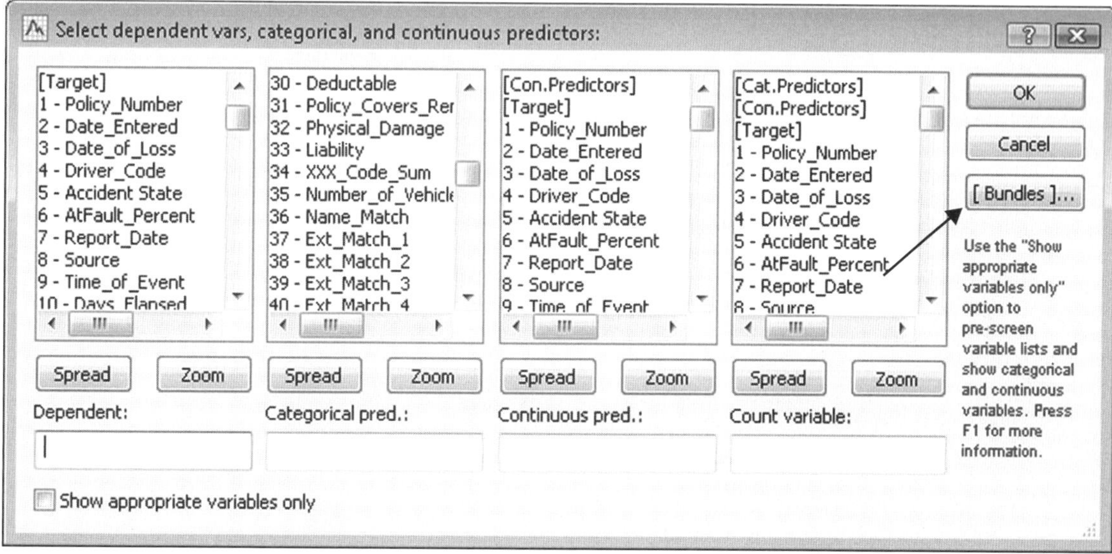

FIGURE C.22

Variable selection. Note that variable names with the square brackets [ ] around them are a "group" of variables called "Bundles of variables." Bundles are very useful to group together variables that are often used together. Here we have bundles for the target variable, Categorical predictors and Continuous predictors. We can create new bundles by clicking on the "[Bundles]..." button.

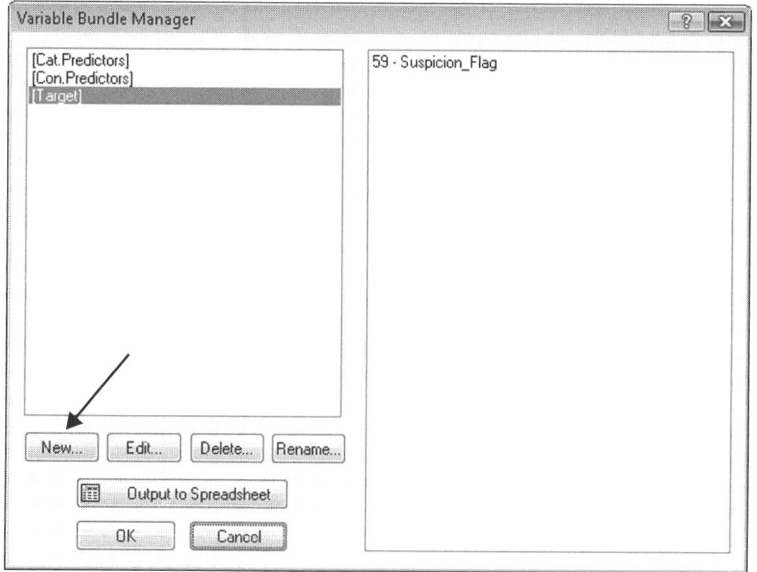

FIGURE C.23

Click on the "New" button to create a new bundle.

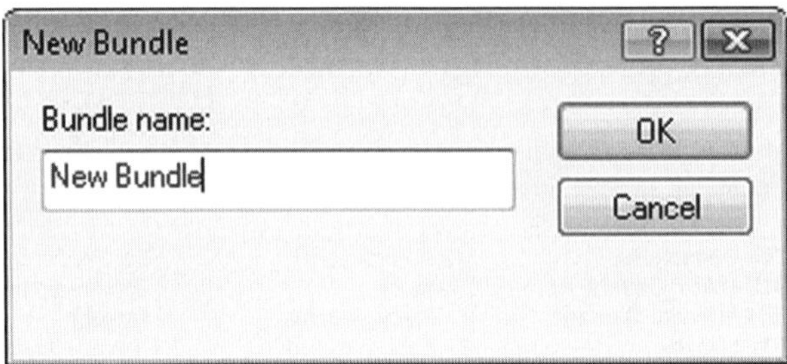

FIGURE C.24
Give the new bundle a name.

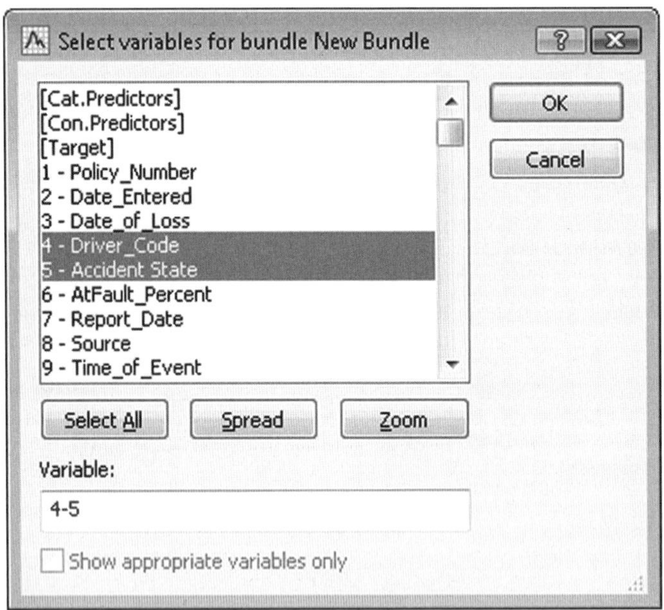

FIGURE C.25
Select variables to be bundled together, and select OK. However, in this tutorial we shall use the three existing bundles (so cancel out of the new bundle).

The next series of figures show how to select the variables for this Boosted Trees computation; see Figures C.26 through C.28.

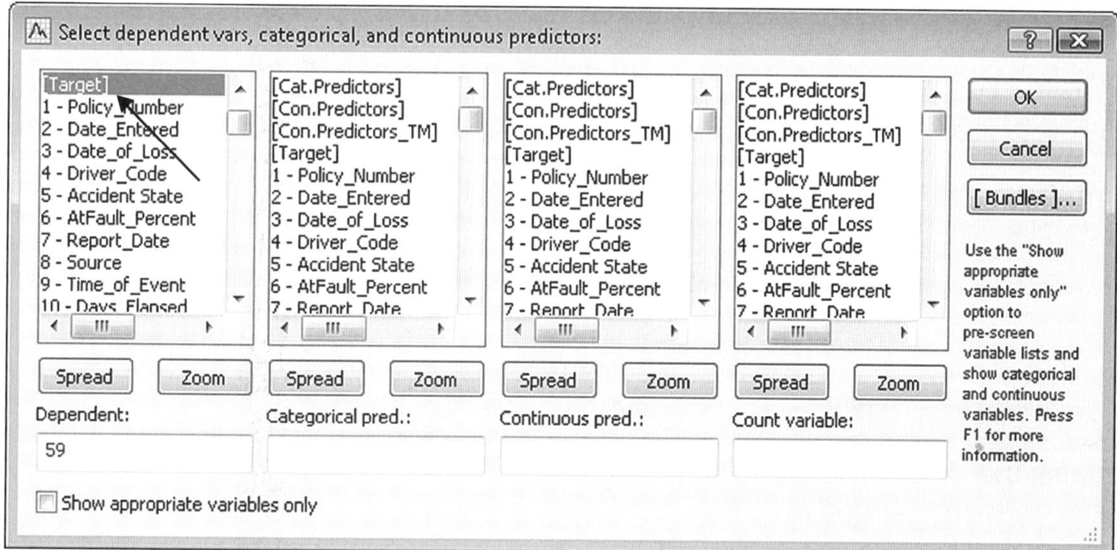

FIGURE C.26
Coming back to the variable selection for the Boosted Trees model. Under "Dependent" select "[Target]."

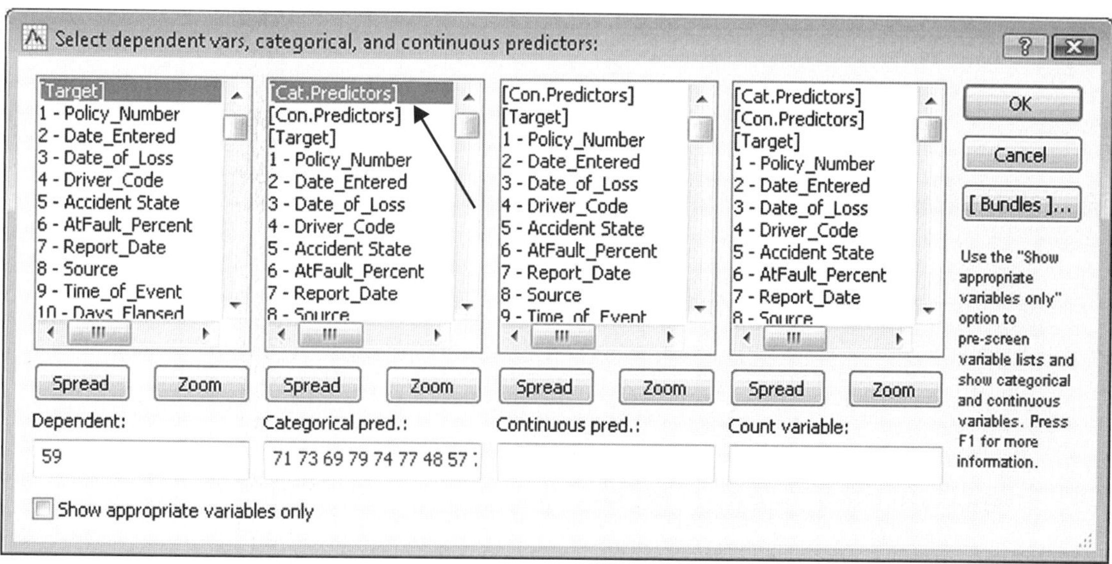

FIGURE C.27
Under "Categorical Pred.:" select "[Cat. Predictors]."

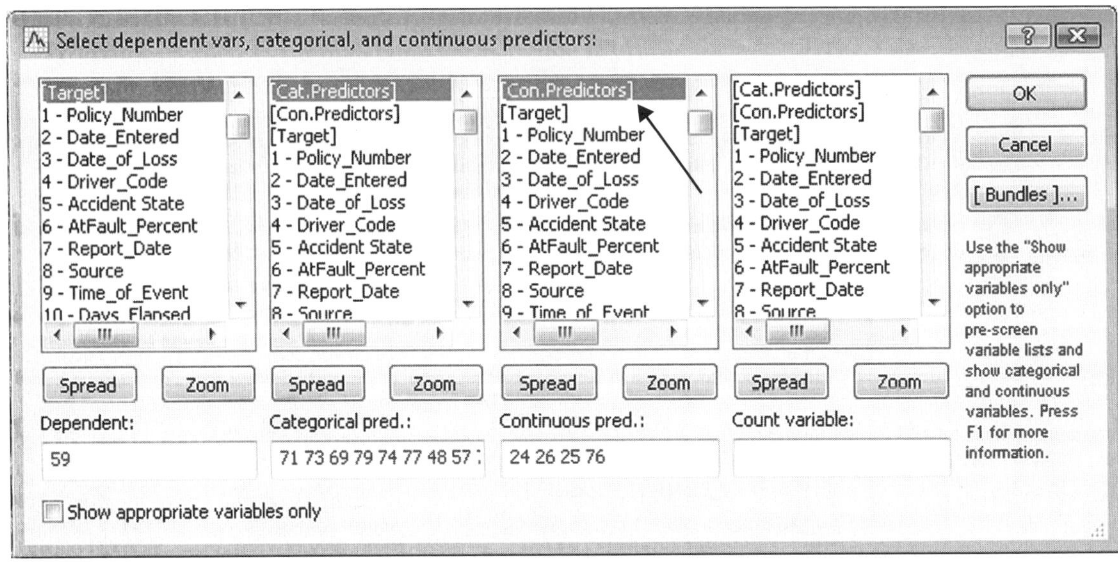

FIGURE C.28

Variable selection. Under "Continuous pred.:" select "[Con. Predictors]."

Then click OK in the diagram of Figure C.28 to start the computation, as illustrated in Figure C.29.

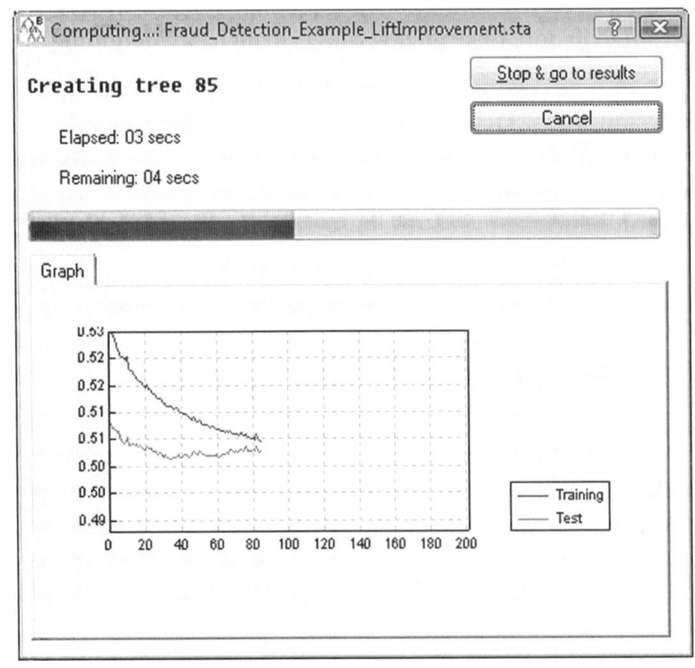

FIGURE C.29

Computation of the "Boosted Trees" is illustrated on the screen. This happens quite rapidly, with the green progress bar moving to the right as the "seconds" go by.

When the computations are completed, the "Boosted Trees Results" dialog will appear, as shown in Figure C.30.

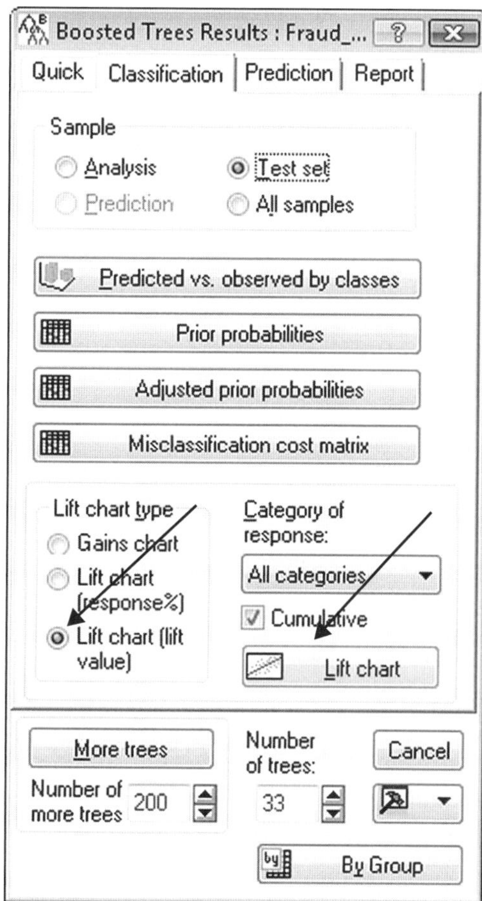

FIGURE C.30
Boosted Trees Results dialog. This dialog contains various options for reviewing the final model. We shall focus mainly on Lift charts, which are used to evaluate and compare the utility of the model for predicting the different categories or classes for the categorical dependent variable. Click on the Classification tab, and select "Test set" under "Sample" to specify the type of sample to use to compute the predicted and residual statistics (classifications). The test set contains all of the observations that were not used to compute the current results but have valid data for all predictor and dependent variables.

Under "Lift chart type" select "Lift Chart (lift value)"and click on the "Lift chart" button to get the lift chart where the vertical (y) axis is scaled in terms of the lift value, expressed as the multiple of the baseline random selection model.

This chart (see Figure C.31) shows that the Boosted Trees model (without text data) that we created is 2.1 times better (more predictive accuracy) than a baseline model (random guess without data mining).

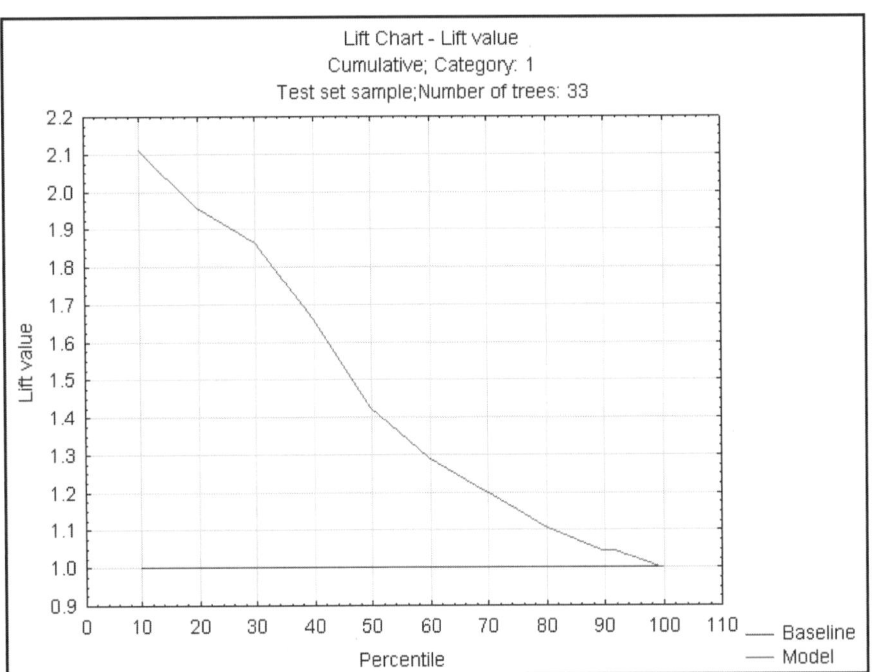

FIGURE C.31
"Cummultive Lift Chart" using Boosted Trees with only the structured data.

BOOSTED TREES ADDING THE TEXT MINING VARIABLES

Now that we have seen that the Boosted Trees models (without text) have added significant lift to our predictions, let us see if including the unstructured data makes our models better. We are now going to repeat the same steps to create a Boosted Trees model, but this time we are going to include the concepts (which we extracted from the unstructured data) as inputs.

Start "Boosted Trees" (see Figures C.32 and C.33) again by selecting it from the Data Mining pull-down menu in classic menus or from the Ribbon Bar.

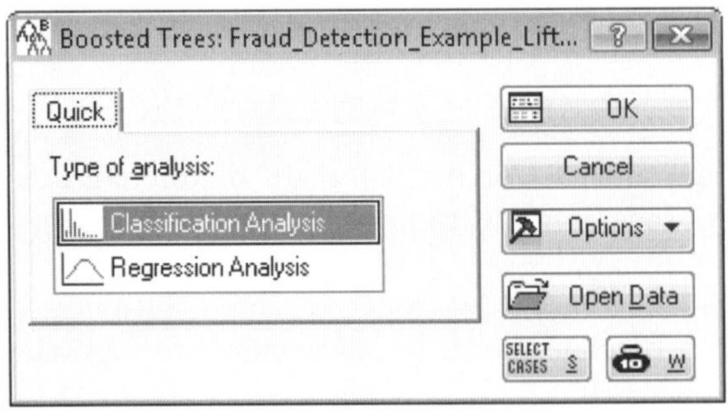

FIGURE C.32
Boosted Trees dialog window. Select "Classification Analysis" under "Type of Analysis," and click OK.

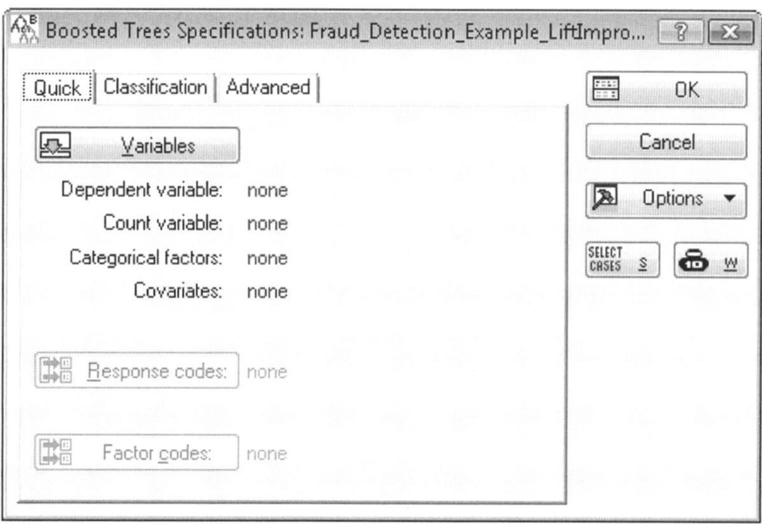

FIGURE C.33
Boosted Trees Specifications window.

The variables, for this combined regular numeric variables plus text variables analysis, are selected as shown in Figure C.34.

Under "Continuousl Pred.:" select "[Con. Predictors]," and also add the concepts (var 87–var 92), which we got from using the text miner on the unstructured data.

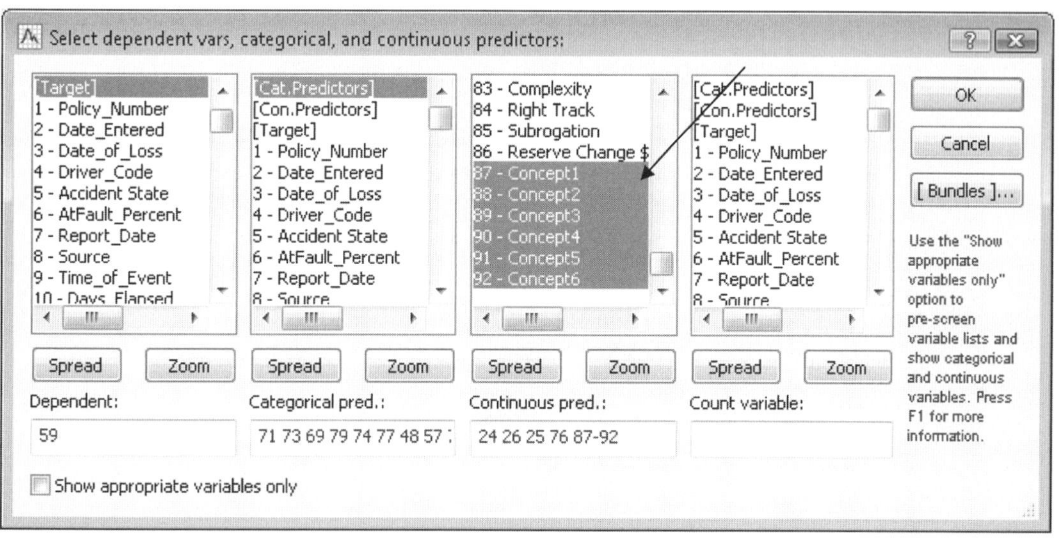

FIGURE C.34
Under "Dependent variable" select "[Target]," and under "Categorical Pred.:" select "[Cat. Predictors]."

Figure C.35 shows the computation process for Boosted Trees; and Figure C.36 shows the final results dialog.

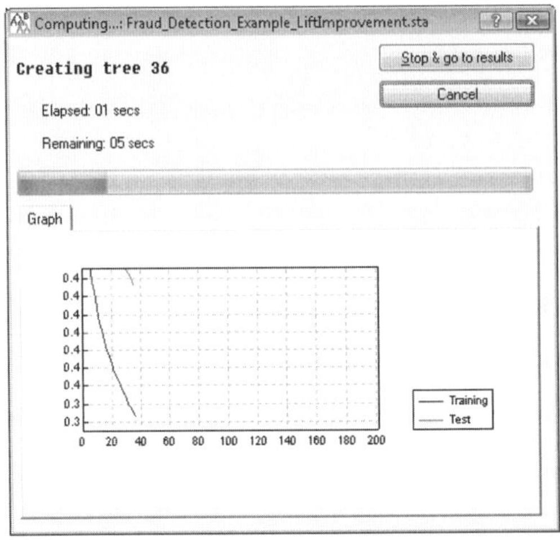

FIGURE C.35

Transient computation window for the Boosted Trees. This computation only takes seconds for a data set of this size.

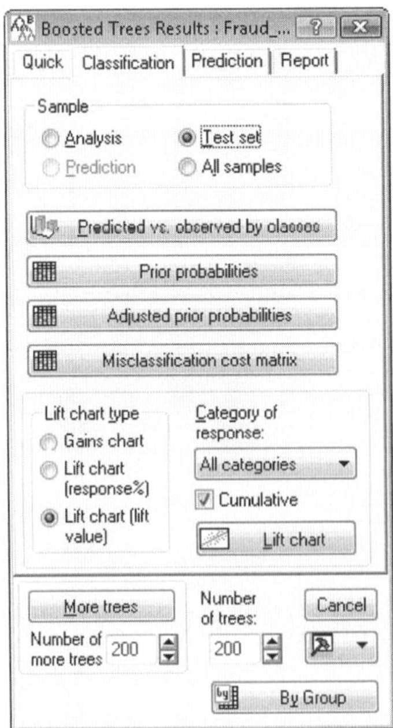

FIGURE C.36

Boosted Trees [with Text Mining variables] Results window.

Click on the Classification tab (as seen in Figure C.36), and select "Test Set" under "Sample." Under "Lift chart type" select "Lift Chart (lift value)," and click on the "Lift chart" button to get the Lift chart as illustrated in Figure C.37.

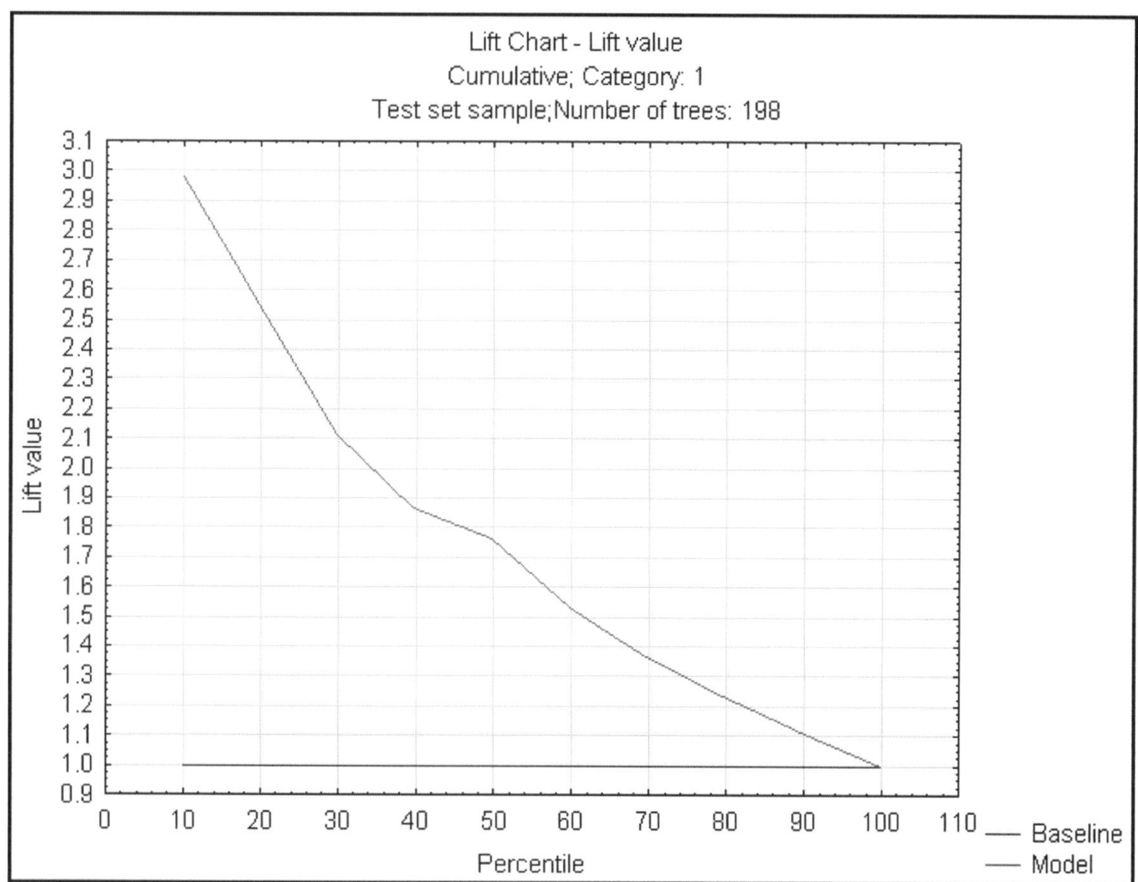

FIGURE C.37
Lift chart for the Boosted Trees with text variables along with numeric variables.

This chart shows that the Boosted Trees model (which includes numeric as well as text data) is three times better (more predictive accuracy) than a baseline model (random guess without data mining). As you can see, that is significantly better than our other model that had only numeric data and no unstructured data.

Now let's "Merge" the two graphs: the graph from the "Boosted Trees numeric variables only" with the "Boosted Trees numeric plus text variables."

HOW TO MERGE GRAPHS

The process of merging graphs and also changing parts of graphs are illustrated in the following set of figures, Figures C.38 through C.45.

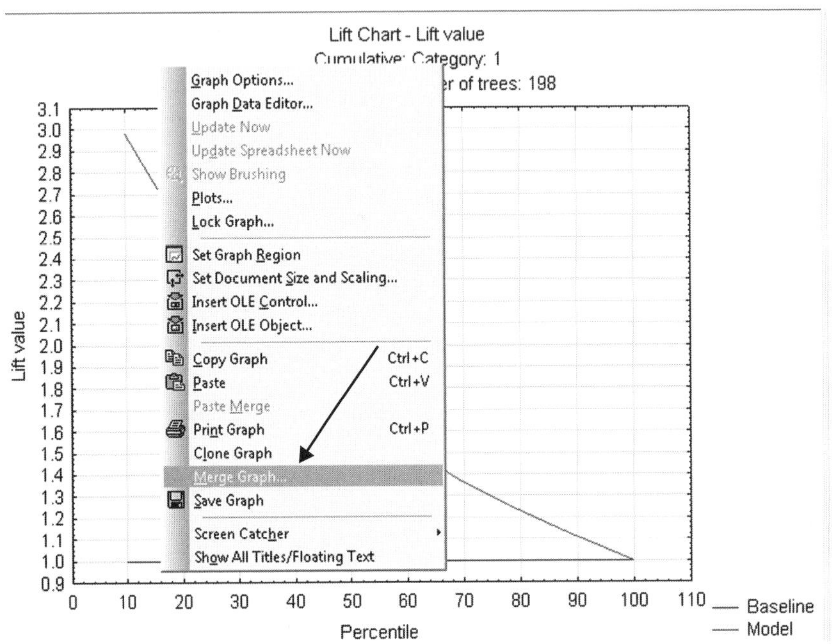

FIGURE C.38

Right click on the Lift chart to get additional graph options. Select "Merge Graph...".

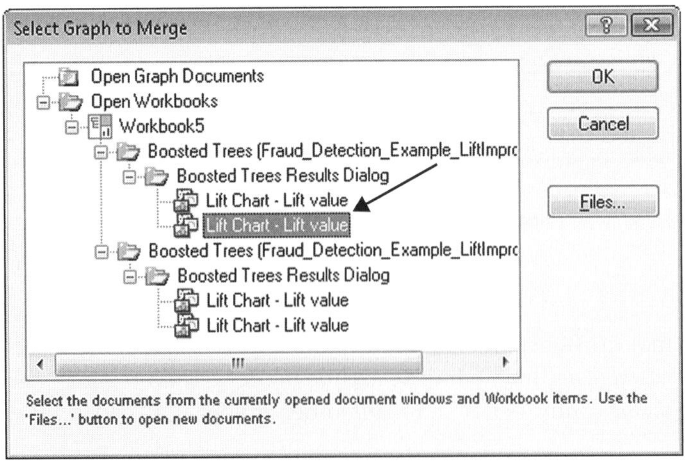

FIGURE C.39

Select the second Lift chart in the previous Boosted Trees model, and click OK.

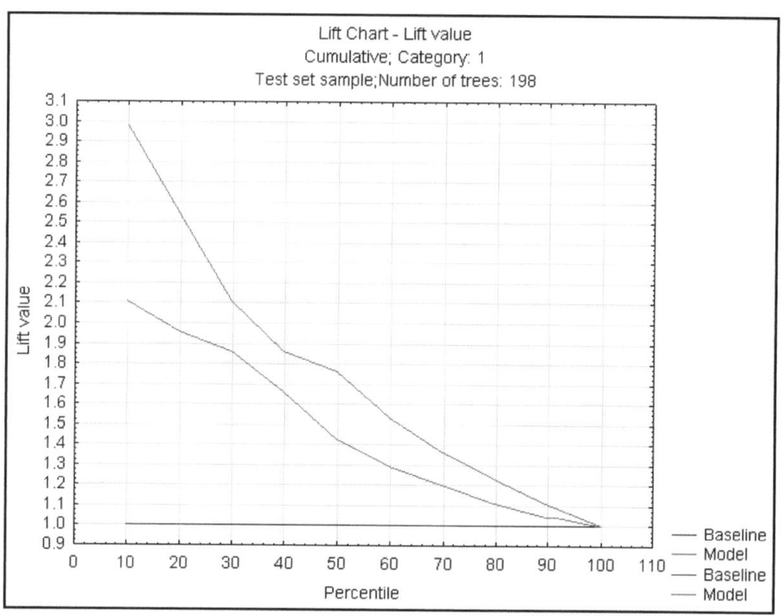

FIGURE C.40

We now have a merged graph that we can use to compare the lifts of different models.

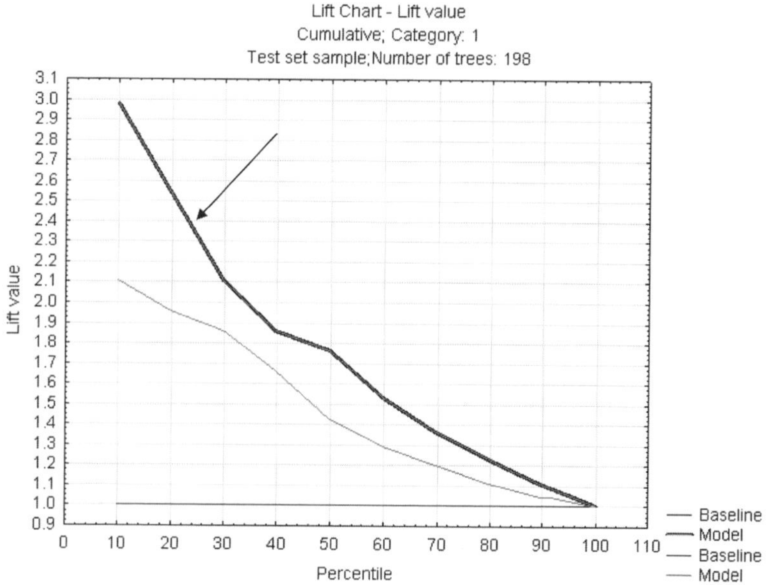

FIGURE C.41

We can now make this graph fancier by giving different colors for different models. Double click on the higher Lift curve (Boosted Trees with Text Mining) to open the graph options dialog.

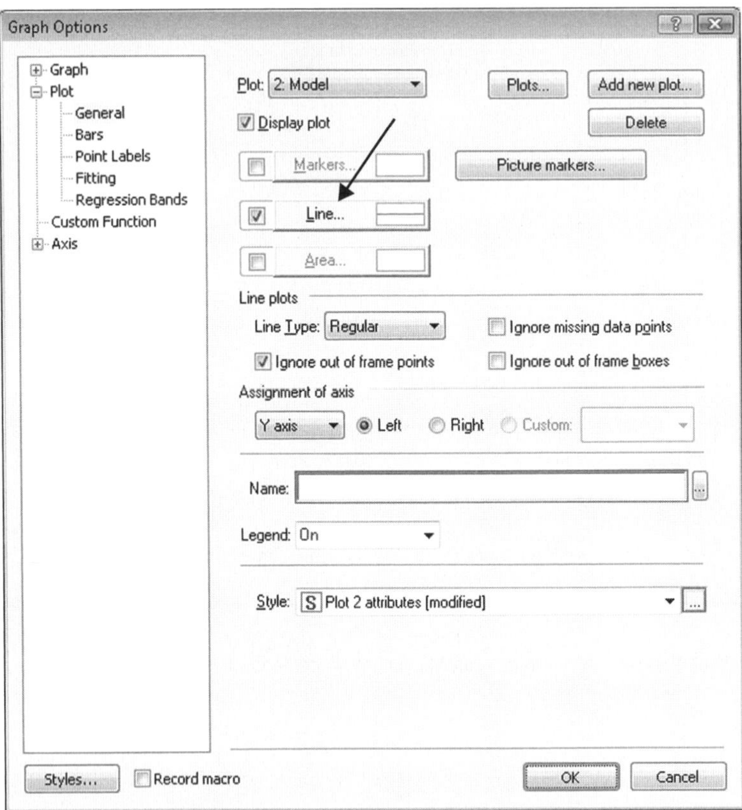

FIGURE C.42
"Graph Options" dialog where the color, line width, and so on of a graph can be changed. Click on "Line."

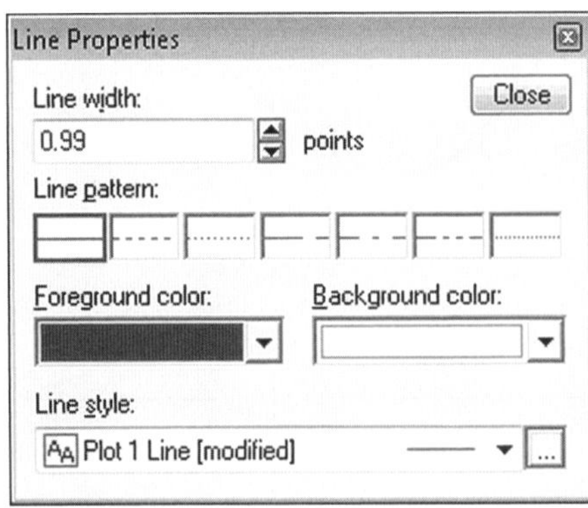

FIGURE C.43
"Line Properties" graph window. Here the pattern, width, Foreground color, Background color, and Line style can be selected. Select a solid line pattern, and change the foreground color to green.

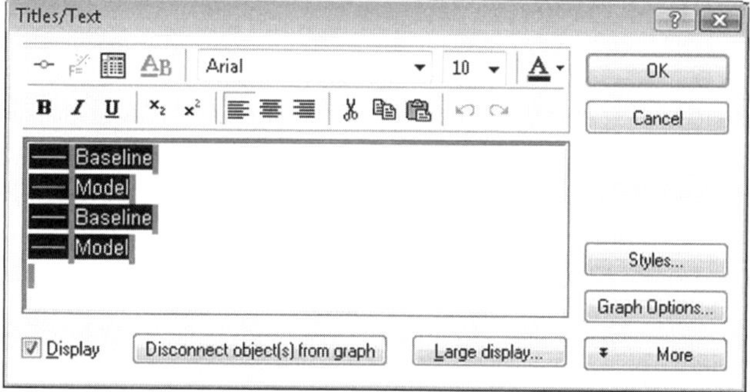

FIGURE C.44

Double click on the Legend to get to the "Titles/Text" window for graphs where the "captions" can be changed to different terms, along with selecting font, color, bolding, and so on. Delete one of the "baselines," and rename the "Model" corresponding to the green line as "Boosted Trees with Text Mining" and the other "Model" as "Boosted Trees without Text Mining." Click OK.

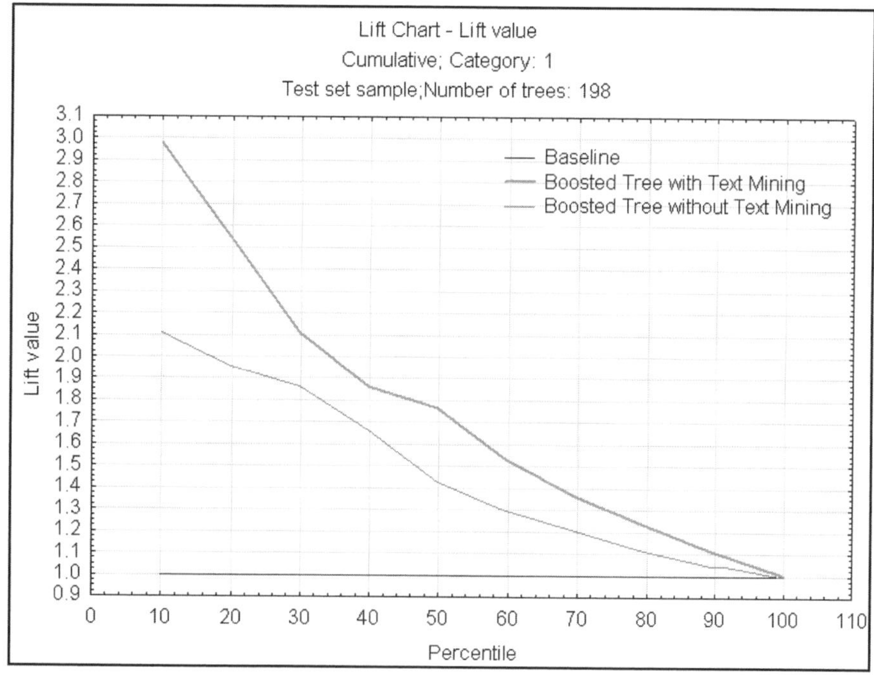

FIGURE C.45

Merged Lift charts.

The legends can be converted into floating text by right clicking on them.

PART B: ENTERPRISE DEPLOYMENT

The purpose of any predictive analytics model is to convert data into knowledge and ultimately help us make well-informed decisions. Especially in the insurance industry, these predictive models need to be deployed in real time (or in batches) to score new claims and predict fraud. The following screen shots illustrate a sample interface that shows how the predictive models that we built in this tutorial may be deployed in real time. Using simple SVB scripts, the models that we created can be deployed to the *STATISTICA* Enterprise platform, where these analyses are automated so, for example, a claims adjuster can rapidly score a new claim.

NOTE: Enterprise deployment is *not* available in the 30-day free trial of *STATISTICA* data miner that is bound with this book. However, for those needing an "enterprise system," the speed and simplicity of deployment in this manner can be very useful, so it is illustrated in Figures C.46 through Figure C.51.

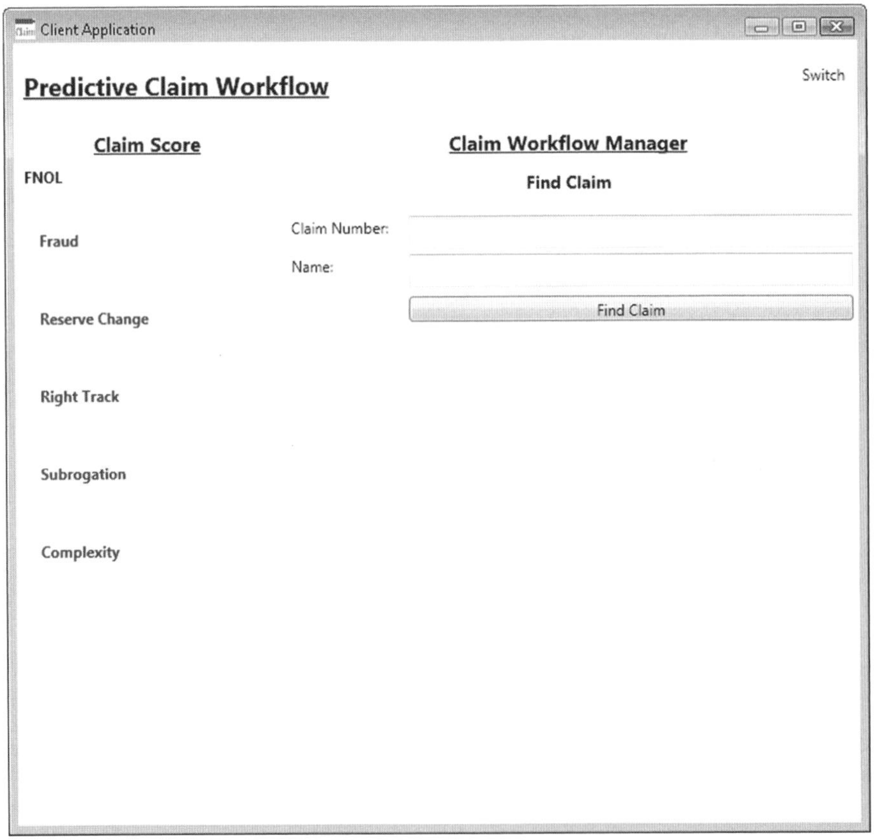

FIGURE C.46

Predictive Claim Workflow. This screen shot illustrates an example user interface that a claims adjuster can use to score new claims to predict fraud.

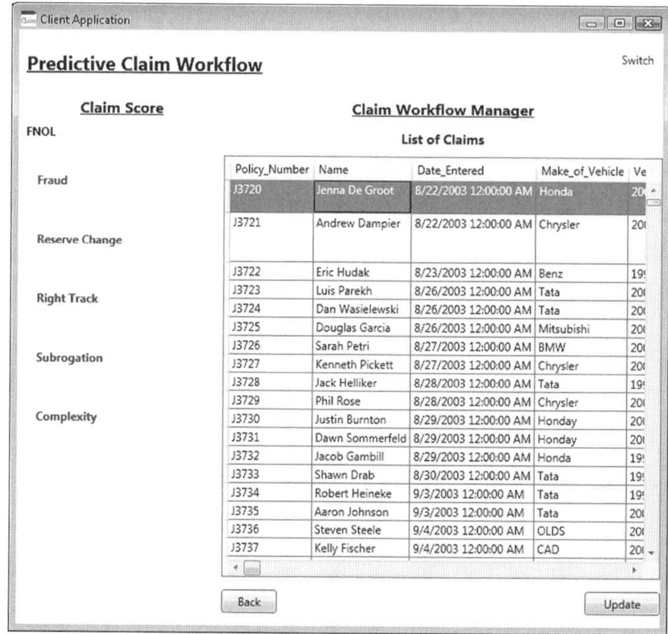

FIGURE C.47

Predictive Claim Workflow with a "list of claims" entered.

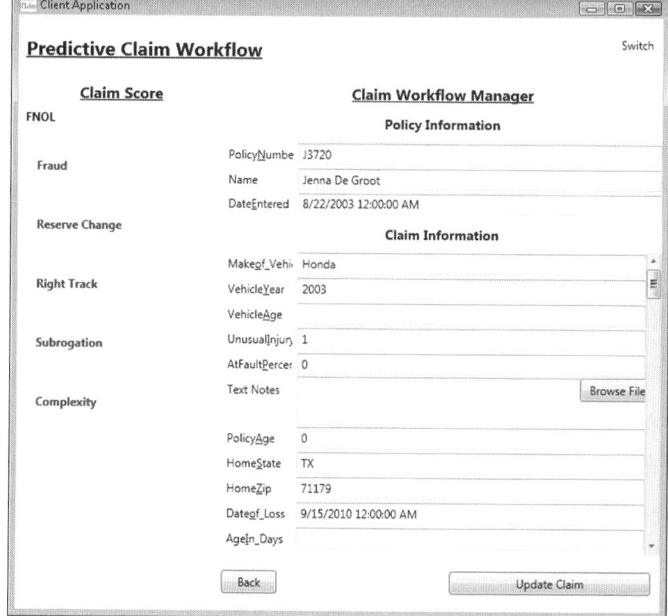

FIGURE C.48

Predictive Claim Workflow with policy and claims data such as policy number, name of claimant, and date entered into the system. NOTE: There are no "text notes" added in this example.

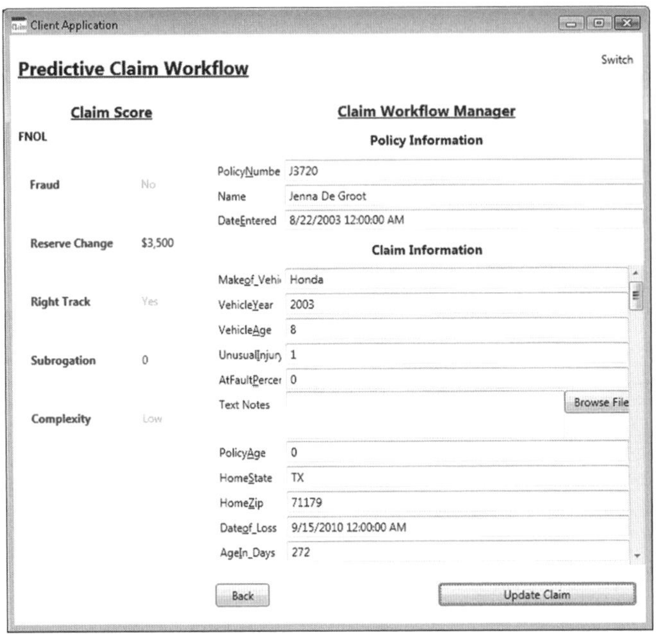

FIGURE C.49

The "Update Claim" button is selected [see lower right of window], and then we can watch as "answers" for Fraud, Reserve Change, Right Track, and so on appear on the left side. "Nonalerting" answers are displayed in green, and an "alert" answer will be displayed in red.

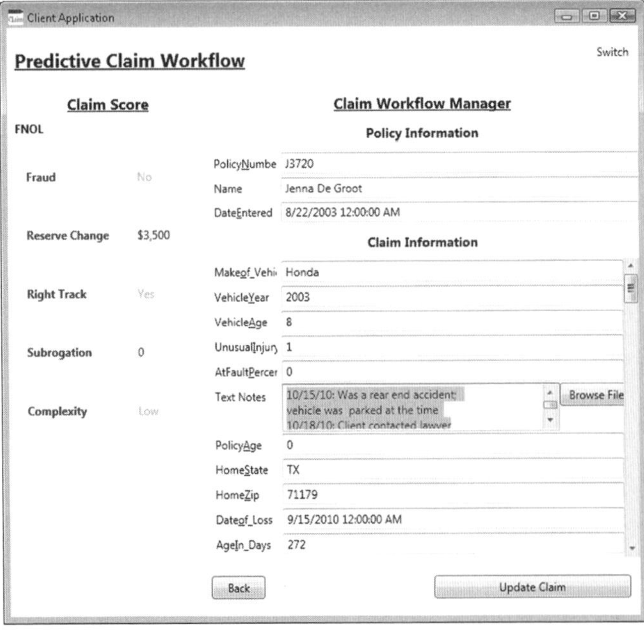

FIGURE C.50

Nothing was displayed in red, indicating no unusual alerts. This means that this could be a very normal claim with nothing suspicious about it. However, note that we have not used any text data yet. Would the models find anything unusual about this claim when we include additional information in the form of text?

We now add the text notes (made by the claims adjuster) for this claim.

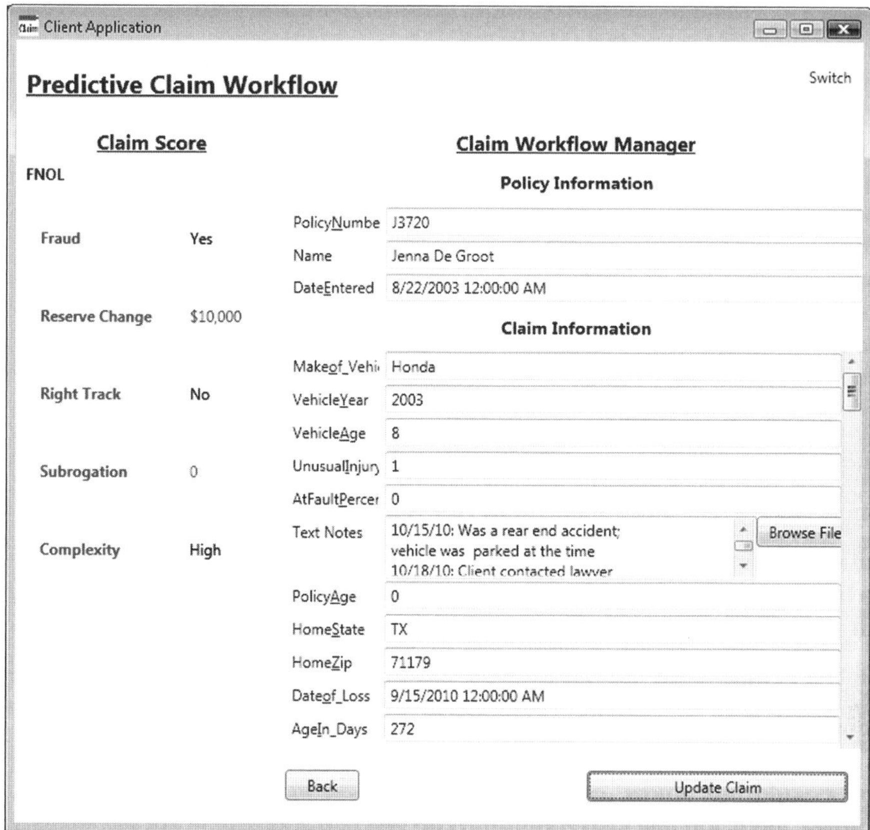

FIGURE C.51

Now, when the Predictive Claim Workflow is rerun, with this text information, we see that FRAUD is highlighted in a red "yes," the "right track" category is given a red "no," and the "complexity" is given a red "high."

SUMMARY

As shown in the first part of this tutorial, we created predictive models (with and without textual data) to predict the probability of fraud for insurance claims, and we saw that the model that included text data along with other numeric data had a significant lift compared to the other models.

Similarly, the real-time deployment example that could not identify a claim of fraud based on structured data alone identified it as a fraudulent claim when we added text notes, thus confirming that using text mining to get information from textual/unstructured data and using it with other structured data is the better way to build robust and powerful predictive models.

TUTORIAL D

Analysis of Survey Data for Establishing the "Best Medical Survey Instrument" Using Text Mining

Jeremy LaMotte, MD
Family Practitioner,
Dadeldhura, Nepal

Ruth Moore, DO
Faculty, In His Image Family Medicine Residency Training Program,
Tulsa, Oklahoma

Sanjay Thomas, MD
Surgery Resident,
Copperston, NY

Chris Jenkins, MD
Faculty, In His Image Family Medicine Residency Training Program,
Tulsa, Oklahoma

Linda A. Miner, PhD
Director of Academic Programs
Southern Nazarene University - Tulsa,
Tulsa, Oklahoma

CONTENTS

Practical Text Mining and Statistical Analysis for Non-structured Text Data Applications. DOI: 10.1016/B978-0-12-386979-1.00012-8

INTRODUCTION

The second-year residents of the In His Image Medical Residency program in Tulsa, Oklahoma, engaged in a research project as part of their requirements. Three of the residents (Drs. LaMotte, Moore, and Thomas) in 2004 and their mentors (Drs. Jenkins and L. Miner) wanted to determine what misconceptions were being taught in the medical schools in Russia. During the Soviet regime, evidence-based medicine was not allowed in the Soviet Republics, and therefore information concerning diseases often was not getting into the medical schools. Human immunodeficiency virus (HIV) was rampant in the countries after the fall of the Soviet Union, and it was presumed that it was due in part to a lack of proper education. The In His Image team wanted to determine what misconceptions about sexual activity were present so the workshops could be designed to eliminate them.

A two-pronged study was initiated. The first stage was open-ended and qualitative and consisted of text mining analyses. The second stage of the project was to use the text mining observations and speculations to formulate hypotheses and to help create closed questions that could be analyzed using more traditional statistical methods. The tutorials in this book involve that qualitative stage of our analysis and include text mining in general and text mining with the "gender" variable as a dependent variable. This tutorial concerns general text mining.

Face-to-face (or sometimes phone) interviews were conducted in which very general questions were asked of physicians and medical students in the former Soviet Republics of Tajikistan, Uzbekistan, Kazakhstan, and Kyrgyzstan. The interviews were conducted by Russian-speaking colleagues who were living in these countries. The answers to the open-ended questions were recorded verbatim, and the responses were translated into English before they were sent to the team. Ninety-six responses were collected from these one-on-one interviews. Not all the questions were asked of each person, so there were incomplete variables for some of the participants. We were not certain whether the questions were always understood, and we certainly had to depend on the translations of the answers. Regardless, this qualitative analysis provided some ideas concerning the various attitudes and conceptualizations about sexual activity and what was being taught in the medical schools. We conducted text mining analysis, first simply on all the answers and then another one using gender as an independent variable. The gender variable was not the gender of the person answering the question; rather the respondent was asked the questions with males in mind or with females in mind. For example, if one were to consider the benefits of being sexually active, there might be a difference in the respondent's mind if the person being active were male versus female. This tutorial concerns the first analysis on the overall comments of the respondents.

THE ANALYSIS

Open the text file, "Russian Study Text File.sta." (See Figure D.1.)

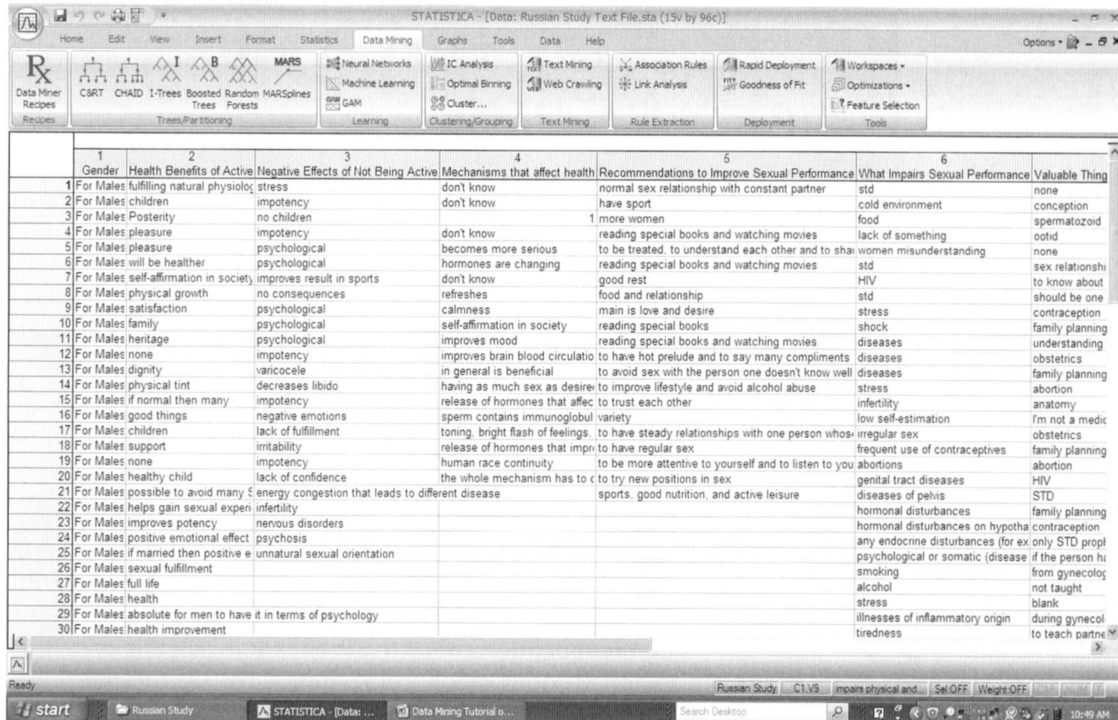

FIGURE D.1
Text File spreadsheet used for analyses in this tutorial, called "Text File.sta."

Go to Data Mining and click on "Text Mining," as in Figure D.2.

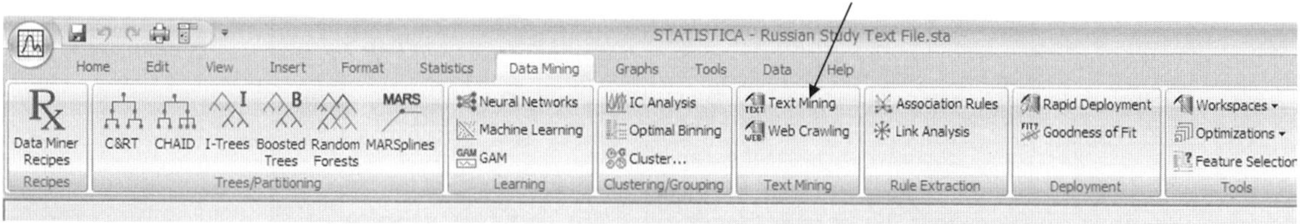

FIGURE D.2
STATISTICA ribbon bar, with "Data Mining" selected (see the large Rx at the very far left) and illustrating where the Text Mining selection is available (see arrow in diagram).

The following will emerge. (See Figure D.3.)

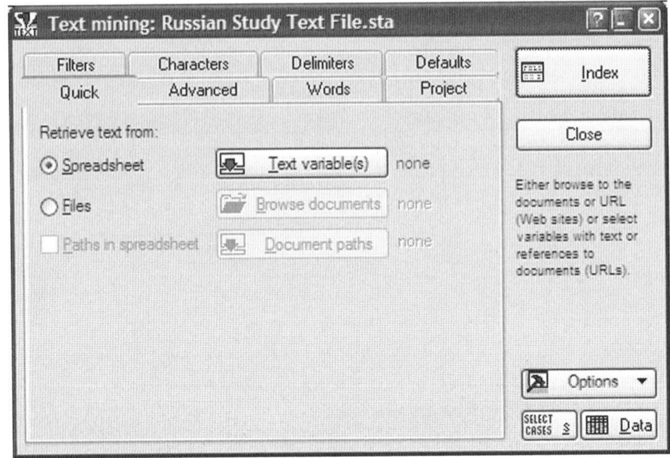

FIGURE D.3
STATISTICA Text Mining main dialog with "Quick" tab selected.

Click on the "Advanced" tab. Then change the percentages from 10 to 80, as in Figure D.4. Words are needed that appear more than just once or twice and less than all the time in order to have any predictive value. Also, set the maximum to a number that seems reasonable. It is possible that 1,000 might be too low a cutoff for words being selected. However, this file is quite small, so the default was left as is. Check the language to make sure it will work. For this study, the words were translated from Russian by translators in Russia. Unfortunately, none of us spoke Russian, or we could have used the original language. We are taking a big chance that meanings are lost by doing so, but that risk was unavoidable.

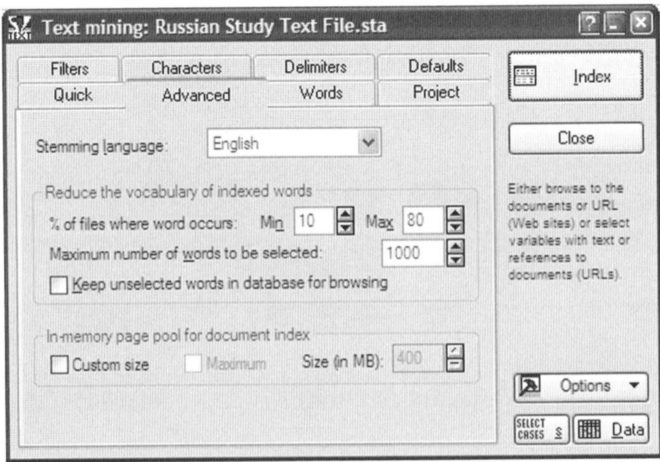

FIGURE D.4
"Advanced" tab of the *STATISTICA* Text Mining windows.

Select the "Words" tab that allows one to filter out words that we do not want the system to identify as important. (See Figure D.5.) There is a default stop file called "EnglishStopList.txt" that filters out very common words. One can also add to the stoplist words that for this particular project would not be meaningful. To save a specialized file, click on the "Save As" button. We did not notice any words in this category. One nice thing about the Russian language is that often articles are not used, leaving only important words.

In our study, however, people often said "STD" (sexually transmitted diseases), or they may have individually identified them as diseases, HIV, and so on. These were placed in the synonyms list. Click on the "Words" tab, as may be seen in Figure D.5, then "Synonyms," and "Edit."

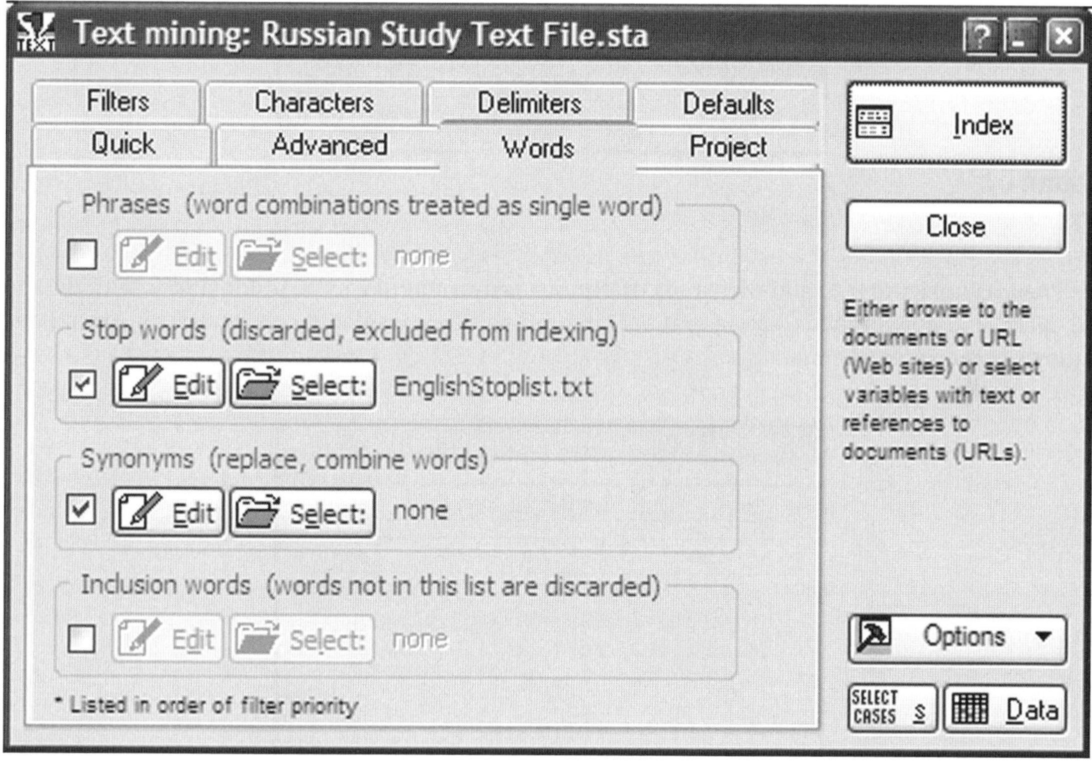

FIGURE D.5
"Words" tab of the *STATISTICA* Text Mining windows.

First, put in the root word *STD*, and click on "Add New Synonym." This puts the word into the root section. Figure D.6 shows what the synonyms window should look like.

FIGURE D.6
"Synonyms" window of *STATISTICA* Text Mining windows.

Then, as in Figure D.7, add the words or phrases you would like to be synonymous with the root, and click "Update synonyms." Note that *STD* and *std* will be considered one by default. Separate the synonyms by using a comma.

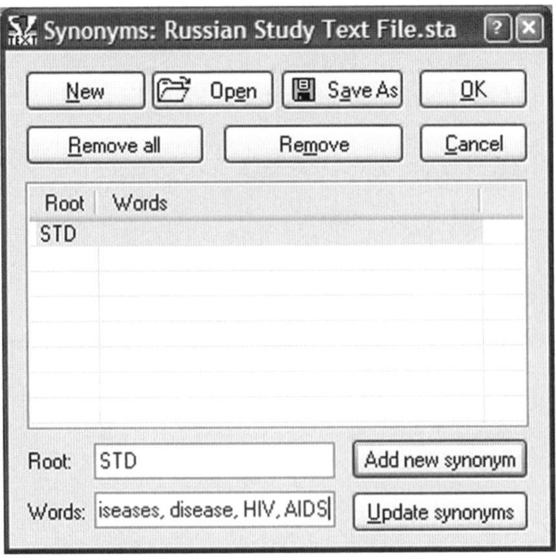

FIGURE D.7
"Synonyms" window with "Words" subwindow updated with the words selected (e.g., typed in) to be synonymous with the root word *STD*.

Update the synonyms, and they appear in the words column. The program will now consider all of those words as one concept. (See Figure D.8 for updating the synonyms.)

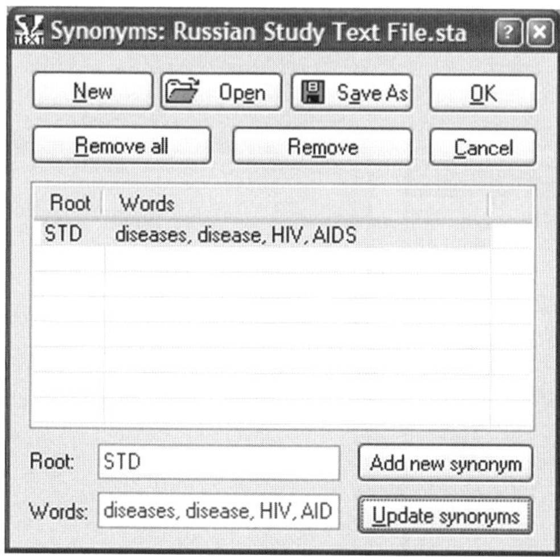

FIGURE D.8
"Synonyms" window with the "Update Synonyms" button having been selected, thus putting these words under "Words" in the main window, just to the right of the "Root" word list.

There can only be one word in the Root column, but many words in the Words column as may be seen in Figure D.9.

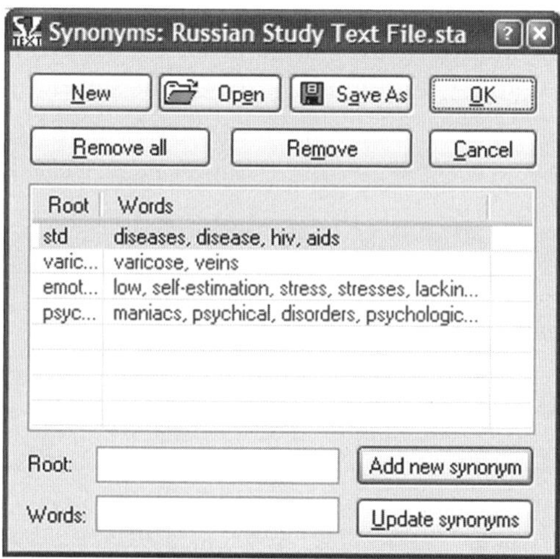

FIGURE D.9
"Synonyms" window illustrating how several "root words" each have several synonym words selected.

We saved this file as the synonyms list in the "Central Asian Study" files. If we liked, we could also select under "Words, Phrases," which would allow us to tell the program to consider certain combinations of words as one word.

We then selected variables to try variables 2–9, and clicked "Index." Under "Text Variables," put in 2–9. (See Figure D.10, consisting of two boxes side by side below.)

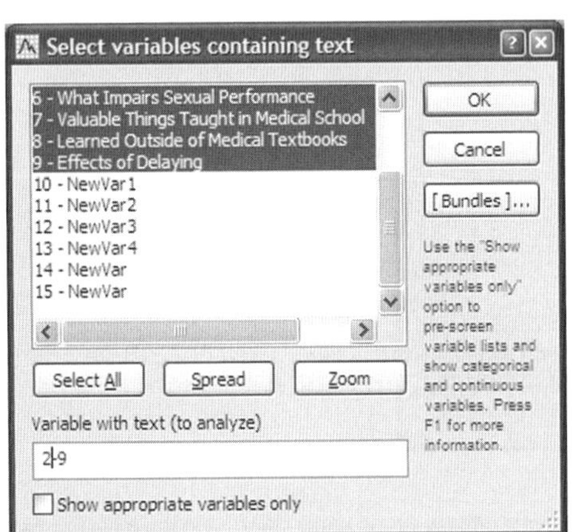

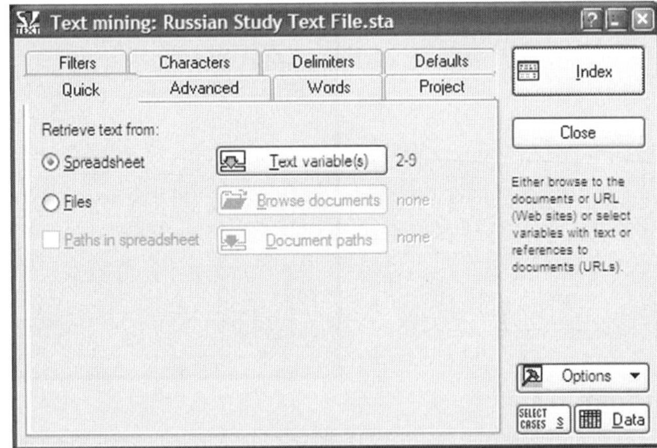

FIGURE D.10

The "Text Variables" button in the "Primary Text Mining Window" was selected, bringing up the "Select Variables containing text" window, where variables 2–9 were selected. Then the "Index" button (upper right-hand corner) of the "Primary Text Mining Window" is selected to perform the "text mining indexing or word counts."

Then click on "Index." Automatically, the program deletes the old file (see Figure D.11 message) and inserts a new one. If the old file is important, it should be saved.

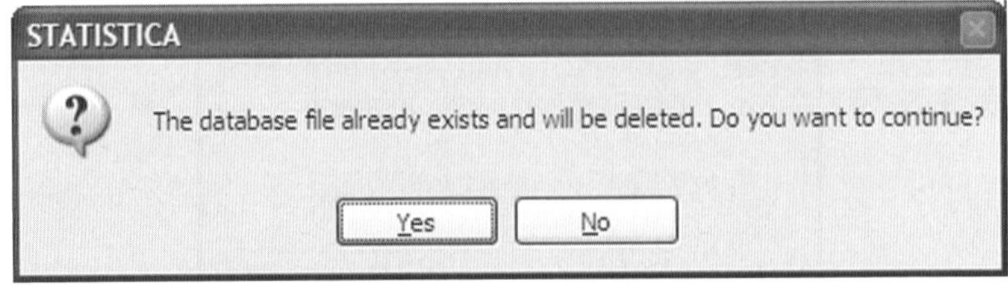

FIGURE D.11

Warning message that appears after "Index button" is selected. Normally one would select "yes," since the data file in use will be used to "add additional variables" that are created in the text mining operations.

The following list emerges. (Three screen shots were necessary for Figure D.12 to show all of the words.)

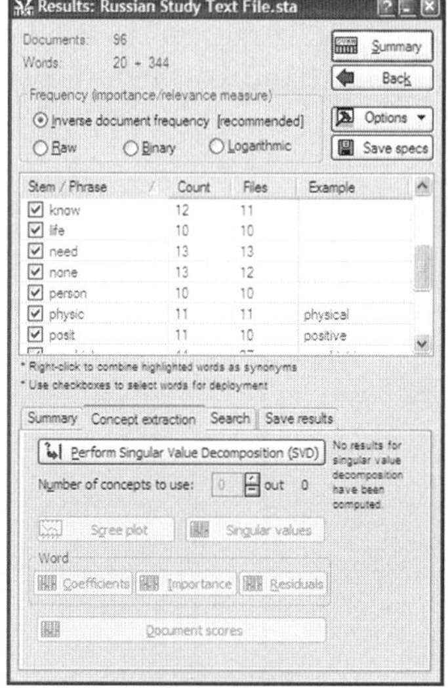

FIGURE D.12

The "Results Dialog" from the *STATISTICA* Text Mining operations. Twenty words were selected, so three screen shots are needed to illustrate all of the words in the windows.

Figure D.13 (two boxes) shows how to deal with two words that should go together. "Don't know" showed up as two concepts: "don" and "know." These two could be combined because they were basically one idea. Highlight each of them, right click, and select "Combine."

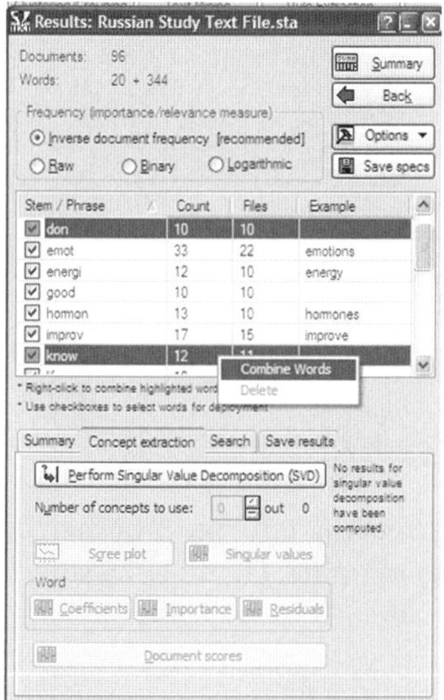

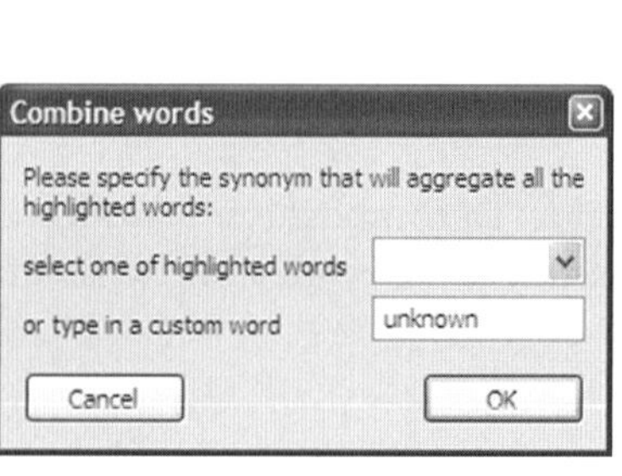

FIGURE D.13

Highlight "don" and "know" to combine as "one idea" or "word." A "Combine Words" dialog will appear where one can select one of these two words as the "new core word," or one can type in another word as the "new core concept."

Then the dialog in Figure D.14, allows one to select a new name for the concept. Write in "unknown," which now appears in the list instead of "don" and "know." One other word, "will," does not seem to be of any particular value, so we can unselect it and remove it from further analysis by unchecking it. Now we can click on "Concept Extraction" to see the singular value decomposition. The two boxes in Figure D.14 demonstrate what happens when the word "will" is unchecked.

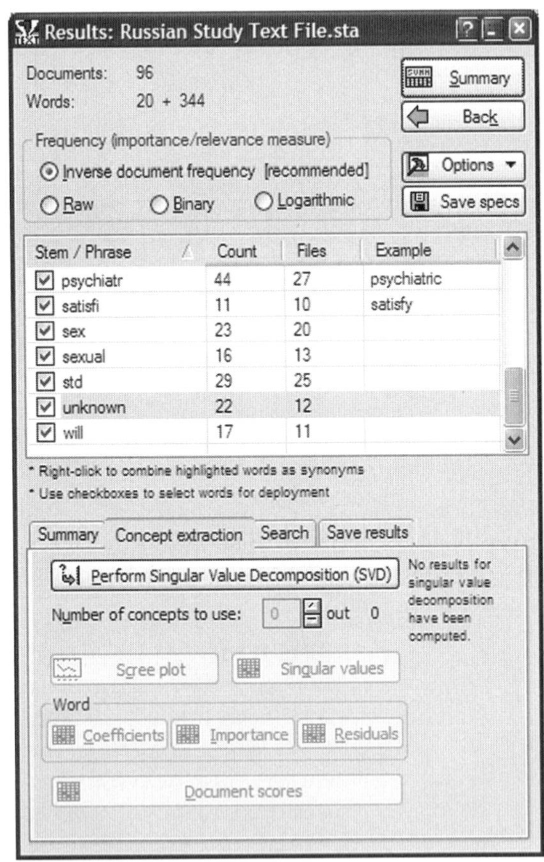

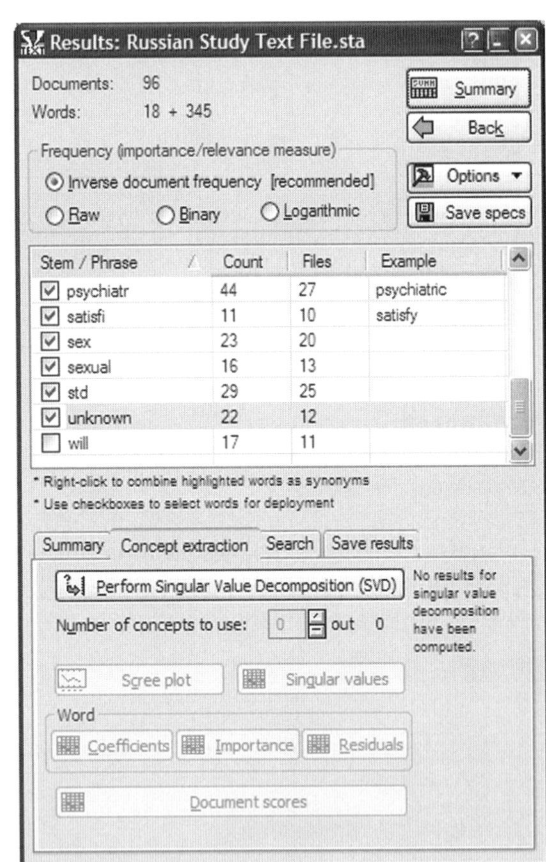

FIGURE D.14
"Concept Extraction" tab highlighted and the "inverse document frequency" radio button selected. The "Perform Singular Value Decomposition" button is then selected to create the "concepts."

Click on "Perform Singular Value Decomposition." Under the selection tab, the number of concepts by default are 18 out of 18. Select "Scree plot" as the first tab to examine. Figure D.15 shows the scree plot.

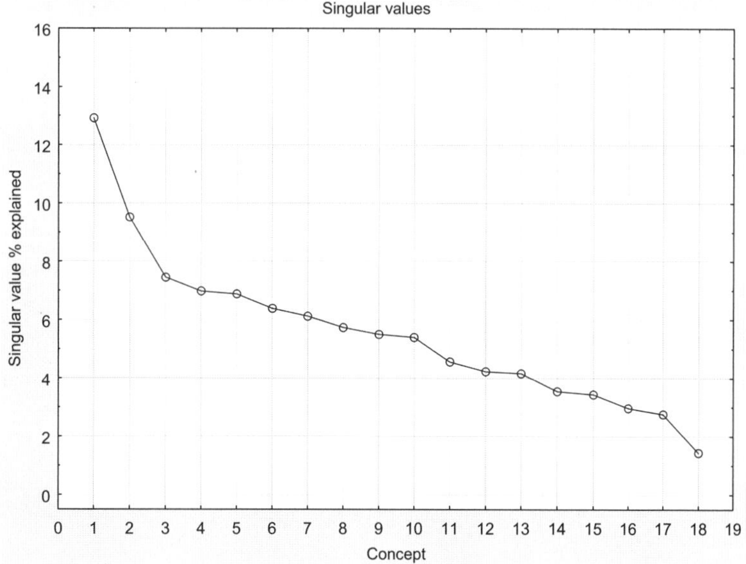

FIGURE D.15
"Scree plot," which is analogous to a "Scree plot in Factor Analysis," is presented when the "Scree plot" button is selected.

Certainly, the first three concepts should be viewed—and perhaps the first six or seven. We should then go back and change the number of concepts to another number. We'll try six. From those six, create a coefficients spreadsheet by clicking on "Coefficients." These can be examined by eyeballing them or by creating scatterplots that show words that are highly related as may be seen in Figure D.16.

	SVD word coefficients (Russian Study Text File.sta)					
	Concept 1	Concept 2	Concept 3	Concept 4	Concept 5	Concept 6
disturb	0.015620	0.024034	0.035015	0.033382	-0.000471	-0.007583
emot	0.017814	-0.007442	-0.041440	0.027405	-0.021569	-0.030834
energi	0.014874	0.014866	-0.043141	0.002007	0.000470	0.030599
good	0.008032	-0.013542	-0.015182	0.012135	0.003950	-0.008357
hormon	0.014609	0.022615	0.043213	0.016556	-0.010608	0.031548
improv	0.010502	-0.000636	0.027764	-0.022975	-0.008354	0.011968
life	0.009991	0.009078	-0.029410	-0.017675	0.001216	0.036198
need	0.011658	0.015194	-0.000130	0.002164	0.051041	-0.028488
none	0.006709	-0.018547	0.009437	-0.002012	0.022648	0.033999
person	0.009845	-0.010419	0.004119	-0.026839	-0.004192	-0.050004
physic	0.009854	0.005172	0.009861	0.006974	0.009859	0.014257
posit	0.015086	0.007038	0.006034	0.002096	-0.044800	-0.020839
psychiatr	0.019345	0.004228	-0.017131	-0.035609	-0.000229	0.042226
satisfi	0.012668	0.019363	-0.011342	0.016631	0.054021	-0.028271
sex	0.016788	-0.004869	0.007856	-0.012819	-0.036922	-0.017700
sexual	0.007648	-0.005328	0.017168	-0.066643	0.026620	-0.019620
std	0.016753	-0.014820	0.000367	-0.006453	-0.009333	-0.012083
unknown	0.015345	-0.054296	0.015713	0.032740	0.022751	0.019067

FIGURE D.16
A coefficients spreadsheet is created by clicking on "Coefficients."

Use the output as an input file so a scatterplot can be created. Right click on the name of the file on the left-hand side, and then click "Use as Active Input." Figure D.17 demonstrates where one can click to use as an active file.

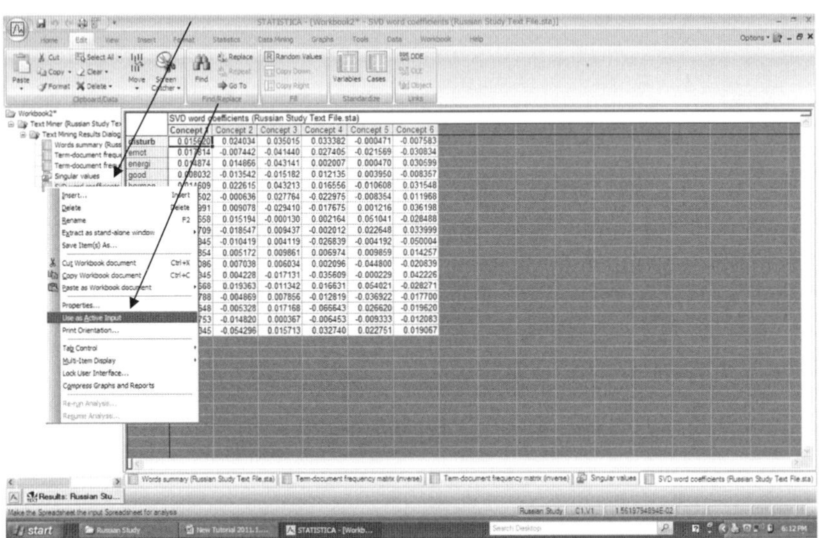

FIGURE D.17

Use the "SVD Word Coefficients" as an "Input File." Right click on the file name, and then select "Use as Active Input."

Select graphs and scatterplot as shown in Figure D.18.

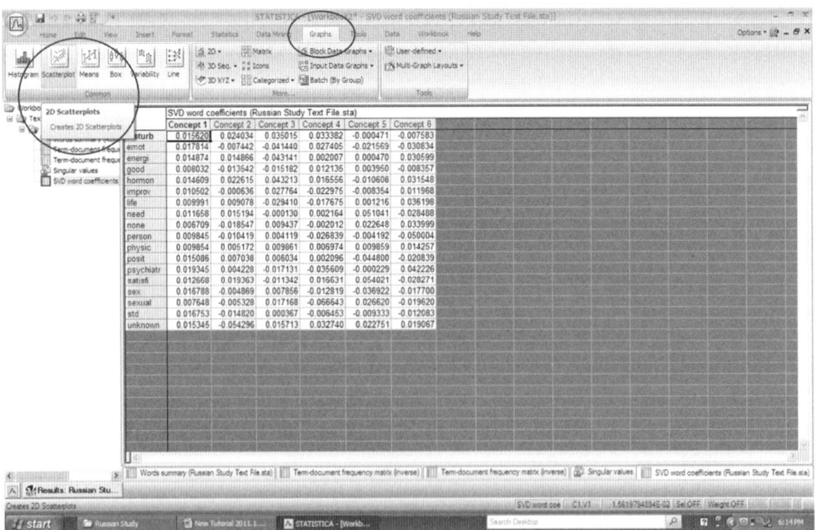

FIGURE D.18

Selecting "Graphs" and "Scatterplot" from the *STATISTICA* ribbon bar menu.

When the scatterplot dialog opens, select Concept 1 and Concept 2 for the graph. Figure D.19 shows the Variables button.

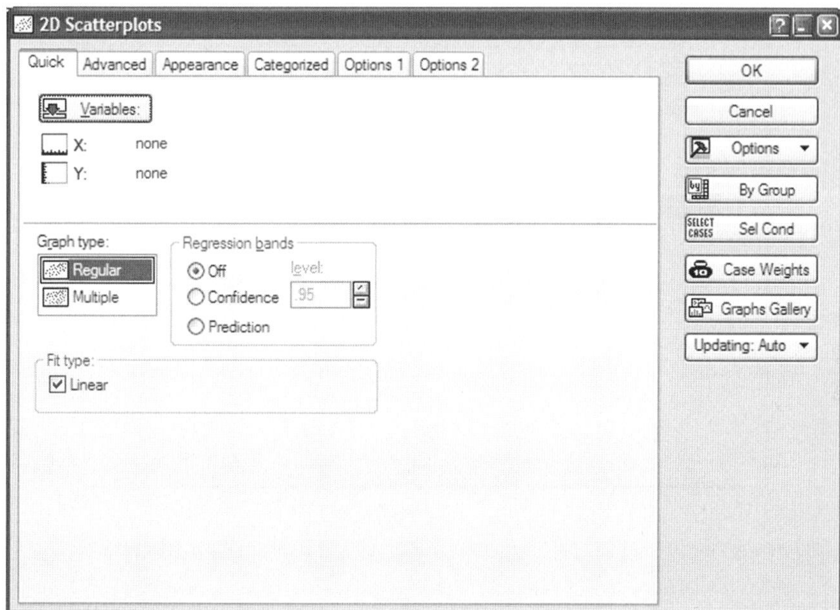

FIGURE D.19
2D scatterplot graphing window. Here the "Variables" button can be selected.

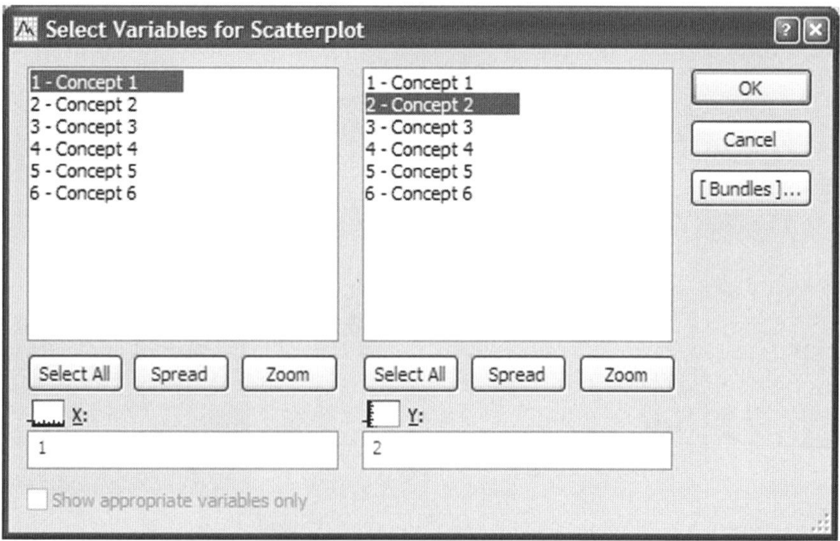

FIGURE D.20
Figure D.20 shows the "Select Variables" window that opens when "Variables" button is selected. Here Concept 1 was selected in the left-hand window, and Concept 2 was selected in the right-hand window.

Click OK and then Figure D.21 will emerge.

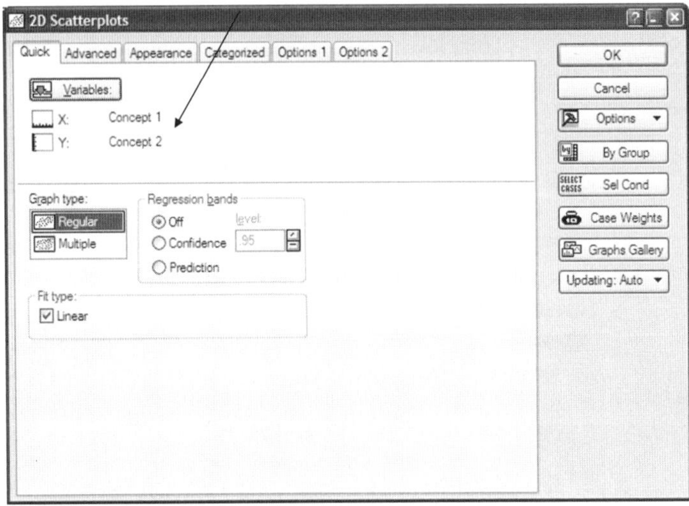

FIGURE D.21
2D scatterplot graphing window, illustrating that now the Concept 1 and Concept 2 variables are selected.

Then click okay to create the scatterplot. Click on the brushing button to pull up the brushing 2D dialog. Figure D.22 shows the resulting scatterplot and where one can click to activate the brushing. Figure D.23 shows the words that then appear when one togles "label" and apply.

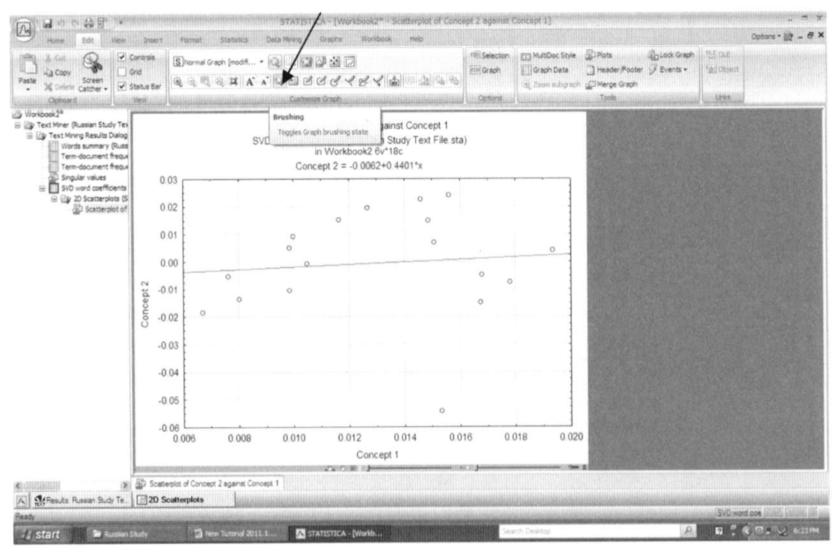

FIGURE D.22
Scatterplot created in statistica workbook window. Then click on the "Brushing" button (see arrow) to get the 2D brushing dialog.

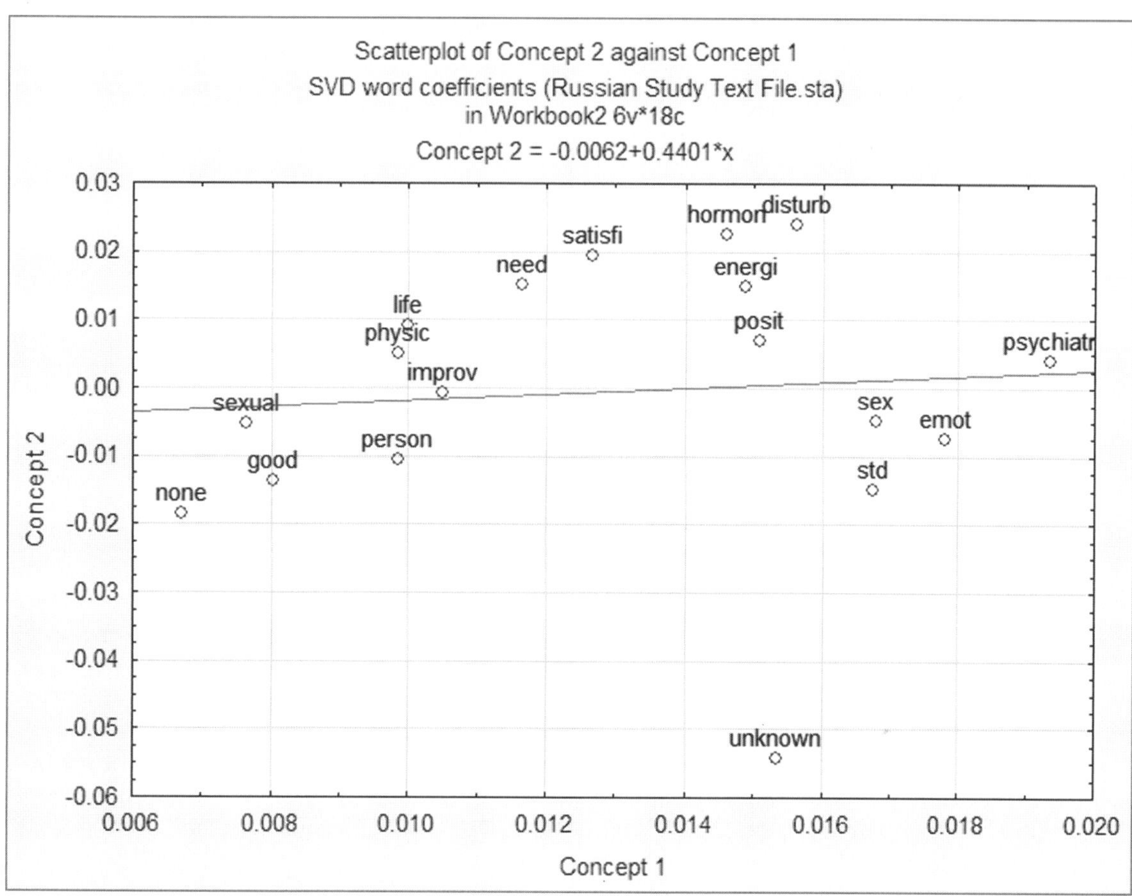

FIGURE D.23

To see the words, click "Toggle Label" and then "Apply."

There could be some interesting concepts that are related in the minds of the respondents. For example, words that are connected with sexually transmitted diseases are associated with sex, emotions, and psychiatric illnesses. Another cluster looks at hormones, energy, disturbances, and positiveness. Need and satisfaction seem to go together, as do improving life, feeling good, and no problems with having sex outside of marriage. These are all themes that could be analyzed by writing questions that elicit feelings concerning these concepts and their relationships.

Next we examined "Word Importance" by going back to the analysis and clicking on "Importance." The following in Figure D.24 emerged.

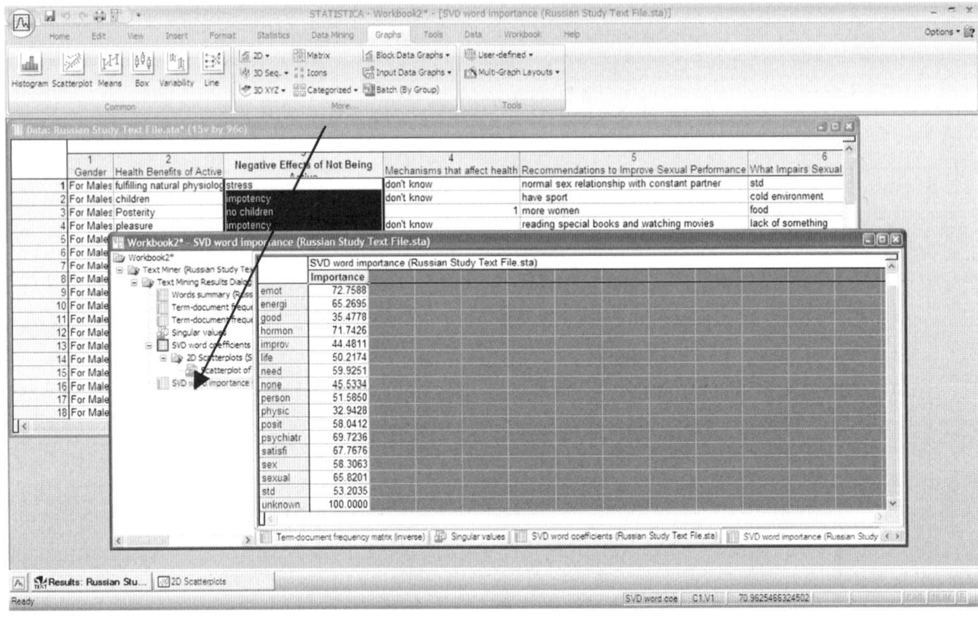

FIGURE D.24
The "Word Importance" table can be obtained by clicking on "Importance" in the table of contents of the workbook.

We ordered the words by importance, descending as in Figure D.25 to see what might be learned. With the "Word Importance" file highlighted, go to "Data" and then "Sort." Add the variable "Importance" by highlighting it and "Add Var(s)." Select "Descending" as the order, and then click OK. Figure D.26 appeared.

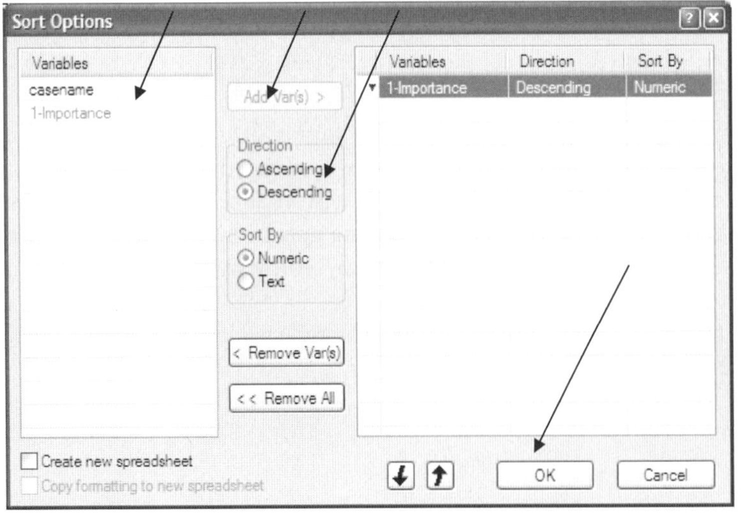

FIGURE D.25
STATISTICA "Sort Options" window, where one can add a variable and then select the order of sorting.

	SVD word importance (Russian Study Text File.sta)
	Importance
unknown	100.0000
emot	72.7588
hormon	71.7426
disturb	70.9625
psychiatr	69.7236
satisfi	67.7676
sexual	65.8201
energi	65.2695
need	59.9251
sex	58.3063
posit	58.0412
std	53.2035
person	51.5850
life	50.2174
none	45.5334
improv	44.4811
good	35.4778
physic	32.9428

FIGURE D.26
The sorted file, sorted by "importance" or the word, can then be inspected for any additional possible meanings.

SUMMARY

"Not known" was the largest response from the SVD Word Importance ordering. This was not surprising because one of the major hypotheses for doing the study in the first place was that while the Soviet Union was in power, there was little information going to that area. Evidence-based medicine was not allowed to penetrate the Iron Curtain. There was a belief, for example, that if a man did not engage in regular sexual activity, he would become impotent. Or if a woman sat on a cold surface, she could become sterile. One of the objectives of the mission trip to the new Russia was to share what was known from research literature concerning sexual activity and the dangers of engaging in unprotected sex.

Emotional issues, disturbances of various kinds, and psychiatric problems were high on the SVD Word Importance list as well. However, it should be noted that one of the "old wives' tales" that had been taught in the medical schools was that unless men had sex on a regular basis, married or not, they would also evidence psychiatric illnesses. We were interested in examining the context of those words to derive the meanings.

The next thing we did with the qualitative part of the study was text mining in a data mining workspace, using gender as the independent variable. We wondered if there were discernible differences when the respondent thought about men versus about women. That analysis is the subject of a separate tutorial.

We decided it would be important to form closed-ended questions to verify if these erroneous "facts" were still believed. The subsequent study, not detailed in this book, did just that and was aided by the initial open-ended, textual, qualitative aspect of the research project.

TUTORIAL E

Analysis of Survey Data for Establishing "Best Medical Survey Instrument" Using Text Mining: Central Asian (Russian Language) Study Tutorial 2: Potential for Constructing Instruments That Have Increased Validity*

Jeremy LaMotte, MD
Family Practitioner,
Dadeldhura, Nepal

Ruth Moore, DO
Faculty, In His Image Family Medical Residency Training Program,
Tulsa, Oklahoma

Sanjay Thomas, MD
Surgery Resident,
Copperston, NY

Chris Jenkins, MD
Faculty, In His Image Family Medical Residency Training Program,
Tulsa, Oklahoma

Linda A. Miner, PhD
Director of Academic Programs,
Southern Nazarene University, Tulsa, Oklahoma

CONTENTS

*Using Data Miner Workspace format in *STATISTICA* (one of three formats for doing predictive analytics in *STATISTICA*).

Practical Text Mining and Statistical Analysis for Non-structured Text Data Applications. DOI: 10.1016/B978-0-12-386979-1.00013-X

INTRODUCTION

This is the second text mining example of the Central Asia (Russian Language) Study data. In this study we wanted to see what meanings would arise when we looked at the words between "for males" and "for females" in the "gender" variable while interviewers asked the open-ended questions. There were 36 cases of "for females" and 39 cases of "for males."

As stated in the first Central Asian Tutorial, one-on-one open-ended interviews were conducted in the Russian language, translated into English, and sent to the research team for analysis. The first tutorial concerned a general text analysis. This analysis involved forming sub–data sets around the "gender" variable. Again, the respondent was asked to answer the questions with either "for a male" or "for a female" in mind. We wanted to see if there were gender-specific hypotheses that could be generated so we could determine closed-ended questions that would be analyzed more in a quantitative manner.

THE ANALYSIS

Open the data files for the Central Asian Russian Language study titled "Central Asia Study Males.sta" and "Central Asia Study Females.sta." Next, click on "Data Mining" and then on "Workspace—All Procedures." Figure E.1 shows this file and the location of the Data Mining tab.

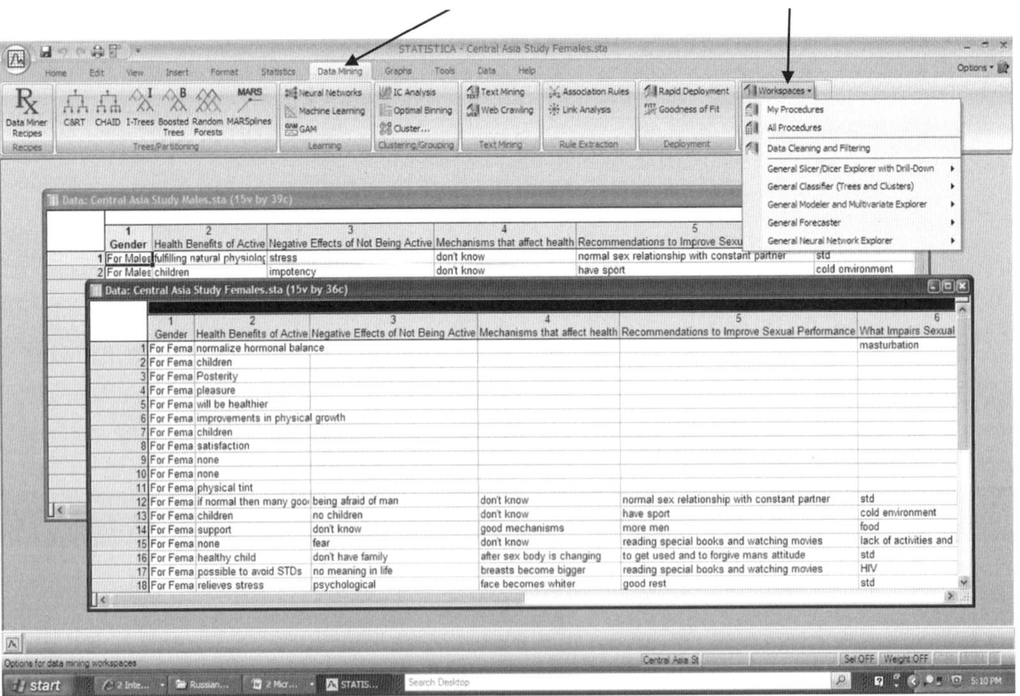

FIGURE E.1

Text file spreadsheet used for analyses in this tutorial called "Central Asia Study Females.sta" and "Central Asia Study Males.sta."

Acquire the data source and select the files that were just opened, one at a time. First open the females file as shown in Figure E.2.

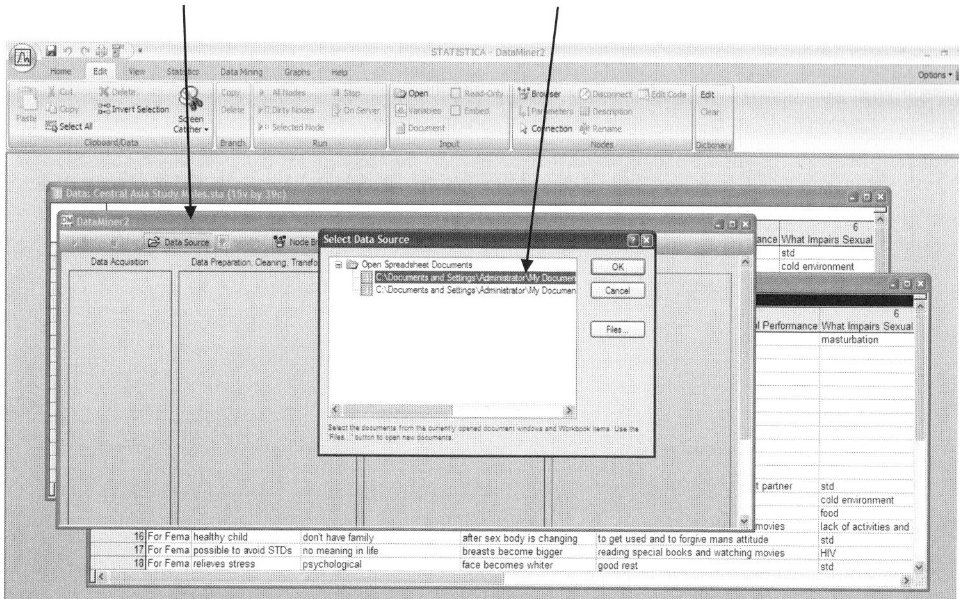

FIGURE E.2

Opening files in data miner workspace.

Next, select the variables of interest from that file. Click on "Variables" and then select 2–9. (See Figures E.3 and E.4.)

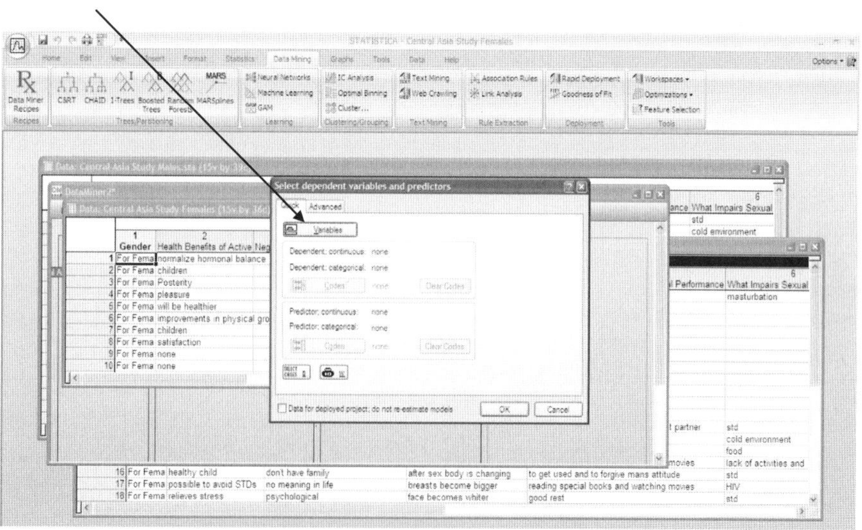

FIGURE E.3

Selecting variables for the data file in the data miner workspace for a specific data file.

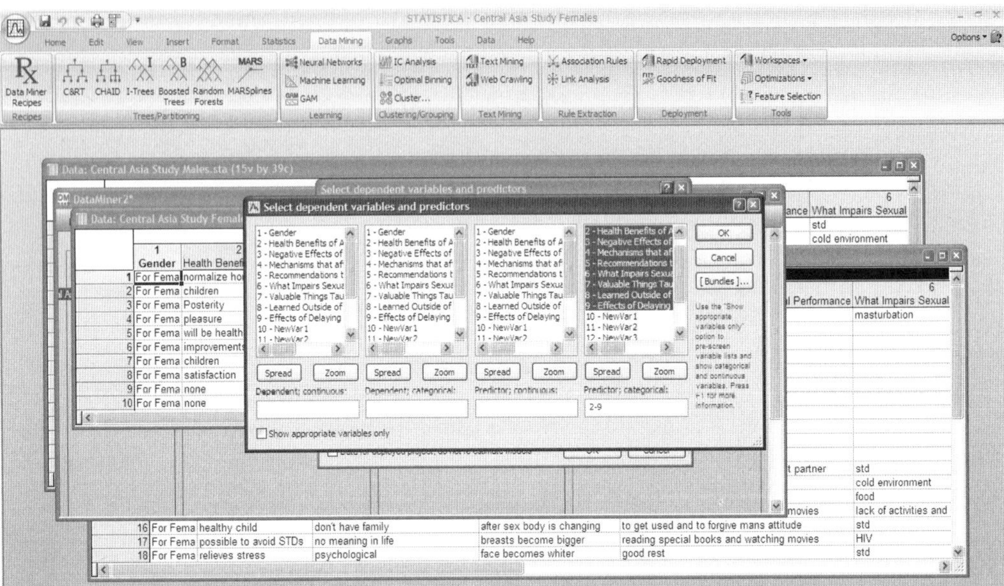

FIGURE E.4
Selecting variables in the data miner workspace for a specific data file.

Click OK and then OK again. Next, bring in the males. Repeat the process. The workspace should look like the following in Figure E.5. Go to the node browser.

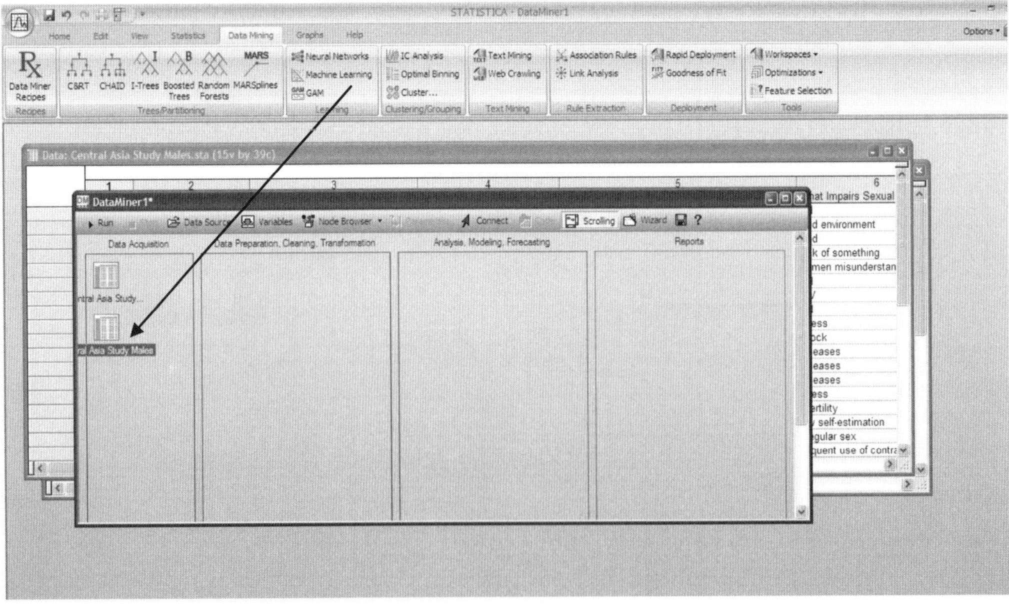

FIGURE E.5

Click on the Node Browser and then on "Text Mining," and then double click "TextMiner" to insert the node into the workspace, or click "insert into workspace" instead of double clicking. See Figure E.6 for location of text miner. Figure E.7 shows the text miner node placed into the workspace.

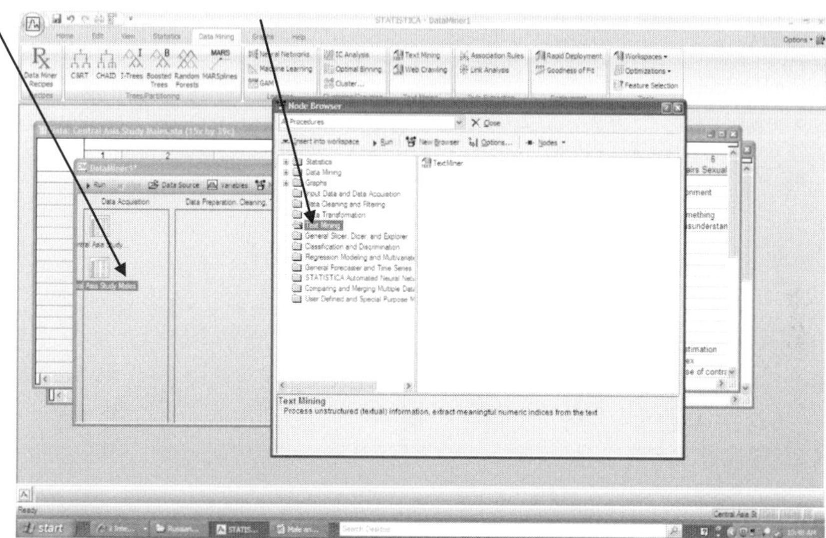

FIGURE E.6
Selecting the text miner within the node browser.

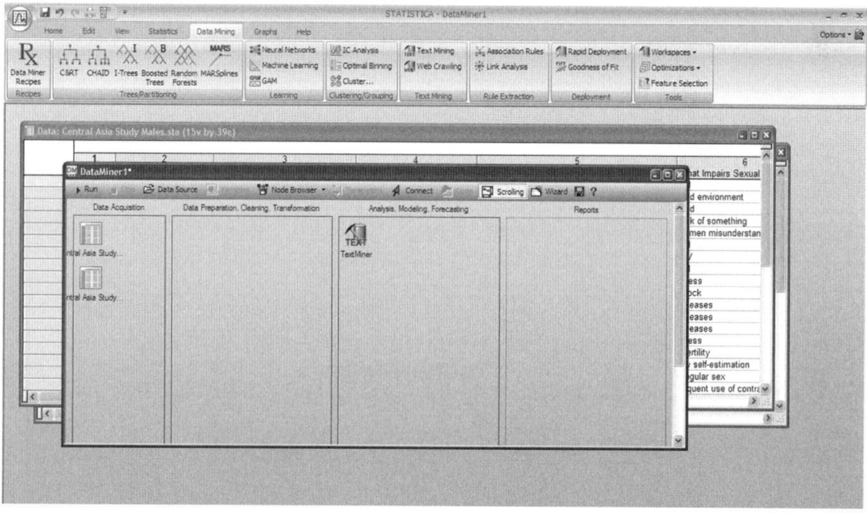

FIGURE E.7
Result of inserting the text miner into the workspace.

Make a connection from the male data to the DataMiner node; click it on the data set, and drag it to the node. Release it so the connection forms. The horizontal arrow in Figure E.8 shows the connection.

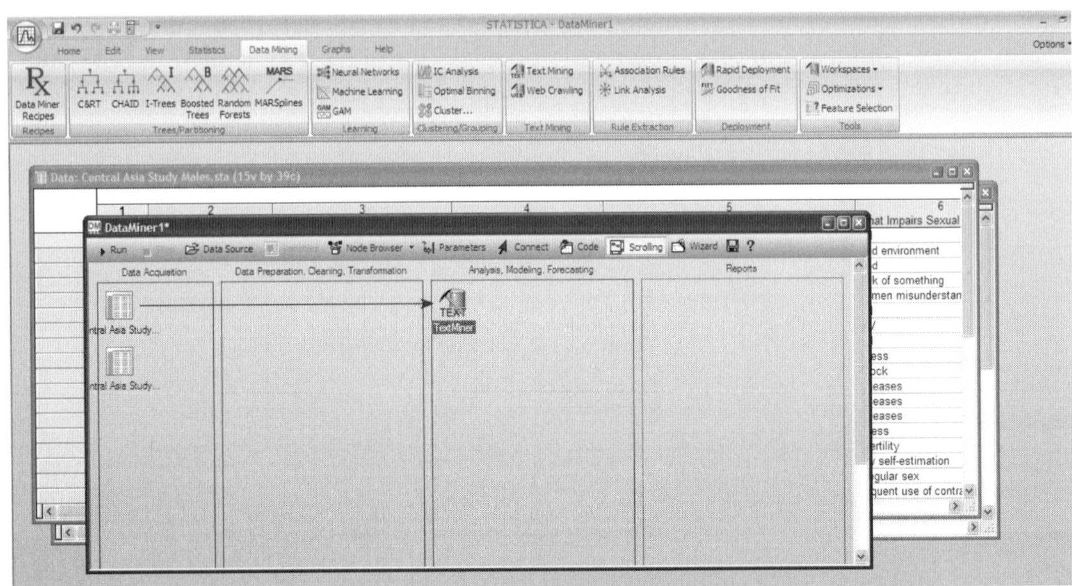

FIGURE E.8
Connecting data to the text miner.

Now we are ready to edit the parameters. Right click on the "DataMiner node," and click on "Edit Parameters." Figure E.9 shows the location of Edit the parameters.

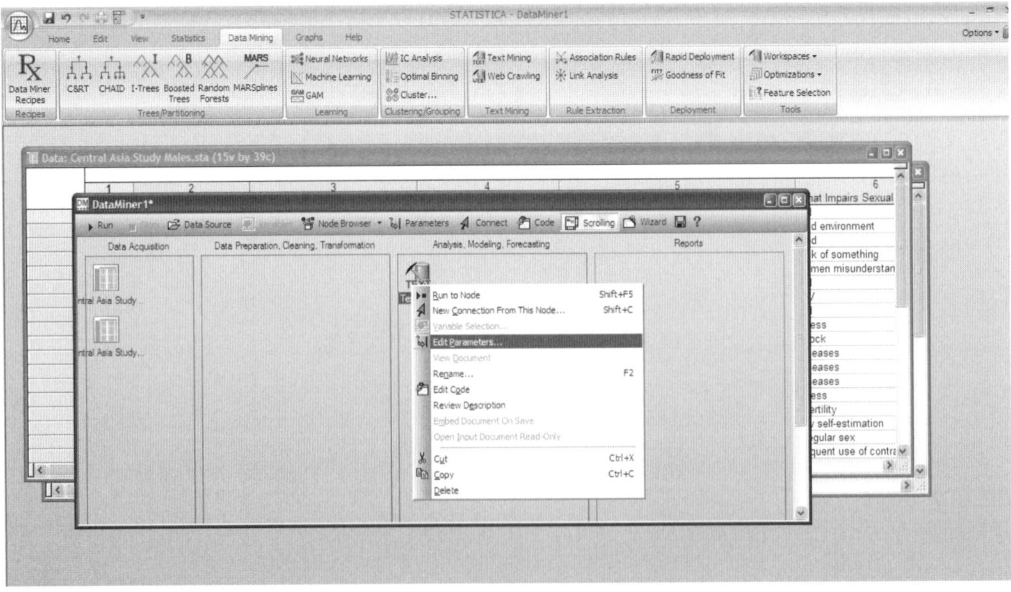

FIGURE E.9
Editing the parameters.

Note in Figure E.10 that the Stop file is included by default.

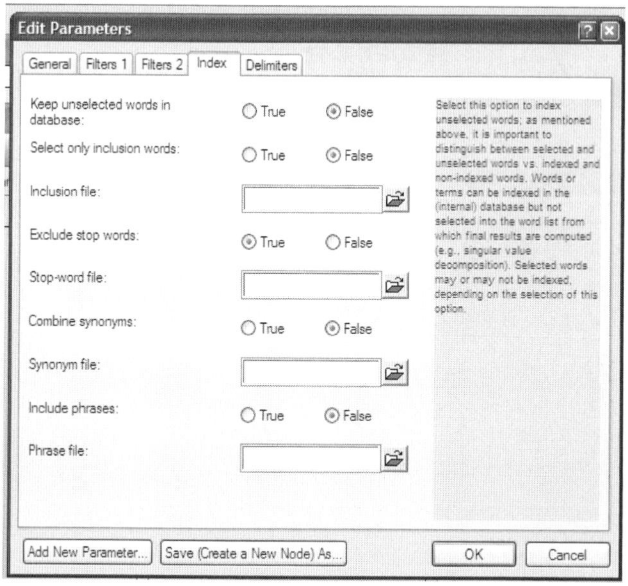

Edit Parameters

General | Filters 1 | Filters 2 | Index | Delimiters

Keep unselected words in database: True False

Select only inclusion words: True False

Inclusion file:

Exclude stop words: True False

Stop-word file:

Combine synonyms: True False

Synonym file:

Include phrases: True False

Phrase file:

Select this option to index unselected words; as mentioned above, it is important to distinguish between selected and unselected words vs. indexed and non-indexed words. Words or terms can be indexed in the (internal) database but not selected into the word list from which final results are computed (e.g., singular value decomposition). Selected words may or may not be indexed, depending on the selection of this option.

Add New Parameter... | Save (Create a New Node) As... | OK | Cancel

FIGURE E.10
The Stop file is automatically selected.

We want the synonym file that we created for the last tutorial. Click on the "Open File" button as in Figure E.11 and select the "Synonyms.txt" file. Click the circle for True, as well.

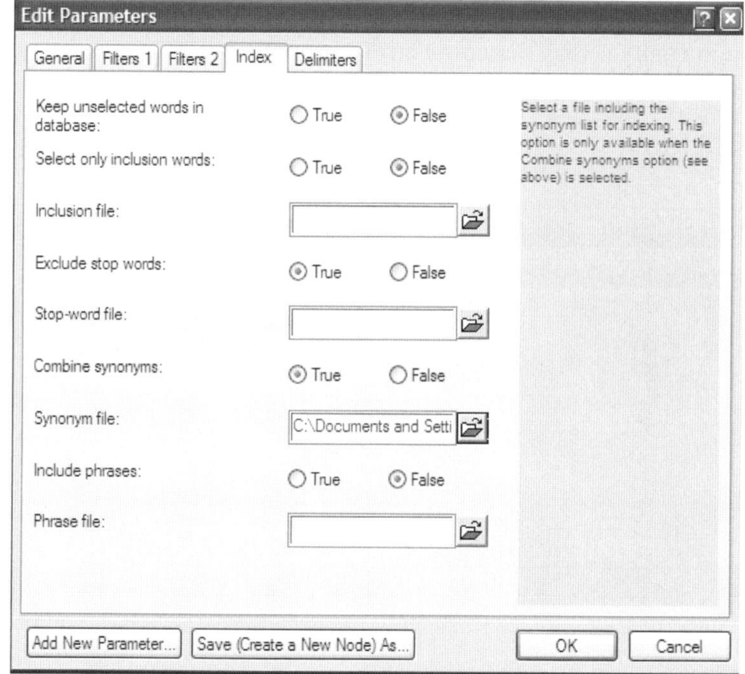

FIGURE E.11
Selecting the "synonyms.txt" file.

Go back to the "General" tab in Figure E.12, and click on "All Results" in "Detail of computed results." Now click OK.

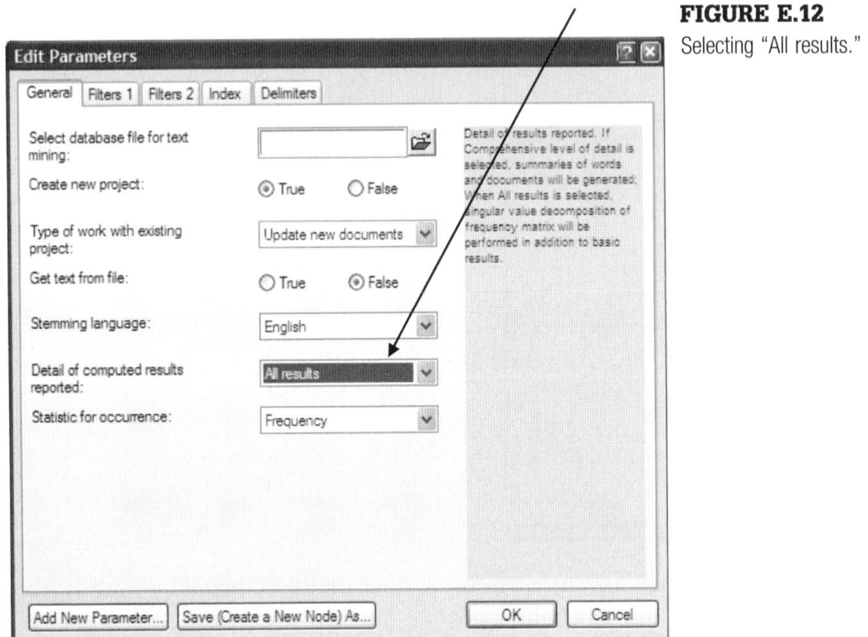

FIGURE E.12
Selecting "All results."

Right click on the "TextMiner node," copy, and then paste into the workspace so there is a node for the female databases as may be seen in Figure E.13. Connect an arrow from the data to the node.

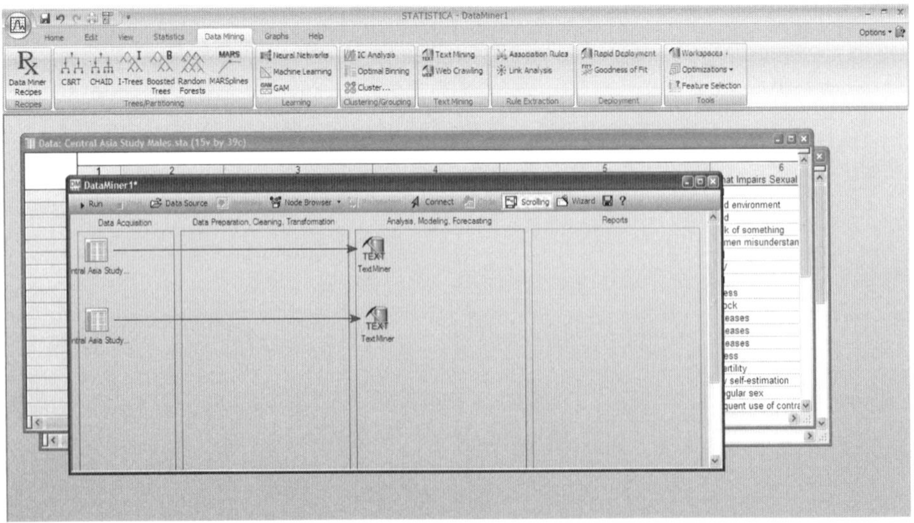

FIGURE E.13
Copy the first TextMiner node for the second database.

Right click on each "TextMiner" and "Run to" node. Right click on each outcome, and check "Embed Document on Save." (Please see Figure E.14.)

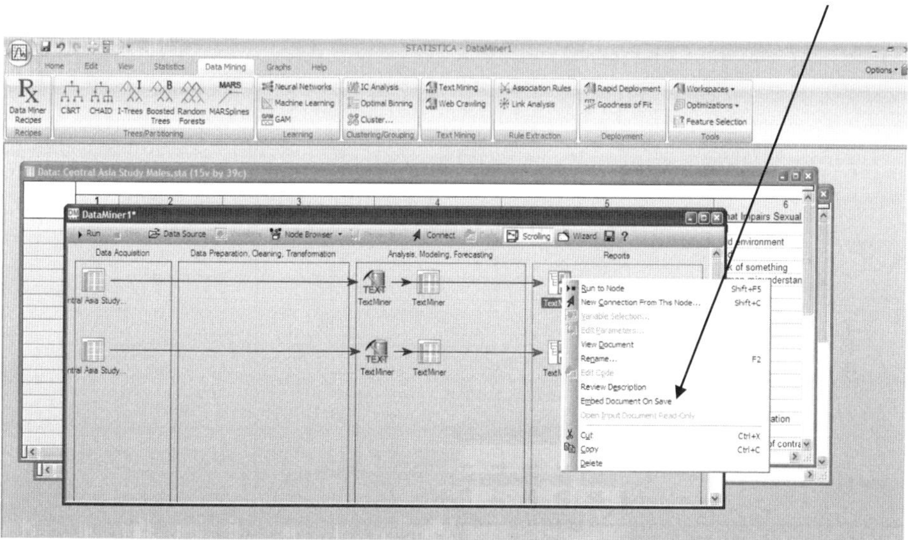

FIGURE E.14
Embedding the document on save.

Double click on the output file, open up the outputs, and examine each of the outcomes generated. For graphs that we would like to compare, we changed the titles to show which group the graph concerned. To do this, simply double click on the title of the graph as in Figure E.15, and the title box will emerge.

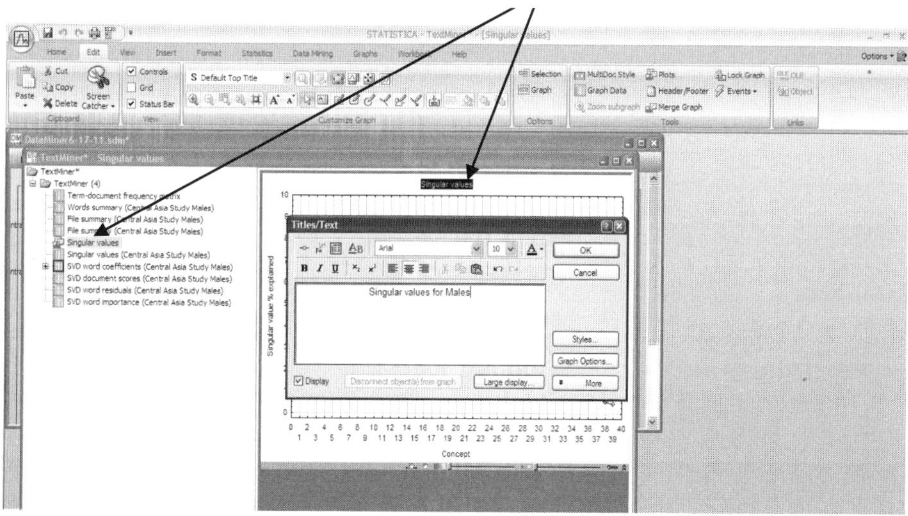

FIGURE E.15
Changing the title of the Scree plot of singular values for the males by double clicking the title on the graph.

The resulting Scree plot in Figure E.16 for the males showed that the first two variables probably were the most important.

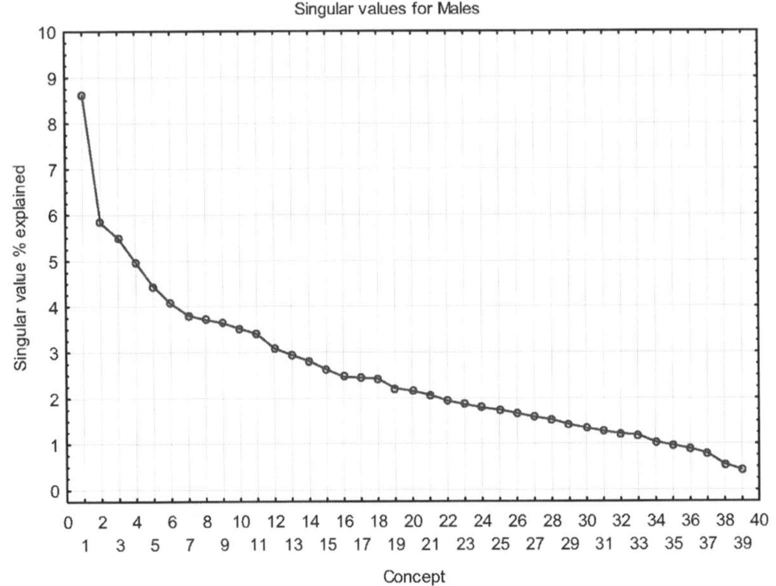

FIGURE E.16
Singular values plot "for males."

Figure E.17 shows the singular values for the females when we repeated the process for the "for females" data file.

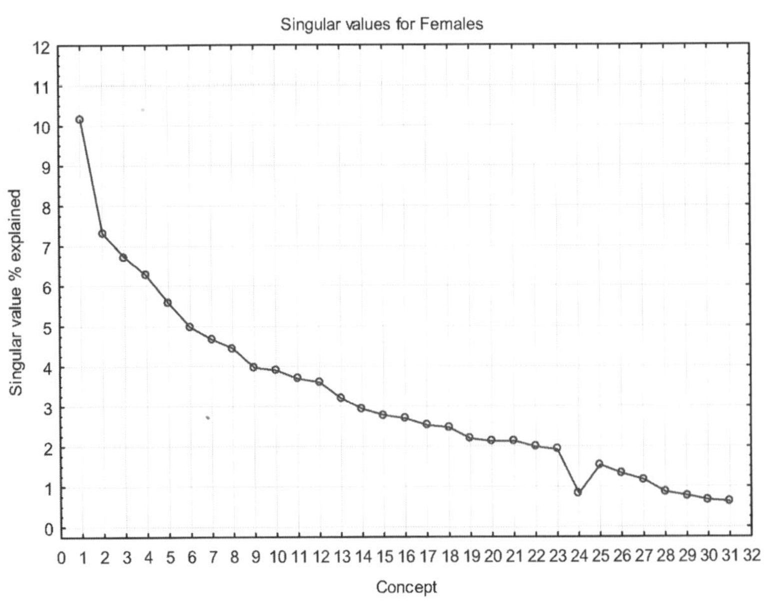

FIGURE E.17
Singular values for the females data file. This is the Scree plot that helps to determine the number of important (predictive) concepts.

Again, it appears that the first two concepts may be important. Later, we will create a scatterplot to examine when we compare the female and the male outcomes.

Next, continuing with the male outcomes, we examined the important words and ordered them, using the data-sort function seen in Figure E.18, descending from most important to least important.

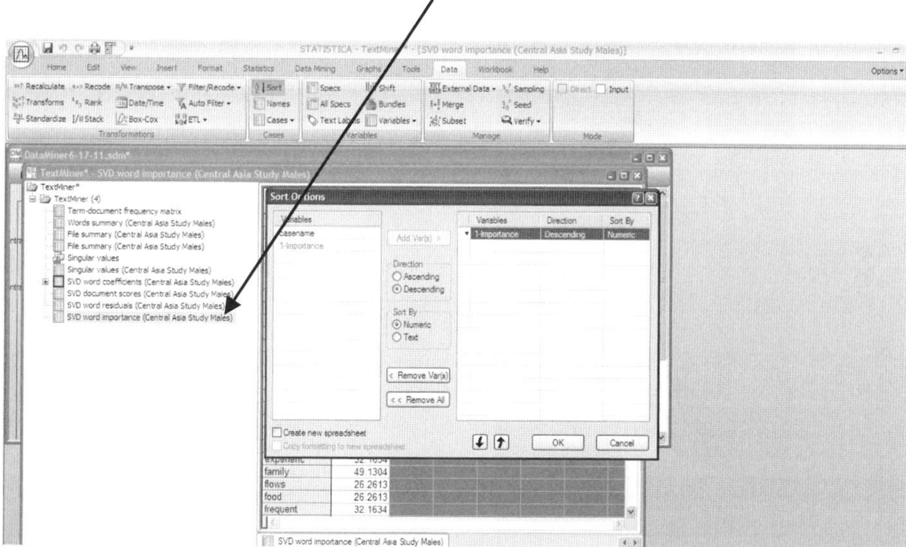

FIGURE E.18

Opening the Word Importance file and sorting the words in descending order.

The resulting list in Figure E.19 showed these as the most important words for the males.

	SVD word importance (Central Asia Study Males)
	Importance
std	100 0000
psychiatric	100 0000
emotion	94 6864
sex	78 7839
wil	74 2781
sexual	61 5882
non	61 5882
need	49 1304
disturbanc	49 1304
family	49 1304
relationship	45 4859
impotency	41 5227
physical	41 5227
book	41 5227
peopl	41 5227
abortion	41 5227
congestion	41 5227
satisfy	41 5227
lead	37 1391
person	37 1391
reading	37 1391
special	37 1391
health	37 1391
childr	37 1391
good	37 1391

FIGURE E.19

Most important to least important ordering of terms for the "for males" data file.

Repeating the process for the "for females" data file yielded the following list in Figure E.20 of important words.

	SVD word importance (Central Asia Study Females)
	Importance
psychiatric	100.0000
unknown	72.2735
disturbanc	63.9565
wil	58.3872
emotion	58.3756
different	47.7086
std	47.5311
sex	45.2077
satisfy	42.6239
level	42.5459
childr	40.5595
hormonal	39.9793
need	36.9466
energy	36.9255
healthy	36.9193
good	36.8968
sexual	36.8407
improv	33.7611
lead	33.6890
physical	32.9842
non	32.7486
becom	30.2441
lif	30.2165
person	26.3908
avoid	26.3225
neurosis	26.1763
book	26.1492
reading	26.1492
special	26.1492

FIGURE E.20

Words ordered by importance for the "for females" data file.

We repeated the sequence with the males to reveal the following outputs. Let's put the two lists together. See Figure E.21 for the two lists.

Females

	SVD word im
	Importance
psychiatric	100.0000
unknown	72.2735
disturbanc	63.9565
wil	58.3872
emotion	58.3756
different	47.7086
std	47.5311
sex	45.2077
satisfy	42.6239
level	42.5459
childr	40.5595
hormonal	39.9793
need	36.9466
energy	36.9255
healthy	36.9193
good	36.8968
sexual	36.8407
improv	33.7611
lead	33.6890
physical	32.9842
non	32.7486
becom	30.2441
lif	30.2165
person	26.3908
avoid	26.3225
neurosis	26.1763
book	26.1492
reading	26.1492
special	26.1492

Males

	SVD word imp
	Importance
std	100.0000
psychiatric	100.0000
emotion	94.6864
sex	78.7839
wil	74.2781
sexual	61.5882
non	61.5882
need	49.1304
disturbanc	49.1304
family	49.1304
relationship	45.4859
impotency	41.5227
physical	41.5227
book	41.5227
peopl	41.5227
abortion	41.5227
congestion	41.5227
satisfy	41.5227
lead	37.1391
person	37.1391
reading	37.1391
special	37.1391
health	37.1391
childr	37.1391
good	37.1391
lif	37.1391
avoid	37.1391
planning	37.1391
energy	37.1391

FIGURE E.21

Examining the important words lists together.

It is interesting to note some of the differences. For example, unknown is second on the female list, but it does not appear in the top group for the males list. Psychiatric is high on the list for both groups and coincides with the idea that if people do not have regular sex, married or not, those people will become mentally unstable, which was one of the "old wives' tales" purported to being passed down in medical schools in the Central Asian countries. STDs are a concern for both groups but higher on the list for the males. Children show up on the female list but not the male list. By looking side by side, we can see what kinds of questions might be asked to differentiate between males and females.

Next, we wanted to see what kinds of scatterplots could be generated and examine those side by side. First, set the male SVD word coefficients as an active file by clicking on "Use as Active Input." (See Figure E.22.)

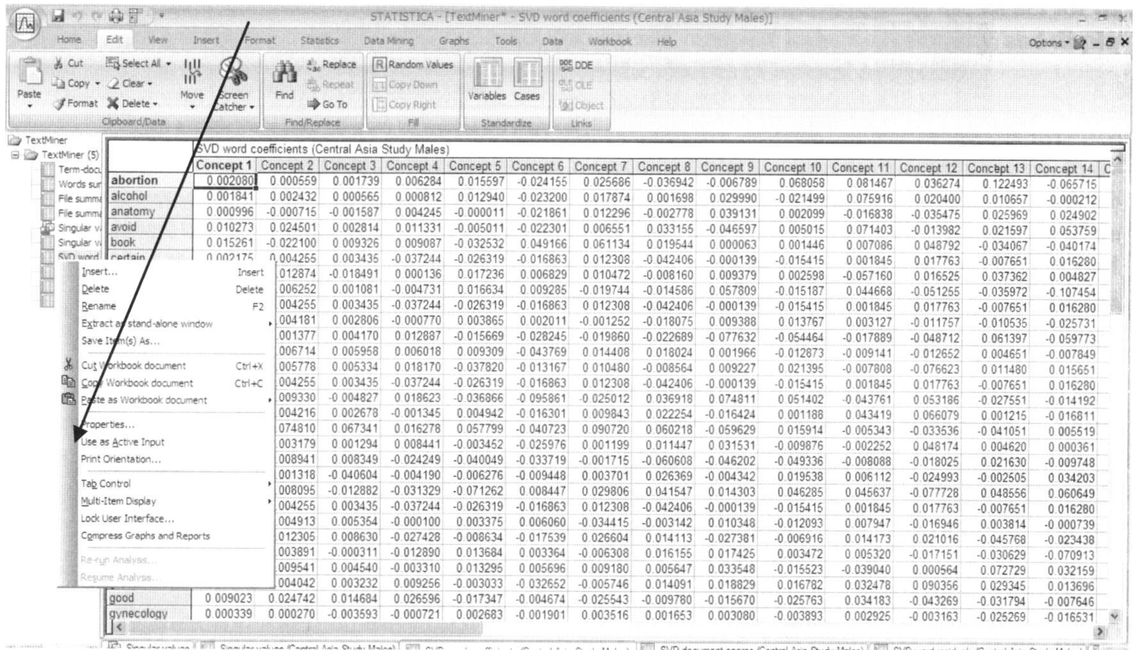

FIGURE E.22
Choosing the SVD word coefficients file as the active input file for the males.

Then, form a scatterplot using the first two concepts (remember that the Scree plots indicated probably no more than two concepts were important). Go to Graphs, 2D, and then Scatterplots as in Figure E.23.

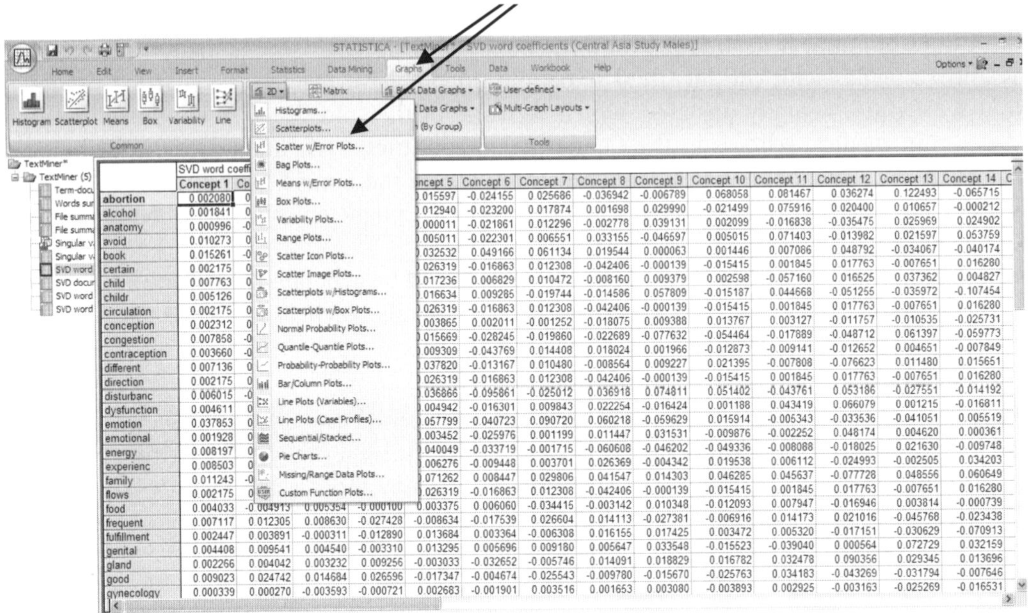

FIGURE E.23

Finding the scatterplots for SVD word coefficients one and two.

Click on the variables tab to select Concepts 1 and 2. Select the variables under the "Quick" tab as may be seen in Figure E.24.

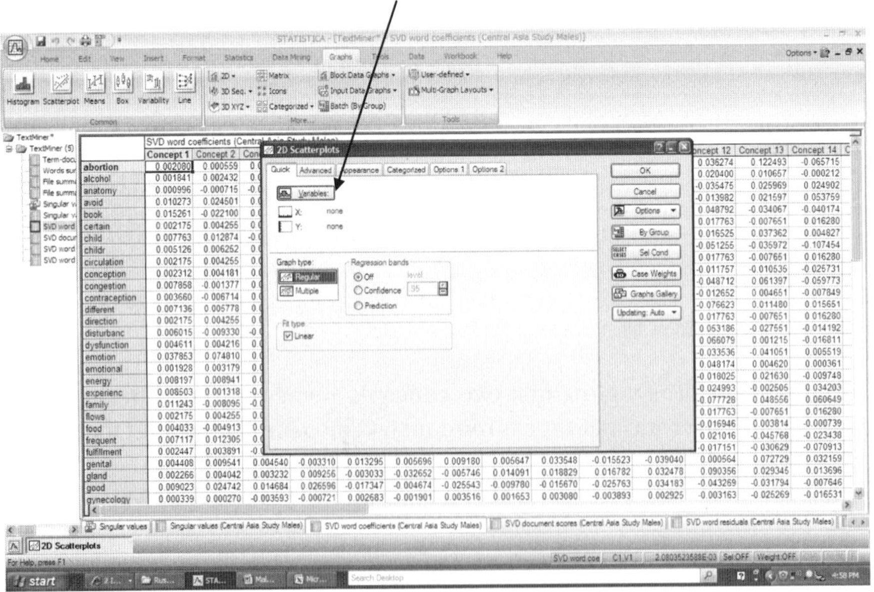

FIGURE E.24

Select Concepts 1 and 2 under the "Variables" tab may be seen in Figure E.25.

Select concept variables 1 and 2. Click OK.

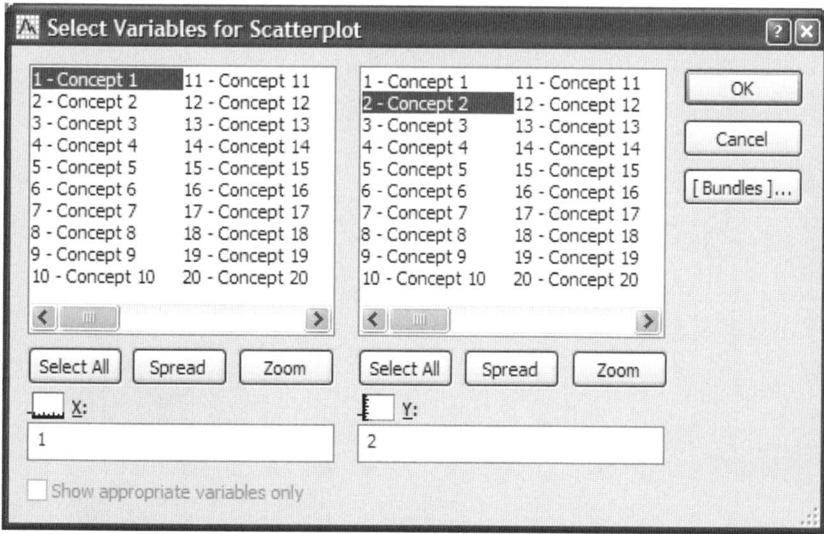

FIGURE E.25
Select Concept 1 on the left and Concept 2 on the right.

Click OK again. Now we can use the brushing tool to look at the words that are highly related to one another, and we can examine the outliers. (The brushing tool is shown in Figure E.26.)

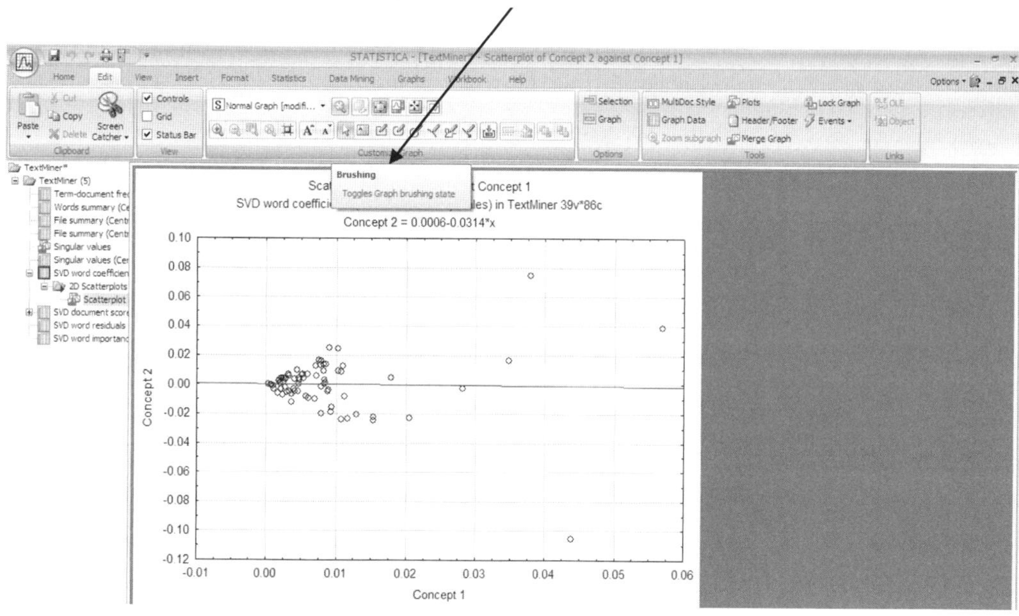

FIGURE E.26
Finding the brushing tool.

Use the brushing tool to click on each circle, and then click "Apply." Figure E.27 shows how words are formed after clicking on each circle and then apply.

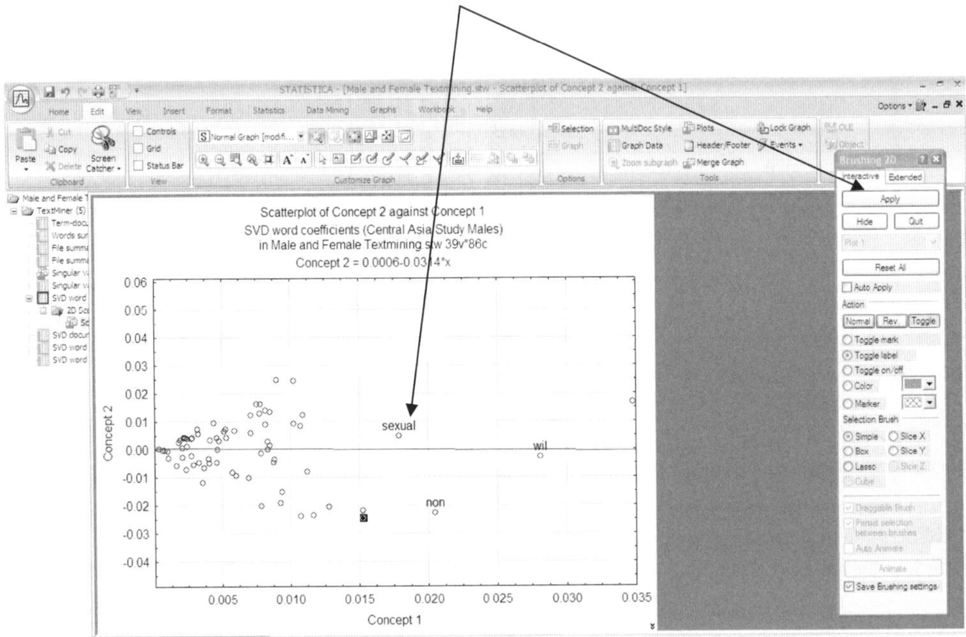

FIGURE E.27
Using the brushing tool.

Or click "Auto Apply" to make the labels work each time you click the brushing tool. Also use the enlarging tool (zoom in) to help break up the clump. (See Figure E.28.)

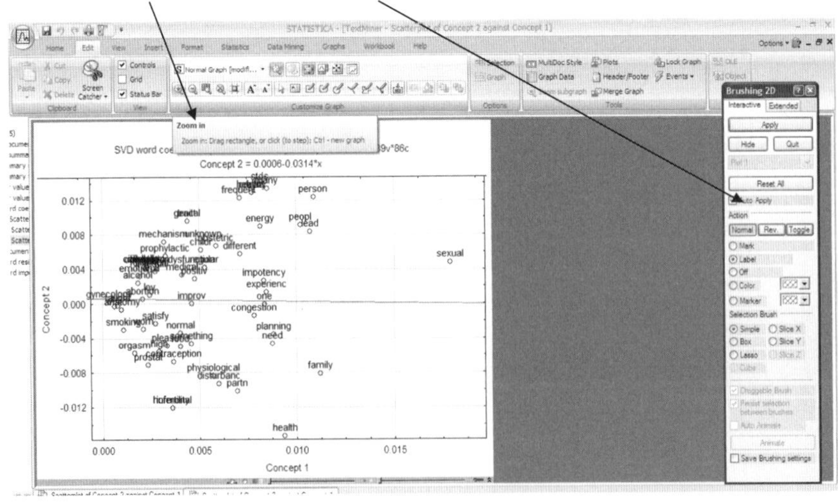

FIGURE E.28
Using the "Zoom in" and "Auto Apply" functions to facilitate the labeling process as may be seen in Figure E.29.

After using the "Zoom in" and "Auto Apply," many of the words are visible in the "clumped" area.

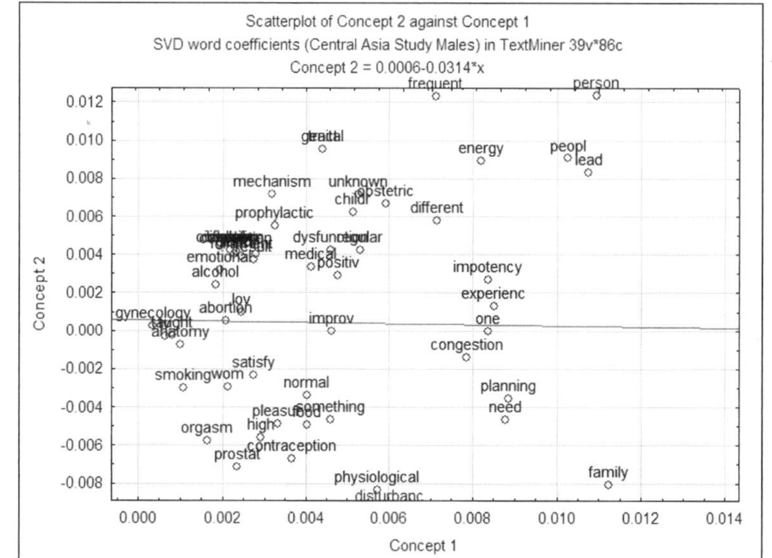

FIGURE E.29
Viewing the result of applying the functions.

Even after enlarging, some of the words are still clumped. One can right click and drag the words to separate them. Note that lines now go to the locations of the words. Or one can further enlarge the area with the "Zoom in" tool. "Unclumping" is demonstrated in Figure E.30.

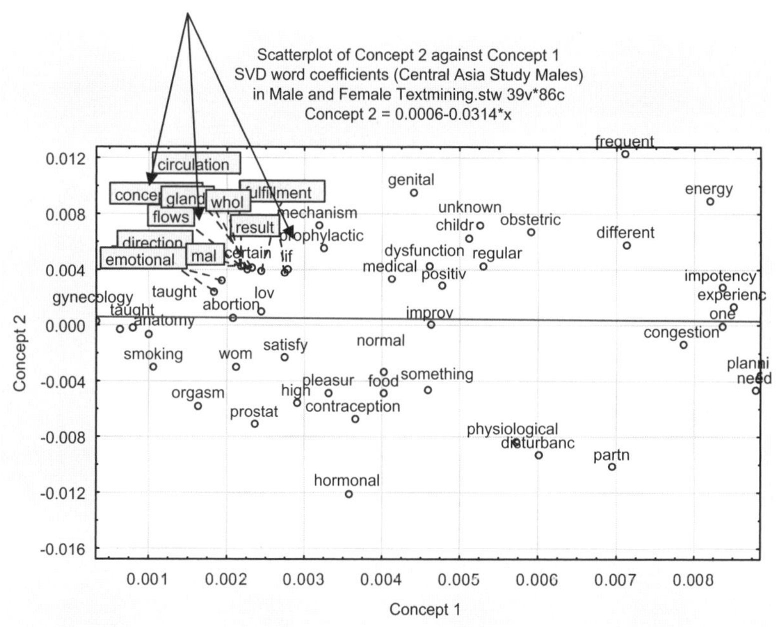

FIGURE E.30
Dragging and separating the words that remain clumped.

For the males, there seems to be a grouping concerning general functioning, with impotence and various dysfunctions concerns. Perhaps closed-ended questions should be formulated reflecting those concerns. The lasso can be used to highlight words that seem to have meaning as seen in Figure E.31.

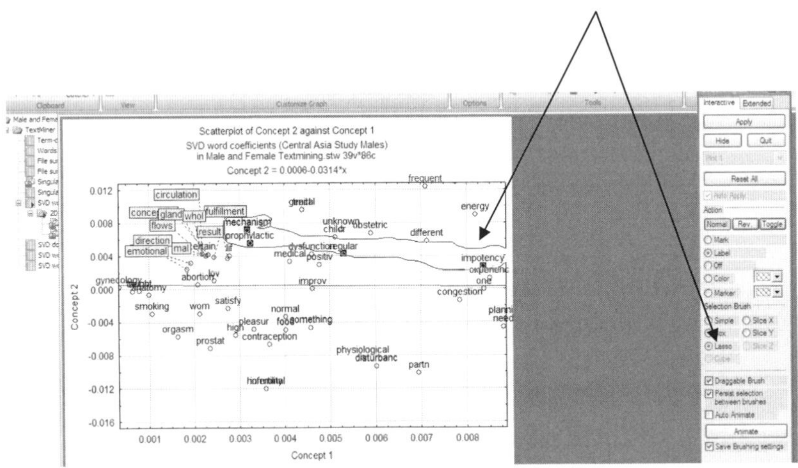

FIGURE E.31
Using the lasso.

There were also some words that grouped together in Figure E.32 concerning pleasure and perhaps, from the male point of view, concerns about satisfaction with respect to frequency. We could go back to the original locations of the terms and see if we were correct. If so, closed-ended questions might be formed around those concepts.

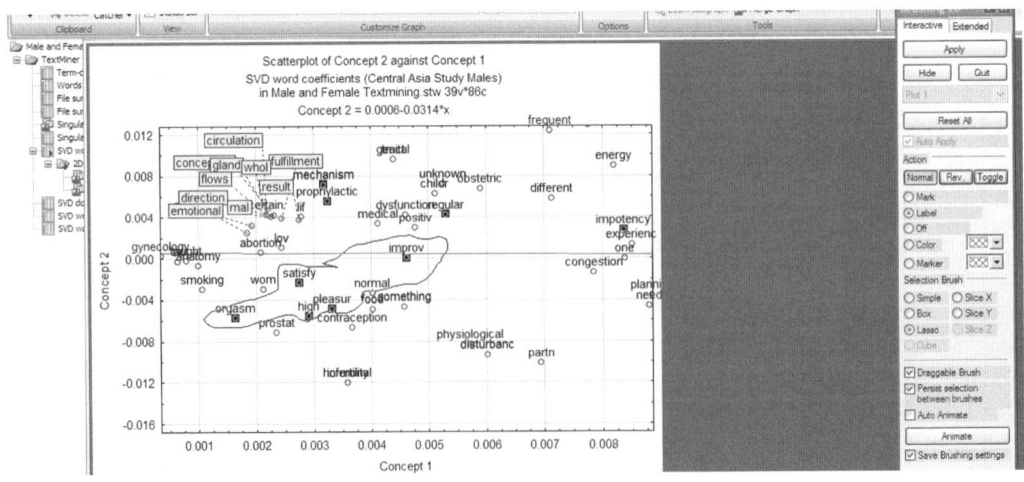

FIGURE E.32
Highlighting other groupings of words.

Now we want to repeat the process with the "for women" concepts. Open the output on the text mining. Set the SVD word coefficients as the active input. See Figure E.33 for repeating the process with women.

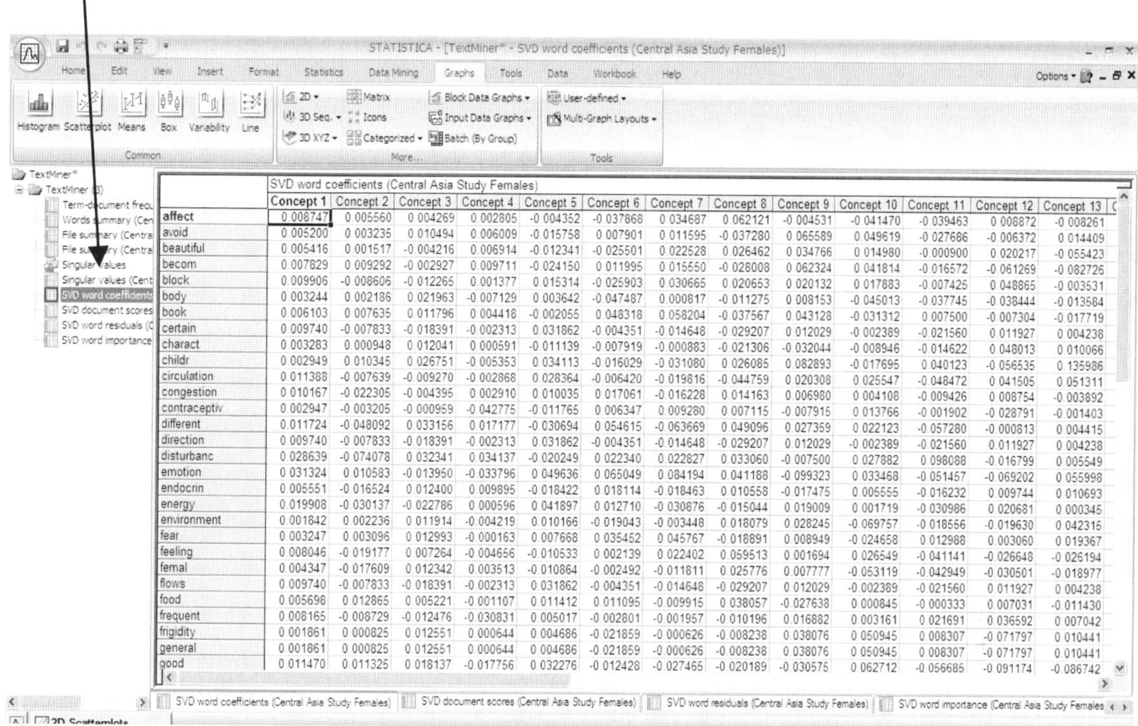

SVD word coefficients (Central Asia Study Females)

	Concept 1	Concept 2	Concept 3	Concept 4	Concept 5	Concept 6	Concept 7	Concept 8	Concept 9	Concept 10	Concept 11	Concept 12	Concept 13
affect	0.008747	0.005560	0.004269	0.002805	-0.004352	-0.037868	0.034687	0.062121	-0.004531	-0.041470	-0.039463	0.008872	-0.008261
avoid	0.005200	0.003235	0.010494	0.006009	-0.015758	0.007901	0.011595	-0.037280	0.065589	0.049619	-0.027686	-0.006372	0.014409
beautiful	0.005416	0.001517	-0.004216	0.006914	-0.012341	-0.025501	0.022528	0.026462	0.034766	0.014980	-0.000900	0.020217	-0.055423
becom	0.007829	0.009292	-0.002927	0.009711	-0.024150	0.011995	0.015550	-0.028008	0.062324	0.041814	-0.016572	-0.061269	-0.082726
block	0.009906	-0.008606	-0.012265	0.001377	0.015314	-0.025903	0.030665	0.020653	0.020132	0.017883	-0.007425	0.048865	-0.003531
body	0.003244	0.002186	0.021963	-0.007129	0.003642	-0.047487	0.000817	-0.011275	0.008153	-0.045013	-0.037745	-0.038444	-0.013584
book	0.006103	0.007635	0.011796	0.004418	-0.002055	0.048318	0.058204	-0.037567	0.043128	-0.031312	0.007500	-0.007304	-0.017719
certain	0.009740	-0.007833	-0.018391	-0.002313	0.031862	-0.004351	-0.014648	-0.029207	0.012029	-0.002389	-0.021560	0.011927	0.004238
charact	0.003283	0.000948	0.012041	0.000591	-0.011139	-0.007919	-0.000883	-0.021306	-0.032044	-0.008946	-0.014622	0.048013	0.010066
childr	0.002949	0.010345	0.026751	-0.005353	0.034113	-0.016029	-0.031080	0.026085	0.082893	-0.017695	0.040123	-0.056535	0.135986
circulation	0.011388	-0.007639	-0.009270	-0.002868	0.028364	-0.006420	-0.019816	-0.044759	0.020308	0.025547	-0.048472	0.041505	0.051311
congestion	0.010167	-0.022305	-0.004395	0.002910	0.010035	0.017061	-0.016228	0.014163	0.006980	0.004108	-0.009426	0.008754	-0.003892
contraceptiv	0.002947	-0.003205	-0.000959	-0.042775	-0.011765	0.006347	0.009280	0.007115	-0.007915	0.013766	-0.001902	-0.028791	-0.001403
different	0.011724	-0.048092	0.033156	0.017177	-0.030694	0.054615	-0.063669	0.049096	0.027359	0.022123	-0.057280	-0.000813	0.004415
direction	0.009740	-0.007833	-0.018391	-0.002313	0.031862	-0.004351	-0.014648	-0.029207	0.012029	-0.002389	-0.021560	0.011927	0.004238
disturbanc	0.028639	-0.074078	0.032341	0.034137	-0.020249	0.022340	0.022827	0.033060	-0.007500	0.027882	0.098088	-0.016799	0.005549
emotion	0.031324	0.010583	-0.013950	-0.033796	0.049636	0.065049	0.084194	0.041188	-0.099323	0.033468	-0.051457	-0.069202	0.055998
endocrin	0.005551	-0.016524	0.012400	0.009895	-0.018422	0.018114	-0.018463	0.010558	-0.017475	0.005555	-0.016232	0.009744	0.010693
energy	0.019908	-0.030137	-0.022786	0.000596	0.041897	0.012710	-0.030876	-0.015044	0.019009	0.001719	-0.030986	0.020681	0.000345
environment	0.001842	0.002236	0.011914	-0.004219	0.010166	-0.019043	-0.003448	0.018079	0.028245	-0.069757	-0.018556	-0.019630	0.042315
fear	0.003247	0.003096	0.012993	-0.000163	0.007668	0.035452	0.045767	-0.018891	0.008949	-0.024658	0.012988	0.003060	0.019367
feeling	0.008046	-0.019177	0.007264	-0.004656	-0.010533	0.002139	0.022402	0.059513	0.001694	0.026549	-0.041141	-0.026648	-0.026194
femal	0.004347	-0.017609	0.012342	0.003513	-0.010864	-0.002492	-0.011811	0.025776	0.007777	-0.053119	-0.042949	-0.030501	-0.018977
flows	0.009740	-0.007833	-0.018391	-0.002313	0.031862	-0.004351	-0.014648	-0.029207	0.012029	-0.002389	-0.021560	0.011927	0.004238
food	0.005698	0.012865	0.005221	-0.001107	0.011412	0.011095	-0.009915	0.038057	-0.027638	0.000845	-0.000333	0.007031	-0.011430
frequent	0.008165	-0.008729	-0.012476	-0.030831	0.005017	-0.002801	-0.001957	-0.010196	0.016882	0.003161	0.021691	0.036592	0.007042
frigidity	0.001861	0.000825	0.012551	0.000644	0.004686	-0.021859	-0.000626	-0.008238	0.038076	0.050945	0.008307	-0.071797	0.010441
general	0.001861	0.000825	0.012551	0.000644	0.004686	-0.021859	-0.000626	-0.008238	0.038076	0.050945	0.008307	-0.071797	0.010441
good	0.011470	0.011325	0.018137	-0.017756	0.032276	-0.012428	-0.027465	-0.020189	-0.030575	0.062712	-0.056685	-0.091174	-0.086742

FIGURE E.33
Examining the "for females" SVD word coefficients. First, set the SVD coefficient as the input file.

Again, form a scatterplot with the first two concepts (graphs—2D—scatterplots). The first graph has no labels, of course, and displays a bit of clumping in Figure E.34. Use the brushing tool and the enlarging tool to examine the words that seem associated (being close together). If one gets out of the brushing tool option, one can also click on words and move them from being on top of one another. One can also always examine outliers.

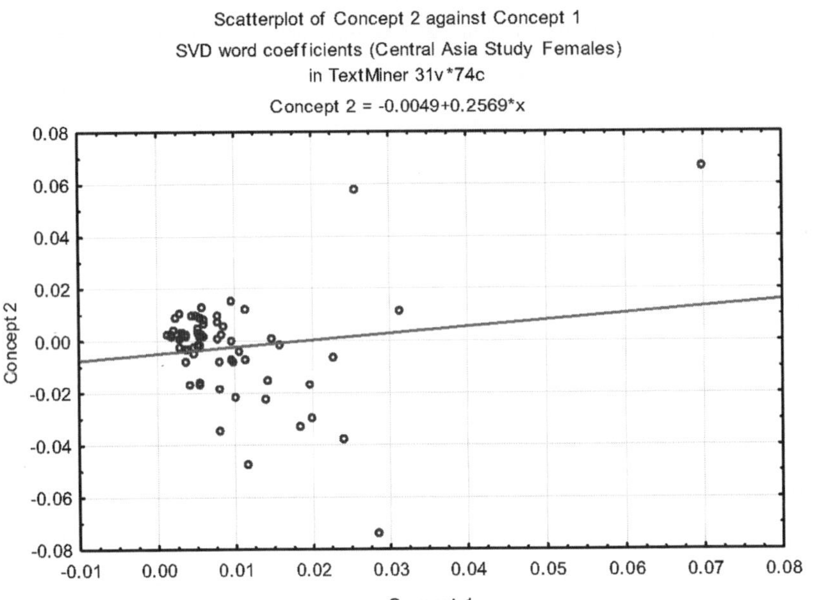

FIGURE E.34
First view of the scatterplot of "for females" words.

Next, the clumps with the words separated may be seen in Figure E.35.

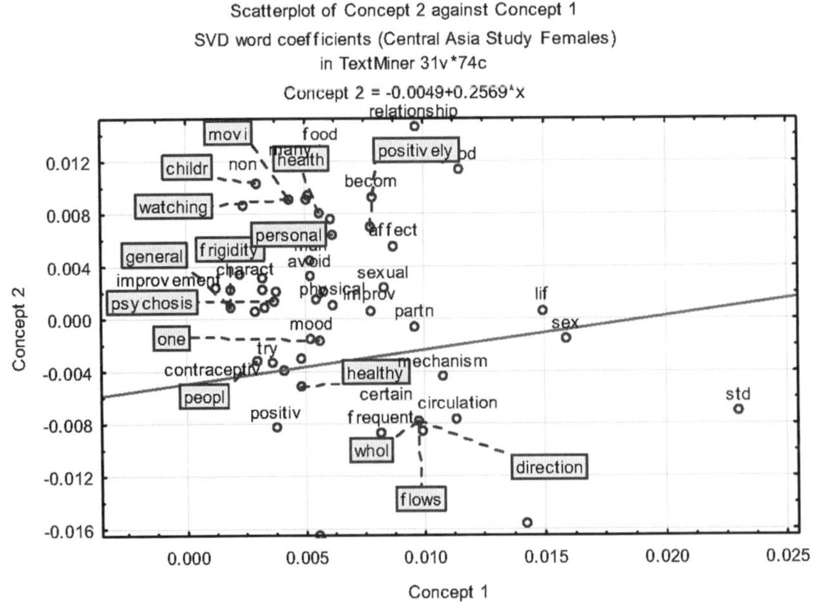

FIGURE E.35
Clump separated and words labeled.

The words in the "for females" seem to have more words centering on family concerns, health, wellness, and relationships (partner, children, personal, people, positive, health, etc.). Given the preceding analyses, it seemed prudent to ask at least some of our questions from the perspective of males versus females in order to obtain more complete results.

SUMMARY

Using the text mining procedure on the data set as a whole, as in the first tutorial, or on the data sets split by a certain variable, as in the "for males" and "for females" data sets, can be one way in which researchers can construct instruments that have increased validity. The process allowed us to reflect upon the kinds of closed-ended questions we wished to ask of the respondents. A qualitative approach allowed us to increase the effectiveness of our subsequent quantitative analysis.

TUTORIAL F

Using eBay Text for Predicting ATLAS Instrumental Learning

Anne Ashby Ghost Bear, EdD
Director, Southern Nazarene University, Tulsa, Tulsa, OK

Linda A. Miner, PhD
Director of Academic Programs, Southern Nazarene University, Tulsa, OK

CONTENTS

INTRODUCTION

The Assessing the Learning Strategies of Adults (ATLAS) instrument was developed for measuring learning strategy preferences in real-life situations. It was created to afford facilitators of adult learning events an easily administered method for discovering the learning strategy preferences of participants in a variety of settings (Conti and Kolody, 1999). Data related to ATLAS were gathered by embedding the instrument into the online questionnaire that placed participants into one of three groups of learners: navigators, problem solvers, or engagers.

EXAMINING THE DATA BY TYPES

The data set comprised 352 cases in which answers to specific questions were given in text form. In addition, the age, gender, educational level, ATLAS category, and Atlas subcategory (learning style) were included. The data are sorted by Atlas category. Please open eBay Database "A Ghost Bear revised.sta." In this database the names were removed and case names were added (variable 43). We were curious as to whether the ATLAS category could be accurately predicted from the kinds of words that people used when they wrote the answers to the questions and, if the subject were technology, if the ATLAS categorizations would still apply in a reliable fashion. Prior to making any kinds of predictive modeling, we decided to randomly separate the larger data set into two randomly generated subsets: a training data

Practical Text Mining and Statistical Analysis for Non-structured Text Data Applications. DOI: 10.1016/B978-0-12-386979-1.00014-1

set and a testing data set. In order to accomplish this goal, we opened a data mining workspace and entered the large data set.

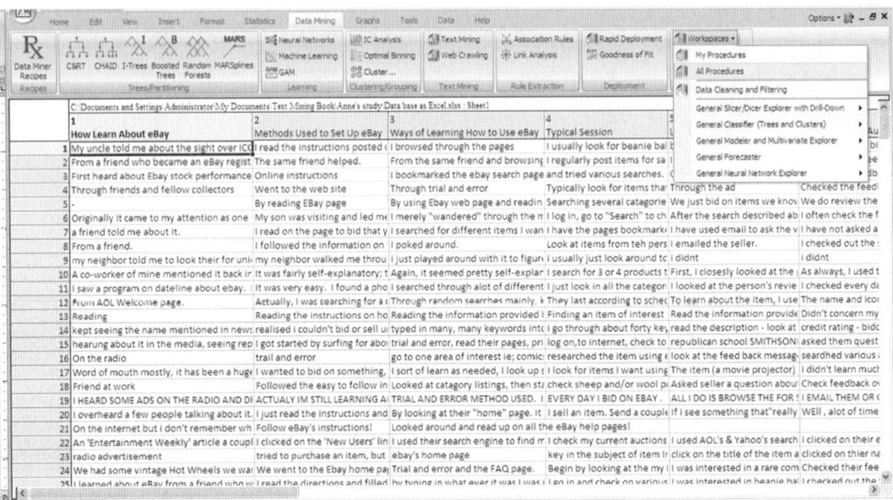

FIGURE F.1
Opening a data mining workspace. Select "All Procedures."

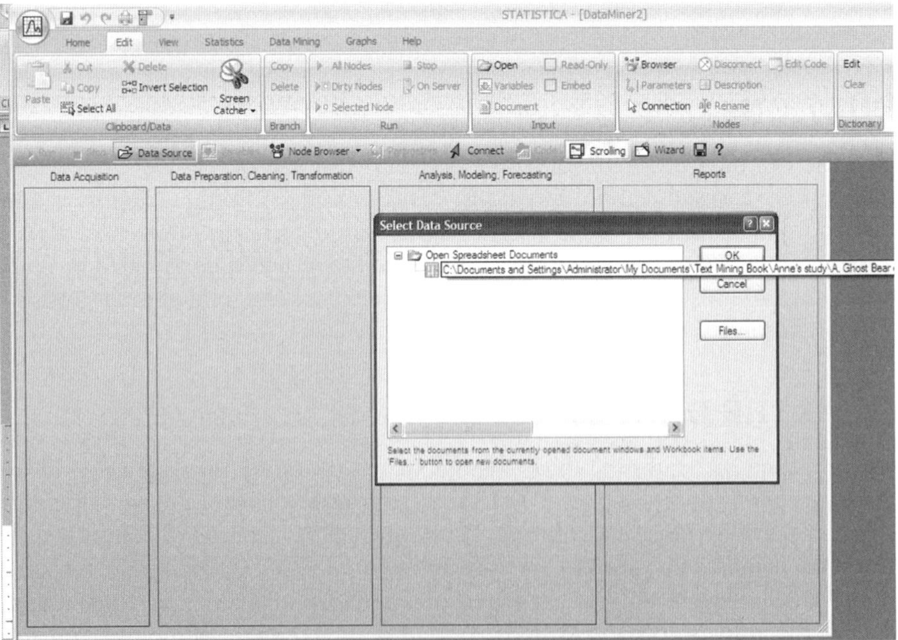

FIGURE F.2
Select the data source that has been opened.

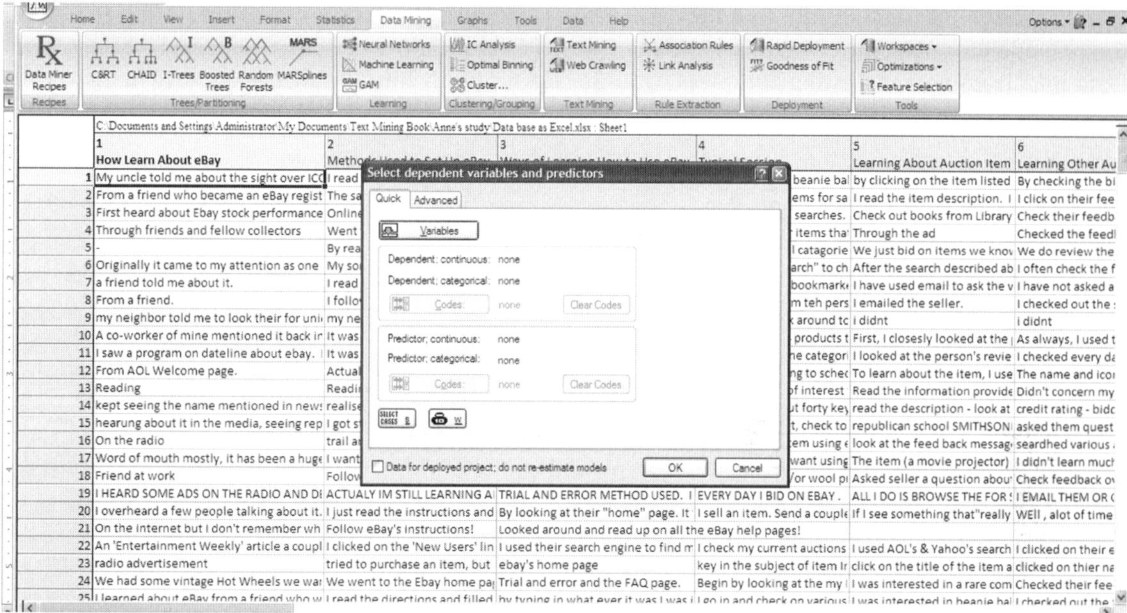

FIGURE F.3

Then select the variables.

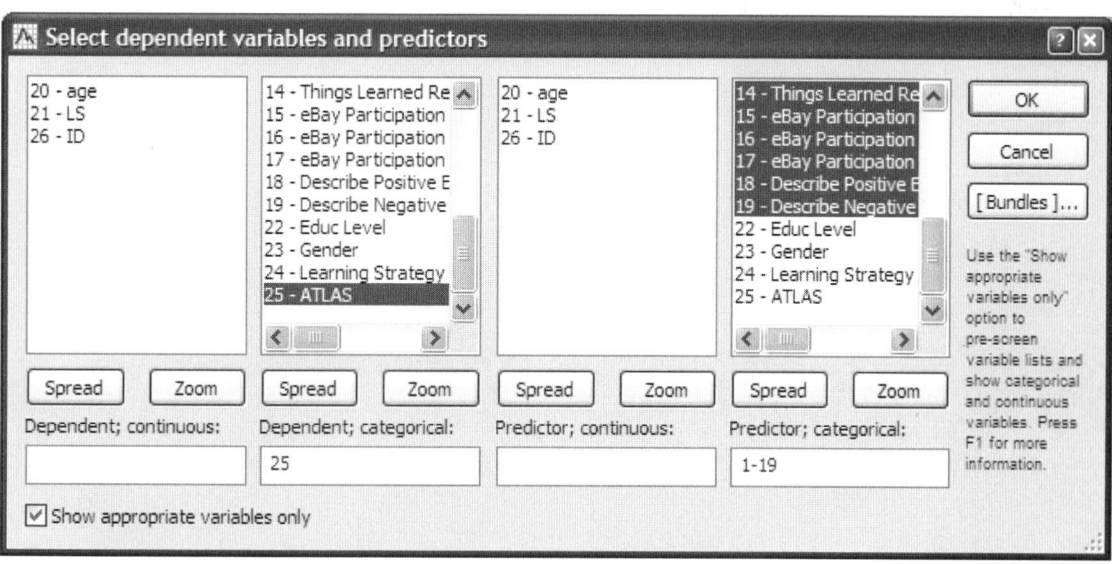

FIGURE F.4

Select ATLAS as the dependent variable and 1–19, the text variables, as the independent predictors.

We would like to randomly split the data into two groups: a training group and a testing group. Go to the Node Browser, select "Graphs," and then go to "Classification and Discrimination." Select "Split Data into Training and Testing Samples." Double click to put it into the workspace (Fig. F.5).

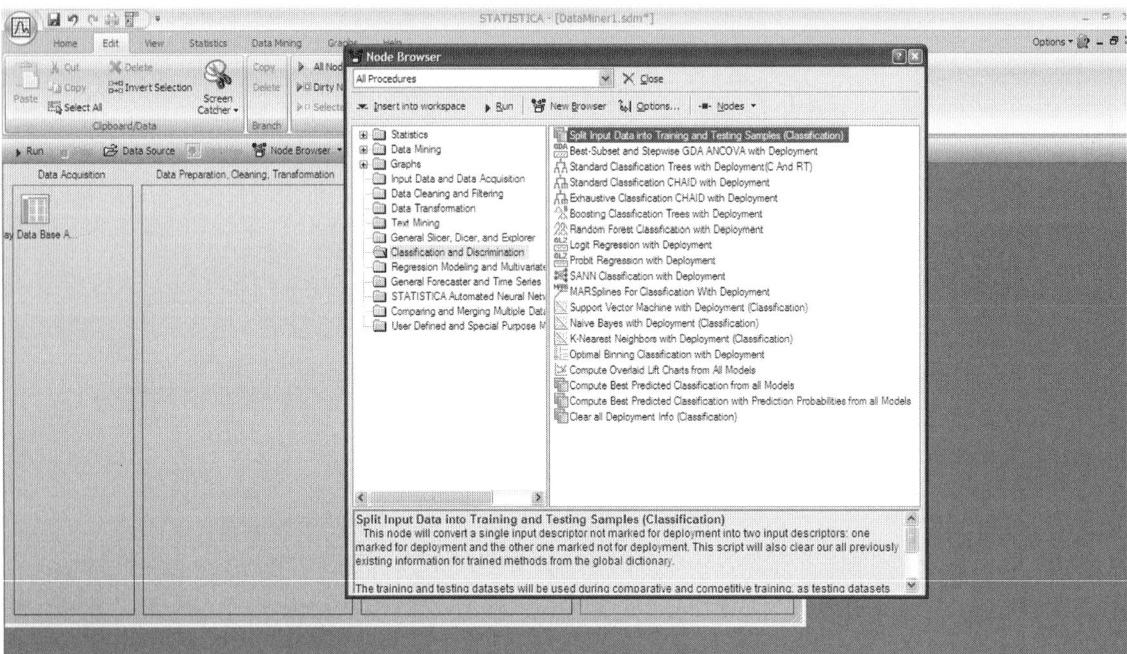

FIGURE F.5
Selecting the procedure from the node browser.

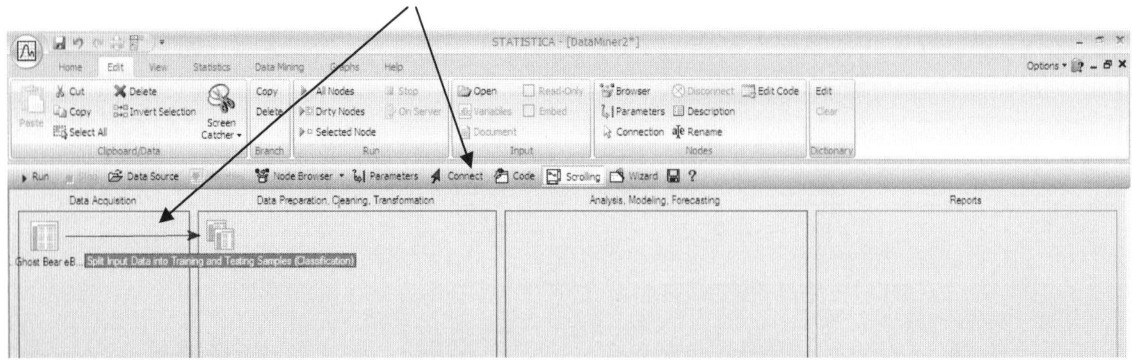

FIGURE F.6
Drag an arrow from the data to the node.

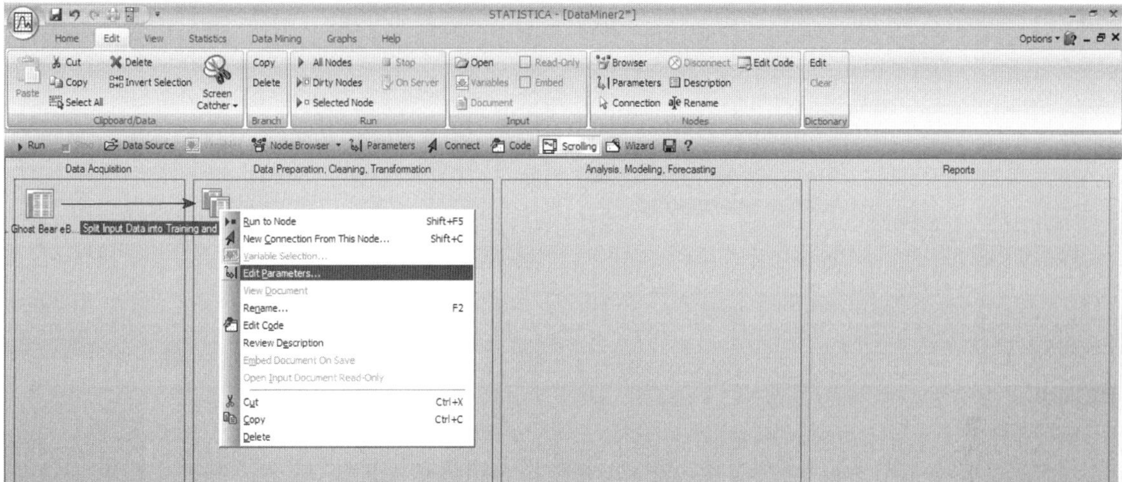

FIGURE F.7

Right click on the node to edit the parameters.

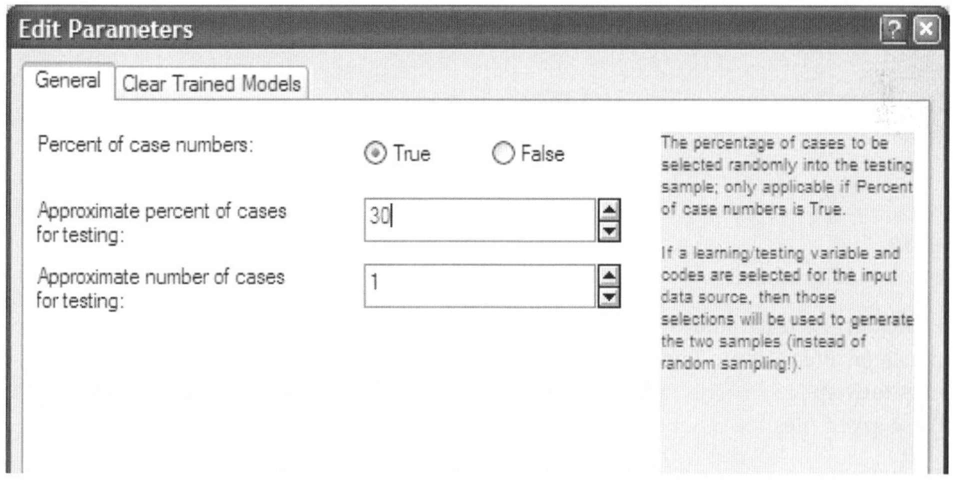

FIGURE F.8

Change the percent of cases from 50 to 30 so we have 30 percent of the cases in the testing database.

There were enough cases that we could use a 70/30 split on the two samples. This will give us 70 percent to play with, and then we can test it on 30 percent. Right click and run the node.

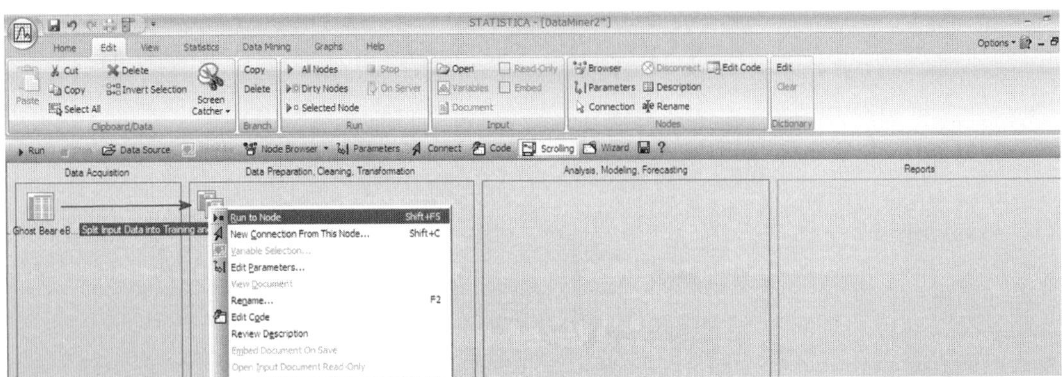

FIGURE F.9
Running the node.

Now we have two data sets that we can use for our analyses. Pull the testing set down a bit so that each can be seen easily (Fig. F.10).

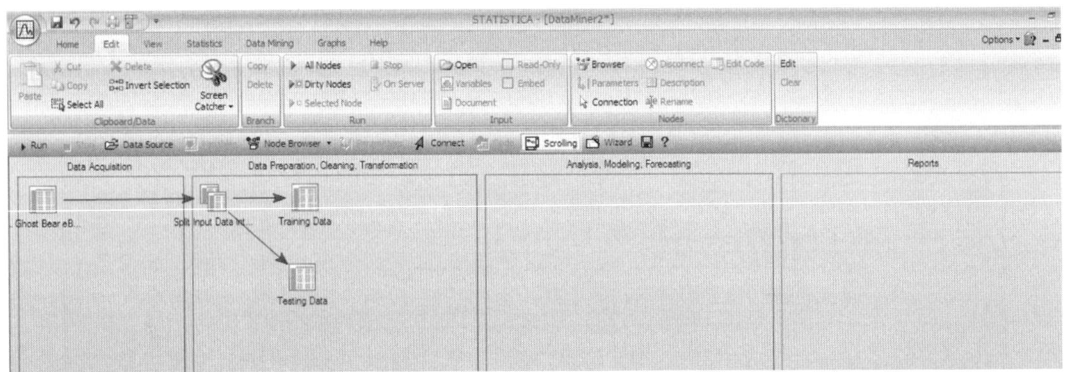

FIGURE F.10
Resulting samples: Training and Testing.

Right click on each file and on "Embed Document on Save." In this way you will be able to open them again later after saving. Right click on the data files, and click on "View Document." After the document comes up, go to "File" and save each. You might save them as training and testing. Later you can use these saved files for other analyses.

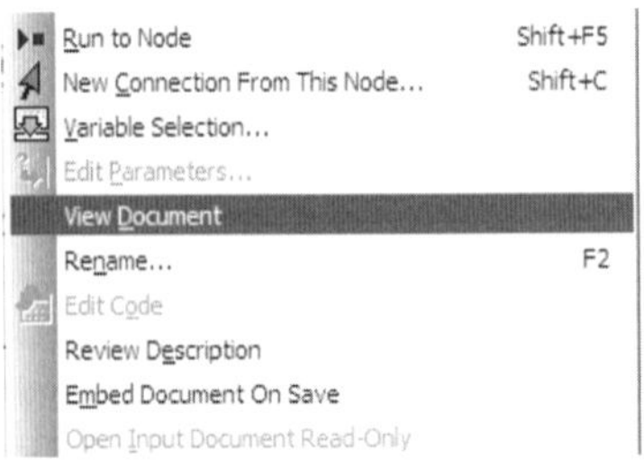

FIGURE F.11
View Document.

Save the data mining file because it will be needed later and close it. Close the full data set for now. Open the training data file that you have just saved. Go to Data Mining and then Text Mining (Fig. F.12).

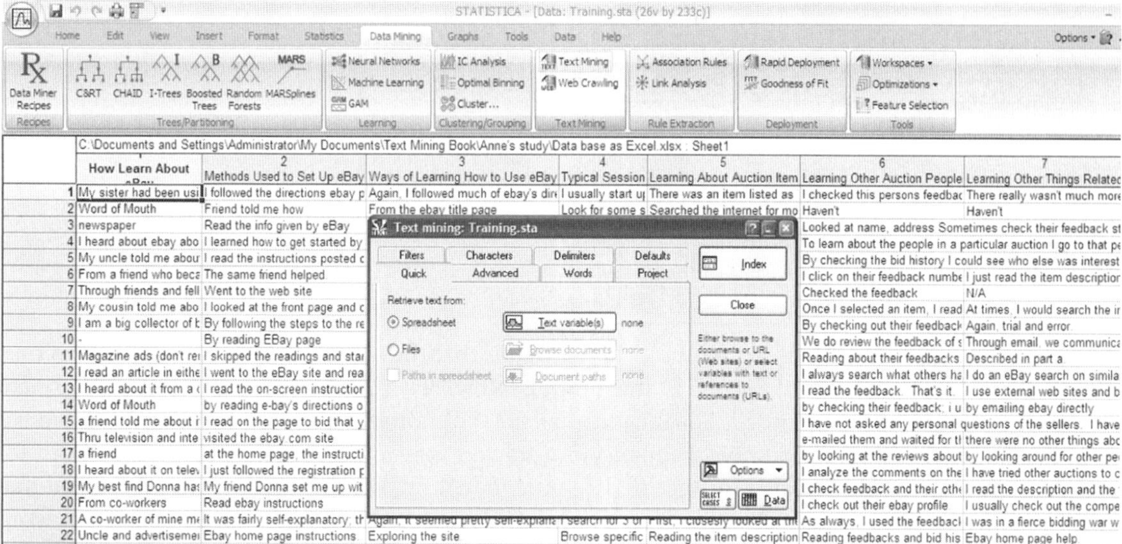

FIGURE F.12
Pulling up Text Mining.

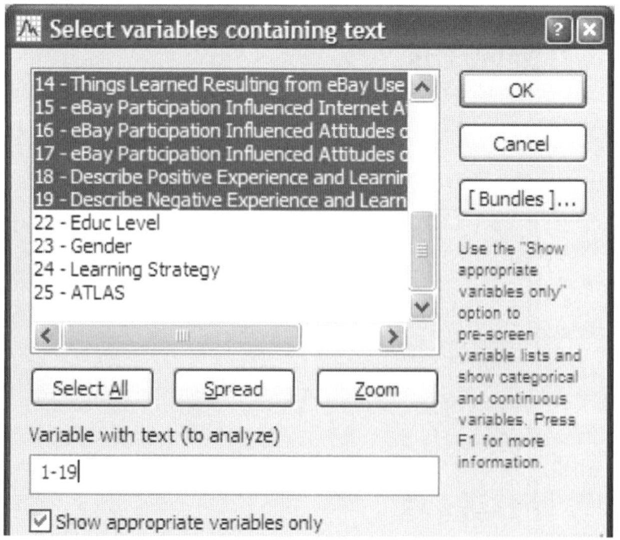

FIGURE F.13
Select "Variables." Choose variables 1–19.

For this analysis we will use no synonyms list or delineators, so simply leave the defaults for now. We do want to change the percentage of words. Click on the "Advanced" tab, change 1 to 10, and change the 100 to 80 (Fig. F.14). Words that nearly never occur will not be useful, nor will words that always show up in all categories.

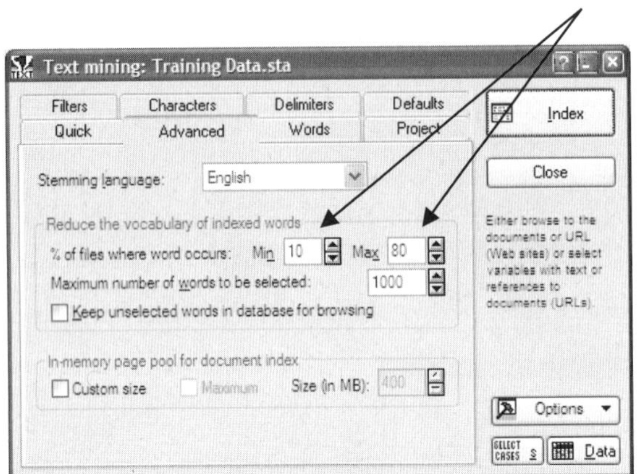

FIGURE F.14

Changing the percentages.

Next, click on "Index." (See Figure F.14.) It will tell us that the old file will be deleted. That is fine. Click with abandon. The next screen is the output. Now, we could click on "Inverse document frequency," which is recommended. For this tutorial, we will not select the recommendation and see what happens. The inverse document frequency is a transformative procedure that accounts both for raw frequencies of words and also for specific semantic meanings. For tutorial F, it was purported that people within their ATLAS categories may use certain words repeatedly. It was felt that simple raw word counts might be the best way to predict the ATLAS categories without regards to their specific contexts.

At another time, you may wish to repeat the tutorial and click inverse document frequency to see if the results change significantly. (To see a tutorial that uses this, go to tutorial R.) Click on "Perform Singular Value Decomposition" under "Concept Extraction." Note that the items underneath are shadowed (See Figure F.15).

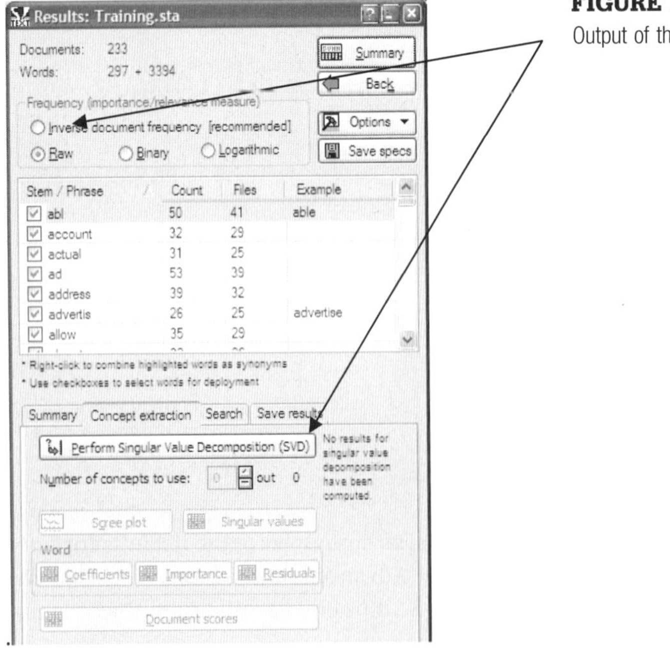

FIGURE F.15

Output of the Indexing.

After clicking, they are no longer shadowed (Fig. F.16).

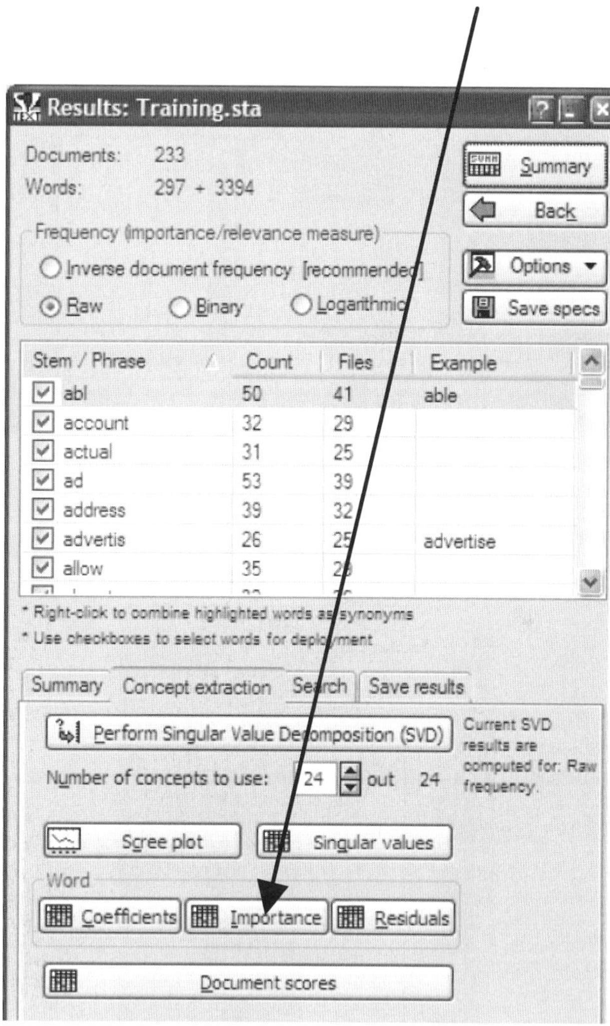

FIGURE F.16
No more shadowing.

Click on "Importance." (See Figure F.16.) A word list will appear. Now choose data, sort, and add the variable importance. Sort in descending order as in Figure F.17.

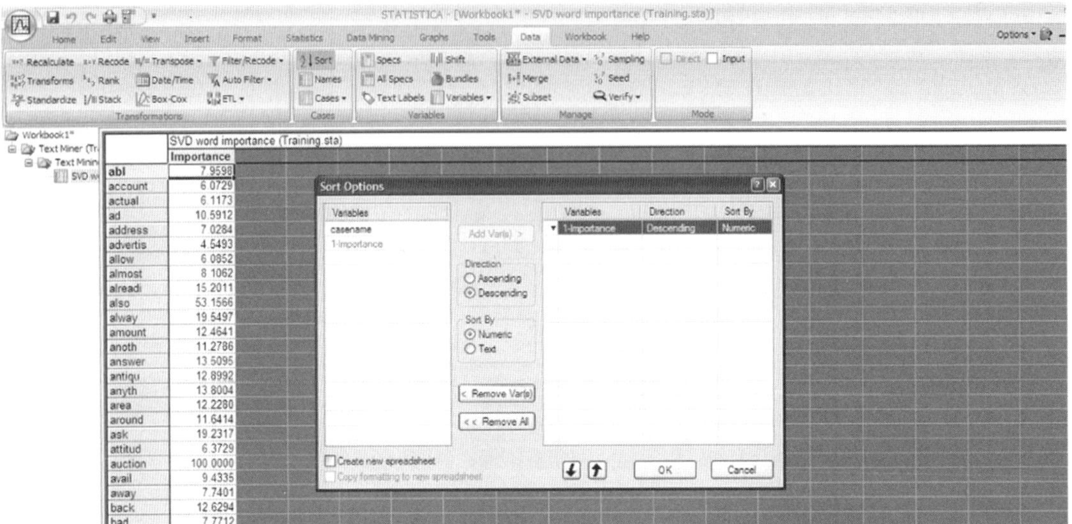

FIGURE F.17
Sorting the variables in descending order.

Click OK (Fig. F.17). The following were the important words as may be seen in Figure F.18.

SVD word importance (Training.sta)

	Importance
auction	100.0000
seller	85.3684
learn	80.8171
use	80.4056
look	74.1540
check	68.2380
peopl	65.8573
get	64.2486
will	63.5602
just	60.3538
can	58.0510
see	57.7256
search	54.3948
interest	54.1301
also	53.1566
time	52.4773
want	52.1184
thing	51.8370
realli	49.9735
internet	49.9180
ve	48.8534
one	48.5000
price	45.7589
sell	45.7558
go	45.6148
like	42.1591
comput	41.9888
read	41.6728
would	40.2619

FIGURE F.18
The important words for training the data set.

Now click on "Saved Results." At this point we want to put all the words into our training data set as separate variables so we can numerically analyze them as predictors. See that there is a "Write back" box (See Figure F.19).

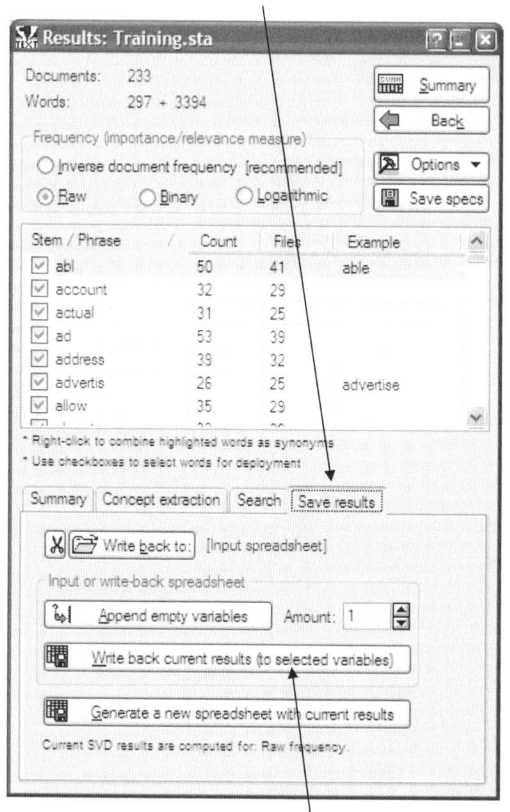

FIGURE F.19
Clicking on "Save results."

Select "Write back to." Write it back to the saved Training Data file. Open it (Fig. F.20).

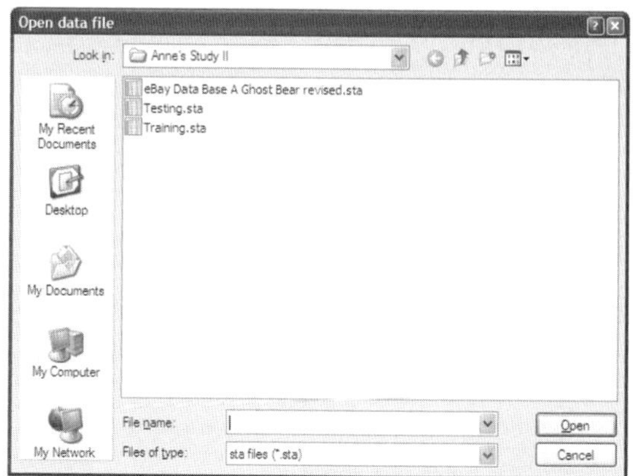

FIGURE F.20
Open the file—in this case "training.sta."

Note there were 297 words selected as seen in Figure F.21.

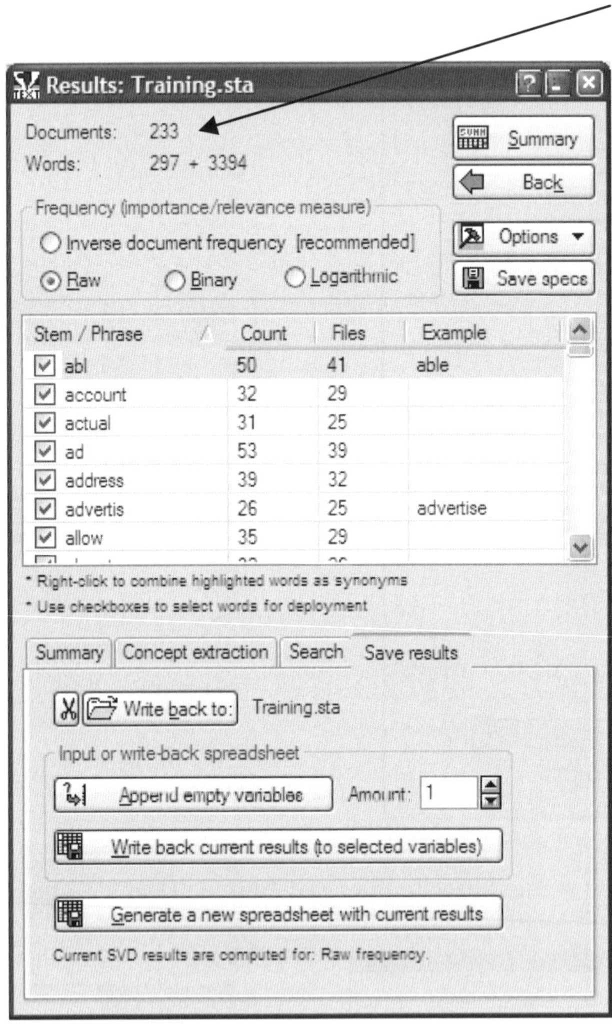

FIGURE F.21
The number of words was 297.

Write that number in the box. (See Figure F.22.)

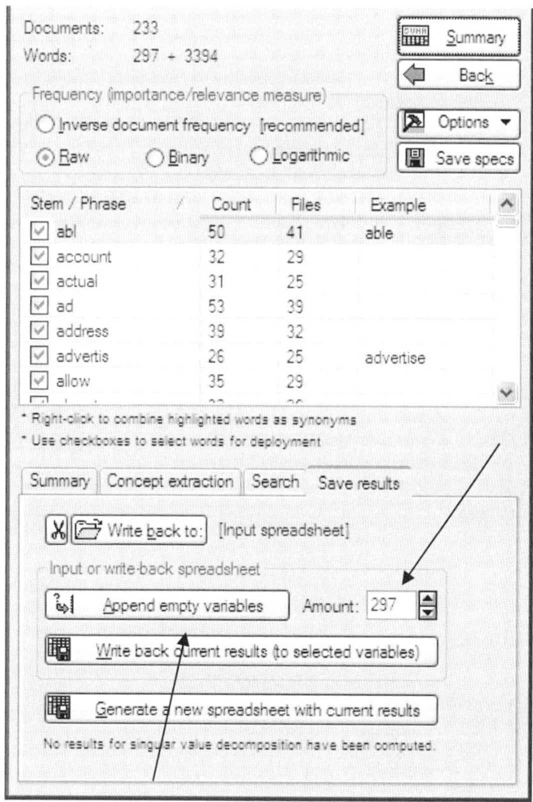

FIGURE F.22
Writing 297 in the Amount box.

Append empty variables as in Figure F.22. Then click OK on writing them into the worksheet.

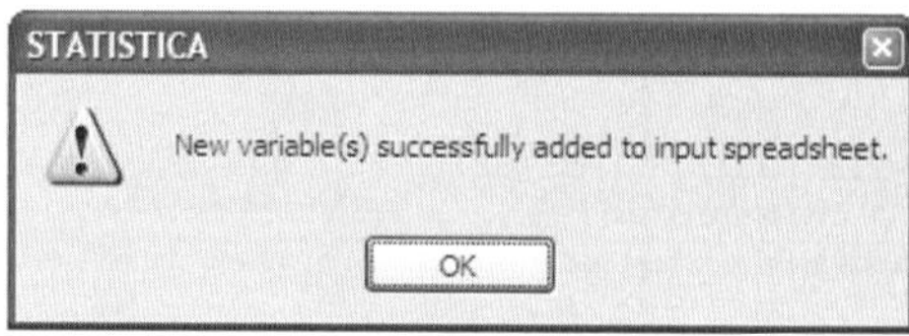

FIGURE F.23
Message that variables were successfully added to input sheet.

Click OK. Note that the original database now has 297 new extra columns that are empty and waiting for the words to be sent over. See Figure F.24.

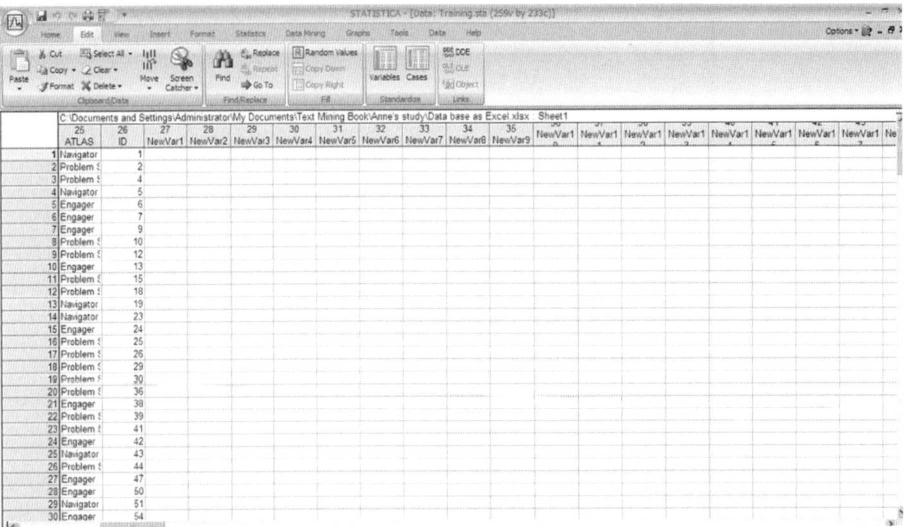

FIGURE F.24
Database now has more variable spaces.

Next write back the current results to selected variables as shown in Figure F.25.

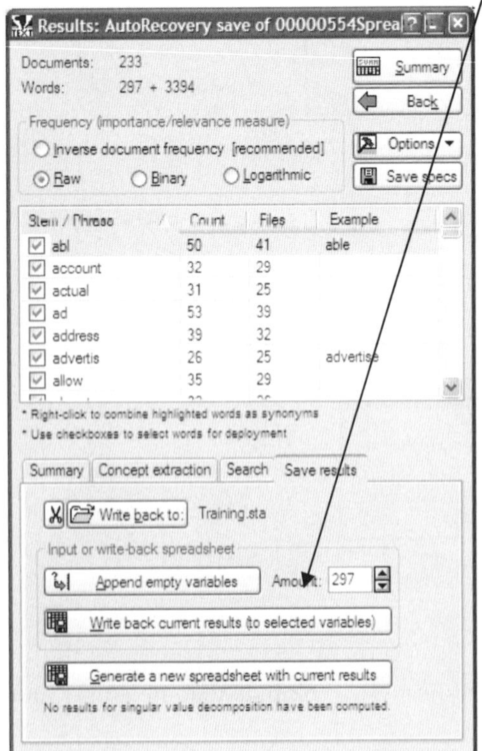

FIGURE F.25
Begin to put the 297 words into the empty variables.

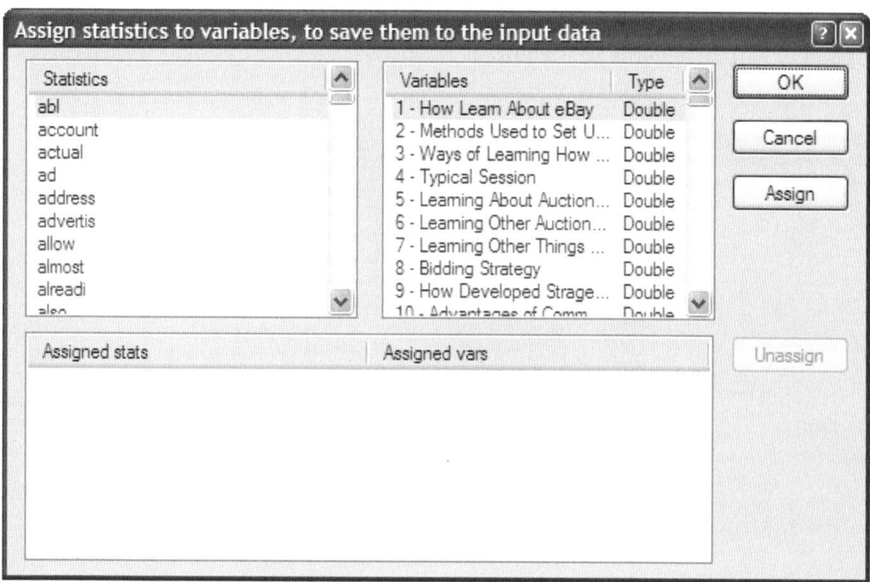

FIGURE F.26
Select them on the left and the right.

Highlight the first word on the left, scroll to the bottom of the list, click Shift, and select all the words as shown in Figure F.27.

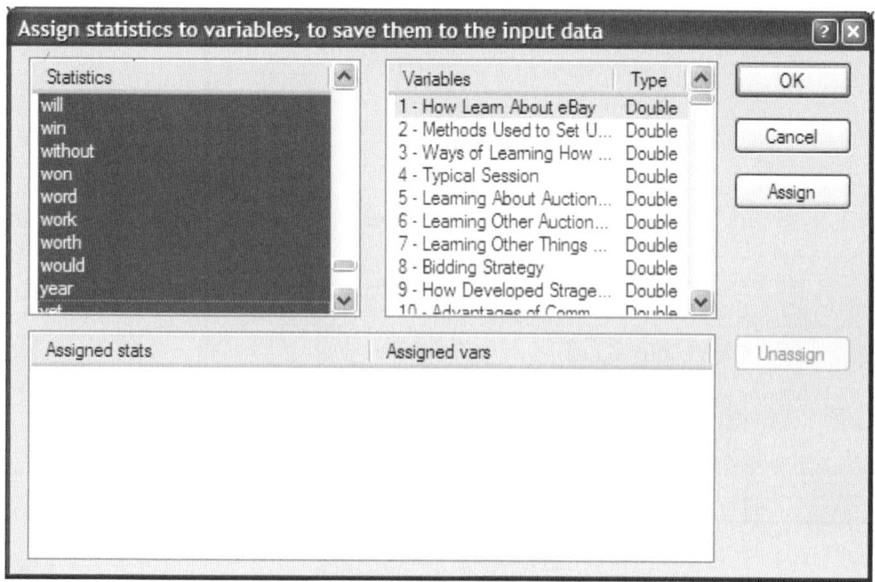

FIGURE F.27
First, select the words on the left.

Now go to the right, click on the first new variable, scroll down to the bottom, click Shift, and highlight all the new variables. Note that Figure F.28 shows the look of both sides; words and variables.

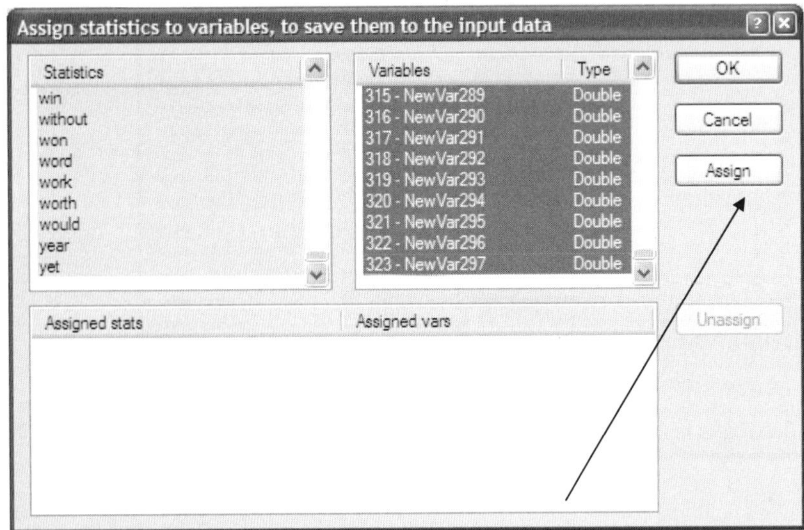

FIGURE F.28
Find the empty variables and select them.

Then click assign button (see arrow Figure F.28). Now they have been assigned (Fig. F.29).

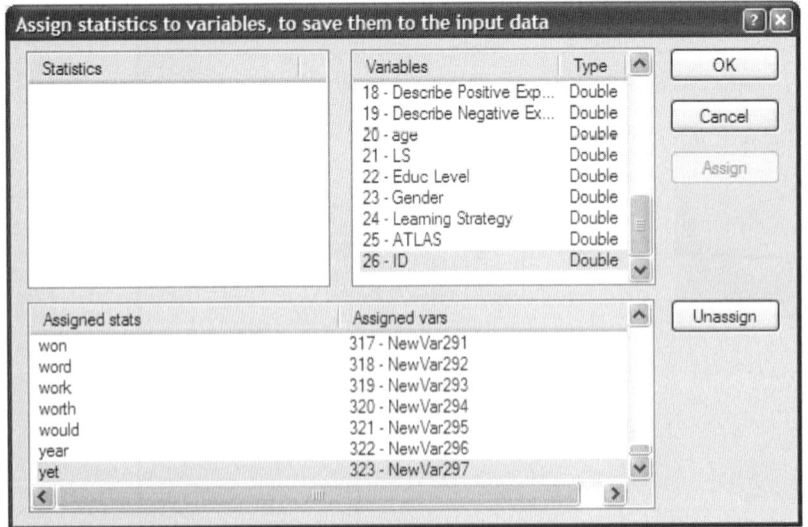

FIGURE F.29
Words assigned to variables.

Click Okay, and they will go into the new columns. The words are recorded as ones, and the empty spots as zeros. (See Fig. F.30.)

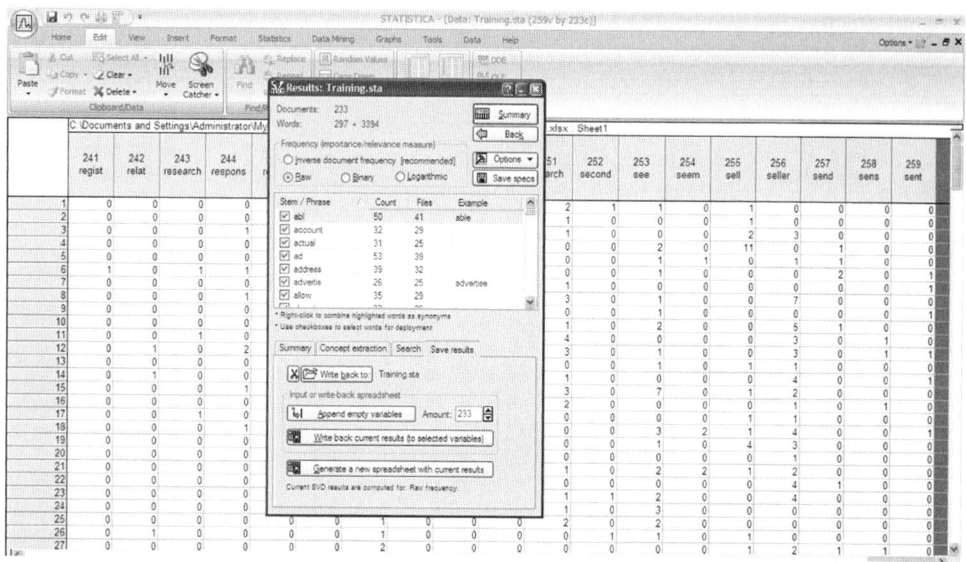

FIGURE F.30

Database now has words in the variables with interval data in each column. This is the quantification of qualitative data.

Click the "Concept extraction" tab to see what else is under there (arrow in Fig. F.31).

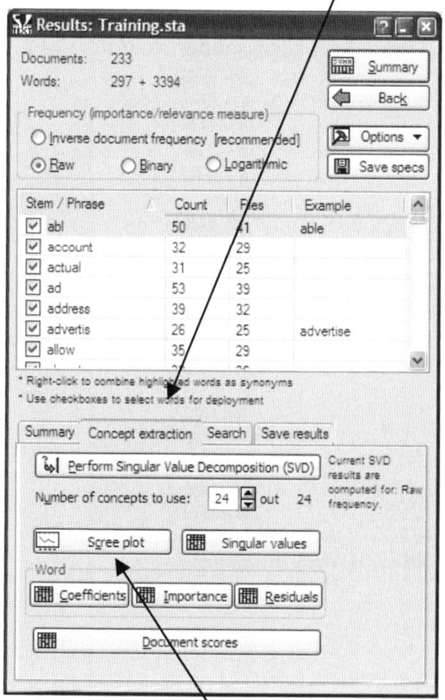

FIGURE F.31

Location of "Concept Extraction" and Scree plot.

Click on the Scree plot (Fig. F.31) to see how many concepts might be significant.

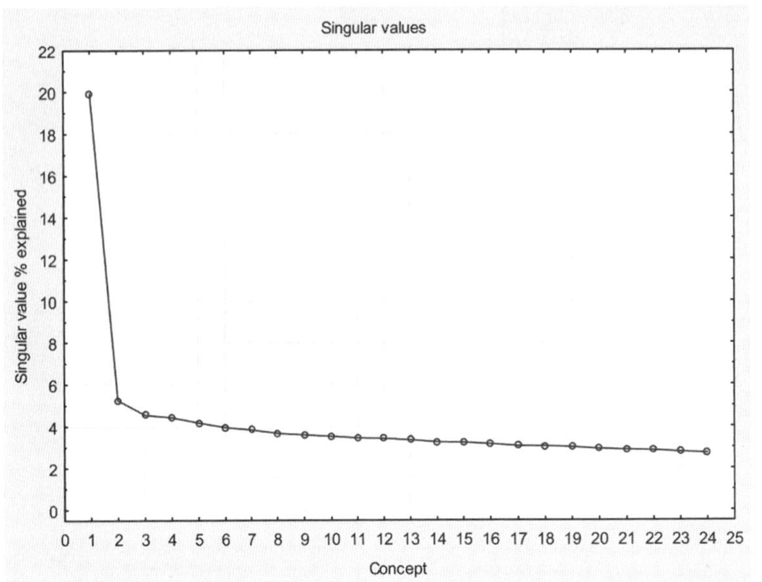

FIGURE F.32
The Scree plot.

The Scree plot in Figure F.32 shows that two main concepts were extracted. We will use those in a scatterplot later. Now click on "Coefficients" (Fig. F.31) to see what the concepts look like. Change the 24 to a 2, since these are the only two that appeared to be significant. (Figure F.33 shows where to enter the 2.)

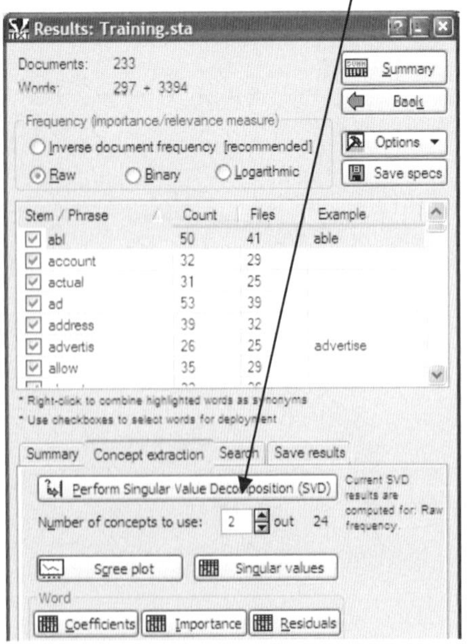

FIGURE F.33
Where to place the 2.

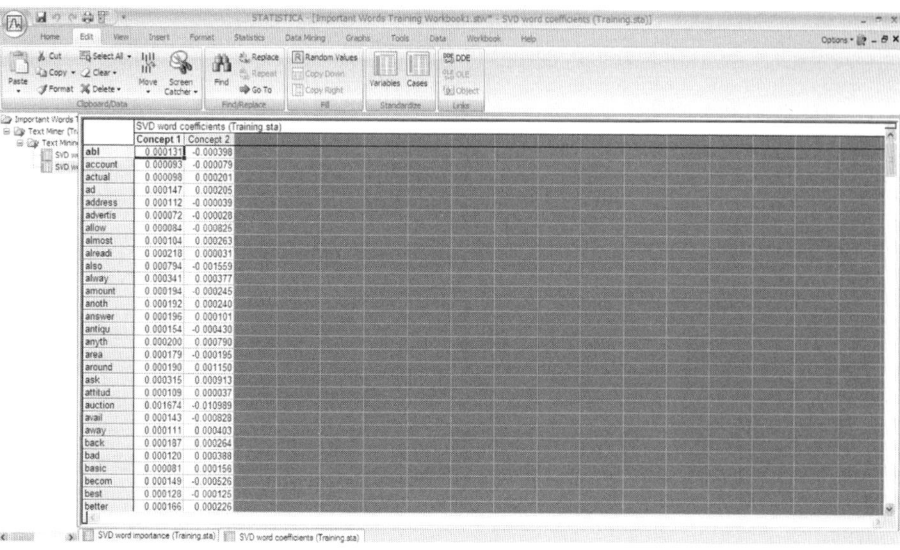

FIGURE F.34
The two concepts that were extracted.

See what they look like in Figure F.34. Click the "Save results" tab. Save the workbook for later use. Go back to the original data mining workspace (Fig. F.35).

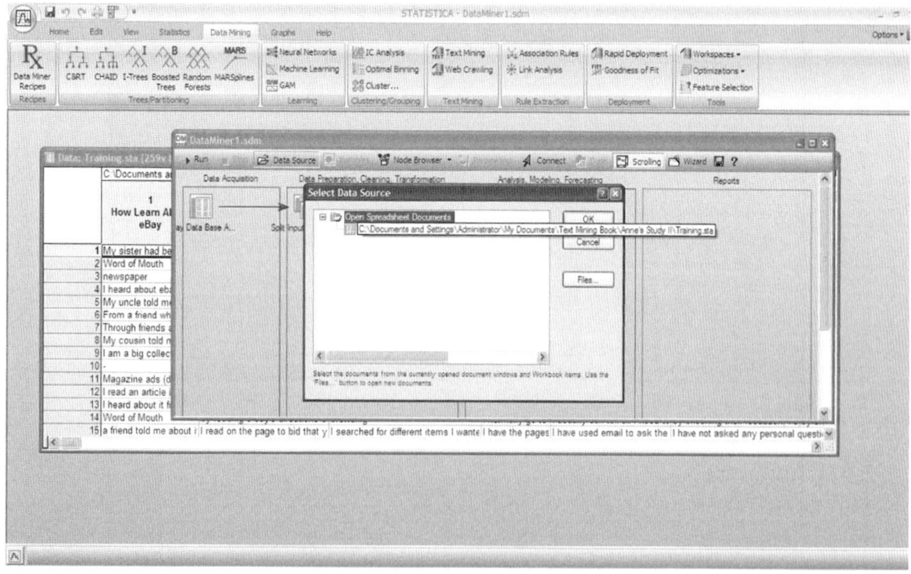

FIGURE F.35
Enter into the data mining workspace the newly saved training data set that now has all the new word variables.

Choose the variables as in Figure F.36. For the dependent categorical variable, choose the ATLAS categories, v25, and for the independent variables, choose the new word variables, v27 through v259.

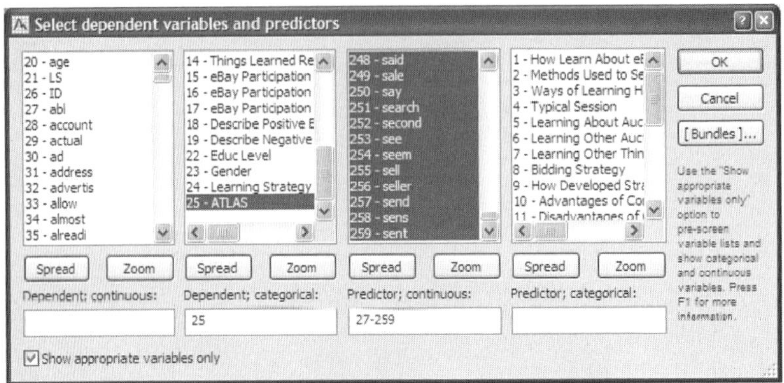

FIGURE F.36

Choosing the variables.

Click OK. Now go to the node browser and select "Feature Selection." Figure F.37 shows selecting the node. Figure F.38 shows connecting the arrow from the data to the node in the data mining workspace.

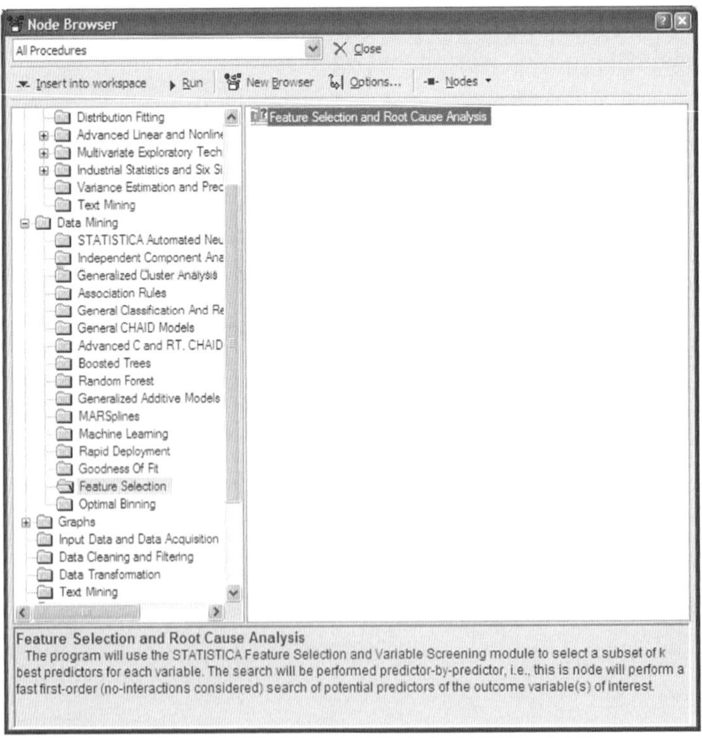

FIGURE F.37

Finding the feature selection and root cause analysis.

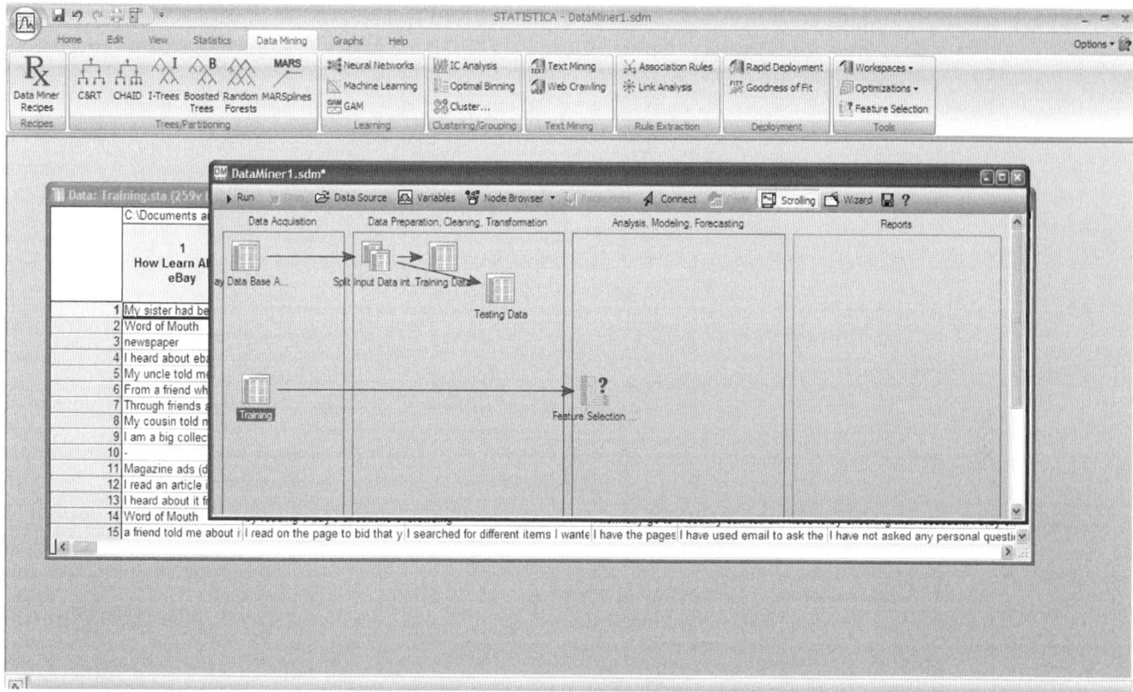

FIGURE F.38

Connect the node to the data using an arrow.

Edit the parameters to include all of the results and base the results upon the p-value (Fig. F.39).

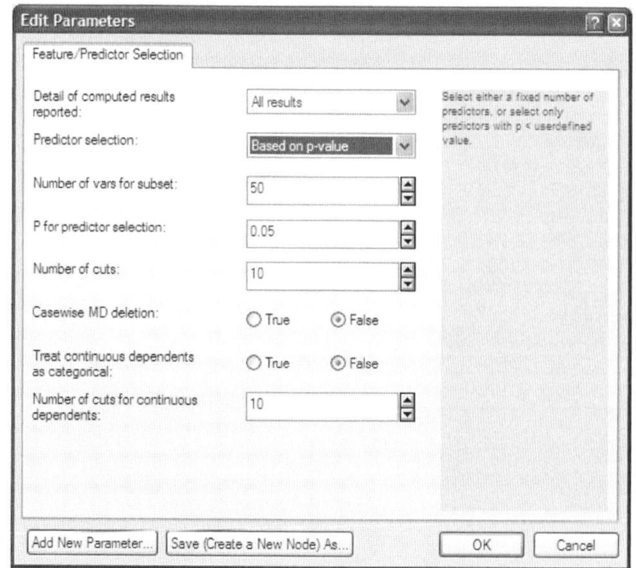

FIGURE F.39

Choose "Based on p-value."

Click OK and then right click on the "Feature" node and run it. Figure F.40 shows selecting running the node, while F.41 shows embedding the output.

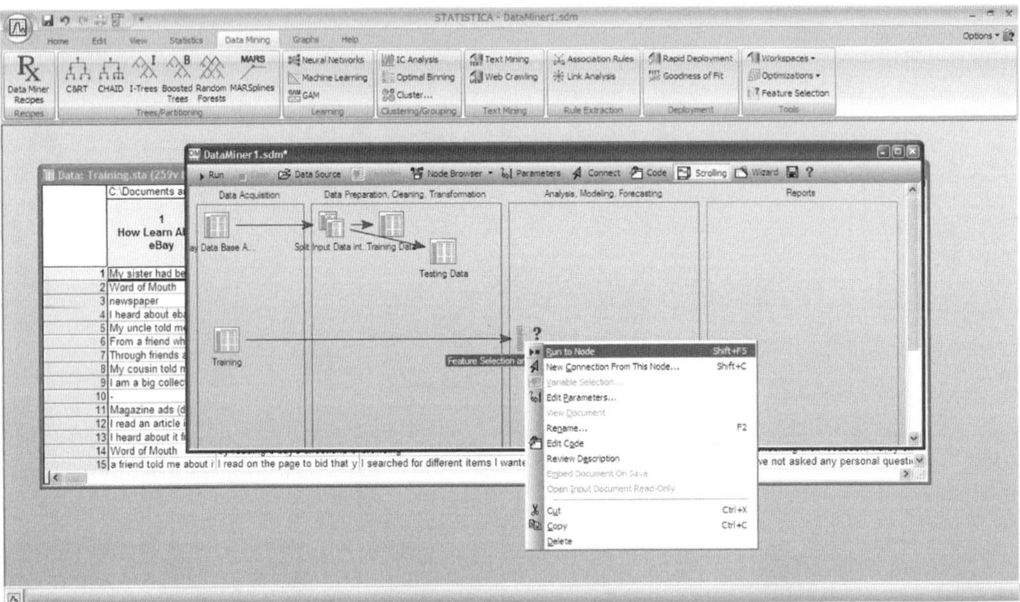

FIGURE F.40

Run the node.

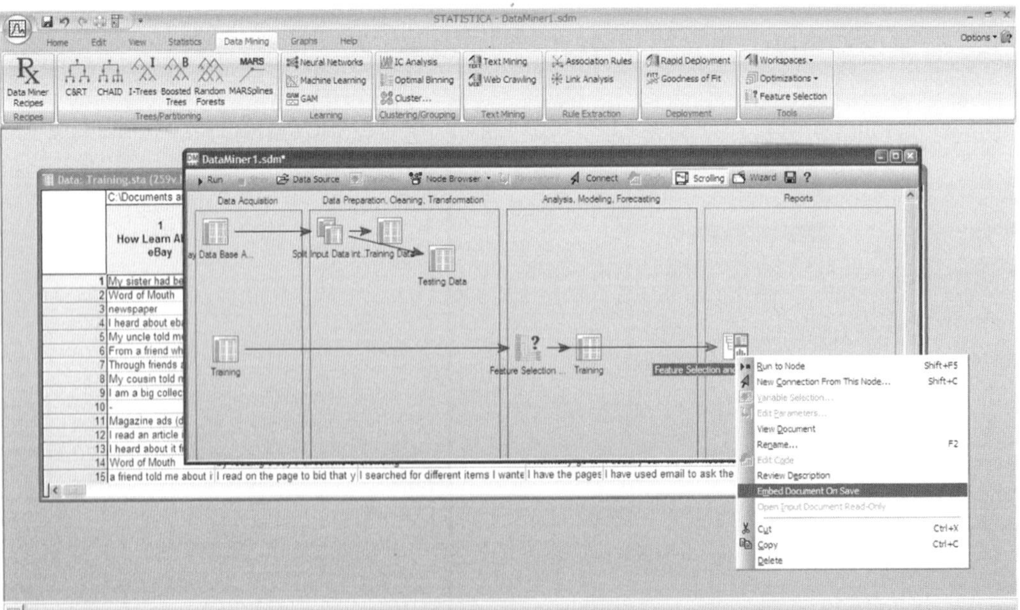

FIGURE F.41

Right click on the output and "Embed Document on Save." Save the data mining project.

Right click on the output and "View the Document," or simply double click on the output. Scroll down to see the graph, which is called the importance plot. The following importance plot resulted (Fig. F.42).

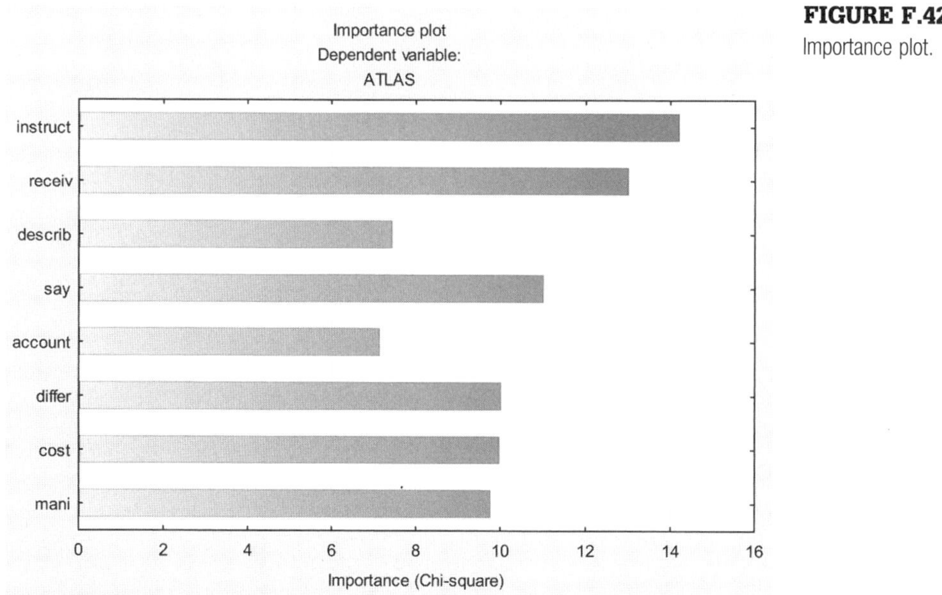

FIGURE F.42
Importance plot.

Eight words were found to be important. Go back to the output and right click on it to rename it (Fig. F.43).

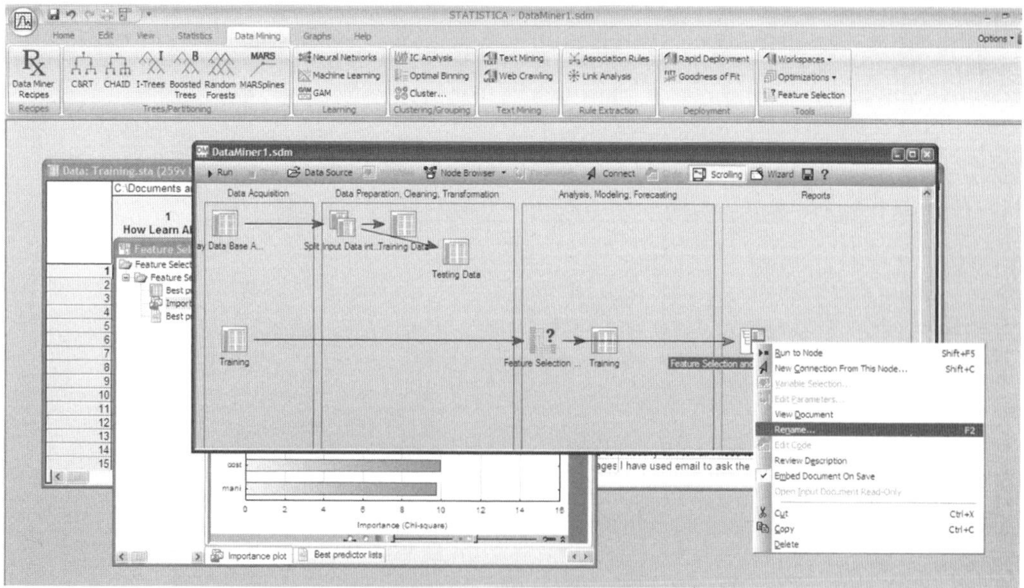

FIGURE F.43
Renaming the output.

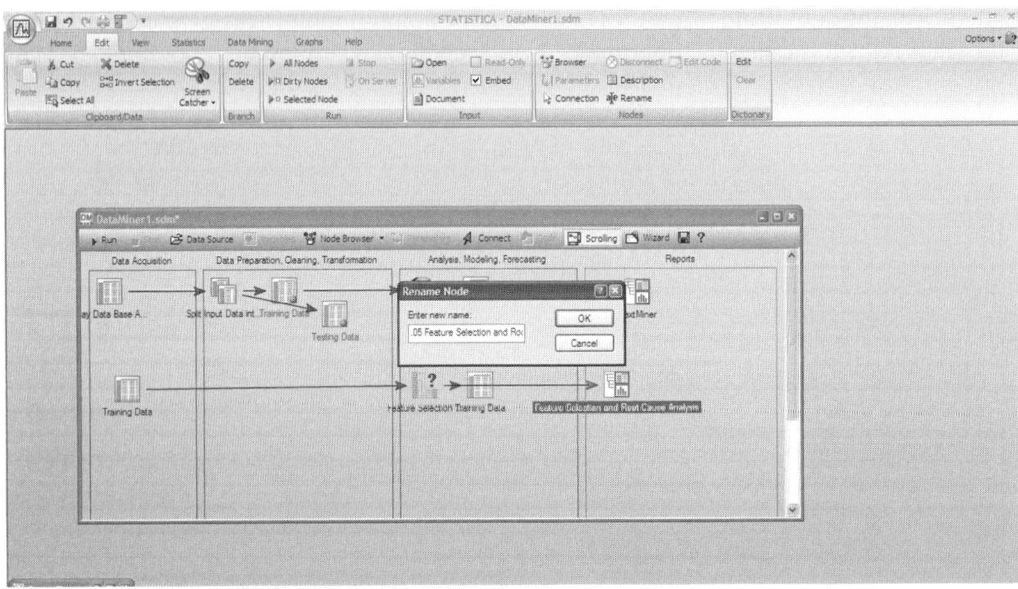

FIGURE F.44
Continuing renaming the output.

Put .05 in the name to remember that this output used the p-value method (Fig. F.44). Now we will go back and repeat the process with different parameters. First, delete the arrow on this output and pull it aside (Fig. F.45).

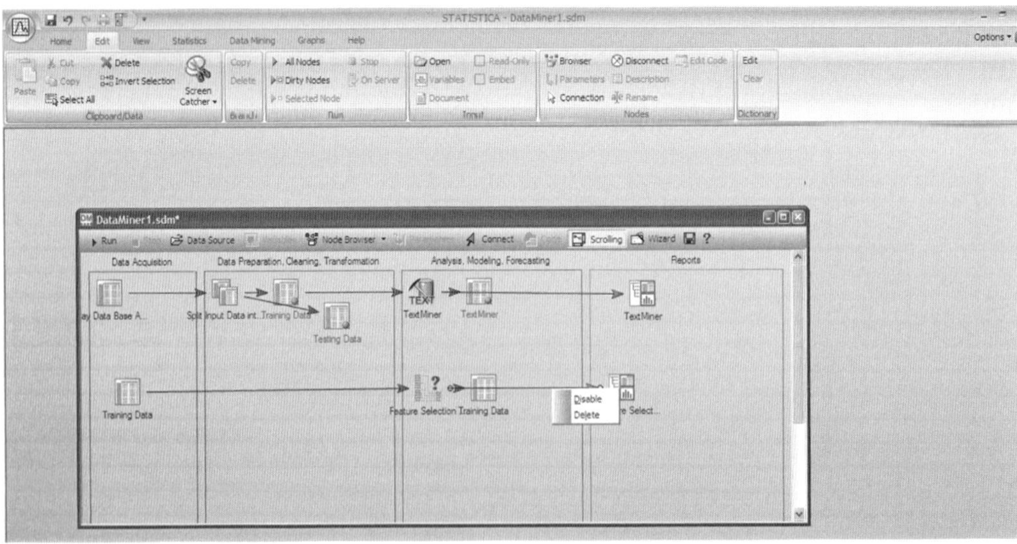

FIGURE F.45
Delete the arrow.

This time in editing the parameters of the node, select "Fixed Number" and 15. Because the p-value gave us 8, 15 at this point should be sufficient to look at. Click OK, run the node, and embed the output for saving. Save the data mining workspace. Open the output. Figure F.46 shows the importance plot based on finding the top 15 predictors.

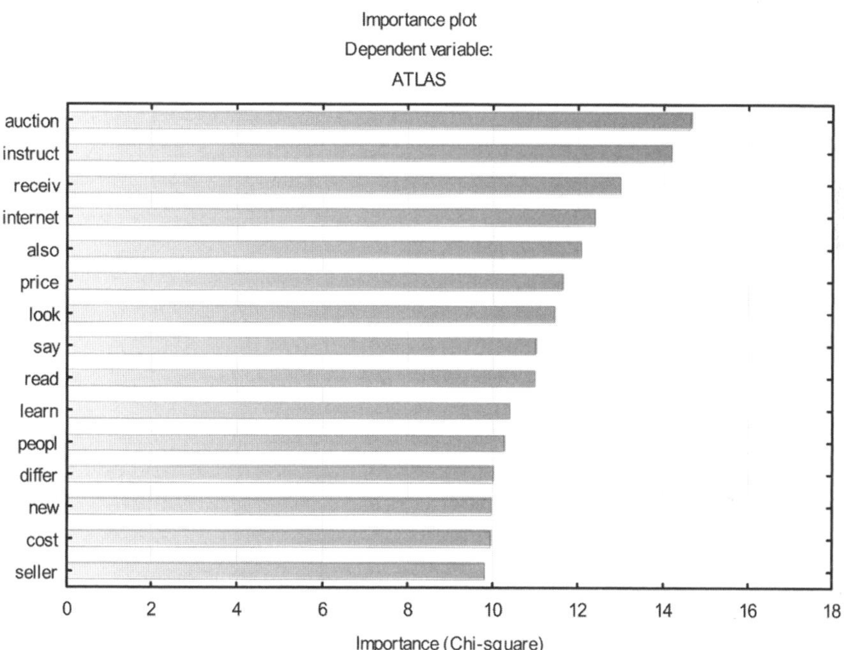

FIGURE F.46
Top 15 predictors.

For this output, the following variables ended up in the top 15 for separating the three groups:

47 151 239 153 36 224 173 250 235 162 214 94 194 82 256

For the p-value analysis, the variables were:

151 239 89 250 28 94 82 182

Common to both are 151, 239, 250, 94, and 82. In other words, the common elements are:

Instruct, receiv, say, differ, cost

Now, to see how these variables might be associated with the ATLAS categories, we did ANOVA procedure (main effects ANOVA in Figure F.47). The graph produced is displayed in Figure F.48. Figures F.49 through F.51 splits out fewer variables in each graph.

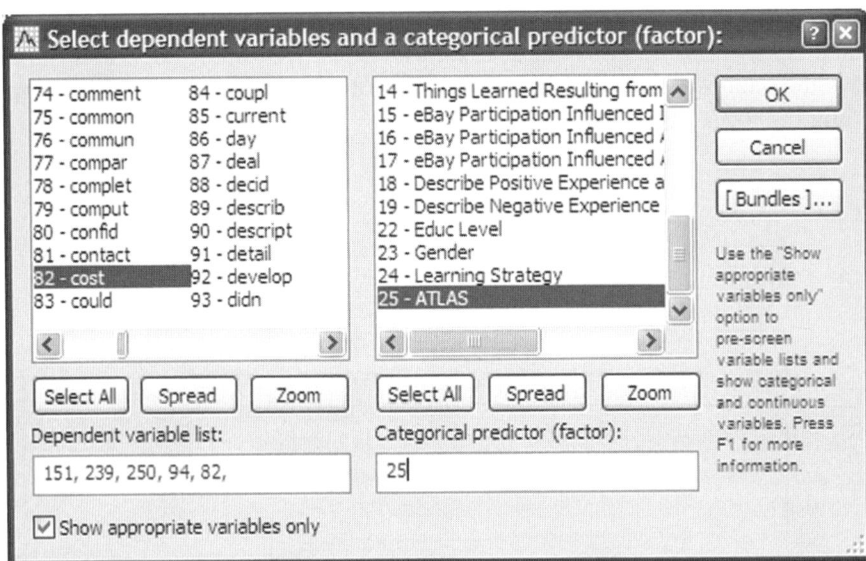

FIGURE F.47
Selecting variables for ANOVA.

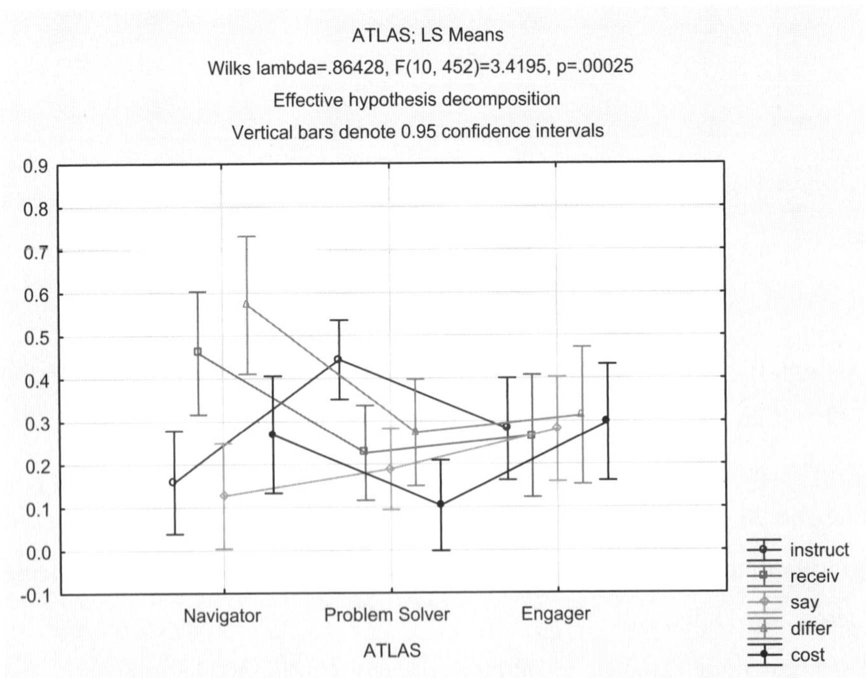

FIGURE F.48
ANOVA graph that is a bit confusing, but the overall F is significant ($p = .00025$).

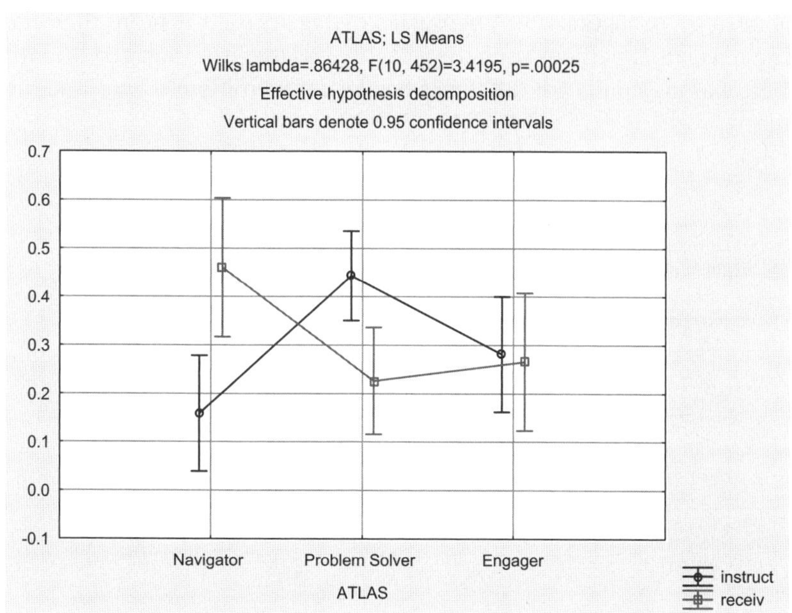

FIGURE F.49

Taking the variables just two at a time: *instruct* and *receiv*.

In Figure F.49 we see that words with the stem *receiv* separated the navigators from the problem solvers and the engagers. Navigators used that word more. However, the navigators used the word *instruct* fewer times than the problem solvers. There probably was not a significant difference between the problem solvers and the engagers with respect to *instruct*.

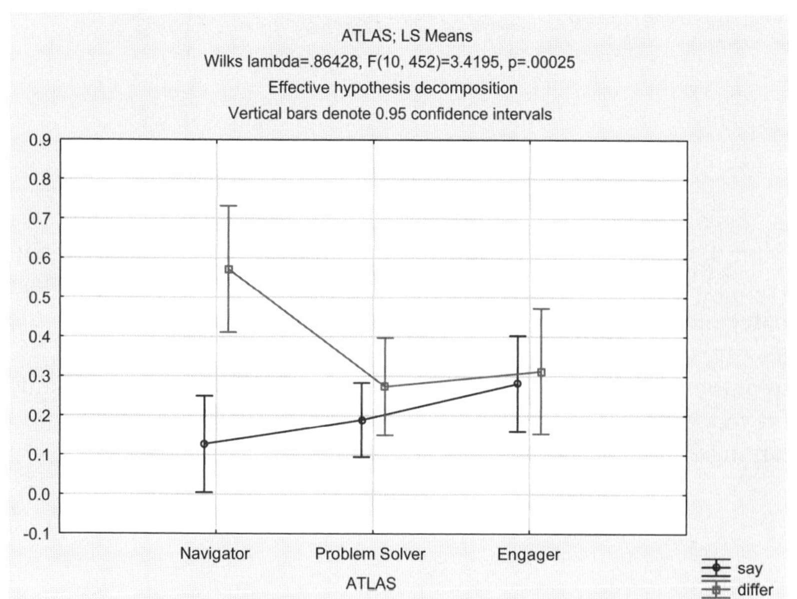

FIGURE F.50

Selecting the words *say* and *differ*.

With respect to the word stem *differ* in Figure F.50, the navigators again used it more than the problem solvers and engagers. The navigators perhaps used the stem *say* less than the engagers but not less than the problem solvers. There was no significant difference between the problem solvers and the engagers for either of those words.

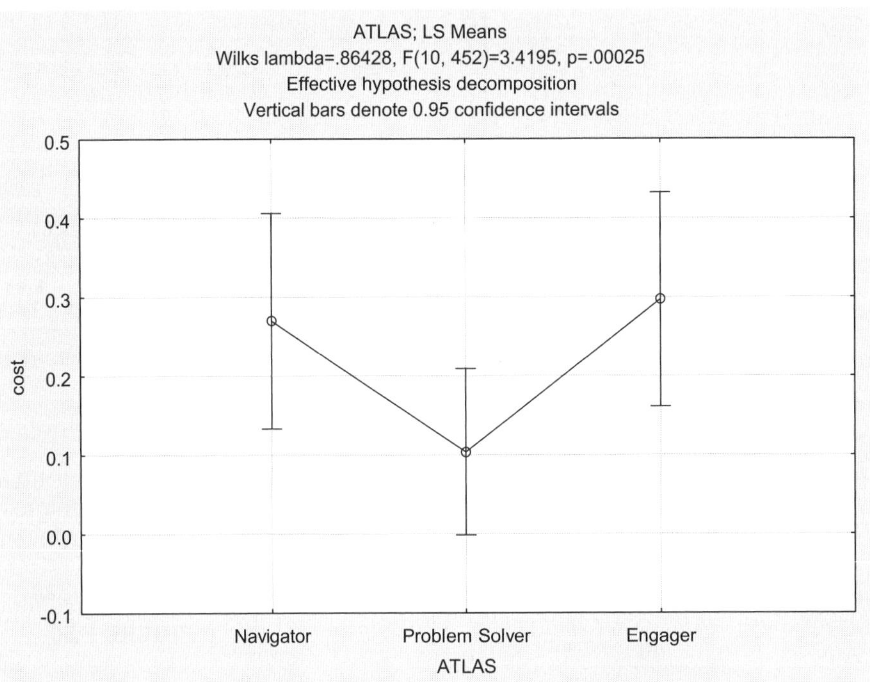

FIGURE F.51
The last word stem: *cost.*

For the last word, *cost* in Figure F.51, if there was a difference (and there was a lot of overlap between all the confidence intervals, so there may not be any difference at all), perhaps the problem solvers used the word *cost* less than either the navigators or the engagers. At least for this analysis, the problem solvers and the engagers were the most alike for most of the words.

At this point, we wanted to see if these five words would have a stable predictive value in separating the three groups. We did a support vector machine using the V-fold method. With the data set up, go to "Data Mining" and "Machine Learning." Select "Support Vector Machine." (See Figures F.52 and F.53.)

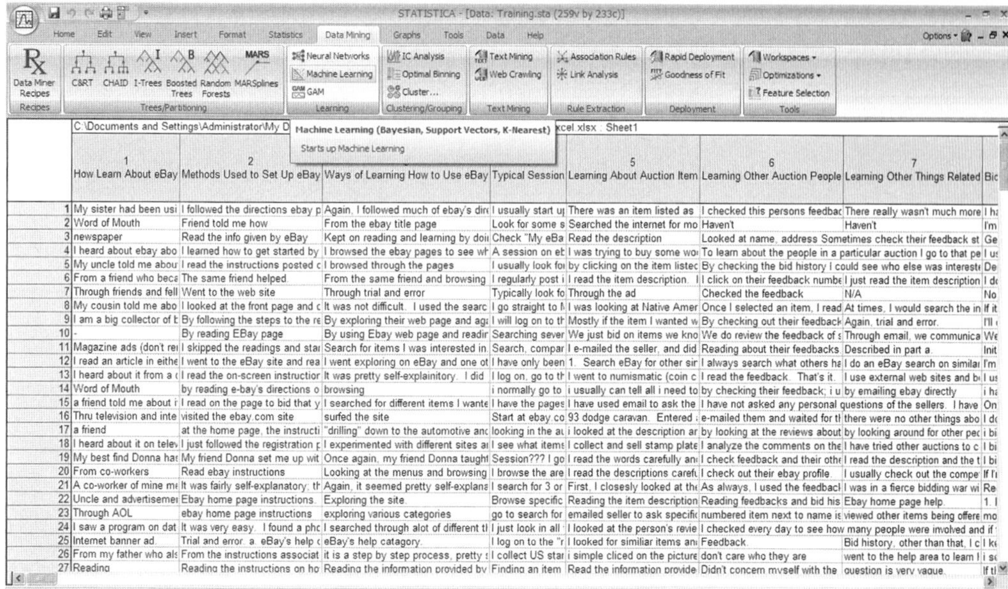

FIGURE F.52
Finding "Support Vector Machine."

FIGURE F.53
Selecting "Support Vector Machine."

Click OK. Select the variables in Figure F.54.

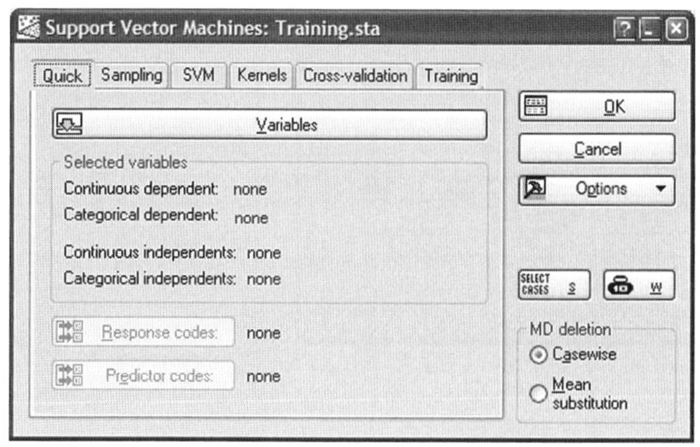

FIGURE F.54
Finding the "Variables" button.

Put in the ATLAS (v25) as the dependent categorical variable and the words (151 239 250 94 82) in as continuous predictors (Fig. F.55).

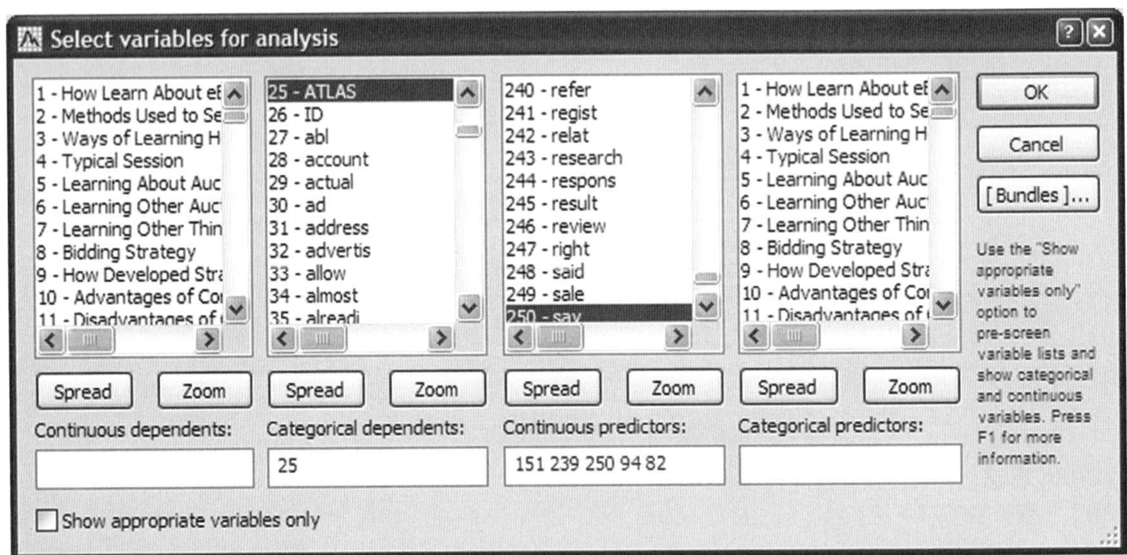

FIGURE F.55

Choosing the dependent and independent predictor factors.

Click OK. Under the "Cross-validation" tab (Fig. F.56), click the box next to "Apply v-fold validation."

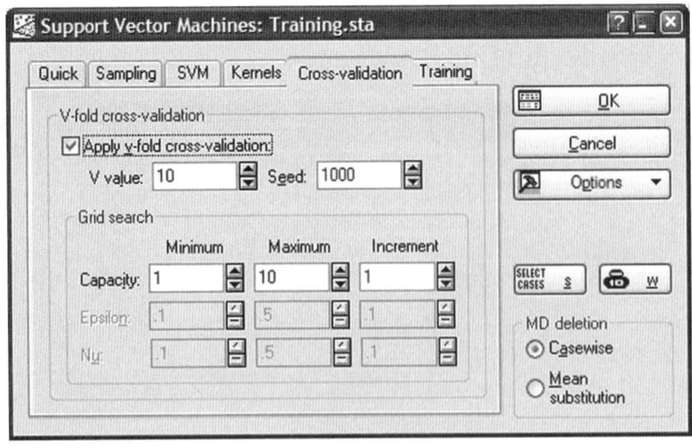

FIGURE F.56

Be sure to "Apply v-fold cross-validation."

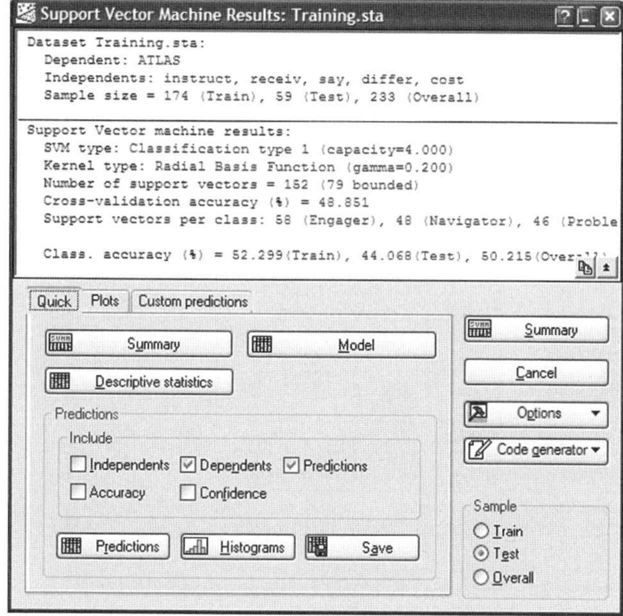

FIGURE F.57
The output revealed.

There appears to be some stability in the prediction model as may be seen in Figure F.57, with the cross-validation accuracy, the training accuracy, and testing accuracy fairly constant. However, the prediction accuracy is terrible. Try again with all 15 variables to see if we can get a better prediction (See Figure F.58).

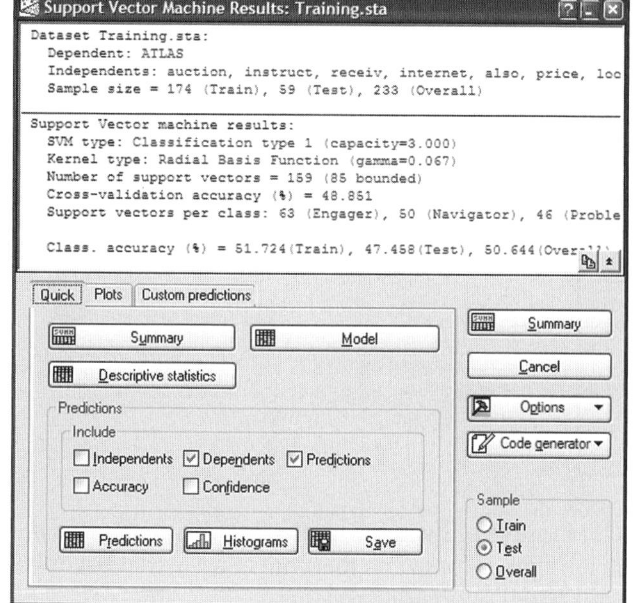

FIGURE F.58
The output using all 15 variables.

The prediction was no better in Figure F.58. Let's try taking away variable 82 in which there was no real difference in the groups. (See Figures F.59 and F.60.)

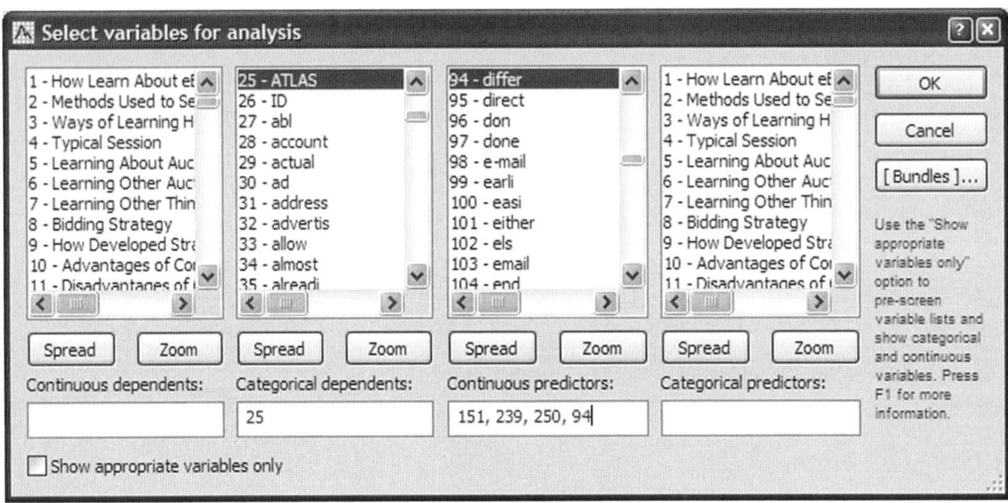

FIGURE F.59

The variable selection looked like this.

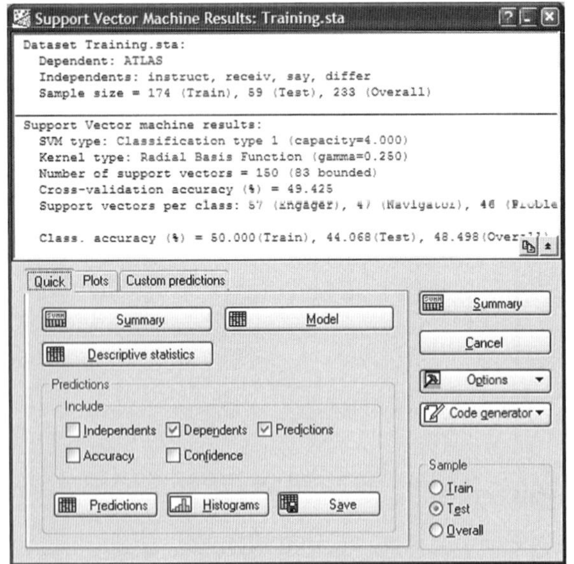

FIGURE F.60

The output, again, using the V-fold method was as follows.

The result in Figure F.60 revealed perhaps a bit more stability in the model, but this certainly is not a good prediction model. We cannot really predict the groups based on the words they used most often. There must be other ways of separating the groups based on their comments.

Now we will move to the graph of the two SVD Concepts 1 and 2. Open the data mining workbook, click on the SVD coefficients, and click "set as the active input" as in Figure F.61.

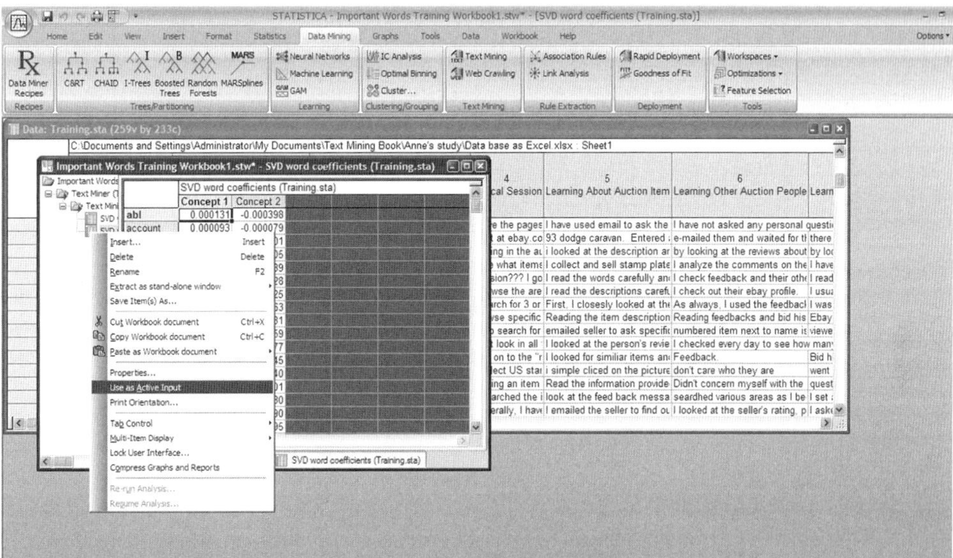

FIGURE F.61
Opening the data mining workbook and setting the SVD coefficients as the active input.

Click on "Graphs and Scatterplots." (See Figures F.62 to F.64.) Click on the variable selection to put in the two concepts.

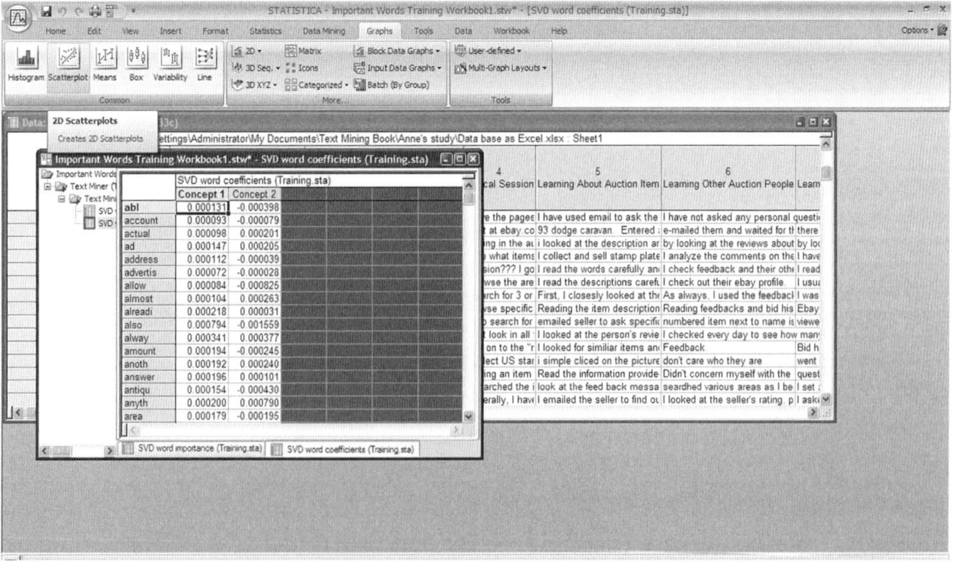

FIGURE F.62
Finding scatterplots.

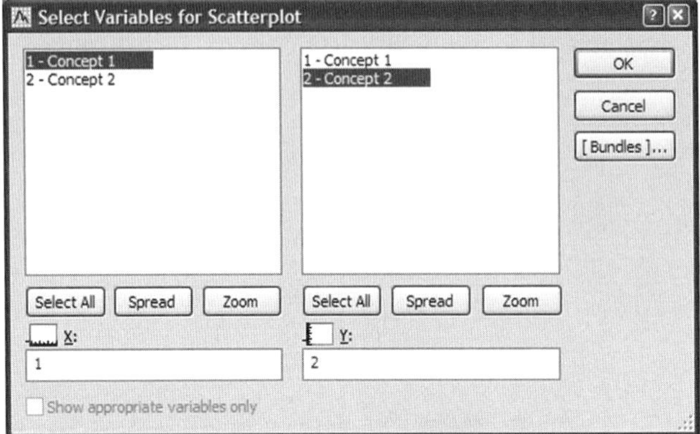

FIGURE F.63
Selecting variable 1 in the left-hand side and variable 2 in the right.

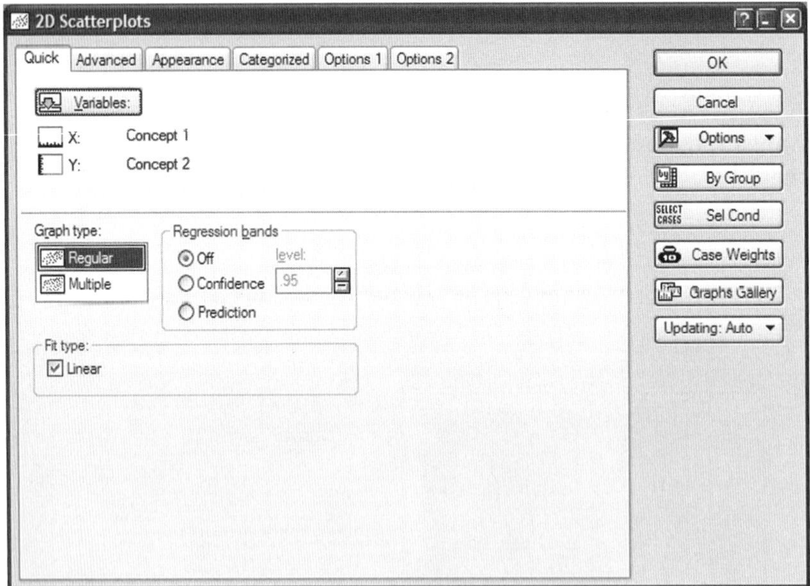

FIGURE F.64
Keeping all the defaults and clicking OK.

Note in Figure F.65 that there is a clump of words and then outliers. We will examine the outliers first using the lasso. Click the brushing tool to the right as in F.66. Click on lasso and also on label.

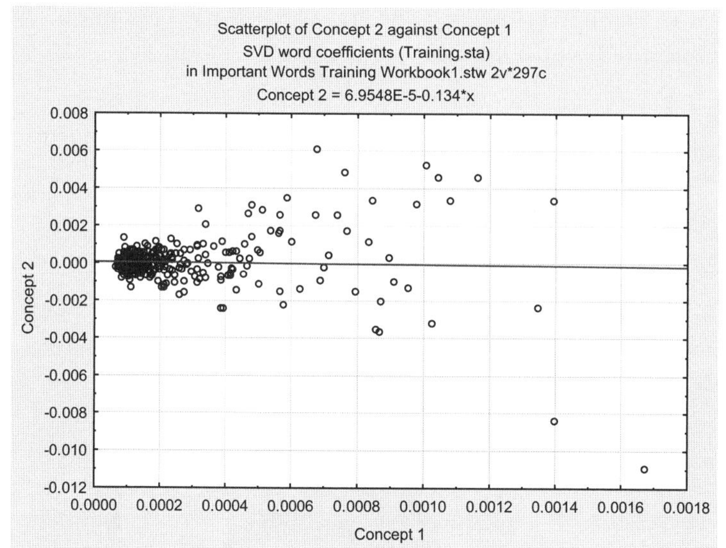

FIGURE F.65
Graph of Concepts 1 and 2.

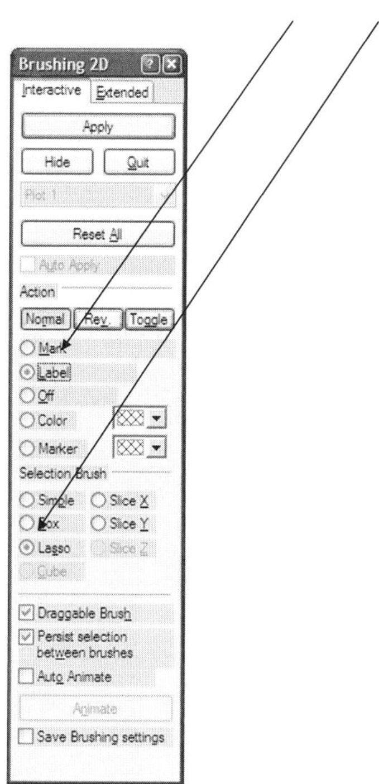

FIGURE F.66
Brushing tool. Click on "Label" and on "Lasso."

Then make a lasso around some of the grouped outliers as in Figures F.67 and F.68 and click "Apply." One may run into a warning, but simply click through it.

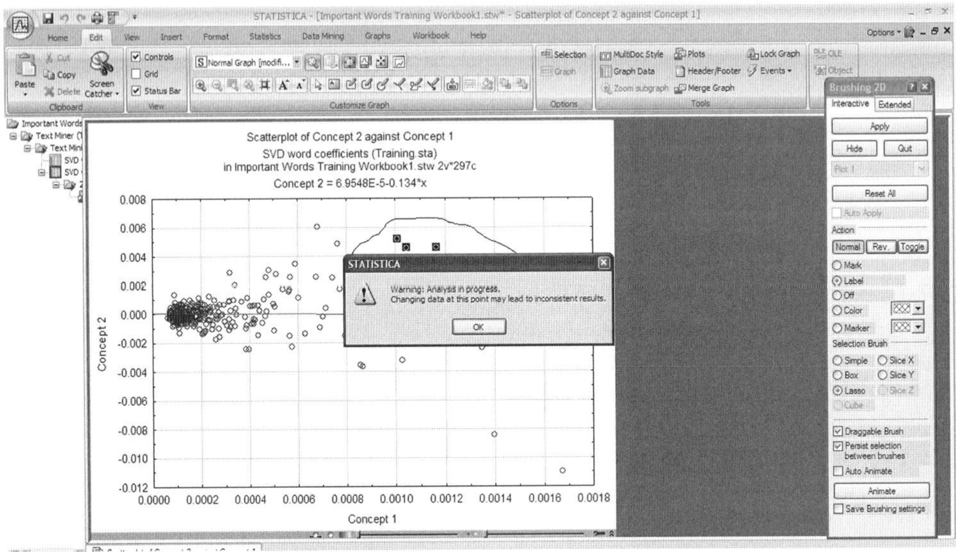

FIGURE F.67

Observe the warning. Click OK and ignore the warning.

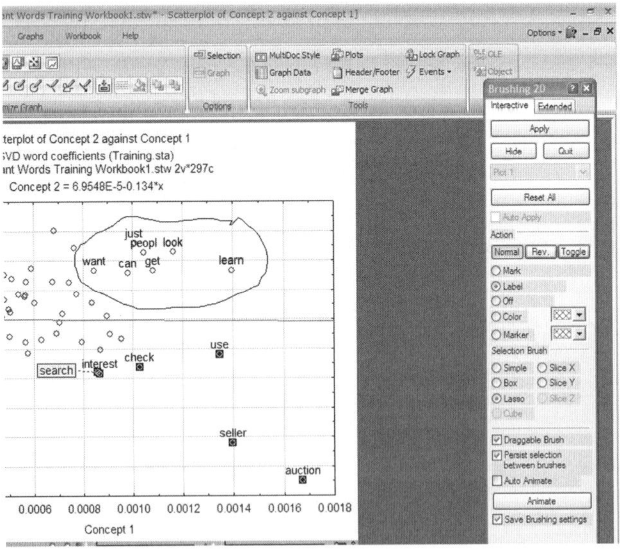

FIGURE F.68

Clicking "Apply" and observing the words.

The words *can, get, want, people, look, learn,* and *just* were often found together (Fig. F.69). This is not surprising because people are involved, and people often just look to learn to get what they want. The second lasso yielded these words.

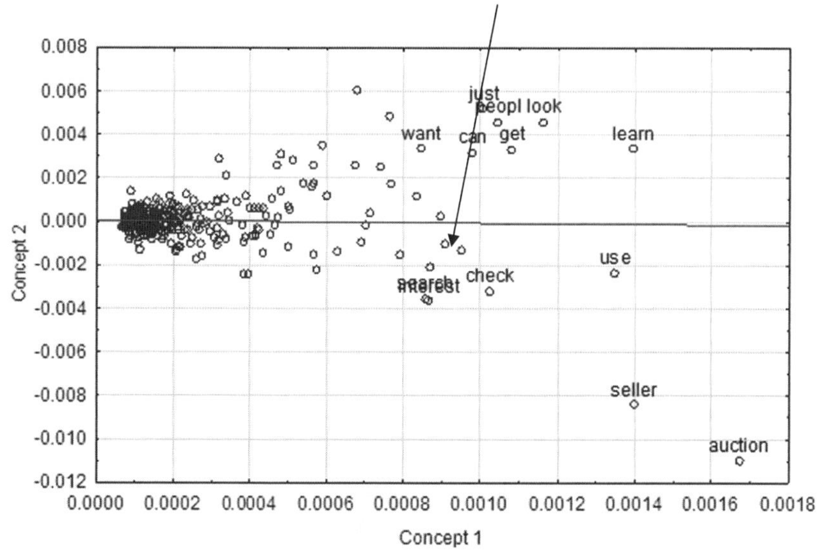

FIGURE F.69
Using the lasso a second time: words overlapping.

We will have to drag the overlapping words apart to see them well. Click on them and try dragging them to the side to separate them as in Figure F.70.

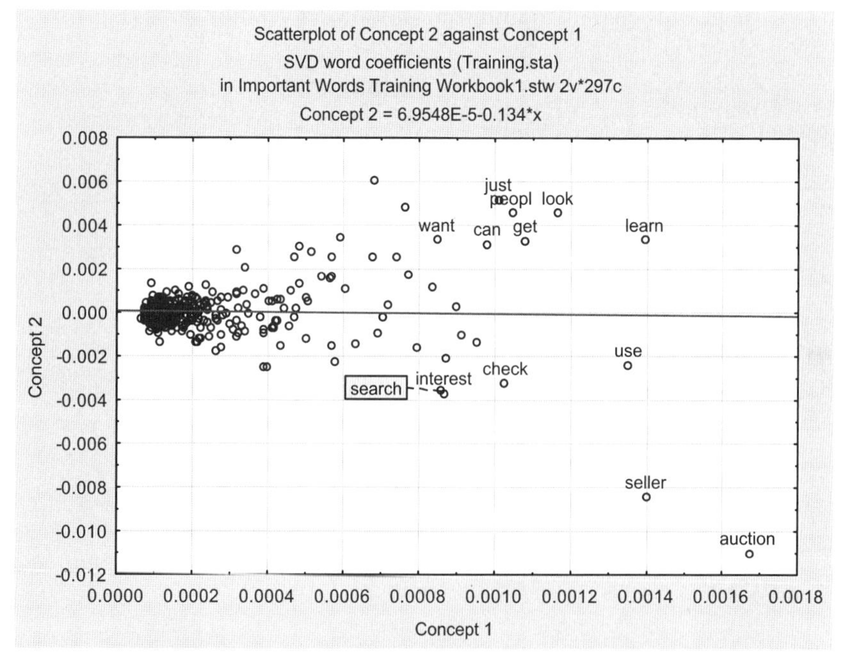

FIGURE F.70
Separating words.

So one might search with interest, checking for useful items that a seller had up for auction. We can also examine the large clump by using the "Zoom in" tool. Then one can again use the lasso and label technique. I copied the scatterplot to do this in an effort not to lose the earlier lasso information. Use the "Zoom in" tool on the second plot in Figure F.71. After just one click in the middle of the clump, this emerged.

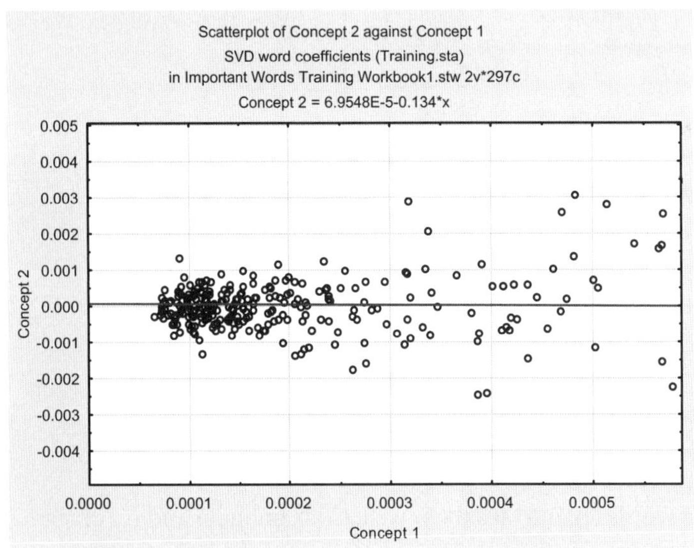

FIGURE F.71

Using the "Zoom in" tool to separate the clump.

Now we can use the lasso tool, drag, and separate to find some additional relationships (Fig. F.72).

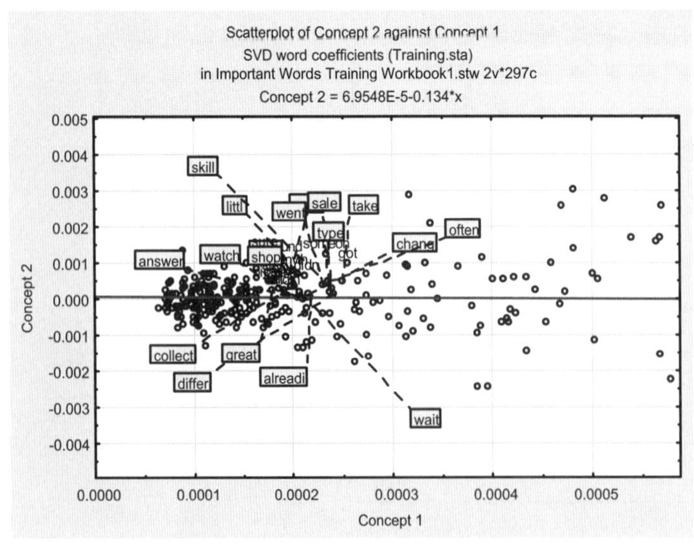

FIGURE F.72

Separating words and examining the cluster.

The process can yield some interesting relationships. One way to move the screen is to use the little hand that shows up when the cursor is held over an axis. One might try looking just at word pairs using the lasso. Figure F.73 illustrates further attempts at examining words that show up together.

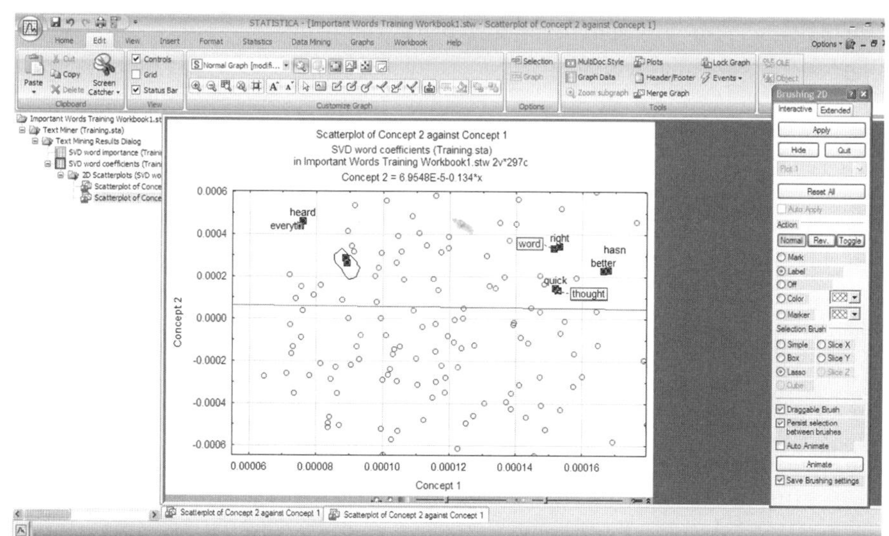

FIGURE F.73 Examining pairs of words that tend to be used together by many people.

These words presumably share semantic space. This certainly seems true when we look at the pairs in Fig. F.74.

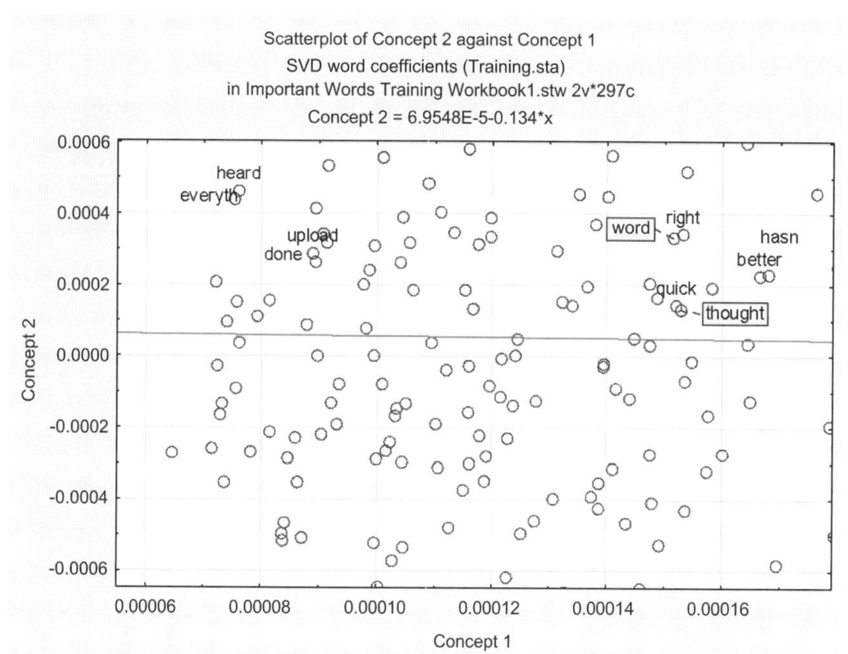

FIGURE F.74 Viewing the selected pairs of words.

Some pairs are "heard everything," "upload done," "right word," and "quick thought." We made three subdatabases to examine the important side by side words by group. We opened the training data, copied each of the groupings, and made three separate data files called "Engager Training," "Navigator Training," and "Problem Solver Training." Pull up each of the databases into *STATISTICA* as we use each below. We will start with the navigators. After opening the database, open Text Mining. (See Figure F.75.)

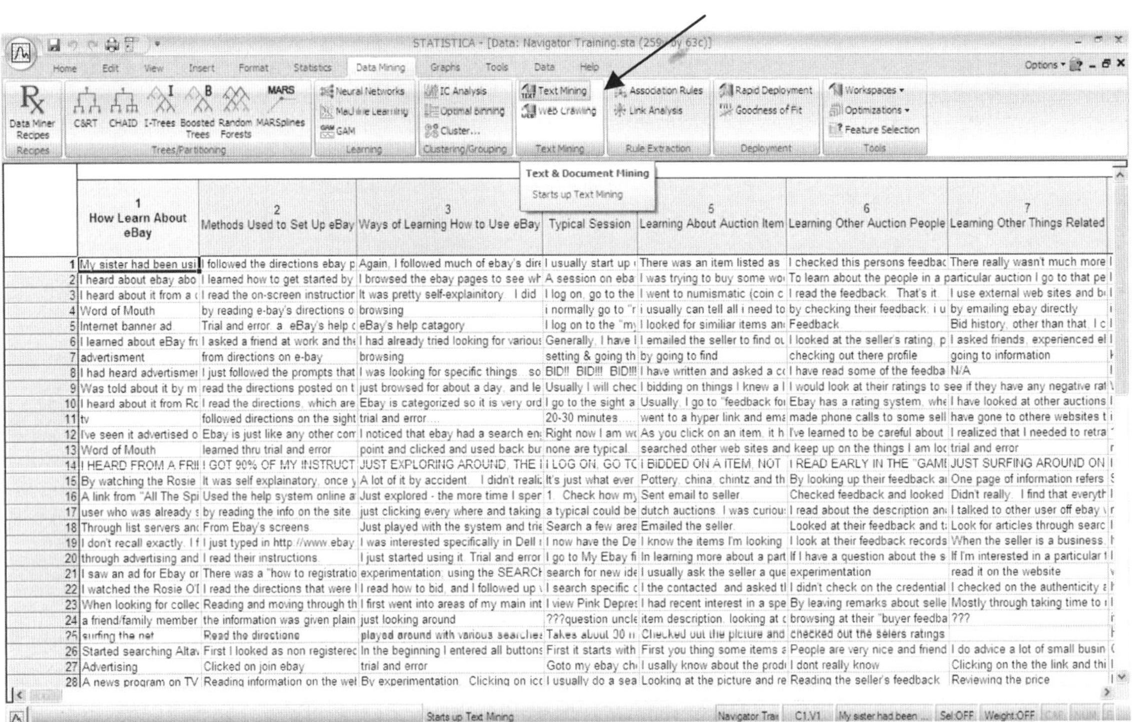

FIGURE F.75

Opening "Navigator Training" and "Text Mining."

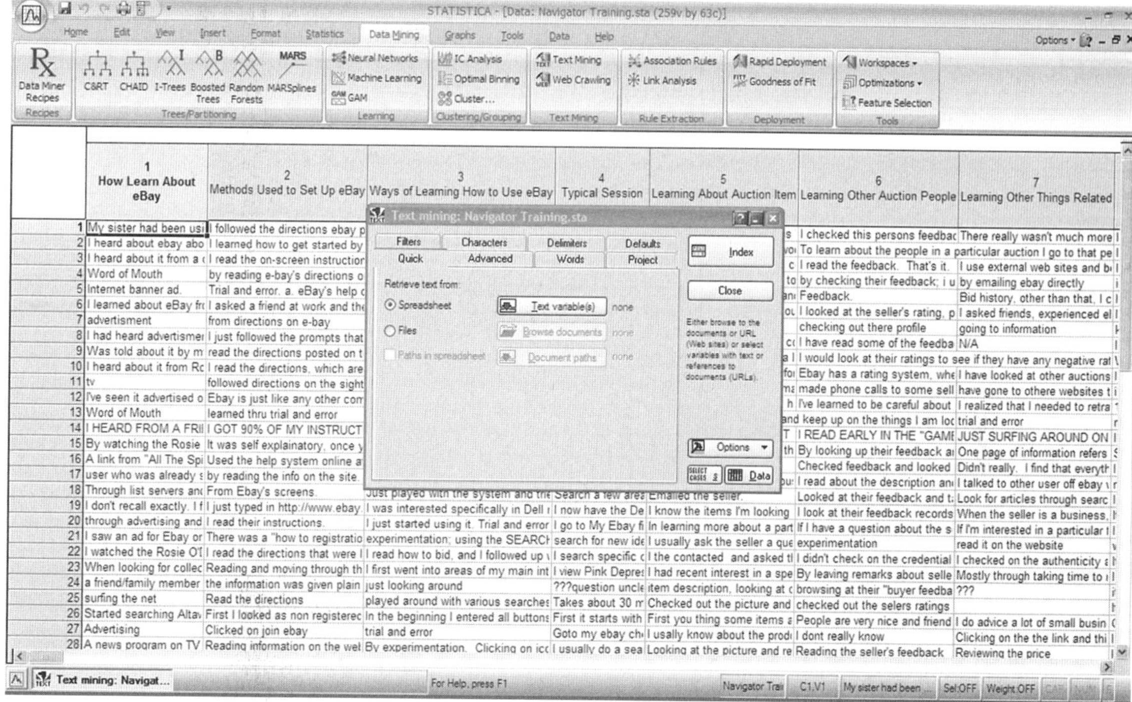

FIGURE F.76
Selecting the variables.

Select the variables (1–19) as in Figure F.76. Set the minimum at 10 and the maximum at 80 as in Figure F.77.

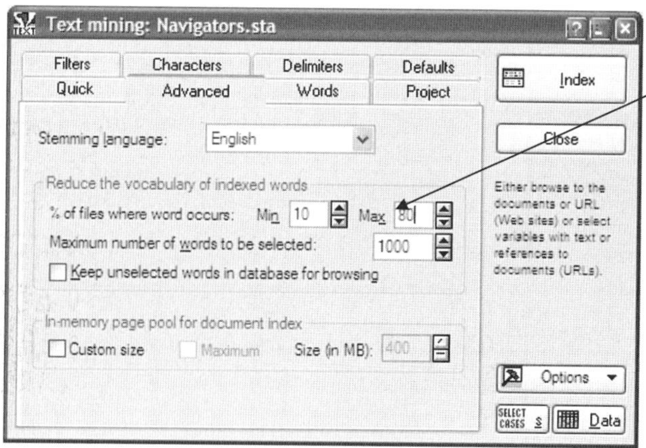

FIGURE F.77
Under advanced, select 10 to 80.

Click "Index." Let the existing database be eliminated. Figure F.78 shows there were 287 words identified.

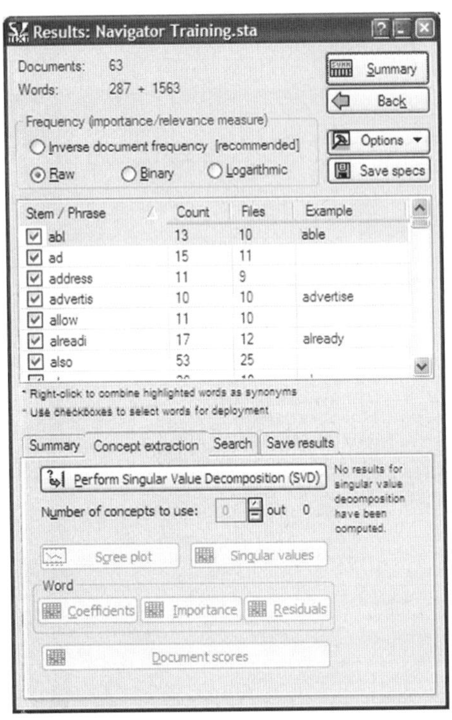

FIGURE F.78

The output. Note there were 287 important words identified.

Click the singular value decomposition shown in Figure F.79, and then order the words using sort. Repeat this process for the engagers and the problem solvers. Obtain the word list for each group as demonstrated in Figure F.80.

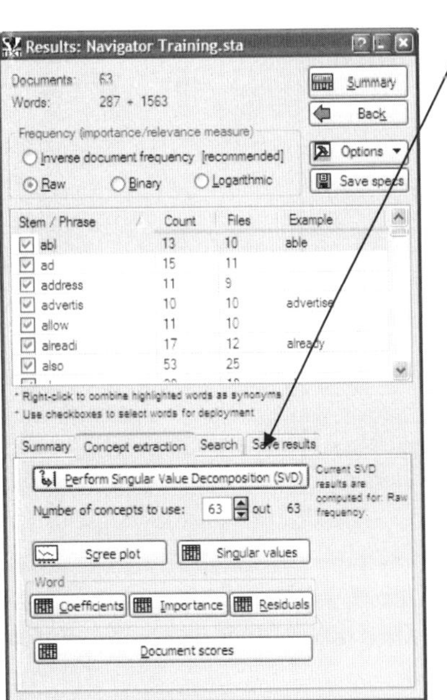

FIGURE F.79

Clicking "Perform Singular Value Decomposition (SVD)."

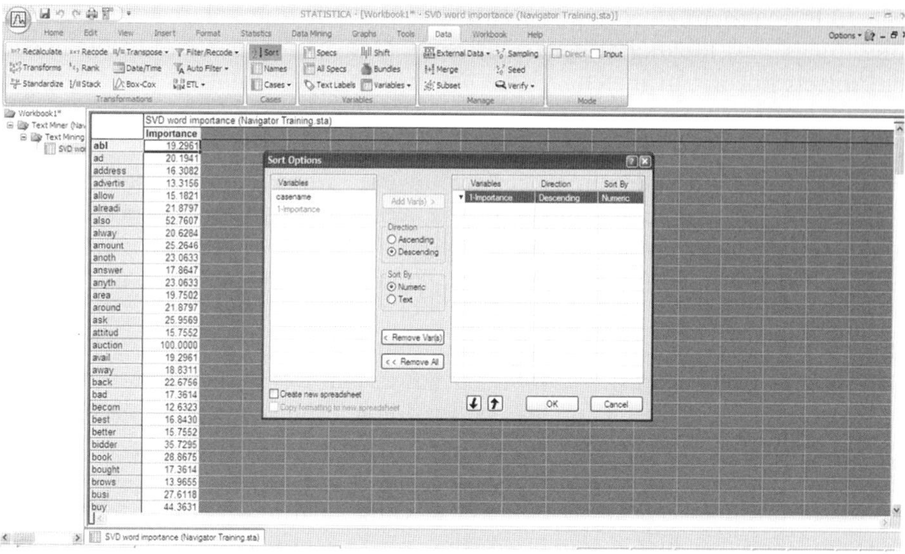

FIGURE F.80
Click “Sort,” add variable, and click “Descending” to obtain an ordered word list.

Use the screen catcher to copy the top importance words for the navigators. Repeat the preceding directions for the engagers and the problem solvers. In the Figure F.81 we put the words side by side.

Navigators		Engagers		Problem Solvers	
	SVD word impo		SVD word imp		SVD word impc
	Importance		Importance		Importance
auction	100.0000	auction	100.0000	auction	100.0000
use	93.3992	seller	85.8484	seller	86.8847
look	89.4229	check	78.7364	use	83.4356
seller	83.3688	see	70.5060	learn	74.2915
learn	81.2142	get	69.3696	look	73.5641
get	80.6665	want	68.1082	peopl	64.5302
peopl	78.4378	will	67.7892	will	62.6302
check	71.2104	can	67.6825	just	61.8122
sell	69.9544	use	65.5127	check	61.4665
internet	68.5462	look	65.0700	interest	58.9956
search	65.3686	just	63.1541	get	57.0041
will	65.2328	peopl	60.5848	thing	56.7398
realli	64.6869	ve	57.7767	search	56.6372
time	64.2745	go	56.7674	time	56.4163
just	62.7389	would	56.3843	also	56.3582
can	62.5974	like	53.4909	can	54.1965
price	61.3096	also	53.0841	see	53.5453
one	60.4358	much	53.0841	ve	53.2932
see	60.4358	feedback	52.8111	one	51.8507
interest	58.3460	interest	52.3991	comput	50.1417
thing	57.4271	internet	51.9838	realli	48.2257
know	55.8620	realli	51.2841	want	48.1335
go	55.3839	purchas	50.5747	price	47.9409
want	54.7399	good	50.5747	read	46.9415
good	53.4285	search	50.2882	internet	46.7244
also	52.7607	well	49.7101	like	45.6792
feedback	51.3989	thing	48.8302	go	42.3353
site	49.8224	time	48.3843	e-mail	42.2026
comput	49.6441	email	47.0211	buy	41.8060
read	48.1943	know	46.2464	would	40.2269

FIGURE F.81
Word lists side by side.

People in all three groups all seem to be checking, looking, learning, mentioning people, listing wants, timing things, and looking at what they want. "Price" is in the top 29 for both navigators and problem solvers but not in the engagers. One interesting difference is that feedback occurs in the top 30 for navigators and engagers but down the line further for the problem solvers. The navigators were interested in getting what they wanted and thus giving them a natural interest in learning about eBay. It is interesting to speculate, however, when words occur a bit more often for one group than for another. Qualitative research is, after all, an exploratory process, and all ideas are valid as suggestions for hypotheses. One thing that we can do is perform a feature selection on specific questions to see if questions that differ conceptually might yield important predictors describing the groups.

Open the training data set. Again, we used 10 and 80 percent as our percentage limits. This time, we developed a synonyms list. We thought that certain words seemed to go together such as trial and error and hunting and pecking; learned by doing, exploring, and experimentation seemed go together, as did followed instructions, followed directions, and reading directions. First, we pulled up navigator database. We selected only variable 2 as our focus. This question involved how the people first set up their website. The ATLAS was intended to separate people based on their learning strategies when they first encountered something new to learn. We would expect that their strategies would differ with respect to the question "What methods did you use to set up your eBay account?" This question involved a beginning process, and ATLAS was designed to separate people's learning style when they first encountered new information or new learning possibilities. We selected variable 2 (Fig. F.82).

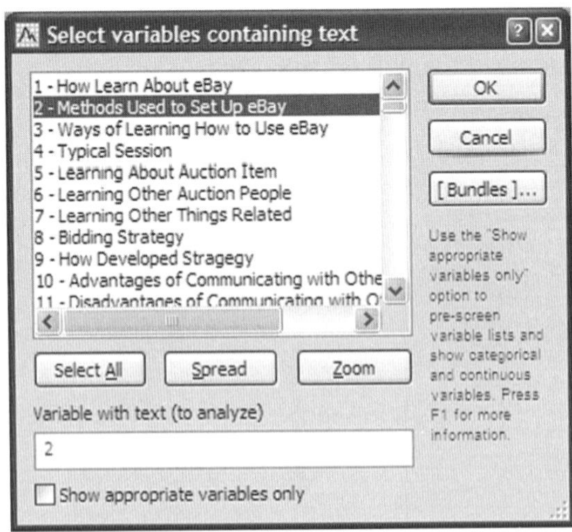

FIGURE F.82
Selecting the variable containing the text.

Under the words tab in Figure F.83, find the spot where one can develop a synonym file.

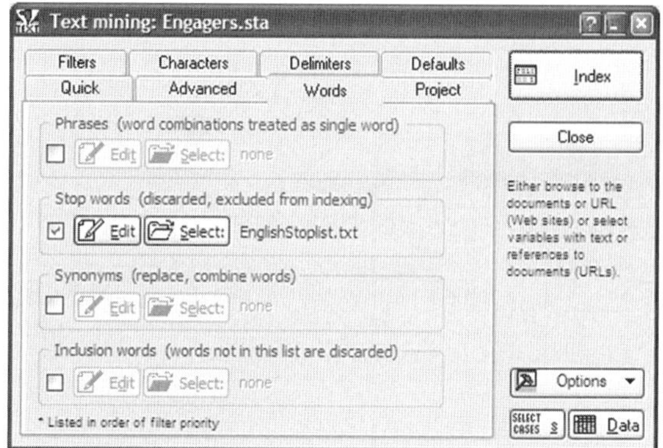

FIGURE F.83
Clicking on the "Words" file.

Note that the "Stop" file is selected by default (Figs. F.84 & F.85). Click on the box under "Synonyms."

FIGURE F.84
Click on the "Synonyms" box.

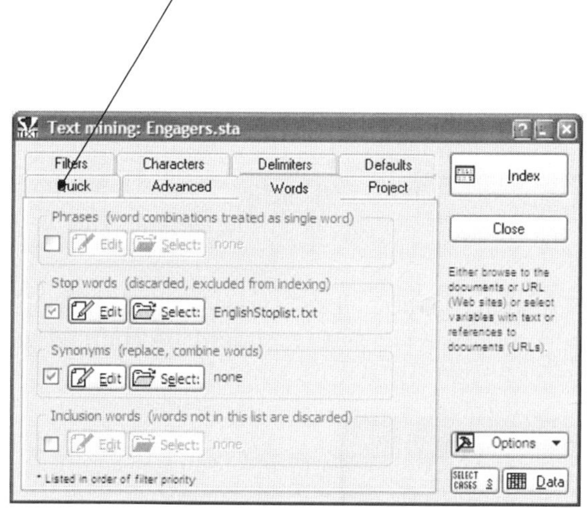

FIGURE F.85
Now click on the "Edit" box.

Here is where we can add new synonyms. First in Figure F.86, we add the root and then we update the synonyms.

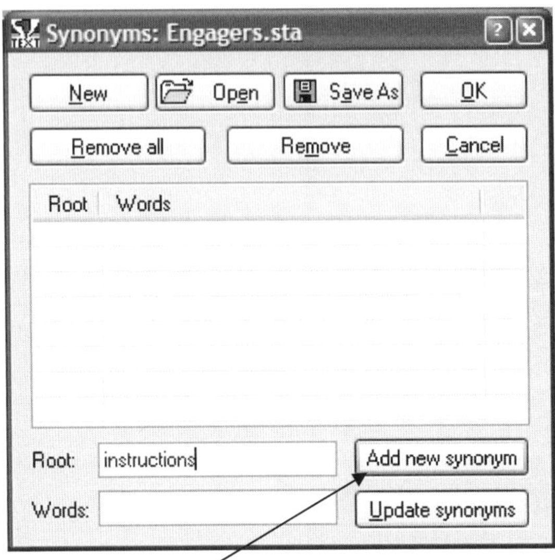

FIGURE F.86
Put the word "instructions" in the root box.

Click on "Add new synonym." (See Figure F.86.) Note that the root, "instr" shows up in the "Root" column in Figure F.87. Figure F.88 shows how to add multiple words that should be counted as synonyms of the root word.

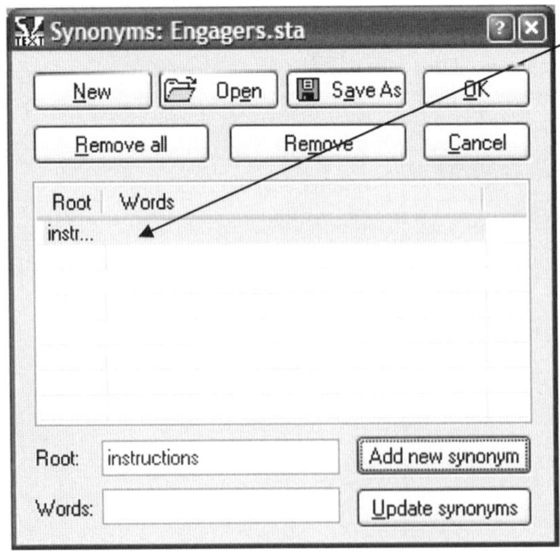

FIGURE F.87
The root of instructions goes into the "Root" box. Highlight the root.

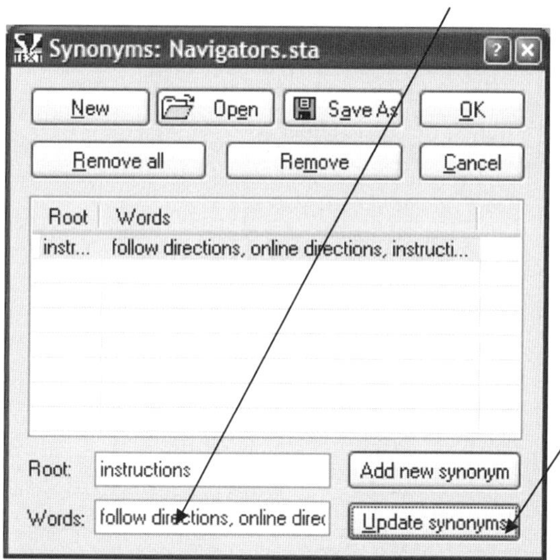

FIGURE F.88

Add the synonym words using a comma between each synonym, and then update the synonyms.

Add the next root word "exploring." (See Figure F.89.)

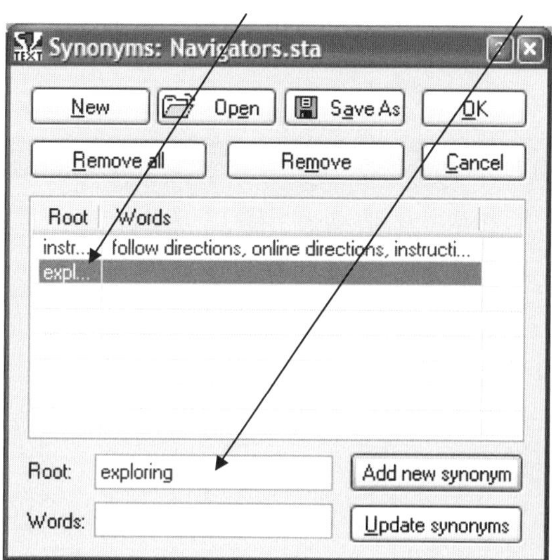

FIGURE F.89

The root *expl* comes from the word "exploring."

Highlight the root. Then add the synonyms by updating them (experimenting, experimentation, exploring, browsing, browsed). Figure F.90 illustrates the result.

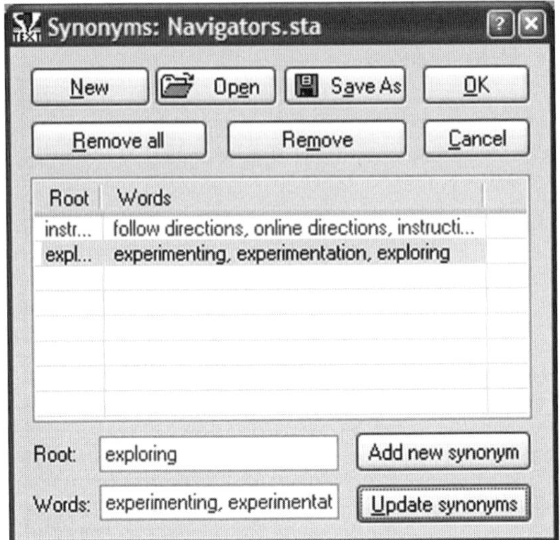

FIGURE F.90
Adding the synonyms for root *expl*.

Finally add "hunt and peck," which goes in as "hunt." Then highlight that root and add the synonyms (learn by doing, trial and error, hunt and peck). (See Figure F.91.)

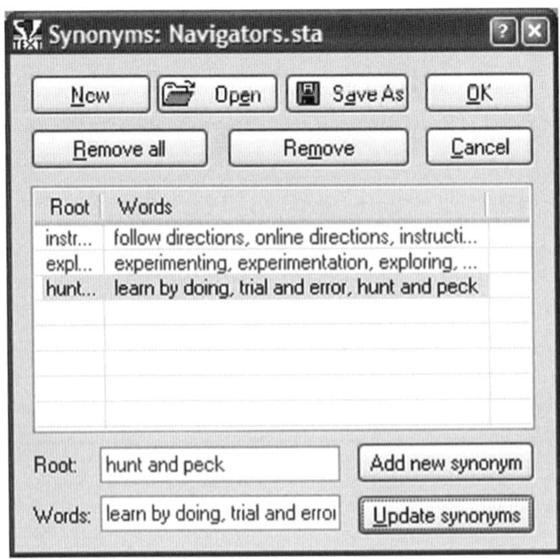

FIGURE F.91
Adding "hunt and peck" as a root.

Click somewhere else so that hunt is not highlighted. Save this list in the same folder as the others for this tutorial. Click on "Edit" to make sure all the roots are saved. Figure F.92 demonstrates saving the synonym list.

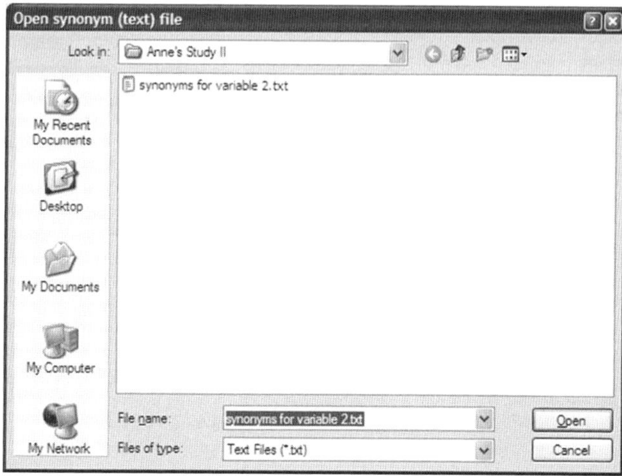

FIGURE F.92
Saving the synonyms list, which we called synonyms for variable 2.

Click OK, and now Index the words. Again, we wish to perform singular value decomposition and so make sure that the synonyms list has been opened (Fig. F.93).

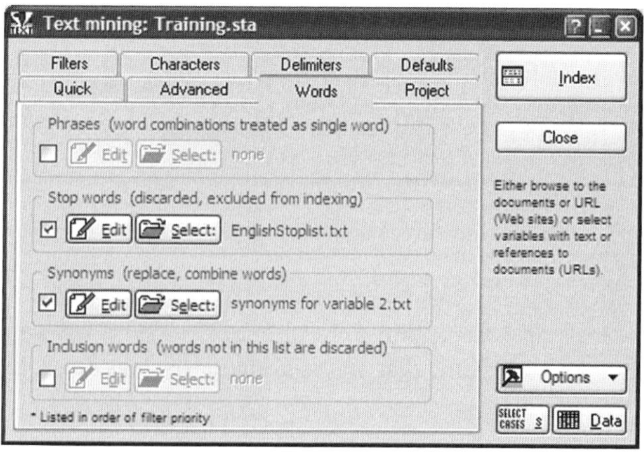

FIGURE F.93
Results of opening the synonyms list by selecting them.

Once again, click on "Singular Value Decomposition" so we can save the words to our data set and then use them for our feature selection analysis. Figure F.94 shows the resultant six important words.

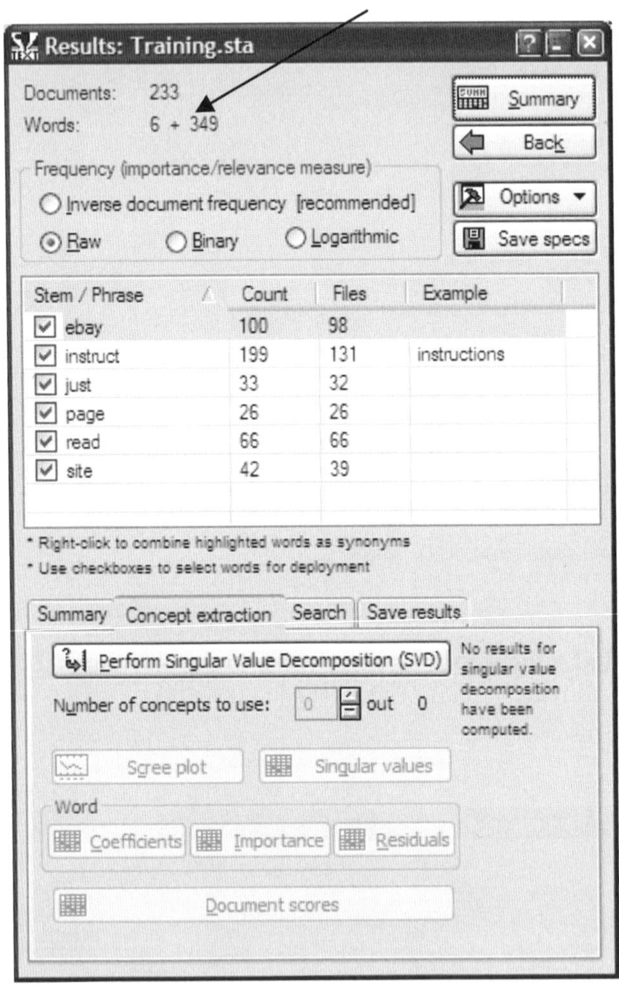

FIGURE F.94
Note there are now six important words.

Click on "Save results," and then put 6 in the box. Again, we wish to perform singular value decomposition and so make sure that the synonyms list has been opened. Then, index the words (Fig. F.95).

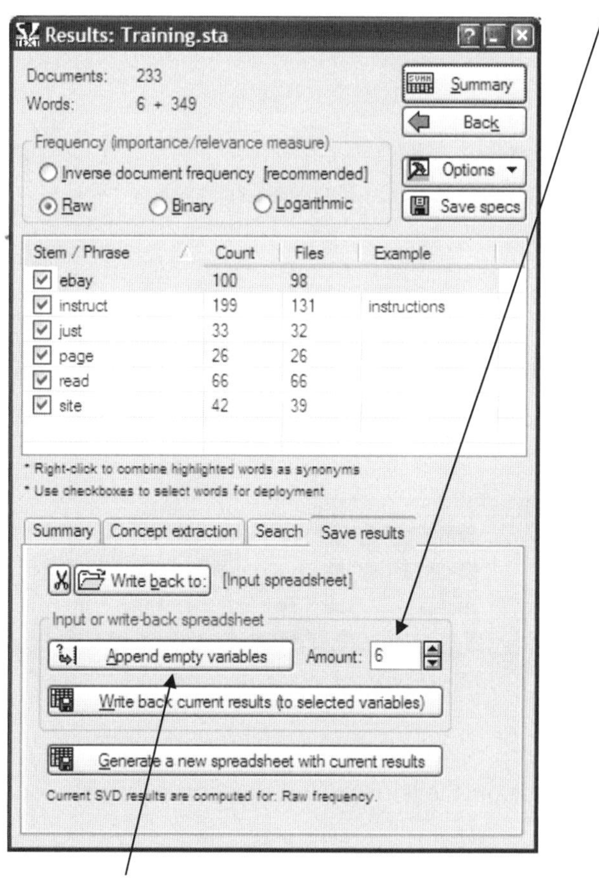

FIGURE F.95

Words have been indexed. The number 6 has been placed in the amount box.

Next append empty values. Figure F.96 shows the box that emerges telling us the new variables have been added to the spreadsheet.

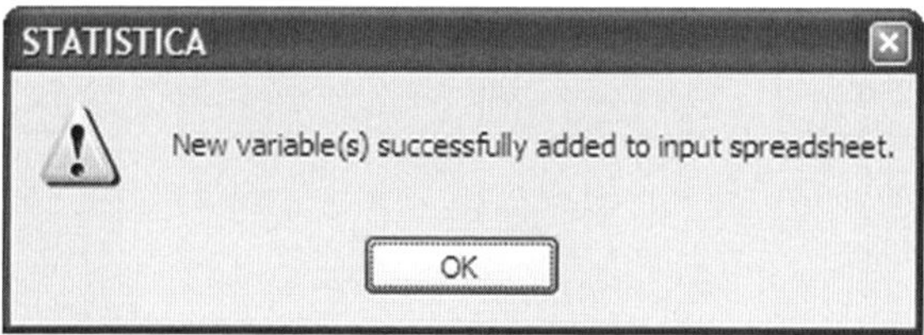

FIGURE F.96

Notice that the new variables have been added to the spreadsheet.

Click OK, and then "Write back current results to selected variables." (See Figure F.97.)

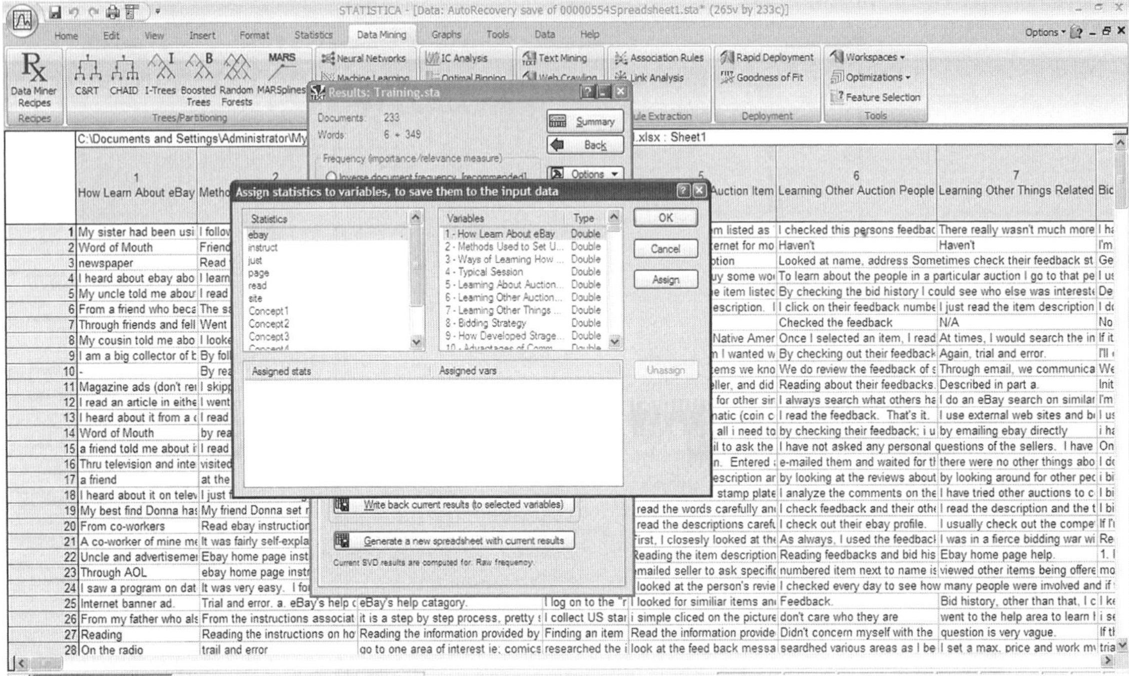

FIGURE F.97
Result of clicking writing back.

Select on each side – words on the left side and the empty variables on the right side as seen in Figure F.98.

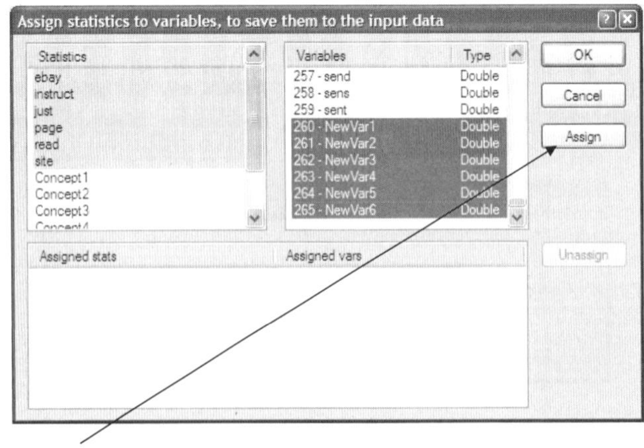

FIGURE F.98
Selecting the words on the left and the new variables on the right.

Assign them as in Figure F.99.

Assign statistics to variables, to save them to the input data

Statistics
Concept1
Concept2
Concept3
Concept4
Concept5
Concept6

Variables	Type
251 - search	Double
252 - second	Double
253 - see	Double
254 - seem	Double
255 - sell	Double
256 - seller	Double
257 - send	Double
258 - sens	Double
259 - sent	Double

Assigned stats	Assigned vars
ebay	260 - NewVar1
instruct	261 - NewVar2
just	262 - NewVar3
page	263 - NewVar4
read	264 - NewVar5
site	265 - NewVar6

OK
Cancel
Assign
Unassign

FIGURE F.99

Result of assigning the words to the variables.

Then click OK. Now the terms are added to the end of the training data set so we can use them in a feature selection. Figure F.100 shows how the data spreadsheet should look.

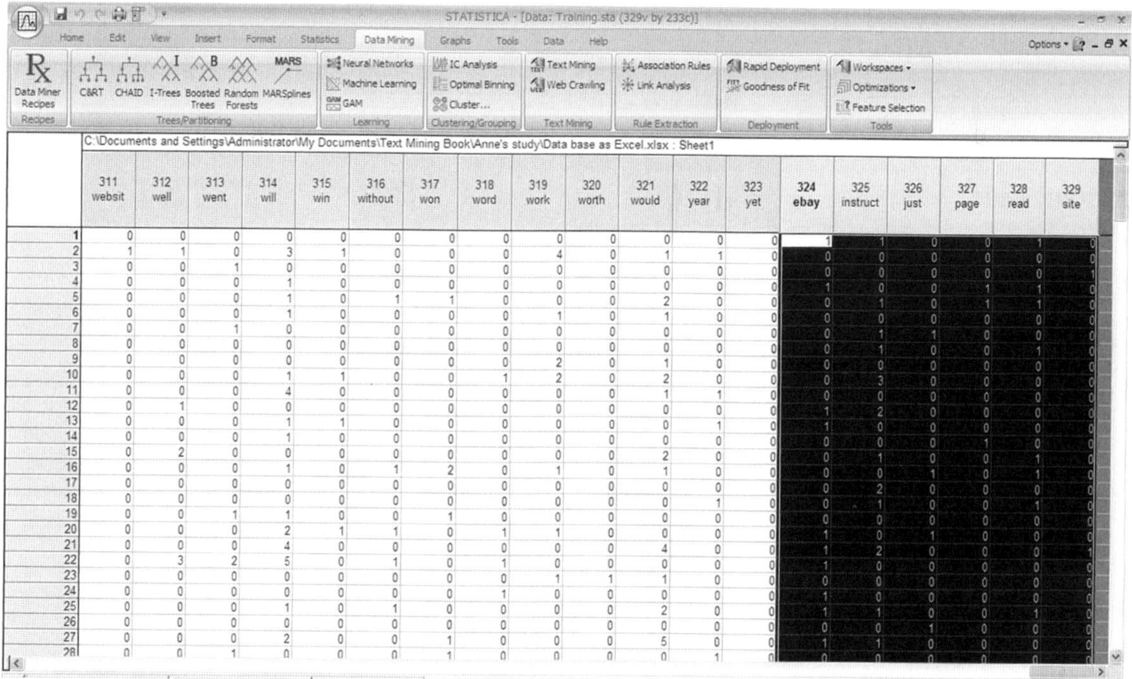

	311 websit	312 well	313 went	314 will	315 win	316 without	317 won	318 word	319 work	320 worth	321 would	322 year	323 yet	324 ebay	325 instruct	326 just	327 page	328 read	329 site
1	0	0	0	0	0	0	0	0	0	0	0	0	0	1	1	0	0	1	0
2	1	1	0	3	1	0	0	0	4	0	1	1	0	0	0	0	0	0	0
3	0	0	1	0	0	0	0	0	0	0	0	0	0	0	0	0	0	0	1
4	0	0	0	1	0	0	0	0	0	0	0	0	0	1	0	0	1	1	0
5	0	0	0	1	0	1	1	0	0	0	2	0	0	0	1	0	1	1	0
6	0	0	0	1	0	0	0	0	1	0	1	0	0	0	0	0	0	0	0
7	0	0	1	0	0	0	0	0	0	0	0	0	0	0	1	1	0	0	0
8	0	0	0	0	0	0	0	0	0	0	0	0	0	0	1	0	0	1	0
9	0	0	0	0	0	0	0	0	2	0	1	0	0	0	0	0	0	0	0
10	0	0	0	1	1	0	0	1	2	0	2	0	0	0	3	0	0	0	0
11	0	0	0	4	0	0	0	0	0	0	1	1	0	0	0	0	0	0	0
12	0	1	0	0	0	0	0	0	0	0	0	0	0	1	2	0	0	0	0
13	0	0	0	1	1	0	0	0	0	0	0	1	0	1	0	0	0	0	0
14	0	0	0	1	0	0	0	0	0	0	0	0	0	0	0	0	1	0	0
15	0	2	0	0	0	0	0	0	0	0	2	0	0	0	1	0	0	1	0
16	0	0	0	1	0	1	2	0	1	0	1	0	0	0	0	1	0	1	0
17	0	0	0	0	0	0	0	0	0	0	0	0	0	0	2	0	0	0	0
18	0	0	0	0	0	0	0	0	0	0	0	1	0	0	1	0	0	1	0
19	0	0	1	1	0	0	1	0	0	0	0	0	0	0	0	0	0	0	0
20	0	0	0	2	1	1	0	1	1	0	0	0	0	1	0	1	0	0	0
21	0	0	0	4	0	0	0	0	0	0	4	0	0	1	2	0	0	0	1
22	0	3	2	5	0	1	0	1	0	0	0	0	0	1	0	0	0	0	0
23	0	0	0	0	0	0	0	0	1	1	1	0	0	0	0	0	0	0	0
24	0	0	0	0	0	0	0	1	0	0	0	0	0	1	0	0	0	0	0
25	0	0	0	1	0	1	0	0	0	0	2	0	0	1	1	0	0	1	0
26	0	0	0	0	0	0	0	0	0	0	0	0	0	0	0	1	0	0	0
27	0	0	0	2	0	0	1	0	0	0	5	0	0	1	1	0	0	0	0
28	0	0	1	0	0	0	1	0	0	0	0	1	0	0	0	0	0	0	0

FIGURE F.100

Note new variables at the end of the database spreadsheet.

The number of times each of the words was used by each case becomes interval data that can be analyzed quantitatively. Save the data set.

Open up the data mining workbook and add the newly saved training data set. (See Figure F.101.) We can now see if these words seem to separate the ATLAS categories.

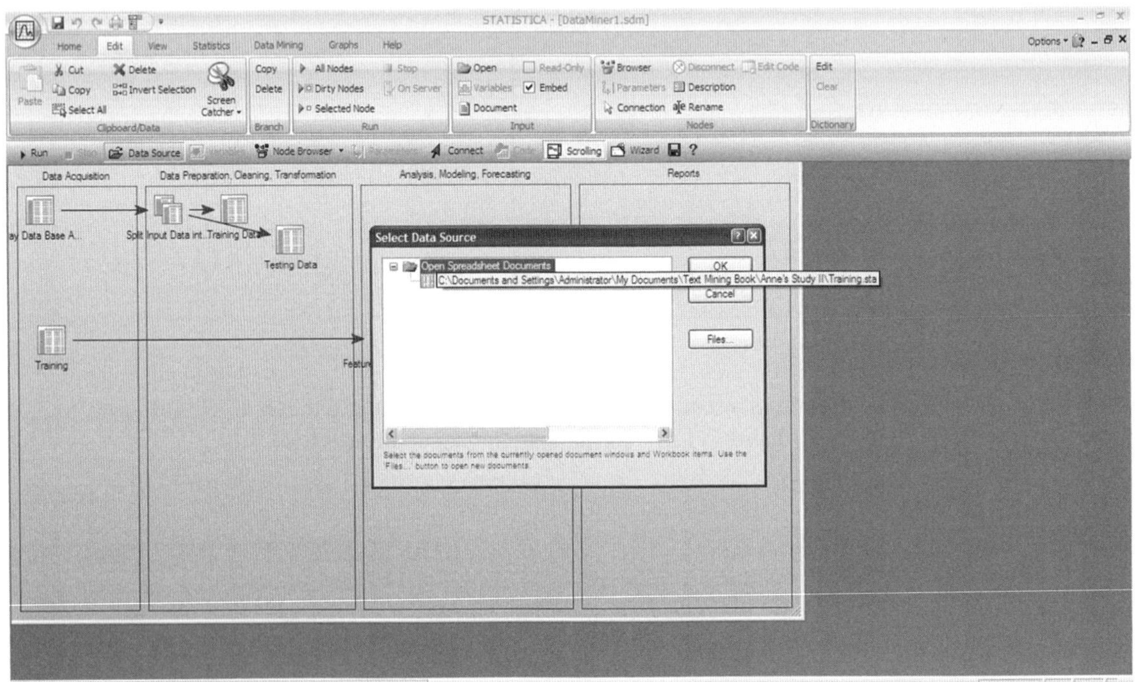

FIGURE F.101
Opening the newly saved training data set in the data mining workspace.

Select the variables associated with question 2 as in Figure F.102.

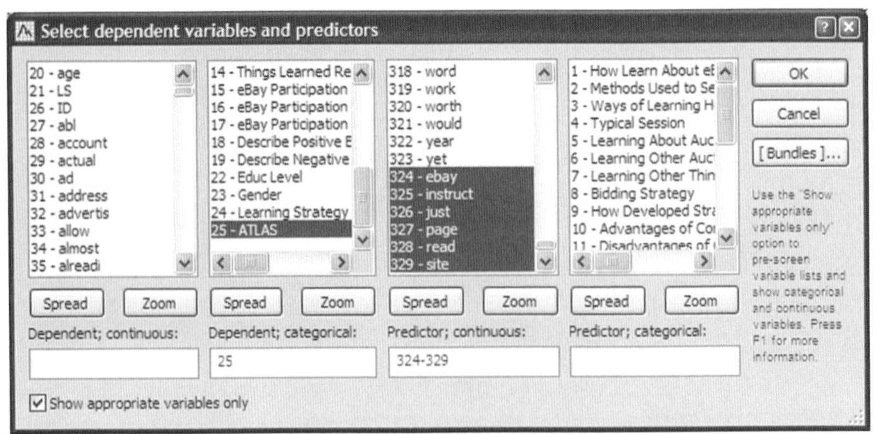

FIGURE F.102
Choose ATLAS as the dependent variable and the six new variable words as the independent variables.

Click OK. Pull the data set down a bit so it is easy to see. Copy the feature selection and paste it into the workspace. Get an arrow and run it from the data set to the procedure (Fig. F.103).

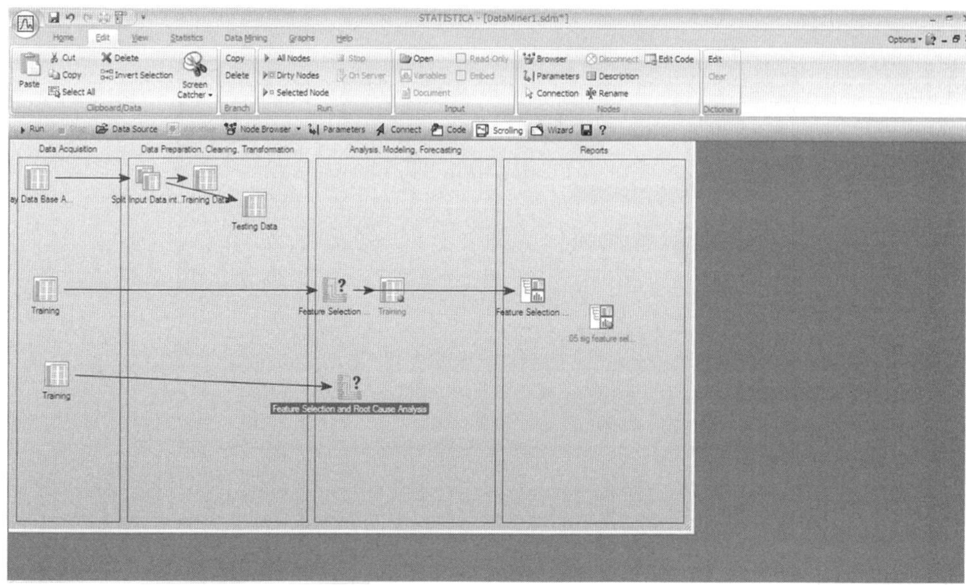

FIGURE F.103

Copy the feature selection and paste it in. Connect an arrow.

The one that was copied above in Figure F.103 used the fixed number procedure. Run the node and embed the output. Open the output. The following importance plot was created (Fig. F.104).

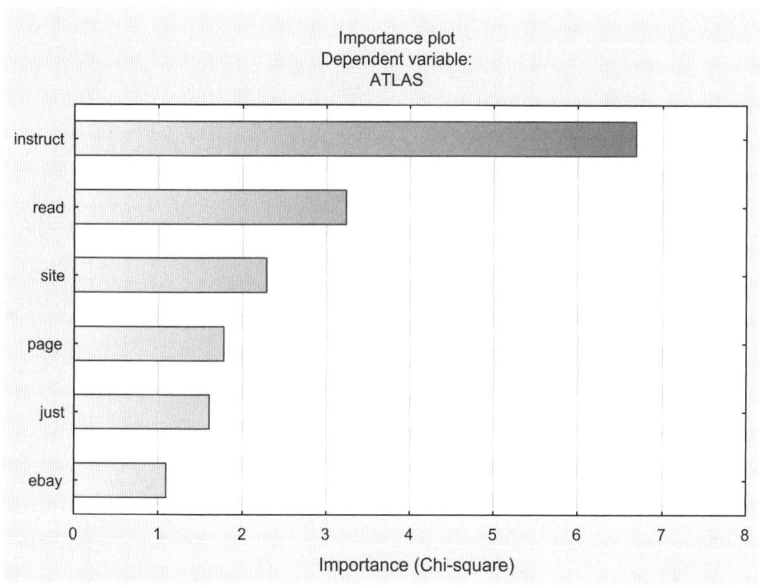

FIGURE F.104

Resulting importance plot.

The "Instruct" root seemed to be the most important predictor, followed by "Read." We used those two in an ANOVA to see what the relationships might be. Pull up a one-way ANOVA and select the variables as in Figure F.105.

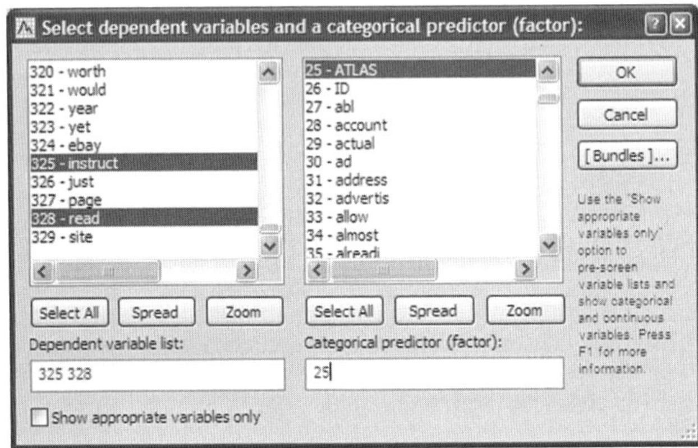

FIGURE F.105
Performing a one-way ANOVA.

The relationships are not strong, but there may be a slight tendency for the problem solvers to use those words more than the other two types.

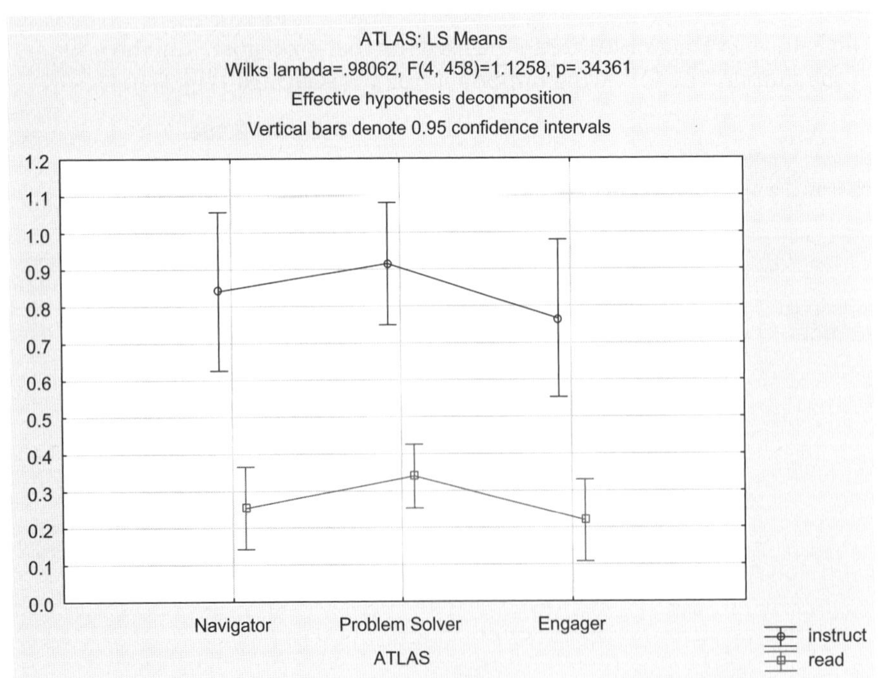

FIGURE F.106
Observing possible relationships.

Again, the relationship lacks significance. It is doubtful that answers to question 2 would be able to separate and predict the groups, but it might be possible that problem solvers tend to read the instructions more than the navigators or the engagers.

One thing we had not done was to use nontext variables to predict the ATLAS. Just to see if there were any significant predictions, we used a feature selection for education and for age to see if either of those would predict the ATLAS. We found that education was a possible predictor, but age was not.

	Best predictors for categorical dependent var: ATLAS (Training)	
	Chi-square	p-value
Educ Level	17.47580	0.064477
age	19.52212	0.360356

FIGURE F.107
Predictions for age and education level.

The ANOVA revealed how the education level may have varied with the categories, with the highest education level predicting the problem solvers. It certainly made sense that if problem solvers tend to get new information by reading instructions or by reading, that schooling might be a place where that strategy is rewarded.

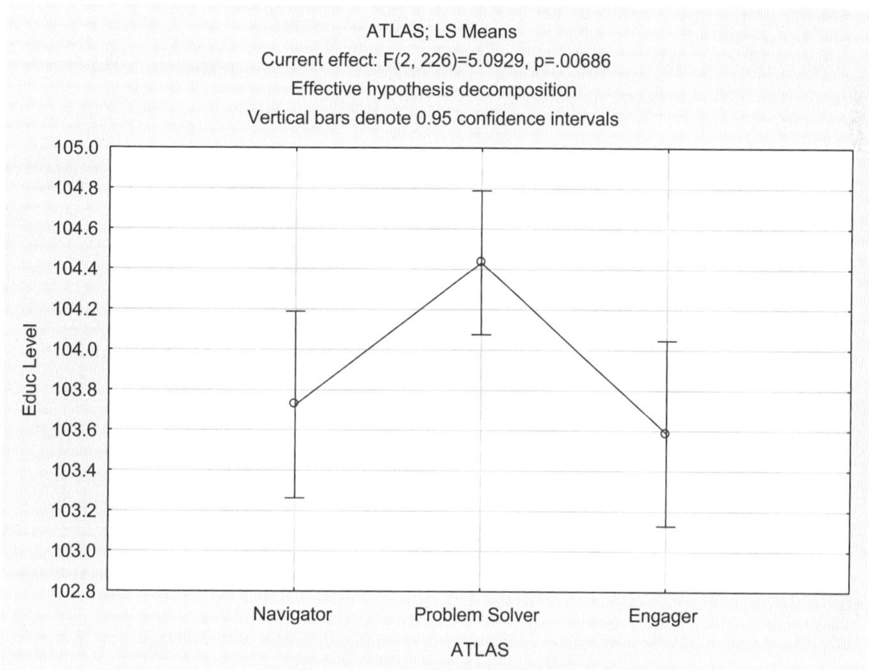

FIGURE F.108
One-way ANOVA for education level. The higher the education level, the more educated the individual.

Perhaps when we try to finalize a support vector machine prediction, we should include education level. We then started looking at another question. We selected question 3 because it also concerns what one would do to learn something for the first time. The question concerned ways of learning how to use eBay. We followed the preceding text mining procedures and appended the training data set with the new words. For this question, there were 16 words (See Figure F.109).

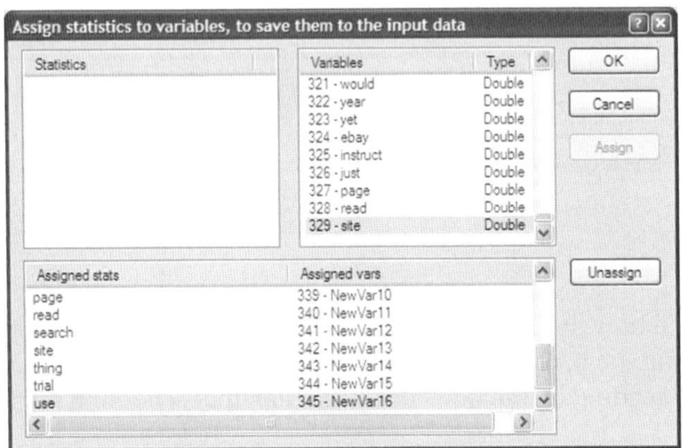

FIGURE F.109

Appending the data set for question 3 using the previous steps.

Once the words have been placed into the training database, save the file and open it in the data mining space. Choose the new variables to predict the ATLAS.

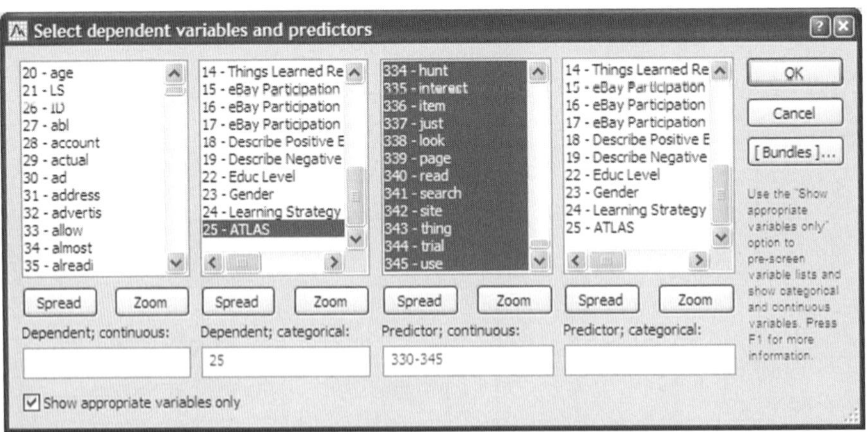

FIGURE F.110

Feature selection for training data after amending the database with the important words from variable 3.

Use the fixed number once again, producing the following importance plot which may be viewed in Figure F.111.

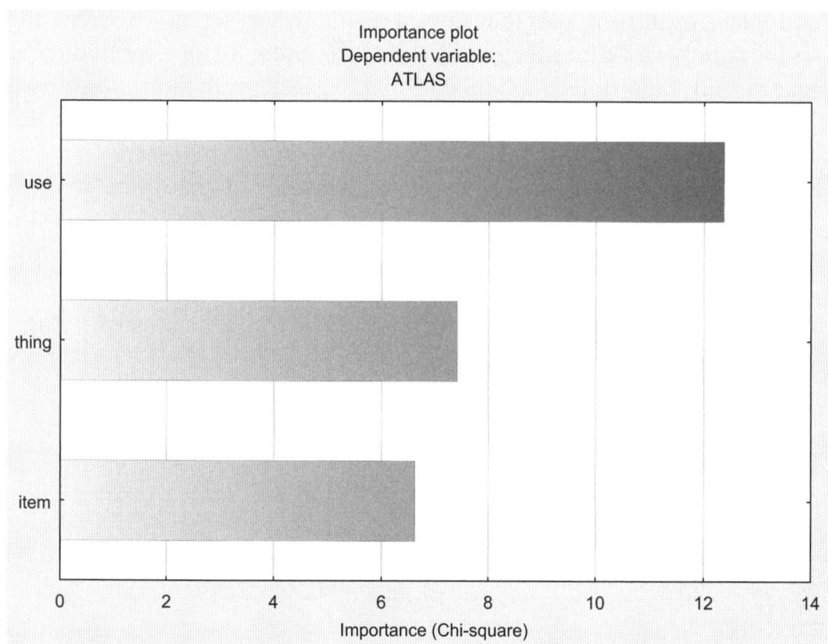

FIGURE F.111
Importance plot.

We decided to use the three words *use, thing,* and *item* in an ANOVA to see what the relationships might be (Fig. F.112).

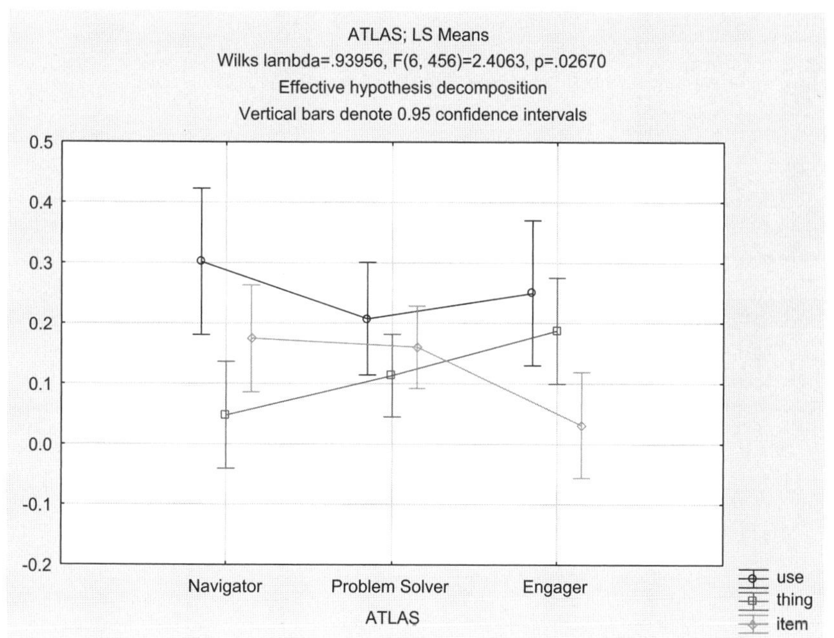

FIGURE F.112
ANOVA for words from importance plot for variable 3.

The only relationship that looked promising was that there was likely a difference between the navigators and the engagers in the number of times they used the word "thing." There seemed to be a difference between the navigators and the problem solvers with the engagers in the times they used the word "item." Navigators also may have used the word "use" more than problem solvers but not more than the engagers.

We decided to do a support vector machine to see if these words could be used in a consistent manner in predicting the groups. First, go to data mining and machine learning as seen in Figures F.113 to F.115.

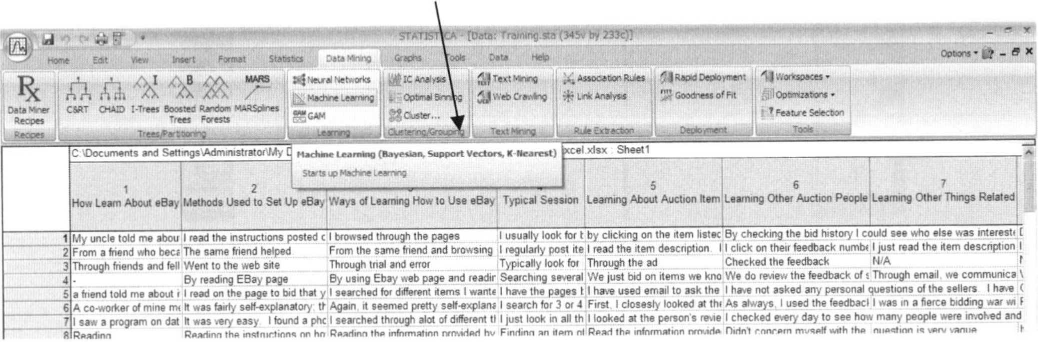

FIGURE F.113
Finding support vector machine.

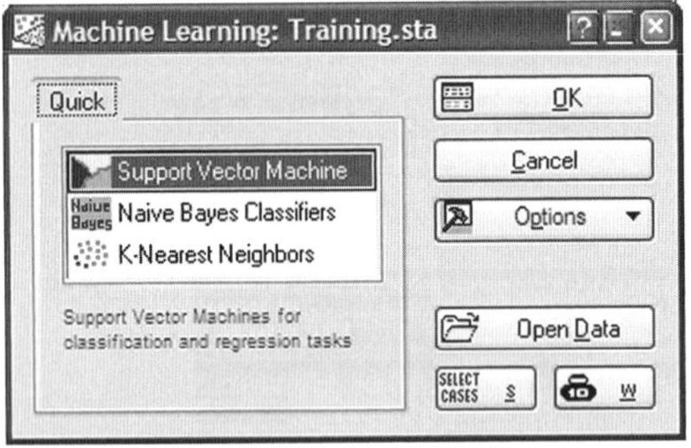

FIGURE F.114
Select "Support Vector Machine."

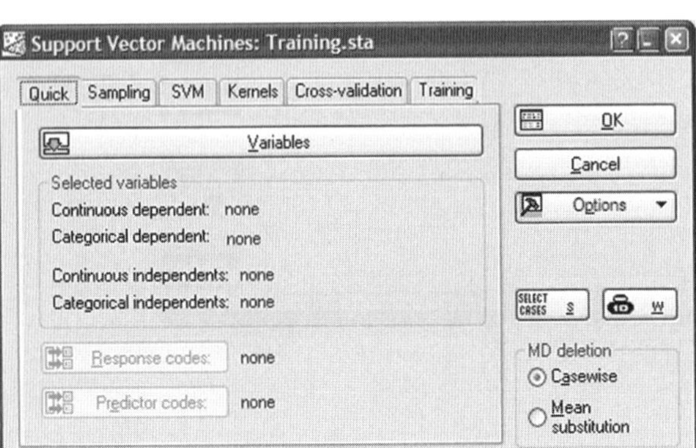

FIGURE F.115
Click the "Variables" button.

The variables put into the support vector machine model were these in Figure F.116.

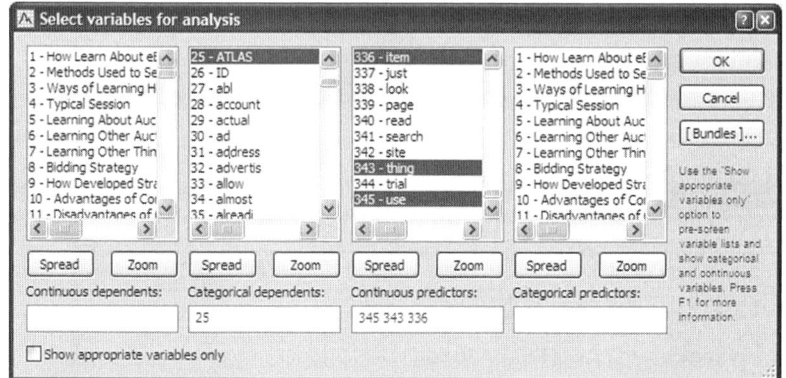

FIGURE F.116
Choosing the appropriate variables.

The V-fold method was selected. These words failed to provide a good prediction model. The ANOVA had a significant F score, but it was not a dependable relationship when subjected to the v-fold method, which is a measure of how the system would work in repeated analyses (Fig. F.117). The groups were predicted accurately less than half the time.

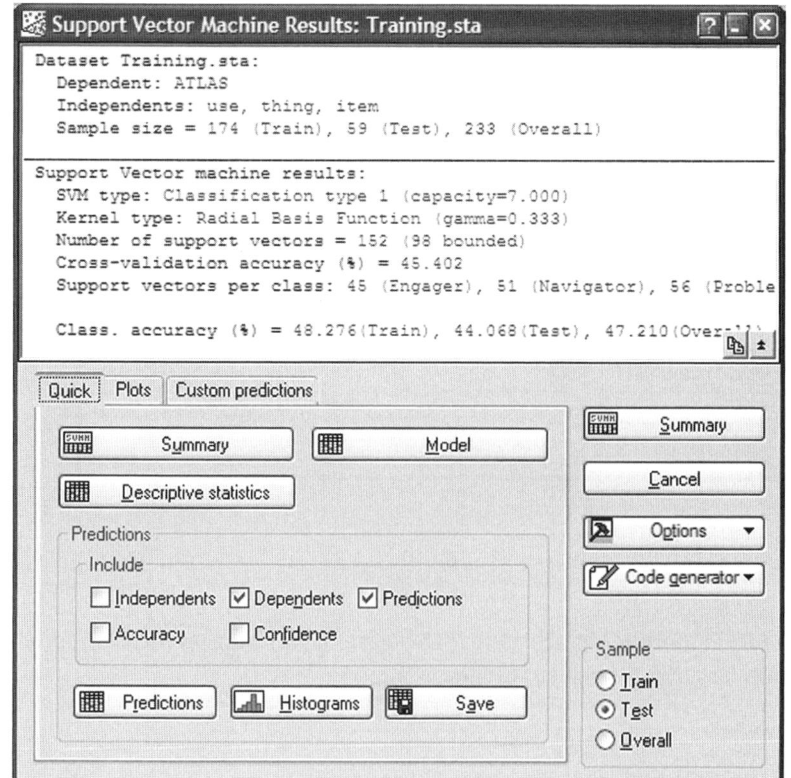

FIGURE F.117
Support vector machine results.

We decided to add education level to see if we could increase the prediction. This was not much better (Fig. F.118).

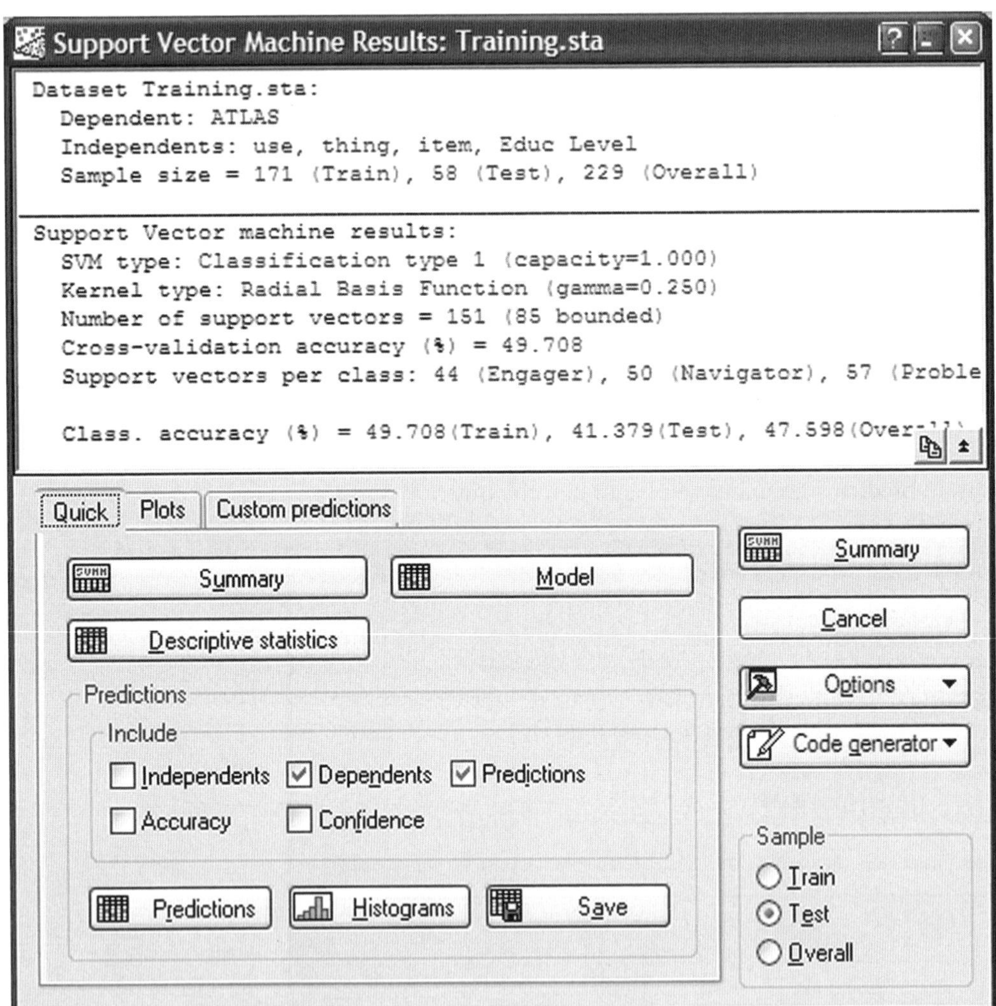

FIGURE F.118
Support vector machine after adding education level.

We decided to use education level and the word "thing." This gave us our best prediction model, but it still was not good, with only about a 50 percent accuracy rate (Fig. F.119).

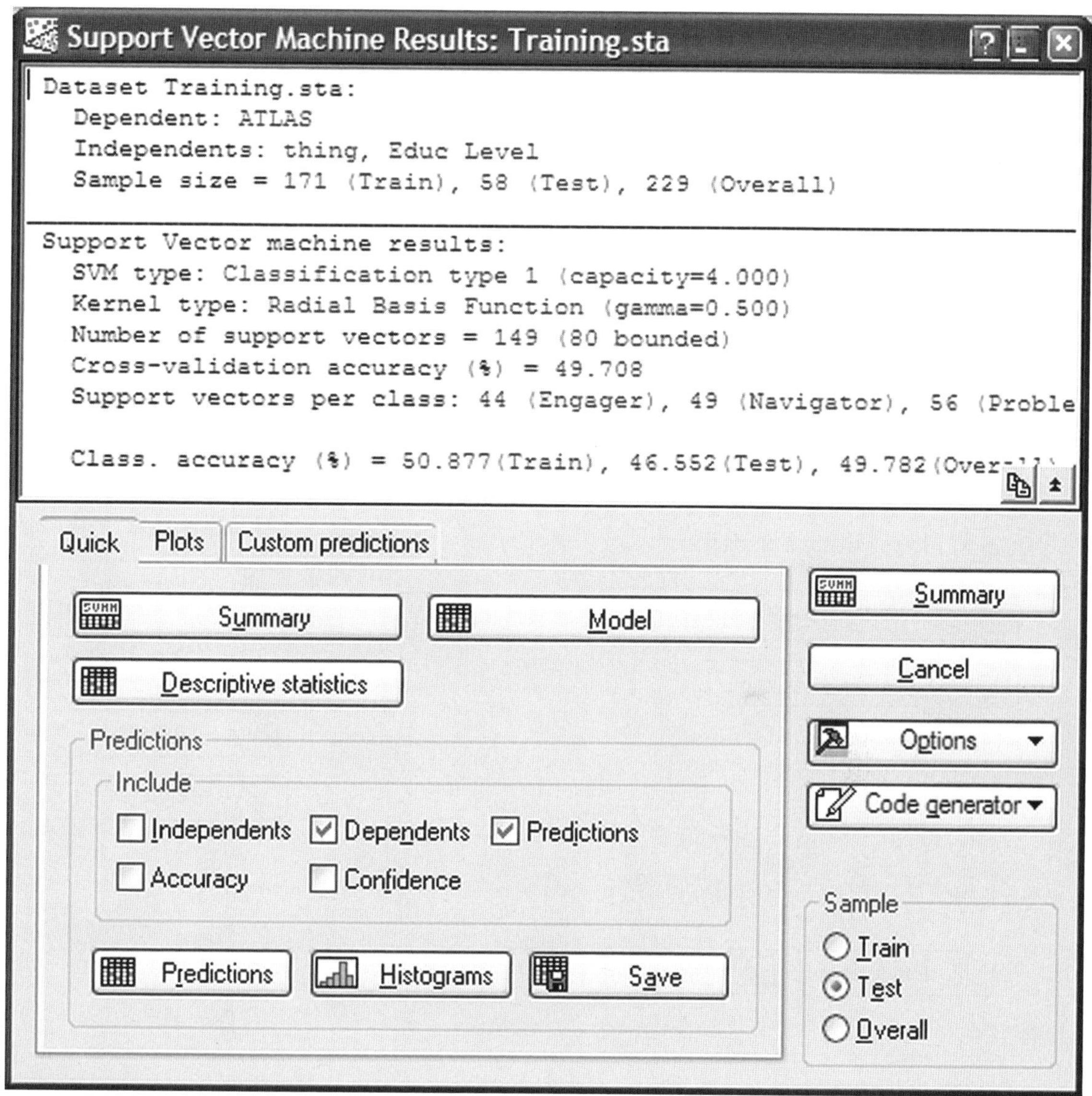

FIGURE F.119
Another attempt to make a good prediction model adding education.

So far, there was nothing that we could really take to the testing sample from feature selection. We tried adding all the best predictors from variables 2 and 3 (Fig. F.120).

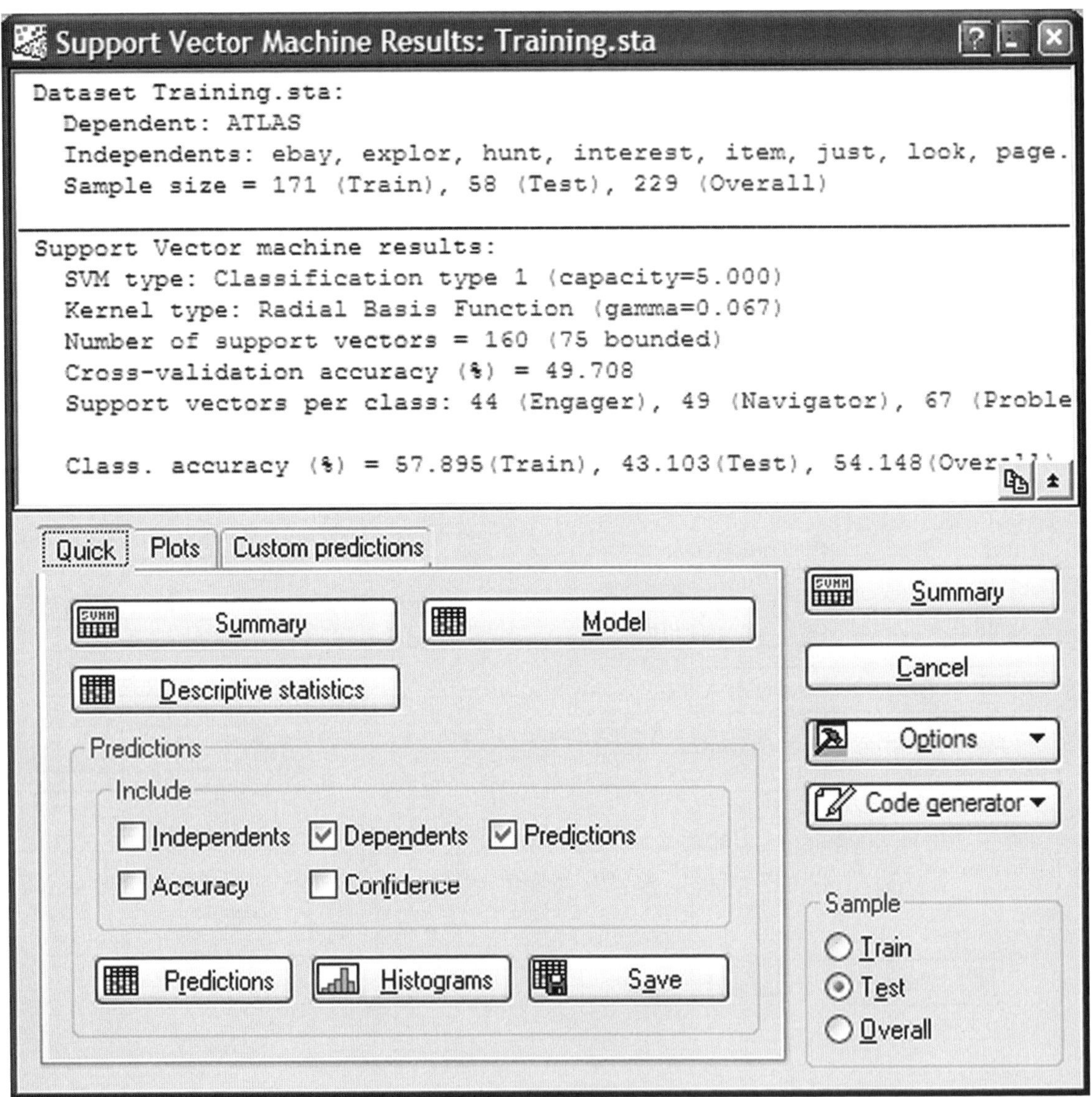

FIGURE F.120
Last attempt combining words from variables 2 and 3 almost got us to 50 percent correct prediction.

Interestingly that seemed to work better. If we can predict 50 percent, that is better than the 33 percent one would expect by chance, given the three categories. If we were to keep at this process and run all the questions and continue to add words, perhaps we would come up with a consistent and good prediction model. However, in an effort to finish this tutorial, we decided to revert back to the words that showed up given all 19 variables with the separated databases repeating the preceding steps.

Navigators

	SVD word impo
	Importance
auction	100.0000
use	93.3992
look	89.4229
seller	83.3688
learn	81.2142
get	80.6665
peopl	78.4378
check	71.2104
sell	69.9544
internet	68.5462
search	65.3686
will	65.2328
realli	64.6869
time	64.2745
just	62.7389
can	62.5974
price	61.3096
one	60.4358
see	60.4358
interest	58.3460
thing	57.4271
know	55.8620
go	55.3839
want	54.7399
good	53.4285
also	52.7607
feedback	51.3989
site	49.8224
comput	49.6441
read	48.1943

Engagers

	SVD word imp
	Importance
auction	100.0000
seller	85.8484
check	78.7364
see	70.5060
get	69.3696
want	68.1082
will	67.7892
can	67.6825
use	65.5127
look	65.0700
just	63.1541
peopl	60.5848
ve	57.7767
go	56.7674
would	56.3843
like	53.4909
also	53.0841
much	53.0841
feedback	52.8111
interest	52.3991
internet	51.9838
realli	51.2841
purchas	50.5747
good	50.5747
search	50.2882
well	49.7101
thing	48.8302
time	48.3843
email	47.0211
know	46.2464

Problem Solvers

	SVD word impc
	Importance
auction	100.0000
seller	86.8847
use	83.4356
learn	74.2915
look	73.5641
peopl	64.5302
will	62.6302
just	61.8122
check	61.4665
interest	58.9956
get	57.0041
thing	56.7398
search	56.6372
time	56.4163
also	56.3582
can	54.1965
see	53.5453
ve	53.2932
one	51.8507
comput	50.1417
realli	48.2257
want	48.1335
price	47.9409
read	46.9415
internet	46.7244
like	45.6792
go	42.3353
e-mail	42.2026
buy	41.8060
would	40.2269

FIGURE F.121

Important words from all 19 questions by ATLAS groups.

As in Figure F.121, we should look at those same lists with testing data to see if there were any similarities. Or we could look at lists from important words from one of the variables as an example. We will choose variable 3 for this analysis. The three resulting important word lists are as follows.

Navigators

	SVD word imp
	Importance
search	100.0000
hunt	92.2958
ebay	88.1917
use	83.8870
explor	79.3492
look	74.5356
just	72.0082
categori	69.3889
click	69.3889
item	69.3889
interest	63.8285
trial	54.4331
site	50.9175
page	50.9175

Problem Solvers

	SVD word impo
	Importance
search	100.0000
hunt	86.3191
use	79.2118
ebay	74.0959
interest	74.0959
explor	72.7607
categori	68.5994
read	62.6224
look	61.0368
site	61.0368
page	61.0368
item	61.0368
trial	57.7350
just	54.2326

Engagers

	SVD word import
	Importance
search	100.0000
hunt	73.5215
use	69.7486
just	67.7834
look	63.6715
page	61.5125
ebay	59.2749
thing	56.9495
explor	56.9495
interest	54.5250
read	43.4959
click	43.4959
trial	43.4959
start	43.4959

FIGURE F.122

Important words from variable 3.

Examining Figure F.122, it was noted that there was no "category" in engagers, no "click" in problem solvers, no "item" in engagers, and no "site" in engagers. Engagers used "thing" instead of "item." There was no "read" in navigators, and engagers used "start," but navigators and engagers did not. It was kind of odd that the problem solvers used the word "interest" more than the engagers, since engagers like to be inherently interested in something before they will explore it. "Explore" and "interest" were both higher for problem solvers than engagers. On the other hand, the word "categories" was found more with navigators and problem solvers, both of whom presumably would be more left-brained as they approach new learning, whereas engagers tend to be more intuitive and right-brained when encountering new learning.

We separated the testing data set into the three ATLAS groups as well and then pulled up their important words for question 3, which involved ways of learning about eBay.

Navigator

	SVD word import
	Importance
search	100.0000
page	86.6025
hunt	77.0552
use	75.0000
item	63.7377
trial	58.6302
ebay	50.0000
just	50.0000
interest	43.3013
click	43.3013
read	39.5285
differ	39.5285

Problem Solver

	SVD word impor
	Importance
hunt	100.0000
search	83.0455
differ	69.4808
ebay	66.9534
click	64.3268
site	64.3268
use	64.3268
start	61.5882
page	61.5882
item	58.7220
explor	58.7220
interest	58.7220
read	55.7086
just	55.7086
look	52.5226
section	52.5226
trial	52.5226
see	49.1304
categori	49.1304
various	41.5227

Engager

	SVD word import
	Importance
search	100.0000
hunt	81.6497
just	79.3492
page	60.8581
read	54.4331
item	50.9175
explor	50.9175
look	50.9175
start	43.0331
around	43.0331
trial	43.0331
use	38.4900
interest	38.4900
need	38.4900
site	38.4900
engin	38.4900

FIGURE F.123

Testing data, variable 3.

"Item" in Figure F.123 is now in all three groups, "site" is now in the engager list but not in the navigator list, "start" is now in the engagers but not in the navigators, "interest and explore" are still used more by the problem solvers than for either the navigators or the engagers and the navigators are not using "explore" at all, and "categories" was used by problem solvers but not by either navigators or engagers. Everyone searched and hunted, everyone read, and the navigators and problem solvers clicked, whereas the engagers did not. One would be hard pressed to notice large differences between the three groups based on the words that the three groups used the most. However, to see if an objective observer could characterize the groups by the words used (testing lists), we asked a colleague to look at the words without the labels. She was unfamiliar with the categories. She said, "Group #1 has a more direct approach. Clearly more people in all these groups are shoppers. Group #2 is more detailed in looking, and more used "explore," which Group #1 did not use at all. Close to 55 percent in Groups #2 and #3 used "read," whereas it is the next to the last in Group #1. No sellers in these groups" (Personal communication with Linda Lauhon). This statement was what one might expect in describing the three groups. Navigators might be expected to be more direct, and problem solvers might be expected to be more explorative in their approaches to new learning. The problem was there was not a consistent relationship with any of the words between training and testing except for the words "interest" and "explore."

Using all the variables these lists emerged in the testing data may be viewed in Figure F.124.

Navigators	SVD word import Importance	Problem Solvers	SVD word impc Importance	Engagers	SVD word impc Importance
search	100.0000	seller	100.0000	seller	100.0000
seller	94.2445	ve	80.9831	hunt	98.9071
use	93.5937	get	80.8156	auction	98.5401
hunt	87.0298	can	80.1422	search	85.3382
just	84.6433	look	79.8033	look	80.0815
time	80.7960	see	79.4630	time	78.9423
see	80.4378	peopl	77.9132	just	77.5532
look	77.3933	go	76.8626	check	77.0846
get	75.9440	check	75.0790	peopl	76.8492
can	75.2677	want	70.8064	realli	74.6975
want	75.0631	internet	68.8662	get	73.7210
interest	74.9481	will	68.2733	want	67.2977
auction	72.6838	search	67.8752	sell	63.4172
check	72.6151	inform	67.4748	will	63.4172
internet	71.4982	someth	67.4748	see	62.5543
email	71.4913	click	65.2284	comput	61.9724
go	68.1702	interest	65.0203	interest	60.4931
realli	67.4915	one	63.7577	much	58.0480
much	67.4829	time	63.7577	thing	57.4204
one	66.3362	thing	63.5448	like	56.4660
sell	66.1206	email	62.6860	can	56.1442
find	63.3779	just	62.0342	email	54.1736
thing	62.5900	price	61.8154	page	52.4750
like	62.2733	find	60.4858	instruct	52.1286
instruct	62.0047	realli	59.8100	would	50.3610
peopl	60.9981	comput	58.8968	find	50.3610
price	59.2417	like	58.4349	internet	48.5291
explor	59.2298	sell	58.4349	product	48.5291
don	58.2666	instruct	57.9692	ask	48.1543
someth	58.1890	feedback	56.0681	don	47.7767

FIGURE F.124
Testing data of all variables by ATLAS categories.

Let's examine the lists side by side, training to testing by group (Figures F.125 to F.127).

Navigators Training

	SVD word impo
	Importance
auction	100.0000
use	93.3992
look	89.4229
seller	83.3688
learn	81.2142
get	80.6665
peopl	78.4378
check	71.2104
sell	69.9544
internet	68.5462
search	65.3686
will	65.2328
realli	64.6869
time	64.2745
just	62.7389
can	62.5974
price	61.3096
one	60.4358
see	60.4358
interest	58.3460
thing	57.4271
know	55.8620
go	55.3839
want	54.7399
good	53.4285
also	52.7607
feedback	51.3989
site	49.8224
comput	49.6441
read	48.1943

Navigators Testing

	SVD word import
	Importance
search	100.0000
seller	94.2445
use	93.5937
hunt	87.0298
just	84.6433
time	80.7960
see	80.4378
look	77.3933
get	75.9440
can	75.2677
want	75.0631
interest	74.9481
auction	72.6838
check	72.6151
internet	71.4982
email	71.4913
go	68.1702
realli	67.4915
much	67.4829
one	66.3362
sell	66.1206
find	63.3779
thing	62.5900
like	62.2733
instruct	62.0047
peopl	60.9981
price	59.2417
explor	59.2298
don	58.2666
someth	58.1890

FIGURE F.125
Navigators training and testing.

A majority of the words in the first list are in the second in Figure F.125.

Problem Solving Training

	SVD word imp
	Importance
auction	100.0000
seller	85.8484
check	78.7364
see	70.5060
get	69.3696
want	68.1082
will	67.7892
can	67.6825
use	65.5127
look	65.0700
just	63.1541
peopl	60.5848
ve	57.7767
go	56.7674
would	56.3843
like	53.4909
also	53.0841
much	53.0841
feedback	52.8111
interest	52.3991
internet	51.9838
realli	51.2841
purchas	50.5747
good	50.5747
search	50.2882
well	49.7101
thing	48.8302
time	48.3843
email	47.0211
know	46.2464

Problem Solving Testing

	SVD word impc
	Importance
seller	100.0000
ve	80.9831
get	80.8156
can	80.1422
look	79.8033
see	79.4630
peopl	77.9132
go	76.8626
check	75.0790
want	70.8064
internet	68.8662
will	68.2733
search	67.8752
inform	67.4748
someth	67.4748
click	65.2284
interest	65.0203
one	63.7577
time	63.7577
thing	63.5448
email	62.6860
just	62.0342
price	61.8154
find	60.4858
realli	59.8100
comput	58.8968
like	58.4349
sell	58.4349
instruct	57.9692
feedback	56.0681

FIGURE F.126
Problem solving training and testing.

Although most of the words agree, it is interesting that the most important word for the training data—"auction"—did not appear in the testing data. (See Figure F.126.)

Engagers Training	SVD word impc Importance	Engagers Testing	SVD word impc Importance
auction	100.0000	seller	100.0000
seller	86.8847	hunt	98.9071
use	83.4356	auction	98.5401
learn	74.2915	search	85.3382
look	73.5641	look	80.0815
peopl	64.5302	time	78.9423
will	62.6302	just	77.5532
just	61.8122	check	77.0846
check	61.4665	peopl	76.8492
interest	58.9956	realli	74.6975
get	57.0041	get	73.7210
thing	56.7398	want	67.2977
search	56.6372	sell	63.4172
time	56.4163	will	63.4172
also	56.3582	see	62.5543
can	54.1965	comput	61.9724
see	53.5453	interest	60.4931
ve	53.2932	much	58.0480
one	51.8507	thing	57.4204
comput	50.1417	like	56.4660
realli	48.2257	can	56.1442
want	48.1335	email	54.1736
price	47.9409	page	52.4750
read	46.9415	instruct	52.1286
internet	46.7244	would	50.3610
like	45.6792	find	50.3610
go	42.3353	internet	48.5291
e-mail	42.2026	product	48.5291
buy	41.8060	ask	48.1543
would	40.2269	don	47.7767

FIGURE F.127
Engagers training and testing.

The second most used word in the testing—"hunt"—did not appear in the training. However, most words did appear in both lists as may be viewed in Figure F.127.

There was consistency between the training and testing databases, but there seemed to be little difference between the groups in the words that they used the most. We tried one last procedure to see if we could somehow separate the groups based on the words that they used. We used the entire database before dividing into training and testing. We then created variables based upon text mining all the words used, for all variables, in the same way that we did previously. First, we used text mining, then singular value decomposition, assigned words to new variables, and so on. Then we used the resulting database to perform a cluster analysis to see if the words would separate into three groups. We found 290 words, and the database was appended. Next, k-means cluster analysis was selected (Figures F.128 and F.129).

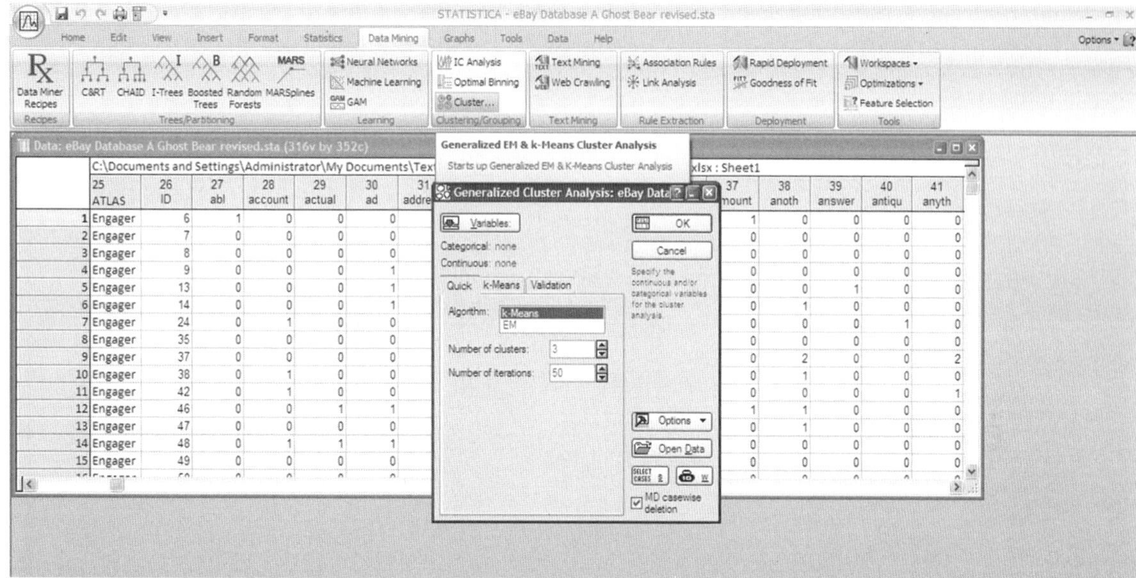

FIGURE F.128
Finding k-means cluster analysis.

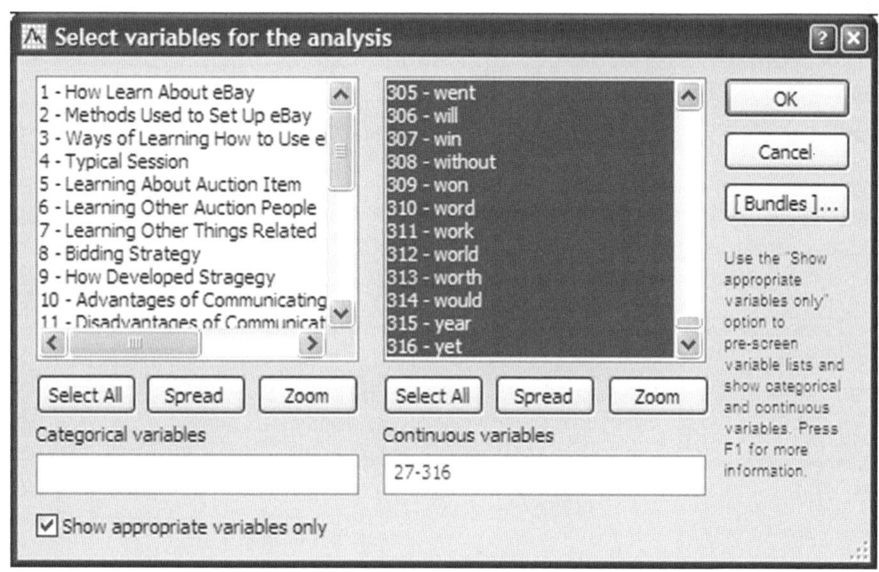

FIGURE F.129
Select the variables.

When run (Fig. F.130), there were two clusters, basically, that came out. A third cluster contained about 1 percent of the cases. One can click on any of the output buttons. First, the cluster means.

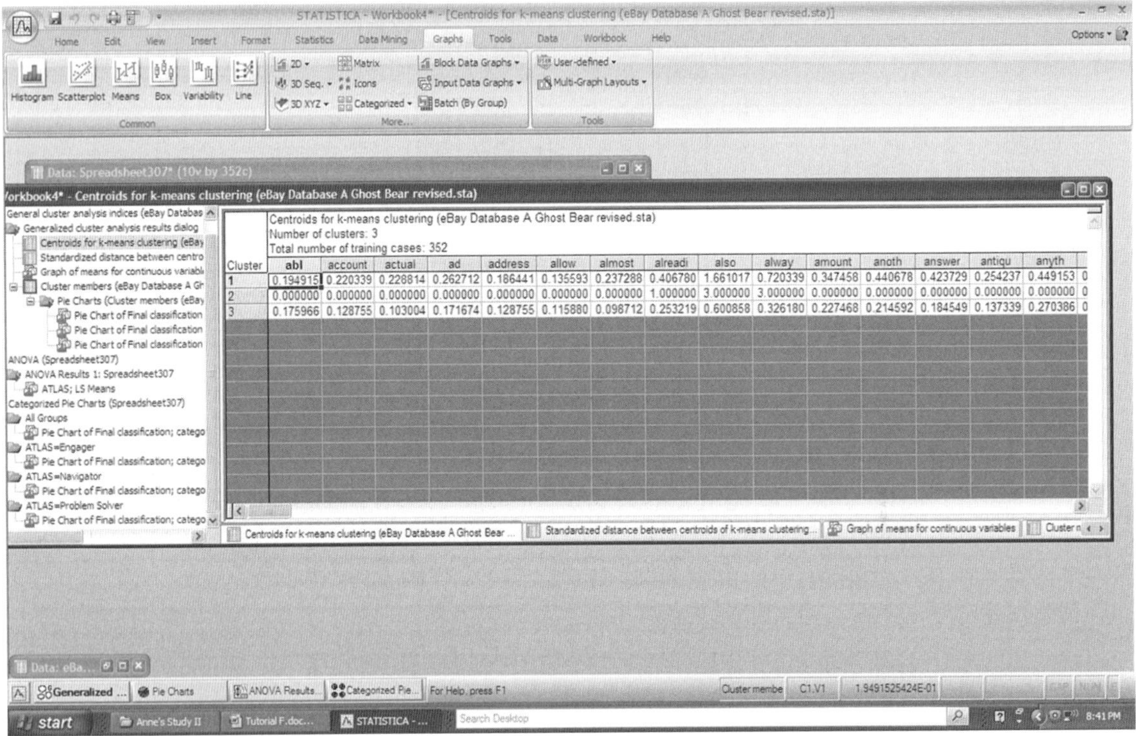

Centroids for k-means clustering (eBay Database A Ghost Bear revised.sta)
Number of clusters: 3
Total number of training cases: 352

Cluster	abl	account	actual	ad	address	allow	almost	alreadi
1	0.194915	0.220339	0.228814	0.262712	0.186441	0.135593	0.237288	0.406780
2	0.000000	0.000000	0.000000	0.000000	0.000000	0.000000	0.000000	1.000000
3	0.175966	0.128755	0.103004	0.171674	0.128755	0.115880	0.098712	0.253219

Cluster	also	alway	amount	anoth	answer	antiqu	anyth
1	1.661017	0.720339	0.347458	0.440678	0.423729	0.254237	0.449153
2	3.000000	3.000000	0.000000	0.000000	0.000000	0.000000	0.000000
3	0.600858	0.326180	0.227468	0.214592	0.184549	0.137339	0.270386

FIGURE F.130
Cluster means output.

Then, the standardized distances were examined (one of the bottom tabs on F.130) under cluster distances (distances seen in Figure F.131).

Standardized distance between centroids of k-means clustering (eBay Database A Ghost Bear revised.sta)
Number of clusters: 3

	Cluster 1	Cluster 2	Cluster 3
Cluster 1	0.000000	3.563812	1.231789
Cluster 2	3.563812	0.000000	3.686091
Cluster 3	1.231789	3.686091	0.000000

FIGURE F.131
Cluster distances.

Finally, the members and distances, which gave the cluster for each case.

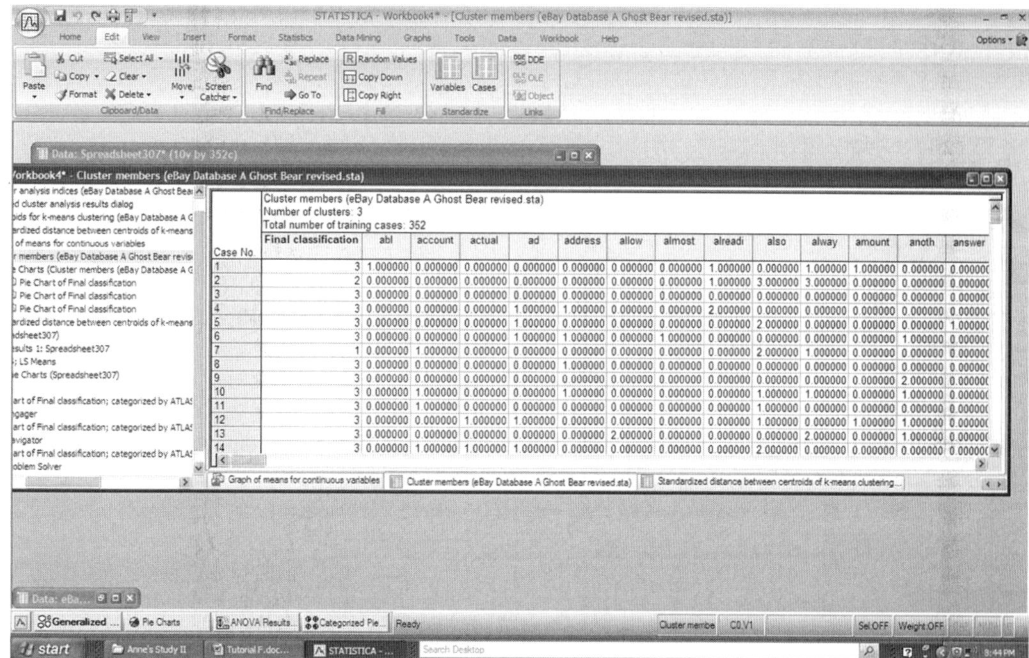

FIGURE F.132
Cluster predictions.

Figure F.132 shows that the members in the data base spreadsheet were in the same order as the original database in which the ATLAS members had been grouped. So the first 95, for example, were engagers in the original data base. These should have all been one number if the procedure was predicting the ATLAS categories. In order to see what was happening graphically, the cluster members were set to the input file (Figure F.133).

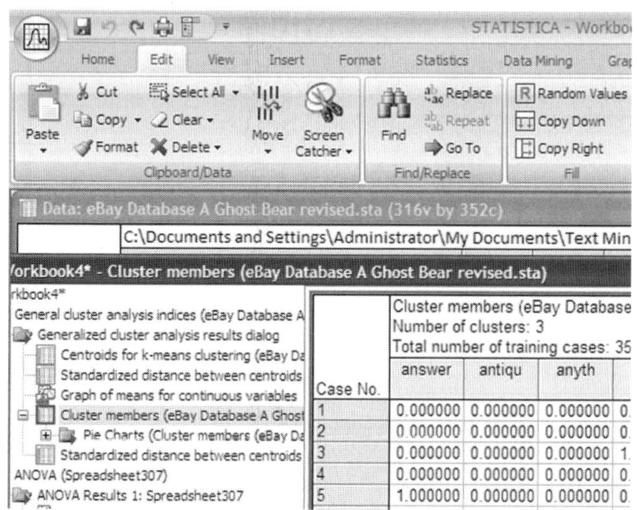

FIGURE F.133
Setting to the input file.

Next, copy with headers the first variable. We will be forming pie charts out of the first variable: final classification. (See Figure F.134 to see what variable needs to be used for copying to the new spreadsheet.)

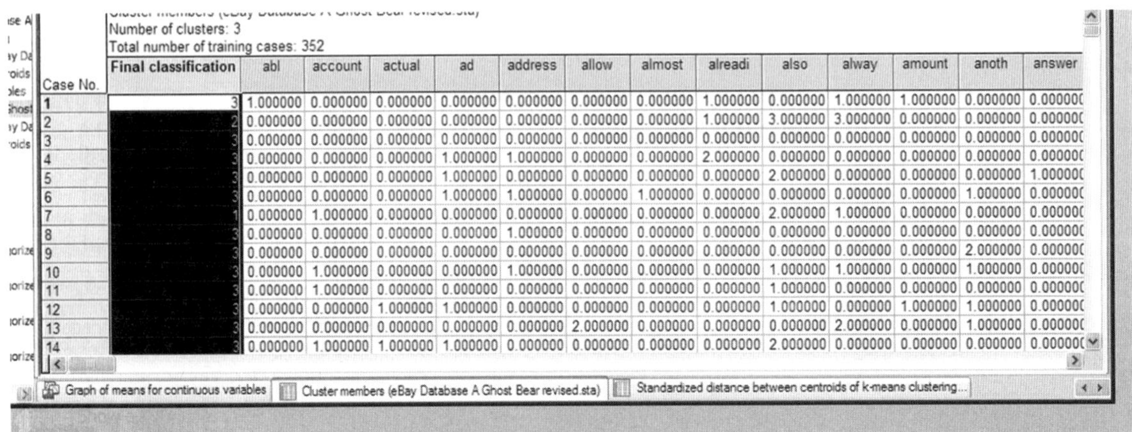

FIGURE F.134
Final classification variable.

Open a new document and paste in this variable. Then, copy the ATLAS categories from the total database and place it next to the final classification. (See Figure F.135.) Copy both variables to the new spreadsheet. Figure F.136 shows the new spreadsheet with the two variables next to each other. Figure F.137 shows how to do a categorized pie chart.

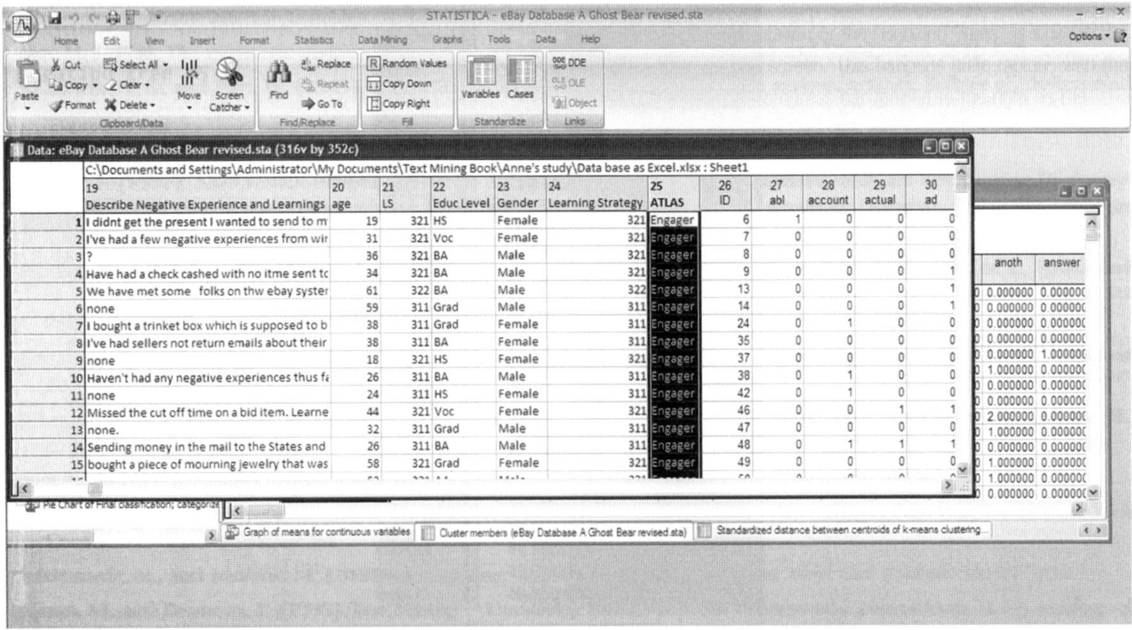

FIGURE F.135
Copying the ATLAS categories for the new spreadsheet.

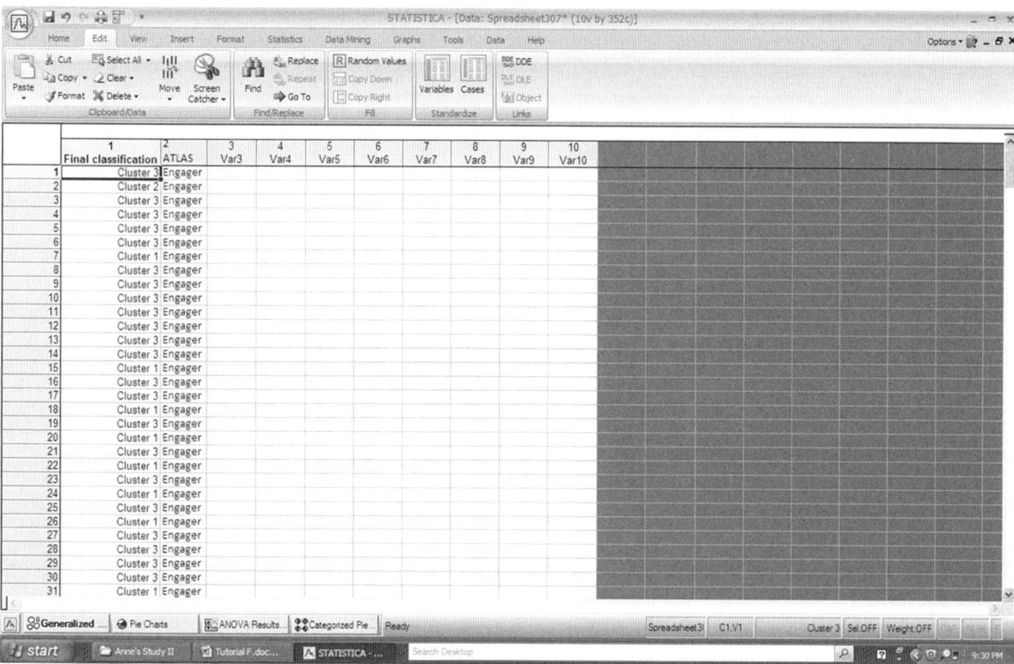

FIGURE F.136
New database with cluster designations and ATLAS categories.

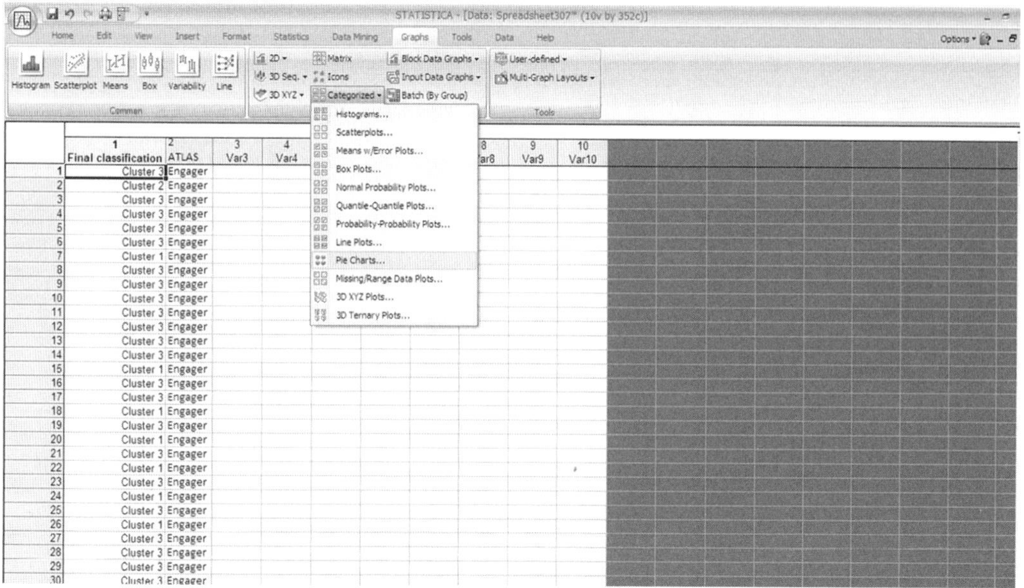

FIGURE F.137
Then to make categorized pie charts, go to graphs, categorized, and pie charts.

Put in the final classification as the variable to be graphed in Figure F.138.

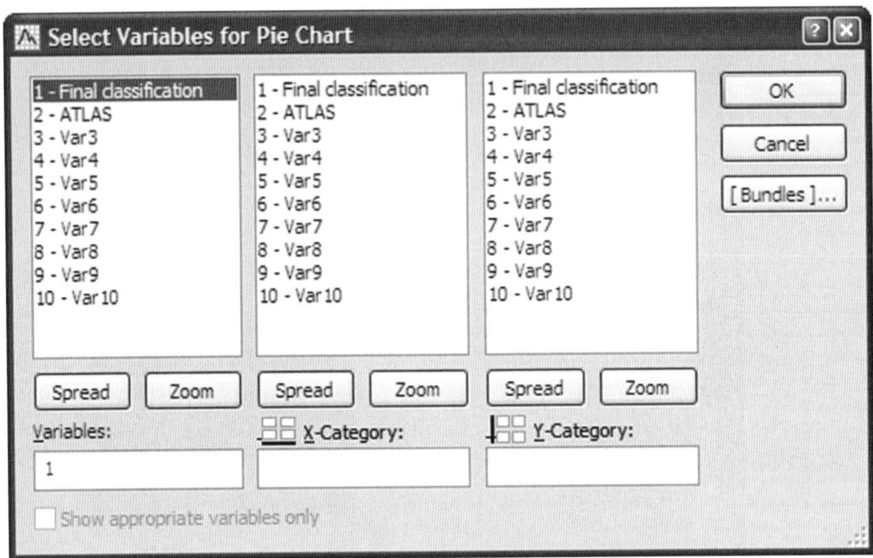

FIGURE F.138
Selecting the variable for pie charts.

Then, when you click OK, you select your categorization variable, which is ATLAS (Figure F.139).

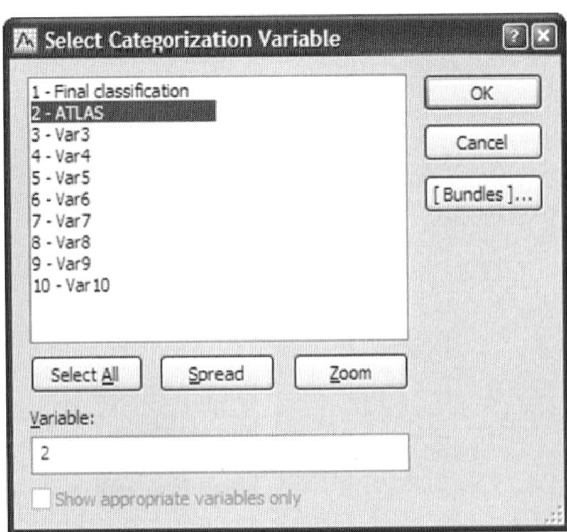

FIGURE F.139
Selecting ATLAS as the categorical variable.

Basically, only two of the clusters came up for the Engagers (Fig. F.140), Problem Solvers (Fig. F.141), and Navigators (Fig. F.142). There was a third cluster that was for only 1 percent of the data within the engagers.

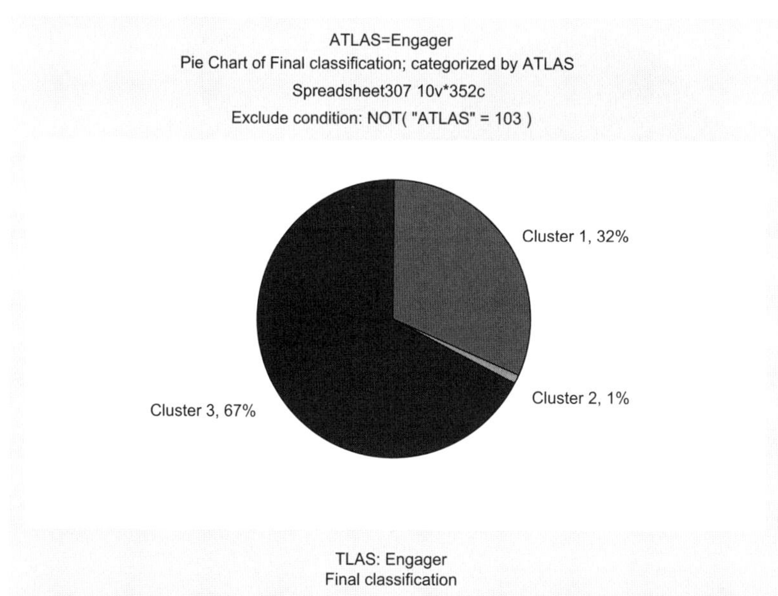

FIGURE F.140
Categories for engagers.

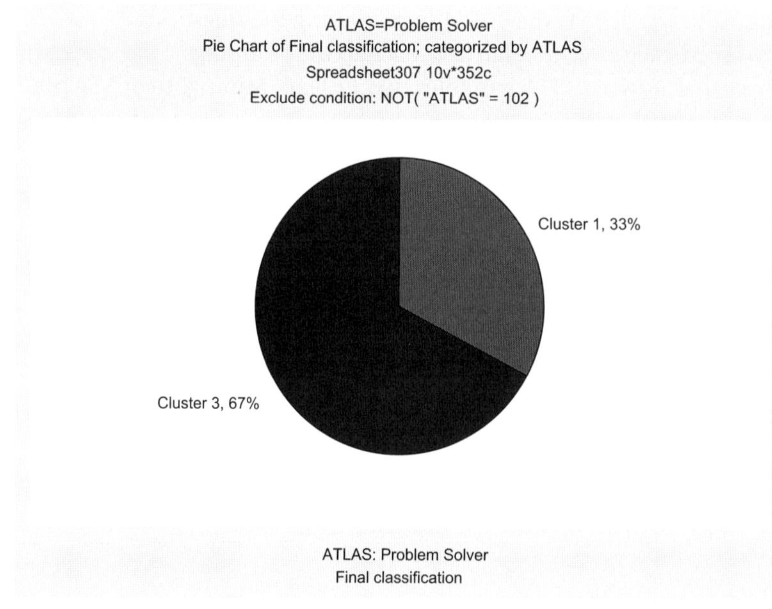

FIGURE F.141
Problem solvers.

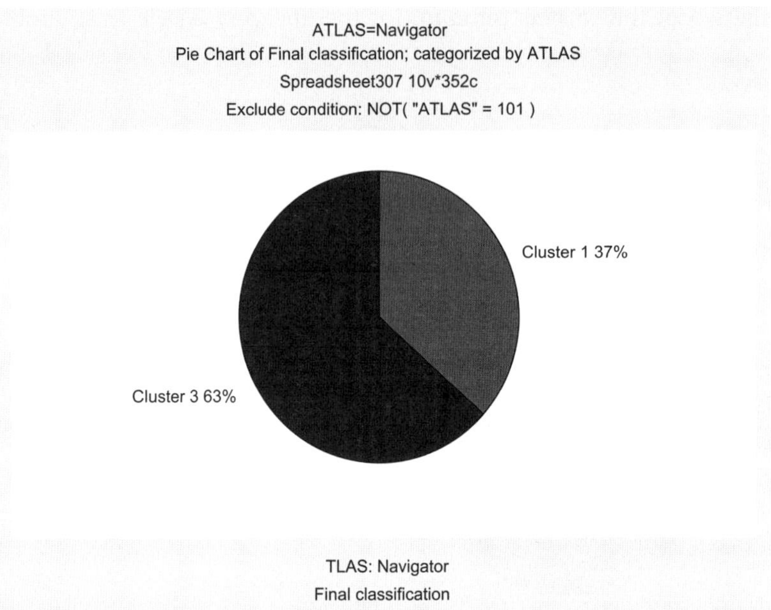

FIGURE F.142
Navigators.

In all ATLAS categories, the words that were most used separated each group into two parts. Basically, two-thirds of each group's words clustered into one group, and one-third of the group's words clustered into another group. According to the cluster analysis, it appears that there are no differences between the three groups based upon the words that they used in answering the questions.

Not to give up, last thing to try would be a more Bayesian approach to the groupings. First, open up the revised eBay database spreadsheet that now has 316 variables. Go to Data Mining, then Machine Learning, and finally Naïve Bayes Classifiers (Fig. F.143).

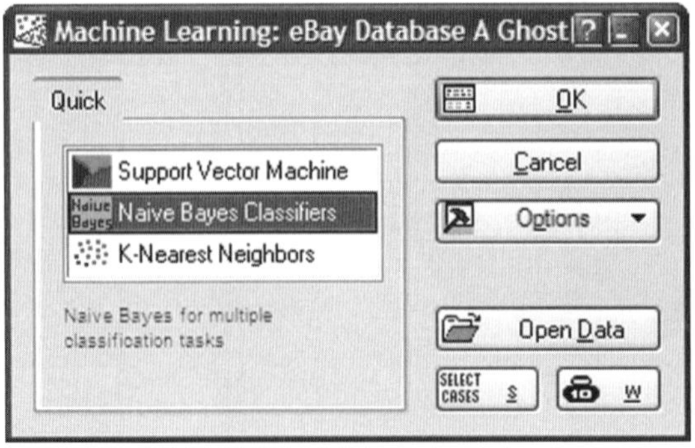

FIGURE F.143
Finding Naïve Bayes Classifiers.

Next, select the variables as in Figure F.144.

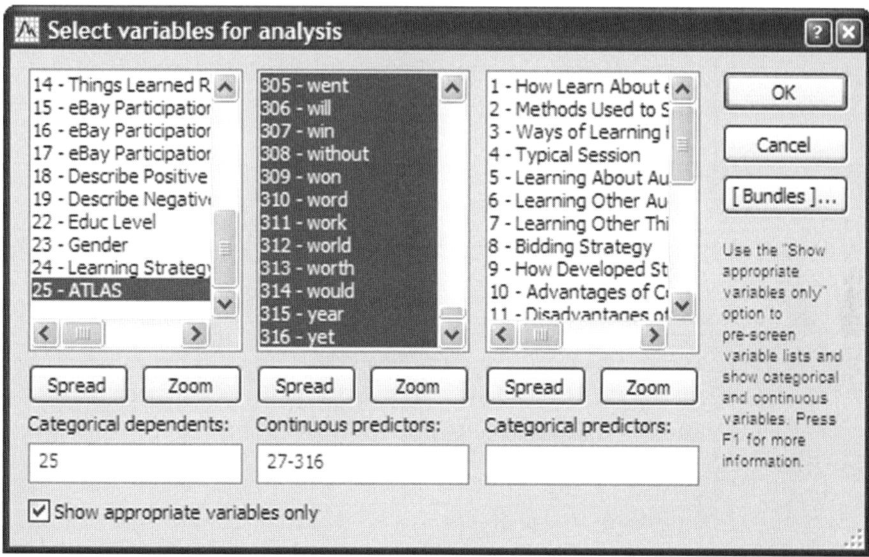

FIGURE F.144

Use ATLAS for the dependent variable and 27 to 316 as the predictors.

We will leave all the defaults as they are, but know that under memory usage, one might want to limit the amount of space that the program will consume. Click OK (Figure F.145).

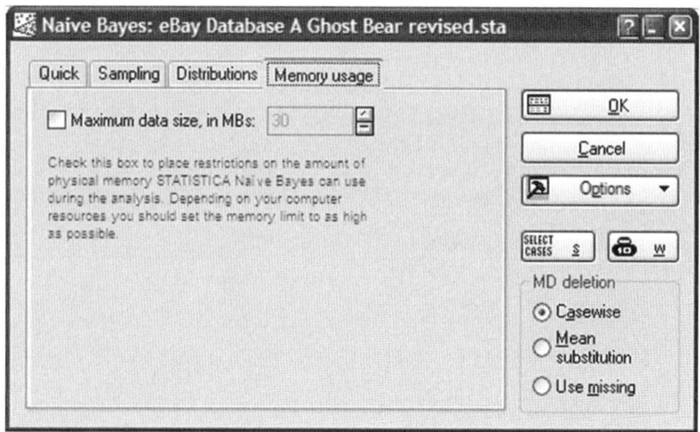

FIGURE F.145

Memory usage consideration.

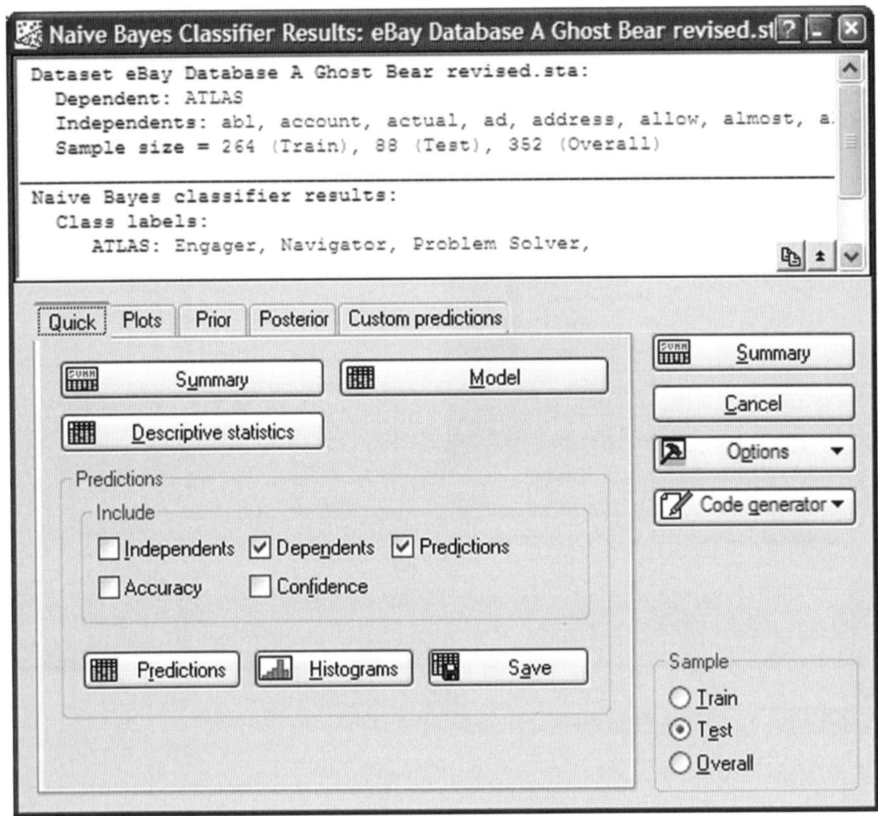

FIGURE F.146

Output screen.

Let's look and see what came out after clicking "model" in Fig. F.146.

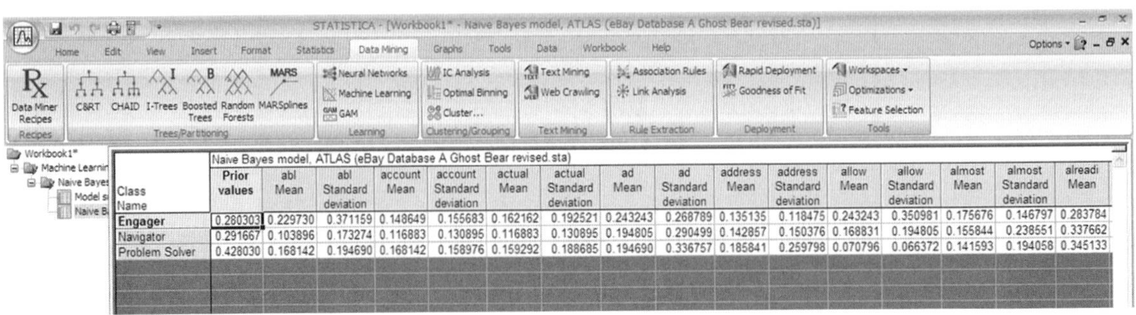

Naive Bayes model, ATLAS (eBay Database A Ghost Bear revised.sta)

Class Name	Prior values	abl Mean	abl Standard deviation	account Mean	account Standard deviation	actual Mean	actual Standard deviation	ad Mean	ad Standard deviation	address Mean	address Standard deviation	allow Mean	allow Standard deviation	almost Mean	almost Standard deviation	alreadi Mean
Engager	0.280303	0.229730	0.371159	0.148649	0.155683	0.162162	0.192521	0.243243	0.268789	0.135135	0.118475	0.243243	0.350981	0.175676	0.146797	0.283784
Navigator	0.291667	0.103896	0.173274	0.116883	0.130895	0.116883	0.130895	0.194805	0.290499	0.142857	0.150376	0.168831	0.194805	0.155844	0.238551	0.337662
Problem Solver	0.428030	0.168142	0.194690	0.168142	0.158976	0.159292	0.188685	0.194690	0.336757	0.185841	0.259798	0.070796	0.066372	0.141593	0.194058	0.345133

FIGURE F.147

The output when one selects "model."

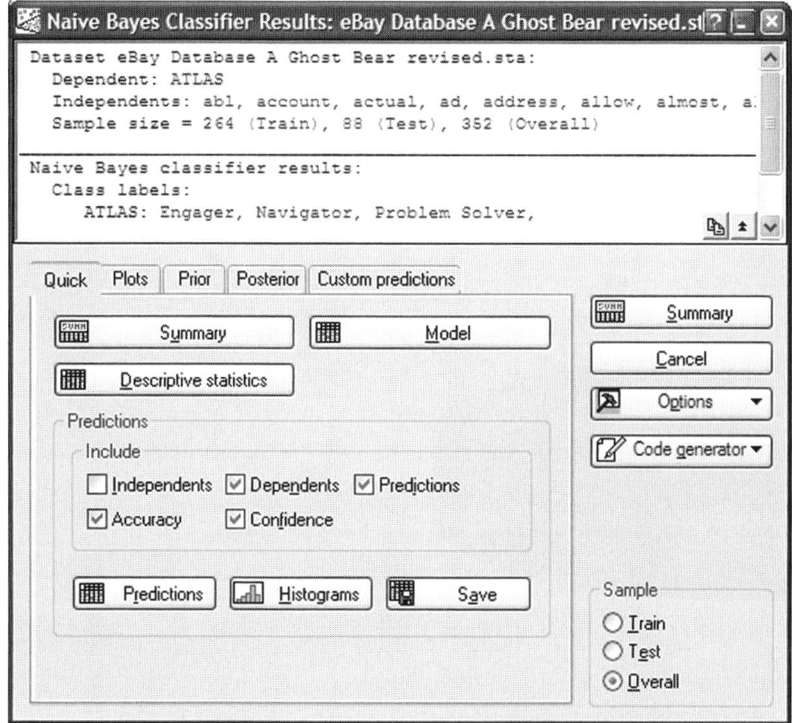

FIGURE F.148
Click on dependents, predictions, accuracy, and confidence.

Then click on predictions to yield the following spreadsheet (Fig. F.149).

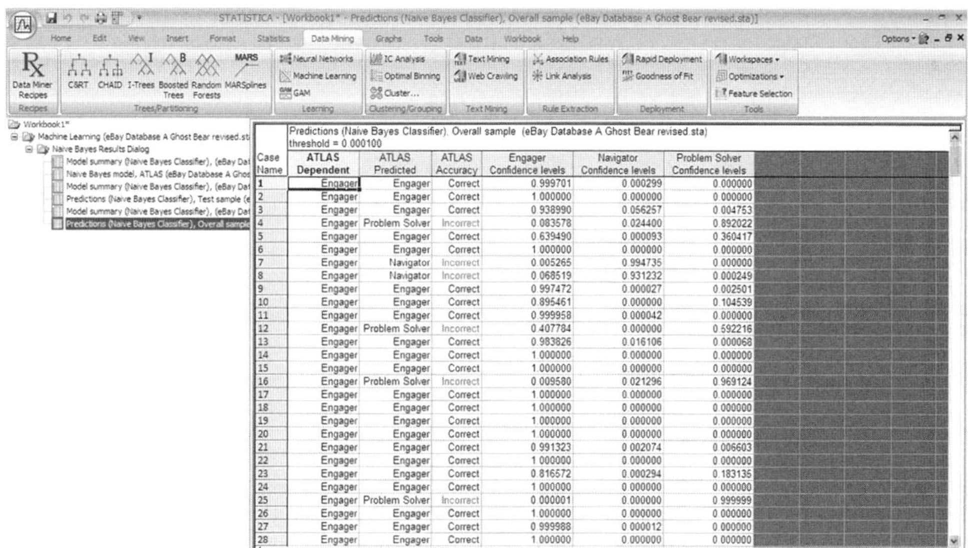

Predictions (Naive Bayes Classifier), Overall sample (eBay Database A Ghost Bear revised.sta)
threshold = 0.000100

Case Name	ATLAS Dependent	ATLAS Predicted	ATLAS Accuracy	Engager Confidence levels	Navigator Confidence levels	Problem Solver Confidence levels
1	Engager	Engager	Correct	0.999701	0.000299	0.000000
2	Engager	Engager	Correct	1.000000	0.000000	0.000000
3	Engager	Engager	Correct	0.938990	0.056257	0.004753
4	Engager	Problem Solver	Incorrect	0.083578	0.024400	0.892022
5	Engager	Engager	Correct	0.639490	0.000093	0.360417
6	Engager	Engager	Correct	1.000000	0.000000	0.000000
7	Engager	Navigator	Incorrect	0.005265	0.994735	0.000000
8	Engager	Navigator	Incorrect	0.068519	0.931232	0.000249
9	Engager	Engager	Correct	0.997472	0.000027	0.002501
10	Engager	Engager	Correct	0.895461	0.000000	0.104539
11	Engager	Engager	Correct	0.999958	0.000042	0.000000
12	Engager	Problem Solver	Incorrect	0.407784	0.000000	0.592216
13	Engager	Engager	Correct	0.983826	0.016106	0.000068
14	Engager	Engager	Correct	1.000000	0.000000	0.000000
15	Engager	Engager	Correct	1.000000	0.000000	0.000000
16	Engager	Problem Solver	Incorrect	0.009580	0.021296	0.969124
17	Engager	Engager	Correct	1.000000	0.000000	0.000000
18	Engager	Engager	Correct	1.000000	0.000000	0.000000
19	Engager	Engager	Correct	1.000000	0.000000	0.000000
20	Engager	Engager	Correct	1.000000	0.000000	0.000000
21	Engager	Engager	Correct	0.991323	0.002074	0.006603
22	Engager	Engager	Correct	1.000000	0.000000	0.000000
23	Engager	Engager	Correct	0.816572	0.000294	0.183135
24	Engager	Engager	Correct	1.000000	0.000000	0.000000
25	Engager	Problem Solver	Incorrect	0.000001	0.000000	0.999999
26	Engager	Engager	Correct	1.000000	0.000000	0.000000
27	Engager	Engager	Correct	0.999988	0.000012	0.000000
28	Engager	Engager	Correct	1.000000	0.000000	0.000000

FIGURE F.149
Spreadsheet of correct and incorrect classifications. Set the spreadsheet to active input.

Next, make a pie chart (as in Figure F.150) of the incorrect versus correct decisions to see if the overall procedure was effective at predicting the ATLAS groups. This time, the prediction was much better, yielding 71 percent accuracy in the predictions (Fig. F.150). Figure F.151 shows how to obtain a breakdown by group.

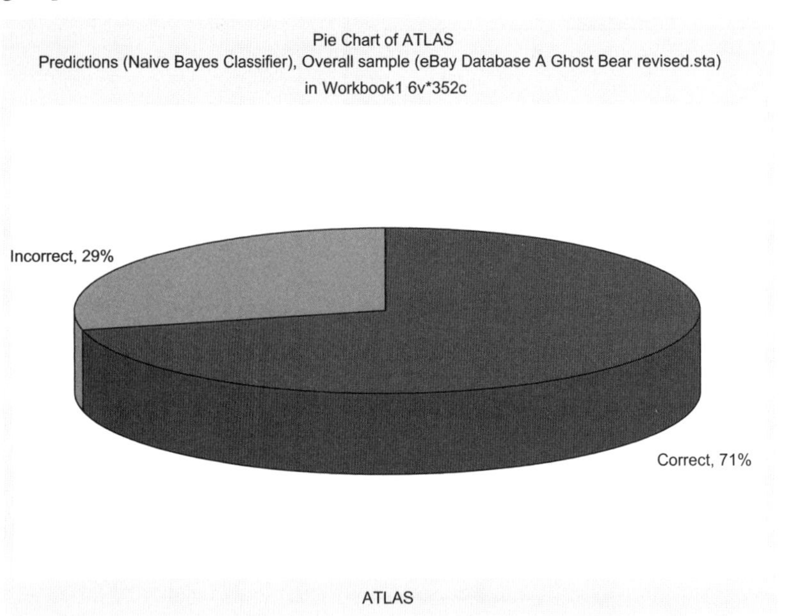

FIGURE F.150
Pie Chart of Correct and Incorrect Predictions for ATLAS groups.

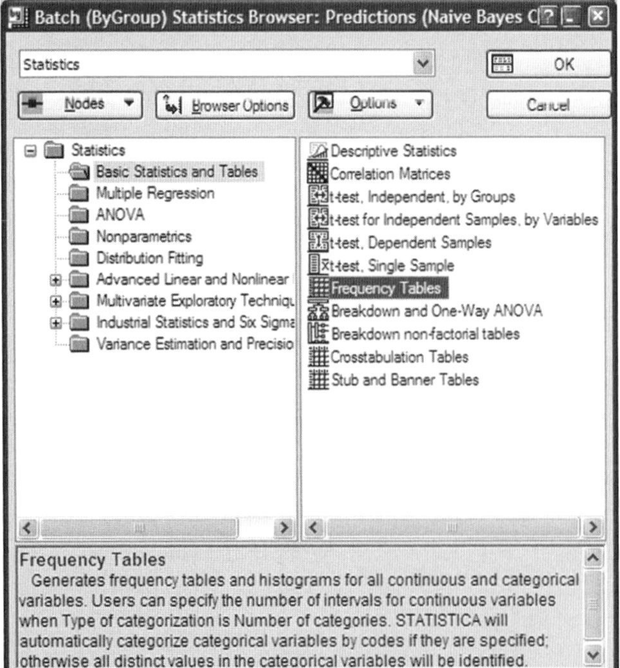

FIGURE F.151
To see the breakdown, we went to "Statistics" and "Batch by group." We then used frequency tables.

Engagers

Category	Frequency table: ATLAS: Accur		
	Count	Cumulative Count	Percent
Correct	68	68	71.57895
Incorrect	27	95	28.42105
Missing	0	95	0.00000

Navigators

Category	Frequency table: ATLAS: Accur		
	Count	Cumulative Count	Percent
Correct	68	68	67.32673
Incorrect	33	101	32.67327
Missing	0	101	0.00000

Problem Solvers

Category	Frequency table: ATLAS: Accur		
	Count	Cumulative Count	Percent
Correct	115	115	73.71795
Incorrect	41	156	26.28205
Missing	0	156	0.00000

FIGURE F.152
Percentages of correctly predicted ATLAS categories.

There was not a significant difference in the percentages of correct predictions (Fig. F.152), but it may have been that the problem solvers were slightly more accurate in being predicted and were the navigators. In any case, this last procedure seemed to have the most accuracy in prediction. We were happy.

SUMMARY

We tried many ways of predicting ATLAS group membership from the words that the cases used the most while answering questions about strategies they used initiating eBay. We used text mining in various ways and combined the procedures with quantitative data mining methods. We were able to make some guesses and hypotheses about the differences in words that the three types were most likely to use. The best predictions came from doing a more Bayesian procedure, in which overall 71 percent of the predictions were correct.

Perhaps to more correctly categorize the types, however, one needed to take a more holistic approach by viewing the sentence structures. Perhaps sentiment analysis, parts of speech tagging, and entity extraction could pull out the more known subtle meanings. These could be done by subjecting this data set to GATE or other freeware tools, in which one projects the structure into the analysis. We suggest that the reader take the data set to try that type of analysis to see if better results could be achieved.

Reference

Conti, G. J., and Kolody, R. C. (1999). *Guide for using ATLAS.* Stillwater, OK: Oklahoma State University.

TUTORIAL G

Text Mining for Patterns in Children's Sleep Disorders Using *STATISTICA* Text Miner

Jennifer Thompson
StatSoft, Inc., Tulsa, OK

Karen Spruyt, PhD
The University of Chicago, Department of Pediatrics, Section of Pulmonary Medicine and Cystic Fibrosis Center, Chicago, IL

CONTENTS

Pediatric sleep disorders are highly underrecognized, but the field is rapidly growing. Sleep-disordered breathing (SDB) is a common and highly prevalent condition in the pediatric age range, affecting up to 27 percent of children, with a median in the 10 percent to 12 percent range. Rather than being an "all or none" condition, SDB is traditionally perceived as encompassing a wide spectrum of clinical severity. Indeed, SDB has been suggested to range from habitual snoring to obesity hypoventilation syndrome, while also including upper airway resistance syndrome and obstructive sleep apnea syndrome (OSAS) of varying severity.

Researchers of children's sleep disorders have a variety of resource material on various studies and findings in their diagnosis and treatment. These resources come from medical journals and other publications. In general, the articles discuss studies on treatment and outcomes of SDB. Two hundred forty-seven articles are available in text format.

The research team is interested in exploring these texts to reveal patterns. They are interested to see if logical patterns emerge that may give insight into research trends. These trends could lead the research team toward areas that need further study. With the large number of articles, a manual exploration of the texts would be tedious and time-consuming. With text mining in *STATISTICA*, the documents can be quickly indexed, revealing keywords. With singular value decomposition (SVD), patterns in the text are revealed.

Practical Text Mining and Statistical Analysis for Non-structured Text Data Applications. DOI: 10.1016/B978-0-12-386979-1.00015-3

SETTING UP THE ANALYSIS

This example uses 247 text files and a *STATISTICA* spreadsheet that contains the names of the files. To begin the analysis, open *text file list.sta* in *STATISTICA* (see Figure G.1).

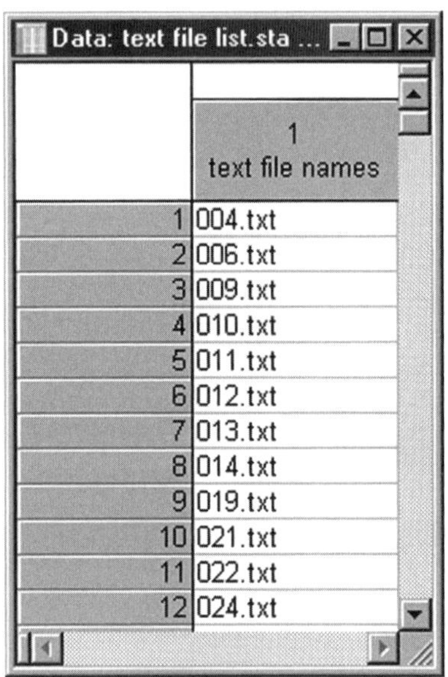

FIGURE G.1
Text data file for this tutorial.

From the *Data Mining* menu, select *Text Mining* to open the *Text mining* dialog (see Figure G.2).

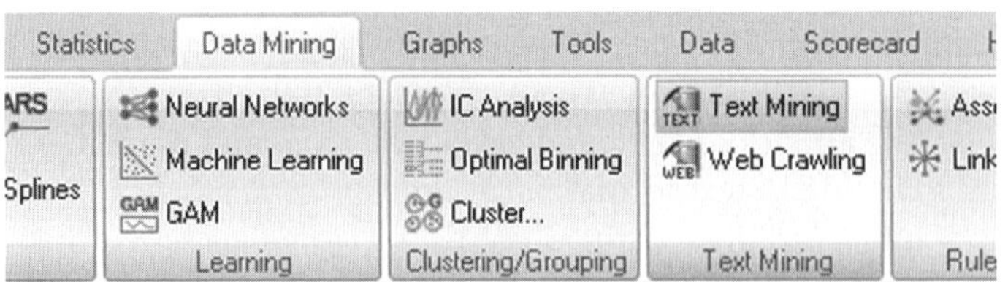

FIGURE G.2
Text Mining selection on ribbon bar.

On the *Quick* tab, select *Files* for the *Retrieve text from* option. Check the *Paths in spreadsheet* option. Select *Document paths* to open the *Select a variable containing file paths* dialog. Select the variable, *text file names*, and click OK. Now the Text Miner tool can access the text files for analysis (see Figures G.3 and G.4).

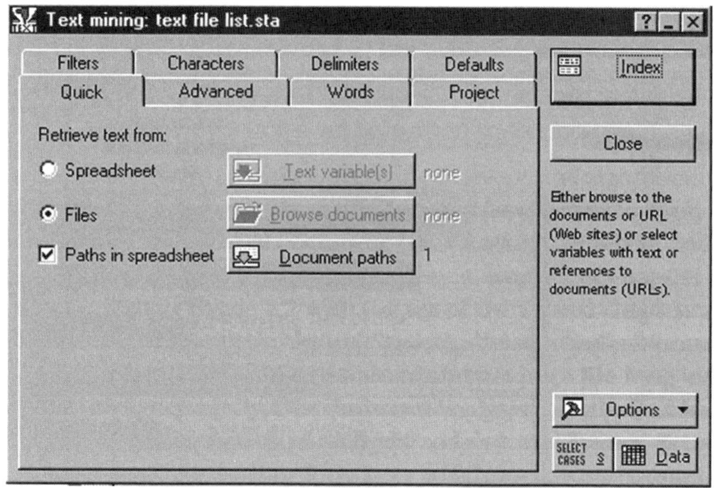

FIGURE G.3
Text mining dialog, Quick tab displayed.

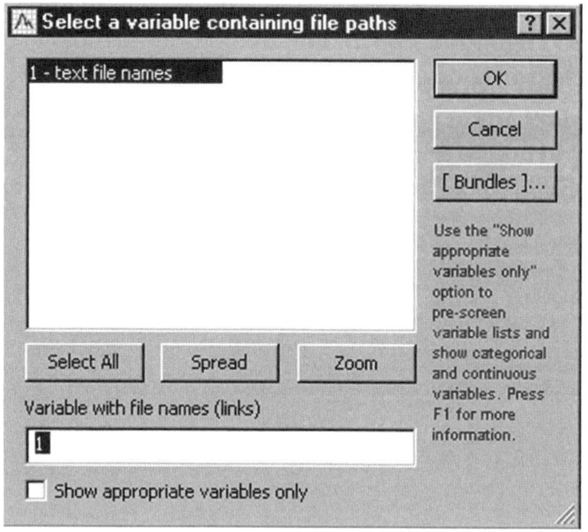

FIGURE G.4
Select a variable containing file paths dialog.

On the *Advanced* tab (see Figure G.5), we have options to select the language and percent of word occurrence in the files. This example uses English as the *Stemming language*. Change the *percent of files where word occurs Min* and *Max* values to 10 and 70, respectively. A word that occurs either in every file or in only a small percentage will not add much value to the analysis. These words will not contribute to concepts extracted from the text. Also increase the *Maximum number of words to be selected* to 5,000.

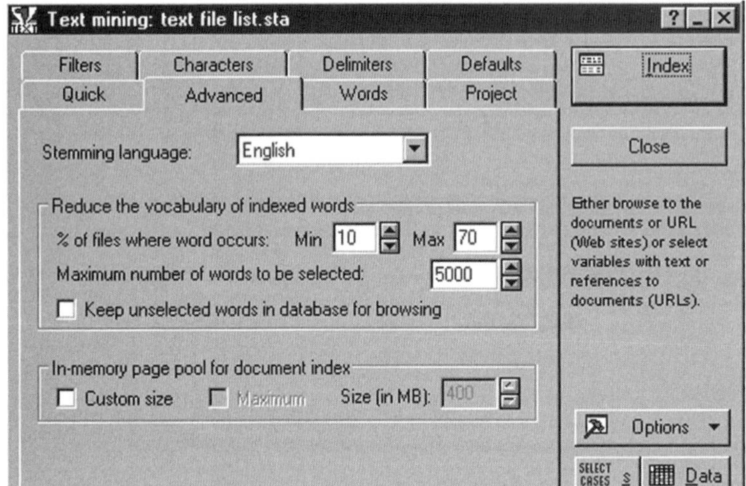

FIGURE G.5
Advanced tab of the Text mining dialog window.

Select the *Words* tab (see Figure G.6). Here, you have control over what words should be ignored with the *Stopwords* list and *Phrases* and *Synonyms* to look for, or you can look only for specific words with an *Inclusion words* list.

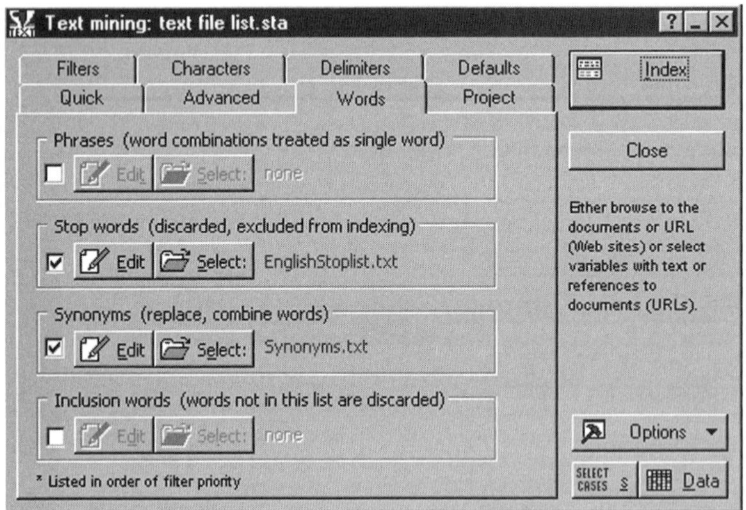

FIGURE G.6
Words tab of the Text mining dialog.

By default, the *EnglishStopList.txt* file is selected, and it lists common words that are not likely to be meaningful in text mining analysis. Often it is a good idea to add words based on your project, which should also be ignored. For this example, add the words "author" and "manuscript" to the stoplist because these words appear frequently but do not have meaning in this study. To do so, select the *Edit* button in the *Stopwords* field to open the *Stopword editor* dialog (see Figure G.7). Add the words to be excluded on separate lines to the end of the list. Likely, you will need to use the Save As button to save the stop list.

Additional words that occur often in articles but do not pertain to the actual content include summary, introduction, methods, results, discussion, objective, aim, figure, table, design, and statistical analysis. These terms may be good to add to the stoplist as well.

FIGURE G.7
Stop-word editor dialog.

This example also has some synonyms that should be specified. Continuous positive airway pressure (CPAP) is a treatment for some sleep disorders. Sometimes it is abbreviated as PAP, so these two acronyms should be synonyms for each other. Click *Edit* in the *Synonyms* field to display the *Synonyms* dialog (see Figure G.8). Enter "CPAP" in the *Root* field and "PAP" in the *Words* field. Click *Add new synonym*. Repeat the process to denote PSG and polysomnography as synonyms.

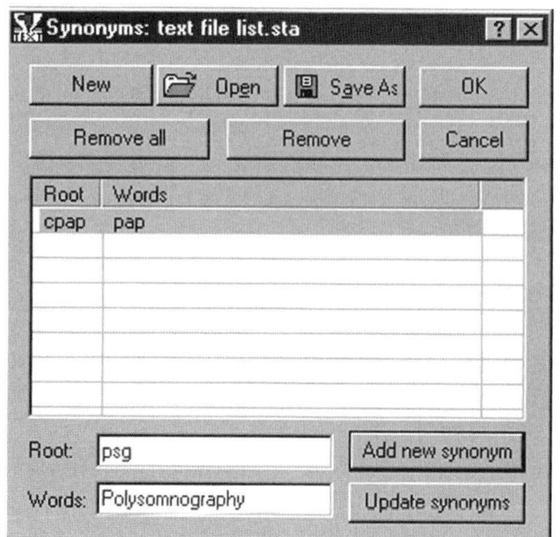

FIGURE G.8
Synonyms dialog.

On the *Filters* tab (see Figure G.9), you can control things like word length and specifics about word construction. The defaults settings here are appropriate.

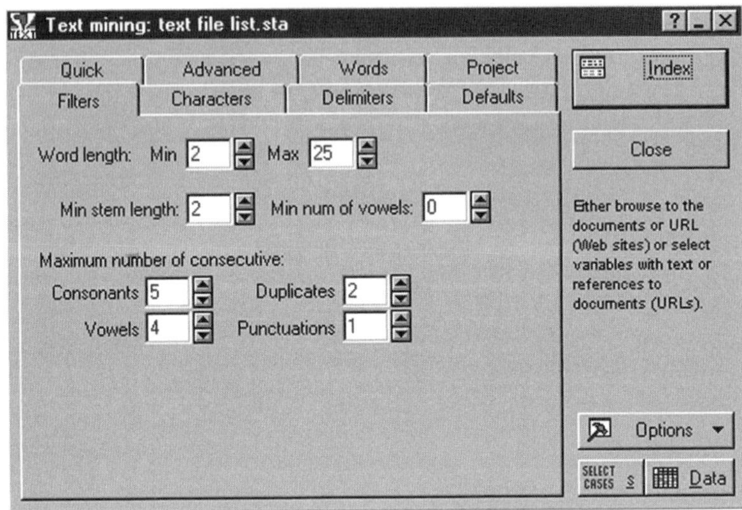

FIGURE G.9
Filters tab.

On the *Characters* tab (see Figure G.10), you can specify characters that are appropriate to form words. By default, numbers 0–9 are allowed for both the *Body* and *Last letter*. Remove the numbers from these character lists.

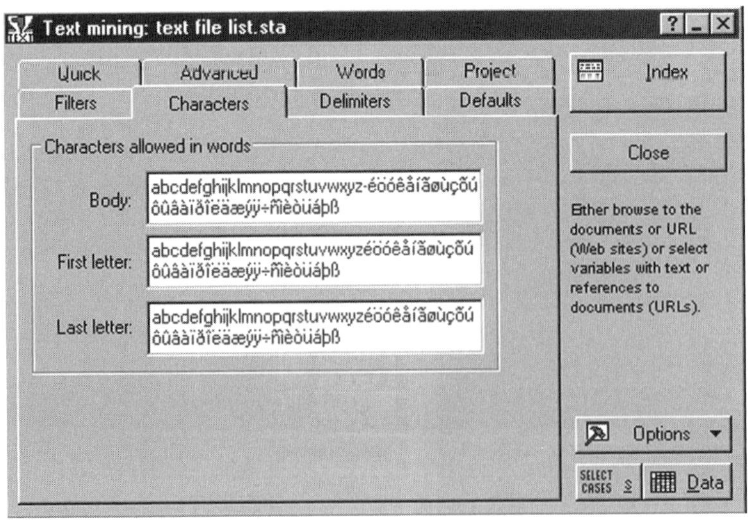

FIGURE G.10
Characters tab dialog.

REVIEWING RESULTS

Accept the other default settings, and select *Index* to begin indexing the set of text files. This process will complete quickly, and a red progress bar at the bottom of the screen shows progress. When complete, the *Results* dialog will display (see Figure G.11). The indexed words are shown in this dialog.

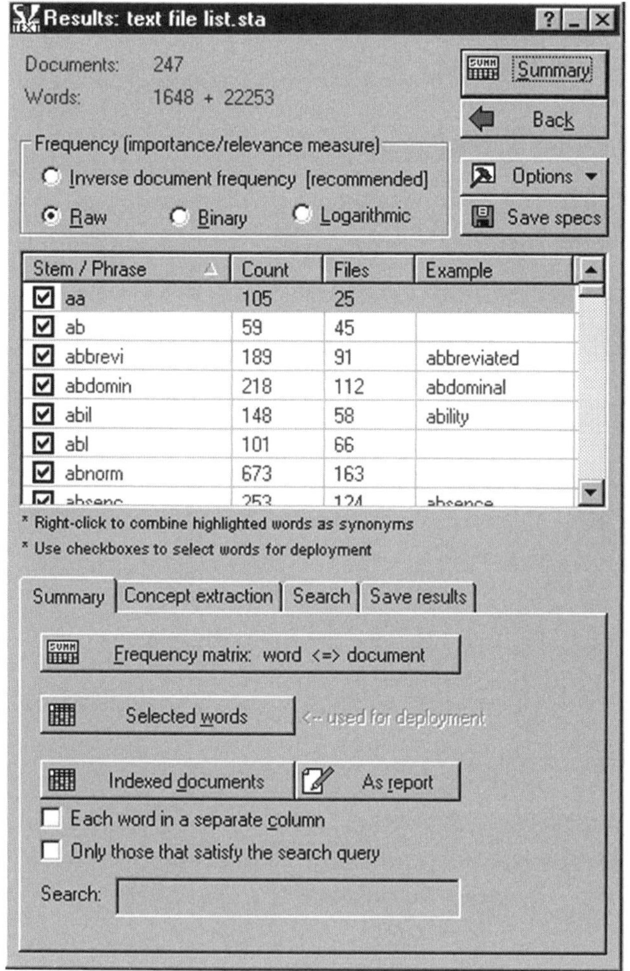

FIGURE G.11
Results dialog following indexing the words; note that the Raw frequency is displayed as the default result.

In practice, care should be taken when reviewing the list of words. Additional synonyms can be combined by highlighting them, right clicking, and selecting *Combine Words*. "Abil" is the stem word for "ability," and "abl" is the stem for "able." These words can be combined as synonyms. Highlight them both, right click, and select *Combine Words* (see Figure G.12).

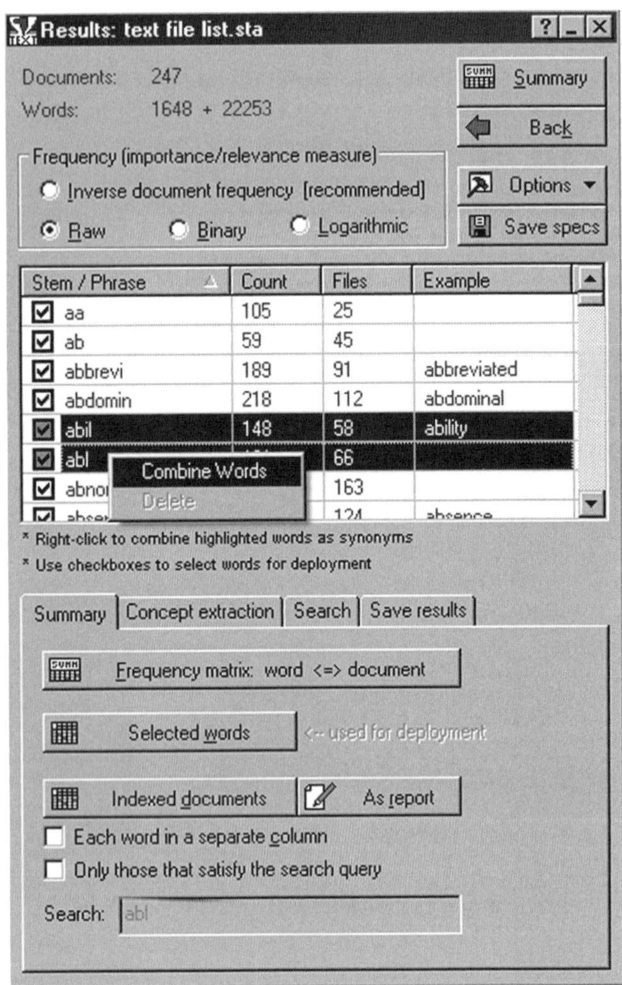

FIGURE G.12
Combining words.

This will open the *Combine words* dialog (see Figure G.13). Select *abl* from the *select one of the highlighted words* drop-down menu and click OK. The two words will be combined into one.

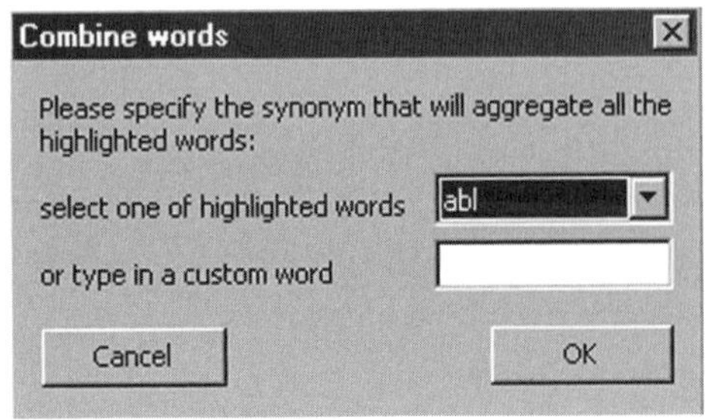

FIGURE G.13
Combine words dialog.

This process should be repeated for all appropriate synonyms to give the best results. Change the *Frequency (importance/relevance measure)* option to *Inverse document frequency*. On the *Concept extraction* tab, select *Perform Singular Value Decomposition (SVD)* (see Figure G.14). The progress bar will again show the progress of the analysis in red.

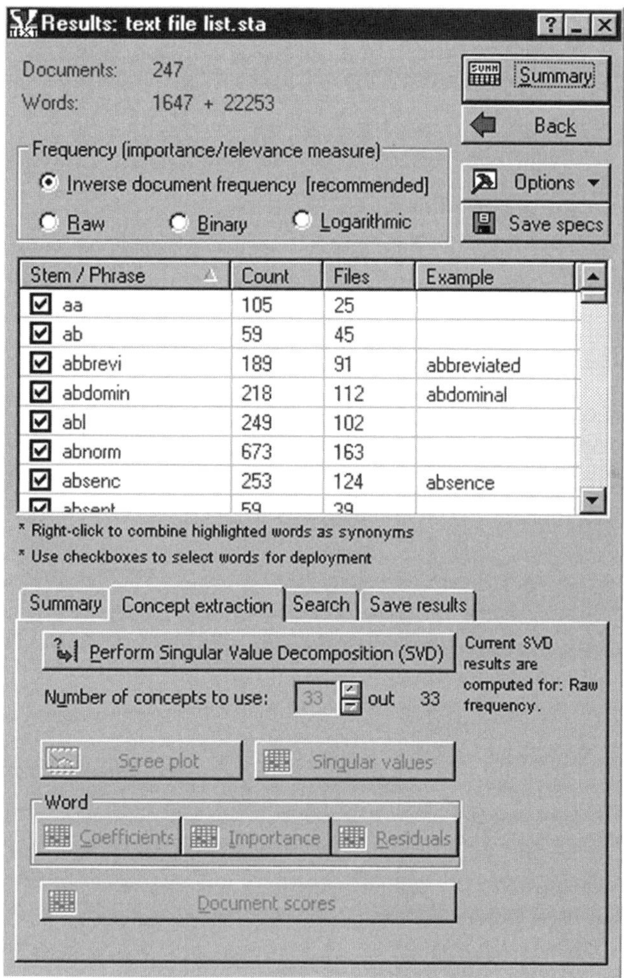

FIGURE G.14

Setting the Inverse document frequency radio button on the Text Mining results dialog.

When complete, the other options on this tab will become active. First make the *Scree plot* (see Figure G.15). With this plot, we can see which extracted concepts will be most informative.

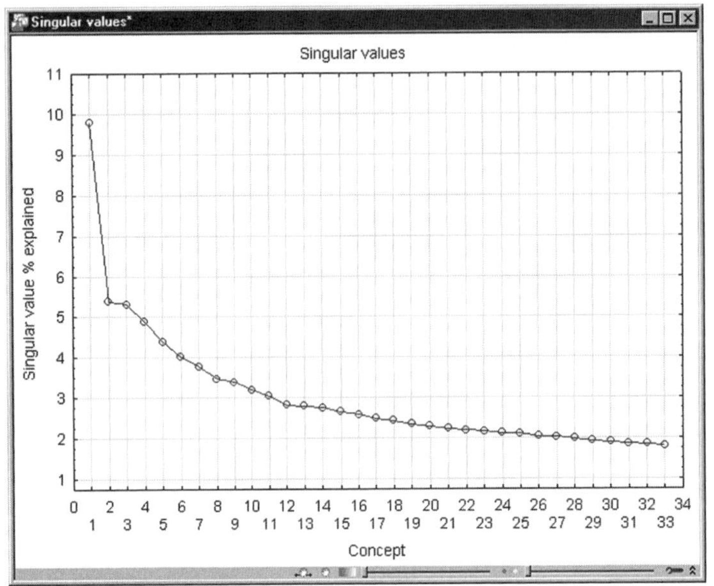

FIGURE G.15

Scree type plot, selected by the *Scree plot* button on the results dialog, following the SVD computation.

From the first plot, it can be seen that the first three concepts extracted will be the most informative and should certainly be reviewed. The next few may also provide some insight.

Change the *Number of concepts to use* to 5, and create the *Coefficients* spreadsheet (see Figures G.18 and G.19).

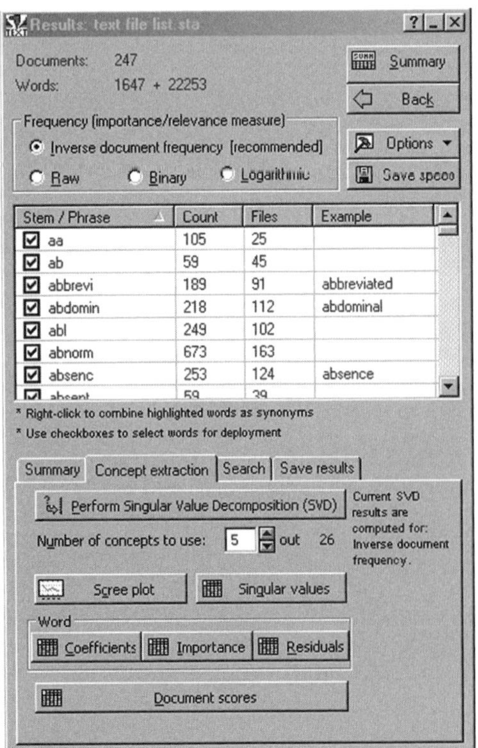

FIGURE G.16

Results dialog following selecting the *Perform Singular Value Decomposition (SVD)*.

Data: SVD word coefficients (text file list)*

	SVD word coefficients (text file list.sta)				
	Concept 1	Concept 2	Concept 3	Concept 4	Concept 5
aa	0.000065	-0.000147	-0.000078	-0.000112	0.000329
ab	0.000057	-0.000264	-0.000077	-0.000197	0.000088
abbrevi	0.000088	-0.000185	-0.000016	0.000224	-0.000380
abdomin	0.000086	-0.000257	-0.000185	0.000117	0.000228
abl	0.000093	0.000264	0.000136	0.000115	0.000232
abnorm	0.000081	-0.000145	0.000017	0.000020	-0.000051
absenc	0.000085	-0.000112	-0.000112	-0.000017	0.000065
absent	0.000066	0.000126	0.000012	0.000056	-0.000049
abstract	0.000071	0.000015	-0.000132	0.000242	0.000001
ac	0.000057	0.000086	-0.000032	0.000042	0.000340
acad	0.000083	0.000366	-0.000327	0.000418	-0.000029

FIGURE G.17
SVD word coefficients results spreadsheet.

This output can then be designated in an input spreadsheet for graphing in a scatterplot to reveal interesting patterns. On the *Data* menu, check the *Input* option. Then on the *Graphs* menu, select *Scatterplot*. The *2D Scatterplots* dialog will open (see Figure G.18).

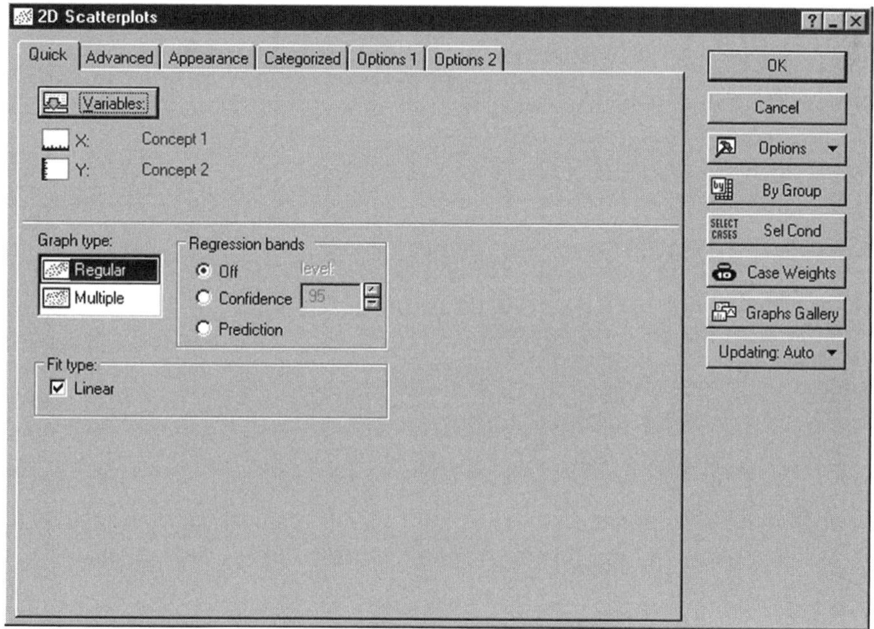

FIGURE G.18
Scatterplot 2D dialog window.

Select *Variables* to open the *Select Variables* for scatterplot dialog. Select *Concept 1* as X and *Concept 2* as Y (see Figure G.19).

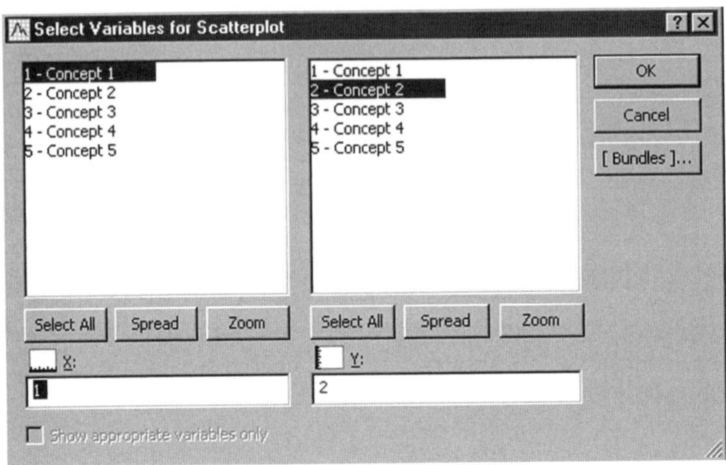

FIGURE G.19
Select Variables dialog for the 2D scatterplot.

Click OK on both dialogs to create the scatterplot (see Figure G.20).

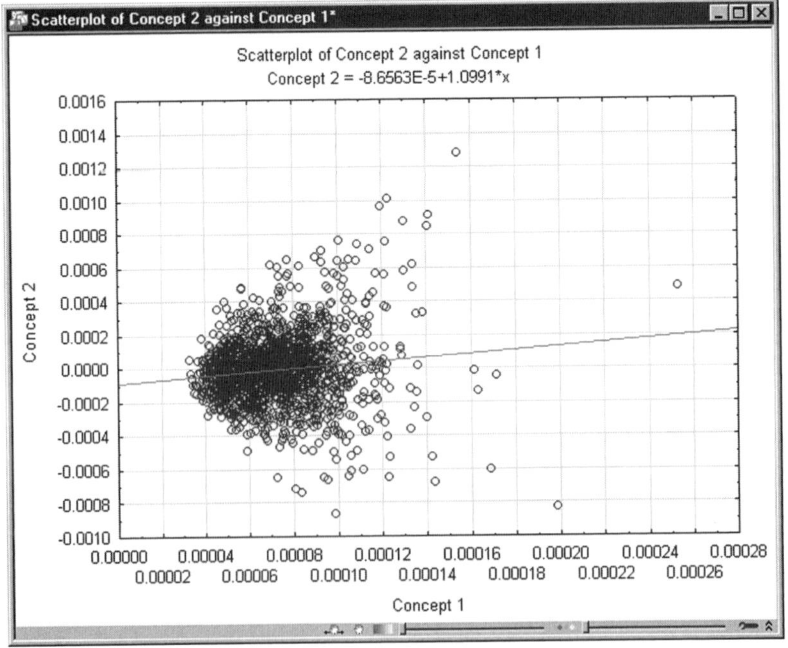

FIGURE G.20
2D scatterplot of Concept 2 versus Concept 1.

On the *Edit* menu, in the *Customize Graph* section, select the graph *Brushing* button (see Figure G.21). It is the one that looks like a magnifying glass with crosshairs, and it is the first icon on the left on the top row.

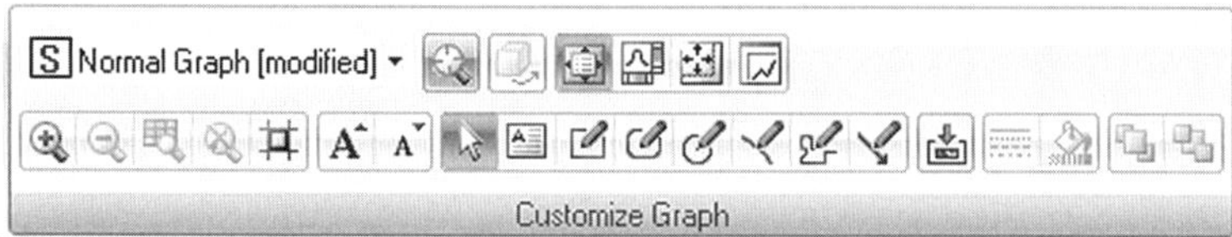

FIGURE G.21
Selecting the Brushing button in the Edit menu.

This will open the *Brushing 2D* dialog window (see Figure G.22).

FIGURE G.22
Brushing 2D dialog window.

Change the *Action* to *Label* and the *Selection Brush* to *Lasso*. This will allow you to draw a "lasso" around the outlier points of this plot and label them with the word the point represents. Draw the lasso around the desired points (see Figure G.23).

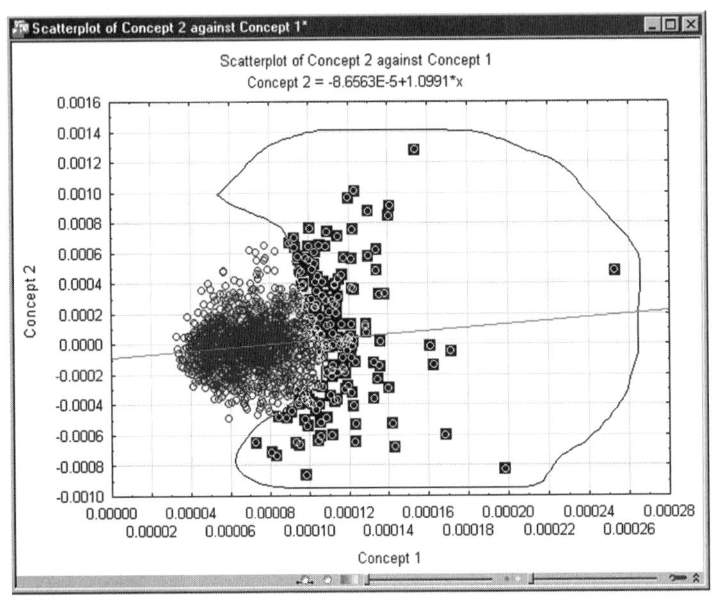

FIGURE G.23

Lassoed section in the 2D scatterplot.

Select *Apply* to add the word labels (see Figure G.24).

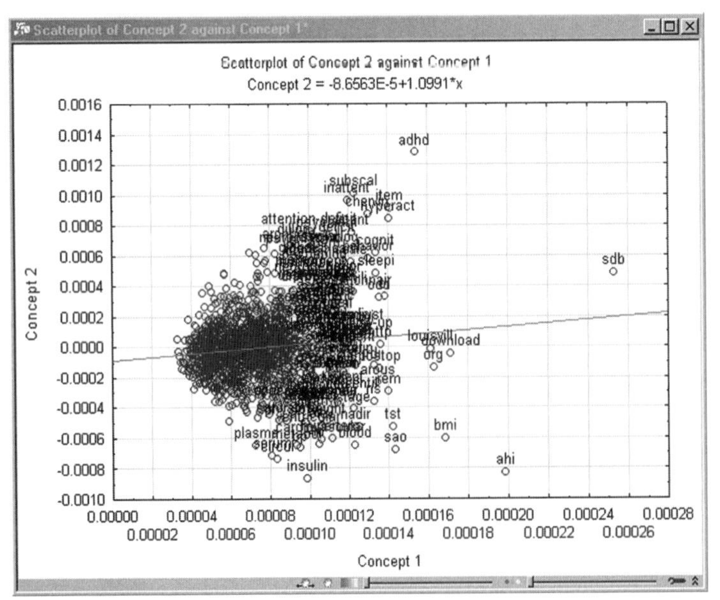

FIGURE G.24

Lassoed points are now labeled with the word these points represent.

This plot shows a lot of words, most of which have labels that overlap. In this initial exploration, we can use the zoom tools to push in on portions of the graph and reveal interesting relationships. When it comes to presenting our findings, we can remove the extra labels that are not needed for display and give a cleaner, easier to follow result.

For example, the upper portion of the graph shows the word "ADHD." The remaining words are difficult to read because of overlap (see Figure G.25). Use the *Zoom* tool, and from the *Edit* menu, *Customize Graph* section. Its icon is on the far left of the second row and looks like a magnifying glass with a plus symbol.

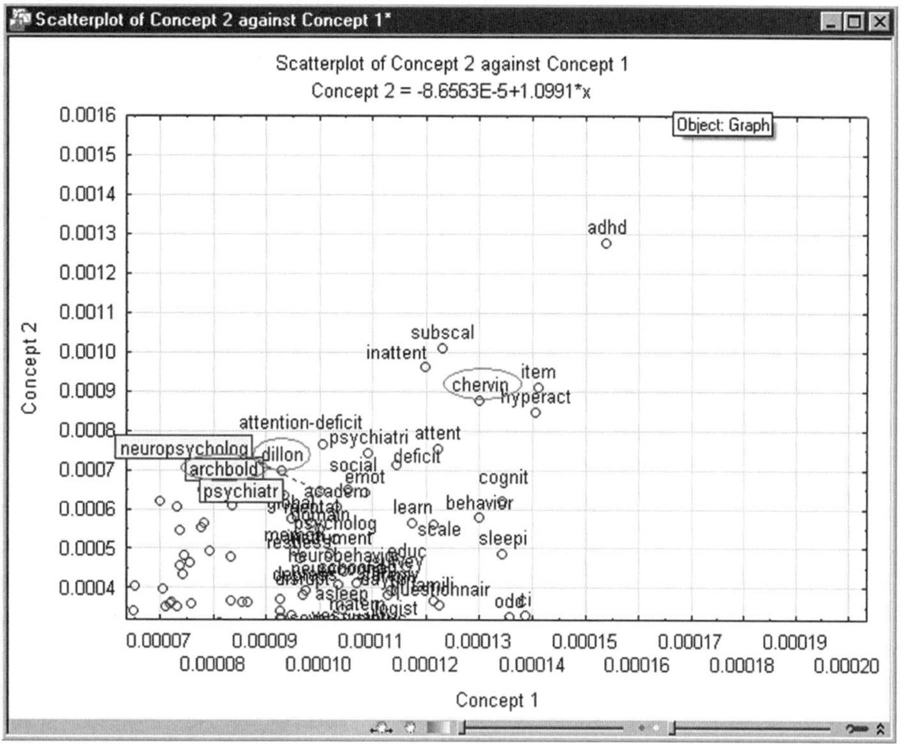

FIGURE G.25

Brushed area zoomed in to enlarge for a better view of each word.

This plot reveals some interesting patterns. Many of the words in this portion of the graph seem to relate to attention deficit hyperactivity disorder (ADHD). Some of the stem words we see are adhd, inattent, hyperact, attent, deficit, cognit, behavior, emot, and attention-deficit. The words circled in red are Chervin, Dillon, and Archbold. These are researchers that focused on ADHD in SDB kids. Dillon and Archbold, in addition to being researchers, are also names of sleep study facilities. So we have revealed an interesting pattern here. Articles that talk about ADHD and related terms also reference Chervin, Dillon, and/or Archbold. We know this because of the clustering of these terms in the concept scatterplot.

Using the same process, the scatterplot of Concepts 2 and 3 shows interesting clusters as well (see Figure G.26). The following graph is a labeled scatterplot of Concept 2 versus Concept 3. Words like bleed, pain, distress, complic, and compromise cluster with words like tonsil, tonsillectomy, adenoidectomi, oper, surgery, and anesthesia. Some articles are discussing reasons for surgery and possible complications from tonsillectomies and adenoidectomies. Mitchel and Rosenfeld are researchers who have published papers centering on quality of life after a tonsillectomy and adenoidectomy.

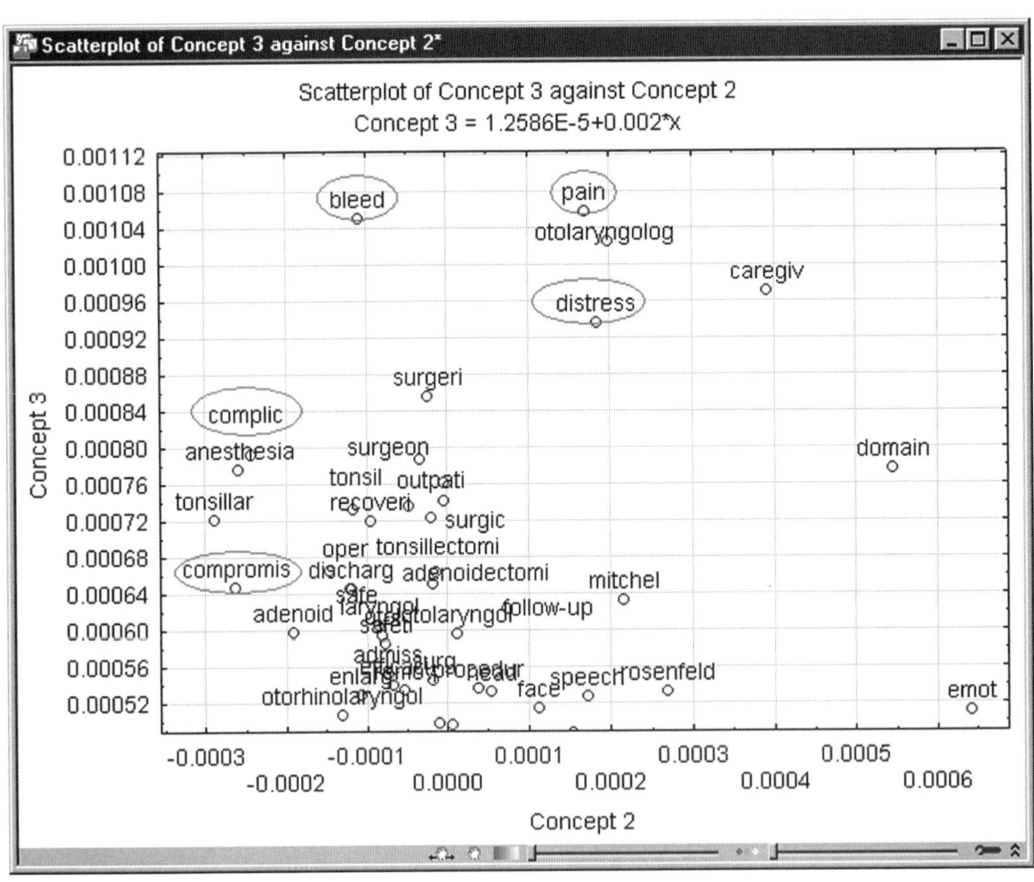

FIGURE G.26
Scatterplot of Concept 3 versus Concept 2.

Another interesting piece of text mining output is *Word Importance*. Create this output from the text mining *Results* dialog. Next, sort the terms by their importance. From the *Data* menu, select *Sort* to open the *Sort Options* dialog. Highlight *Importance* and select *Add Var(s)*. Change the *Direction* to *Descending* and select OK (see Figure G.27).

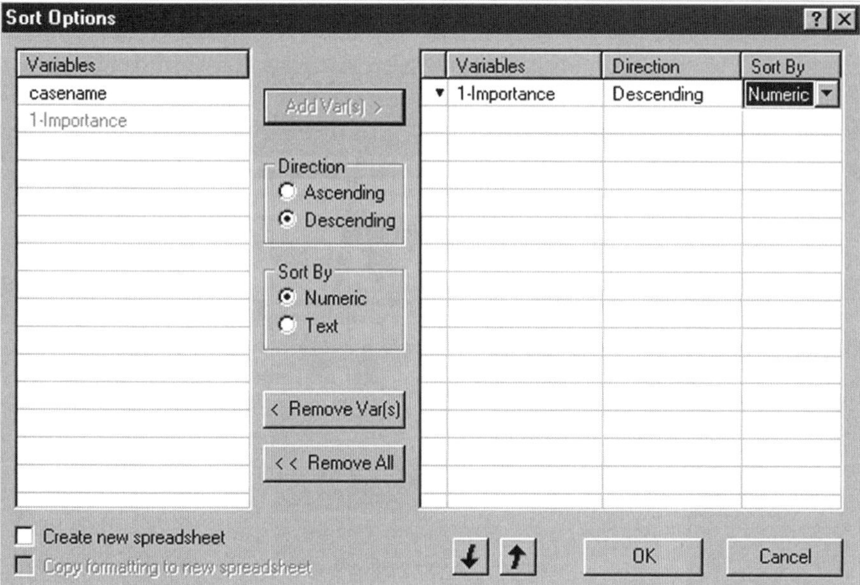

FIGURE G.27
Sort Options dialog.

The terms will then be sorted with the highest importance first. This shows which terms contributed most to the SVD analysis (see Figure G.28).

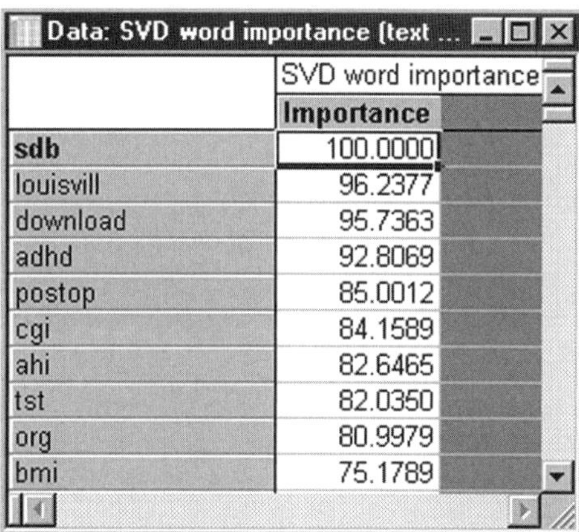

FIGURE G.28
Words sorted in order of decreasing importance.

SUMMARY

Further exploration of the concept plots can reveal more interesting relationships in these published papers. Some relationships may be expected, while others are not. These unexpected relationships may help researchers to form new hypotheses to test. The unstructured learning gained from text mining could lead to new discoveries in children's sleep disorders by pointing research in the right direction. Text mining can find patterns in text that would otherwise be too complex to discover.

Extracting Knowledge from Published Literature Using RapidMiner

Dursun Delen, PhD
Spears School of Business, Oklahoma State University

CONTENTS

INTRODUCTION

The main purpose of this tutorial is to illustrate the text mining capabilities of RapidMiner's text analytics extension using an easily understandable example data set. In fact, the details of the data set, which was used to conduct a similar study with a different set of software tools, can be found in Delen and Crossland (2008).

MOTIVATION

Researchers as well as practitioners conducting reviews of the existing body of knowledge published in literature are facing an increasingly complex and voluminous task. With the increasing wealth of potentially significant research reported in ever increasing numbers of publication outlets (sometimes in related fields and sometimes in what is traditionally deemed as "nonrelated" fields of particular domain of study), the researcher's task is ever more daunting if a thorough job is desired.

In this tutorial, we illustrate a method to assist and enhance the efforts of researchers in this situation by enabling a semiautomated analysis of large volumes of unstructured data (in the form of published journal articles) through the application of text mining. By accessing the extensive number of

Practical Text Mining and Statistical Analysis for Non-structured Text Data Applications. DOI: 10.1016/B978-0-12-386979-1.00016-5

abstracts that are available online, herein we detail how one can use the text mining capabilities of RapidMiner, a free and open source data mining software, to analyze related research.

Using standard digital libraries and online publication search engines, we downloaded and collected all of the available articles for the three major journals in the field of information systems: *MIS Quarterly* (MISQ), *Information Systems Research* (ISR), and the *Journal of Management Information Systems* (JMIS). For each article, we extracted its title, abstract, author list, published keywords, volume, issue number, and year of publication. Also included in the data set was a field that designated the journal type of each article to serve for future pattern analysis. At the end, 901 articles were included in the corpus of their study.

In our text mining part of the analyses, we chose to use only the abstract of an article as the only source of information. We have not included the title or the keywords of the article for two main reasons: Under normal circumstances, the abstract would already include the listed keywords, and therefore inclusion of the listed key words for the analysis would mean repeating the same information and potentially giving them unmerited weight; and the listed keywords may be terms that authors would like their article to be associated with (as opposed to what is really contained in the article), therefore potentially introducing unquantifiable bias to the analysis of the content. We adopted the three-step process (see Figure H.1) of text mining (described in detail in Chapter 5) to execute the text mining project explained in this tutorial.

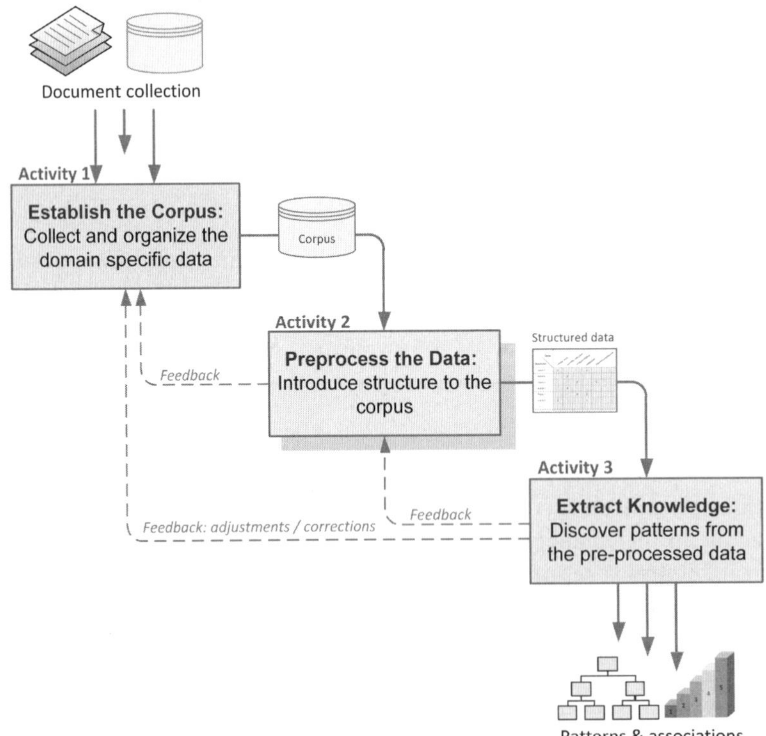

FIGURE H.1

The three-step process for text mining.

A BRIEF INTRODUCTION TO RAPIDMINER

RapidMiner is a free of charge, open source software tool for data and text mining. In addition to Windows operating systems, RapidMiner also supports Macintosh, Linux, and Unix systems. It is available as a stand-alone application for data/text analysis and as a data/text mining engine for the integration into your own products. Thousands of applications of RapidMiner in more than 40 countries are successfully developed to give its users a competitive edge.

The RapidMiner software tool, along with its extensions (including text analytics extension) and documentation, can be found and downloaded from www.rapid-i.com. Once the proper version of the tool is downloaded and installed, it can be used for a variety of data and text mining projects. Its graphical user interface is a little different from the ones we often see in other commercial data mining tools, such as IBM SPSS Modeler, SAS Enterprise Miner, and *STATISTICA* Data Miner. Such differences may lead to a longer learning curve, but once understood it is quite logical and informative.

When the RapidMiner tool is first started, the user is asked to specify a repository; either connect to a remote repository or create a new local repository (see Figures H.2 and H.3). A repository in RapidMiner is a central storage mechanism for all project-related files (processes, models, outputs, etc.).

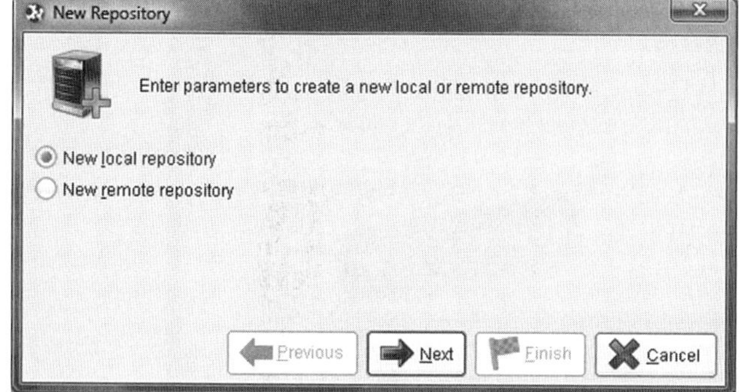

FIGURE H.2
Specifying a repository for managing a project in RapidMiner.

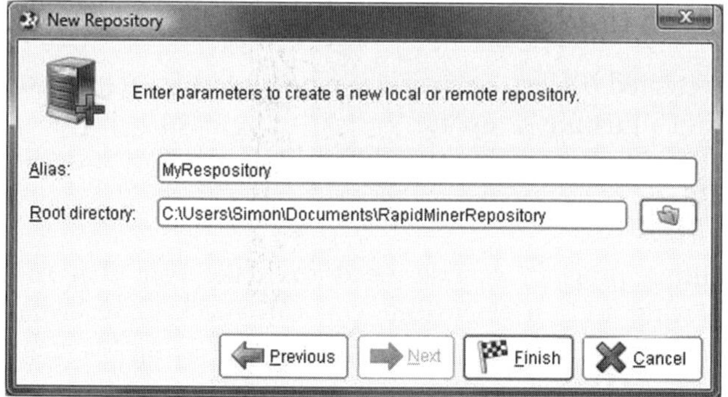

FIGURE H.3
Creating a local repository on your computer the first time you start the program.

Once the local repository and the project name are both specified, you will be forwarded into the so-called Welcome window (Figure H.4). There, the lower section shows current news about RapidMiner if you have an Internet connection. The list in the center of the window shows the analysis processes recently worked on. Users can choose to open one of the recent processes or create a brand new one.

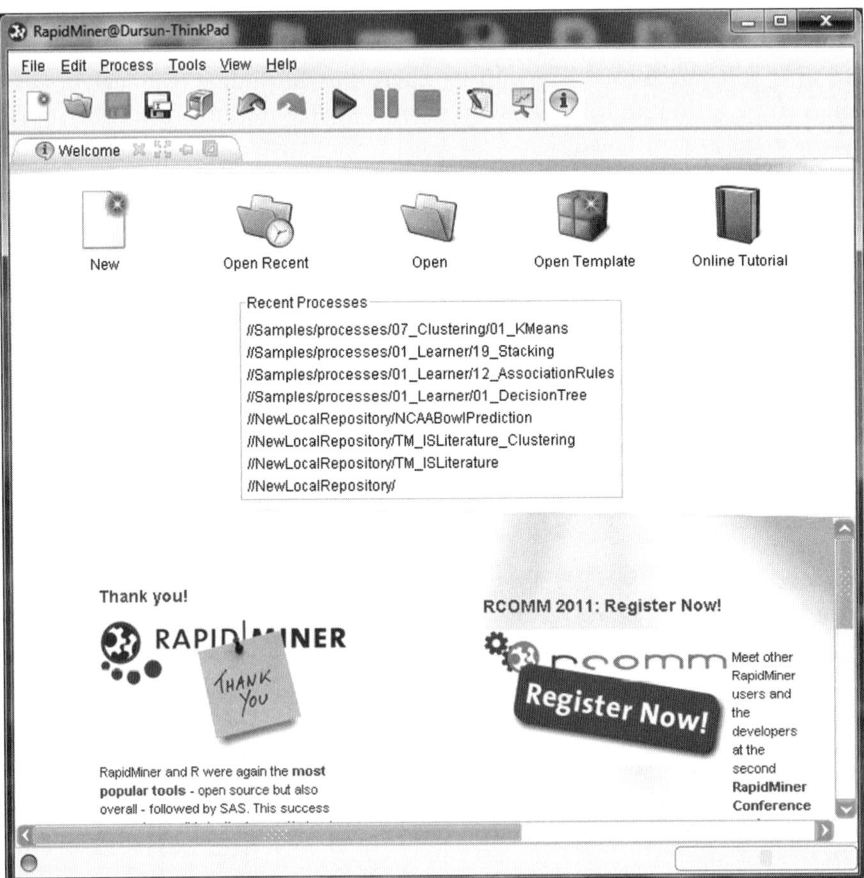

FIGURE H.4
RapidMiner's Welcome window.

In this window the user can open a recently created project, open an existing project from a file, or start a new project either from scratch or by using a prebuilt template.

TEXT ANALYTICS IN RAPIDMINER

In order to use RapidMiner for this text mining project, we first need to make sure that our version of RapidMiner includes the Text Processing add-on extension. You can check to see what extensions are already installed on your RapidMiner by clicking "Help" and then selecting "Manage

Extensions" (Figure H.5). In the pop-up window you will see all of the installed extensions with their current versions (Figure H.6). In this interface you can deactivate or uninstall already installed extensions.

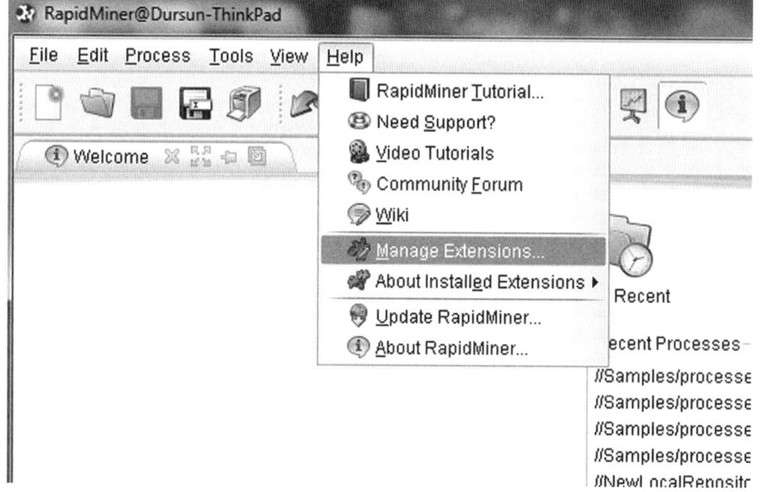

FIGURE H.5
Adding Text Mining extension to the RapidMiner tool.

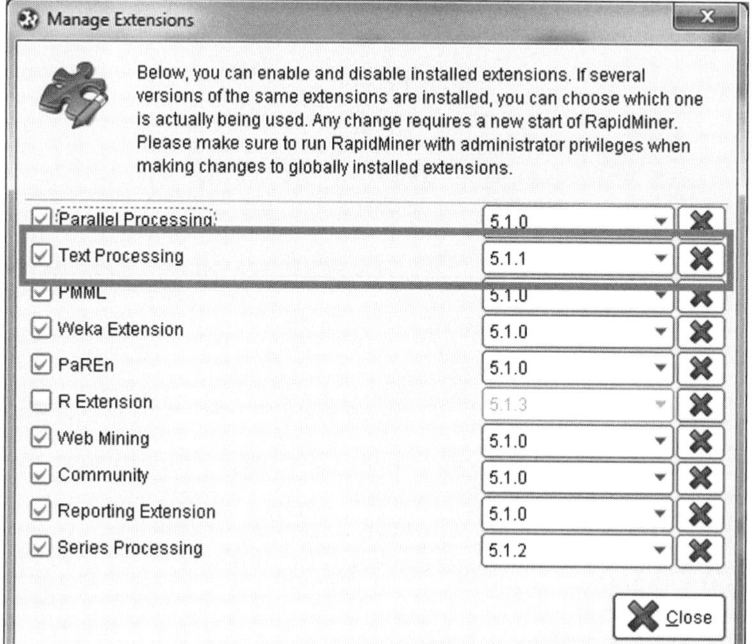

FIGURE H.6
Managing extensions.

If you don't see Text Processing listed, then it is not installed. All you have to do is go to Help, click on Update RapidMiner, and select and install Text Processing (Figure H.7).

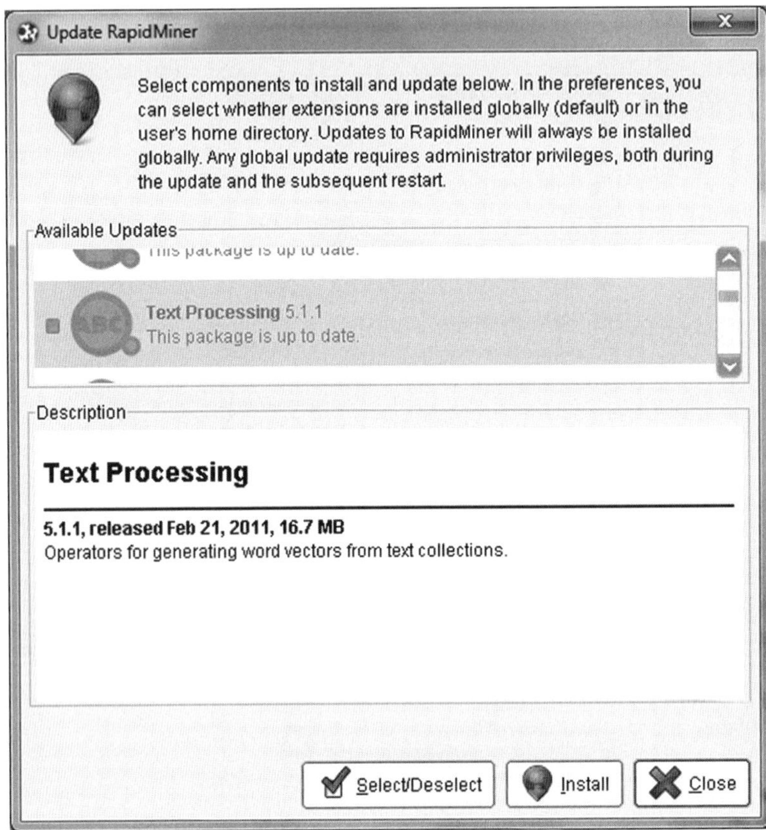

FIGURE H.7
Installing Text Processing extension (in this figure Text Processing is grayed out because it is already installed with the latest version).

STARTING A NEW PROCESS

In the Welcome window, click on New to start a new process. Then select a repository to store the contents of the new project, give it a name, and click OK (Figure H.8).

To load documents into RapidMiner for text mining, you have a number of different options (Figure H.9). You can process documents from data, create documents, read documents, or load the documents from file or mail.

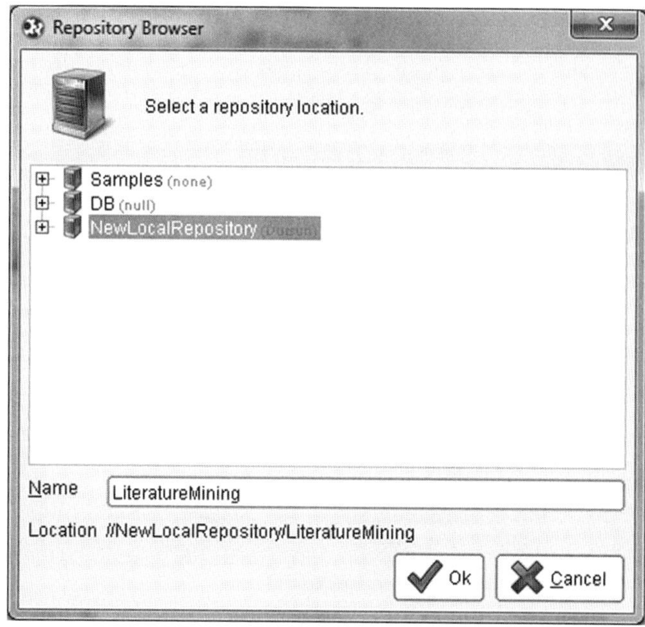

FIGURE H.8
Selecting a repository and naming the new project.

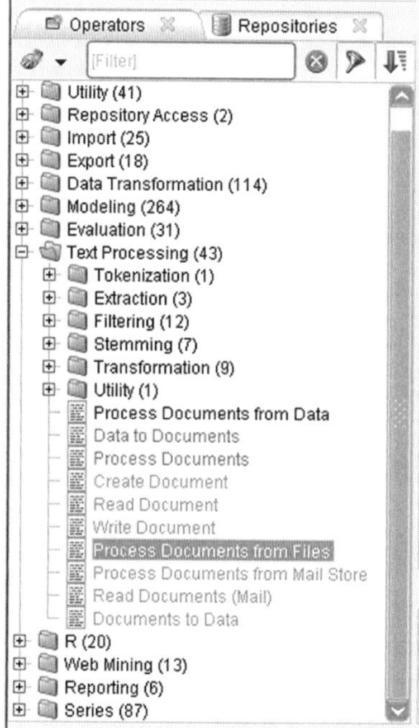

FIGURE H.9
Options to read documents for text processing.

If your documents are saved as individual text files in a folder, then you need to select Process Documents from Files, and in the pop-up dialog box specify the folder (or folders) where the text files are stored (Figure H.10).

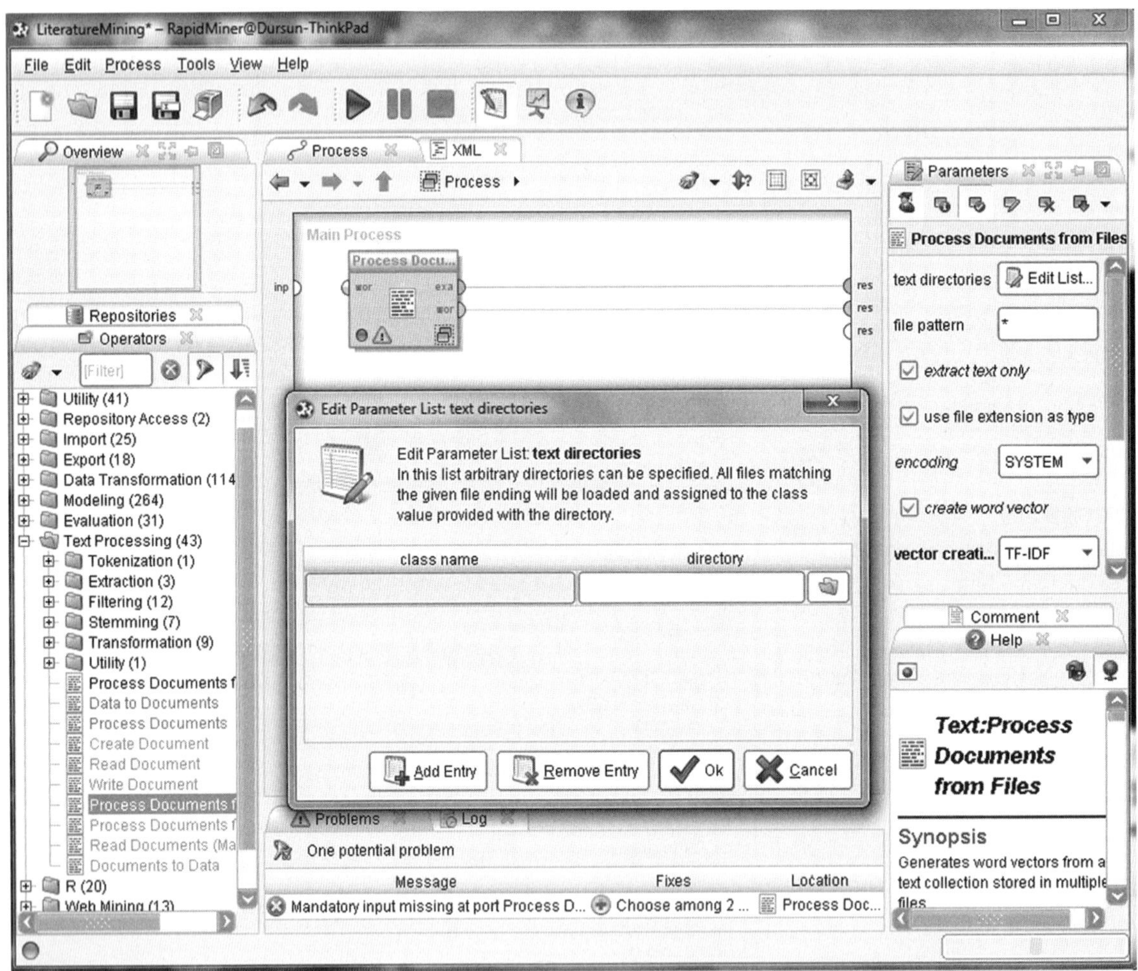

FIGURE H.10

Loading documents from files stored in one or more folders.

In our case, our text data are stored in an Excel file (Figure H.11).

In order to read an Excel file into RapidMiner, you can use the Read Excel operator (i.e., process node). You can find this operator under Import>Data>Read Excel in the Operators tree pane. A quick way to find an operator is to use the Filter (or Search) function, which is located on the top of the Operators pane. There, if you type "Excel," you would get all of the operators that include the word "Excel" in their name (Figure H.12).

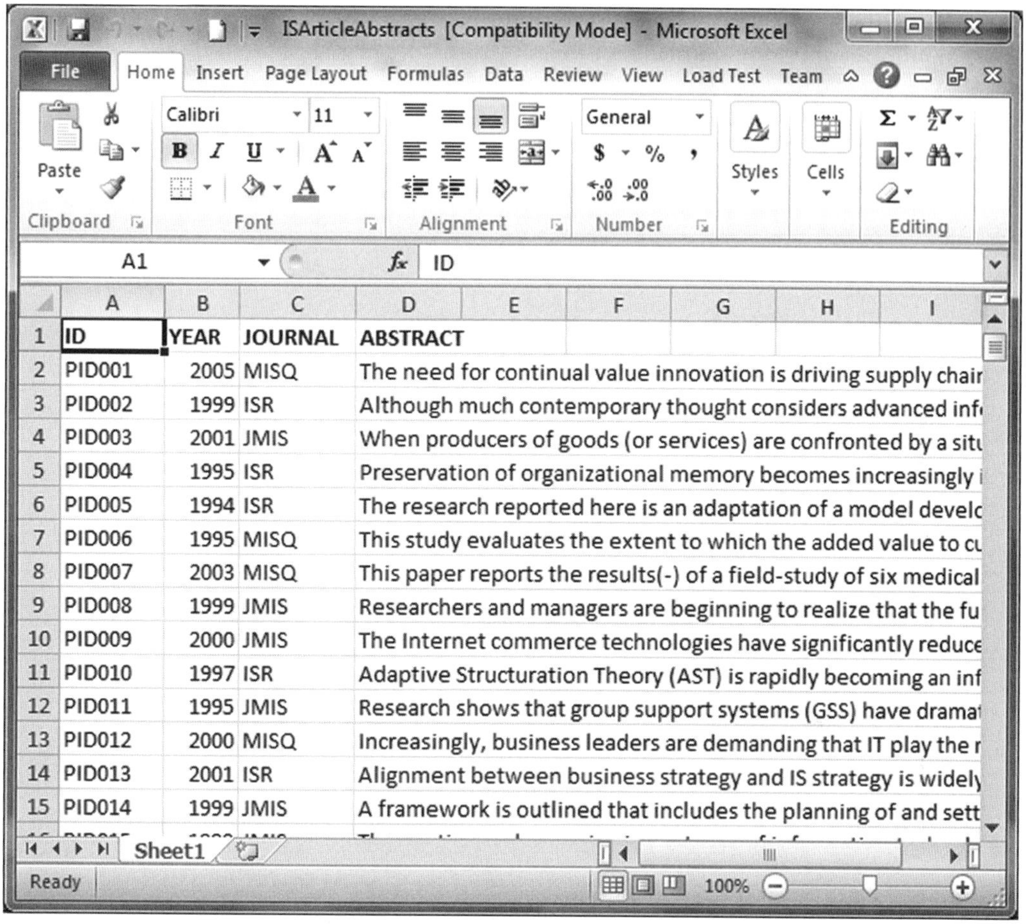

	A	B	C	D
1	ID	YEAR	JOURNAL	ABSTRACT
2	PID001	2005	MISQ	The need for continual value innovation is driving supply chair
3	PID002	1999	ISR	Although much contemporary thought considers advanced inf
4	PID003	2001	JMIS	When producers of goods (or services) are confronted by a situ
5	PID004	1995	ISR	Preservation of organizational memory becomes increasingly i
6	PID005	1994	ISR	The research reported here is an adaptation of a model develc
7	PID006	1995	MISQ	This study evaluates the extent to which the added value to cu
8	PID007	2003	MISQ	This paper reports the results(-) of a field-study of six medical
9	PID008	1999	JMIS	Researchers and managers are beginning to realize that the fu
10	PID009	2000	JMIS	The Internet commerce technologies have significantly reduce
11	PID010	1997	ISR	Adaptive Structuration Theory (AST) is rapidly becoming an inf
12	PID011	1995	JMIS	Research shows that group support systems (GSS) have drama
13	PID012	2000	MISQ	Increasingly, business leaders are demanding that IT play the r
14	PID013	2001	ISR	Alignment between business strategy and IS strategy is widely
15	PID014	1999	JMIS	A framework is outlined that includes the planning of and sett

FIGURE H.11

Textual data (article abstract) stored in an MS Excel file.

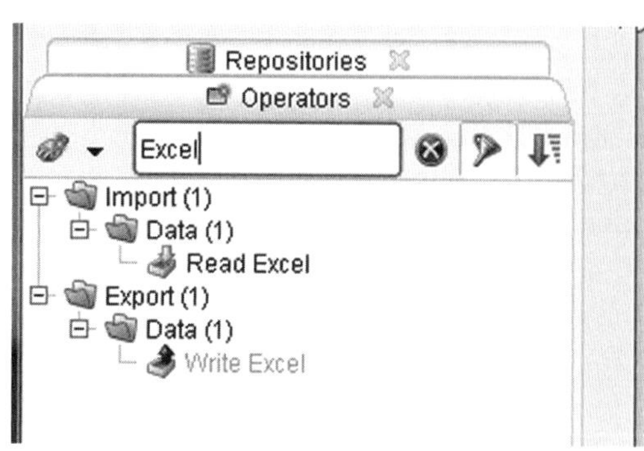

FIGURE H.12

Searching for an operator to manage Excel files.

You can drag and drop the read Excel operator into the process map. Once the operator is selected, its properties are shown on the right side of the screen (Figure H.13).

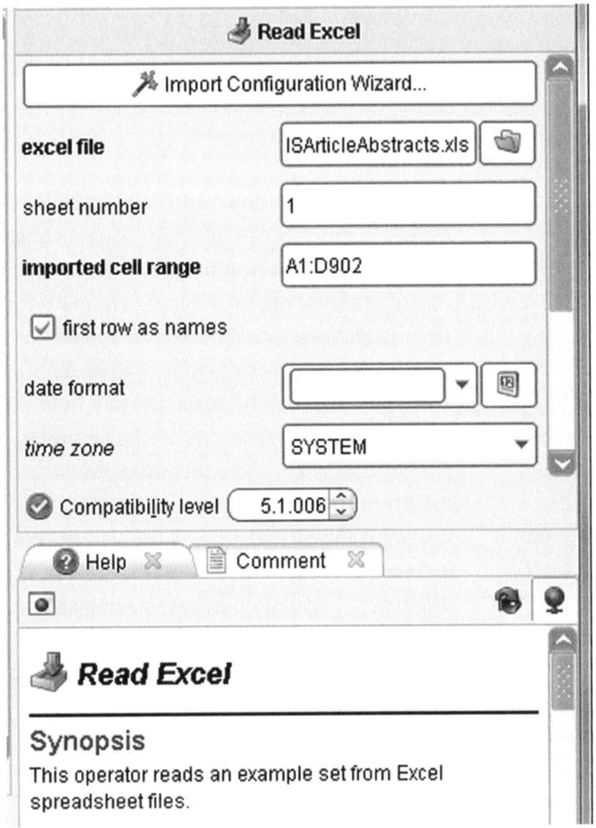

Read Excel	
Import Configuration Wizard...	
excel file	ISArticleAbstracts.xls
sheet number	1
imported cell range	A1:D902
☑ first row as names	
date format	
time zone	SYSTEM
Compatibility level	5.1.006

Help | Comment

Read Excel

Synopsis

This operator reads an example set from Excel spreadsheet files.

FIGURE H.13
Properties for Read Excel operator.

A quick way to specify the file name and various details about the variables in the data set is to use the Import Configuration Wizard (located on the top of the Read Excel property pane). This wizard takes you through a few steps to set the file and variable specific parameters (see Figures H.14, H.15, H.16, and H.17).

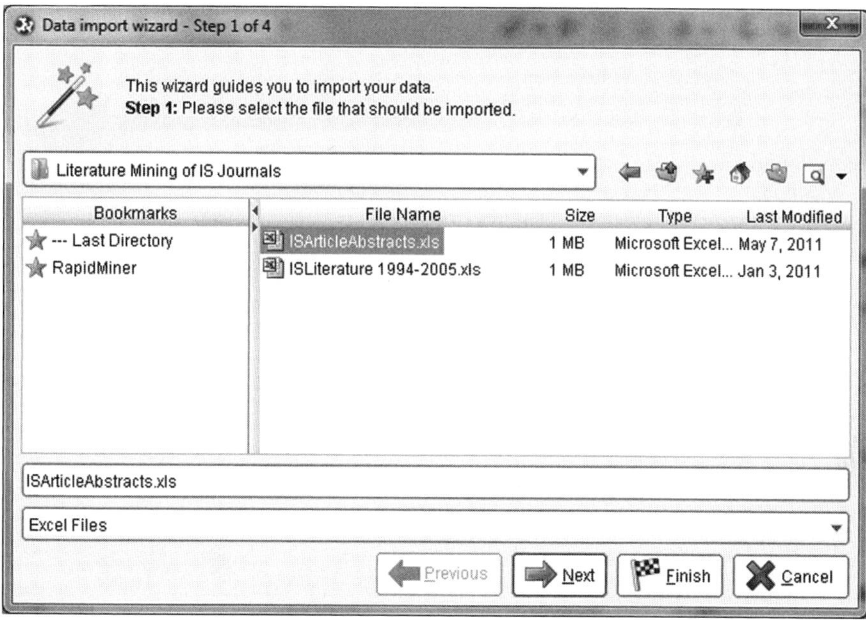

FIGURE H.14
Selecting the Excel file to read.

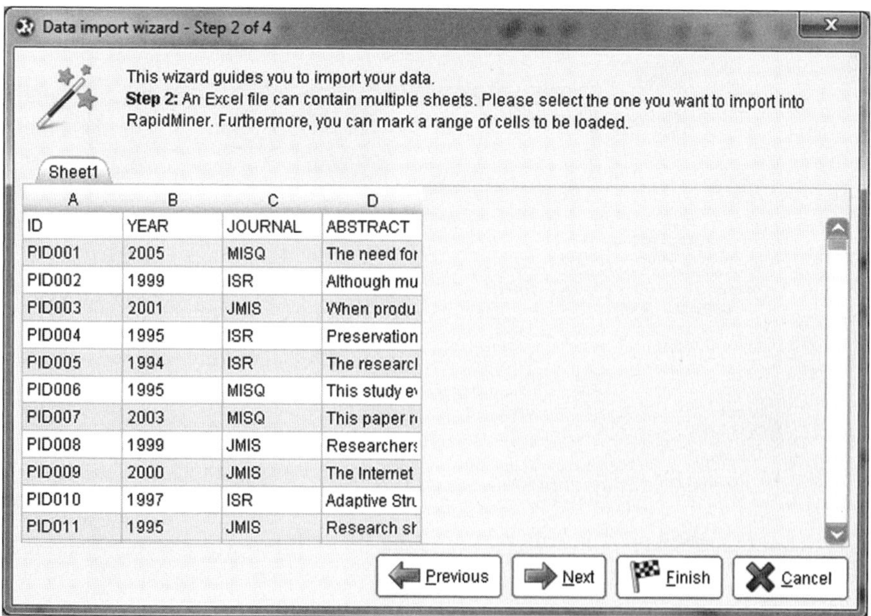

FIGURE H.15
Specifying the sheet within the selected Excel file to process.

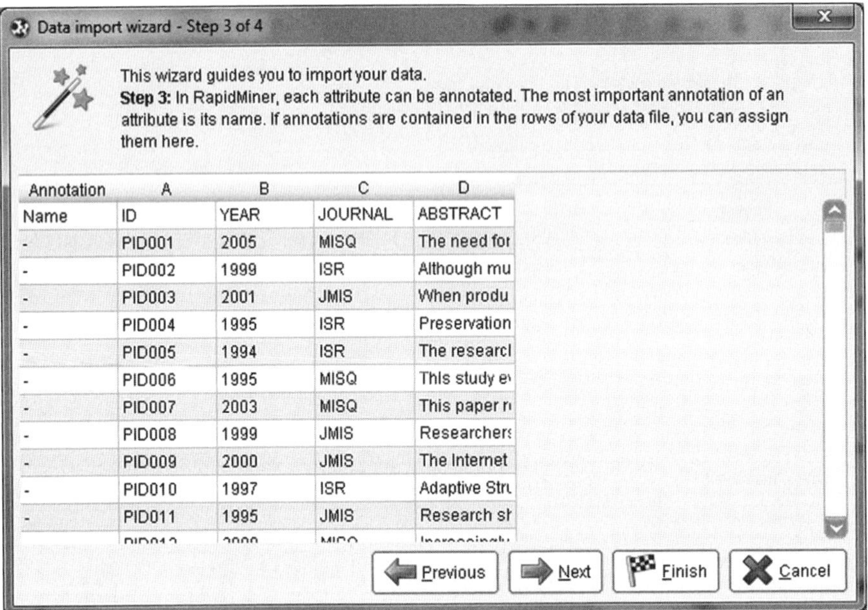

FIGURE H.16

Annotating the attributes (if so desired) interface.

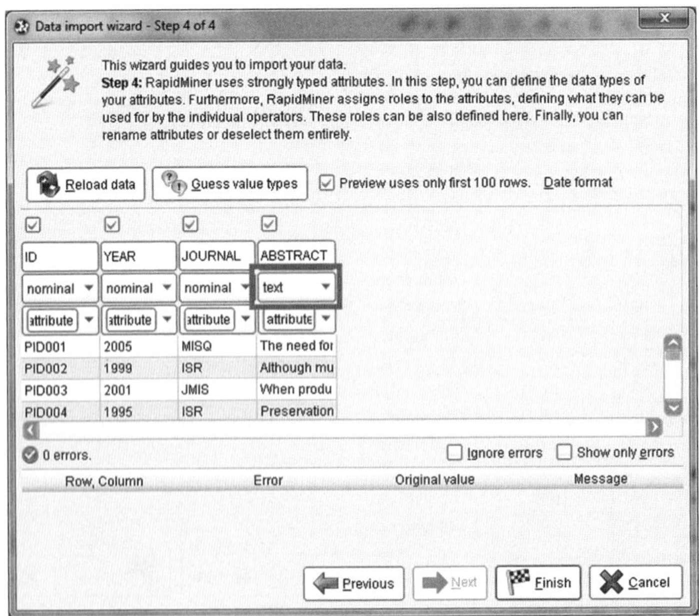

FIGURE H.17

Specifying the types for attributes.

You should make sure the variable types are set properly. The text variable (in this case the ABSTRACT variable) especially needs to have the type specified as *text*. Then you can click Finish to return back to the process window.

Once the data are loaded, you can use the proper operators to process the text. First, you need to drop Process Documents from the Data operator into the process map and connect it to the Read Excel operator (Figure H.18).

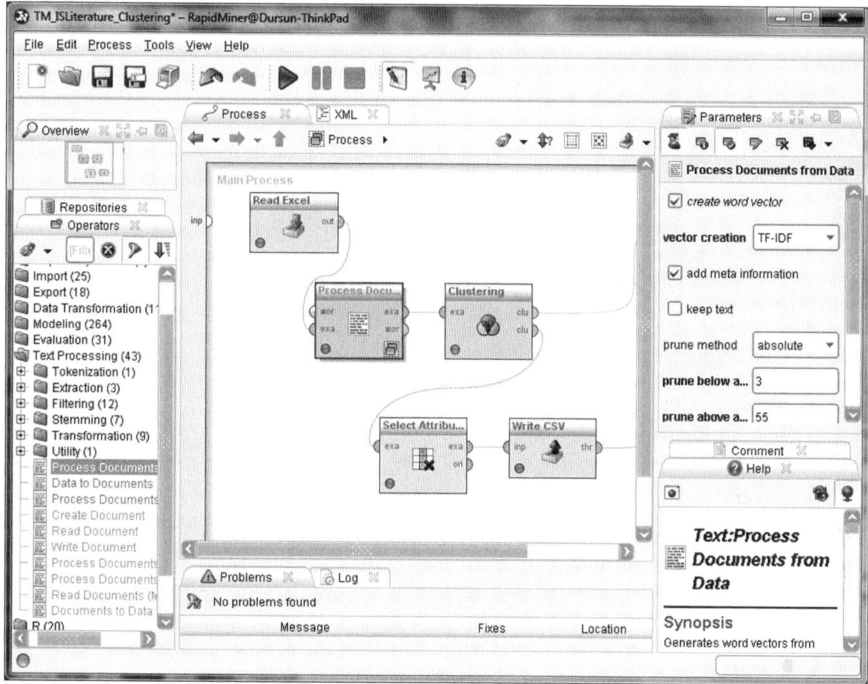

FIGURE H.18
The complete screen shot for the text mining and clustering process.

Figure H.18 shows the complete process where the Process Documents from Data operator is selected. In the RapidMiner window, specific properties of the selected operator are shown on the right side of the window. In Figure H.18, the right side of the window pane shows the properties of the Process Documents from the Data operator. Specific attention is to be paid to the property named "vector creation." Here the user has four different options (Figure H.19), each of which represents the relationship between the words/terms and the documents with different numbers:

1. **Binary Term Occurrence** places 1 in the intersection cell between a document (row) and a word/term (column) if the word/term occurs at least once in that document and places 0 otherwise. The number of occurrences in the document is ignored in this measure.
2. **Term Occurrence** places the exact number of occurrences of a word/term in the intersection cell between the document (row) and the word/term (column). If the word/term does not occur in that document, 0 is placed in the intersection cell.

3. **Term Frequency** places the relative frequency of the word/term in the document in the intersection cell. This measure is calculated by dividing the number of occurrences of a word/term into the number of total words in that document.
4. **TF-IDF** stands for *Term Frequency—Inverse Document Frequency*. It is arguably the most commonly used numerical representation in text mining. It calculates a numerical value that emphasizes both the frequency of the term in a document (more is better) and the rareness of the same term in the collection of all documents (less is better).

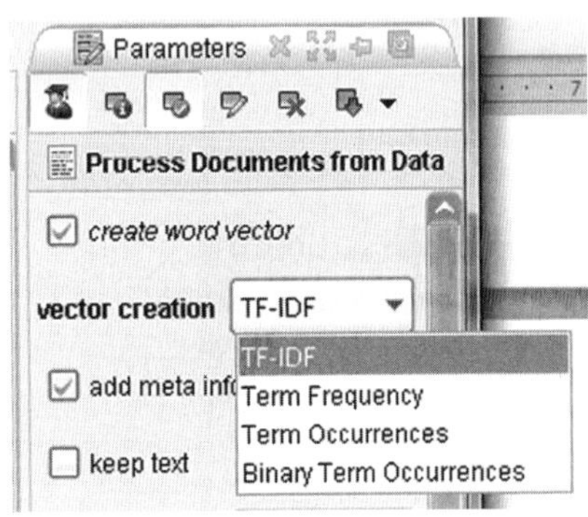

FIGURE H.19
Options to select for vector creation.

The Process Documents from the Data operator has a subprocess where the user is expected to specify the details of how the text should be mined. The operators (like Process Documents from Data) that have an icon that looks like overlapping windows in the lower right corner indicates that there is a subprocess that needs to be specified by the user. User can double click on the parent process to go into the subprocess. In this example the subprocess includes a number of operators (Figure H.20) to convert unstructured text (i.e., article abstracts) into a structured data file (i.e., a term-document matrix).

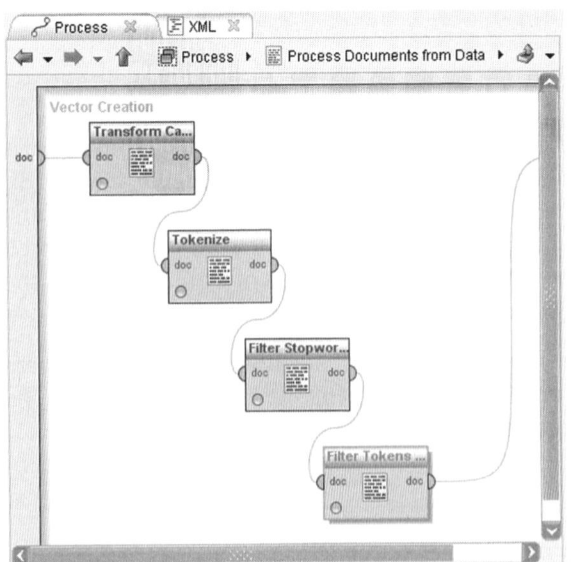

FIGURE H.20
Subprocess under Process Documents from Data.

In this specific subprocess, we used four operators. First, we used a Transform Cases operator to transform all of the characters in the document collection to either lower case (or upper case, if preferred) so the identification of the same words/terms is not biased by the capitalization. Following the transformation, a tokenization operation is applied. Tokenization operator splits the text of documents into a sequence of tokens. There are several options on how to specify the splitting points. You may use all nonletter characters, which is the default setting and works well with English text. This process would result in tokens consisting of one single word/term. Next we applied a Filter Stopwords operator. This operator filters English stopwords from a document collection by removing every token that equals a stopword from the built-in stopword list. Please note that for this operator to work properly, every token should represent a single English word only. The list of stopwords includes words/terms that are commonly found in most text documents (e.g., a, an, is, am, are, the, etc.) that have no contribution to either identification or discrimination of documents from each other. Lastly, we applied the Filter Tokens (by Length) operator to filter out the tokens based on their length (i.e., less than two characters long).

Now we go back to the parent process by clicking on the up arrow at the top of the process window. The output of the Process Documents from Data operator is a term-document matrix where the relationships between the documents and the words/terms are represented with numerical indices—in this case TF-IDF measures (Figure H.21). These structured data are then fed into the Clustering operator.

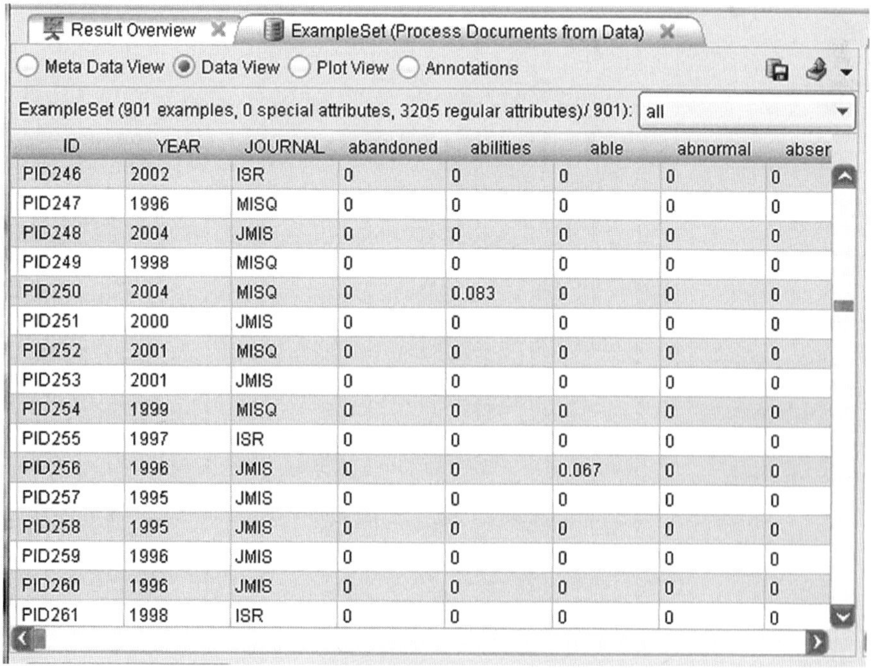

Result Overview | ExampleSet (Process Documents from Data)

Meta Data View | Data View | Plot View | Annotations

ExampleSet (901 examples, 0 special attributes, 3205 regular attributes)/ 901): all

ID	YEAR	JOURNAL	abandoned	abilities	able	abnormal	abser
PID246	2002	ISR	0	0	0	0	0
PID247	1996	MISQ	0	0	0	0	0
PID248	2004	JMIS	0	0	0	0	0
PID249	1998	MISQ	0	0	0	0	0
PID250	2004	MISQ	0	0.083	0	0	0
PID251	2000	JMIS	0	0	0	0	0
PID252	2001	MISQ	0	0	0	0	0
PID253	2001	JMIS	0	0	0	0	0
PID254	1999	MISQ	0	0	0	0	0
PID255	1997	ISR	0	0	0	0	0
PID256	1996	JMIS	0	0	0.067	0	0
PID257	1995	JMIS	0	0	0	0	0
PID258	1995	JMIS	0	0	0	0	0
PID259	1996	JMIS	0	0	0	0	0
PID260	1996	JMIS	0	0	0	0	0
PID261	1998	ISR	0	0	0	0	0

FIGURE H.21
Output of Process Documents from Data operator (a term-document matrix).

For clustering we selected the k-means clustering algorithm (a popular statistical technique to find natural groupings of records using a simple multidimensional distance measure). In k-means, the user is expected to specify the number of clusters that he or she would like to have. In this tutorial, after some experimentation, we chose to set the number of clusters to 9. We kept the rest of the parameters of Clustering operator as their default values. Figure H.22 shows a screen shot of one of the interesting output reports generated by the Clustering operator. In this two-pane window, all nine clusters are shown in a tree view (left-hand side), and the list of documents that belong to each of the nine clusters is shown in the list view (right-hand side). In this dynamic view, the list of documents changes according to the selection of the cluster number in the tree view.

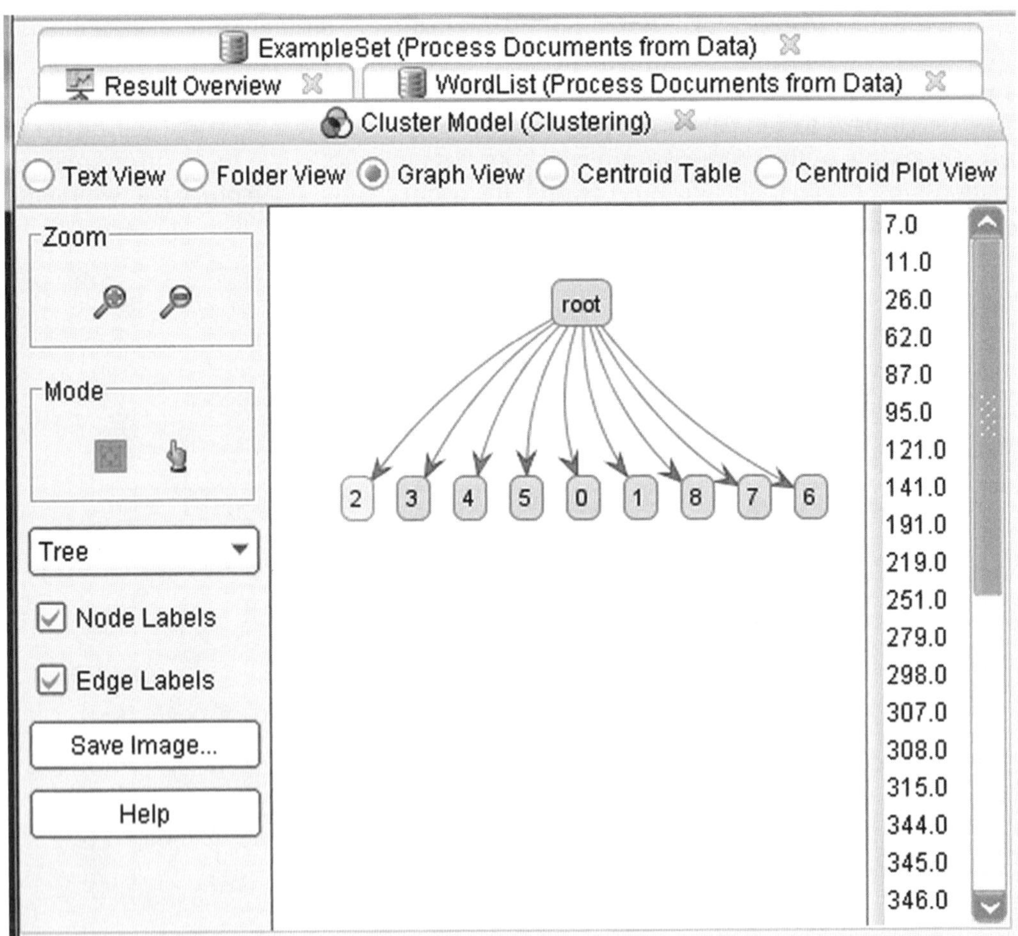

FIGURE H.22

Output of the clustering operator. When one of the nine clusters is selected in the tree view, the document numbers/IDs that fall into that cluster are shown in a list view on the right side.

Next, we wanted to export the clustering results into an easily importable file format; for that we choose the comma separated values (CSV) file format. In this file, in addition to clusters, we also included the original variables (YEAR and JOURNAL). The CSV file is then imported in the *STATISTICA* Data Miner software tool for further graphical reporting (Figure H.23).

Data: ClusterData* (3v by 901c)

	1 JOURNAL	2 YEAR	3 CLUSTER
1	MISQ	2005	3
2	ISR	1999	8
3	JMIS	2001	3
4	ISR	1995	6
5	ISR	1994	0
6	MISQ	1995	4
7	MISQ	2003	5
8	JMIS	1999	6
9	JMIS	2000	4
10	ISR	1997	0
11	JMIS	1995	5
12	MISQ	2000	6
13	ISR	2001	0
14	JMIS	1999	6
15	JMIS	1999	6
16	MISQ	1994	1
17	ISR	1996	3
18	JMIS	1996	7
19	JMIS	1997	5
20	ISR	2002	0
21	JMIS	2005	8

FIGURE H.23
Clustering results in the CSV file are imported into the *STATISTICA* Data Miner tool.

Using these simple data, we created two graphical reports:

1. A histogram (per cluster) that shows time-dependent changes in the number of articles (i.e., documents) for each of the nine clusters (Figure H.24). With this graphical report, one can deduce the increasing/decreasing popularity of topics (represented as clusters) over the 12-year period. Topic specification can be heuristically/annually determined by looking at the dominant terms that define each of the nine clusters.

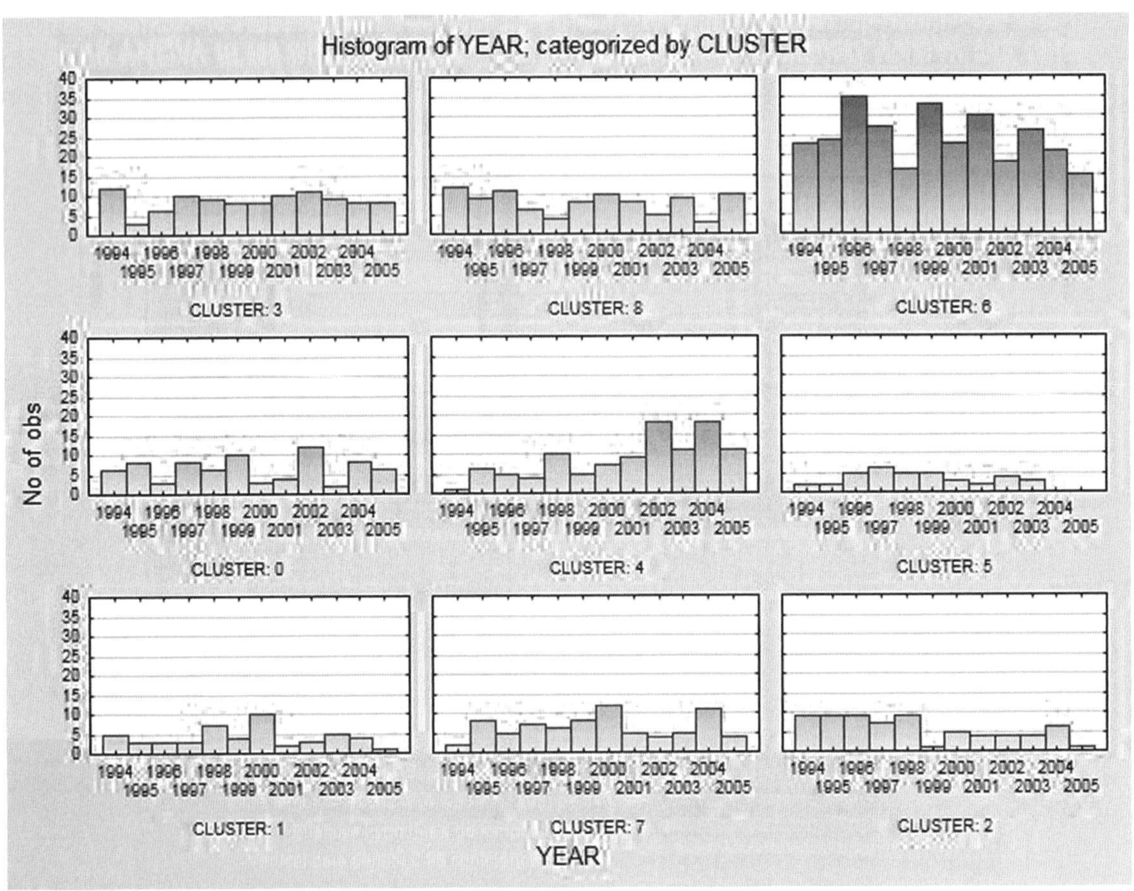

FIGURE H.24
Time-dependent changes in the number of articles published in each of the nine clusters.

2. A histogram (per cluster) that shows the number of articles coming from each of the three journal types (Figure H.25). Such a report could be used to determine if any of the three journals are more inclined to publish certain topics (i.e., clusters).

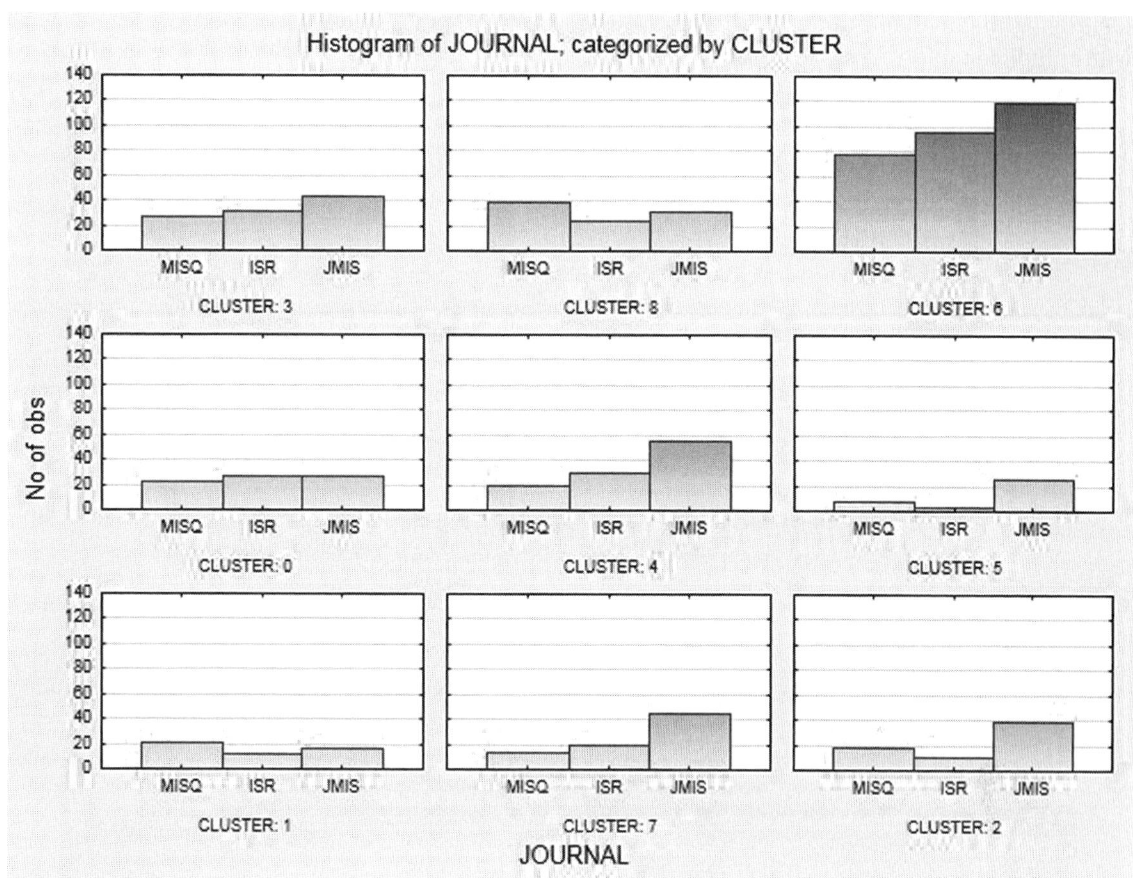

FIGURE H.25

Numerical representation of each of the three journals in each of the nine clusters.

SUMMARY

The amount of unstructured data collected and stored in databases is increasing at a higher rate than any traditional, mostly manual method can keep up with. As the digitized data (either structured or unstructured) become more widely available and accessible, tools that allow us to extract information and knowledge from this mountain of data with ease (e.g., data mining and text mining) are likely to become more valuable. Even though text mining is a relatively new technology, its applications and benefits have already been realized in such fields as medicine, health care, homeland security, law, education, and customer relationship management.

This tutorial showed that it is relatively straightforward to apply text mining techniques to a readily available set of data in the form of downloaded journal article abstracts. A total of 901 articles from three major MIS journals were downloaded and analyzed using text mining, with the objective of identifying major themes of research and how the major themes may have varied over time, both within individual journals and within the entire set.

The field of text mining is presently in a growth phase in the research literature. Researchers may apply it to a wide spectrum of unstructured, textual data, from genetic sequence analyses to business email. It makes sense to find ways to use this powerful tool to help analyze the status and direction of the research itself.

Reference

Delen, D., and M. Crossland (2008). "Seeding the Survey and Analysis of Research Literature with Text Mining," *Expert Systems with Applications*, 34 (3), 1707–1720.

Text Mining Speech Samples: Can the Speech of Individuals Diagnosed with Schizophrenia Differentiate Them from Unaffected Controls?

Jennifer Thompson
StatSoft, Inc.

Susan L. Trumbetta, PhD
Vassar College

Gary D. Miner, PhD
StatSoft, Inc.

Irving I. Gottesman, PhD
University of Minnesota - Bernstein Professor in Adult Psychiatry & Senior Fellow, Department of Psychology, University of Minnesota & Sherrell J. Aston Professor of Psychology Emeritus, University of Virginia

CONTENTS

INTRODUCTION

From the time of Freud, clinicians and scientists have used transcripts of clinical interviews and other speech samples to gain a better understanding of the mechanisms of psychopathology. Because speech provides a window into human thought, it may be particularly helpful to our understanding of thought disorders such as schizophrenia. Schizophrenia symptoms typically affect both receptive and expressive language, and speech samples have long informed scientific investigations of the disorder. Bleuler (1950) noted that in schizophrenia, speech became more adversely affected with increasing symptom

Practical Text Mining and Statistical Analysis for Non-structured Text Data Applications. DOI: 10.1016/B978-0-12-386979-1.00017-7

severity; the milder the schizophrenia symptoms, the less difference there was between schizophrenic and nonschizophrenic speech. Investigators since then have developed typologies of speech anomalies associated with schizophrenia (Andreasen, 1986; Docherty et al., 1996); anomalies that also are more common in the speech of first-degree unaffected relatives of people with schizophrenia than in the speech of general population controls (Docherty et al., 2004).

In schizophrenia, different types of symptoms seem to affect speech differently. Individuals with prominent negative symptoms typically display alogia or poverty of speech or poverty of content of speech (Liddle, 1987), so one would expect that their speech samples would generally be briefer and less varied in topic and vocabulary than ordinary speech.

Referential errors also occur in schizophrenia (Docherty et al., 2003) and may be more common among individuals with prominent positive symptoms (Allen and Allen, 1985). Unclear referents may be manifest in various forms, including *nonsequiturs* (Barch and Berenbaum, 1996) and ambiguous pronomial referents (Caplan et al.,1992; Harvey and Brault, 1986; Harvey and Serper, 1990). These types of unclear referents seem to be based on the speaker's inadequate communication of his thought context to his listener, which seems to fall into the category of missing information references (Docherty et al., 2004). Although *nonsequiturs* are currently difficult to detect solely by computer, the emerging technology of syntactical programming, exemplified in the performance of Watson, the computer contestant on the game show *Jeopardy*, may one day may improve detection of these patterns. However, one might expect that a relatively simple scan for pronoun usage frequency might differentiate schizophrenia-affected from unaffected speech, since a greater frequency of pronouns relative to other forms of words could indicate a greater likelihood of unclear referents. Our tutorial presents an analysis of pronoun frequency in clinical interviews of twins with schizophrenia and with their illness-concordant or -discordant cotwins.

Of course, schizophrenia is a heterogeneous illness, both across and within affected individuals. Symptoms of the illness may change in the person over time and with different medication. With those symptom changes, speech patterns also may change. Speech samples from individuals with schizophrenia will likely reflect variable patterns, and data mining may provide data for further analysis in the development of typologies of speech characteristic of different types, phases, and levels of neurocognitive anomalies associated with the disorder.

OBJECTIVES

In this tutorial, we use data mining to study speech patterns in a sample of individuals with schizophrenia and their identical twins from Gottesman and Shields's (1972) historic study. These speech samples represent twins' responses to standard questions from an interviewer (Gottesman). For illustrative purposes, this data mining tutorial focuses on two types of speech patterns in schizophrenia. First, we focus on the relative brevity of patient discourse in response to the interviewer's questions. Shorter responses would be more likely in the presence of prominent negative symptoms. Second, we focus on pronoun usage as a somewhat indirect index of unclear or ambiguous referents, which is more common to individuals with positive symptoms of schizophrenia.

CASE STUDY: THE STEPS USED TO PREPARE THE DATA

The current study examines clinical transcripts from Gottesman and Shields's (1972) foundational study of twins with schizophrenia. For the purpose of demonstrating psychiatric applications of data mining techniques, we selected the fairly simple task of comparing pronoun frequency across the speech of individuals either affected or unaffected with schizophrenia. We de-identified clinical transcripts to protect the confidentiality of the participants in the Gottesman and Shields (1972) study, replacing the names of people and places with a single initial followed by four dashes. The initials are fictitious, assigned alphabetically in order, according to the first appearance of each name as mentioned in a given speech sample. We also have fictionalized months, dates, and years, as well as some occupational references, numbers of children, siblings, and the like. In order to examine the content of the speech, we also regularized the presentation of some words (*going* instead of *goin'* or *myself* instead of *meself*, for example) for comparability across dialects. What remains for each interview is an otherwise verbatim transcript. The data set used for this tutorial is called *MZ_Maudsley.sta* and contains paths to the 46 interview transcripts as well as additional information about the patient. A portion of the data is seen here (see Figure I.1).

Data: MZ _Maudsley.sta* (7v by 46c)

	1 MZ Twin ID #	2 Twin A or B	3 MZ Text File names	4 Sex	5 Sc DX - yes or no	6 interview ?	7 Dominant psych. symptoms
1	1	A	MZ-1-A_anon_interview.txt	Male	Sc Dx	0	Schizophrenia
2	1	B	MZ-1-B_anon_interview.txt	Male	Sc Dx	0	Schizophrenia
3	2	A	MZ-2-A_anon_interview.txt	Male	Sc Dx	1	Schizophrenia
4	2	B	MZ-2-B_anon_interview.txt	Male	Sc Dx	1	Schizophrenia
5	3	A	MZ-3-A_anon_interview.txt	Male	Sc Dx	1	Schizophrenia
6	3	B	MZ-3-B_anon_interview.txt	Male	No SC Dx	0	
7	4	A	MZ-4-A_anon_interview.txt	Female	Sc Dx	1	Schizophrenia
8	4	B	MZ-4-B_anon_interview.txt	Female	No SC Dx	1	
9	5	A	MZ-5-A_anon_interview.txt	Female	Sc Dx	1	Schizophrenia
10	5	B	MZ-5-B_anon_interview.txt	Female	No SC Dx	1	

FIGURE I.1
Data file.

RESULTS AND ANALYSIS

Frequency of Pronoun Use

From the *STATISTICA* Data Mining menu, select *Text Mining* to open the Text Mining dialog (see Figure I.2).

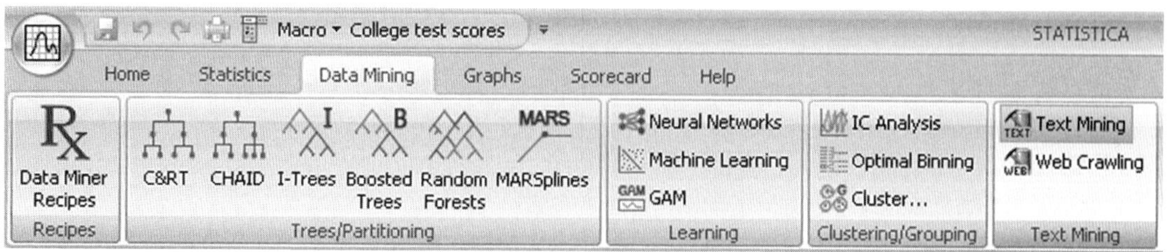

FIGURE I.2
Selecting Text Mining from the *STATISTICA* ribbon bar menu.

On the Quick tab (see Figure I.3), change the Retrieve text from option to Files and check the option Paths in spreadsheet. Select Document paths to open the Select a variable containing file paths dialog.

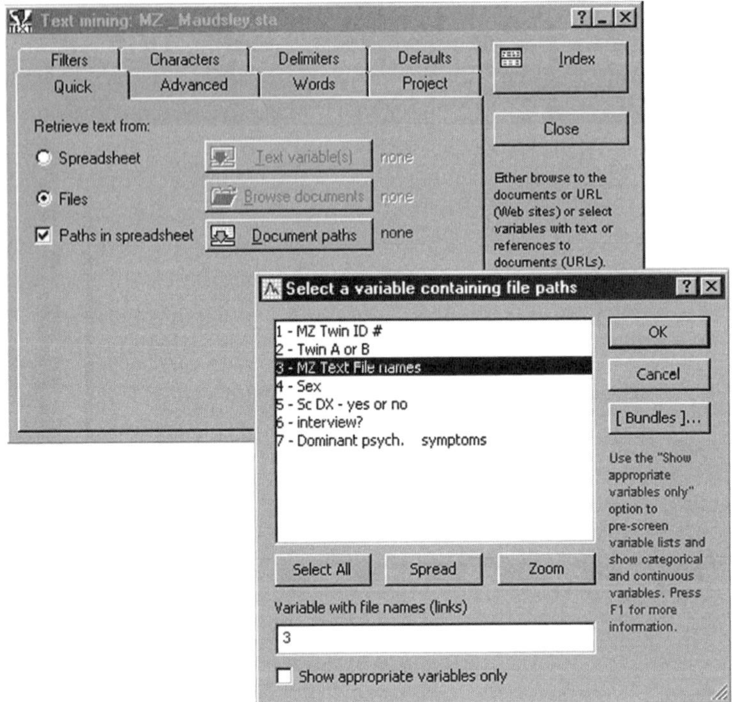

FIGURE I.3
Selecting text file variable for the text mining main dialog window.

Select variable 3 (see Figure I.3), *MZ Text File names*, and click OK to update the selection. This tells *STATISTICA* to retrieve the text from the paths given in this variable. Select the *Words* tab (see Figure I.4).

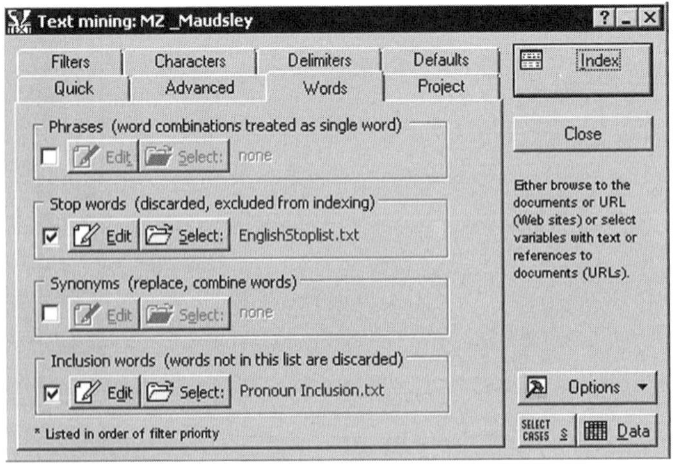

FIGURE I.4
Words tab of the text mining dialog.

Since we are interested in exploring pronoun use, only pronouns should be indexed in this analysis. Check the option for *Inclusion words* to activate this option. Use the *Select* button in this field to browse to and select the *Pronoun Inclusion.txt* file. This is a list of pronouns, and they are the only words that will be indexed in the analysis. Select *Edit* to review the list (see Figure I.5).

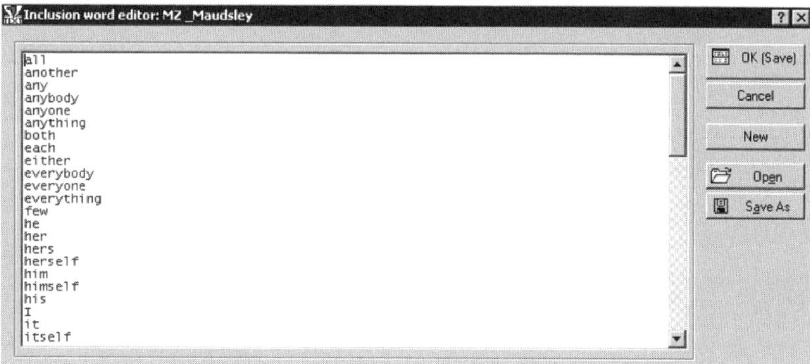

FIGURE I.5
Inclusion word editor.

Click OK on the *Inclusion word editor* to close the dialog box, and click *Index* to begin indexing the interview transcripts. When indexing completes, the *Results* dialog will display (see Figure I.6).

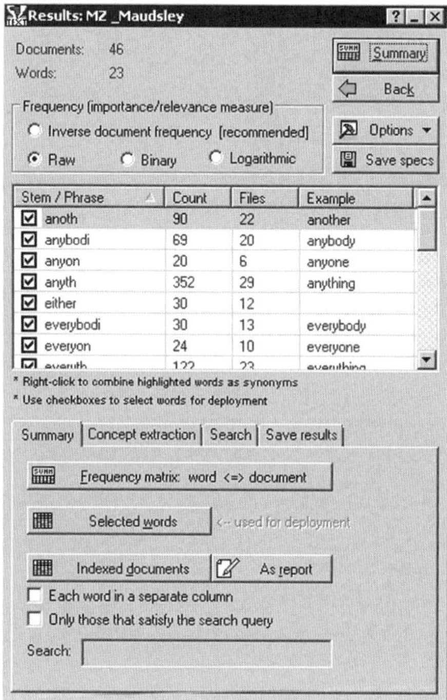

FIGURE I.6
Results of Indexing the words with the text miner.

The results show that 46 documents were indexed, one for each of 23 sets of twins. Twenty-three words were found from the inclusion list provided. Click *Selected words* to output a list of all the selected words, their frequency, and the number of files in which they occurred (see Figure I.7).

Data: Words summary (MZ _Maudsle...

	Words summary (MZ _Maudsle		
Stem / Phras	Count	Number of documents	Example
anoth	90	22	another
anybodi	69	20	anybody
anyon	20	6	anyone
anyth	352	29	anything
either	30	12	
everybodi	30	13	everybody
everyon	24	10	everyone
everyth	122	23	everything
littl	128	27	little
mani	95	26	many
mine	9	6	
much	343	29	
neither	2	2	
nobodi	17	8	nobody
none	6	5	
noth	109	22	nothing
one	547	29	
sever	10	8	several
somebodi	82	19	somebody
someon	25	12	someone
someth	306	29	something
us	116	24	
whatev	15	9	whatever

FIGURE I.7
Word summary list.

The word "one" occurred most often—547 times. "Anything," "much," and "something" all occurred over 300 times.

Select the *Save results* tab (see Figure I.8) so that the indexed terms can be written back to the input spreadsheet. Change the *Frequency* option to *Inverse document frequency*. In the *Input or write-back spreadsheet* field, enter 23 next to *Amount*. Click *Append empty variables*. This will add 23 new, empty variables to the input spreadsheet. These variables will be filled in with inverse document frequencies for the 23 indexed words.

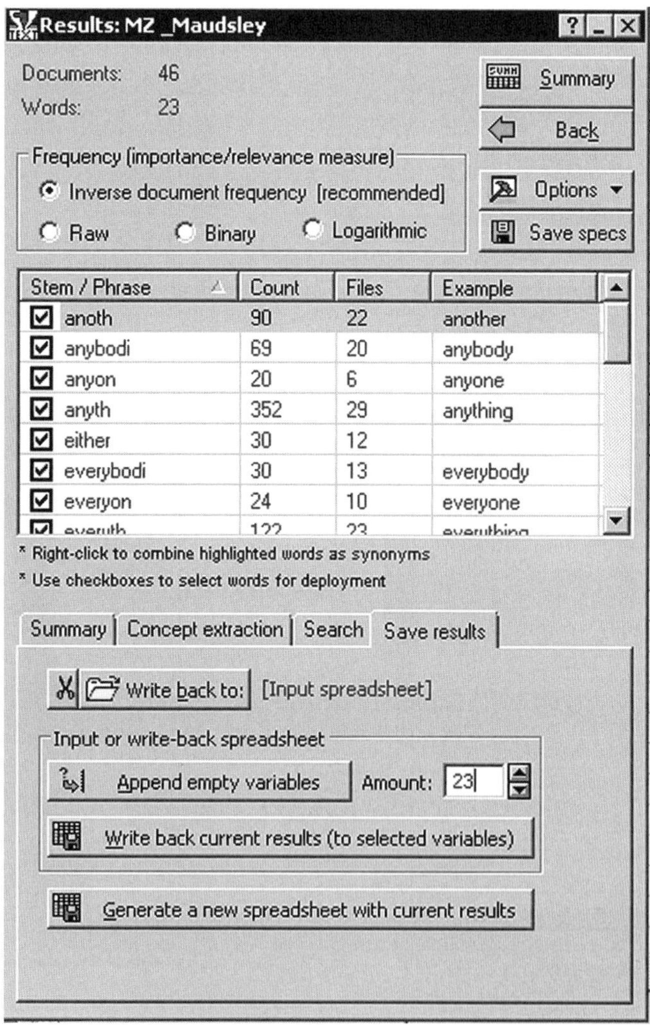

FIGURE I.8

Inverse document frequency selected as the importance/relevance measure for word frequency.

A dialog will display to indicate that the variables were successfully added (see Figure I.9).

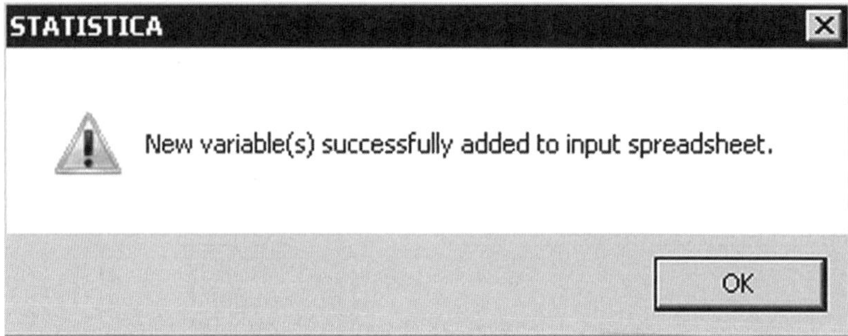

FIGURE I.9
Message alerting that the "new variable(s)" were successfully added to the data spreadsheet.

Select *Write back current results (to selected variables)* to display the *Assign statistics to variables to save them to the input data* dialog box (see Figure I.10).

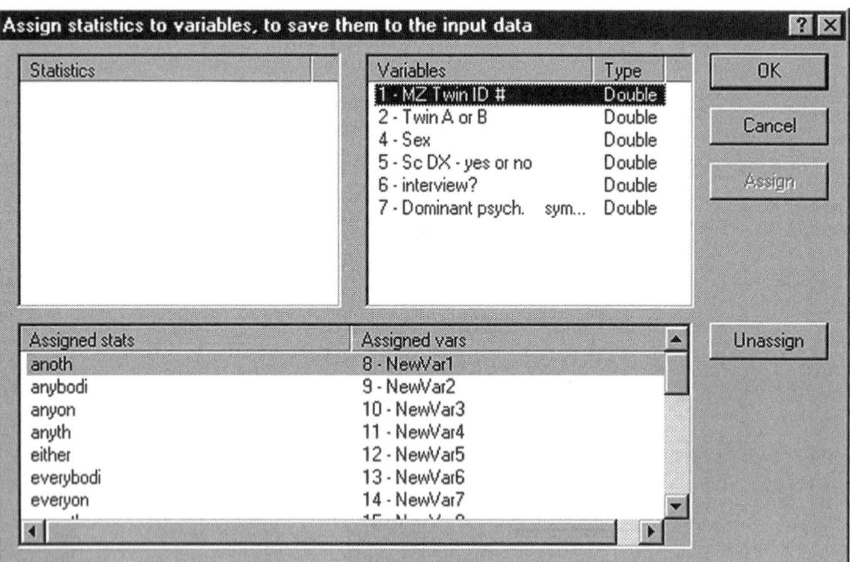

FIGURE I.10
Assign statistics to variables dialog window.

In the *Statistics* field, select the first indexed term, *anoth*. In the *Variables* field, select the first newly created destination variable, *NewVar1*. With both items selected, click *Assign* to move the pair to the *Assigned stats* field below. Repeat this process until all of the indexed terms are assigned to destination variables (see Figure I.10). Click OK to write the statistics to the input spreadsheet (see Figure I.11).

Data: MZ _Maudsley.sta* (30v by 46c)

	1 MZ Twin ID #	2 Twin A or B	3 MZ Text File names	4 Sex	5 Sc DX - yes or no	6 interview ?	7 Dominant psych. symptoms	8 anoth	9 anybodi	10 anyon	11 anyth	12 either
1	1	A	MZ-1-A_anon_interview.txt	Male	Sc Dx	0	Schizophrenia					
2	1	B	MZ-1-B_anon_interview.txt	Male	Sc Dx	0	Schizophrenia					
3	2	A	MZ-2-A_anon_interview.txt	Male	Sc Dx	1	Schizophrenia		0.832909		1.475025	
4	2	B	MZ-2-B_anon_interview.txt	Male	Sc Dx	1	Schizophrenia	0.737599			1.768436	2.819978
5	3	A	MZ-3-A_anon_interview.txt	Male	Sc Dx	1	Schizophrenia				0.461346	
6	3	B	MZ-3-B_anon_interview.txt	Male	No SC Dx	0						
7	4	A	MZ-4-A_anon_interview.txt	Female	Sc Dx	1	Schizophrenia		0.832909		1.567604	
8	4	B	MZ-4-B_anon_interview.txt	Female	No SC Dx	1					1.740467	
9	5	A	MZ-5-A_anon_interview.txt	Female	Sc Dx	1	Schizophrenia	1.547934	1.747953		1.203853	
10	5	B	MZ-5-B_anon_interview.txt	Female	No SC Dx	1		1.760128	0.832909		1.644674	3.206547

FIGURE I.11

Data spreadsheet with the pronoun frequencies being added as variables.

These newly created pronoun frequency statistics can now be used to build predictive models.

Length of Responses

Another statistic that likely varies between individuals with and without schizophrenia is the number of words found in the interviews—document length. To add this information to the data set, return to the *Summary* tab. Select *Indexed documents*.

The output gives the document length and the number of words indexed and then lists the indexed words (see Figure I.12). Copy the *Document length* variable.

Data: File summary (MZ _Maudsley)*

	File summary (MZ _Maudsley)		
	Document length	Number of words	Indexed text
MZ-1-A_anon_interview.txt	0	0	
MZ-1-B_anon_interview.txt	0	0	
MZ-2-A_anon_interview.txt	12183	38	much someth anybodi anyth anyth someth anyth anyth one anyth one much much much much
MZ-2-B_anon_interview.txt	18437	65	anyth sever much someon one someth someth much much anyth anyth one anyth much anyth
MZ-3-A_anon_interview.txt	2951	6	one much anyth someth much much
MZ-3-B_anon_interview.txt	0	0	
MZ-4-A_anon_interview.txt	11838	41	one one someth someth much anyth anyth much anyth much much one anyth us us us us muc
MZ-4-B_anon_interview.txt	14325	47	littl anyth anyth anyth much one someth someth everyth littl anyth littl much much anyth anyth
MZ-5-A_anon_interview.txt	16283	52	one anoth anyth mani mani mani mani mani anoth much us us anybodi littl much anyth much o
MZ-5-B_anon_interview.txt	26455	61	one one littl anyth either one either anyth whatev someth mani us someth anyth much littl anyth
MZ-6-A_anon_interview.txt	9653	14	anoth one anyth someth littl much littl anyth littl someth much anyth anyth much

FIGURE I.12

Text file summary giving document length and other statistics.

With the input spreadsheet, *MZ_Maudsley.sta*, active, select the *Data* menu. Select *Add* (see Figure I.13) from the *Variables* drop-down menu.

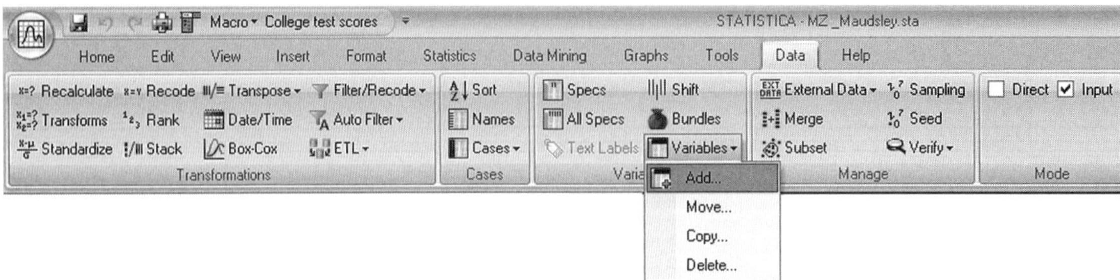

FIGURE I.13
Adding a new variable to a data spreadsheet.

On the *Add variables* dialog box (see Figure I.14), make the following specifications: *How many* = 1, *After* = *whatev*, *Name* = *Document length*. Select OK to insert the new variable.

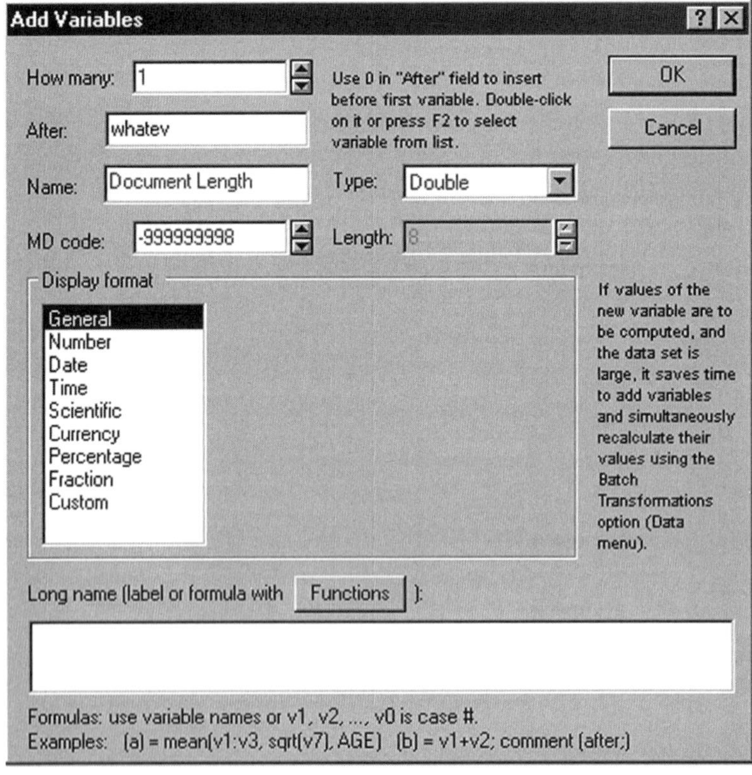

FIGURE I.14
Add variable dialog.

Paste the copied *Document length* variable to the newly created variable in the input spreadsheet (see Figure I.15).

Data: MZ _Maudsley.sta* (34v by 46c)

	22 much	23 neither	24 nobodi	25 none	26 noth	27 one	28 sever	29 somebodi	30 someon	31 someth	32 us	33 whatev	34 Document Length
1													0
2													0
3	1.5676				1.2489	1.1009		0.8842		0.7811			12183
4	1.6077				1.5479	1.475	1.7492		1.3437	1.475	0.6506	1.6314	18437
5	0.9682					0.4613				0.4613			2951
6													0
7	1.475					1.1009				0.7811	1.8163		11838
8	1.5236					0.9682		0.8842		1.288	1.1015		14325
9	1.3591		1.7492		1.5479	1.6789	1.7492	0.8842		0.4613	1.3653		16283
10	1.5676		1.7492			1.5236				1.288	1.1015	1.6314	26455

FIGURE I.15
Document length, a new variable, added to the data spreadsheet.

Now, the data set has pronoun usage variables found from indexing the interview transcripts and the overall length of the interview. Using these variables, along with the gender variable, predictive models can be built to classify patients as having a schizophrenia diagnosis or not.

Interviews were not available for some of the participants. The indicator variable *Interview?* specifies those where an interview was not available with 0 and 1 when the interview was available. Using *Case selection conditions*, only the cases with interview (and therefore indexed pronouns) can be used. From the *Tools* menu, select *Edit* from the *Selection Conditions* drop-down menu (see Figure I.16).

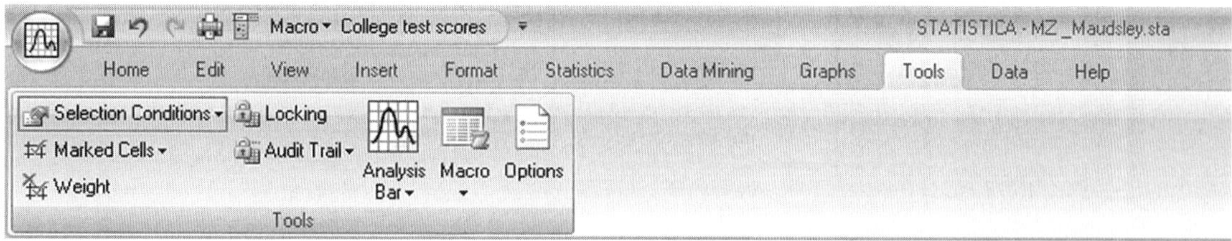

FIGURE I.16
Accessing the Selection Conditions from the Tools menu.

On the *Spreadsheet Case Selection conditions* dialog (see Figure I.17), check *Enable Selection Conditions.* In the *Include cases* field, select *Specific, selected by:* and enter the expression: *"interview?" = 1.*

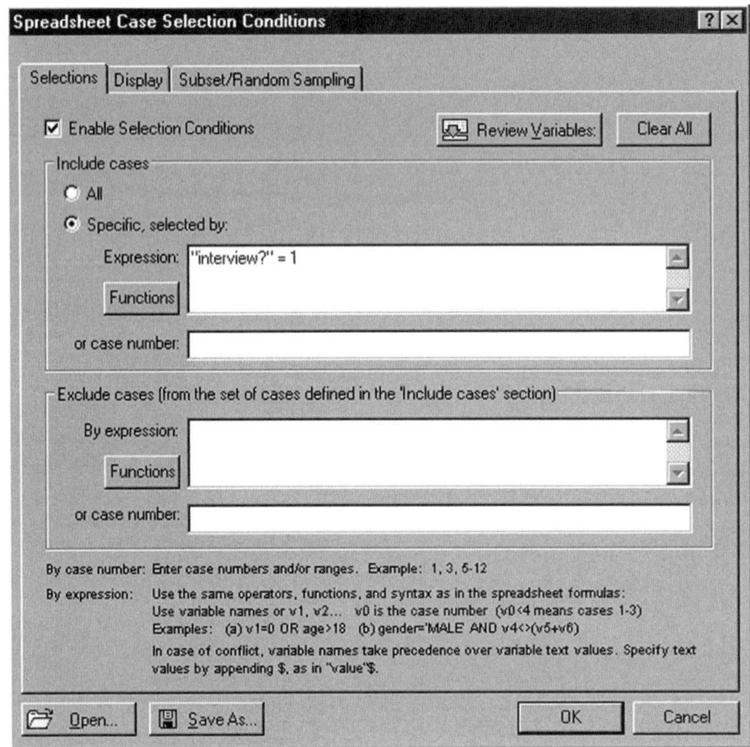

FIGURE I.17
Case Selection dialog.

Click OK to update the spreadsheet with the case selection conditions. The cases that will be used for analysis are highlighted (by default) in light gray (colored light green on the computer screen) (see Figure I.18).

Data: MZ _Maudsley.sta* (30v by 46c)

	1 MZ Twin ID #	2 Twin A or B	3 MZ Text File names	4 Sex	5 Sc DX - yes or no	6 interview ?	7 Dominant psych. symptoms	8 anoth	9 anybodi	10 anyon	11 anyth	12 either
1	1	A	MZ-1-A_anon_interview.txt	Male	Sc Dx	0	Schizophrenia					
2	1	B	MZ-1-B_anon_interview.txt	Male	Sc Dx	0	Schizophrenia					
3	2	A	MZ-2-A_anon_interview.txt	Male	Sc Dx	1	Schizophrenia		0.832909		1.475025	
4	2	B	MZ-2-B_anon_interview.txt	Male	Sc Dx	1	Schizophrenia	0.737599			1.768436	2.819978
5	3	A	MZ-3-A_anon_interview.txt	Male	Sc Dx	1	Schizophrenia				0.461346	
6	3	B	MZ-3-B_anon_interview.txt	Male	No SC Dx	0						
7	4	A	MZ-4-A_anon_interview.txt	Female	Sc Dx	1	Schizophrenia		0.832909		1.567604	
8	4	B	MZ-4-B_anon_interview.txt	Female	No SC Dx	1					1.740467	
9	5	A	MZ-5-A_anon_interview.txt	Female	Sc Dx	1	Schizophrenia	1.547934	1.747953		1.203853	
10	5	B	MZ-5-B_anon_interview.txt	Female	No SC Dx	1		1.760128	0.832909		1.644674	3.206547

FIGURE I.18
The cases selected for further data analysis are highlighted in gray (colored light green on the computer screen).

From the *Data Mining* menu, select *Neural Networks* to open the *SANN—New Analysis/Deployment* dialog (see Figure I.19). (For the reader not familiar with Neural Networks, please refer to the discussion of this algorithm in *Handbook of Statistical Analysis & Data Mining Applications,* by Nisbet, Elder, and Miner, 2009.)

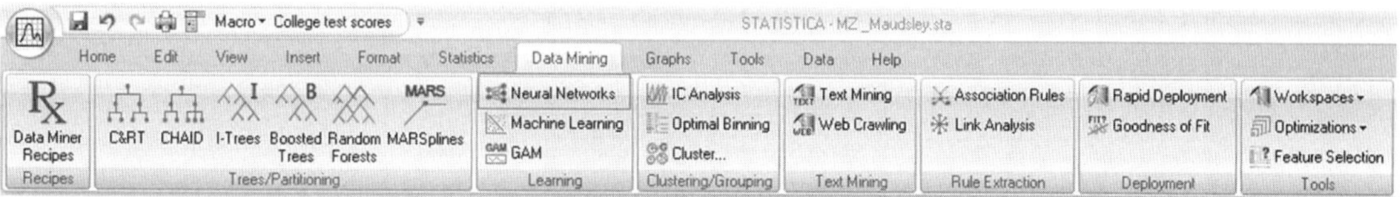

FIGURE I.19

Selecting Neural Networks from the data mining menu.

In the *New analysis* field (see Figure I.20), select *Classification* as the target variable, *Sc DX—yes or no,* is categorical in nature.

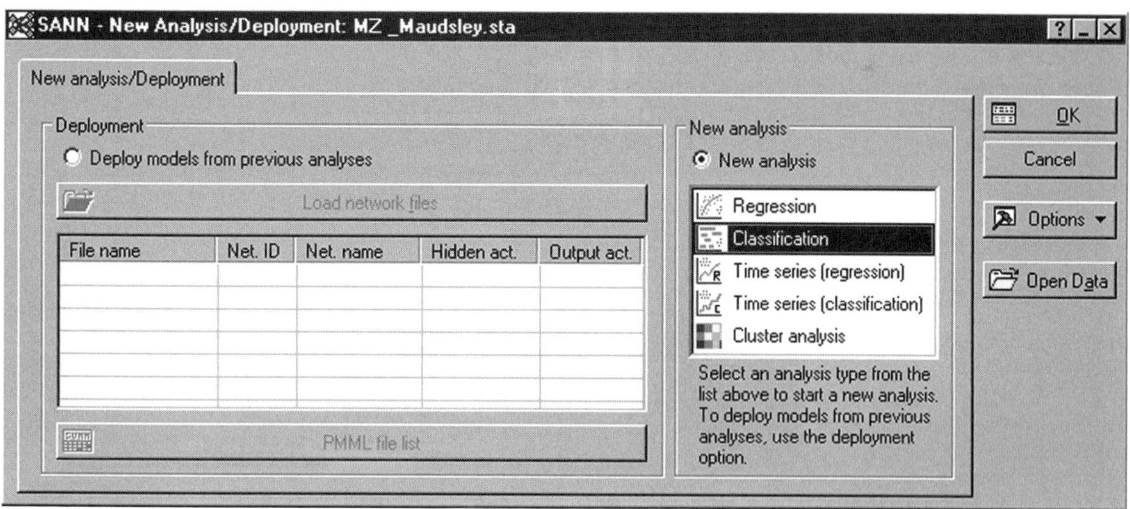

FIGURE I.20

STATISTICA Automated Neural Networks (SANN) opening dialog.

Click OK to advance to the *SANN—Data selection* dialog. Click *Variables* to open the *Select variables for analysis* dialog box. Select *Sc SX—yes or no* as the *Categorical target,* and select the indexed pronoun variables and the *document length* variable as *Continuous inputs* (see Figure I.21).

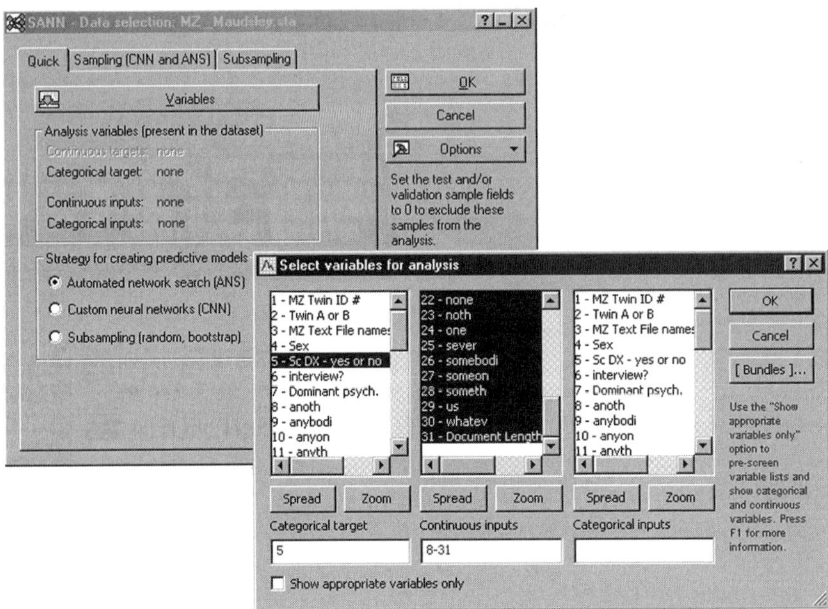

FIGURE I.21
Variable selection dialog.

Accept the other default settings and advance to the *SANN—Automated Network Search (ANS)* dialog. Increase the number of *Networks to train* to 500 (see Figure I.22).

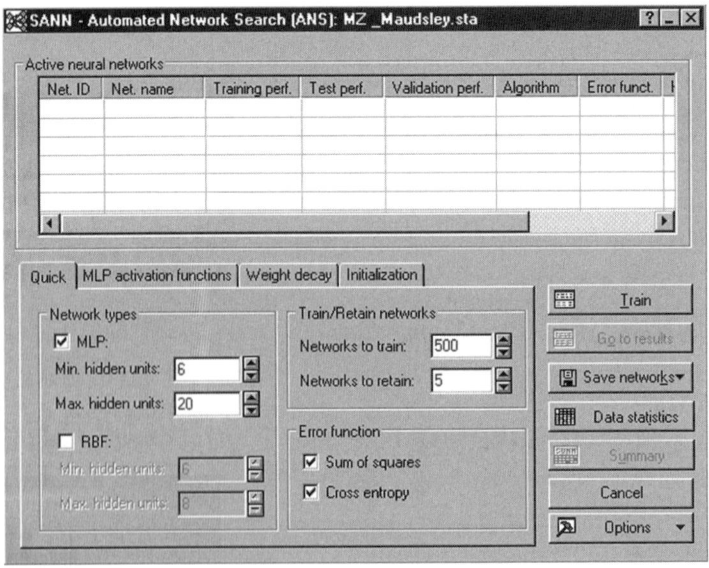

FIGURE I.22
Automated Network Search for the best neural network models.

The networks train quickly. Accept the other default settings and select *Train*.

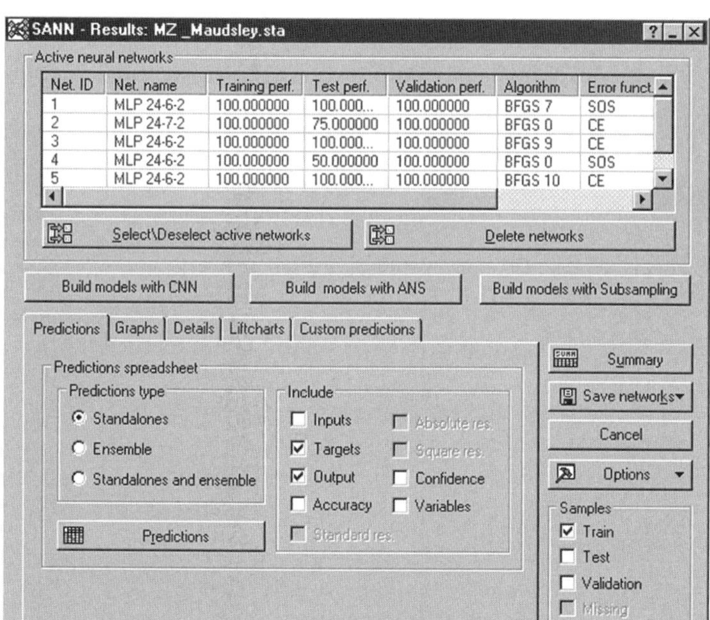

FIGURE I.23
When the search for the best networks is completed, the top five networks will be displayed in the SANN dialog window.

On the *SANN–Results* dialog, the five best networks are listed (see Figure I.23). Select the *Liftcharts* tab (see Figure I.24).

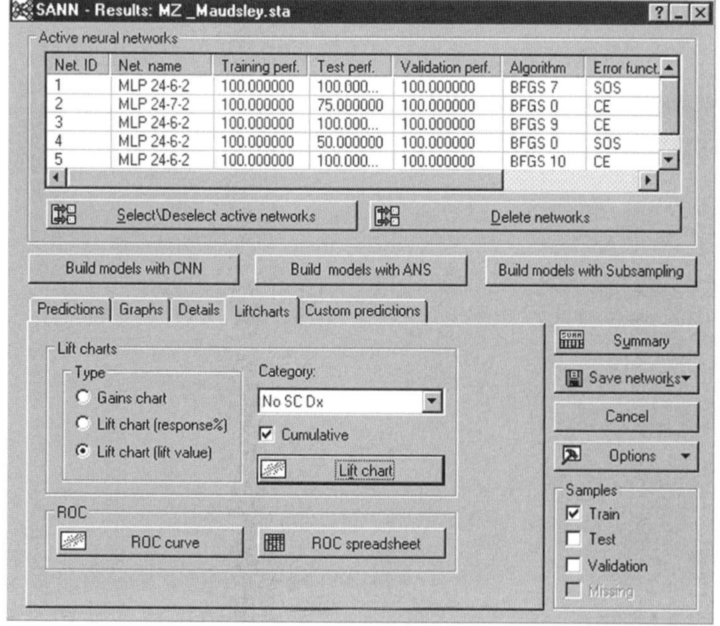

FIGURE I.24
Liftcharts tab of the SANN dialog window.

Change the *Type* to *Lift chart (lift value)* and check the Cumulative option. Select *Lift chart* to create the lift charts for the five selected networks.

Here is one such lift chart (see Figure I.25), showing the added utility gained from using the SANN predictive model, compared to the baseline model, which is a random choice. Select the *Details* tab to explore the model further (see Figure I.26).

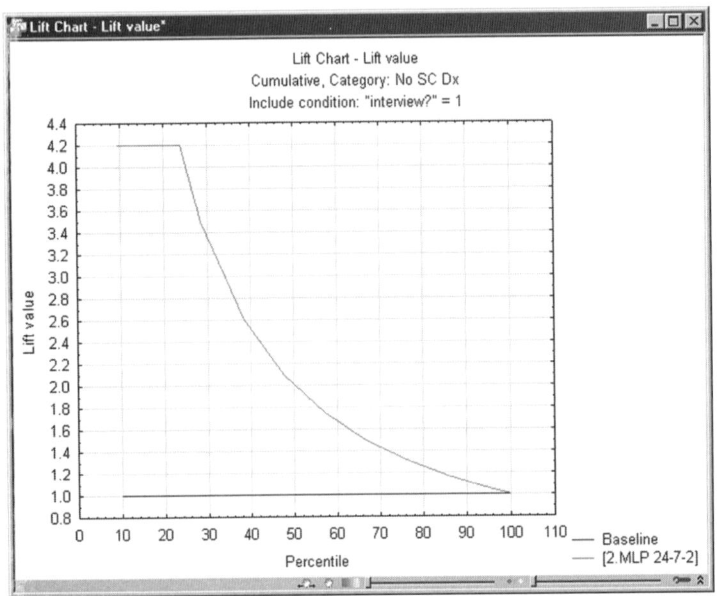

FIGURE I.25
Lift chart.

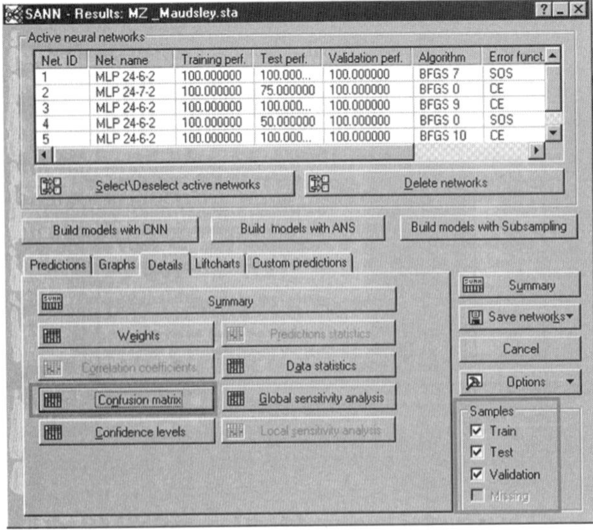

Net. ID	Net. name	Training perf.	Test perf.	Validation perf.	Algorithm	Error funct.
1	MLP 24-6-2	100.000000	100.000...	100.000000	BFGS 7	SOS
2	MLP 24-7-2	100.000000	75.000000	100.000000	BFGS 0	CE
3	MLP 24-6-2	100.000000	100.000...	100.000000	BFGS 9	CE
4	MLP 24-6-2	100.000000	50.000000	100.000000	BFGS 0	SOS
5	MLP 24-6-2	100.000000	100.000...	100.000000	BFGS 10	CE

FIGURE I.26
Details tab of the SANN dialog window; selecting the "confusion matrix" output using Train, Test, and Validation samples of the data set.

In the *Samples* field, check all three samples: *Train, Test,* and *Validation*. Click *Confusion matrix* to output the table comparing *observed vs predicted* values (see Figure I.27).

Data: Sc DX - yes or no (Classification summary) (MZ _Maudsley)*

Sc DX - yes or no (Classification summary) (MZ _Maudsley.sta
Samples: Train, Test, Validation
Include condition: "interview?" = 1

		Sc DX - yes or no-No SC Dx	Sc DX - yes or no-Sc Dx	Sc DX - yes or no-All
1.MLP 24-6-2	Total	7.0000	22.0000	29.0000
	Correct	7.0000	22.0000	29.0000
	Incorrect	0.0000	0.0000	0.0000
	Correct (%)	100.0000	100.0000	100.0000
	Incorrect (%)	0.0000	0.0000	0.0000
2.MLP 24-7-2	Total	7.0000	22.0000	29.0000
	Correct	6.0000	22.0000	28.0000
	Incorrect	1.0000	0.0000	1.0000
	Correct (%)	85.7143	100.0000	96.5517
	Incorrect (%)	14.2857	0.0000	3.4483
3.MLP 24-6-2	Total	7.0000	22.0000	29.0000
	Correct	7.0000	22.0000	29.0000
	Incorrect	0.0000	0.0000	0.0000
	Correct (%)	100.0000	100.0000	100.0000
	Incorrect (%)	0.0000	0.0000	0.0000
4.MLP 24-6-2	Total	7.0000	22.0000	29.0000
	Correct	6.0000	21.0000	27.0000
	Incorrect	1.0000	1.0000	2.0000
	Correct (%)	85.7143	95.4545	93.1034
	Incorrect (%)	14.2857	4.5455	6.8966
5.MLP 24-6-2	Total	7.0000	22.0000	29.0000
	Correct	7.0000	22.0000	29.0000
	Incorrect	0.0000	0.0000	0.0000
	Correct (%)	100.0000	100.0000	100.0000
	Incorrect (%)	0.0000	0.0000	0.0000

FIGURE I.27
Classification summary for the five best neural networks models.

The overall correct percentages are quite high for each of the five networks. Numbers 1, 3, and 5 had 100 percent accuracy in the *Train, Test,* and *Validation* samples. The remaining two models had predictive accuracy in the 90 percent range.

This example used a fairly small sample size, with only 29 patients interviewed. But the predictive model seems to perform quite well for the validation sample, which was not used to build the models.

SUMMARY

It is interesting that we got such a high accuracy percentage in our neural networks modeling. On the surface we have to be concerned that this does not represent "overfitting," some artifact, or a biased sample selection/data preparation procedure. In general statistical analysis we need to remember that

90 to 99 percent predictive accuracy *can* mean that one variable (or phenomenon) explains practically all variance of the target variable. In social (as opposed to natural) sciences, the pure error of every measurement (e.g., every questionnaire or the like) is always or usually very large. For example a test-retest reliability of 70 to 80 percent of any measure in social sciences is often considered almost "perfect," practically eliminating the possibility that such a variable could ever reliably correlate at a 90 percent (let alone 99 percent) level with anything else. Generally reporting something like this without qualifiers and an explanation of why you believe that this "unbelievable" result is still "believable" can raise a red flag. However, in the predictive analytics method used here, neural networks, we selected a validation sample (also known as a "holdout sample") because this is necessary to guard against overfitting. It is always possible that the "devil was in the design," and there is some artifact or biased sample selection contributing to these unusually high predictive results. Nonetheless, this data set has been well studied over the years, is well organized more like a biological model, and thus we can place credibility in the result using modern statistical learning theory methods. Of course, a replication of these results in another setting would provide extra confirmation. The test sample and the validation samples used in predictive analytics, in effect, are virtual replications of the experiment multiple times, and when the validation accuracy is high, this indicates a stable result.

The heterogeneity of this historic twin sample includes a range of illness stages, prominent symptoms, and types and levels of, and responsiveness to, psychotropic medications. Some twins were interviewed while in remission, including some who enjoyed freedom from all but relatively minor symptoms and whose functional capacities included steady work in their communities and relatively stable family lives. Others were interviewed in the hospital during an acute phase of their illness. Despite this heterogeneity and the variability it may contribute to the data, we obtained results that demonstrated very high accuracy, above 90 percent, which compares well with longitudinal predictions of schizophrenia from psychometric data obtained years before diagnosis (Bolinskey et al., 2010).

References

Andreasen, N.C. (1986). Scale for the Assessment of Thought, Language, and Communication (TLC). *Schizophrenia Bulletin, 12,* 473–482.

Barch, D.M., & Berenbaum, H. (1996). Language production and thought disorder in schizophrenia. *Journal of Abnormal Psychology, 105,* 81–88.

Bleuler, E. (1950). *Dementia Praecox or the Group of Schizophrenias.* Translated by Zinkin, J. New York: International Universities Press.

Docherty, N.M., Cohen, A.S., Nienow, T.M., Dinzeo, T.J., & Dangelmaier, R.E. (2003). Stability of formal thought disorder and referential communication disturbances in schizophrenia. *Journal of Abnormal Psychology, 112,* 469–475.

Docherty, N.M., DeRosa, M., & Andreasen, N.C. (1996). Communication disturbances in schizophrenia and mania. *Archives of General Psychiatry, 53,* 358–364.

Docherty, N.M., Gordinier, S.W., Hall, M.J., & Dombrowski, M.E. (2004). Referential communication distrubances in the speech of nonschizophrenic siblings of schizophrenia patients. *Journal of Abnormal Psychology, 113,* 399–405.

Harvey, P., & Brault, J. (1986). Speech performance in mania and schizophrenia: The association of positive and negative thought disorders and reference failures. *Journal of Communication Disorders, 19,* 161–173.

Harvey, P., & Serper, M. (1990). Linguistic and cognitive failures in schizophrenia. A multivariate analysis. *Journal of Nervous and Mental Disease, 178,* 487–493.

Liddle, P.F. (1987). The symptoms of chronic schizophrenia. A re-examination of the positive-negative dichotomy. *British Journal of Psychiatry, 151,* 145–151.

Text Mining Using STM™, CART®, and TreeNet® from Salford Systems: Analysis of 16,000 iPod Auctions on eBay

Dan Steinberg
dans_salford@yahoo.com

Mykhaylo Golovnya
golomi@salford-systems.com

Ilya Polosukhin
ilyap@salford-systems.com

CONTENTS

This tutorial is based on a data mining competition held over the past few years. For the original DMC2006 competition website, visit http://www.data-mining-cup.de/en/review/dmc-2006/. We recommend that you visit this site for information only. The URLs for data and tools for preparing that data are available in this tutorial.

Text mining is an important and fascinating area of modern analytics. On the one hand, text mining can be thought of as just another application area for powerful learning machines. On the other hand, text mining is a distinct field with its own dedicated concepts, vocabulary, tools, and techniques.

In this tutorial we aim to illustrate some important analytical methods and strategies from both perspectives on data mining by introducing tools that are specific to the analysis text and deploying general machine learning technology.

The Salford Text Mining (STM) utility is used in this tutorial for the text processing system that prepares data for machine learning analytics. This is followed by predictive analytics modeling using the Salford Systems CART® decision tree and stochastic gradient boosting TreeNet®. (Evaluation copies of the proprietary technology in CART and TreeNet, as well as the STM are available from http://www.salford-systems.com.)

To follow along with this tutorial, you may want to have the analytical tools being demonstrated installed on your computer. Everything you need may already be on a CD disk containing this tutorial

Practical Text Mining and Statistical Analysis for Non-structured Text Data Applications. DOI: 10.1016/B978-0-12-386979-1.00018-9

and analytical software, but if not, you can use the following link. Create an empty folder on your computer hard drive named "**stmtutor**." This is the root folder where all of the work files related to this tutorial will reside. You may also use the following link to download the SPM: http://www.salford-systems.com/dist/SPM/SPM680_Mulitple_Installs_2011_06_07.zip.

After downloading the package, unzip its contents into "**stmtutor**," which will create a new folder named "**SPM680_Mulitple_Installs_2011_06_07**." The Salford Systems software you've just downloaded needs to be both installed and licensed. Free license codes for a 30-day period are available on request to visitors of this tutorial. (Be aware, however, that Salford Systems reserves the right to decline to offer the free license at its discretion.) To do this, double click on the "**Install_a_Transform_SPM.exe**" file located in the "**SPM680_Mulitple_Installs_2011_06_07**" folder to install the specific version of SPM used in this tutorial. Follow the simple installation steps on your screen.

For the original DMC2006 competition website, visit http://www.data-mining-cup.de/en/review/dmc-2006/. We recommend that you visit this site for information only; data and tools for preparing that data are available at the URL listed following. For the STM package, prepared data files, and other utilities developed for this tutorial, please visit http://www.salford-systems.com/dist/STM.zip. After downloading the archive, unzip its contents into the "**stmtutor**" folder.

When you launch the SPM, you will be greeted with a License dialog containing information needed to secure a license via email (Figure J.1).

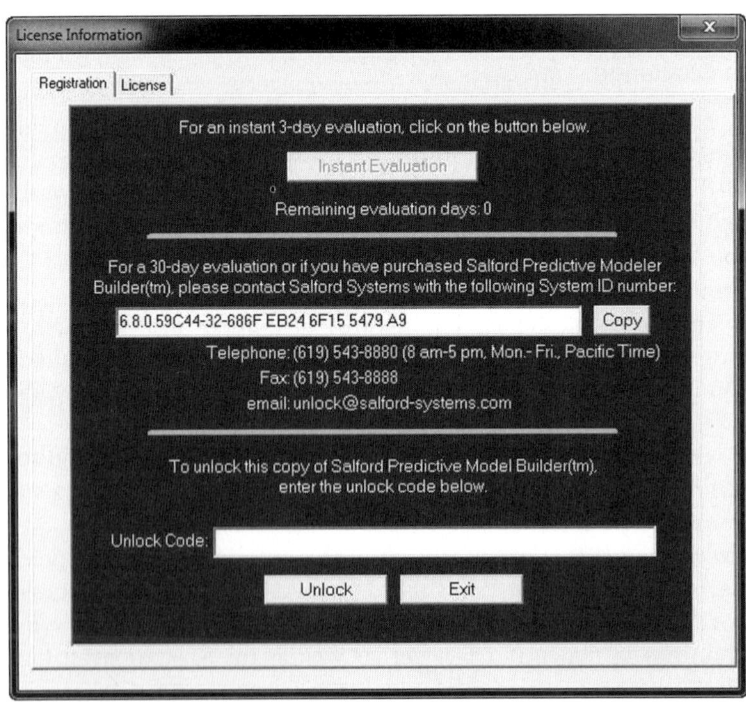

FIGURE J.1
License dialog.

Send the necessary information to Salford Systems to secure your license by entering the "Unlock Code," which will be emailed back to you. The software will operate for three days without any licensing, but you can secure a 30-day license on request.

INSTALLING THE SALFORD TEXT MINER

In addition to the Salford Predictive Modeler (SPM), you will also work with the Salford Text Miner (STM) software. No installation is needed, and you should already have the "**stm.exe**" executable in the "**stmtutor\STM\bin**" folder from unzipping the "**STM.zip**" package. STM builds upon the Python 2.6 distribution and the NLTK (Natural Language Tool Kit) but makes text data processing for analytics very easy to conduct and manage. Expect to see several folders and a large number of files located under the "**stmtutor\STM**" folder. It is important to leave these files in the location to which you have installed them. In order for the software to work properly, please *do not move or alter* any of the installed files other than those explicitly listed as user-modifiable.

NOTE: "**stm.exe**" will expire in the middle of 2012. Please contact Salford Systems to get an updated version beyond this date (http://www.salford-systems.com/). Their phone number is 619-543-8880, and their email address is info@salford-systems.com).

The best examples are drawn from real-world data sets, and we were fortunate to locate data publicly released by eBay. Good teaching examples also need to be simple. Unfortunately, real-world text mining could easily involve hundreds of thousands, if not millions, of features characterizing billions of records. Professionals need to be able to tackle such problems, but to learn, we need to start with simpler situations. Fortunately, there are many applications in which text is important but the dimensions of the data set are radically smaller, either because the data available are limited or because a decision has been made to work with a reduced problem. In this tutorial, we use our simpler example to illustrate many useful ideas for beginning text miners, while pointing the way to working on larger problems.

In 2006 the DMC data mining competition (restricted to student competitors only) introduced a predictive modeling problem for which much of the predictive information was in the form of unstructured text. These data sets can be downloaded from http://www.data-mining-cup.de/en/review/dmc-2006/. For your convenience, however, we have repackaged these data and made it somewhat easier to work with. This repackaged data are included in the STMU package described at the beginning of this tutorial. The data summarize 16,000 iPod auctions held on eBay from May 2005 through May 2006 in Germany. Each auction item is represented by a text description written by the seller (in German), as well as a number of flags and features available to the seller at the time of the auction. Auction items were grouped into 15 mutually exclusive categories based on distinct iPod features: storage size, type (regular, mini, nano), and color. The competition's goal was to predict whether the closing price would be above or below the category average.

COMMENTS ON THE CHALLENGE

One might think that a challenge written in German might not be of general interest outside of Germany. However, working with a language that is essentially unfamiliar to any member of the analysis team helps to illustrate one important point:

> Text mining via tools that have no "understanding" of the language can be strikingly effective.

We have no doubt that dedicated tools that embed knowledge of the language being analyzed can yield predictive benefits. We also believe we could have gained further valuable insight into the data if any of the authors spoke German! *But our performance without this knowledge is still impressive.* In contexts where simple methods can yield more than satisfactory results, or in contexts where the same methods must be applied uniformly across multiple languages, the methods described in this tutorial can provide excellent guidance.

- Figure J.2 summarizes the positioning of the four basic models made in this tutorial with respect to the 173 official competition entries.
- The TN model with text mining processing is among the top 10 winners!

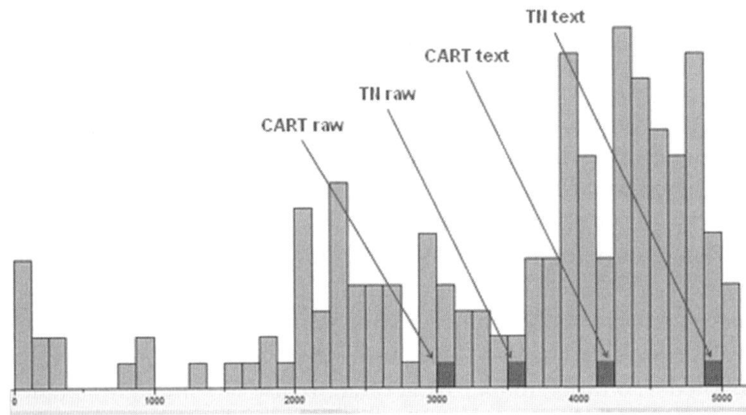

FIGURE J.2
Four models.

Now to work through this tutorial, please go to the DVD packaged with this book and find "Tutorial J" in the electronic folder, and then find the PowerPoint of over 100 slides. This tutorial is presented as a PPT. This PPT is extremely thorough, taking you through all of the steps needed to learn how to use Salford Systems and their new text mining module.

TUTORIAL K

Predicting Micro Lending Loan Defaults Using SAS® Text Miner

Richard Foley
SAS, Raleigh, NC, USA

CONTENTS

Practical Text Mining and Statistical Analysis for Non-structured Text Data Applications. DOI: 10.1016/B978-0-12-386979-1.00019-0

In this section, we use SAS® Text Miner, a plug-in to SAS® Enterprise Miner, to combine text mining techniques with standard data mining techniques, thereby predicting loan failures of microloans. Microfinance is a general term to describe financial services to low-income individuals or to those who do not have access to typical banking services.

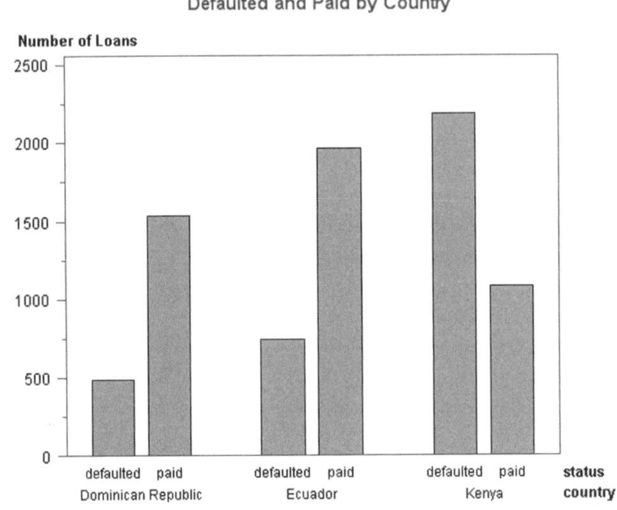

Kiva is an online lending platform that allows individuals to make small loans to borrowers around the world. For more information on Kiva and micro-lending please go to http://www.kiva.org.

For the purposes of this demo we will only look at countries with a good mix of defaulted and paid loans. As you can see Ecuador, Kenya, and the Dominican Republic have a good sampling of defaulted loans with respect to paid loans.

We compare different decision trees from structure data only, unstructured data only, and the combination of structured and unstructured data.

INTRODUCTION

SAS® Text Miner uses a latent semantics analytics (LSA) technique to generate topics. First, Natural Language Processing is used to do the following:

- Parse
- Tokenize
- Stemming
- Part of speech tagging
- Spell checking

Single value decomposition (SVD) is used to cluster the documents. Each SVD vector is then rotated around the document cutoff value maximizing the sum of the least squares. The rotation moves low-weight terms to zero and elevates the important terms, thus providing knowledge of the terms within the category.

These rotated SVDs can then be combined with structured data to enhance analytical models as well as provide more information to help make a decision on whether or not to fund a loan.

ABOUT SAS® TEXT MINER

SAS® Text Miner is a plug-in for the SAS® Enterprise Miner environment, which provides a rich set of data mining tools that facilitate the prediction aspect of text mining. The integration of SAS® Text Miner within SAS® Enterprise Miner combines textual data with traditional data mining variables. The integration provides the ability to add text mining nodes into SAS® Enterprise Miner process flow diagrams. SAS® Text Miner encompasses the parsing and exploration aspects of text mining and prepares data for predictive mining and further exploration using other SAS® Enterprise Miner nodes. SAS® Text Miner supports various sources of textual data: local text files, text as observations in SAS® data sets or external databases, and files on the web.

As part of SAS® Enterprise Miner, SAS® Text Miner, follows the concept of nodes, where each node provides a customizable task. SAS® Text Miner consists of four of these highly customizable nodes.

PROJECT OVERVIEW

This tutorial takes a step-by-step approach in showing the text mining techniques to build the flow diagram (Figure K.1).

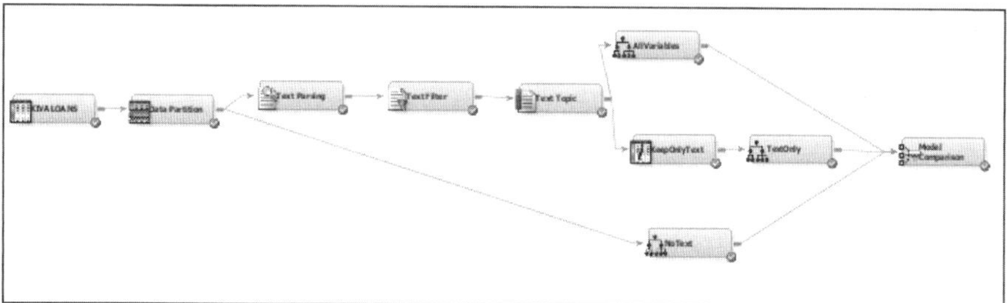

FIGURE K.1
SAS® Text Mining flow.

Much of the loan information is translated from other languages and contains verbiage unique to a region and industry. Because of the language issues generating stoplists, synonyms and multiterm words will play an extremely important part in the use of text mining. This project will demonstrate every stage of the process of determining stopwords, finding multiterm words, and finding synonyms in order to build a better model.

We will then create the topics through the SAS® Text Topic node and create three decision trees:

- Relational data only
- Text data only
- Combined relational and text data

We will import the complete data flow, run it, look at each decision tree, and compare their performance. We will look at each tree and compare the models' performance.

FIGURE K.2
Login screen.

PREPARING THE DATA AND SETTING UP THE DIAGRAM

Before we begin, copy the directory on the CD to the C-drive. This will create a directory called "TMProjects" with two subdirectories, "Data" and "KivaLoans." This folder will be used to store the metadata and data found on the CD provided in this book.

CREATING A NEW PROJECT

Log in to SAS® Enterprise Minerselect New Project in order to create a new project.

Select the "New Project" link.
A new window will appear.

First select Next. This brings up step 2 of the Creating New Project wizard. You will be asked to fill in a Project Name and an SAS® server directory.

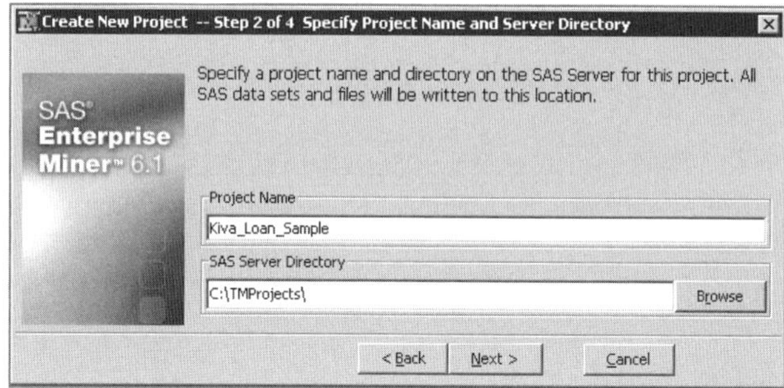

In the Project Name enter "Kiva_Loan_Sample." This will define the project name.

In the SAS® Server Directory, enter a directory name—for this example, "c:\TMProjects\." This directory will be the location of the SAS® data sets and metadata pertaining to the your text mining projects. Click Next until you finish the wizard. When finished, you will be taken to the newly created Kiva_Loan_Sample project.

First, we will define the location of the Kiva data. Select Library from the New icon. It looks like a blank sheet of paper with a corner folder over in the icon toolbar drop-down list.

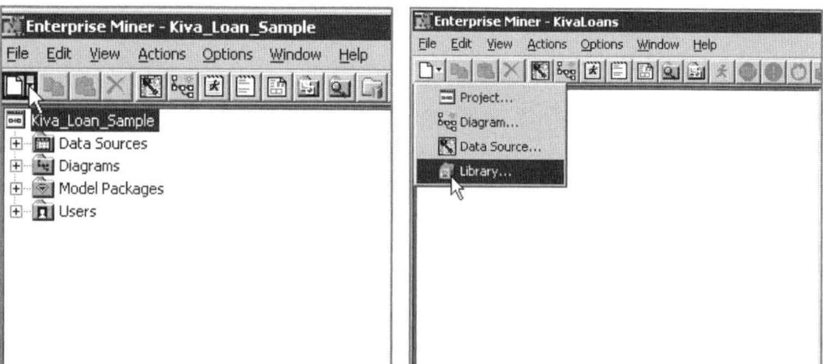

The Library wizard window will appear. Click Next to go to step 2. In the Name window, type in "Kiva." Keep the Engine to base. In the path, type in the name of the location of the data set: "c:\tmprojects\data\." Leave the options text box blank.

Click Next until you are finished. You have just registered the library where the data are stored.

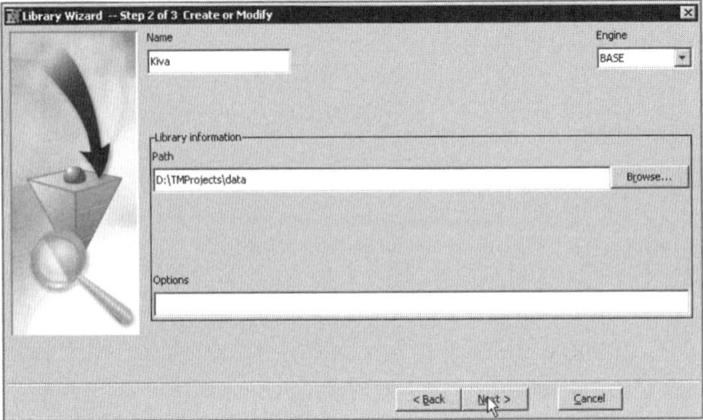

REGISTERING THE TABLE

Go back to the New icon and select Data Source. The Metadata source window appears. In the drop-down box make sure the Source is SAS® Table. Click Next.

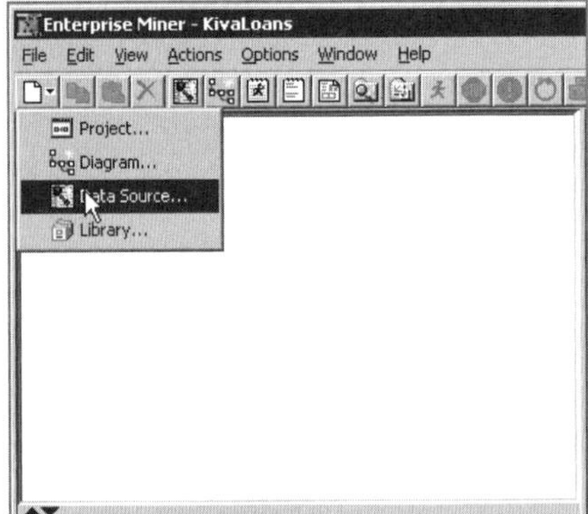

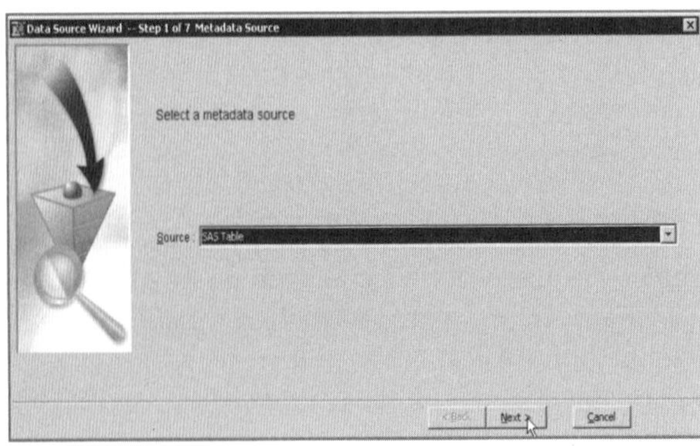

In step 2 select browse.

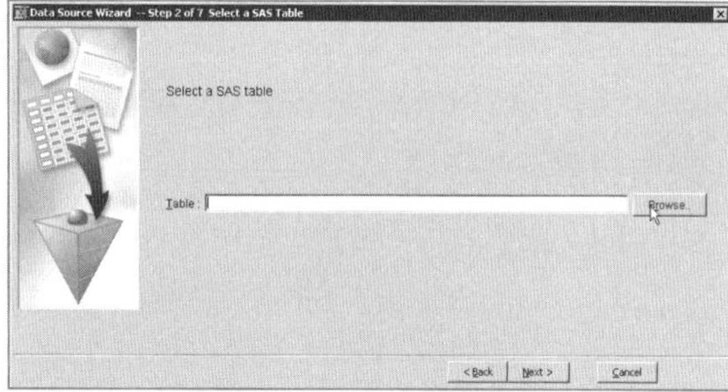

Select the Kiva library. Then select "Kivaloans," and press OK.

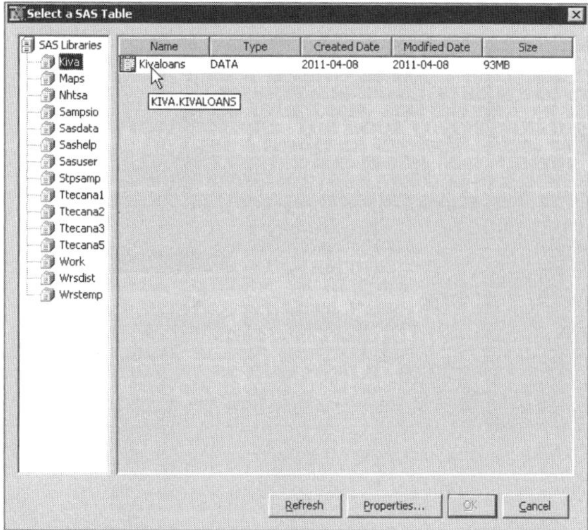

Select Next. Take the defaults for the rest of the way.

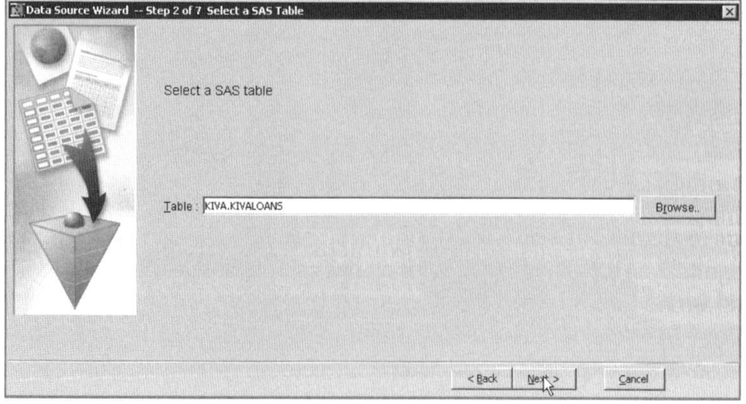

CREATING A NEW DIAGRAM

Select the "New" icon, and in the drop-down box select "Diagram."

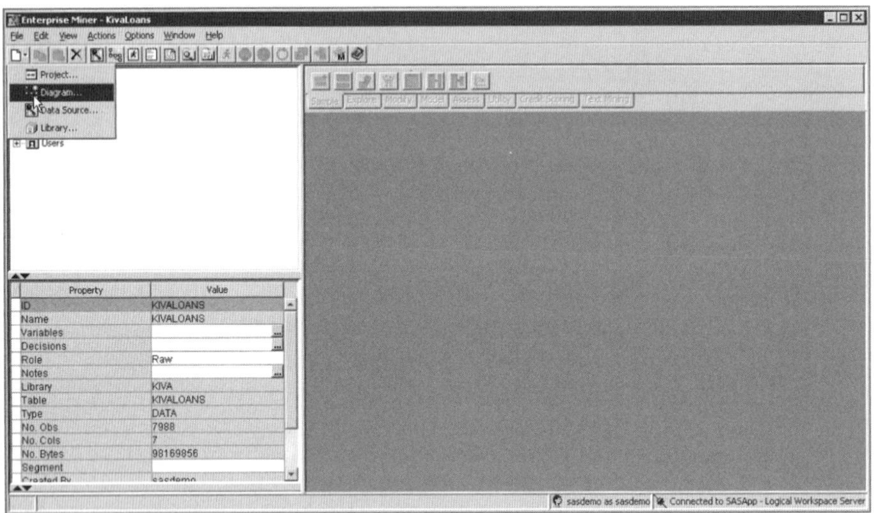

Type in "KivaLoans" in the text box. Click OK.

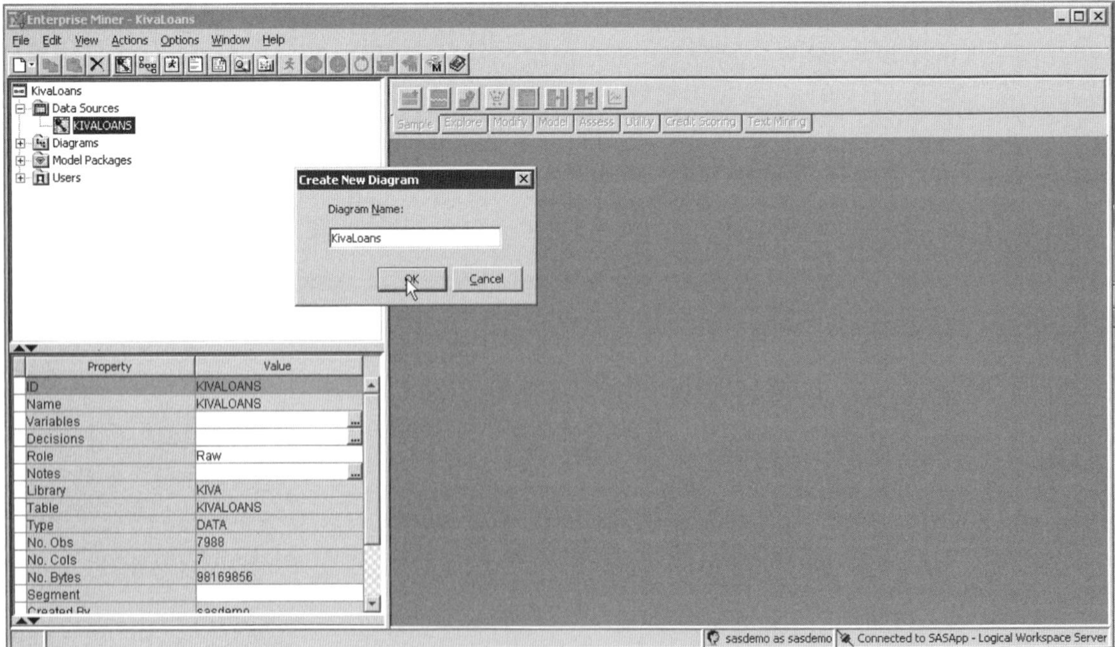

In the Diagrams folder, you will see KivaLoans. Double click the KivaLoans diagram. This will open the KivaLoans diagram, which will be the workspace to build the text and data mining project.

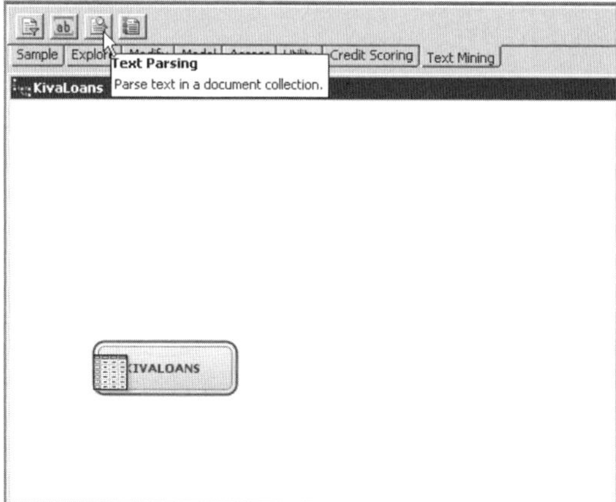

The Text Parsing node isolates the parsing to a single node and gives us control over many natural language parsing options, including the following:

- Stoplist
- Start list
- Synonym list
- Multiterm list
- Concept extraction

TEXT FILTER NODE

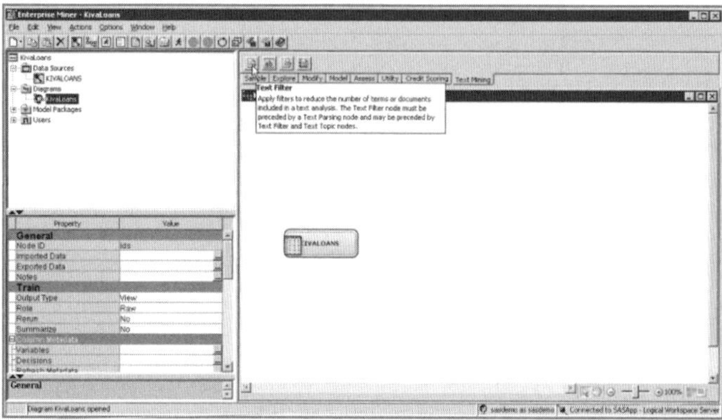

The Text Filter node is the node used for term understanding. In this node you can perform:

- advanced searches to understand use of words within the actual text
- concept linking to understand terms in context that allows you to identify unexpected multiword terms and synonyms
- spell checking
- filtering of documents based on terms
- filtering of documents based on relational data

TEXT TOPIC NODE

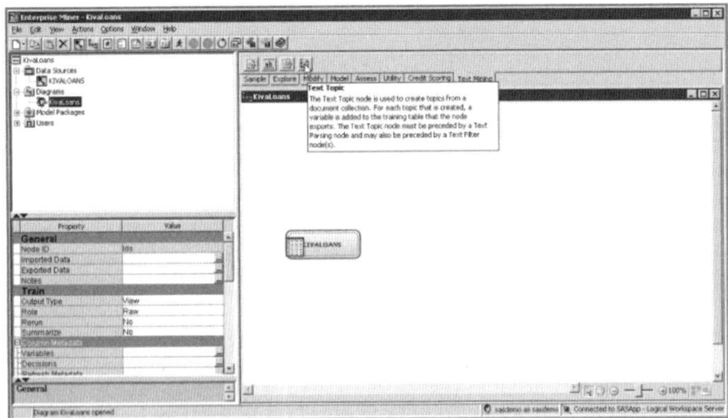

The Text Topic node uses a rotated SVD methodology to determine its topics. However, there will be many SVD vectors generated, and it is important to keep only the vectors that are far enough away from one another to make distinct enough document clusters. Other things done in the Text Topic node are the ability to:

- change term cutoff values
- change document cutoff values
- view documents in a cluster
- rename clusters to provide more meaningful titles

CREATING THE TEXT MINING FLOW

In this section we will create the text mining flow, adding the data, the Text Parsing node, the Text Filter node, and finally the Text Topic node. We will (1) run the Text Parsing node to discover term frequencies and improve our stoplist; (2) run the Text Filter node to determine synonyms and multiword terms; and (3) make the proper corrections and run the Text Topic node, which will provide us with the information we need to start our predictive models.

INSERTING THE DATA

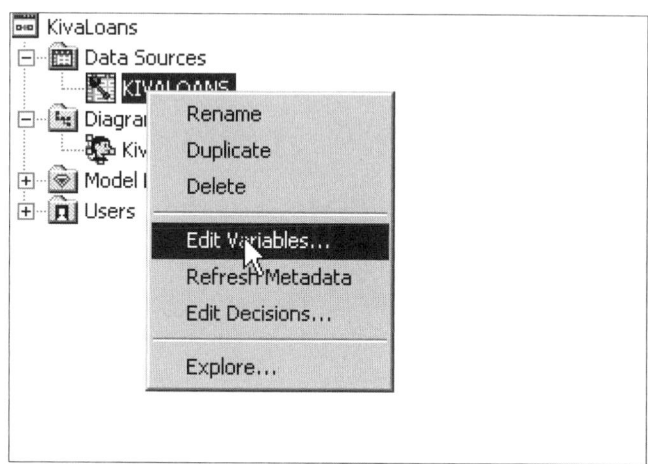

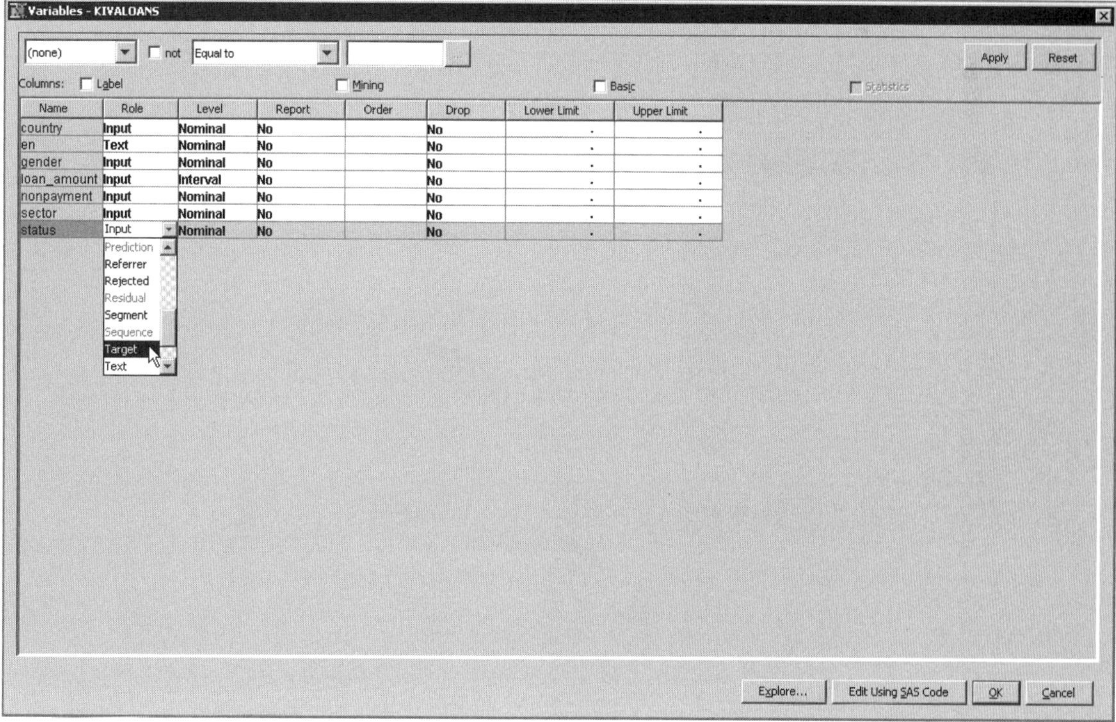

"Status" is the variable that we want to use as our target in our predictive model. Select the row with "Status" in the Name field. Change the Role to "Target."

Drag the KIVALOANS data set in the "Data Sources" folder onto the KivaLoans pallet.

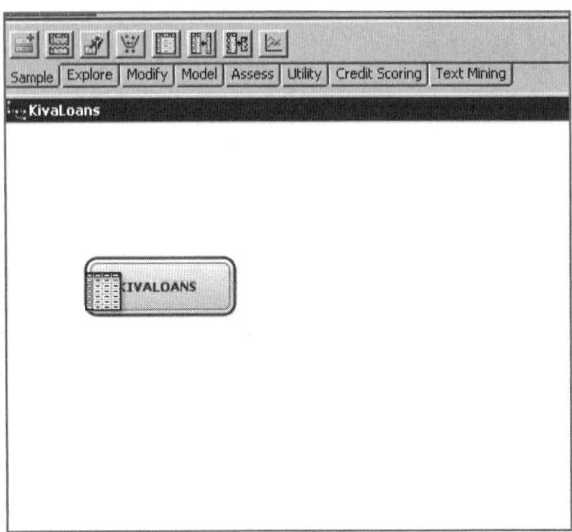

Select the Sample Tab.

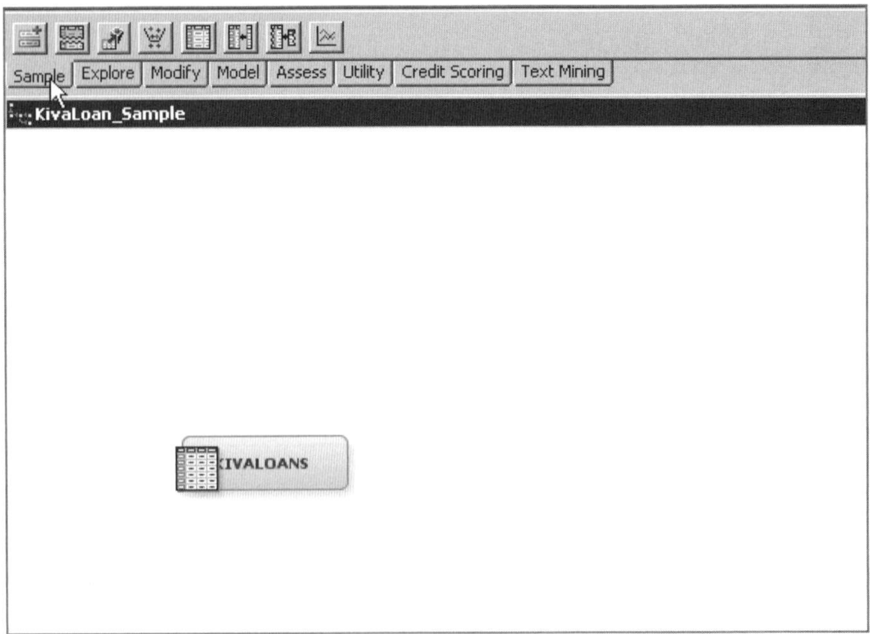

Select the Partition icon, and drag it onto the pallet. Then connect the KivaLoans data set node with the Partition node as in Figure K.3.

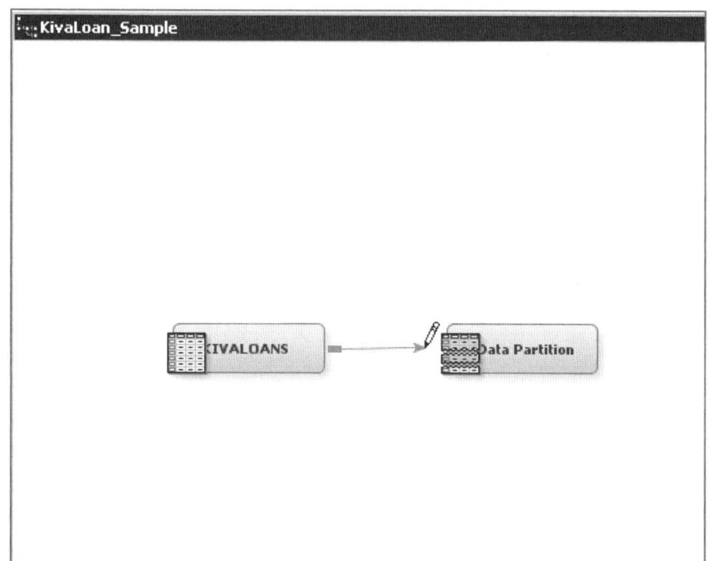

FIGURE K.3
Connecting the data set with the Partition node.

Select the Text Mining Tab.

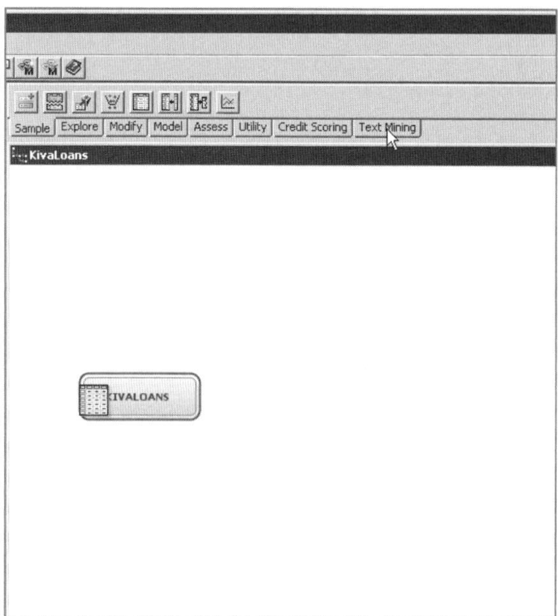

Select the Text Parsing icon.

Drag the Text Parsing node on the KivaLoans pallet. Connect the KivaLoans data set with the Text Parsing node as shown in the figure below.

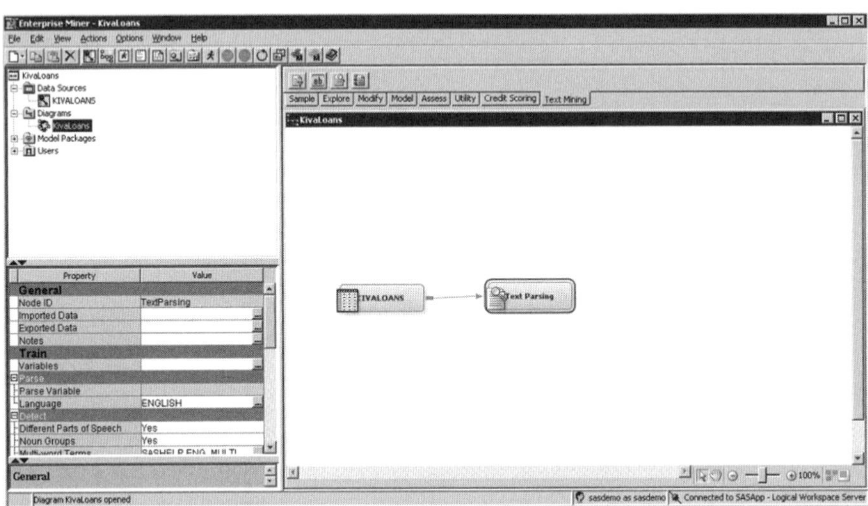

Drag the Filter node onto the KivaLoans pallet. Connect the Text Parsing node to the Text Filter node, as shown in the figure below. Then drag the Text Topic node onto the KivaLoans pallet and connect the Text Topic node to the Text Filter node as shown below.

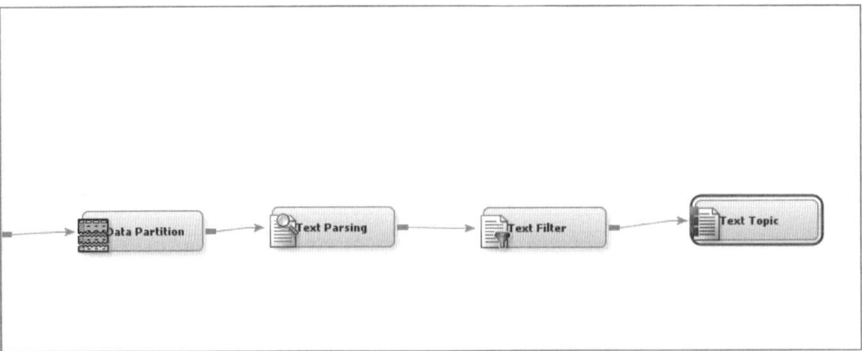

We are now ready to begin our process.

UNDERSTANDING TEXT PARSING

Select the Text Parsing node, and then click on the "Running Person" icon in the toolbar, as shown in figure below. The "Running Person" icon executes the flow up to the selected node.

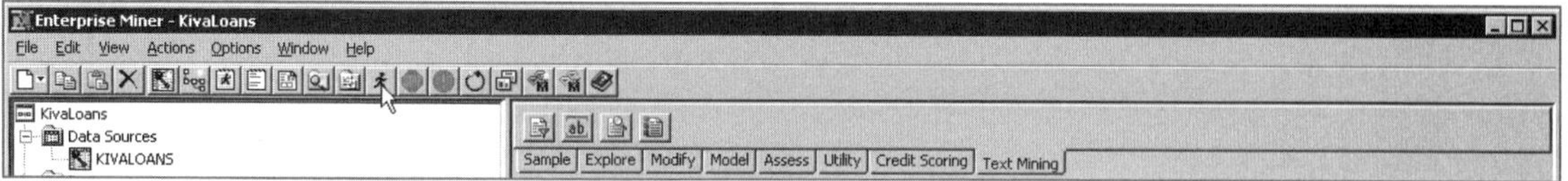

When the process is finished you will see a Run Completed box. Select the Results button.

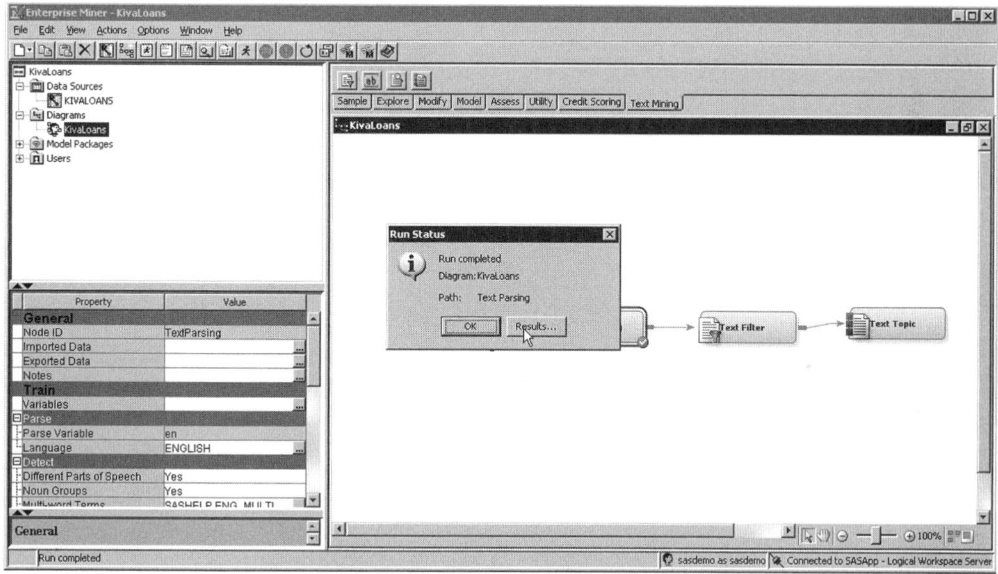

The Text Parse node results provide very useful statistics that can be used to help create your stoplist. A stoplist is useful to help filter out noisy terms. Noisy terms are terms that appear in documents with great frequency.

The results from the Text Parsing node show graphs and tables with information on the frequency of the terms, how many documents the terms are in, and a count of the various entities. Figure K.4 shows outliers from the main cluster. The largest outliers with the greatest frequency and number of documents are the verbs "be" and "have," which are already dropped, as you can see from the table below the Number of Documents by Frequency diagram. The highest-frequency term is "business." Since the documents are about business loans, the fact that the word "business" is in the document does not add to the information and will hide important information that is contained in the documents. However, you may want to keep a word like "sell," since not every business sells products; some businesses may be taxi drivers, farmers, or mechanics.

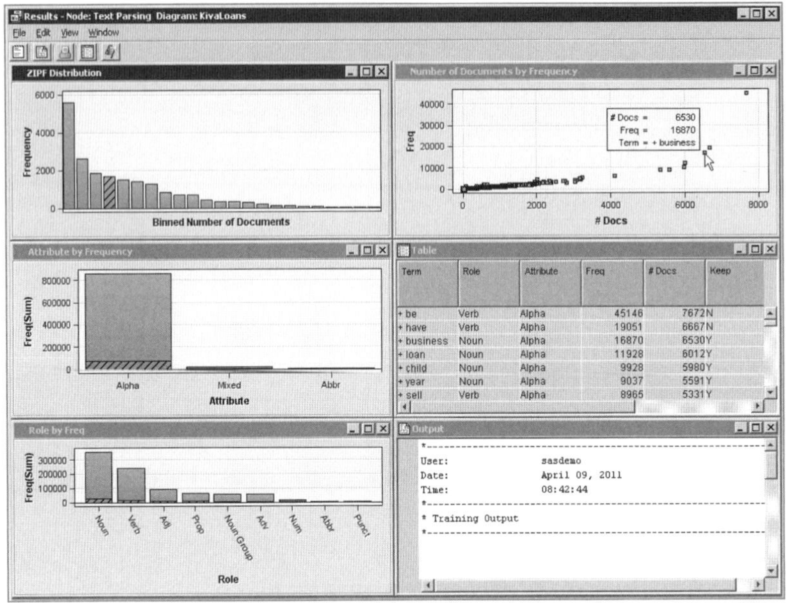

FIGURE K.4
Text Parsing node results.

Now that we have identified a stopword, we need to add the word to our stoplist. Close the results window and select the Text Parse node from the KivaLoan pallet. On the left you will see the properties sheet. Scroll down until you see the stoplist. Click on the ellipses.

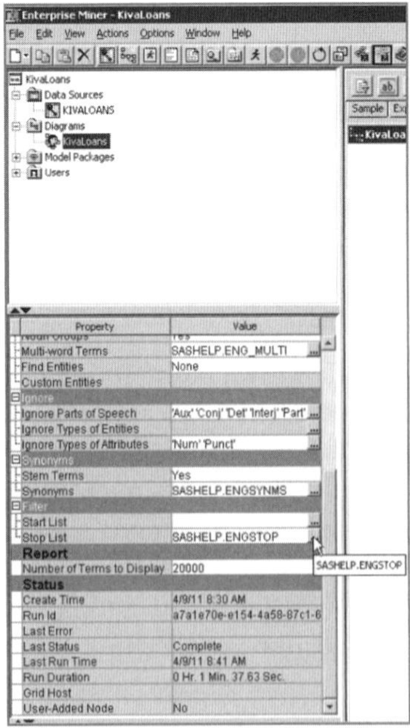

The editor for the TextParsing stoplist appears. The editor allows you to add, modify, and delete stopwords. Select the folder paper and add the icon to add a new term, as shown in Figure K.5.

FIGURE K.5
Stoplist.

Scroll down to the bottom, and enter the term "business," as shown in Figure K.6.

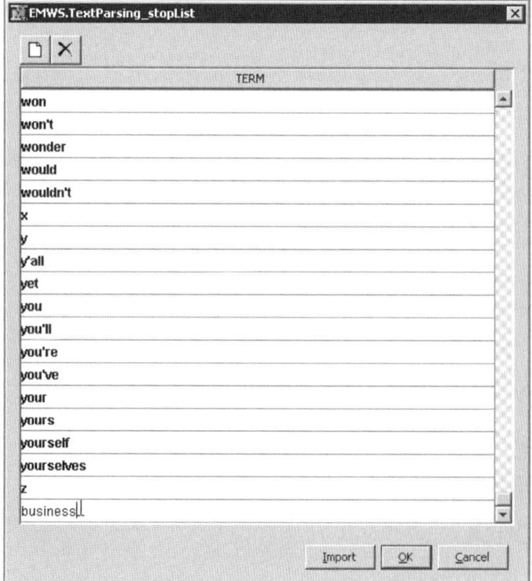

FIGURE K.6
Adding a term to the stoplist.

Run the Text Parse node again, and you will see that the term "business" is now listed as N in "Keep," as shown in Figure K.7.

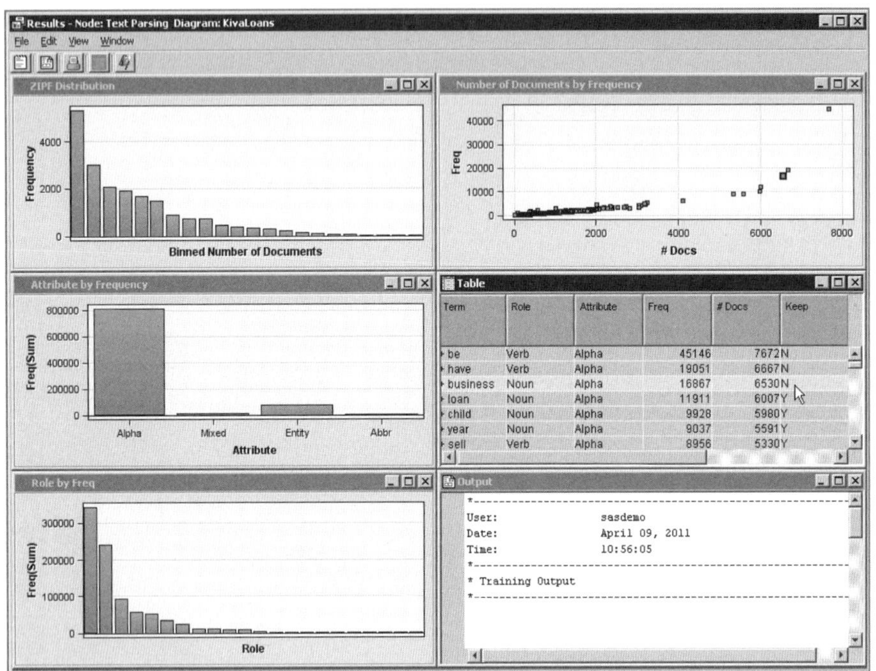

FIGURE K.7

Text Parsing node results modified.

SYNONYMS AND MULTITERM WORDS

Next, we will examine various words and the context in which they appear. This will help find synonyms and multiterm words, as well as terms that we may want to drop. Select the Text Filter node. In the property sheet on the right, scroll down until you see the heading Spelling. Under the "Check Spelling" heading, select "Yes" in the drop-down box (Figure K.8). This will find misspelled terms.

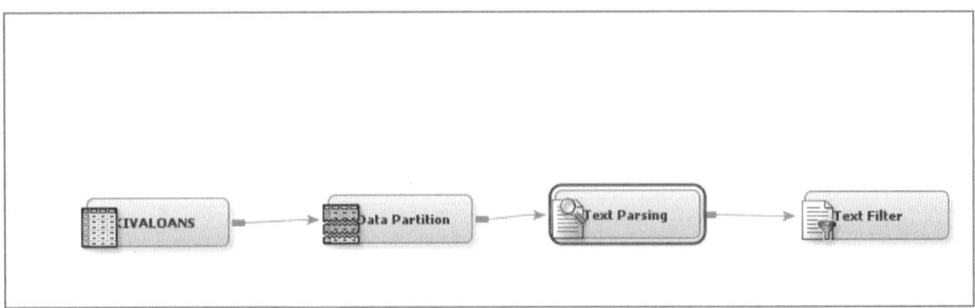

FIGURE K.8

Adding the Text Filter node.

Run the Text Filter node, and then click "No" on the results window. Then go to the property sheet on the right and scroll down until you see the Interactive Filter heading. Click on the ellipses to open the Interactive Filter. The Interactive Filter Viewer appears. The window is broken up into two parts. The top part contains the search box where you can enter in search filters as well as each document with its associated relational information.

The bottom section contains a list of terms with the term's statistics. In the "Keep" column a check means the term is used in the analysis. You can uncheck the term, meaning the term will be dropped in the analysis. Select the checkbox for the term "loan" and uncheck it, as shown in Figure K.9, thus dropping the term "loan."

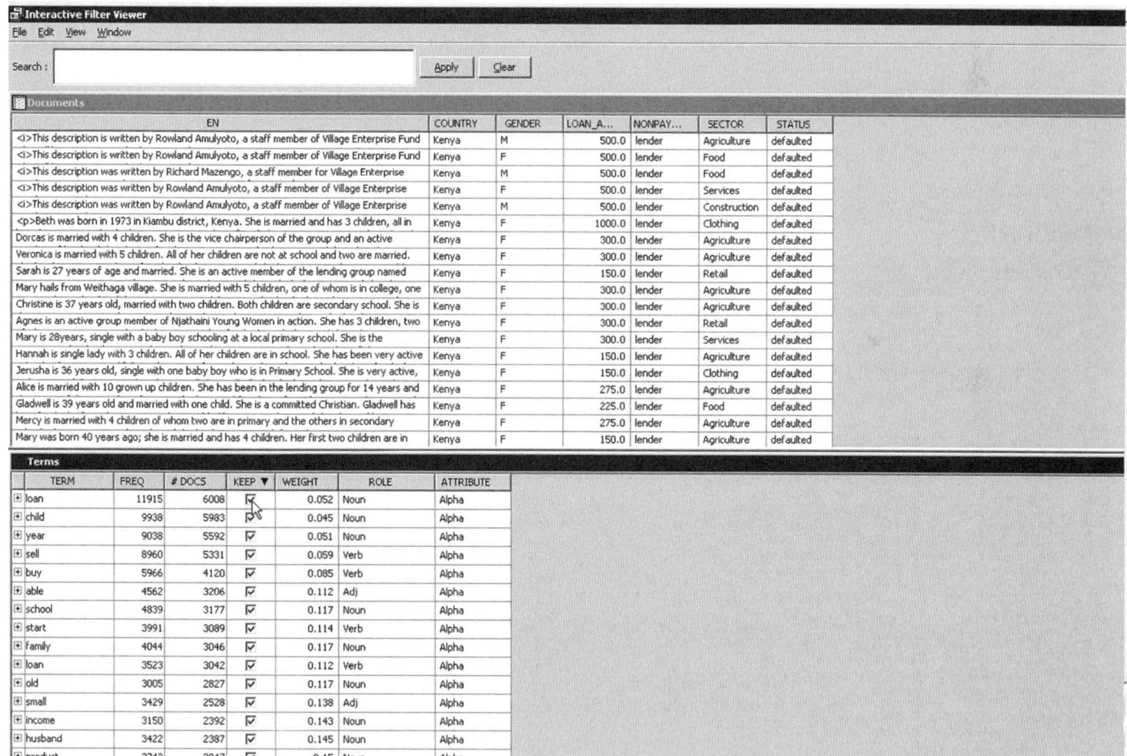

EN	COUNTRY	GENDER	LOAN_A...	NONPAY...	SECTOR	STATUS
<i>This description is written by Rowland Amulyoto, a staff member of Village Enterprise Fund	Kenya	M	500.0	lender	Agriculture	defaulted
<i>This description is written by Rowland Amulyoto, a staff member of Village Enterprise Fund	Kenya	F	500.0	lender	Food	defaulted
<i>This description was written by Richard Mazengo, a staff member for Village Enterprise	Kenya	M	500.0	lender	Food	defaulted
<i>This description was written by Rowland Amulyoto, a staff member of Village Enterprise	Kenya	F	500.0	lender	Services	defaulted
<i>This description was written by Rowland Amulyoto, a staff member of Village Enterprise	Kenya	M	500.0	lender	Construction	defaulted
<p>Beth was born in 1973 in Kiambu district, Kenya. She is married and has 3 children, all in	Kenya	F	1000.0	lender	Clothing	defaulted
Dorcas is married with 4 children. She is the vice chairperson of the group and an active	Kenya	F	300.0	lender	Agriculture	defaulted
Veronica is married with 5 children. All of her children are not at school and two are married.	Kenya	F	300.0	lender	Agriculture	defaulted
Sarah is 27 years of age and married. She is an active member of the lending group named	Kenya	F	150.0	lender	Retail	defaulted
Mary hails from Weithaga village. She is married with 5 children, one of whom is in college, one	Kenya	F	300.0	lender	Agriculture	defaulted
Christine is 37 years old, married with two children. Both children are secondary school. She is	Kenya	F	300.0	lender	Agriculture	defaulted
Agnes is an active group member of Njathaini Young Women in action. She has 3 children, two	Kenya	F	300.0	lender	Retail	defaulted
Mary is 28years, single with a baby boy schooling at a local primary school. She is the	Kenya	F	300.0	lender	Services	defaulted
Hannah is single lady with 3 children. All of her children are in school. She has been very active	Kenya	F	150.0	lender	Agriculture	defaulted
Jerusha is 36 years old, single with one baby boy who is in Primary School. She is very active,	Kenya	F	150.0	lender	Clothing	defaulted
Alice is married with 10 grown up children. She has been in the lending group for 14 years and	Kenya	F	275.0	lender	Agriculture	defaulted
Gladwell is 39 years old and married with one child. She is a committed Christian. Gladwell has	Kenya	F	225.0	lender	Food	defaulted
Mercy is married with 4 children of whom two are in primary and the others in secondary	Kenya	F	275.0	lender	Agriculture	defaulted
Mary was born 40 years ago; she is married and has 4 children. Her first two children are in	Kenya	F	150.0	lender	Agriculture	defaulted

TERM	FREQ	# DOCS	KEEP ▼	WEIGHT	ROLE	ATTRIBUTE
loan	11915	6008	☑	0.052	Noun	Alpha
child	9938	5983	☑	0.045	Noun	Alpha
year	9038	5592	☑	0.051	Noun	Alpha
sell	8960	5331	☑	0.059	Verb	Alpha
buy	5966	4120	☑	0.085	Verb	Alpha
able	4562	3206	☑	0.112	Adj	Alpha
school	4839	3177	☑	0.117	Noun	Alpha
start	3991	3089	☑	0.114	Verb	Alpha
family	4044	3046	☑	0.117	Noun	Alpha
loan	3523	3042	☑	0.112	Verb	Alpha
old	3005	2827	☑	0.117	Noun	Alpha
small	3429	2528	☑	0.138	Adj	Alpha
income	3150	2392	☑	0.143	Noun	Alpha
husband	3422	2387	☑	0.145	Noun	Alpha
product	3743	2347	☑	0.15	Noun	Alpha

FIGURE K.9
Interactive Filter viewer.

Click in the lower window to activate the terms window. Place your cursor on the term "loan," and select it, making sure the row is highlighted. Next, a drop-down list appears. Select "Find," as shown in Figure K.10. In the box that appears, type in "motorcycle" and Enter. This will take you to the term "motorcycle."

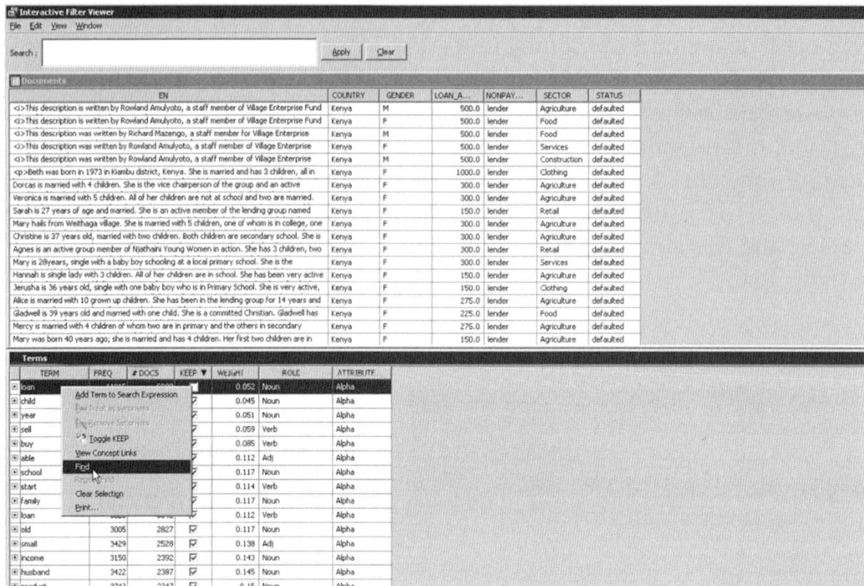

FIGURE K.10
Text Filter node results.

Right click the mouse again, and this time select "View Concept Links."

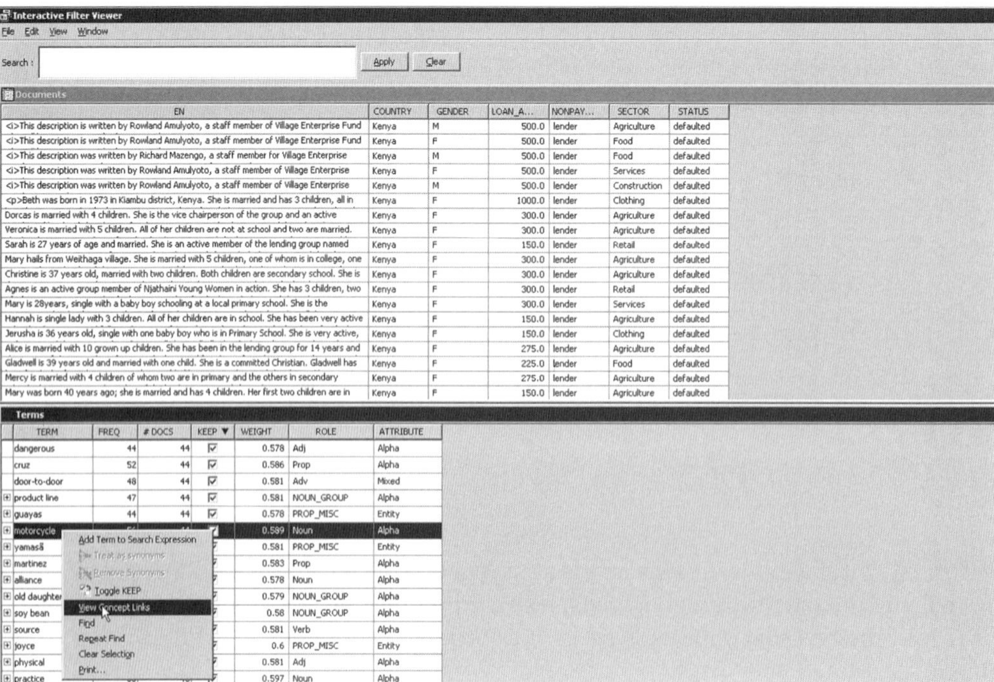

The diagram below Figure K.10 is the concept link viewer. "Motorcycle" is the center term, and the darker the lines, the stronger the relationship is to those terms. Notice that "motorcycle" and "taxi" have a strong relationship, which could mean that people are looking for loans for motorcycle taxis. Another thing to notice is the relationship between motorcycle and spare parts shops, as well as motorcycle repair shops. This could be a possible synonym.

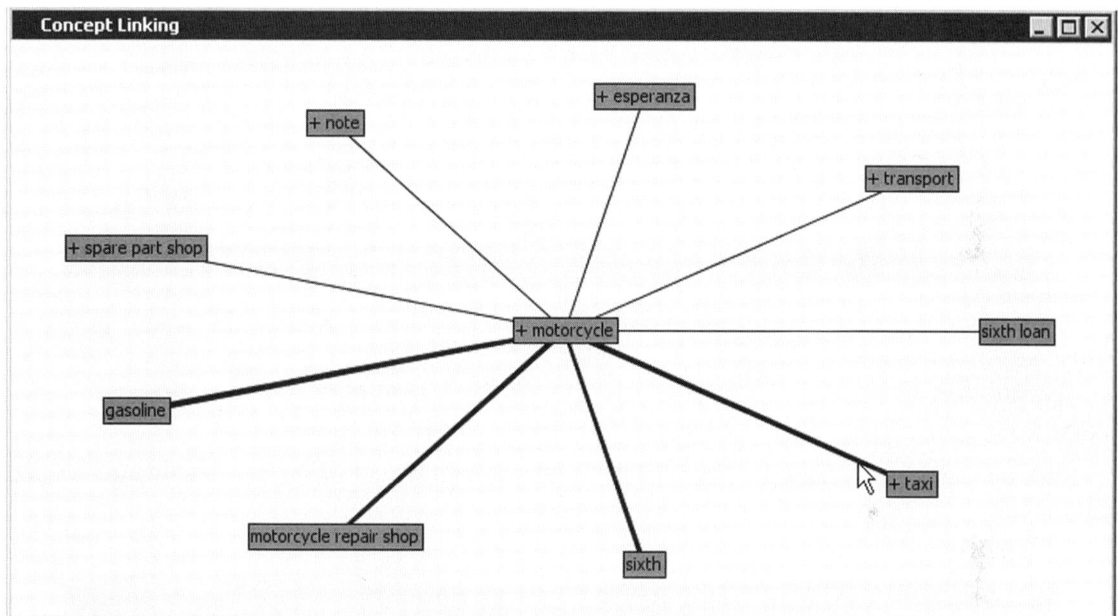

FIGURE K.11
Concept linking.

In order to verify the relationship between "motorcycle" and "taxi," right click on "motorcycle" in the terms window, but this time select Search. This will add "motorcycle" to the search term. The "+motorcycle" means return all values of motorcycle regardless of spelling or stemming. In the search box after "motorcycle." type in "&taxi." This will get all of the values of "motorcycle" wherever "taxi" is mentioned. Hit Enter.

Examine the documents, and you will see, as shown in Figure K.11 that motorcycle taxi businesses are listed. Thus, "motorcycle taxi" needs to be called a multiword term.

Select the Term Window again. This time we want to sort the terms alphabetically. Click on the Term column, and this will sort the terms alphabetically. Find the "motorcycle" term again.

Terms

TERM ▲	FREQ	# DOCS	KEEP	WEIGHT	ROLE	ATTRIBUTE
motor taxi service	3	3	☐	0.0	NOUN_GROUP	Alpha
⊞ motor vehicle	7	6	☑	0.805	NOUN_GROUP	Alpha
motor vehicle engineering	1	1	☐	0.0	NOUN_GROUP	Alpha
motor vehicle garage	1	1	☐	0.0	NOUN_GROUP	Alpha
motor vehicle spare	3	2	☐	0.0	NOUN_GROUP	Alpha
motor water	1	1	☐	0.0	NOUN_GROUP	Alpha
⊞ motor water pump	2	2	☐	0.0	NOUN_GROUP	Alpha
motor-driven	1	1	☐	0.0	Adj	Mixed
motor-driven sprayer	1	1	☐	0.0	NOUN_GROUP	Mixed
motorbike business	1	1	☐	0.0	NOUN_GROUP	Alpha
motorbike repair shop	1	1	☐	0.0	NOUN_GROUP	Alpha
⊞ motorbike spare part	1	1	☐	0.0	NOUN_GROUP	Alpha
⊞ motorcycle	92	59	☑	0.567	Noun	Alpha
⊞ motorcycle	7	7	☑	0.783	Verb	Alpha
⊞ motorcycle driver	1	1	☐	0.0	NOUN_GROUP	Alpha
motorcycle maintenance	1	1	☐	0.0	NOUN_GROUP	Alpha
motorcycle repair shop	5	5	☑	0.821	NOUN_GROUP	Alpha

Click on the plus sign next to "motorcycle," and you will see all of the possible misspellings of "motorcycle."

Terms

TERM ▲	FREQ	# DOCS	KEEP	WEIGHT	ROLE	ATTRIBUTE
motor vehicle engineering	1	1	☐	0.0	NOUN_GROUP	Alpha
motor vehicle garage	1	1	☐	0.0	NOUN_GROUP	Alpha
motor vehicle spare	3	2	☐	0.0	NOUN_GROUP	Alpha
motor water	1	1	☐	0.0	NOUN_GROUP	Alpha
⊞ motor water pump	2	2	☐	0.0	NOUN_GROUP	Alpha
motor-driven	1	1	☐	0.0	Adj	Mixed
motor-driven sprayer	1	1	☐	0.0	NOUN_GROUP	Mixed
motorbike business	1	1	☐	0.0	NOUN_GROUP	Alpha
motorbike repair shop	1	1	☐	0.0	NOUN_GROUP	Alpha
⊞ motorbike spare part	1	1	☐	0.0	NOUN_GROUP	Alpha
⊟ motorcycle	92	59	☑	0.567	Noun	Alpha
motorbikes	6	4			Noun	
motorbike	35	22			Noun	
motor cycle	2	1			NOUN_GROUP	
motorbicycle	2	1			Noun	
motorcycles	13	11			Noun	
motorcycle	34	28			Noun	

You will also notice many variations of motorcycle taxi service.

Terms

TERM ▲	FREQ	# DOCS	KEEP	WEIGHT	ROLE	ATTRIBUTE
⊞ motorbike spare part	1	1	☐	0.0	NOUN_GROUP	Alpha
⊞ motorcycle	92	59	☑	0.567	Noun	Alpha
⊞ motorcycle	7	7	☑	0.783	Verb	Alpha
⊞ motorcycle driver	1	1	☐	0.0	NOUN_GROUP	Alpha
motorcycle maintenance	1	1	☐	0.0	NOUN_GROUP	Alpha
motorcycle repair shop	5	5	☑	0.821	NOUN_GROUP	Alpha
⊞ motorcycle taxi	12	10	☑	0.0	NOUN_GROUP	Alpha
⊞ motorcyclist	2	2	☐	0.0	Noun	Alpha
⊞ motorize	4	4	☑	0.845	Verb	Alpha
motorola	1	1	☐	0.0	PROP_MISC	Entity
motto	2	2	☐	0.0	Noun	Alpha
mould	1	1	☐	0.0	Noun	Alpha
⊞ mould	3	3	☐	0.0	Verb	Alpha
moulding	1	1	☐	0.0	Noun	Alpha
moumounan	1	1	☐	0.0	PROP_MISC	Entity
moumounan	1	1	☐	0.0	Prop	Alpha
moumounan aristil	1	1	☐	0.0	PROP_MISC	Entity

Select "motorcycle taxi," and then hold down the Control key, and then select the following terms:

- Motorbike taxi business
- Motorbike transportation
- Motorbike transportation business
- Motorcycle taxi – noun
- Motorcycle taxi – Noun_Group
- Motorcycle taxi business

Right click and select "Treat as synonym." Select the second "motorcycle taxi" term; this will be the noun version of this term, and all of the other terms will now roll up into the term "motorcycle taxi."

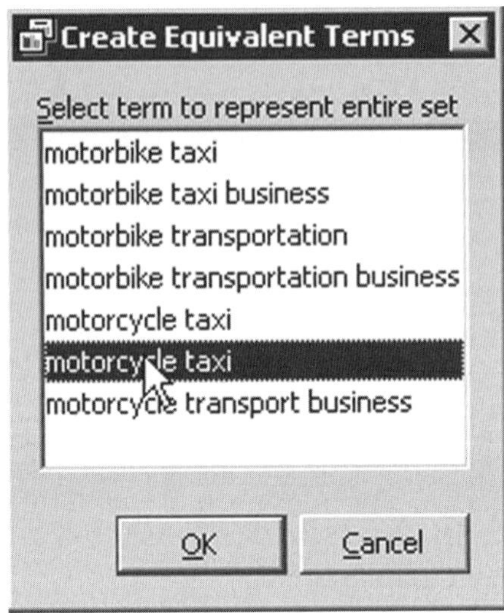

Make sure the "Keep" box is checked

DEFINING TOPICS

Select the Text Topic node and click the "Run" icon. This will define the topics for this flow. Go to the property sheet, and scroll down to Topic Viewer and click on the ellipses.

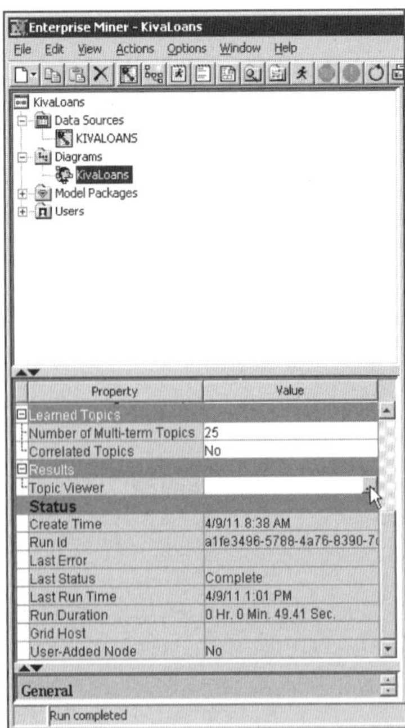

The Interactive Topic Viewer appears.

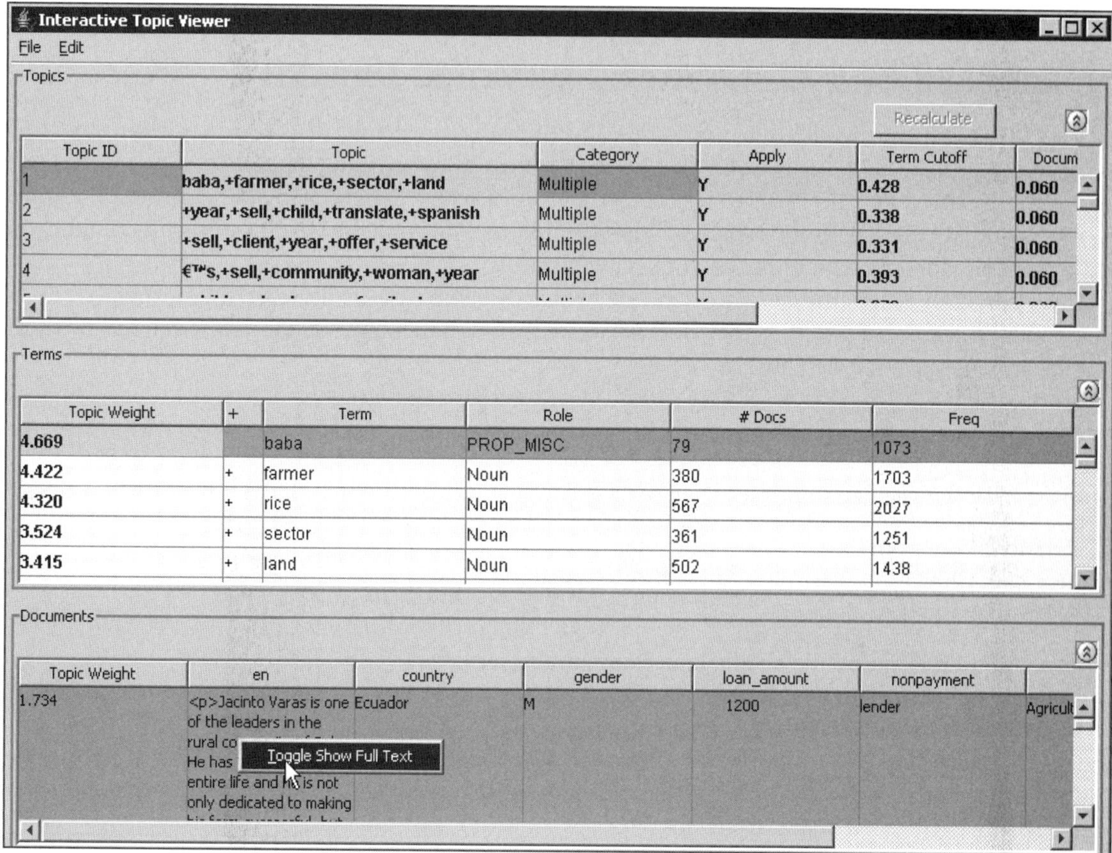

The Topic Viewer shows the topics that the SVD revealed, and, as you can see, some of the topics have terms that need more investigation, such as baba. If you select the Documents section and right click on the end column, you will be able to toggle between showing the full text or just showing a snippet of the text.

There are also bad character terms in the data set that are causing issues. These characters are because of the non-English characters and terms that are in the data set. (The data are translated from the original language of the region, and many of the words have no English equivalent or the translator left in the original spelling or symbol of the word.)

By showing the full text of documents in a topic, you can get an idea of the topic classification. As you can see, the first classification is about rice farming. However, we want to remove or change how some of the terms are used, such as define a multiterm word or add to a synonym list. This will allow for better classification.

Our text mining process is done. We have created topics and classified the documents. As you can see, text mining is an extremely iterative process. You can analyze the terms and the concepts looking for information that adds noise to the data, is a synonym, or is a multiword term.

OTHER USES OF THE INTERACTIVE TOPIC VIEWER

We can lower the term weights in each topic by selecting the Topic in the term section and then changing the topic weight. We can also change the term cutoff by selecting the term cutoff and changing the weight. By increasing the term weight cutoff, you are effectively decreasing the number of terms in the topics and also changing the weights of the documents. Affecting the weights of the documents can add or remove documents within the Topic classification.

You can also change the document cutoff, which will increase or decrease the number of documents in that topic. To simplify the process, we will not change any of the term weights or cutoff weights and just take the defaults.

MAKING THE PREDICTIVE MODEL

In this section we use decision trees to make a predictive model. We will make three decision trees:

- Relational data only
- Text data only
- Relational and Text data only

We will then examine the results and compare the models.

Relational Decision Tree

The first decision tree we create is the relational decision tree. Based on Figure K.13 it seems that Country will play an important role in the predictive model, but we'll also want to see what else there could be so we can make better decisions.

Select the Model tab.

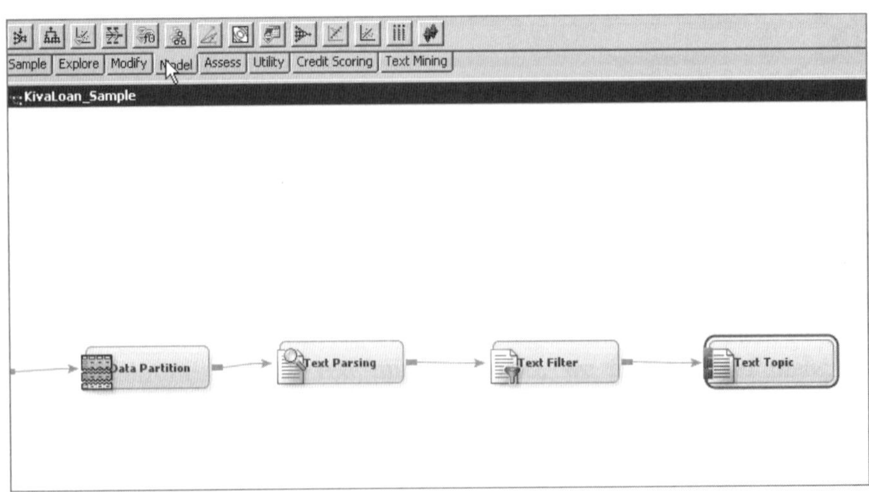

Select the Decision Tree icon, and drag the icon onto the pallet to create a Decision Tree node.

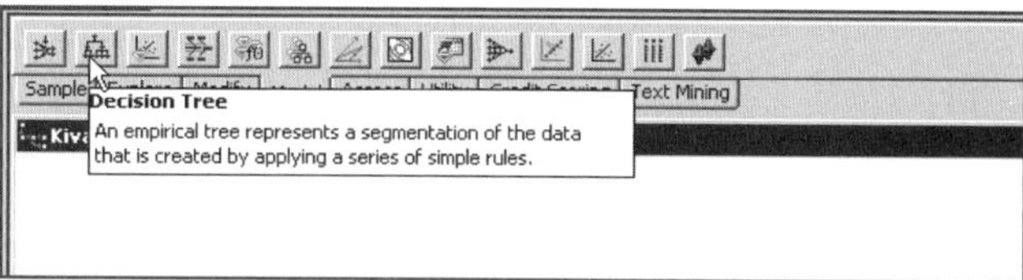

Connect the KivaLoans data set icon to the Decision Tree, as shown in Figure K.12. Doing so eliminates all of the information obtained through text mining, treating all the information as if it were relational.

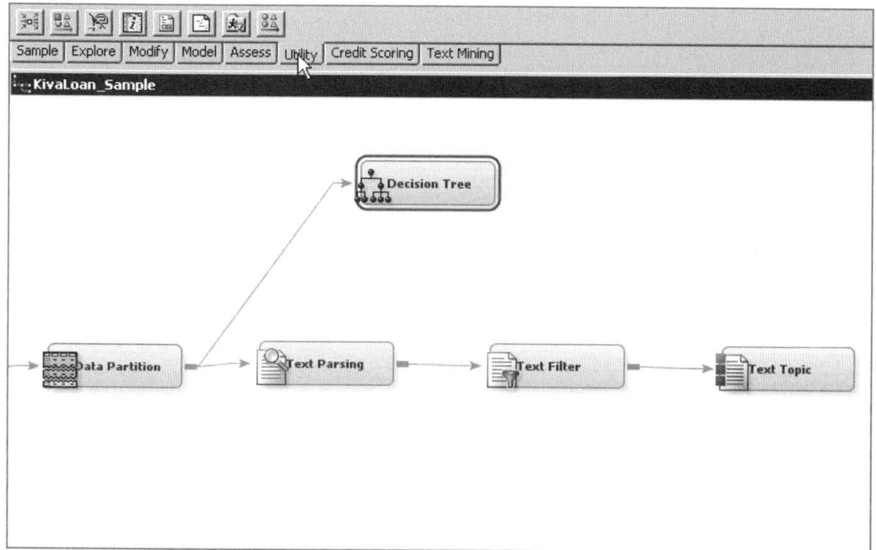

FIGURE K.12
Decision Tree.

Run the Decision Tree node by selecting the Decision Tree node and clicking on the "Running Person" as before.

When the run is finished, view the results. The diagram comes up. Click on the Tree Diagram as shown in Figure K.13.

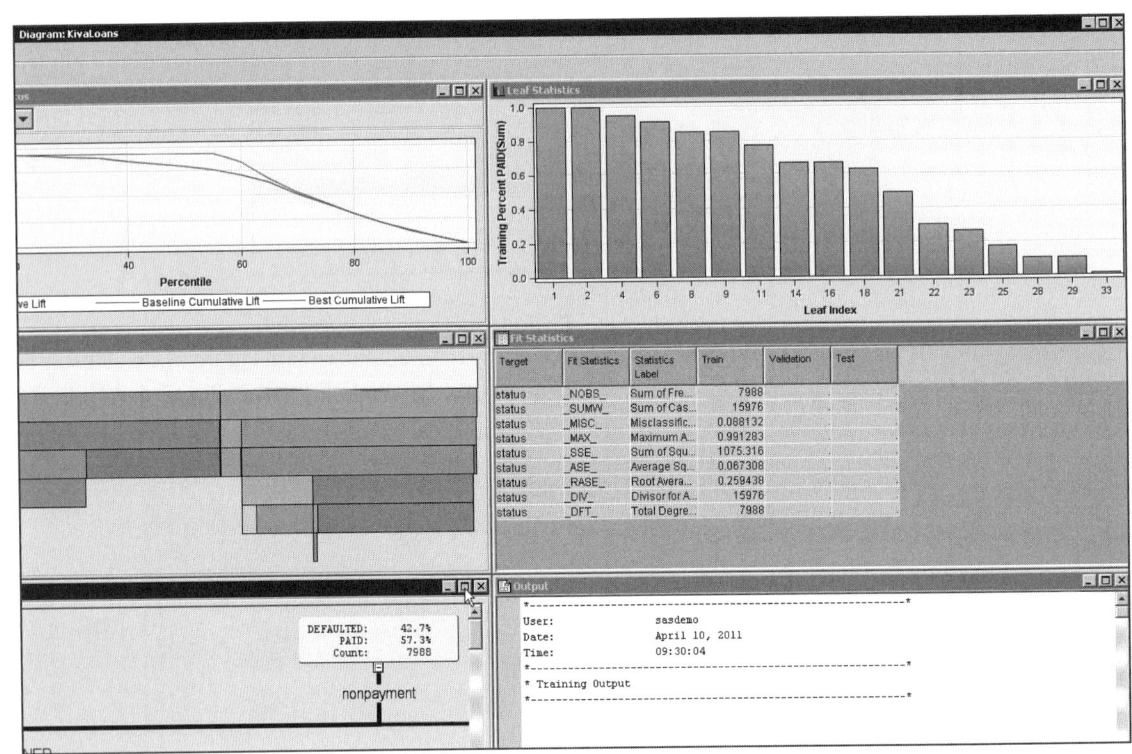

Target	Fit Statistics	Statistics Label	Train	Validation	Test
status	_NOBS_	Sum of Fre...	7988	.	.
status	_SUMW_	Sum of Cas...	15976	.	.
status	_MISC_	Misclassific...	0.088132	.	.
status	_MAX_	Maximum A...	0.991283	.	.
status	_SSE_	Sum of Squ...	1075.316	.	.
status	_ASE_	Average Sq...	0.067308	.	.
status	_RASE_	Root Avera...	0.259438	.	.
status	_DIV_	Divisor for A...	15976	.	.
status	_DFT_	Total Degre...	7988	.	.

FIGURE K.13

Decision Tree Viewer.

As you can see from Figure K.14, the primary factors for loan default or risk are:

- Nonpayment: Who takes on the risk—the lender/bank or a partner organization? The lender takes on more risk.
- Country: As stated before, the country is a strong indicator of risk.
- Loan amount: Larger loans tend to have a better chance of being paid than smaller loans.

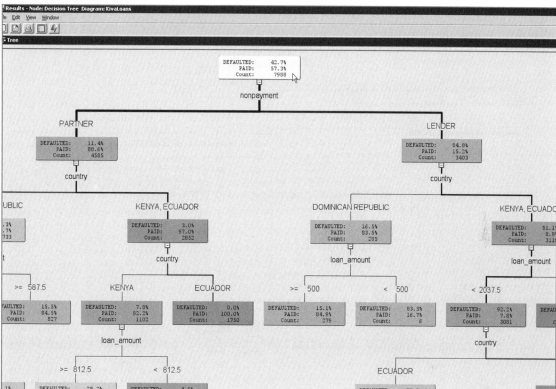

FIGURE K.14
Relational Decision Tree results.

Now we will look at how text alone predicts. This analysis will not use any relational data, so first we will remove all the relational data (you can do this from the Decision Tree node, but this method highlights the metadata change in the data mining flow).

First select the Utility tab as seen in Figure K.15. Then select the Metadata icon—the icon with an italic *i*. The Metadata node lets you change variable information or variable metadata.

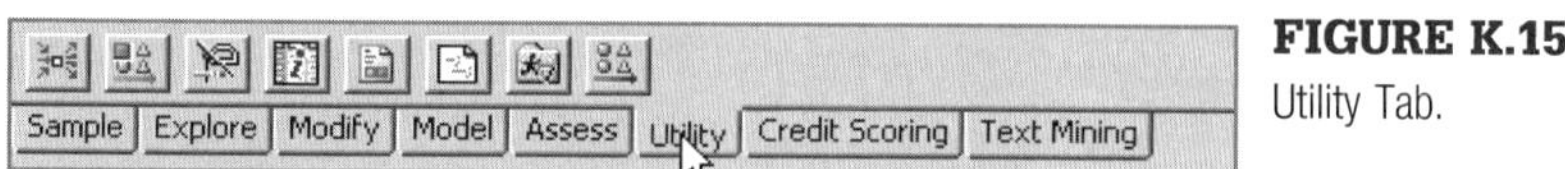

FIGURE K.15
Utility Tab.

Then drag the icon on to the palette and connect the Metadata node to the Text Topic node as shown in Figure K.16.

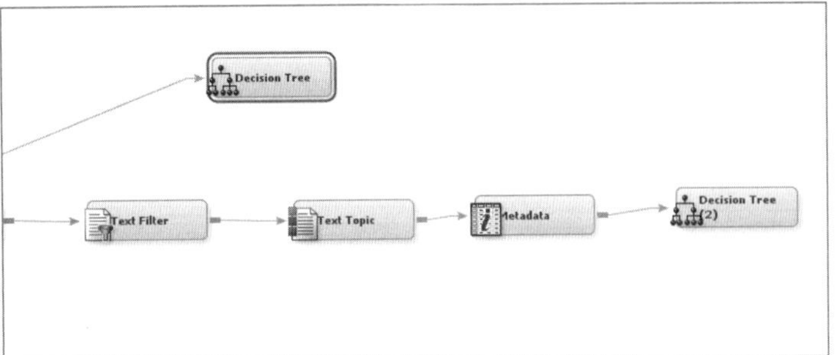

FIGURE K.16
Text Mining flow with Metadata node.

In the property sheet on the left, select the ellipses associated with Train. In the Meta Viewer highlight the following variables:

- Country
- En
- Gender
- Loan_amount
- Nonpayment
- Sector

Then click the drop-down box in the New Role column and change Default to Rejected as in Figure K.17.

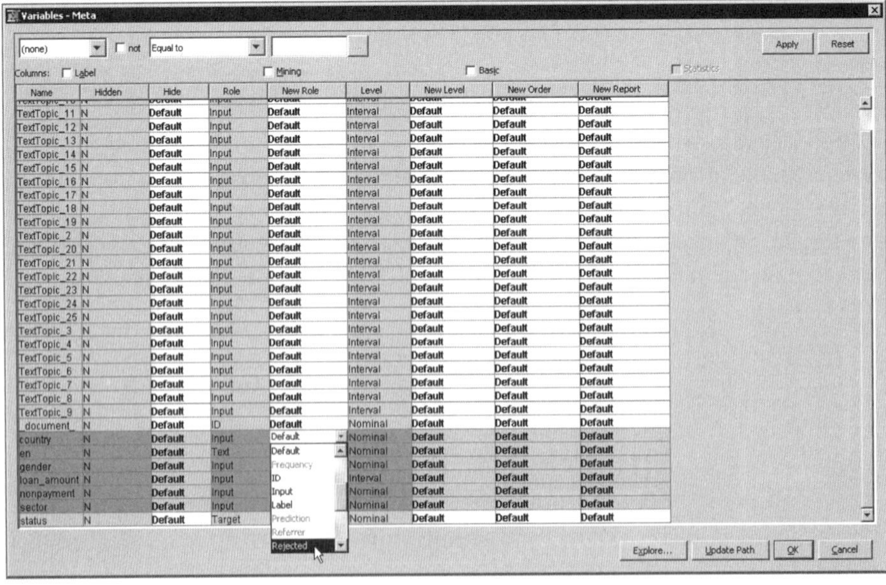

FIGURE K.17
Changing Metadata.

This will remove all the structured values from the analysis and allow us to only look at the Text Mining results.

Next, drag a Decision Tree onto the pallet and connect the Metadata node with the new Decision Tree node so the diagram looks like Figure K.18.

You can run the node and view the results as before. However, there is a lot more text that needs to be put in the stoplist and terms that need to be treated as multiword terms or used as synonyms. We will save the results until later in this section.

Next, drop a new Decision Tree onto the pallet and connect the Text Topic node to the New Decision Tree Node. Because we are not removing any of the variables from the analysis, this decision tree will use both structured and unstructured information in the analysis.

If you right click in any of the nodes, you can rename the node to something more descriptive. This has been done in Figure K.14.

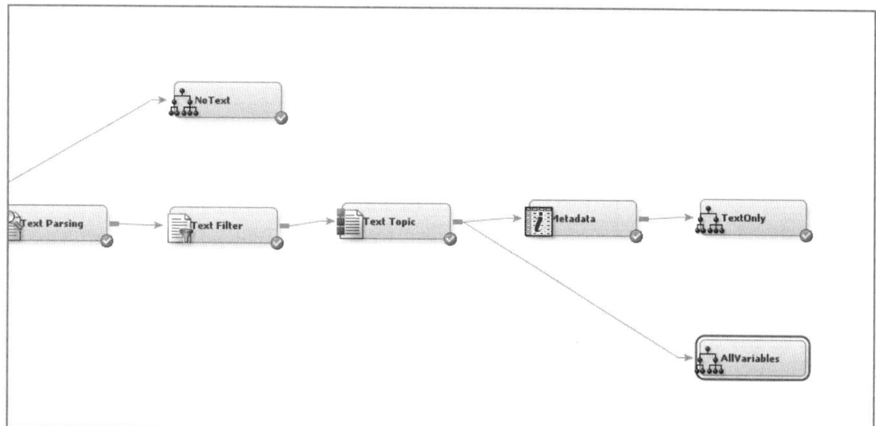

FIGURE K.18
Text Mining flow with Decision Trees.

Next we want to compare the models to see which model is the champion. Select the Assess tab as shown in Figure K.19.

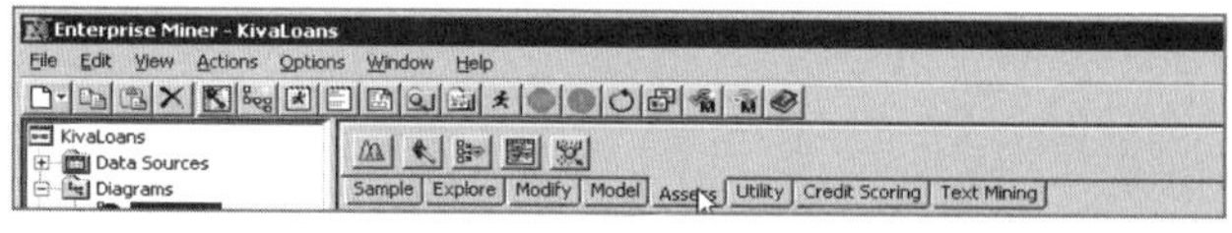

FIGURE K.19
Assess tab.

Next select the Model Comparison icon as shown in Figure K.20.

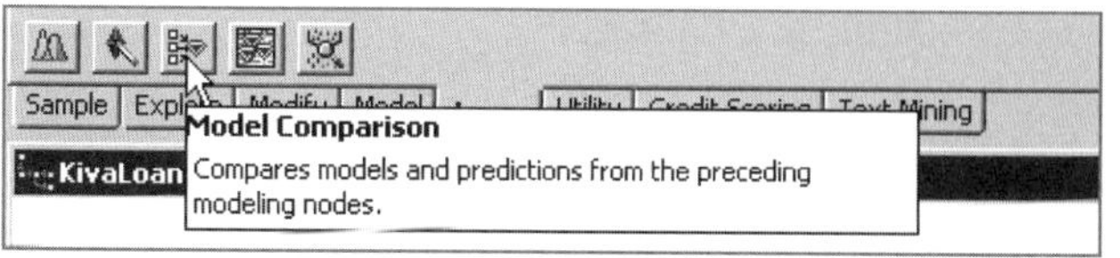

FIGURE K.20
Model Comparison node.

Then drag the icon onto the pallet and connect all of the decision trees to the model comparison node, as shown in Figure K.21. You will have the finished Data Mining flow diagram.

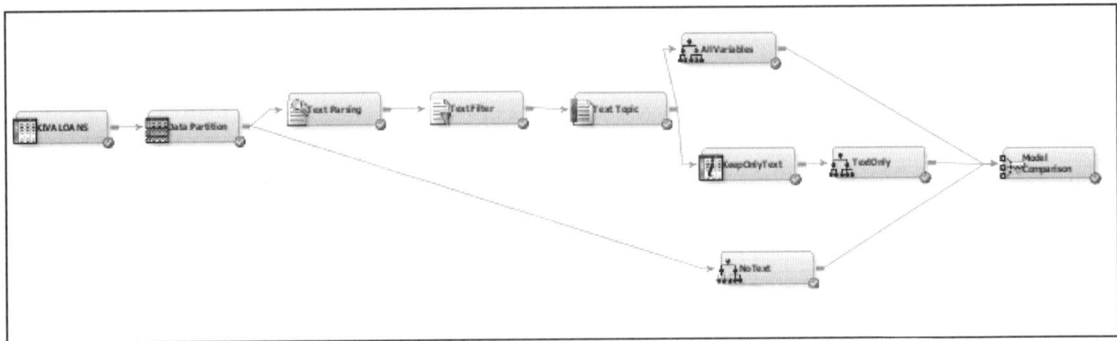

FIGURE K.21
Completed Text/Data Mining Flow Diagram.

FINAL RESULTS

As stated earlier, other terms needed to be added to the stoplist or used as multiword terms or used as synonyms to improve the model. A SAS® Text Mining flow has already been done. In this section we will load the prerun node and look at the results.

Adding the Prebuilt Model

In the first section, you copied the TMProjects folder into your C-drive. In the TMProjects folder, there is a subdirectory called "KivaLoans." Make sure this folder is there. If the folder is not there, find the folder and copy it to the TMProjects directory.

Next, Open SAS® Enterprise Miner, and, as before, create a new project. This time when asked to enter the project name on the second screen, type in "KivaLoans," and point the SAS® Server Directory to c:\TMProjects\. SAS® Enterprise Miner will recognize that there is already a current project located at C:\TMprojects\ with the name KivaLoans.

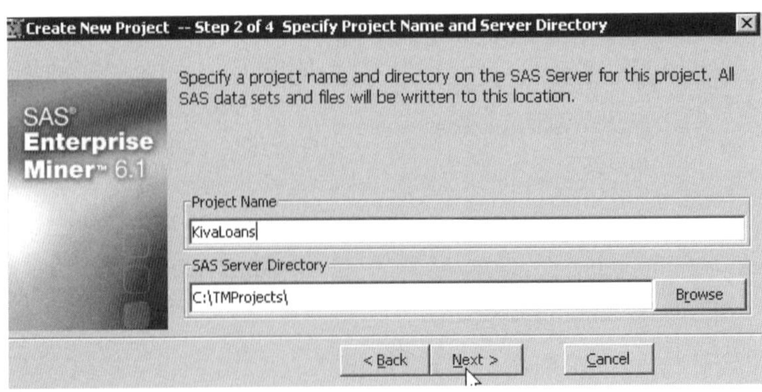

Click Next. You will see a window notifying you that the project already exists and asking if you would like to continue.

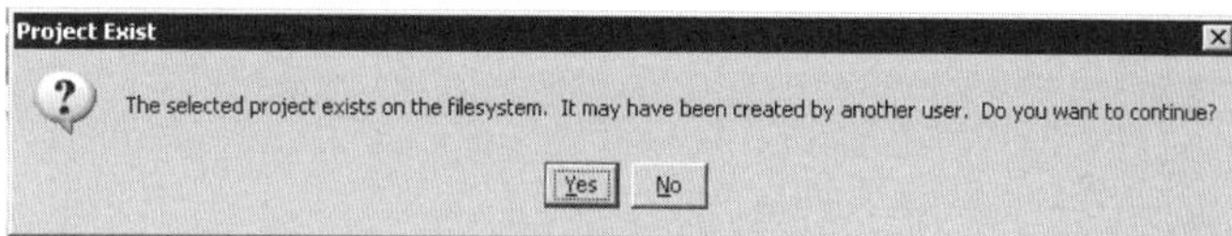

Click Yes, and finish by selecting all of the defaults. KivaLoans is now a project in your SAS® Enterprise Miner. The next step is to validate the data source location.

First, click on the new icon as if you are about to create a new SAS® Library. Select Library.

You will be taken to the Library Wizard. Find the Kiva Library and make sure it is pointing to c:\TMProjects\data. If not, then select the Modify Library option and click Next.

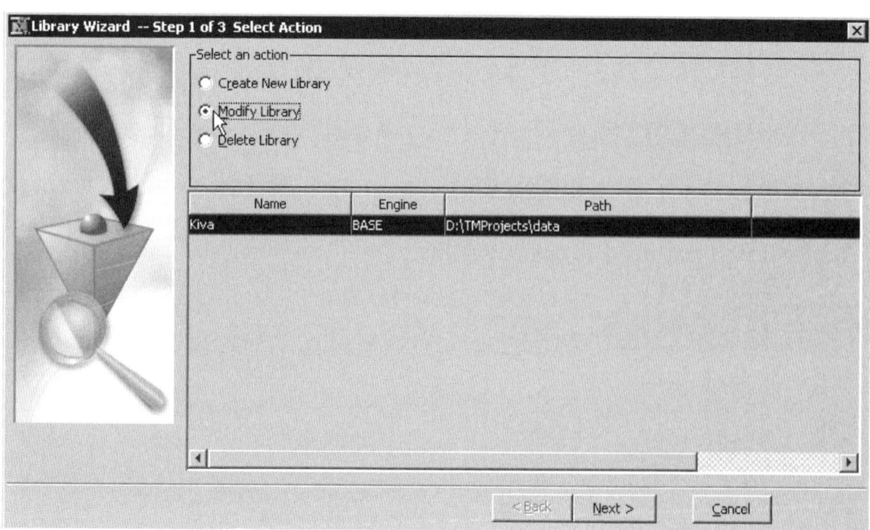

Change the path to the location of your TMProjects\data directory. In this case, it is C:\TMProjects\data.

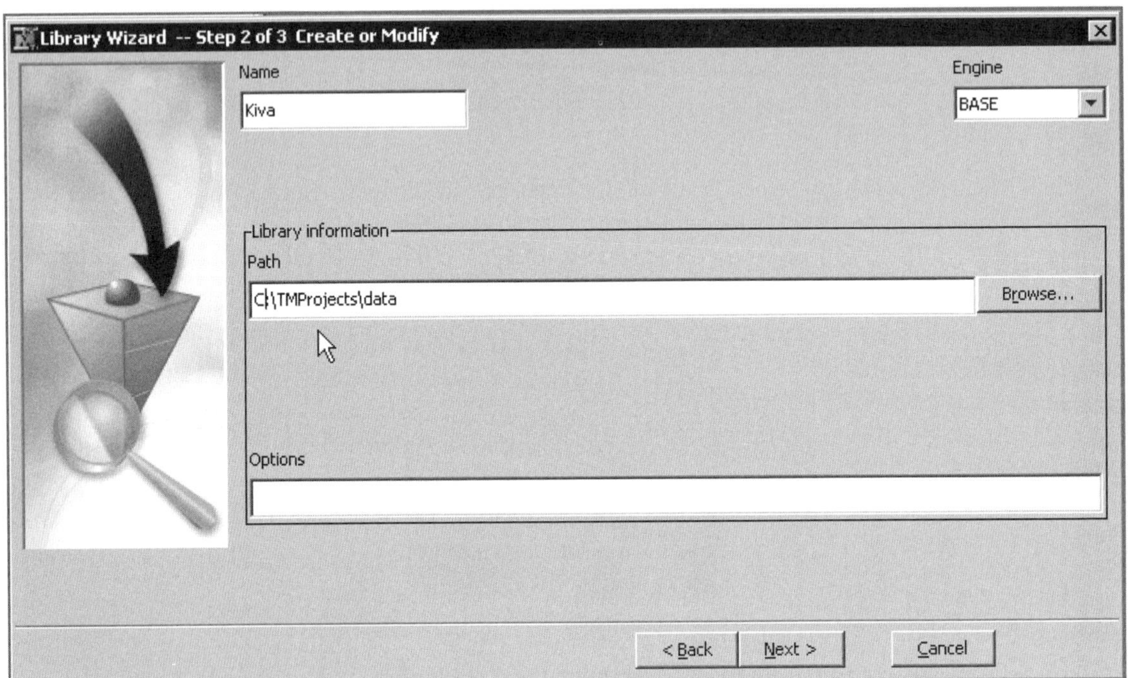

Click Next and select defaults. The flow is now ready to be modified and/or run.

VIEWING THE REPORTS

In this section we will examine the results of the decision trees. The flow was already run when it was loaded, so there is no need to run it again. If you right click on any of the decision trees, a drop-down window appears, and you just need to select Results to view the analysis.

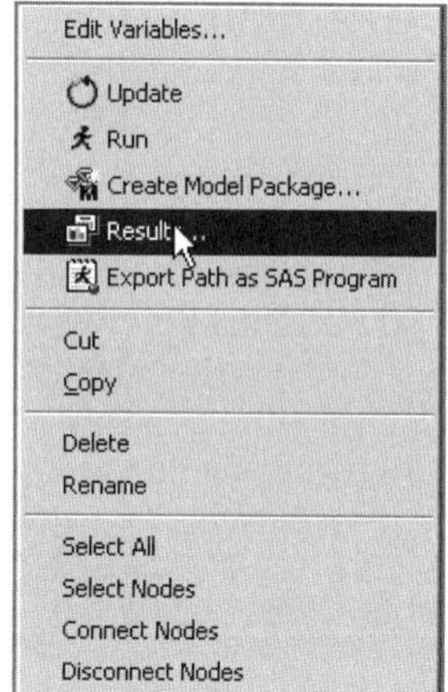

The Analysis

In order to simplify the flow, a different format was used, which changes the raw SVD values to binary SVD values. A "1" represents documents that are in that category, and a "0" represents documents that are not in the category. For the decision tree numbers, less than .5 are in the category numbers, and more than .5 are not in the category. This also affects the model somewhat. You get better performance if the raw SVD numbers are used. I will show the comparison at the end. In the next release of SAS® Text Miner, SAS® Text Miner 5.1, SAS® defaults to the raw SVD values for this type of analysis.

TEXT ONLY DECISION TREE

In the Text Only Decision Tree (Figure K.22), we see that there are not large differentiators. The School split had just over 60 Pay if they are in the category, and

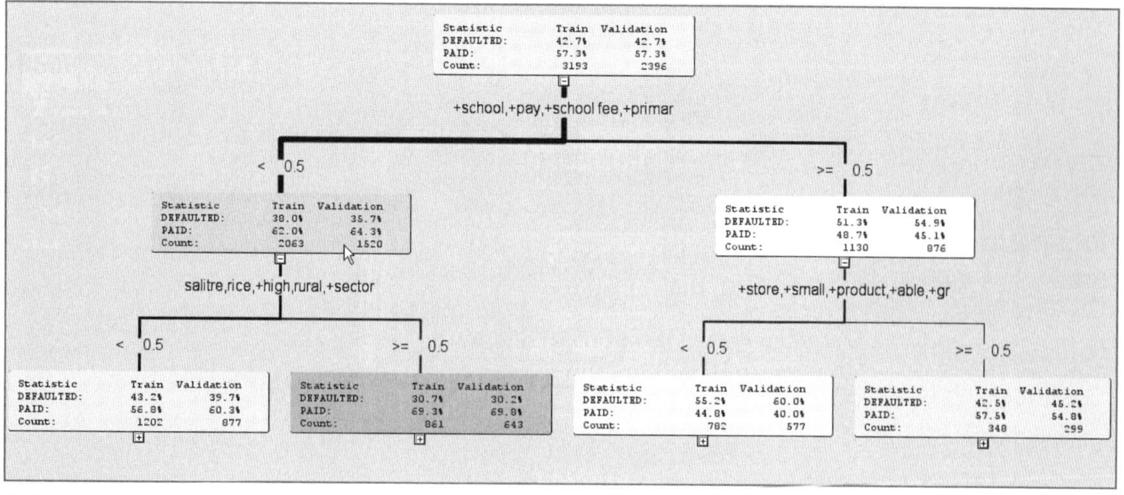

FIGURE K.22
Text Only Decision Tree.

ALL VARIABLE TEXT AND RELATIONAL

Text does start to effect the model. People in the Purchase Machine category start to have a higher propensity to pay (Figure K.23).

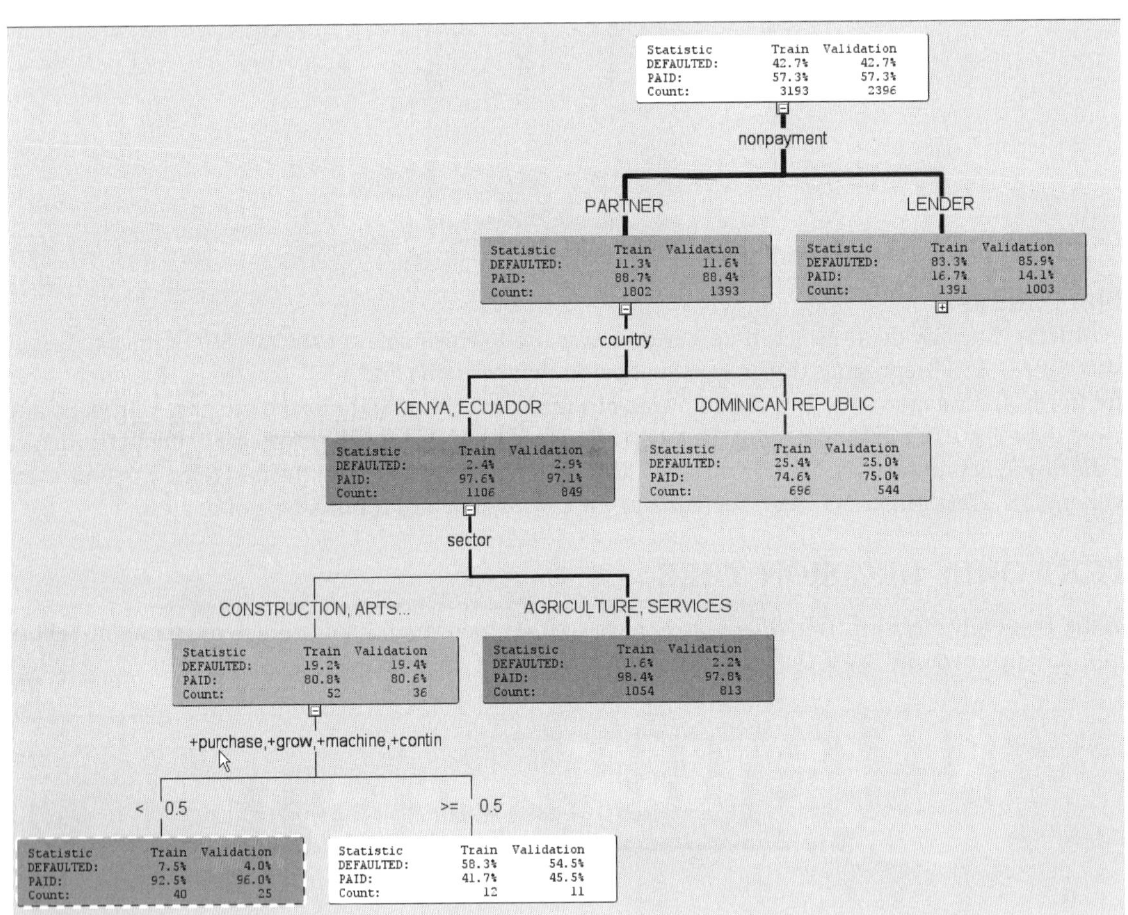

FIGURE K.23
Structured and Unstructured Decision Tree.

When we look at the cumulative effect, AllVariables and NoText perform about the same as you would expect; however, more information is provided with AllVariables than with NoText. This extra information can help make better decisions, as shown in Figure K.24.

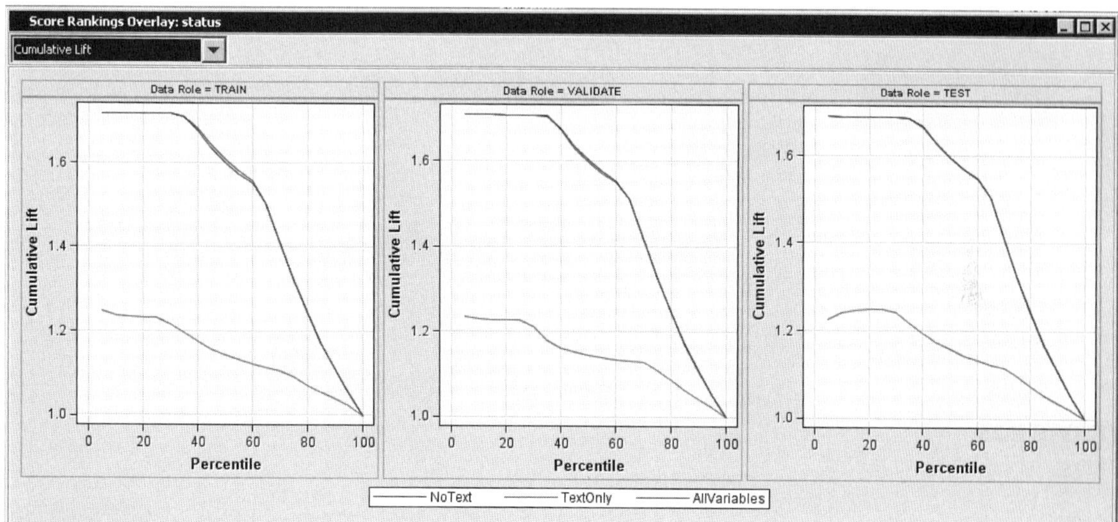

FIGURE K.24
Model comparison.

The raw values of the rotated SVD do no rounding of the SVD values. When using the raw values, the SVD vector does not have a simple binary split of 1's and 0's. The data are a set of rational numbers that represent the cluster. The cluster is still represented correctly, but it is just that the raw values provide better weighting of the terms and documents, whereas the binary provides a simpler representaion of the data.

By looking at the Decision Tree in Figure K.25, we see that Text is a very strong predictor of what type of loan will be paid or default. The other surprising thing to note is that Country is not a factor; the text data dominate the model.

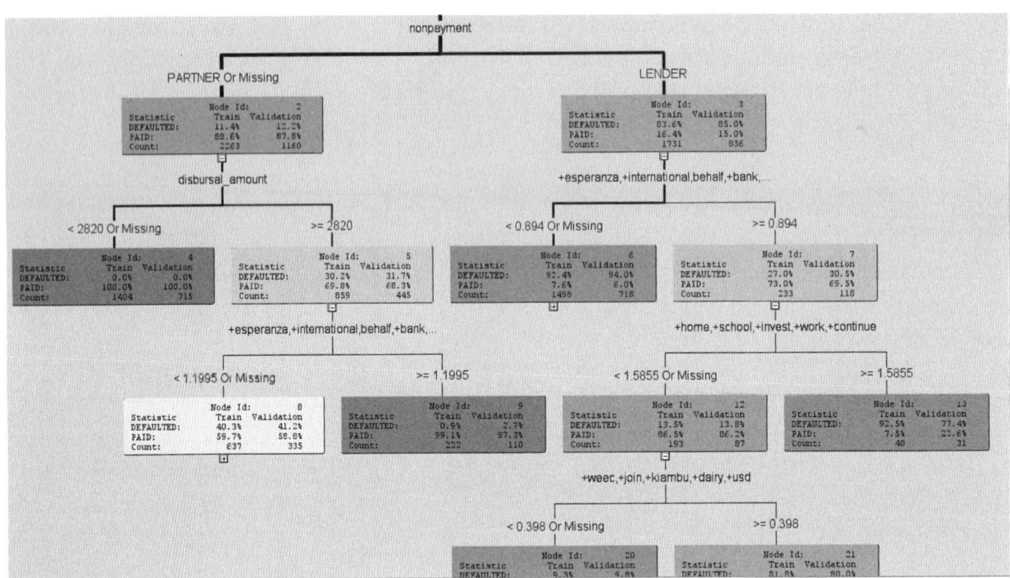

FIGURE K.25
Unstructured and Structured raw SVD.

When we compare the models between the Decision Trees (Figure K.26):

- Text only performs the worst.
- Structured only performs better.
- Structured and Unstructured outperform them all.

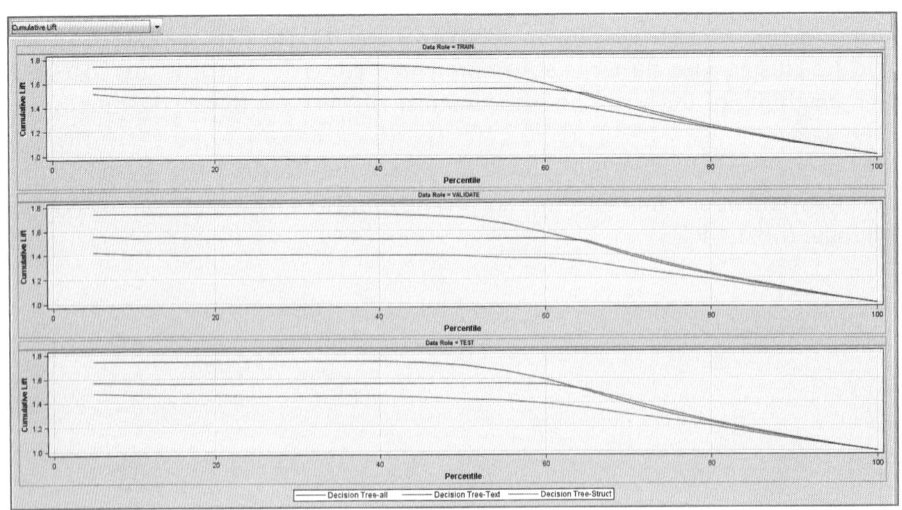

FIGURE K.26
Model comparison of raw SVD values.

CONCLUSION

The analysis of unstructured data requires a lot of iterations to completely filter out the information. There is a lot of noise in data, as shown in the SAS® Text Parsing node. There are concepts and synonyms that need to be addressed to make the topics and classifications more accurate and meaningful, as discovered in the SAS® Text Filter node.

The proper use of unstructured data can uncover meaningful information where there was no insight before. Then, when combined with relational data, it can add more detailed information as well as improve predictive models.

We were able to show what type of business people are investing in and what information can predict if a loan will default or be paid. This is useful information for anyone making a loan, even if it is only a $25 loan.

Opera Lyrics: Text Analytics Compared by the Composer and the Century of Composition—Wagner versus Puccini

Gary Miner

Opera lyrics are fairly simple in one sense. They all seem to have to do with love, remorse, trust, and similar interpersonal communications set in a scene from the sea, to a Paris neighborhood, to a serious Wagnerian German scenario. Yet, intuitively we'd expect that the lyrics to operas could be quite different among different centuries of the writing and also among different composers. Most composers chose lyricists to write the words, but some composers also wrote the words to their music. With this intriguing idea, we set out to see if we could generate patterns of words from different composers.

For this tutorial we only had time to compare three of Wagner's operas with three of Puccini's. But you, the reader, after going through this tutorial, will understand the process and could extend the number of composers and centuries into more interesting studies. These could also be extended to the languages in which the opera was written and usually the language of a performance.

To generate the text data for this tutorial, the words in English of selected operas were typed into txt documents (we used Notepad, found as an "accessory tool" in Microsoft Office software, to store the txt files). In this study we had to actually type the words into the txt documents, as the originals were not in a format easily scanned into a PDF or text document. Only the first acts of the operas were used. Each opera was put into one txt document. Thus, since we used only two composers and three operas each in this study, there are only six txt documents. These documents are given a name, and this name is the "handle" used in the data spreadsheet for the variable column "Name of Text File," as illustrated in Figure L.1. When the Text Mining program performs its computations, it uses this text file name to go to the computer folder where it is stored and bring the entire text into the analysis.

In Figure L.1, you will note seven variable columns: Name of Opera, Composer, Act, Lyricist, Year Written, Century Written, and Name of Text File. The Lyricist column is empty of data because we decided not to use this for the tutorial. However this variable, and possibly many more "numeric type variables" that you, the reader, can think of, could be added to this type of study.

To begin this tutorial, put the DVD that came with this book in your DVD drive and find the dataset in the "Tutorial L" folder. The dataset has the name "Opera—Wagner vs Puccini_TEXTMINING.sta." This is a *STATISTICA* data set. If you need to do this tutorial in another data analysis program, you can install the *STATISTICA* program, also found on the DVD, then open this file, and then do a "Save As" to another of the many other data file types available for other software programs. Then you will have the data set available for analysis in this other program.

Practical Text Mining and Statistical Analysis for Non-structured Text Data Applications. DOI: 10.1016/B978-0-12-386979-1.00020-7

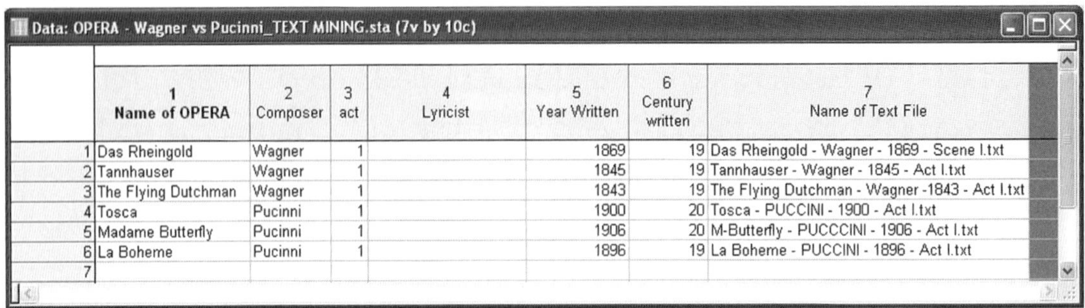

Data: OPERA - Wagner vs Pucinni_TEXT MINING.sta (7v by 10c)

	1 Name of OPERA	2 Composer	3 act	4 Lyricist	5 Year Written	6 Century written	7 Name of Text File
1	Das Rheingold	Wagner	1		1869	19	Das Rheingold - Wagner - 1869 - Scene I.txt
2	Tannhauser	Wagner	1		1845	19	Tannhauser - Wagner - 1845 - Act I.txt
3	The Flying Dutchman	Wagner	1		1843	19	The Flying Dutchman - Wagner -1843 - Act I.txt
4	Tosca	Pucinni	1		1900	20	Tosca - PUCCINI - 1900 - Act I.txt
5	Madame Butterfly	Pucinni	1		1906	20	M-Butterfly - PUCCCINI - 1906 - Act I.txt
6	La Boheme	Pucinni	1		1896	19	La Boheme - PUCCINI - 1896 - Act I.txt
7							

FIGURE L.1
Original data set before Text Mining.

In Figure L.2 we see that the "Text & Document Mining" program is found in the Data Mining pull-down menu.

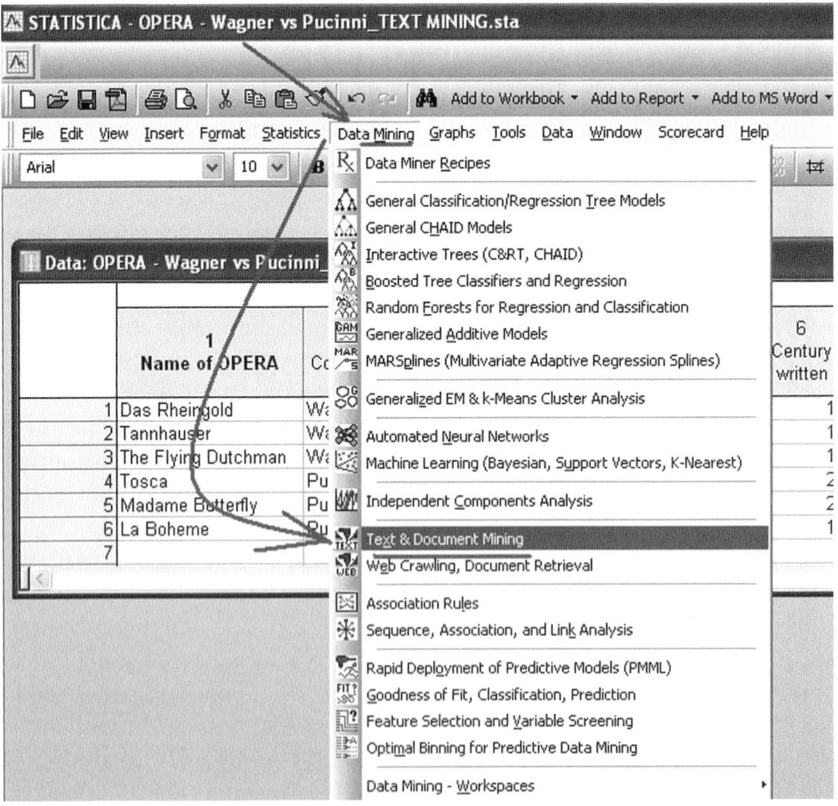

FIGURE L.2
STATISTICA main window with the "Data Mining" pull-down menu selected and the "Text & Document Mining" selection highlighted.

Clicking on this highlighted selection, and the Text Mining window will open. This is illustrated in Figure L.3.

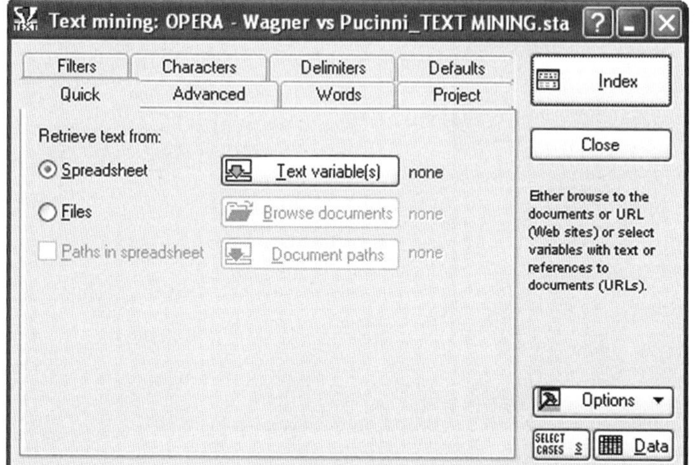

FIGURE L.3
Opening window of the Text Mining application.

Next, we need to select the text data files for the analysis. The first part of this is illustrated in Figure L.4. There are three ways to get text data into the analysis: a spreadsheet, where the text material is actually typed or pasted into the cells of the data sheet; the "files" selection, which when selected allows you to browse your computer folder for the files needed (in this tutorial the six opera txt files); or paths in spreadsheets where we direct the path to a folder containing the files we need.

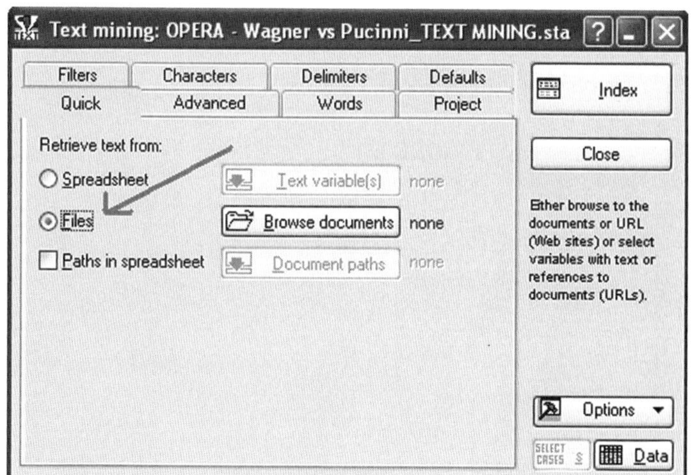

FIGURE L.4
The Quick tab of the Text Miner dialog window with the "Files" radio button selected.

Then click on "Browse documents."

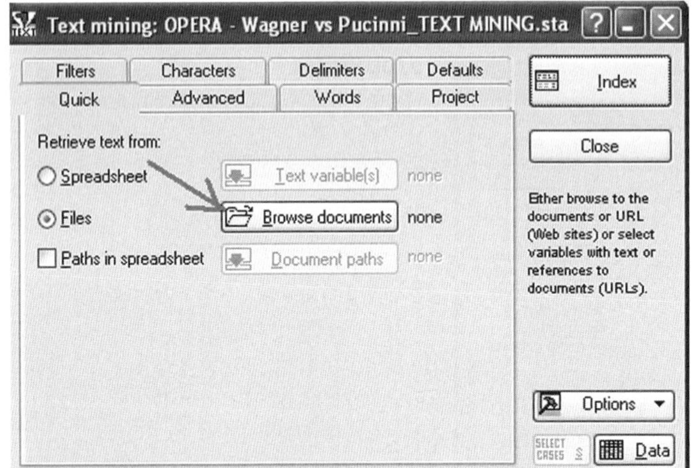

FIGURE L.5
"Browse documents" button on the Quick tab.

This brings up the "Open Document Files" window.

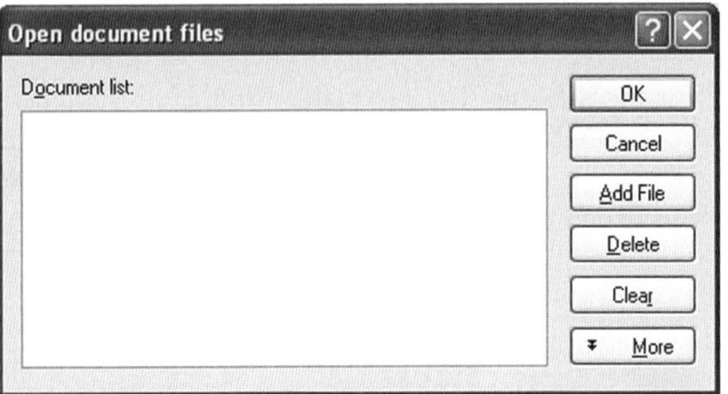

FIGURE L.6
Open document files dialog window.

Then click the "Add File" button in the "Open Document Files" window.

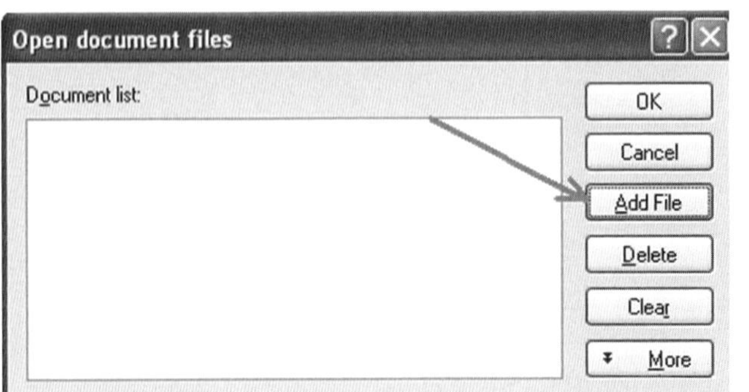

FIGURE L.7
Add File button.

This opens up the actual files on your computer hard drive folder.

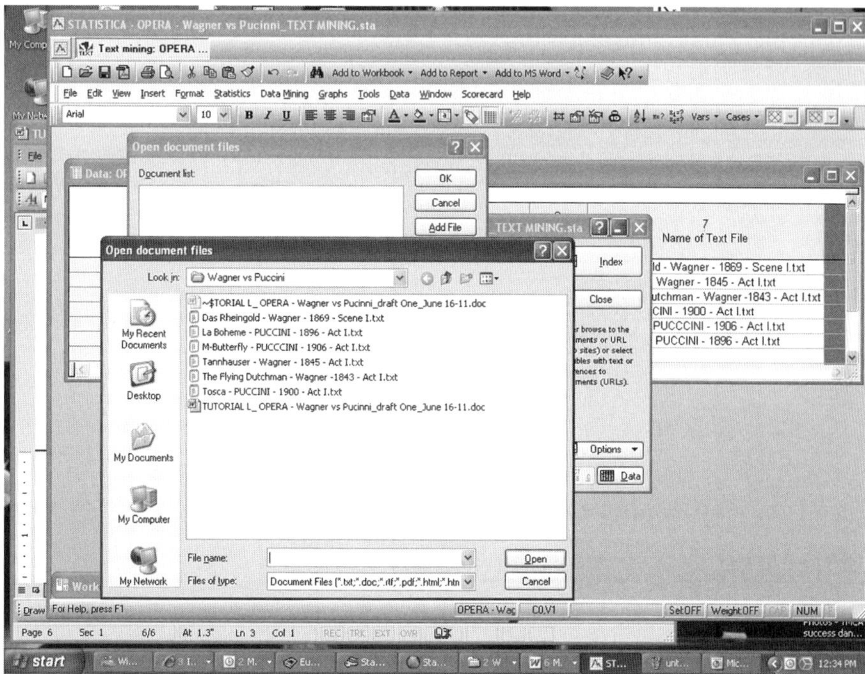

FIGURE L.8

Text files needed for your analysis are in the top window.

Figure L.9 shows the folder's files more closely.

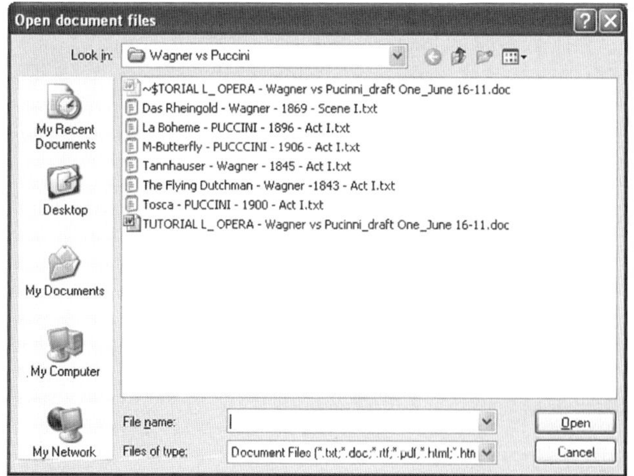

FIGURE L.9

The names of the six text files to be used in the text mining analysis.

Note in Figure L.9 that the "order" of the txt files for these opera lyrics is not in the same order as the "case numbers 1–6" in the data file. Thus, to make it easier to import these files in the correct order for Variable 7 in the spreadsheet, we can add a number to the beginning of each opera in the order of the "cases" in the data file. This is illustrated in Figure L.10.

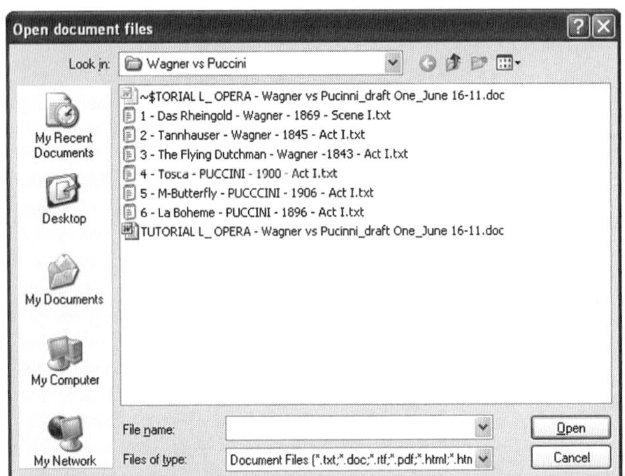

FIGURE L.10
Rearranged order of the six opera act I lyrics to be consistent with how these are ordered in the *STATISTICA* data spreadsheet.

Select the text files "1–6."

Then click on Open (lower right-hand corner of Figure L.11). That will get you to the "Text Miner Open document files window," shown in Figure L.12.

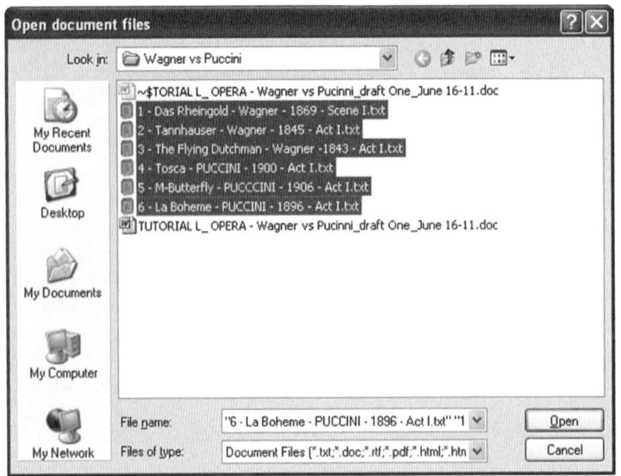

FIGURE L.11
The six text files needed are highlighted.

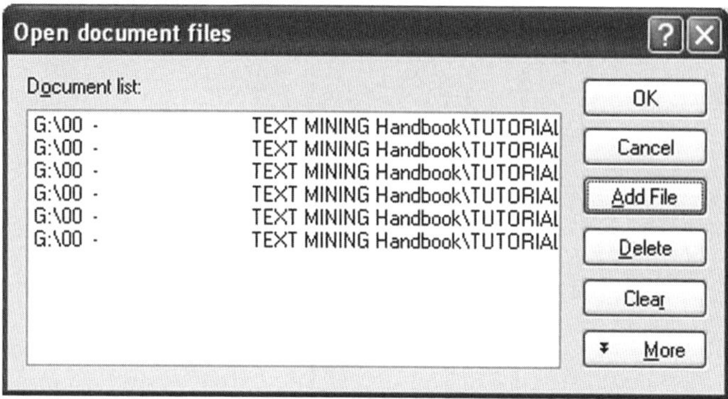

FIGURE L.12
Files inserted into the "Open document files" window.

Now click the "OK" button in the upper right-hand corner of the "Open document files" window, and the documents will be seen as "selected" in the *STATISTICA* Text Miner main dialog window (see Figure L.13).

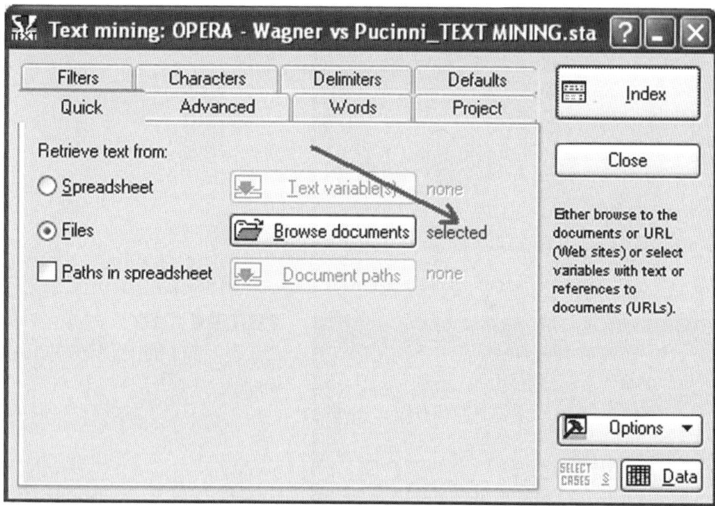

FIGURE L.13
Documents now selected for text analysis.

At this point we'll leave the remaining "tabs" of the Text Miner dialog windows at their defaults, but each are illustrated in the next seven figures so we can all understand what they contain (see Figures L.14 through L.20).

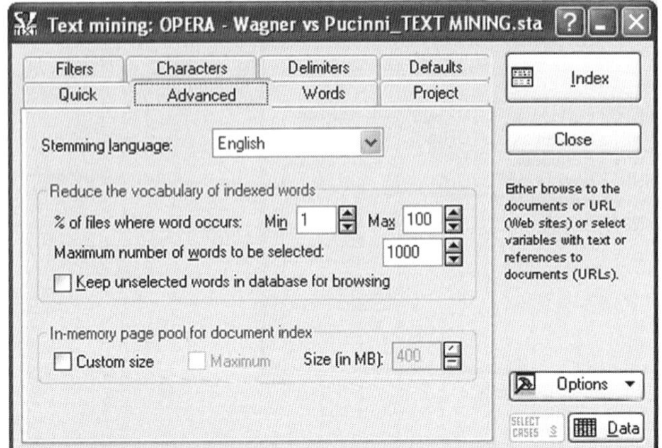

FIGURE L.14
Advanced Tab of the *STATISTICA* Text Miner main dialog window.

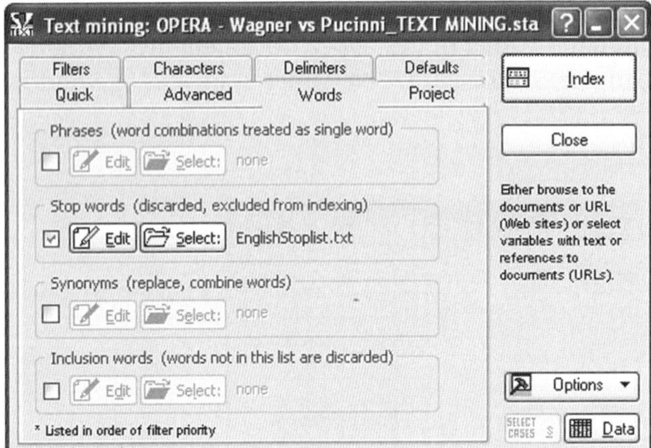

FIGURE L.15
Words tab of the *STATISTICA* Text Miner main dialog window.

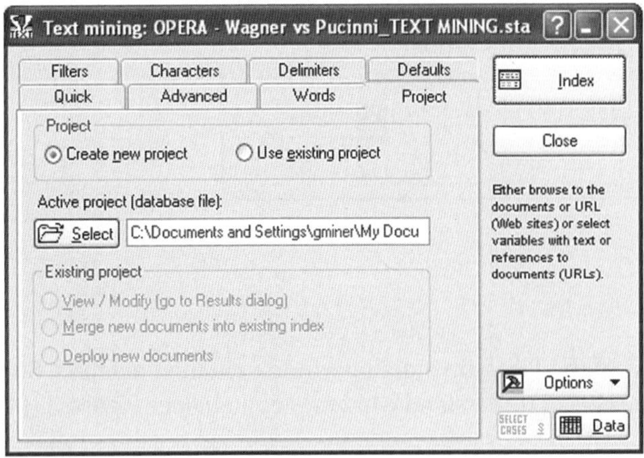

FIGURE L.16
Project tab of the *STATISTICA* Text Miner main dialog window.

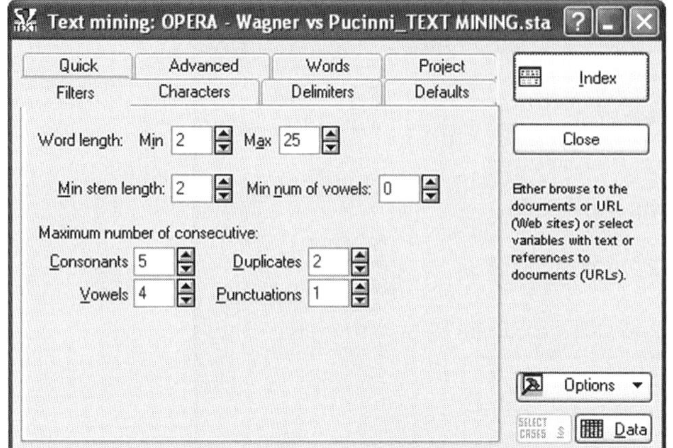

FIGURE L.17

Filters tab of the *STATISTICA* Text Miner main dialog window.

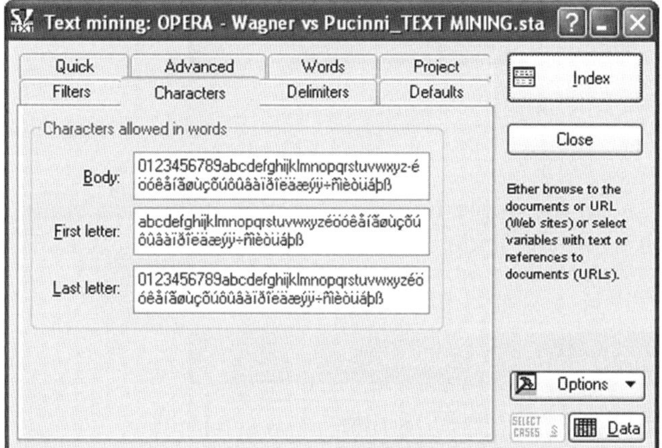

FIGURE L.18

Characters tab of the *STATISTICA* Text Miner main dialog window.

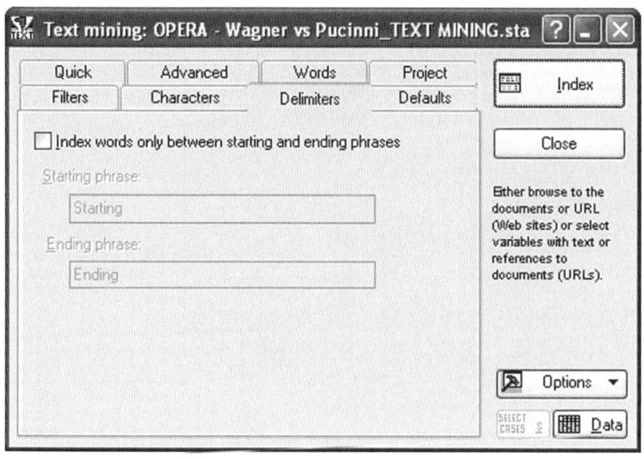

FIGURE L.19

Delimiters tab of the *STATISTICA* Text Miner main dialog window.

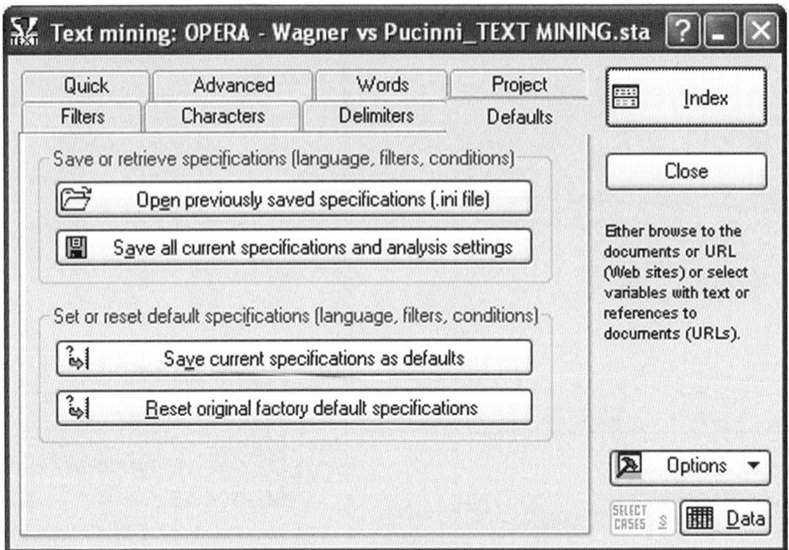

FIGURE L.20
Defaults tab of the *STATISTICA* Text Miner main dialog window.

Now, click the Index button (see Figure L.21) in the upper right-hand corner of the Text Mining main dialog to compute the frequencies of each word (or "index" each word).

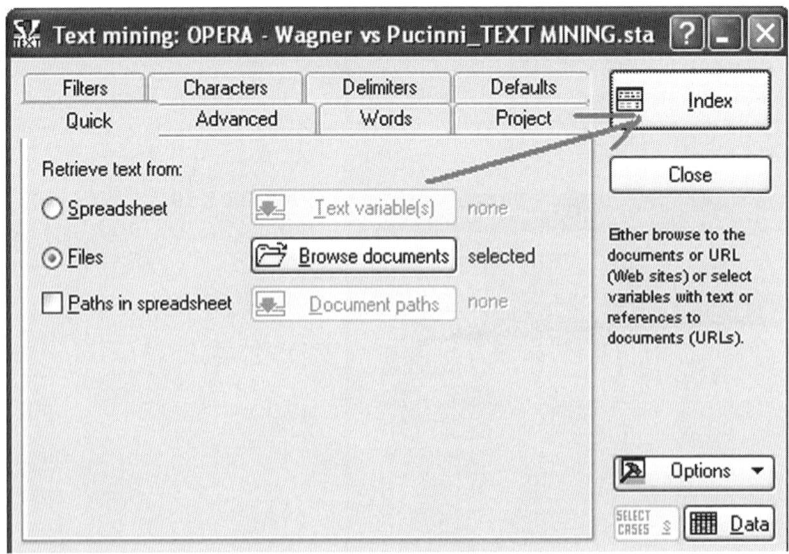

FIGURE L.21
Index button is clicked to compute the word index, where each word's frequency will be counted and put into a results output.

When we click the Index, the next thing we see is a "warning dialog." (see Figure L.22)

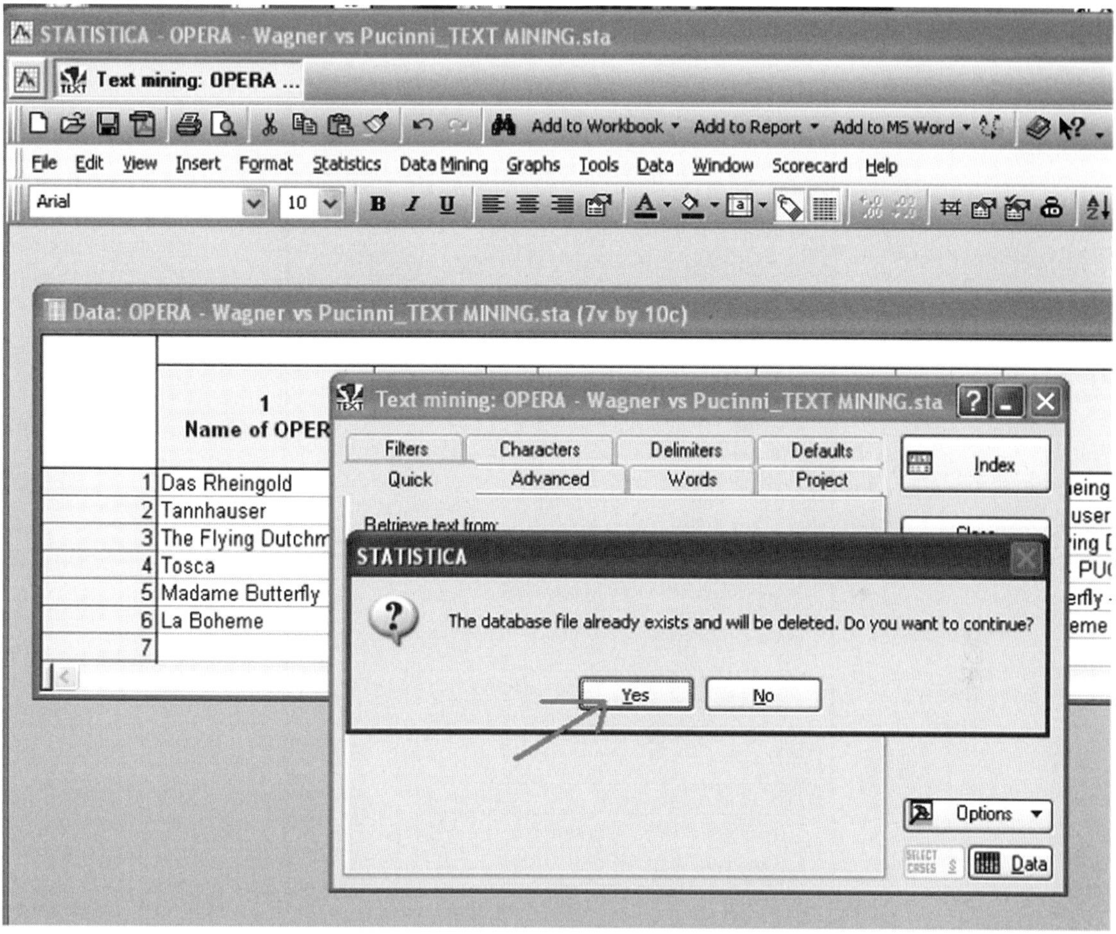

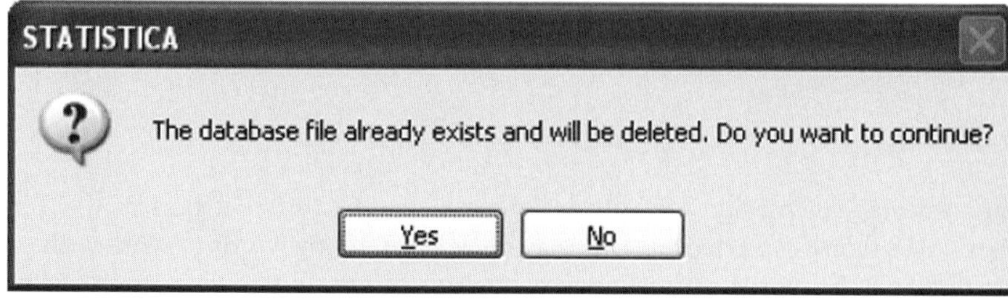

FIGURE L.22
Warning dialog box asking if you want to continue.

If you do *not* want the index words to be placed in your original data file, then click "No." Then you'll have to do a "Save As" for your data file, giving it a different name, like "Data file_original.sta."

But most of the time, you don't have to worry about this, so click "Yes" to compute the index/counts of words. After a few seconds, the Results dialog appears (see Figure L.23).

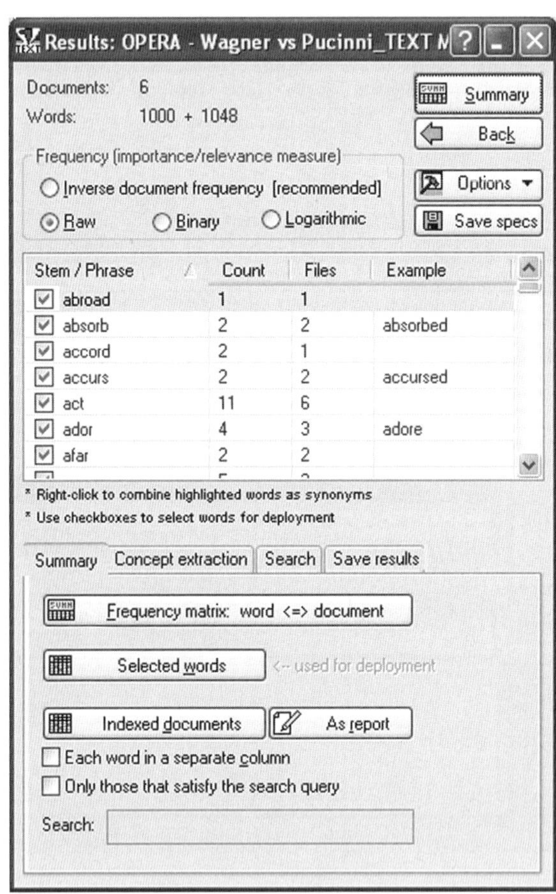

FIGURE L.23
Results dialog from *STATISTICA* Text Mining.

There are 1,000 words selected, using our "default parameters" on the different tabs of the Text Miner dialog, with another 1,048 words *not* selected. This may be way too many words to work with, so eventually we will need to reduce these words. This is most easily done by shoring up the parameters on the "Text Miner main dialog tabs" and rerunning the indexing. But for now, let's see if we can get any information out of this plethora of 1,000 words.

Select the Save tab on the Results dialog (see Figure L.24).

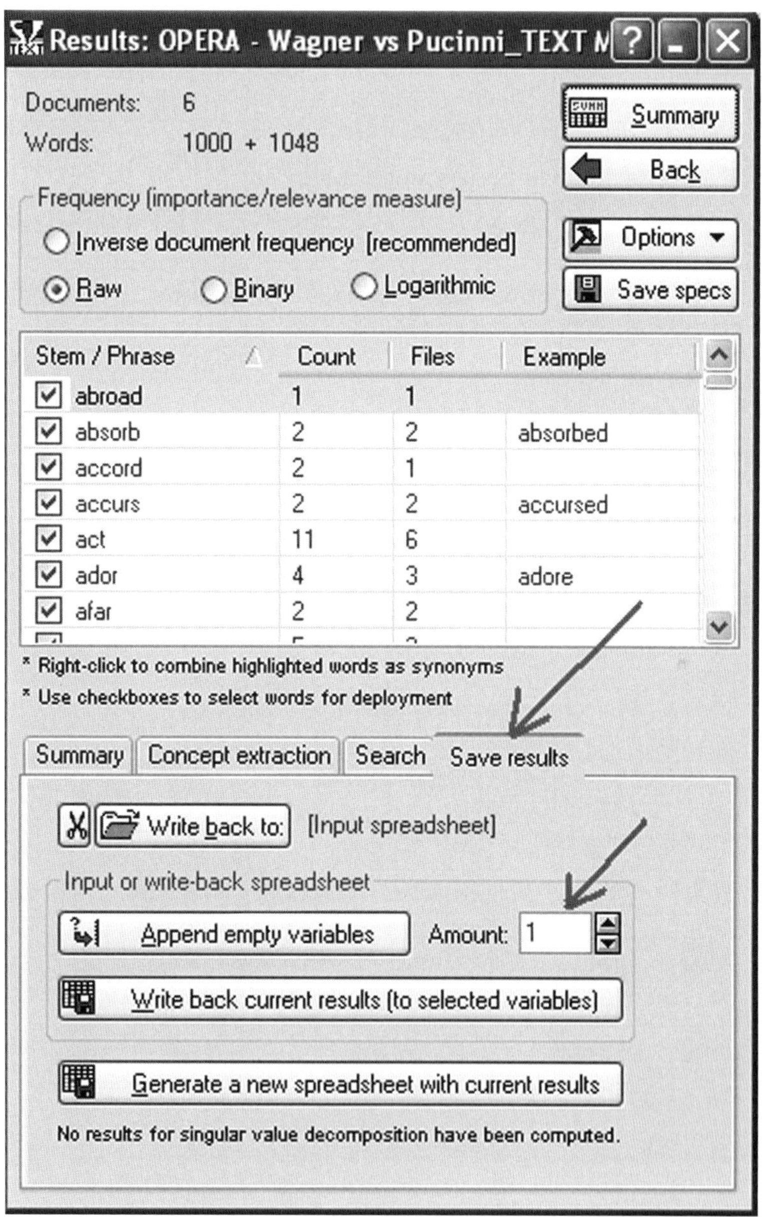

FIGURE L.24
Save tab on the results dialog selected.

And then type "1,000" for the Amount of words/variables (see Figure L.25).

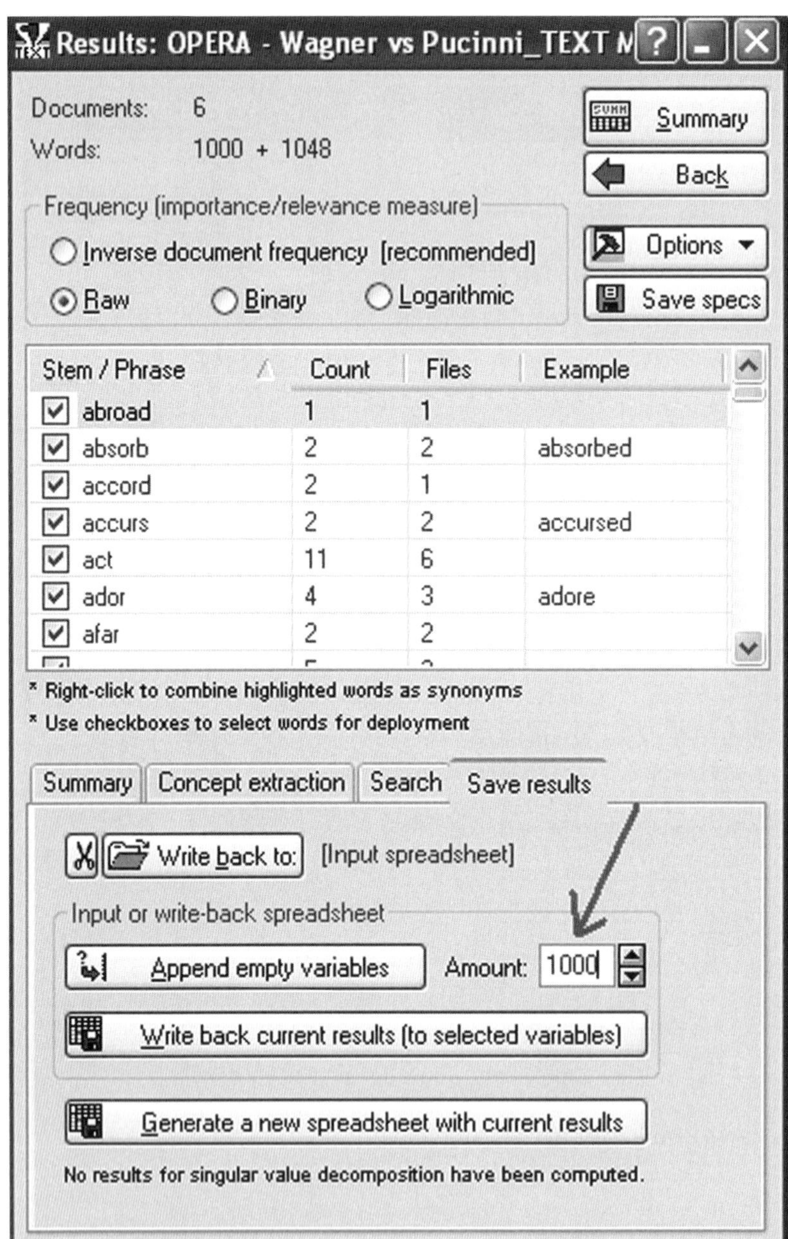

FIGURE L.25
1,000 variables or "words" are selected. This will create 1,000 new columns in our spreadsheet, so the frequencies of these words in each of the six opera documents can be placed into these new variable columns.

Next, click the "Write back to" button to select a spreadsheet to add these word frequencies (see Figure L.26).

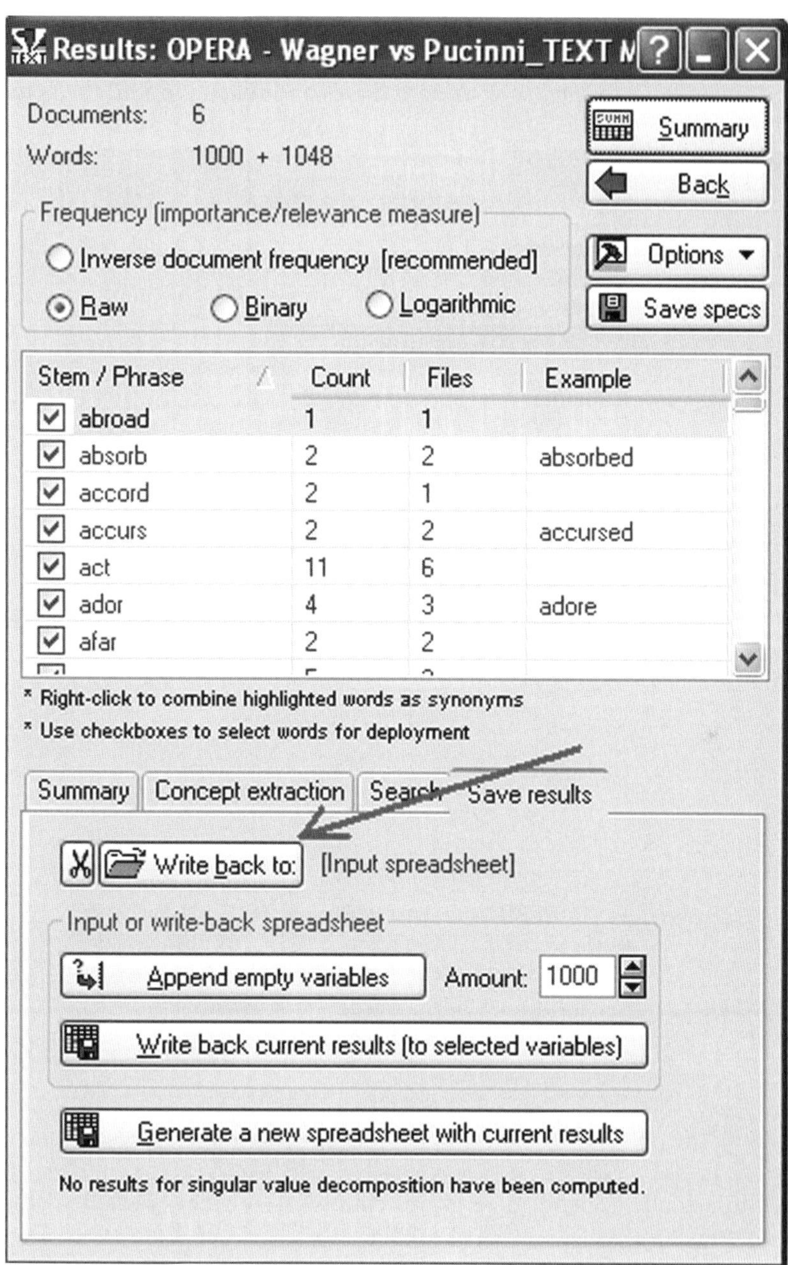

FIGURE L.26

"Append empty variables" selection on the Results dialog.

This will open up the window "Open data file" (see Figure L.27). Here, the data file to which we want the word frequencies immediately pops open, since it is the data spreadsheet we started with. But in other circumstances, you might have to browse your hard drives and folders to find the spreadsheet needed.

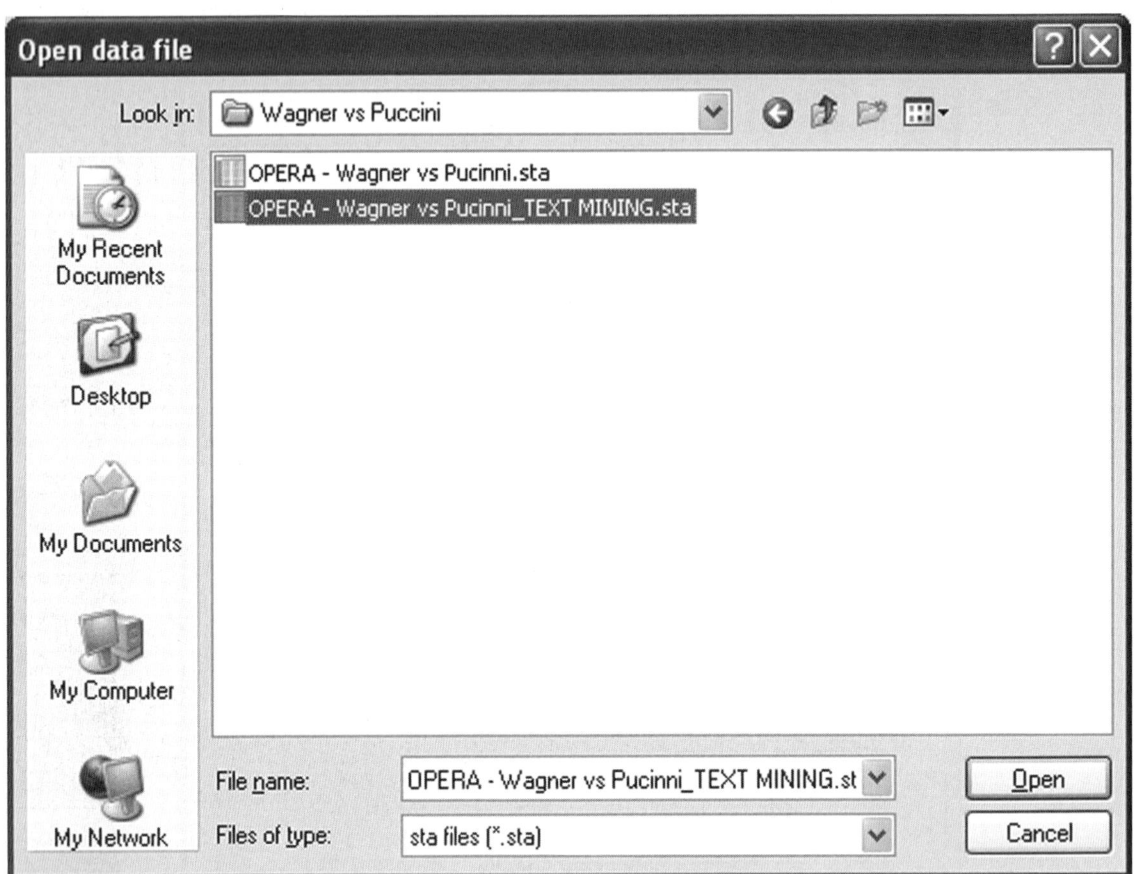

FIGURE L.27
Open data file window.

Click Open, and the Results dialog will show that you have selected this data file source (see Figure L.28).

Next, we need to actually append these "word counts" to our new variable columns in the data spreadsheet (see Figure L.29).

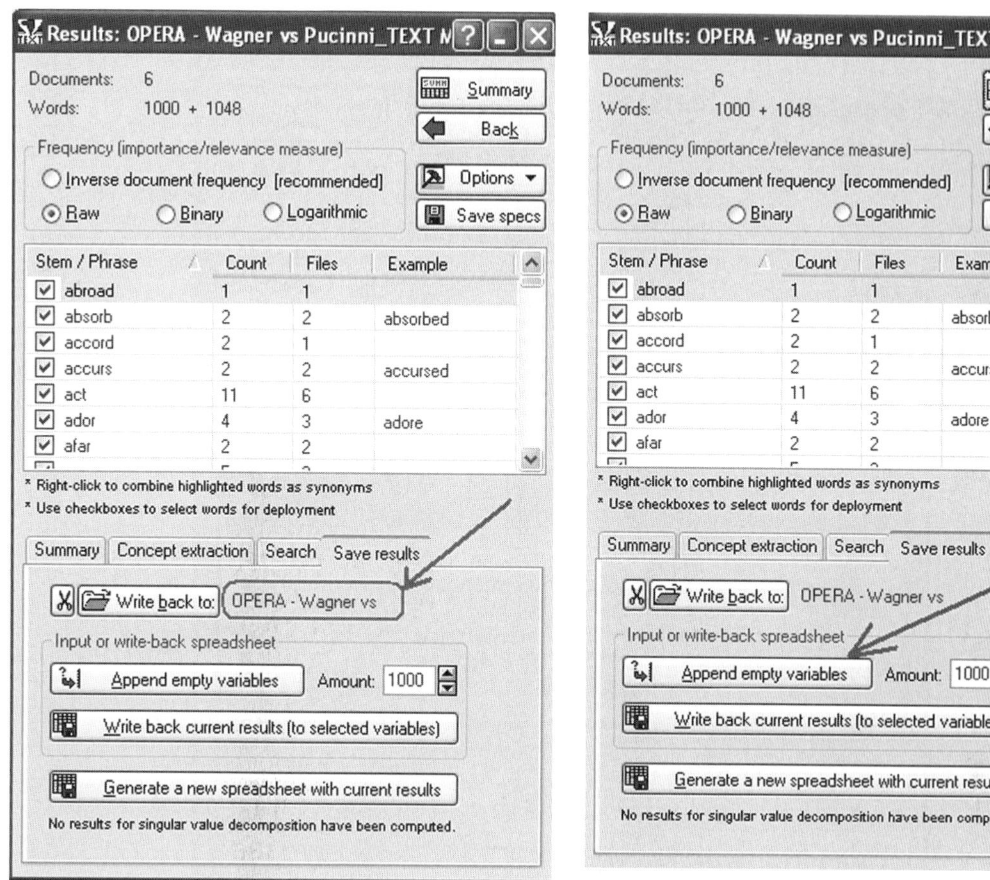

FIGURE L.28
Results dialog.

FIGURE L.29
"Append empty variables" button selected to actually write "word frequencies" back to the data spreadsheet.

A message will pop up, letting the user know that these variables have successfully been added to the data sheet (see Figure L.30).

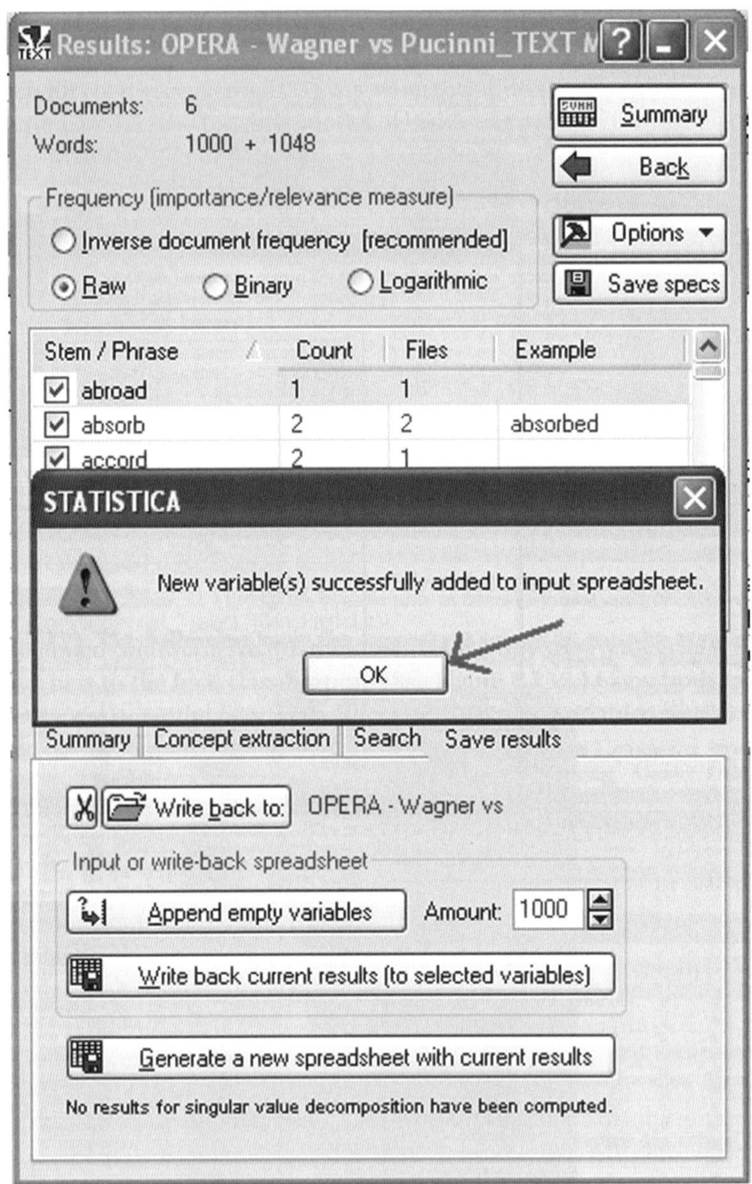

FIGURE L.30
Variables successfully added to spreadsheet pop-up dialog.

Click on the OK button to accept.

Now if you scroll the *STATISTICA* data spreadsheet to the right, you will see that these "variable columns" have been added but are missing data at this point (see Figure L.31).

Data: OPERA - Wagner vs Pucinni_TEXT MINING.sta (1007v by 10c)

	5 Year Written	6 Century written	7 Name of Text File	8 NewVar1	9 NewVar2	10 NewVar3	11 NewVar4	12 NewVar5	13 NewVar6	14 NewV
1	1869	19	Das Rheingold - Wagner - 1869 - Scene I.txt							
2	1845	19	Tannhauser - Wagner - 1845 - Act I.txt							
3	1843	19	The Flying Dutchman - Wagner -1843 - Act I.txt							
4	1900	20	Tosca - PUCCINI - 1900 - Act I.txt							
5	1906	20	M-Butterfly - PUCCCINI - 1906 - Act I.txt							
6	1896	19	La Boheme - PUCCINI - 1896 - Act I.txt							
7										

FIGURE L.31
Spreadsheet with NewVars added, one for each of the word frequencies selected.

When the *STATISTICA* data set slider bar was selected (to slide to the right to expose the "new variables" created by the TM process), the TM Results Dialog "jumped" to the top tool bar on the installation being used for writing this tutorial. The default position for "minimized dialogs" is the bottom of the *STATISTICA* screen, but by "grabbing" the minimized screen at the default bottom, one can drag it and put it to the top of the screen, as illustrated here (see Figure L.32).

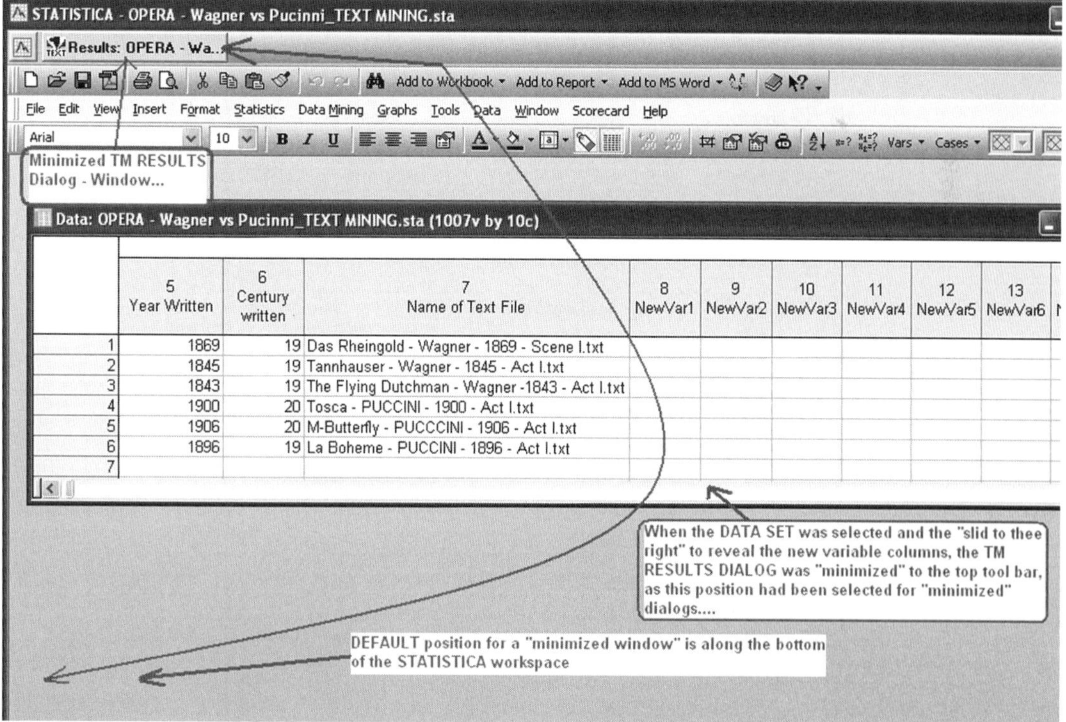

FIGURE L.32
Minimized dialog windows can be set to either go to the bottom of the *STATISTICA* workspace, or to a "tool bar" at the top of the workspace. The top is selected for the PC installation used for this tutorial.

Next, we need to do *singular value decomposition*, a procedure analogous to factor analysis or principal components analysis, to see if we can derive what in Text Mining are called "concepts" out of all the words.

To do this, we need to select the Concepts tab (see Figure L.33) and then select the radio button for "Inverse document frequency." (We won't go into the reasons for selecting "Inverse document frequency" in this tutorial, but you can go to the Online Help in *STATISTICA* by clicking on the "?" mark that is in the upper right-hand corner of the TM Results dialog, which will take you to the section that deals with this dialog, including all the buttons and selections.)

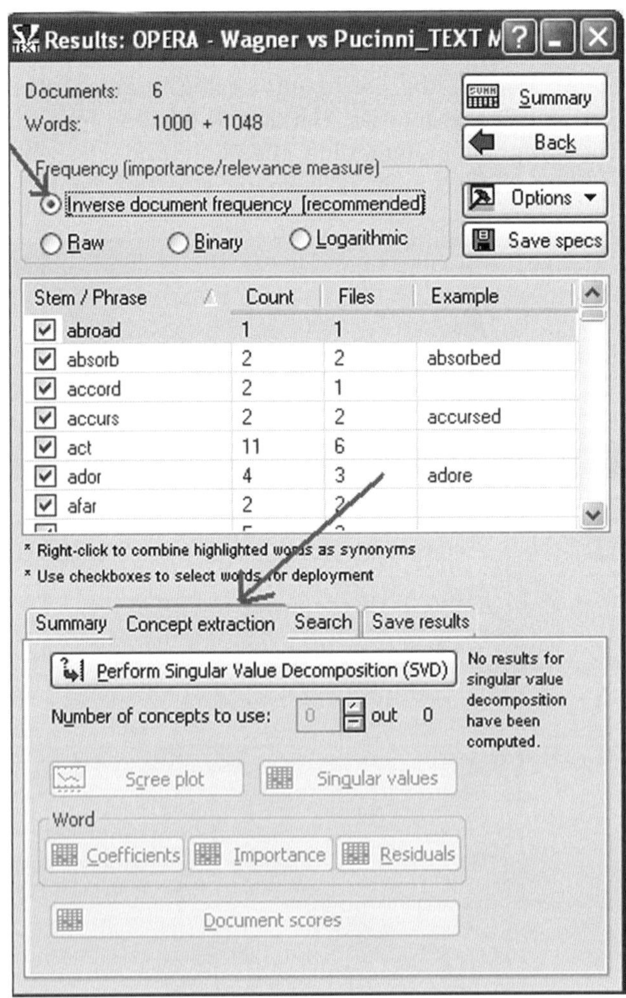

FIGURE L.33

"Inverse document frequency" button and the Concept Extraction tab in the *STATISTICA* Text Mining Results dialog.

Now, click the button to "Perform Singular Value Decomposition (SVD)" (note that all of the buttons below this are "grayed out"—that is, they are not active at this point).

Following the "Singular Value Decomposition" computation, the buttons below this are no longer grayed out (are now active) (see Figure L.34).

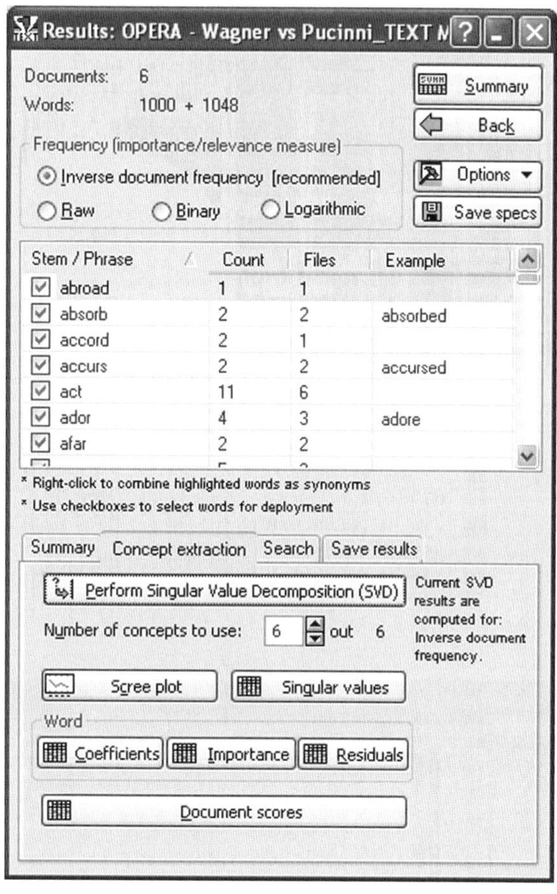

FIGURE L.34

TM Results dialog showing that SVD has been computed, since the buttons below SVD are not active (e.g., are grayed in).

We note that 6 concepts were found in this data set. Now we have a choice of selecting the number of concepts we want to look at further among these 6 concepts. In many data sets, you'd get 10 to 20 or more concepts, so in those cases, you need to reduce the number. As in principal component analysis in traditional statistics, you would look for the inflection point in a Scree plot, and then usually end up with just 2, 3, or 5 "components" or "factors" to explain the data.

Next, we need to look at the Scree plot to see the curve of these components. To do this, click on the "Scree plot" button in the TM Results dialog window (see Figures L.35 through L.38).

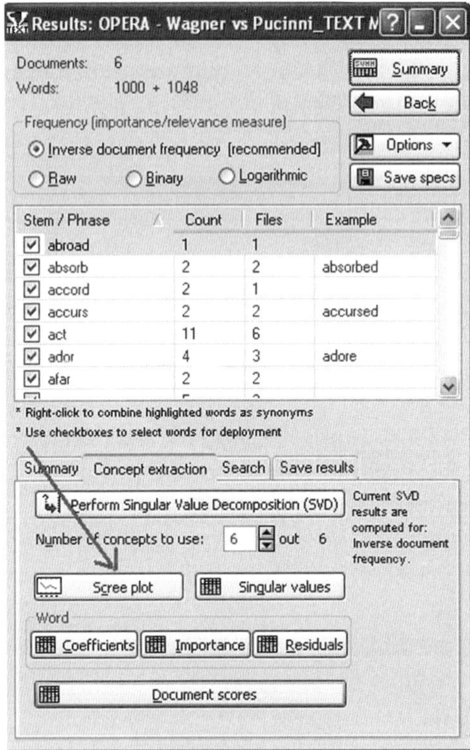

FIGURE L.35

Scree plot selection following SVD.

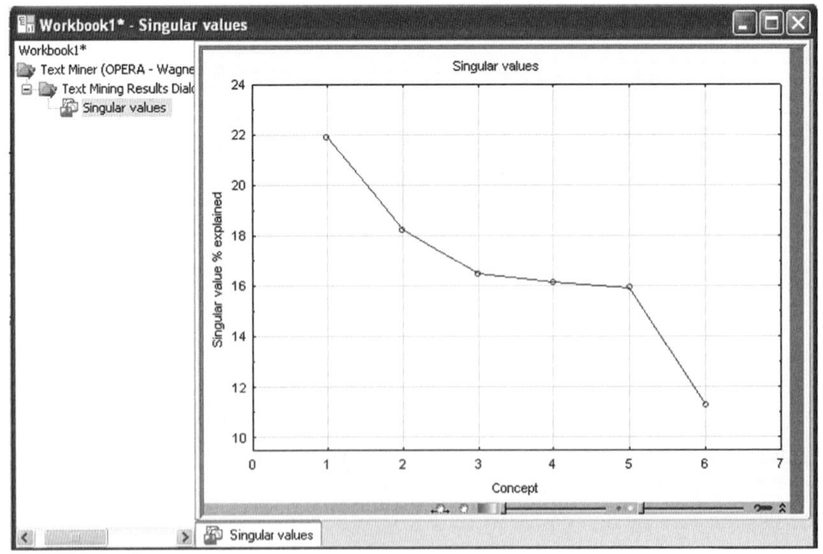

FIGURE L.36

Results workbook containing the Scree plot.

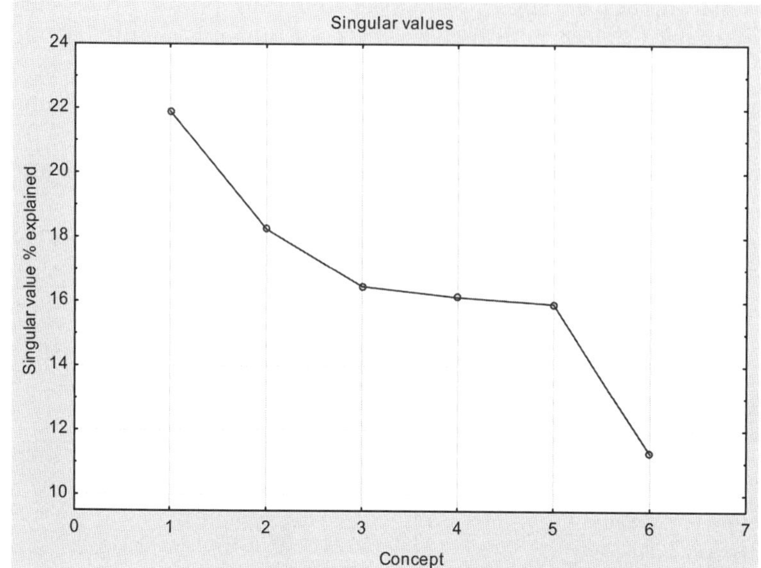

FIGURE L.37

Scree plot extracted as a stand-alone window.

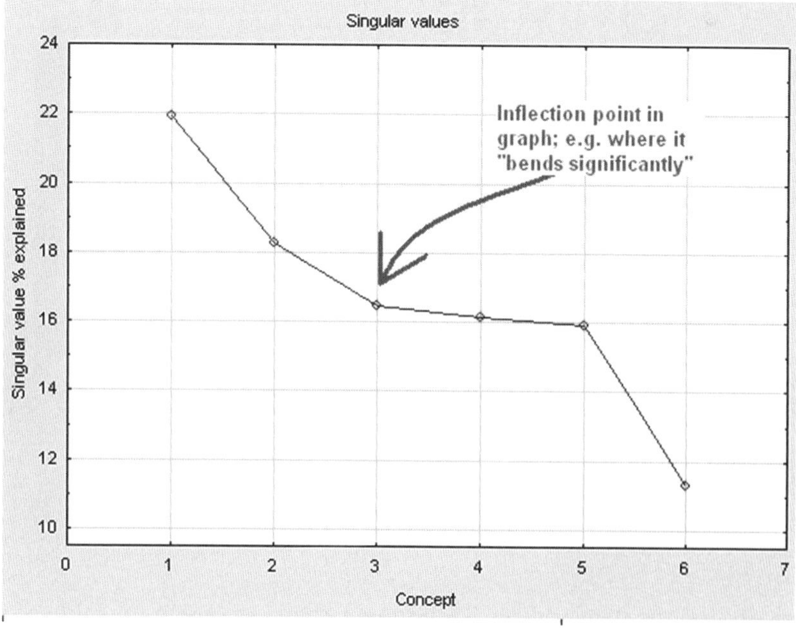

FIGURE L.38

The "inflection point" in a Scree plot is where the curve bends significantly, as shown in the figure. Usually the data points to the left are considered the most significant, and the data points to the right of this bend are considered much less significant for a data set.

Since the Scree plot "levels off" starting at Concept 3, Concepts 1 and 2 are probably the most important and may be the only ones we need to look at further; however, we probably should include Concept 3 as well. We might even want to consider Concepts 1 through 5, but obviously Concept 6 drops off significantly from the others, so Concept 6 is probably not of much value in understanding the lyrics of these operas.

We can also print out the relative importance of these "concepts" by clicking the "Importance" button.

You can note from Figure L.39 that the "SVD word importance" is alphabetized rather than sorted in "order of importance." It might be more useful for us to see the order by importance.

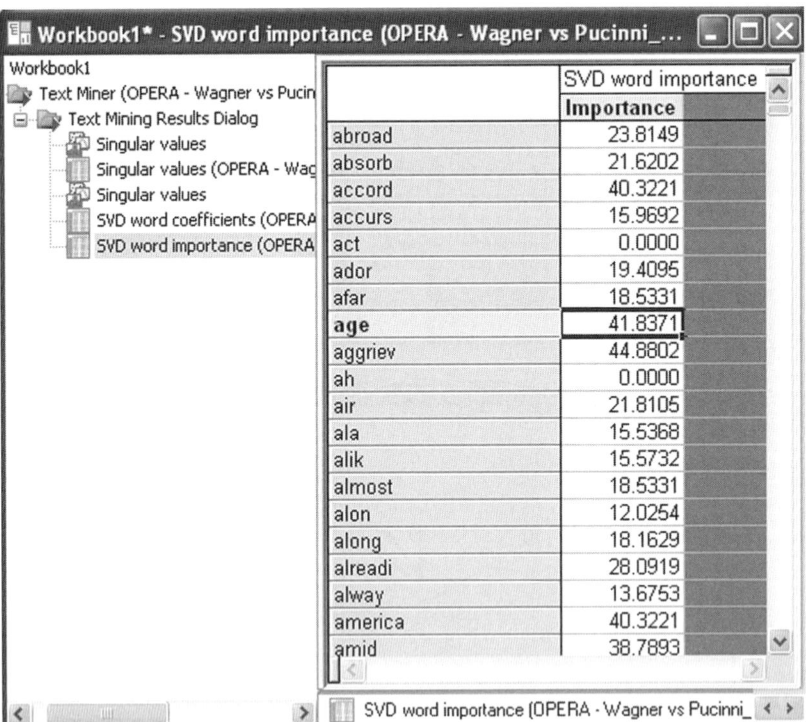

	SVD word importance
	Importance
abroad	23.8149
absorb	21.6202
accord	40.3221
accurs	15.9692
act	0.0000
ador	19.4095
afar	18.5331
age	41.8371
aggriev	44.8802
ah	0.0000
air	21.8105
ala	15.5368
alik	15.5732
almost	18.5331
alon	12.0254
along	18.1629
alreadi	28.0919
alway	13.6753
america	40.3221
amid	38.7893

FIGURE L.39
Word relative importance.

We can right click on the name of this "importance chart" (name in the table of contents on the left side of the workbook), and then select "Use as active input." This will cause the results table to become a data spreadsheet, and then we can manipulate it any way a data file can be used in *STATISTICA* (see Figure L.40).

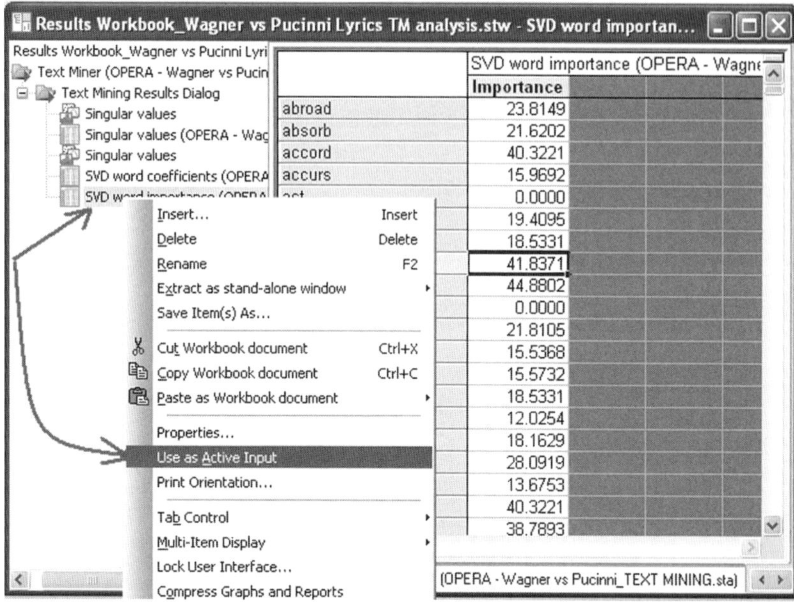

FIGURE L.40

Process to give "active input data sheet" to the Importance chart, so we can "sort" by importance rather than by alphabet.

When you make the Importance chart the "active input data set," a red border is put around the icon in the workbook that represents this workbook file (see Figure L.41).

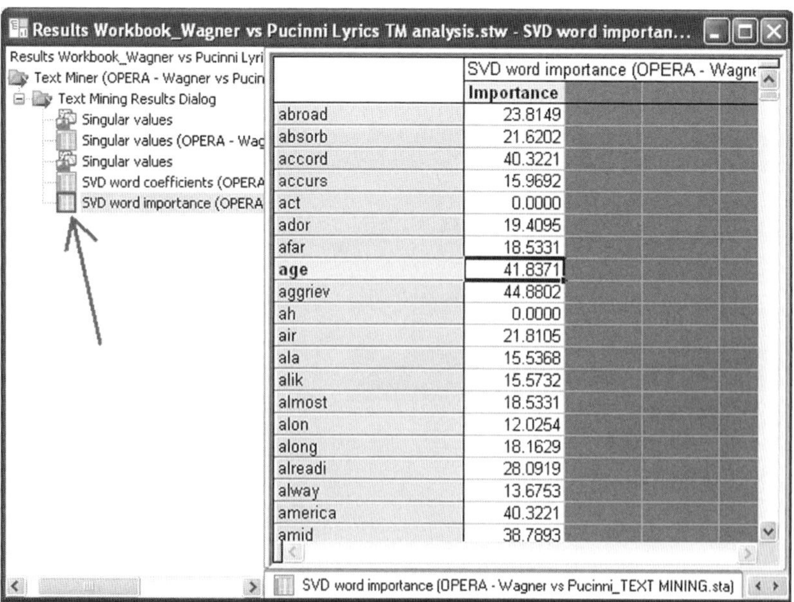

FIGURE L.41

Red border around the new "Active Input" spreadsheet.

Click on the "Variable" label for importance to highlight the entire column (see Figure L.42).

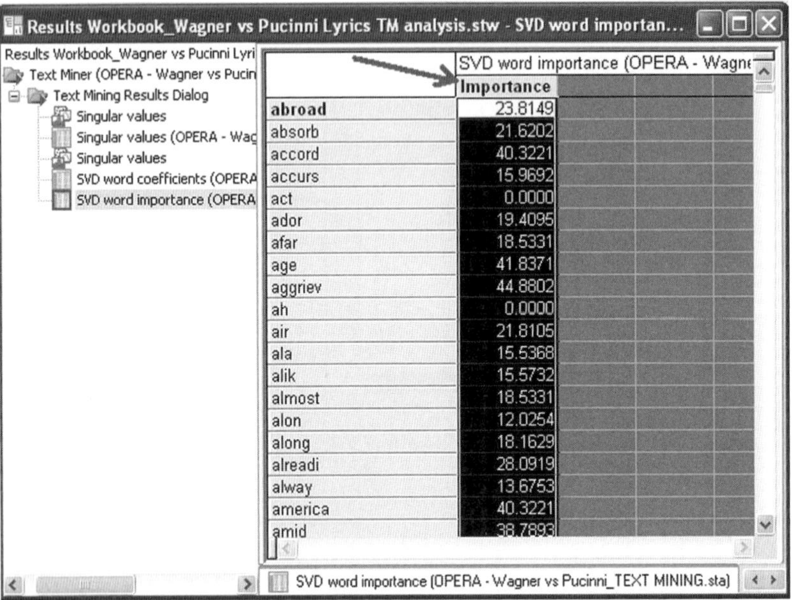

FIGURE L.42
The importance scores for the words are highlighted by clicking on the "Variable" label for the column.

Then click on the "Data" pull-down menu (see Figure L.43), and select "Sort."

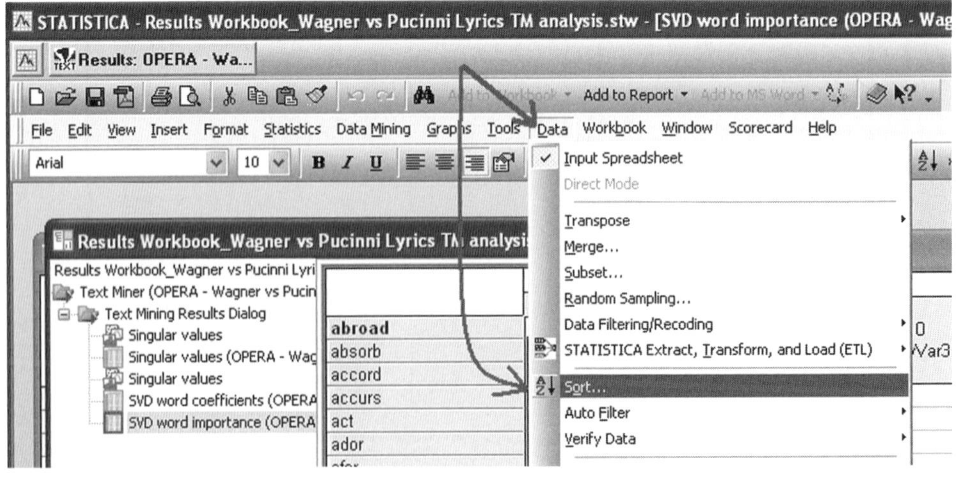

FIGURE L.43
Using the "Data" pull-down menu to select "Sort" so the word importance can be sorted in order of importance.

Click on the "Add Var(s) >" button in Figure L.44 to add the variable to the right-hand window (see Figure L.45).

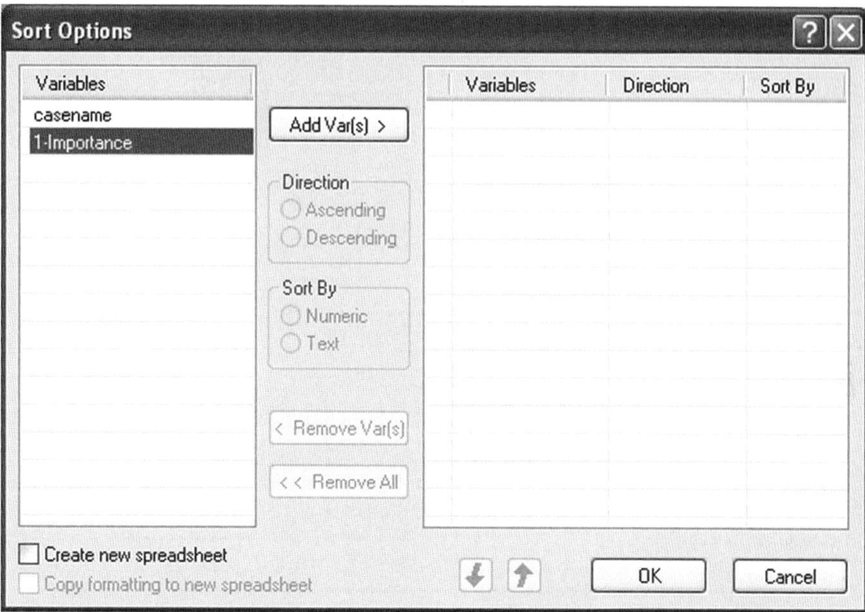

FIGURE L.44
Sort "Options" dialog.

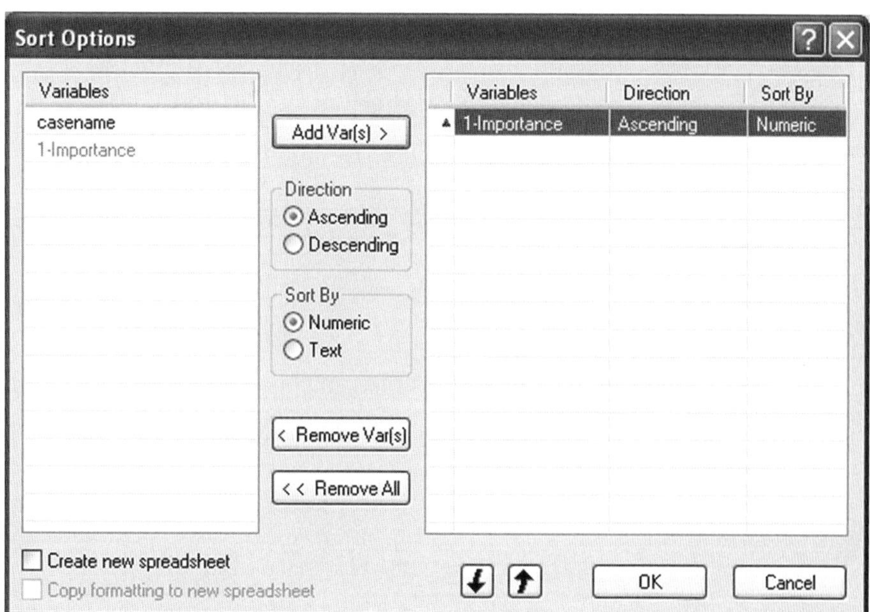

FIGURE L.45
Ascending order is the default, but we'll want to change this to descending order.

Using the ascending pull-down tab, select descending order (see Figures L.46 and L.47).

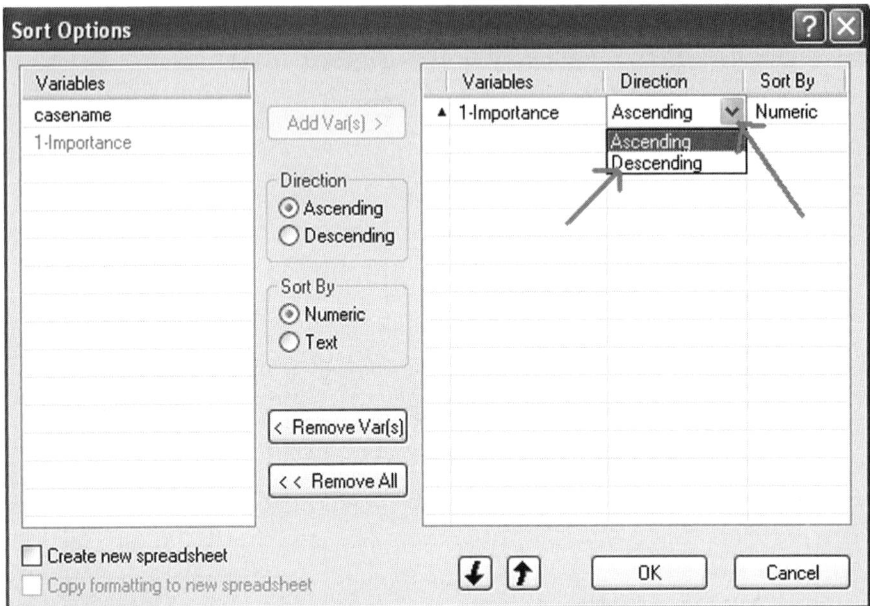

FIGURE L.46
Selecting the descending sort order.

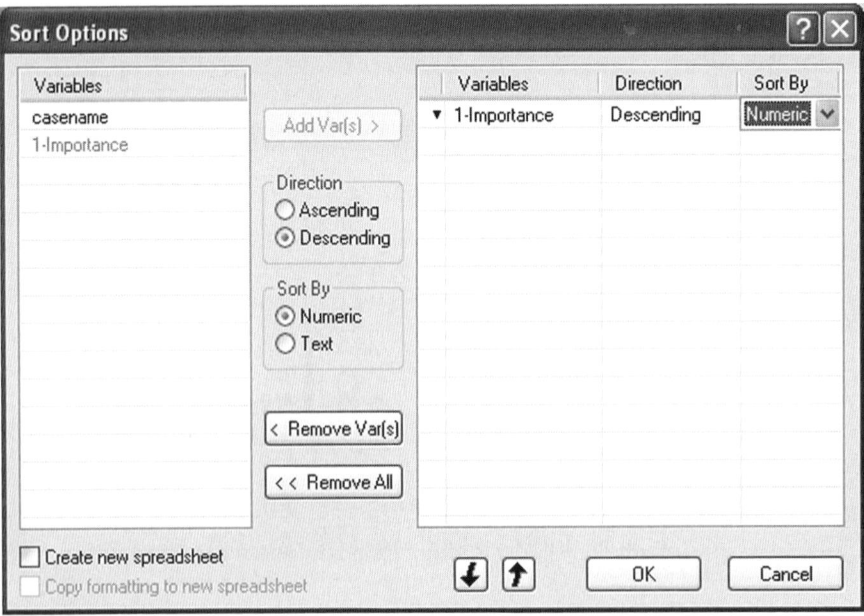

FIGURE L.47
Descending sort order selected.

Now click OK to sort the words based on descending order of the "importance score." (see Figure L.48)

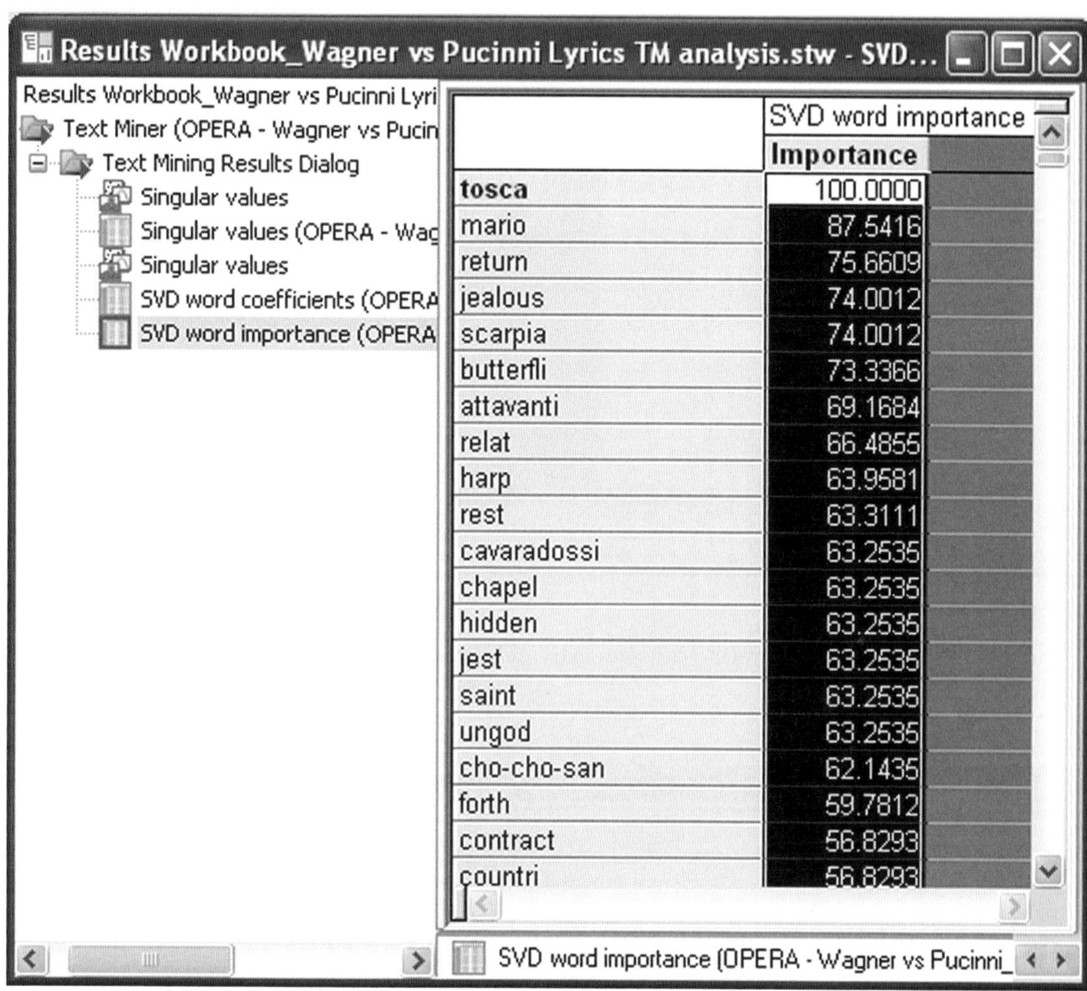

FIGURE L.48
Results workbook showing the SVD word importance spreadsheet open.

Note that "tosca" is the most important word (see Figure L.48 above), followed by "mario," "return," and "jealous."

And also note that small words like "us," "ve," and "will" are at the bottom of the list of importance (see Figure L.49). However, words like "heart" and "love" are also at the bottom of the list, which is a surprise, since many of the opera lyrics talk about "love," "heart," and such matters (but maybe love and heart are "inferred" more than actually used as words in the lyrics).

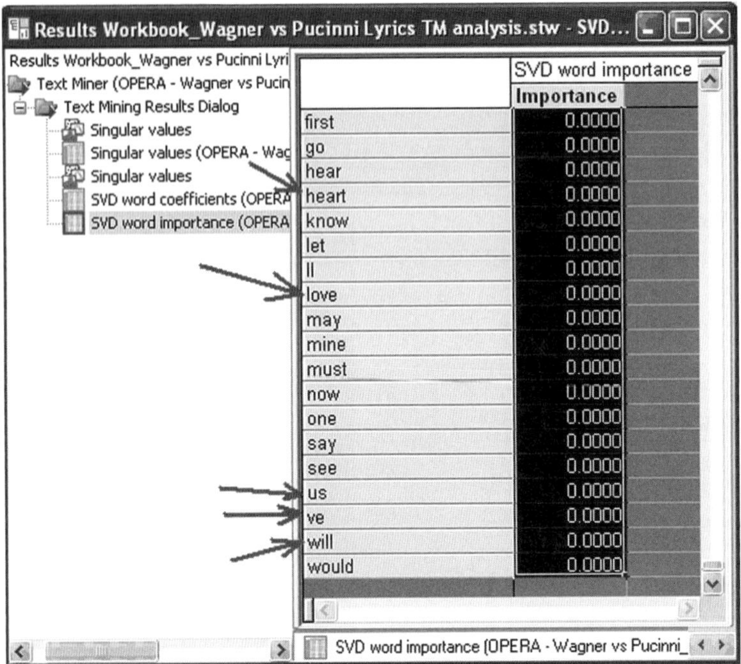

FIGURE L.49
Least important words.

Looking at "word importance" can be extremely valuable in further analysis of text data (see Figure L.50).

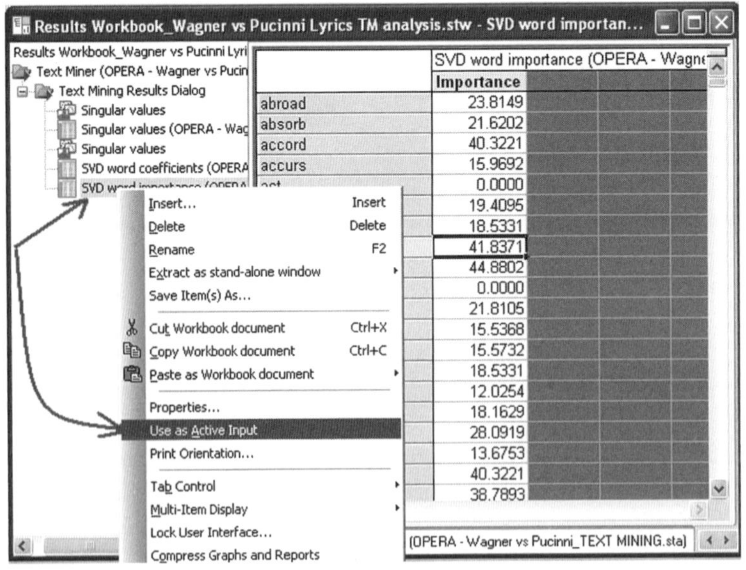

FIGURE L.50
Selecting "word importance" as active input so this column can be used for further data analysis, like making a bar graph of the most important words so we have a visual representation.

And we can also get the "word coefficients" if wanted for any other computations (see Figure L. 51).

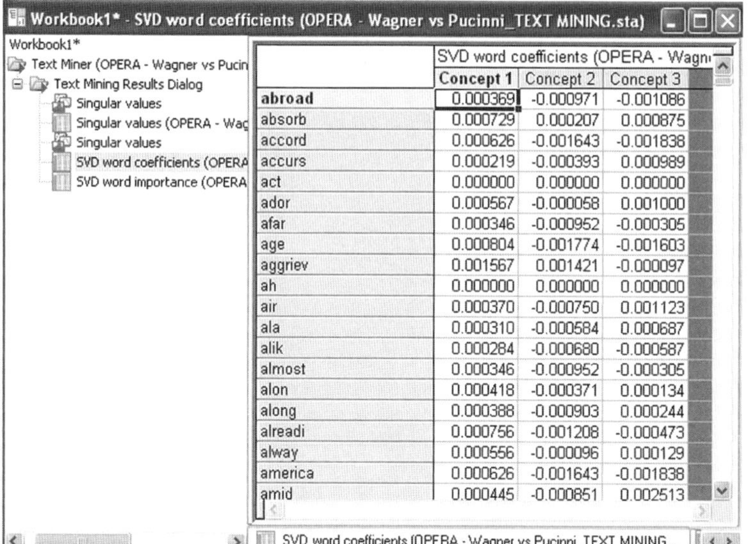

FIGURE L.51
Word coefficients.

For our tutorial, we'll select three concepts (see Figure L.52).

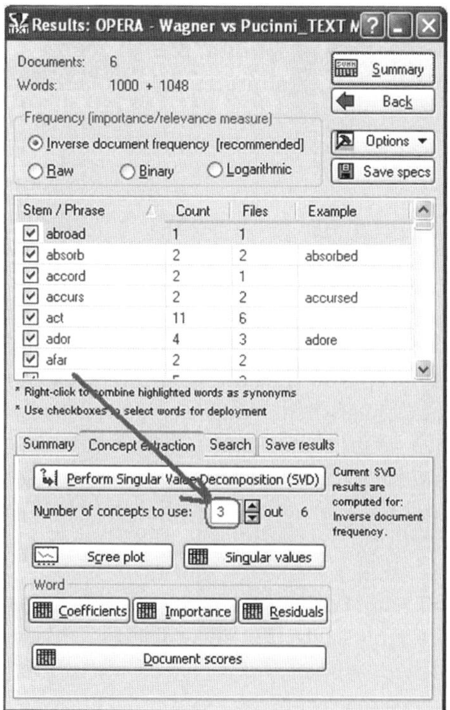

FIGURE L.52
Three concepts selected for further analysis.

If we now click on the Scree plot button, we will see that only three points are plotted (see Figure L.53).

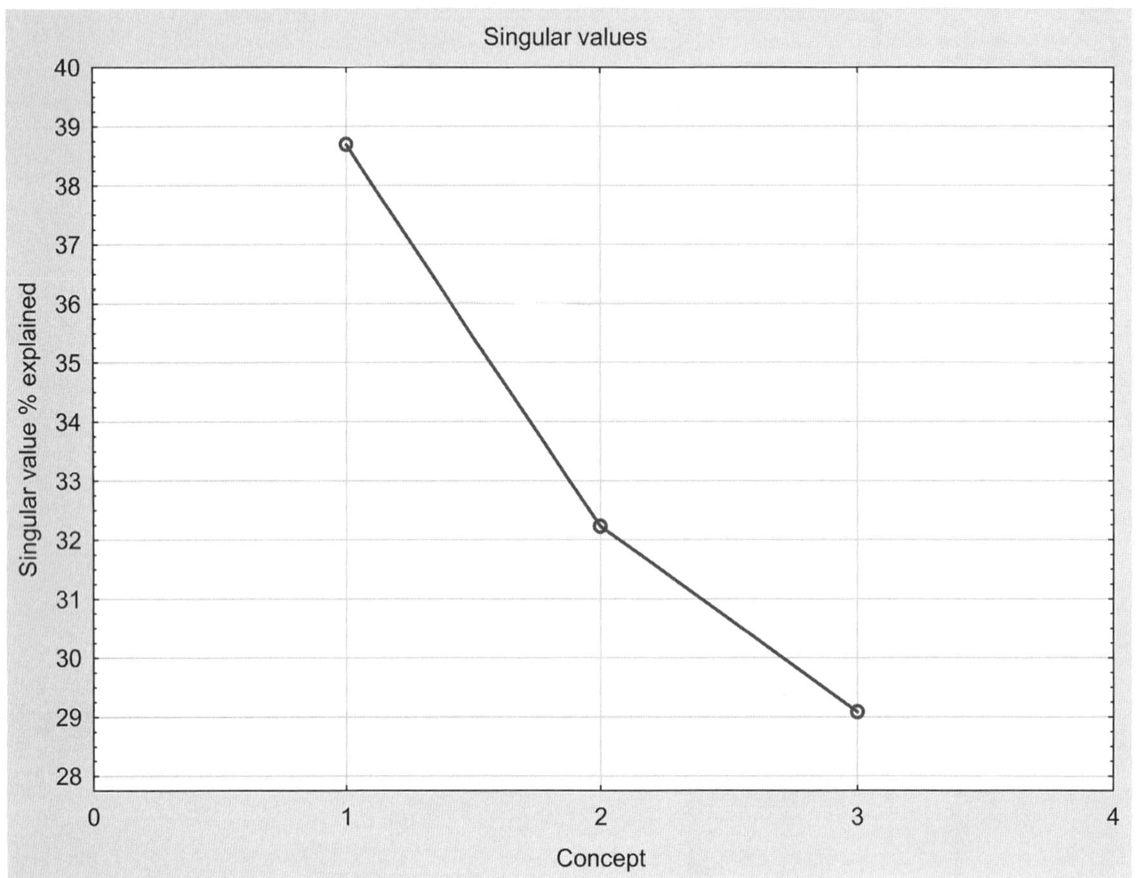

FIGURE L.53
Scree plot with only three concepts selected.

I'm going to change my mind and actually look at all six concepts so that the values of these six concepts can be added as six new variables at the end of our data spreadsheet. After we add these new variables, we can do various exploratory things with them, such as graphing, and it may be that we really do want to look at all six concepts, since so few were found for these data.

So change the number of concepts to look at back to 6, and then select the Save tab on the TM Results dialog (see Figures L.54 through L.58).

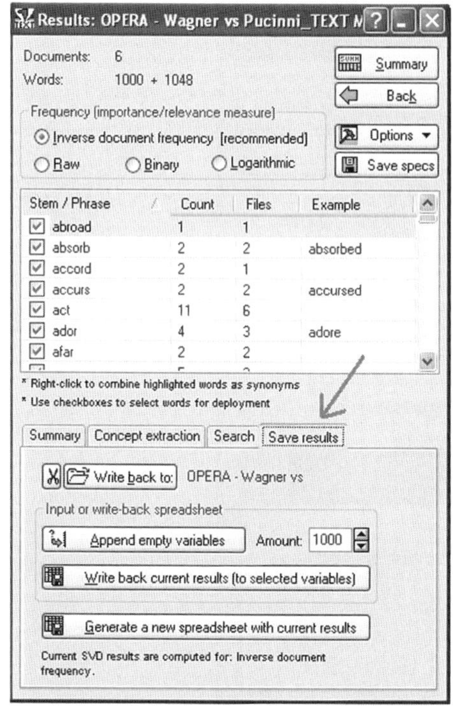

FIGURE L.54

Save tab on TM Results dialog selected so we can save back to the spreadsheet the values for the six concepts.

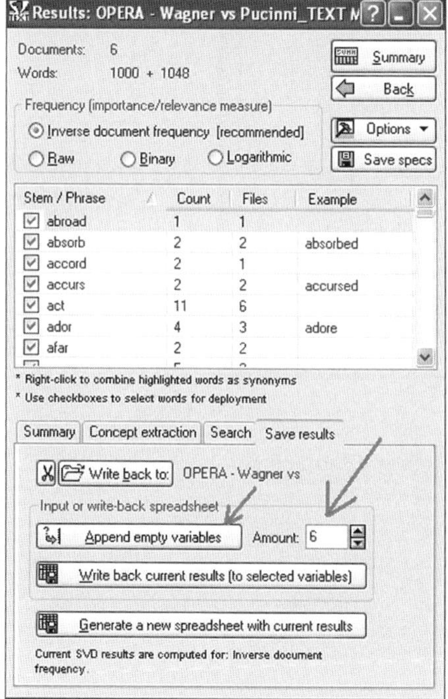

FIGURE L.55

Selecting the concepts to add to 6 in the "Amount" box and clicking on the "Append variables" button to make room for these extra 6 concepts in the data spreadsheet.

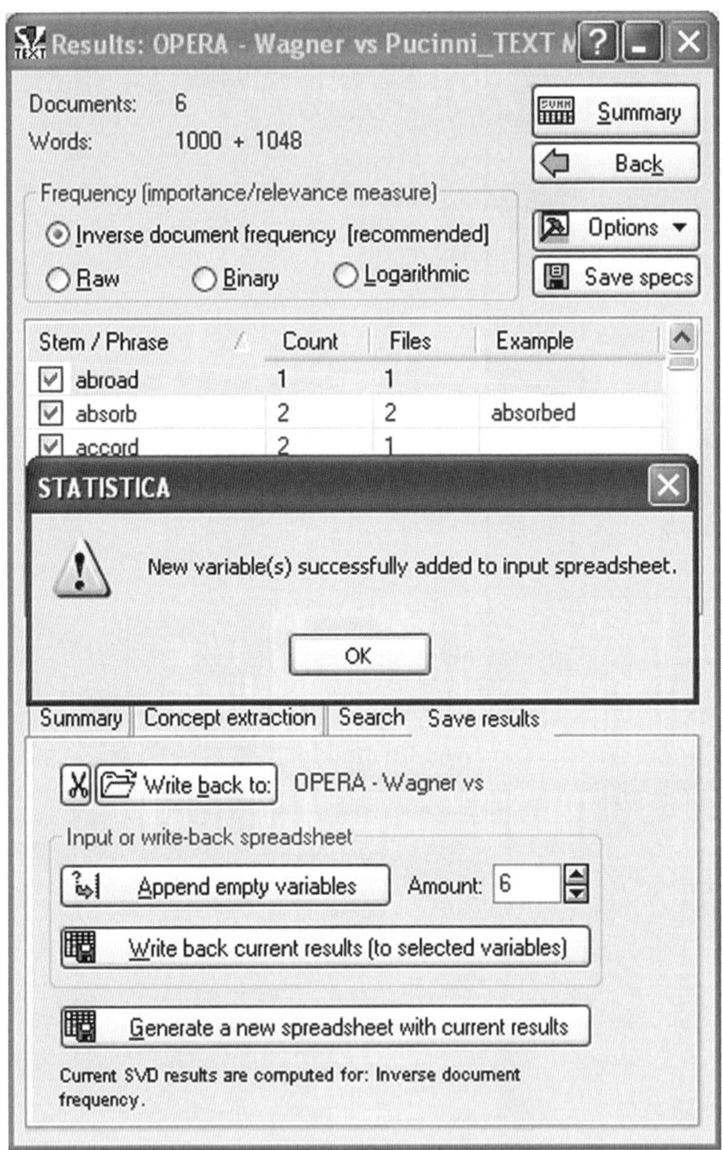

FIGURE L.56
Pop-up window telling us that the new variables were added to the input master data file.

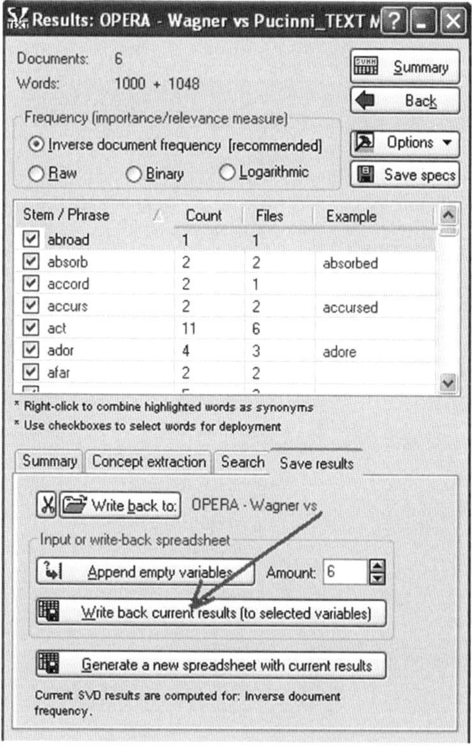

FIGURE L.57
"Write back to current results" is needed to copy the values for the 6 concepts to the data spreadsheet.

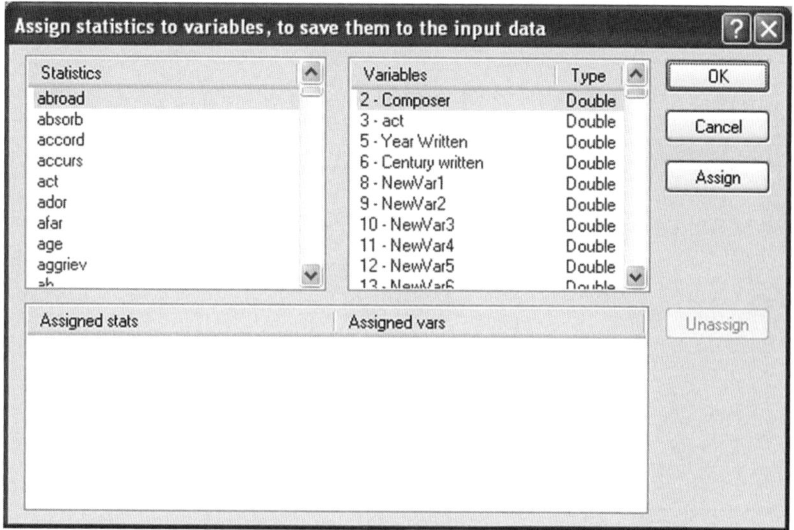

FIGURE L.58
Clicking the "Write back current results" for adding the values for the 6 concepts produces this "Assign statistics to variables" window.

So the concepts need to be selected on the left side, and the six new variables on the right side (see Figures L.59c through L.62).

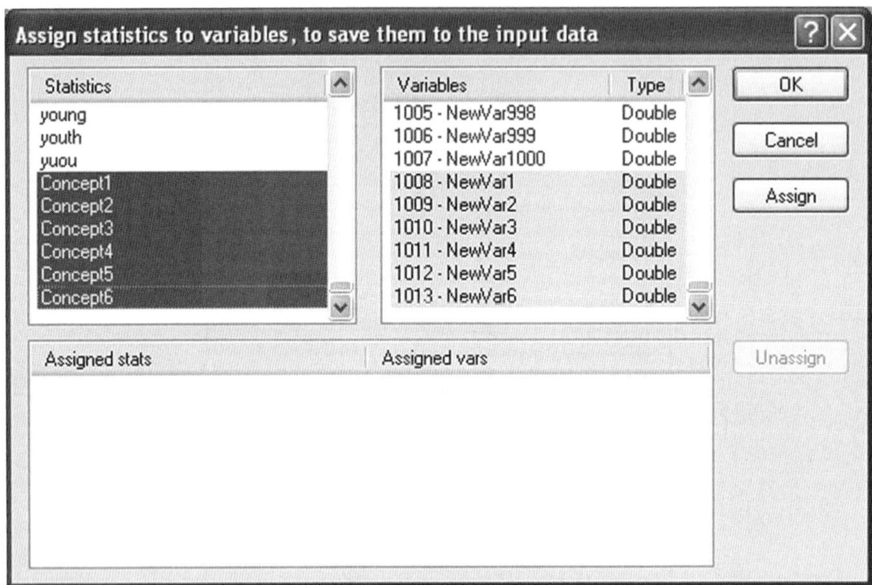

FIGURE L.59
"Assign statistics to variables" dialog window.

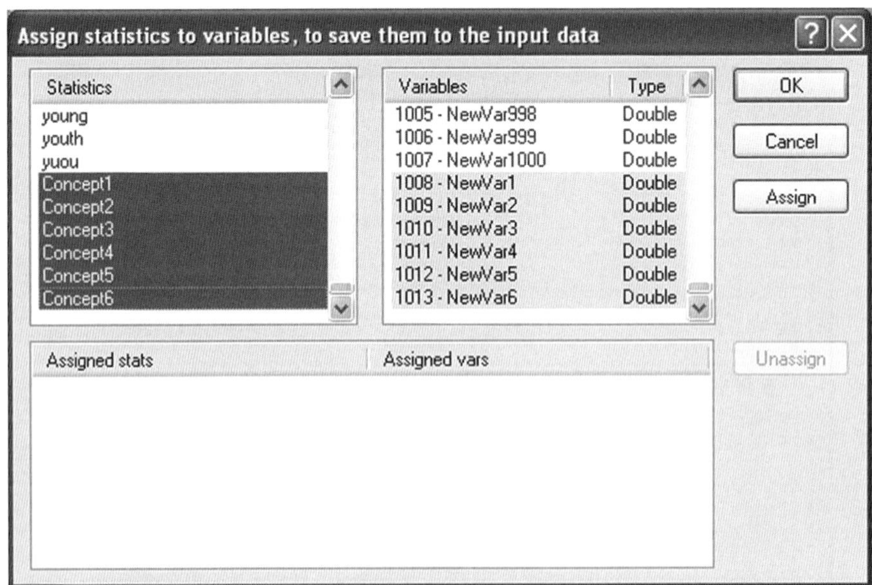

FIGURE L.60
Click "Assign" to assign the values to the variables.

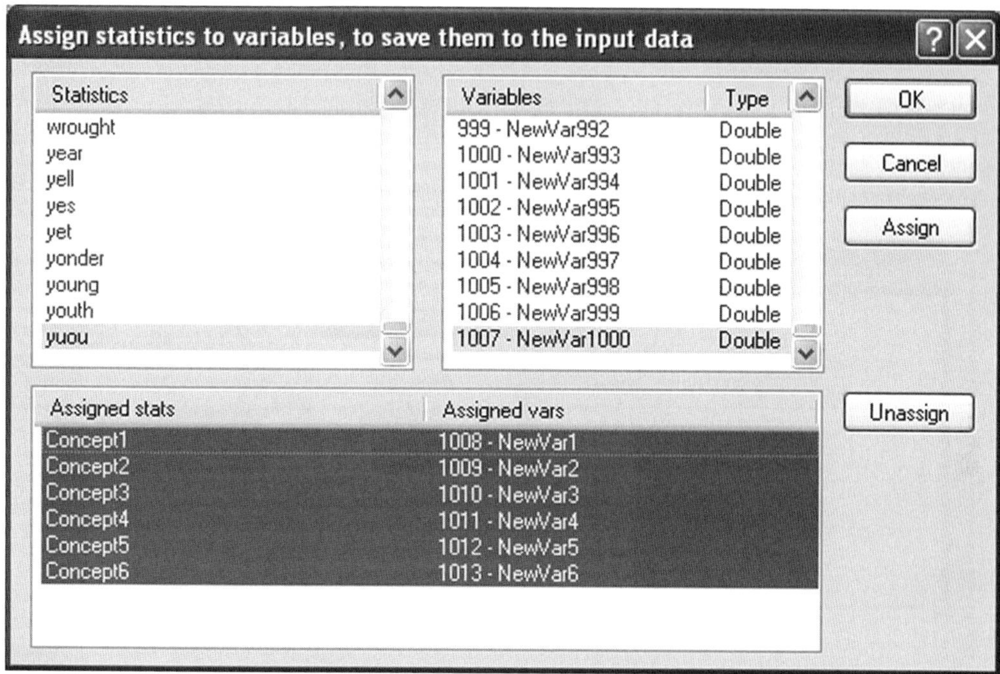

FIGURE L.61

Then click OK to have the values of the six concepts written in the data spreadsheet.

Data: OPERA - Wagner vs Pucinni_TEXT MINING.sta (1013v by 10c)

	1006 NewVar999	1007 NewVar1000	1008 Concept1	1009 Concept2	1010 Concept3	1011 Concept4	1012 Concept5	1013 Concept6
1			0.084874	-0.0876	0.066097	0.023275	0.012156	-0.98998
2			0.239206	-0.31689	0.762716	0.20814	-0.45558	0.098772
3			0.176554	-0.36708	0.302449	-0.02952	0.857733	0.07765
4			0.841794	0.529454	-0.02938	0.073818	0.063871	0.025877
5			0.335959	-0.61228	-0.55785	0.427259	-0.12469	0.054251
6			0.287958	-0.31929	-0.10178	-0.87594	-0.19227	0.023191
7								

FIGURE L.62

Looking at the data spreadsheet, by scrolling all the way to the right, we now see that the values of these six concepts have been added.

Now let's make some scatterplots of Concept 1 compared to Concept 2, and so on (see Figures L.63 through L.66).

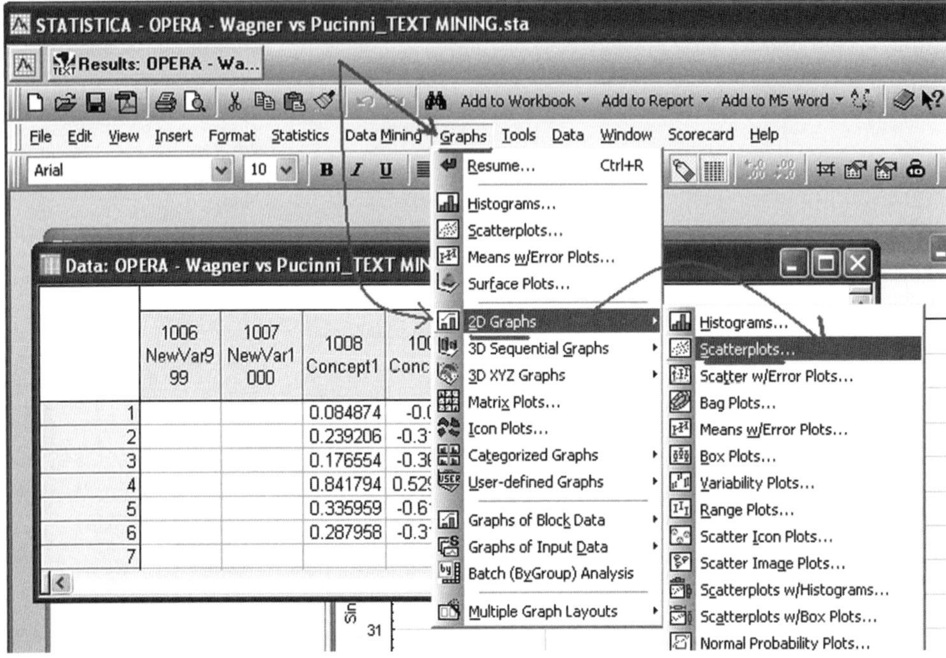

FIGURE L.63
Process of steps/clicks to select the 2D Scatterplot dialog.

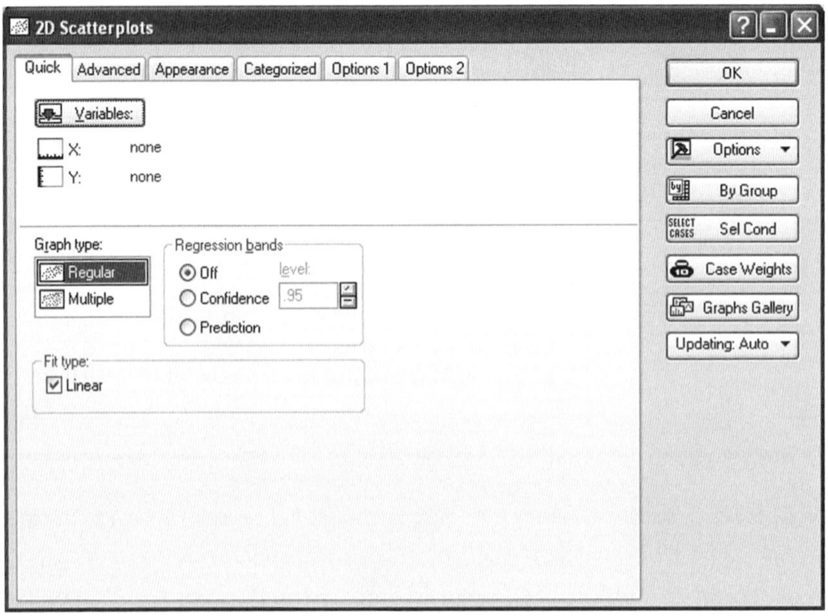

FIGURE L.64
2D Scatterplots dialog window.

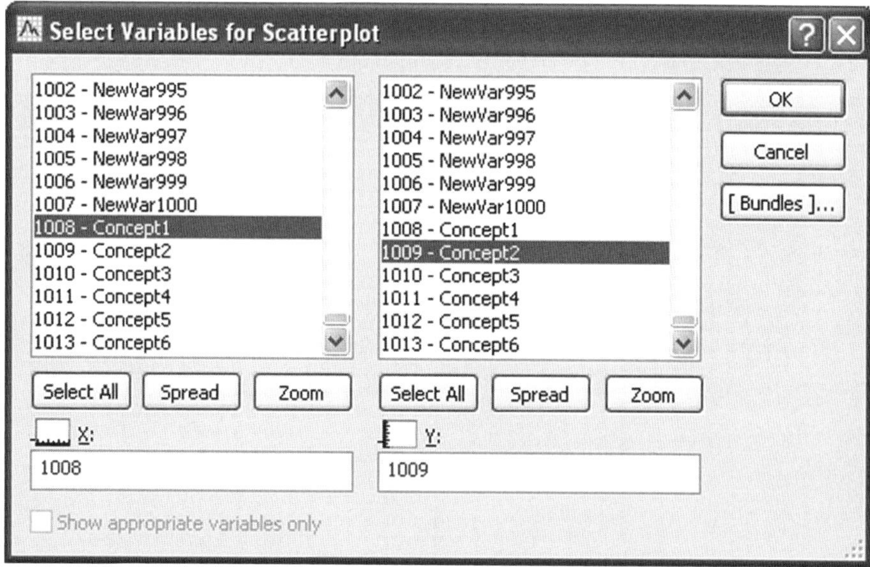

FIGURE L.65

Select Concept 1 in the left panel and Concept 2 in the right panel; then click OK, and then OK again on the 2D Graphs dialog, and the graph will be produced.

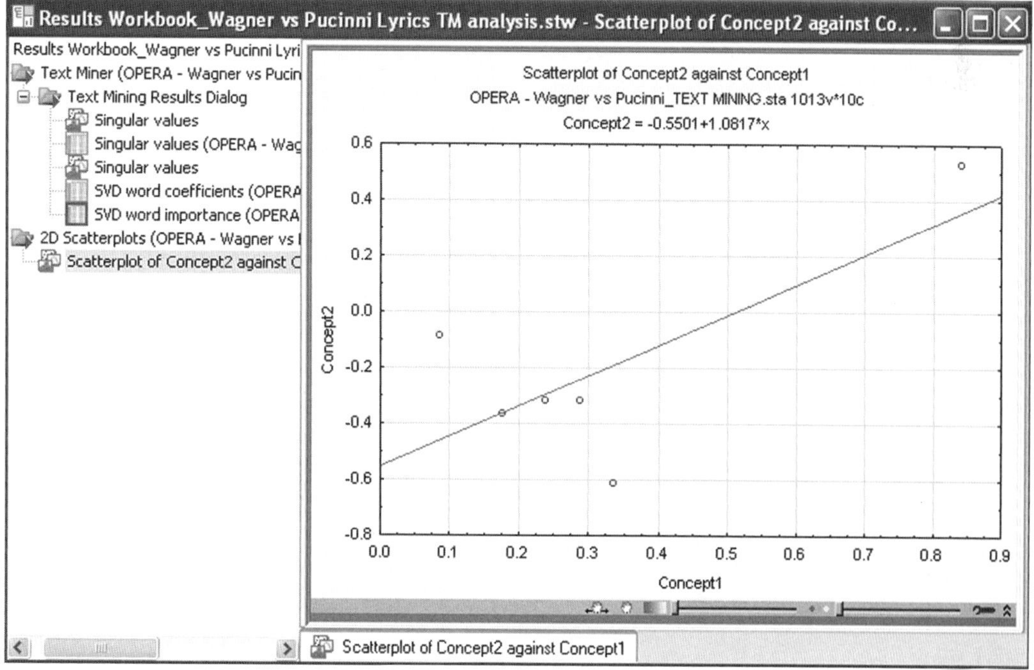

FIGURE L.66

2D Scatterplot of Concept 1 compared to Concept 2.

I'm not sure if the preceding is too much information to absorb. So let's try another approach (see Figures L.67 through L.69).

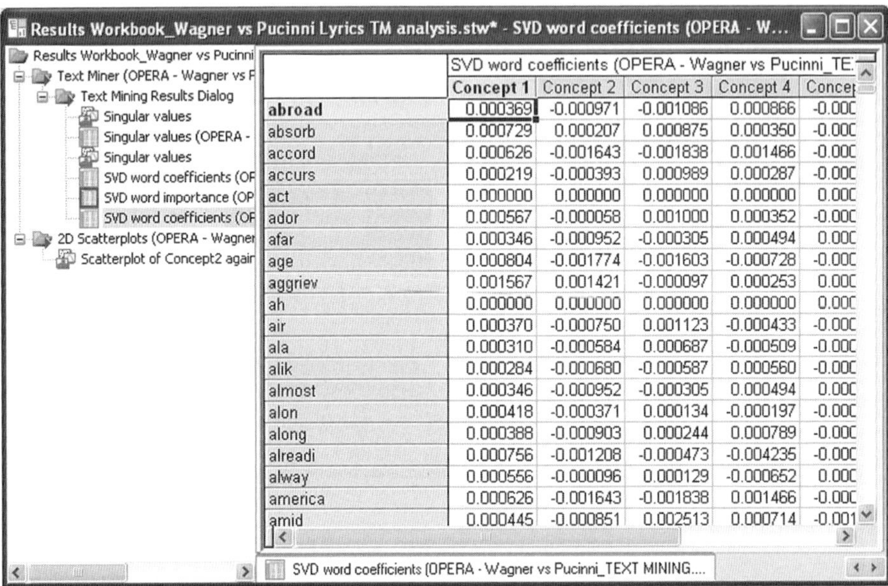

	Concept 1	Concept 2	Concept 3	Concept 4	Concep
abroad	0.000369	-0.000971	-0.001086	0.000866	-0.00C
absorb	0.000729	0.000207	0.000875	0.000350	-0.00C
accord	0.000626	-0.001643	-0.001838	0.001466	-0.00C
accurs	0.000219	-0.000393	0.000989	0.000287	-0.00C
act	0.000000	0.000000	0.000000	0.000000	0.00C
ador	0.000567	-0.000058	0.001000	0.000352	-0.00C
afar	0.000346	-0.000952	-0.000305	0.000494	0.00C
age	0.000804	-0.001774	-0.001603	-0.000728	-0.00C
aggriev	0.001567	0.001421	-0.000097	0.000253	0.00C
ah	0.000000	0.000000	0.000000	0.000000	0.00C
air	0.000370	-0.000750	0.001123	-0.000433	-0.00C
ala	0.000310	-0.000584	0.000687	-0.000509	-0.00C
alik	0.000284	-0.000680	-0.000587	0.000560	-0.00C
almost	0.000346	-0.000952	-0.000305	0.000494	0.00C
alon	0.000418	-0.000371	0.000134	-0.000197	-0.00C
along	0.000388	-0.000903	0.000244	0.000789	-0.00C
alreadi	0.000756	-0.001208	-0.000473	-0.004235	-0.00C
alway	0.000556	-0.000096	0.000129	-0.000652	0.00C
america	0.000626	-0.001643	-0.001838	0.001466	-0.00C
amid	0.000445	-0.000851	0.002513	0.000714	-0.001

FIGURE L.67

SVD Word Coefficients for the six concepts.

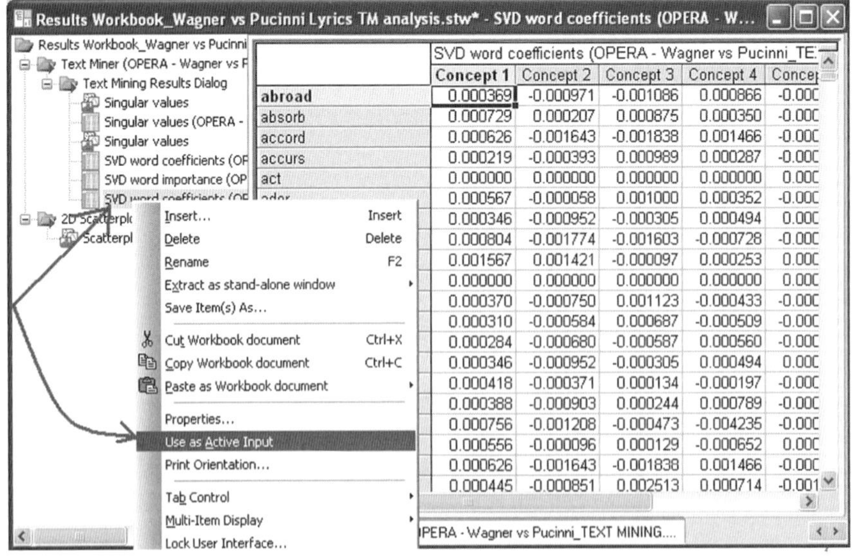

FIGURE L.68

Now let's use the "SVD word coefficients" as the "Active Input" (first, we'll have to "deselect" the SVD word importance as the active input).

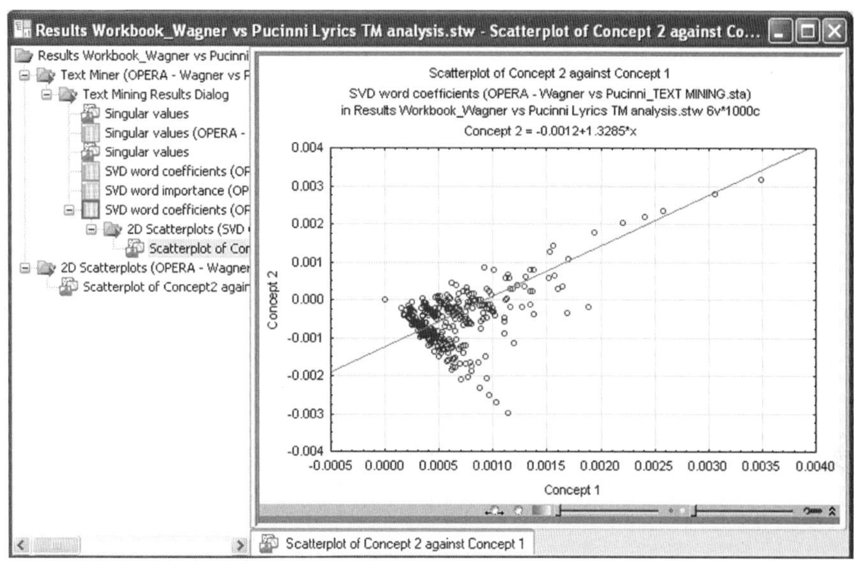

FIGURE L.69

2D Scatterplot of Concept 2 compared to Concept 1.

Then let's look at the graph more closely (see Figures L.70 through L.74).

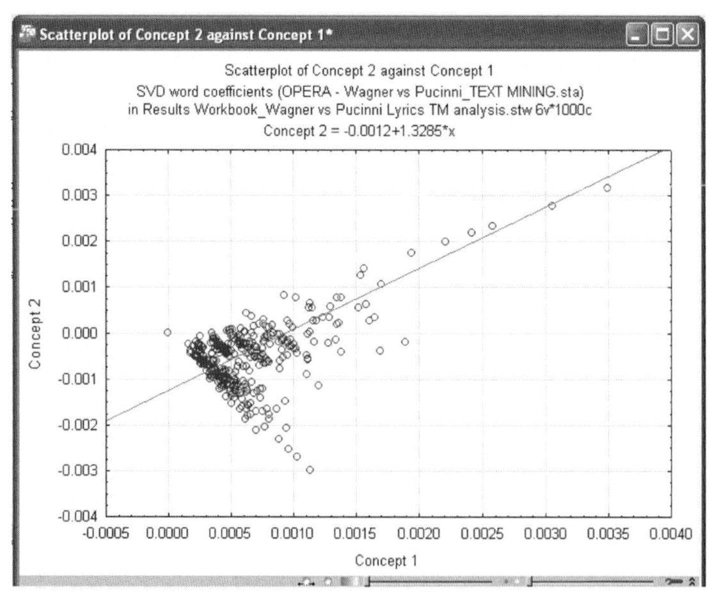

FIGURE L.70

Expanded view of Concept 2 compared to Concept 1.

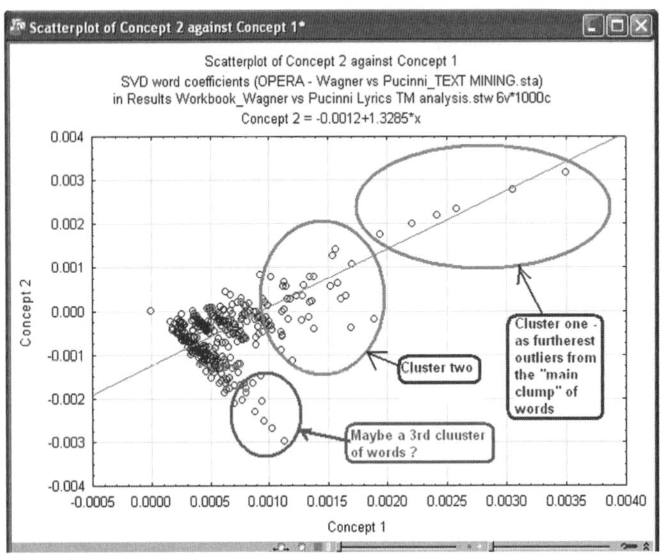

FIGURE L.71
Visual examination for "clusters" of words in the Concept 1 versus Concept 2 scatterplot.

Then apply Brushing as before (see Figures L.72 through L.73).

FIGURE L.72
Choosing both "Lasso" and "Label" on the Brushing window.

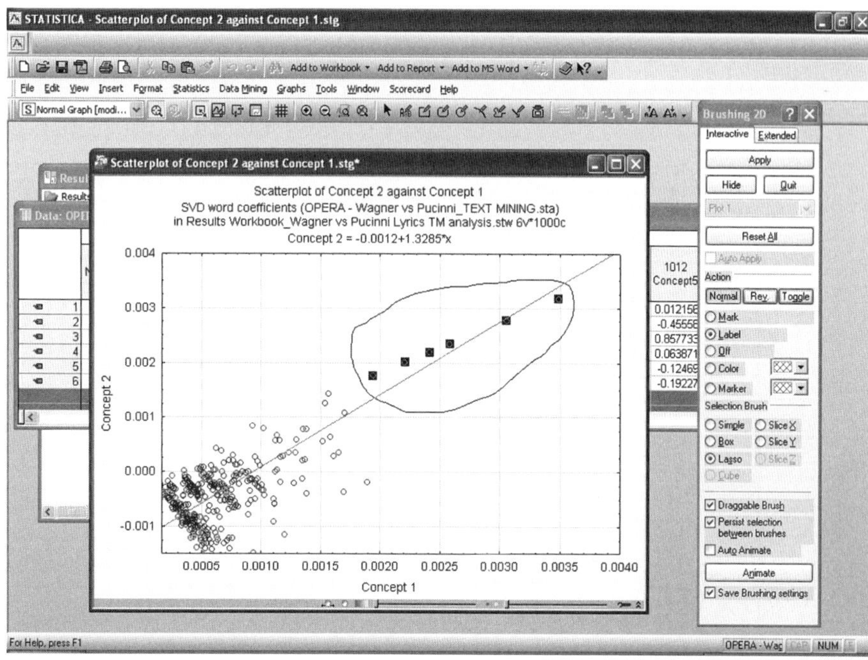

FIGURE L.73

Using "Lasso" and "Label" (see selections on Figure L.72), we can draw a circle (or "lasso") around what appear to be "clusters" of words on the scatterplot, and then the labels will be revealed.

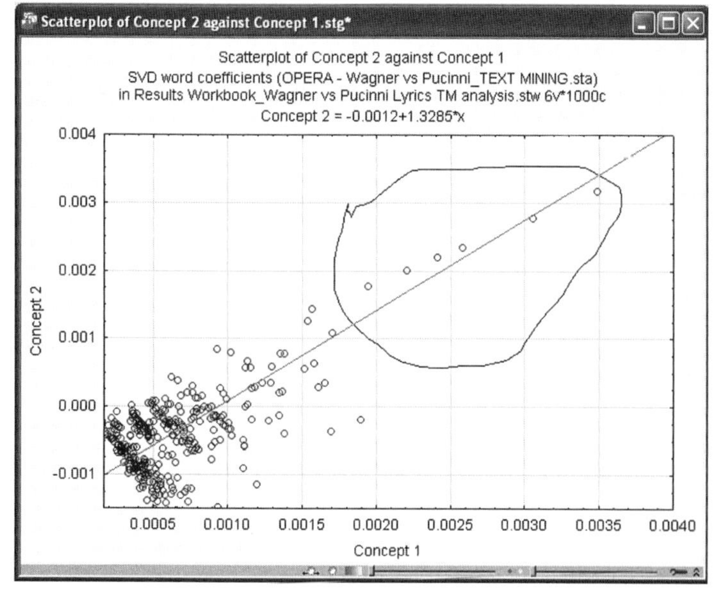

FIGURE L.74

Close-up view of the previous figure, showing just the graph.

Then by hitting the "Apply" button on the Brushing dialog window, the words for each "point" in the lassoed area appear (see Figure L.75).

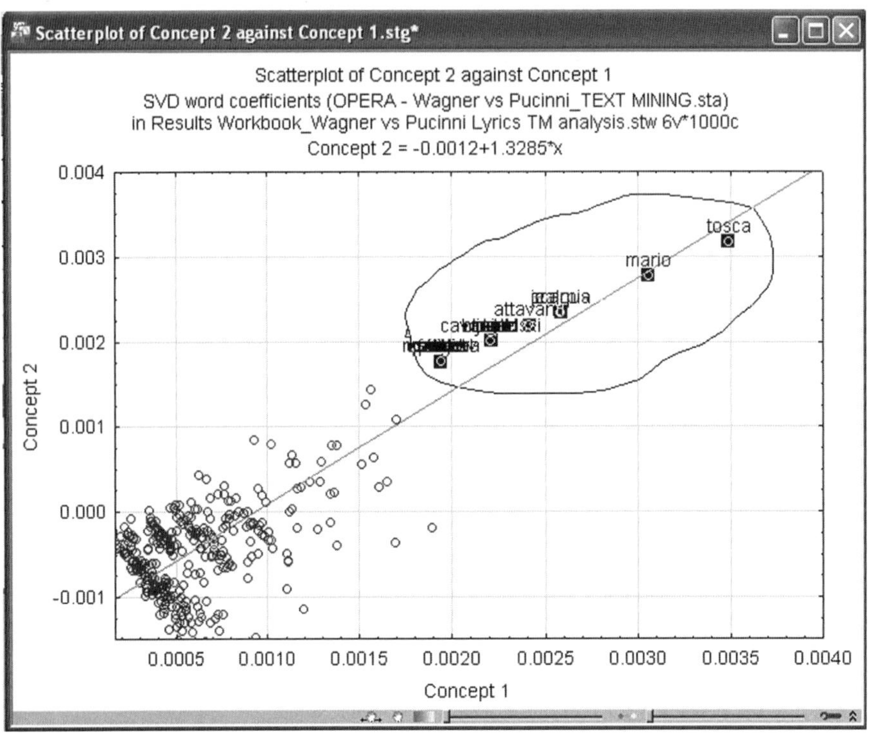

FIGURE L.75
Words in the lassoed area or cluster in the upper right area of the 2D scatterplot.

Now let's use the "Zoom" tool to get closer to this cluster and see if we can separate some of the "overlapping" words so they can be read easily (see Figures L.76 through L.78).

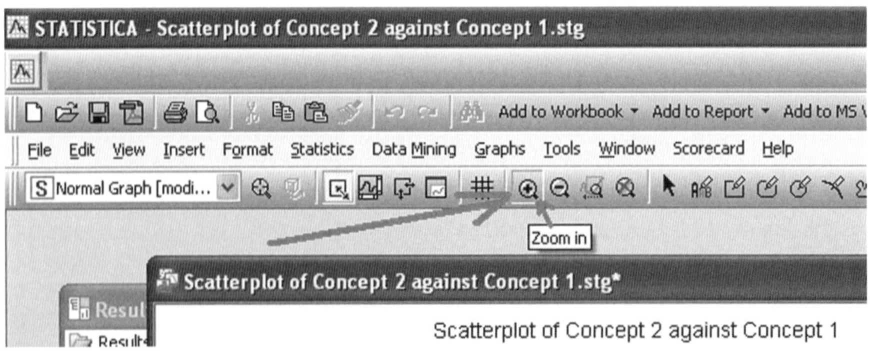

FIGURE L.76
Using the "Zoom in" tool will make the area larger and separate some of the overlapping word labels.

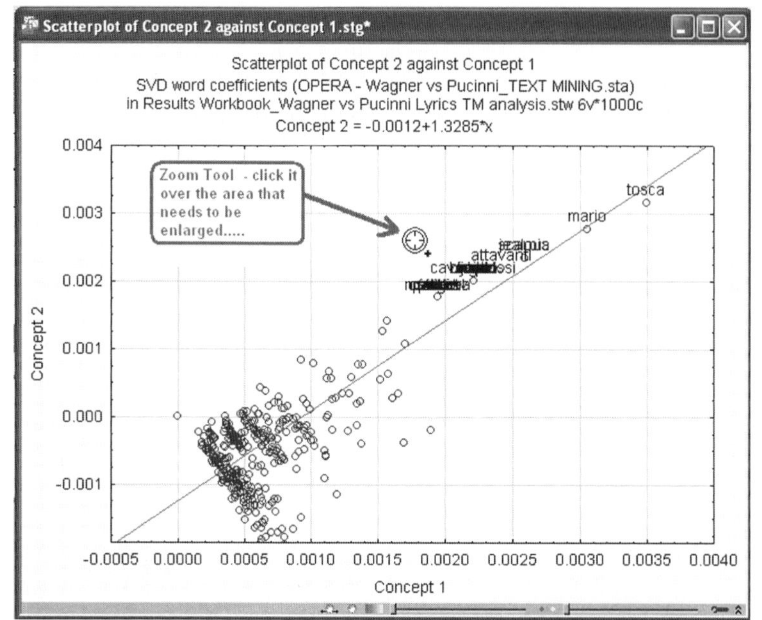

FIGURE L.77
How to use the "Zoom in" tool.

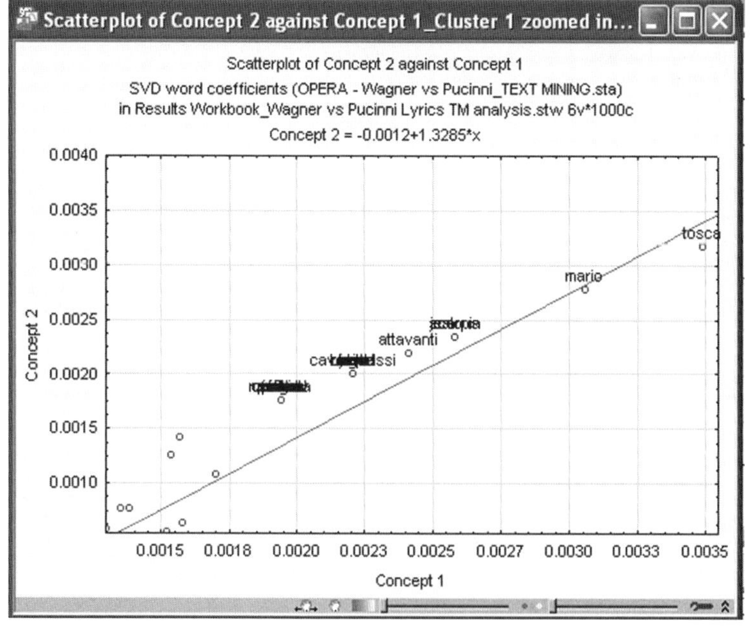

FIGURE L.78
Zoomed in view of Cluster One of the Opera Lyrics Concept 1 versus Concept 2.

Note that some of the data points for what we are calling "Cluster One" have more than one word for that point (see Figures L.79 through L.80). Thus, the word labels are on top of each other, causing a "black blur." These overlapping labels can be clicked on with the regular cursor (not the brushing cursor), so we need to exit the Brushing window to return to the regular cursor. Then, by clicking on the "overlapping" area and holding the mouse and dragging a label out, they can be separated.

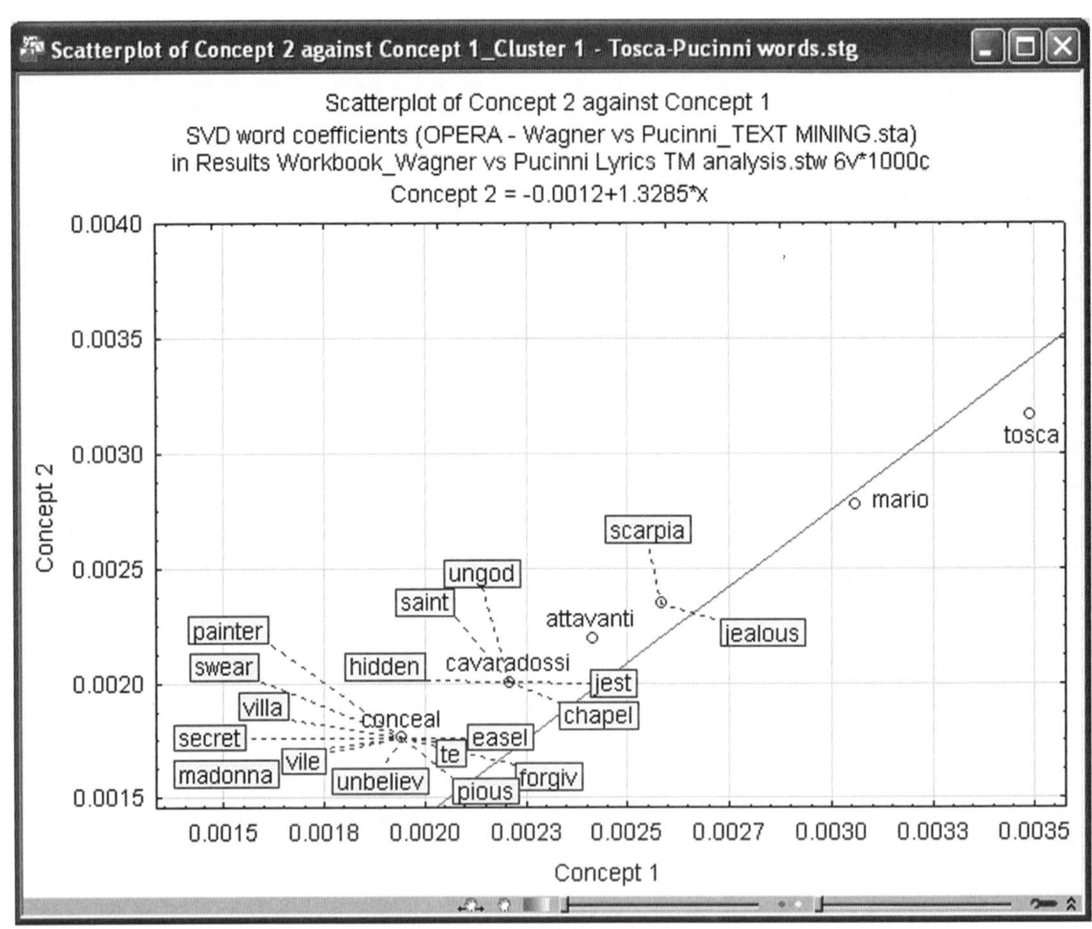

FIGURE L.79

Pulling the "overlapping words" apart. This is done by "exiting" the Brushing window, thus getting back regular cursors, and then clicking on the overlapping words and dragging each out from the "data point." If the words are dragged away from the data point far enough, a dotted line (will be coloured blue on the computer screen) will be drawn from the data point to the word.

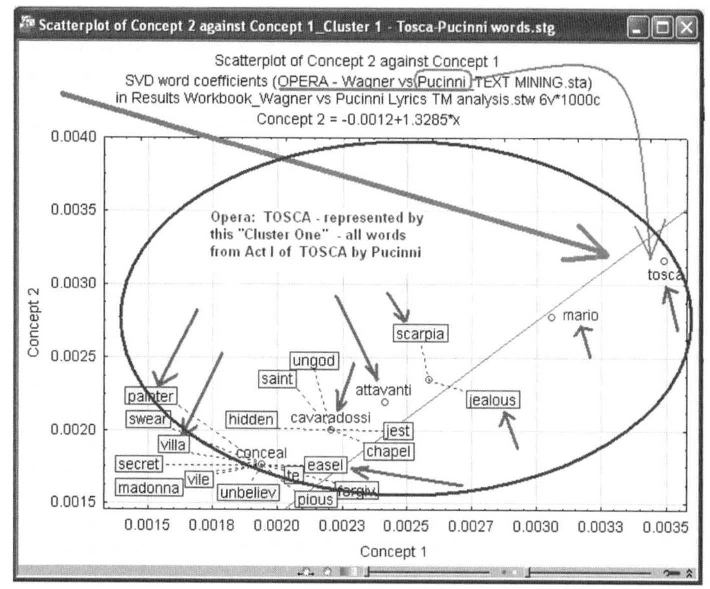

FIGURE L.80
The opera *Tosca* by Puccini seems to be primarily represented in "Cluster One" of the word coefficients of Concept 1 plotted versus Concept 2.

Now, using the Brushing tool again, let's "capture" "Cluster Two" on the 2D scatterplot of Concept 1 versus Concept 2 (see Figures L.81 through L.83).

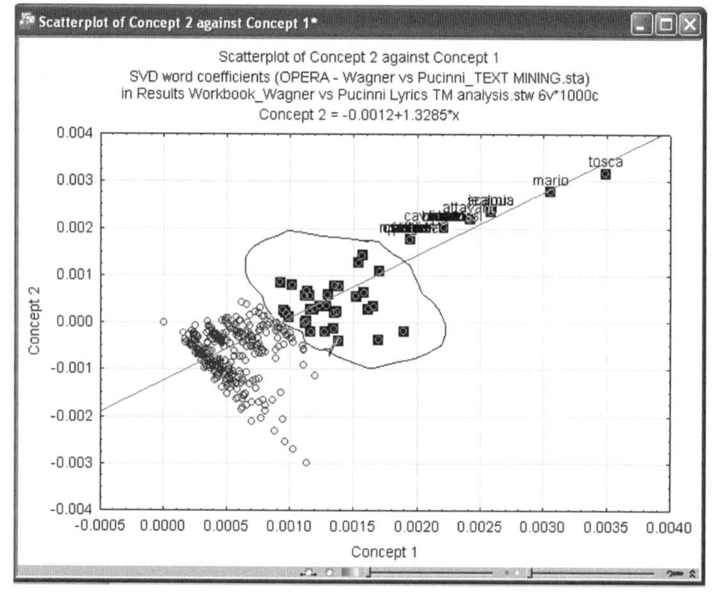

FIGURE L.81
"Cluster Two" on the 2D scatterplot of Concept 1 versus Concept 2 "lassoed" and ready to have the words exposed.

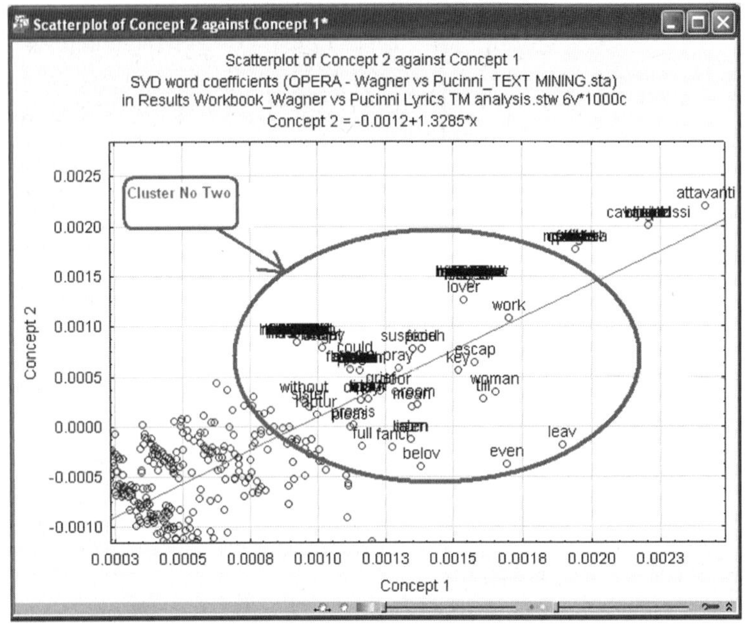

FIGURE L.82

"Cluster Two" of the words. Again, some of these overlapping words need to be pulled apart.

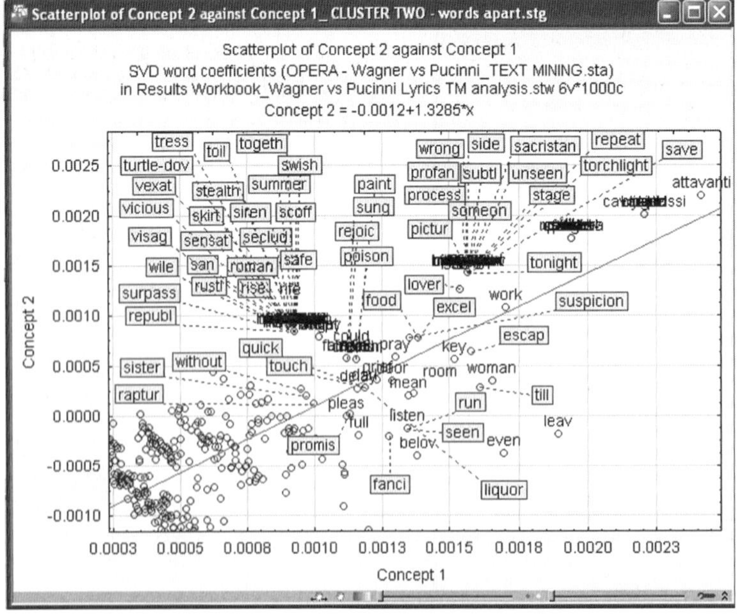

FIGURE L.83

Most of the overlapping words on "Two" pulled apart.

Using just a cursory glance at the words in "Cluster Two," it is difficult immediately to jump to a conclusion that this represented Wagner or Puccini. Many of these words seem to be more of the "dark side" seen in Wagnerian operas as compared to the "lighter – frilly – fun" operas of Puccini. But to be sure, we'd have to check to see if any of these words come from both Puccini and Wagner operas in this study.

Now let's take a quick look at "Cluster Three (see Figures L.84 through L.86)."

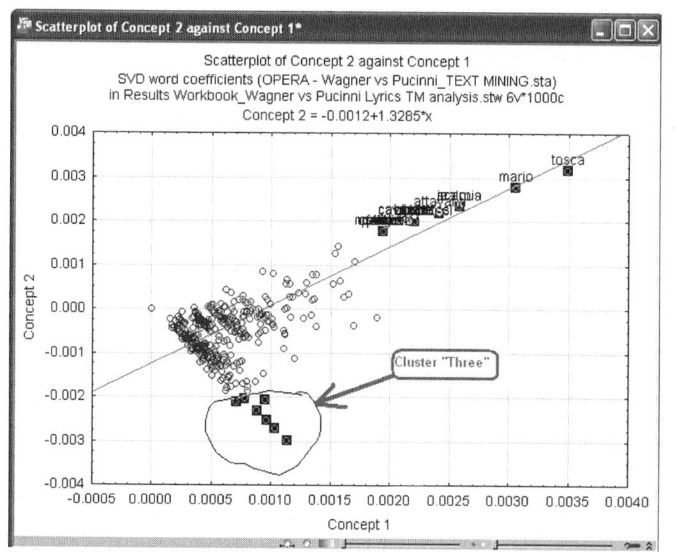

FIGURE L.84

Cluster Three.

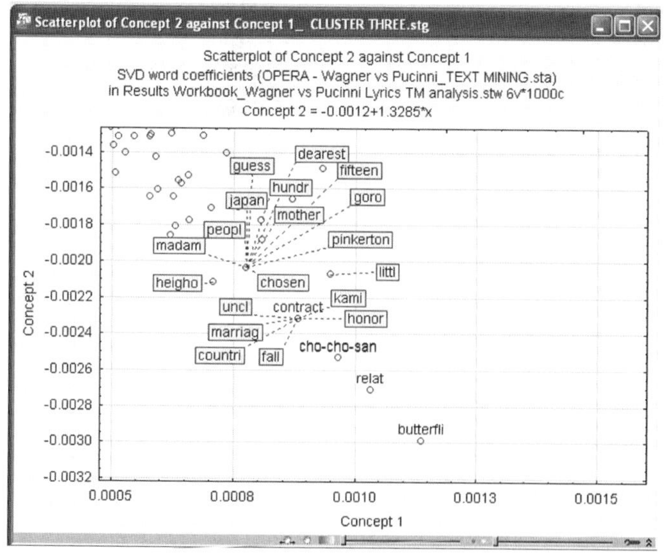

FIGURE L.85

"Cluster Three" overlapping words separated.

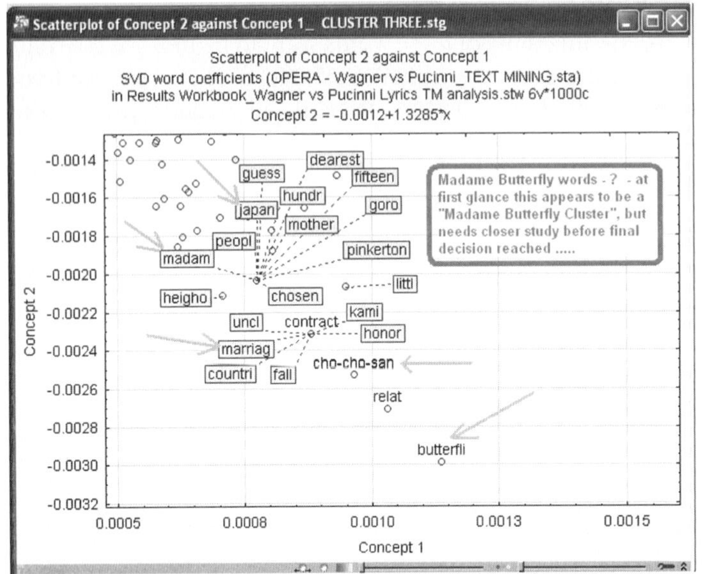

FIGURE L.86
Cluster Three words that appear to be related to Puccini's *Madame Butterfly*.

My suspicion at this point is that most of the Wagner words are in the "heavy clustering of words" at the lower left of the 2D scatterplot (see Figures L.87 through L.89).

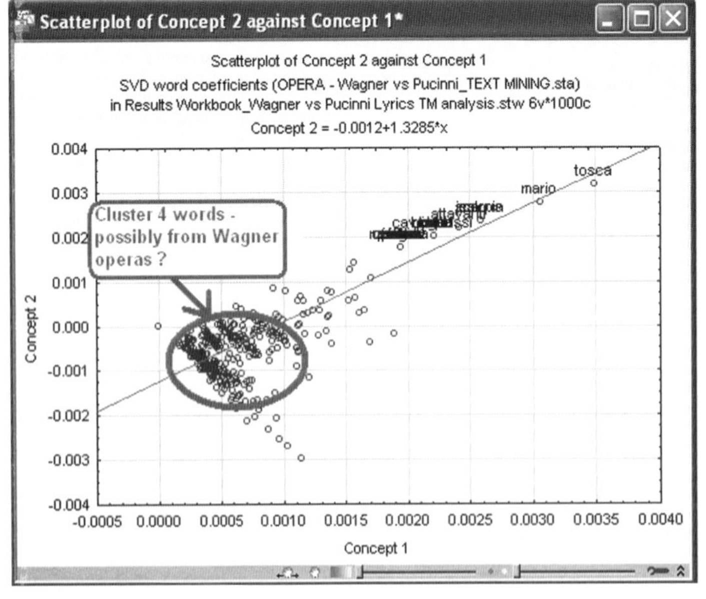

FIGURE L.87
Possible "Cluster Four," where the Wagner opera words may be clustered, Wagner lyrics written by himself, and more "words" and "darker side of life" type of words.

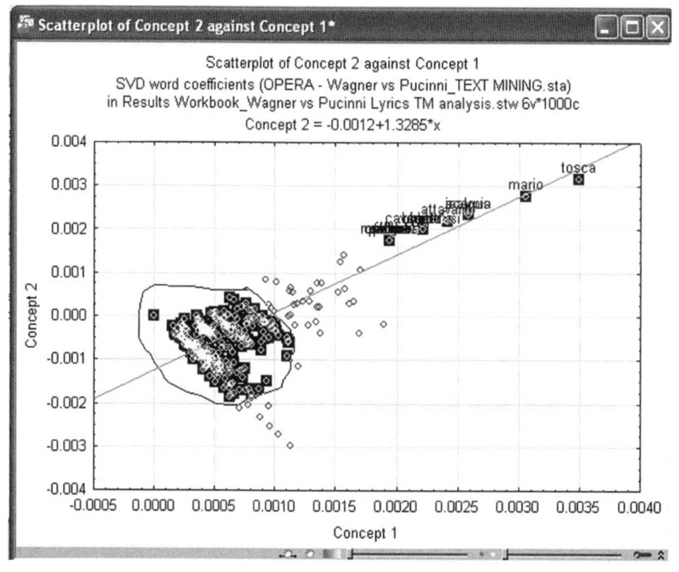

FIGURE L.88

Selecting the Cluster Four words with the Brushing Lasso tool so they can be labeled and some separations made.

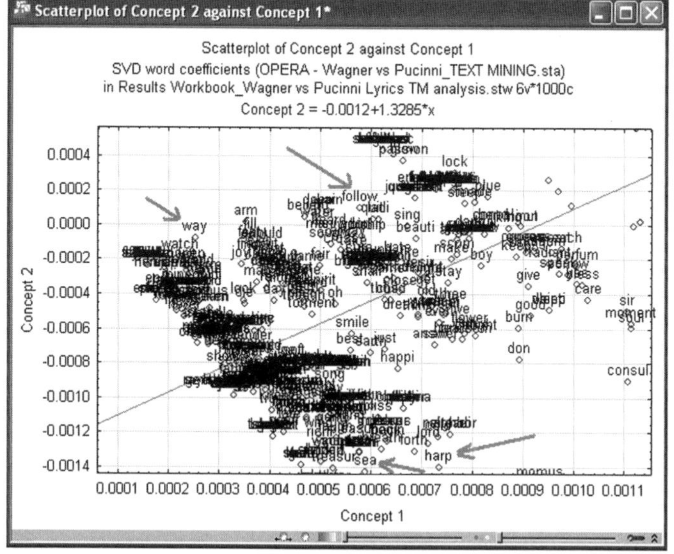

FIGURE L.89

Expanded (zoomed in on) Cluster Four prior to separating any "overlapping words." As you can see, there are so many overlapping words that this may be a difficult operation, at least for one window. We may have to take "pieces" of Cluster Four and do a separation on each if we really want to dissect and find out which words come from Puccini or Wagner, or both sets of operas.

But a cursory glance of Cluster Four does have words like "sea," "harp," "way," and "follow," which, as I recall, all have to do with the two ships in one of the Wagner operas. So this is a "hint" that many of these words may be related to Wagner's operas. Careful dissection will be needed, however, to come to that conclusion. I suspect that the conclusion will be that (1) some of the words are a Wagner cluster, and (2) other words in this cluster are more "generic" and can be found in both Wagner and Puccini operas. At this point we leave it up to you, the reader, to separate the words in these "four clusters" further to see if all or most of the words belong to the Wagner opera.

CASE STUDY: Sentiment-Based Text Analytics to Better Predict Customer Satisfaction and Net Promoter® Score Using IBM®SPSS® Modeler

Olivier Jouve, Eric Martin, and Marie-Claude Guerin
IBM Corp.

CONTENTS

INTRODUCTION

Net Promoter Score (NPS) is a customer loyalty metric developed and registered by Satmetrix. The NPS is a convenient and simple way to analyze customer overall satisfaction. Customers rate their satisfaction on a 0 to 10 scale about a company, a service, or a product they use. Results are used to segment customers into three groups: Promoters, Passives, and Detractors. The percentage of Detractors is then subtracted from the percentage of Promoters to obtain a Net Promoter score. Data are collected on typical customer satisfaction surveys, along with some other customer attributes like additional geo-demographic data (i.e., age, gender, zip code, etc.), behavioral data (i.e., "How often do you visit our store?"), and attitudinal data (i.e., customers wishes, likes and dislikes). Attitudinal data have been increasingly used to better understand the reasons why customers belong to one of those three groups. Marketers use open-ended questions because they don't frame customer responses and help marketers in detecting new trends, and they don't treat customers as an "answering machine" and leave some room to personal comments and motivations. These free text comments usually enlighten the reasons of a poor NPS. For example, customers who give a low score to a mobile Telco operator quality of service are likely to complain about some specific points (i.e., coverage, call

Practical Text Mining and Statistical Analysis for Non-structured Text Data Applications. DOI: 10.1016/B978-0-12-386979-1.00021-9

centers, audio quality, etc.), which may or may not have been addressed by previous questions. Hence marketers use Text Analytics techniques to analyze customers' free text feedback so they can quantify topics and related opinions from the customers' own words. The full insight comes then when structured and unstructured data are brought together to better understand customers' behaviors and detect complex patterns from the data. Survey data can deliver more value than just insight from the restricted (even if significant) set of polled customers. Customers' comments and attributes can be used as inputs to model NPS through data mining techniques. Those predictive models can then be used and deployed on some other customers to predict their NPS status (Promoter, Passive Detractor). So eventually all customers' satisfaction groups can be inferred from a customer satisfaction survey. People who have similar attributes, issues, and opinions are likely to be in the same NPS group.

BUSINESS OBJECTIVES

The data collected in this study are very similar to what mobile operators do on a regular basis to measure NPS. Each NPS segment can be addressed differently by marketing operations:

- *Detractors* are derived from customers with customer satisfaction between 0 and 6. They are unsatisfied customers. Many of them are likely to churn. They can be enrolled in retention programs or follow-up actions to fix their issues and turn them into satisfied customers. Their feedback is valued to better understand what's going wrong and to better understand the key criteria associated with quality of service.
- *Passives* correspond to customer satisfaction that equals 7 or 8. They are supposed to be neutral. They are supposed to remain active customers as long as they don't experience any major issue.
- *Promoters* correspond to customer satisfaction that equals 9 or 10. They are very satisfied customers likely to spread positive word-of-mouth around them. Companies can better understand the reasons for their satisfaction and replicate that across different geographies or markets. They can also serve as targets for viral marketing campaigns.

The main issue when analyzing customer surveys is to properly and consistently analyze customers' own words on a reasonable scale. Market researchers, for example, often categorize or "code" free-text responses in surveys. Because people are good at understanding text, this approach is quite accurate, but it is time-consuming and expensive. In addition, a manual approach cannot offer guidance in identifying relationships or trends in the information analyzed. With the immense volume of text now available, often in multiple languages, other approaches are needed.

A second approach is to employ automated solutions based on statistics. Some of these, however, simply count the number of times terms occur and calculate their proximity to related terms. Because they cannot factor in the ambiguities in human languages, relevant relationships may be hidden in masses of irrelevant findings—or missed altogether. Some of these statistics-based solutions compensate by providing ways for analysts to create rule books that help suppress irrelevant results. But these rule books need to be created and continually updated by analysts, which adds cost and complexity.

Linguistics-based text analytics offer the speed and cost-effectiveness of statistics-based systems but offer a far higher degree of accuracy. Linguistics-based text analytics is based on the field of study known as natural language processing (NLP). The understanding of language that is possible with the NLP approach cuts through the ambiguity of text, making linguistics-based text analytics the most accurate possible approach. Initially, linguistics-based solutions may require some human intervention in developing dictionaries for a particular industry or field of study. But the benefit obtained from these efforts is significant. The results are more accurate, and the techniques involved are more transparent, meaning that they can be modified by users to further increase the accuracy of the results. Sentiment Analysis is at the cutting edge of NLP techniques because it can discriminate mixed opinions within the same sentence or response, providing a high level of accuracy while structuring unstructured data. For instance, in a sentence like "I'm fine with my plan, but it takes way too long to get connected to the help desk," the technology must be smart enough to recognize that there are two different opinions in the same sentence and to properly link satisfaction to the "plan" and dissatisfaction to the "help desk."

Another Text Analytics issue is that all results need to be aggregated at a higher level to be exploitable. Similar issues and topics need to be grouped together in order to work with a limited but meaningful set of new customer attributes derived from text.

CASE STUDY

The goal of this tutorial is to demonstrate the value of the use of unstructured data in better understanding the reasons for customer (dis-)satisfaction and improving the accuracy of predictive models to predict NPS from a sample of customer data.

Data

The survey data set consists of 1,340 records, each corresponding to a mobile phone customer interview. The survey fields correspond to customer ID, marital status, gender, age, number of children, estimated income, ownership of a car (yes or no), education level, employment status, satisfaction score, and comments related to the provided level of satisfaction (as free text).

Stream Creation

IBM SPSS Modeler contains several nodes to analyze unstructured data. Among them, the Text Mining modeling node is used to access data and extract concepts in a stream. You can use any source node to access data, such as a Database node, Var. File node, Web Feed node, a Fixed File node, or an Excel File node. For text that resides in external documents, a File List node can be used. In this application, we will use an Excel File node as the source and connect it to the Text Mining node.

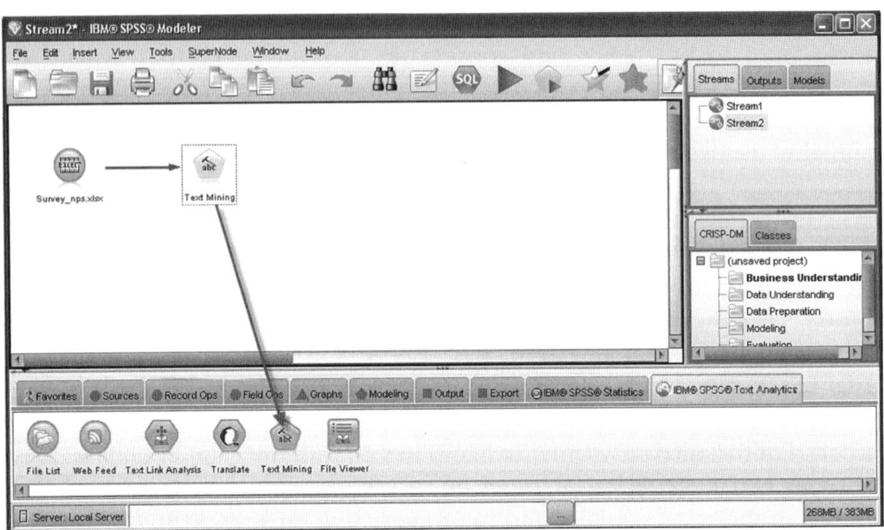

FIGURE M.1

Creating a stream.

Before executing the stream, we need to identify which part of the file to analyze; this is done from the Fields tab in the Text Mining node.

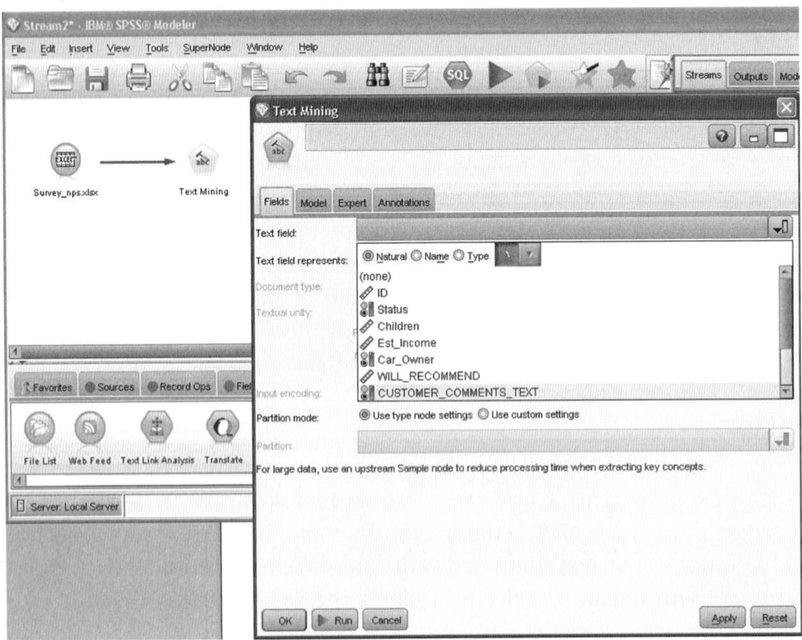

FIGURE M.2

Text field selection.

We also need to specify the type of model to be built and the linguistic resources (a.k.a. dictionaries) and/or sets of categories to be used. When the Build interactively (category model nugget) is selected, there are three different types of outputs:

1. Using extraction results to build categories
2. Exploring text link analysis (TLA) results
3. Analyzing coword clusters

Text link analysis is a pattern-matching technology that enables you to use existing pattern rules; define your own, if needed; and compare these to actual extracted concepts and relationships found in your text. This is particularly suited to extract opinions on products or services.

IBM SPSS Text Analytics rapidly and accurately captures and extracts key concepts and relationships from text data. This extraction process relies heavily on linguistic resources to dictate how to extract information from text data. When you install the software, you also get a set of specialized resources that you can fine-tune in the Resource Editor view, if needed. These resources come in various forms, and each can be used in your session. Resources can be found in the following:

- **Resource templates**. Templates are made up of a set of libraries, types, and some advanced resources, which together form a specialized set of resources adapted to a particular domain or context, such as product opinions.
- **Text analysis packages (TAPs)**. In addition to the resources stored in a template, TAPs also bundle together one or more specialized category sets generated using those resources so that both the categories and the resources are stored together and reusable.

Selecting a resource template or a TAP will automatically define the language of the application and load the relevant language-based extraction dictionaries and grammars.

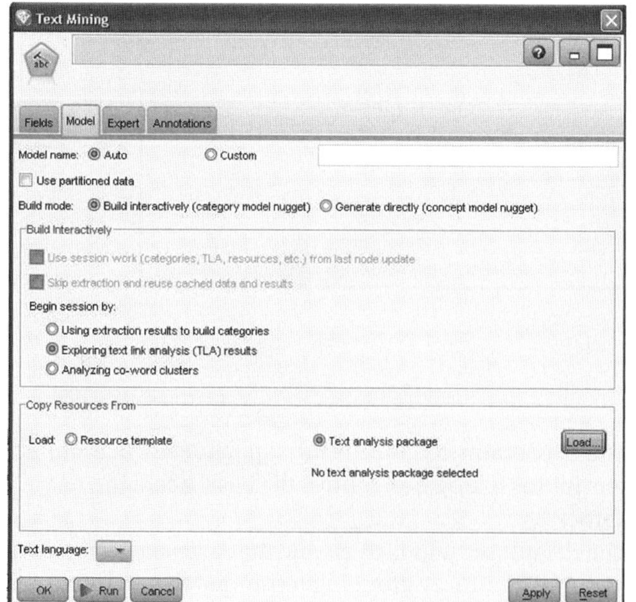

FIGURE M.3

Select the model and linguistic resources to be used.

The Customer Satisfaction TAP contains a set of predefined categories on opinions about products and services. Since the goal is to identify reasons for dissatisfaction, it is recommended to use the set of categories on negative opinions.

FIGURE M.4
Selecting a TAP and the right set of categories.

Expert Tab

The third tab of the Text Mining node, the Expert tab, contains certain advanced parameters that impact how text is extracted and handled. The parameters in this dialog box control the basic behavior, as well as a few advanced behaviors, of the extraction process.

The only parameter that we will change in the default settings is to enable the *Accommodate spelling for a minimum root character limit of: 5*. This option applies a fuzzy grouping technique that helps group commonly misspelled words or closely spelled words under one concept. The fuzzy grouping algorithm temporarily strips all vowels (except the first one) and strips double/triple consonants from extracted words and then compares them to see if they are the same so that *modeling* and *modelling* would be grouped together.

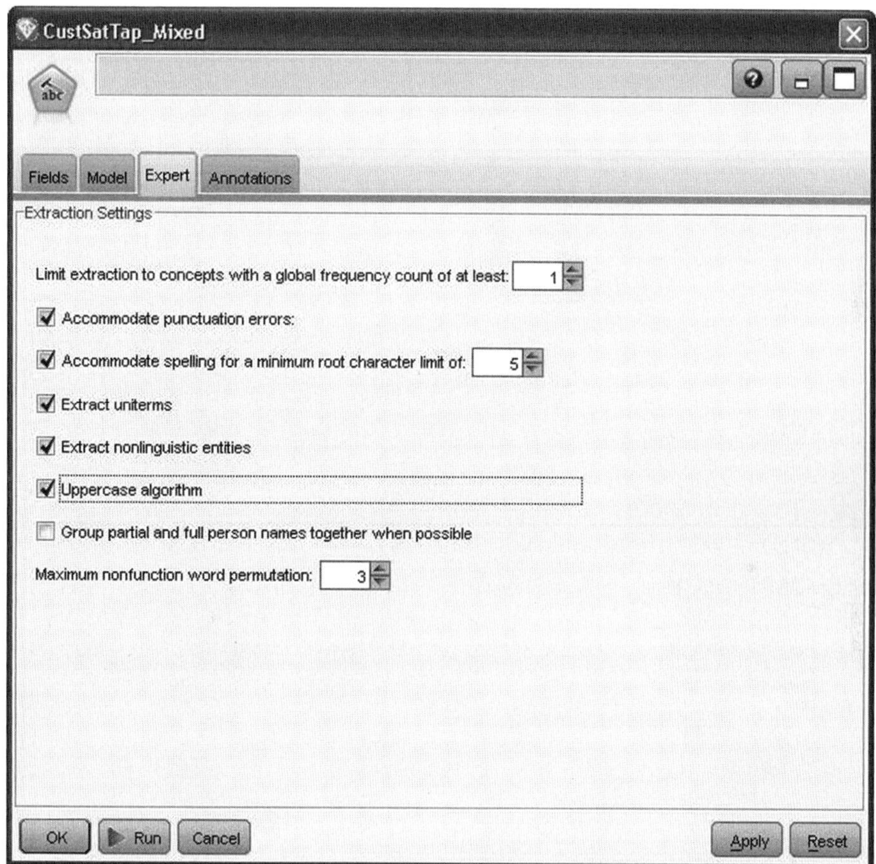

FIGURE M.5
Expert option tab.

From the Text Mining node, click *Run* to execute the stream. Since the model chosen was *Exploring text link analysis (TLA results)*, results are displayed directly in this pane. We will discuss each of the three other panes later.

The Text Link Analysis View

When you select a type pattern, the corresponding concept patterns are displayed in the lower part of the pane. This is also shown in the Concept Web pane, on the right upper part of the panel. The Concept Web pane provides a graphical overview on how concepts are connected with one another. In Figure M.7, you can see that the strongest link is between *calls* and *dropped*.

If you click on *Display*, the corresponding documents or records will appear in the right lower part of the panel.

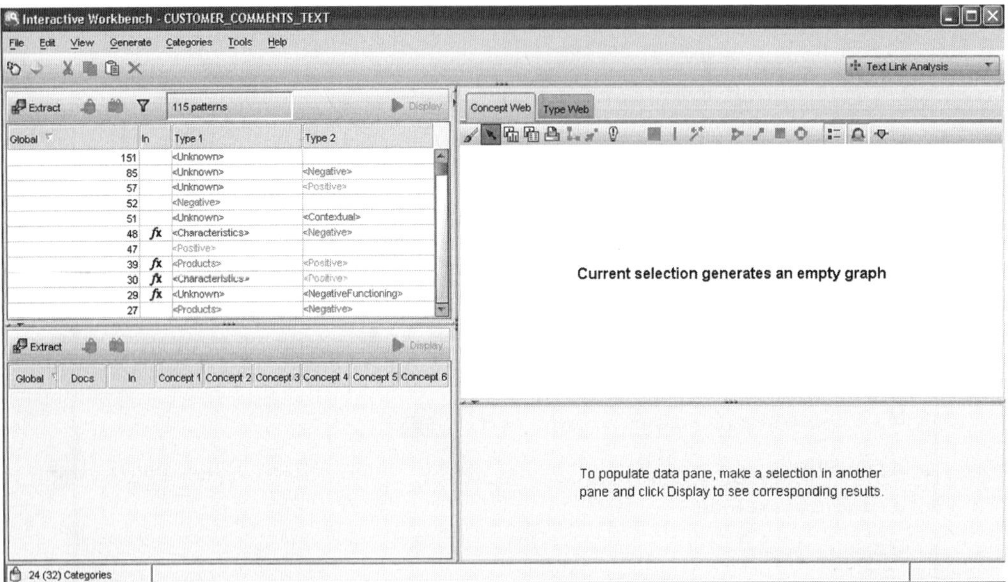

FIGURE M.6
Results on the Text Link Analysis (TLA) pane.

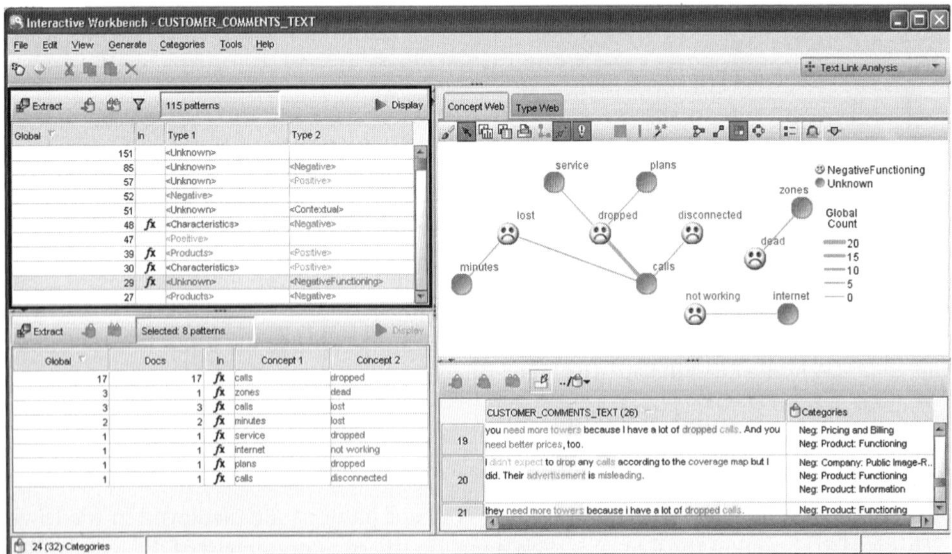

FIGURE M.7
Select a type pattern to display concepts and source text.

The Categories and Concepts View

The Categories and Concepts view is the window in which you can create and explore categories, as well as explore and tweak the extraction results. **Categories** refers to a group of closely related ideas and patterns to which documents and records are assigned through a scoring process, while **concepts** refer to the most basic level of extraction results available to use as building blocks, called descriptors, for your categories.

Categories Pane

Located in the upper left corner, this area presents a table in which you can manage any categories you build. To run this application, remember that we have selected a TAP; this explains why this pane is already populated with categories (top and subcategories) and descriptors.

If you double click a category name, the Category Definitions dialog box opens and displays all of the descriptors that make up its definition, such as concepts, types, and rules.

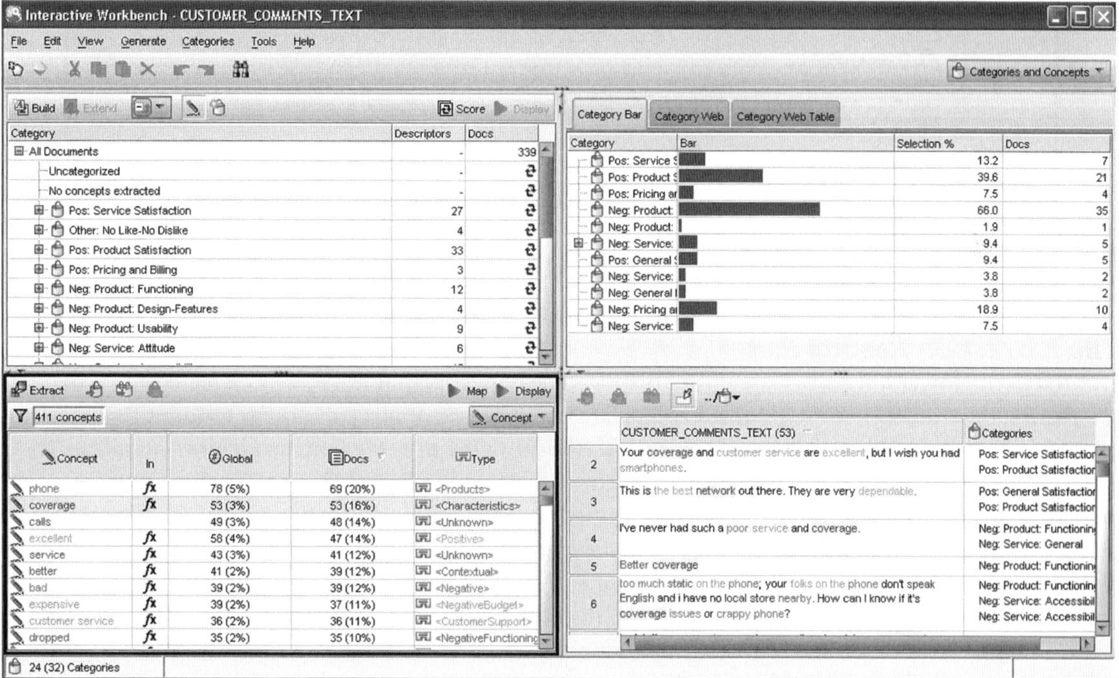

FIGURE M.8

The four panes of the *Categories and Concepts* view. Category definitions (top left), Category bar charts (top right), Concepts (bottom left), and original text with categorization results (bottom right).

IBM SPSS Text Analytics supports hierarchical categories. In Figure M.9, *Neg: Service: Accessibility* is a top category, which contains some subcategories such as *Language Issues, Wait Time*, and *Phone Support.*

[no answer] or *[* interaction & (<Negative> | better | no)]* are descriptors.

Build Extend Score Display

<table>
<tr><th>Category</th><th>Descriptors</th><th>Docs</th></tr>
<tr><td>Pos: Service Satisfaction</td><td>27</td><td>59</td></tr>
<tr><td>Pos: General Satisfaction</td><td>35</td><td>48</td></tr>
<tr><td>Neg: Service: Accessibility</td><td>45</td><td>43</td></tr>
<tr><td>fx [no answer]</td><td></td><td>2</td></tr>
<tr><td>fx [* support * & (<Negative> | better | no | greater) & !(<Busine</td><td></td><td>2</td></tr>
<tr><td>fx [* people * & (no | more)]</td><td></td><td>1</td></tr>
<tr><td>fx [* interaction & (<Negative> | better | no)]</td><td></td><td>0</td></tr>
<tr><td>fx [* answer * & (<Negative> | no | better) & !(too long | * time)]</td><td></td><td>0</td></tr>
<tr><td>fx [* automated *]</td><td></td><td>0</td></tr>
<tr><td>fx [* access * & (<Negative> | <NegativeFunctioning> | better | r</td><td></td><td>0</td></tr>
<tr><td>fx [* machine* * & more]</td><td></td><td>0</td></tr>
<tr><td>Language Issues</td><td>3</td><td>14</td></tr>
<tr><td>Wait Time</td><td>7</td><td>12</td></tr>
<tr><td>Phone Support</td><td>3</td><td>11</td></tr>
</table>

FIGURE M.9

The Categories pane. The category "negative opinions about accessibility to the service" is shown with its descriptors.

The Extraction Results Pane

Located in the lower left corner, this pane displays the concepts that have been extracted from the data set. During the extraction process, the text data are scanned and analyzed in order to identify interesting or relevant single words (such as *coverage* or *calls*) and word phrases (such as *rollover minutes*). These words and phrases are collectively referred to as *terms*. Using the linguistic resources, the relevant terms are extracted, and then similar terms are grouped together under a lead term called a **concept**. By default, those concepts are displaying by decreasing number, but you can also choose to sort them alphabetically.

Each concept has a Global count (the total number a concept occurs in the data set), a Docs count (which may be lower than the Global count, because if a concept occurs twice in the same document, it will be counted only once here), and a Type.

You can see the set of underlying terms for a concept by hovering your mouse over the concept name. Doing so will display a tool tip showing the concept name and up to several lines of terms that are grouped under that concept. These underlying terms include the synonyms defined in the linguistic resources (regardless of whether or not they were found in the text), as well as the any extracted plural/singular terms, permuted terms, terms from fuzzy grouping, and so on.

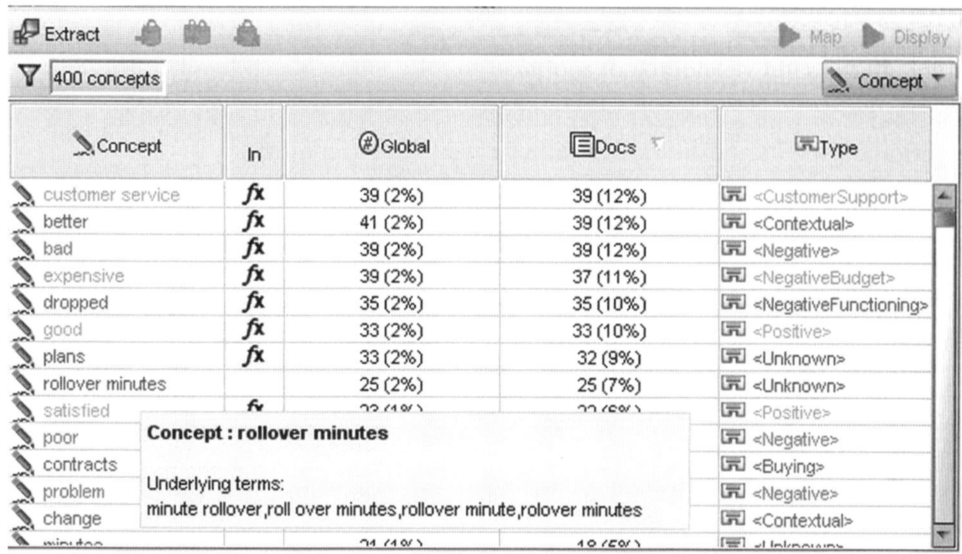

FIGURE M.10
The Extraction Results pane.

The Data Pane

You can also select a concept, click Display, and see the corresponding source text in the Data pane, located in the lower right corner.

FIGURE M.11
The Data pane.

The Vizualization Pane

Located in the upper right corner, this area presents multiple perspectives on the commonalities in document/record categorization. Each graph or chart provides similar information but presents it in a different manner or with a different level of detail. These charts and graphs can be used to analyze your categorization results and aid in fine-tuning categories or reporting.

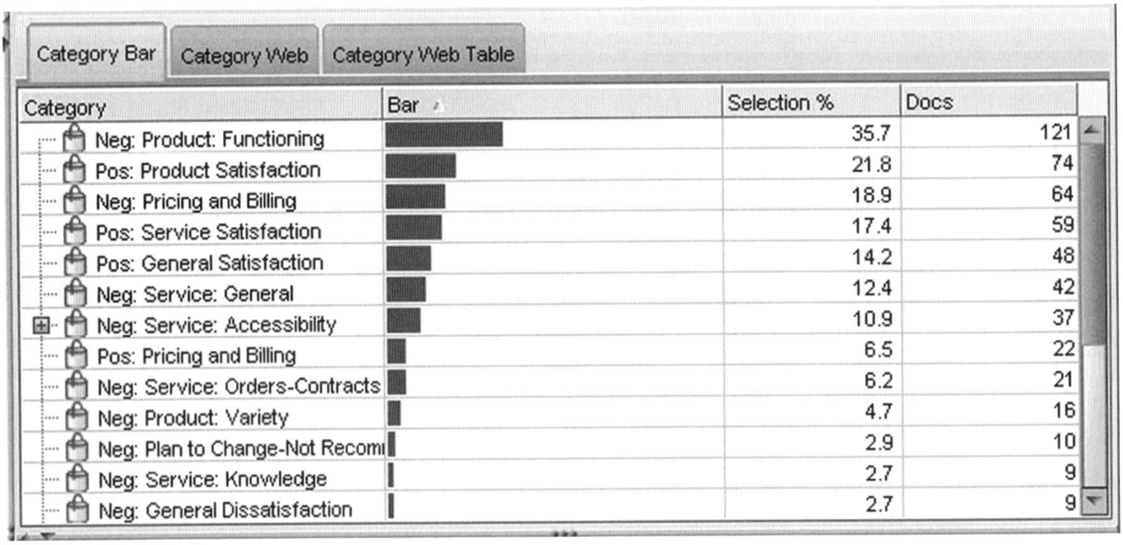

Category Bar | Category Web | Category Web Table

Category	Bar	Selection %	Docs
Neg: Product: Functioning		35.7	121
Pos: Product Satisfaction		21.8	74
Neg: Pricing and Billing		18.9	64
Pos: Service Satisfaction		17.4	59
Pos: General Satisfaction		14.2	48
Neg: Service: General		12.4	42
Neg: Service: Accessibility		10.9	37
Pos: Pricing and Billing		6.5	22
Neg: Service: Orders-Contracts		6.2	21
Neg: Product: Variety		4.7	16
Neg: Plan to Change-Not Recomi		2.9	10
Neg: Service: Knowledge		2.7	9
Neg: General Dissatisfaction		2.7	9

FIGURE M.12
The Vizualization pane.

The Clusters View

When you first access the Clusters view, no clusters are visible. You can build the clusters through the menus (*Tools* > *Build Clusters*) or by clicking the *Build...* button on the toolbar. This action opens the Build Clusters dialog box in which you can define the settings and limits for building your clusters.

A cluster is a grouping of related concepts generated by clustering algorithms based on how often these concepts occur in the document/record set and how often they appear together in the same document, also known as co-occurrence. Each concept in a cluster co-occurs with at least one other concept in the cluster.

Since we have decided to use a TAP for this application, we do not need to build clusters, but this is a very useful way to discover what is in your data. In Figure M.13, we notice that the strongest link is between *calls* and *dropped*, then *lost*.

Each time a category is created or updated, the documents or records can be scored by clicking the *Score* button to see whether any text matches a descriptor in a given category. If a match is found, the document or record is assigned to that category. The end result is that most, if not all, of the documents or records are assigned to categories based on the descriptors in the categories.

Click the *Score* button on top of the left part of the panel. When you click twice on the *Docs* column, categories are sorted by decreasing number of documents/records matched. The first category on this set of documents is by far *Neg: Product Functioning*. Expand this category to see the different descriptors and their corresponding scores; there are no subcategories in this category.

Select the second descriptor fx *reception * | * coverage * | * signal* *) & (<Negative> | better | no)]* and click Display to see the corresponding documents.

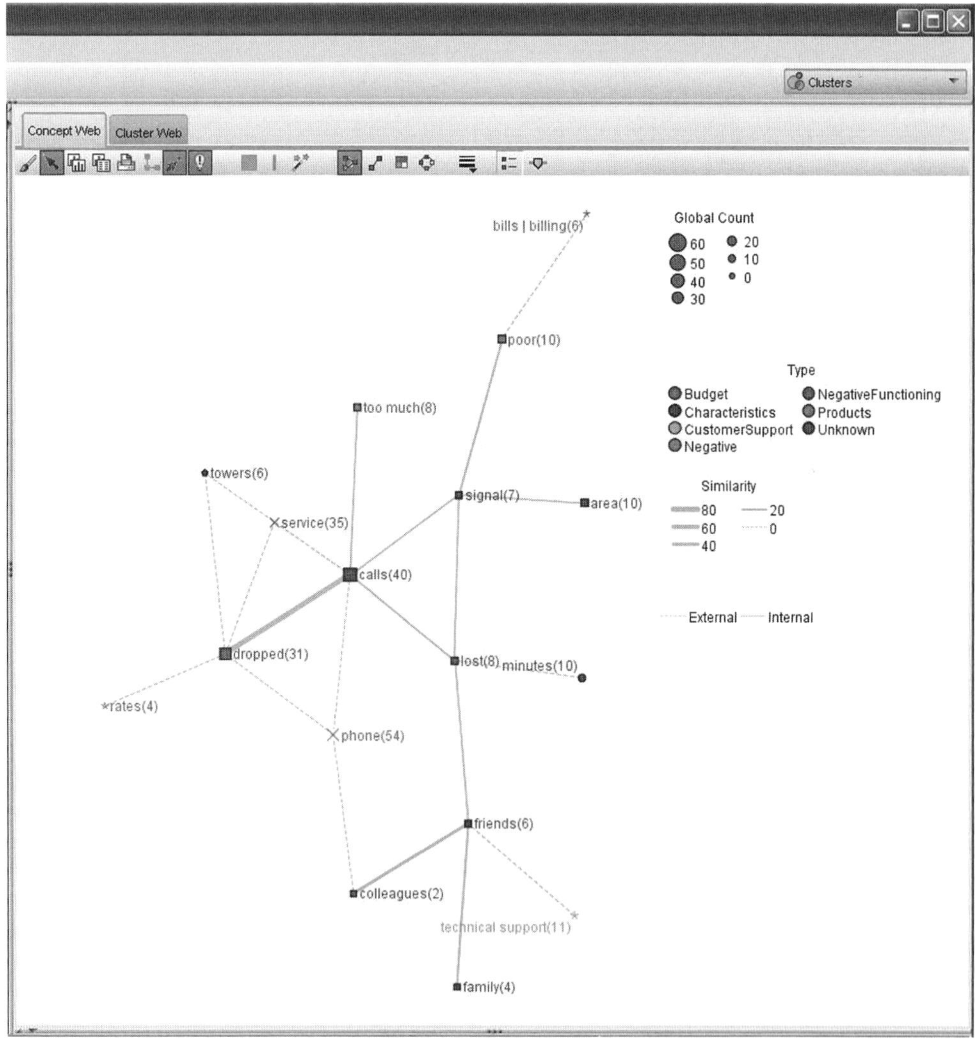

FIGURE M.13
Cluster "calls."

	CUSTOMER_COMMENTS_TEXT (26)	Categories
1	I have a lot of problems with the reception; And i don't like your billing process;;.	Neg: Pricing and Billing Neg: Product: Functioning
2	There should be more towers where we live. Coverage is poor.	Neg: Product: Functioning
3	I've never had such a poor service and coverage.	Neg: Product: Functioning Neg: Service: General
4	Better coverage	Neg: Product: Functioning
5	The coverage is not really good in my area.	Neg: Product: Functioning
6	I don't get service in my house. Poor reception around my house	Neg: Product: Functioning Neg: Service: General
7	Change your plans. I don't need text messages, but i'd like better phone coverage.	Neg: Product: Functioning
8	Improve your network: dead zones, dead zones, dead zones;..	Neg: Product: Functioning
9	crappy phones. Poor reception. Outragous bills	Neg: Product: Functioning Neg: Pricing and Billing
10	I wish there was a tower where we live. Apart from coverage issues, I have no problem with your services.	Neg: Product: Functioning Pos: Product: Functioning
11	Nothing special, except that i'd like better coverage. I drop calls too often.	Neg: Product: Functioning Other: Don't Know
12	No coverage where I live. No problem at my girlfriend's place	Neg: Product: Functioning
13	cheaper plans. More rollover minutes. Better coverage in some areas.	Neg: Pricing and Billing Neg: Product: Functioning
14	the phone I got is great, but unfortunately I live in the mountains and reception is awful. Most calls are lost.	Neg: Product: Functioning Pos: General Satisfaction
15	lower minute prices and improve coverage.	Neg: Pricing and Billing Neg: Product: Functioning
16	improve coverage in some places, especially in the moiuntains.	Neg: Product: Functioning
17	cheaper phone cards, more ring tones, better coverage in certain areas.	Neg: Pricing and Billing Neg: Product: Functioning
18	Poor signal. Nobody here gets signal.	Neg: Product: Functioning
19	the signal here is very poor. I lose most of my calls.	Neg: Product: Functioning
20	Increase the size of the network to reduce roaming charges	Neg: Pricing and Billing Neg: Product: Functioning
21	you probably need more coverage	Neg: Product: Functioning
22	make your coverage better; I live in the mountains and sometimes I can't get calls.	Neg: Product: Functioning
23	better signal in some areas; sometimes, you don't get calls.	Neg: Product: Functioning
24	I have several reasons to be dissatisfied. 1. too many dropped calls. Poor coverage where i live. 2. rates are too expensive. 3. billing issues. 4. don't know about technical support because I've never contacted them.	Neg: Pricing and Billing Neg: Product: Functioning Neg: Service: Accessibility

FIGURE M.14
Documents scored by descriptor reception * | * coverage * | * signal* *) & (<Negative> | better | no)].

CREATING NEW CATEGORIES AND ADDING MISSING DESCRIPTORS

The results obtained with the Customer Satisfaction TAP seem correct and cover most of the records. However, some of them are not yet assigned to categories. Looking at Uncategorized records is helpful.

Make sure that the category model is scored (otherwise, click on the *Score* button), select *Uncategorized,* and click on *Display*. Any record with no category will display in the Data pane.

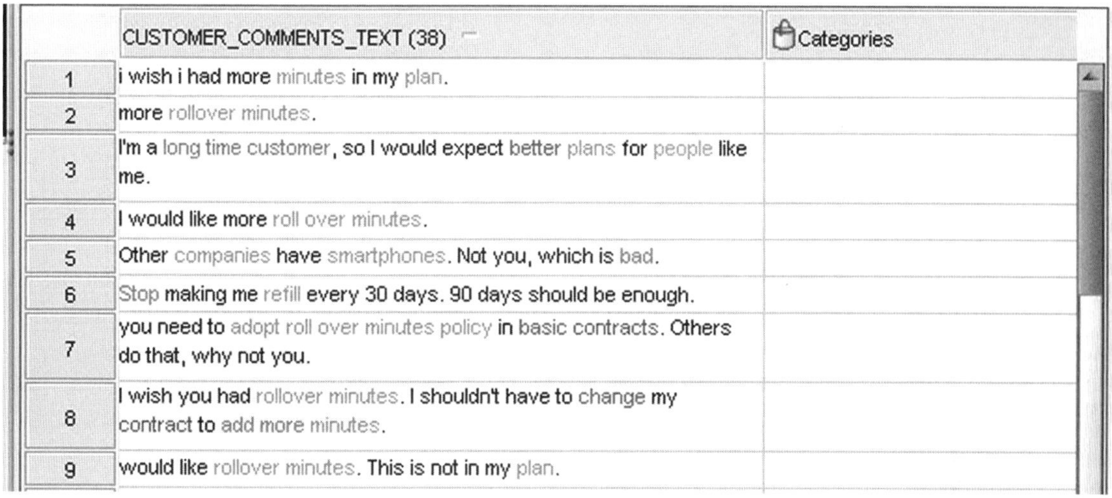

	CUSTOMER_COMMENTS_TEXT (38)	Categories
1	i wish i had more minutes in my plan.	
2	more rollover minutes.	
3	I'm a long time customer, so I would expect better plans for people like me.	
4	I would like more roll over minutes.	
5	Other companies have smartphones. Not you, which is bad.	
6	Stop making me refill every 30 days. 90 days should be enough.	
7	you need to adopt roll over minutes policy in basic contracts. Others do that, why not you.	
8	I wish you had rollover minutes. I shouldn't have to change my contract to add more minutes.	
9	would like rollover minutes. This is not in my plan.	

FIGURE M.15
Example of uncategorized records.

What appears here is that there is no category for *rollover minutes*. So let's create one. From the Categories and Concepts pane, select *All Documents* and right click. A submenu pops up and contains a *Create Empty Category*... option. A window called *Category Properties* opens; you can change the name of the category and add a *Label* and/or *Annotations*, if needed. Annotations may be helpful because you can store a definition or examples, and when you hover the mouse over a category, you will see this information without having to open the category.

Let's call this category *Neg: Plans*. Select the concept *rollover minutes*, right click, and *Add to Category*... The whole list of available categories (top and subcategories) appears, and you just have to select the relevant one. Since we have just created the category *Neg: Plans*, this category appears first in the list, but you can also choose to display categories by alphabetical order.

You will notice that the list of concepts has been updated in the Extraction Results pane to indicate that *rollover minutes* is now a concept that is used as a descriptor.

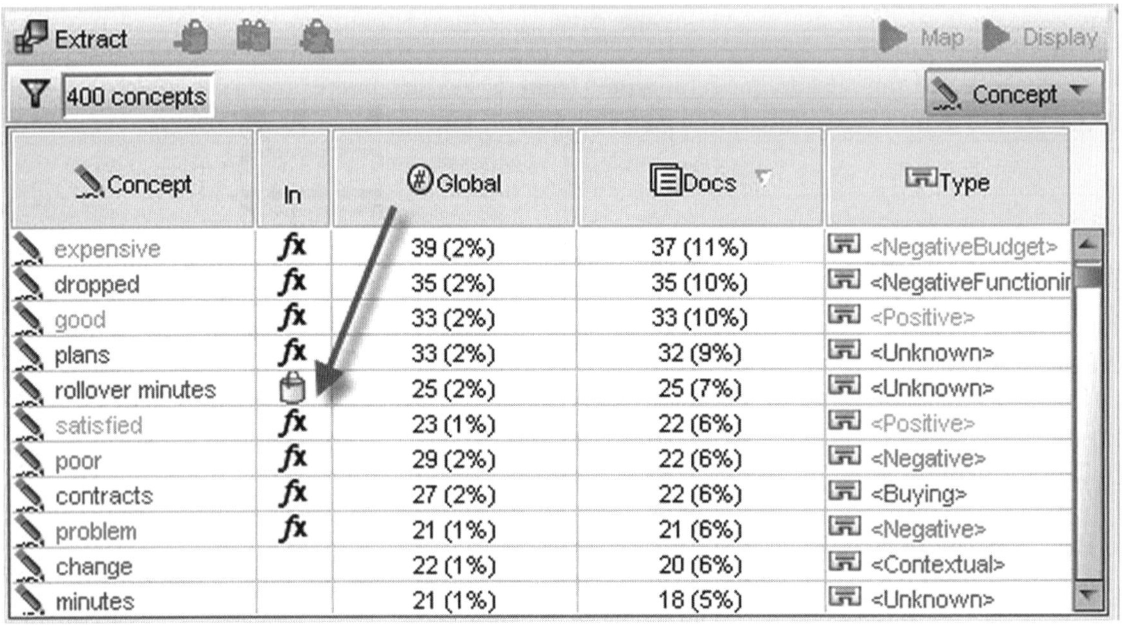

Extract | Map | Display

400 concepts | Concept

Concept	In	Global	Docs	Type
expensive	fx	39 (2%)	37 (11%)	<NegativeBudget>
dropped	fx	35 (2%)	35 (10%)	<NegativeFunctionir
good	fx	33 (2%)	33 (10%)	<Positive>
plans	fx	33 (2%)	32 (9%)	<Unknown>
rollover minutes		25 (2%)	25 (7%)	<Unknown>
satisfied	fx	23 (1%)	22 (6%)	<Positive>
poor	fx	29 (2%)	22 (6%)	<Negative>
contracts	fx	27 (2%)	22 (6%)	<Buying>
problem	fx	21 (1%)	21 (6%)	<Negative>
change		22 (1%)	20 (6%)	<Contextual>
minutes		21 (1%)	18 (5%)	<Unknown>

FIGURE M.16
Descriptor rollover minutes.

You can add some other descriptors—concepts, TLA results, or business rules—depending on the context, to complete the model.

Reorganizing Categories

The model used contains 26 top categories and 34 including subcategories. In this application we are only interested in the reasons for dissatisfaction, so we will group all Positive and Other categories into a single one.

We will see how easy it is to change the structure of a category model. Click on the top of the column to have the categories sorted by alphabetic order. Select all the relevant categories (i.e., all the categories starting with *Pos* and *Other*), right click, and *Move to Category*.

As previously done for the *Neg: Plan* category, a window will pop up, and you can create a new category. Let's call it *Pos&Other*. Note that only the top structure (the depth) has been modified. Top categories have now become subcategories, but their internal structure has been preserved, so a category with subcategories will still have subcategories.

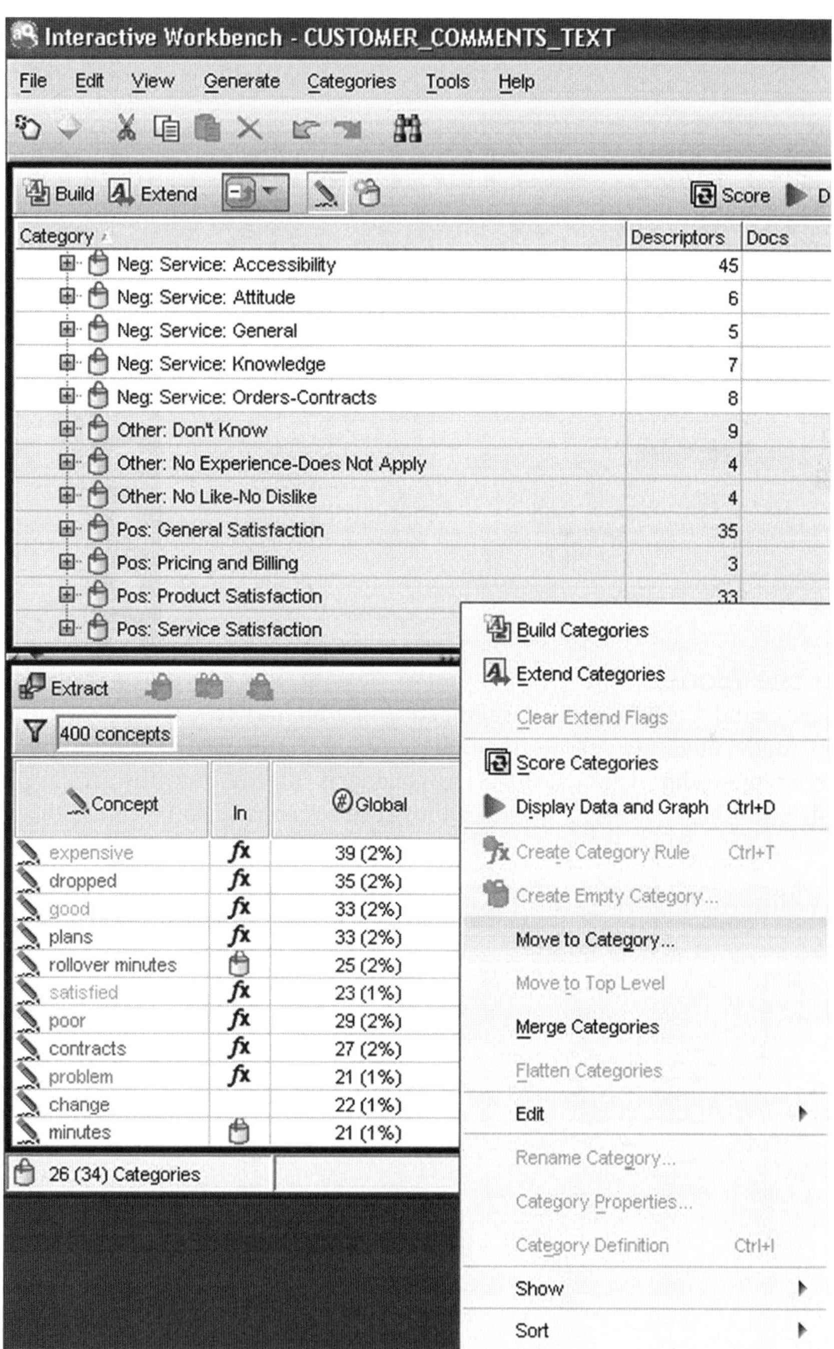

FIGURE M.17
Moving categories.

Validating the Category Model

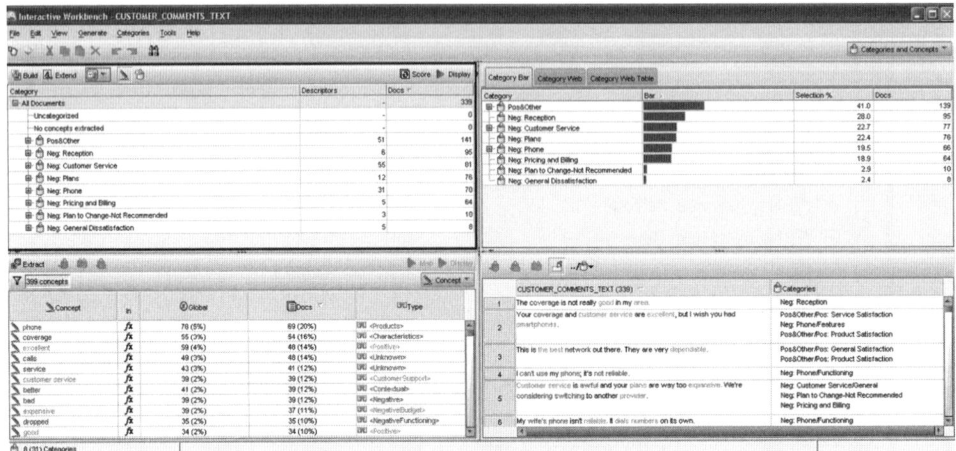

FIGURE M.18
Final category model.

Generating the Model

When the category model is finished and validated in the interactive session, you can generate a text mining category model nugget (a yellow diamond) to use it to score new data. From the menus, choose **Generate > Generate Model**. A model nugget is generated directly onto the Model palette with the default name. You just have to drag and drop it into your stream and connect it to the relevant nodes.

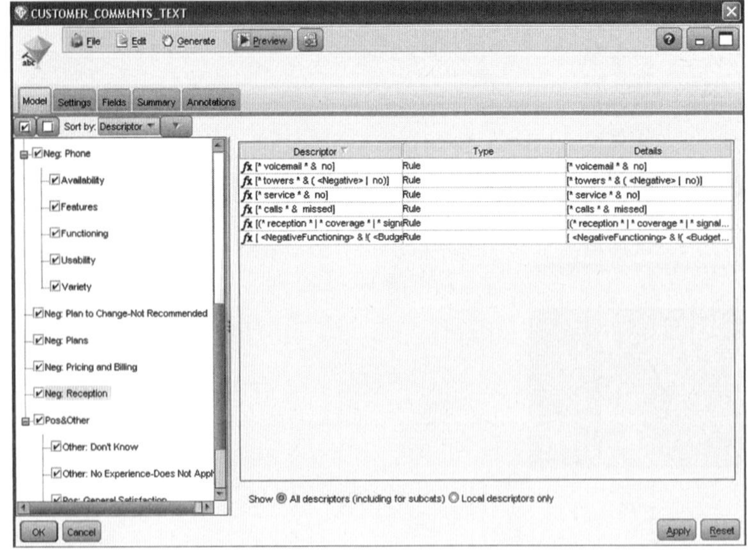

FIGURE M.19
Deployed model.

There are two scoring modes available: *Categories as fields* and *Categories as records*. With *Categories as fields*, there are just as many output records as there were in the input. However, each record now contains one new field for every category that was selected (using the check mark) on the Model tab. For each field, enter a flag value for **True** and for **False**, such as *Yes/No, True/False, T/F,* or *1* and *2*. The storage types are set automatically to reflect the values chosen. For example, if you enter numeric values for the flags, they will be automatically handled as an integer value. The storage types for flags can be string, integer, real number, or date/time.

With *Categories as records*, a new record is created for each category, document pair. Typically, there are more records in the output than there were in the input. In addition to the input fields, new fields are also added to the data, depending on what kind of model it is.

RESULTS AND ANALYSIS

Once the categorization model has been created, it is then easy to score all the records against the final set of categories and create some reports looking at several attributes at the same time.

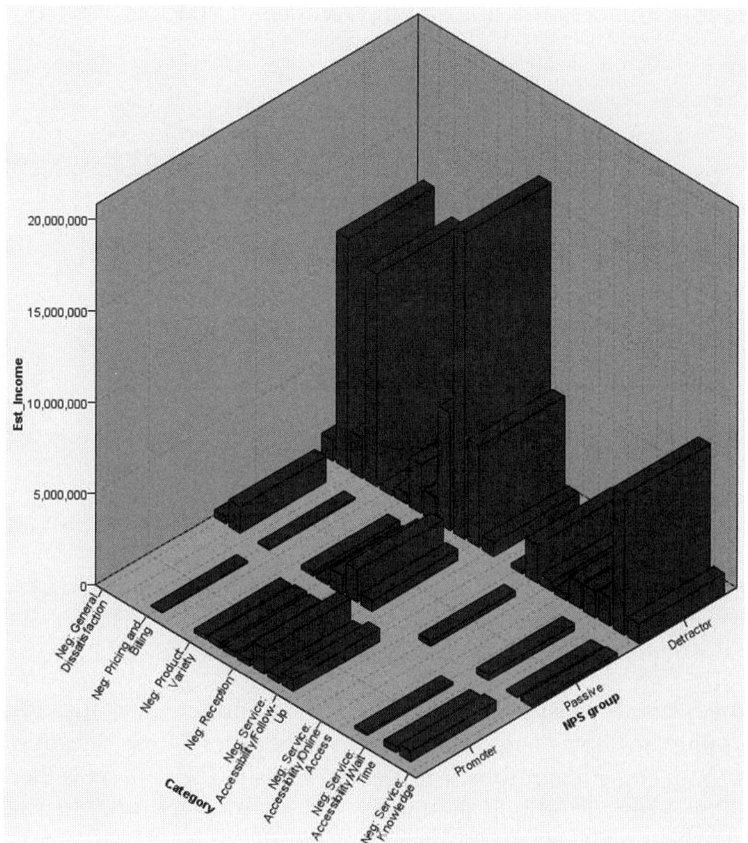

FIGURE M.20
3D bar chart based on categorization results (x: categories derived from text comments; y: NPS group; z: estimated customer income).

In Figure M.20, a graph board node has been used to create a 3D bar chart combining categorization results with NPS group and customer estimated income. For instance it clearly appears that "detractors" with the highest salaries mainly complain about "reception," "pricing & billing," and "plan." Many similar reports can be built in IBM SPSS Modeler. Alternatively, results can be directly passed to IBM® Cognos™ or any other Business Intelligence solution to publish dashboards to reflect the survey results for a large audience. Such reports do provide a lot of value to marketers or customer-related departments.

Customer satisfaction can also be inferred, like any other behavioral or attitudinal data. People who share the same pattern are likely to behave or even "think" the same way. Inferring customer satisfaction can be used on surveys to predict answers from nonrespondents and more widely to address the whole customer database. In this case, survey data help in building up a predictive model that is going to be then applied to a larger set of customers. We're going to use data mining techniques to try to predict the customer satisfaction group. In the first case, we're going to build up a model based on customer-structured data only. A second model integrating all survey data—with categorized responses as seen above—is going to be built in parallel to show any additional value.

CHAID, or Chi-squared Automatic Interaction Detection, is a classification method for building decision trees by using Chi-square statistics to identify optimal splits. Figure M.21 shows the process to create two CHAID models to predict the customer satisfaction group (Promoters, Detractors, Passives).

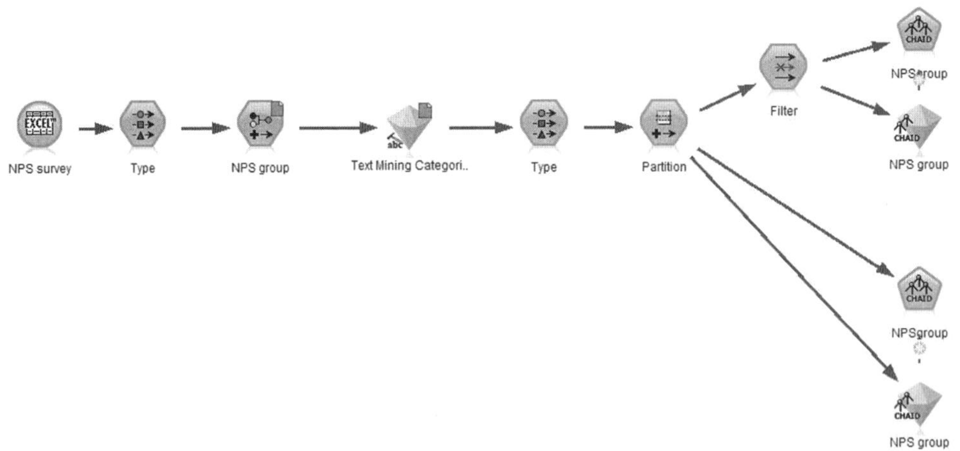

FIGURE M.21

Stream to create two CHAID models to predict the customer satisfaction group, one based on customer-structured data only (top right yellow nugget) and the other based on all available customer data (bottom right yellow nugget).

The NPS group is derived from the customer satisfaction score. Then the text mining categorization model is applied to create new customer attributes based on matching categories. These categories reflect customers' negative opinions about some major topics, as seen previously. The data set is then partitioned into two subsets: a training set used to build the model and a testing set used then to compare results from the model with actual customer data.

In the first model, all text mining categories are filtered out first to make sure the model is not based on any text mining input. In the second model, all customer data are used as input. In both models the variable "NPS group" is defined as target, and CHAID node default settings are used.

Once decision trees have been generated, Figure M.22 shows the set of rules derived to predict the NPS group based on structured fields only.

```
EMPLOYMENT in [ "Employed full-time" "Employed part-time, Stay at home parent" "U" ] [ Mode: Detractor ]
    Children <= 0 [ Mode: Detractor ]
        EDUCATION in [ "Assoc Degree" "Bachelors Degree" "High School Grad" "No High School diploma" "Some College" ] [ Mode: Detractor ]
            Est_Income <= 8,359 [ Mode: Detractor ] ⇨ Detractor
            Est_Income > 8,359 and Est_Income <= 69,486 [ Mode: Detractor ] ⇨ Detractor
            Est_Income > 69,486 and Est_Income <= 81,641 [ Mode: Detractor ] ⇨ Detractor
            Est_Income > 81,641 [ Mode: Detractor ] ⇨ Detractor
        EDUCATION in [ "Grad / Post-grad degree" "U" ] [ Mode: Detractor ] ⇨ Detractor
    Children > 0 [ Mode: Detractor ] ⇨ Detractor
EMPLOYMENT in [ "Employed full-time, Student" "Employed part-time, Student" "Retired" ] [ Mode: Detractor ]
    Status in [ "M" ] [ Mode: Detractor ]
        EDUCATION in [ "Assoc Degree" "Bachelors Degree" "Grad / Post-grad degree" "High School Grad" "Some College" ] [ Mode: Detractor ]
            Children <= 0 [ Mode: Detractor ] ⇨ Detractor
            Children > 0 [ Mode: Detractor ] ⇨ Detractor
        EDUCATION in [ "No High School diploma" ] [ Mode: Promoter ] ⇨ Promoter
    Status in [ "S" ] [ Mode: Promoter ] ⇨ Promoter
EMPLOYMENT in [ "Employed part-time" "Not currently employed" "Stay at home parent" "Student" ] [ Mode: Detractor ] ⇨ Detractor
```

FIGURE M.22

Rules of the CHAID model to predict the NPS group based on customer data without text mining.

Figure M.23a shows a new set of rules once the text mining categorization results have been used as predictors within the CHAID model, whereas Figure M.23b shows the list of predictors, by decreasing importance, of the same model.

(a)

```
Category_Neg: Reception in [ "F" ] [ Mode: Detractor ]
    Category_Neg: Plans in [ "F" ] [ Mode: Detractor ]
        Category_Neg: Service: General in [ "F" ] [ Mode: Promoter ]
            Category_Neg: Pricing and Billing in [ "F" ] [ Mode: Promoter ]
                Category_Neg: Product Features in [ "F" ] [ Mode: Promoter ] ⇨ Promoter
                Category_Neg: Product Features in [ "T" ] [ Mode: Detractor ] ⇨ Detractor
            Category_Neg: Pricing and Billing in [ "T" ] [ Mode: Detractor ]
                Car_Owner in [ "N" ] [ Mode: Detractor ] ⇨ Detractor
                Car_Owner in [ "Y" ] [ Mode: Detractor ] ⇨ Detractor
        Category_Neg: Service: General in [ "T" ] [ Mode: Detractor ]
            EMPLOYMENT in [ "Employed full-time" "Employed full-time, Student" "Retired" "Stay at home parent" "Student" "U" ] [ Mode: Detractor ] ⇨ Detractor
            EMPLOYMENT in [ "Not currently employed" ] [ Mode: Promoter ] ⇨ Promoter
    Category_Neg: Plans in [ "T" ] [ Mode: Detractor ] ⇨ Detractor
Category_Neg: Reception in [ "T" ] [ Mode: Detractor ]
    EMPLOYMENT in [ "Employed full-time" "Employed full-time, Student" "Employed part-time, Student" "Retired" ] [ Mode: Detractor ] ⇨ Detractor
    EMPLOYMENT in [ "Employed part-time" "Not currently employed" ] [ Mode: Detractor ]
        Est_Income <= 14,578 [ Mode: Detractor ] ⇨ Detractor
        Est_Income > 14,578 and Est_Income <= 30,763 [ Mode: Passive ] ⇨ Passive
        Est_Income > 30,763 [ Mode: Detractor ] ⇨ Detractor
    EMPLOYMENT in [ "Stay at home parent" "Student" ] [ Mode: Passive ] ⇨ Passive
```

FIGURE M.23a

Rules of the CHAID model to predict the NPS group based on all available customer data.

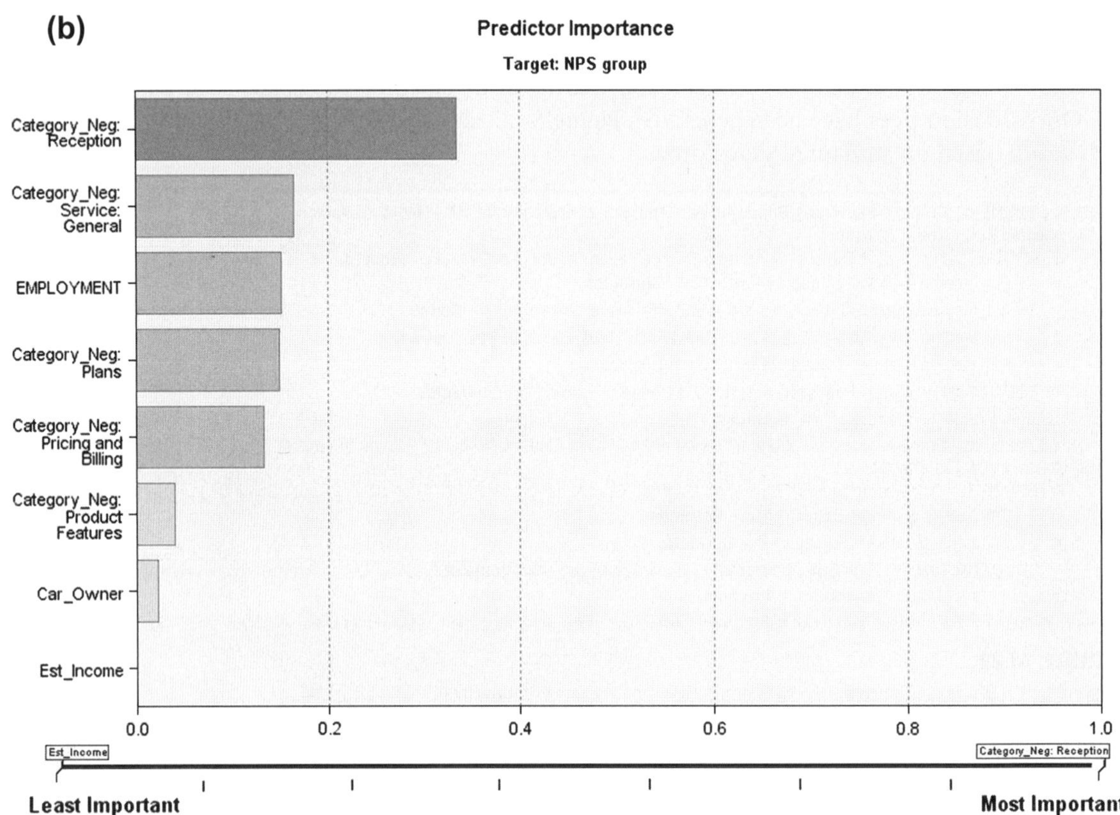

FIGURE M.23b

Importance Plot: this graph shows the relative importance of each of the text mining categories, thus "Neg_Reception" has the greatest importance relatives to "Est_Income" which is the least important as a predictor in the CHAID model.

It is noteworthy that analyzing the customers' own words through text mining categorization brings in richer rules combining customers' traditional attributes with sentiment-based categories they have been scored against. When comparing both models, "Employment" is the major predictor in the first one, whereas it's only ranked third in the second one. It turns out for mobile customers that complaining about the quality of the reception is the top predictor when predicting the NPS group.

Both models are then compared by applying them on the testing data set. Gain charts are used to effectively illustrate how widely you need to cast the net to capture a given percentage of all of the hits in the tree. In Figure M.23, the gain chart compares both predictive models. The blue line represents the performance of the first CHAID model (structured data only), and the red line represents the performance of the text mining–enabled CHAID model. The chart shows that the first model can accurately predict the NPS group in 60 percent of cases by analyzing only 50 percent of the customers, whereas the second model is 70 percent accurate, bringing a gain of 10 percent. This clearly demonstrates the value of sentiment-based text mining categories as predictors in inferring customer satisfaction.

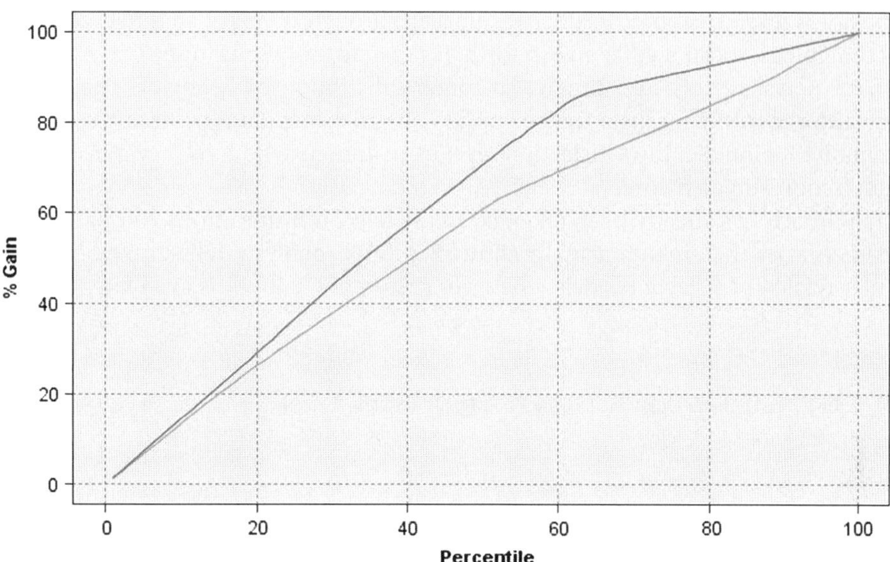

FIGURE M.24
Gain chart comparing both predictive models on the testing data set with no text mining (light gray curve) or with text mining (dark gray curve).

SUMMARY

Customer satisfaction surveys are regularly used by marketers to "feel the pulse" of the client database. NPS is a basic metric that helps in better understanding customer satisfaction at a macro level. The trouble is that NPS is only derived from factual answers that do not reflect the reasons why people think or behave this way. Using free-text customer feedback enables organizations to really listen to the "voice of the customer." Text Analytics allows looking at such comments to understand and quantify what customers are complaining about.

This tutorial shows step by step how text mining can be efficiently applied to structure unstructured data for business purposes. The challenge of a good text mining process requires properly capturing customers' sentiments and their related topics, aggregating the results at a less granular level (a.k.a. categories) so it's usable for further modeling, and keeping the categories meaningful enough for business users.

IBM SPSS Modeler Premium has been designed to help analysts in addressing all of the preceding points. Features and functions go from unstructured data discovery up to creating relevant categorization models. Combining text analytics results with the rest of customer data delivers very valuable new insights. In this case, we showed, for instance, that many customers complained about a poor quality of reception. Among them, many high salaries were detractors. Many other analyses can be easily performed to better understand customer feedback. But capturing customers' emotions and feelings also brings in a new dimension in predictive modeling. In this example, the use of text analytics as a new input for a CHAID model results in more accurate predictions to assess customer satisfaction.

At the business level, customer feedback is tracked through different channels as free text: emails, surveys, call center logs, social media, and so forth. It's therefore obvious to see Text Analytics becoming more and more pervasive to better predict customer satisfaction or behavior. The predictive model created in this tutorial could, for instance, be applied on call center data (including comments from the sales representative) to predict the satisfaction level of customers who have never been targeted by any customer satisfaction survey before.

Such an approach has already been taken by many telecommunications companies to proactively contact customers at risk before they actually churn. Sentiment analysis delivers tangible additional ROI to all customer-related predictive models whenever customer feedback is available.

References

Fred Reichheld (2006). *The Ultimate Question: Driving Good Profits and True Growth.* Harvard Business Press, Cambridge, MA.

Nucleus Research (2007). *Guidebook: SPSS Text analytics.* Document H99. Wellesley, MA.

SPSS Inc. (2008). *Mastering New Challenges in Text Analytics.* Chicago, IL.

CASE STUDY: Detecting Deception in Text with Freely Available Text and Data Mining Tools

Christie Fuller, PhD
College of Business, Louisiana Tech University

Dursun Delen, PhD
Spears School of Business, Oklahoma State University

CONTENTS

INTRODUCTION

In this tutorial we explain in a step-by-step fashion an interesting study where we analyzed person-of-interest statements completed by people involved in crimes on military bases in order to develop classification models to predict deceptive statements using a variety of text and data mining tools and techniques. In these statements, suspects and witnesses are required to write their recollection of the event in their own words. Base law enforcement (LE) personnel searched archival data for statements that they could conclusively identify as being truthful or deceptive. These decisions were made on the basis of corroborating evidence and case resolution (i.e., not just the personal opinion of LE personnel). The definition of *deception* relies on an intentional communication of false information; therefore, statements where a person-of-interest was simply mistaken in their recall of events were not labeled as deceptive. Once labeled as truthful or deceptive, the law enforcement personnel removed identifying information and gave the statements to the research team. In total, 366 statements were used in our analysis, and 79 of these statements were deceptive. The statements were from many different types of crimes, such as traffic infractions, shoplifting, assault, and arson. All statements were provided by adults.

Practical Text Mining and Statistical Analysis for Non-structured Text Data Applications. DOI: 10.1016/B978-0-12-386979-1.00022-0

We underscore the importance of this real-world data set. Unlike many past studies that used data collected from student groups conducting mock lies or deceptions, the individuals involved in these cases faced severe consequences for lying on an incident statement. Military members could face penalties up to and including court martial for creating a false official statement. Civilians could face disbarment from the base, or in the case of DoD employees, termination of employment. These penalties are, of course, in addition to those that the person may be facing due to conviction for involvement in the crime.

The methodology employed in this study is based on a process known as message feature mining (MFM). This process relies on elements of text mining and data mining techniques.

Message Feature Mining Steps

Traditionally, data mining analyzes categorical or numerical variables to find meaningful patterns in a large volume of structured/tabular data. Text mining also seeks to find meaningful patterns in data, though the data usually originate as unstructured text. Often, this text must be transformed into some structured format prior to further analysis. Figure N.1 illustrates this step-by-step process. Even though no backtrackings/redoes/corrections are shown in Figure N.1, as is the case in any data and/or text mining project, they are a common part of these experimental studies.

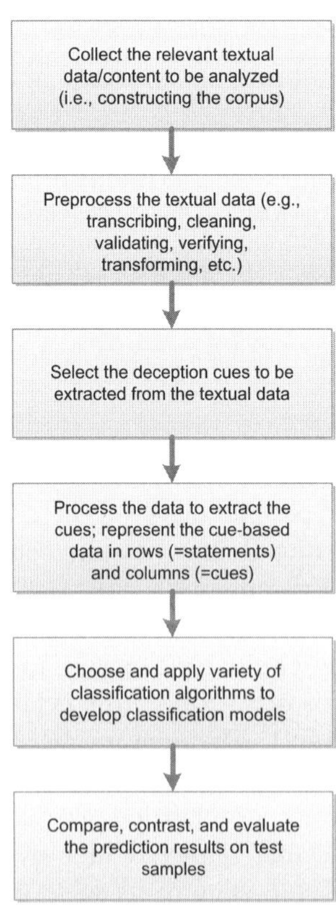

FIGURE N.1
A generic deception detection process map.

Step 1: Collect the Relevant Textual Data/Content

The overall process begins with collecting and organizing the data in a single file or file location (a folder). In text mining, this activity is commonly referred to as constructing the corpus. In this study, a number of person-of-interest statements are collected and digitized for further processing.

Step 2: Data Preprocessing

The main idea in this step is to clean and transform the data into a machine processable form. That is, if the text is not in a soft format (but in a piece of paper), it needs to be transcribed, validated/verified, and transformed into a specific format. The data for this project were photocopies of handwritten statements. The statements were transcribed using Notepad and saved as text files. Any identifying information, such as names, Social Security numbers, phone numbers, and so on was blacked out by the law enforcement agency that supplied the statements. To preserve overall word counts and capture as much information as possible, this information was replaced with generic data in the same format as the information that was omitted. For example, males were given the name of John Doe, and Social Security numbers were replaced with 111-11-1111. A naming standard was incorporated for the text files. The naming standard indicates the name of the transcriber, the gender of the statement author, the output class (truthful or deceptive) of the statement, and a number for the statement.

Step 3: Select Deception Cues

A number of linguistic cues have been previously identified that may be used to classify data for deception detection (Fuller et al., 2009; Hancock et al., 2005; Zhou and Zhang, 2006). Many of the cues coincide with the variables included in the Linguistic Inquiry and Word Count (LIWC) tool (Pennebaker and Francis, 2001). General Architecture for Text Engineering (GATE) (Cunningham, 2002; Cunningham et al., 2005) was originally configured for use in deception detection by the University of Arizona (Cao et al., 2003; Zhou et al., 2003). In addition to using standard features of LIWC and GATE, custom dictionaries have also been added to the software packages. GATE is an open source software tool, while LIWC must be purchased for a small fee.

Step 4: Process Data to Extract the Cues

After transcription, the statements were processed to extract the levels of the desired cues from each statement using GATE and LIWC. In this step, the software tokenizes the text files and identifies the words matching each cue. After this identification, the total of words matching each cue is calculated for each statement. For example, a part-of-speech tagger identifies the verbs in a statement and then sums the number of verbs. The process varies somewhat by software, and the steps for both GATE and LIWC are provided following. The output of this step is a representation of the text in columns and rows, where each row represents an individual statement and each column represents a cue. This output serves as the input to classification models.

GENERAL ARCHITECTURE FOR TEST ENGINEERING

The first step in GATE is to upload the text to be processed. In GATE, this text is referred to as a language resource. The language resources will be loaded into a corpus, or collection of documents. In GATE, you will first create and name a new corpus, and then load the documents into this corpus. During this process you will specify the location of the text and the file extension of the text.

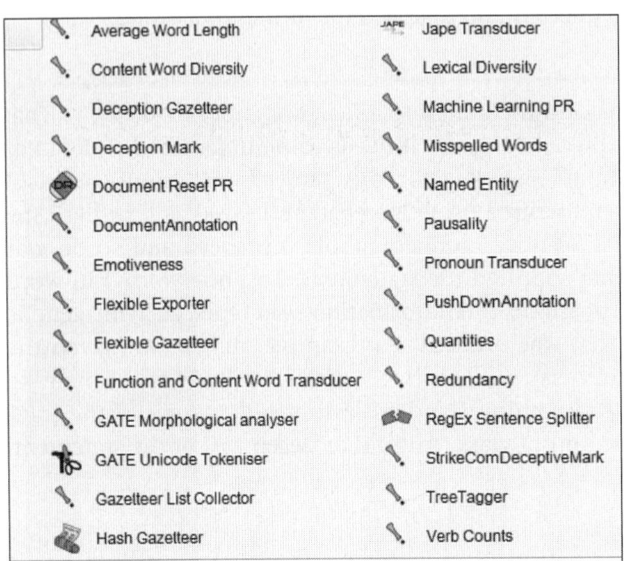

FIGURE N.2

Select GATE processing resources.

After creating the corpus, you will specify which of the available processing resources should be used to process the text (Figure N.2). Here we show both default and custom resources.

After loading both the processing and language resources, a corpus pipeline is created to combine the two. In the corpus pipeline interface, the desired corpus to be processed is selected if more than one corpus is available. From the processing resources that were previously loaded, all or a subset can be selected for processing the corpus. The processing resources can also be arranged in the desired order (Figure N.3). For example, the tokenizer must be run before the sentence splitter and the part of speech tagger. In this example, GATE has been configured to be the output results in .arff format for use in WEKA. The WEKA output resource would last on the list of selected processing resources. After the corpus and processing resources are selected, the application is run and the results are exported.

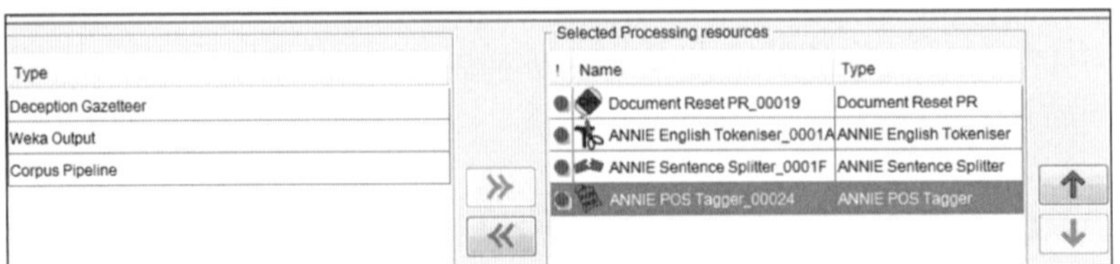

FIGURE N.3

GATE Corpus Pipeline.

LINGUISTIC INQUIRY AND WORD COUNT

The first step in using LIWC is to choose the variables that should be extracted from the text. LIWC includes 74 standard variables across five categories. As shown in Figure N.4, you can choose to extract all of the variables or one of the five subsets. You can also choose to load a custom dictionary using the dictionary menu.

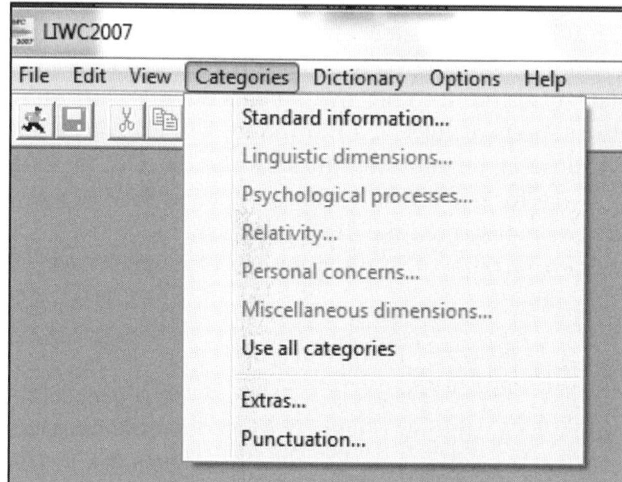

FIGURE N.4
Selection of LIWC variables.

The next step is to choose the file or files to process. In Figure N.5, all of the text files have been loaded into a single folder and the Select All option can be used. The location for results to be saved to and the desired format for the results should also be chosen. For example, the results may be saved as a Microsoft Excel spreadsheet.

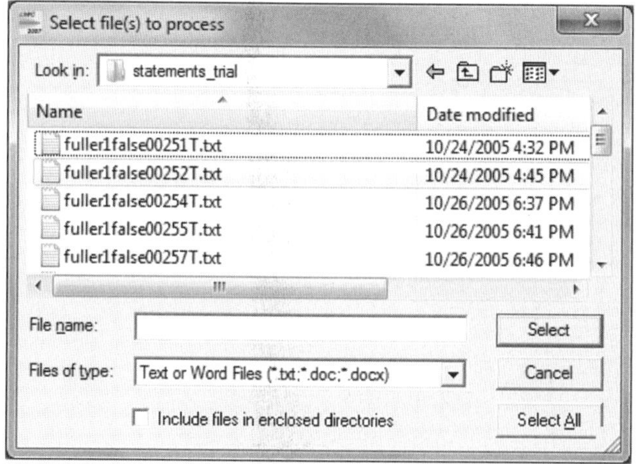

FIGURE N.5
LIWC Selection of Text.

WORKING WITH GENERAL ARCHITECTURE FOR TEST ENGINEERING AND LINGUISTIC INQUIRY AND WORD COUNT OUTPUT

After the text has been processed through LIWC and GATE, the next step is to merge the results into a single spreadsheet. Once the results have been combined, you can choose to proceed directly to the selected data mining program for classification. Here, Waikato Environment for Knowledge Analysis (WEKA) will be used for the classification step (Witten and Frank, 2000). It may be desirable to do some further processing of the data before using WEKA. While the desired cues were defined in Step 3, the data may need to be edited at this point to narrow the data down to just these variables. For example, since LIWC extracts entire subsets of cues, at a minimum, it is likely that not all variables extracted by LIWC are useful, and those extraneous variables can be deleted. If using WEKA, the spreadsheet will need to be saved in csv format before it can be opened in WEKA. As noted previously, only about one-quarter of the data represented the deceptive class. Therefore, prior to classification, the data were balanced to achieve a ratio of approximately 50 percent each of deceptive and truthful cases.

Step 5: Classification

The first step in the classification process is to use WEKA explorer (Figure N.6) to open the data file. In the explorer, the distribution of the individual variables can be viewed (Figure N.7). The attributes to be used for classification can be chosen using the preprocessing tab or the Select Attributes tab. Any extraneous cues that were not removed in the previous step can be eliminated. For some classification models, such as neural networks, it may be useful to further limit the number of cues used in model development or to try different combinations of cues. The Select Attributes tab can be used to algorithmically select attributes, as demonstrated in Fuller and colleagues (2008).

FIGURE N.6
WEKA Explorer.

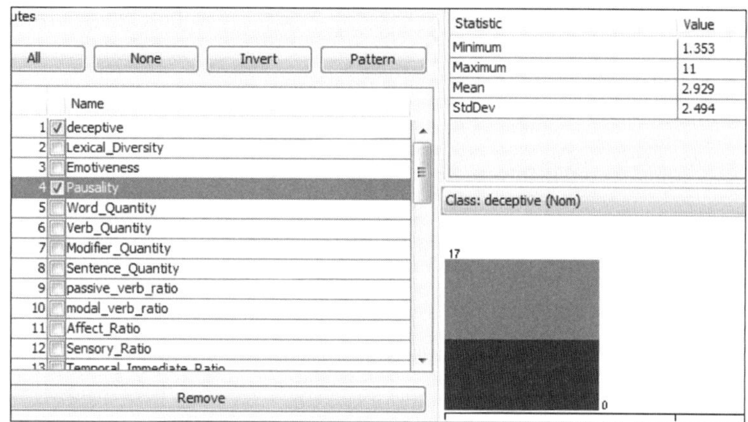

FIGURE N.7
WEKA preprocessing tab.

After selecting the attributes, proceed to the classify tab (Figure N.8). First, indicate the output variable using the drop-down menu. Additional options such as 10-fold cross validation can also be specified on this screen.

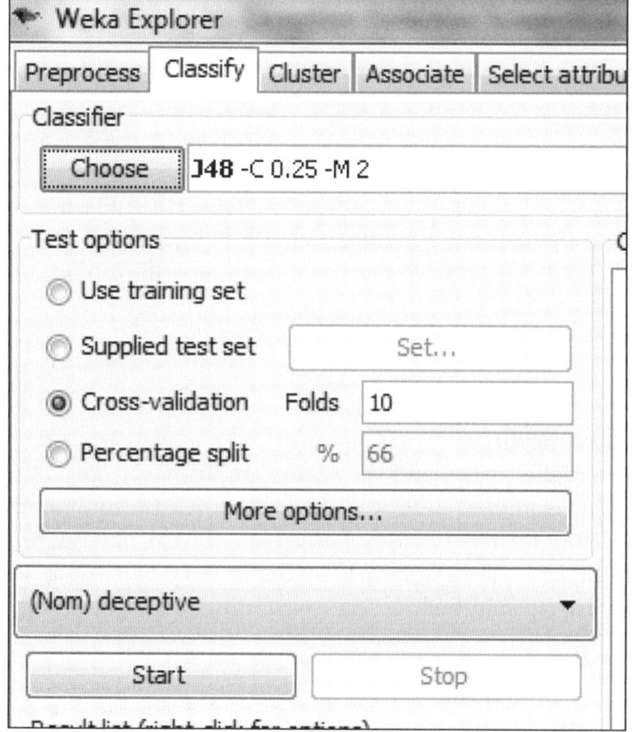

FIGURE N.8
WEKA Classify tab.

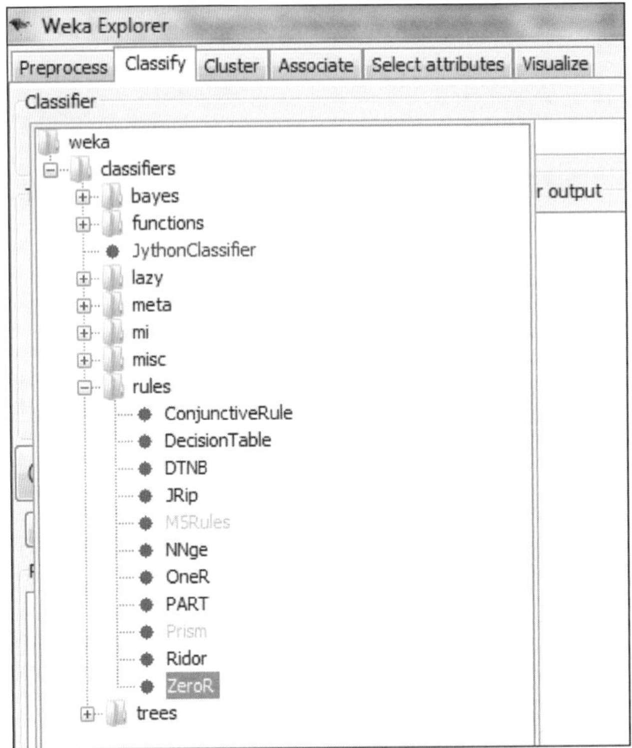

FIGURE N.9
Available classifiers.

Next, choose the desired classifiers (Figure N.9). The classifiers that can be selected (in black, not grayed out) depends on the type of dependent variable (numeric or nominal). Common classifiers include the multilayer perceptron, simple logistic regression, and J48, an implementation of the C4.5 algorithm.

Step 6: Compare, Contrast, and Evaluate Results

After choosing all the appropriate options, the classifier is run and the results are displayed for interpretation (Figure N.10). The results across various classifiers and/or cue sets can then be compared. Both the overall classification accuracy percentage and the rate of false positives are of particular interest in this data set. In addition, the variables retained for models such as decision trees or ranking of variable importance for neural networks might also be analyzed.

```
=== Summary ===

Correctly Classified Instances          12               63.1579 %
Incorrectly Classified Instances         7               36.8421 %
Kappa statistic                          0.2652
Mean absolute error                      0.4034
Root mean squared error                  0.5822
Relative absolute error                 80.6725 %
Root relative squared error            116.2926 %
Total Number of Instances               19

=== Detailed Accuracy By Class ===

               TP Rate   FP Rate   Precision   Recall  F-Measure   ROC Area  Class
                 0.667     0.4        0.6        0.667     0.632      0.606    true
                 0.6       0.333      0.667      0.6       0.632      0.606    false
Weighted Avg.    0.632     0.365      0.635      0.632     0.632      0.606

=== Confusion Matrix ===

 a b   <-- classified as
 6 3 | a = true
 4 6 | b = false
```

FIGURE N.10

A sample screen shot of WEKA classification results.

SUMMARY

A maximum accuracy rate of just over 76 percent, with false positives around 19 percent, has been achieved using this data set and technique (Fuller et al., 2008; Fuller et al., 2009) across several different cue sets and classification models. Once models are built, this technique can be used for real-time deception detection on unseen data. While this accuracy rate is certainly not sufficient to convict someone of a crime, it is sufficient to aid law enforcement in investigations. These results are comparable to those found in field investigations of the well-known polygraph (Honts and Raskin, 1988; Raskin, 1987). Unlike the polygraph, this method of deception detection does not require a highly trained operator or subject cooperation.

The technique described here can be extended to a wide variety of real-world text needing to be classified. There are a few drawbacks that may need to be addressed. Ideally, data and text mining techniques are applied to data sets containing thousands or even millions of records. In this case, fewer than 400 records were available that could be confidently identified as truthful or deceptive by law enforcement personnel. Due to this, the selection of algorithm and cues must be done with care to accommodate the limited data. The problem of attaining large, real-world data sets certainly extends beyond the context of deception. Further, in these real-world scenarios, the situation may arise, as it did here, where the classes are represented in unbalanced proportions. This must also be addressed so that models are not biased toward the class with more representation. Otherwise, models may not perform well on unseen data. Similarly, the overall classification accuracy must be balanced against the rate of false positives. Despite these limitations, these readily available tools can be used effectively to classify text as truthful or deceptive.

References

J. Cao, J. M. Crews, M. Lin, J. Burgoon, and J. F. Nunamaker, "Designing Agent99 Trainer: A Learner-Centered, Web-Based Training System for Deception Detection," in Lecture Notes in Computer Science: Proceedings of Intelligence and Security Informatics: First NSF/NIJ Symposium, ISI 2003, Tucson, AZ, June 2–3, 2003.

H. Cunningham, "GATE, a General Architecture for Text Engineering," *Computers and the Humanities*, vol. 36, pp. 223–254, 2002.

H. Cunningham, D. Maynard, K. Bontcheva, V. Tablan, C. Ursu, M. Dimitrov, M. Dowman, N. Aswani, and I. Roberts, "Developing Language Processing Components with GATE Version 3 (a User Guide), http://gate.ac.uk/sale/tao/index.html#x1–1710008.4." vol. 2006, 2005.

C. Fuller, D. Biros, and D. Delen, "Exploration of Feature Selection and Advanced Classification Models for High-Stakes Deception Detection," in *41st Hawaii International Conference on System Sciences*, 2008.

C. Fuller, D. Biros, and R. Wilson, "Decision support for determining veracity via linguistic-based cues," *Decision Support Systems*, vol. 46, p. 695, 2009.

J. T. Hancock, L. Curry, S. Goorha, and M. Woodworth, "Automated Linguistic Analysis of Deceptive and Truthful Synchronous Computer-Mediated Communication," 2005, p. 22c.

C. R. Honts and D. C. Raskin, "A field study of the validity of the directed lie control question," *Journal of Police Science and Administration*, vol. 16, pp. 56–61, 1988.

J. W. Pennebaker and M. E. Francis, *Linguistic Inquiry and Word Count: LIWC 2001*, Mahwah, NJ: Erlbaum Publishers, 2001.

D. C. Raskin, "Methodological issues in estimating polygraph accuracy in field applications," *Canadian Journal of Behavioral Science*, vol. 19, pp. 389–404, 1987.

I. H. Witten and E. Frank, *Data Mining: Practical Machine Learning Tools and Techniques with Java*. San Francisco: Morgan Kaufmann, 2000.

L. Zhou, D. P. Twitchell, T. T. Qin, J. K. Burgoon, and J. F. Nunamaker, Jr., "An exploratory study into deception detection in text-based computer-mediated communication," in *36th Annual Hawaii International Conference on System Sciences*, Big Island, Hawaii, 2003.

L. Zhou and D. S. Zhang, "A comparison of deception behavior in dyad and triadic group decision making in synchronous computer-mediated communication," *Small Group Research*, vol. 37, pp. 140–164, 2006.

TUTORIAL O

Predicting Box Office Success of Motion Pictures with Text Mining

Dursun Delen, PhD
Department of Management Science and Information Systems, Spears School of Business, Oklahoma State University, Tulsa, Oklahoma

CONTENTS

INTRODUCTION

Predicting the financial success of a movie prior to its production cycle is arguably one of the most challenging yet essential tasks for decision makers in the motion picture industry. If done accurately, such information could allow decision makers to optimally allocate their resources (financial and otherwise) to maximize their RIO. In-depth knowledge about the factors affecting the financial success of a movie would be of great use in making project selection, investment, and production-related decisions. However, forecasting financial success (box office receipts) of a particular motion picture is considered a very difficult (often impossible) problem. Most domain experts think that "Hollywood is the land of hunches and wild guesses" due largely to the uncertainty associated with predicting the product demand. Jack Valenti, long-time president and CEO of the Motion Picture Association of America, once said, "No one can tell you how a movie is going to do in the marketplace … not until the film opens in a darkened theater and the sparks fly up between the screen and the audience." Journals and trade magazines of the motion picture industry have been full of examples, statements, and experiences that support this claim.

The difficulty associated with the perceived unpredictable nature of the problem has intrigued researchers and practitioners to develop "models" for understanding and hopefully forecasting the financial success of motion pictures. Most analysts have used a combination of numerical and nominal variables (e.g., MPAA rating, genre, star power, time of release, special effects, etc.) to

Practical Text Mining and Statistical Analysis for Non-structured Text Data Applications. DOI: 10.1016/B978-0-12-386979-1.00023-2

predict the box office receipts of motion pictures after a movie's initial theatrical release. Because they attempt to determine how a movie is going to do based on the early financial figures, the results were quite accurate, but they are not helpful for making investment and production-related decisions, which are to be made during the planning phase. Some studies have attempted to forecast the performance of a movie before it is released but had only limited prediction success. These previous studies, which are either good for predicting the financial success of a movie after its initial theatrical release or are not accurate enough predictors for decision support, leave us with an unsatisfied need for a forecasting system capable of making a prediction prior to a movie's theatrical release. Our ongoing research aims to fill this need by developing and embedding sophisticated forecasting models into a web-based decision support system that can be readily accessed and used by Hollywood managers.

In this tutorial, we used a text mining approach. Using five years of movie data (story lines for over 1,000 movies produced and launched between 2002 and 2006), along with RapidMiner's Text Processing extension, we tried to predict the financial success of movies. Following the format used by Sharda and Delen (2006) and Delen and colleagues (2007), we discretized the dependent variable into nine classes based on the following breakpoints:

Table O.1 Breakdown of the Dependent Variables into Nine Classes

Class No.	1	2	3	4	5	6	7	8	9
Range (in Millions)	<1 (Flop)	>1 <10	>10 <20	>20 <40	>40 <65	>65 <100	>100 <150	>150 <200	>200 (Blockbuster)

In summary, we used the textual variable of "story line" as the input to the prediction system and used the class numbers (nine-value nominal-ordinal variable as shown above) as the output of our system. Once such a prediction model is trained and deployed, the user can input the story line of a hypothetical movie and find out what success class it belongs to.

ANALYSIS

Text Processing is an extension to the RapidMiner data mining software tool. In order to use RapidMiner for this text mining project, we must first make sure that our version of RapidMiner includes the Text Processing extension. You can check to see what extensions are already installed on your RapidMiner by clicking Help and then selecting Manage Extensions (Figure O.1). In the pop-up window you will see all of the installed extensions with their current versions (Figure O.2). In this interface you can deactivate or uninstall already installed extensions.

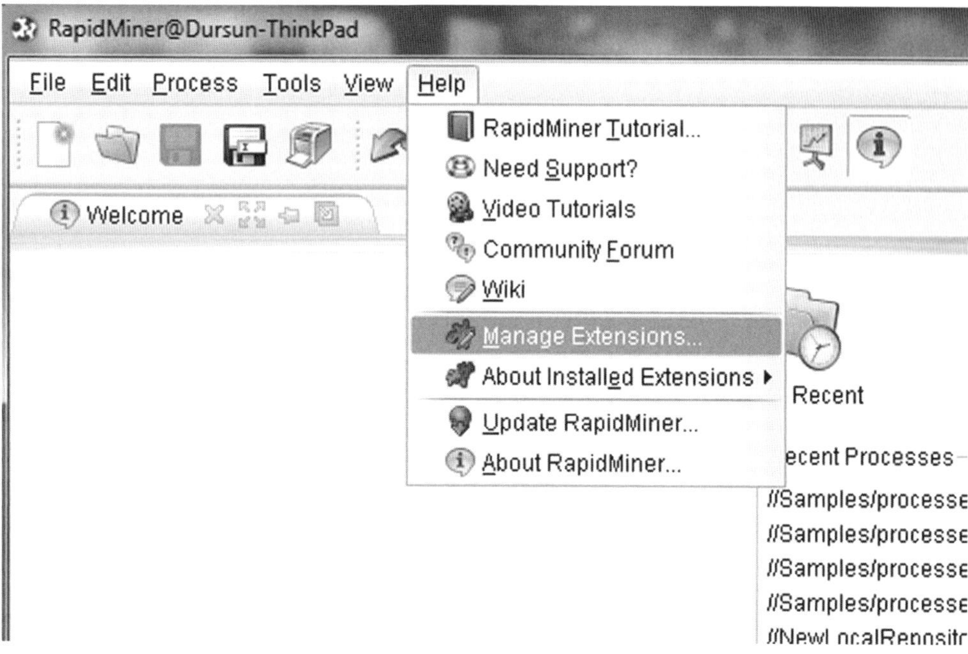

FIGURE O.1

Visualizing the already installed extensions in RapidMiner software tool.

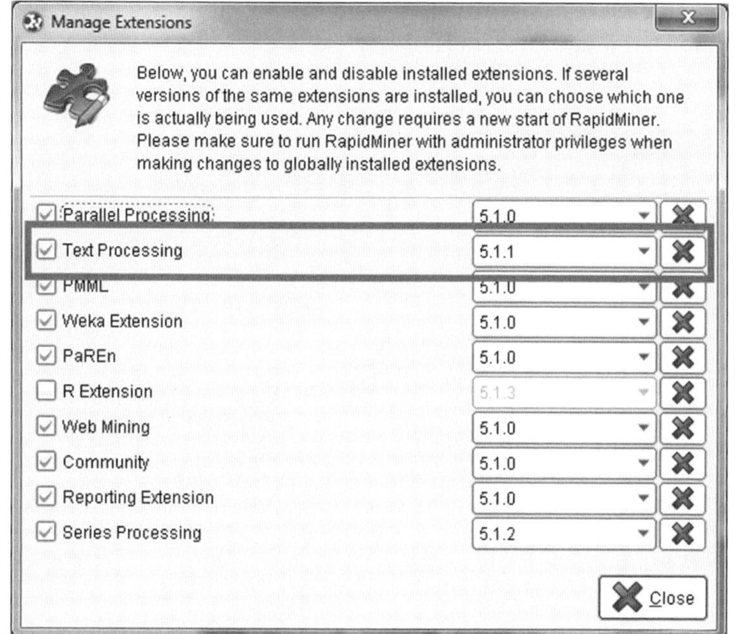

FIGURE O.2

Managing extensions in RapidMiner.

If you don't see Text Processing listed, then it is not installed. All you have to do is go to Help, click on Update RapidMiner, and select and install Text Processing (Figure O.3).

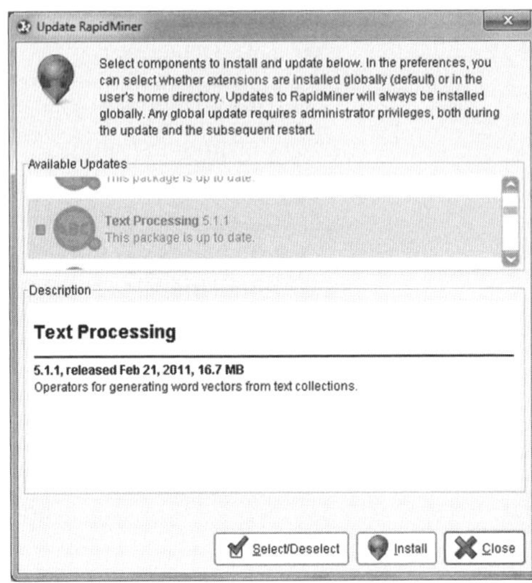

FIGURE O.3
Installing the Text Processing extension. Text Processing is grayed out because it is already installed with the latest version.

Once we have confirmed the existence of the Text Processing extension in RapidMiner, we can start building our process. For that we need to start a new process (Figure O.4).

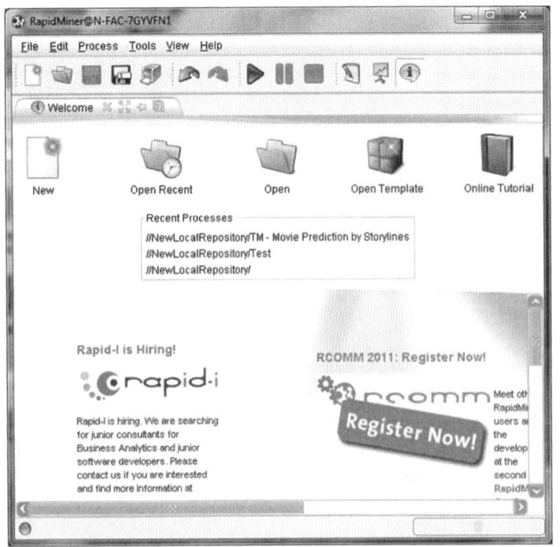

FIGURE O.4
Starting a new process in RapidMiner.

As soon as you click on a new process, RapidMiner asks you to specify the repository location. To put it simply, a repository in RapidMiner is a centralized location to collect and organize all project related files. In this interface (Figure O.5), you can select an existing repository or to create a new repository.

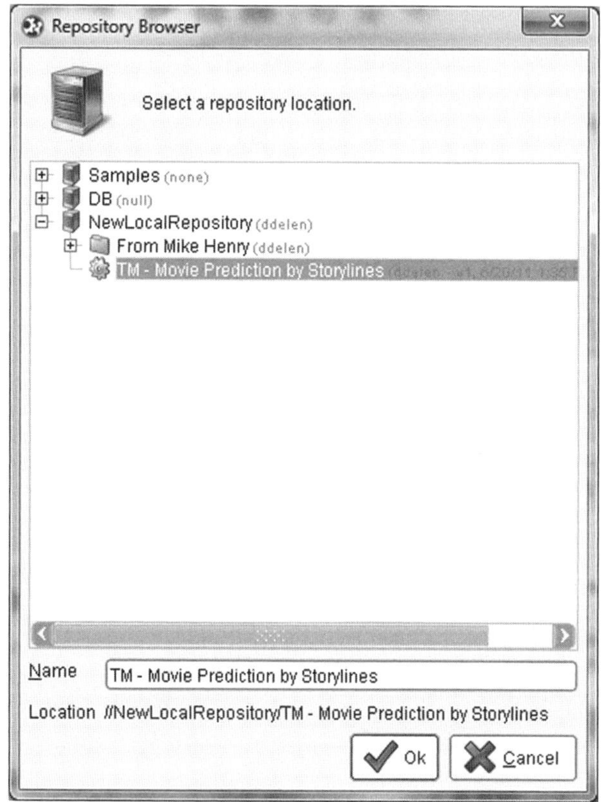

FIGURE O.5
Selecting an existing project or creating a new project under a repository.

Once the repository is specified, you will be directed to the process window (Figure O.6). There you can browse, select, or search Operators to drop into the Project workspace. Once located, you can simply drag and drop it into the workspace. Based on your selection, the connections between the input/output ports (and ports between the Operators) can be made automatically or manually (by clicking and dragging the connection from the source port and clicking and dropping it into the destination port).

In this project, we will read an Excel file as our input file. Therefore, we search for the Read Excel operator (step "1"). Once located, we can drag and drop it into the Process workspace (step "2"). We can select the Read Excel process in the Process workspace and see its properties in the pane on the right-hand side. There we can click on Import Configuration Wizard to guide us through locating and specifying the properties of our input data file (Figure O.6).

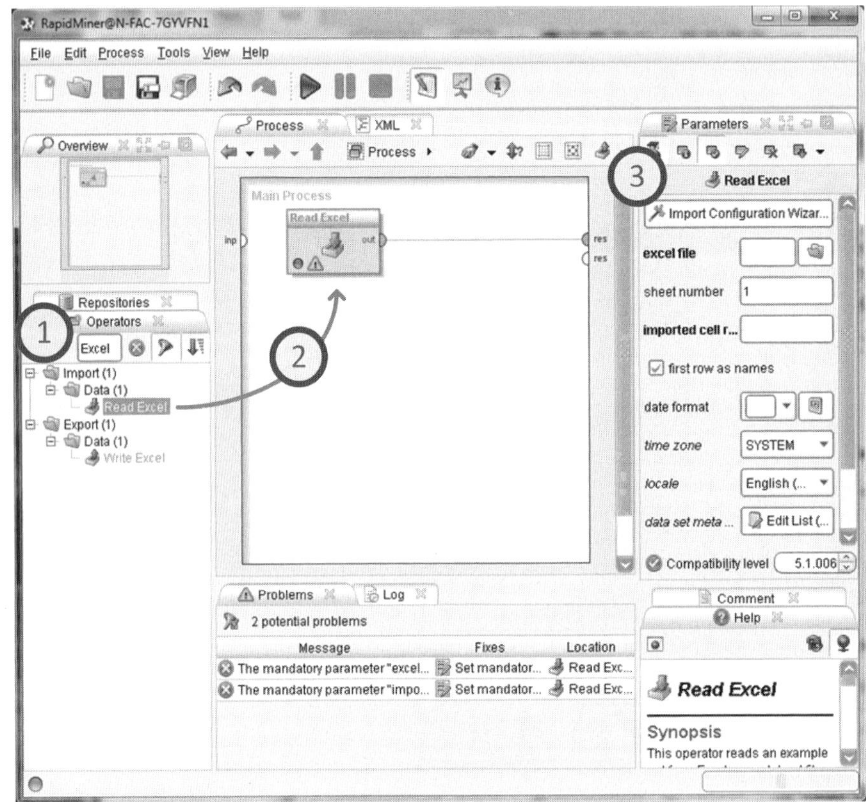

FIGURE O.6
Connecting to an Excel file in the Process workspace.

Once the Import Configuration Wizard has started, we will be asked to go through four simple steps:

Step 1. Locate and select your Excel file. A small snapshot of the Excel file is shown in Figure O.7.

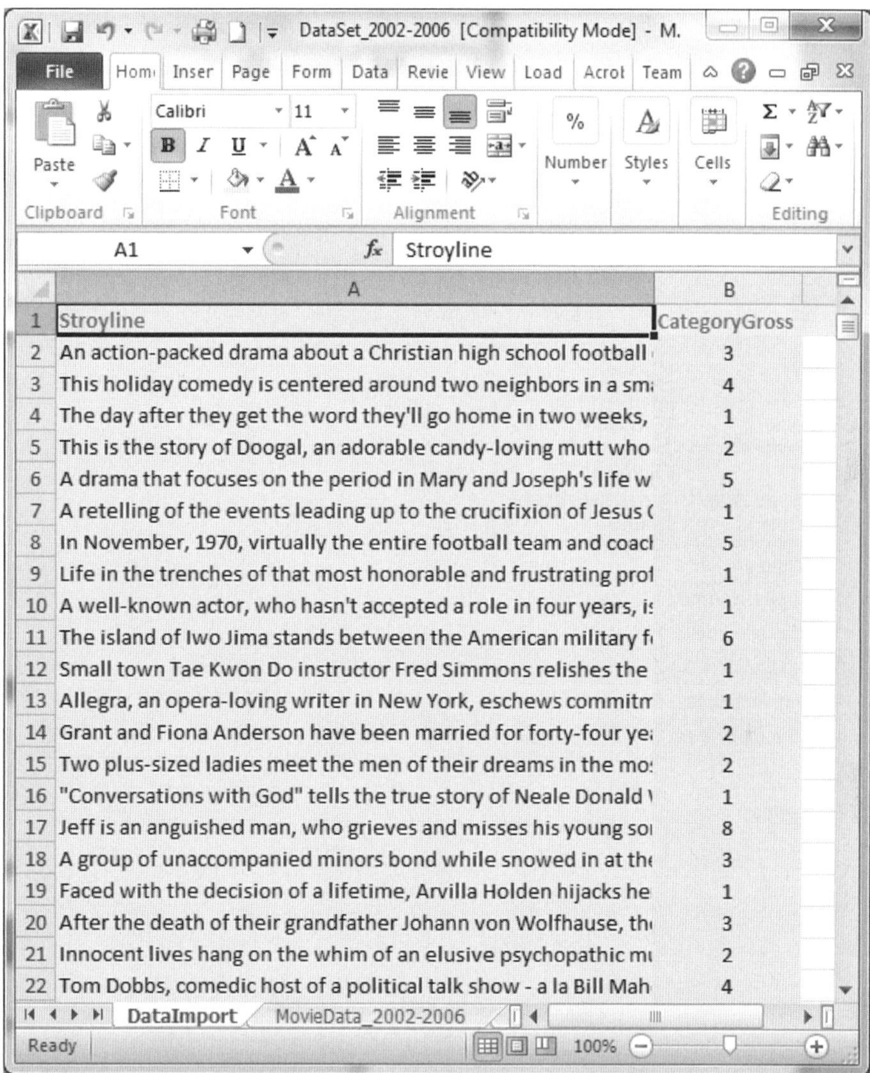

FIGURE O.7
A snapshot for the data set in Microsoft Excel.

Step 2. Select your Excel sheet within the selected Excel file.
Step 3. Make annotations to your cases (if you so desire).
Step 4. Specify the data types for your variables (e.g., numeric, nominal, text, etc.) and the role that they will play (e.g., attribute, label, ID, etc.). The final step in the Import Configuration Wizard is shown in Figure O.8. Then click Finish to save and return to the text mining process window.

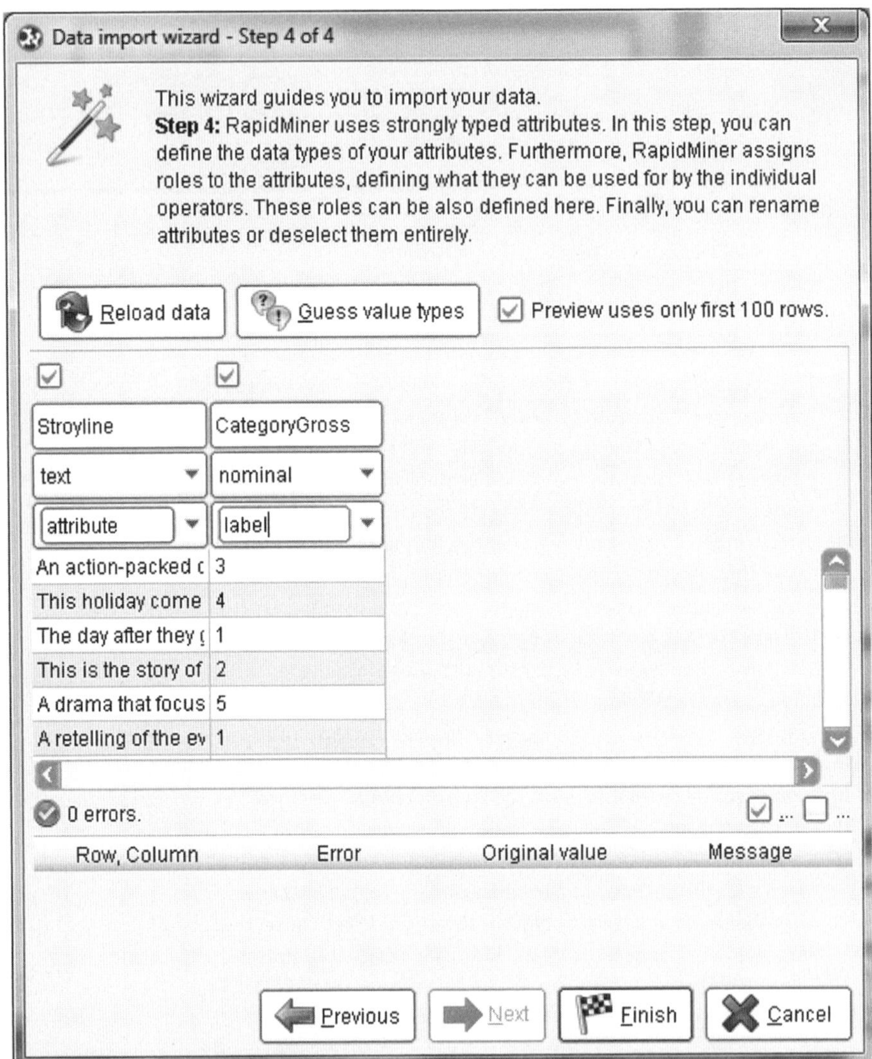

FIGURE O.8
Step 4 in the Import Configuration Wizard process.

Once the input data are specified, we can locate and place other operators into the process workspace (Figure O.9). As shown in Figure O.9, immediately following Read Excel operator, we used a Process Documents from Data operator. This operator has a subprocess underneath, where we provide additional operators to convert textual data into a structured form (essentially into a matrix where the rows represent the documents/movies and the columns represent the unique terms/words identified from the collection of story lines). The relationships between the rows and columns are represented by some sort of indices in the intersecting cells.

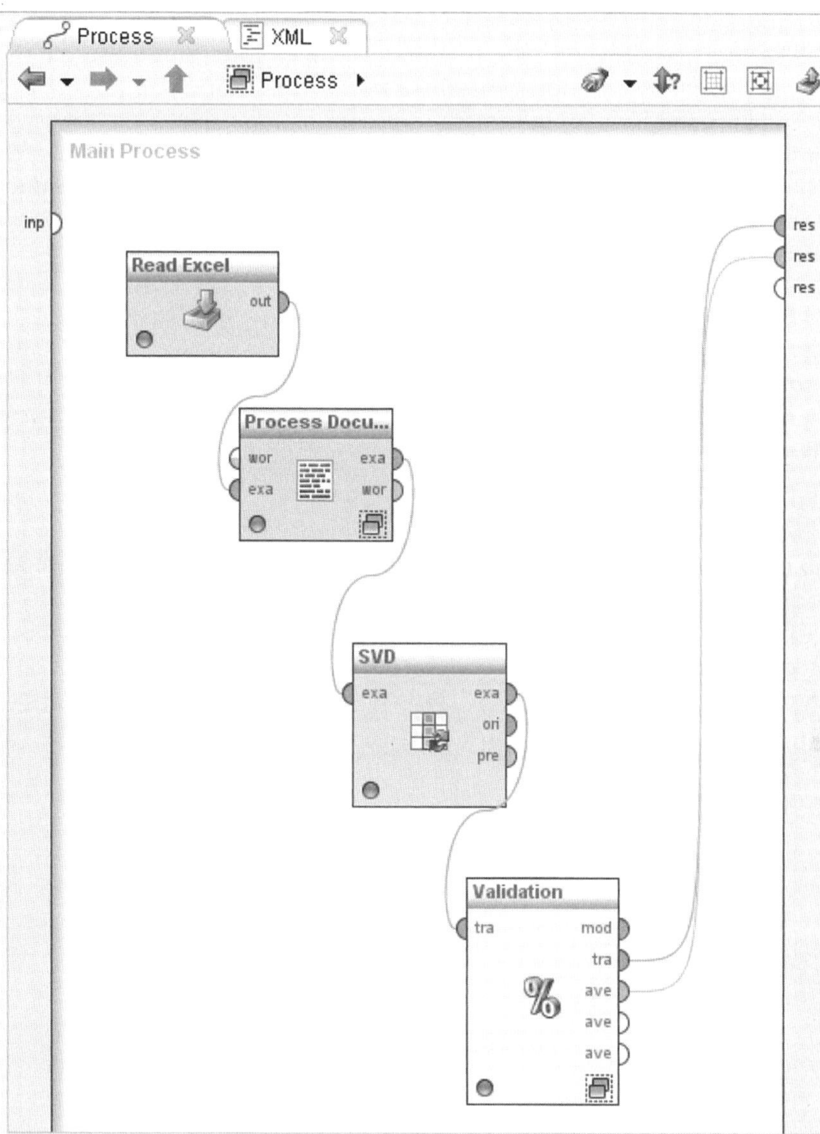

FIGURE O.9
The main process for the text mining project.

Figure O.10 shows the subprocess under the Process Documents from Data operator. As can be seen, we used five Operators to convert unstructured text (in the form of story lines) into a matrix. The first Operator (i.e., Transform Cases) converts the text into a single case type (either all lower case or all upper case) so that the case differences between the terms/words would not incorrectly lead to confusion in term/word identification. The next operator (i.e., Tokenize) takes the input of the previous operator and separates the terms/words from each other. Following the tokenization, comes the Filter Stopwords (English) operator. This operator filters out the most commonly observed words in English from the term/word list. The assumption is that these words (e.g., *a, an, is, am, are, the,* etc.) repeat many times in text documents and have no information/discrimination value. Next, we used another operator (i.e., Filter Tokens by Length) to remove words/terms that are less than three characters long. Finally, we used an operator (i.e., Generate n-Grams) to identify two- and three-word terms and make them a part of the term-document matrix.

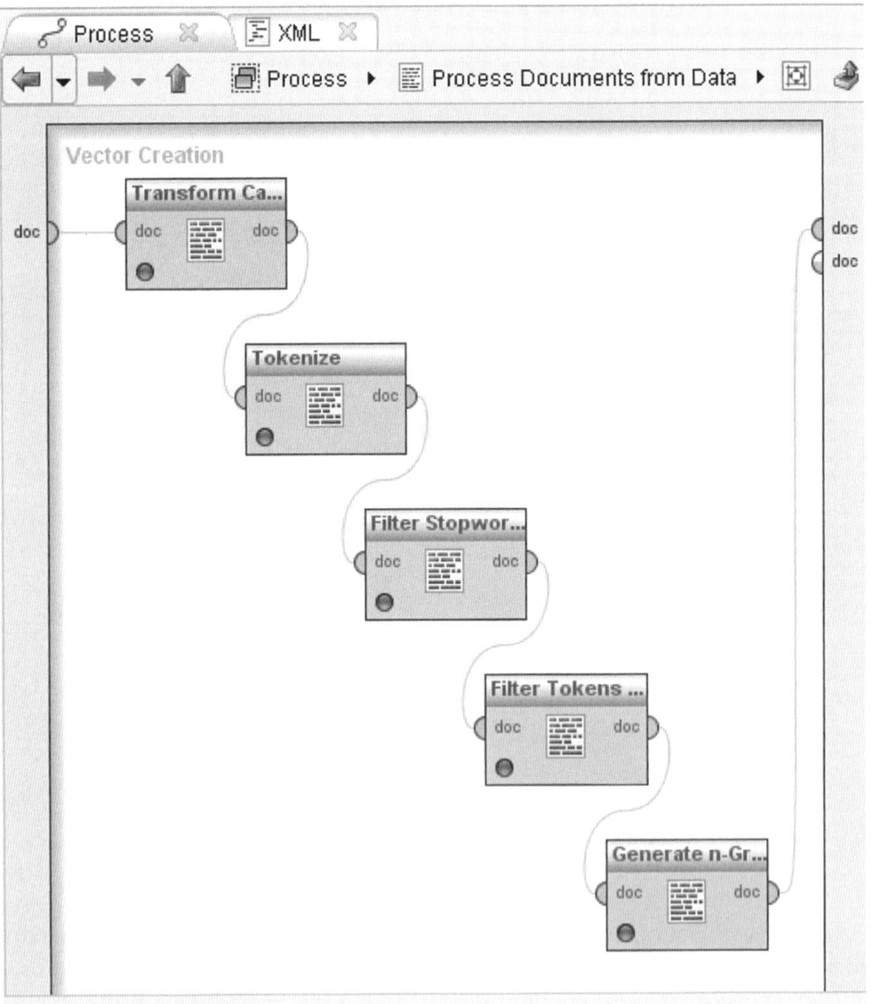

FIGURE O.10

Subprocess of the Process Documents from Data operator.

In the main process (see Figure O.9) following the Process Documents from Data operator, we used a data reduction operator called singular value decomposition (SVD). This operator takes the input data and generates a much smaller number of pseudo-variables (often called the variable reduction/transformation process). The output of the Process Documents from Data operator is a large matrix, more than 1,000 rows and 2,095 columns (variables—words/terms that are created after the five-step text processing). This matrix is shown in Figure O.11. As can be seen, this is a sparse matrix with most of the cells filled with zero values. For classification (which is the case in this study) or clustering type data mining tasks, such large and sparse matrixes are often converted into a smaller size while maintaining underlying patterns. In test mining this reduction process is often performed using SVD. In this example, we reduced 2,095 variables into 5 pseudo-variables (singular values) for further processing.

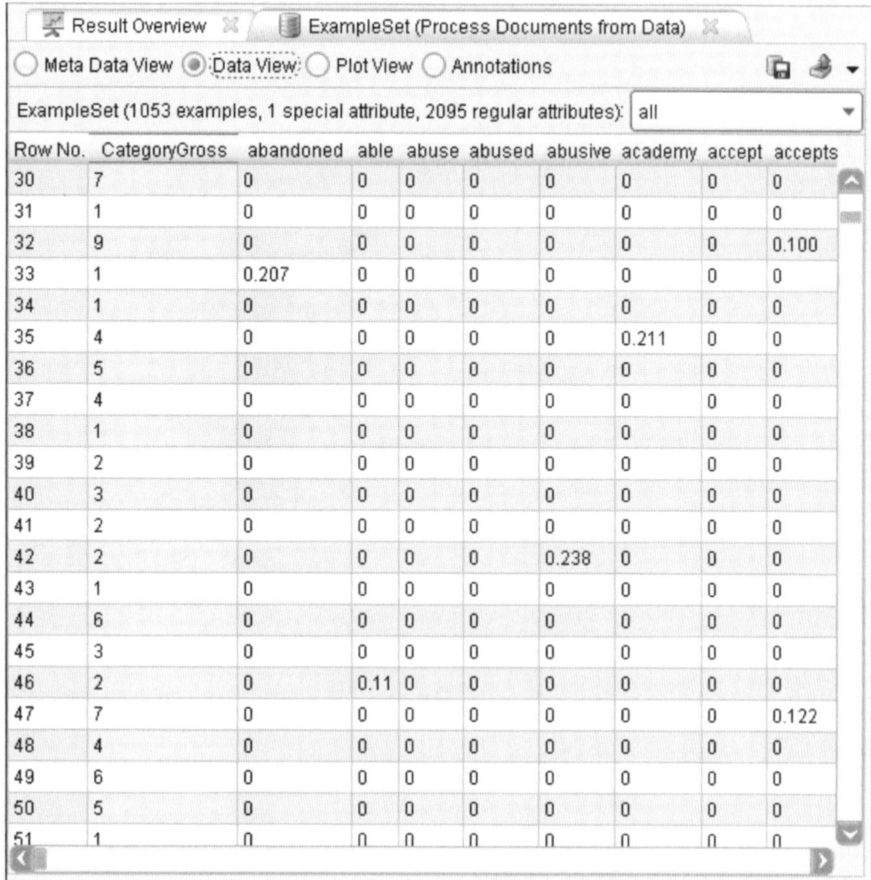

Result Overview | ExampleSet (Process Documents from Data)

Meta Data View | Data View | Plot View | Annotations

ExampleSet (1053 examples, 1 special attribute, 2095 regular attributes): all

Row No.	CategoryGross	abandoned	able	abuse	abused	abusive	academy	accept	accepts
30	7	0	0	0	0	0	0	0	0
31	1	0	0	0	0	0	0	0	0
32	9	0	0	0	0	0	0	0	0.100
33	1	0.207	0	0	0	0	0	0	0
34	1	0	0	0	0	0	0	0	0
35	4	0	0	0	0	0	0.211	0	0
36	5	0	0	0	0	0	0	0	0
37	4	0	0	0	0	0	0	0	0
38	1	0	0	0	0	0	0	0	0
39	2	0	0	0	0	0	0	0	0
40	3	0	0	0	0	0	0	0	0
41	2	0	0	0	0	0	0	0	0
42	2	0	0	0	0	0.238	0	0	0
43	1	0	0	0	0	0	0	0	0
44	6	0	0	0	0	0	0	0	0
45	3	0	0	0	0	0	0	0	0
46	2	0	0.11	0	0	0	0	0	0
47	7	0	0	0	0	0	0	0	0.122
48	4	0	0	0	0	0	0	0	0
49	6	0	0	0	0	0	0	0	0
50	5	0	0	0	0	0	0	0	0
51	1	0	0	0	0	0	0	0	0

FIGURE O.11
The term-document matrix generated by the Process Documents from Data operator.

Once the SVD operator completed its processing of the data, we can observe the relationships between the five dimensions in a scatterplot (Figure O.12). By doing so, we can go back and either increase or reduce the dimensions to "optimize" the number of new variables to be used for the classification task that follows.

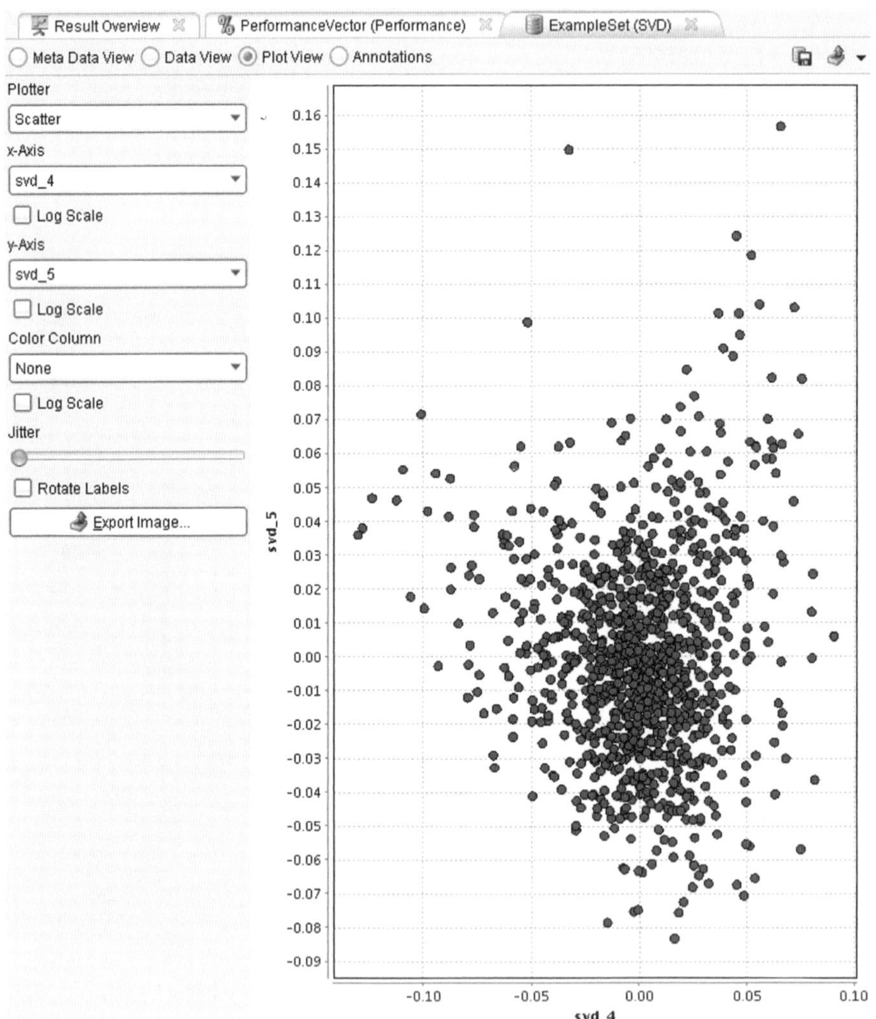

FIGURE O.12
A scatterplot showing the relationship between SVD dimensions 4 and 5.

The last operator in the main process is a subprocess called Validation. This operator is used for classification and regression tasks where the data are automatically split into *x* number of folds and the experimentation is repeated for *x* times, each time one fold of the data is used as the test sample (holdout sample). Figure O.13 shows the subprocess used under the Validation operator.

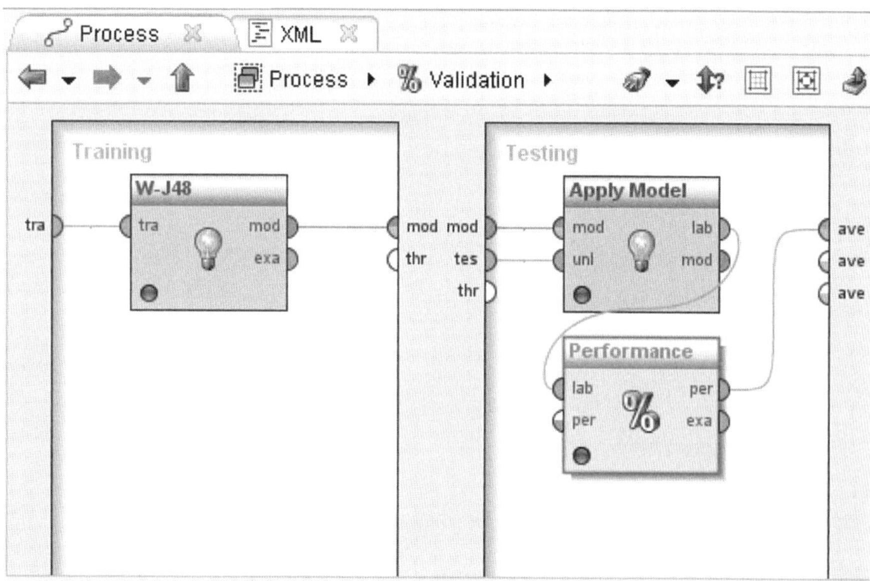

FIGURE O.13
The subprocess under the Validation operator.

As can be seen in Figure O.13, there are two distinct panes in this subprocess: a model training pane and a model testing pane. In this example, in the model training pane, we used an operator called W-J48. This operator is a part of the Weka library. Weka is another freely downloadable open source data mining tool (www.cs.waikato.ac.nz/ml/weka/index_downloading.html). In addition to its own algorithms, RapidMiner also seamlessly incorporates most of Weka's data mining algorithms. J48 is essentially the Java implementation of Quinlan's famous C4.5 decision tree generation algorithm.

On the right-hand side of this subprocess, the trained models are tested and assessed. First, the developed model is applied to the test data (as specified by the current fold) using Apply Model operator, and then its accuracy is assessed using Performance operator.

The classification results of the X-Validation (also known as k-fold cross validation) are shown in Figure O.14 in a confusion matrix format. Herein, all of the folds are combined and aggregated into a single confusion matrix. In this confusion matrix, the columns represent the actuals, while rows represent the predictions. For instance, the very first cell says that 30 of the Class-3 records in the test data sets are accurately predicted as Class-3, accounting for 21.28 percent classification accuracy for Class-3.

As the top of the window shows, the overall accuracy on the test data set is only 15.77. That is, out of all the movies used in this study, this model accurately predicted only 15.77 percent of them correctly. This is not a great prediction accuracy. Considering that the random change of accurately assigning a movie into one of nine classes is roughly 11 percent (i.e., 1/9), 15.77 percent accuracy is only marginally better than random chance but obviously not accurate enough for decision making.

Multiclass Classification Performance   Annotations

Table View   Plot View

accuracy: 15.77% +/- 2.42% (mikro: 15.76%)

	true 3	true 4	true 1	true 2	true 5	true 6	true 8	true 7	true 9	class precis
pred. 3	30	29	27	32	19	16	8	11	4	17.05%
pred. 4	12	21	24	24	13	23	5	8	13	14.69%
pred. 1	30	27	44	37	22	15	6	20	12	20.66%
pred. 2	30	24	46	35	18	21	5	11	12	17.33%
pred. 5	10	13	10	18	11	13	3	6	4	12.50%
pred. 6	11	15	9	18	5	6	2	7	9	7.32%
pred. 8	7	10	2	5	4	1	2	3	2	5.56%
pred. 7	6	4	9	8	4	5	5	9	4	16.67%
pred. 9	5	10	7	4	7	9	5	4	8	13.56%
class recall	21.28%	13.73%	24.72%	19.34%	10.68%	5.50%	4.88%	11.39%	11.76%	

FIGURE O.14
Classification results on holdout samples represented in a confusion matrix.

SUMMARY

In this tutorial, we used textual data (i.e., story lines) of over 1,000 movies launched between 2002 and 2006 to predict the financial success. As the results indicated, textual information in a story line may not be enough to accurately predict how a movie is going to do at the box office. A short paragraph (a story line) about a movie may not have enough information content to explain potential box office success. At least that is what we deduce from the prediction models developed for this tutorial. It may, however, provide additional variables (words/terms) that can complement other, more traditional data mining-based prediction models.

References

Dursun Delen, Ramesh Sharda, and P. Kumar, 2007, "Movie Forecast Guru: A Web-Based DSS for Hollywood Managers," *Decision Support Systems*, 43(4), 1151–1170.

Ramesh Sharda, and Dursun Delen, 2006, "Predicting Box Office Success of Motion Pictures with Neural Networks," *Expert Systems with Applications*, 30, 243–254.

A Hands-On Tutorial of Text Mining in PASW: Clustering and Sentiment Analysis Using Tweets from Twitter

Raja Kakarlapudi
Student, Department of Marketing, Oklahoma State University

Satish Garla
Student, Department of Marketing, Oklahoma State University

Dr. Goutam Chakraborty
Professor, Department of Marketing, Oklahoma State University

CONTENTS

INTRODUCTION

Unstructured textual data such as customer interactions with call center support staff, product reviews given by customers on companies' websites, customer responses from online surveys, and customer comments about products/services in blogs or on Twitter are being increasingly collected and stored. Most analysts and researchers are used to analyzing numerical data, but they usually find handling and analyzing textual data problematic. Thus, often such textual data are not analyzed. However, if analyzed properly, such unstructured data can often lead to valuable information about a company's own products or services, as well as competitors' products and services. We illustrate two approaches to analyzing such textual data: clustering of text data into natural groupings and sentiment analysis that identifies if customers are generally happy or unhappy with a company's

Practical Text Mining and Statistical Analysis for Non-structured Text Data Applications. DOI: 10.1016/B978-0-12-386979-1.00024-4

products or services. We use the context of consumer postings on Twitter to demonstrate these two approaches.

OBJECTIVE

During September 2010, Twitter changed its website interface and launched a new interface with more features and controls. This tutorial demonstrates how to use text mining and sentiment analysis using the PASW® Text Analytics tool to evaluate user acceptance of the new Twitter web interface. We also explore the reasons for users' satisfaction or disappointment due to the change.

CASE STUDY

We use the PASW® Text Analytics add-on in PASW® Modeler 14 to perform this analysis. The PASW Text Analytics package has strong natural language processing (NLP) and advanced linguistics capabilities to process a large variety of unstructured textual data and extract key concepts. In this tutorial we demonstrate concept extraction, category creation, cluster analysis, and text link analysis. The demonstration is intended to give an overview of such tools that can be used. For more details or technical background behind such analyses, interested readers should refer to relevant chapters in this text as well as the help file from PASW.

Data

The data we use for this analysis are the tweets (posting by users) from Twitter. To understand user reaction to the new interface, the tweets about the new interface are an ideal source of information. After using the new website, most of the Twitter users commented on what and how they felt about the change. Information on Twitter related to users or comments made by the users are publicly available unless the user chooses not to disclose any of his or her information. These data were collected using an SAS macro program (%GetTweet macro). For more details on using this macro, refer to http://support.sas.com/resources/papers/proceedings11/324-2011.pdf.

The macro we used was configured to collect tweets that contained the key words *New Twitter*, *Twitter*, *New Interface*, or the hash tag *#newtwitter*. Total collected tweets using this program during first ten days of September are around 8,600. The variables in the dataset include:

Id: Unique ID generated for each tweet
Text: The actual user tweet (the analysis variable)
Pubdate: Date of the tweet

The Analysis Process

While there are various ways to access different types of data, the steps explain how to use the SAS File node in PASW to read SAS data sets. Figure P.1 shows the starting screen in PASW Modeler.

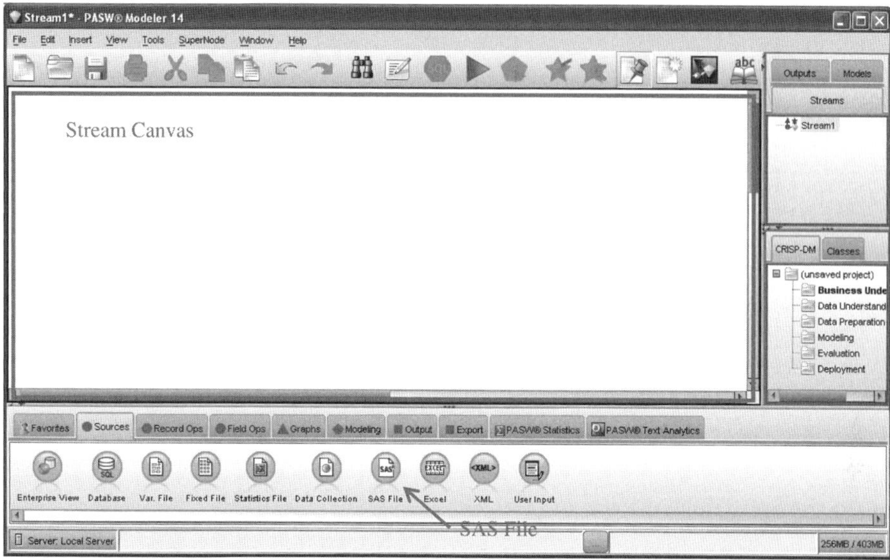

FIGURE P.1
IBM-SPSS Modeler starting screen.

1. Drag and drop **SAS File** node onto the stream canvas.
2. Double click the SAS File node on the canvas.
3. Select data using the browse button next to **Import File** text field (see Figure P.2).

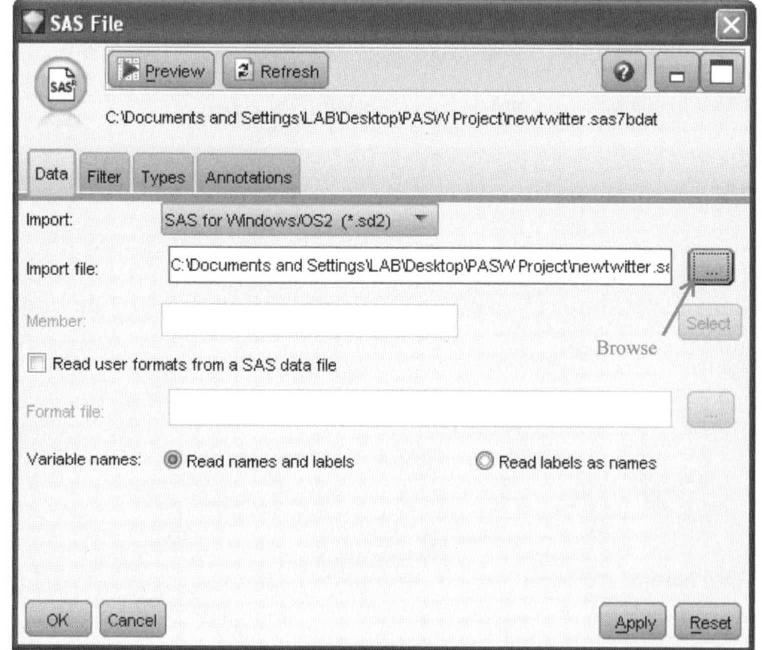

FIGURE P.2
SAS file dialog where the SAS file can be selected for importing into IBM-SPSS Modeler.

4. Click on the **Types** tab (see Figure P.3), and click **Read Values** and select **OK** if there is any pop-up message.

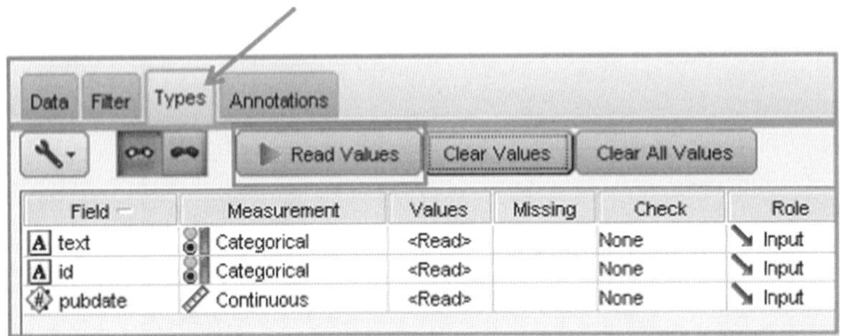

FIGURE P.3
The Types Tab in the SAS Data Miner and Text Miner software.

5. Click **OK** on the SAS File window.

Note: During this step, you can click on the Preview button to view sample data to make sure the input data are being read correctly.

At the end of this step, the data on the stream canvas can be used for further analysis. For demonstrating text mining, we will be using the Text Mining node available in the PASW Text Analytics tab (see Figure P.4).

FIGURE P.4
The PASW Text Analytics set of nodes, including the Text Mining node.

6. Drag the **Text Mining** node onto the stream canvas.
7. Right click on the **SAS File** node and click **Connect**. Then select the **Text Mining** node to join both nodes (see Figure P.5).

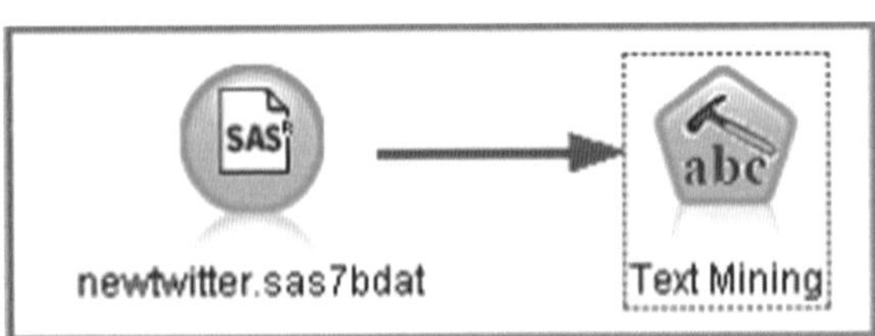

FIGURE P.5
Connecting the SAS node and the Text mining node.

8. Double click on the **Text Mining** node. Under the **Fields** tab, select the variable that contains the text data for the **Text Field**.
9. Click on the **Model** tab and select "Build interactively" (category model nugget) for the **Build mode**.

Note: Two different model build modes are available in this tool. Interactive mode allows you to retain all extracted concepts, whereas the build model mode keeps only the concepts used by the model. There are various model fine-tuning options available with interactive mode.

10. From the Copy Resources from Section, click "Load" to load the linguistic resource library.
11. From the Load Resource Template, select "Customer Satisfaction Opinions (English)" and then click OK (see Figure P.6).

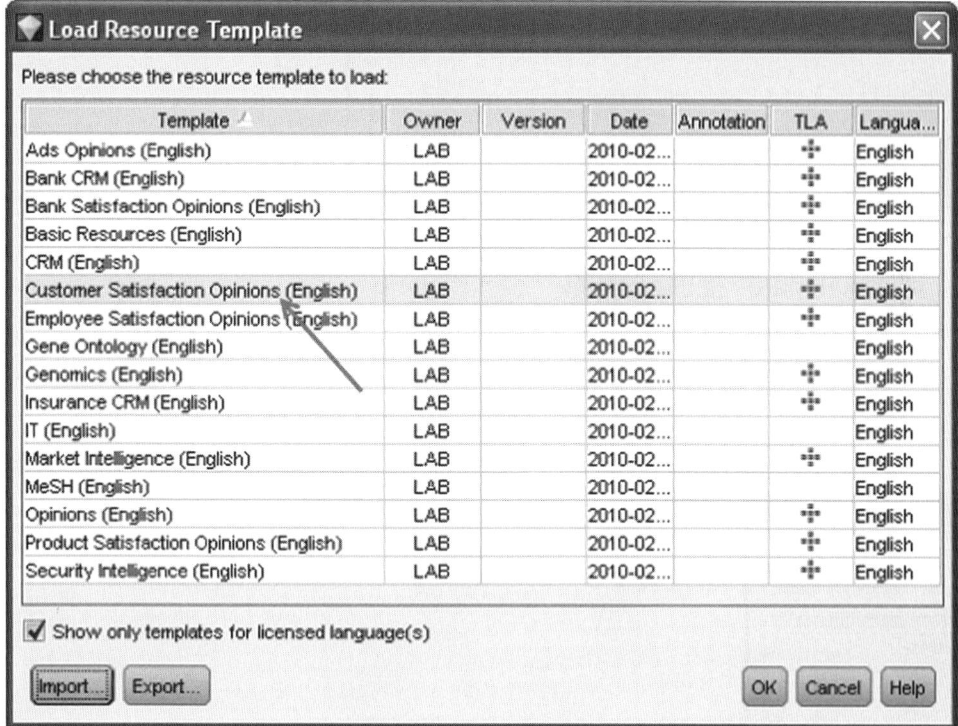

FIGURE P.6
The Load Resource Template.

12. Go to the **Expert** tab and enter 2 for **Limit Extraction to concepts with a global frequency count of at least:** Leave other values to default (see Figure P.7).

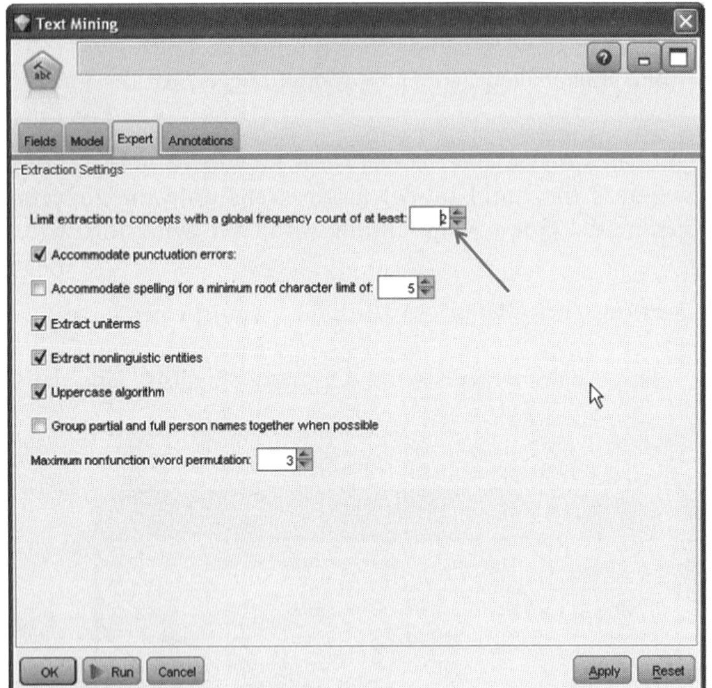

FIGURE P.7
The Expert Tab in the Text Mining window.

13. Click **Run** on the Text Mining window.

The first step in the Text Mining process, Extraction, is initiated. You will see a small message window (Extracting) with the message "Please wait, extraction in progress." You will encounter similar status windows during the course of your analysis for most of the tasks (see Figure P.8).

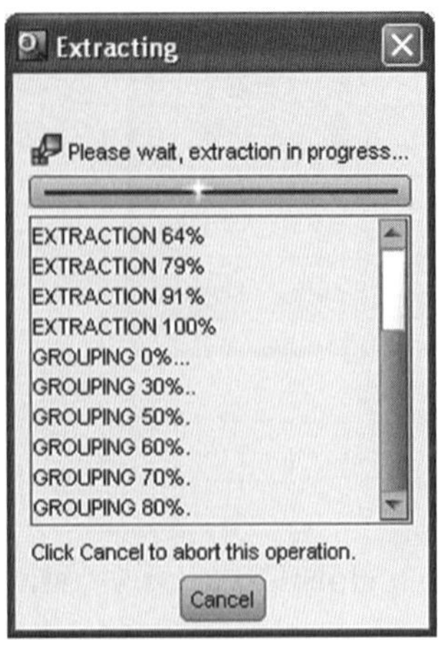

FIGURE P.8
The Extracting window.

After the stream is executed, an advanced interface is launched, through which you can perform exploratory analysis, refine extraction results, create categories, explore text link patterns, and view results in graphs (see Figure P.9).

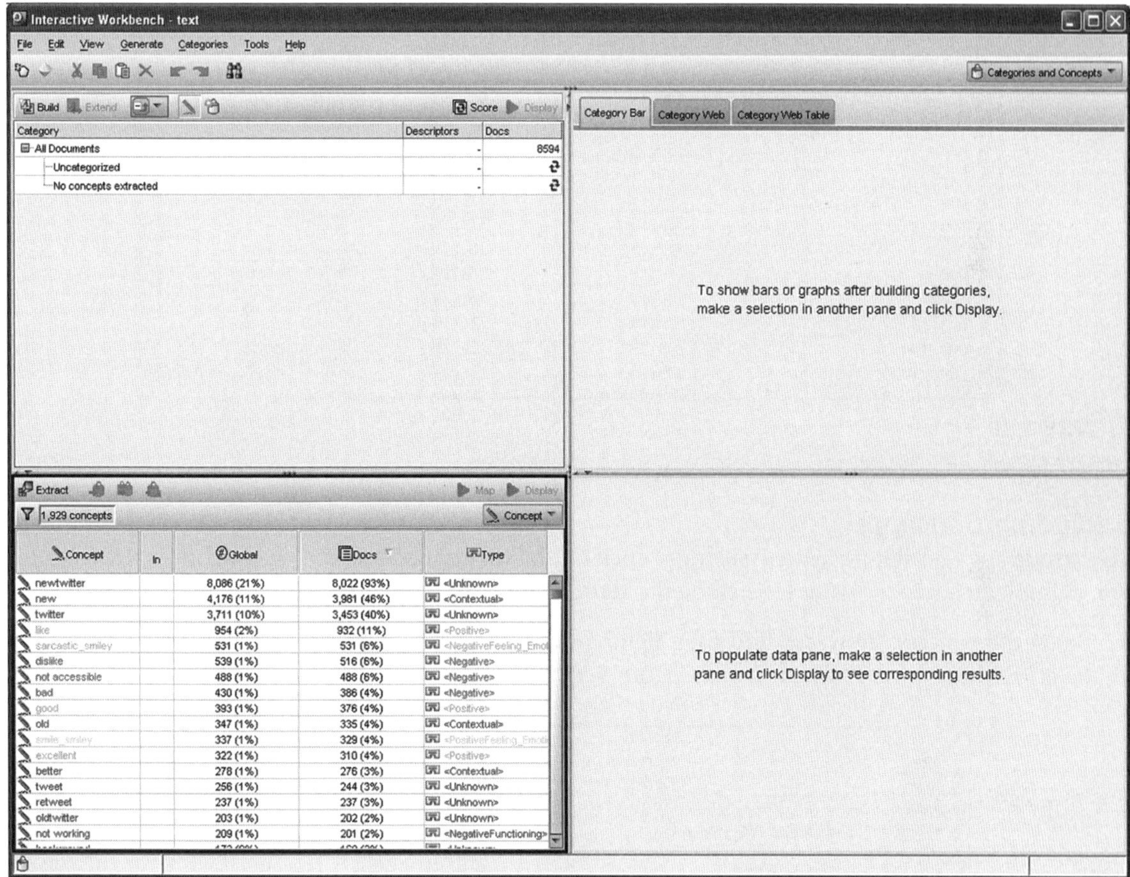

FIGURE P.9
The PASW Advanced Window where various computations and comparisons can be explored.

Concept Extraction

Figure P.10 shows the concepts extracted from the data set based on the default settings and the selected language resource library. For better results, we need to fine-tune the extracted concepts by excluding concepts, combining concepts to form synonyms, and assigning concepts to types.

Extract · Map · Display
1,929 concepts · Concept

Concept	In	Global	Docs	Type
newtwitter		8,086 (21%)	8,022 (93%)	<Unknown>
new		4,176 (11%)	3,981 (46%)	<Contextual>
twitter		3,711 (10%)	3,453 (40%)	<Unknown>
like		954 (2%)	932 (11%)	<Positive>
sarcastic_smiley		531 (1%)	531 (6%)	<NegativeFeeling_Emo
dislike		539 (1%)	516 (6%)	<Negative>
not accessible		488 (1%)	488 (6%)	<Negative>
bad		430 (1%)	386 (4%)	<Negative>
good		393 (1%)	376 (4%)	<Positive>
old		347 (1%)	335 (4%)	<Contextual>
smile_smiley		337 (1%)	329 (4%)	<PositiveFeeling_Emoti
excellent		322 (1%)	310 (4%)	<Positive>
better		278 (1%)	276 (3%)	<Contextual>
tweet		256 (1%)	244 (3%)	<Unknown>
retweet		237 (1%)	237 (3%)	<Unknown>
oldtwitter		203 (1%)	202 (2%)	<Unknown>
not working		209 (1%)	201 (2%)	<NegativeFunctioning>

FIGURE P.10

Concepts extracted from the data.

Excluding Concepts

To exclude any concept from analysis, right click on the concept and select "Exclude from Extraction." For example, assume we want to exclude the term *retweet* from analysis. From the Extract pane:

1. Select the concept "retweet," and then right click on the concept.
2. Select **Exclude from Extraction** (see Figure P.11).

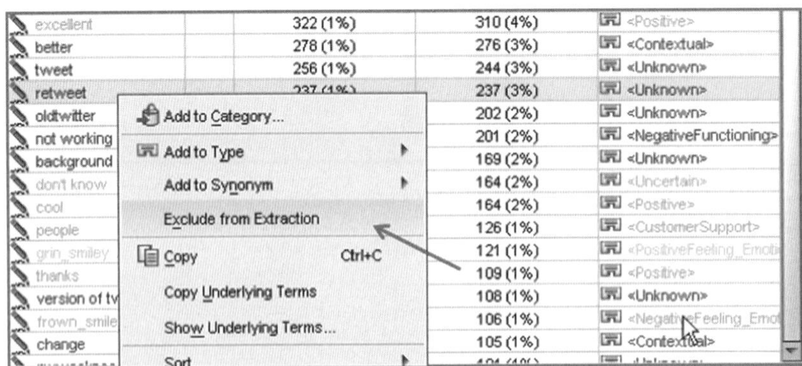

FIGURE P.11

Exclude from Extraction from the flying menu dialog.

Repeat the same task for all of the terms that you think are not relevant to your analysis. After making any change, the Extract window turns yellow, indicating it needs to be rerun to apply the changes (see Figure P.12).

better		278 (1%)	276 (3%)	<Contextual>
tweet		256 (1%)	244 (3%)	<Unknown>
retweet		237 (1%)	237 (3%)	<Unknown>
oldtwitter		203 (1%)	202 (2%)	<Unknown>
not working		209 (1%)	201 (2%)	<NegativeFunctioning>
background		172 (0%)	169 (2%)	<Unknown>
don't know		165 (0%)	164 (2%)	<Uncertain>
cool		167 (0%)	164 (2%)	<Positive>

FIGURE P.12
Retreat selected by the Exclude from extraction, ready to be re-run.

Combining Terms

Multiple terms can be combined to form synonyms with a parent term. This reduces the number of extracted concepts and generally produces better results. For example, assume we want to combine the terms *twitter* and *twitter.com*:

1. Select both terms by holding the **Ctrl** key and right click on any of these terms.
2. Click on **Add to Synonym** and then click **New** (see Figure P.13).

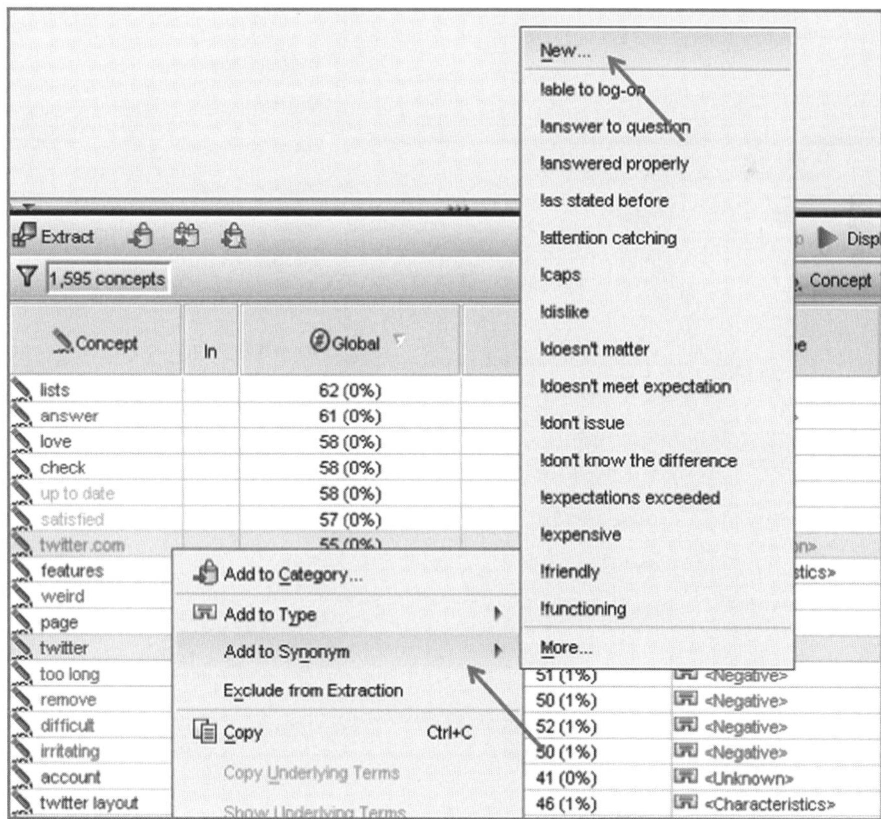

FIGURE P.13
Selecting "Add to Synonym".

3. Enter a parent name for both the terms in the **Target** field. You can add more terms to this list, separating terms with a comma.
4. Click **OK**.

Repeat the same task for all the terms that you think should be grouped together.

Assigning Terms to New Type

We can assign new types to terms that are identified as belonging to Unknown type. Good domain knowledge will help to correctly identify types. For example, in this analysis Twitter is the product being studied. By default, the term *Twitter* is not identified as a product. To assign a new type to the term *Twitter*, do the following:

1. Select the concept from the extract window.
2. Right click and select **Add to Type**.
3. If you do find the type you are looking for, click on **More** (see Figure P.14).

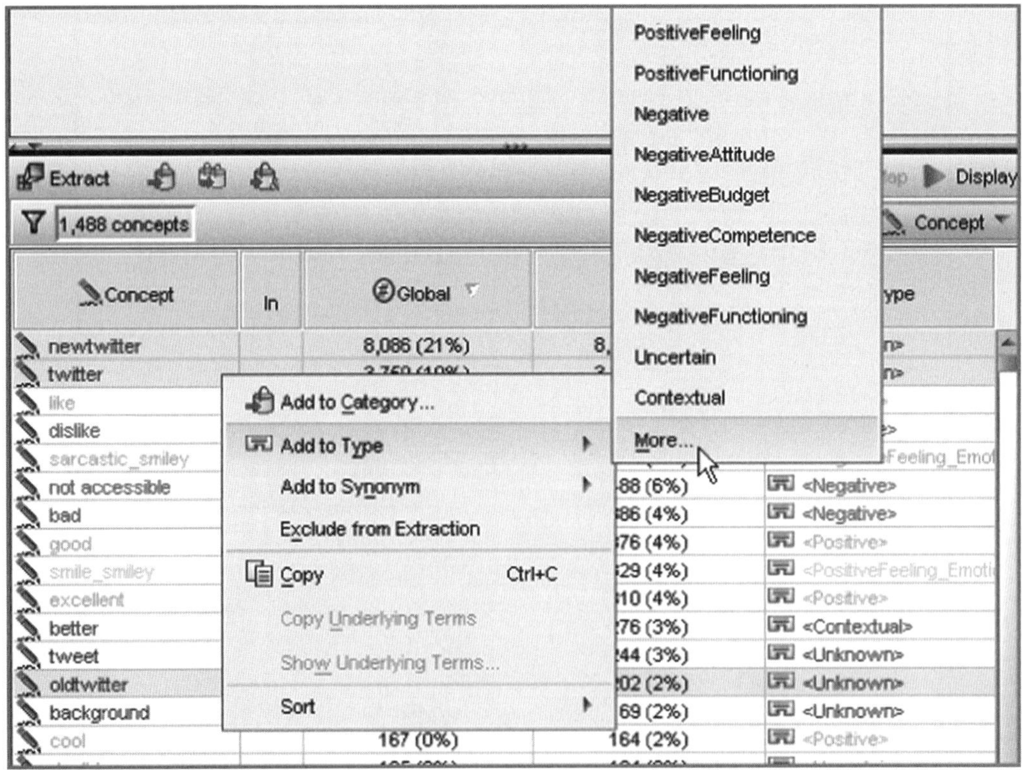

FIGURE P.14
Clicking on the MORE selection in the Extract dialog.

4. From the All Types window, select **Product** and click **OK** (see Figure P.15).

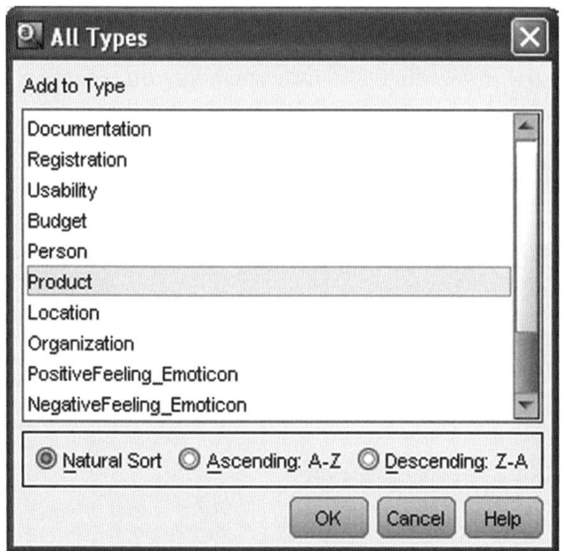

FIGURE P.15
Selecting "Product" from the All Types dialog.

Often in writing tweets, users use shortcuts or abbreviations that need to be put in types. For example, we may want to assign terms such as *lovin* and *luv* as positive and terms like *psssst* and *grrrr* as negative. This task should be repeated for all the terms that you believe are wrongly classified or classified as unknown (see Figure P.16).

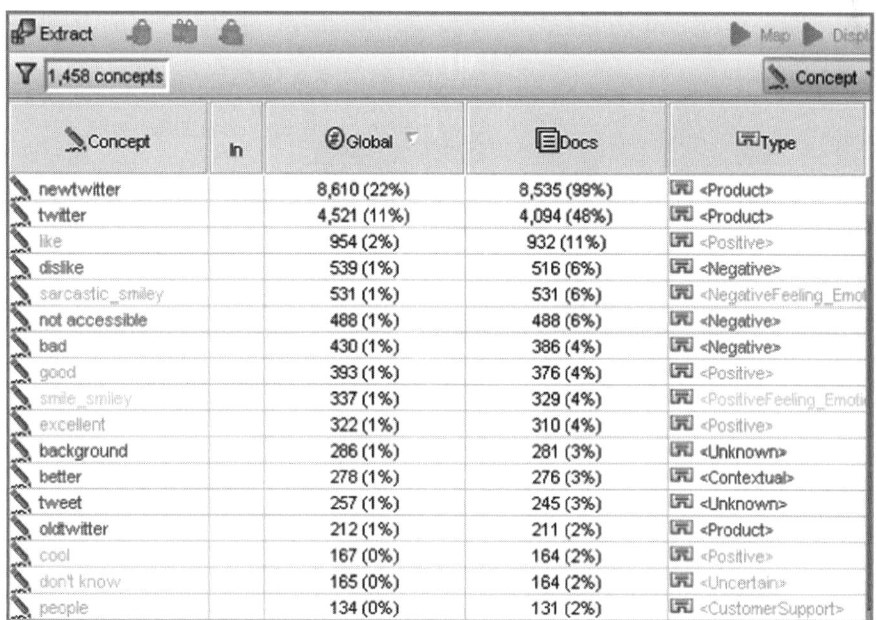

Concept	In	Global	Docs	Type
newtwitter		8,610 (22%)	8,535 (99%)	<Product>
twitter		4,521 (11%)	4,094 (48%)	<Product>
like		954 (2%)	932 (11%)	<Positive>
dislike		539 (1%)	516 (6%)	<Negative>
sarcastic_smiley		531 (1%)	531 (6%)	<NegativeFeeling_Emo
not accessible		488 (1%)	488 (6%)	<Negative>
bad		430 (1%)	386 (4%)	<Negative>
good		393 (1%)	376 (4%)	<Positive>
smile_smiley		337 (1%)	329 (4%)	<PositiveFeeling_Emoti
excellent		322 (1%)	310 (4%)	<Positive>
background		286 (1%)	281 (3%)	<Unknown>
better		278 (1%)	276 (3%)	<Contextual>
tweet		257 (1%)	245 (3%)	<Unknown>
oldtwitter		212 (1%)	211 (2%)	<Product>
cool		167 (0%)	164 (2%)	<Positive>
don't know		165 (0%)	164 (2%)	<Uncertain>
people		134 (0%)	131 (2%)	<CustomerSupport>

FIGURE P.16
Extract dialog; assigning abbreviated terms to "positive" or "negative".

CATEGORIZATION

After identifying the concepts in the data, we can create categories that represent a high-level topic that capture key ideas or sentiment expressed in the text. Categories can be created using several different methods. The Categories pane is the area in the Interactive Workbench window where you create and manage categories. In this demonstration we will show how to create categories automatically as well as manually.

Creating Categories Automatically

1. From the Menu bar, go to **Categories** → **Build Categories**.
2. Click **Edit** to change the default settings (see Figure P.17).

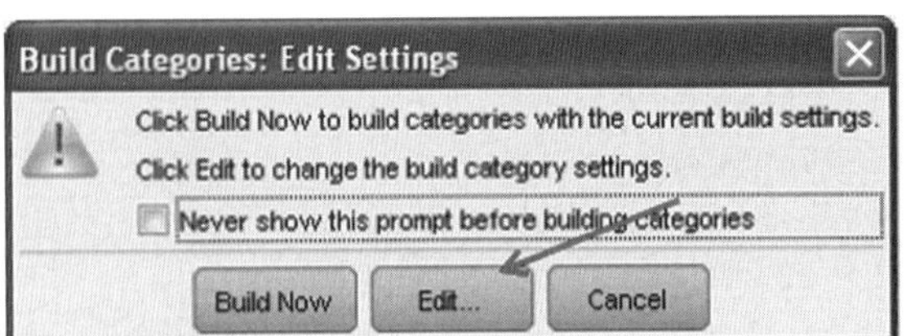

FIGURE P.17
The Build Categories - Edit Settings dialog.

3. In the **Inputs** section make sure **Types** is selected in the **Build Categories from:** drop-down.
4. Select all the Types related to Positive and Negative, like **<Positive>, <Positive Feeling>**, ... and **<Negative>, <Negative Emoticon>**,.... and deselect all other types (see Figure P.18).

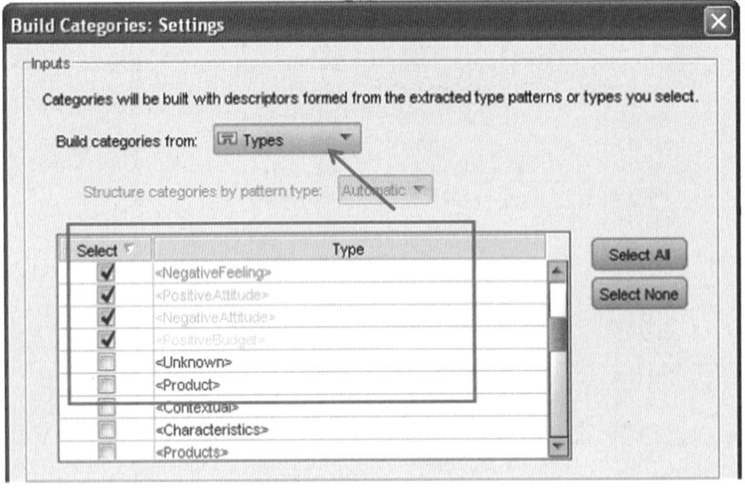

FIGURE P.18
The Build Categories - Settings dialog.

5. Click the **Advanced Settings** button in the **Techniques** section (see Figure P.19).

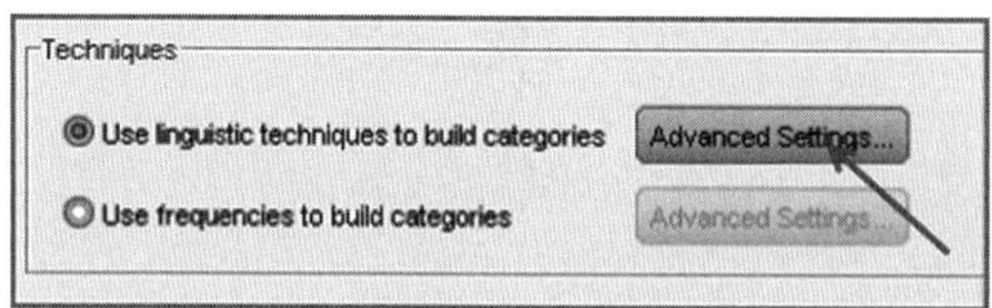

FIGURE P.19
The Advanced Settings button in the Techniques section.

6. Select **All Extraction Results** from the **Input and Output** section (see Figure P.20).

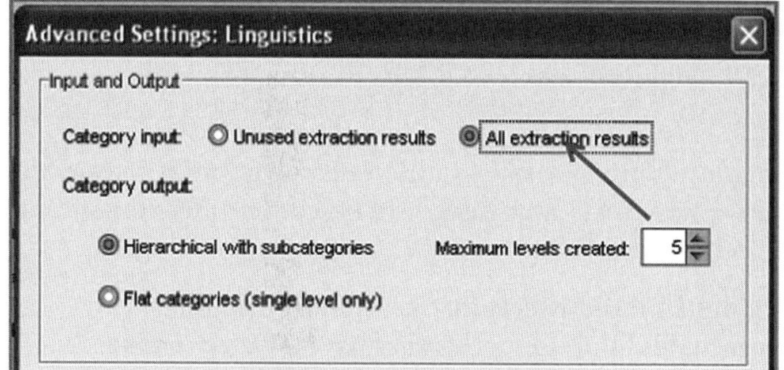

FIGURE P.20
Selecting the All Extraction Results button from the Input and Output section.

7. Leave all other settings to default and click **OK**.
8. Click **Build Categories** Build Categories on the Build Categories: Settings window. The categories pane shows all the categories that were generated based on your settings (see Figure P.21).

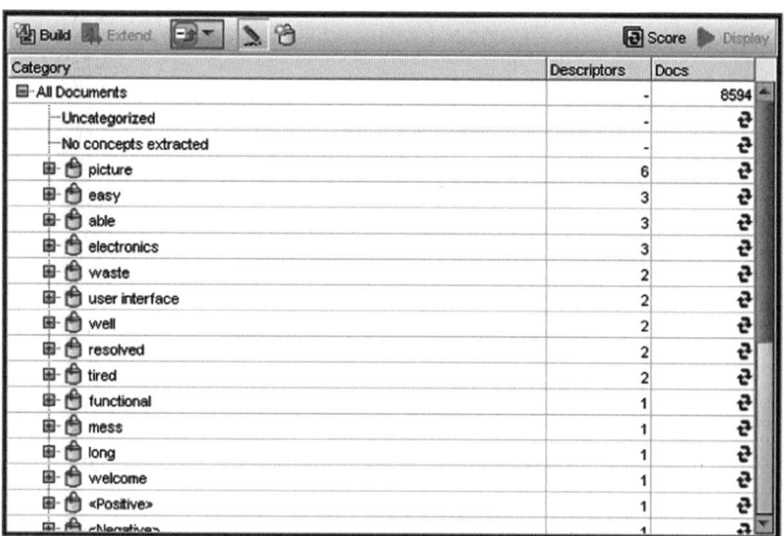

FIGURE P.21
The Categories pane.

9. To score the documents with the categories, click on the Score button on the top right corner of the Category pane (see Figure P.22).

FIGURE P.22
The Score button in the Category pane.

To delete categories, highlight the category and press the **Delete** key on your keyboard.

Creating Categories Manually

If the categorization results created automatically are not satisfactory, it is advisable to create the categories manually. In this demonstration we will create four categories based on different concepts emphasizing Positive, Negative, Usability, and Features.

New Twitter is great: This category includes all those tweets that have a positive tone.
New Twitter is not great: This category includes all those tweets that have a negative tone.
Why is New Twitter great?: This category includes all those tweets that are both positive in tone and talk about features or usability issues.
Why is New Twitter bad?: This category includes all those tweets that are both negative in tone and talks about features and usability issues.

In order to create these categories, we need to define rules that satisfy the category definition. However, the first two categories can be created easily by simply assigning Types to the category.

Creating the New Twitter Is Great Category

1. From the **Menu** bar, click **Categories,** then select **Create Empty Category** (see Figure P.23).

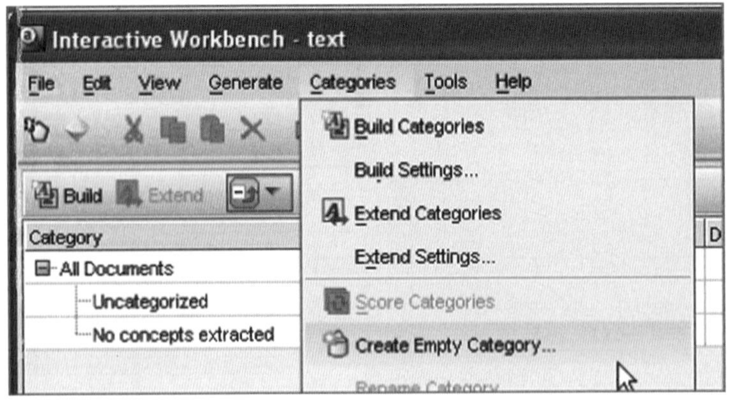

FIGURE P.23
Selecting the Create Empty Category from the Interactive Workbench dialog.

2. Enter *New Twitter is great* for the category **Name** and click **OK** (see Figure P.24).

FIGURE P.24
Entering the name of a new category.

3. From the Extract pane select Type in the drop-down (see Figure P.25).

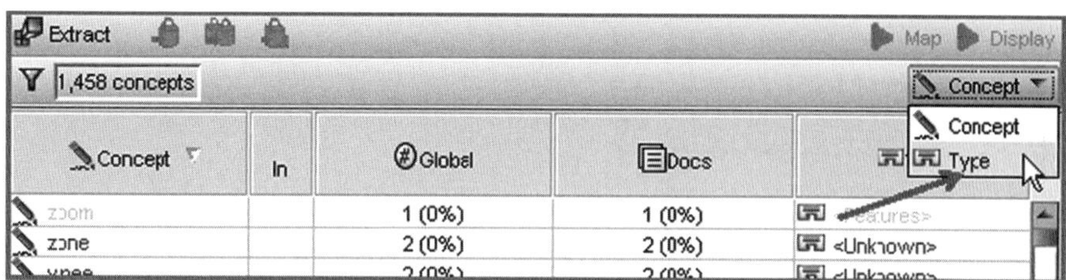

FIGURE P.25
Selecting Type in the Extract pane.

4. Click on the **Type** column header to sort the Types in A to Z order (see Figure P.26).

FIGURE P.26
Clicking on the Type column to sort the types into alphabetical order.

5. From the **Type** column scroll down to the types starting with the letter **P**.
6. Highlight all the Types related to **Positive**.
7. Right click and select **Add to Category** (see Figure P.27).

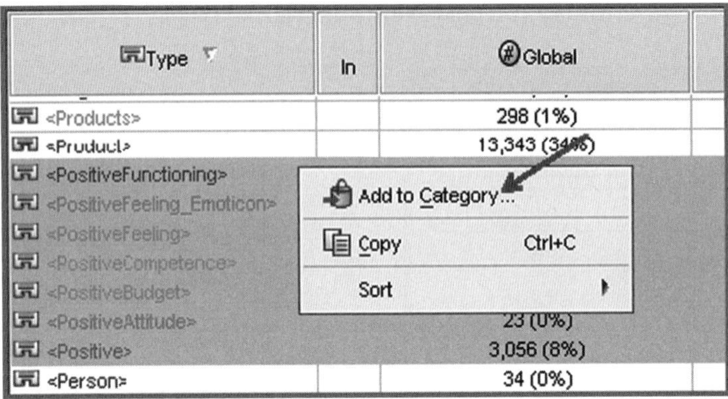

FIGURE P.27
Selecting Add to Category.

8. In the All Categories window, select **New Twitter is great** and click **OK** (see Figure P.28).

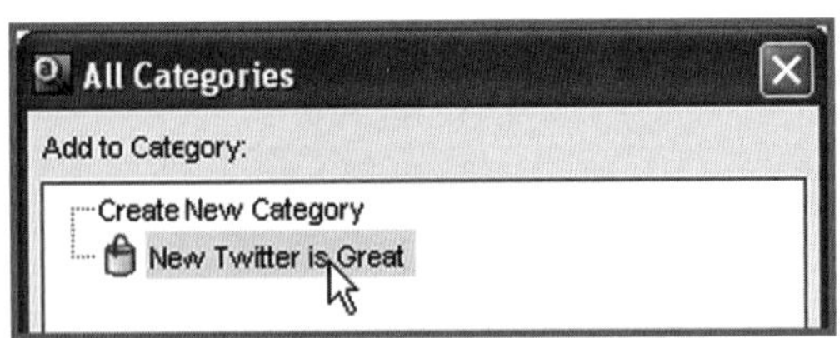

FIGURE P.28
Selecting 'New Twitter is Great' category.

You will now find all these categories added to the *New Twitter is great* category in the Category pane. In the same way, create the category *New Twitter is not great* (see Figure P.29).

Category	Descriptors	Docs
All Documents	-	8594
Uncategorized	-	
No concepts extracted	-	
New Twitter is Great	7	
<PositiveFunctioning>		
<PositiveFeeling_Emoticon>		
<PositiveFeeling>		
<PositiveCompetence>		
<PositiveBudget>		
<PositiveAttitude>		
<Positive>		
New Twitter is not Great	6	
<NegativeFunctioning>		
<NegativeFeeling_Emoticon>		
<NegativeFeeling>		
<NegativeCompetence>		
<NegativeAttitude>		
<Negative>		

FIGURE P.29
All these categories are now added to the "New Twitter is Great" category.

Creating the Why Is New Twitter Great? Category

1. Create an empty category called **Why is New Twitter great?**
2. From the Category pane toolbar, click on the **Collapse all** icon (see Figure P.30).

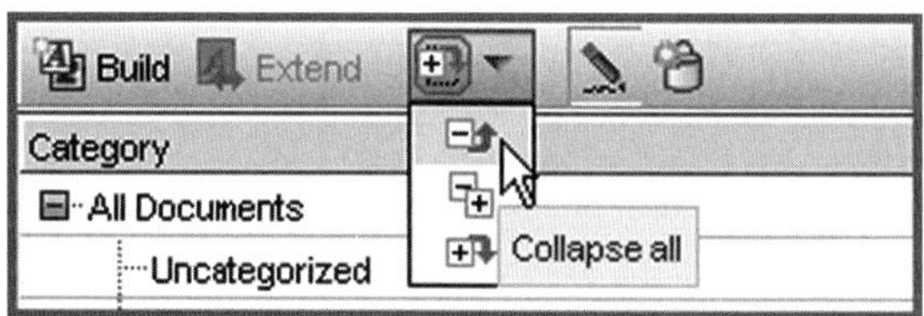

FIGURE P.30
Clicking on the "Collapse all" from the Category pane toolbar.

3. Highlight the new category, then right click and select **Create Category Rule** (see Figure P.31).

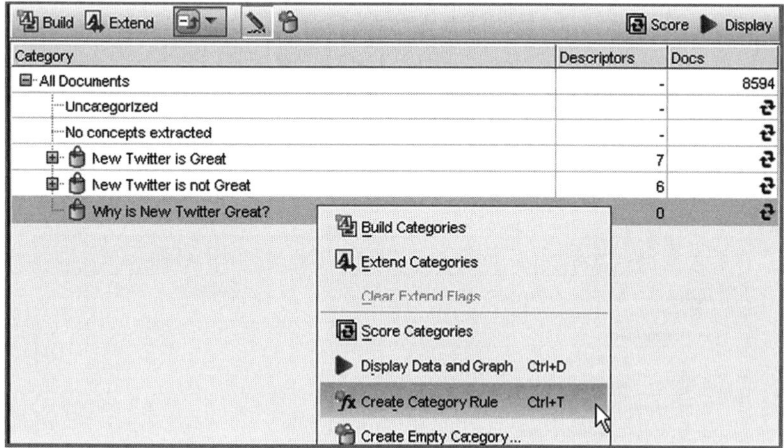

FIGURE P.31
Selecting 'Create Category Rule' after highlighting the new category.

4. In the rules editor window, give the **Rule Name** as ***Great*** and enter the following rule (see Figure P.32).
   (<Positive> | <PositiveFunctioning> | <PositiveFeeling> | <PositiveAttitude>) &
   (<Usability> | <Performance> | <Features> | <characteristics>)

This rule filters all the tweets that were classified as both positive and contain any of the words related to performance, usability, features, and so on.

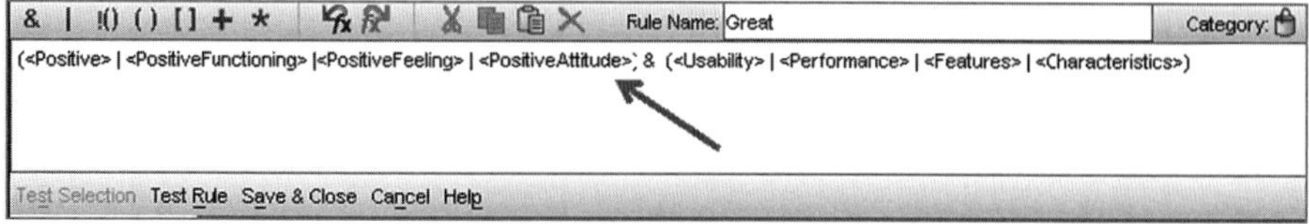

FIGURE P.32
Rules editor window.

5. Test the rule by clicking on the **Test Rule** Test Rule button (see Figure P.33). You will find that 361 comments satisfied this rule.

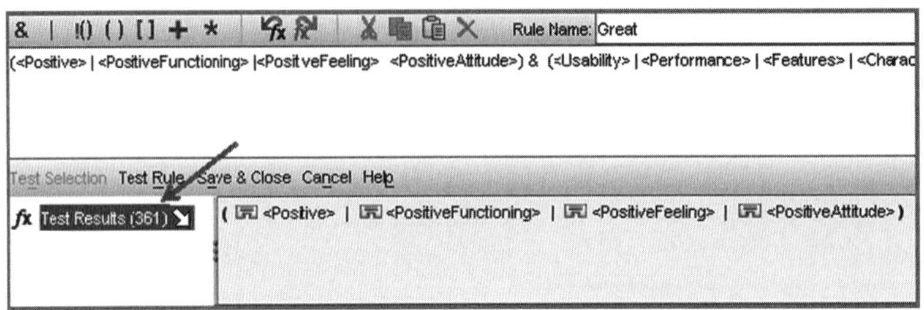

FIGURE P.33
Clicking on the 'Text Rule' button.

6. Click on the **Save & Close** button on the rule editor window pane.
Similarly create a category **Why is New Twitter bad?** with **Rule Name** as ***Not Great*** defined as:
(<Negative>| <NegativeFeeling>| <NegativeAttitude>| <NegativeFunctioning> | <NegativeCompetence>) & (<Usability> | <Performance> | <Features> | <characteristics>)

With all four categories created, the category pane should look like Figure P.34.

Build Extend Score Display

Category	Descriptors	Docs
All Documents	-	8594
Uncategorized	-	
No concepts extracted	-	
New Twitter is Great	7	
New Twitter is not Great	6	
Why is New Twitter Great?	1	
Why is New Twitter bad?	1	

FIGURE P.34
The 'catetory pane' after all four categories have been created.

7. Click on **Score** to score all the comments with these categories.

After scoring, the category panes should look as follows with the document frequency identified for each category (see Figure P.35).

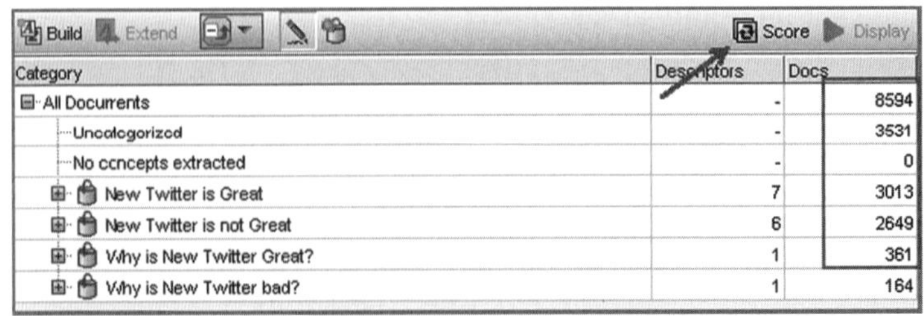

Build Extend Score Display

Category	Descriptors	Docs
All Documents	-	8594
Uncategorized	-	3531
No concepts extracted	-	0
New Twitter is Great	7	3013
New Twitter is not Great	6	2649
Why is New Twitter Great?	1	361
Why is New Twitter bad?	1	164

FIGURE P.35
After hitting the 'Score' button, the category window will show the document frequencies in the right most column.

8. To visually explore the comments in each category, highlight the category in the category pane and click on **Display**. In this example, highlight **Why is New Twitter great?** and click **Display** (see Figure P.36).

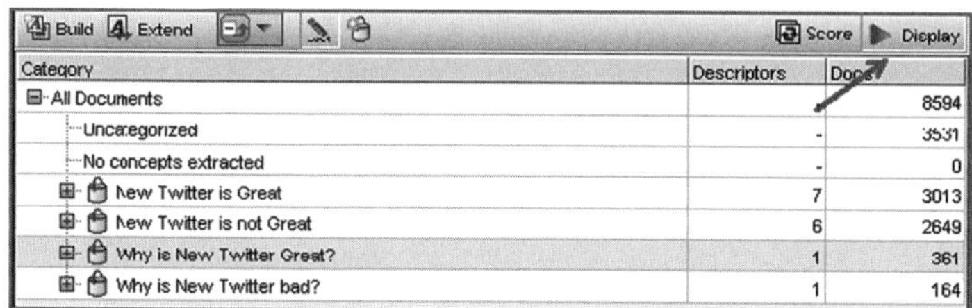

FIGURE P.36

Clicking 'Display' button after highlighting the category 'Why is New Twitter great'.

9. You will find the display results in the two right panels of the interactive workbench. The top section in Figure P.37 shows the overlap in the classification of all the 361 comments of "Why is New Twitter great?" with other categories. The bottom window shows the actual tweets with color coding. Similarly you can highlight other categories and explore those comments.

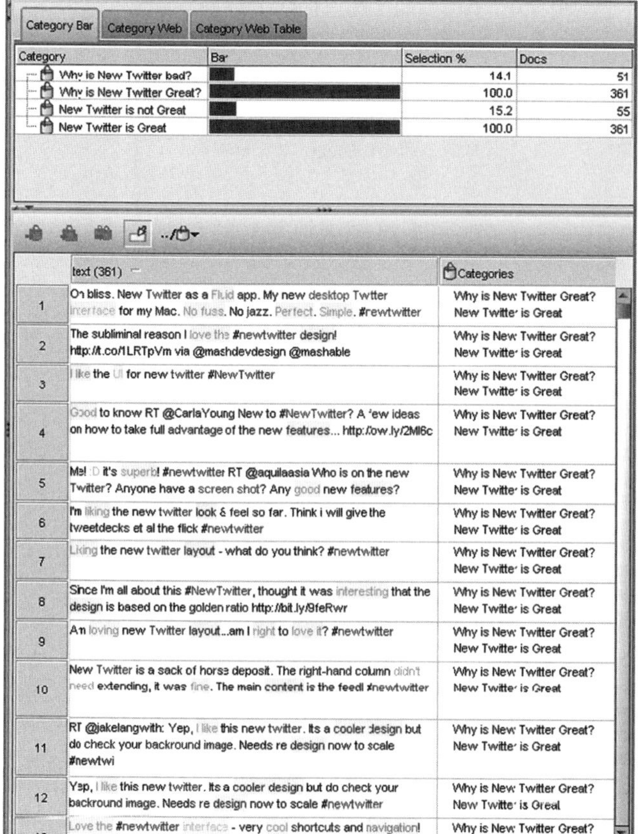

FIGURE P.37

The results for categories displayed in the "interactive workbench" window.

On the top right pane, the **Category Web** and **Category Web Table** tabs provide more details on the overlapping of the comments between the categories (see Figures P.38 and P.39).

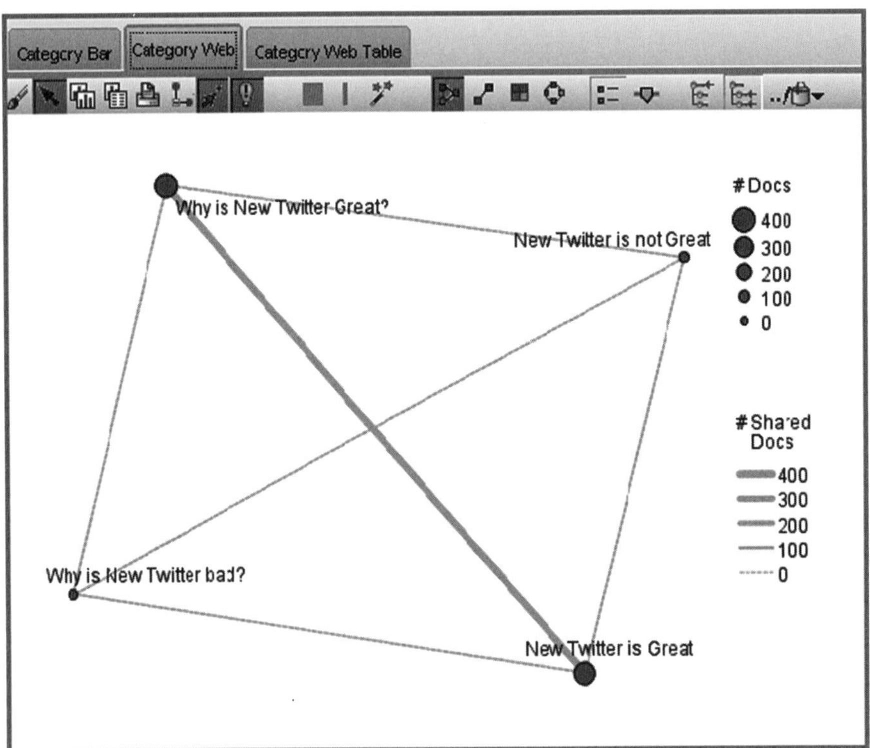

FIGURE P.38
The Category Web button clicked giving the Category Web graph.

Category Bar | Category Web | Category Web Table

Count	Category 1	Category 2
361	Why is New Twitter Great?(361)	New Twitter is Great(361)
55	Why is New Twitter Great?(361)	New Twitter is not Great(55)
55	New Twitter is Great(361)	New Twitter is not Great(55)
51	Why is New Twitter Great?(361)	Why is New Twitter bad?(51)
51	New Twitter is Great(361)	Why is New Twitter bad?(51)
51	New Twitter is not Great(55)	Why is New Twitter bad?(51)

FIGURE P.39
The Category Web Table selected.

CLUSTER ANALYSIS

We can build and explore clusters that are natural groupings of concepts, whereas categories are groupings of documents or records. Using clusters we can understand the co-occurrence of concepts.

To access the clusters window, select **Clusters** in the drop-down present at the top right corner of the Interactive Workbench window (see Figure P.40).

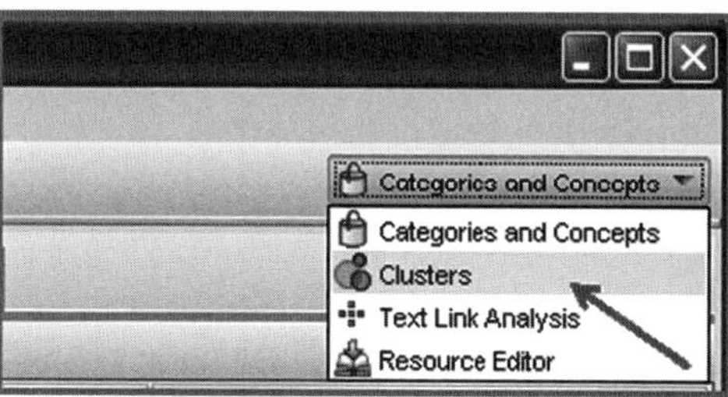

FIGURE P.40
Selecting 'Clusters' from the drop down menu found at the top right corner of the Interactive Workbench window.

To build clusters:

1. Click on the **Build** button on the toolbar, or go to **Tools → Build**.
2. In the **Input** section of the Cluster Settings window, select the **Select None** button to deselect all the types.
3. Select all the types that we used in building categories and also select **<Product>** type in order to see the co-occurrence of the concepts **<Product>** with all the **<Positive>** or **<Negative>** concepts.
4. For the **Maximum number of documents to use** field, enter "8600" to include all the documents for clusters analysis (see Figure P.41).

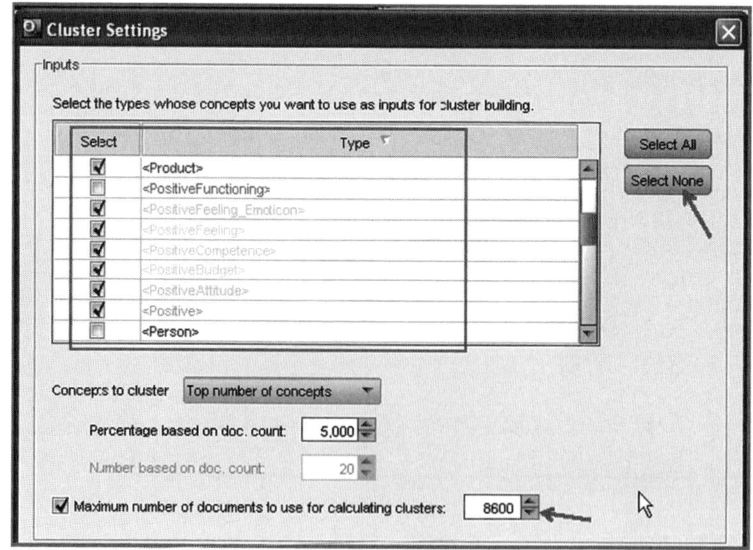

FIGURE P.41
The 'Cluster Settings' window.

5. In the **Output** section, change the **Minimum link value** to 2 (see Figure P.42).

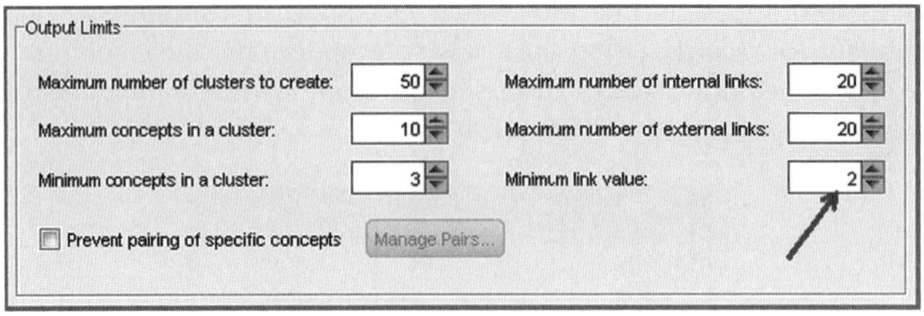

Output Limits

Maximum number of clusters to create: 50

Maximum number of internal links: 20

Maximum concepts in a cluster: 10

Maximum number of external links: 20

Minimum concepts in a cluster: 3

Minimum link value: 2

Prevent pairing of specific concepts

Manage Pairs...

FIGURE P.42

The Output section of the Interactive Window.

6. Click on the **Build Clusters** [Build Clusters] button

You can see the clusters created in the left panel of the Interactive Workbench window. Select the cluster to view the Concept web for the corresponding clusters in the right panel (see Figure P.43).

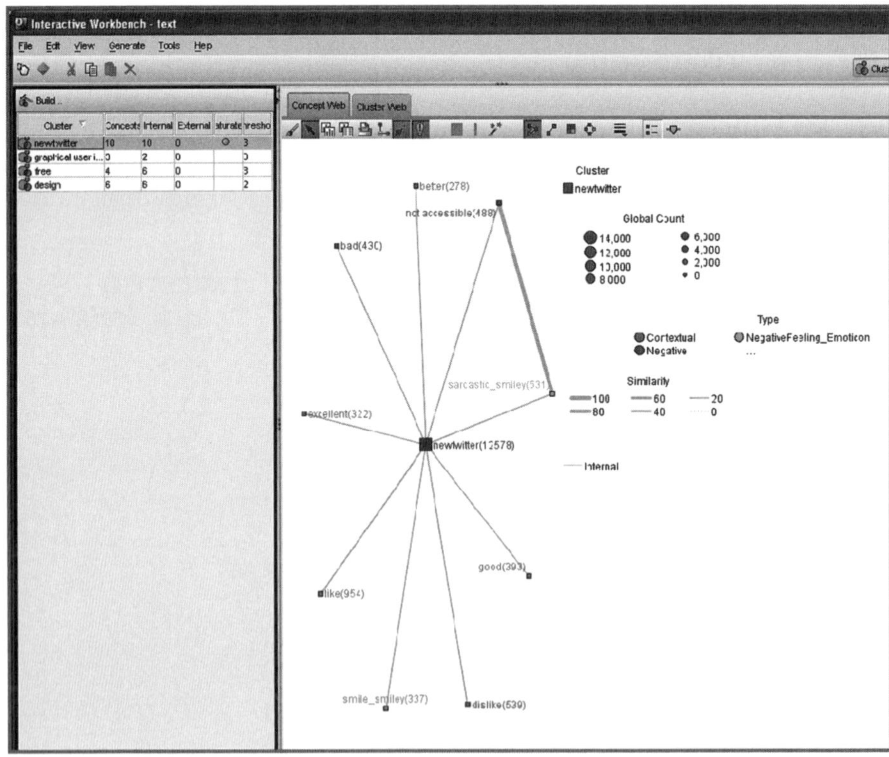

FIGURE P.43

Select the 'Cluster to view' in the Interactive Workbench window.

For example, when you select the "newtwitter" cluster in the left panel, you can see the concept web for this term in the right panel. The concept web shows the concepts that are closely related to the selected concept. As you can see, the concept "newtwitter" is associated more with the terms like "good" than with terms like "bad" or "dislike." We also see strong association between "newtwitter" and the words "not accessible." This is likely due to the venting of frustration by a large number of users who did not have access to New Twitter.

ANALYZING TEXT LINKS

Text Link Analysis (TLA) includes a pattern-matching alogrithm that enables you to define pattern rules and compare these to actual extracted concepts. To access the TLA view, select **Text Link Analysis** from the drop-down menu on the top right corner of the Interactive Workbench window (see Figure P.44).

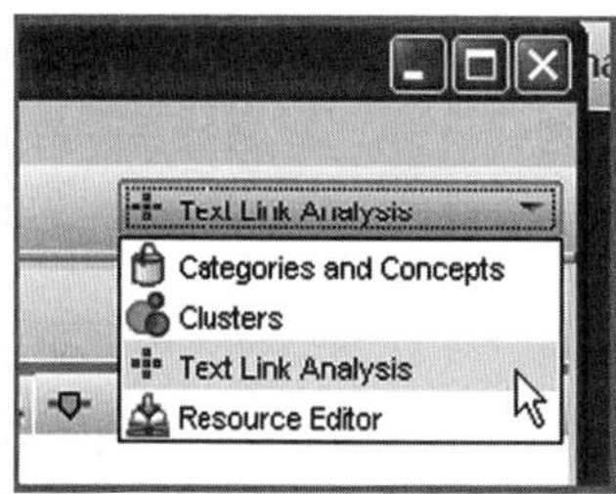

FIGURE P.44
Text Link Analysis (TLA) found in the top right corner of Interactive Workbench window.

To extract patterns, click on the **Extract** button from the toolbar. The pattern extraction results panel is organized into four panes as shown in Figure P.45.

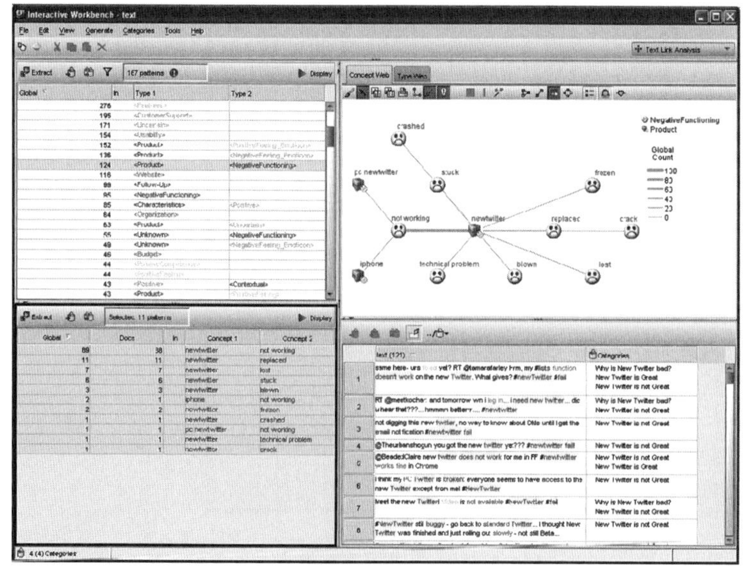

FIGURE P.45
Pattern extraction results window.

The top left pane displays all the significant combinations of concept types. The data and graphs in the top right window and bottom left window are displayed when a row is selected in the top left window—for example, select <Product> and <Negative Functioning> in the top left window (see Figures P.46 and P.47).

Extract | 167 patterns | Display

Global	In	Type 1	Type 2
276		<Features>	
195		<CustomerSupport>	
171		<Uncertain>	
154		<Usability>	
152		<Product>	<PositiveFeeling_Emoticon>
136		<Product>	<NegativeFeeling_Emoticon>
124		<Product>	<NegativeFunctioning>
116		<Website>	
99		<Follow-Up>	
95		<NegativeFunctioning>	
85		<Characteristics>	<Positive>
64		<Organization>	
63		<Product>	<Uncertain>
55		<Unknown>	<NegativeFunctioning>
49		<Unknown>	<NegativeFeeling_Emoticon>
46		<Budget>	
44		<PositiveCompetence>	
44		<PositiveFeeling>	
43		<Positive>	<Contextual>
43		<Product>	<PositiveFeeling>

FIGURE P.46
Significant combination of concept types window.

This action will display the link patterns in the top right pane and the bottom left pane.

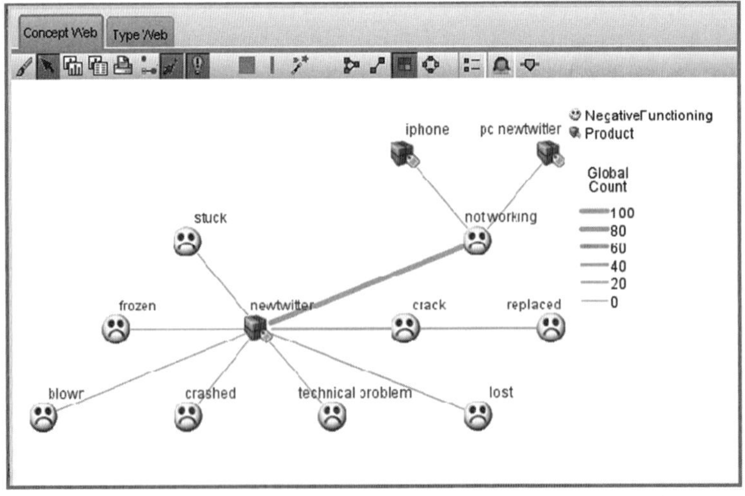

FIGURE P.47
Link patterns graph.

To view actual tweets containing this pattern, click on the **Display** button on the toolbar of the top left pane. This will populate the bottom right pane with the actual text data (see Figure P.48).

	text (121)	Categories
1	same here- urs fixed yet? RT @tamarafarley Hrm, my #lists function doesn't work on the new Twitter. What gives? #newTwitter #fail	Why is New Twitter bad? New Twitter is Great New Twitter is not Great
2	RT @meetkochar: and tomorrow wn i log in... i need new twitter... did u hear that???... hmmmm betterrr.... #newtwitter	Why is New Twitter bad? New Twitter is not Great
3	no: digging this new twitter, no way to know about DMs until I get the email notification #newtwitter fail	New Twitter is not Great
4	@Theurbanshogun you got the new twitter yet??? #newtwitter fail!	New Twitter is not Great
5	@BeadedClaire new twitter does not work for me in FF #newtwitter works fine in Chrome	New Twitter is not Great New Twitter is Great
6	I think my PC Twitter is broken: everyone seems to have access to the new Twitter except from me! #NewTwitter	New Twitter is not Great

FIGURE P.48
View of actual text of tweets contained in a pattern.

In a similar way, you can view patterns for various concept combinations. For example, when you select **<Positive>** and **<contextual>**, the link patterns that are displayed are shown as in Figure P.49.

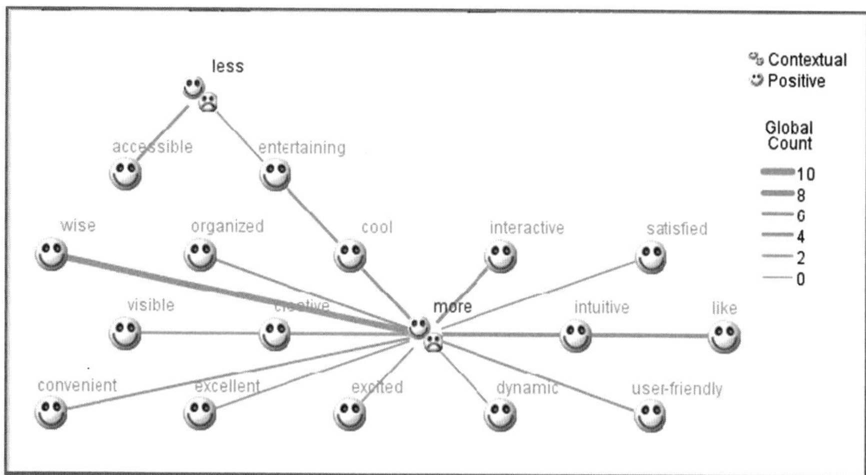

FIGURE P.49
Link patterns from selecting 'Positive' and 'contextual'.

ADDITIONAL SETTINGS

We can improve this analysis further by controlling and managing the resource libraries that are used. The default customer satisfaction library is built to support extracting and categorizing concepts correctly for a small number of categories. We can control all the language capabilities from the

Resource Editor window (see Figure P.51). To access this window, select **Resource Editor** from the Interactive Workbench drop-down (see Figure P.50).

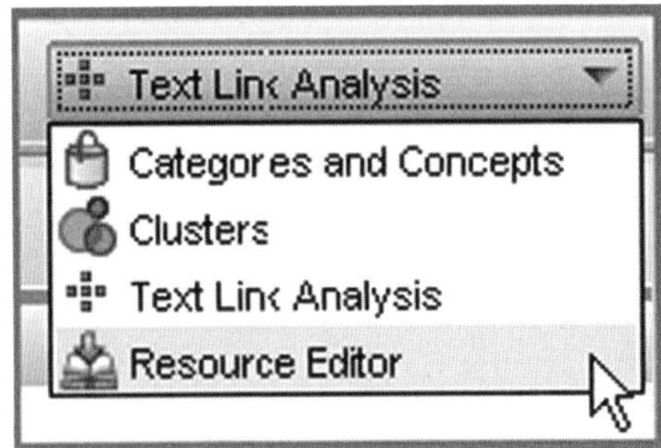

FIGURE P.50
Selecting the Resource Editor window.

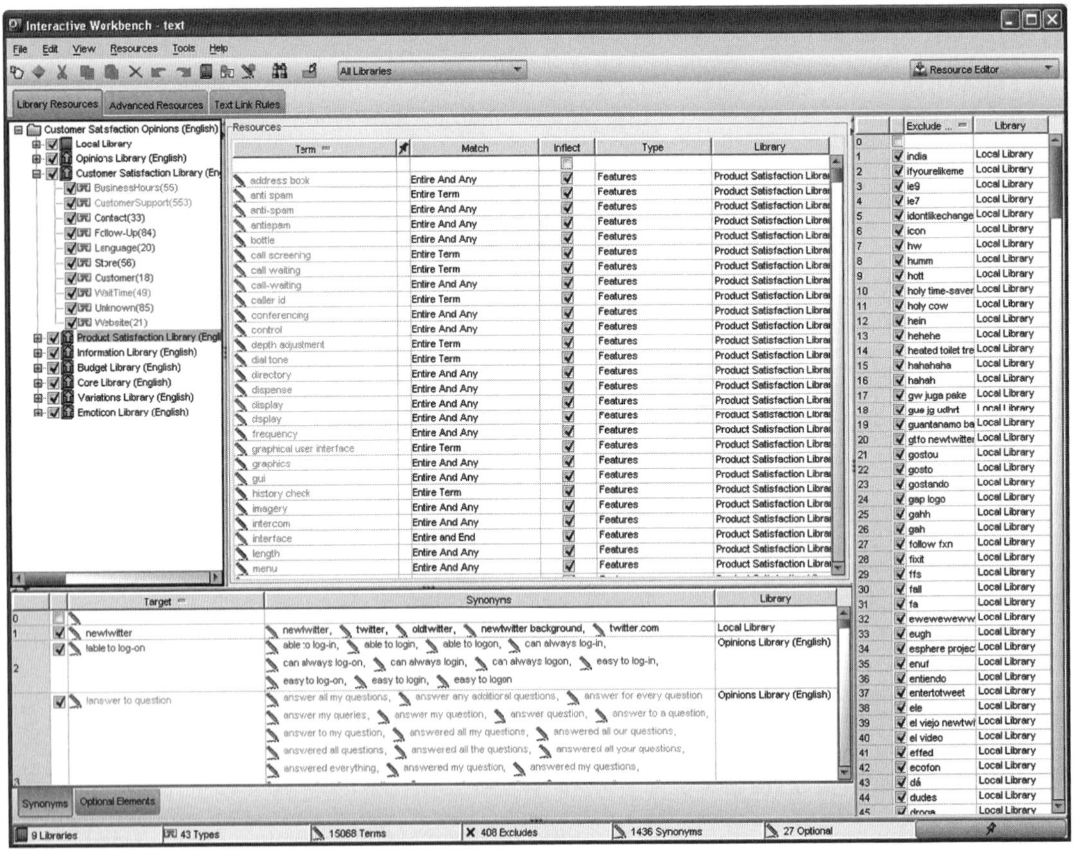

FIGURE P.51
Resource Editor window.

SUMMARY

All the steps described in this tutorial outline the high-level activities that need to be performed for text mining using the PASW Text Analytics tool. Most of the activities are done in an iterative fashion to fine-tune the model and achieve better results. The PASW Text Analytics tool has a lot more features to offer for text mining and sentiment analysis that are beyond the scope of this tutorial.

From this text mining analysis, we can fairly say that most of the users are happy with the change in the Twitter interface. The primary reasons for such happiness are availability of features such as instant video streaming, layout of the interface, user-friendly design, and creative and convenient organization. A significant number of users were unhappy with the change, and the reasons were technical issues like browser noncompatibility, confusing layout, frustrating design, bad color, bugs in the website, and so on. Users complaining about such issues are normal reactions to any new application launch. This text mining analysis will help the Twitter development team to identify how the users react to their new upgrades to the website. Tweets are valuable sources of information because they are quick and voluntary reactions of customers using their own words/terms and are therefore not biased by what questions were asked, such as in a customer survey.

A Hands-On Tutorial on Text Mining in SAS®: Analysis of Customer Comments for Clustering and Predictive Modeling

Maheshwar Nareddy
Graduate Student, Email: maheshn@ostatemail.okstate.edu

Dr. Goutam Chakraborty
Professor, Department of Marketing, Oklahoma State University, Email: Goutam.chakraborty@okstate.edu

CONTENTS

INTRODUCTION

In order to retain existing customers and attract new ones, many companies use customer loyalty programs. This practice is very common among retail businesses, where it has been popular for quite some time. Periodic evaluation of these loyalty programs through customer surveys helps in identifying the important updates that need to be done to the programs to keep them effective.

Customer surveys typically contain closed-end questions that generate structured numerical data and open-ended questions or comments that generate unstructured textual data. Survey analysis so far has been confined mostly to structured quantitative data analysis. According to Dobson (2010), many companies fail to analyze the patterns in their large sets of data. However, textual comments and responses provided in these surveys often contain a wealth of information (Boire, 2009; Dobson, 2010). When these textual comments are further explored using text mining techniques, they have the potential to increase the predictive and/or explanatory ability of models (Sullivan and Ellingsworth, 2003). Kleij and Musters (2003) demonstrate how text analysis of open-ended survey responses can complement preference mapping.

Practical Text Mining and Statistical Analysis for Non-structured Text Data Applications. DOI: 10.1016/B978-0-12-386979-1.00025-6

OBJECTIVE

In this tutorial, we illustrate how textual data can be grouped together to generate insights into customers' expectations and how such groupings can be used as input variables to build better predictive models than models based on numerical data alone in a retail context. The data used for this demonstration were collected via a survey administered at a national conference by a client company (that wishes to remain anonymous) to determine whether the company has the best loyalty program in the industry. This company provides both B2B and B2C products and services in the retail sector.

The textual responses from the customers are first clustered using the text mining node in SAS® Enterprise Miner™ 6.2, and later these clusters are used as inputs to predict the target variable. The text miner node of SAS® can be customized to run different types of analyses. Interested readers are referred to the help menu in SAS® Enterprise Miner™ 6.2 for more details. We also note that this tutorial is based on a paper that was published in SAS® Global Forum 2011 (http://support.sas.com/resources/papers/proceedings11/223-2011.pdf).

CASE STUDY

In this case study we will look at both B2B and B2C products and services, illustrating how text can generate insights into customer's perceptions and expectations and thus build better models than could be obtained with just numerical data only.

Data Description

For this demonstration, we will use two data sets: *survey_textual* and *survey_text_numeric*. We use eight numerical questions and four open-ended textual questions to predict the target variable. The target variable (best_loyalty_card_bin) captures customers' responses to the question of whether the client company has the best loyalty program. Of the 315 respondents, 55.56 percent felt that the client company has the best loyalty program in the industry.

The textual comments comprised responses to open-ended questions such as the following:

1. Why does the respondent feel that a company has the best loyalty program? (Variable: Why_Best_Lylty_Card)
2. Why does the respondent feel that a company has the worst loyalty program? (Variable: Why_Worst_Lylty_Card)
3. Why does the respondent feel that a store is his/her favorite stop to shop? (Variable: Why_Fav_Stop)
4. Why does the respondent feel that a store is his/her least favorite stop to shop? (Variable: Why_Least_Fav_Stop)

Typical surveys contain a lot of missing values, as does this survey data. Since the objective of this tutorial is on how to use the textual comments in building predictive models, we have not emphasized on how to deal with missing data.

Methodology

Text mining techniques were applied on the textual data set (*survey_textual*), and then the resulting textual data set with clusters was merged with the numeric data set. The merged data set (*survey_text_numeric*) has both numeric data and cluster memberships. This data set will be used in the second phase of this tutorial: predictive modeling.

Text Mining

The textual data set (*survey_textual*) has four variables that capture textual responses. Each of these variables was used to cluster the respondents using the text mining node in SAS® Enterprise Miner™ 6.2. These are the steps to mine text data:

1. Create the Project

Log on to SAS® Enterprise Miner™ 6.2, create a new project, and name it **TextMine**, as shown in Figure Q.1.

FIGURE Q.1
The SAS Enterprise Miner opening dialog.

Then click on Next and finally click on Finish.

2. Create the Diagram

Once the project window opens, we must create a new diagram. This diagram is used to build the text mining model. Right click on Diagrams and select Create Diagram, as shown in Figure Q.2.

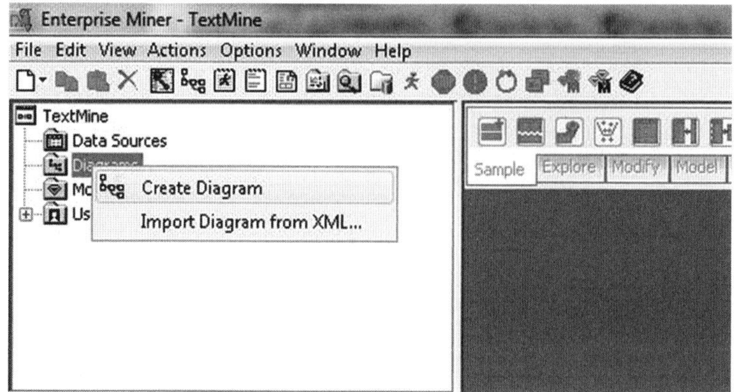

FIGURE Q.2
Creating a new diagram.

Name the diagram **TextMine_1,** as shown in Figure Q.3.

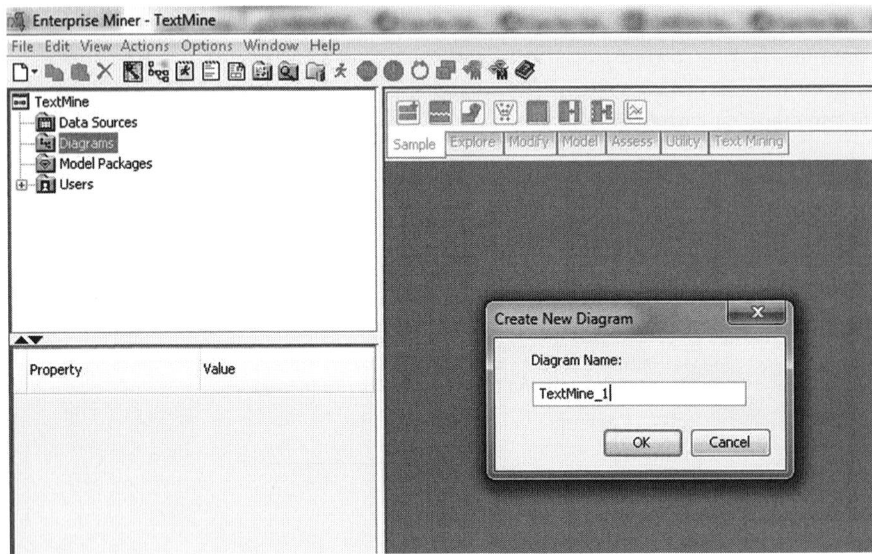

FIGURE Q.3
Naming the new diagram.

3. Create the Library

Once this diagram is created, we will create a library and then a data source. In order to create a library, click on the drop-down menu just below the file menu and select Library, as shown in Figure Q.4. Enter the library name as **srvydata** and browse the path to the location on your computer where the data sets are located.

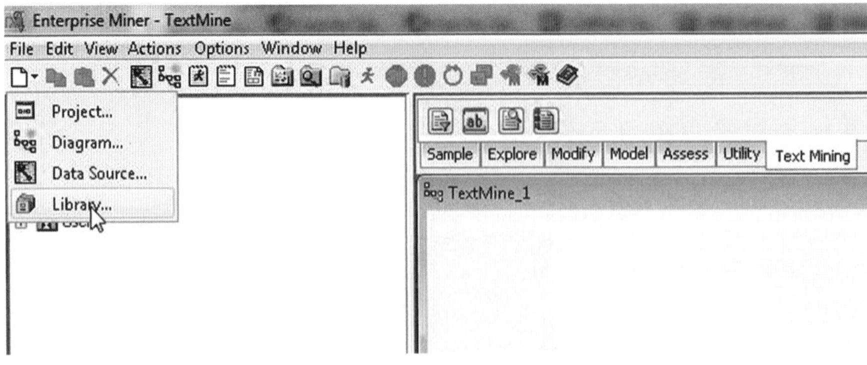

FIGURE Q.4
Creating a library name and a data source.

4. Create the Data Source

In the project panel, right click on Data Sources, and select Create Data Source, as shown in Figure Q.5. Select SAS Table as the source and click on Next. In the pop-up window, click on **Browse** and double click on the library **srvydata.** Select the **survey_textual** data set and click on **OK,** as shown in Figure Q.6.

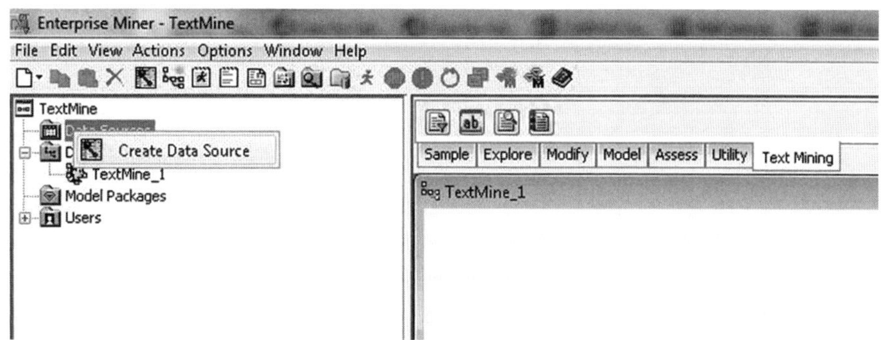

FIGURE Q.5
Selecting "Create Data Source".

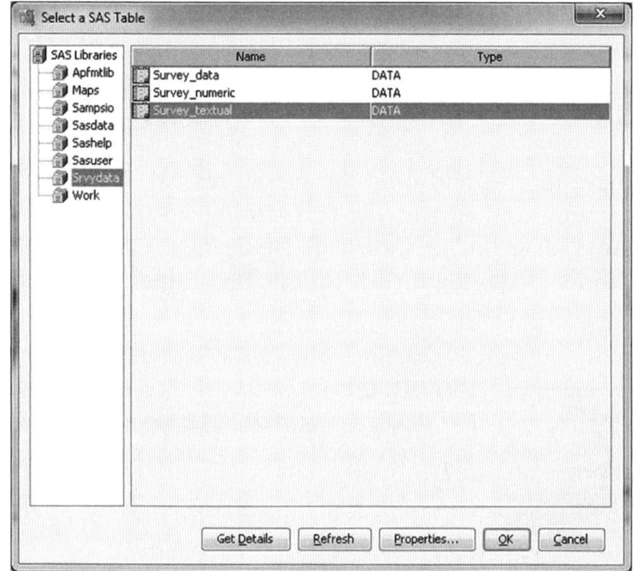

FIGURE Q.6
Selecting the "survey_textual" data set.

Keep clicking **Next** until you get the **Column Metadata** window, where we define roles and levels. Assign the roles and levels as shown in Figure Q.7. Click on **Next** until you get the **Data Source Attributes** window. Select the role as **Raw** and click on **Next**. Click on **Finish** in the last step of the data source creation wizard.

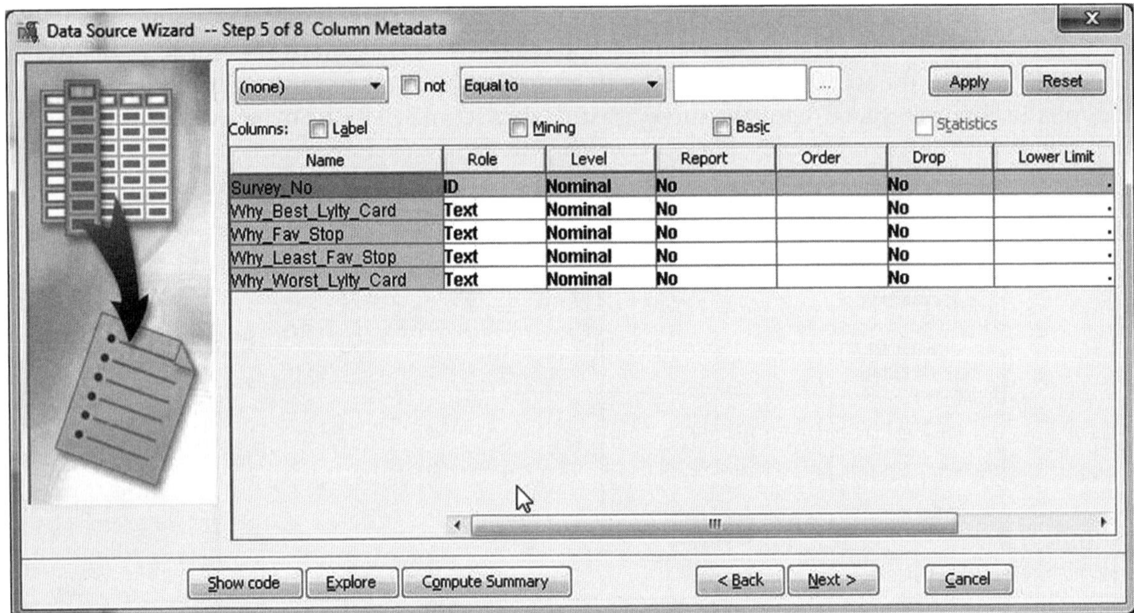

FIGURE Q.7
Roles and Levels are provided in this Data Source Wizard.

We have now successfully created the data source for our text mining purposes.

5. Text Mining

In order to build the text mining model, we have to drag and drop the **survey_textual** data source from the project panel onto the **TextMine_1** diagram. Now, click on the **Text Mining** tab and drag and drop the **Text Miner** node onto the **TextMine_1** diagram, as shown in Figure Q.8. Connect the **survey_textual** data source node to the **Text Miner** node. Right click on the **Text Miner** node and rename it **Why_Best_Loyalty_Card,** as shown in Figure Q.9.

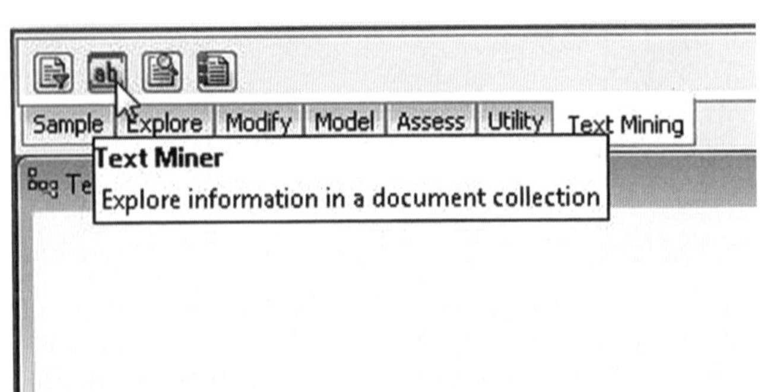

FIGURE Q.8
After clicking on the text mining tab, drag and drop the Text Miner node onto TextMine_1 diagram.

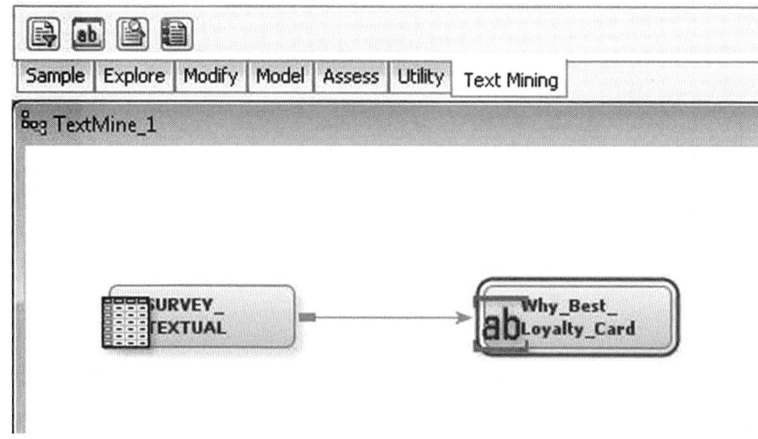

FIGURE Q.9
Renaming the Text Miner node as Why_Best_Loyalty_Card.

Now, right click on the Text Miner node and select **Edit Variables**. Select the variables and set the **Use** value to **No**, as shown in Figure Q.10. Make sure that the value of **Use** is set to **Yes** for the variable **Why_Best_Lylty_Card.** We will text mine one variable at a time. By setting the value of **Use** to **No**, we are excluding those variables in our text mining process at this stage. Now click on **OK.** We will change the values of **Use** for each of these variables in later stages of this case study.

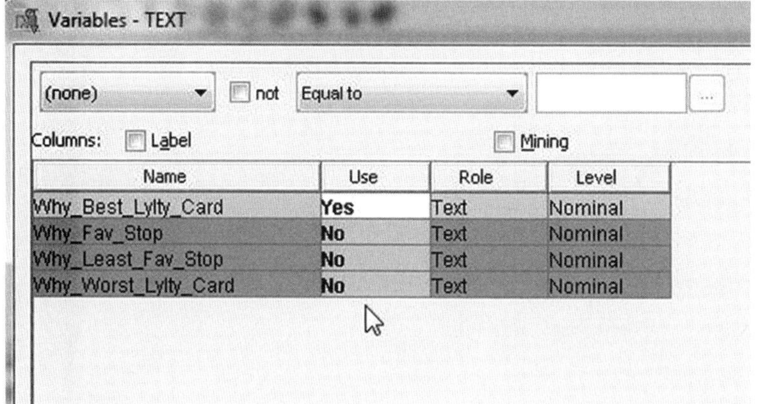

FIGURE Q.10
Setting the value of USE to NO.

6. Changing Properties

Now select the **Why_Best_Loyalty_Card** Text Miner node on the diagram. On the left-side pane, all the properties of this node are displayed, as shown in Figure Q.11. We can change the values of these properties according to our requirements. Click on the ellipsis next to **Ignore Parts of Speech** property. Now select all the parts of speech you want to ignore in this case, as shown in Figure Q.12, and click on **OK**.

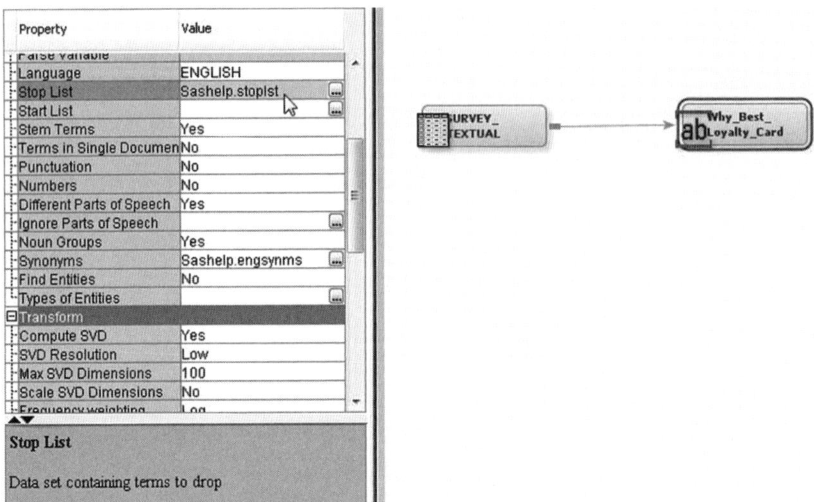

FIGURE Q.11
On the left side pane, all the properties of this node are displayed.

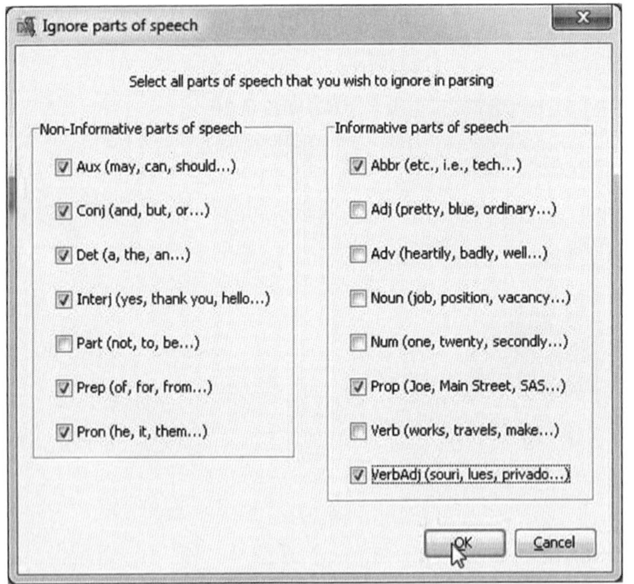

FIGURE Q.12
Selecting all the parts of speech you want to ignore in this case.

Now, go to the **Cluster** section in the properties pane and change the properties of **Automatically Cluster, Exact of Maximum Number,** and **Number of Clusters,** as shown in Figure Q.13. For demonstration purposes, we have used ten clusters. We can use numbers other than ten, depending on our requirements. Text Miner node has other properties such as **Start List, Stop List, Punctuation, Numbers,** and **Stem Terms** that can be used to fine-tune the results. For more details, see the help menu in SAS®.

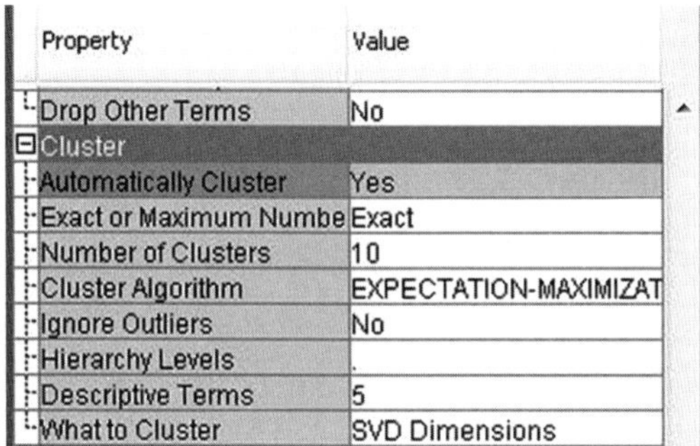

FIGURE Q.13

Here we can change the properties of Automatically Cluster, Exact of Maximum Number, and Number of Clusters.

Start List and **Stop List** are used to control the terms that we use for text mining. A stoplist contains terms we want to drop from our analysis, whereas a start list contains terms we want to keep for our analysis. Text Miner node uses these lists to cluster documents. Again, these can be customized based on our requirements. In this demonstration we are using default lists.

7. Results

Select the **Why_Best_Loyalty_Card** Text Miner node on the diagram. Right click on it and select **Run**. In the confirmation pop-up window, select **Yes**. Once the run is completed, click on **OK.** Now go to the **Train** section of the properties panel of the Text Miner node and click on the ellipsis against the **Interactive** property, as shown in Figure Q.14.

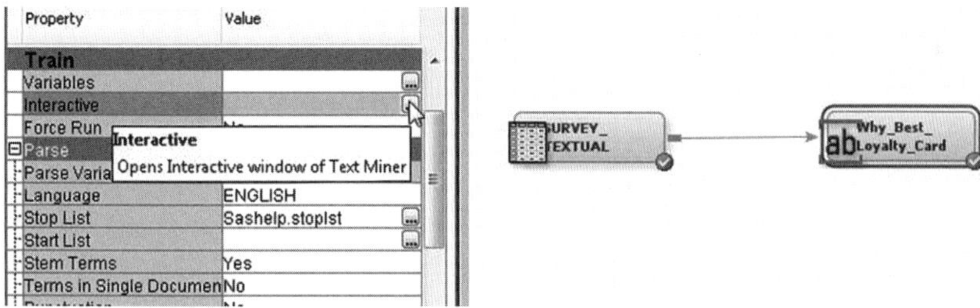

FIGURE Q.14

The Train section of the properties panel.

A window pops up with windows related to **Clusters, Terms,** and **Documents**. The Documents window displays the cluster membership of each document (observation). The other important windows in which we are interested are **Terms** and **Clusters**. The **Terms** table provides a list of terms used in text mining. For each of the terms, the frequency of occurrence and also the number of documents in which the term appears are displayed in the **Terms** table, as shown in Figure Q.15. We can either keep or drop terms at our discretion based on domain expertise. The **Cluster** window displays the descriptive terms of each of the clusters, as shown in Figure Q.16. We notice that there are 99 observations in cluster 1 but no descriptive term. All the observations with missing comments fall into this cluster (while we retained the cluster with missing values for this demonstration, in practice, these missing observations are often removed before running text mining clusters). We also notice that the descriptive terms such as double points, platinum program, general merchandise, and so on describe the other clusters. These terms are very specific to the industry in our case.

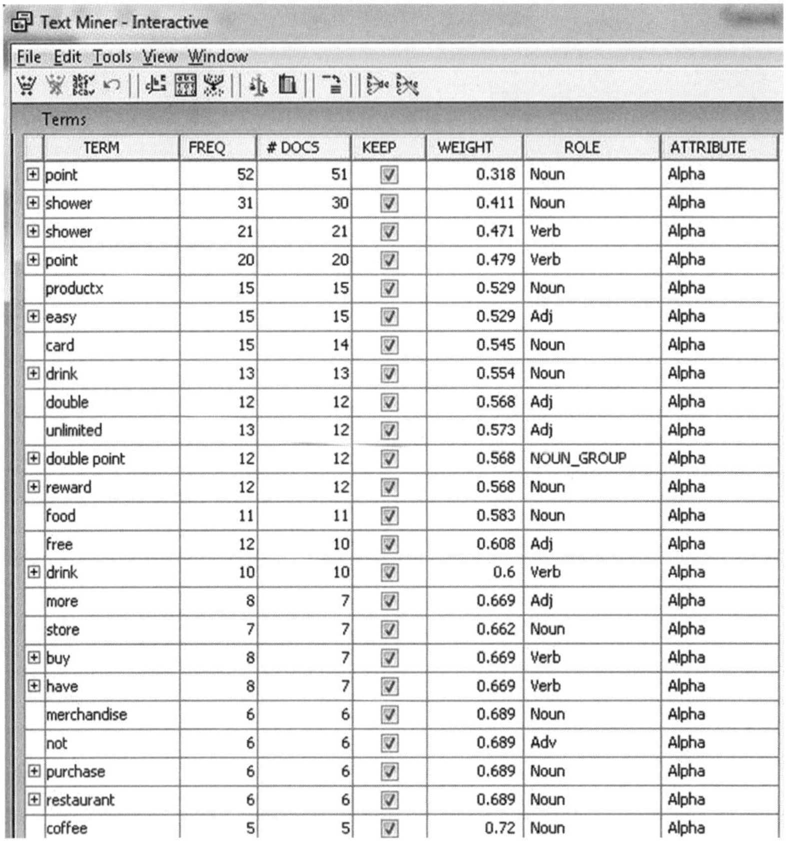

Text Miner - Interactive

File Edit Tools View Window

Terms

	TERM	FREQ	# DOCS	KEEP	WEIGHT	ROLE	ATTRIBUTE
⊞	point	52	51	☑	0.318	Noun	Alpha
⊞	shower	31	30	☑	0.411	Noun	Alpha
⊞	shower	21	21	☑	0.471	Verb	Alpha
⊞	point	20	20	☑	0.479	Verb	Alpha
	productx	15	15	☑	0.529	Noun	Alpha
⊞	easy	15	15	☑	0.529	Adj	Alpha
	card	15	14	☑	0.545	Noun	Alpha
⊞	drink	13	13	☑	0.554	Noun	Alpha
	double	12	12	☑	0.568	Adj	Alpha
	unlimited	13	12	☑	0.573	Adj	Alpha
⊞	double point	12	12	☑	0.568	NOUN_GROUP	Alpha
⊞	reward	12	12	☑	0.568	Noun	Alpha
	food	11	11	☑	0.583	Noun	Alpha
	free	12	10	☑	0.608	Adj	Alpha
⊞	drink	10	10	☑	0.6	Verb	Alpha
	more	8	7	☑	0.669	Adj	Alpha
	store	7	7	☑	0.662	Noun	Alpha
⊞	buy	8	7	☑	0.669	Verb	Alpha
⊞	have	8	7	☑	0.669	Verb	Alpha
	merchandise	6	6	☑	0.689	Noun	Alpha
	not	6	6	☑	0.689	Adv	Alpha
⊞	purchase	6	6	☑	0.689	Noun	Alpha
⊞	restaurant	6	6	☑	0.689	Noun	Alpha
	coffee	5	5	☑	0.72	Noun	Alpha

FIGURE Q.15

The Terms table.

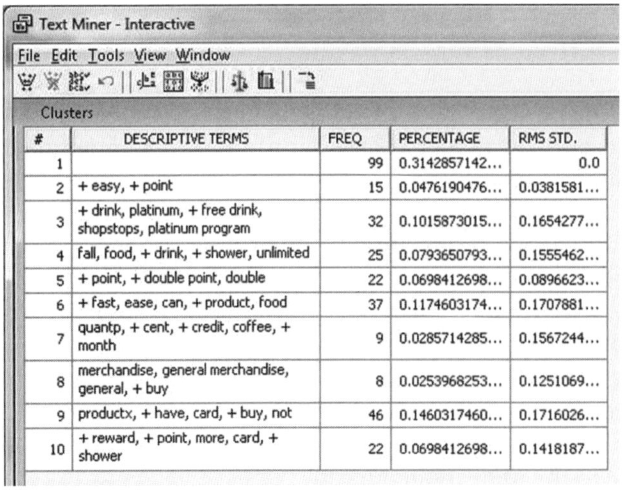

Text Miner - Interactive

File Edit Tools View Window

Clusters

#	DESCRIPTIVE TERMS	FREQ	PERCENTAGE	RMS STD.
1		99	0.3142857142...	0.0
2	+ easy, + point	15	0.0476190476...	0.0381581...
3	+ drink, platinum, + free drink, shopstops, platinum program	32	0.1015873015...	0.1654277...
4	fall, food, + drink, + shower, unlimited	25	0.0793650793...	0.1555462...
5	+ point, + double point, double	22	0.0698412698...	0.0896623...
6	+ fast, ease, can, + product, food	37	0.1174603174...	0.1707881...
7	quantp, + cent, + credit, coffee, + month	9	0.0285714285...	0.1567244...
8	merchandise, general merchandise, general, + buy	8	0.0253968253...	0.1251069...
9	productx, + have, card, + buy, not	46	0.1460317460...	0.1716026...
10	+ reward, + point, more, card, + shower	22	0.0698412698...	0.1418187...

FIGURE Q.16

The Cluster window.

We can export the data set with cluster memberships to our library and later use it for predictive modeling. The exported data set has a variable named _CLUSTER_, which has the information related to the cluster membership for each of the observations. This variable is renamed **best_loyalty_clust**.

So far, we have used responses to only one open-ended question. There are four open-ended questions in this data set. We have to repeat the text mining process mentioned in step 5 through step 7 above for the other three variables. Drag and drop the Text Miner nodes onto the diagram and connect the survey_textual data set to each of the nodes, as shown in Figure Q.17. The variable _CLUSTER_ in the output data sets (that denotes cluster membership for each observation) is renamed **fav_stop_clust, least_fav_stop_clust, worst_loyalty_clust** to represent the cluster memberships for the three remaining textual variables.

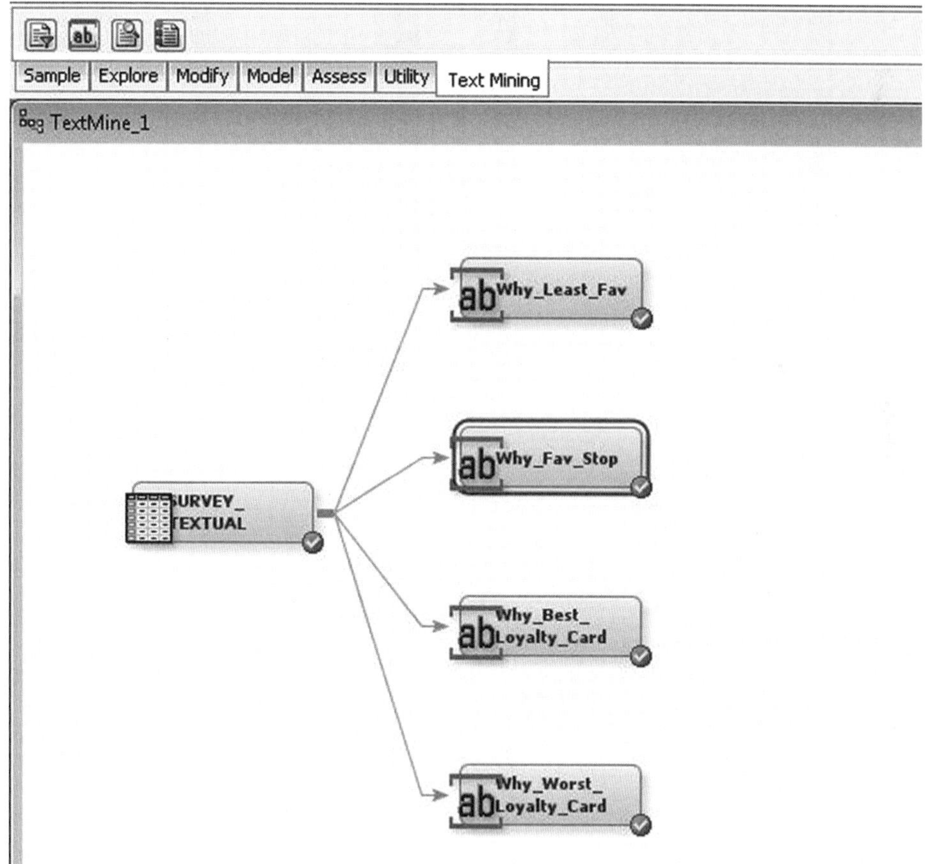

FIGURE Q.17
Dragging and dropping the Text Miner nodes onto the diagram and connect to the survey_textual data.

For the sake of convenience, we have provided the data set (survey_text_numeric) with the cluster memberships for all four textual variables as well as the numeric variables.

Predictive Modeling

We will use the survey_text_numeric data set for building the predictive models. Create a new data source survey_text_numeric by following the steps mentioned earlier with the roles and levels as shown in Figure Q.18. Then create a new diagram and rename it **PredictiveModeling,** as shown in Figure Q.19.

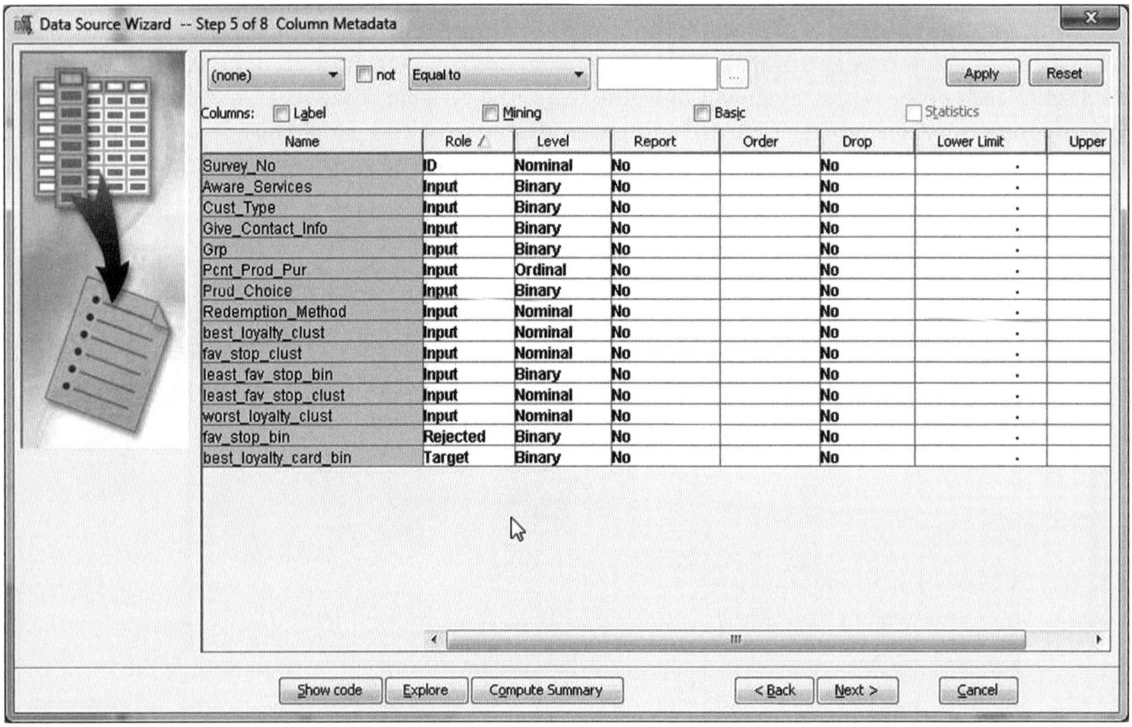

Data Source Wizard -- Step 5 of 8 Column Metadata
(none)
not
Equal to
Apply
Reset
Columns:
Label
Mining
Basic
Statistics
Name
Role
Level
Report
Order
Drop
Lower Limit
Upper
Survey_No ID Nominal No No
Aware_Services Input Binary No No
Cust_Type Input Binary No No
Give_Contact_Info Input Binary No No
Grp Input Binary No No
Pcnt_Prod_Pur Input Ordinal No No
Prod_Choice Input Binary No No
Redemption_Method Input Nominal No No
best_loyalty_clust Input Nominal No No
fav_stop_clust Input Nominal No No
least_fav_stop_bin Input Binary No No
least_fav_stop_clust Input Nominal No No
worst_loyalty_clust Input Nominal No No
fav_stop_bin Rejected Binary No No
best_loyalty_card_bin Target Binary No No
Show code
Explore
Compute Summary
< Back
Next >
Cancel

FIGURE Q.18

Creating a new data source survey_text_numeric with the roles and levels.

Enterprise Miner - TextMine
File Edit View Actions Options Window Help
TextMine
Data Sources
SURVEY_TEXT_NUMERIC
SURVEY_TEXTUAL
Diagrams
PredictiveModeling
TextMine_1
Model Packages
Users

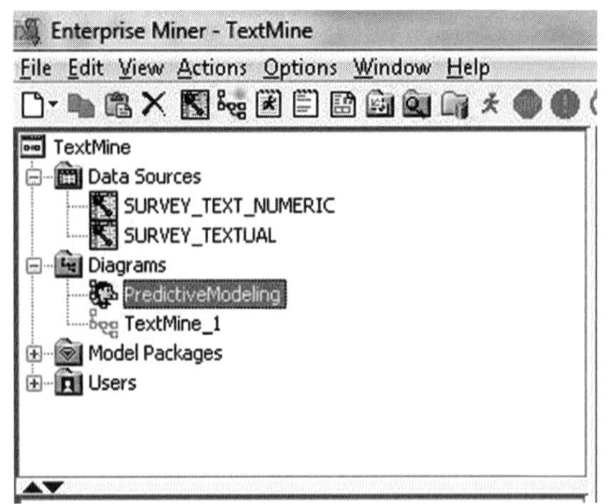

FIGURE Q.19

Creating a new diagram and renaming it PredictiveModeling.

Drag and drop the survey_text_numeric data source on to the diagram **PredictiveModeling**. Before applying the models, a tree imputation method can be used to impute the missing values of all *numerical variables*. In order to do so, connect an Impute node to the survey_text_numeric data source. We can find the **Impute** node from the **Modify** tab, as shown in Figure Q.20.

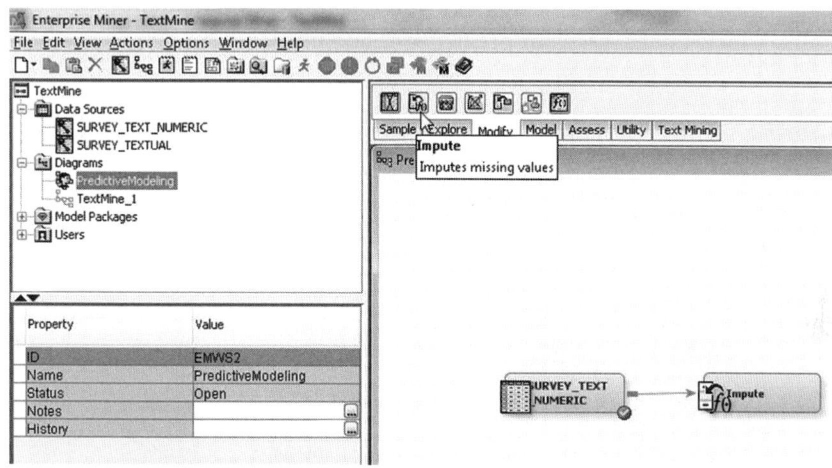

FIGURE Q.20
Finding the Impute node from the Modify tab.

Go to the properties panel of the **Impute** node and change the values for **Default Input Method** for both class and interval variables to **None** and **Type of Indicator Variables** to **Unique,** as shown in Figure Q.21. Setting the indicator variables to **Unique** creates new unique variable for every variable that is imputed.

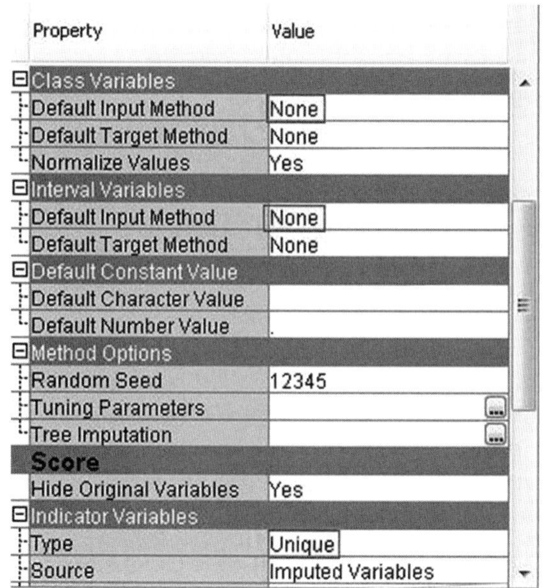

FIGURE Q.21
The properties panel of the Impute node where values can be changed.

Right click on the **Impute** node and select **Edit Variables**. Change the value of **Use** and **Use Tree** to **Yes** and **Method** to **Tree** for the variables highlighted in Figure Q.22. The missing values for these *numeric* variables are imputed using the tree imputation method. There are many other imputation methods available in SAS, and interested readers should consider SAS help and documentation. Click on **OK** and run the **Impute** node. The results in Figure Q.23 show the new imputed variables and the indicator variables, which can be used for predictive modeling.

Variables - Impt

(none) | not | Equal to

Columns: Label | Mining | Basic

Name	Use	Method	Use Tree	Role	Level
Aware_Services	Yes	Tree	Yes	Input	Binary
Cust_Type	Yes	Tree	Yes	Input	Binary
Give_Contact_Info	Yes	Tree	Yes	Input	Binary
Grp	Yes	Tree	Yes	Input	Binary
Pcnt_Prod_Pur	Yes	Tree	Yes	Input	Ordinal
Prod_Choice	Yes	Tree	Yes	Input	Binary
Redemption_Method	Yes	Tree	Yes	Input	Nominal
best_loyalty_card_bin	Default	Default	Default	Target	Binary
best_loyalty_clust	Default	Default	Default	Input	Nominal
fav_stop_bin	Default	Default	Default	Rejected	Binary
fav_stop_clust	Default	Default	Default	Input	Nominal
least_fav_stop_bin	Yes	Tree	Yes	Input	Binary
least_fav_stop_clust	Default	Default	Default	Input	Nominal
worst_loyalty_clust	Default	Default	Default	Input	Nominal

FIGURE Q.22

Changing the value of Use and Use Tree of the highlighted variables.

Imputation Summary

Variable Name	Impute Method	Imputed Variable	Indicator Variable	Impute Value	Role	Measurement Level	Label	Number of Missing for TRAIN
Aware_Services	TREE	IMP_Aware_Services	M_Aware_Services	.	INPUT	BINARY		3
Cust_Type	TREE	IMP_Cust_Type	M_Cust_Type	.	INPUT	BINARY		2
Give_Contact_Info	TREE	IMP_Give_Contact_Info	M_Give_Contact_Info	.	INPUT	BINARY		15
Grp	TREE	IMP_Grp	M_Grp	.	INPUT	BINARY		4
Pcnt_Prod_Pur	TREE	IMP_Pcnt_Prod_Pur	M_Pcnt_Prod_Pur	.	INPUT	ORDINAL		3
Prod_Choice	TREE	IMP_Prod_Choice	M_Prod_Choice	.	INPUT	BINARY		2
Redemption_Method	TREE	IMP_Redemption_Method	M_Redemption_Method	.	INPUT	NOMINAL		14
least_fav_stop_bin	TREE	IMP_least_fav_stop_bin	M_least_fav_stop_bin	.	INPUT	BINARY		30

FIGURE Q.23

New imputed variables and the indicator variables, which can be used for predictive modeling.

After imputation, a stratified partitioning method is applied to the data set to split it into training (80 percent) to build predictive models and validation (20 percent) to fine-tune and test the models. This can be achieved by dragging and dropping a **Data Partition** node under the **Sample** tab onto the **PredictiveModeling** diagram. Connect the **Impute** node to this **Data Partition** node and change the **Partitioning Method** to **Stratified**, Training data set to **80%**, and Validation data set to **20%**, as shown in Figure Q.24. After changing the properties, **Run** the node and check the results for confirmation.

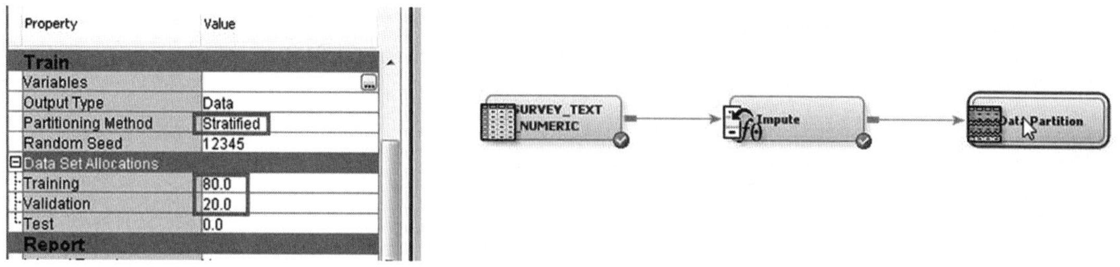

FIGURE Q.24

Change the Partitioning Method to Stratified, Training data set to 80%, and Validation data set to 20%.

In this case, two different types of models will be built. In the first model type, only numerical (quantitative) responses were considered as inputs for prediction purposes. In the second model type, the clusters representing the textual comments were added, along with the numerical responses as inputs to predict the target. For demonstration, we build logistic regression (predictive) models. Again, SAS provides many options for building predictive models such as Decision Trees and Neural Net, and interested readers should refer to the SAS documentation.

Go to the **Model** tab and then drag and drop a **Regression** node onto the **PredictiveModeling** diagram. Rename the node **Num Inputs Log Reg** and connect this node to the **Data Partition** node. Right click on the **Num Inputs Log Reg** regression node and select **Edit Variables**. Set the values of **Use** to **No** for cluster variables for this regression model, as shown in Figure Q.25 and click on **OK**. Go to the properties panel of this regression node and change the properties under the **Model Selection** section. Set the **Selection Model** to **Stepwise** and **Selection Criterion** to **Validation Error,** as shown in Figure Q.26.

Variables - Reg

(none) ☐ not Equal to

Columns: ☐ Label ☐ Mining

Name	Use	Report	Role	Level
IMP_Aware_Services	Default	No	Input	Binary
IMP_Cust_Type	Default	No	Input	Binary
IMP_Give_Contact_Info	Default	No	Input	Binary
IMP_Grp	Default	No	Input	Binary
IMP_Pcnt_Prod_Pur	Default	No	Input	Ordinal
IMP_Prod_Choice	Default	No	Input	Binary
IMP_Redemption_Method	Default	No	Input	Nominal
IMP_least_fav_stop_bin	Default	No	Input	Binary
M_Aware_Services	Default	No	Rejected	Binary
M_Cust_Type	Default	No	Rejected	Binary
M_Give_Contact_Info	Default	No	Rejected	Binary
M_Grp	Default	No	Rejected	Binary
M_Pcnt_Prod_Pur	Default	No	Rejected	Binary
M_Prod_Choice	Default	No	Rejected	Binary
M_Redemption_Method	Default	No	Rejected	Binary
M_least_fav_stop_bin	Default	No	Rejected	Binary
best_loyalty_card_bin	Yes	No	Target	Binary
best_loyalty_clust	No	No	Input	Nominal
fav_stop_bin	Default	No	Rejected	Binary
fav_stop_clust	No	No	Input	Nominal
least_fav_stop_clust	No	No	Input	Nominal
worst_loyalty_clust	No	No	Input	Nominal

FIGURE Q.25

Setting the values of Use to No for cluster variable.

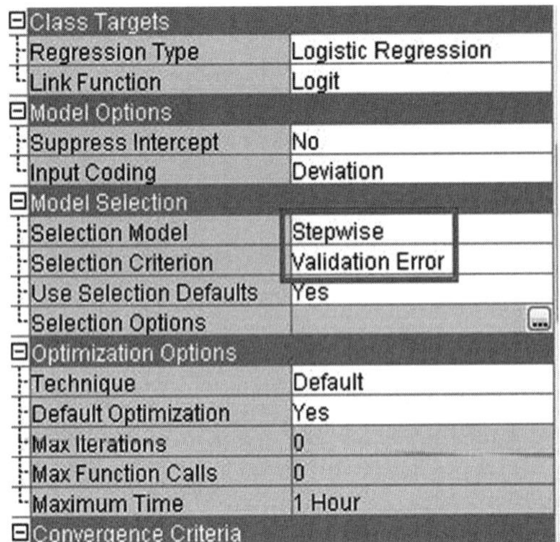

FIGURE Q.26

Setting the Selection Model to Stepwise and Selection Criterion to Validation Error.

Similarly, connect another Regression node to the Data Partition node and rename it **Num+Text Inputs Log Reg.** Now, change the properties of this node as we did for the previous model. However, the values for **Use** for the cluster variables should be set either to **Yes** or **Default,** as shown in Figure Q.27. This time we are including the cluster membership variables as inputs to our model.

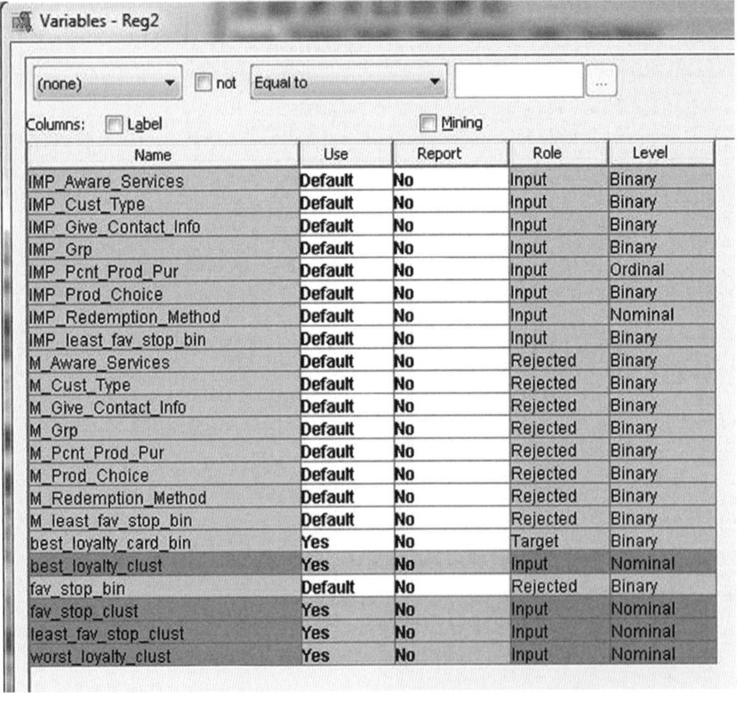

Name	Use	Report	Role	Level
IMP_Aware_Services	Default	No	Input	Binary
IMP_Cust_Type	Default	No	Input	Binary
IMP_Give_Contact_Info	Default	No	Input	Binary
IMP_Grp	Default	No	Input	Binary
IMP_Pcnt_Prod_Pur	Default	No	Input	Ordinal
IMP_Prod_Choice	Default	No	Input	Binary
IMP_Redemption_Method	Default	No	Input	Nominal
IMP_least_fav_stop_bin	Default	No	Input	Binary
M_Aware_Services	Default	No	Rejected	Binary
M_Cust_Type	Default	No	Rejected	Binary
M_Give_Contact_Info	Default	No	Rejected	Binary
M_Grp	Default	No	Rejected	Binary
M_Pcnt_Prod_Pur	Default	No	Rejected	Binary
M_Prod_Choice	Default	No	Rejected	Binary
M_Redemption_Method	Default	No	Rejected	Binary
M_least_fav_stop_bin	Default	No	Rejected	Binary
best_loyalty_card_bin	Yes	No	Target	Binary
best_loyalty_clust	Yes	No	Input	Nominal
fav_stop_bin	Default	No	Rejected	Binary
fav_stop_clust	Yes	No	Input	Nominal
least_fav_stop_clust	Yes	No	Input	Nominal
worst_loyalty_clust	Yes	No	Input	Nominal

FIGURE Q.27

The values for Use for the cluster variables should be set either to Yes or Default.

A model comparison node will be used to evaluate the performance of each model against the other model using the validation error criteria. Go to the **Assess** tab, drag and drop the **Model Comparison** node on to the diagram, and connect it to both the regression nodes, as shown in Figure Q.28. Go to the properties pane and set the **Selection Statistic** to **Average Squared Error**. The **Selection Table** is set to **Train** for demonstration purposes. This can be changed to Validation also. Because of the limited number of observations in the validation data set, we chose to compare the models using the Average Squared Error in Training data set for this demonstration. Run the Model Comparison node and check the results.

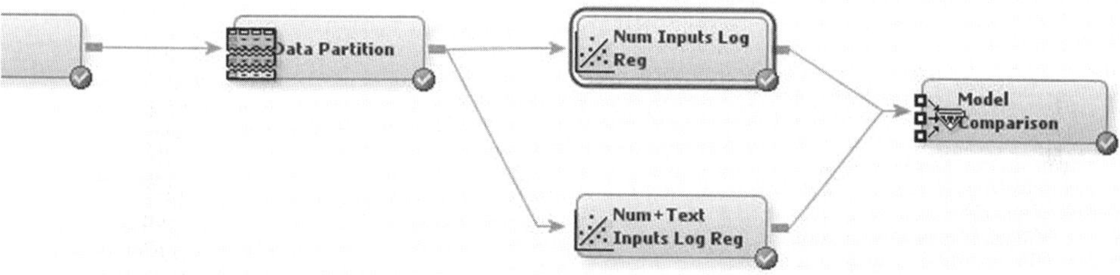

FIGURE Q.28
Using the Assess tab, drag and drop the Model Comparison node on to the diagram, and connect it to both the regression nodes.

Results

The grouping of the textual data in the text mining phase of this case revealed several important factors (as shown in Figure Q.16) for each of the four textual questions relating to best/worst loyalty programs and most/least favorite stop to shop. These factors are inferred based on our subjective interpretation of terms in each of the clusters for each question. These factors are summarized in Table Q.1.

Table Q.1 Summary of the Important Factors for each of the four textual questions.

Customers' perception of favorite stop to shop	Customers' perception of least favorite stop to shop	Customers' perception of best loyalty program	Customers' perception of worst loyalty program
Faster customer service	Slow customer service & bad service experience	Free drink refill	Mailing the redemption coupons
Large & convenient parking	Dirty restrooms & facilities	Free showers & unlimited showers	
Free coffee & drinks	Long queues/lines	Double reward points per dollar spent	
	Rude employees	Redemption points expire late	
		Instant redemption of points	

Comparison of the logistic regression model using numerical data only (referred to as NumInputs Log Reg) with the logistic regression model using numerical plus textual data (referred to as Num+TextLog Reg) clearly shows the utility of using textual data in addition to the numerical data. From the results of the model comparison node shown in Figure Q.29, we notice that the average squared error (0.1483) of the model with both numeric and cluster membership inputs is smaller compared to that of the model with just numeric inputs alone (0.1714). In addition to that, the misclassification rate in the training data sets has decreased from 25.9 percent (Num Inputs model) to 20.3 percent (Num+Text Inputs model). While the efficacy of using textual comments in conjunction with numeric data is quite clear in the training data, the results are less impressive for the validation data. We suspect this has happened because of the very small number of observations in the validation data and our decision to retain the cluster with missing textual comments.

Results - Node: Model Comparison Diagram: PredictiveModeling

File Edit View Window

Fit Statistics

Selected Model	Predecessor Node	Model Node	Model Description	Target Variable	Train: Average Squared Error	Train: Misclassification Rate	Valid: Average Squared Error	Valid: Misclassification Rate
Y	Reg2	Reg2	Num+Text Inputs Log Reg	best_loyalty_card_bin	0.148324	0.203187	0.170093	0.265625
	Reg	Reg	Num Inputs Log Reg	best_loyalty_card_bin	0.171432	0.258964	0.173948	0.265625

FIGURE Q.29

Results of the model comparison node - summary statistics.

The ROC charts in Figure Q.30 and the summary statistics reported in Figure Q.29 provide more diagnostics that show the superiority of the **Num + Txt** model over the **Num** model. The area under the ROC curve of the model with numerical and textual inputs (red curve) is more compared to that of the model with numerical inputs alone (green curve). Again, the effect is more evident in the training data than the validation data.

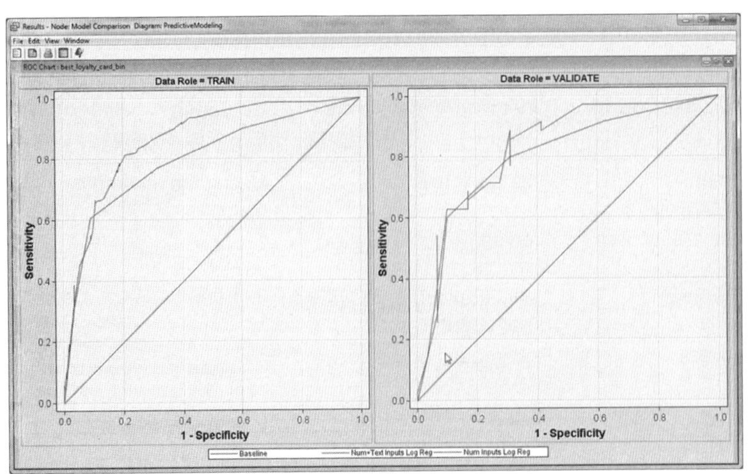

FIGURE Q.30

The ROC charts from the analyses.

SUMMARY

As shown in Table Q.1, customers' perceptions of "least favorite stop to shop" consist of comments such as rude employee behavior, long lines, bad/slow service experience, high prices, and unclean facilities. On the other hand, a customer's perception of his/her "favorite stop to shop" consists of comments such as faster customer service, free coffee/drinks, and large and convenient parking. Customers' perceptions of "best or worst loyalty program" capture comments such as late expiration of loyalty points, free/unlimited refills, double reward points per dollar spent, and instant redemption of points rather than mailing back redemption coupons. These findings provide actionable diagnostics for the client company to work on improving the image of their retail shops.

From the ROC chart (refer to Figure Q.30), we can conclude that the winning logistic regression model (with numerical and textual inputs) fares better than the logistic regression model (with numerical inputs only) in terms of predicting whether a customer feels a company's loyalty program is the best compared to its competitors.

Periodic evaluation of loyalty programs through survey analysis not only helps companies understand customers' ever-changing needs but also provides insightful information about where they stand in providing service relative to their competitors. When analyzed together, unstructured textual comments along with the quantitative data from the customer surveys can create a wealth of information to better understand customers' expectations and predict customers' satisfaction levels. This in turn will assist companies in designing and revamping their loyalty programs, thus allowing them to position themselves well ahead of the competition.

References

Boire, Richard. 2009. "Data Mining For Customer Loyalty." *Direct Marketing*. (see: http://www.boirefillergroup.com/pdf/DataMiningforCustLoyalty.pdf).

Dobson, David. 2010. "Segmenting Textual Data for Automobile Insurance Claims." *Proceedings of the SAS® Global 2010 Conference*. Available at http://support.sas.com/resources/papers/proceedings10/125-2010.pdf.

Kleij, Frederieketen and Musters, Pieter A.D. 2003. "Text Analysis of Open-Ended Survey Responses: A Complementary Method to Preference Mapping." *Food Quality and Preference*, Vol.14 (1). pp. 43–52.

Sullivan, Dan, and Ellingsworth, Marty. 2003. "Text Mining Improves Business Intelligence and Predictive Modeling in Insurance." *DM Review Magazine*. Available at http://www.information-management.com/issues/20030701/6995-1.html.

SAS® Enterprise Miner™ 6.2 help documentation at http://support.sas.com/documentation/onlinedoc/miner/index.html.

Scoring Retention and Success of Incoming College Freshmen Using Text Analytics

Linda Miner, PhD and Gary D. Miner, PhD
8110 S. Florence Place Tulsa, OK 74137

Mary Jones, PhD
Provost and Chief Academic Officer Acting Dean of the College of Business, Education and Kinesiology Southern Nazarene University

CONTENTS

INTRODUCTION

Because of the difficulty of getting a good set of university "institutional research data," this tutorial uses a hypothetical data set that is biased toward giving an accurate predictive model with just the numeric data. However, in a second analysis a text variable is added to see if we can enhance the "accuracy score" of the predictive models. The data are biased with an assumption that half the incoming traditionally aged students come from families that have been long-time local residents who think they "know" the strengths and weaknesses of this small state-run college. The other half of the traditionally aged students are Hispanic from newly emigrated families. They are thus not as familiar with the college's strengths. These students, to a large extent, come from close-knit family units that are hard-working and that want their children to acquire a college education. Many of the students, regardless of background, are attracted to the college because it is conveniently located, which allows them to commute.

We have chosen a hypothetical college in a Mexico—United States border town. This college has about 50 percent of its student population coming from the long-term residents of this community and the other 50 percent coming from emigrated families. The four-year college is strong in math and the sciences, with its graduates having almost 100 percent success in entering Ivy League medical, dental, and engineering graduate schools. But the arts are weak, including both music and the visual arts.

Figure R.1 shows the beginning data set of 1,000 cases (students). The numeric variables are age, sex, population (long-term residents and new residents), proposed major, left college early, or graduated. The one text variable was the written answer to the question "Statement by student."

Practical Text Mining and Statistical Analysis for Non-structured Text Data Applications. DOI: 10.1016/B978-0-12-386979-1.00026-8

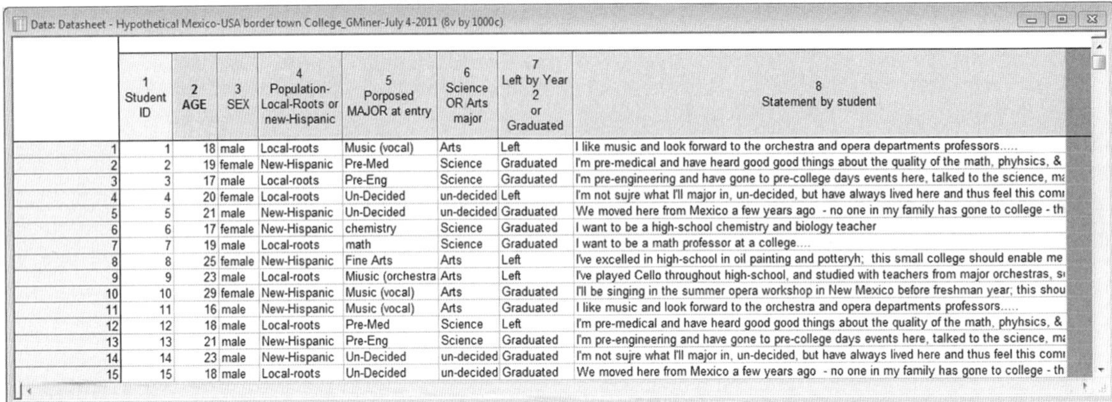
Data: Datasheet - Hypothetical Mexico-USA border town College_GMiner-July 4-2011 (8v by 1000c)

	1 Student ID	2 AGE	3 SEX	4 Population-Local-Roots or new-Hispanic	5 Porposed MAJOR at entry	6 Science OR Arts major	7 Left by Year 2 or Graduated	8 Statement by student
1	1	18	male	Local-roots	Music (vocal)	Arts	Left	I like music and look forward to the orchestra and opera departments professors.....
2	2	19	female	New-Hispanic	Pre-Med	Science	Graduated	I'm pre-medical and have heard good good things about the quality of the math, phyhsics, &
3	3	17	male	Local-roots	Pre-Eng	Science	Graduated	I'm pre-engineering and have gone to pre-college days events here, talked to the science, ma
4	4	20	female	Local-roots	Un-Decided	un-decided	Left	I'm not sujre what I'll major in, un-decided, but have always lived here and thus feel this comı
5	5	21	male	New-Hispanic	Un-Decided	un-decided	Graduated	We moved here from Mexico a few years ago - no one in my family has gone to college - th
6	6	17	female	New-Hispanic	chemistry	Science	Graduated	I want to be a high-school chemistry and biology teacher
7	7	19	male	Local-roots	math	Science	Graduated	I want to be a math professor at a college....
8	8	25	female	New-Hispanic	Fine Arts	Arts	Left	I've excelled in high-school in oil painting and potteryh; this small college should enable me
9	9	23	male	Local-roots	Miusic (orchestra	Arts	Left	I've played Cello throughout high-school, and studied with teachers from major orchestras, s
10	10	29	female	New-Hispanic	Music (vocal)	Arts	Graduated	I'll be singing in the summer opera workshop in New Mexico before freshman year; this shou
11	11	16	male	New-Hispanic	Music (vocal)	Arts	Graduated	I like music and look forward to the orchestra and opera departments professors.....
12	12	18	male	Local-roots	Pre-Med	Science	Left	I'm pre-medical and have heard good good things about the quality of the math, phyhsics, &
13	13	21	male	New-Hispanic	Pre-Eng	Science	Graduated	I'm pre-engineering and have gone to pre-college days events here, talked to the science, ma
14	14	23	male	New-Hispanic	Un-Decided	un-decided	Graduated	I'm not sujre what I'll major in, un-decided, but have always lived here and thus feel this comı
15	15	18	male	Local-roots	Un-Decided	un-decided	Graduated	We moved here from Mexico a few years ago - no one in my family has gone to college - th

FIGURE R.1

The data spreadsheet used for this tutorial. This is a "made-up" data set, purposely fashioned to provide very high accuracy with the numeric data. Thus, we will see if the small quantity of text data can improve the accuracy of a predictive model.

PART I. PREDICTIVE MODELING USING ONLY THE NUMERIC VARIABLES

The first thing we will do is use the *STATISTICA* Data Miner Recipe (DMRecipe) for a quick exploration of the numeric data to see what variables best predict *retention* of the student.

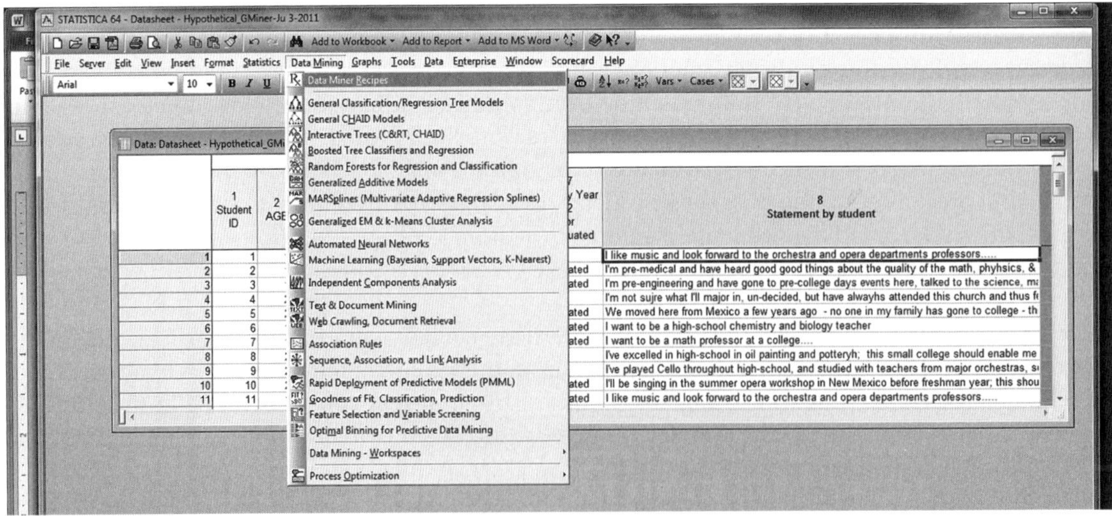

FIGURE R.2

Selecting the *STATISTICA* Data Miner Recipe (DMRecipe) from the Data Mining pull-down menu.

Click on the Data Miner Recipes selection (see Figure R.2). This opens the Data Miner Recipes window, where we can either open a previously made data miner recipe project or we can select the "New" button (see Figure R.3) in order to create a new "data miner recipe" project (see Figure R.4) for this new data set.

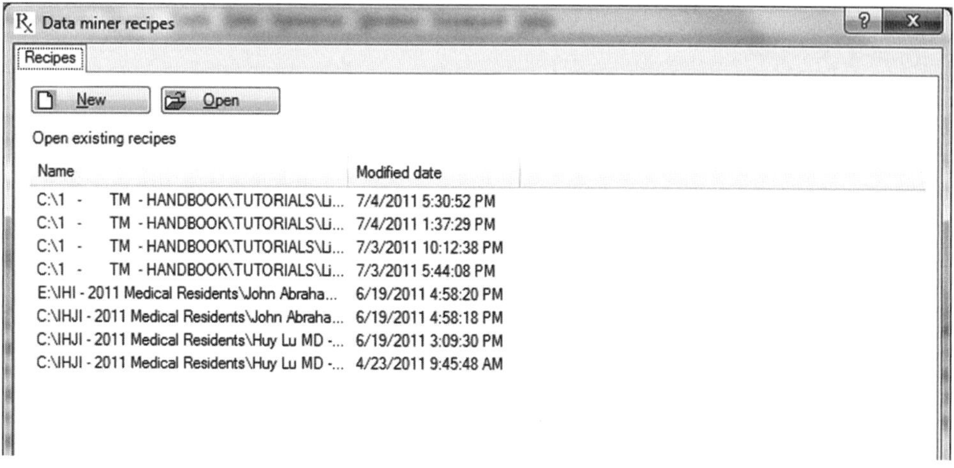

FIGURE R.3
The DMRecipe dialog window. To start a new DMRecipe project, select the "New" button.

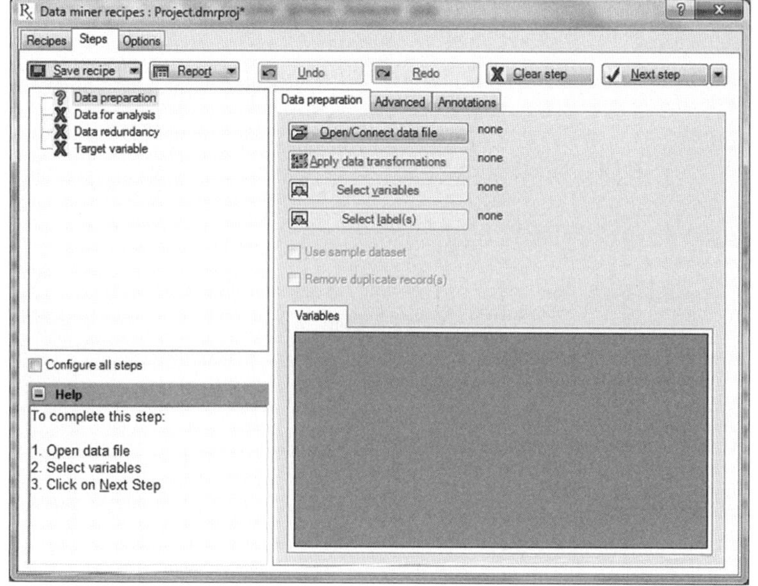

FIGURE R.4
The DMRecipe new dialog window, where one can make a few "clicks" to enter the new data set, select variables, and change any of the parameters from their default, if desired.

Select the "Open/Connect data file" button on the DMRecipe new project dialog window to bring up the "select Data Source" window (see Figure R.5).

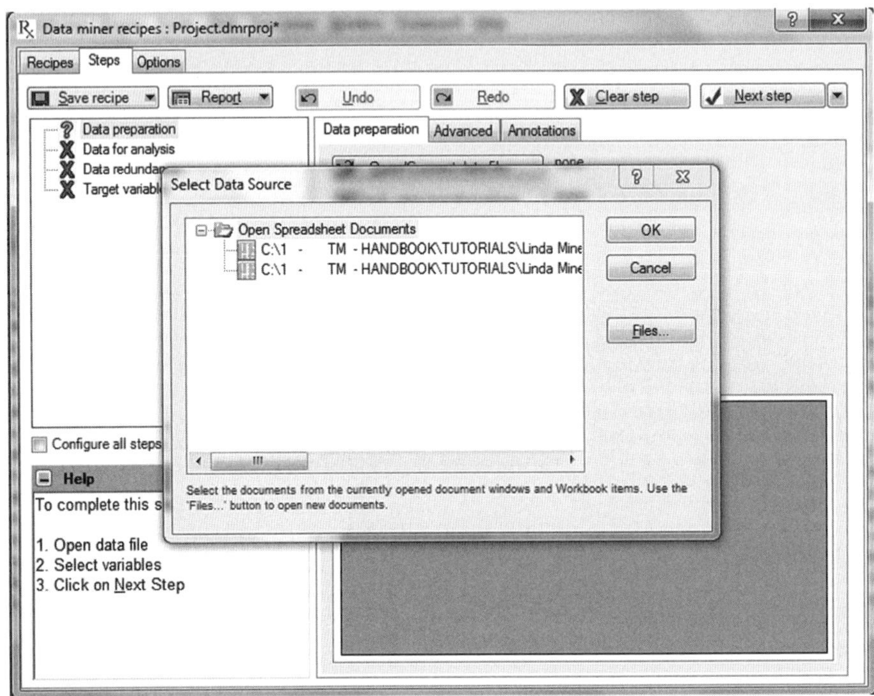

FIGURE R.5
DMRecipe Select Data Source window.

Select the correct data file from the two listed in the Select Data Source window (see Figure R.6).

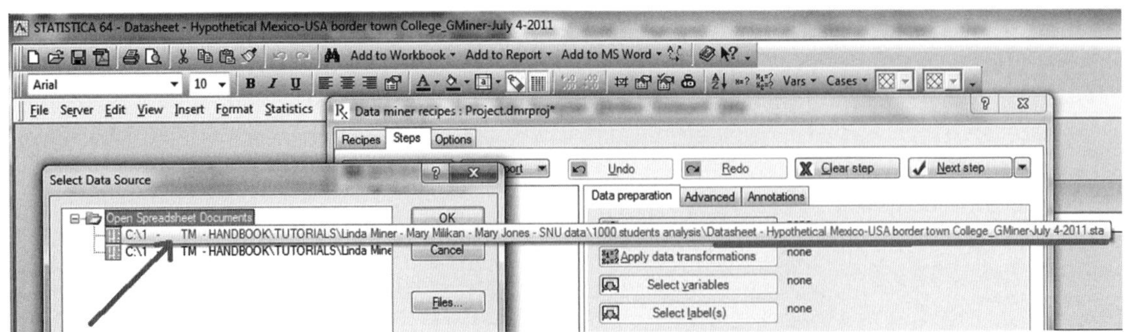

FIGURE R.6
When there is more than one data set (or can also be workbooks or other windows open on the screen), select the data set that is needed; in this example, two data sets are being used—one with only the numeric data plus the original text variable and one with the "text mined" word frequencies and concepts added to the database. At this point we only need the original data set prior to text mining additions.

Figure R.7 illustrates that the Dataset of interest has been selected, and has been entered into the DMRecipe dialog window.

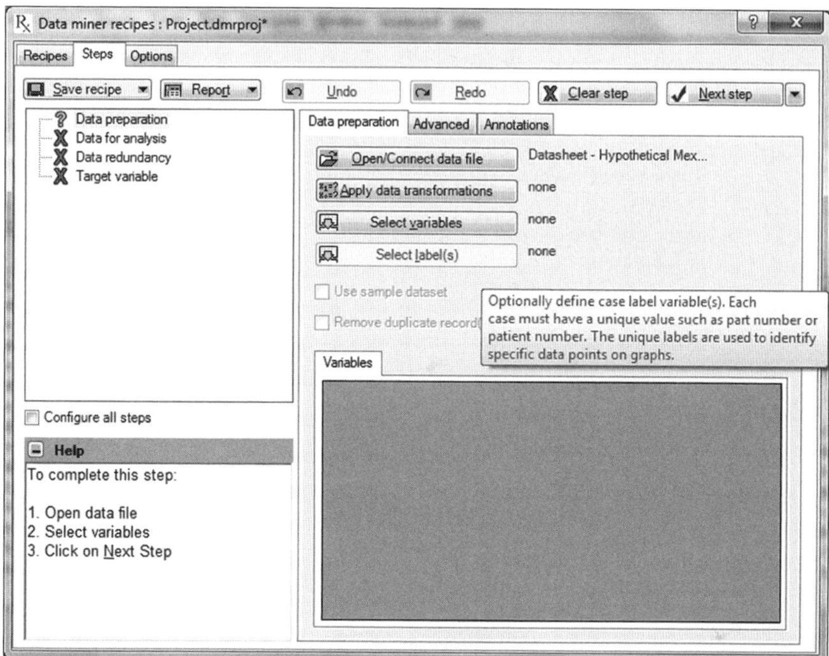

FIGURE R.7

DMRecipe dialog showing that the data set has been selected.

Now we need to select the variables (see Figure R.8).

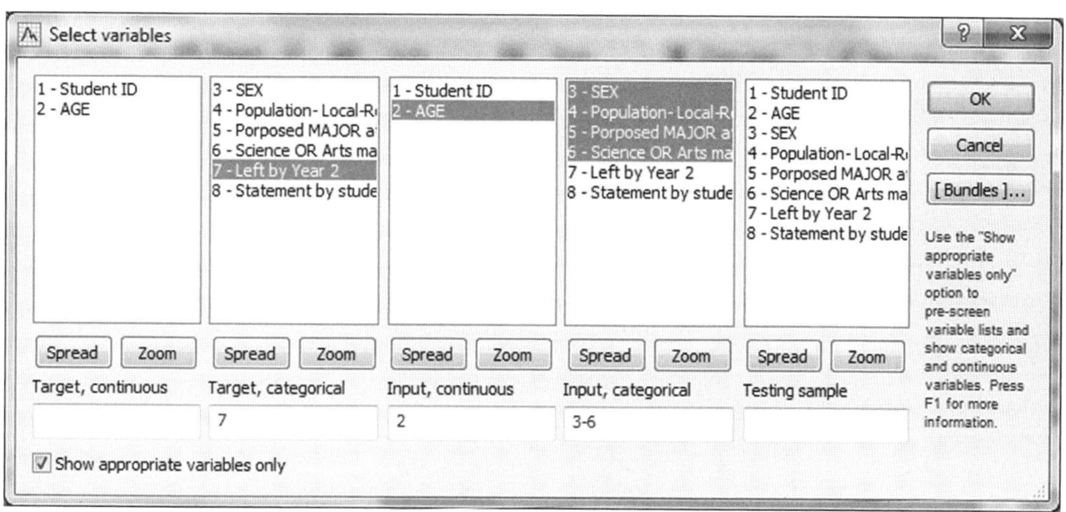

FIGURE R.8

Selecting the variables for the "numeric variables only" DMRecipe computations.

Click OK (see Figure R.8) on the data selection window to place these selections into the DMRecipe project. Now we see that the variables selected are indicated on the DMRecipe dialog (see Figure R.9).

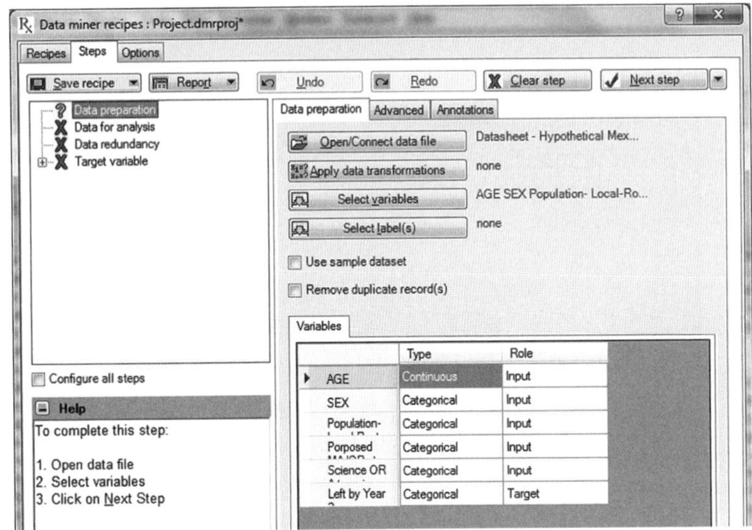

FIGURE R.9

Variables selected on the DMRecipe window.

Next, check the box on the left center of the DMRecipe dialog which says: "Configure all steps". This turns the icons for the menu tree to a different shape and also colored purple, as illustrated in Figure R.10. Then click the button labeled "Select testing sample" (Figure R.10).

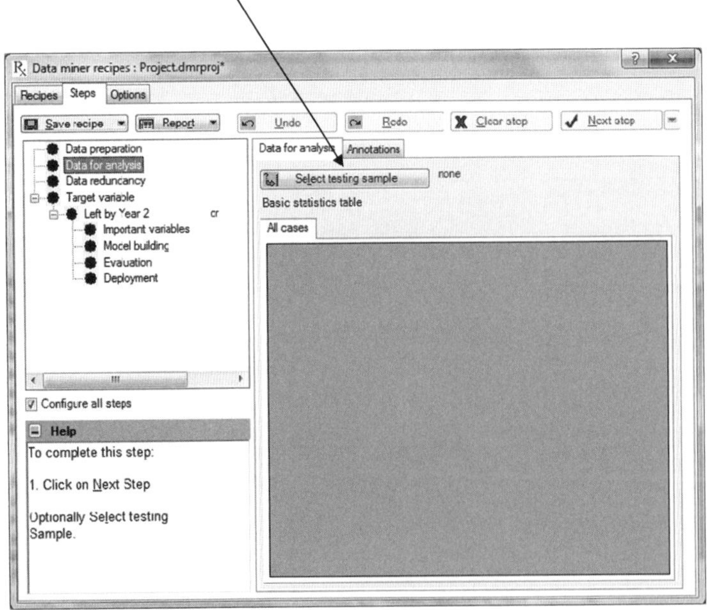

FIGURE R.10

The *STATISTICA* Data Miner Recipe dialog window with the Configure All Steps checked brings purple (will be colored purple on the computer screen) stars as the icons for each of the TOC tree contents. "Data for analysis" is selected, so that the "Select testing sample" button can be checked in order to make Train and Test subsamples from the data set.

We will set the Training sample at 66 percent of the cases in the data set. To do this, see Figure R.11.

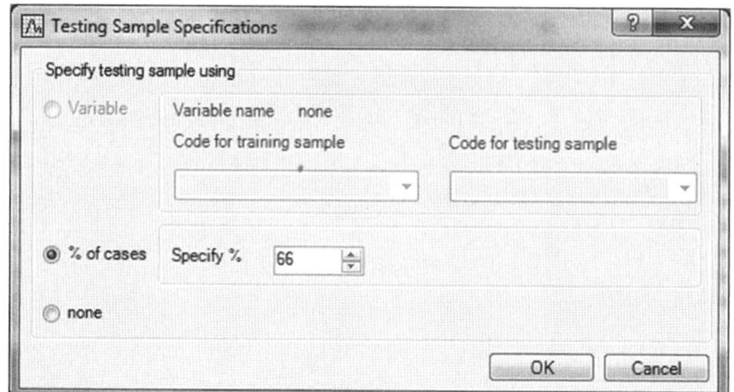

FIGURE R.11
Testing Sample Specifications window.

We need to select "Fast predictor screening" to screen out any of the predictor variables that don't have enough to contribute to our target variable, as illustrated in Figure R.12 below.

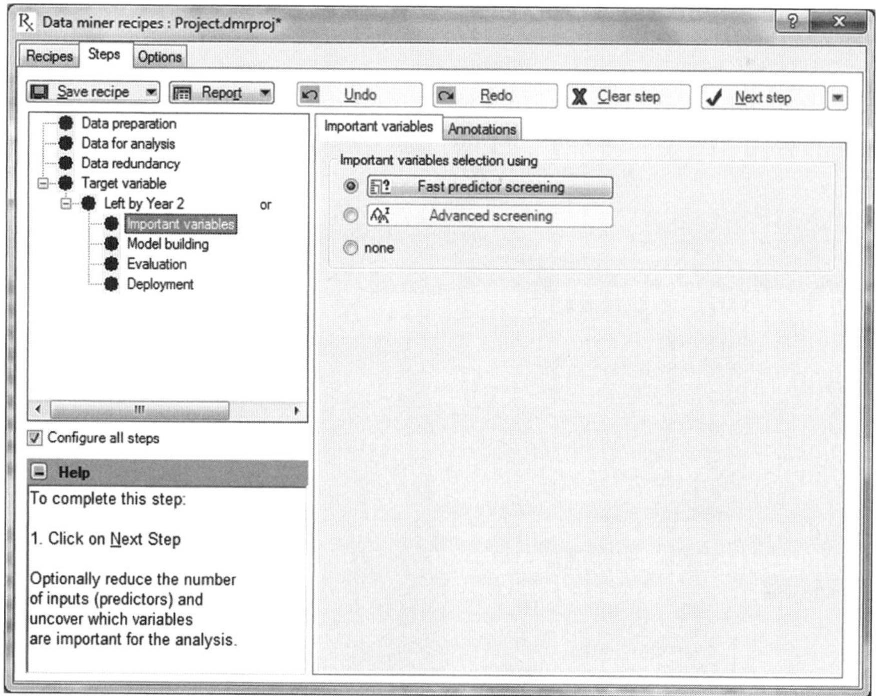

FIGURE R.12
Important Variables selection on the TOC tree in the left panel. "Fast predictor screening" is selected in the right-hand window by clicking and thus highlighting the radio button.

Next, we need to check the Model building window (see Figure R.13) to ensure that all of the data mining models are selected for the competitive comparison that the DMRecipe will compute.

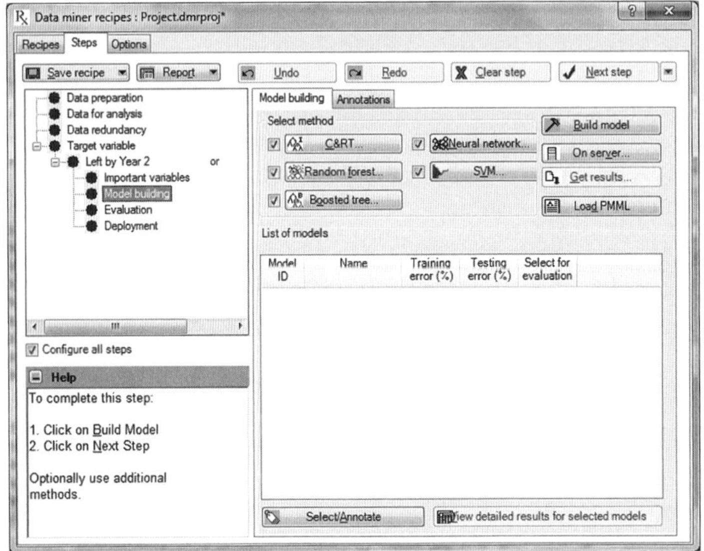

FIGURE R.13
Model building selection on the TOC tree (left panel). All of the five predictive analytical algorithm methods are selected.

Then uncheck "Configure all steps," which will return the DMRecipe window to the "red icons" state, in which we can let the analysis be computed (see Figure R.14).

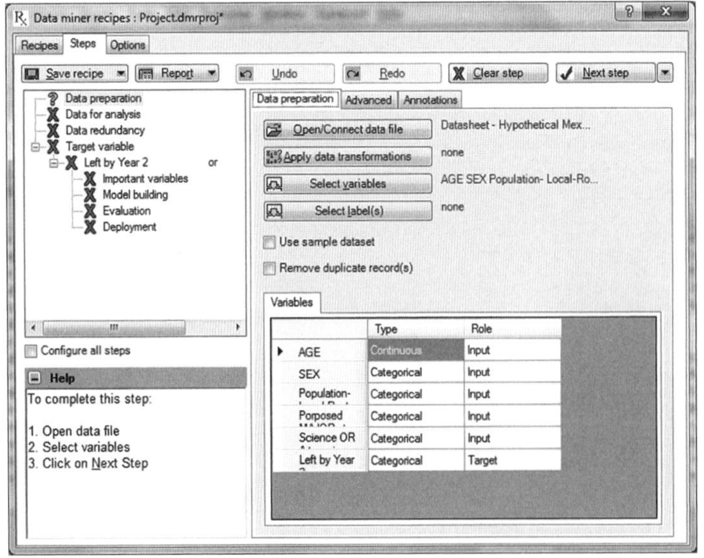

FIGURE R.14
The DMRecipe action window with all selections and adjustments to any parameters made, thus ready for the computations.

The final step to get the DMRecipe format to run the computations is to click the "down arrow" in the upper right-hand corner of the dialog window (see Figure R.15). When the "run to completion" is selected, the DMRecipe will do all the computations of all the algorithms and make competitive comparisons among all the algorithms automatically, generating many results windows for each algorithm, in addition to summary/comparison results windows and a final window that gives a summary tabulation of algorithms and the "error score" or "accuracy score" for each (see Figures R.17 and R.18).

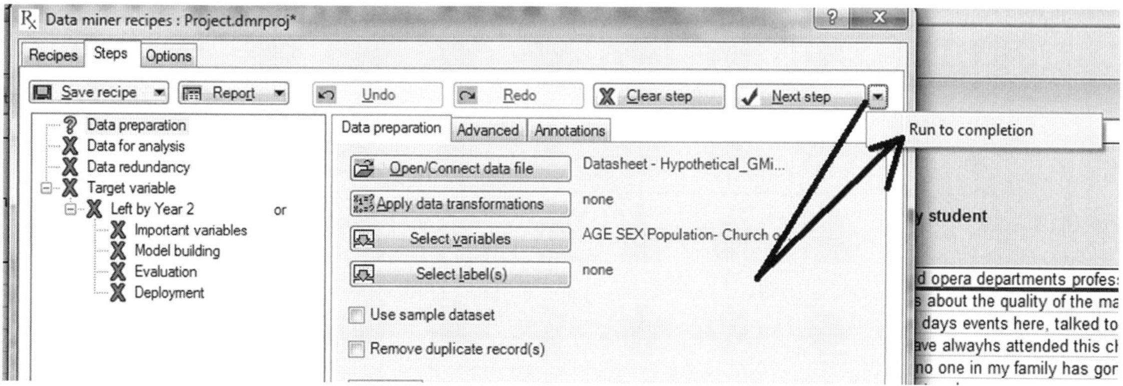

FIGURE R.15
The upper right down arrow of the DMRecipe window is selected, and then "Run to completion" can be clicked to cause all the different modeling algorithms to be computed at once, with the final result being a "competitive comparison chart" telling which algorithm models the data at the highest accuracy rate.

During the DMRecipe computation phase, various progress bars will appear to let the user know the stage of the project's computation (see Figure R.16).

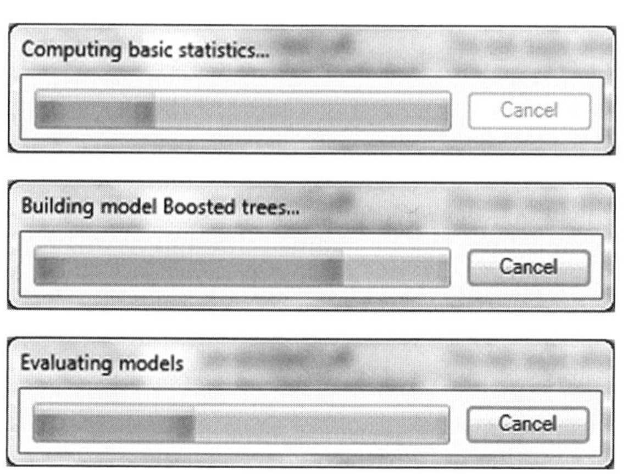

FIGURES R.16
Progress bars for the various computation components of the DMRecipe.

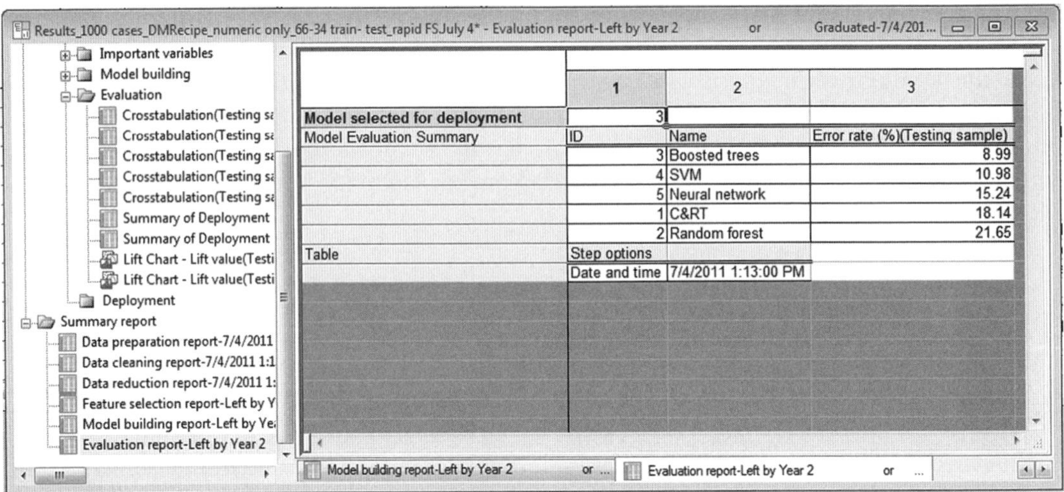

FIGURE R.17
DMRecipe results for using only the numeric predictors.

	1	2	3
Model selected for deployment	3		
Model Evaluation Summary	ID	Name	Error rate (%)(Testing sample)
	3	Boosted trees	8.99
	4	SVM	10.98
	5	Neural network	15.24
	1	C&RT	18.14
	2	Random forest	21.65
Table	Step options		
	Date and time	7/4/2011 1:13:00 PM	

FIGURE R.18
Close-up view of the summary evaluation report of the DMRecipe, numeric variables only used to predict the target variable (stayed in college through graduation or left prior to graduation).

Boosted Trees gave the best model, with only an 8.99 percent error rate (see Figure R.18), or this equals a 91.11 percent accuracy rate in predicting if an incoming freshman student at this particular Mexico–United States border town community four-year college will be retained through graduation.

PART II. TEXT MINING AND TEXT VARIABLES' WORD FREQUENCIES AND CONCEPTS

We would like to find out if text mining could increase the accuracy of our Boosted Trees prediction model. First, we need to select Text Mining in the Data Mining pull-down menu (see Figure R.19).

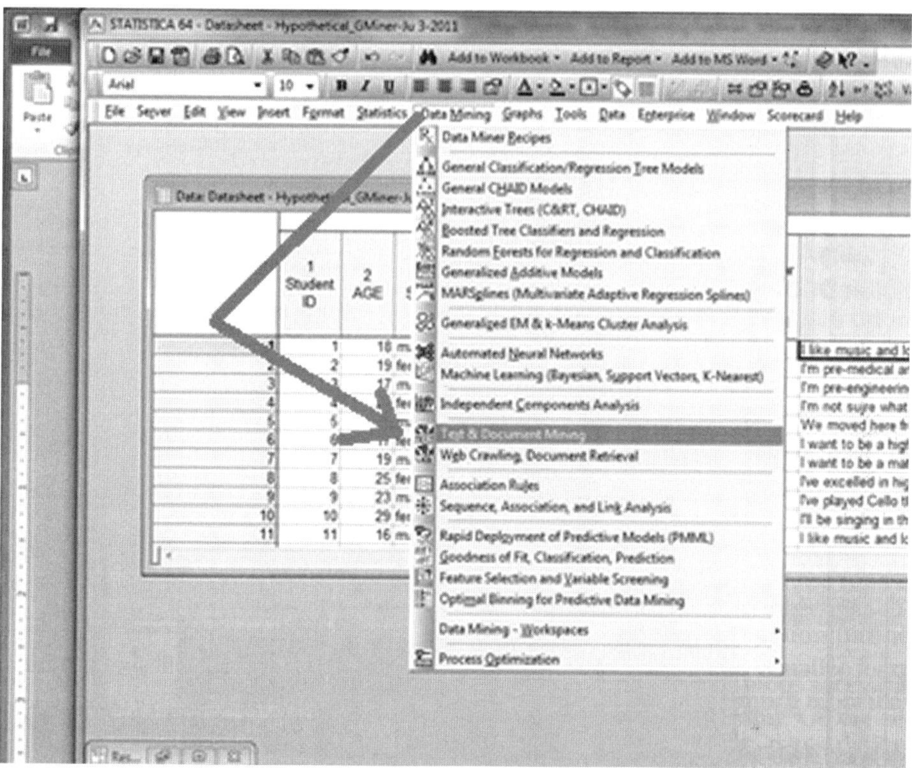

FIGURE R.19
Selecting the Text Mining module in the *STATISTICA* main window.

This brings up on your computer screen the main Text Mining window, with the Quick tab selected by default (see Figure R.20).

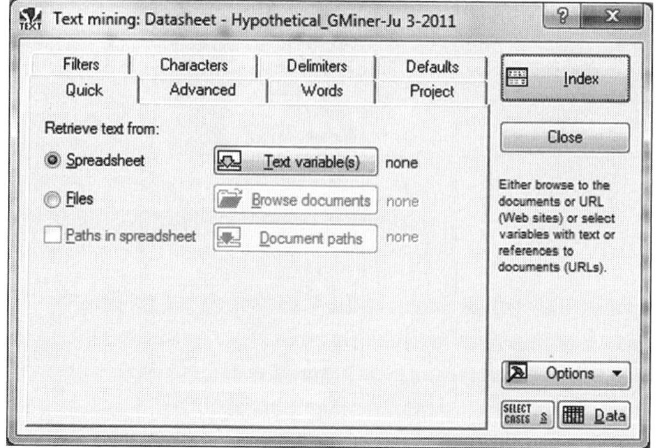

FIGURE R.20
STATISTICA Text Mining main dialog, quick tab.

Click the "Text variable(s)" button (see Figure R.20) to select this variable for text mining (see Figure R.21 below).

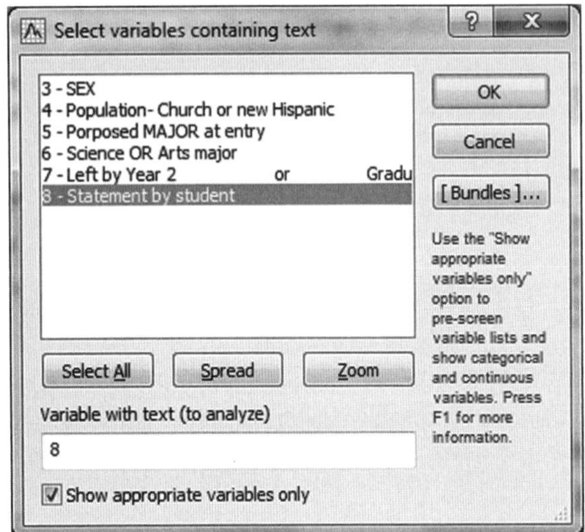

FIGURE R.21
Selecting the variable containing text window.

Select variable 8, the text variable (see Figure R.21), and click OK on the "Select variables containing text" window. Then the text variable, No. 8, will show in the Quick Tab window of the Text Miner, as illustrated below in Figure R.22.

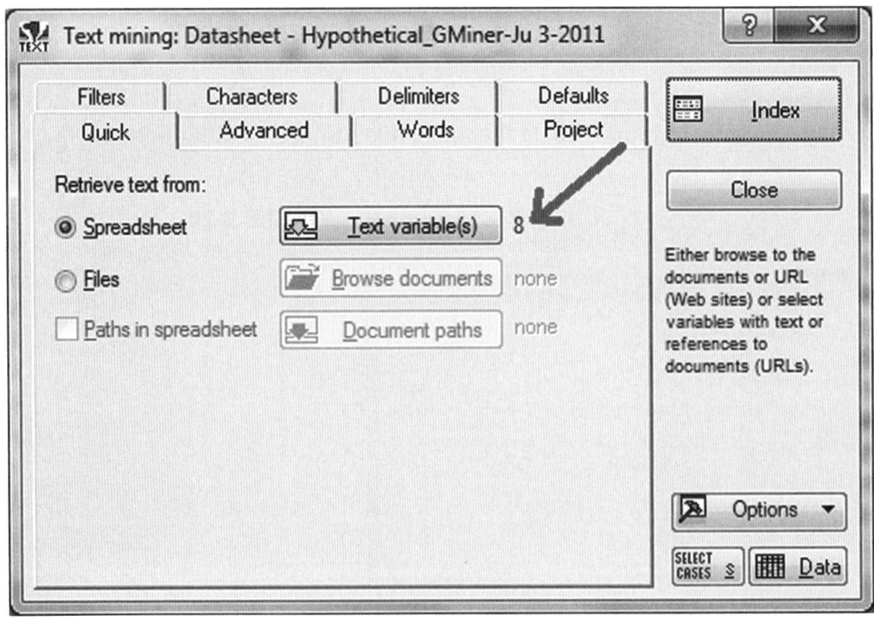

FIGURE R.22
Text variable selected.

Click the "Index" button (upper right corner of the Text Mining dialog as shown in Figure R.22). This will start the computation of word frequencies for the text selected.

When the results dialog appears (see Figure R.23), we see that of the 1,000 documents (e.g., cases), 81 words were selected to be counted. These words are presented in alphabetical order. If you wish to see the order by decreasing word frequency, all one has to do is click on the heading "Count" column," and the words will be so arranged (see Figure R.24).

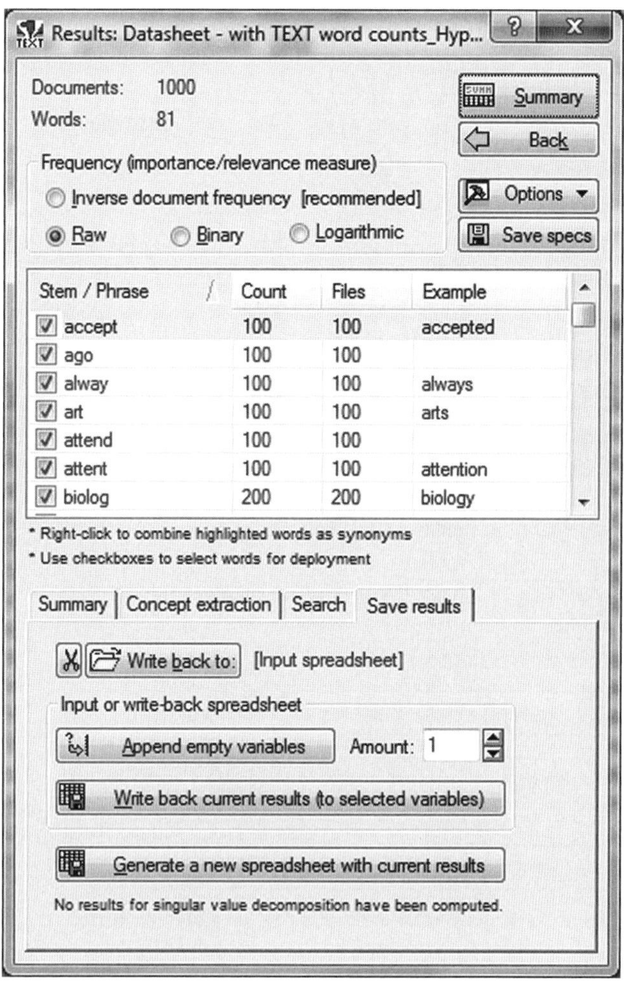

FIGURE R.23
Results of the text mining dialog: words in alphabetical order.

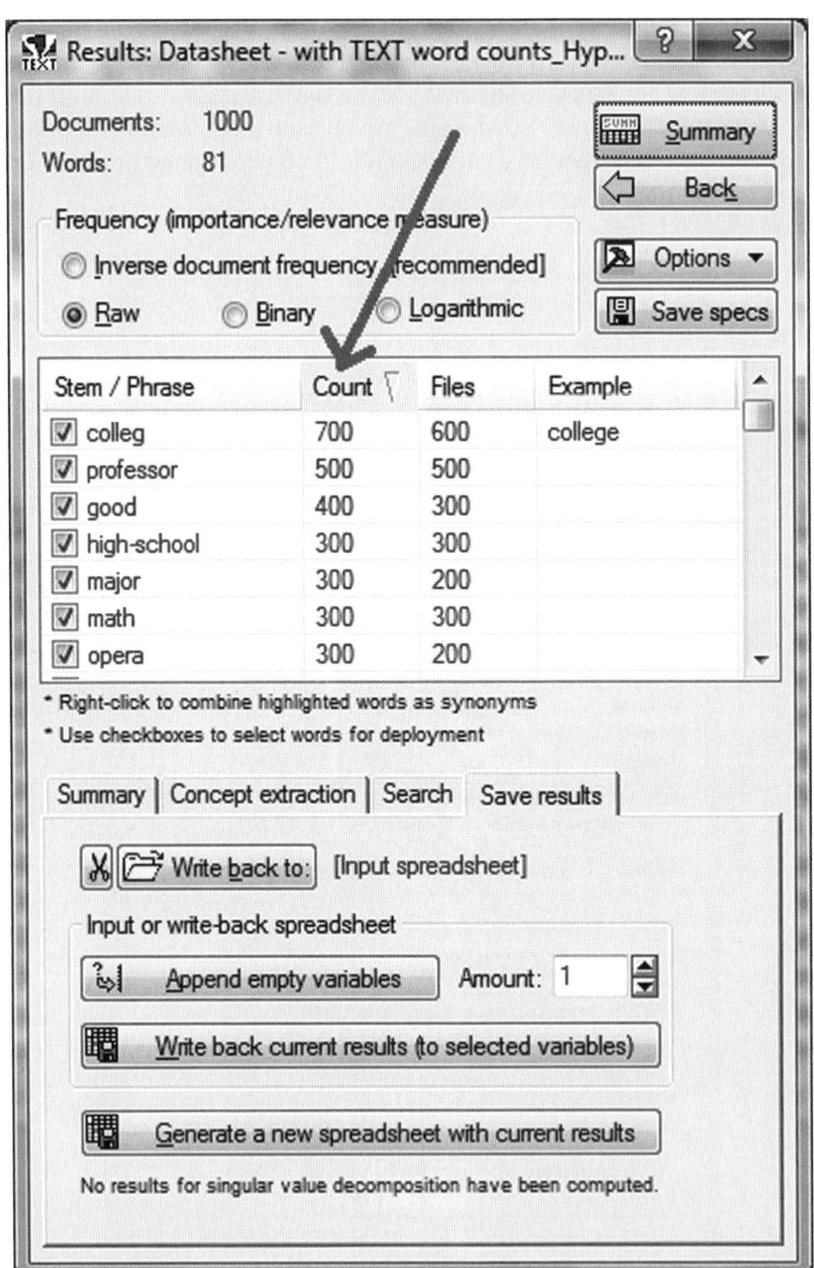

FIGURE R.24
Looking at descending order by frequency of words, obtained by clicking on the "Count" heading in the Indexing Results dialog.

Next, let's do a Save Results. Click on the "Save results" tab (see Figure R.25).

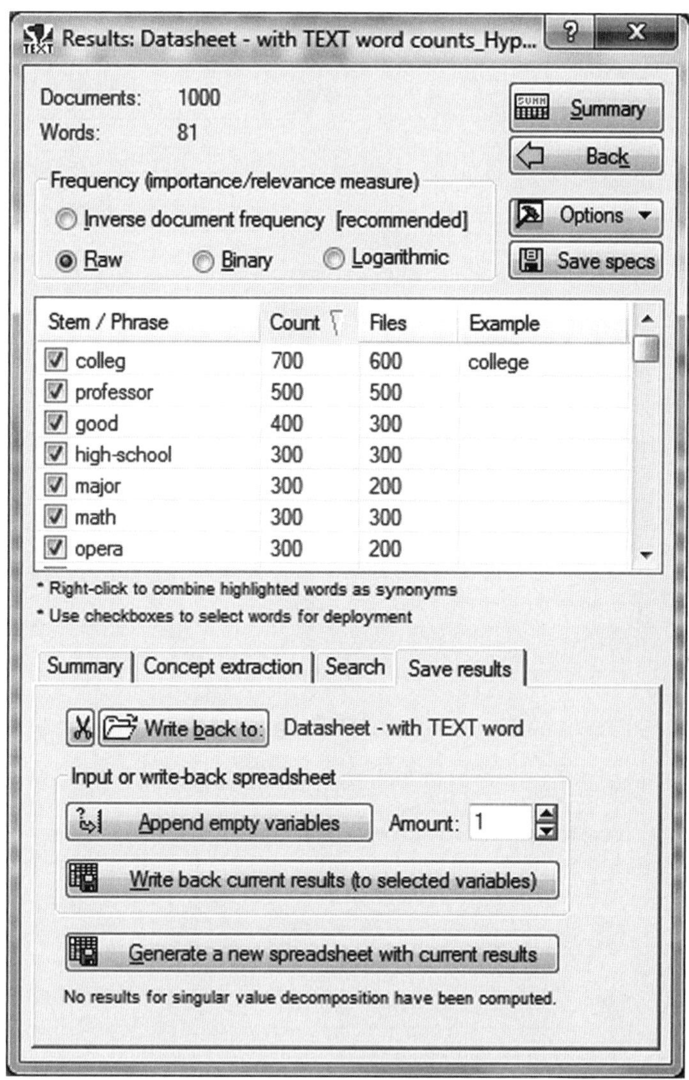

FIGURE R.25
Selecting the "Save results" tab of the Text Miner window brings up new selections that allow the saving of the word frequencies to our original or another data set.

Click on the "Write back to" button (see above in Figure R.25), and select the spreadsheet we are currently using after renaming it "Datasheet - with TEXT word counts_Hypothetical Mexico-USA border town College_GMiner-July 4-2011."

Next, we need to set the "Amount" to 81, since there are 81 words of concern that have been selected by the text mining process; this needs to be done on the "Save results" tab (see Figure R.26).

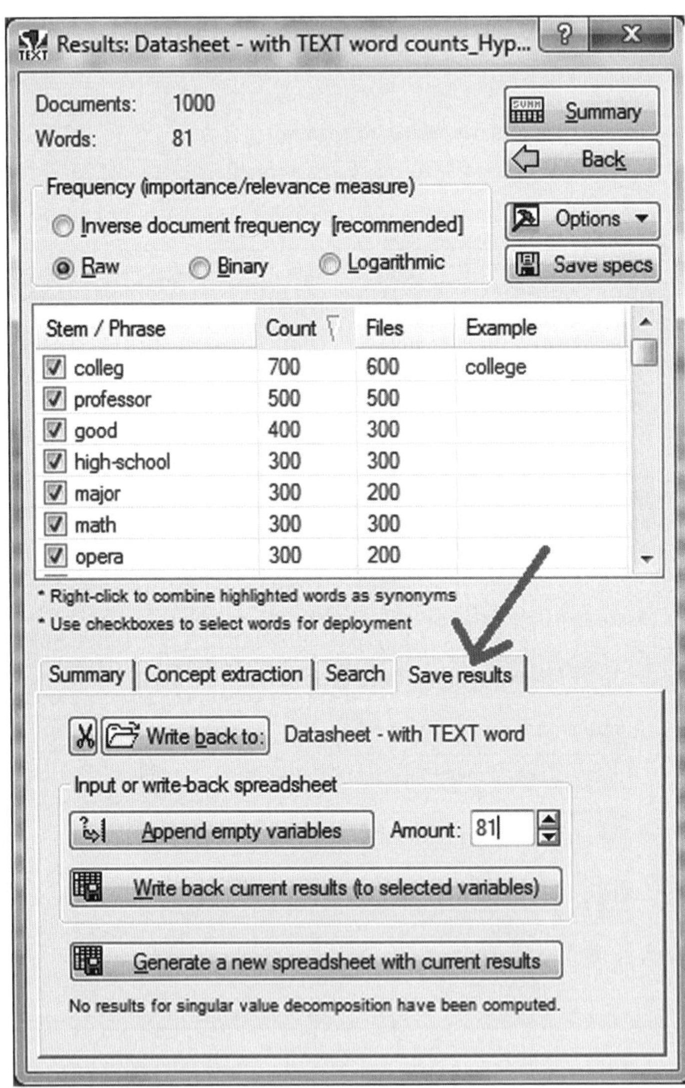

FIGURE R.26
"Save results" tab of the text mining results dialog.

Click the "Append empty variables" button (see Figure R.26 above), and then look at the spreadsheet to confirm visually that 81 new empty variable columns have been added (see Figure R.27).

Data: Datasheet - with TEXT word counts_Hypothetical Mexico-USA border town College_GMiner-July 4-2011 (89v by 1000c)

	8 Statement by student	9 NewVar	10 NewVar1	11 NewVar2	12 NewVar3	13 NewVar4	14 NewVar5	15 NewVar6	16 NewVa
1	I like music and look forward to the orchestra and opera departments professors.....								
2	I'm pre-medical and have heard good good things about the quality of the math, phyhsics, & I								
3	I'm pre-engineering and have gone to pre-college days events here, talked to the science, ma								
4	I'm not sujre what I'll major in, un-decided, but have always lived here and thus feel this comn								
5	We moved here from Mexico a few years ago - no one in my family has gone to college - thi:								
6	I want to be a high-school chemistry and biology teacher								
7	I want to be a math professor at a college....								
8	I've excelled in high-school in oil painting and potteryh; this small college should enable me t								
9	I've played Cello throughout high-school, and studied with teachers from major orchestras, so								
10	I'll be singing in the summer opera workshop in New Mexico before freshman year; this shoul								
11	I like music and look forward to the orchestra and opera departments professors.....								
12	I'm pre-medical and have heard good good things about the quality of the math, phyhsics, & I								
13	I'm pre-engineering and have gone to pre-college days events here, talked to the science, ma								
14	I'm not sujre what I'll major in, un-decided, but have always lived here and thus feel this comn								
15	We moved here from Mexico a few years ago - no one in my family has gone to college - thi:								

Data: Datasheet - with TEXT word counts_Hypothetical Mexico-USA border town College_GMiner-July 4-2011 (89v by 1000c)

	74 NewVar65	75 NewVar66	76 NewVar67	77 NewVar68	78 NewVar69	79 NewVar70	80 NewVar71	81 NewVar72	82 NewVar73	83 NewVar74	84 NewVar75	85 NewVar76	86 NewVar77	87 NewVar78	88 NewVar79	89 NewVar80
1																
2																
3																
4																
5																
6																
7																
8																
9																
10																
11																
12																
13																
14																
15																

FIGURE R.27
Spreadsheet appended with the new columns to which we'll write back the current word counts to these variables.

Next, we need to assign the word frequency variables to the new variable columns in the spreadsheet. To do this, click on the "Write back to current results (to selected variables)" button on the "Save results" tab (see Figure R.28), and follow through as indicated in the following few figures starting with Figure R.28.

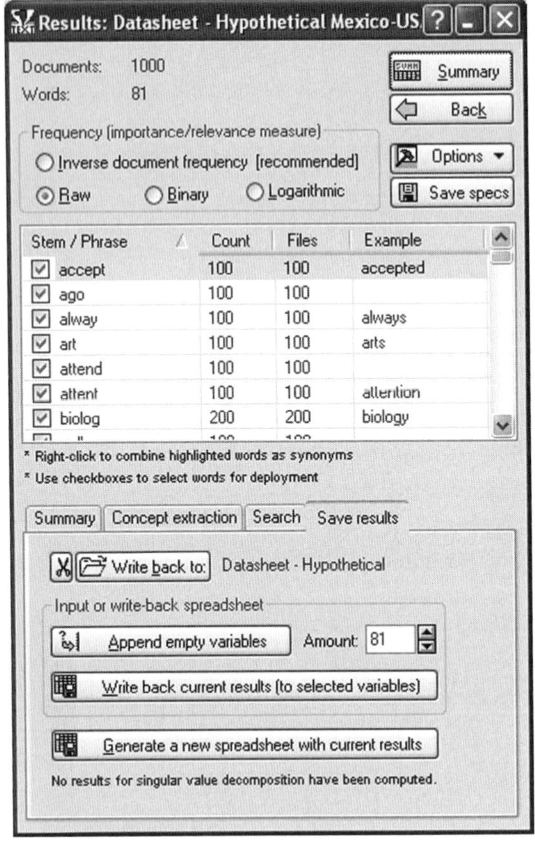

FIGURE R.28
Assigning variables to the new word frequency columns in the data sheet. Follow the next few diagrams for the process. In this first in the series of figures, click on the "Write back current results (to selected spreadsheet)" button (the second button from the bottom of the Results dialog in this figure).

Highlight all of the words you want to retain in the left upper panel. In this case we will highlight all 81 words (see Figures R.29 and R.30).

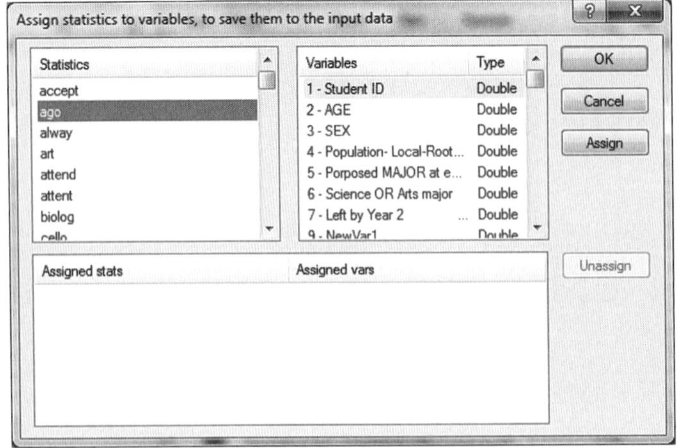

FIGURE R.29
Selecting the first word in the left panel.

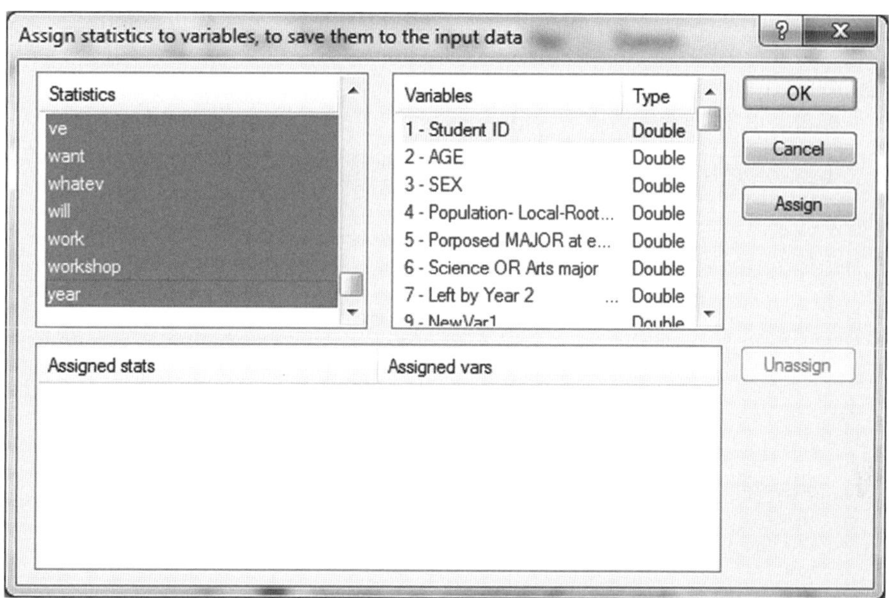

FIGURE R.30
Selecting the last word with the Shift key held down, and thus the entire 81 words.

Then highlight all of the new variables in the new spreadsheet in the right panel (see Figure R.31). Then click the "Assign" button (see Figure R.31).

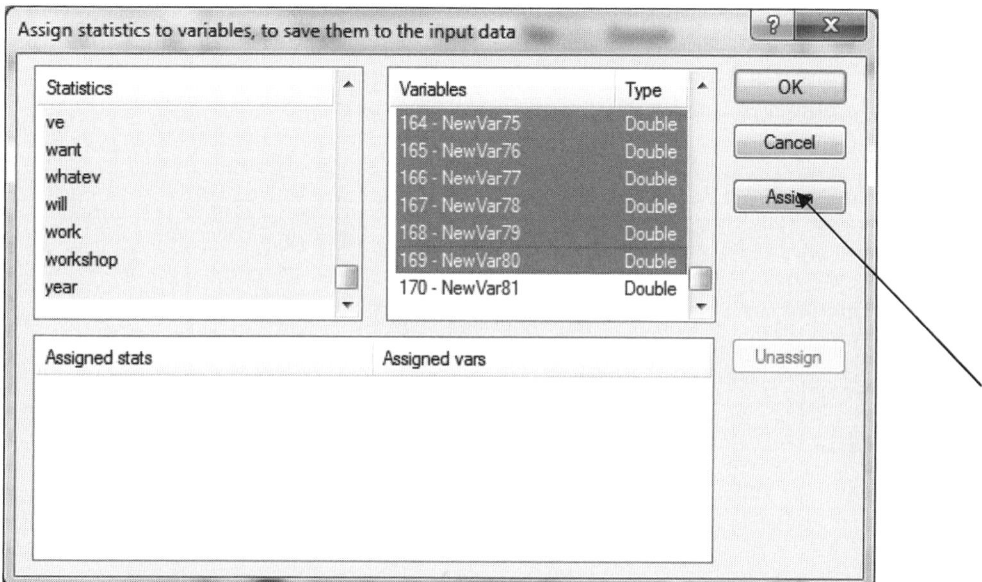

FIGURE R.31
Variable columns to which the word counts will be placed are highlighted in the upper right panel.

Note that all of the words and variables are now put into the bottom panel of this dialog (see Figure R.32). When the OK button is clicked, these word frequency counts will be added to the new variable columns in the spreadsheet (see Figure R.33).

Assign statistics to variables, to save them to the input data

Statistics

Variables	Type
1 - Student ID	Double
2 - AGE	Double
3 - SEX	Double
4 - Population- Local-Root...	Double
5 - Porposed MAJOR at e...	Double
6 - Science OR Arts major	Double
7 - Left by Year 2 ...	Double

OK
Cancel
Assign

Assigned stats	Assigned vars
whatev	85 - NewVar76
will	86 - NewVar77
work	87 - NewVar78
workshop	88 - NewVar79
year	89 - NewVar80

Unassign

FIGURE R.32
Assigned variables and the words in the bottom half panel.

Click OK, and the word frequencies are written into the columns in the data spreadsheet (see Figure R.33). Check to see that the process worked and that the new variables contain interval data. The data represent the number of times each person used each word.

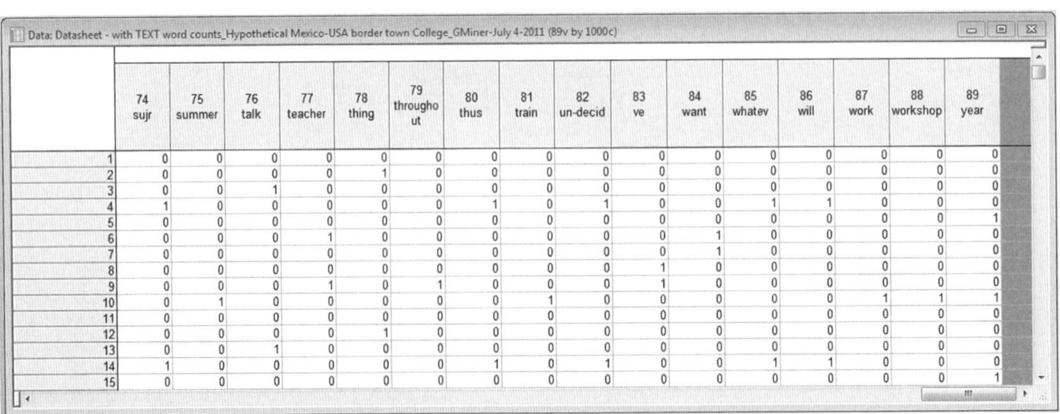

Data: Datasheet - with TEXT word counts_Hypothetical Mexico-USA border town College_GMiner-July 4-2011 (89v by 1000c)

	74 sujr	75 summer	76 talk	77 teacher	78 thing	79 throughout	80 thus	81 train	82 un-decid	83 ve	84 want	85 whatev	86 will	87 work	88 workshop	89 year
1	0	0	0	0	0	0	0	0	0	0	0	0	0	0	0	0
2	0	0	0	0	1	0	0	0	0	0	0	0	0	0	0	0
3	0	0	1	0	0	0	0	0	0	0	0	0	0	0	0	0
4	1	0	0	0	0	0	1	0	1	0	0	1	1	0	0	0
5	0	0	0	0	0	0	0	0	0	0	0	0	0	0	0	1
6	0	0	0	1	0	0	0	0	0	0	1	0	0	0	0	0
7	0	0	0	0	0	0	0	0	0	0	1	0	0	0	0	0
8	0	0	0	0	0	0	0	0	0	1	0	0	0	0	0	0
9	0	0	0	1	0	1	0	0	0	1	0	0	0	0	0	0
10	0	1	0	0	0	0	0	1	0	0	0	0	0	1	1	1
11	0	0	0	0	0	0	0	0	0	0	0	0	0	0	0	0
12	0	0	0	0	1	0	0	0	0	0	0	0	0	0	0	0
13	0	0	1	0	0	0	0	0	0	0	0	0	0	0	0	0
14	1	0	0	0	0	0	1	0	1	0	0	1	1	0	0	0
15	0	0	0	0	0	0	0	0	0	0	0	0	0	0	0	1

FIGURE R.33
Word frequencies written to the new variable columns.

Next we'll perform SVD (Singular Value Decomposition), setting the radio button on the Indexing Results dialog to "Inverse document frequency" (see Figure R.34), a procedure analogous to principal components analysis, which will make "concepts" out of combinations of words. In Tutorial W, we did not select the inverse document frequency, since the raw numbers of the words were thought to be more important in predicting accurately. However, in this example, we are selecting the SVD option.

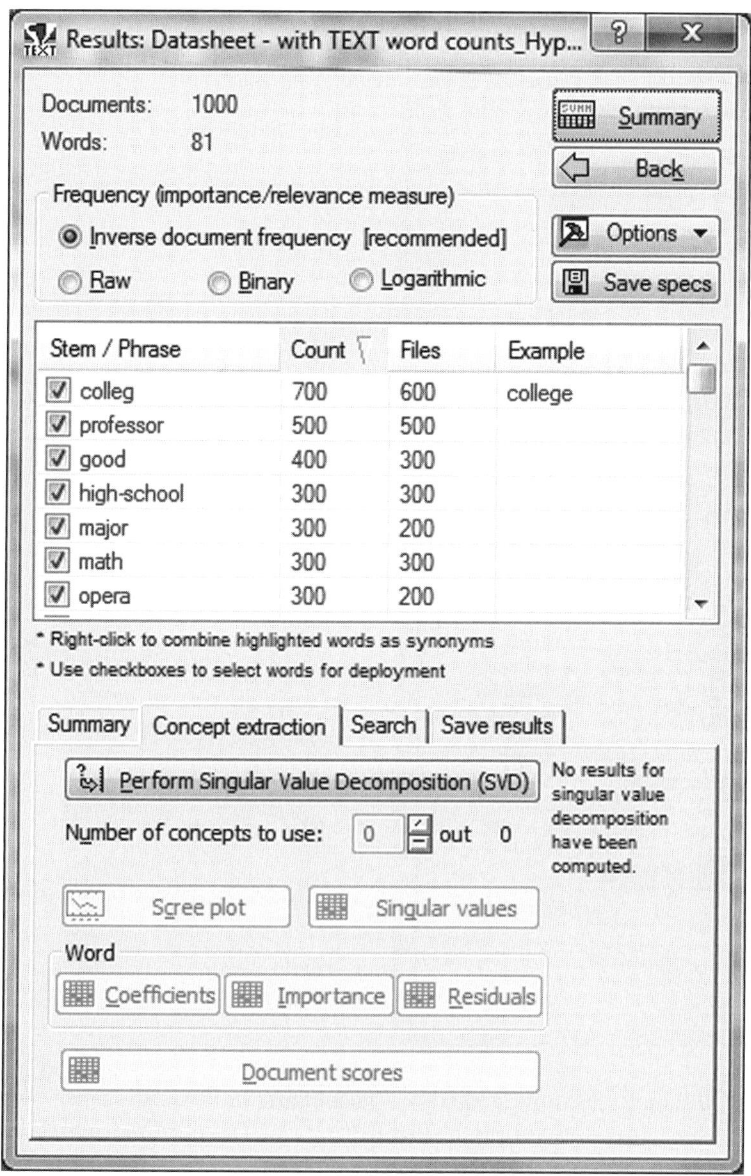

FIGURE R.34
The "Inverse document frequency" radio button is selected prior to performing SVD.

Then click on the "Perform Singular Value Decomposition" button (see Figure R.35) to let the concepts be computed. In Figure R.35 we see the program computed 10 concepts.

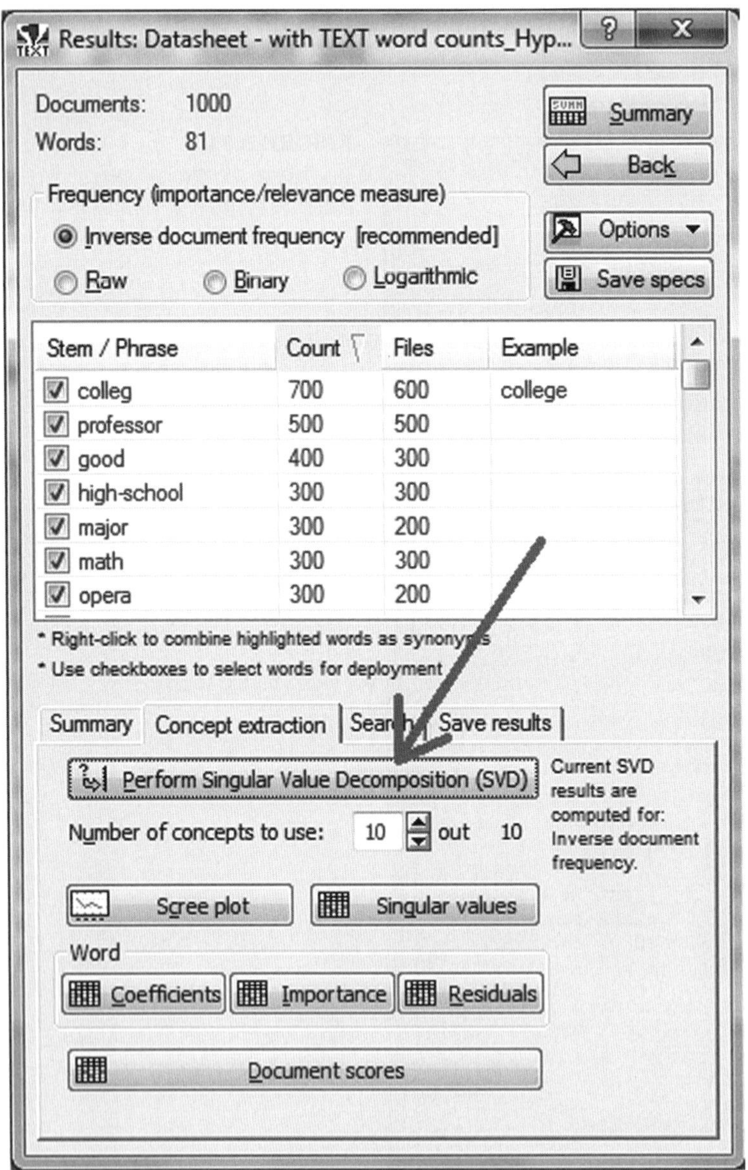

FIGURE R.35
"Perform Singular Value Decomposition" button.

And note that the bottom part of the dialog is now "grayed in," (see Figure R.35) so the Scree plot and other buttons will now function, since these are for the results from the SVD computations. Click on the Scree plot (see Figure R.35 above).

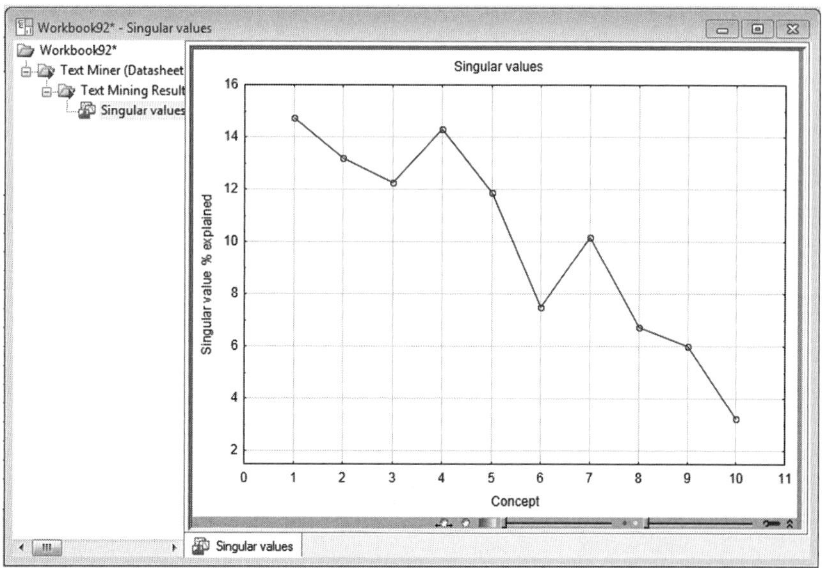

FIGURE R.36
Scree plot result from SVD of the text mining word frequencies.

We see that this is not the usual Scree plot (see Figure R.36), which normally continues to go right across the x-axis, usually with the first 3 points being the most important, and after that points 4 and 5 level off near the x-axis. But here the concepts "jump around" as to the percentage of the "singular value explained." Point one explains 15 percent, 2 explains 13 percent, 3 explains just slightly more than 12 percent, and so on. We suspect this phenomenon happened because the data were made up and repeated through the spreadsheet. The relative values of the concepts are presented in Figure R.37.

	Singular values
	Value
Concept 1	91.15569
Concept 2	81.70127
Concept 3	75.96483
Concept 4	88.48584
Concept 5	73.44306
Concept 6	46.36339
Concept 7	63.10297
Concept 8	41.80420
Concept 9	37.19350
Concept 10	19.98779

FIGURE R.37
Numeric relative values of each of the concepts in spreadsheet format.

Word Importance is another relative value calculated with the SVD, as illustrated in Figures R.38 and R.39. The order of importance can be put in either ascending or descending order by clicking on the header and then selecting the order.

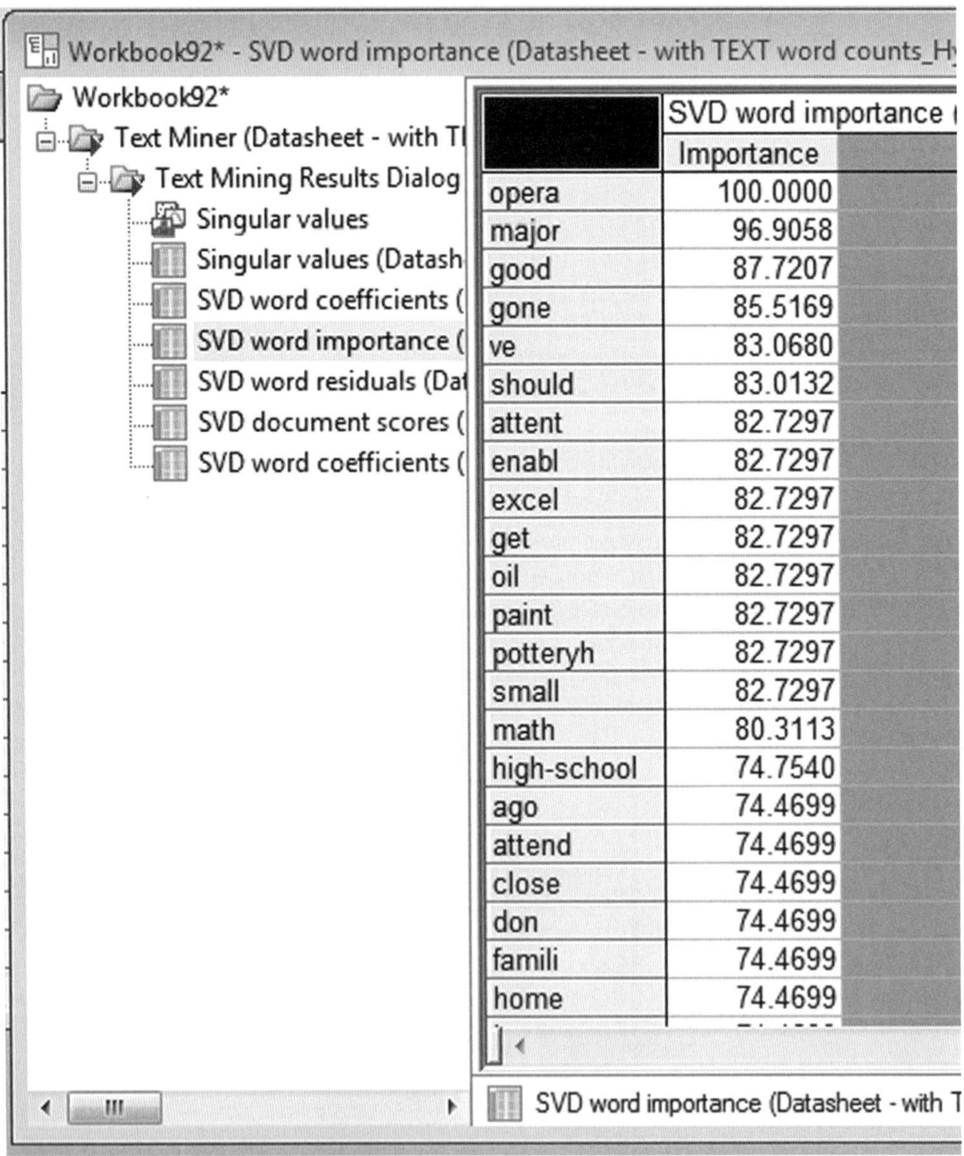

FIGURE R.38
Sorting the words in order value of their importance. This is done by clicking on the heading for the words and then selecting the descending order of importance sort.

	SVD word importance (Datasheet - with TEXT word counts Hypothetical Mexico-USA border town College GMiner-July 4-2011)
	Importance
opera	100.0000
major	96.9058
good	87.7207
gone	85.5169
ve	83.0680
should	83.0132
attent	82.7297
enabl	82.7297
excel	82.7297
get	82.7297
oil	82.7297
paint	82.7297
potteryh	82.7297
small	82.7297
math	80.3113
high-school	74.7540
ago	74.4699
attend	74.4699
close	74.4699
don	74.4699
famili	74.4699
home	74.4699
leav	74.4699
move	74.4699
one	74.4699
freshman	74.1265
new	74.1265
sing	74.1265
summer	74.1265
train	74.1265
work	74.1265
workshop	74.1265
cello	72.8947
play	72.8947

FIGURE R.39

Complete SVD word important value listing, obtained by right clicking on the "SVD Word Importance" listing in the TOC — tree of results in the left panel and then selecting "Copy document."

Now we need to save the concepts to the spreadsheet, just in case we want to work with them further in predictive modeling. There are only 10 concepts, so we will change Amount to 10 on the Indexing dialog.

Click the "Append" button, and then the variable columns are added to the spreadsheet (see Figure R.40 below).

Data: Datasheet - with TEXT word counts_Hypothetical Mexico-USA border town College_GMiner-July 4-2011* (99v by 1000c)

	84 want	85 whatev	86 will	87 work	88 workshop	89 year	90 NewVar1	91 NewVar2	92 NewVar3	93 NewVar4	94 NewVar5	95 NewVar6	96 NewVar7	97 NewVar8	98 NewVar9	99 NewVar10
1	0	0	0	0	0	0										
2	0	0	0	0	0	0										
3	0	0	0	0	0	0										
4	0	1	1	0	0	0										
5	0	0	0	0	0	1										
6	1	0	0	0	0	0										
7	1	0	0	0	0	0										
8	0	0	0	0	0	0										
9	0	0	0	0	0	0										
10	0	0	0	1	1	1										
11	0	0	0	0	0	0										
12	0	0	0	0	0	0										
13	0	0	0	0	0	0										
14	0	1	1	0	0	0										
15	0	0	0	0	0	1										

FIGURE R.40
New variables added to the spreadsheet for the 10 new concepts.

Then select the "Write back to current spreadsheet" button, and arrange for these 10 new concepts to be placed in our spreadsheet. Ten new columns will be needed. Do this in the same sequence of steps as we did for the word frequencies (see Figures R.41 and R.42).

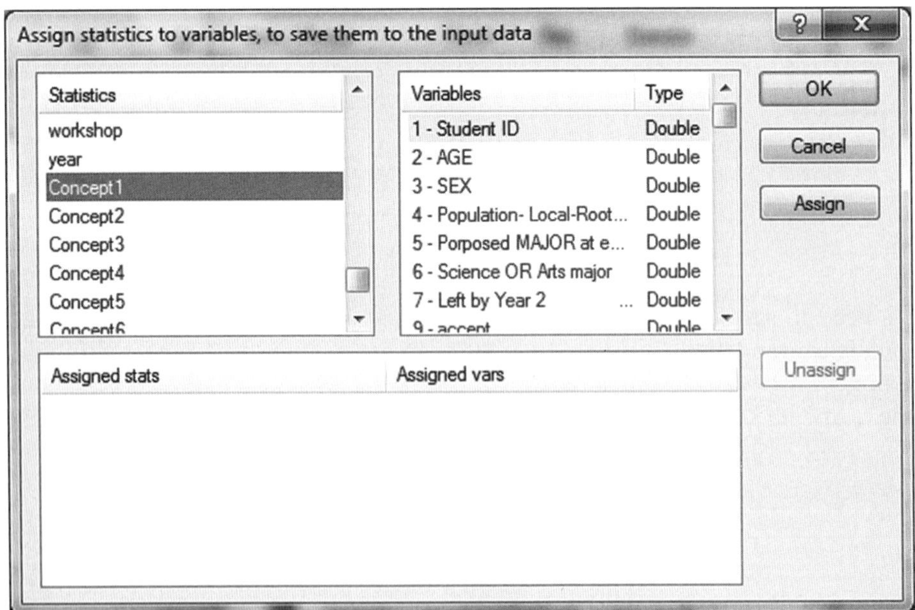

FIGURE R.41
Assign statistics to the variables window. Concepts 1—10 are being assigned to 10 new variable columns in the data spreadsheet.

Data: Datasheet - with TEXT word counts_Hypothetical Mexico-USA border town College_GMiner-July 4-2011 (99v by 1000c)

	84 want	85 whatev	86 will	87 work	88 workshop	89 year	90 Concept1	91 Concept2	92 Concept3	93 Concept4	94 Concept5	95 Concept6	96 Concept7	97 Concept8	98 Concept9	99 Concept10
1	0	0	0	0	0	0	0.003689	-0.01545	0.00381	0.007053	0.00686	-0.00841	0.03257	0.086442	0.032538	0.00368562
2	0	0	0	0	0	0	0.003647	-0.03345	0.073917	-0.03588	0.010799	-0.02504	-0.00303	-0.00195	-0.0077	0.00045436
3	0	0	0	0	0	0	0.008534	-0.01692	0.008404	-0.05315	0.011677	-0.05004	0.004988	-0.00351	-0.00203	0.00851916
4	0	1	1	0	0	0	0.097139	0.020536	0.000013	-0.00176	-0.00048	-0.00119	-0.00942	0.002699	0.001918	0.00019007
5	0	0	0	0	0	1	0.006142	-0.04352	-0.05337	-0.04454	-0.0443	-0.04016	-0.00035	0.002374	0.000671	0.00085196
6	1	0	0	0	0	0	0.000952	-0.00309	0.006713	0.001774	-0.00479	0.005335	0.012181	-0.03911	0.087436	0.02444911
7	1	0	0	0	0	0	0.00093	-0.003	0.00238	-0.00324	0.000205	-0.00811	0.0012	-0.00623	0.024098	-0.0964383
8	0	0	0	0	0	0	0.004867	-0.02734	0.033346	0.048469	-0.07095	0.064611	-0.02934	0.006374	-0.00005	-0.0003813
9	0	0	0	0	0	0	0.011472	-0.00684	0.009406	0.018774	-0.02267	0.020562	0.088071	-0.0255	-0.02493	-0.0038226
10	0	0	0	1	1	1	0.016416	-0.0723	-0.01874	0.033702	0.046547	0.022195	-0.00701	-0.01548	-0.00514	0.00002739
11	0	0	0	0	0	0	0.003689	-0.01545	0.00381	0.007053	0.00686	-0.00841	0.03257	0.086442	0.032538	0.00368562
12	0	0	0	0	0	0	0.003647	-0.03345	0.073917	-0.03588	0.010799	-0.02504	-0.00303	-0.00195	-0.0077	0.00045436
13	0	0	0	0	0	0	0.008534	-0.01692	0.008404	-0.05315	0.011677	-0.05004	0.004988	-0.00351	-0.00203	0.00851916
14	0	1	1	0	0	0	0.097139	0.020536	0.000013	-0.00176	-0.00048	-0.00119	-0.00942	0.002699	0.001918	0.00019007
15	0	0	0	0	0	1	0.006142	-0.04352	-0.05337	-0.04454	-0.0443	-0.04016	-0.00035	0.002374	0.000671	0.00085196

FIGURE R.42

Concepts 1–10 written back to the spreadsheet.

Now, let's go back to the Workbook and make the "SVD word coefficients" the "Input Spreadsheet." Do this by right clicking on the "SVD word coefficients" file in the directory (on left panel in Figure R.43) and then accepting "Make active input."

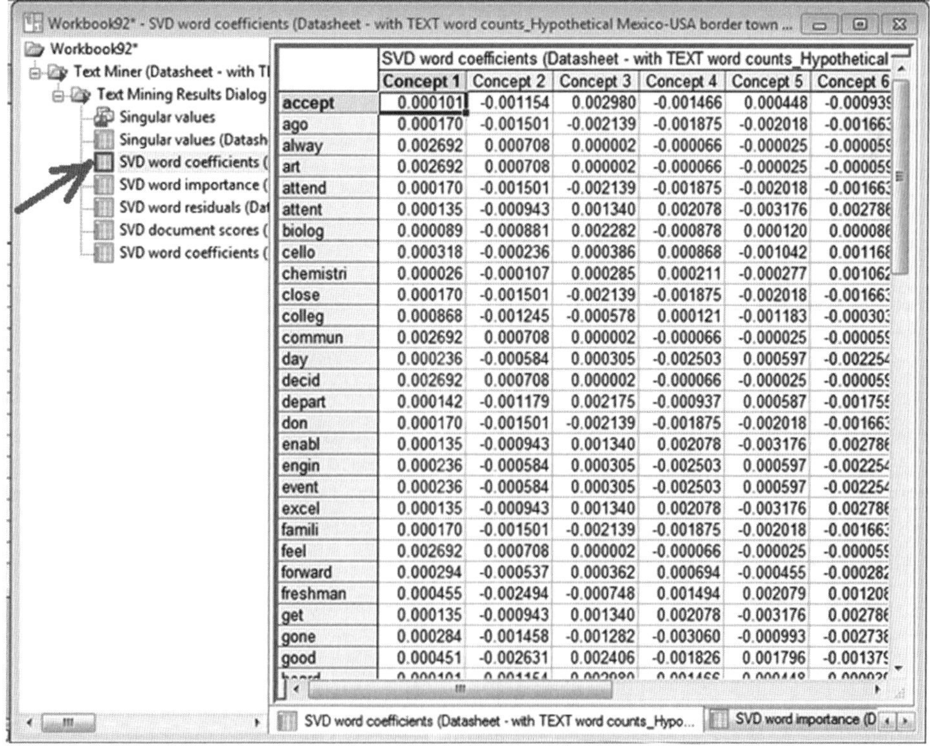

SVD word coefficients (Datasheet - with TEXT word counts_Hypothetical

	Concept 1	Concept 2	Concept 3	Concept 4	Concept 5	Concept 6
accept	0.000101	-0.001154	0.002980	-0.001466	0.000448	-0.00093
ago	0.000170	-0.001501	-0.002139	-0.001875	-0.002018	-0.00166
alway	0.002692	0.000708	0.000002	-0.000066	-0.000025	-0.00005
art	0.002692	0.000708	0.000002	-0.000066	-0.000025	-0.00005
attend	0.000170	-0.001501	-0.002139	-0.001875	-0.002018	-0.00166
attent	0.000135	-0.000943	0.001340	0.002078	-0.003176	0.00278
biolog	0.000089	-0.000881	0.002282	-0.000878	0.000120	0.00008
cello	0.000318	-0.000236	0.000386	0.000868	-0.001042	0.00116
chemistri	0.000026	-0.000107	0.000285	0.000211	-0.000277	0.00106
close	0.000170	-0.001501	-0.002139	-0.001875	-0.002018	-0.00166
colleg	0.000868	-0.001245	-0.000578	0.000121	-0.001183	-0.00030
commun	0.002692	0.000708	0.000002	-0.000066	-0.000025	-0.00005
day	0.000236	-0.000584	0.000305	-0.002503	0.000597	-0.00225
decid	0.002692	0.000708	0.000002	-0.000066	-0.000025	-0.00005
depart	0.000142	-0.001179	0.002175	-0.000937	0.000587	-0.00175
don	0.000170	-0.001501	-0.002139	-0.001875	-0.002018	-0.00166
enabl	0.000135	-0.000943	0.001340	0.002078	-0.003176	0.00278
engin	0.000236	-0.000584	0.000305	-0.002503	0.000597	-0.00225
event	0.000236	-0.000584	0.000305	-0.002503	0.000597	-0.00225
excel	0.000135	-0.000943	0.001340	0.002078	-0.003176	0.00278
famili	0.000170	-0.001501	-0.002139	-0.001875	-0.002018	-0.00166
feel	0.002692	0.000708	0.000002	-0.000066	-0.000025	-0.00005
forward	0.000294	-0.000537	0.000362	0.000694	-0.000455	-0.00028
freshman	0.000455	-0.002494	-0.000748	0.001494	0.002079	0.00120
get	0.000135	-0.000943	0.001340	0.002078	-0.003176	0.00278
gone	0.000284	-0.001458	-0.001282	-0.003060	-0.000993	-0.00273
good	0.000451	-0.002631	0.002406	-0.001826	0.001796	-0.00137

FIGURE R.43

SVD word coefficients.

Now, let's make a 2D scatterplot of Concept 1 versus Concept 2. By making the "SVD word coefficients" the Active Datasheet, this is easily done. Right click on the name of the SVD word coefficients in the Text Mining results workbook (see Figure R.44) and then select "Make input data."

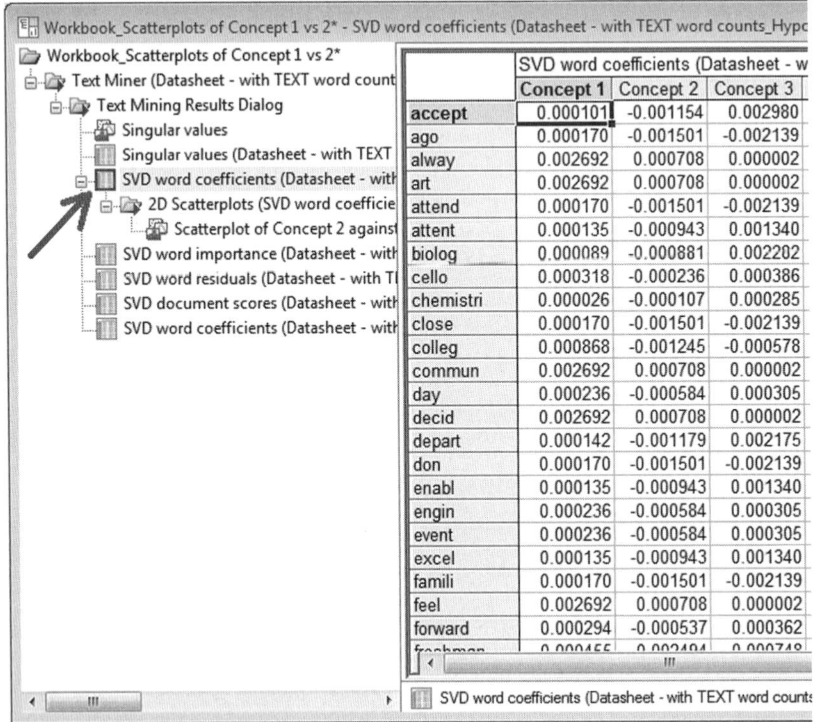

FIGURE R.44
Making the SVD word coefficient datasheet the "Input spreadsheet."

The next two figures, Figures R.45 and R.46, illustrate how to make a scatterplot graph using the GRAPHS pull down menu.

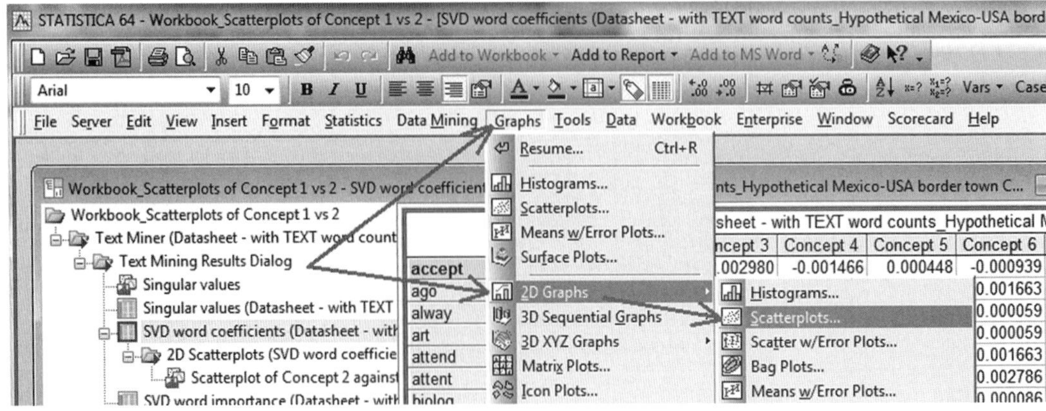

FIGURE R.45
Selecting the Scatterplot graph from the Graphs pull-down menu.

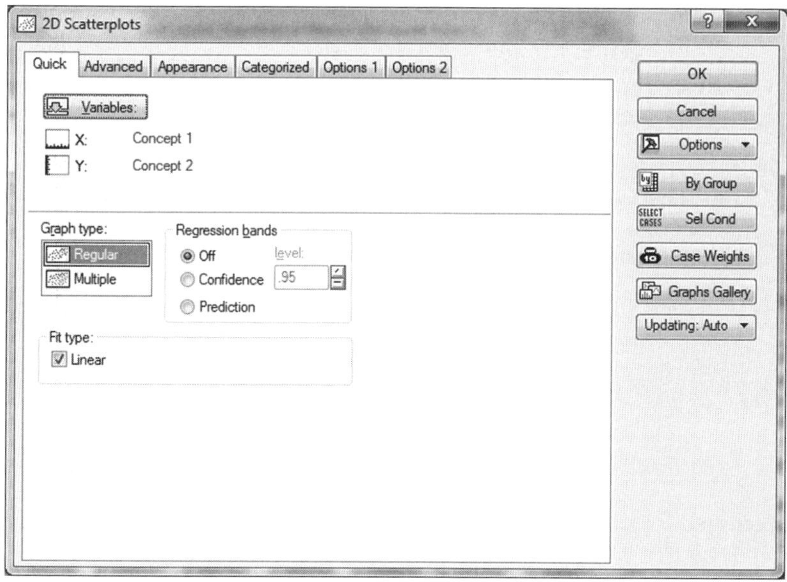

FIGURE R.46

Scatterplot graph menu. Concept 1 and Concept 2 are selected as the two variables to graph against each other.

Clicking OK on the 2D Scatterplot dialog (see Figure R.46 above) produces the graph in Figure R.47.

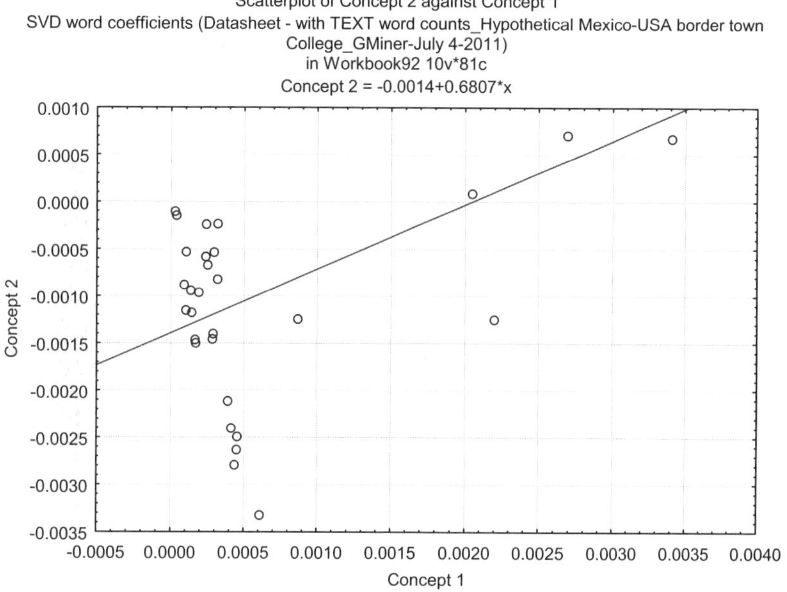

FIGURE R.47

2D Scatterplot of Concept 1 versus Concept 2.

Now do a Feature Selection using the Workspace (see Figure R.48) to see which of the Words and Concepts are the "most important."

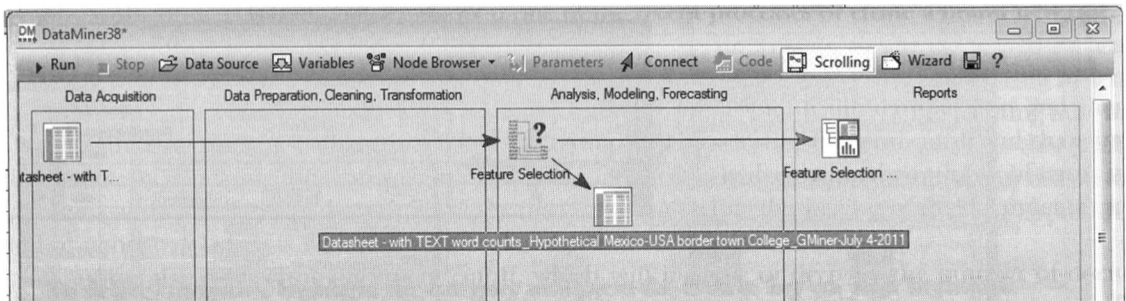

FIGURE R.48

Data Miner Workspace with the spreadsheet icon entered in the left data set panel and the feature selection entered in the middle panel.

The variables selected for this Feature Selection computation are highlighted in Figure R.49 below.

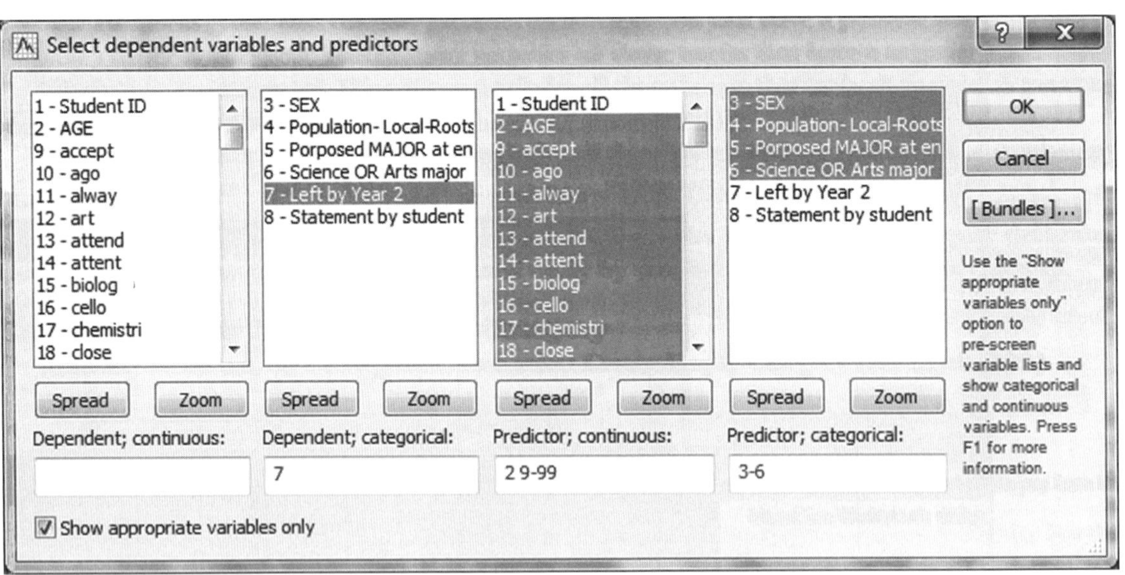

FIGURE R.49

Variables selected to be put through the feature selection process.

After the feature selection is computed, you can double click on the "green" results icon (see Figure R.48 above) in the right panel to get a workbook of results (see Figure R.50). Included in this is an "Importance Plot" and also a listing of the variable numbers that have been selected (such that those wanted can easily be copied and then pasted into subsequent variable selection dialogs in other data analysis modules).

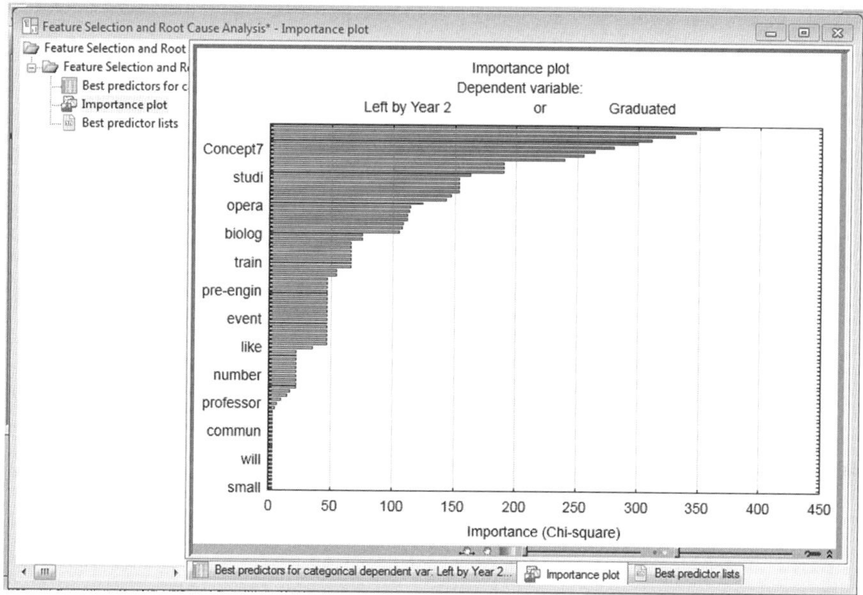

FIGURE R.50

Importance Plot of variables selected by "feature selection."

We can see from Figures R.51 and R.52 (a zoomed in view) that the concepts are mostly represented by the more important variables selected by feature selection.

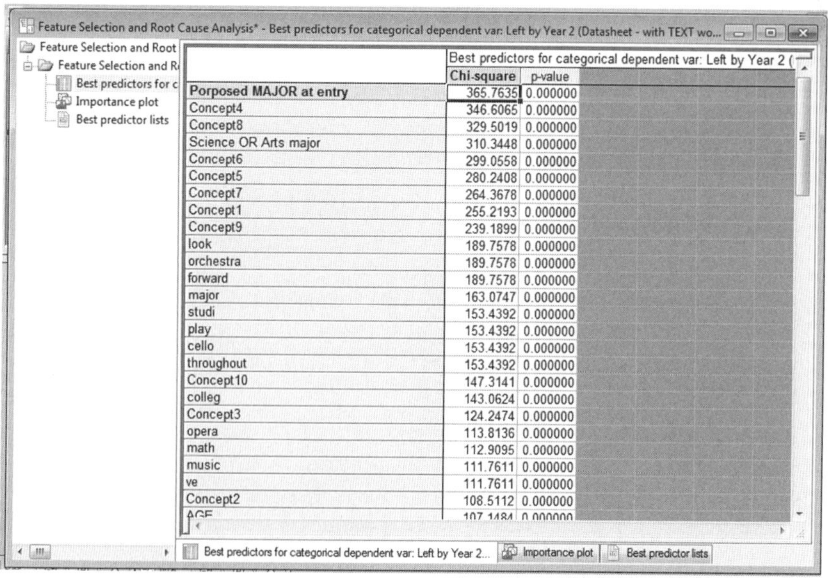

Feature Selection and Root Cause Analysis* - Best predictors for categorical dependent var: Left by Year 2 (Datasheet - with TEXT wo...

Best predictors for categorical dependent var: Left by Year 2 (

	Chi-square	p-value
Porposed MAJOR at entry	365.7635	0.000000
Concept4	346.6065	0.000000
Concept8	329.5019	0.000000
Science OR Arts major	310.3448	0.000000
Concept6	299.0558	0.000000
Concept5	280.2408	0.000000
Concept7	264.3678	0.000000
Concept1	255.2193	0.000000
Concept9	239.1899	0.000000
look	189.7578	0.000000
orchestra	189.7578	0.000000
forward	189.7578	0.000000
major	163.0747	0.000000
studi	153.4392	0.000000
play	153.4392	0.000000
cello	153.4392	0.000000
throughout	153.4392	0.000000
Concept10	147.3141	0.000000
colleg	143.0624	0.000000
Concept3	124.2474	0.000000
opera	113.8136	0.000000
math	112.9095	0.000000
music	111.7611	0.000000
ve	111.7611	0.000000
Concept2	108.5112	0.000000
AGE	107.1484	0.000000

FIGURE R.51

Variable importance listing from feature selection.

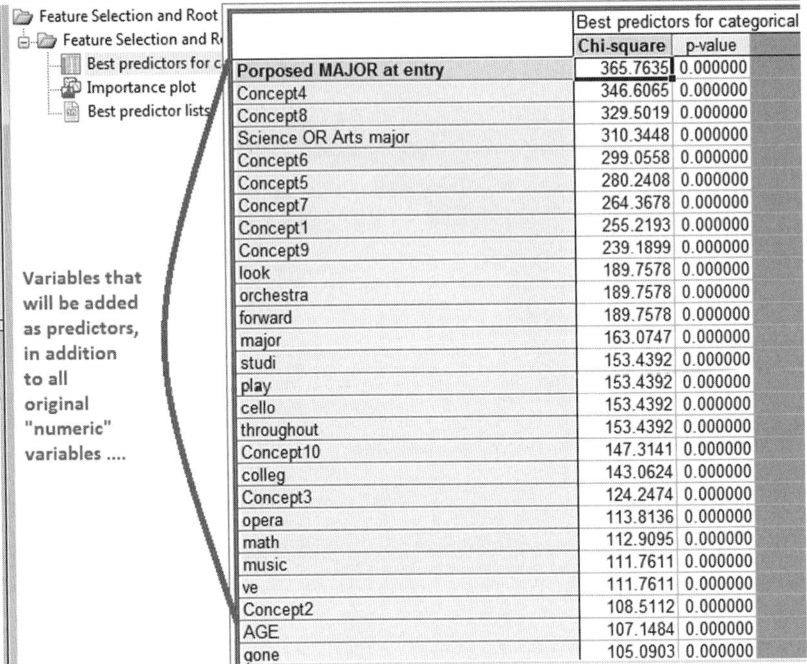

	Best predictors for categorical	
	Chi-square	p-value
Porposed MAJOR at entry	365.7635	0.000000
Concept4	346.6065	0.000000
Concept8	329.5019	0.000000
Science OR Arts major	310.3448	0.000000
Concept6	299.0558	0.000000
Concept5	280.2408	0.000000
Concept7	264.3678	0.000000
Concept1	255.2193	0.000000
Concept9	239.1899	0.000000
look	189.7578	0.000000
orchestra	189.7578	0.000000
forward	189.7578	0.000000
major	163.0747	0.000000
studi	153.4392	0.000000
play	153.4392	0.000000
cello	153.4392	0.000000
throughout	153.4392	0.000000
Concept10	147.3141	0.000000
colleg	143.0624	0.000000
Concept3	124.2474	0.000000
opera	113.8136	0.000000
math	112.9095	0.000000
music	111.7611	0.000000
ve	111.7611	0.000000
Concept2	108.5112	0.000000
AGE	107.1484	0.000000
gone	105.0903	0.000000

FIGURE R.52
The list of variables that are being selected to put into "Predictive Analytic Models."

We can select from the list outputted by just highlighting the top ones, down to AGE (Variable # 2) and then copy and paste into the Variable selection dialog (see Figure R.53) when we do the next DMRecipe.

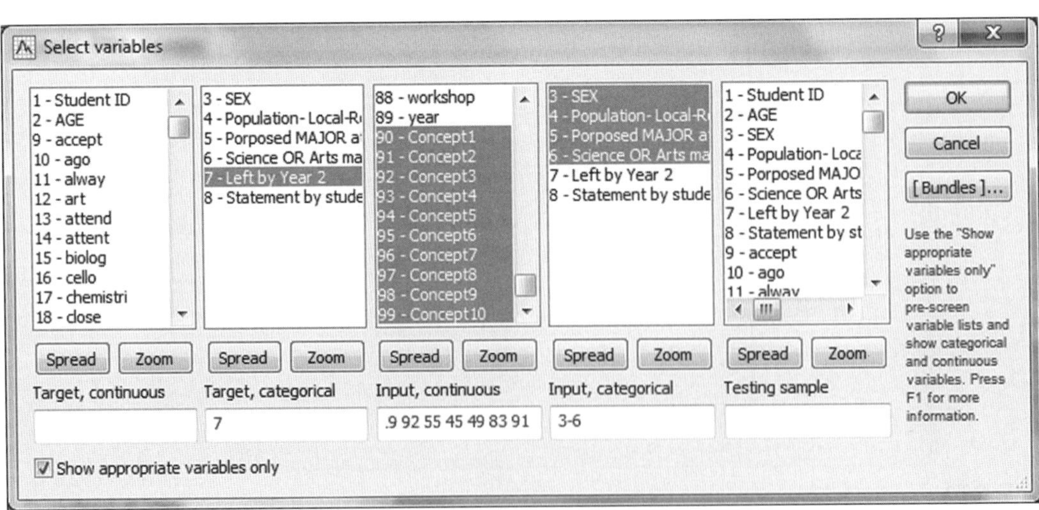

FIGURE R.53
Variable selection for data mining modeling of both the numeric variables and the text mining variables.

The DMRecipe process is illustrated in the next set of figures: Figure R.54 through the last figure in this tutorial, Figure R.63.

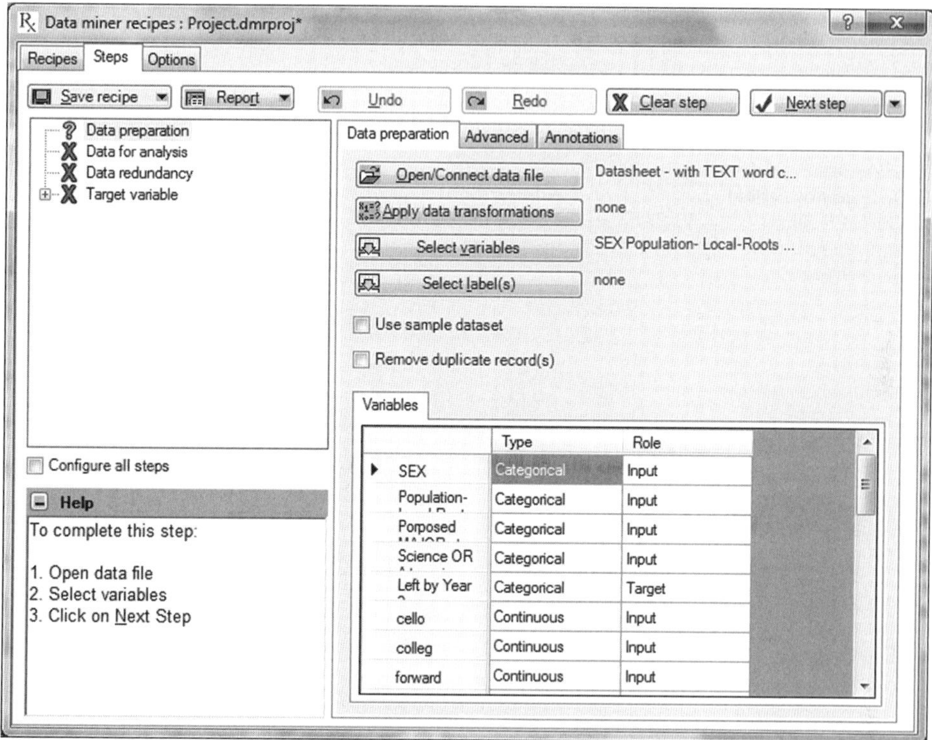

FIGURE R.54

DMRecipe computation for text variables added to the numeric variables.

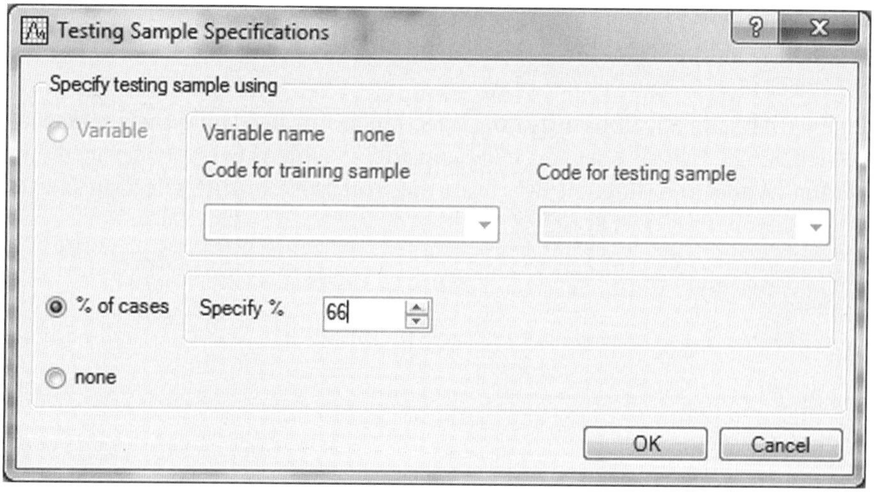

Testing Sample Specifications

Specify testing sample using

Variable

Variable name none

Code for training sample

Code for testing sample

% of cases

Specify %

66

none

OK

Cancel

FIGURE R.55

Specifications for the Train and the Test sample percentages.

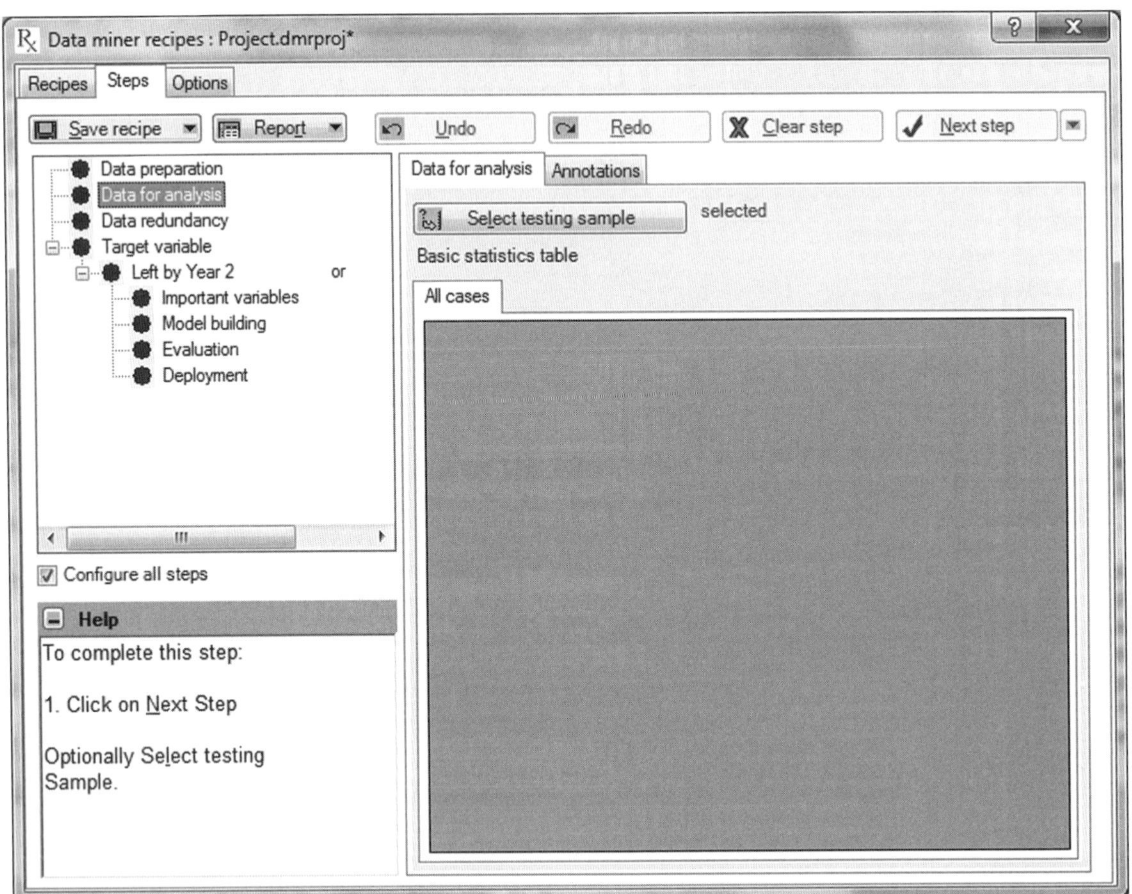

FIGURE R.56
DMRecipe main dialog showing that the Train/Test sample has been selected.

Even though we have already done a "feature selection" using the data miner workspace, let's set the "fast predictor screening," just as a second check on the variable previously selected, since this "fast predictor screening" works in a different way from the workspace feature selection (see Figure R.57).

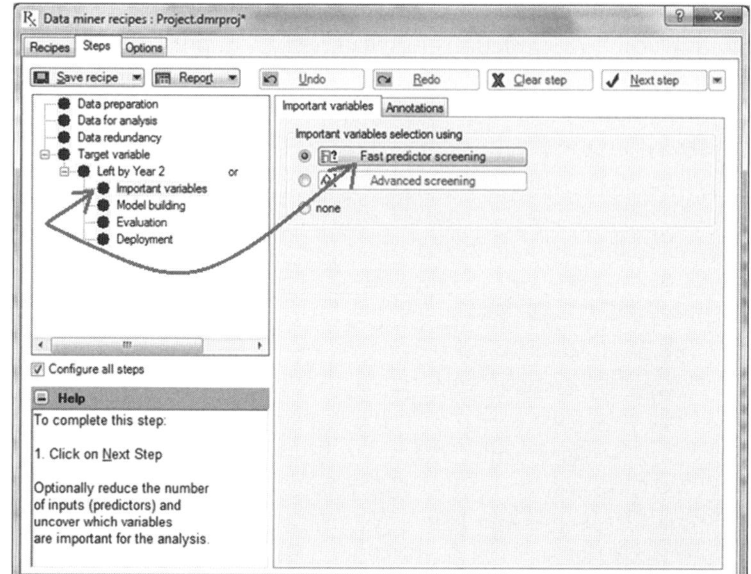

FIGURE R.57
Setting the "Fast predictor screening" in the DMRecipe.

And finally we want to use all of the models available in the DMRecipe for the competitive evaluation of models to see which can give the highest accuracy prediction of our target variable (see Figure R.58).

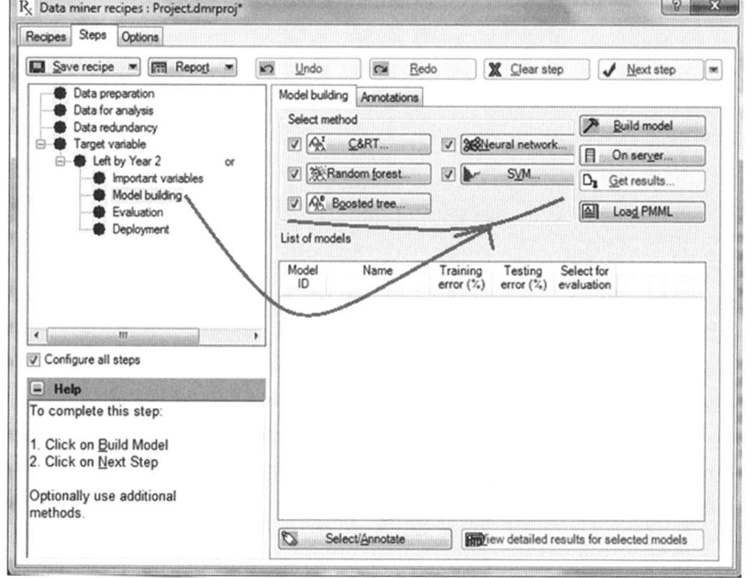

FIGURE R.58
Selecting all of the data mining algorithms in the "Model building" window.

Now we need to "uncheck" the "Configure all steps" radio button, taking us back to the primary DMRecipe window, and then we will select "Run to completion" (see Figures R.59 and R.60) so that all these algorithms will compute automatically, simultaneously with distributed parallel processing, and automatically spit out an entire array of results worksheets and graphs, with the last one being a summary evaluation showing which algorithm models this data set best.

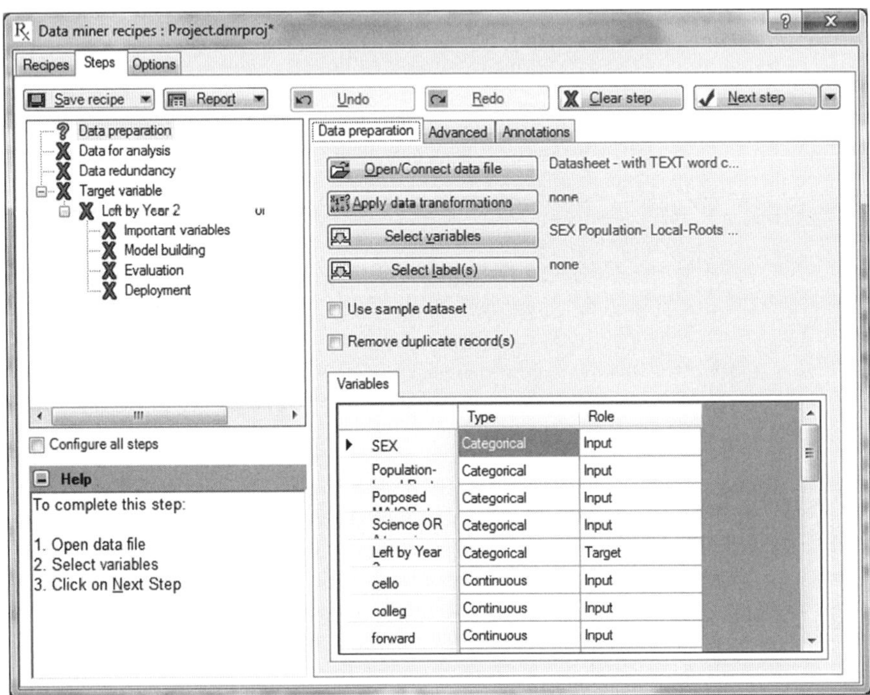

FIGURE R.59

DMRecipe dialog, parameters all set, ready to run the computations.

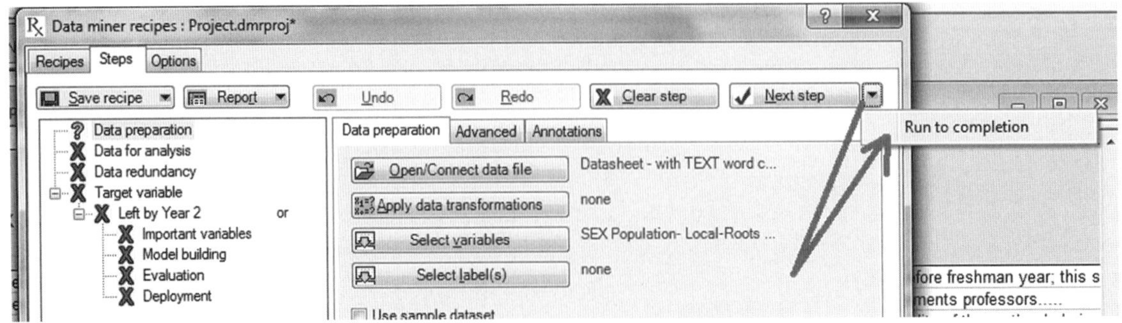

FIGURE R.60

By clicking the "down arrow" in the upper right, the "Run to completion" selection appears. Select this to have the entire set of computations run automatically.

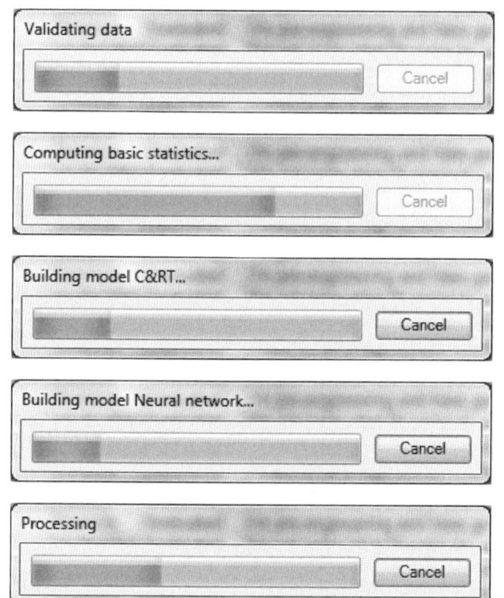

FIGURE R.61
Progress bars that appear for the different algorithms and processes during the DMRecipe computations.

Finally, the Results Workbook is displayed (see Figure R.62), with the summary evaluation tabulated as the top window.

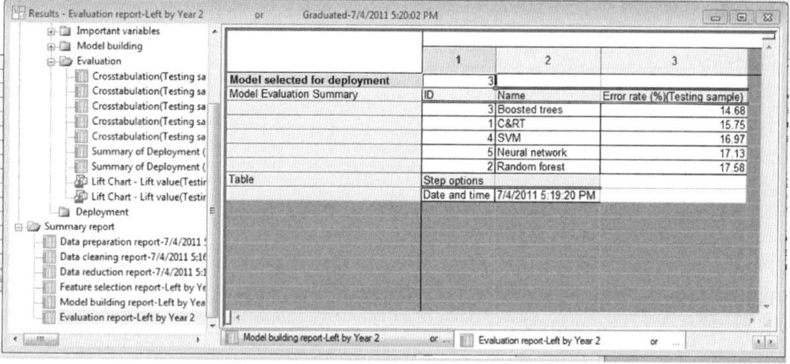

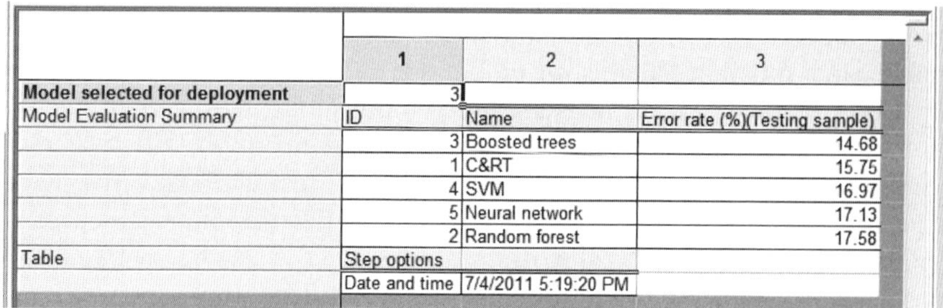

	1	2	3
Model selected for deployment	3		
Model Evaluation Summary	ID	Name	Error rate (%)(Testing sample)
	3	Boosted trees	14.68
	1	C&RT	15.75
	4	SVM	16.97
	5	Neural network	17.13
	2	Random forest	17.58
Table	Step options		
	Date and time	7/4/2011 5:19:20 PM	

FIGURE R.62
Results workbook for the combined "text + numeric" predictor variables.

If we look at the Lift chart (see Figure R.63), we see visually how Boosted Trees compared to the other algorithms.

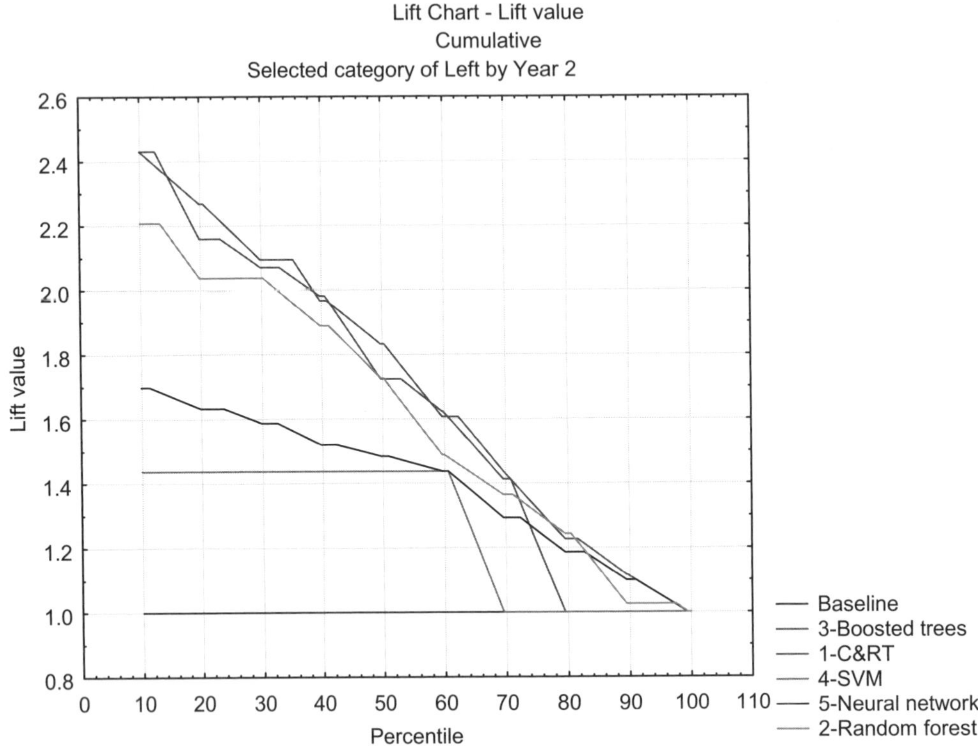

FIGURE R.63
Lift chart visually illustrating that Boosted Trees is overall the best algorithm to model this data.

We see that with all these extra variables, we did *not* get a higher accuracy, but Boosted Trees is still the best algorithm for modeling this data. However, some of the other algorithms seem to have a higher accuracy than with just the numeric variables.

We should note, however, that the variables in this tutorial are a hypothetical (made-up) data set, where the "major study" was selected to "graduate" if in the math and sciences and mostly set to "leave the college before graduation" if the major was in the arts. (Thus, "Declared major" was the most important variable.) In a situation like this, we would not necessarily find that the text would add that much lift. In the real situation, where things are much more subtle, we'd expect that a "text variable" would add lift—for example, add accuracy to predictions of retention. This was demonstrated in Tutorial C on insurance fraud and other tutorials with real or "real-like" data. The Retention Data set used here was purposely biased to make "declared major" the determining factor. This variable would probably not be that important in the real world.

Having an accurate handle on whether a prospective college applicant will succeed and stay through to graduation at a college is an important admission office skill. Scoring new applicants by a predictive analytics model and accepting those who have a high accuracy of retention will allow a college to be more efficient and cost-effective, and will avoid wasted time by students dropping out to find another college.

The reader may be interested in a link that has great variables identified for retention and persistence: http://www.act.org/research/policymakers/pdf/college_retention.pdf. The predictive data help to inform university administrators regarding more effective uses of resources, marketing strategies, and admissions guidelines. Even a few points' increase in an institution's retention rate can make a huge difference in revenues. Scoring new applicants by using a predictive analytical model can enable universities to better use resources and assist students and admissions counselors in making college admissions decisions.

TUTORIAL S

Searching for Relationships in Product Recall Data from the Consumer Product Safety Commission with *STATISTICA* Text Miner

Jennifer Thompson
Woodward, OK, USA

CONTENTS

The Consumer Product Safety Commission (CPSC) is the government agency responsible for protecting the public from products that may be potentially harmful. The agency evaluates the risk of injury and even death from the use of consumer products and works with the makers of those products in recalling dangerous ones. The CPSC gives information about product recalls on its website, www.CPSC.gov. This information is a brief synopsis of the recalled products and issues behind the recall. The recall information is broken down by month.

Products can be recalled for a variety of reasons. It is interesting to explore what types of products are recalled and the reasoning behind it. Using the information available on the CPSC website and *STATISTICA* Text Miner, patterns and relationships can be discovered. These relationships can provide insights about the nature of product recalls.

This example is particularly useful in helping to understand the purpose of unstructured text mining. Unstructured text mining is used to find relationships in text to gain understanding. Product recall information is widely available, and the public has a basic understanding of it. We know the main reasons for recalls, which include dangers to children such as choking, suffocation, falling, lead paint, and so on, and dangers to everyone, such as catching fire, among other things. In this example, we will see patterns emerge that are in line with our expectations. This will help you to understand how text mining can be used and how to apply it to your areas of interest.

Recall information is available from the web. Using a *STATISTICA* spreadsheet containing the web addresses, the Text Mining tool can access the text on these pages and process it for analysis.

SPECIFYING THE ANALYSIS

To begin, open the *STATISTICA* spreadsheet, *Product Recall web addresses.sta*. The first column in the data is web URLs for the CPSC website recall pages (see Figure S.1). The dates represented in these pages range from January 2000 to December 2010.

Practical Text Mining and Statistical Analysis for Non-structured Text Data Applications. DOI: 10.1016/B978-0-12-386979-1.00027-X

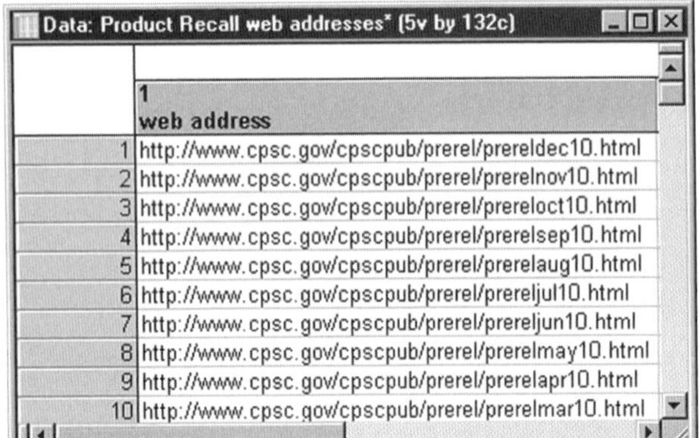

FIGURE S.1
Product Recall web address spreadsheet data file.

From the *Data Mining* menu, select *Text Mining* to open the *Text mining* dialog. In the *Retrieve text from* section, select *Files* and check the option *Paths in spreadsheet* (see Figure S.2).

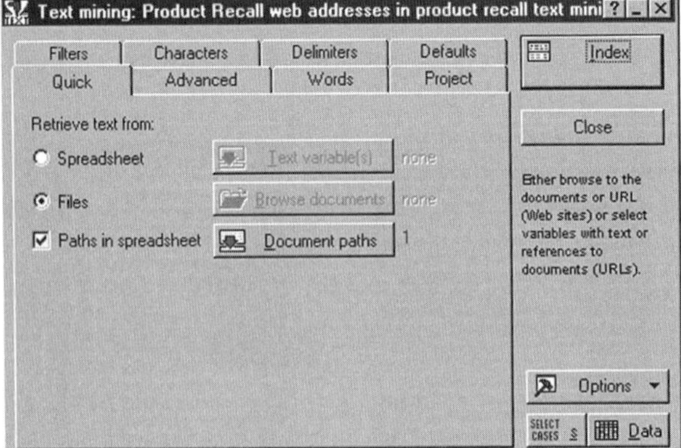

FIGURE S.2
Text Mining dialog window.

Select *Document paths* to open the *Select a variable containing file paths*. Select variable 1 and click OK (see Figure S.3).

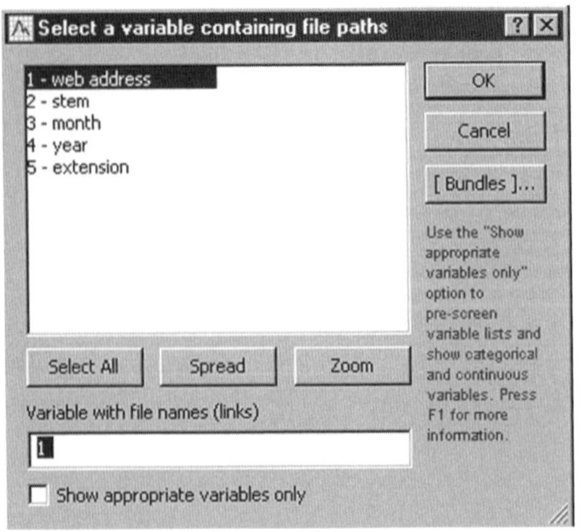

FIGURE S.3
Select variable with file path for web address dialog.

Select the *Advanced* tab. Change the *Reduce the vocabulary of indexed words* option, *% of files where word occurs: Min* and *Max* to 15 and 85, respectively (see Figure S.4). This will exclude the words occurring in most all or only one or two of the pages.

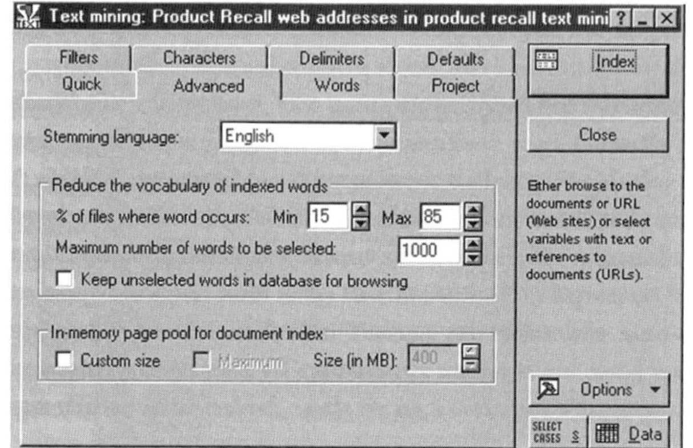

FIGURE S.4

Advanced tab of text mining dialog.

Next, select the *Words* tab and check the *Synonyms* option (see Figure S.5). Add the synonym Root *children* and the *Words childrenâ*, using the *Synonyms* dialog. In several places throughout the product recall page, the word *children's* is inadvertently shown as *childrenâ€™s*. The apostrophe is not rendered properly. In indexing of the words, this mistake is taken to be a separate word when it is not. Specifying the synonym allows us to avoid this issue.

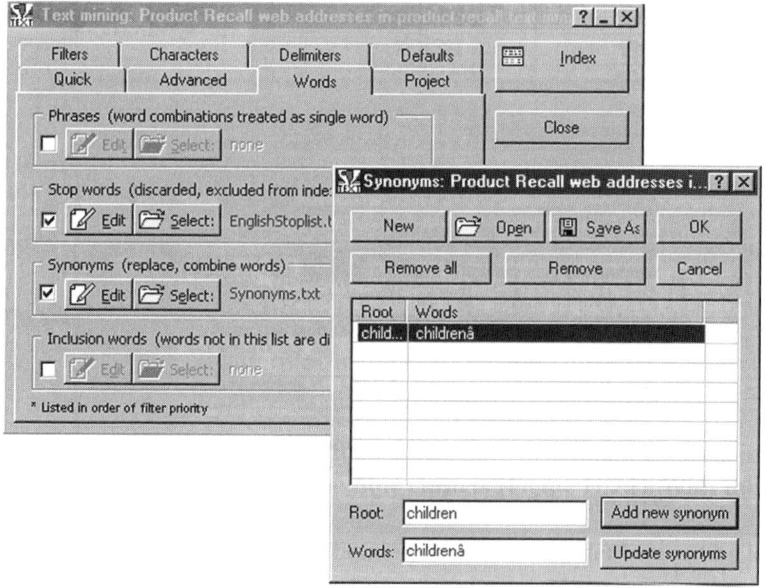

FIGURE S.5

Synonyms window from the "Words" tab of text mining dialog.

Accept all other default settings and select *Index* to begin indexing the web pages.

REVIEWING THE RESULTS

After a brief time of processing, the *Results* dialog is displayed (see Figure S.6).

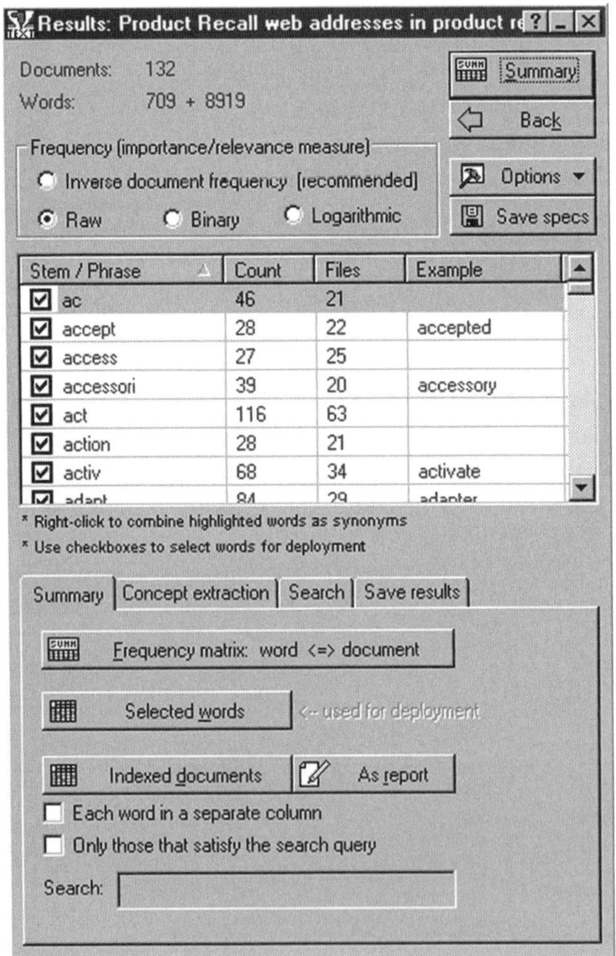

FIGURE S.6
Results dialog.

Click *Frequency matrix: word <=> document*. This creates the *Term document frequency matrix* spreadsheet (see Figure S.7). This spreadsheet gives the frequency of terms on each of the web pages and can be used in other analysis tools for further analysis.

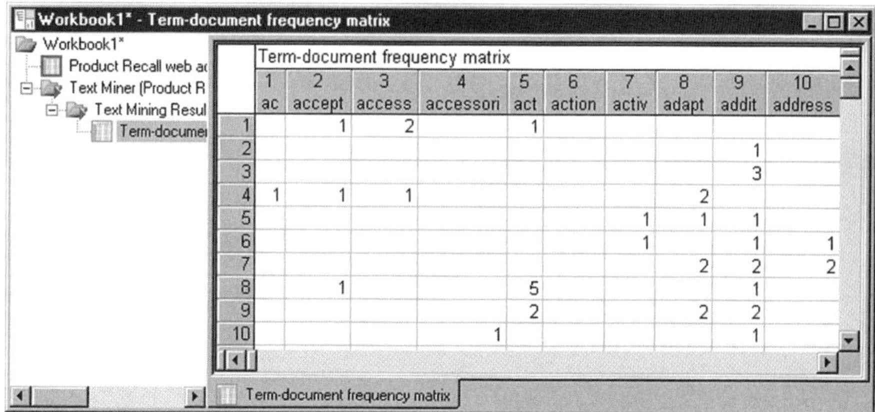

FIGURE S.7
Term document frequency spreadsheet.

Select the *Concept extraction* tab and click *Perform Singular Value Decomposition (SVD)*. After a short time processing, the options on this tab will become active (see Figure S.8).

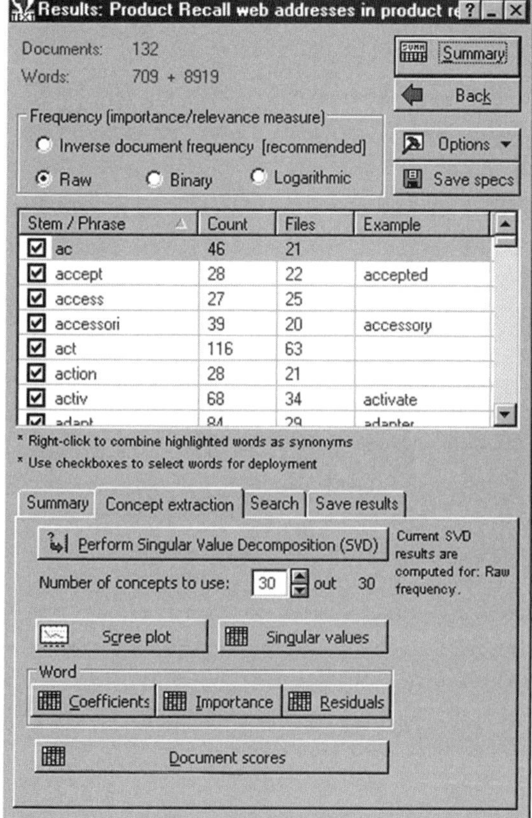

FIGURE S.8
Results of indexing words.

Click *Scree plot* to create the Scree plot of the concepts extracted from the text (see Figure S.9). This plot shows what percent singular value is explained by each of the extracted concepts. With this plot, you can determine which concepts will be most useful in exploring the textual relationships.

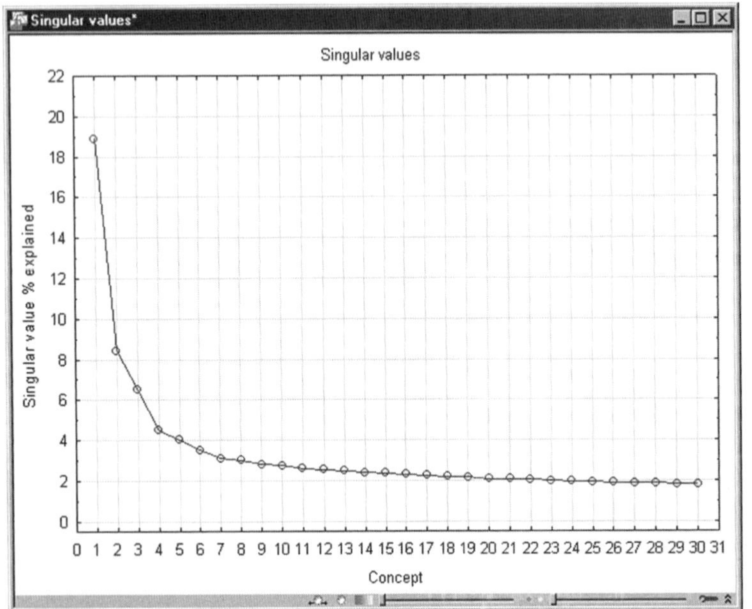

FIGURE S.9
Scree plot of the concepts created by the SVD process.

The scree plot begins to level off after Concept 4, so the main focus should be on these first four concepts.

Select *Importance* to create the SVD importance output. From the *Data* menu, select *Sort* to open the *Sort Options* dialog. Add the variable *Importance* to the sort list and change the direction to *Descending* (see Figure S.10).

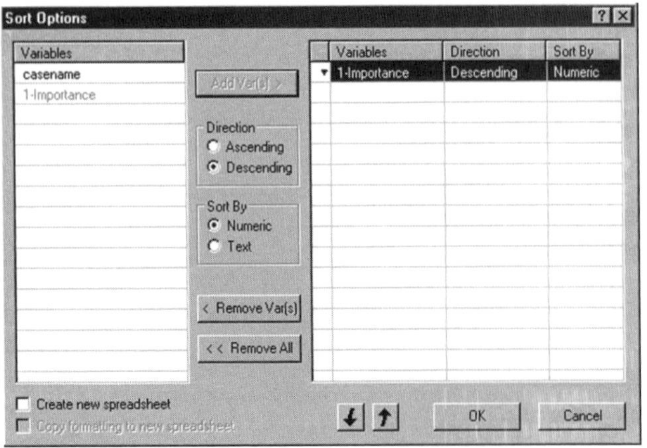

FIGURE S.10
Sort Options dialog window.

Click OK to sort the data. The output shows the words, ranked by importance, that contribute most to the SVD analysis (see Figure S.11). These words are the ones that will show us textual patterns in the data.

Data: SVD word importanc...

	SVD word importance
	Importance
due	100.0000
lead	76.4214
paint	55.0458
standard	41.0946
violat	36.5862
strangul	34.1336
bicycl	29.1695
contain	28.5539
crib	27.7758
drawstr	26.9390
hood	25.3127
feder	25.0889
light	25.0310
level	23.7540
gas	23.7180
batteri	23.3696

FIGURE S.11
SVD word importance spreadsheet.

Select *Coefficients* to create the spreadsheet of output for SVD (see Figure S.12). This output can be used in scatterplots to show relationships in the text. On the *Data* menu, check *Input* in the *Mode* section to allow this output spreadsheet to be used as input for graphical analysis.

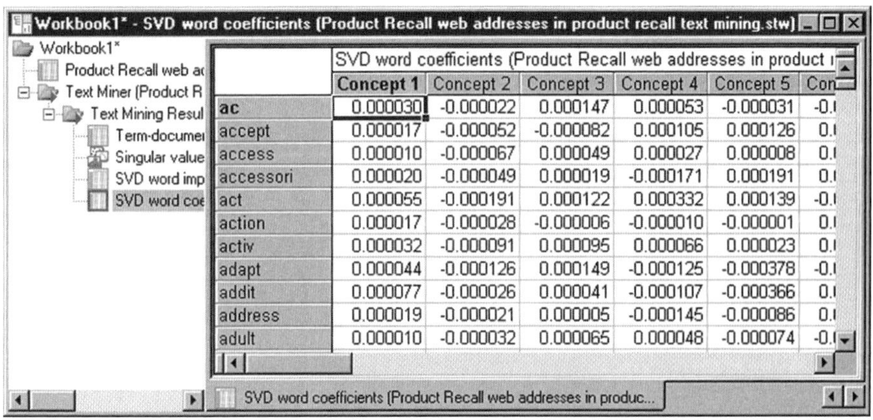
Workbook1* - SVD word coefficients (Product Recall web addresses in product recall text mining.stw)

Workbook1*
Product Recall web ad
Text Miner (Product R
Text Mining Resul
Term-documen
Singular value
SVD word imp
SVD word coe

	SVD word coefficients (Product Recall web addresses in product r					
	Concept 1	Concept 2	Concept 3	Concept 4	Concept 5	Con
ac	0.000030	-0.000022	0.000147	0.000053	-0.000031	-0.(
accept	0.000017	-0.000052	-0.000082	0.000105	0.000126	0.(
access	0.000010	-0.000067	0.000049	0.000027	0.000008	0.(
accessori	0.000020	-0.000049	0.000019	-0.000171	0.000191	0.(
act	0.000055	-0.000191	0.000122	0.000332	0.000139	-0.(
action	0.000017	-0.000028	-0.000006	-0.000010	-0.000001	0.(
activ	0.000032	-0.000091	0.000095	0.000066	0.000023	0.(
adapt	0.000044	-0.000126	0.000149	-0.000125	-0.000378	-0.(
addit	0.000077	-0.000026	0.000041	-0.000107	-0.000366	0.(
address	0.000019	-0.000021	0.000005	-0.000145	-0.000086	0.(
adult	0.000010	-0.000032	0.000065	0.000048	-0.000074	-0.(

SVD word coefficients (Product Recall web addresses in produc...

FIGURE S.12
Workbook of the text mining containing result spreadsheets and graphs. Shown here are the SVD word coefficients.

From the *Graphs* menu, select *2D Scatterplot* to open the *2D Scatterplots* dialog. Select *Variables* to open the *Select Variables for Scatterplot* dialog. Choose *Concept 1* as *X* and *Concept 2* as *Y* (see Figure S.13).

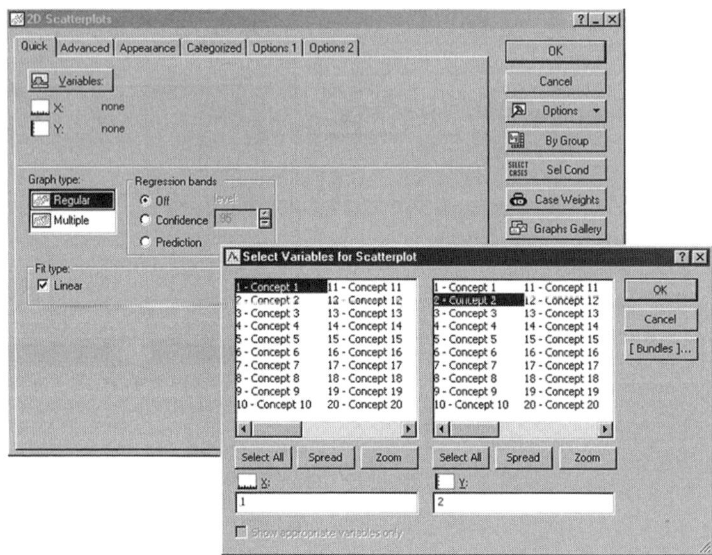

FIGURE S.13
Select variables window for the 2D Scatterplot graph dialog.

Click OK on both the *Select variables for Scatterplot* and *2D Scatterplots* dialogs to create the plot. The resulting plot shows points that represent each of the indexed words from text mining (see Figure S.14).

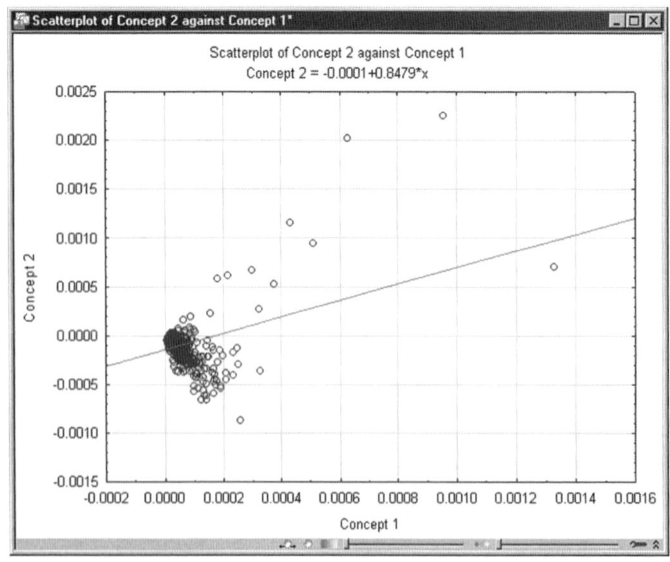

FIGURE S.14
2D Scatterplot of Concept 2 versus Concept 1.

On the *Edit* menu, select the *Brushing* tool in the *Customize Graph* section. The icon is a magnifying glass with crosshairs and is located on the top row on the left (see Figure S.15).

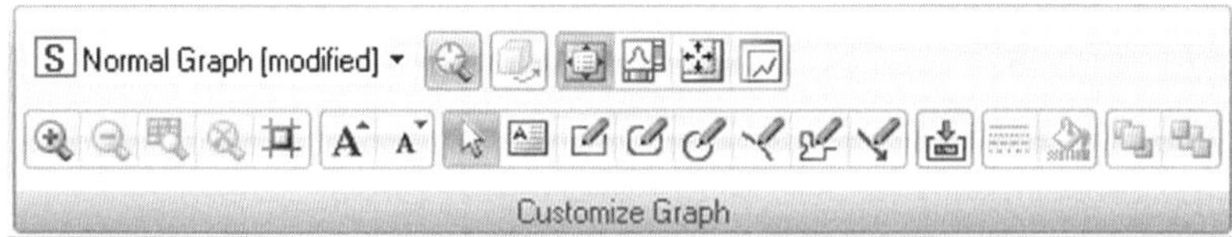

FIGURE S.15
Selecting the Brushing tool from the Edit menu.

This will open the *Brushing 2D* dialog. Change the *Selection Brush* option to *Lasso* and the *Action* to *Label* (see Figure S.16).

FIGURE S.16
Brushing dialog.

On the scatterplot, use the Brushing tool to draw a loop around the points you would like to label with their corresponding text value (see Figure S.17).

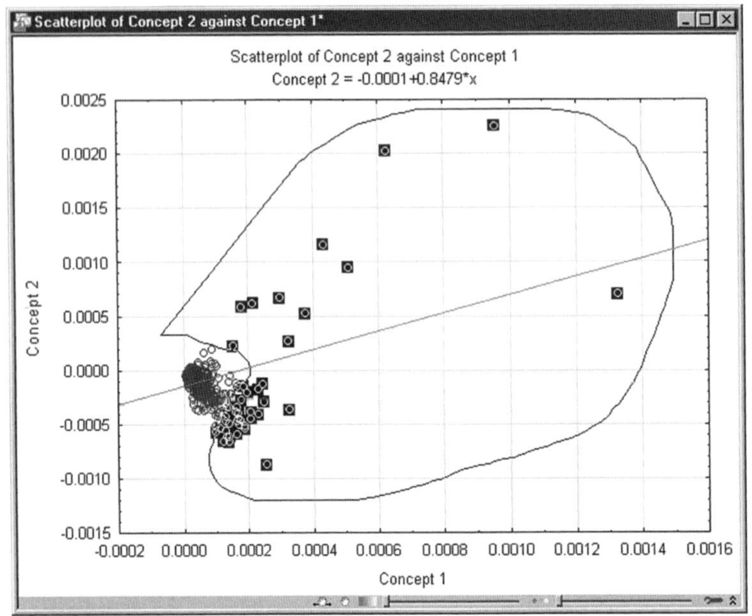

FIGURE S.17
Scatterplot of Concept 2 versus Concept 1.

Then select *Label* to add the text labels (see Figure S.18).

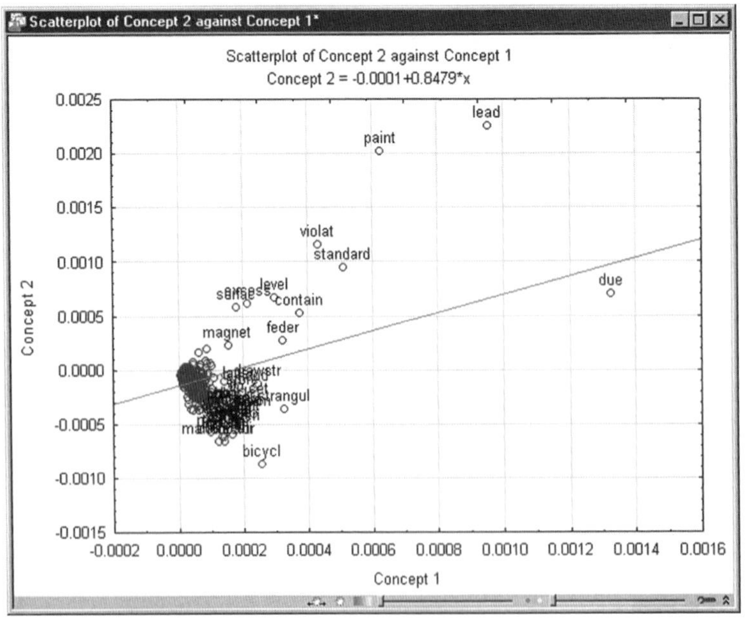

FIGURE S.18
Scatterplot with data points labeled with words.

Right away, a relationship we might expect in product recall information becomes apparent. The words *lead* and *paint* are shown together in the plot, indicating that these words occur together in various monthly recall listings. Also close by are the stem words for *violate, standard, level, contain,* and *federal*. This clustering of terms seems appropriate, since lead paint is a common reason for product recalls.

Using the Zoom tool, also found on the *Edit* menu, we can find other interesting results. The *Zoom* icon is found on the second row on the left: a magnifying glass with a plus symbol.

Looking at a zoomed in portion of the *Concept 1* versus *Concept 3* scatterplot, you can see the pattern resulting from the recall of drop-side cribs related to infant deaths (see Figure S.19).

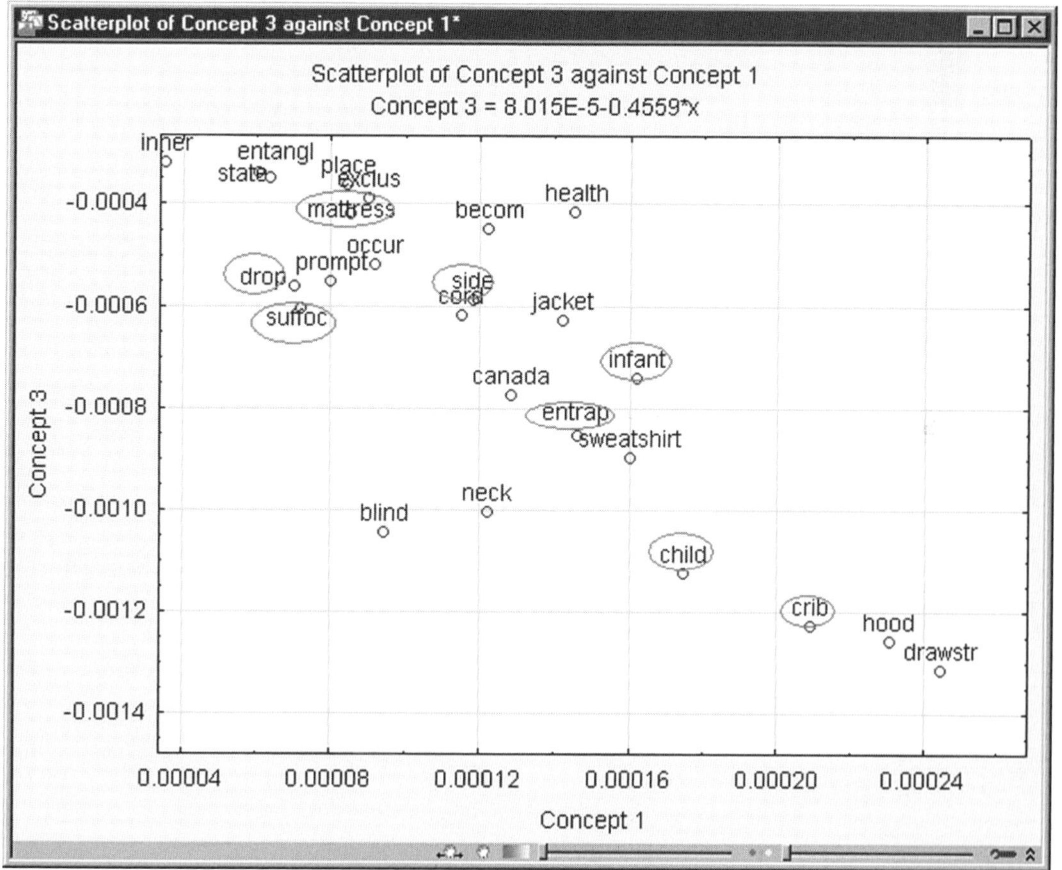

FIGURE S.19

Concept 3 versus Concept 1 plotted and words separated.

In this same plot, words including *blinds* and *cord* were seen with *neck* and *strangle* (see Figure S.20).

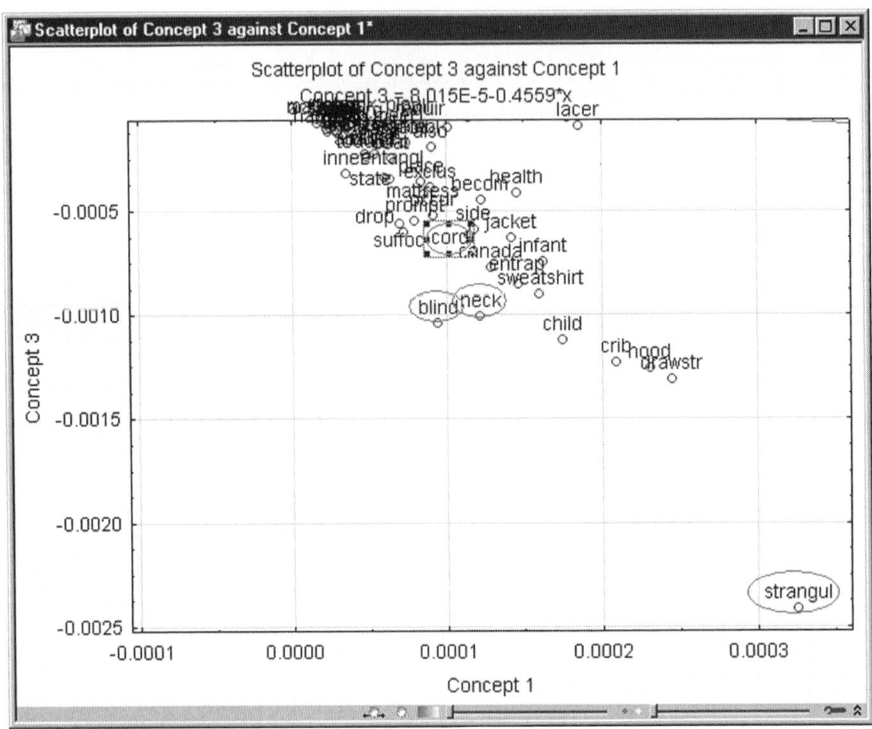

FIGURE S.20

Concept 3 versus Concept 1 graphed with some words separated.

Continued exploration of this text mining project could reveal more patterns in the product recall information.

Potential Problems That Can Arise in Text Mining: Example Using NALL Aviation Data

Jennifer Hunter Thompson

CONTENTS

INTRODUCTION

When working with unstructured text, a variety of issues can arise. Some of these issues are more easily fixed than others. Awareness is always the first step to a resolution. This tutorial will explore some potential issues you may run into while exploring unstructured text. Suggestions for solutions will also be explored.

SPELLING ERRORS

Some sources of text may present an issue of spelling errors in the text. This will be particularly true when text comes from a web form where spell-check is not available or emails where slang may be used. Even website content has the potential for spelling errors. Most professional reports will not present this issue.

A good text mining tool will be able to accurately detect and index spelling mistakes automatically. Depending on the spelling error made, text mining tools may not detect a misspelled word as the same as the correct word. *STATISTICA* Text Miner actually has the ability to discern many misspellings correctly. At times the misspelling may be so great or may be so similar to a different word that the tool

Practical Text Mining and Statistical Analysis for Non-structured Text Data Applications. DOI: 10.1016/B978-0-12-386979-1.00028-1

cannot correctly classify it. So when looking at the indexed text, you may see multiple spellings of what is meant to be one word. For example, *calendar* is a commonly misspelled word. When misspelled, it is typically written *calander* or *calender*.

Examining the indexed words is one way to find these spelling mistakes. Additionally, subject matter experts likely have an idea of the most common spelling mistakes they will encounter. Even before the text mining project begins, a list may be available to ensure correct indexing of misspellings. This misspelling information can be used as a synonym list to properly combine words with their misspellings.

EXAMPLE: FINDING SPELLING ERRORS IN TEXT MINER

The Excel data, DW review.xls (see Figure T.1), contains free-form text reviews of various models of dishwashers from a variety of manufacturers. The text review variable comes from an online review forum from the www.viewpoints.com website.

Some of the manufacturer brands are commonly misspelled. Some of these misspellings still allow for proper indexing of terms. For instance, *Whirlpool* is a dishwasher manufacturer found in this data set. One of the reviews spells the name *Wirlpool*. This spelling error is seen in red (will be colored "red" on the computer screen) in the following review.

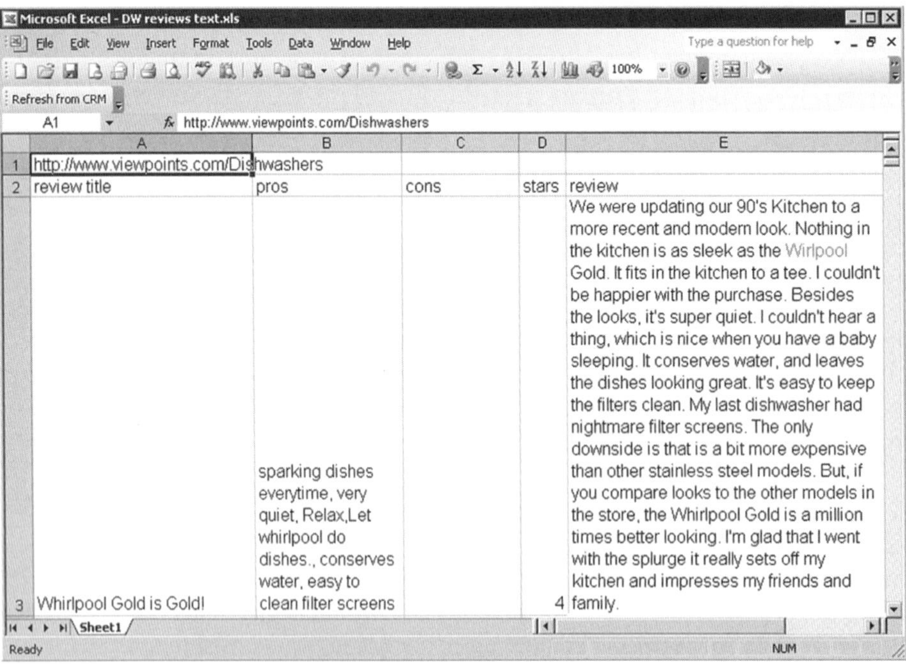

FIGURE T.1.

Microsoft Excel spreadsheet used for this tutorial, "DW reviews text.xls".

Another spelling mistake occurs with the manufacturer *Bosch* (see Figure T.2). One reviewer writes *Bosh*, as seen in red (will be colored "red" on the computer screen) in the following review.

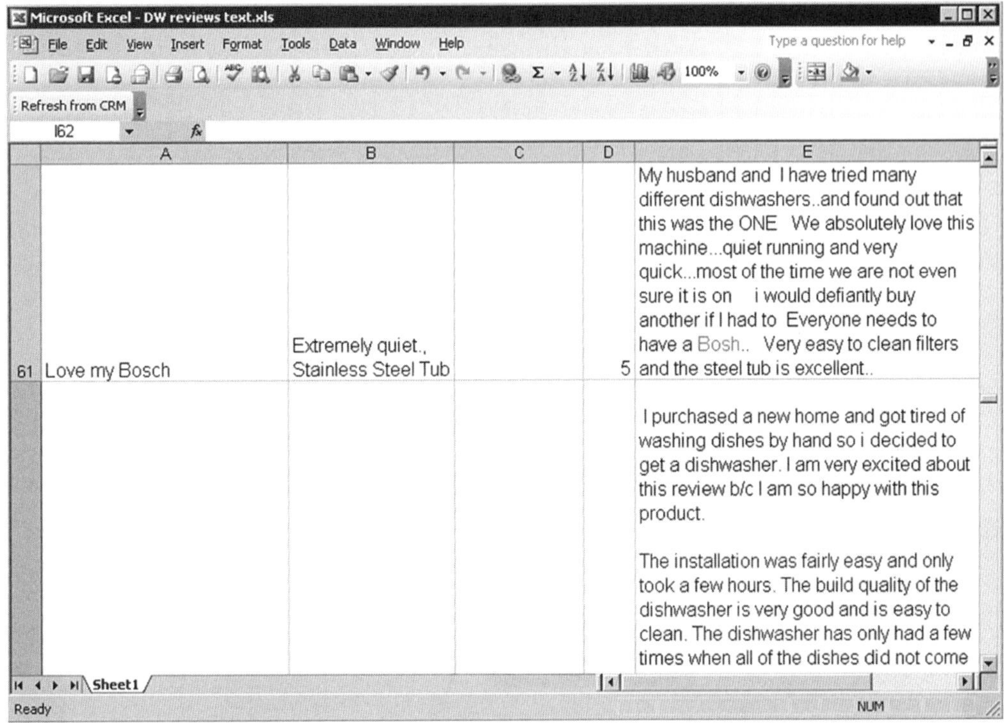

FIGURE T.2.
DW review text.xls showing that spelling mistakes occurs with the manufacturer Bosch.

STATISTICA Text Miner is able to detect some spelling errors as the proper term. After indexing the documents, a brief review will reveal spelling errors that were not combined.

The data are stored in an Excel file. In *STATISTICA*, on the *Home* menu, select *Open* (see Figure T.3) to open the *Open* dialog.

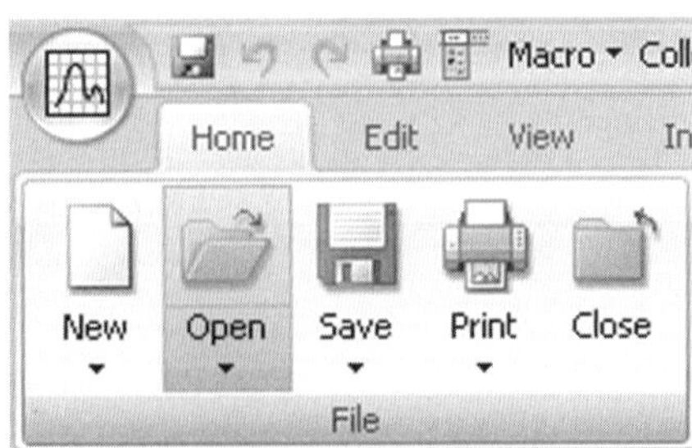

FIGURE T.3.
Selecting 'Open' from the "Home" tab on the ribbon bar format of STATISTICA.

Browse to the file, *DW reviews.xls*, and click *Open* (see Figure T.4).

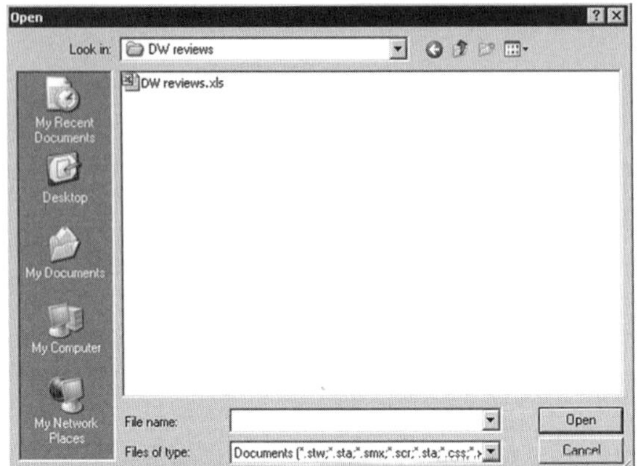

FIGURE T.4.
Open window where the dataset desired can be selected.

The *Opening file* dialog gives options for how to import the Excel file into *STATISTICA* (see Figure T.5).

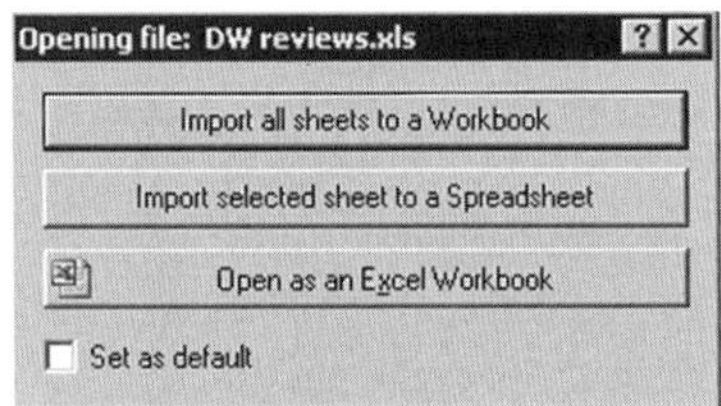

FIGURE T.5.
Opening files dialog giving options on how to import a spreadsheet.

Select *Open as an Excel Workbook* (see Figure T.5). This will open the file as an Excel document (see Figure T.6), giving access to both *STATISTICA* data analysis and graphing tools, as well as Excel tools. Notice in the screen shot below, the Excel file is open in the *STATISTICA* application. The menu items *Statistics, Data Mining,* and *Graphs* are those from *STATISTICA*. The other menu items belong to Excel.

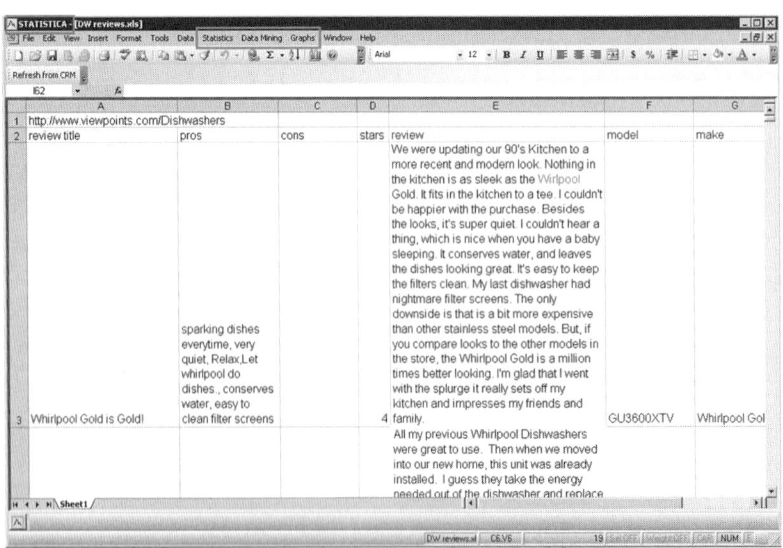

FIGURE T.6.
An Excel Workbook opened within the STATISTICA application.

From the Data Mining menu, select Text & Document Mining (see Figure T.7) to open the Select Excel range for the analysis dialog.

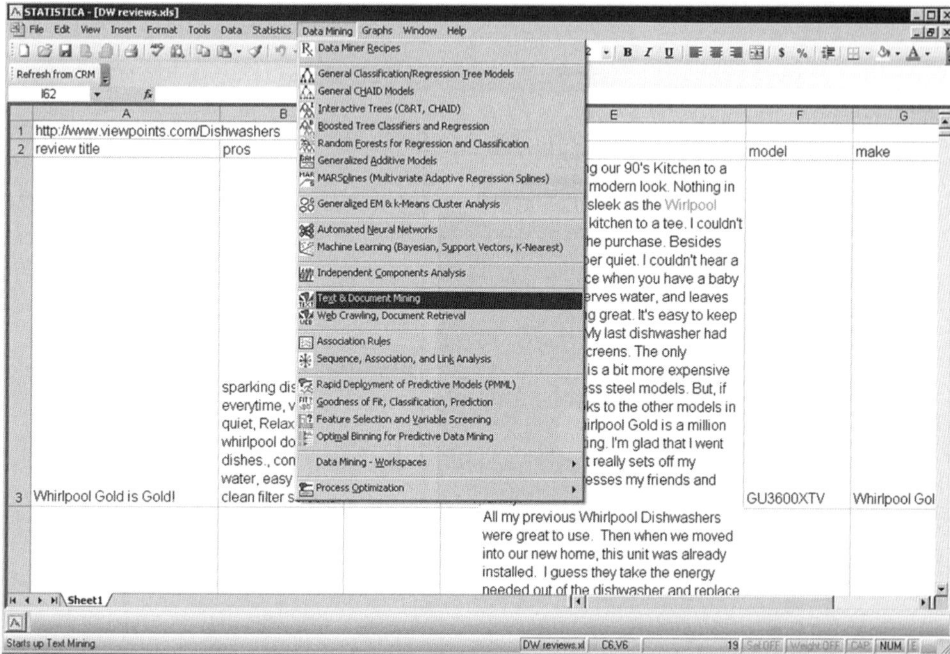

FIGURE T.7.
Selecting the 'Text & Document Mining' from the 'Data Miner' pull down menu of STATISTICA.

Before analysis can begin, the data range must be specified. The data start in row 3 and column A. Variable names are found in row 2. Make these specifications as in the following screen shot (see Figure T.8).

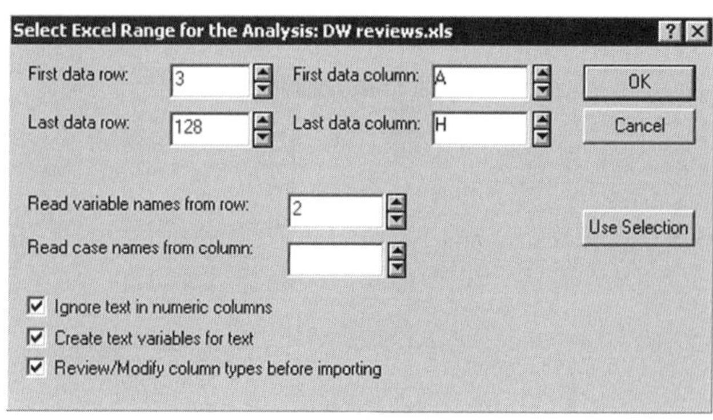

FIGURE T.8.
The 'Select Excel Range for the Analysis' dialog.

Click OK to open the *Review/Edit Column Types* dialog (see Figure T.9). The column types are appropriate for each variable. Click OK to proceed with *Text Mining*.

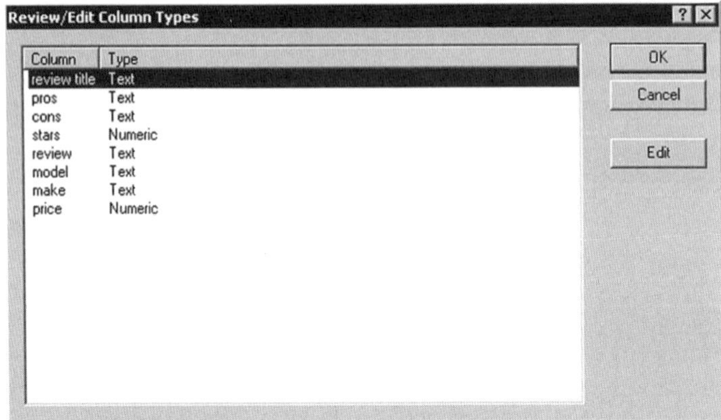

FIGURE T.9.
The 'Review / Edit Column Types' dialog.

On the Text Mining dialog, Quick tab, select Text variable(s) to open the Select variables containing text dialog. Select review as the Variable with text (to analyze) (see Figure T.10). Click OK.

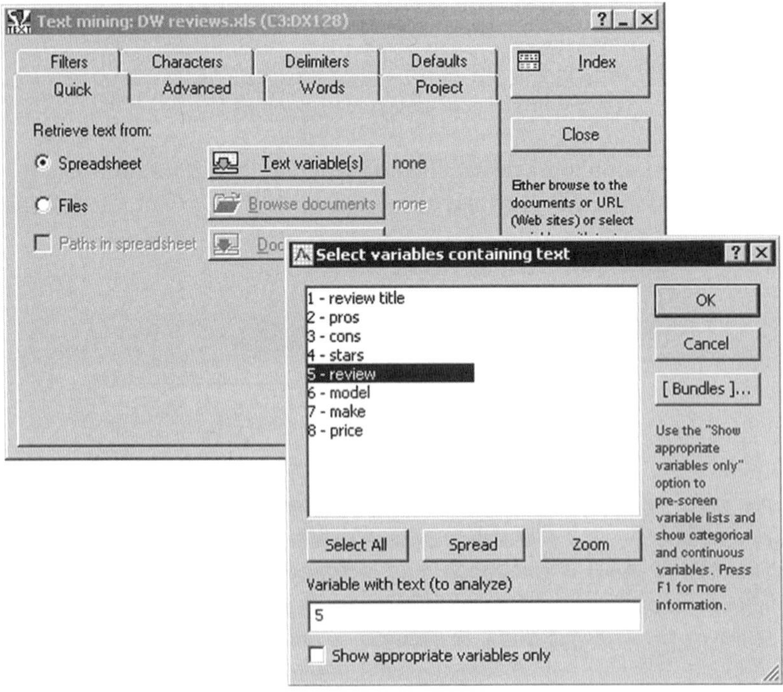

FIGURE T.10.
Quick Tab of the text mining dialog, with 'Text Variables' button selected so that variables for analysis can be selected.

Select *Index* to accept all default settings and begin indexing the text. The document indexing will complete quickly and display the *Results* dialog (see Figure T.11). Scroll down through the *Stem/Phrase* list to find *Bosch* and *Bosh*.

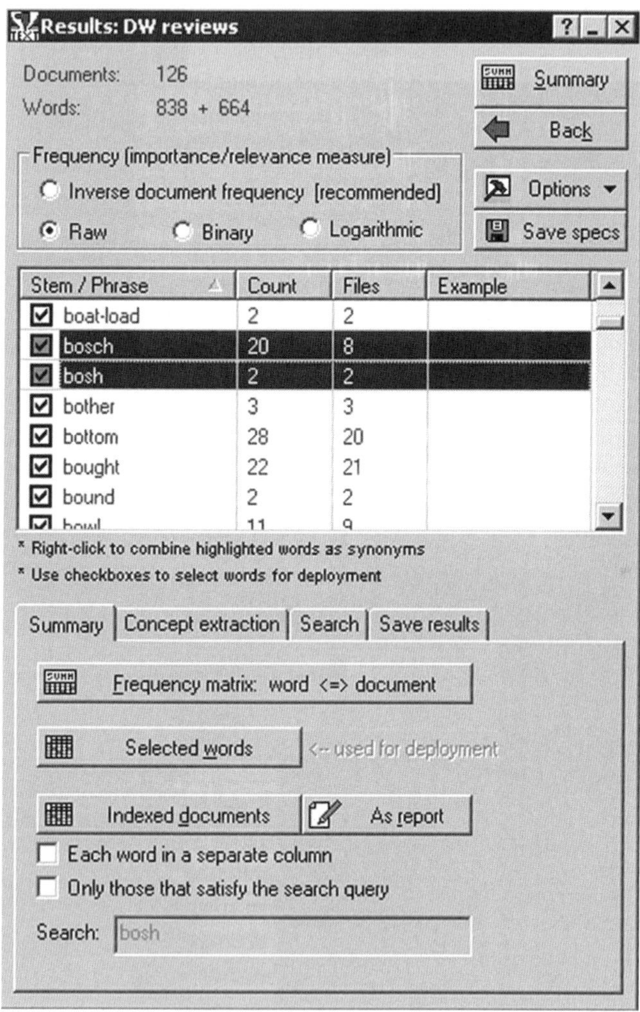

FIGURE T.11.
Results of text mining dialog, following the indexing of frequencies of the words.

In this particular study, the term *Bosh* is a misspelling of *Bosch*. These two terms should be combined. Two options exist for combining these terms into one. On the *Results* dialog, you can *combine words*. If the misspelling will occur often in your *Text Mining* projects, it likely makes more sense to add these to the *Synonyms list*.

Before combining these terms, let's take a look at the other known spelling error in this example. *Whirlpool* was also written as *Wirlpool*. As seen in the following results, this spelling error was correctly indexed as the intended term *Whirlpool* (see Figure T.12). (The next term after *wipe* is *wish*. *Wirlpool* would come just before it, if present.) There is no need to do anything for proper indexing of this term.

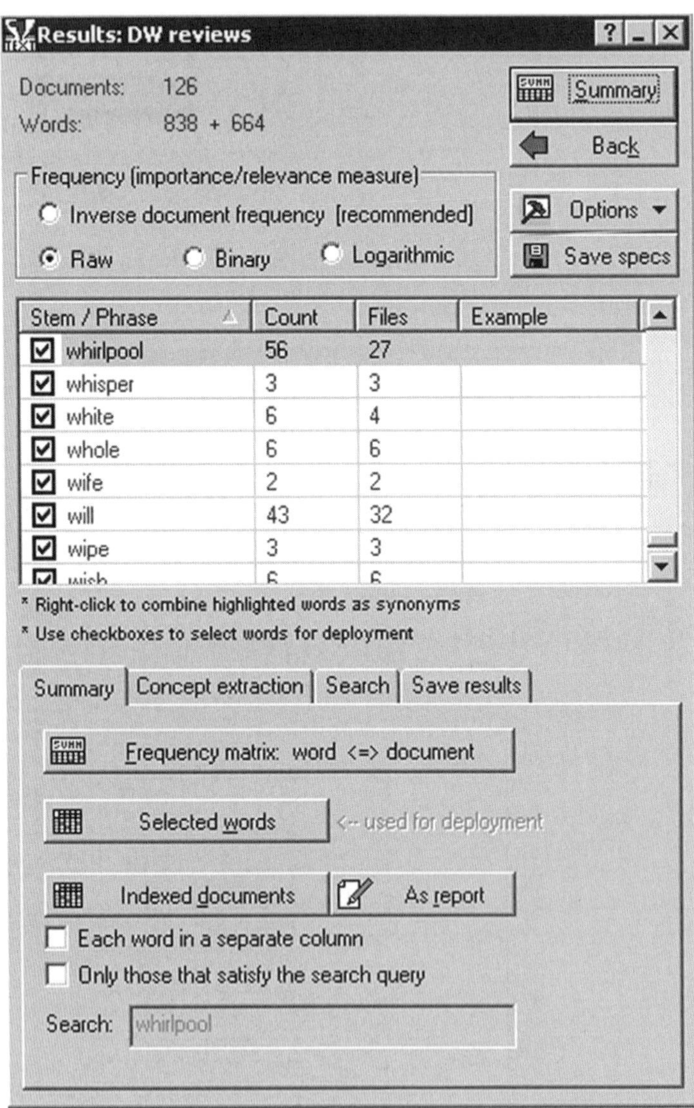

FIGURE T.12.
Results dialog showing that 'whirlpool' and 'wirlpool' were automatically recognized as the same word.

COMBINE WORDS

Highlight the terms *Bosch* and *Bosh*. Then right click and select *Combine words* (see Figure T.13) to combine the words and continue to move forward with the analysis.

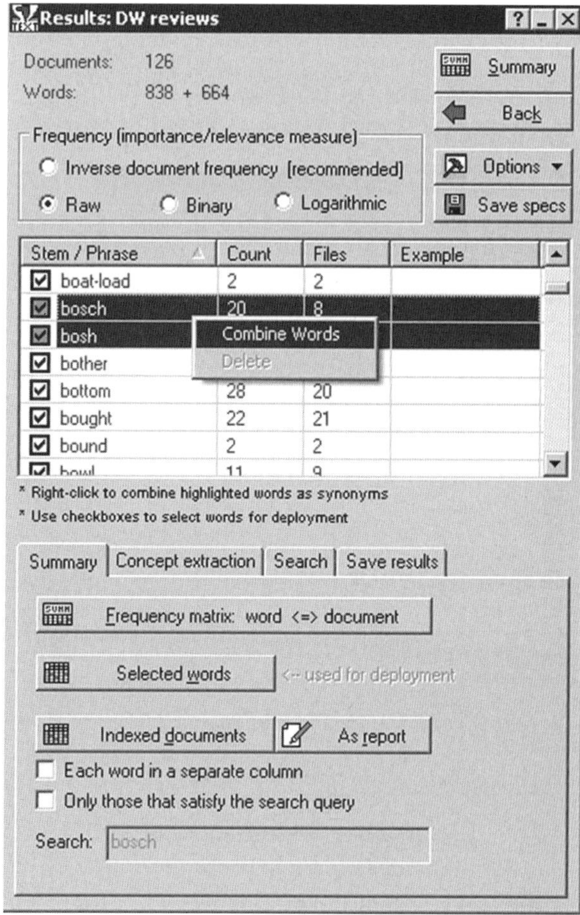

FIGURE T.13.
Right-clicking on the 2 highlighted words allows the combination of these words to be seen as the same.

On the *Combine words* dialog, select the correct spelling *Bosch* in the field, and select one of the highlighted words (see Figure T.14). Click OK to complete the operation.

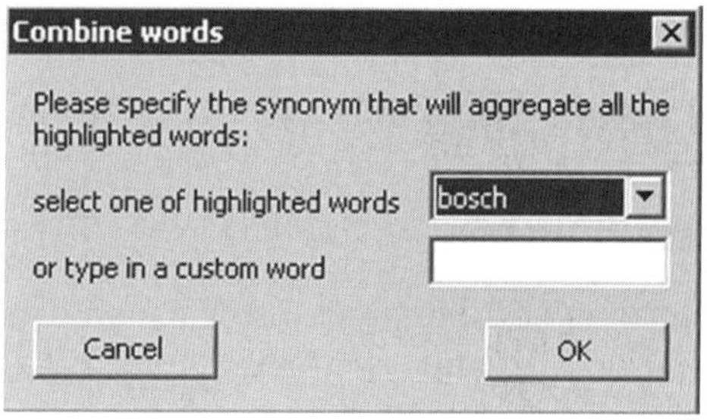

Combine words
Please specify the synonym that will aggregate all the highlighted words:
select one of highlighted words bosch
or type in a custom word
Cancel
OK

FIGURE T.14.
Combine words dialog.

The *Results* dialog will update, combining these two terms (see Figure T.15). Before, *Bosch* was found 20 times in 8 reviews. *Bosh* was found twice in 2 reviews. Combined, the terms are found 22 times in 9 reviews.

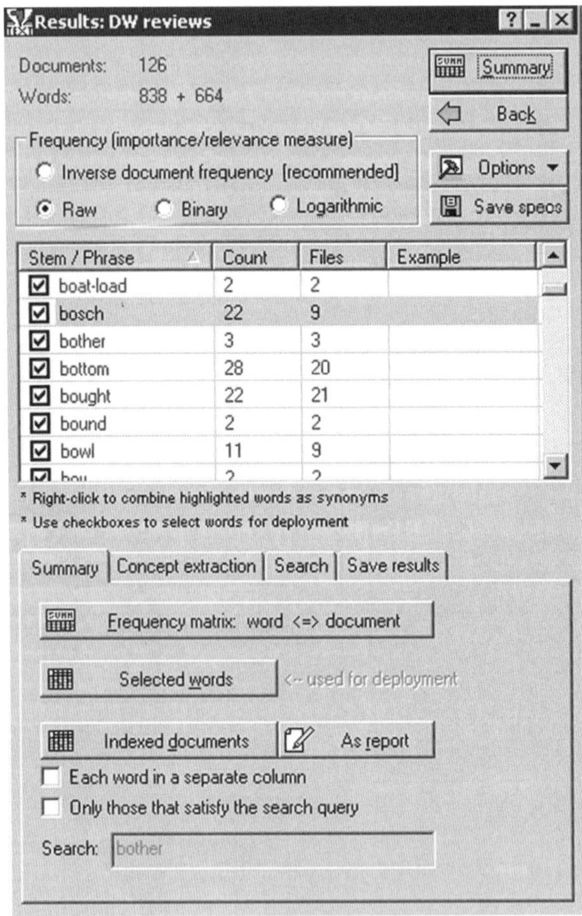

FIGURE T.15.
Results dialog updated with 'bosch' and its misspelled 'bosh' combined into one word, 'bosch'.

This is a quick way to combine words and move forward with the analysis.

MISSPELLINGS AS SYNONYMS

If the same issue will come up again in future analysis, using a synonyms list for common misspellings can save a lot of time. Click *Back* on the *Results* dialog to return to the *Text Mining* dialog. Select the *Words* tab (see Figure T.16).

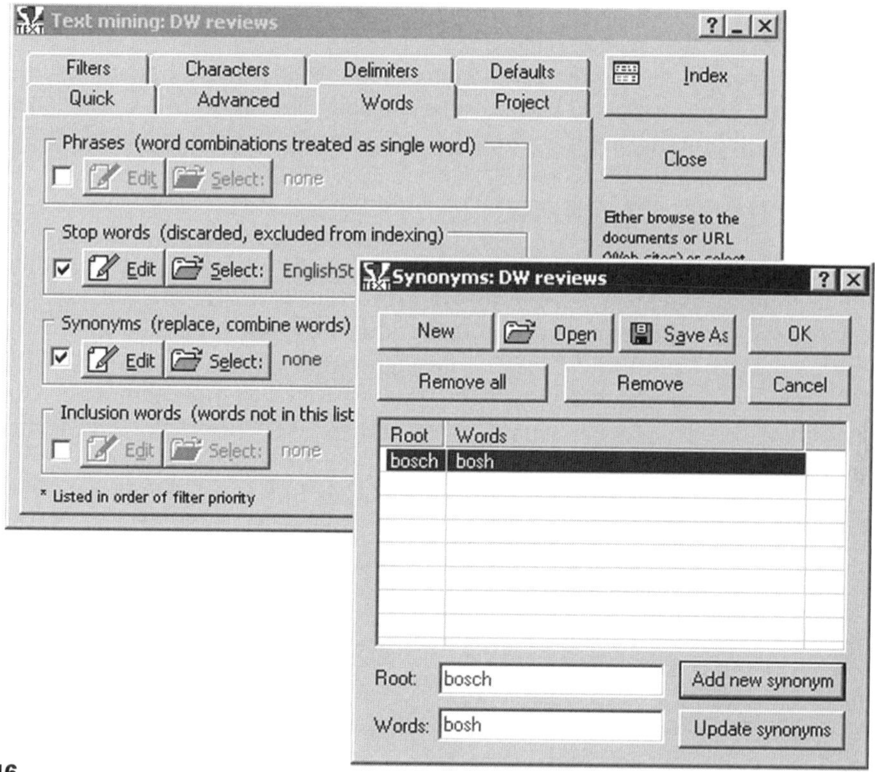

FIGURE T.16.
Adding words to a 'Synonyms list'.

Check the *Synonyms (replace, combine words)* option and select *Edit* to open the *Synonyms* dialog (see Figure T.16). The words selected previously to combine are automatically entered in the synonyms list, since they were combined as one word on the results dialog. Additional synonyms can be entered at this time. Click *Save As* to save the synonyms file for future use (see Figure T.16).

UNEXPECTED TERMS

While reviewing the indexed terms, you may find some unexpected text. This may be words that you don't expect or character strings that are not words. When this happens, it may be helpful to explore possible causes of the unexpected text. Or it may be sufficient to simply exclude the terms that are not relevant. The unexpected terms can be excluded from analysis using the *Exclusion* list.

Published papers will often give figures and tables a title and description. One author may write out *Figure,* while another will abbreviate it as *Fig.* This may lead to the terms *Figure* and *Fig.* having high importance in the singular value decomposition analysis. The terms actually do not have anything to do with the actual content of the paper, but more to do with the publisher or author. These types of terms can be found by reviewing the indexed terms and the importance output of SVD.

Text mining of web content may show some unexpected terms coming from the website structure. Often web pages have a set of links like *Home, Contact Us, Login, Sign in, Register*, and so on. These terms are not related to the content of the web pages but are terms likely to be encountered on web pages.

No matter the source of unexpected terms, understanding their cause or developing a plan to work with those terms is an important step in gaining the most insight from unstructured text information.

EXAMPLE: FINDING UNEXPECTED TERMS

The Air Safety Institute releases annual reports called NALL reports. These reports outline trends in aviation accidents. The reports are available in PDF form from their website, aopa.org. These reports outline what happened in a given year in aviation accidents. We are interested in finding patterns in these reports that might give additional insight into flight safety trends.

While the reports are given in PDF format, say that we were given the reports in text file format. From the *Data Mining* menu, select *Text Mining* to open the *Text Mining* dialog (see Figure T.17).

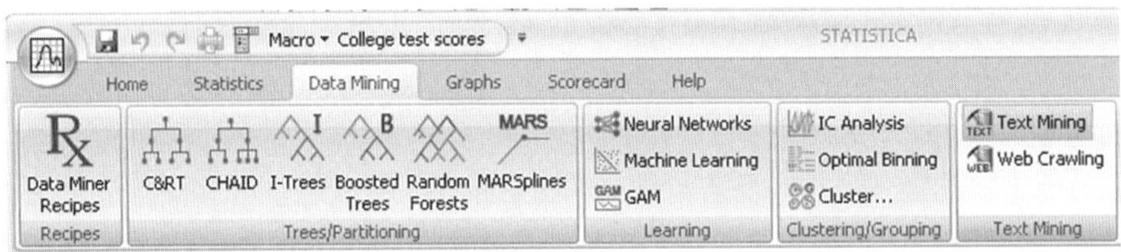

FIGURE T.17.
Data Mining ribbon bar, where 'Text Mining' can be selected in the upper right hand corner.

On the *Quick* tab, select the *Files* option and click *Browse documents* to open the *Open document files* dialog (see Figure T.18).

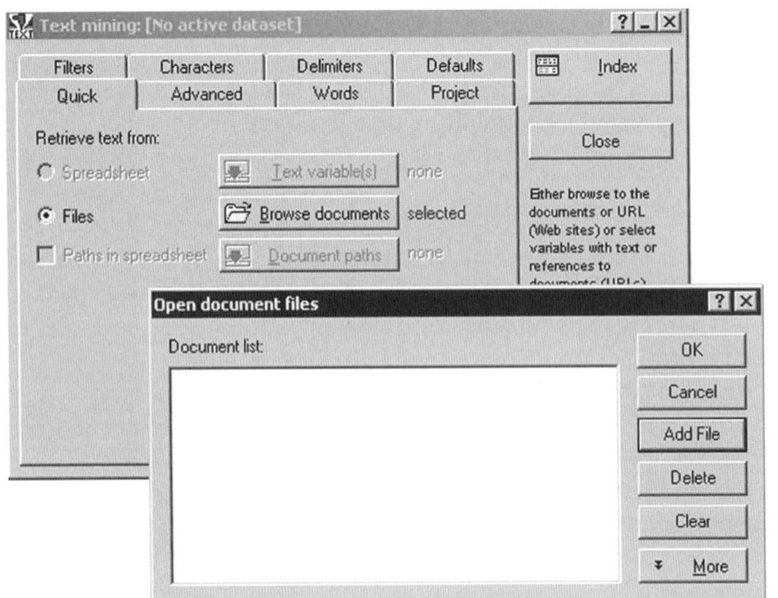

FIGURE T.17.
Open Document Files window obtained from Quick Tab and the selecting Files and selecting the button "Browse documents".

Select *Add file* (see Figure T.17) to open the *Open document files* dialog (see Figure T.18).

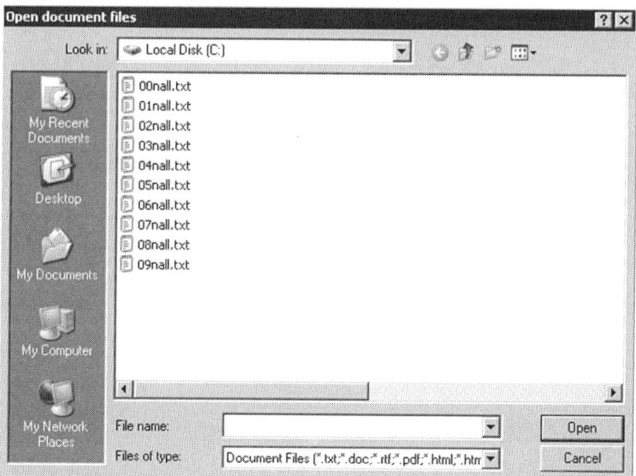

FIGURE T.18.
Open document files dialog.

Browse to the NALL report Text Files. Select all 10 files and click *Open* (see Figure T.19).

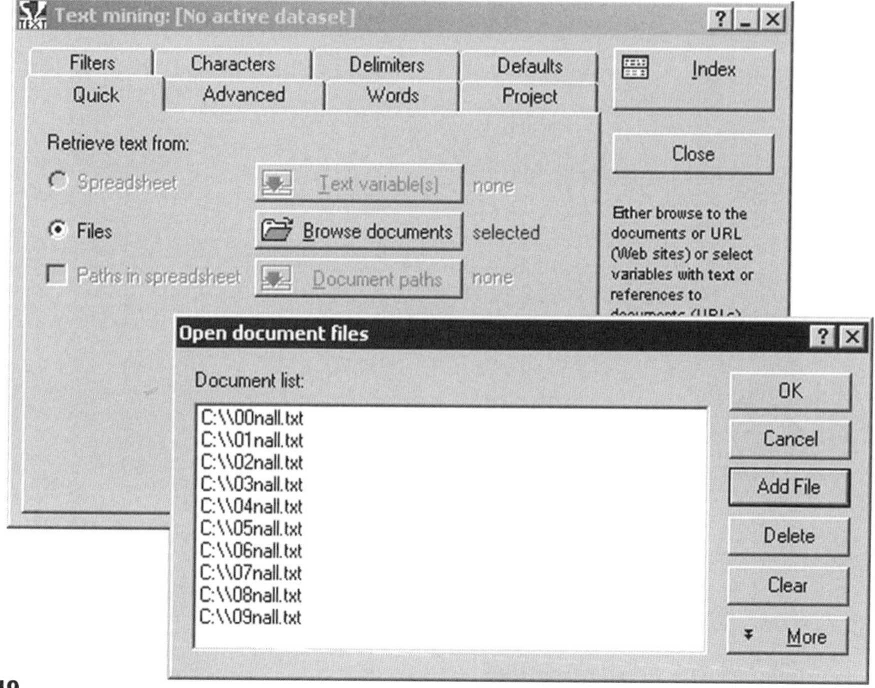

FIGURE T.19.
NALL report text files; all 10 will be selected.

The *Document list* is updated on the *Open document files* dialog. Click OK to continue.

Accept the default settings and click *Index* to begin indexing the 10 text files. When indexing completes, the *Results* dialog is displayed (see Figure T.20). Scroll through the indexed words to find *ing*.

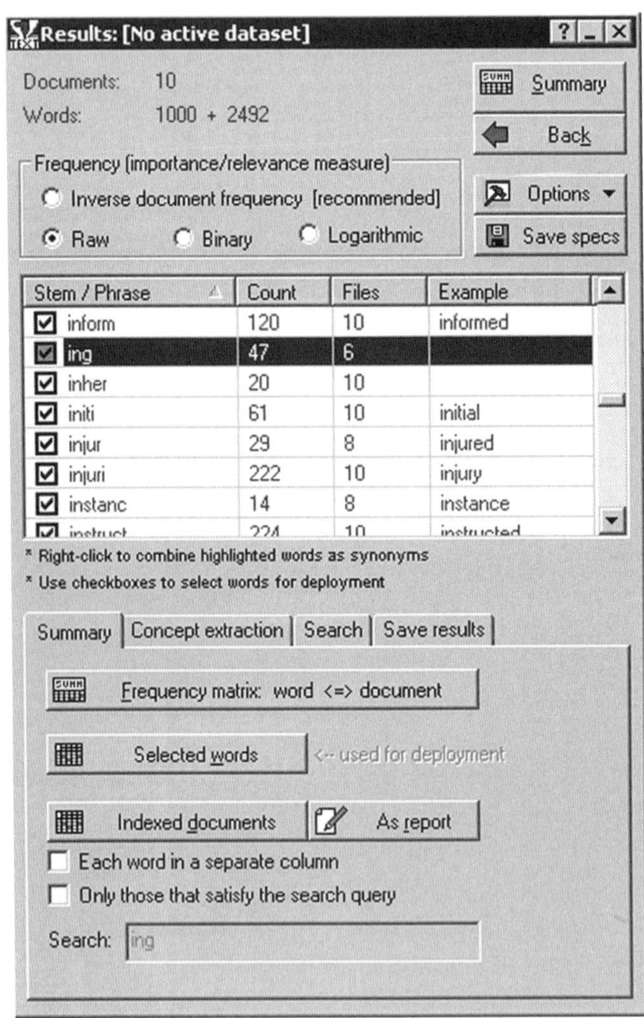

FIGURE T.20.
Results of indexing the words, with 'ing' highlighted.

This term *ing* is not a word and not something that we would expect to find in this study. However, the text string occurs 47 times throughout 6 files.

Next scroll down to the term *tion* (see Figure T.21).

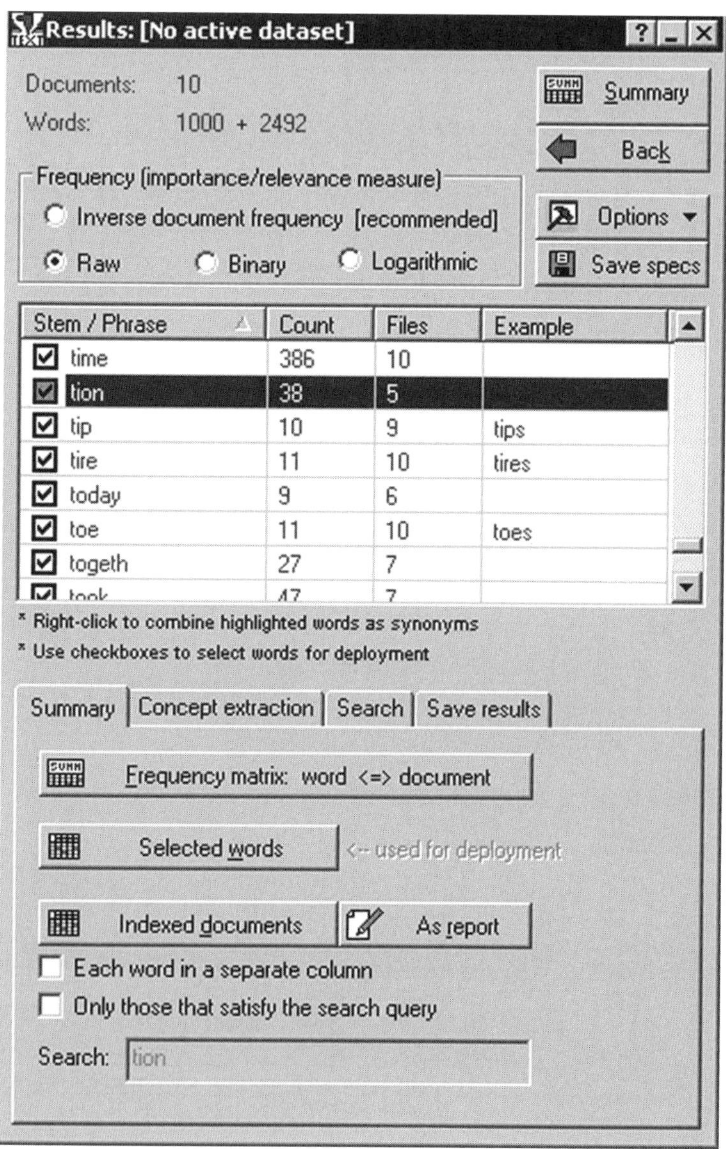

FIGURE T.21.
Results of indexing words with 'tion' selected.

This is another nonword, unexpected text string. This one occurs 38 times across 5 documents. Select the *Search* tab and type in *ing* (see Figure T.22) into the *Words*: field.

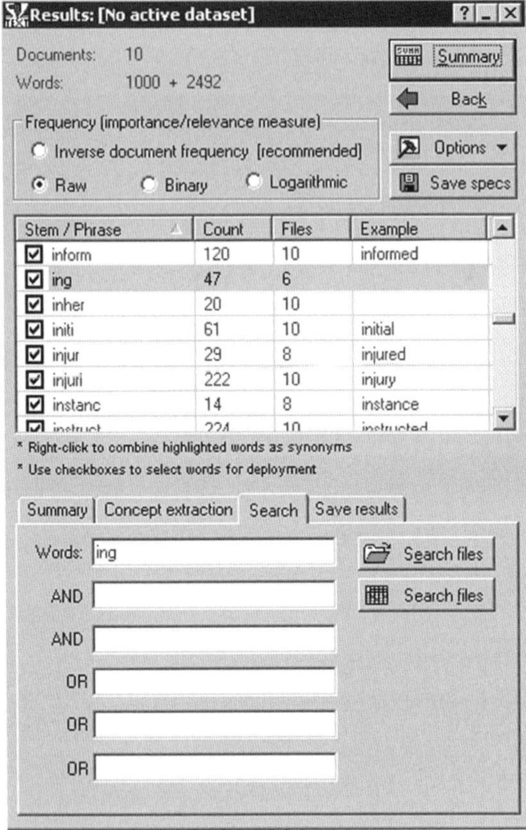

FIGURE T.22.

The 'ing' is put into the "words" list.

Click the *Search files* button (see Figure T.22) with the spreadsheet icon to produce a spreadsheet (see Figure T.23) listing the 6 files containing the text string *ing*.

Data: File search summary ([No active dataset])*

File search summary ([No active dataset])

	Document length	Number of words	Indexed text
00nall.txt	90625	77	ıe sometim three year effort provid current safe
01nall.txt	139675	76	aviat aircraft weigh less pound provid current sa
02nall.txt	94751	79	ear report dedic memori joe nall ntsb board me
07nall.txt	84238	78	:tor dedic joseph nall report aopa air safeti foun
08nall.txt	131542	79	pa air safeti foundat annual review general aviat
09nall.txt	196924	78	ıal review general aviat aircraft accid occur prev

FIGURE T.23.

File search summary spreadsheet listing the files that contain the text string 'ing'.

The file *00nall.txt* contains this text string. Open this text file and use the *Find* tool (see Figure T.24) to find the text string *ing*.

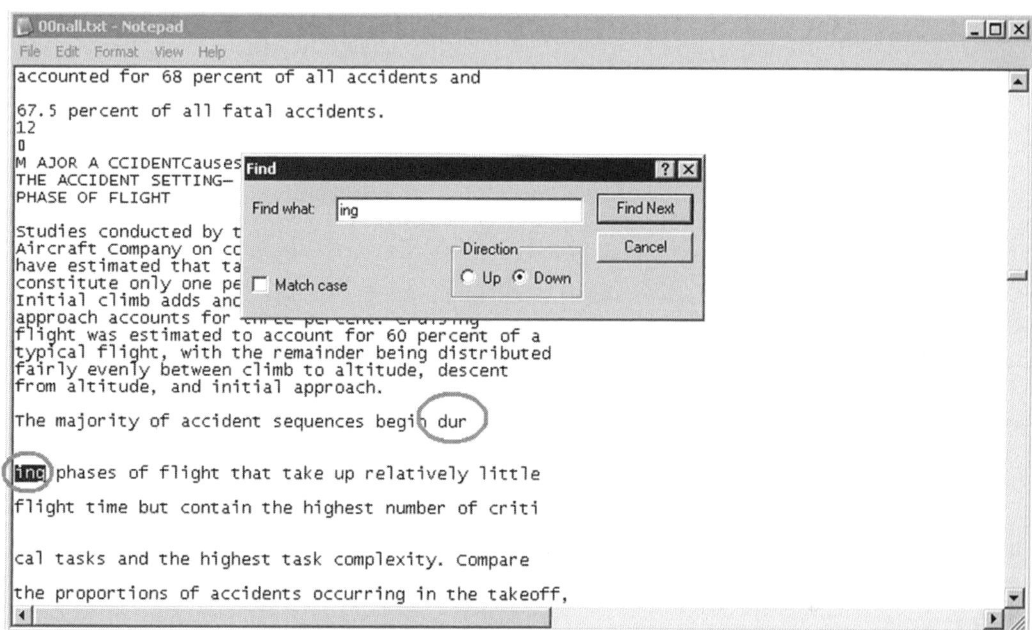

FIGURE T.24.
Using the 'Find tool' of Notepad, for finding the text string 'ing' in a text file.

This reveals the source of the issue. The word *during* is split between two lines. In the text file, the second half of this word is recognized as a different term, where the two pieces actually make one term.

Looking a few lines down, the term *critical* is also split in this way. The way the document was written presents a challenge for indexing the text. Typically when words are split between lines, the stem of the word will be given before the split. If this is the case, the terms will be correctly indexed. Terms like *ing, tion,* and so on can be ignored by unselecting them or listing them on the exclusion list.

On the *Text Mining Results* dialog, find the terms *ing* and *tion*. Uncheck the box next to these terms to unselect them for analysis (see Figure T.25). They will now be ignored.

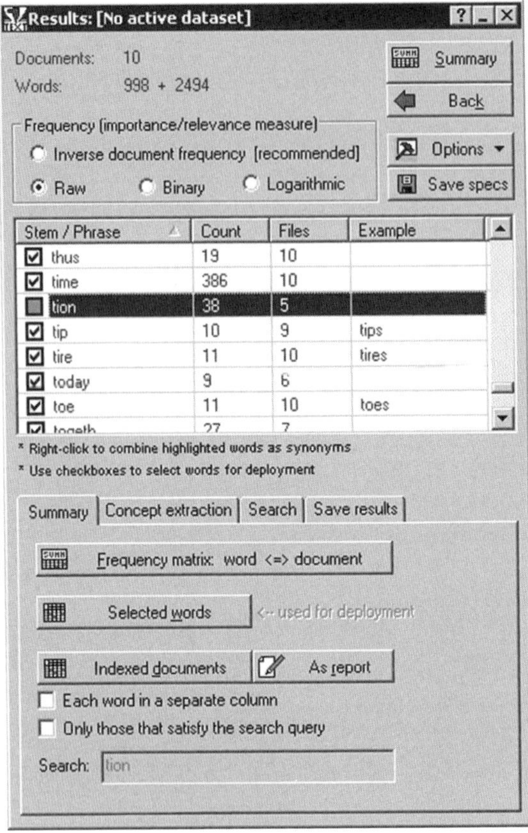

FIGURE T.25.
Un-ckecking the "checkbox" next to 'tion' so that this term will be ignored in future analyses.

If these types of terms will occur often in analysis, adding them to the *Exclusion* list will be most efficient. Click *Back* to return to the *Text Mining* dialog. Select the *Words* tab (see Figure T.26).

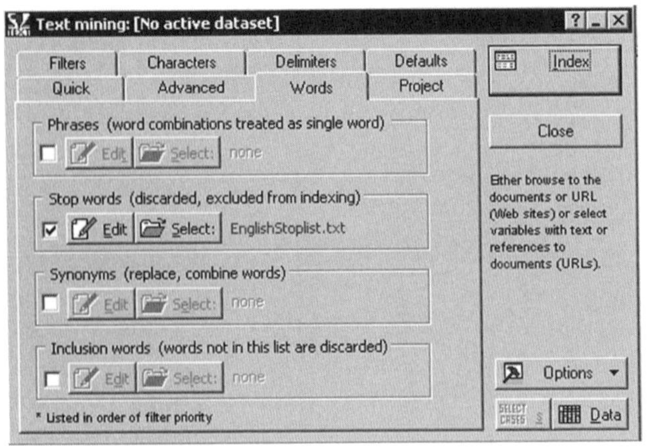

FIGURE T.26.
Word tab where the 'Stop words' is checked to allow adding words to the Exclusion list.

On the *Quick* tab, select the *Files* option and click *Browse documents* to open the *Open document files* dialog (see Figure T.31).

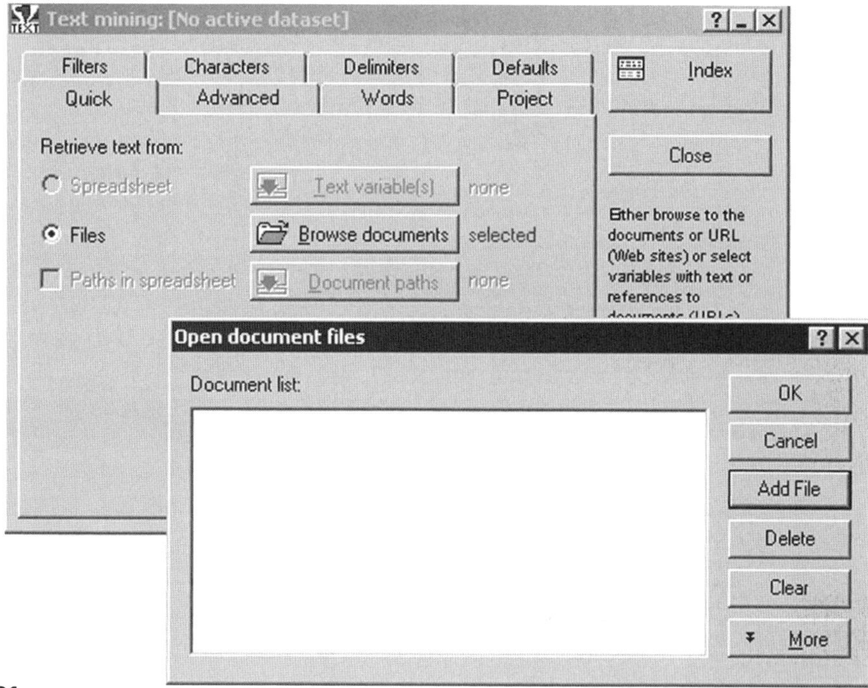

FIGURE T.31.
Starting a new text mining analysis by finding new documents by clicking on the button called "Add File" in the "Open document files" window.

Select *Add file* to open the *Open document files* dialog (see Figure T.32).

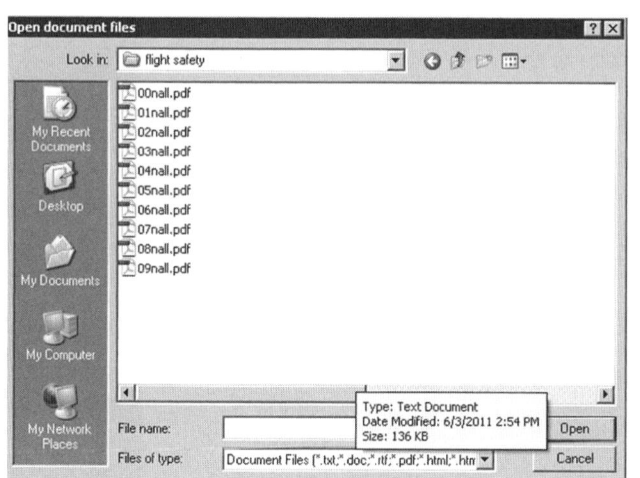

FIGURE T.32.
Finding specific pdf files to submit to a new text mining analysis.

Browse to the nall report PDF Files. Select all 10 files and click Open (see Figure T.33).

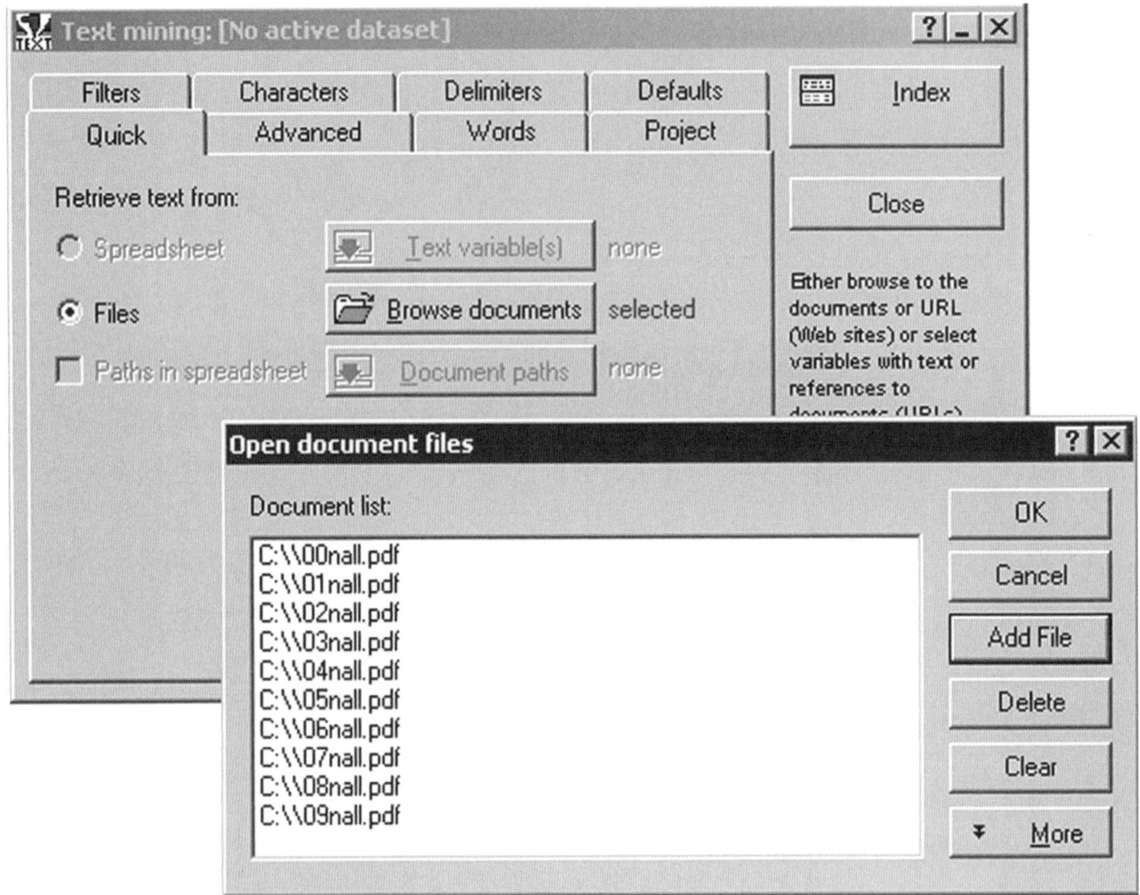

FIGURE T.33.
Select all of the 10 'nall.pdf' files and then click "OK" to add to the analysis.

The *Document list* is updated on the *Open document files* dialog. Click OK to continue.

Accept the default settings and click *Index* to begin indexing the 10 text files. When indexing completes, the *Results* dialog is displayed (see Figure T.34). Scroll through the indexed words to find *ing*.

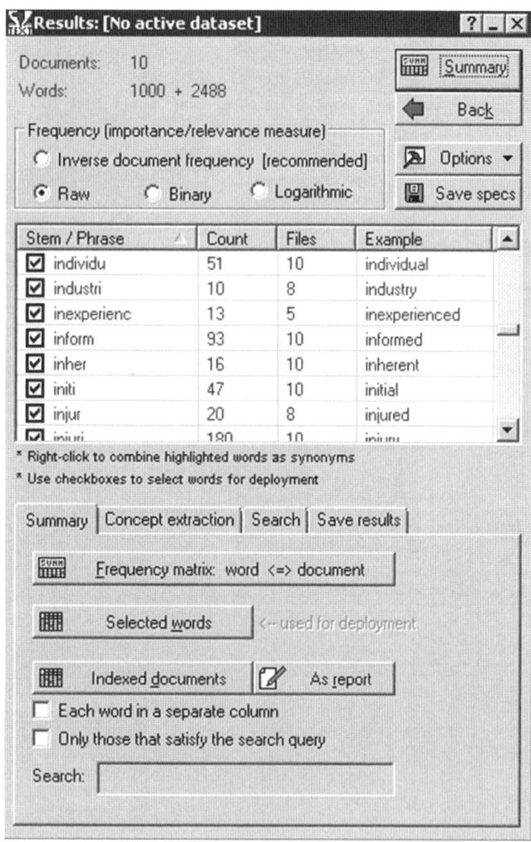

FIGURE T.34.
Results dialog of this new text mining analysis.

The term *ing* is not present in the indexed text for the PDF NALL reports. The PDF files correctly associate the word *during*, which was split between 2 lines, as one term.

This exercise illustrates that we must keep in mind the quality of the materials available. Understanding the source of the data can aid in the analysis. It can help in decision making for cleaning the results.

SUMMARY

The potential hidden in unstructured text can be immense. That potential far outweighs any snags you may run into while processing the text. Typically, effort put into cleaning the results, handling various issues such as spelling differences or unexpected terms, can pay off.

Exploring the Unabomber Manifesto Using Text Miner

Jennifer Thompson
Woodward, OK

Susan L. Trumbetta, PhD
Vassar College

Gary D. Miner, PhD
StatSoft, Inc.

Irving I. Gottesman, PhD
University of Minnesota-Bernstein Professor in Adult Psychiatry and Senior Fellow, Department of Psychology, University of Minnesota & Sherrell J. Aston Professor of Psychology Emeritus, University of Virginia

CONTENTS

INTRODUCTION

The written communications of individuals implicated in criminal activity sometimes provide clues for law enforcement and may lead to an arrest. This was true in the case of Theodore "Ted" Kaczynski, also known as the Unabomber. Over a 20-year span Kaczynski mailed 16 packages containing bombs. These mail bombs killed 3 and injured 23 people. He was a well-educated child prodigy who was accepted to Harvard University at the age of 16. After earning an undergraduate degree there, he studied mathematics at the University of Michigan, earning a PhD.

As an adult, he moved to a secluded cabin in Montana, where he began his terror campaign, mailing bombs to universities and airlines as a response to the destruction of the wilderness around him. On April 24, 1995, Kaczynski sent a letter to the *New York Times*. The letter promised "to desist from terrorism" if his manifesto, "Industrial Society and Its Future," was published in either the *Times* or the *Washington Post*. The manifesto's publication led to Kaczynski's arrest, when family members recognized his unique language patterns.

Practical Text Mining and Statistical Analysis for Non-structured Text Data Applications. DOI: 10.1016/B978-0-12-386979-1.00029-3

All of this occurred through human recognition, and yet, the more objective method of text mining may give law enforcement additional empirical clues, which may help to reveal important aspects of the mental state of a wanted individual. The Unabomber text is nearly 35,000 words and states Kaczynski's views on society and strategies for the future. The intention of the document is to persuade others to his line of thinking. For comparison, we used a set of *New York Times* opinion editorials from the same year. These editorials are also examples of writings intended to persuade, typically written by individuals as highly educated as Kaczynski and usually represent similarly strong beliefs or convictions.

The Unabomber manifesto, TK_manifesto.doc, and the following *New York Times* opinion editorials: *NYT OP-ED Abroad at Home, Back to McCarthy.doc; NYT OP-ED Abroad at Home, Bare Ruined Choirs.doc; NYT OP-ED at Home Abroad, The New South Africa.doc; NYT OP-ED Batter Up, Already.doc; NYT OP-ED Get Rid of Corporate Welfare.doc; NYT OP-ED Greenspan's Error.doc; NYT OP-ED Let Affirmative Action Die.doc; NYT OP-ED Life and Liberty.doc; NYT OP-ED New York's desperate hours.doc; NYT OP-ED Observer, Art, Debt, Slumber.doc; NYT OP-ED Observer, Eight Easy Pieces.doc; NYT OP-ED on My Mind.doc; NYT OP-ED on My Mind, Cover-Up Chronology.doc; NYT OP-ED on My Mind, Jihad in America.doc, NYT OP-ED on My Mind, The Document of May 30.doc; NYT OP-ED Preparing Students for the World of Jobs.doc; NYT OP-ED Return to Saigon, Tet 1995.doc; NYT OP-ED Take the Lead, Mr. Clinton. doc; and NYT OP-ED The Good Earth Looks Better.doc,* can be indexed using *STATISTICA* Text Miner. The text mining tool can be used to find the most frequently used words. This information then can be used to gain a basic overview of the themes of the text without reading all 64 pages of the complete manifesto. Additionally, the patterns of text usage can be compared across the set of persuasive writings.

SUMMARIZING THE TEXT

From the *STATISTICA Data Mining* menu, select *Text Mining* (see Figure U.1) to open the *Text Mining* dialog (see Figure U.2).

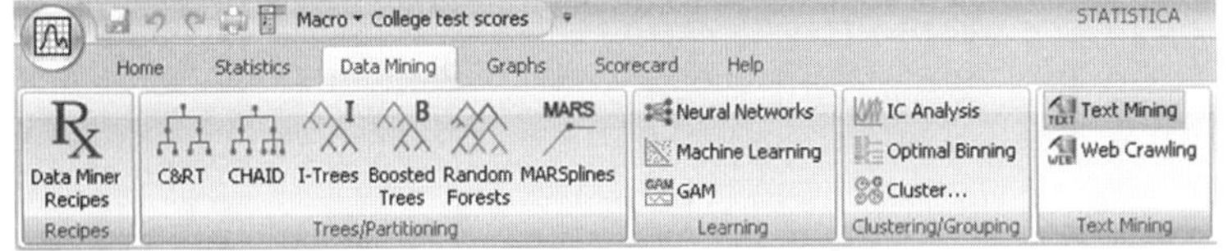

FIGURE U.1
The ribbon bar data mining menu.

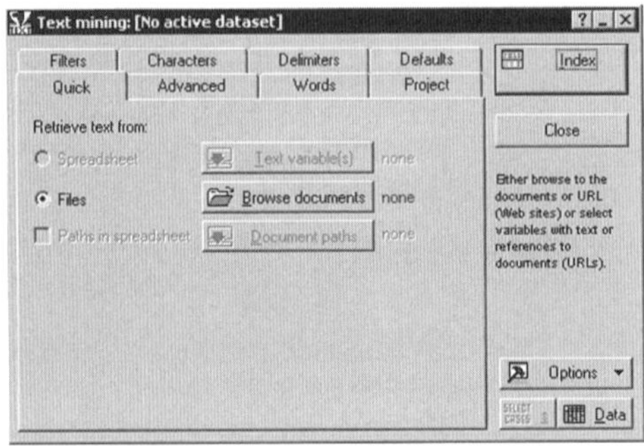

FIGURE U.2
The Text Mining dialog in STATISTICA.

On the *Quick* tab, select *Browse documents* to open the *Open document files* dialog.

Select *Add Files* to open the *Open document files* dialog (see Figure U.3).

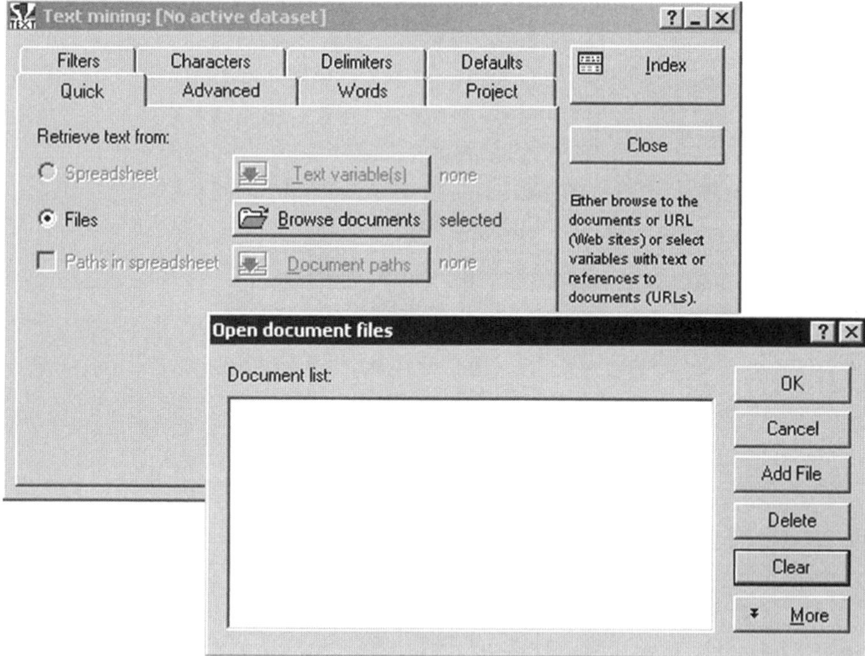

FIGURE U.3
The OPEN DOCUMENTS FILES dialog.

Browse to the *TK_manifesto.doc* file and select *Open* (see Figure U.4).

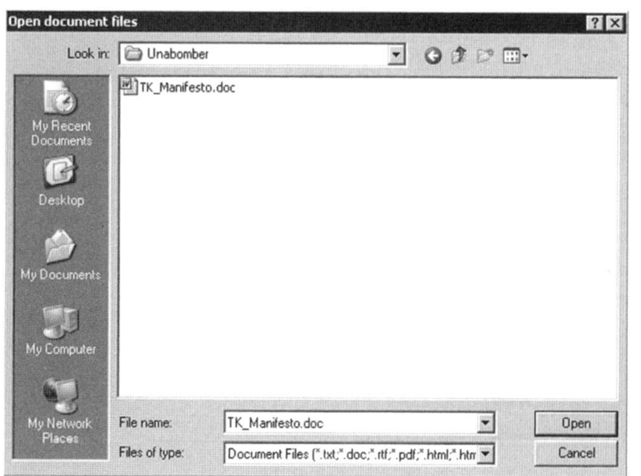

FIGURE U.4
Selecting the text document for the text mining process.

The *Open document files* dialog will update with the selected document (see Figure U.5).

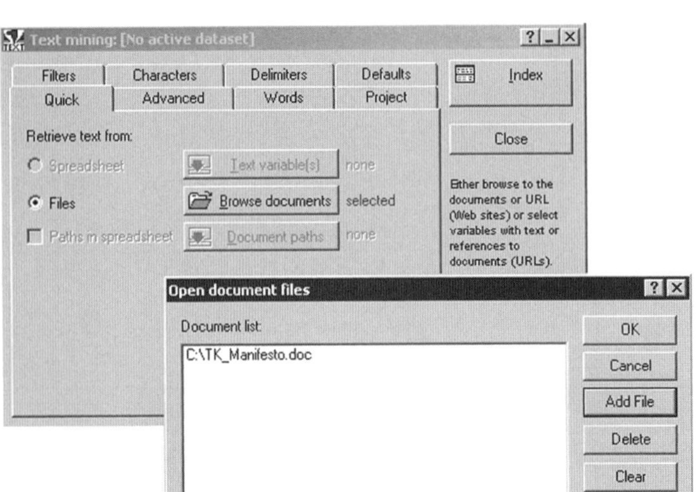

FIGURE U.5
The document of interest is showing in the Open Document File dialog.

Select OK on the *Open document files* dialog and then *Index* on the *Text mining* dialog. After the document is indexed, the *Results* dialog is displayed (see Figure U.6).

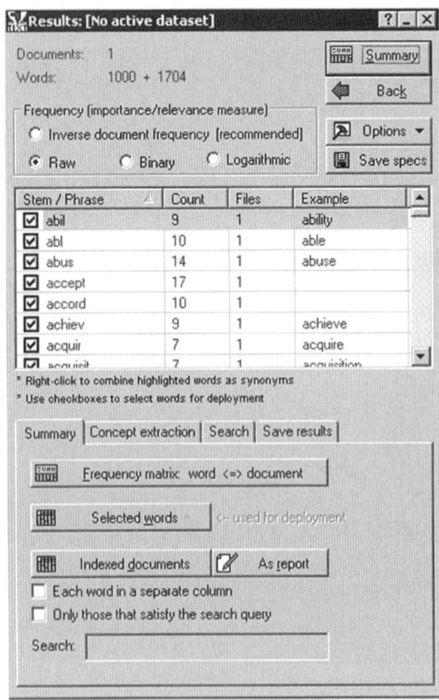

FIGURE U.6
The text mining results dialog, following the indexing process.

Click *Selected words* to output the word summary in a spreadsheet (see Figure U.7).

Data: Words summary ([No active dataset])*

Stem / Phras	Count	Number of documents	Example
abil	9	1	ability
abl	10	1	able
abus	14	1	abuse
accept	17	1	
accord	10	1	
achiev	9	1	achieve
acquir	7	1	acquire
acquisit	7	1	acquisition
act	9	1	
action	14	1	
activ	59	1	active
activist	14	1	
actual	7	1	
address	3	1	
adequ	5	1	adequate
adjust	9	1	
admit	5	1	
advanc	33	1	advance
advantag	10	1	advantage
advertis	17	1	advertising

FIGURE U.7
Word Summary spreadsheet.

The most frequently used terms may reveal key themes, so the document should be sorted by the count. From the *Data* menu, select *Sort* to open the *Sort Options* dialog (see Figure U.8).

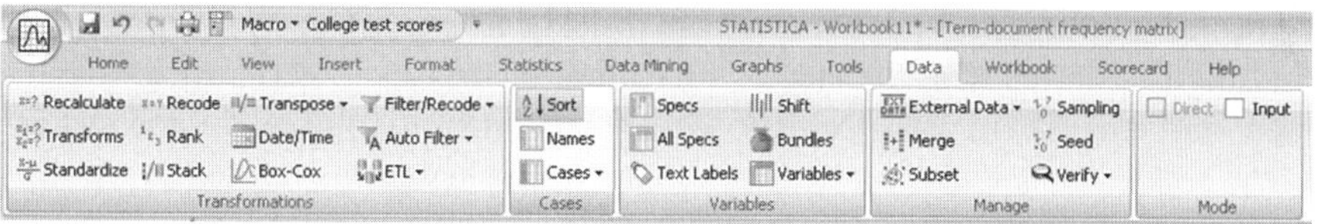

FIGURE U.8
Selecting the "sort options" dialog.

Select the variable containing document frequencies, *Count,* and select *Add var(s)*. Change the direction to *Descending* (see Figure U.9).

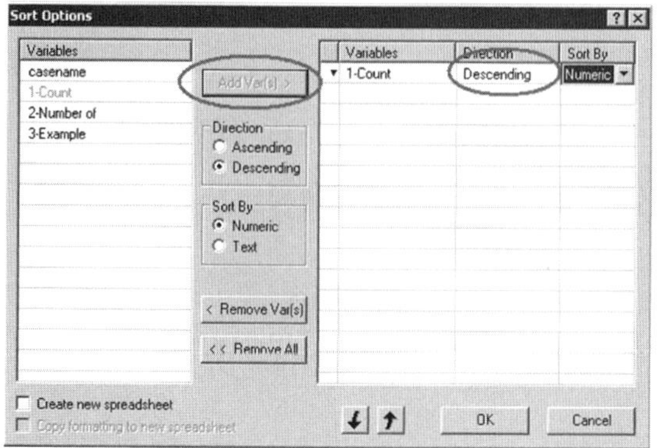

FIGURE U.9
The "sort options" dialog.

Click *OK* to sort the data. Now the most frequently used terms are seen first (see Figure U.10).

Data: Words summary ([No active dataset])*

Stem / Phras	Words summary ([No active datase		
	Count	Number of documents	Example
societi	267	1	societ
will	237	1	
system	236	1	
peopl	226	1	peopl
power	177	1	
technolog	176	1	technolog
would	154	1	
human	149	1	
one	148	1	
leftist	126	1	
need	121	1	
social	119	1	
can	108	1	
may	102	1	
make	100	1	
mani	97	1	man
modern	95	1	
freedom	93	1	
problem	87	1	
use	87	1	

FIGURE U.10
Words summary spreadsheet, with words assorted in descending order of frequency.

Without reading the 64-page document, we can quickly see the topics of this manifesto from its list of most frequently used terms. The list suggests frequent references to the future with the words *will, would, need, can,* and *may*. There also seems to be a focus on humanity as an abstract entity through terms like *society, people, human,* and *social*. We also see some key terms referencing politics, including *power, leftist,* and *freedom*. A very simple generalization we can take away from this is that the text speaks of a need for action in the political arena.

SEARCHING FOR TRENDS WITH PRONOUNS

One area of research in schizophrenia is in the use of pronouns. The evidence is mixed between spoken and written discourse, with some research suggesting that pronouns without clear referents occur more frequently in the speech of schizophrenia patients relative to controls (Dougherty et al., 2003; Harvey and Brault, 1986; Harvey and Serper, 1990), and yet the writings of schizophrenic patients in another study showed fewer pronouns than the writings of well controls (Lee et al., 2007).

Because Ted Kaczynski, the Unabomber, was diagnosed with paranoid schizophrenia, it is interesting to explore the use of pronouns in his manifesto. Did his use of pronouns differ from that of his similarly educated peers? For the purposes of this study, we used as a proxy of peer writings some opinion editorials from the *New York Times* of 1995, the same year Kaczynski's "Industrial Society and Its Future" appeared. The authors of these opinion editorials are diverse, and yet, they may typify "average" persuasive writing by individuals whose education was relatively similar to that of the Unabomber.

Analyzing Pronoun Use

From the *STATISTICA data mining* menu, select *text mining* to open the *text mining* dialog (see Figure U.11).

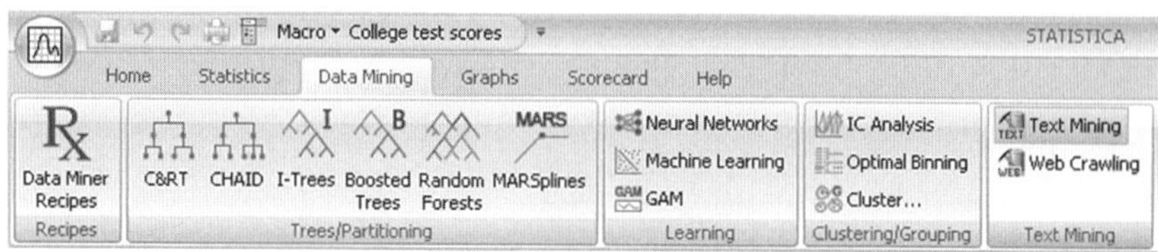

FIGURE U.11
Selecting text mining from the data mining ribbon bar.

On the *Quick* tab, select *Browse documents* to open the *Open document files* dialog (see Figure U.12).

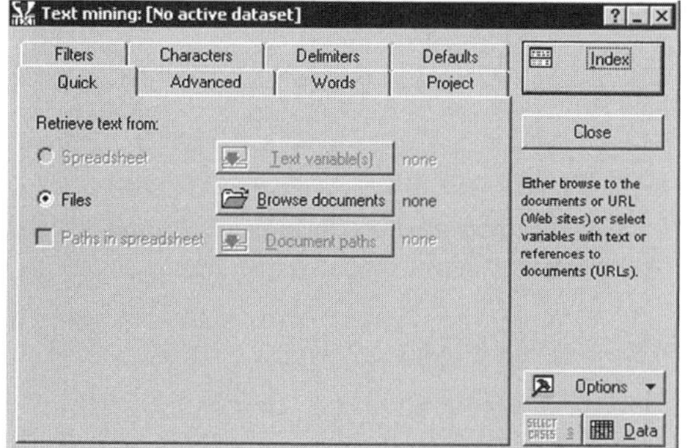

FIGURE U.12
Text Mining main dialog.

Select *Add Files* to open the *Open document files* dialog (see Figure U.13).

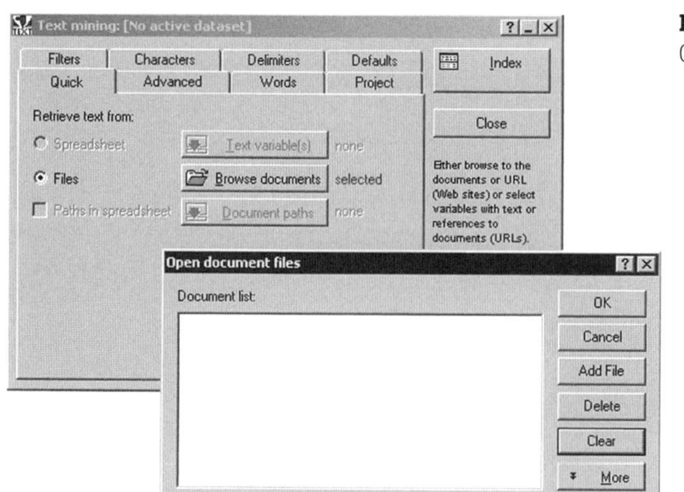

FIGURE U.13
Open document files dialog.

Browse to the set of files including *TK_manifesto.doc* and the *New York Times* Opinion Editorial pages (see Figure U.14). Select *Open*.

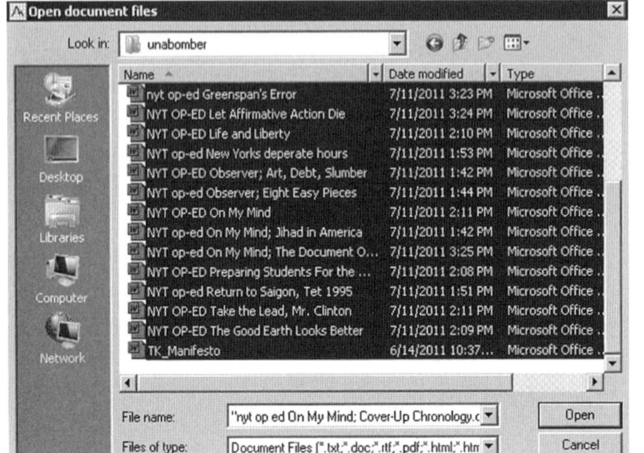

FIGURE U.14
Documents to be used in text mining are all selected.

The *Open document files* dialog will update with the selected document.

Select OK on the *Open document files* dialog to update the *Text Mining* dialog (see Figure U.15). Select the *Words* tab. Uncheck the *Stop words* option. Select the *Inclusion words* option.

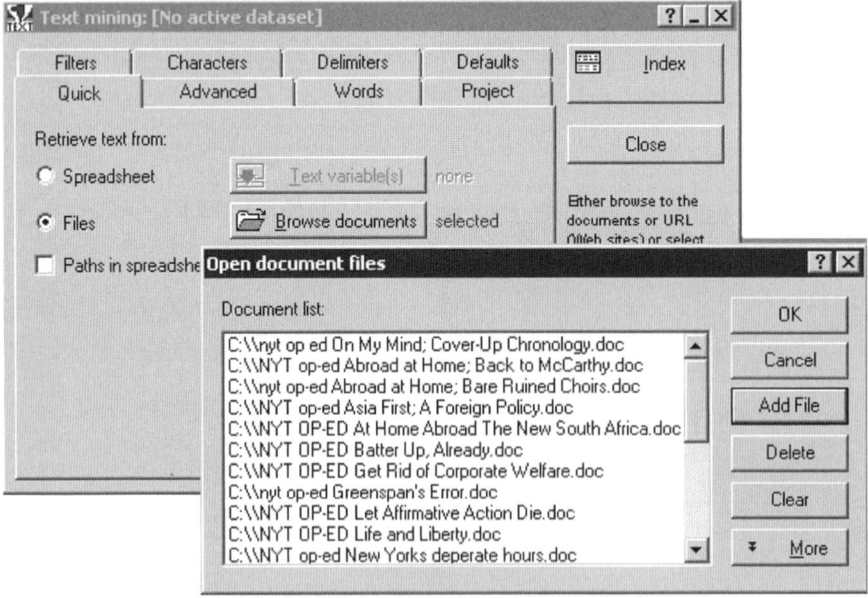

FIGURE U.15
Select OK on the "open document files" dialog to update the documents selected in the Text Mining main dialog.

Select *Edit* to open the *Inclusion word editor* dialog (see Figure U.16).

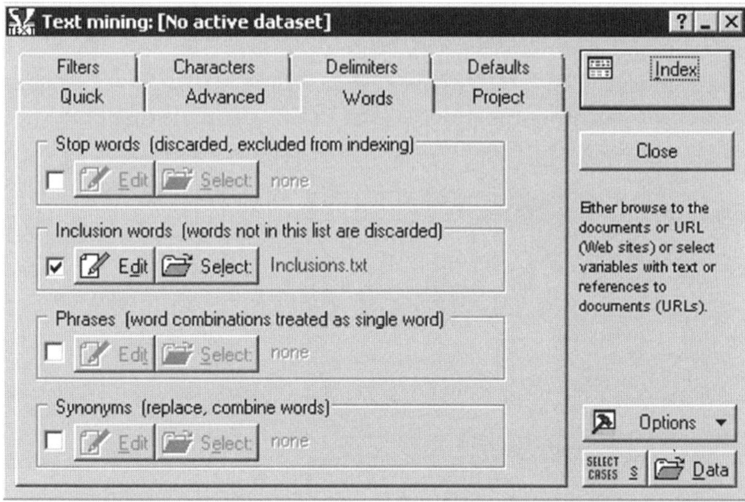

FIGURE U.16
Words tab selected in Text Mining dialog.

FIGURE U.17
Inclusion word editor dialog.

Enter a list of pronouns to index in the selected documents (see Figure U.17). Each term on the inclusion list should start a new line. Only words on the list will be indexed. (Alternatively, select the *Pronoun Inclusion.txt* file as the inclusion list.) The following pronouns are listed: *all, another, any, anybody, anyone, anything, both, each, either, everybody, everyone, everything, few, he, her, hers, herself, him, himself, his, I, it, its, itself, little, many, me, mine, more, most, much, my, myself, neither, nobody, none, nothing, one, other, others, our, ours, ourselves, several, she, some, somebody, someone, something, that, their, theirs, them, themselves, these, they, this, those, us, we, what, whatever, which, whichever, who, whoever, whom, whomever, whose, you, your, yours, yourself, yourselves.*

After entering the list, select OK (*Save*) to save the list. Select *Index* on the *Text mining* dialog to begin indexing the texts. After the document is indexed, the *Results* dialog is displayed (see Figure U.18).

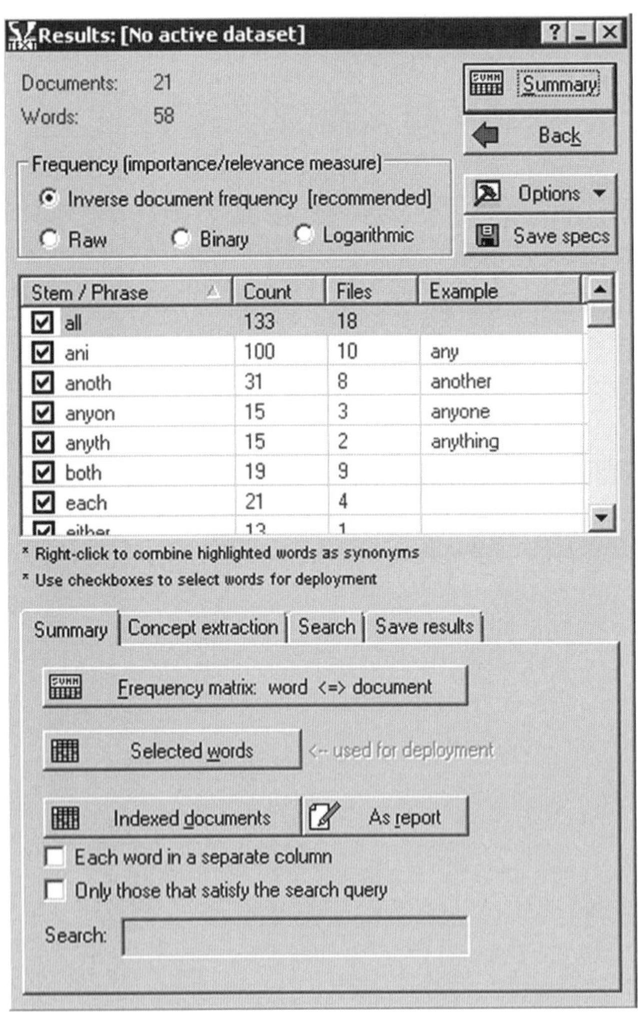

FIGURE U.18

Results dialog after indexing the frequency of the words.

The documents are of varying lengths, so using the *Inverse document frequency* will take into account not only the number of times the word occurs but also the document length. Effectively, the word occurrence is adjusted for document length.

On the *Concept extraction* tab, select *Perform Singular Value Decomposition (SVD)*. When complete, the options on this tab will become active (see Figure U.19).

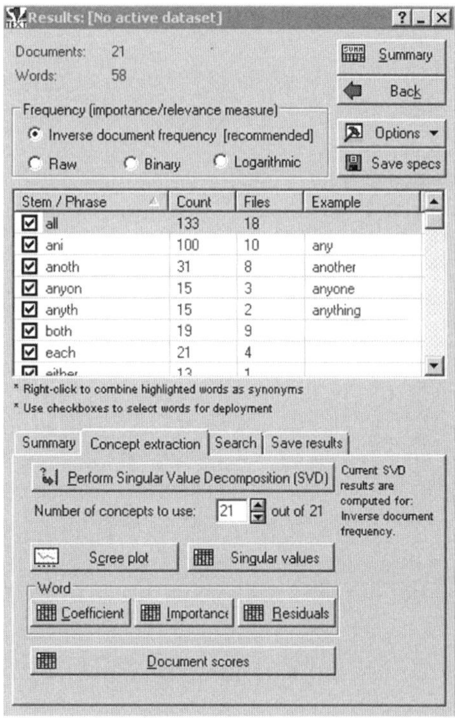

FIGURE U.19

Concept extraction tab selected so SVD can be computed.

The interest is with the documents themselves. Does Ted Kaczynski's manifesto follow the same patterns in terms of pronoun use as opinion editorials from the *New York Times* from the same year? We can explore this with the *Document scores* output (see Figure U.20).

Data: SVD document scores ([No active dataset])* (21v by 21c)

	SVD document scores ([No active dataset])			
	Concept 1	Concept 2	Concept 3	Concept 4
nyt op ed On My Mind; Cover-Up Chronology.doc	0.053450	-0.222675	-0.367770	0.028695
NYT op-ed Abroad at Home; Back to McCarthy.doc	0.053952	-0.185190	0.026235	0.214723
nyt op-ed Abroad at Home; Bare Ruined Choirs.doc	0.027563	-0.155031	0.206607	0.022478
NYT op-ed Asia First; A Foreign Policy.doc	0.051993	-0.266866	0.114243	-0.037666
NYT OP-ED At Home Abroad The New South Africa.doc	0.045047	-0.295601	0.203900	-0.041528
NYT OP-ED Batter Up, Already.doc	0.046855	-0.109345	0.040079	-0.340888
NYT OP-ED Get Rid of Corporate Welfare.doc	0.016249	-0.056887	0.023248	0.025150
nyt op-ed Greenspan's Error.doc	0.024208	-0.136764	-0.060114	-0.052329
NYT OP-ED Let Affirmative Action Die.doc	0.072052	-0.358579	-0.073900	0.110710
NYT OP-ED Life and Liberty.doc	0.020395	-0.127427	0.046343	0.037391
NYT op-ed New Yorks deperate hours.doc	0.027284	-0.097182	-0.033331	0.000019
NYT OP-ED Observer; Art, Debt, Slumber.doc	0.050904	-0.377854	-0.353032	0.079555
NYT op-ed Observer; Eight Easy Pieces.doc	0.058132	-0.279472	-0.195692	0.205384
NYT OP-ED On My Mind.doc	0.042336	-0.237581	-0.236510	0.162576
NYT op-ed On My Mind; Jihad in America.doc	0.047244	-0.146298	-0.211955	-0.848947
NYT op-ed On My Mind; The Document Of May 30.doc	0.014960	-0.129916	-0.037680	-0.011443

FIGURE U.20

SVD document scores.

Designate the SVD Document Scores output as input by selecting the *Data* menu and checking *Input* in the *Mode* field (see Figure U.21).

FIGURE U.21
Making the "document scores output" as the input spreadsheet from the DATA menu.

From the *Graphs* menu, select *Scatterplot* to open the *2D Scatterplots* dialog (see Figure U.22).

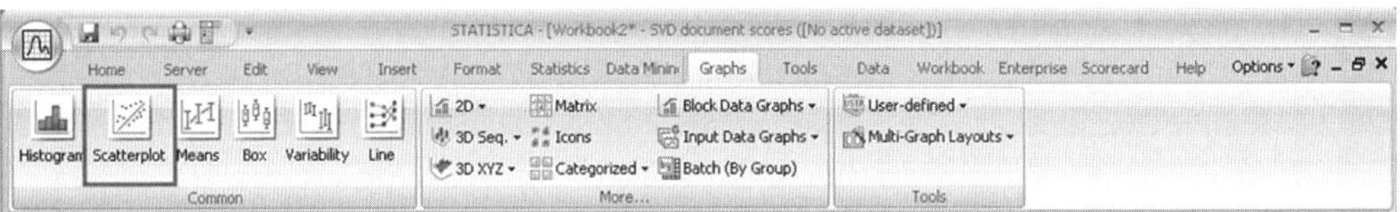

FIGURE U.22
Using Graphs menu to select 2D Scatterplots dialog.

Select *Concept 1* and *Concept 2* as *X* and *Y*, respectively. Turn off the *Fit type* (see Figure U.23).

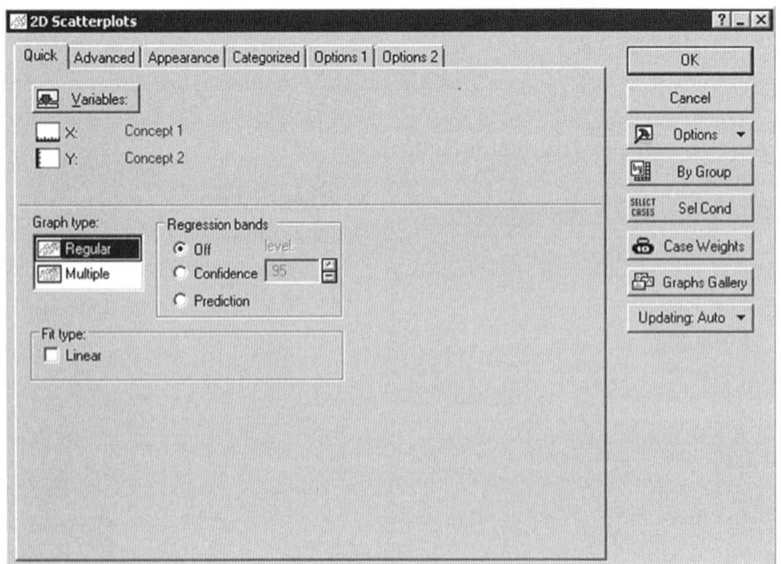

FIGURE U.23
2D Scatterplots graph dialog.

Select OK to create the plot.

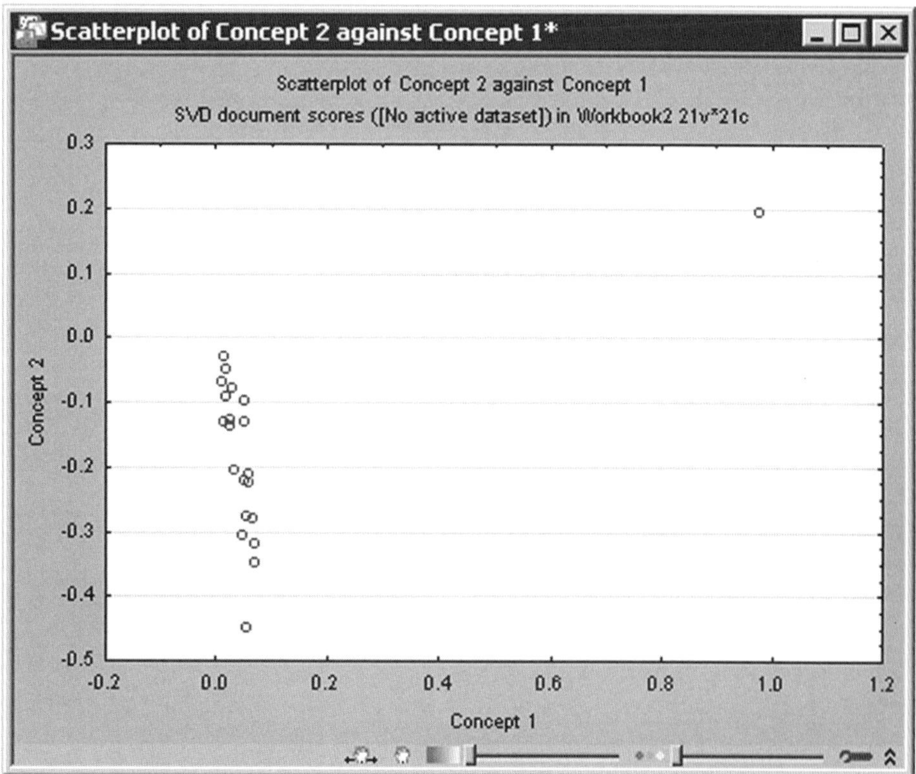

FIGURE U.24
2D Scatterplot of Concept 2 vs Concept 1.

Each point in the scatterplot (see Figure U.24) represents one of the documents in this study. One document stands alone, while the others cluster together. To reveal the document represented by the outlying point, it should be labeled.

From the *Edit* menu, select *Brushing* (see Figure U.25). The icon is in the *Customize Graph* section and looks like a magnifying glass with crosshairs.

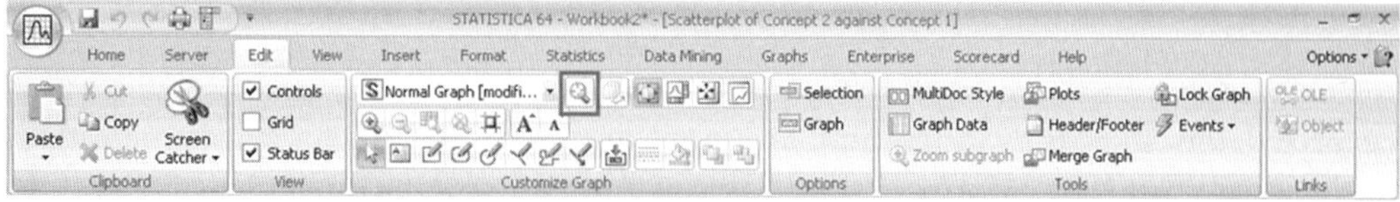

FIGURE U.25
Selecting Brushing from the Edit menu.

Use the *Brushing 2D* tool to select the outlying point (see Figure U.26). Change the *Action* to *Label* and click *Apply*.

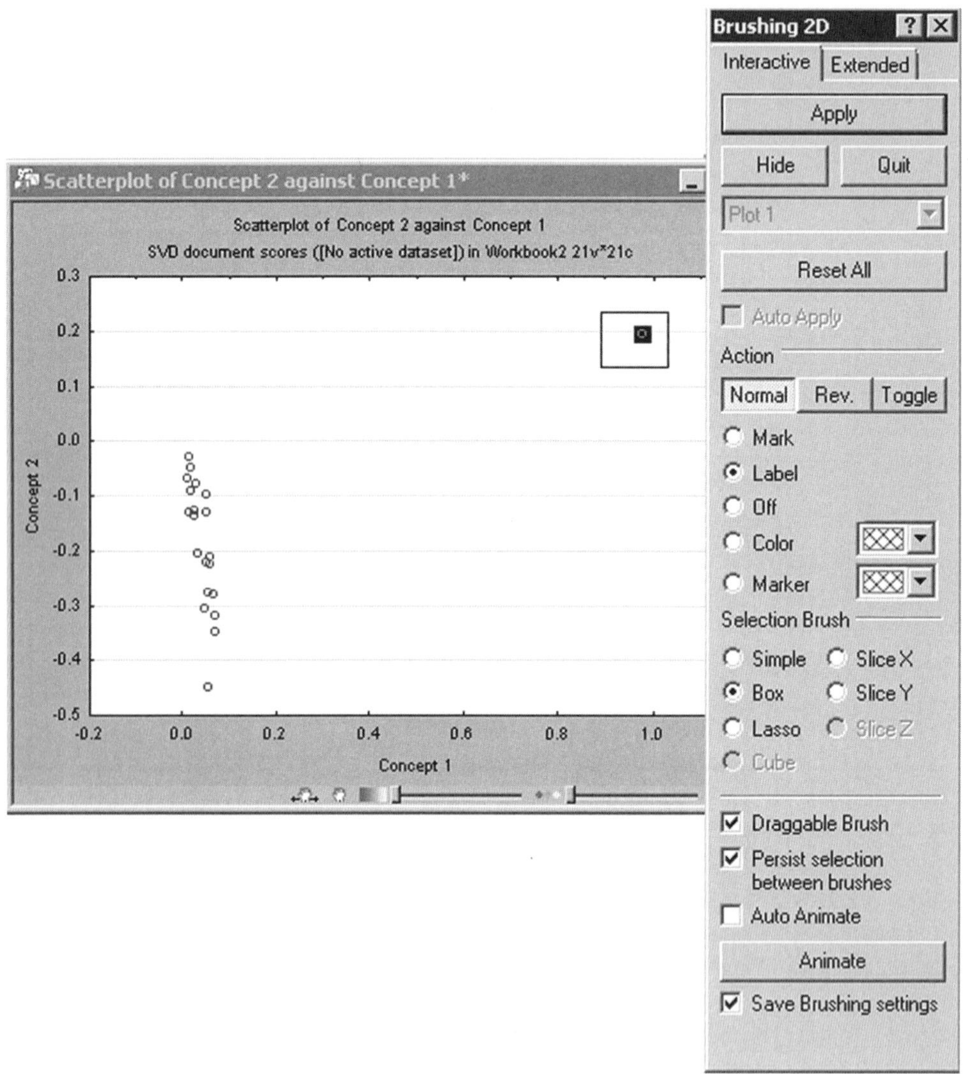

FIGURE U.26
Using the brushing tool.

As expected, the "Industrial Society and Its Future" document, the Ted Kaczynski manifesto, is the outlying document (see Figure U.27). This tells us that this particular document was different from the set of documents in terms of its use of pronouns. Pronouns were the only terms indexed in this analysis, so they are the only type of words considered here.

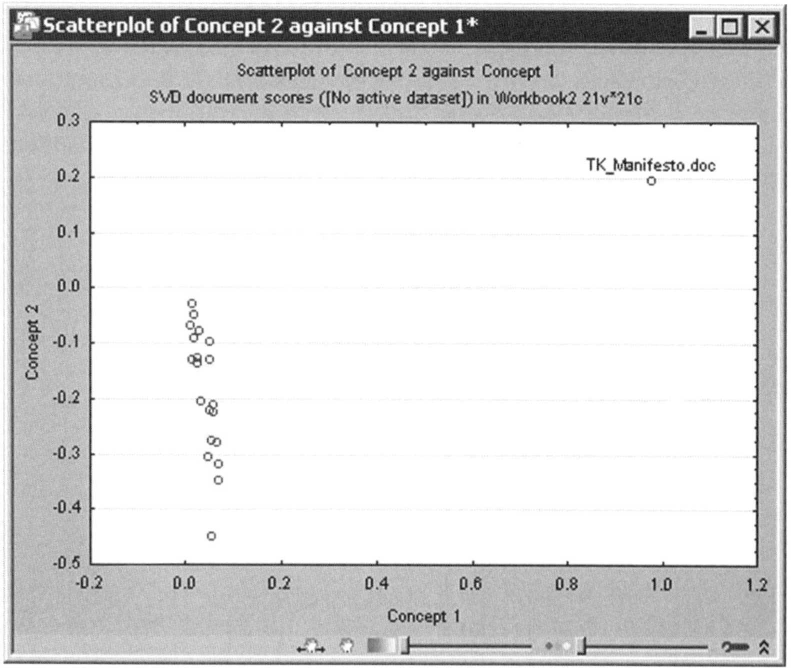

FIGURE U.27

The TK_Manifesto document is observed as an 'outlier" among the clustering of documents.

Return to the *Summary* tab of the *Results* dialog. Select *Indexed documents* to create output showing word counts for each document (see Figure U.28).

Data: File summary ([No active dataset])*

	File summary ([No active dataset])		
	Document length	Number of words	
nyt op ed On My Mind; Cover-Up Chronology.doc	4208	50	my all it that that that that anoth m
NYT op-ed Abroad at Home; Back to McCarthy.doc	4351	66	one this which their who them wha
nyt op-ed Abroad at Home; Bare Ruined Choirs.doc	4433	52	that they our our we our what our i
NYT op-ed Asia First; A Foreign Policy.doc	7585	98	much neither we other we it most i
NYT OP-ED At Home Abroad The New South Africa.doc	4119	61	mani who he him who it who their
NYT OP-ED Batter Up, Already.doc	4177	42	it some they littl they most his sev
NYT OP-ED Get Rid of Corporate Welfare.doc	3583	38	their each much more it it that one
nyt op-ed Greenspan's Error.doc	3683	39	one that one anoth those more we
NYT OP-ED Let Affirmative Action Die.doc	7450	115	it which it it it it this that he his his
NYT OP-ED Life and Liberty.doc	4043	54	more our it it who some it both whi
NYT op-ed New Yorks deperate hours.doc	7627	82	this that it that this it all that that it
NYT OP-ED Observer; Art, Debt, Slumber.doc	4244	86	my which some it you you this it w
NYT op-ed Observer; Eight Easy Pieces.doc	4273	66	this some it our which they this it
NYT OP-ED On My Mind.doc	4278	75	my that that one other that this all
NYT op-ed On My Mind; Jihad in America.doc	4293	49	my that someth one their that it it
NYT op-ed On My Mind; The Document Of May 30.doc	4349	55	my which it them this it their that i
NYT OP-ED Preparing Students For the World Of Jobs.doc	5713	67	our who it more our them that it th
NYT op-ed Return to Saigon, Tet 1995.doc	7110	110	it his other they their he he he thei
NYT OP-ED Take the Lead, Mr. Clinton.doc	3658	35	that it this that most it much that c
NYT OP-ED The Good Earth Looks Better.doc	6687	66	you one our it it that that some mc
TK_Manifesto.doc	220939	116	us it they those us who they it it it

FIGURE U.28

Word counts for each document.

The first column (see Figure U.28) of output is the *Document length*, which shows that the Unabomber manifesto is quite long in comparison to the other documents. This is why the inverse document frequency is a better selection in this case. It controls for that discrepancy in document length.

The next column is *Number of Words*. Again, this is the number of pronouns found in each text. In order to examine pronoun use relative to document length, begin with the *Data* menu and select *Add* from the *Variables* drop-down menu (see Figure U.29).

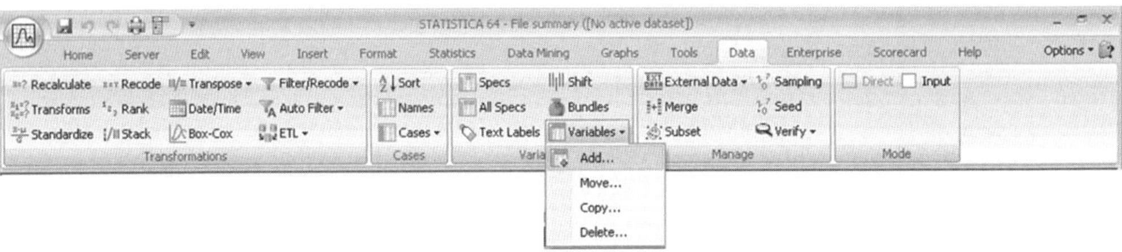

FIGURE U.29
Selecting "add" from the Variables drop-down menu.

On the *Add Variables* dialog (see Figure U.30), make the following settings: *How many* = 1, *After* = *Number of*, Name = *Rate*, *Display format* = *Percentage*, *Decimal places* = 2, and *Long name (label or formula with Functions)* = "=v2/v1."

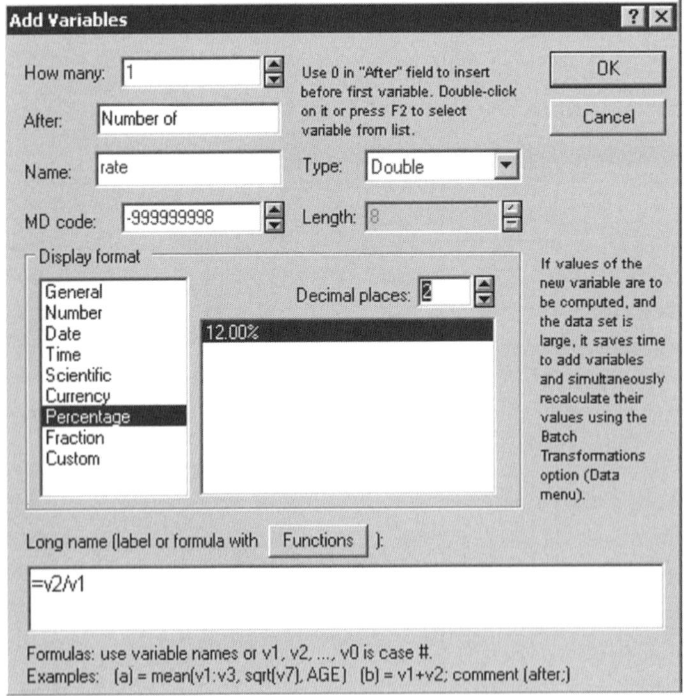

FIGURE U.30
Add Variables dialog.

Click OK to create the new variable.

Data: File summary ([No active dataset])*

	File summary ([No active dataset])			
	Document length	Number of words	rate =v2/v1	
nyt op ed On My Mind; Cover-Up Chronology.doc	4208	50	1.19%	my all it that that that th
NYT op-ed Abroad at Home; Back to McCarthy.doc	4351	66	1.52%	one this which their whc
nyt op-ed Abroad at Home; Bare Ruined Choirs.doc	4433	52	1.17%	that they our our we our
NYT op-ed Asia First; A Foreign Policy.doc	7585	98	1.29%	much neither we other v
NYT OP-ED At Home Abroad The New South Africa.doc	4119	61	1.48%	mani who he him who it
NYT OP-ED Batter Up, Already.doc	4177	42	1.01%	it some they littl they m
NYT OP-ED Get Rid of Corporate Welfare.doc	3583	38	1.06%	their each much more it
nyt op-ed Greenspan's Error.doc	3683	39	1.06%	one that one anoth thos
NYT OP-ED Let Affirmative Action Die.doc	7450	115	1.54%	it which it it it it this that
NYT OP-ED Life and Liberty.doc	4043	54	1.34%	more our it it who some
NYT op-ed New Yorks deperate hours.doc	7627	82	1.08%	this that it that this it all
NYT OP-ED Observer; Art, Debt, Slumber.doc	4244	86	2.03%	my which some it you y
NYT op-ed Observer; Eight Easy Pieces.doc	4273	66	1.54%	this some it our which tl
NYT OP-ED On My Mind.doc	4278	75	1.75%	my that that one other t
NYT op-ed On My Mind; Jihad in America.doc	4293	49	1.14%	my that someth one the
NYT op-ed On My Mind; The Document Of May 30.doc	4349	55	1.26%	my which it them this it
NYT OP-ED Preparing Students For the World Of Jobs.doc	5713	67	1.17%	our who it more our ther
NYT op-ed Return to Saigon, Tet 1995.doc	7110	110	1.55%	it his other they their he
NYT OP-ED Take the Lead, Mr. Clinton.doc	3658	35	0.96%	that it this that most it r
NYT OP-ED The Good Earth Looks Better.doc	6687	66	0.99%	you one our it it that tha
TK_Manifesto.doc	220939	116	0.05%	us it they those us who

FIGURE U.31

Data file summary of the New York Times articles.

The *New York Times* opinion editorial articles (see Figure U.31) all have anywhere from .96% pronouns to 2.03%. The Unabomber manifesto has .05% pronouns. This large difference partly explains why this text stood out in the scatterplot earlier. Vague and abstract language may be more typical of a treatise or manifesto than of opinion editorials, and this may account, in part, for Kaczynski's less frequent use of pronouns, and yet, it also is true, however, that schizophrenia sometimes manifests itself in similarly vague and abstract language (Dougherty et al., 2004). In some ways, the manifesto demonstrates two dimensions of speech observed in schizophrenia: verbosity and discontinuity (Berenbaum et al., 1995).

Masculine and Feminine Pronoun use

Do any other interesting patterns exist in the pronoun usage of Ted Kaczynski in his manifesto? To explore only this text, select *Back* on the *Results* dialog to return to the *Text Mining* dialog. On the *Quick* tab, select *Browse documents* to open the *Open document files* dialog. *Delete* all but the file, *TK_manifesto.doc* (see Figure U.32).

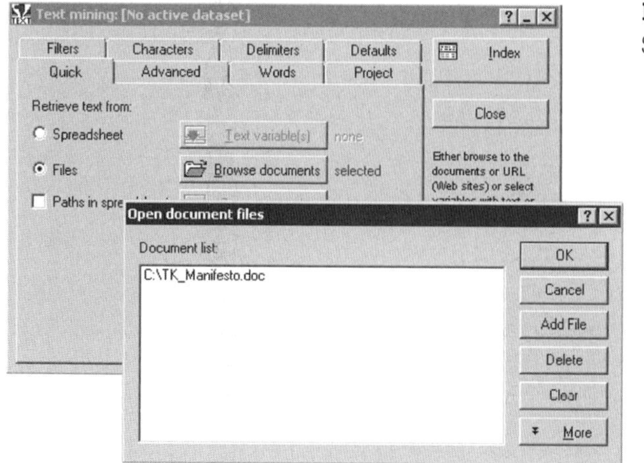

FIGURE U.32

Selecting only the TK_manifesto.doc.

Click OK to close the *Open document files* dialog.

On the *Words* tab, select *Edit* in the *Inclusion words* field (see Figure U.33). This will allow us to narrow the inclusion list for simplicity.

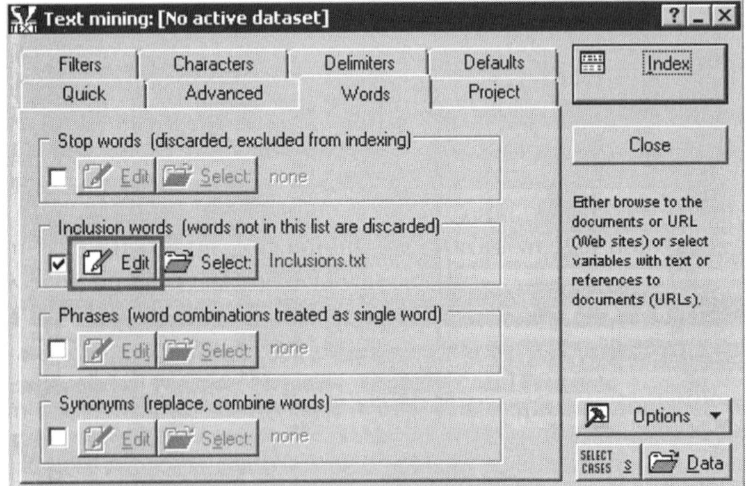

FIGURE U.33
Inclusion words editing.

In the *Inclusion word editor* (see Figure U.34), change the inclusion list to the following: *he, her, herself, him, himself, his, I, it, me, mine, my, myself, our, ourselves, she, their, them, themselves, they, us, we, you, your, yourself, yourselves.*

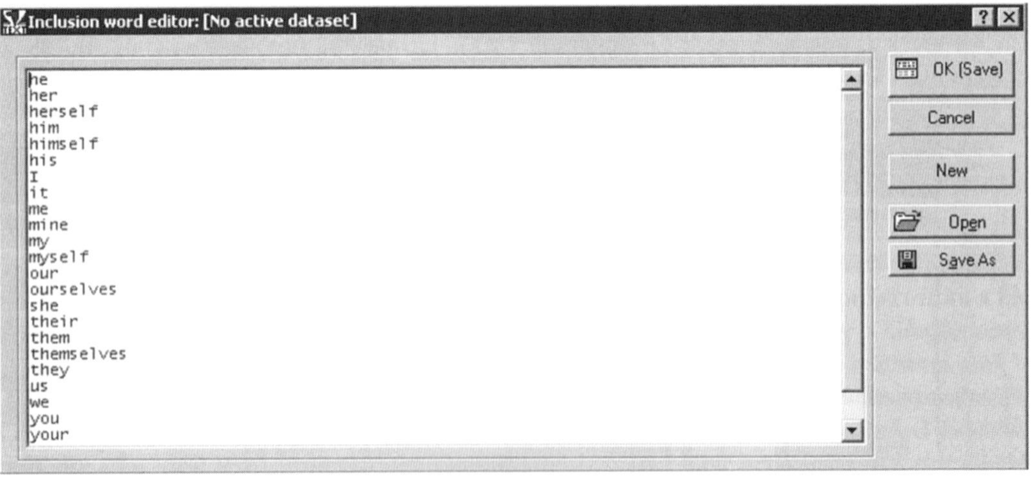

FIGURE U.34
Inclusion word editor dialog.

Select OK to update the list. Click *Index* to begin indexing the one file.

On the *Results* dialog, we see that only one document was indexed and 17 words were selected (see Figure U.35).

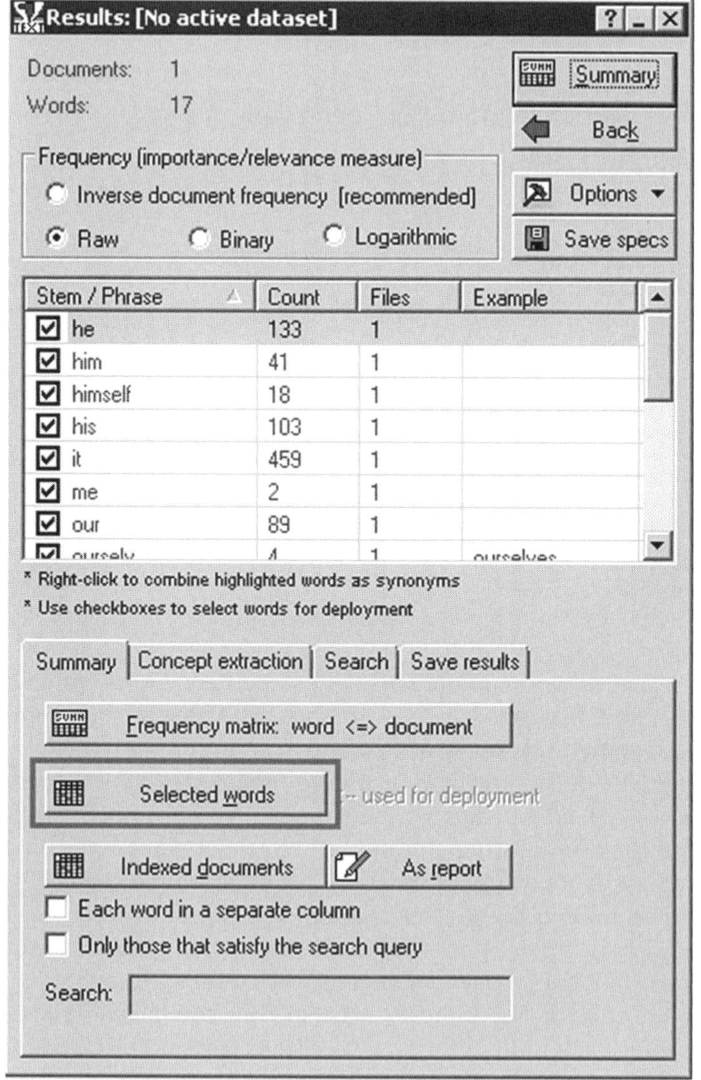

FIGURE U35
Results dialog with only 1 document having been selected.

Click *Selected words* to create the *Word summary* output, showing the counts for each of the selected words (see Figure U.36).

	Words summary ([No active data		
Stem / Phras	Count	Number of documents	Example
he	133	1	
him	41	1	
himself	18	1	
his	103	1	
it	459	1	
me	2	1	
our	89	1	
ourselv	4	1	ourselves
their	192	1	
them	82	1	
themselv	38	1	themselves
they	253	1	
us	25	1	
we	210	1	
you	33	1	
your	7	1	
yourself	3	1	

FIGURE U.36
Document words summary spreadsheet.

Masculine pronouns, such as *he, him, himself,* and *his,* are used with some frequency. Feminine pronouns, such as *she, her, herself,* and *hers,* are completely absent from the text. Although one might infer from his exclusive use of masculine pronouns that Kaczynski had certain attitudes about the roles of males and females in society, the manifesto's exclusive use of masculine pronouns may reflect a more gender-neutral meaning, in which a term like *him,* intended to represent the human individual, carries more universal connotations. In English, this use of masculine pronouns for gender-neutral meaning also may reflect highly formal writing, perhaps more typical of treatises than of opinion editorials.

This tutorial has illustrated the use of text mining as a method that can summarize key themes of a large document and compare texts in ways that may yield clues about the writer's thoughts relative to those of other writers. With further empirical development, text mining may be useful in situations like law enforcement or mental health, where information about the likely perspective and mental state of an individual may contribute to better-informed and more appropriate decisions that may involve him or her.

References

Berenbaum, H., Oltmanns, T.F., Gottesman, I.I. (1985). Formal thought disorder in schizophrenics and their twins. *Journal of Abnormal Psychology, 94,* 3–16.

Chase, A. (2003). *Harvard and the Unabomber: The Education of an American Terrorist.* New York: Norton.

Docherty, N.M., Cohen, A.S., Nienow, T.M., Dinzeo, T.J., and Dangelmaier, R.E. (2003). Stability of formal thought disorder and referential communication disturbances in schizophrenia. *Journal of Abnormal Psychology, 112*, 469–475.

Harvey, P., and Brault, J. (1986). Speech performance in mania and schizophrenia: The association of positive and negative thought disorders and reference failures. *Journal of Communication Disorders, 19*, 161–173.

Harvey, P., & Serper, M. (1990). Linguistic and cognitive failures in schizophrenia. A multivariate analysis. *Journal of Nervous and Mental Disease, 178*, 487–493.

Kaczynski, T. (October 2008). *The Road to Revolution: The Complete Writings of Theodore J. Kazcynski*. Vevey, Switzerland: Xenia Books.

Lee, Chang H., Sungwoo Ahn, Myungju Lee, & Kim, Kyungil (2007). Preliminary analysis of language styles in a sample of schizophrenics. *Psychological Reports, 101*, 392–394.

Baker, Russell. "Observer; Art, Debt, Slumber." *New York Times*, January 10, 1995, Opinion Editorial.

Baker, Russell. "Observer; Eight Easy Pieces." *New York Times*, February 4, 1995, Opinion Editorial.

Easterbrook, Gregg. "The Good Earth Looks Better." *New York Times*, April 21, 1995, Opinion Editorial.

Galbraith, James K. "Greenspan's Error." *New York Times*, July 11, 1995, Opinion Editorial.

Kasich, John. "Get Rid of Corporate Welfare." *New York Times*, July 9, 1995, Opinion Editorial.

Lewis, Anthony. "Abroad at Home; Back to McCarthy." *New York Times*, February 24, 1995, Opinion Editorial.

Lewis, Anthony. "Abroad at Home; Bare Ruined Choirs." *New York Times*, July 10, 1995, Opinion Editorial.

Lewis, Anthony. "Abroad at Home; The New South Africa." *New York Times*, January 9, 1995, Opinion Editorial.

Lind, Michael. "Asia First: A Foreign Policy." *New York Times*, April 18, 1995, Opinion Editorial.

Miller, Matthew. "Take the Lead, Mr. Clinton." *New York Times*, May 16, 1995, Opinion Editorial.

O'Cleireacain, Carol. "New York's Desperate Hours." *New York Times*, March 14, 1995, Opinion Editorial.

Rosenthal, A.M. "On My Mind; Cover-Up Chronology." *New York Times*, April 4, 1995, Opinion Editorial.

Rosenthal, A.M. "On My Mind; Jihad in America." *New York Times*, February 3, 1995, Opinion Editorial.

Rosenthal, A.M. "On My Mind: The Document of May 30." *New York Times*, July 25, 1995, Opinion Editorial.

Rosenthal, A.M. "On My Mind." *New York Times*, July 4, 1995, Opinion Editorial.

Schumer, Charles E. "Life and Liberty." *New York Times*, April 28, 1995, Opinion Editorial.

Smith, Hedrick. "Preparing Students for the World of Jobs." *New York Times*, April 20, 1995, Opinion Editorial.

Stone, Robert. "Return to Saigon, Tet 1995." *New York Times*, March 12, 1995, Opinion Editorial.

Sullivan, Andrew. "Let Affirmative Action Die." *New York Times*, July 23, 1995, Opinion Editorial.

Zimbalist, Andrew. "Batter Up, Already." *New York Times*, February 14, 1995, Opinion Editorial.

Text Mining PubMed: Extracting Publications on Genes and Genetic Markers Associated with Migraine Headaches from PubMed Abstracts

Nephi Walton, MS, MD, and PhD
Candidate University of Utah, Departments of Medical Informatics and School of Medicine

Vladimir Rastunkov, PhD
Data Mining Consultant, StatSoft, Tulsa, OK

Gary Miner, PhD
Senior Statistician and Data Mining Consultant, StatSoft, Tulsa, OK

Not very long ago, it was feasible for clinicians and biomedical researchers to keep informed of developments in their respective areas of specialization by subscribing to a few select pertinent journals and periodically browsing a few more in the library. As the number of papers, journals, and publishers has increased dramatically in the past few decades, along with the increasing availability of such information from other countries around the world, this has become an almost insurmountable task. This is particularly true in the field of genetic research, where there are often thousands of papers published for any given disease. Wading through this information can be cumbersome and time-consuming, taking hours to days just to review the abstracts of all the available publications on a given topic.

Using text mining we can attempt to predict the articles that are most relevant to what we are looking for and at the same time look for key common concepts in those articles to help formulate hypotheses. In this tutorial we look for genes or genetic markers that are related to migraine headaches. Doing a search for genetics and migraine retrieves over 1,700 papers at the time this tutorial was written, and this number will increase dramatically over time. A similar search for breast cancer genetics brings back over 43,000 publications! This number is growing every day. There are more targeted searches that could be executed. However, those searches may pass over an article that focuses on another disease but whose conclusion or discussion suggests that the gene may be involved in the pathogenesis of migraine headaches based on their findings and the pathway that the gene is involved in.

Practical Text Mining and Statistical Analysis for Non-structured Text Data Applications. DOI: 10.1016/B978-0-12-386979-1.00030-X

Perhaps by analyzing these abstracts and extracting pertinent phrases and then creating prediction models based on these results, we can then deploy predictive models to score abstracts automatically to determine which papers are relevant to our research needs. This tutorial will take you through the process of bringing large numbers of abstracts from PubMed into *STATISTICA* for analysis with Text Miner. Your particular research question may be very different from the question posed here, but the process of sorting through the data should be the same.

The first step in this tutorial is to create a folder on your C: drive to store your files. You will need to know the absolute path to these files later when we use a visual basic macro to import the abstracts. The files created at each stage of the process are included on the DVD provided with this text.

Once the folder has been created, go to the PubMed website and retrieve the abstracts for migraine genetics. To do this open your web browser and type in the following address: http://www.ncbi.nlm.nih.gov/pubmed (see Figure V.1).

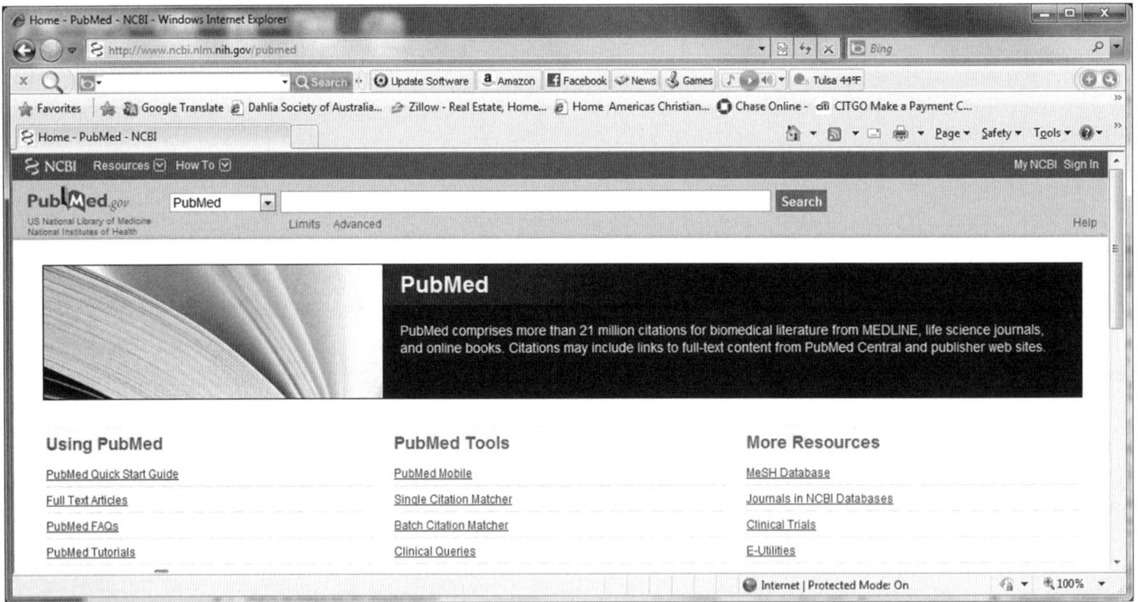

FIGURE V.1

Pub-Med web site; where you start with this tutorial.

Once the PubMed website appears, type "migraine genetics" in the search box and click on "Search." This brings up the results you see in Figure V.2.

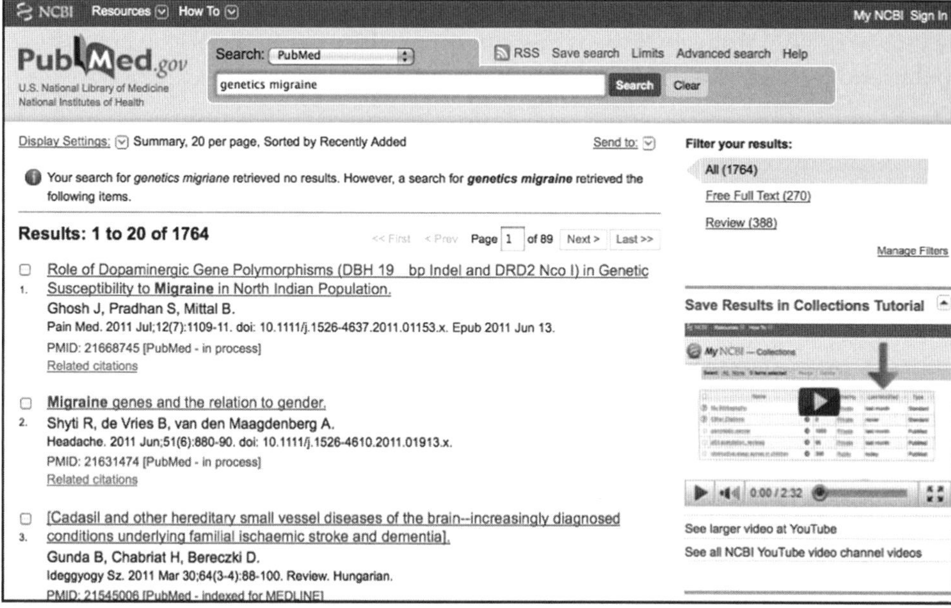

FIGURE V.2
The results of the search on "genetics migraine" in Pub-Med web site.

After completing the search, the abstracts retrieved need to be stored in a text file. To do this click on "Send to:" on the right-hand side below the Search button (see Figure V.2 above in the upper right hand side of figure to find "search button"). This gives you a list of destinations to which you can export your results (see Figure V.3 below). Click on the "File" radio button (see Figure V.3).

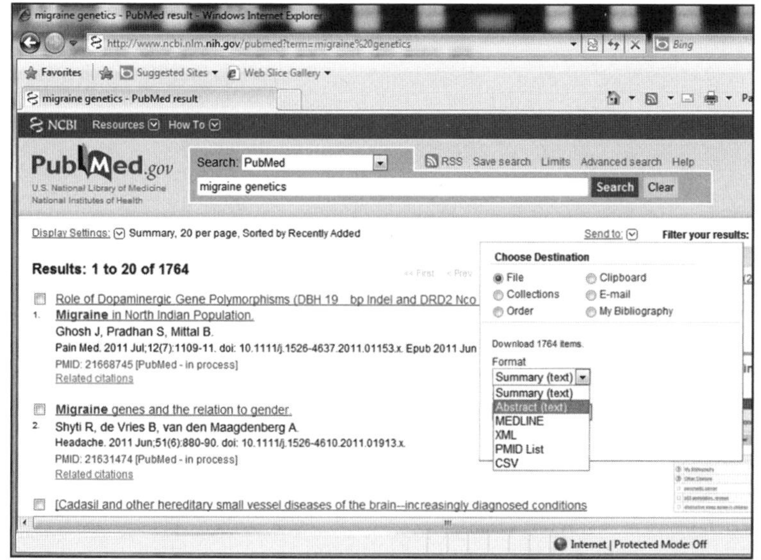

FIGURE V.3
Exporting the results of the search to a destination of your choice.

Once you click on the "File" radio button, you are given additional options. From the "Format" drop-down menu select "Abstract (text)." This selection enables output of all the abstracts for all the selected publications (see Figure V.4).

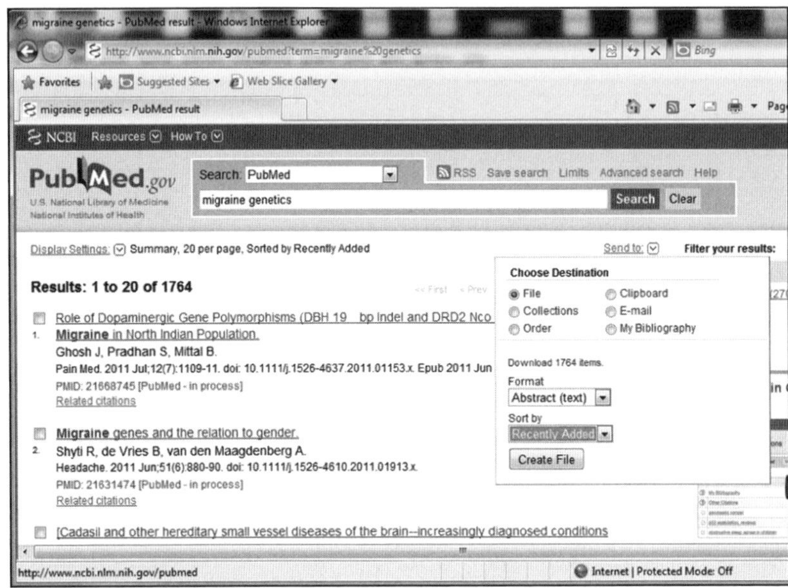

FIGURE V.4
Use the 'Format' drop down menu to select 'Abstract(text)'.

The "Sort by" drop-down enables you to sort the records by various parameters. We will leave the setting to the default of sorting by the order the abstracts were added to PubMed. Now click on the "Create File" button (see Figure V.5).

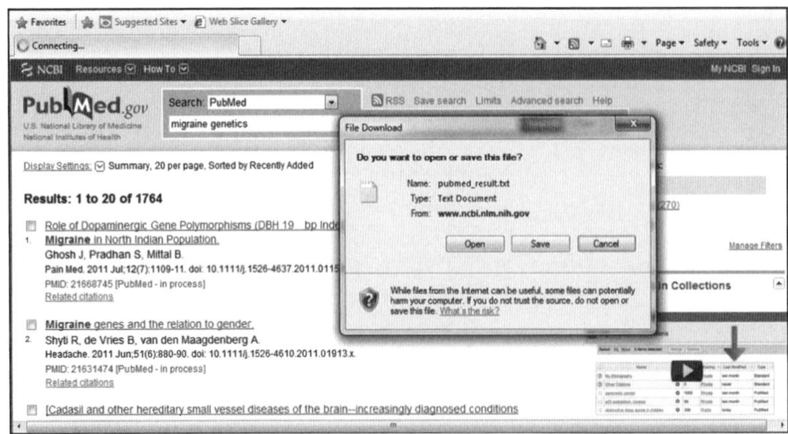

FIGURE V.5
Clicking on the 'Create File' button.

Afer you click on "Create File," a dialog appears asking you if you want to save or open the file. Save the file to the folder you created earlier on your C: drive. The file downloaded from PubMed has been included with this tutorial and is named "pubmed_result.txt."

Now that you have the abstracts from PubMed, we need to process them so that we can use them in *STATISTICA*. You can do this in any word processor, but in this tutorial we will use Microsoft Word. For this tutorial we are going to analyze or train our model on only the first 100 abstracts in the text file. To do this, open a new document in Microsoft Word. Now open the text file from PubMed in a text editor or Microsoft Word, and copy and paste the first 100 abstracts from the text file to your new Microsoft Word document.

Now that you have the first 100 abstracts, we need to create separators in between the abstracts so they can be independently analyzed. To do this, we can use the "Find/Replace" feature in your word processor (see Figure V.6). We will use the terms "MyStart" and "MyEnd" to denote the start and end of each abstract record, respectively. To do this, it is necessary to replace reliably placed phrases within the abstract file with these terms. The following phrases need to be replaced:

```
"[PubMed - indexed for MEDLINE]"
"[PubMed - in process]"
"[PubMed - as supplied by publisher]"
"[PubMed]"
"[PubMed - OLDMEDLINE]"
```

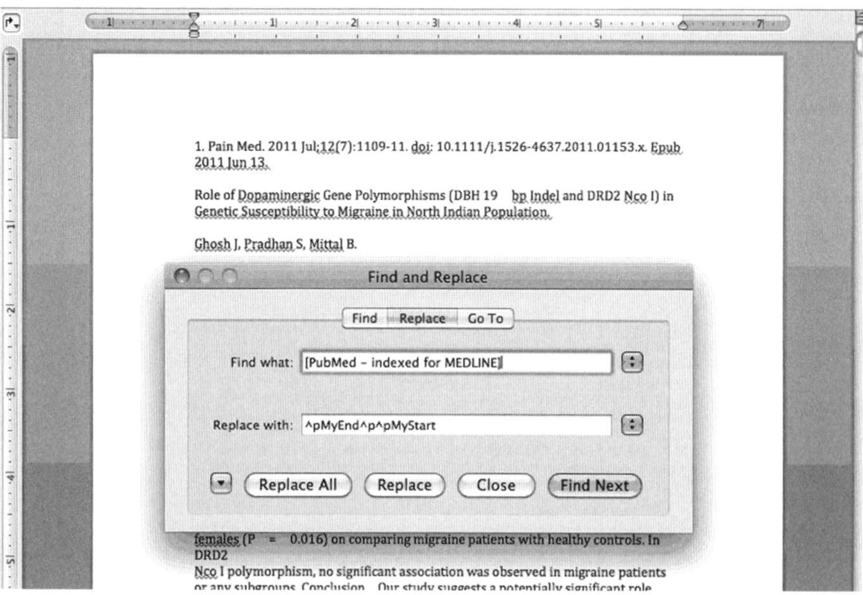

FIGURE V.6
Using the "Find/Replace" feature in Microsoft Word.

We will replace them with the following string: "^pMyEnd^p^pMyStart" as shown above in Figure V.6. In Microsoft Word, "^p" denotes a paragraph separation or return. By replacing all the phrases above, we place separators between each abstract that allow us to import the data into

STATISTICA. Once you have entered the phrase to replace and the replacement phrase, click on "Replace All" to replace every occurrence in the document.

After replacing all the phrases above, you must modify the beginning and ending of the document by adding "MyStart" at the very top of the document and removing the "MyStart" from the end of the document, where it is not followed by an abstract or a "MyEnd" (see Figures V.7 and V.8 below).

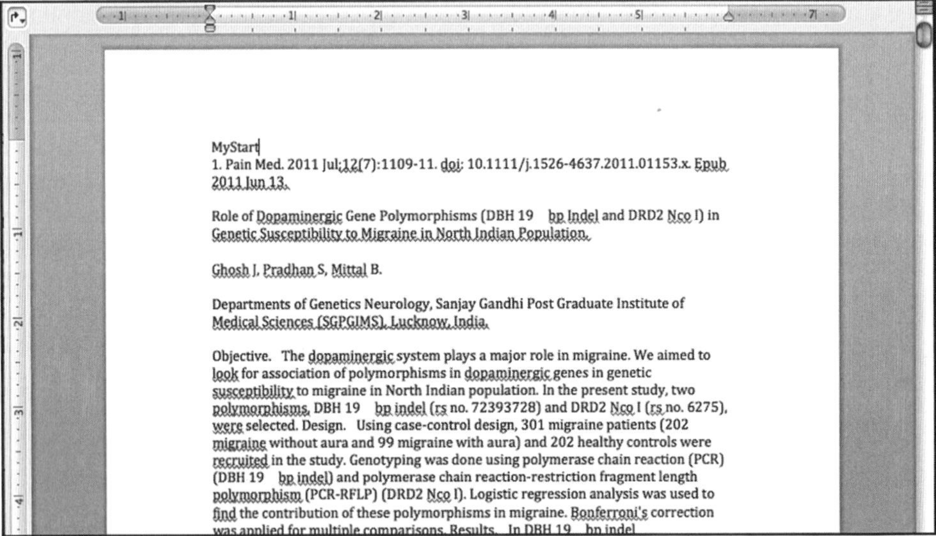

MyStart
1. Pain Med. 2011 Jul;12(7):1109-11. doi: 10.1111/j.1526-4637.2011.01153.x. Epub 2011 Jun 13.

Role of Dopaminergic Gene Polymorphisms (DBH 19 bp Indel and DRD2 Nco I) in Genetic Susceptibility to Migraine in North Indian Population.

Ghosh J, Pradhan S, Mittal B.

Departments of Genetics Neurology, Sanjay Gandhi Post Graduate Institute of Medical Sciences (SGPGIMS), Lucknow, India.

Objective. The dopaminergic system plays a major role in migraine. We aimed to look for association of polymorphisms in dopaminergic genes in genetic susceptibility to migraine in North Indian population. In the present study, two polymorphisms, DBH 19 bp indel (rs no. 72393728) and DRD2 Nco I (rs no. 6275), were selected. Design. Using case-control design, 301 migraine patients (202 migraine without aura and 99 migraine with aura) and 202 healthy controls were recruited in the study. Genotyping was done using polymerase chain reaction (PCR) (DBH 19 bp indel) and polymerase chain reaction-restriction fragment length polymorphism (PCR-RFLP) (DRD2 Nco I). Logistic regression analysis was used to find the contribution of these polymorphisms in migraine. Bonferroni's correction was applied for multiple comparisons. Results. In DBH 19 bp indel

FIGURE V.7

Adding "MyStart" at the top of a document.

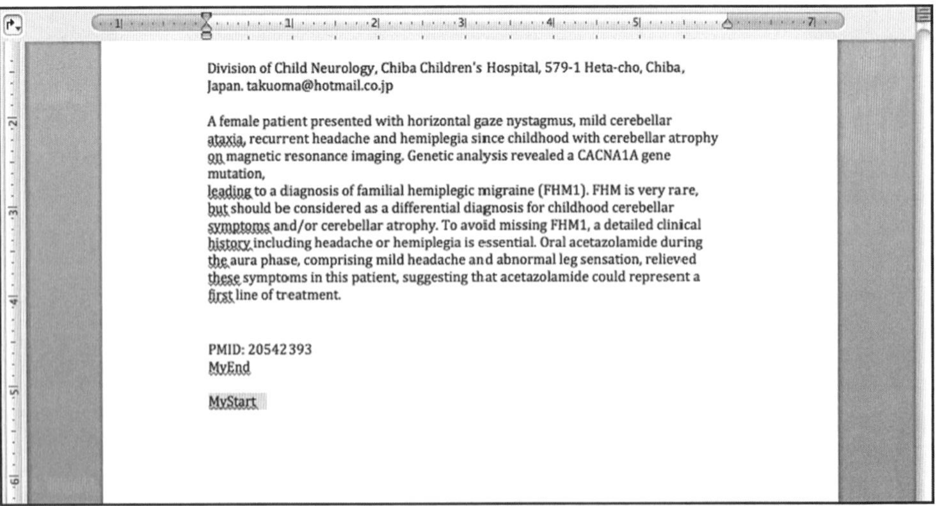

Division of Child Neurology, Chiba Children's Hospital, 579-1 Heta-cho, Chiba, Japan. takuoma@hotmail.co.jp

A female patient presented with horizontal gaze nystagmus, mild cerebellar ataxia, recurrent headache and hemiplegia since childhood with cerebellar atrophy on magnetic resonance imaging. Genetic analysis revealed a CACNA1A gene mutation,
leading to a diagnosis of familial hemiplegic migraine (FHM1). FHM is very rare, but should be considered as a differential diagnosis for childhood cerebellar symptoms and/or cerebellar atrophy. To avoid missing FHM1, a detailed clinical history including headache or hemiplegia is essential. Oral acetazolamide during the aura phase, comprising mild headache and abnormal leg sensation, relieved these symptoms in this patient, suggesting that acetazolamide could represent a first line of treatment.

PMID: 20542393
MyEnd

MyStart

FIGURE V.8

"MyStart" must be removed from the end of the document, where it is not followed by an abstract or a "MyEnd".

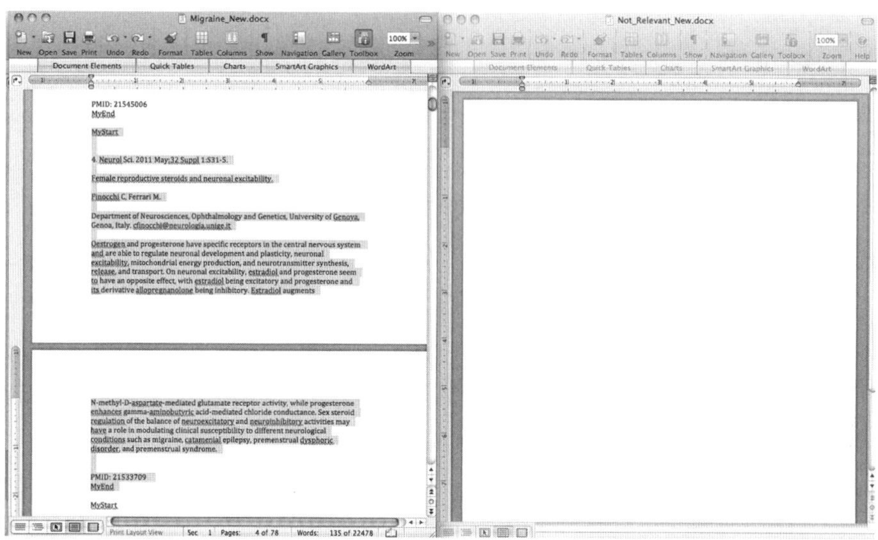

FIGURE V.9

Manual sorting of the 100 abstracts into two categories: Genetic relevant VS Not Genetic relevant.

Once the abstracts have been formatted as described above, we begin the manual process of classifying the first 100 abstracts into two categories: those abstracts showing relevance to genetic markers in migraines and those abstracts that are not relevant. To do this, open another Word document and place side by side with your document containing the 100 abstracts (see Figure V.9 above). Save the document with the 100 abstracts as "Migraine," and save the blank document as "Not_Relevant."

Now manually review each of the 100 abstracts. If the abstract does not contain information on or mention a specific gene or genetic marker related to migraine, then cut and paste the abstract from the "Migraine" document into the "Not_Relevant" document; otherwise leave it in the Migraine document. It is important to cut the entire abstract from the MyStart to MyEnd text markers so that the abstracts can be read into *STATISTICA*. It is also very important when you do this for your own project that you define a very explicit set of rules for document classification before starting this step. If you change your rules as you classify, the accuracy of your class prediction will not be as good because you may have similar abstracts in each file. In this case the paper must mention specific genes or genetics markers related to migraines. The main topic of the paper does not need to be migraines, but the paper must have genes or genetic markers that have been suspected or associated with migraine headaches.

After a quick manual sort, we get 38 abstracts that are not relevant and 62 that are relevant. Be sure to save your files when you have completed this step. Now do "Save As" and save both files as text (".txt") files in your project folder. So you should have a file titled "migraine.txt" and a file titled "not_relevant.txt."

Now to import the files into *STATISTICA*, we need to create a macro. The visual basic macro to do this has been included in this tutorial and is titled "Macro_to_separate_docs.svb." You can run or build this macro in three different ways. You can quickly open the macro by double clicking it, you can open the macro in a text editor and copy the contents and then paste them into a new macro, or you can just manually type the contents of the macro.

The following steps show you how to manually create a macro or create a new macro into which you can paste the contents of the "Macro_to_separate_docs.svb." As an alternative you can double click the icon for the "Macro_to_separate_docs.svb" file, and it will automatically open in *STATISTICA* and you can skip these steps. Going through the following steps, however, will show you how to create a macro from scratch (see Figure V.10).

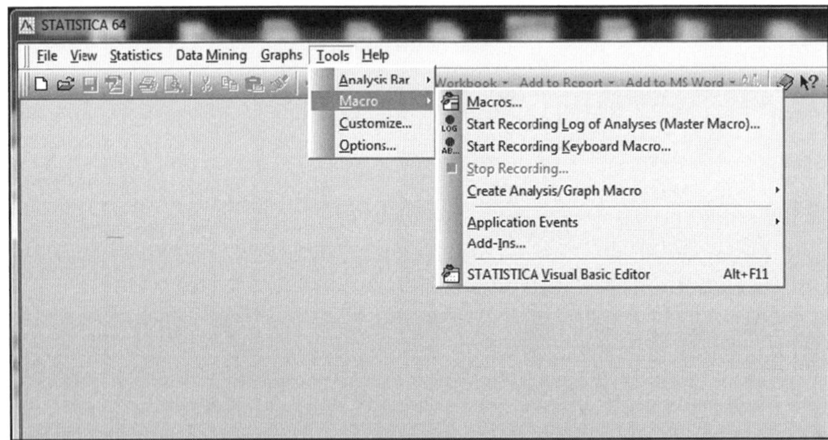

FIGURE V.10
Process to start a new Macro.

From the "Tools" menu select "Macro" and then select "(see Figure V.10) Macros…". This will pull up the Macros dialog box (see Figure V.11a).

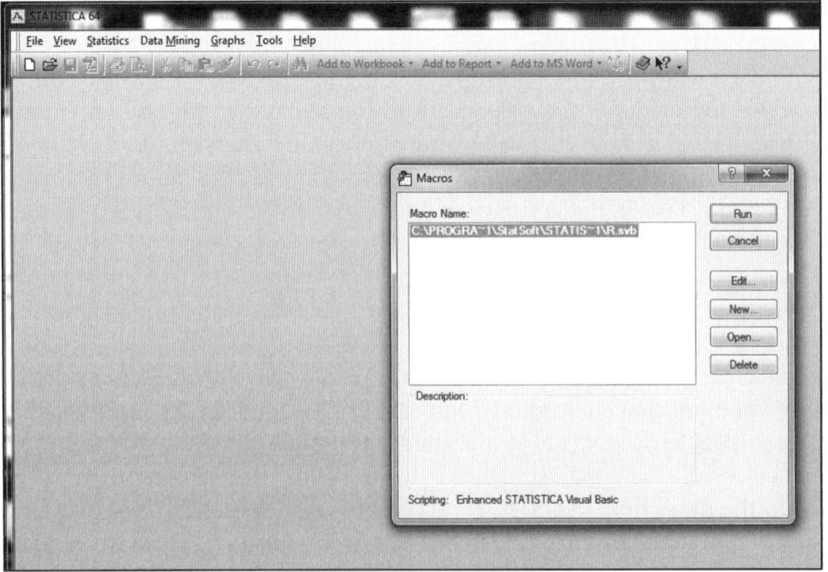

FIGURE V.11a
Macro dialog window.

Once the Macros dialog box appears, click on the "New…" button.

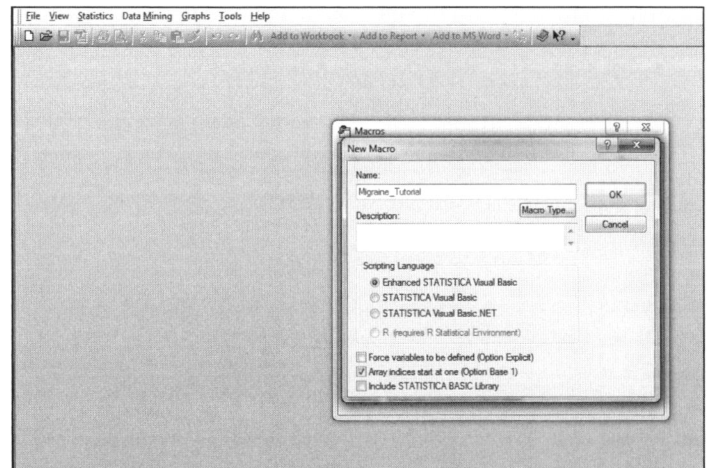

FIGURE V.11b

New Macro window.

The New Macro dialog then appears (see Figure V.11b). Type "Migraine_Tutorial" as the name of the macro. The name is arbitrary; you can name the macro whatever you like, but just make it something that will help you remember its function. Now click OK, and the macro edit screen appears with some text prepopulated:

```
'#Language "WWB-COM"
Option Base 1

Sub Main

End Sub
```

This is the skeleton code for a macro. You can either fill in the missing information by manually typing the rest of the macro as shown below (see Figure V.12) or you can copy and paste the text from the "Macro_to_separate_docs.svb" file.

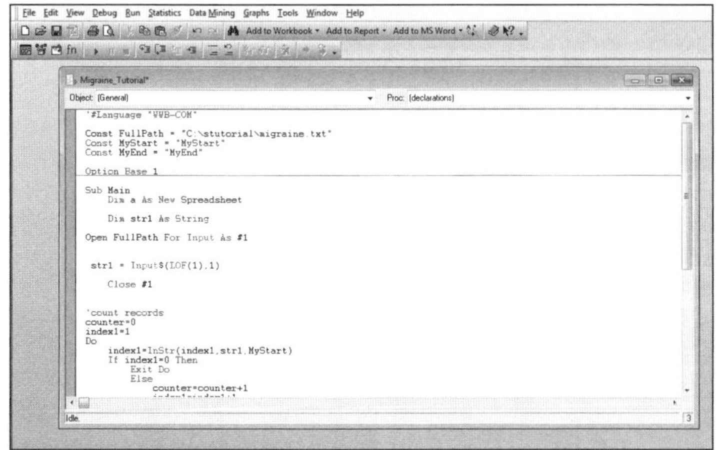

FIGURE V.12

Filling in the lines of the Macro.

The contents of the macro are shown below:

```
'#Language "WWB-COM"

Const FullPath = "C:\stutorial\not_relevant.txt"
Const MyStart = "MyStart"
Const MyEnd = "MyEnd"

Option Base 1

Sub Main
     Dim a As New Spreadsheet

     Dim str1 As String

Open FullPath For Input As #1


 str1 = Input$(LOF(1),1)

   Close #1


'count records
counter=0
index1=1
Do
     index1=InStr(index1,str1,MyStart)
     If index1=0 Then
          Exit Do
          Else
               counter=counter+1
               index1=index1+1
     End If
Loop While True
'MsgBox(CStr(counter))

a.SetSize(counter,1)
index1=1
For i=1 To counter
     index1=InStr(index1,str1,MyStart)

a.Value(i,1)=Mid(str1,index1+Len(MyStart),InStr(index1+1,str1,MyEnd)-index1-Len(MyEnd)-2)
     index1=index1+1
```

```
Next
a.Visible=True
End Sub
```

This macro essentially takes all the text in between the separators "MyStart" and "MyEnd" and puts them in separate cells of a spreadsheet within *STATISTICA*. Since this tutorial is not meant to be a full tutorial on creating macros in visual basic, we will concern ourselves with only three lines in this macro. They are as follows:

```
Const FullPath = "C:\stutorial\not_relevant.txt"
Const MyStart = "MyStart"
Const MyEnd = "MyEnd"
```

The first line contains the absolute path to our text file that we want to import. You can save this file wherever you like, but just make sure you know the absolute path to put into this macro. The second two lines specify the delimiters used to separate abstracts. You can really use any term or set of characters you want for these delimiters. The important thing is *not* to use a term or set of characters that might be found in an abstract or abstract record. For example, using the word "genetic" would be a bad idea because it would be found in multiple abstracts and would create abstract separations in the middle an abstract.

Run the macro in the *STATISTICA* program, putting the individual documents into a spreadsheet called "not_relevant.sta." Do this by Selecting "Run Macro" from the "Run" menu of *STATISTICA* (see Figure V.13).

Once this spreadsheet is formed, we need to either add the "Migraine.txt" documents to the *STATISTICA* spreadsheet2 or make a separate spreadsheet of the Migraine documents and then later merge the two spreadsheets, but, very important, by adding a target variable of "Migraine = Yes or 1" and "Not relevant = No or 2."

To get the Migraine documents, we must modify the line shown below to contain the path to your "migraine.txt" file in the folder you saved it in.

```
Const FullPath = "C:\stutorial\migraine.txt"
```

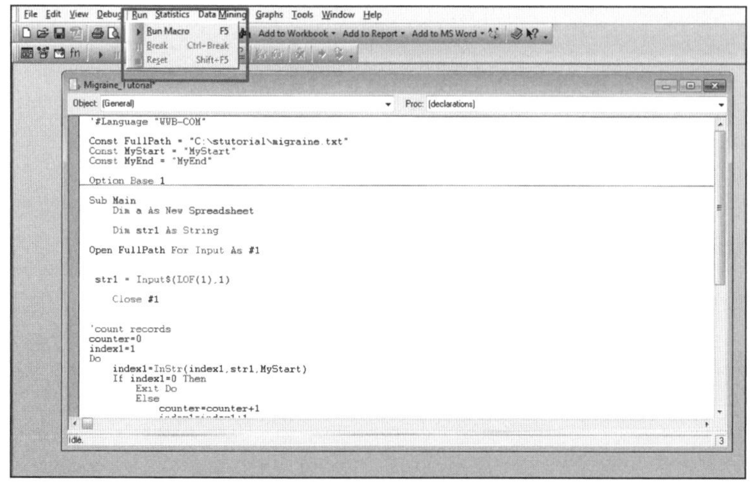

FIGURE V.13
Selecting 'Run Macro' from the 'Run' pull down menu in *STATISTICA*.

Once you have modified the line above, you can run the macro and import the data by Selecting "Run Macro" from the "Run" menu, as shown above.

This will put the "Migraine.txt" into a *STATISTICA* spreadsheet (see Figure V.14), with each separate original PubMed document being in a separate line (case) of the spreadsheet that we can call "Migraine_Genetics.sta."

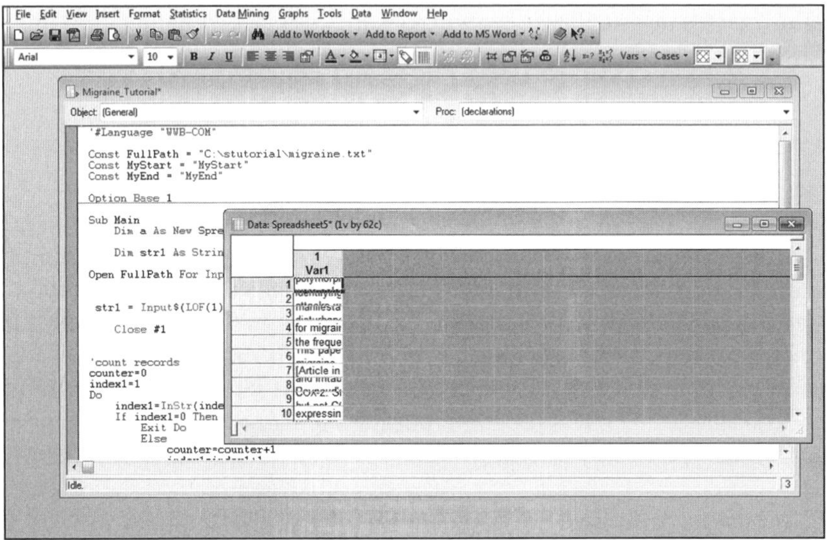

FIGURE V.14

Spreadsheet being formed from running the Macro.

After running the macro, a spreadsheet will appear with one column; each cell in the column contains a separate abstract (see Figure V.15). These abstracts are now ready to be mined with text miner. But first we must merge the two data sets. A "Merge" function is available in *STATISTICA*, but we'll use a simple copy and paste routine here. The two spreadsheets are illustrated in the following two figures, Figure V.15 and Figure V.16).

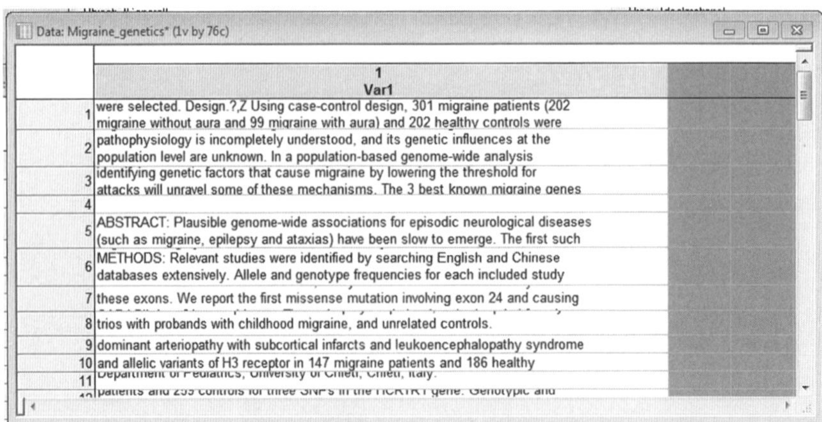

FIGURE V.15

Spreadsheet "Migraine_genetics.sta."

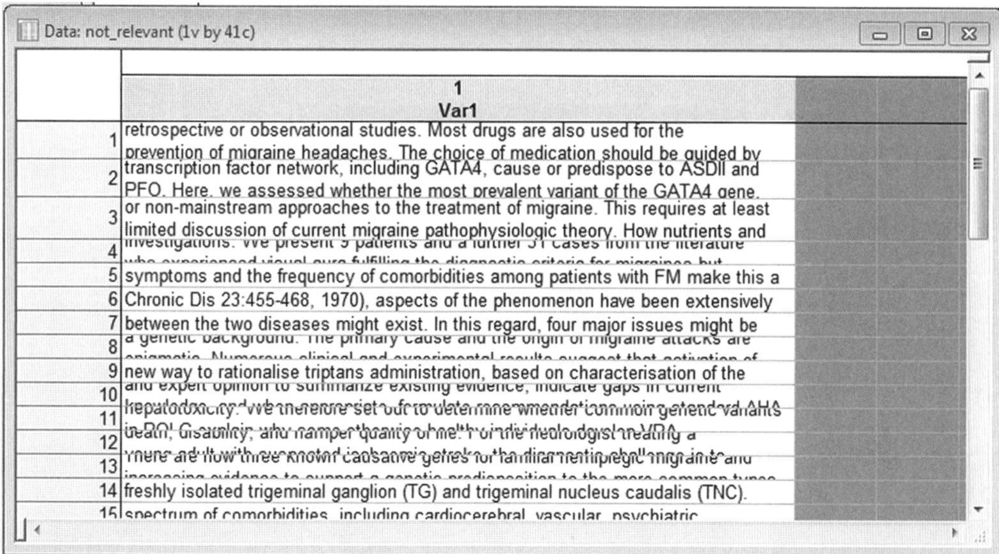

FIGURE V.16
Spreadsheet "not_relevant.sta."

Let's make the "Migraine_genetics.sta" our base spreadsheet. (Please follow Figures V.17 through V.26 for this entire process of merging two spreadsheets into one.) Since we may want to refer to this original spreadsheet later, we'll do a Save As and make a second copy that we'll call "Migraine PUB-MED Text Mining.sta." (see Figure V.17)

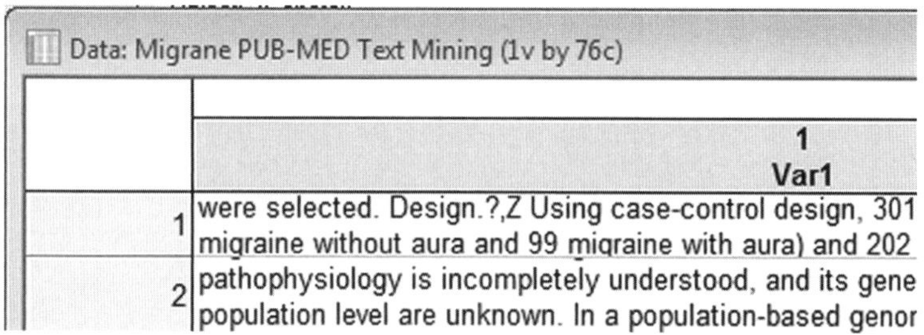

FIGURE V.17
New name for our spreadsheet that will combine both the "Migraine_genetics.sta" and the "not_relevant.sta" text files.

Now, by going to the Data pull-down menu in *STATISTICA*, we will release on "Cases" and then select "Add Cases." (see Figure V.18)

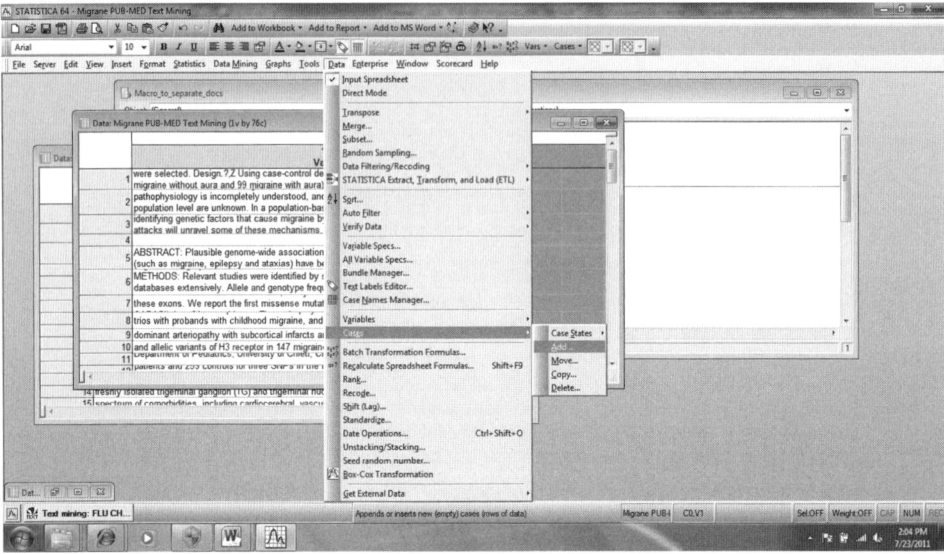

FIGURE V.18

Data pull-down menu with "Cases" and "Add" selected.

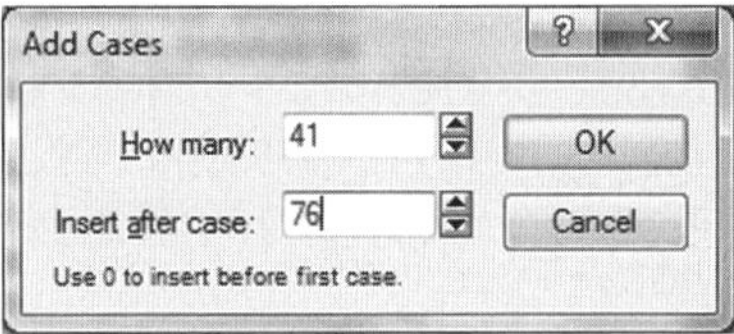

FIGURE V.19

Window asking how many cases to add, and then which numbered case.

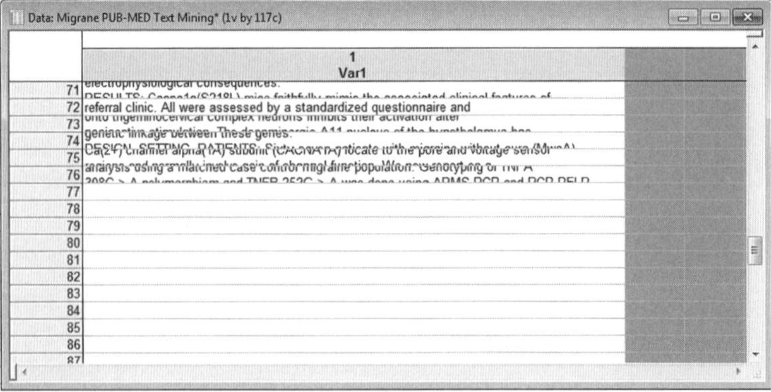

FIGURE V.20

Enlarged spreadsheet with 41 new rows to accept the data from the "not_relevant.sta" data file.

Highlight the text variable column (see Figure V.21) in the "not_relevant.sta" data set by clicking on the "header" for this variable.

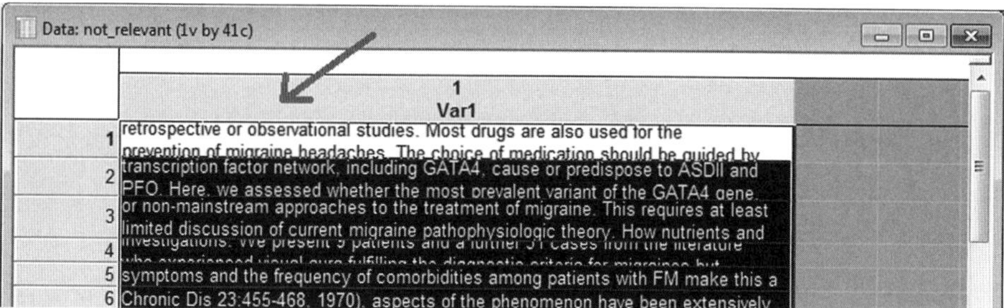

FIGURE V.21
Selecting the text variable column in the "not_relevant.sta" spreadsheet.

Right click on the darkened highlighted area of the text column of data, and select "Copy" from the pop-up menu (see Figure V.22).

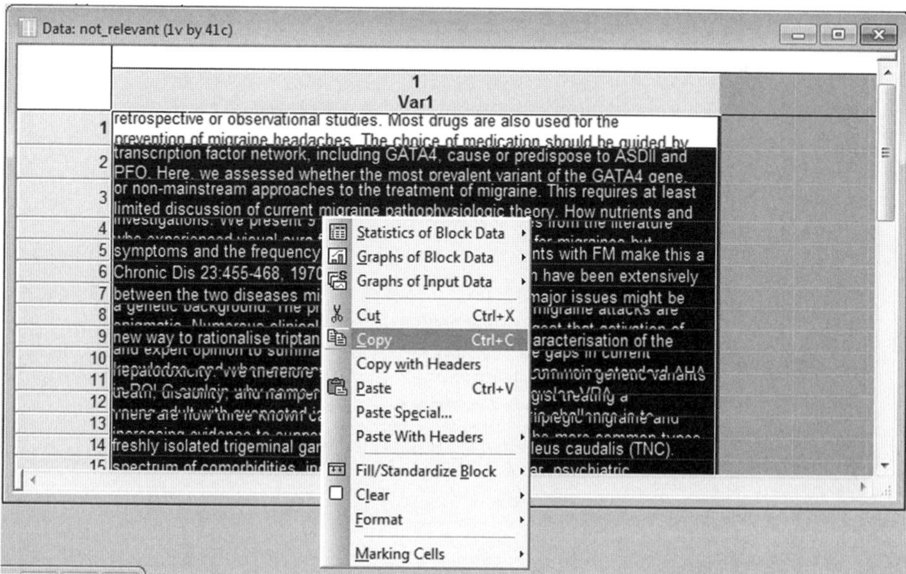

FIGURE V.22
Selecting the "not_relevant.sta" text data to copy into our master data file.

Next, we need to click on to "select" Case No 77 in the master data file. This will allow copying of the "not_relevant.sta" text data into case row 77 and the remaining rows of the master data file.

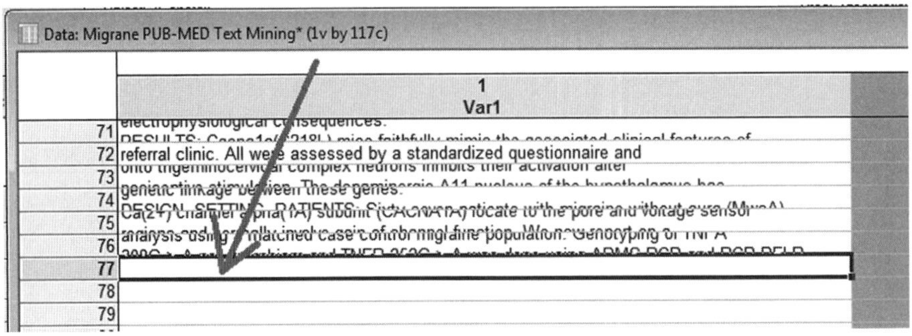

FIGURE V.23
Selecting the row start point in which to paste in another set of data.

Next, right click on case row #77 in Figure V.24, and from the pop-up menu select "Paste."

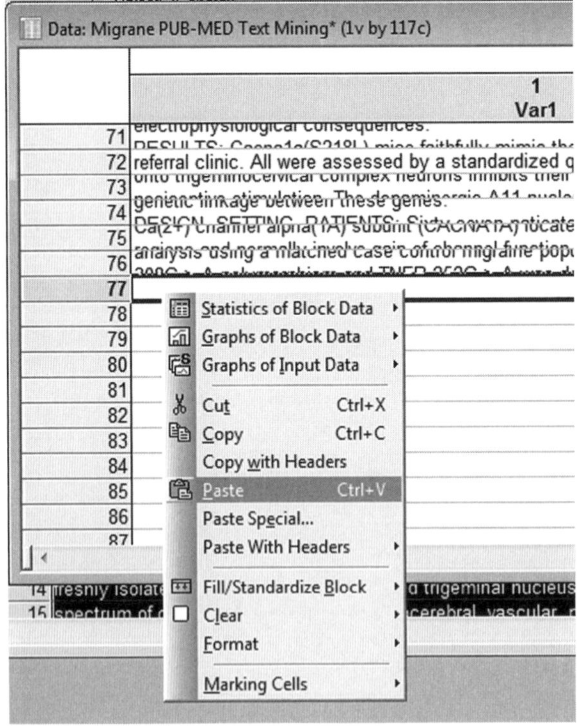

FIGURE V.24
Pasting a data column into master data file.

Next, paste the two data files together in one master data file as illustratted in Figure V.25. This will be the "master data file" that we'll use for the text mining in the remaining part of this tutorial.

FIGURE V.25

The two separate data files are now "pasted together" in one master file, the one we'll use for text mining in the rest of this tutorial.

Now we need to add a new variable to our master data file (follow through on the next 9 figures for this entire process; e.g. Figure V.26 – Figure V.34). This will be the "target variable" that can be used for predictive data analysis following the text mining part. This variable will be coded "0" for the "non_relevant" documents—that is, Cases 77–117. Cases 1–76 will be coded "1" for "Migraine_genetic" documents (e.g., documents that contain both references to "migraine headaches" and also "genetics markers").

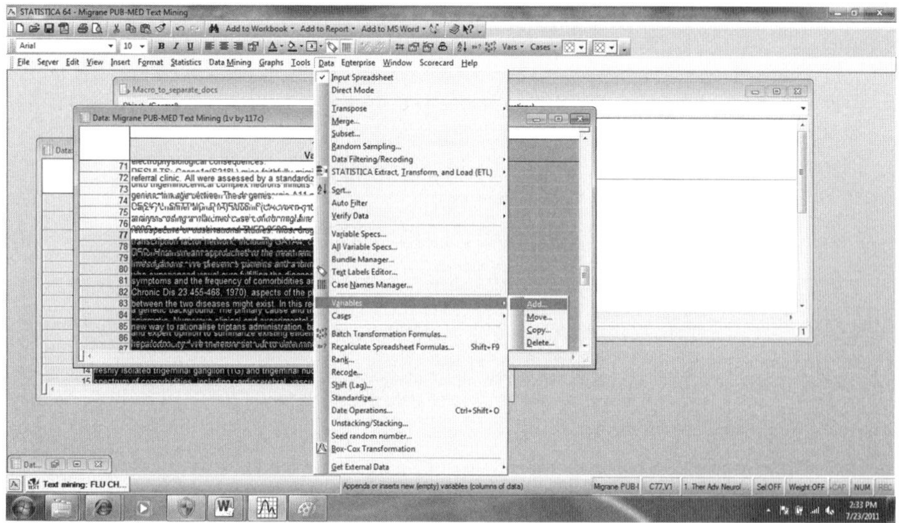

FIGURE V.26

Process to add a new variable to the data file.

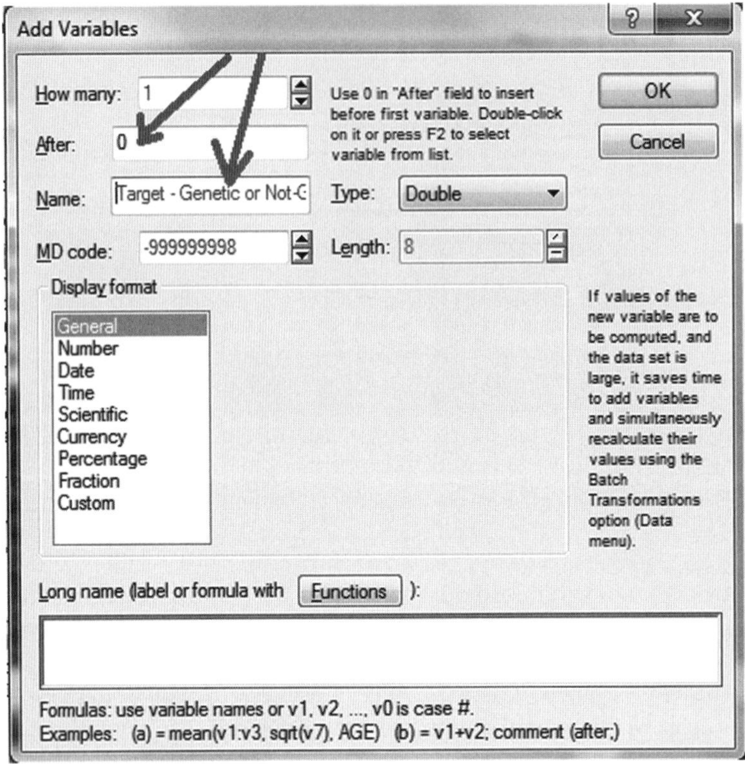

FIGURE V.27
Dialog for adding new variable. The variable will be added after the "0" variable—for example, just before the text variable.

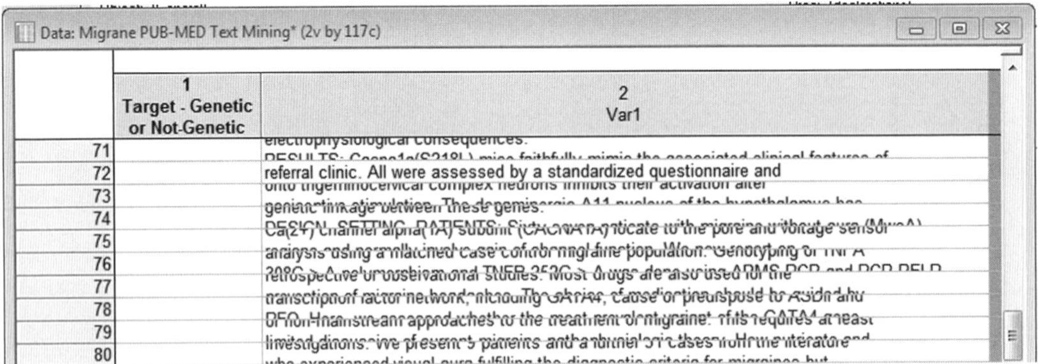

FIGURE V.28
New Variable Column in the Var 1 spot. The text variable has moved to the Var 2 position.

Now we need to select the "codes" for Var 1, the Target variable. We will do this through the "variable specs" dialog obtained by double clicking on the Var 1 header.

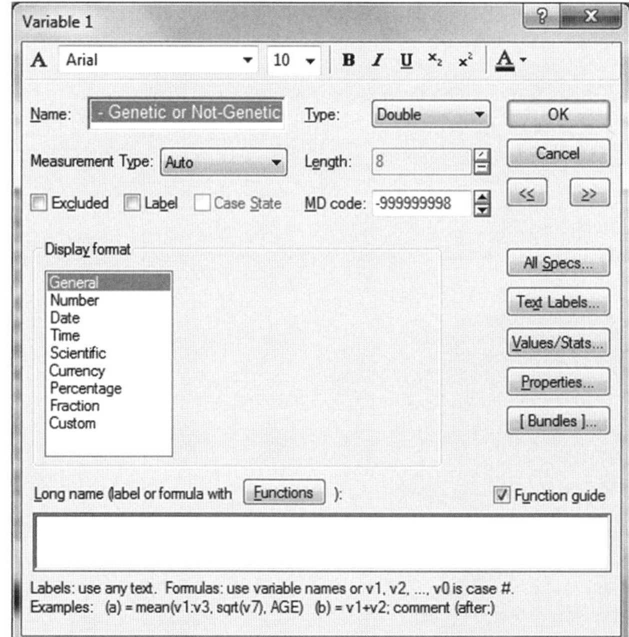

FIGURE V.29

The Variable specs dialog.

Select the "Text Labels" button on the center right side of the "Variable Specs" dialog and then the "Text Labels Editor" dialog will appear as shown in the next figure, Figure V.30.

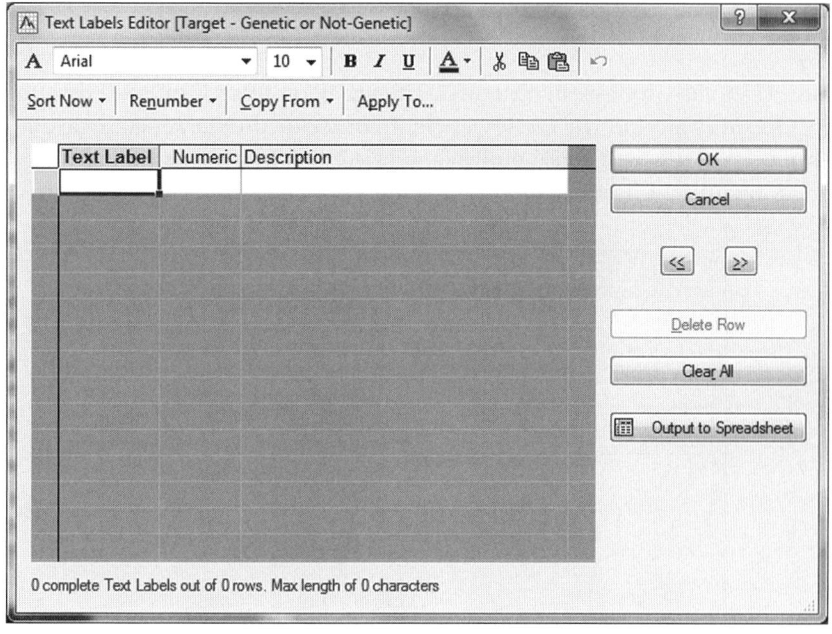

FIGURE V.30

Text Labels Editor dialog.

Then type in 0 (zero) and 1 for the numeric, and "Non-relevant" and "Genetic" for the Text Label.

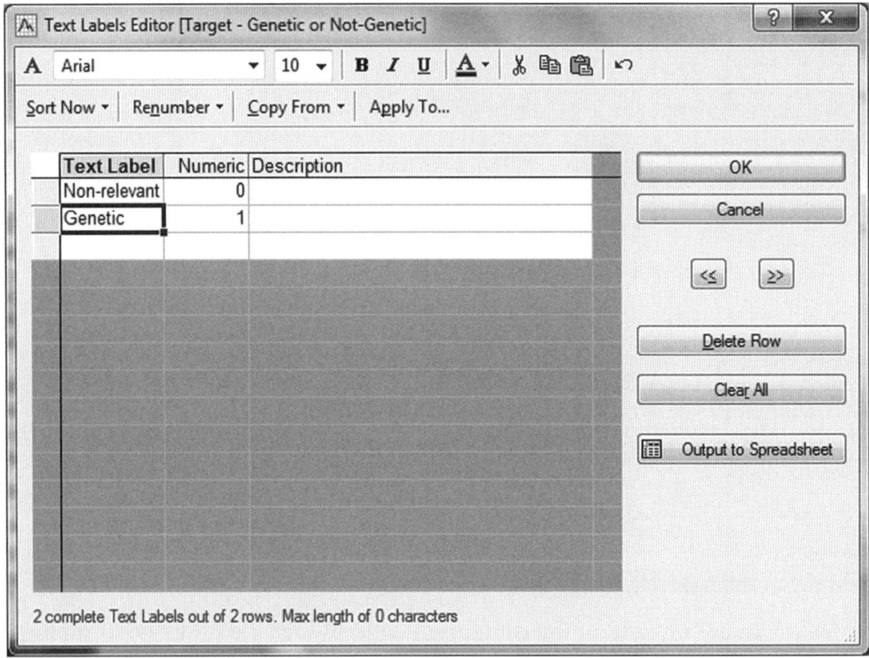

FIGURE V.31

Text Labels and corresponding numeric selected.

Select OK.

Then type in "1" in Var 1 for cases 1, 2, and 3. Then click on Case 1 and pull down to Case 3; then, using the cursor, move it around on the lower right corner of the darkened area until a "+" cursor appears. Then hold down the left mouse button and select this new + cursor and drag it down to case #76 (the Genetic Migraine documents are cases 1–76).

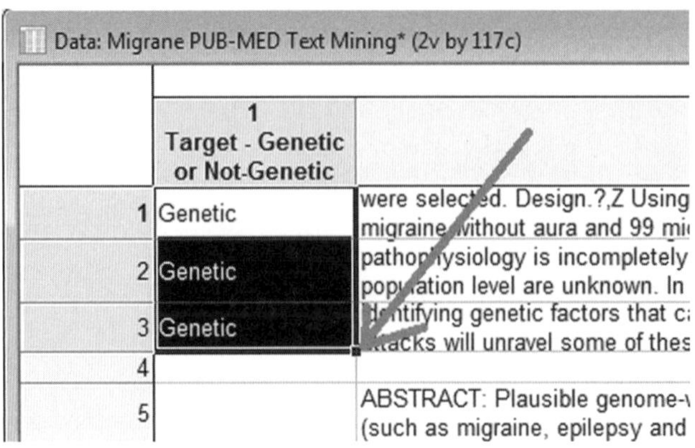

FIGURE V.32

Filling Var 1 with "1" for Genetic for cases 1–76.

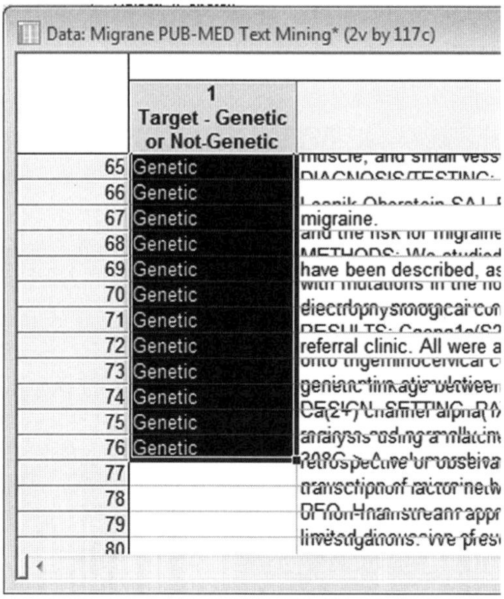

FIGURE V.33
Genetic pasted in for Var 1 through case 76.

Now, do the same for "0" for Cases 77, 78, and 79, and pull down to the bottom of the master data file.

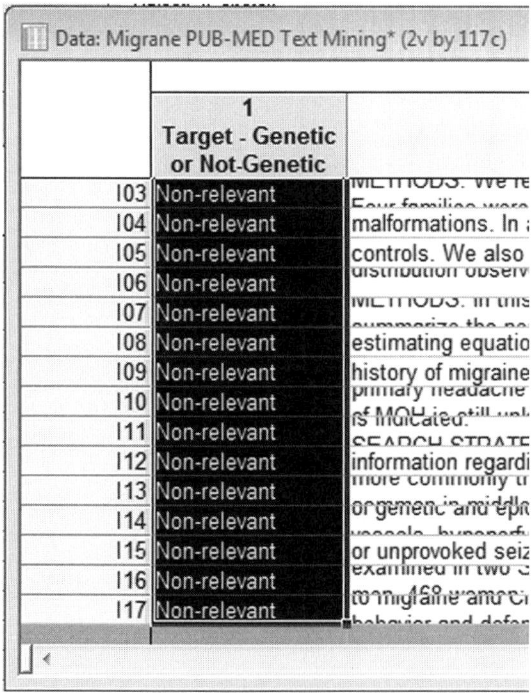

FIGURE V.34
Variable 1 fully defined for the target as "Genetic" or "Non-relevant."

Now the Text Mining will begin: Select the "Text and Document Mining" module from the Data Mining pull-down menu (see Figure V.35).

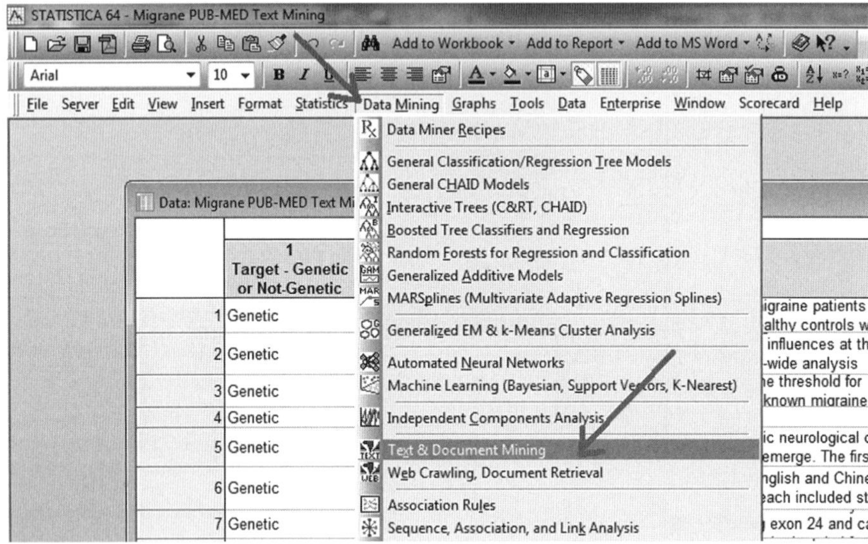

FIGURE V.35

Selecting the "Text Mining" module in *STATISTICA*.

Click on the Text Variable button (see Figure V.36), and then select the "text variable" in our master spreadsheet (see Figure V.37 and Figure V.38).

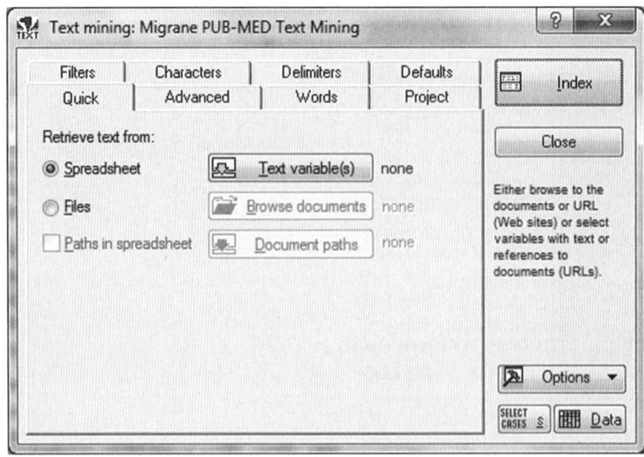

FIGURE V.36

The Text Mining opening dialog window.

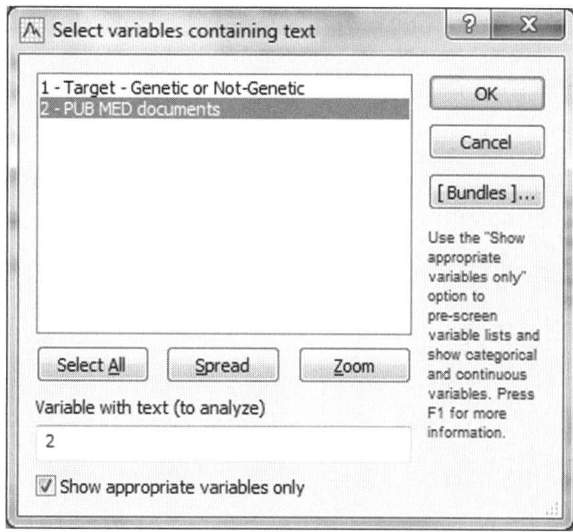

FIGURE V.37

Variable selection window pops up after "Text Variable" button is selected; PUB MED Documents, Var 2, selected.

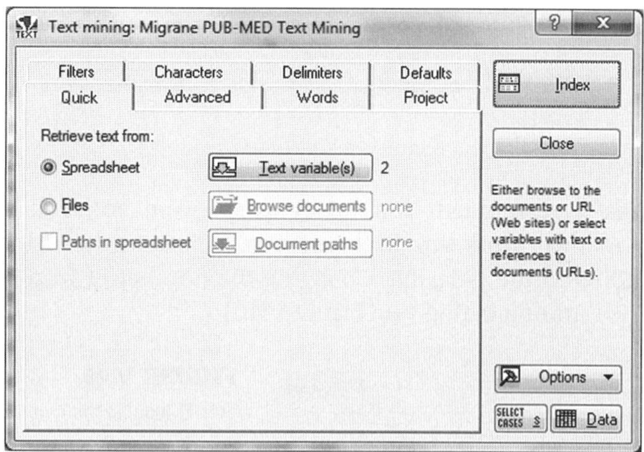

FIGURE V.38

Text Mining dialog showing that the text variable is selected.

We could add other parameters like including more words to the default "word exclusion list" or making a "word inclusion list" and selecting other delimiters and filters, but for now we will just accept all defaults.

Click the Index button in the upper right-hand corner of the Text Mining dialog, and the word frequency indexing will begin. It will take about 25 seconds or less on a 64-bit PC computer to do the indexing of this data set. When completed, a "results" dialog will pop up on the screen (see Figure V.39).

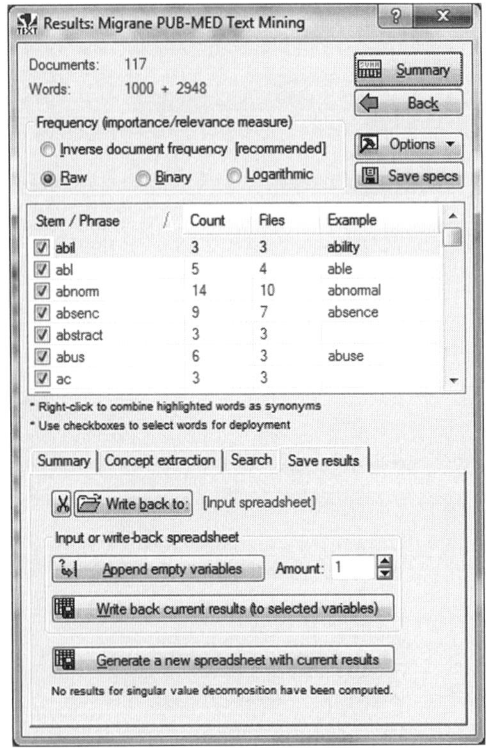

FIGURE V.39
Results of indexing words.

One thousand words were selected, and 2,948 words were not. The default setting is 1,000 words. I think this is too many to work with in this project, so let's set the "selected words" to a smaller number, maybe 300, and rerun the indexing. Changing the number of words selected is done in the "Advanced Tab" of the text mining dialog (see Figure V.40).

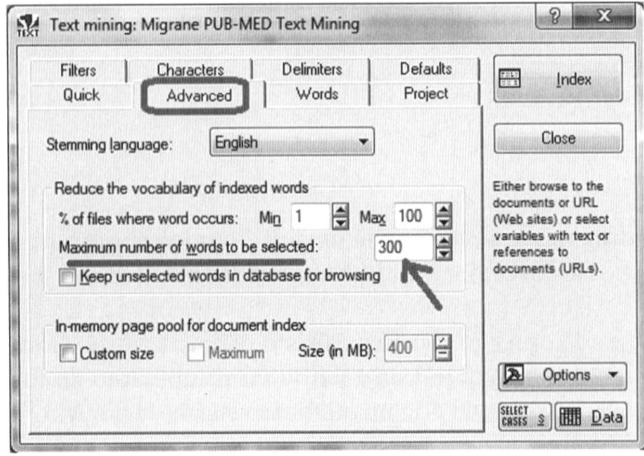

FIGURE V.40
Setting the maximum number of words selected to 300.

When the INDEX button is selected (see Figure V.40 above) the computation of word frequencies will be done; during this process you may see a "Progress bar" as illustrated in Figure V.41.

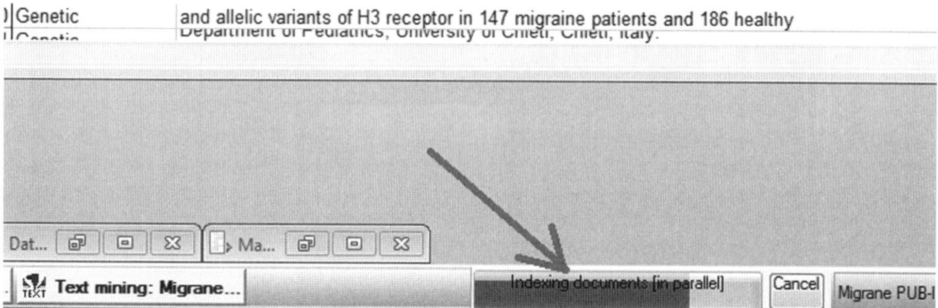

FIGURE V.41
During the indexing process, a green progress bar will be seen at the bottom of the screen.

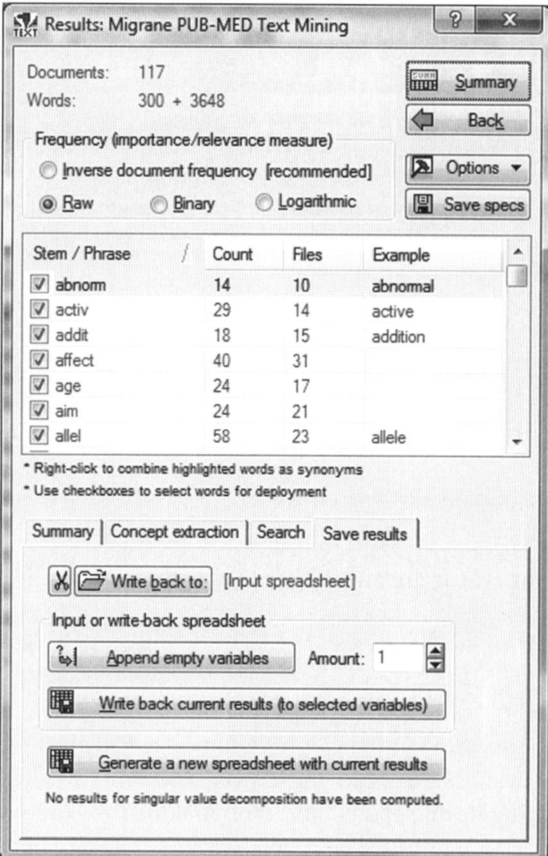

FIGURE V.42
Results of indexing a maximum of 300 words.

We note in the figure above (see Figure V.42) that the words are in alphabetical order. We may wish to see them in order of word frequency, from highest word frequency to lowest. We can do this by clicking on the "count" heading above the words (see Figure V.43).

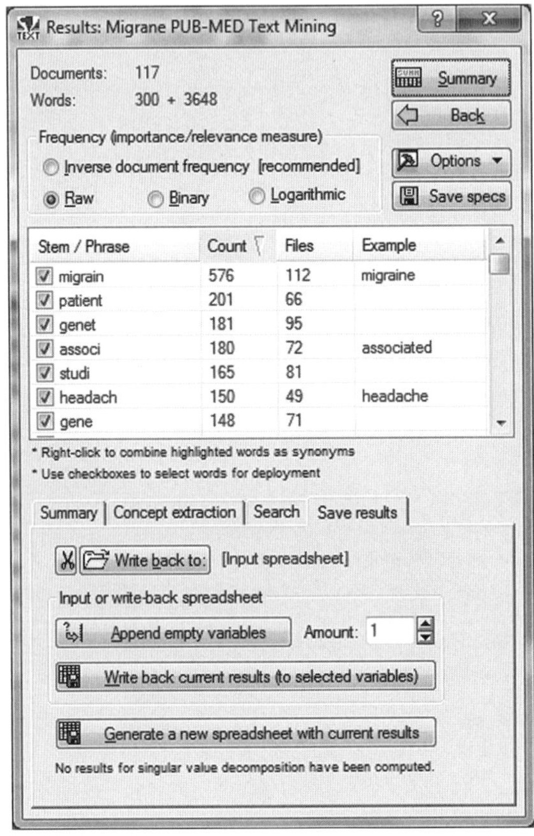

FIGURE V.43

Word frequency count with "count" heading clicked to order words in count frequency instead of alphabetical.

We see that the following words are among those at the top of the list:

a. migraine
b. genetic
c. headache

Next, we need to save these word frequencies back to our spreadsheet. That means we need to make 300 new variable columns available to hold these word counts. This is done by selecting the "Save Results" tab on the Text Mining Results dialog and then making the "Amount" 300, and then selecting "Append Empty Variables." This will bring up a window in which to select the word variables in one window pane and also the new variables numbers (in the data file) to place these word counts. Then, after these arrangements have been made, the "Write back current results" button is selected to finalize the words being put into our master data file (see Figures V.44 through V.51 for the entire process).

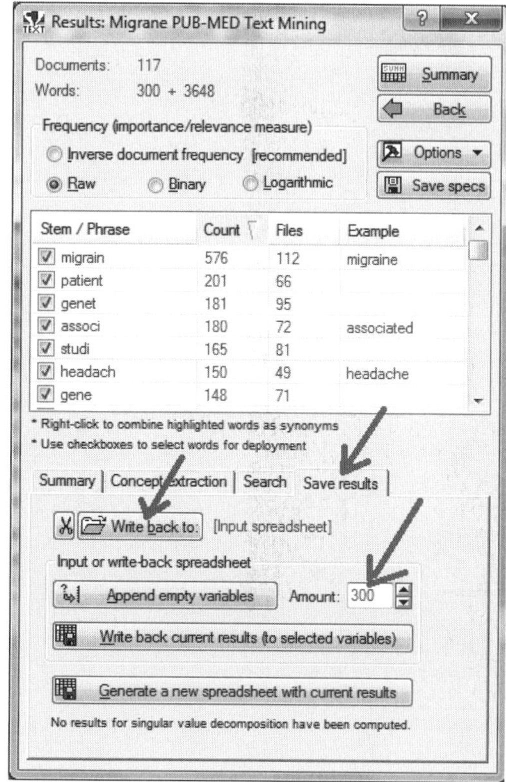

FIGURE V.44
Save Results tab. The amount of words to send to the master data file and the "Write Back" button are used to get the word counts into the data file.

Click the "Write back to" button, and a directory of saved files pops up. At this point, select the "Migraine PUB-MED Text Mining.sta" data file.

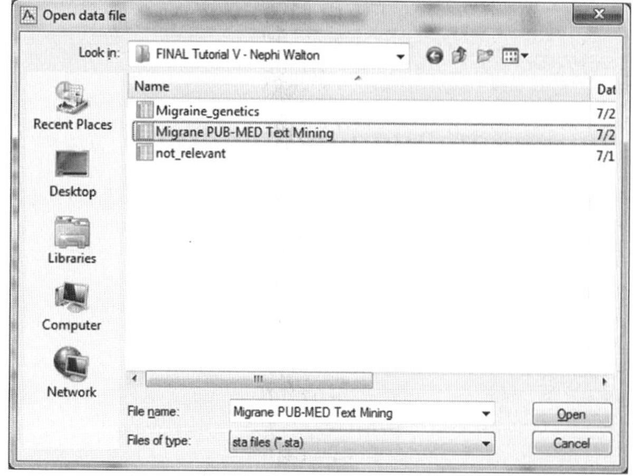

FIGURE V.45
Open data file window where the "master data file" is selected to write back word frequency counts.

Select the "Migraine PUB-MED Text Mining.sta" data file, and click "Open" on the dialog.

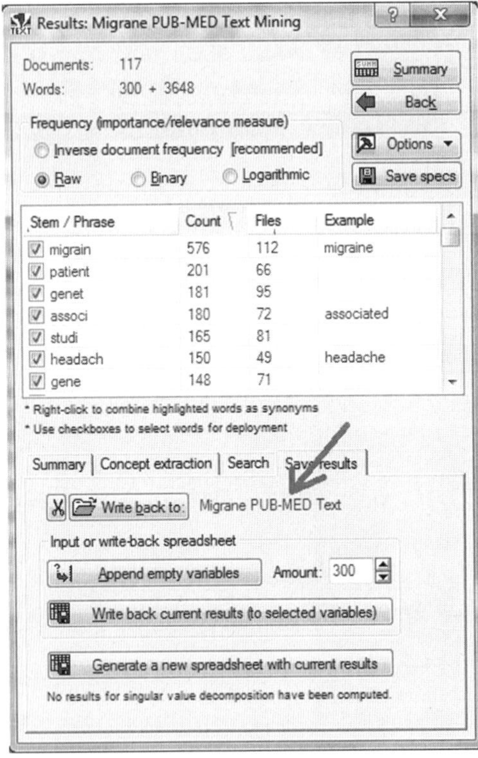

FIGURE V.46

Migraine PUB-MED Text Mining data file selected.

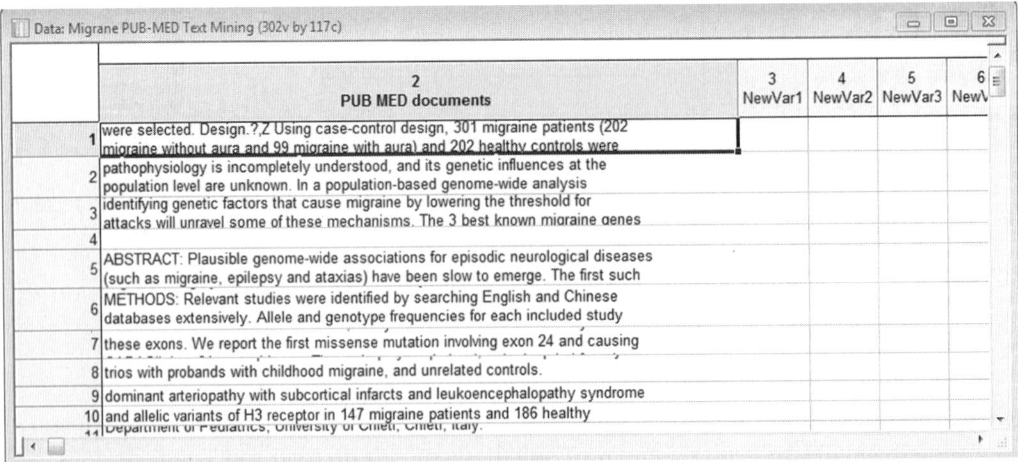

FIGURE V.47

New variable columns created in our master spreadsheet to hold the new indexed word frequencies.

Now click the "Write back current results" button to specify which words go into which variable number columns. A new window will pop up; select the words on the left and the variable columns on the right.

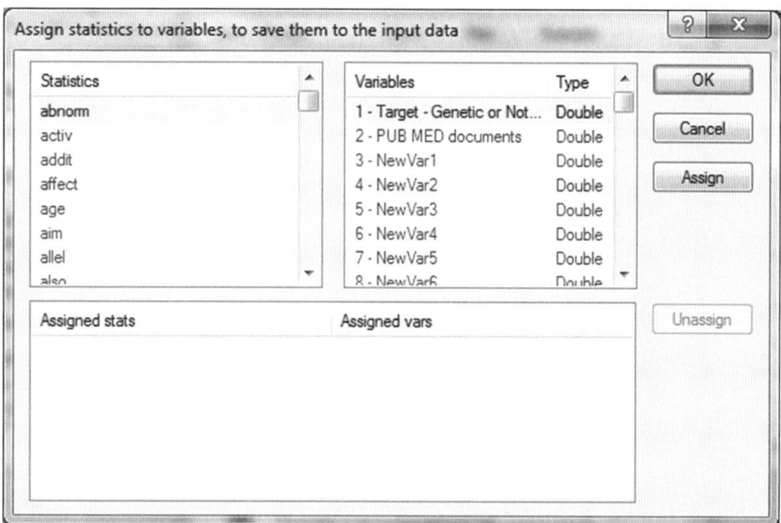

FIGURE V.48
Assign statistics to variables window.

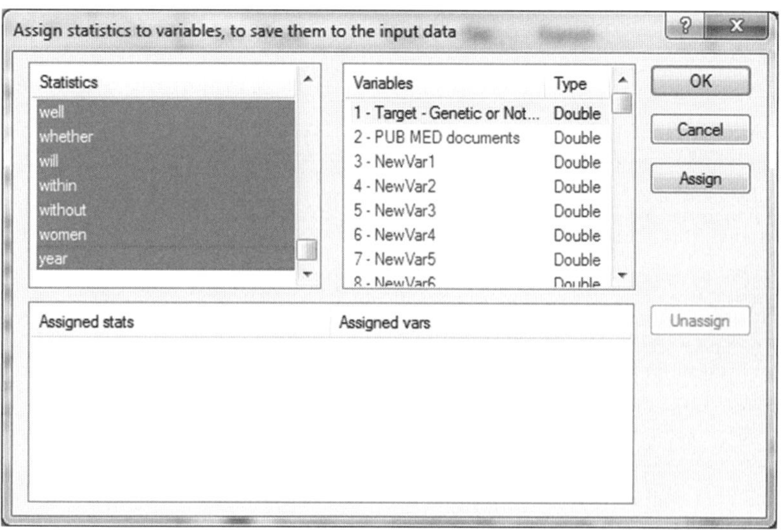

FIGURE V.49
Highlight all the 300 words in the left panel.

Highlight all the new 300 variables in the right-hand panel.

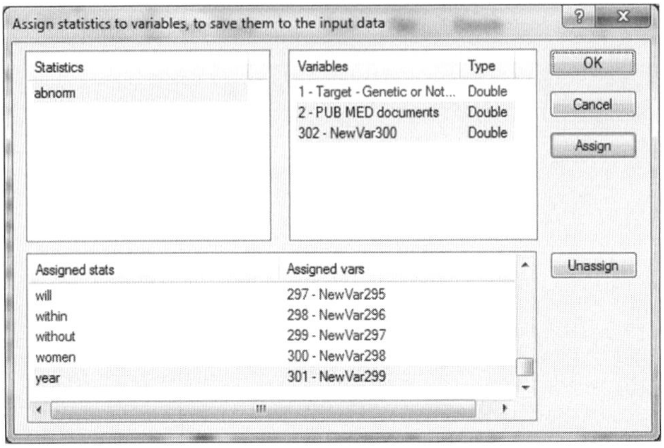

Assign statistics to variables, to save them to the input data

Statistics
abnorm

Variables	Type
1 - Target - Genetic or Not...	Double
2 - PUB MED documents	Double
302 - NewVar300	Double

OK
Cancel
Assign

Assigned stats	Assigned vars
will	297 - NewVar295
within	298 - NewVar296
without	299 - NewVar297
women	300 - NewVar298
year	301 - NewVar299

Unassign

FIGURE V.50
Click "Assign" to assign words to variable numbers.

Variables (words) assigned are now available for inspection in the bottom window of the "Assign statistics to variables" window. We can inspect this to make sure the right words are assigned to the correct variable number.

Now click OK to actually put these word frequencies into the master data file.

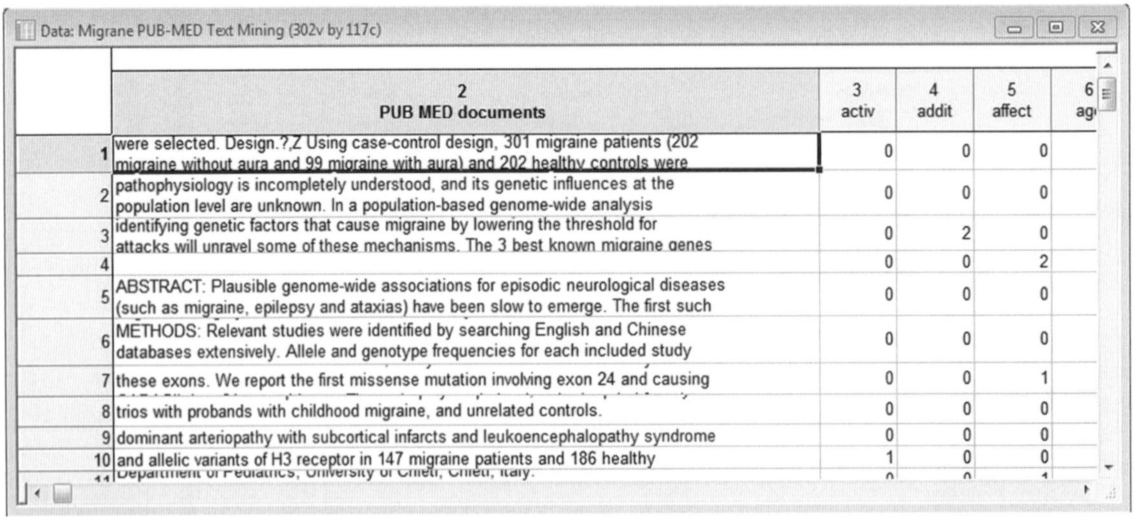

Data: Migrane PUB-MED Text Mining (302v by 117c)

	2 PUB MED documents	3 activ	4 addit	5 affect	6 ag
1	were selected. Design.?,Z Using case-control design, 301 migraine patients (202 migraine without aura and 99 migraine with aura) and 202 healthy controls were	0	0	0	
2	pathophysiology is incompletely understood, and its genetic influences at the population level are unknown. In a population-based genome-wide analysis	0	0	0	
3	identifying genetic factors that cause migraine by lowering the threshold for attacks will unravel some of these mechanisms. The 3 best known migraine genes	0	2	0	
4		0	0	2	
5	ABSTRACT: Plausible genome-wide associations for episodic neurological diseases (such as migraine, epilepsy and ataxias) have been slow to emerge. The first such	0	0	0	
6	METHODS: Relevant studies were identified by searching English and Chinese databases extensively. Allele and genotype frequencies for each included study	0	0	0	
7	these exons. We report the first missense mutation involving exon 24 and causing	0	0	1	
8	trios with probands with childhood migraine, and unrelated controls.	0	0	0	
9	dominant arteriopathy with subcortical infarcts and leukoencephalopathy syndrome	0	0	0	
10	and allelic variants of H3 receptor in 147 migraine patients and 186 healthy	1	0	0	
11	Department of Pediatrics, University of Chieti, Chieti, Italy.	0	0	1	

FIGURE V.51
Looking at Var 3, Var 4, Var 5, and so on, we can see that the word frequencies have been put into the new columns for each "case" or "each document."

Now we want to do one more thing to prepare all the information we can use from the word counts prior to doing predictive modeling. This is performing a singular value decomposition (SVD) using the "inverse document frequency" form of the indexed words (see Figure V.52).

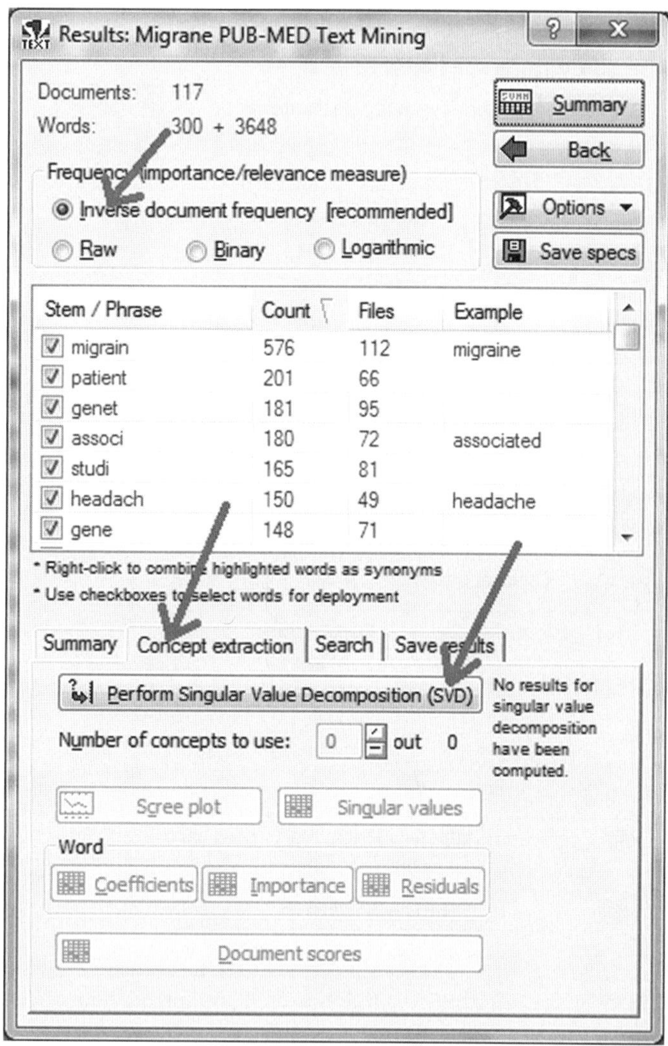

FIGURE V.52

"Inverse document frequency" selected, "Concept extraction" tab selected, and SVD button will be selected to perform the computations.

Notice that the bottom $^1/_3$ of the Results dialog is "grayed out" when the "Concept extraction" tab is selected. Now select the "Summary" button (upper right-hand corner).

After selecting "Perform Singular Value Decomposition," we see that the bottom $^1/_3$ of the dialog is grayed in and that 28 Concepts have been created (see Figure V.53). Concepts are like Principal Components in PCA: there are "new variables" composed of portions of words from the original 300 words we selected.

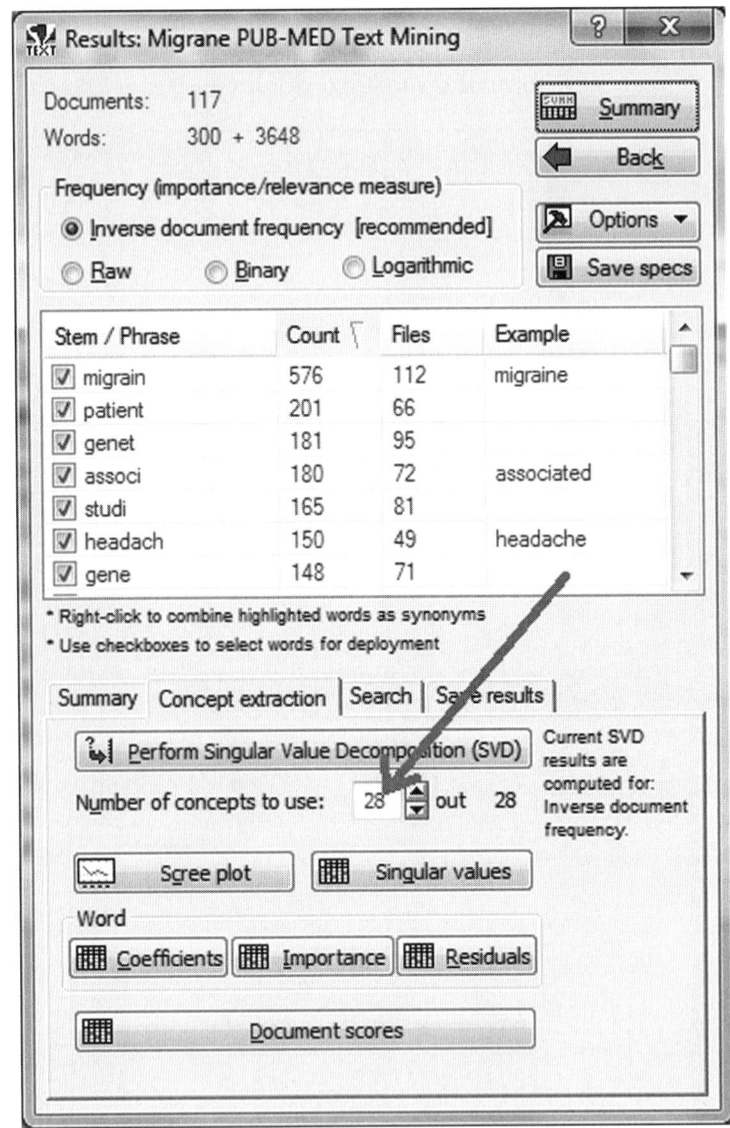

FIGURE V.53

Concepts are now created, 28 in number, and the bottom 1/4 of this tab is now grayed in, so we can select various result buttons and send this information to a workbook.

Now let's look at the "Scree plot" (see Figure V.54) and also click on the other results buttons to put them into a workbook in case we need any for futher exploration or predictive modeling.

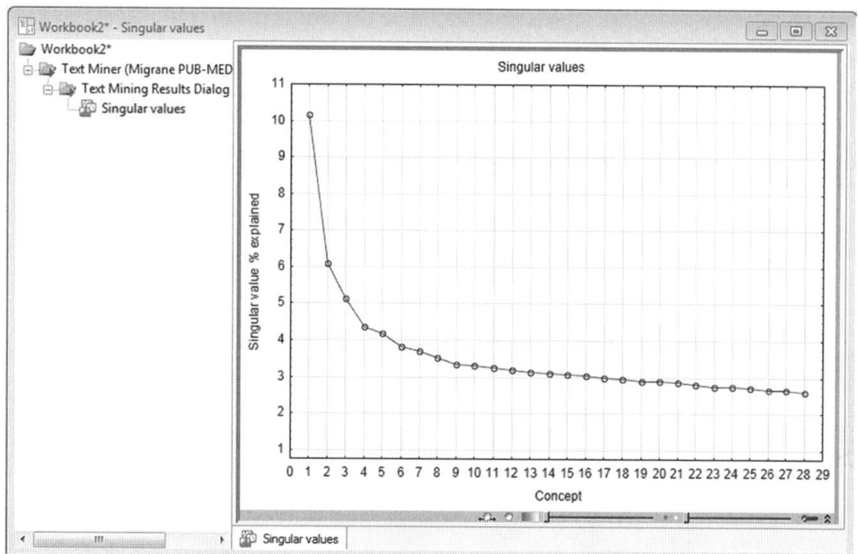

FIGURE V.54

Scree plot of the 28 concepts extracted from the 300 PUB-MED documents.

The Scree plot information is used just as we would in factor analysis or principal components analysis. The new variables, called concepts (see Figure V.55), that are most important are those to the left of any "inflection" point in the curve. Thus, in this case, the first 4 concepts are clearly important, but we may also want to look at 5 and 6 because the curve does not fully begin to level off until we reach concept number 6.

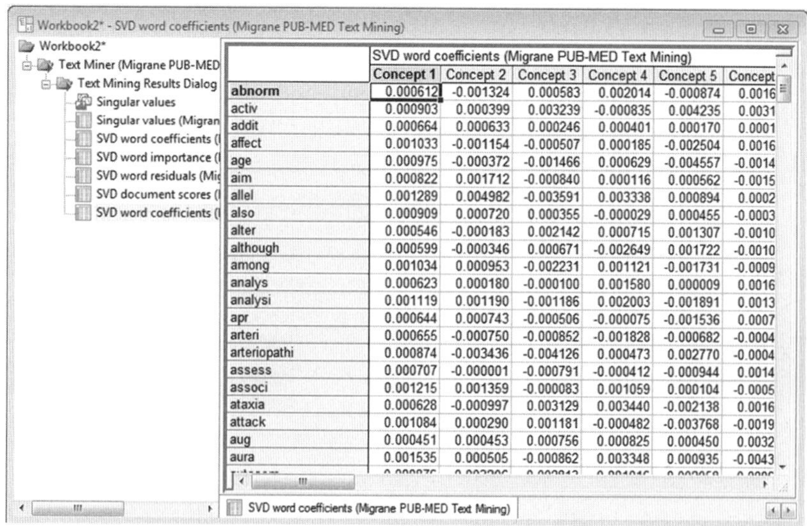

SVD word coefficients (Migrane PUB-MED Text Mining)

	Concept 1	Concept 2	Concept 3	Concept 4	Concept 5	Concept
abnorm	0.000612	-0.001324	0.000583	0.002014	-0.000874	0.0016
activ	0.000903	0.000399	0.003239	-0.000835	0.004235	0.0031
addit	0.000664	0.000633	0.000246	0.000401	0.000170	0.0001
affect	0.001033	-0.001154	-0.000507	0.000185	-0.002504	0.0016
age	0.000975	-0.000372	-0.001466	0.000629	-0.004557	-0.0014
aim	0.000822	0.001712	-0.000840	0.000116	0.000562	-0.0015
allel	0.001289	0.004982	-0.003591	0.003338	0.000894	0.0002
also	0.000909	0.000720	0.000355	-0.000029	0.000455	-0.0003
alter	0.000546	-0.000183	0.002142	0.000715	0.001307	-0.0010
although	0.000599	-0.000346	0.000671	-0.002649	0.001722	-0.0010
among	0.001034	0.000953	-0.002231	0.001121	-0.001731	-0.0009
analys	0.000623	0.000180	-0.000100	0.001580	0.000009	0.0016
analysi	0.001119	0.001190	-0.001186	0.002003	-0.001891	0.0013
apr	0.000644	0.000743	-0.000506	-0.000075	-0.001536	0.0007
arteri	0.000655	-0.000750	-0.000852	-0.001828	-0.000682	-0.0004
arteriopathi	0.000874	-0.003436	-0.004126	0.000473	0.002770	-0.0004
assess	0.000707	-0.000001	-0.000791	-0.000412	-0.000944	0.0014
associ	0.001215	0.001359	-0.000083	0.001059	0.000104	-0.0005
ataxia	0.000628	-0.000997	0.003129	0.003440	-0.002138	0.0016
attack	0.001084	0.000290	0.001181	-0.000482	-0.003768	-0.0019
aug	0.000451	0.000453	0.000756	0.000825	0.000450	0.0032
aura	0.001535	0.000505	-0.000862	0.003348	0.000935	-0.0043

FIGURE V.55

Workbook of results derived from the "Concept" tab of the Text Mining Results dialog.

At this point, before we send the concepts over to our master data file, let's look at a few of the concepts by making scatterplots of one concept versus another. We will start with Concept 1 versus Concept 2. If this gives us information that appears useful, we may stop there, but if not, we may make other plots, like Concept 2 versus Concept 3, or Concept 4 versus Concept 1, and so on.

These scatterplots are made from the "Concept Coefficients," so we will make the Coefficient spreadsheet an "Active data input" spreadsheet by right clicking on the name of this in the left side of the Workbook and then selecting "Use as active input" from the pop-up menu (see Figure V.56 and Figure V.57).

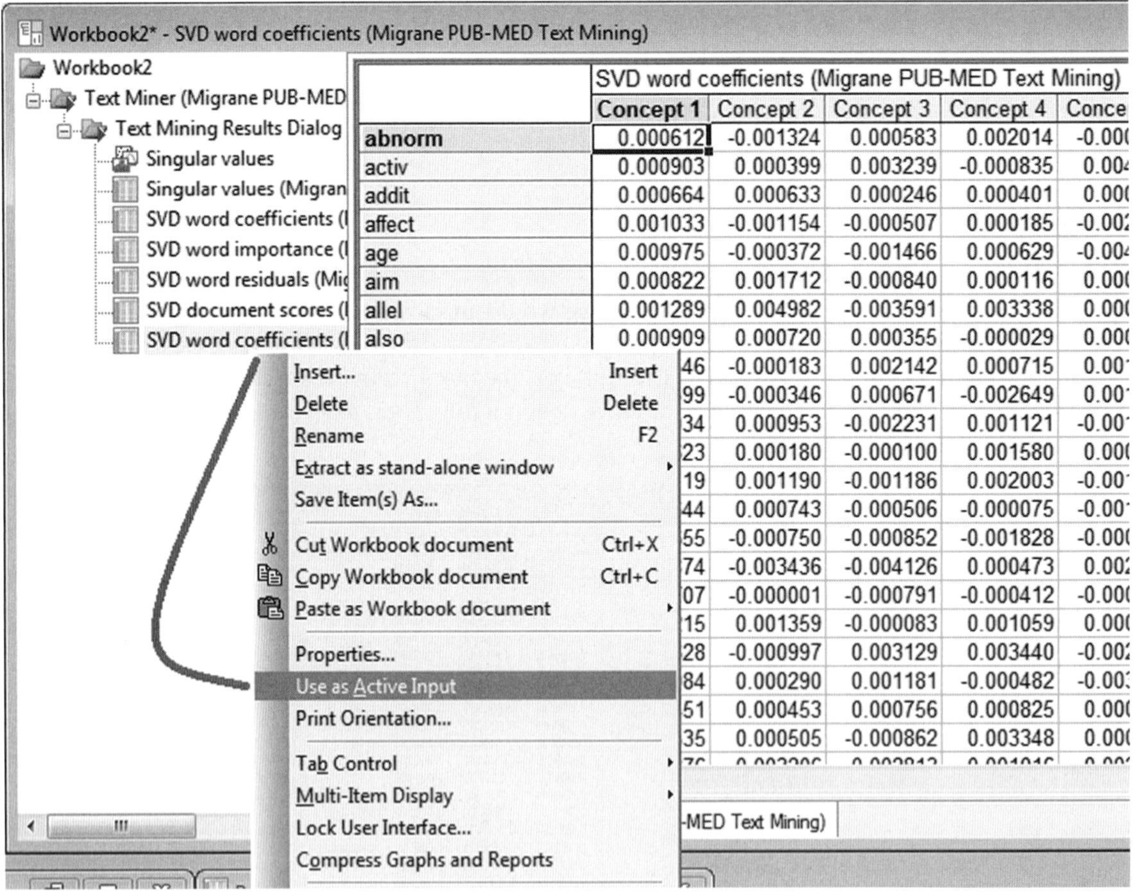

FIGURE V.56

Selecting the Concept Coefficients as the "active input data."

Workbook or Results - PUB MED analysis - SVD word coefficients (Migrane PUB-MED Text M

Workbook or Results - PUB MED a
Text Miner (Migrane PUB-MED
Text Mining Results Dialog
Singular values
Singular values (Migran
SVD word coefficients (
SVD word importance (
SVD word residuals (Mi
SVD document scores (
SVD word coefficients (

	SVD word coefficients (Mi	
	Concept 1	Concept 2
abnorm	0.000612	-0.001324
activ	0.000903	0.000399
addit	0.000664	0.000633
affect	0.001033	-0.001154
age	0.000975	-0.000372
aim	0.000822	0.001712
allel	0.001289	0.004982
also	0.000909	0.000720
alter	0.000546	-0.000183
although	0.000599	-0.000346
among	0.001034	0.000953
analys	0.000623	0.000180
analysi	0.001119	0.001190
apr	0.000644	0.000743
arteri	0.000655	-0.000750

FIGURE V.57
"SVD word coefficients" is now the "active input," as denoted by the red box around the icon to the left of the "SVD word coefficients" name in the "table of contents" in the left panel of the results workbook.

Now let's make the scatterplots (see Figures V.58 and V.59).

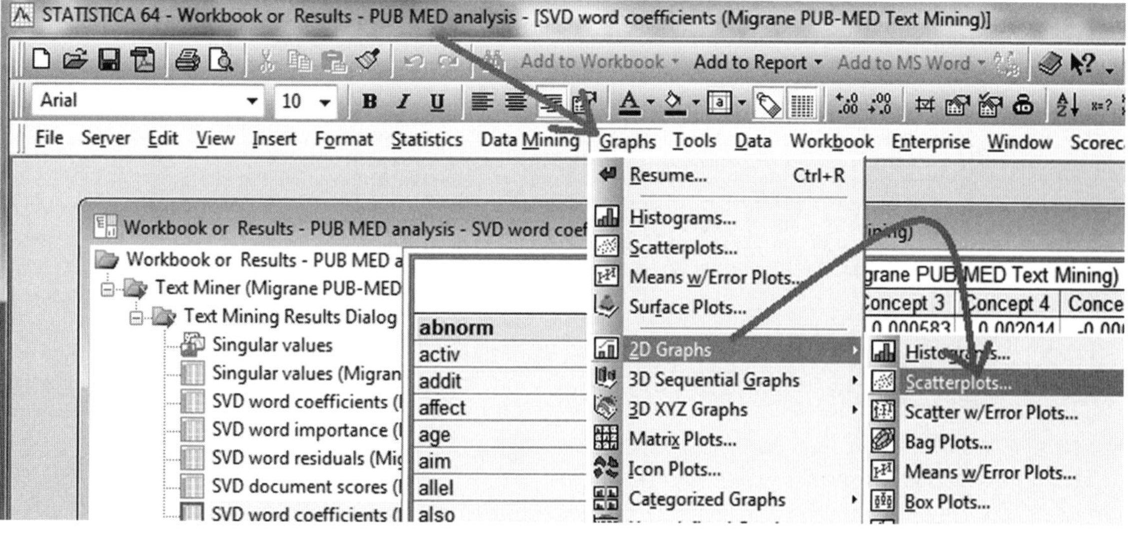

FIGURE V.58
Selecting Graphs → 2D Graphs → Scatterplots for producing a graph of one concept versus another concept.

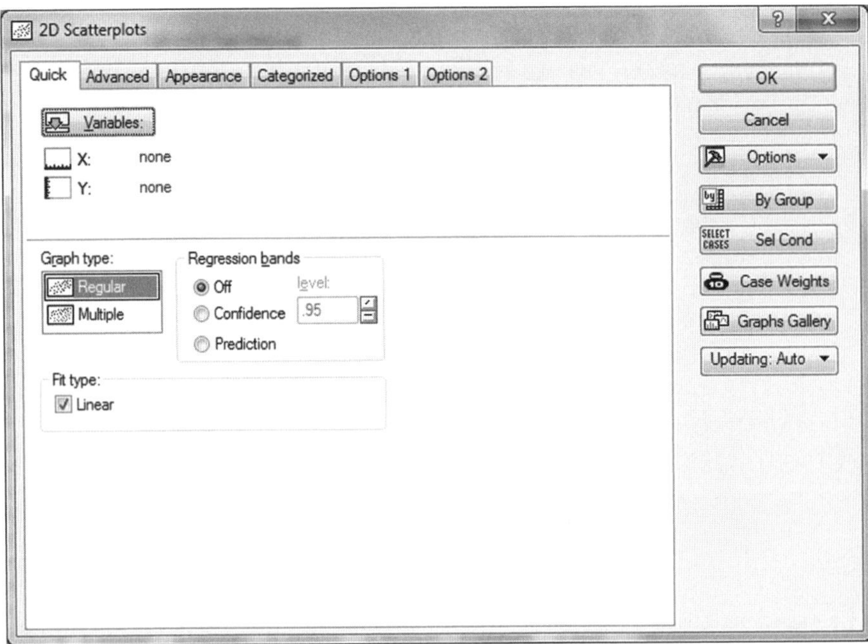

FIGURE V.59
2D Graphs Scatterplots dialog.

When we click on "Variable," the variable selection dialog appears (see Figure V.60).

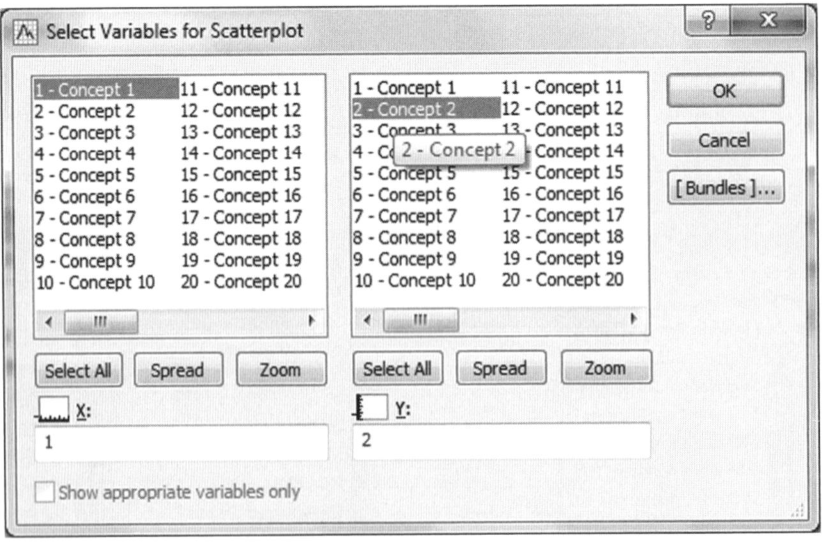

FIGURE V.60
Variable Selection dialog for the 2D Scatterplots.

When we click on OK in the "Select Variables" dialog, the variables are placed into the 2D Scatterplot dialog, and then when we click OK on that dialog, the graphs are produced (see Figure V.61).

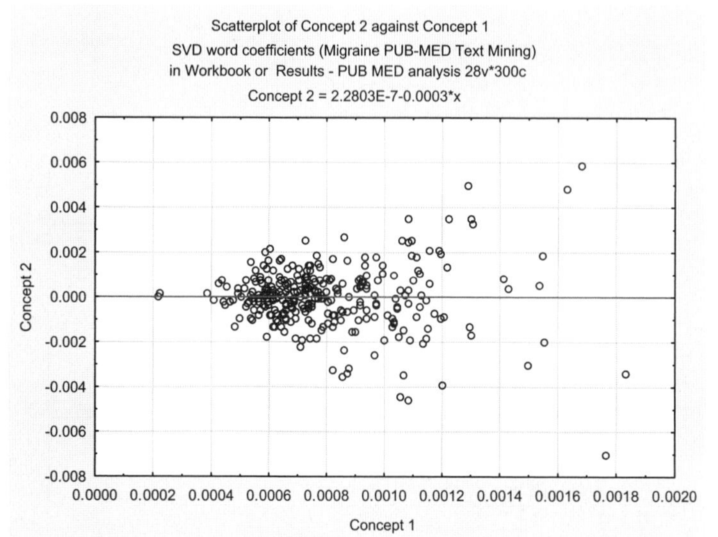

FIGURE V.61
2D Scatterplot of Concept 1 versus Concept 2.

Let's look at a few more concept versus concept comparisons:

Concept 2 versus Concept 3 looks interesting because different clusters poking out in different directions seem to be apparent; see the next figure, Figure V.62.

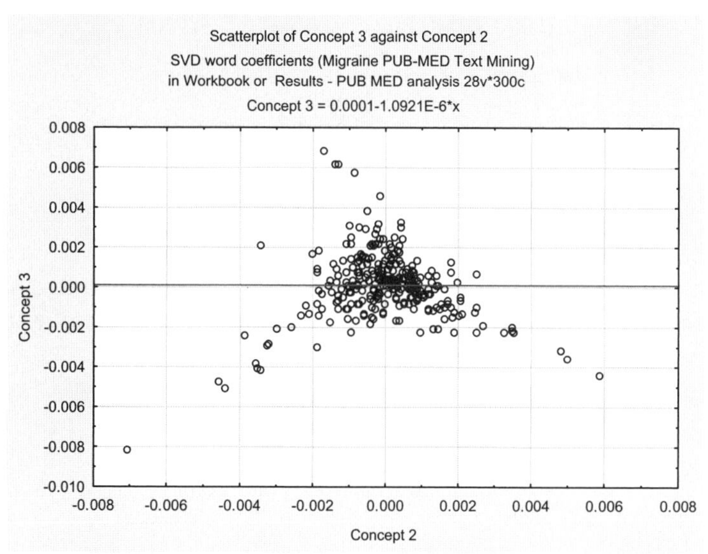

FIGURE V.62
Concept 3 versus Concept 2.

And Concept 5 versus Concept 2 also looks interesting (see Figure V.63).

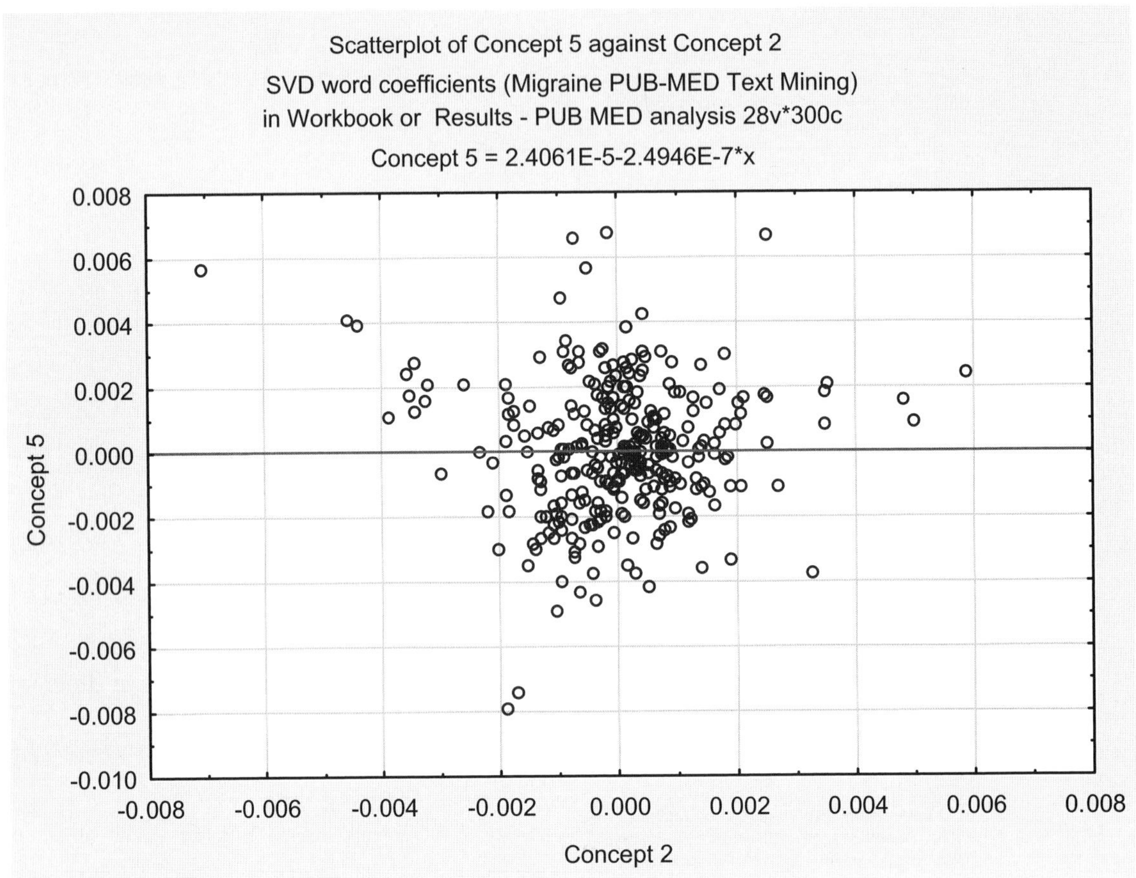

FIGURE V.63
Concept 5 versus Concept 2.

Let's select the Brushing tool in *STATISTICA* to make a "lasso" around different groupings of points in the Concept 5 versus Concept 2 graph and then have them labeled by the word the points represent (see Figures V.64 and V.65). If word names overlap, making them difficult to read, we may have to "pull them apart" and zoom into the area in order to fully comprehend them.

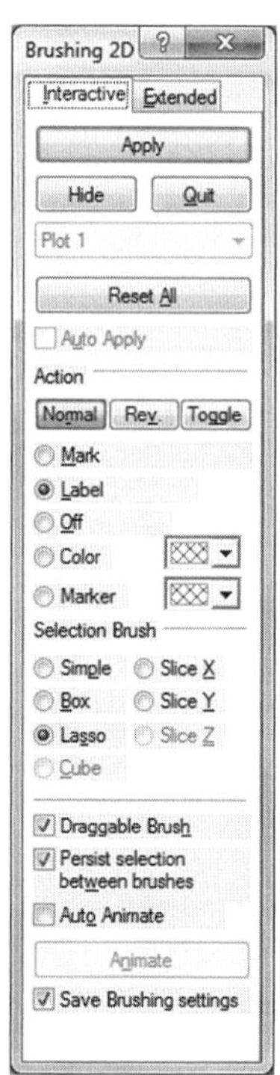

FIGURE V.64

The Brushing dialog, with the "label" and "lasso" selected.

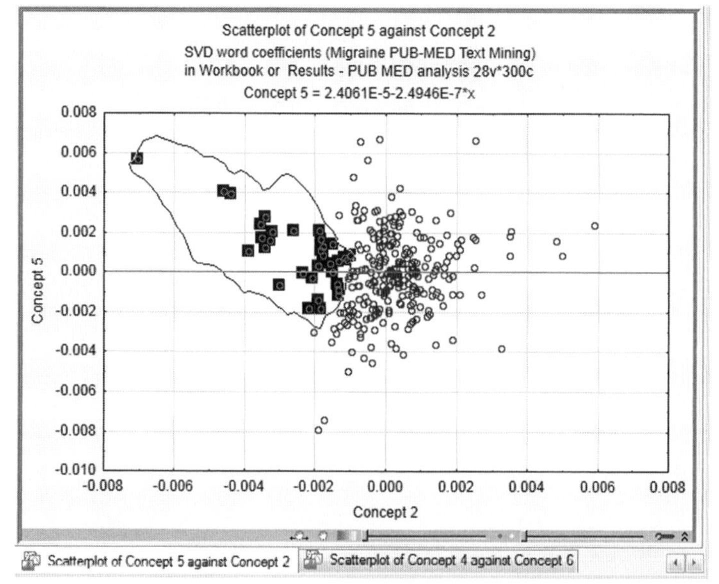

FIGURE V.65

One grouping, or "cluster," of words lassoed.

Then hit "Apply" on the Brushing tool dialog (see Figure V.66).

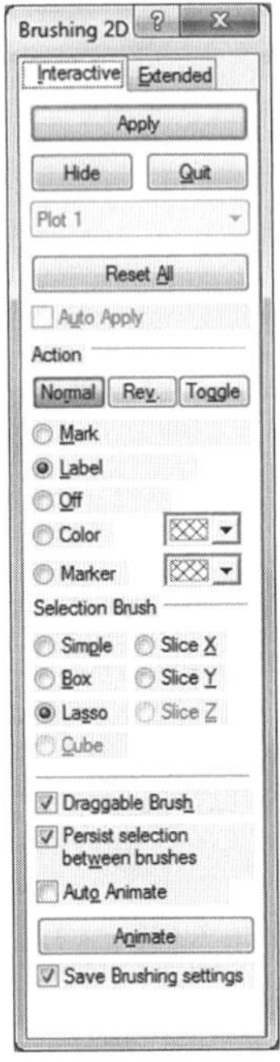

FIGURE V.66
Brushing tool dialog.

The data points in the lassoed cluster are now labeled with their word name as illustrated in Figure V.67.

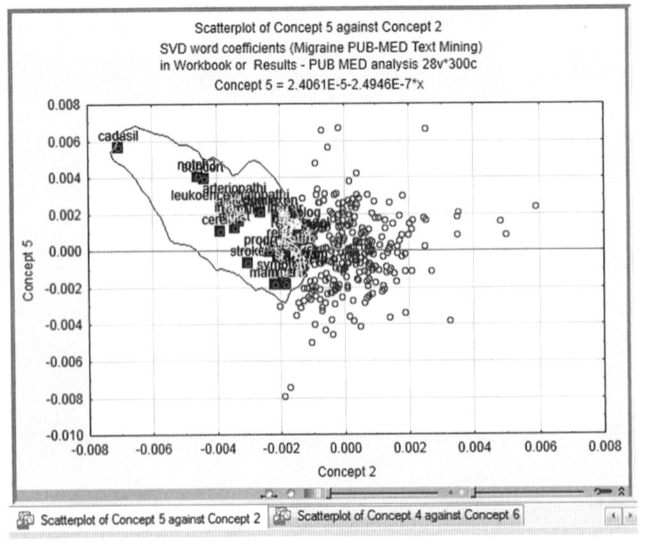

FIGURE V.67
Labeled data point in the first cluster.

We will make lassoes around a couple of other clusters, taking the ones that are way out from the central cluster. Then, by pulling the overlapping words apart, we begin to get an idea of whether or not there are clusters of "like-type words" (see Figure V.68).

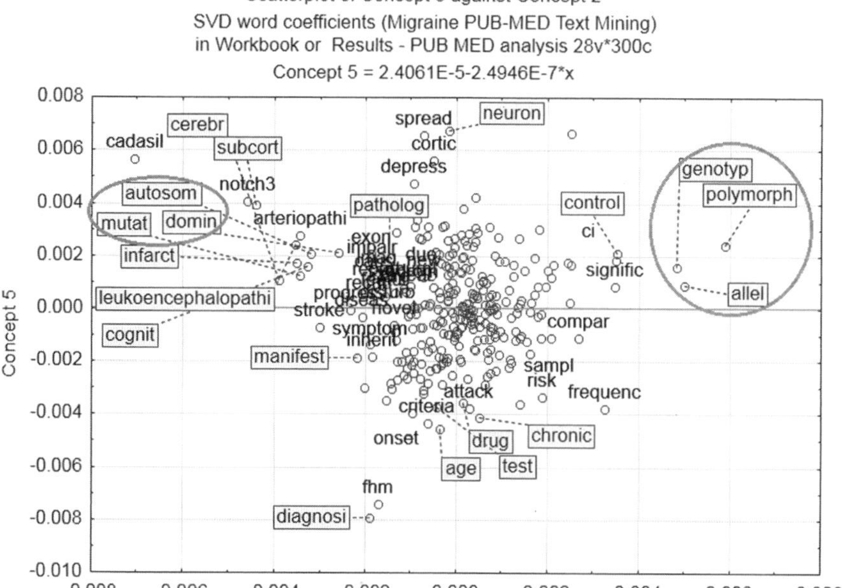

FIGURE V.68
Word groups in Concept 5 versus Concept 2.

Above we see two small clusters of words that relate to genetics. The one on the upper right has *genotype, polymorphism,* and *allele,* all terms from genetics. Also the small group on the left has *autosomal, mutant,* and *dominant*—all genetic terms.

We could go on with this further by pulling apart the words of other clusters and also doing the same for other concept comparisons. However, we will leave this for you to experiment with. The data files are included on the DVD with this book, so you can replicate what is done above but also go off in your own direction.

At this point, let's turn to another avenue of text mining, making predictive models. We'll use the "yes/no" in the "Variable" column 1 for the target that we want to predict, and the text variables, the 300 word frequencies, and the 28 concepts will be our predictor variables.

Let's see if our predictor variables can predict whether a document is Genetic or Not Genetic in its contents. First, we need to add the 28 concepts to the "master PUB MED data file." We'll do this in the same way as we did the word frequencies, only this time selecting 28 as the number of new variables to add (see Figure V.69).

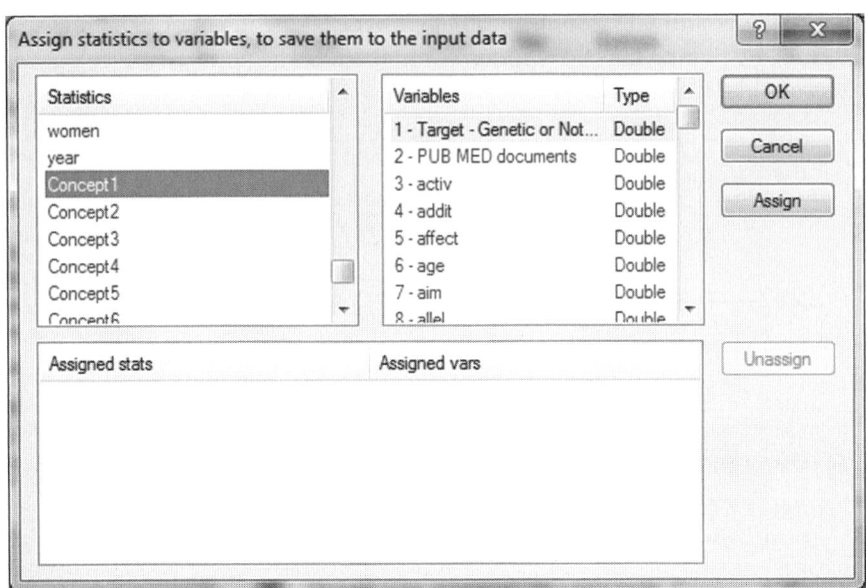

FIGURE V.69
Selecting the concepts to write back to the master data file.

When completed, the master data set will look as in the next figure (Figure V.70).

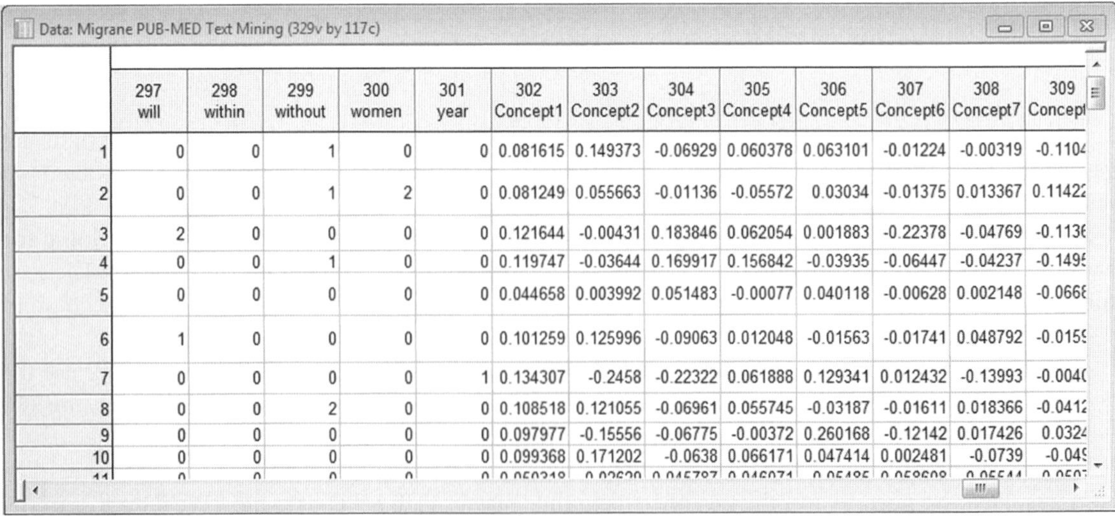
Data: Migrane PUB-MED Text Mining (329v by 117c)

	297 will	298 within	299 without	300 women	301 year	302 Concept1	303 Concept2	304 Concept3	305 Concept4	306 Concept5	307 Concept6	308 Concept7	309 Concept
1	0	0	1	0	0	0.081615	0.149373	-0.06929	0.060378	0.063101	-0.01224	-0.00319	-0.1104
2	0	0	1	2	0	0.081249	0.055663	-0.01136	-0.05572	0.03034	-0.01375	0.013367	0.11422
3	2	0	0	0	0	0.121644	-0.00431	0.183846	0.062054	0.001883	-0.22378	-0.04769	-0.1136
4	0	0	1	0	0	0.119747	-0.03644	0.169917	0.156842	-0.03935	-0.06447	-0.04237	-0.1495
5	0	0	0	0	0	0.044658	0.003992	0.051483	-0.00077	0.040118	-0.00628	0.002148	-0.0668
6	1	0	0	0	0	0.101259	0.125996	-0.09063	0.012048	-0.01563	-0.01741	0.048792	-0.0159
7	0	0	0	0	1	0.134307	-0.2458	-0.22322	0.061888	0.129341	0.012432	-0.13993	-0.0040
8	0	0	2	0	0	0.108518	0.121055	-0.06961	0.055745	-0.03187	-0.01611	0.018366	-0.0412
9	0	0	0	0	0	0.097977	-0.15556	-0.06775	-0.00372	0.260168	-0.12142	0.017426	0.0324
10	0	0	0	0	0	0.099368	0.171202	-0.0638	0.066171	0.047414	0.002481	-0.0739	-0.049

FIGURE V.70
PUB-MED master data file with concepts added.

Next we will use the DATA MINER RECIPE format to make a predictive model; see Figure V.71 for how to select this DMRecipe from the *STATISTICA* window.

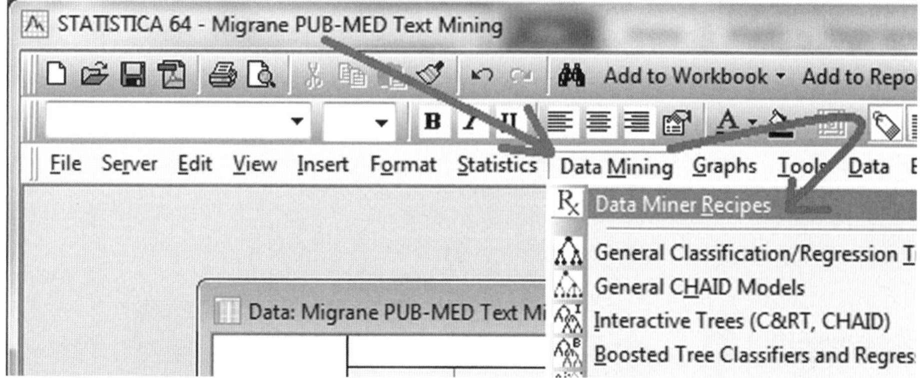

FIGURE V.71
Selecting "Data Miner Recipes" from the Data Miner pull-down menu.

Next, the Data Miner Recipes dialog appears (see Figure V.72). Follow the process of the DMRecipe using Figures V.72 – V.81.

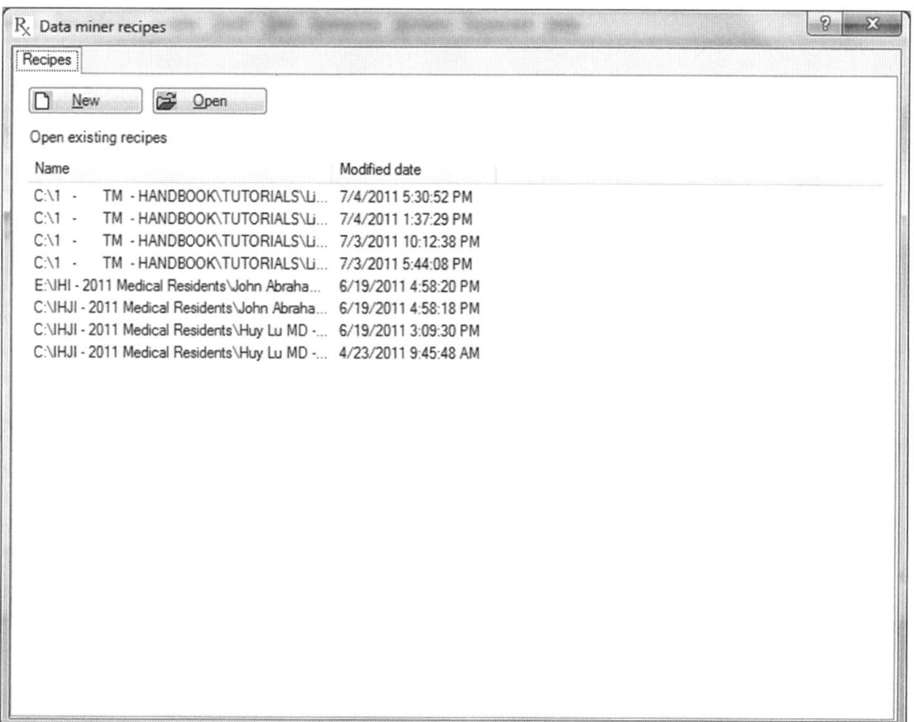

FIGURE V.72
Data Miner Recipes dialog.

Select the "New" button in upper left of the DMRecipe dialog (see Figure V.73).

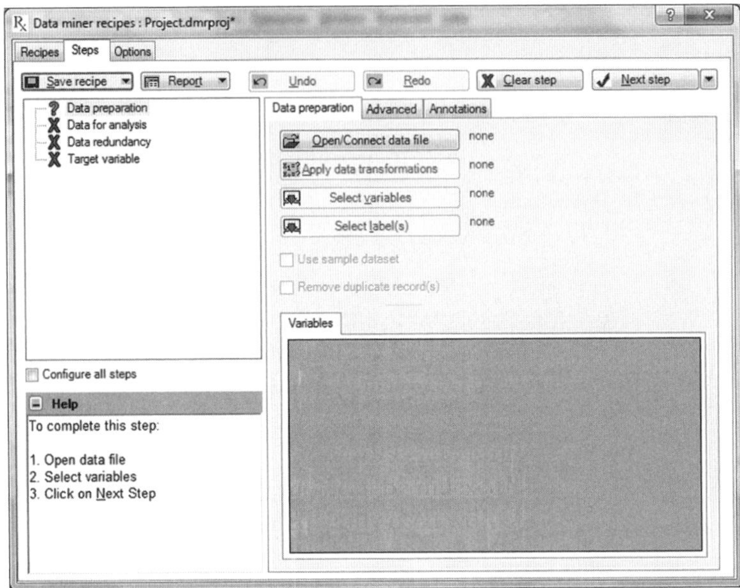

FIGURE V.73
A new Data Miner Recipes dialog, waiting for the data set to be chosen, and then variables selected, and then the process runs.

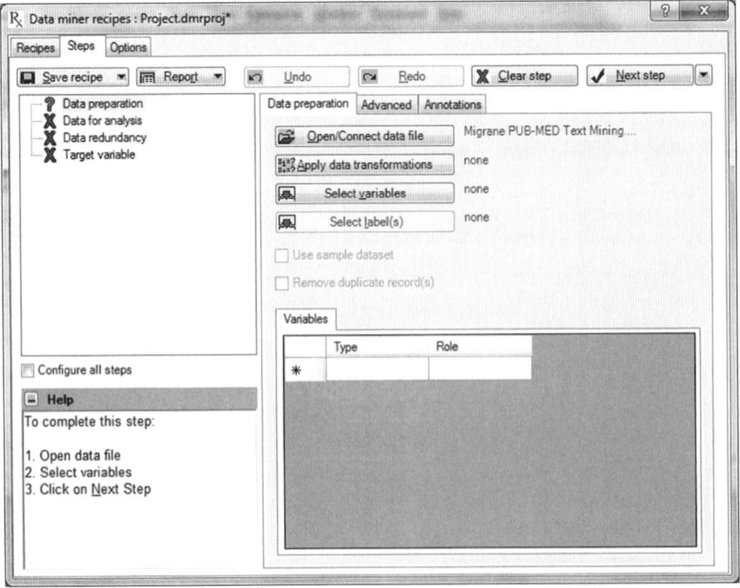

FIGURE V.74
Variables selected.

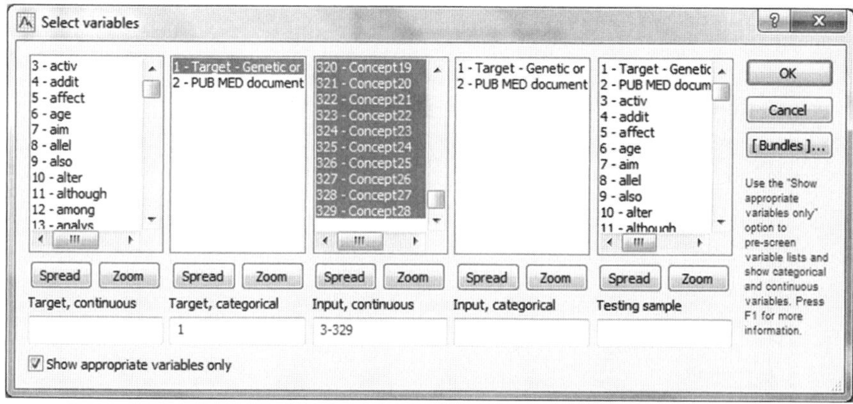

FIGURE V.75
Variables selected for DMRecipe computation.

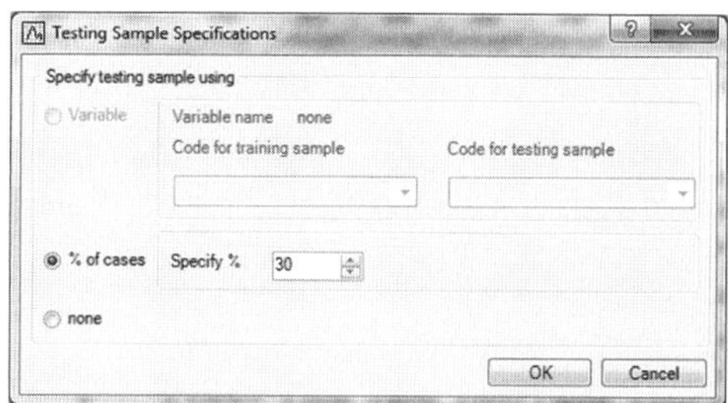

FIGURE V.76
Specifying 30% of the cases to be “held out” for the Text Sample.

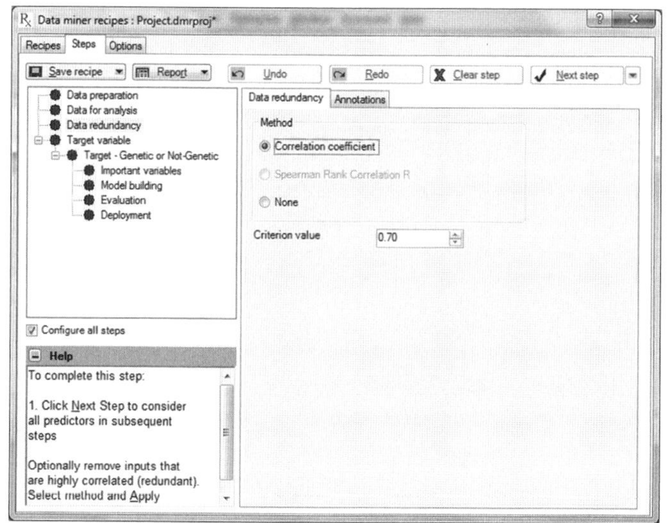

FIGURE V.77
Setting a data redundancy using the correlation coefficient among variables of 70% or greater.

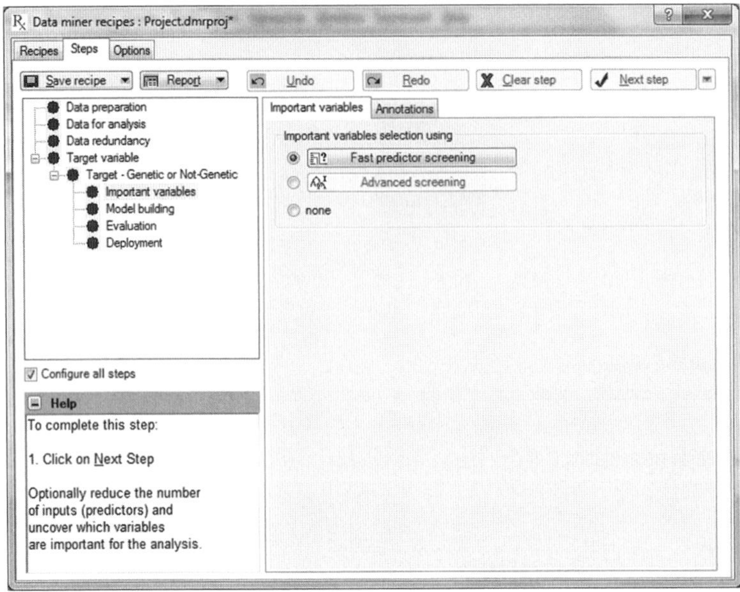

FIGURE V.78

Fast Predictor Screening chosen to sort out and delete the variables without sufficient predictive power so only the best predictors are used in the final modeling.

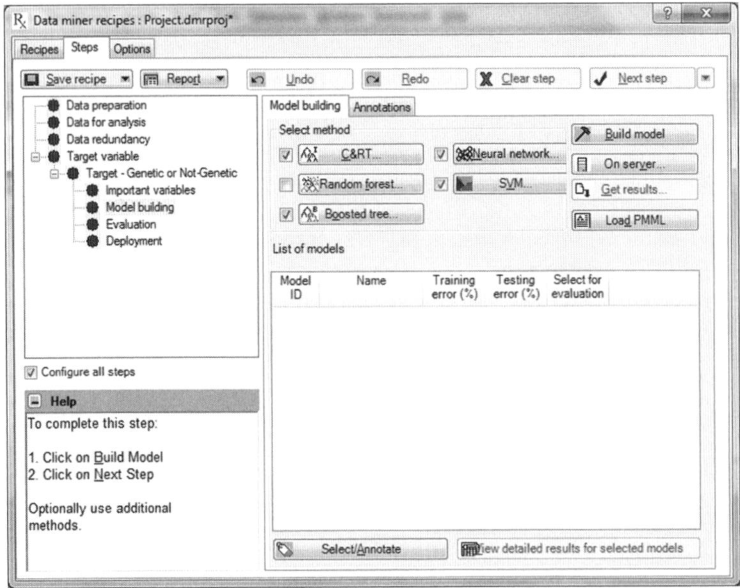

FIGURE V.79

All possible models are chosen for the DMRecipe. The process will competitively compare all algorithms and output a summary tabulation showing the accuracy of each model.

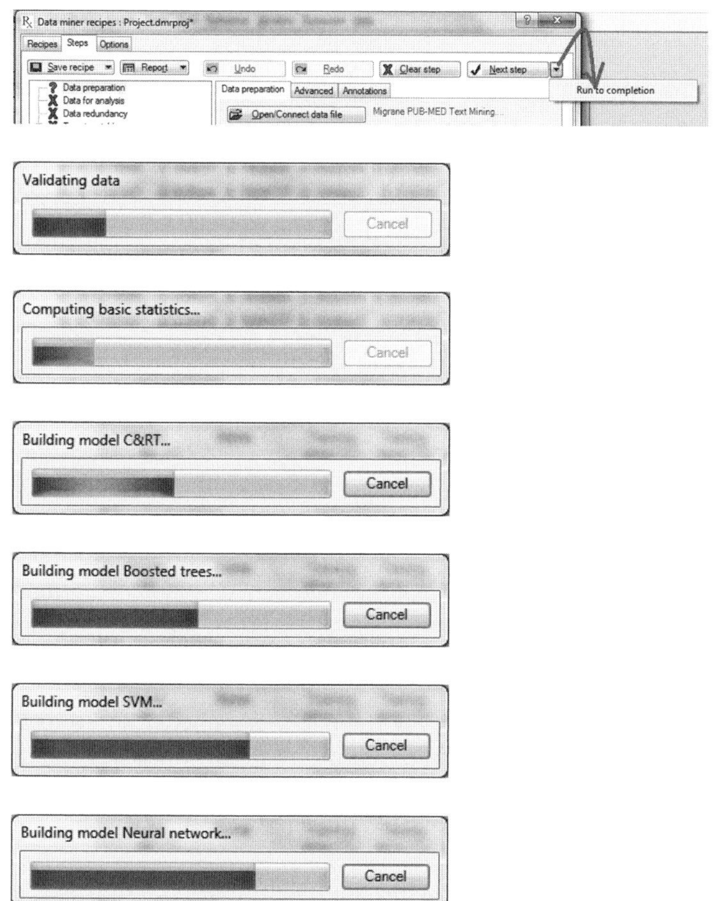

FIGURE V.80

Select "Run to completion," as shown in this figure, to cause all the computations to be performed automatically.

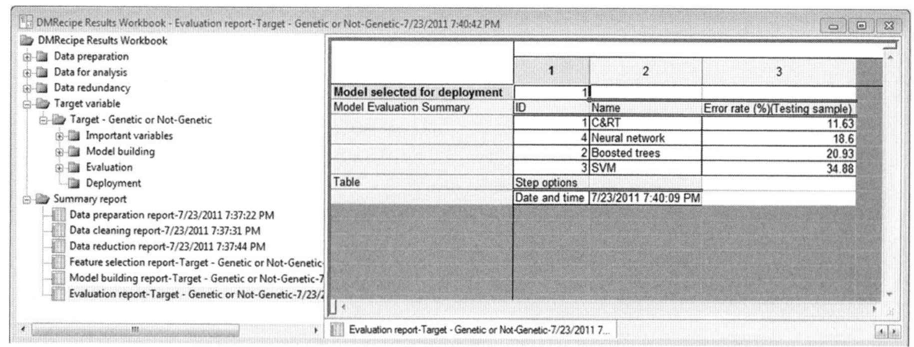

FIGURE V.81

Summary Evalution of all models as shown in the Results Workbook from the DMRecipe computations.

Looking more closely at the summary in Figure V.82:

	1	2	3
Model selected for deployment	1		
Model Evaluation Summary	ID	Name	Error rate (%)(Testing sample)
	1	C&RT	11.63
	4	Neural network	18.6
	2	Boosted trees	20.93
	3	SVM	34.88
Table	Step options		
	Date and time	7/23/2011 7:40:09 PM	

FIGURE V.82
Close-up view of Summary of Models Accuracy.

We can see that C&RT does the most accurate modeling, except that a v-fold cross-validation measure is not run with this. The Accuracies listed above are for the "Test Sample," which was selected at 30% of the total data file. Neural Networks and Boosted trees gave about the same accuracy: about 80% Accurate in predicting if a PubMed document contained genetic marker information or did not. SVM was surprisingly low: only about 65% accurate in predicting.

We could go further with this by running a Feature Selection in the Data Miner Workspace (see other tutorials for how to do this) to reduce the number of predictors used in an "Interactive data mining algorithm model," where we can tweak the parameters better and thus hopefully get a more accurate model, using v-fold cross-validation in addition to both train and text samples.

Another way to explore this data would be to apply one of the models, NN or Boosted trees, to the entire large set of PubMed documents that were obtained in the search (e.g., not just the 117 documents used in the modeling).

We are not going to go further with this in this tutorial, but the preceding suggestions are things that you can try now that you have worked through several tutorials and learned various methods of predictive analysis.

CASE STUDY: The Problem with the Use of Medical Abbreviations by Physicians and Health Care Providers

Mitchell Goldstein, MD
Associate Professor, Pediatrics, Division of Neonatology, Loma Linda University Children's Hospital, Loma Linda, CA

Gary D. Miner, PhD
Tulsa, OK

CONTENTS

Physician Notes Abbreviations: The National Patient Safety Goal requiring accredited organizations to develop and implement a list of "**Preferred Medical Abbreviations**" from The Joint Commission (TJC) List, using Text Analytics to understand current use in one medical setting.

THE PRESENT PROBLEM IN THE USE OF MEDICAL ABBREVIATIONS BY PHYSICIANS AND HEALTH CARE PROVIDERS

In 2001, The Joint Commission (TJC; formerly known as JCAHO: Joint Commission on Accreditation of Healthcare Organizations) issued a Sentinel Event Alert on the subject of medical abbreviations, and just one year later, its Board of Commissioners approved a National Patient Safety Goal requiring accredited organizations to develop and implement a list of abbreviations *not* to use. In 2004 The Joint Commission created its "do not use" list of abbreviations as part of the requirements for meeting that goal (http://www.csahq.org/pdf/bulletin/issue_3/dailey.pdf). In 2010, this was integrated into the

Practical Text Mining and Statistical Analysis for Non-structured Text Data Applications. DOI: 10.1016/B978-0-12-386979-1.00031-1

Information Management standards as elements of performance. The lists presented here are verbatim from the JCAHO website (http://www.coursewareobjects.com/objects/evolve/E2/pdf/SD_JCAHO_DoNotUse_Abbrev.pdf).

TJC (JCAHO) "DO NOT USE" ABBREVIATIONS

Official "Do Not Use" List 1

Applies to all orders and all medication-related documentation that is handwritten (including free-text computer entry) or on preprinted forms.

Do Not Use	*Potential Problem*	Use Instead
U, u (unit)	Mistaken for "0" (zero), the number "4" (four) or "cc"	Write "unit"
IU (International Unit)	Mistaken for IV (intravenous) or the number 10 (ten)	Write "International Unit"
Q.D., QD, q.d., qd (daily) Q.O.D., QOD, q.o.d, qod (every other day)	Mistaken for one another Period after the Q mistaken for "I" and the "O" mistaken for "I"	Write "daily" Write "every other day"
Trailing zero (X.0 mg) Lack of leading zero (.X mg)	Decimal point is missed	Write X mg Write 0.X mg
MS MSO_4 and $MgSO_4$	Can mean morphine sulfate or magnesium sulfate Confused for one another	Write "morphine sulfate" Write "magnesium sulfate"

ADDITIONAL ABBREVIATIONS, ACRONYMS, AND SYMBOLS

For possible future inclusion in the official "Do Not Use" list.

Do Not Use	*Potential Problem*	Use Instead
> (greater than) < (less than)	Misinterpreted as the number "7" (seven) or the letter "L" Confused for one another	Write "greater than" Write "less than"
Abbreviations for drug names	Misinterpreted due to similar abbreviations for multiple drugs	Write drug names in full
Apothecary units	Unfamiliar to many practitioners Confused with metric units	Use metric units
@	Mistaken for the number "2" (two)	Write "at"
cc	Mistaken for U (units) when poorly written	Write "mL" or "ml" or "milliliters" ("mL" is preferred)
µg	Mistaken for mg (milligrams) resulting in 1,000-fold overdose	Write "mcg" or "micrograms"

Additional sites that deal with National Patient Safety Standards are http://www.jointcommission.org/standards_information/npsgs.aspx [Joint 2011 Standards].

Within our practice, we had commonly used the abbreviation "cc" for cubic centimeter as the equivalent of the Joint Commission–preferred "mL" or milliliter. In 2010, coincident with the recommendation of the Commission, we began a process of encouraging our member physicians to switch to using the accepted abbreviations. Compliance was difficult to measure because daily physician notes are written in MS Word 2003 format, which was difficult to parse manually, since some daily notes exceeded 20 pages in length, and over 4,500 such documents were generated over the index period. Furthermore, some physicians might be predicted to be less or more compliant with the group directive. Complicating this further, some notes might contain both accepted and unaccepted abbreviations, and notes from the beginning of the year might not have any accepted abbreviations because of a delay in implementation.

To provide better insight into whether we were improving as a group, we used the text miner capacity of *STATISTICA* to analyze all of the patient care notes generated in the year 2010. Notes were stored on a networked drive and grouped by directory according to the month in which they were created. For our initial data analysis, we decided to look just at the totals of a few selected words over the entire year. The words/abbreviations chosen were "cc" and "ml" instead of "mL" or milliliter; "MS," and "MSO_4" and "$MgSO_4$" instead of morphine sulfate or magnesium sulfate; and "U" or "u" instead of unit."

USING THE "TEXT MINING PROJECT" FORMAT OF *STATISTICA* TEXT MINER

Go to the Data Mining pull-down menu, and release on Text Mining to get the initial Text Mining dialog (see Figure W.1):

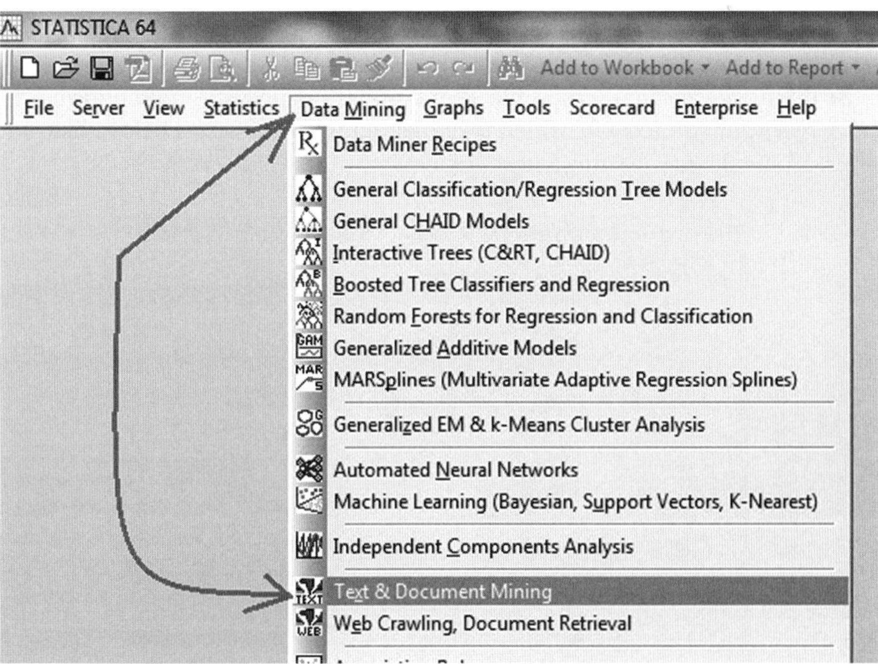

FIGURE W.1
Selecting the "Text & Document Mining" from the "Data Mining" pull down menu of STATISTICA.

Then select the "Project" tab on this dialog (see Figure W.2):

FIGURE W.2

The "Project tab" in the *STATISTICA* Text Miner.

Note that you can either "Create a new project" or "Use an existing project." In this case study, the "new project" was created by Dr. Goldstein by having this dialog reach out to the Networked Server in his medical setting and bring it into the "Text Miner." *STATISTICA* Text Miner crawled through all of the notes for 2010, pulling in all of the text and indexing it according to the frequency of words. Then Dr. Goldstein sent this Text Mining project file (having a *.dbs file ending) to Dr. Miner, who selected the needed results embedded in this project file.

Thus, to use the already created project file, the "Use existing project" radio button was selected on the *STATISTICA* Text Miner project tab, which then highlights the processes below, like "View/modify" or "Merge new documents into existing index," as illustrated in Figure W.3.

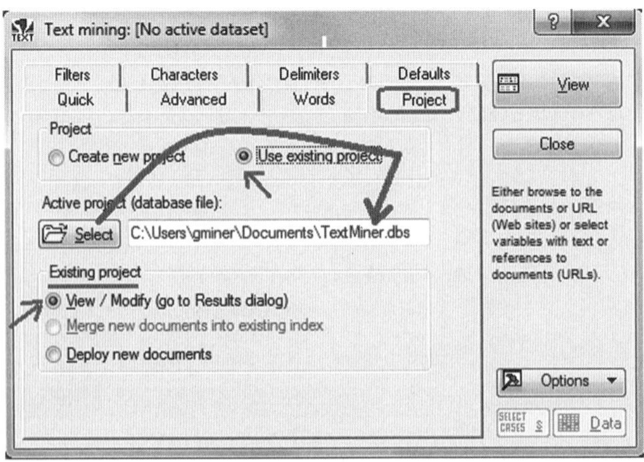

FIGURE W.3

Selecting the "Use existing project", selecting the pathway to the documents of interest, and then selecting "View / Modify" radio button.

But first, let's take an overall look at how this "Project" format works in *STATISTICA* Text Miner, since we have not presented this in any of the other tutorials or case studies in this book.

The following is taken from the *STATISTICA* On-Line Help section on the Text Miner Project Tab.

Text Mining Start-up Panel Project Tab

Selecting the *Project* tab of the *Text mining* Start-up Panel provides options to select an internal database file that will be used for indexing of the documents. For performance reasons, *STATISTICA Text Mining and Document Retrieval* incorporates advanced relational database components to maintain and update the index of words or terms by document. With large document collections and complex text, these databases can become extremely large, so an efficient relational database scheme was chosen to store this information. An additional advantage of this approach is that this database can be stored in a user-defined location on the hard disk and reused. Note that you can use the options on the *Defaults* tab to save or retrieve the settings for these options and to set the defaults for future analyses.

Note: Choosing an existing database; deploying databases: With the options on this tab, you cannot only determine the location where the database containing the index of words/terms and documents is to be stored, but you can also select a specific existing database created during a previous analysis and use the information contained therein.

Databases created with *STATISTICA Text Mining and Document Retrieval* will contain not only the list of indexed words and their frequencies in each document, but also information about which of the indexed words were selected for the analysis (words or terms can be indexed, but not selected and, thus, ignored for subsequent results), as well as results from a singular value decomposition of the frequencies or other derived indices for the selected words. Using the options available on this tab, you can either update an existing database with the words or terms found in new documents or you can index a new set of documents using only the selected terms in an existing database.

This type of indexing based on words selected during previous analyses can be considered a form of "deployment" of the database, in the sense of deployment commonly used in the context of trained models in predictive analytics. Note that the program can also compute word coefficients and document scores based on results from singular value decomposition performed in a prior analysis and stored in the database. Hence, these options enable you to compute scores for new documents based on a previous analysis; this functionality may be critical if you want to use information extracted from text in predictive data mining projects based on numeric indices that were derived during training from unstructured text.

Project Tab. Use the options in this group box either to create a new database for indexing (project) or to select an existing database and index created in previous analyses.

Create new project. Select this option button to create a new index; the index and database will be created in the location indicated in the *Active project (database file)* box, described below.

Use existing project. Select this option button to use an existing database—for example, to deploy an existing database to score new documents based on the information extracted in prior analyses. You can then select the database file using the *Select* button, described below.

Active project (database file). Specify here the name of an existing database or the name for a new database to hold the index and other information computed by *STATISTICA Text Mining and Document Retrieval*. Note that these database files can become quite large (with large collections of complex documents); hence, make sure to store this information on a hard disk with sufficient free space.

Select. Click this button to browse to an existing database file or to specify a new file name and location (depending on *Project* selection). Clicking this button will display a standard file selection dialog.
Existing project. Use the options in this group box to specify how to use the information in an existing database (project); these options are only available if the *Use existing project* option button is selected in the *Project* group box.
View/Modify (go to Results dialog). Select this option button and click the *View* button to go directly to the "Text Mining Results" dialog, where you can review the results from previous analyses as stored in the current database. This option is useful for reviewing the information from previous analyses and to update it by, for example, selecting different words or computing different indices.
Merge new documents into existing index. Select this option button to append the indexing results for a new set of documents to an existing index/database instead of overwriting it.
Deploy new documents. Select this option button to "score" the new documents, using the information contained in the current database and index. This option enables you to "deploy" the information in the existing database, in the sense of this term as it is commonly used in predictive analytics. You can use the information in the database to process the selected new documents, create results based on the previously selected words and terms, and compute word coefficients and document scores based on the singular value decomposition of results for documents used during "training."

The preceding is taken from the *STATISTICA* Online Help section on the Text Miner Project Tab (*STATISTICA*, 2011).

USING TEXTMINER3.DBS

You can use the TextMiner3.dbs "text miner project" format to compile the total 2010 use of three abbreviations in a medical care setting. We will do this later on in this case study using the "TextMiner3.dbs" sent from one institution, where compiled, to another organization, where completed.

But for now, to show you how to make a new "text mining project," go to the *Text Mining* Dialog, and select the default "Create new project" (see Figure W.4).

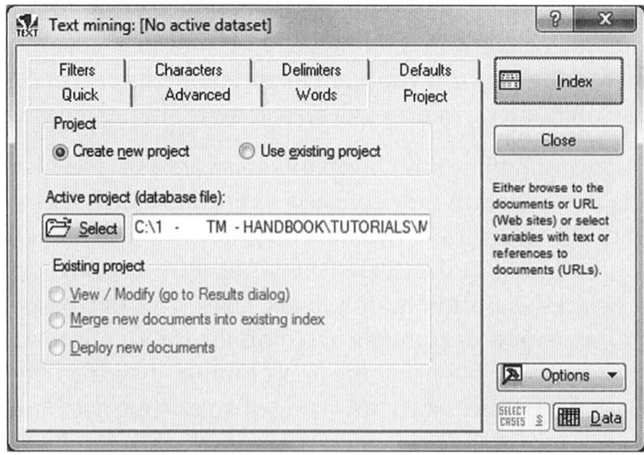

FIGURE W.4
Using "Create new project" in the Text Mining quick tab.

Continue with the Quick Tab as illustrated in Figure W.5, and select "Browse documents."

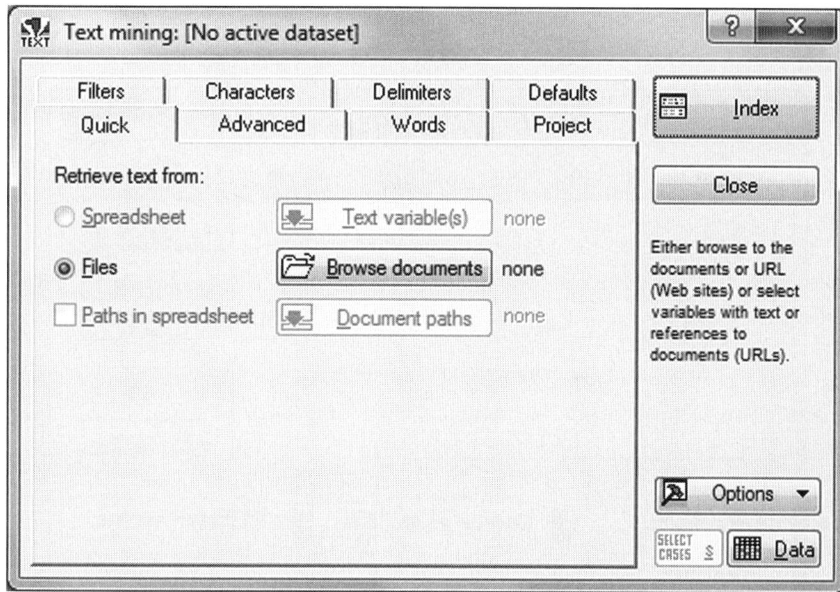

FIGURE W.5
Selecting the "Browse documents" button.

When "Browse documents" is selected, the "Open document files" window appears (see Figure W.6).

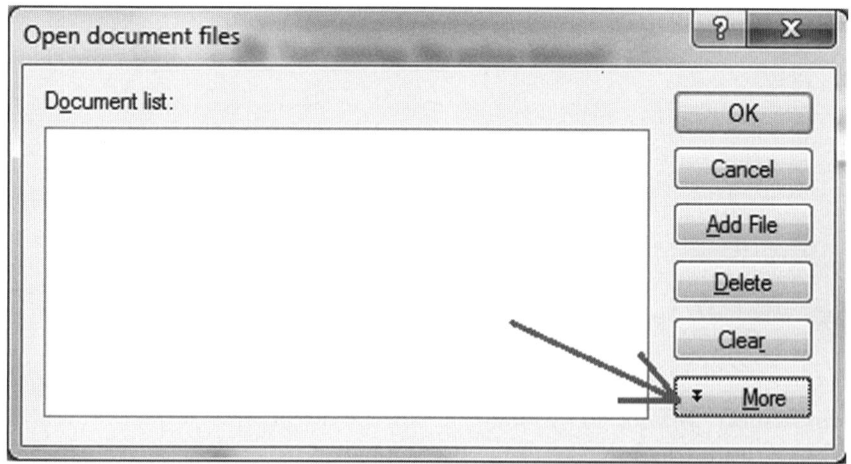

FIGURE W.6
Open document files window.

Click on "More" to bring up a more complete "Open documents files" dialog (see Figure W.7).

FIGURE W.7
The "Open documents" window when the "more" button is selected to open up a larger window.

Then a "destination" is selected (where one has the documents stored, folder, pathway), and the "Add to Crawl" button is selected. This went out to a networked server where the physician patient notes were stored and brought them in by "crawling" this destination, one note at a time. In the following screen shot, each of the notes is listed in the bottom (but blurred out so as not to identify any case), and the exact pathway for each document is placed in the top window (see Figure W.8).

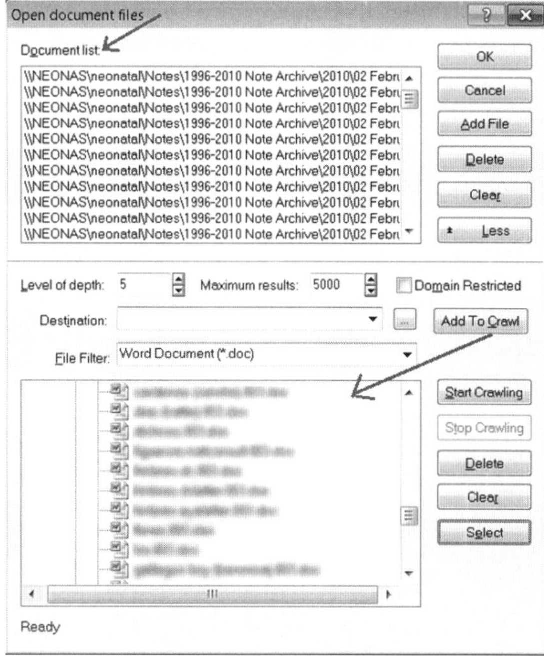

FIGURE W.8
The documents have populated the previously blank window in the "Open document files" window.

If we examine the above screen closely, we see that the physician notes were in a NEONAS root directory, and following the path to the right, we see that the documents showing in the window are from February 2010.

When this selection of documents is completed, we go back to the original Text Miner dialog, but notice that the word "selected" is present, meaning that all the documents that were requested have been brought into the text miner.

At this point we can select the Index button on the dialog, and the words of interest will be selected, as specified on the other "tabs" of the Text Miner dialog (see Figure W.9).

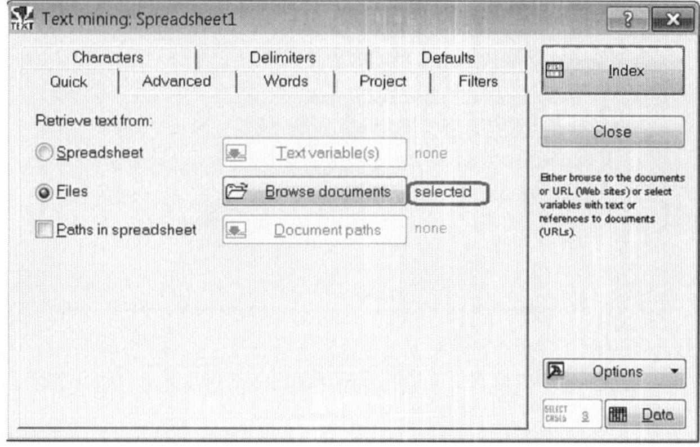

FIGURE W.9
At this point, click on the INDEX button in upper right hand corner, to start the computation of the frequencies of specific words in the documents.

During the indexing of the documents, if the document size is large the process can take considerable time; during this time, if one is doing the indexing on multiple cores, the following progress bar will appear (see Figure W.10). Because of the extensiveness of the documents in this project, even with multiple cores, this process took 8 hours.

FIGURE W.10
Progress bar during the word frequency indexing process.

Then the Indexed Results dialog appears as illustrated in Figure W.11.

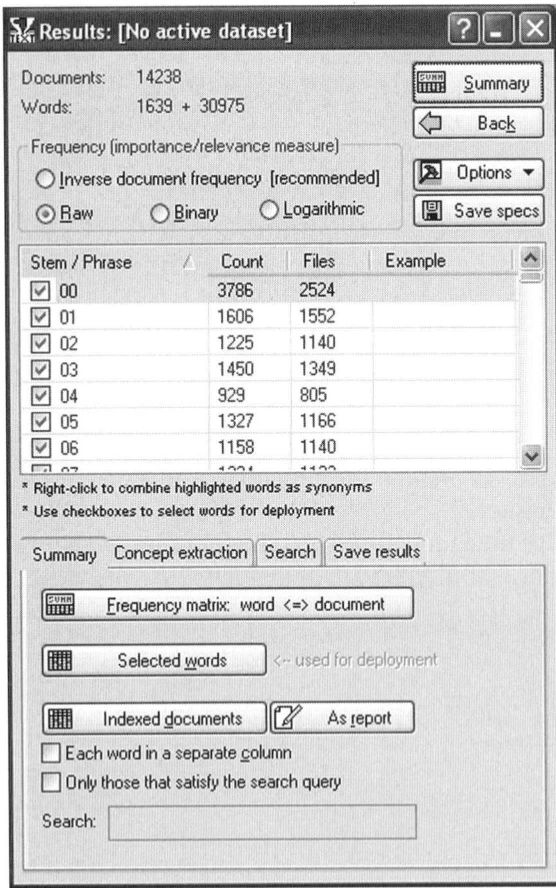

FIGURE W.11
The Results of the indexing of words; note that the button in the upper right hand corner is now termed "Summary" rather than "Index".

At this point one can take this saved project and send to another person for further analysis; this is what we did in this case study.

When the TextMiner3.dbs project was received, it was reopened in the *STATISTICA* Text Miner by selecting it as shown below in the text miner dialog (see Figure W.12).

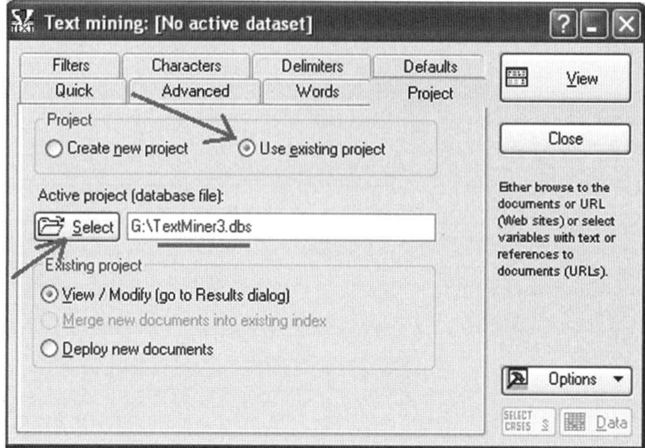

FIGURE W.12

Select the "Use existing project" in order to bring in the "TextMiner3.dbs" file.

Now, selecting the View button in Figure W.12 (upper right-hand corner), the same Results dialog as the one two screen shots above reappears (see Figure W.13).

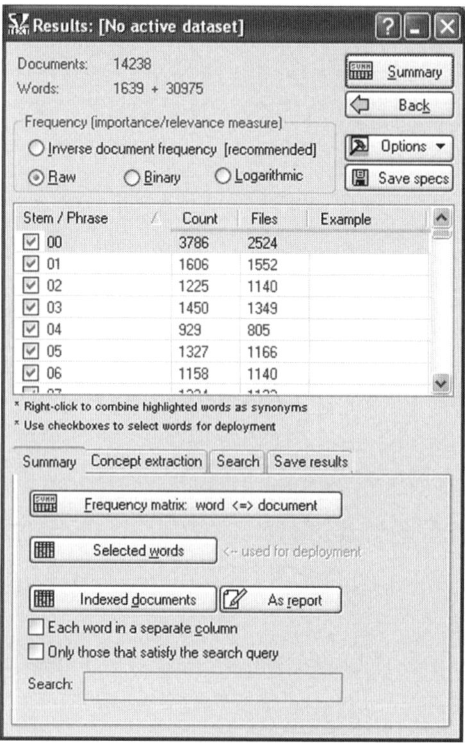

FIGURE W.13

Results dialog from the TextMiner3.dbs saved project.

By selecting some of the various buttons on the different tabs on the lower part of this Results dialog, we can get a workbook of results (see Figure W.14), which, among other things, can contain a "File search summary," where we can see which of the words we wanted indexed are present in selected physician notes, as shown below.

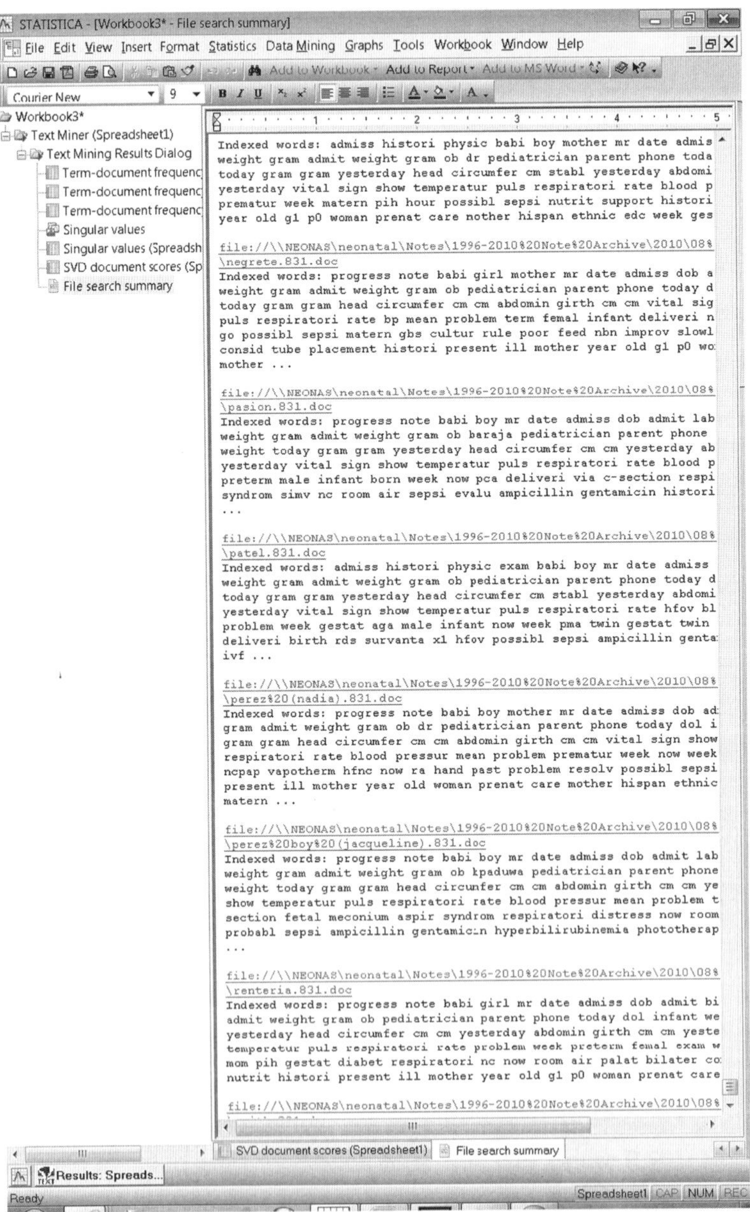

FIGURE W.14

A workbook of results.

However, for our purposes in this case study, we want to look at the entire 2010 year frequencies of a few selected abbreviations/words to see if the "do not use" terms are being used more than the "Use Instead" group.

If we use the *STATISTICA* Text Miner Results dialog to create a spreadsheet of the indexed words, leaving the default percentage of words for selection, 1,639 different words are placed in this spreadsheet for each of the 14,238 documents, as shown below in Figure W.15.

Data: SPREADSHEET _ with all words and concepts_Oct 3-2011 (1671v by 14238c)

	1 Var1	2 Var2	3 Var3	4 Var4	5 00	6 01	7 02	8 03	9 04	10 05	11 06	12 07	13 08	14 09	15 1-2	16 1/32	17 1/35	
1					0	0	0	0	0	0	0	0	0	0	2.415324	0	0	0.
2					0	0	0	2.356551	0	0	0	0	0	0	0	0	0	0.
3					0	0	0	0	2.872827	4.236822	2.524886	0	4.571271	0	0	5.190152	2.682258	1.
4					0	0	0	0	0	0	0	0	0	0	0	0	0	
5					0	0	0	0	0	0	0	0	0	0	0	0	0	0.
6					0	0	0	2.356551	0	0	0	0	0	0	0	0	0	0.
7					0	0	0	0	0	0	0	0	0	0	0	0	0	0.
8					0	0	0	0	0	0	0	0	0	0	0	0	0	0.
9					0	0	0	0	0	0	0	0	0	0	0	0	0	
10					1.73007	0	0	0	0	0	0	0	0	2.396632	0	0	0	0.
11					0	0	0	0	0	0	0	0	0	0	0	0	0	
12					0	0	2.524886	0	0	0	0	0	0	0	0	0	0	
13					0	0	0	0	0	0	0	0	0	0	0	0	0	0.
14					0	0	0	0	0	0	0	0	0	0	0	0	0	0.
15					0	3.752641	0	0	0	0	0	0	0	0	0	0	2.682258	0.
16					0	0	0	0	0	0	0	0	0	0	0	0	0	0
17					0	0	0	0	0	0	0	2.531045	0	0	0	3.065388	0	0.
18					0	0	0	0	0	0	0	0	0	0	2.415324	0	0	0.

FIGURE W.15

Spreadsheet of the indexed word frequencies.

In the above spreadsheet, Var1, Var2, Var3, and Var4 were created as "placeholders" in which to later add in the Physicians and Months, and even break it down by further grouping categories, such as Nurse versus Physician and weeks.

Since we only want to work with U / u / units, cc / ml / mL / milliliter, and MS / MSO_4 / $MgSO_4$ / morphine sulfate / magnesium sulphate in this case study, we'd have to go through all these words in the spreadsheet and just select these few, deleting the others if we wanted only the words of interest to be viewable on the screen at once. It would probably be easier to use the TM Results dialog and just "deselect" all the words, scroll through and select only the words of interest, and then make a new spreadsheet of just these few words. This is what we'll do, and we'll illustrate how in the next series of screen shots.

To deselect checked Indexed words and then reselect only the few needed, the following procedure is used:

Select the top word with the Shift key held down (see Figure W.16).

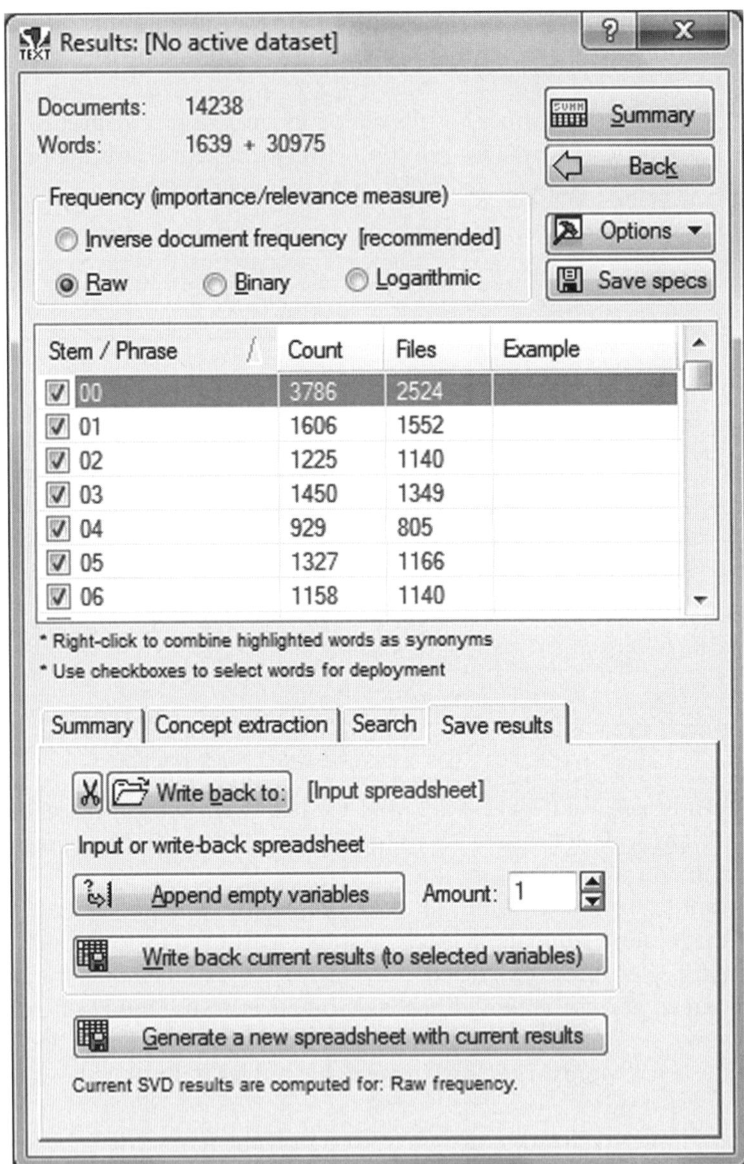

FIGURE W.16
The top most word is highlighted, with Ctl key held down.

Then scroll down to the bottom of the indexed word list, and with the Shift key held down, select "Last one," thus selecting all (see Figure W.17).

Then when all are selected, again click on one of the highlighted words at the bottom, holding down the Shift key, and "decheck" this one word (by unchecking it in the box as shown in Figure W.17), and then wait for all to be unchecked as illustrated in Figure W.18, and then for the highlighting to clear as seen in Figure W.20.

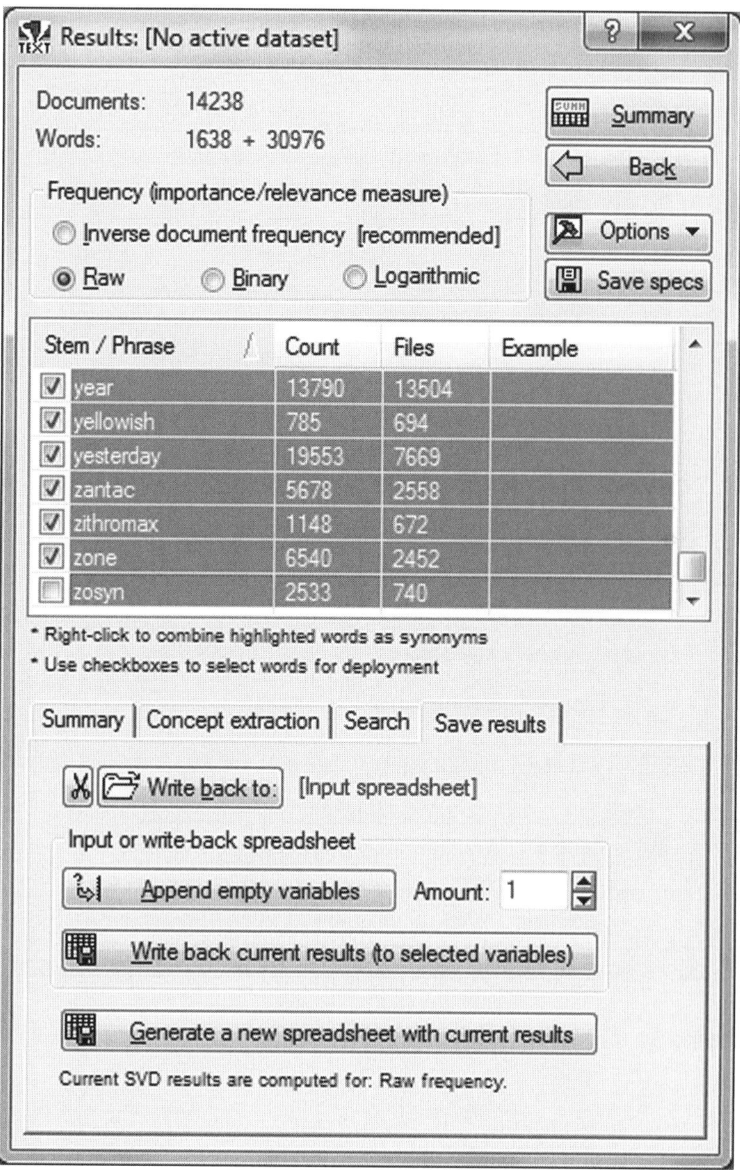

FIGURE W.17

By holding Ctl key down while scrolling to the bottom word in the list and then "selecting" that word with curser, all words are selected.

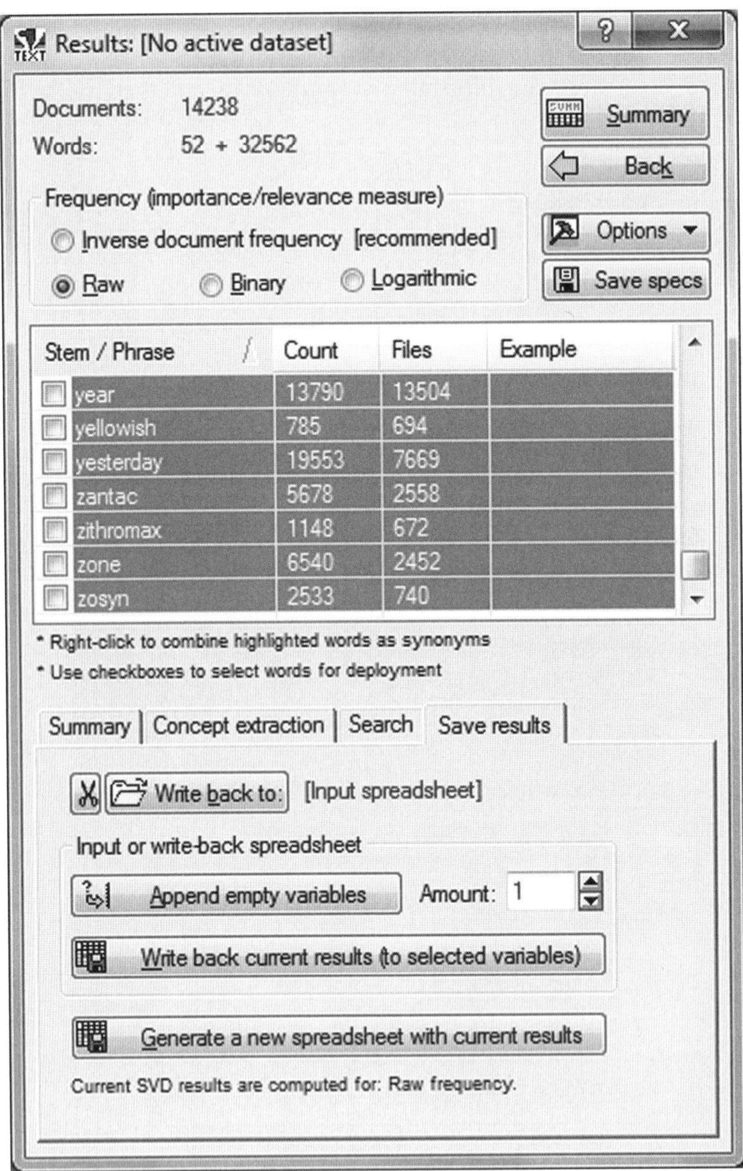

FIGURE W.18

Note that all the words are now "un-checked" in their boxes to left of words.

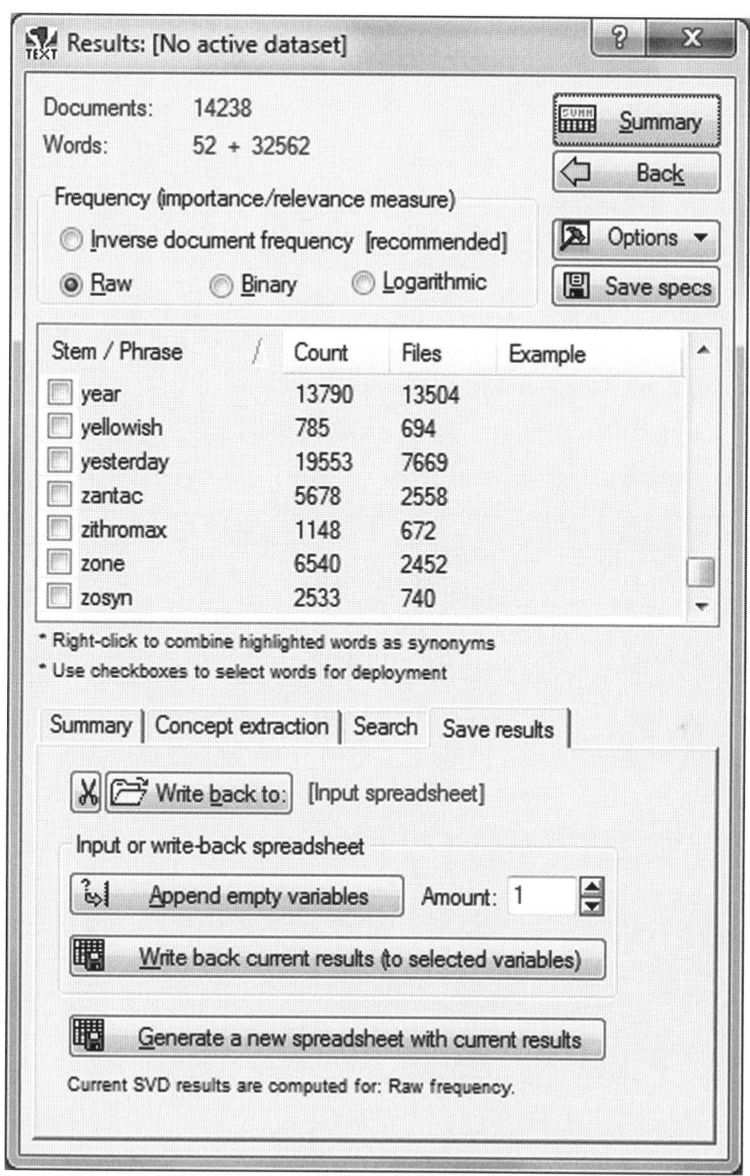

FIGURE W.19

The highlighted words are now both cleared, and also un-checked.

Then you can go back and check just those words wanted in a final spreadsheet. In this example, we will just check the cc / ml / ML / milliter and the other two abbreviation examples as illustrated in Figure W.20.

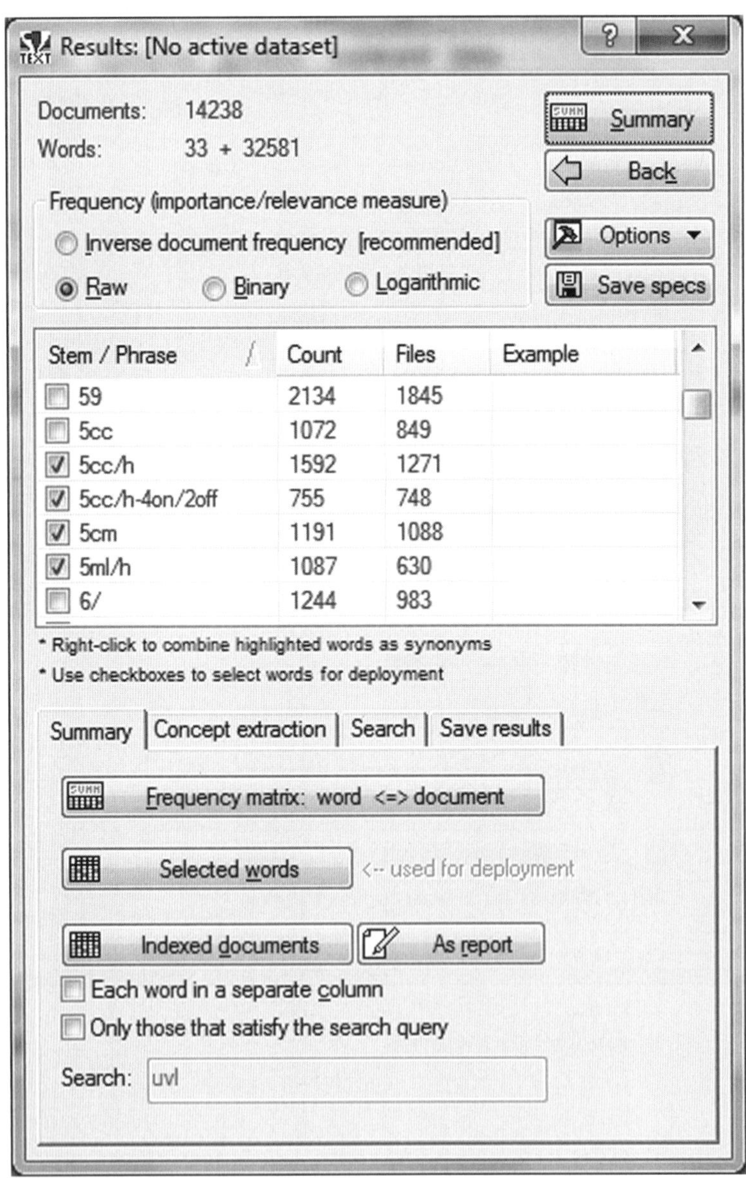

FIGURE W.20

Only the words of interest are now checked.

Then save the results of these 33 selected words (see Figure W.20 and also Figure W.21 below) from the indexed words to put into a new spreadsheet of just these 33 words. (Please refer to earlier tutorials in this book to find out how to use the "Save Results" tab to make this new spreadsheet.)

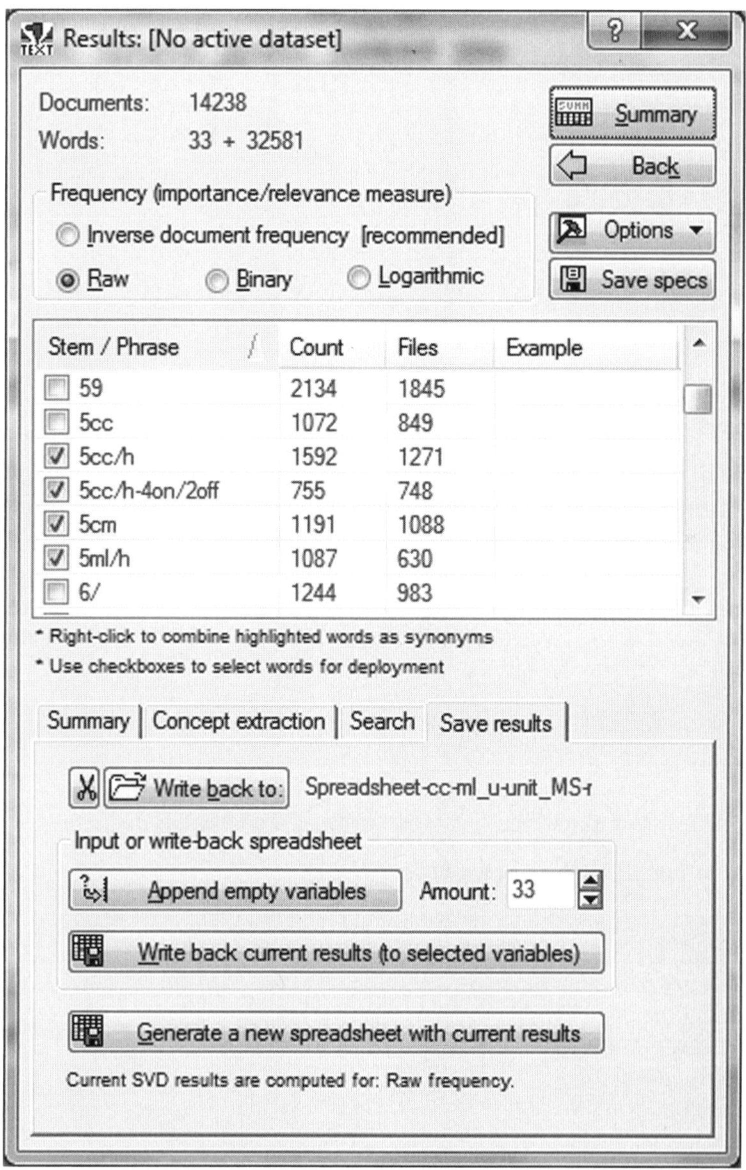

FIGURE W.21
The "Save results" tab is selected; next these 33 words need to be saved.

Figure W.22 shows the spreadsheet with addition of two categorical variables, Var 1 and Var 2. What the spreadsheet looks like after the addition of Var 1 – Month - during year 2010, and Var 2 – Physician making the patient notes.

Data: Spreadsheet-cc-ml_u-unit_MS-morphine-magnesium_Oct8-2011 (36v by 14238c)

	1 Month - during year 2010	2 Physician	3 Var3	4 1cc/h	5 5cc/h	6 5cc/h-4on/2off	7 5cm	8 5ml/h	9 cc	10 cc/feed	11 cc/h	12 cc/hr	13 cc/kg	14 cc/kg/day	15 cc/kg/hr	16 cm
1				0	0	0	0	0	4	0	0	0	0	2	0	4
2				0	0	0	0	0	2	0	0	0	0	2	0	4
3				0	0	0	0	0	5	0	1	0	0	3	1	5
4				0	0	0	0	0	0	0	0	0	0	0	0	4
5				0	0	0	0	0	3	0	0	0	0	2	0	5
6				0	0	0	0	0	1	0	0	0	0	1	0	7
7				0	0	0	0	0	0	0	0	0	0	2	1	3
8				0	0	0	0	0	0	0	0	0	0	2	1	5
9				0	0	0	0	0	1	1	0	0	0	2	1	4
10				0	0	0	0	0	0	0	0	0	0	3	1	5
11				0	0	0	0	0	0	0	0	0	0	4	1	6
12				0	0	0	0	0	1	1	0	0	0	2	0	7
13				0	0	0	0	0	1	0	0	0	0	2	1	7
14				0	0	0	0	0	5	0	0	0	0	2	0	4
15				1	3	1	0	0	6	0	4	3	0	3	0	6
16				0	0	0	0	0	1	0	0	0	1	2	1	4
17				0	0	1	2	0	22	0	2	1	0	3	1	11

FIGURE W.22
Spreadsheet with all variables needed to complete this example.

Highlighting all of the variables of interest (e.g., the cc-mL, MS-morphine or magnesium or $MgSO_4$, or u-unit) and then right clicking on one of the highlighted variable headings, and then from flying menu selecting STATS – sums, we get the following spreadsheet (see Figure W.23).

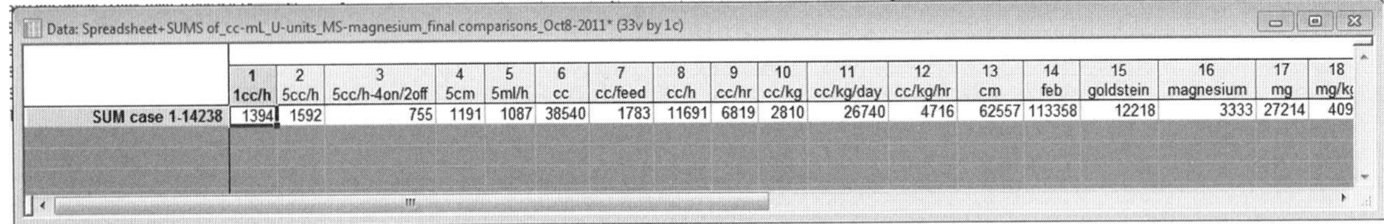

Data: Spreadsheet+SUMS of_cc-mL_U-units_MS-magnesium_final comparisons_Oct8-2011* (33v by 1c)

	1 1cc/h	2 5cc/h	3 5cc/h-4on/2off	4 5cm	5 5ml/h	6 cc	7 cc/feed	8 cc/h	9 cc/hr	10 cc/kg	11 cc/kg/day	12 cc/kg/hr	13 cm	14 feb	15 goldstein	16 magnesium	17 mg	18 mg/kg
SUM case 1-14238	1394	1592	755	1191	1087	38540	1783	11691	6819	2810	26740	4716	62557	113358	12218	3333	27214	409

FIGURE W.23
Results spreadsheet of just the sums of each variable for the entire number of documents examined.

Looking closer, at just the left half (see Figure W.24):

Data: Spreadsheet+SUMS of_cc-mL_U-units_MS-magnesium_final comparisons_Oct8-2011* (33v by 1c)

	1 1cc/h	2 5cc/h	3 5cc/h-4on/2off	4 5cm	5 5ml/h	6 cc	7 cc/feed	8 cc/h
SUM case 1-14238	1394	1592	755	1191	1087	38540	1783	11691

FIGURE W.24
Close up view of this spreadsheet of SUMS of each column.

For this simple case study, the above information will serve our purposes. We can add up all the frequencies of use of "cc" and compare it to the frequencies of "ml," "mL," and "milliliter," if any, and so on for the other two abbreviations.

The next Figure W.25 examines just "cc" and "mL" in another tabulation done separately for just this one abbreviation:

Data: Spreadsheet39* (13v by 1c)

	Term-document frequency matrix												
	1 cc	2 cc/feed	3 cc/h	4 cc/hr	5 cc/kg	6 cc/kg/day	7 cc/kg/hr	8 ml	9 ml/h	10 ml/hr	11 ml/kg	12 ml/kg/day	13 ml/kg/hr
SUM case 1-14238	38535	1777	11673	6805	2811	26733	4709	31746	3447	10071	1290	10903	2432

FIGURE W.25
Sums for "cc" and "mL".

What we are really interested in here is the *proportion* of the "nonapproved abbreviation" cc compared to the "Preferred" abbreviation of mL, of which there are absolutely zero in this data set, and optionally "ml," which is "acceptable" for cc but not preferred.

We see that "cc" has been used 38,536 times in these documents and that "ml" has been used 31,746 times.

CONCLUSION

The abbreviation "cc" is still being about used 50 percent of the time in this data set; mL is not being used at all; ml is being used almost as frequently as "cc"; and the fully written-out word "milliliter," which is preferred over "cc," is also not being used.

We also examined three other abbreviations/words:

1. "cm" versus "centimeter": "cm" was used by all physicians in 2010—62,557 times; "centimeter" was used "zero" times.
2. "U" or "u" for "unit" was used in different formats 22,688 times, and "unit" was used 1,395 times.
3. "$MgSO_4$" was used 1,295 times; "MS" was not used at all; and "MSO_4" was not used at all; "magnesium" was used 3,333 times, and "sulfate" was used 1,569 times; thus, the full spelling of these words, which can be so easily confused with "morphine sulfate," is being done in this medical setting to a higher degree than the other three terms examined in this case study.

INTERVENTION TRAINING NEEDED

It appears that an intervention training of some type is needed. Interventions can take the form of a "special seminar" for all of the hospital staff involved with these patients, or a "poster" can be placed in various locations in the clinic to remind physicians, nurses, and consultants that preferred abbreviations/words are the goal of this clinic. Also, attending physicians can remind staff on all rounds. Whatever the intervention method, it appears that some type of attention is needed if these "do not use" abbreviations are to be eliminated in medical records.

After an intervention is put in place, it must be determined if the desired results are improving over both time and by physician. Thus, the categorical variables of "month" for time and "physician" for doctor will need to be added to the data set. The needed data set will look like this after the indexed words are added (see Figure W.26).

Data: Spreadsheet-cc-ml_u-unit_MS-morphine-magnesium_Oct8-2011* (36v by 14238c)

	1 Month - during year 2012	2 Physician	3 Var3	4 1cc/h	5 5cc/h	6 5cc/h-4on/2off	7 5cm
1				0	0	0	0
2				0	0	0	0
3				0	0	0	0
4				0	0	0	0
5				0	0	0	0
6				0	0	0	0
7				0	0	0	0
8				0	0	0	0
9				0	0	0	0
10				0	0	0	0
11				0	0	0	0
12				0	0	0	0
13				0	0	0	0
14				0	0	0	0
15				1	3	1	0
16				0	0	0	0
17				0	0	1	2

FIGURE W.26
The format of our desired dataset, where month and physician are included as categorical variables.

Once we get the data in this format, we can easily track the "learning curve" in a changeover in physicians, eliminating the "do not use" JCAHO list of medical abbreviations both by physician and time. Possibly a "time series analysis" and whatever other "Statistical Learning Theory" algorithms that are appropriate could be used to get a much better understanding of this issue, and thus in the long run make changes that lead to improved health care delivery, communication, and enhanced patient safety.

Because of the complexity of the model, we will probably need a nonparametric regression procedure that makes no assumptions about the underlying relationship among the dependent and independent variables. We suspect that a Multivariate Adaptive Regression Splines (MARSplines) will be needed to analyze the sample but will submit future data sets to other predictive algorithms to see which will best handle the data.

References

JCAHO "do not use abbreviations" website (2010): http://www.coursewareobjects.com/objects/evolve/E2/pdf/SD_JCAHO_DoNotUse_Abbrev.pdf

(2004) http://www.csahq.org/pdf/bulletin/issue_3/dailey.pdf

(2011) http://www.jointcommission.org/standards_information/npsgs.aspx (Joint 2011 Standards)

StatSoft, Inc. (2011). *STATISTICA* (data analysis software system), version 10. www.statsoft.com. Online Help, section on "Project" Tab of Text Miner.

Classifying Documents with Respect to "Earnings" and Then Making a Predictive Model for the Target Variable Using Decision Trees, MARSplines, Naïve Bayes Classifier, and K-Nearest Neighbors with *STATISTICA* Text Miner

CONTENTS

INTRODUCTION: AUTOMATIC TEXT CLASSIFICATION

This example is based on the "classic" Reuters collection of documents. Specifically, 5,000 documents were selected from the Reuters-21578 database, which is a collection of 21,578 articles from Reuters that appeared on the newswires in 1987. The documents were assembled and indexed with categories by personnel from Reuters Ltd. in 1987. Note that the copyright for these articles resides with Reuters Ltd. and Carnegie Group, Inc., and these files are available for research and demonstration purposes only. You can also review Chapter 16 in Manning and Schütze (2002) to learn more about these documents and the specific types of analyses illustrated in this example. The body of the articles was placed into XML (Extensible Markup Language) files. Following is an example of such a file (see Figure X.1).

Practical Text Mining and Statistical Analysis for Non-structured Text Data Applications. DOI: 10.1016/B978-0-12-386979-1.00032-3

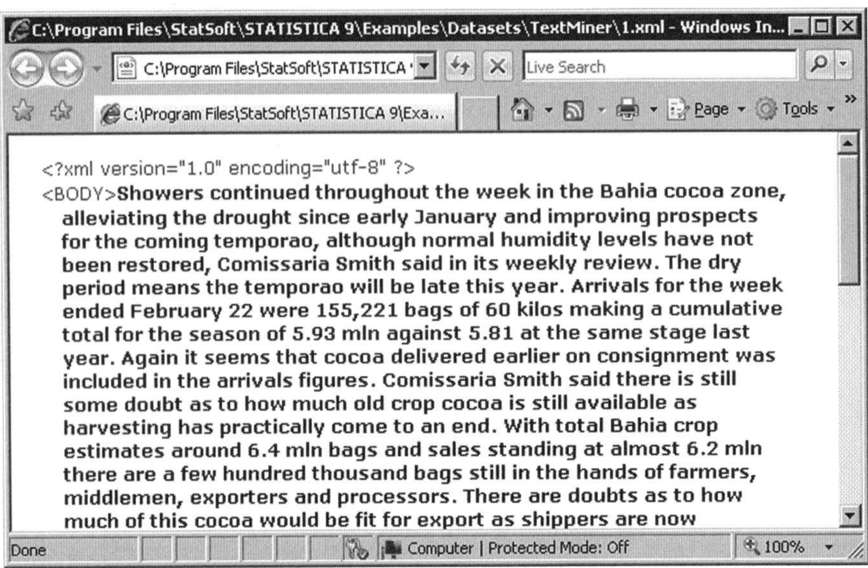

FIGURE X.1
XML (Extensible Markup Files) in a *STATISTICA* text file supplied with the software.

The value of this collection of documents is that it was carefully coded by experts with respect to different content categories. The one of interest for this example is the "Earnings" category—that is, the goal of this text mining project is to derive a simple classifier that enables us to automatically classify the articles as either dealing with earnings or not (see also Manning and Schütze, 2002, p. 579).

Needless to say, the general utility of such methods that enable you to automatically classify large numbers of texts into certain categories (e.g., of interest or not of interest; or categories that allow for automated routing of documents to the appropriate offices, departments, etc.) can be immense. Once a good (accurate) classification method has been determined, hundreds or perhaps thousands of human work hours could be saved by implementing an automated system to perform necessary classifications of documents. (Note that the *STATISTICA* system is ideally suited to implement such systems because it supports *deployment* of *text mining* results and because the system is completely programmable, so it can be seamlessly integrated with existing electronic management systems, such as the *STATISTICA Document Management System*.)

DATA FILE WITH FILE REFERENCES

To reiterate, the purpose of this analysis is to derive a model that will enable us to automatically determine whether a document is relevant to the *Earnings* category. The *STATISTICA Text Mining and Document Retrieval* system includes many options for retrieving documents or references to documents, including web or file crawling. In this case, the example data file *Reuters.sta* will be used (see Figure X.2), which already contains the necessary information to retrieve all documents.

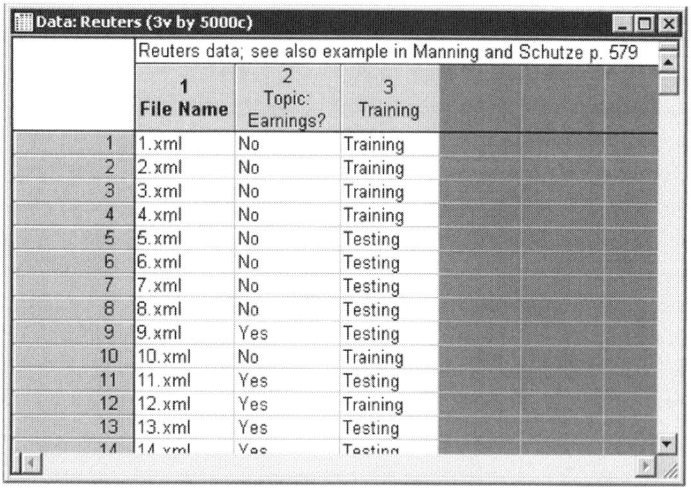

Data: Reuters (3v by 5000c)

Reuters data; see also example in Manning and Schutze p. 579

	1 File Name	2 Topic: Earnings?	3 Training
1	1.xml	No	Training
2	2.xml	No	Training
3	3.xml	No	Training
4	4.xml	No	Training
5	5.xml	No	Testing
6	6.xml	No	Testing
7	7.xml	No	Testing
8	8.xml	No	Testing
9	9.xml	Yes	Testing
10	10.xml	No	Training
11	11.xml	Yes	Testing
12	12.xml	Yes	Training
13	13.xml	Yes	Testing
14	14.xml	Yes	Testing

FIGURE X.2
STATISTICA data file with "Text" column, column No. 1, having the "file name" entered for each cell. When a text mining analysis is executed, each of these text files will be pulled into the analysis from the folder where stored.

The variable *File Name* contains the actual file names to be explored. The second variable, *Topic: Earnings?,* is how the experts classified each document (as relevant or not relevant to *Earnings*). Also, there is a variable called *Training* that will later be used during cross-validation of the final model to evaluate its predictive validity and accuracy.

SPECIFYING THE ANALYSIS

Begin by opening the example data file *Reuters.sta:*

Ribbon bar: Select the *Home* tab. In the *File* group, click the *Open* arrow and select *Open Examples* to display the *Open a STATISTICA Data File* dialog. Open the *Datasets* folder. The *Reuters.sta* data file is located in the *TextMiner* folder.
Classic menus: Select *Open Examples* from the *File* menu to display the *Open a STATISTICA Data File* dialog. Open the *Datasets* folder. The *Reuters.sta* data file is located in the *TextMiner* folder.

Next, launch *STATISTICA* Text Miner:

Ribbon bar: Select the *Data Mining* tab. In the Text Mining group, click Text Mining to display the *Text mining Startup Panel.*
Classic menus: From the *Data Mining* menu, select *Text & Document Mining* to display the *Text mining* Startup Panel.

On the *Quick* tab, we need to specify the source of text data (e.g., from spreadsheet cases, from files, or from a file in locations specified by in a spreadsheet column). Select the *Files* option button, and select the *Paths in spreadsheet* checkbox (see Figure X.3).

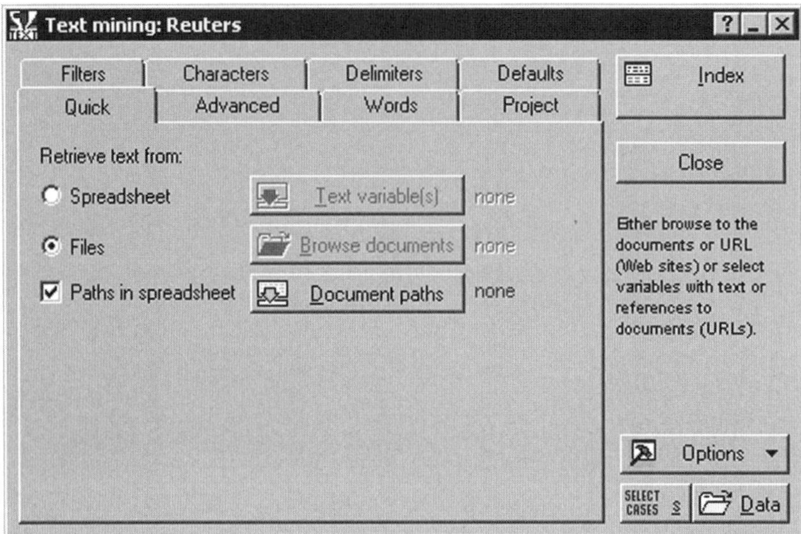

FIGURE X.3
STATISITCA Text Miner "Quick Tab."

Now, click the *Document paths* button to display a variable selection dialog (see Figure X.4) in which you select the variable *File Name* (which is the variable containing the complete references to the input document [XML] files).

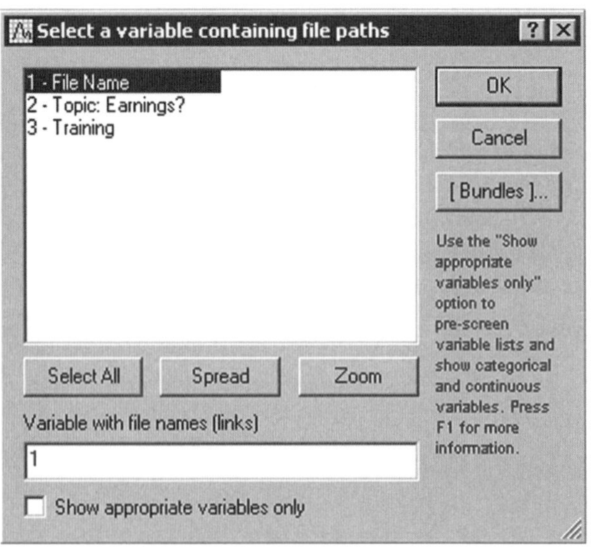

FIGURE X.4
STATISTICA Text Miner "variable selection" dialog.

Click the OK button to return to the Startup Panel (see Figure X.5).

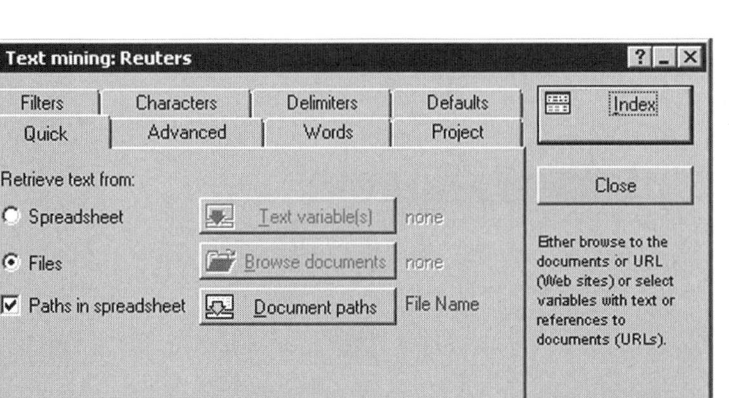

FIGURE X.5
STATISTICA Text Miner Quick Tab dialog showing that the "file name" or text data source has been selected.

Next, select the *Advanced* tab. Change the *% of files where word occurs option* to 3 in order to filter out infrequent words. Now, select the *Words* tab (see Figure X.6), and select the *Stop words (discarded, excluded from indexing)* checkbox. Click the adjacent *Select* button to display the *Open stop-word (text) file* dialog. Browse to the *EnglishStoplist.txt* file (which is in the *TextMiner* subdirectory of the *STATISTICA Text Mining and Document Retrieval* installation).

Click the *Open* button to load that file as the default stop list—that is, the words and terms contained in that stoplist will be excluded from the indexing that occurs during the processing of the documents.

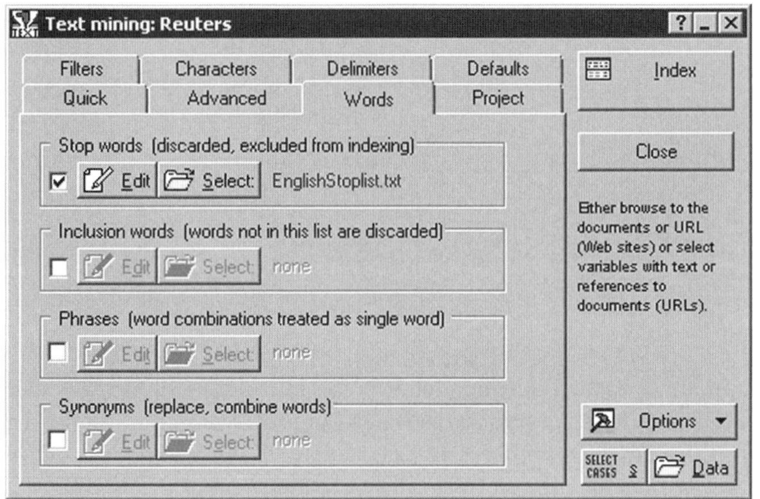

FIGURE X.6
STATISTICA Text Miner "Words tab" dialog, showing that the English stoplist, called "EnglishStoplist.txt," has been selected.

PROCESSING THE DATA ANALYSIS

Next, click the *Index* button in the Startup Panel to begin the processing of the documents. After a few seconds (or minutes, depending on the speed of your computer hardware), the *Results* dialog will be displayed as illustrated in Figure X.7).

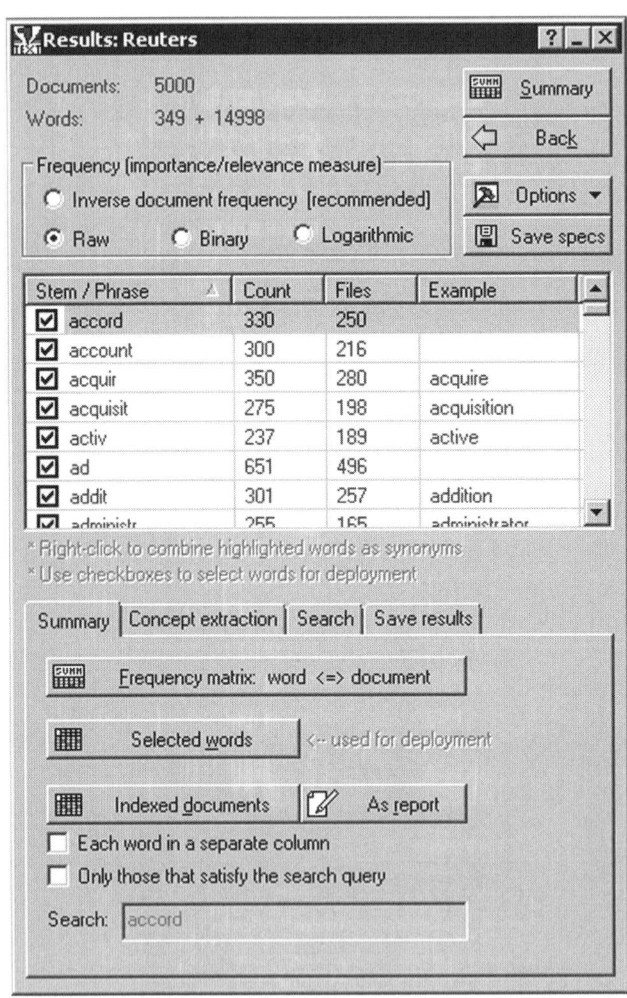

FIGURE X.7
STATISTICA Text Miner "Results" dialog.

The options available at this point are described in some detail in the Introductory Overview (see the ON-LINE HELP which if part of *STATISTICA* Text Miner), as well as in the documentation for the *TM results* dialog. The primary goal of this research is to derive a good classification model for automatically classifying documents (news stories) as relevant or not relevant to *Earnings*.

SAVING THE EXTRACTED WORD FREQUENCIES TO THE INPUT FILE

The next step is to write the extracted word frequencies back to the input file so we can use these frequencies for further analyses. Select the *Save results* tab (see Figure X.8). To write the 349 words that were extracted back into the input file, we need to first "make room" in the data file. To do this, enter 349 into the *Amount* field.

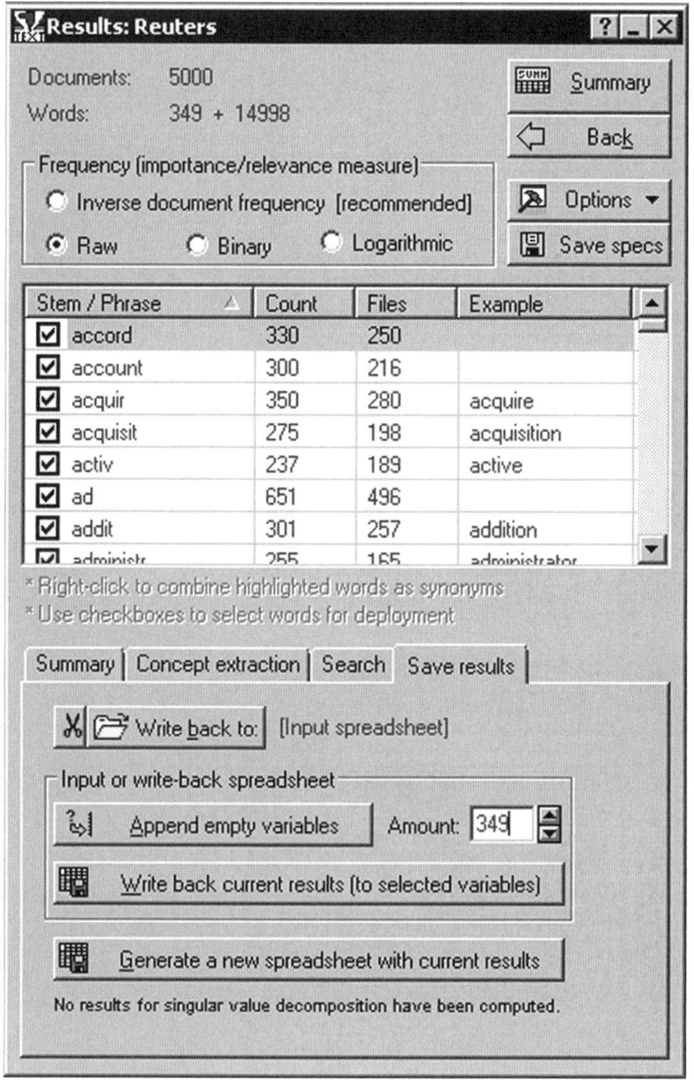

FIGURE X.8

STATISTICA Text Miner "Results Dialog" with "Save Results" tab selected.

Then click the *Append empty variables* button. If *Reuters.sta* was opened as a read-only file, we will be asked to save the file to a different directory (see Figure X.9).

Data: Reuters* (352v by 5000c)

Reuters data; see also example in Manning and Schutze p. 579

	1 File Name	2 Topic: Earnings?	3 Training	4 NewVar1	5 NewVar2	6 NewV
1	1.xml	No	Training			
2	2.xml	No	Training			
3	3.xml	No	Training			
4	4.xml	No	Training			
5	5.xml	No	Testing			
6	6.xml	No	Testing			
7	7.xml	No	Testing			
8	8.xml	No	Testing			
9	9.xml	Yes	Testing			
10	10.xml	No	Training			
11	11.xml	Yes	Testing			
12	12.xml	Yes	Training			
13	13.xml	Yes	Testing			
14	14.xml	Yes	Testing			

FIGURE X.9

STATISTICA Text Miner spreadsheet with the "NewVar1—through NewVar349" added to the spreadsheet, opening up these columns so that the "word counts" of the 349 words selected can be written back to this spreadsheet.

With this operation, 349 blank new variables will be appended to the input file (see Figure X.9). Next, click the *Write back current results (to selected variables)* button to display the *Assign statistics to variables, to save them to the input data* dialog (see Figure X.10). Select all extracted words (variables) in the left pane and all newly created variables in the right pane.

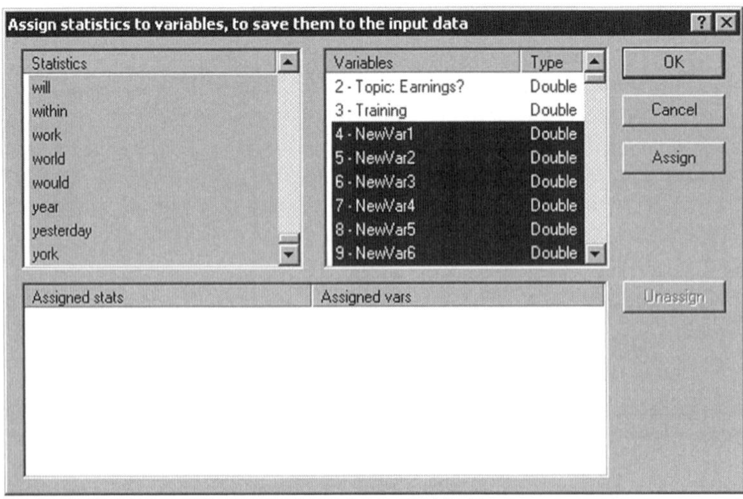

FIGURE X.10

STATISTICA Text Miner "Assign statistics to variables—save to input spreadsheet" dialog.

Then click *Assign* (see Figure X.11).

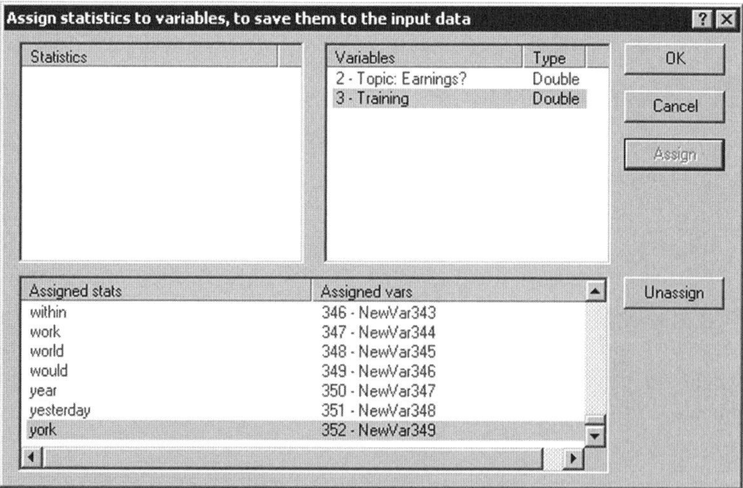

FIGURE X.11

STATISTICA Text Miner "Assign statistics to variables" dialog, after selecting the "assign button." The words assigned and the spreadsheet column to which assigned are indicated in the lower panel of this dialog.

Next, click OK to complete this operation. The newly added variables will automatically be assigned the appropriate variable names to reflect the respective word that was extracted, and the respective frequency counts will automatically be written to the new variables (see Figure X.12).

Data: Reuters.sta* (352v by 5000c)

Reuters data; see also example in Manning and Schutze p. 579

	1 File Name	2 Topic: Earnings ?	3 Training	4 accord	5 account	6 acquire	7 acquisition
1	1.xml	No	Training	0	0	0	0
2	2.xml	No	Training	0	0	0	0
3	3.xml	No	Training	0	0	0	0
4	4.xml	No	Training	0	0	0	0
5	5.xml	No	Testing	0	0	0	0
6	6.xml	No	Testing	0	0	0	0
7	7.xml	No	Testing	0	0	0	1
8	8.xml	No	Testing	0	0	0	0
9	9.xml	Yes	Testing	0	0	0	0
10	10.xml	No	Training	0	0	1	0

FIGURE X.12

STATISTICA Text Miner spreadsheet illustrating that the "word counts" have now been placed into the spreadsheet, following the "assigning of these" in previous "assign dialog."

These simple steps conclude the text mining specific portion of this analysis. What remains is the task to build a good model for predicting the contents (*Earnings - Yes/No*) of the news stories so that we can automatically classify them.

INITIAL FEATURE SELECTION

There are several ways in which we could proceed. As a first step, let's use the powerful and efficient *Feature Selection and Variable Screening* facilities to identify a subset of important predictors from the 349 words that were extracted for inclusion in further model building. Technically, this isn't necessary here because practically all methods for *predictive classification* available in *STATISTICA Data Miner* can handle this many predictors. However, to illustrate how quickly models can be built, let's first use the *Feature Selection and Variable Screening* methods.

Select *Feature Selection and Variable Screening* (see Figure X.13) from the *Data Mining* menu. Then select variable "Topic: Earnings?" as the categorical dependent variable and all variables containing the word counts (which we wrote back to the input data) as continuous predictors (see Figure X.13).

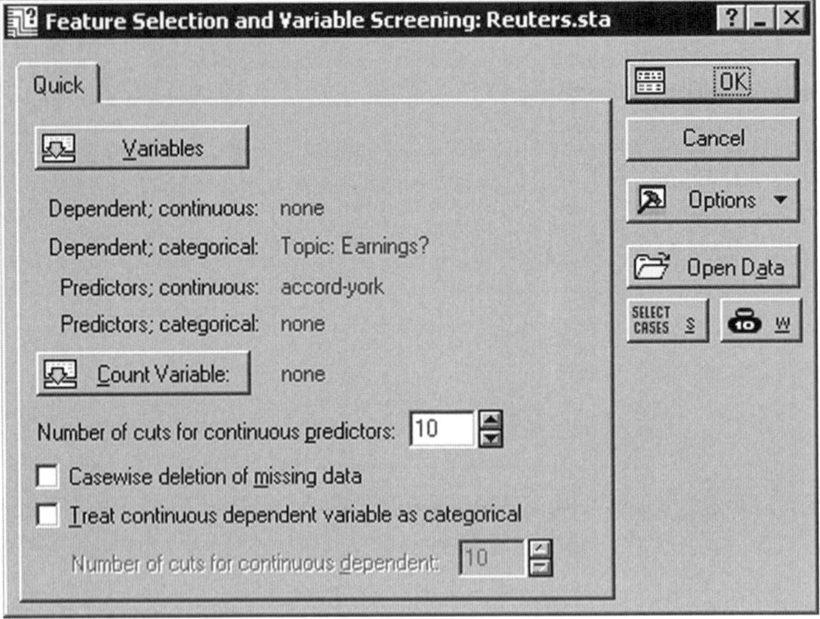

FIGURE X.13
STATISTICA Text Miner "Feature Selection and Variable Screening" dialog.

Then click OK on the Feature Selection and Variable Screening dialog to display the *FSL Results* dialog. Specify to display the best 50 predictors of "Topic: Earnings?" (enter 50 into the Display field) and create the graph of the predictor importance (click the Histogram of importance for best k predictors button as illustrated in Figure X.14).

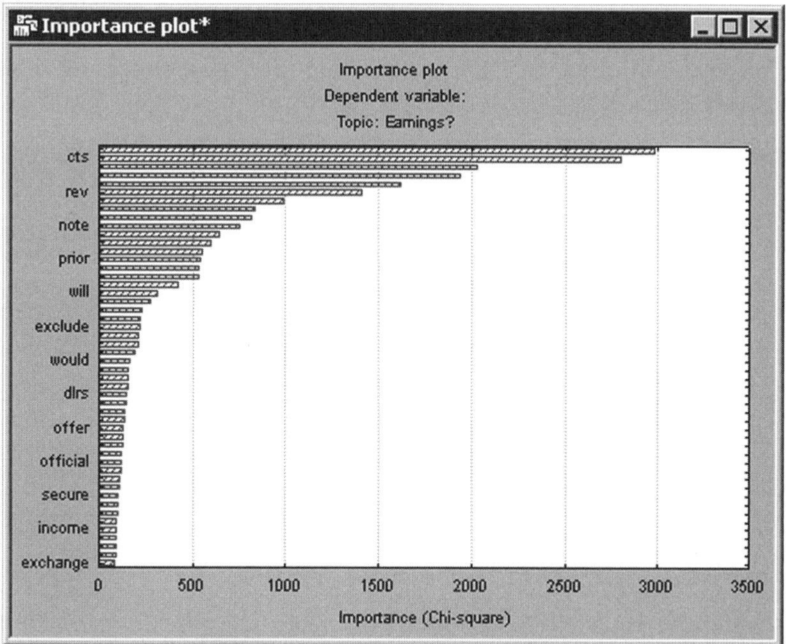

FIGURE X.14
STATISTICA Text Miner "Importance Plot" generated from "Feature Selection and Variable Importance" computations.

Judging from this plot, it may be sufficient to take only the first 20 or so predictors for final modeling (refer also to the Feature Selection and Variable Screening Overviews). We will use the best 20 variables (words) as the predictors (see Figure X.15) for further model building, specifically to use *Classification and Regression Trees* to build a final predictive model.

In the *Display* field, specify to display 20 predictors, and click the *Report of best k predictors (features)* button to display the list of the best predictors in a report. Copy the 20 predictors (see Figure X.15) to the Clipboard to be used in the *General Classification and Regression Trees (GC&RT)* analysis.

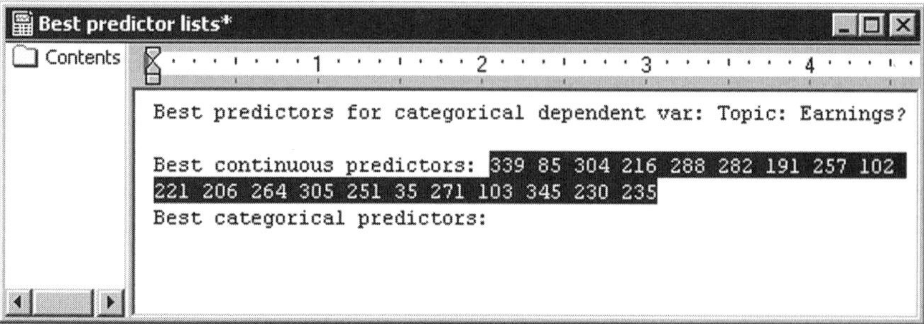

FIGURE X.15
STATISTICA Data Miner and Text Miner "best predictor lists" results from the "Feature Selection" process. These can be highlighted and copy and pasted as the variables to use in additional analyses.

GENERAL CLASSIFICATION AND REGRESSION TREES

Select General Classification/Regression Trees Models from the Data Mining menu. Standard C&RT is selected by default (see Figure X.16).

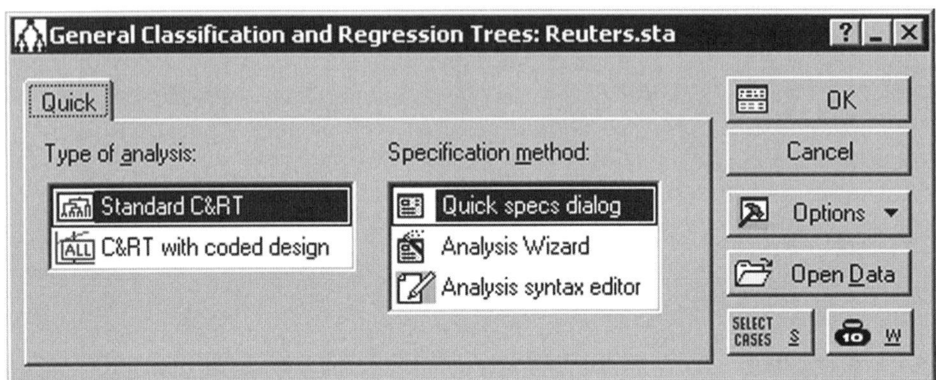

FIGURE X.16
STATISTICA Data Miner "General Classification and Regression Trees" dialog.

Click the OK button to display the *Standard C&RT* dialog, select the *Categorical response (categorical dependent variable)* checkbox, click the *Variables* button, and select variable "Topic: Earnings?" (see Figure X.17) as the *Dependent* variable. As the *Continuous predictors,* select the best 20 predictors (paste them into the variable selection dialog from the Clipboard) derived from the *Feature Selection and Variable Screening* analysis.

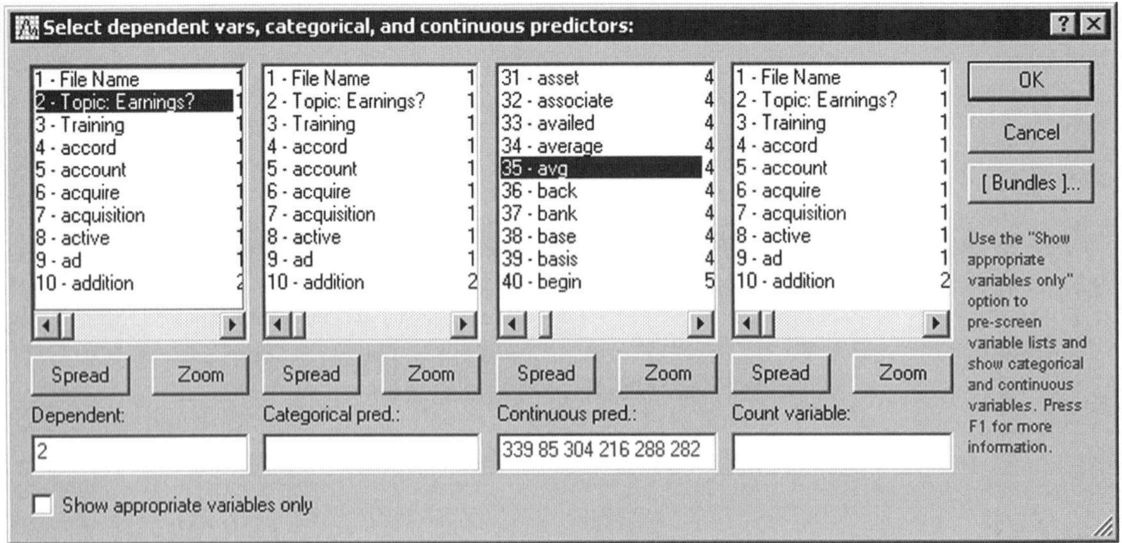

FIGURE X.17
STATISTICA variable selection dialog.

Click OK on the "Variable Selection" dialog. which brings back the "Standard C&RT" window.

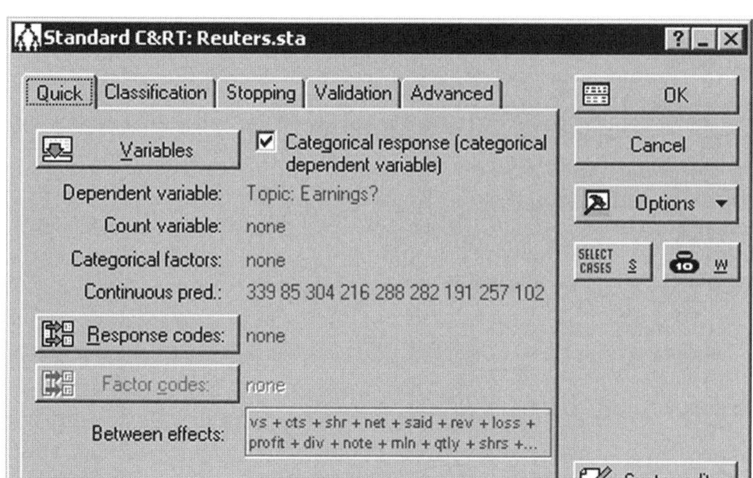

FIGURE X.18

STATISTICA Standard C&RT dialog showing that the variables selected are now placed into the model to be computed when the OK button is selected.

Next, on the *Validation* tab, select the *V-fold cross-validation* checkbox (to automatically select a robust model), and also specify variable *Training* as a *Test sample* variable (see Figure X.19), with the code *Training* as to define the sample from which we will build the model (we will use the remaining cases to test the predictive validity of the model).

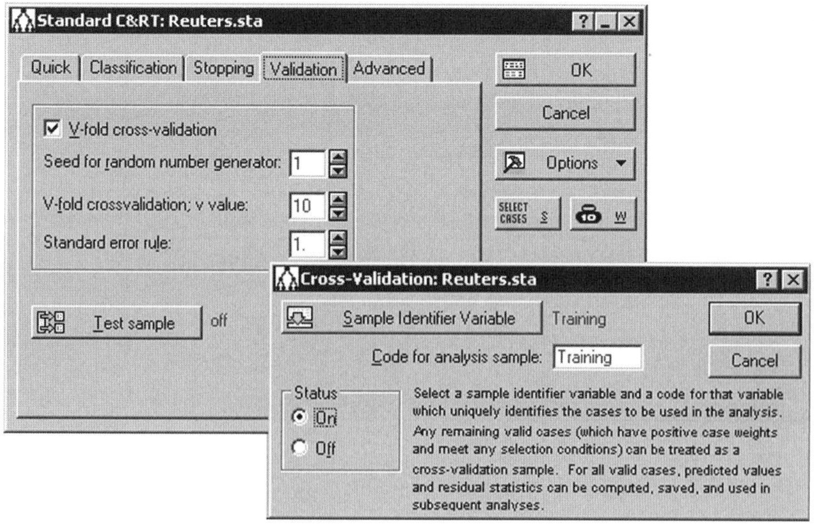

FIGURE X.19

STATISTICA C&RT dialog with "Validation Tab" selected, "V-fold cross-validation" checked, and the "cross-validation" window set for the *Training* sample, which is set to "On" status.

Now click OK in the *Standard C&RT* dialog to begin the analysis. After a few seconds, the *GC&RT Results* dialog will be displayed. Click the *Tree graph* button on the *Summary* tab to review the final tree (see Figure X.20).

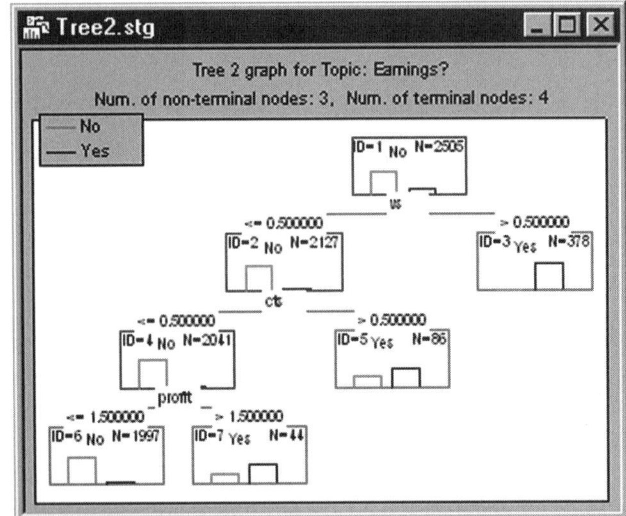

FIGURE X.20
STATISTICA Data Miner C&RT tree graph results.

The final tree is similar, although not identical to that shown in Manning and Schütze (2002, Figure 16.1). Nevertheless, if you select the *Classification* tab of the *GC&RT Results* dialog, select the *Test set* option button to compute the predicted classification for the (holdout) test sample, and click the *Predicted vs. observed by classes* button to get the following confusion (misclassification) matrix as illustrated in Figure X.21.

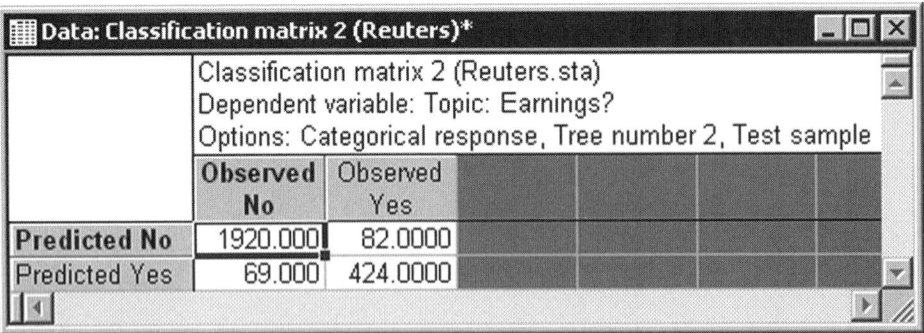

Data: Classification matrix 2 (Reuters)*

Classification matrix 2 (Reuters.sta)
Dependent variable: Topic: Earnings?
Options: Categorical response, Tree number 2, Test sample

	Observed No	Observed Yes
Predicted No	1920.000	82.0000
Predicted Yes	69.000	424.0000

FIGURE X.21
STATISTICA Data Miner C&RT tree "classification matrix" or also known as "Confusion — misclassification — Matrix" results dialog.

This translates into a classification model with a predictive accuracy rate of 94 percent!

MARSplines example: We will use just 100 of the Reuters documents to make this example run faster for use as a teaching example (see Figures X.22 and X.23).

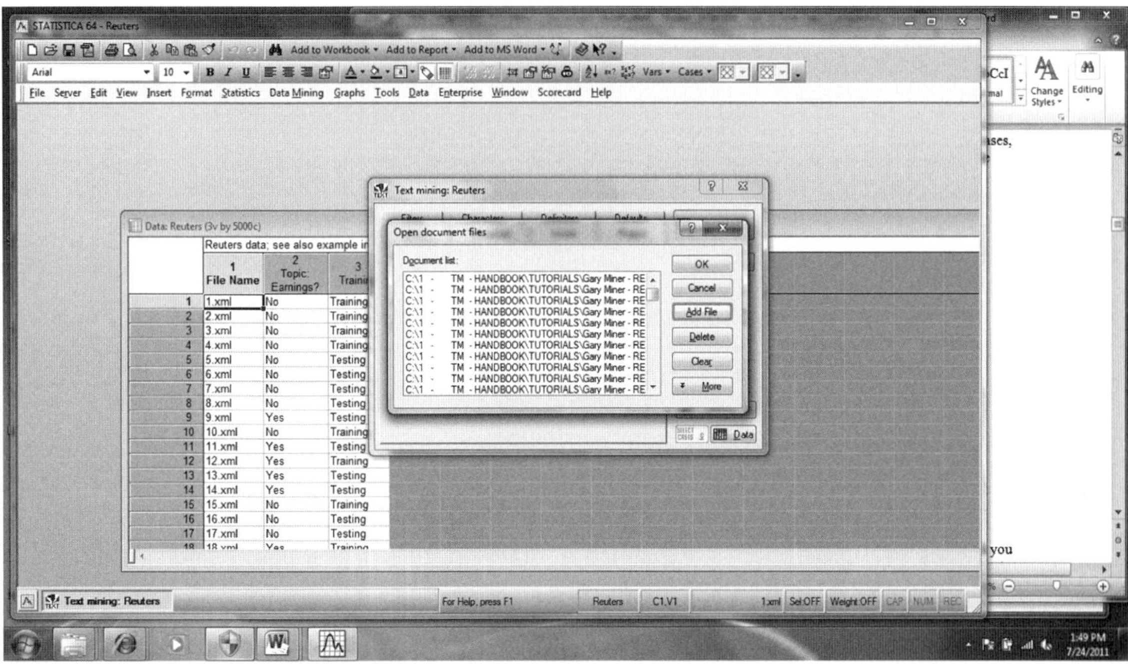

FIGURE X.22
Selection of Reuters text files from the pathway where they reside.

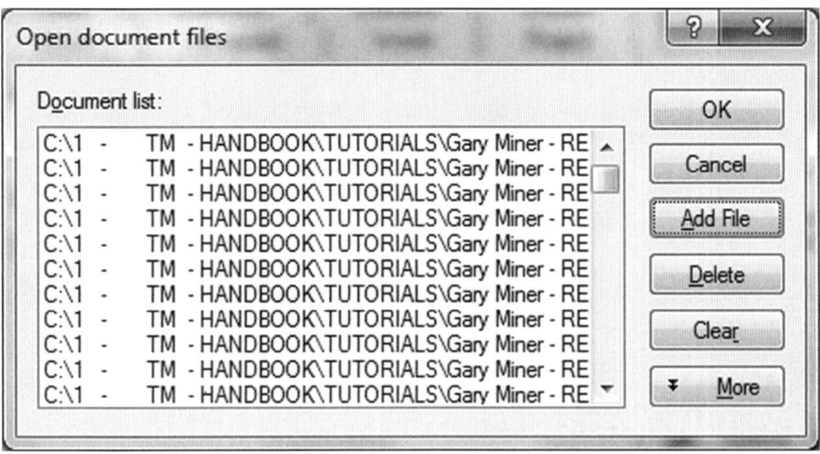

FIGURE X.23
Open document files window.

All parameters of all tabs will be left at their defaults, except the number of words to be selected. We'll change this to 300 words (see Figure X.24).

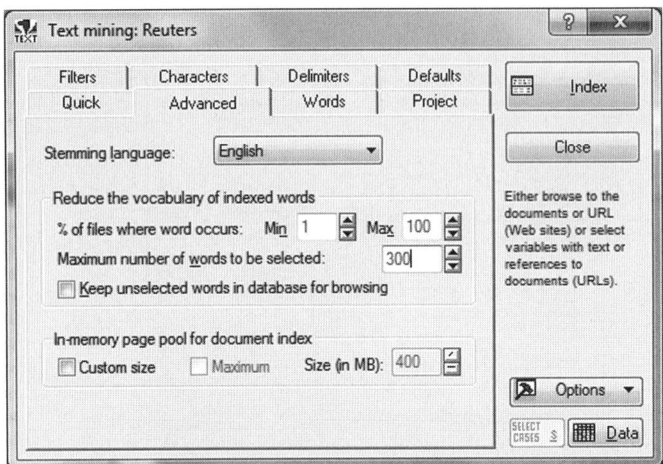

FIGURE X.24

Text Mining dialog "Advanced" tab, where we have selected just 300 words to be returned from the indexing.

Click Index to compute the frequency of words as illustrated in Figure X.25. Then compute Concepts using the SVD procedure (as explained in other tutorials in this book), and save all results back to the master data file. Thirty-six concepts were extracted.

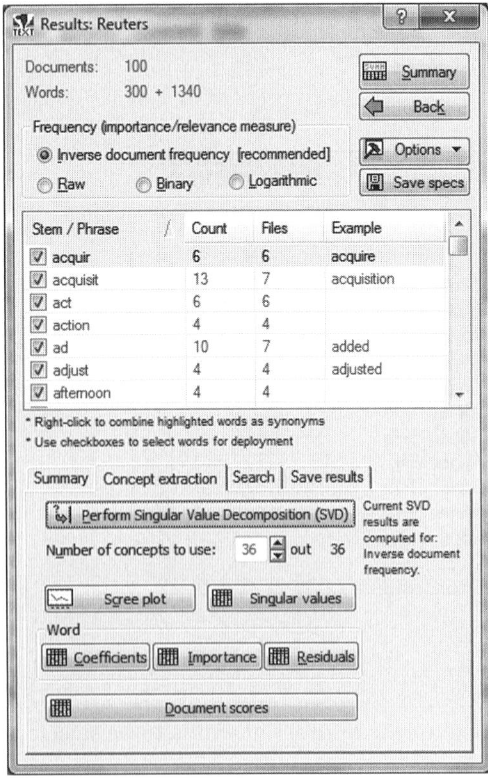

FIGURE X.25

Results of text mining the subset of just 100 documents.

Select "MARSplines" from the Data Mining pull-down menu (see Figure X.26 and X.27).

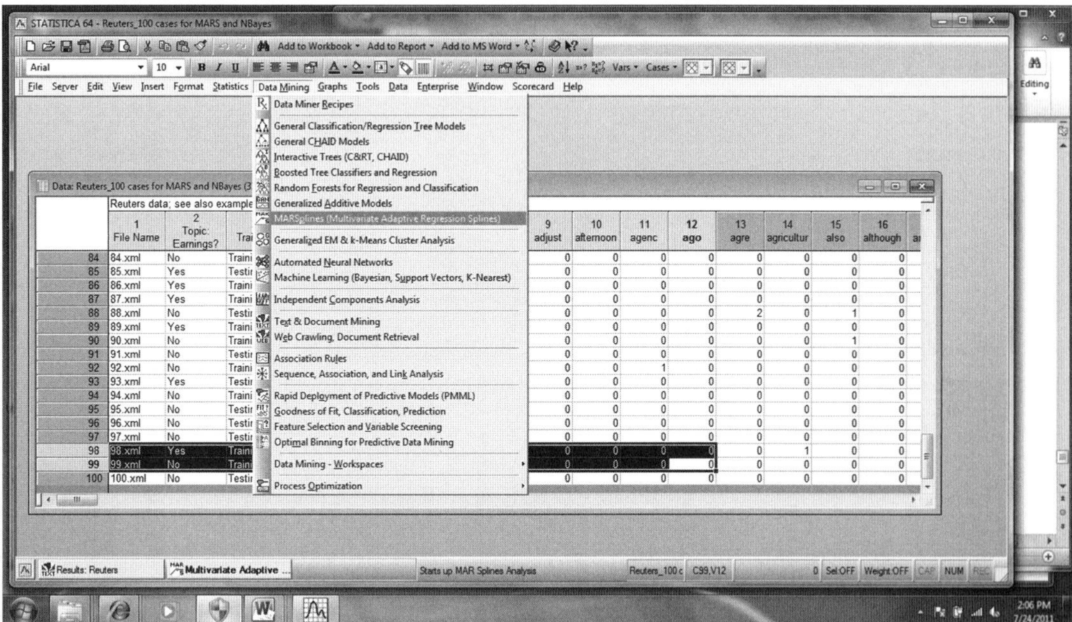

FIGURE X.26

MARSplines algorithm selection from the Data Mining pull-down menu.

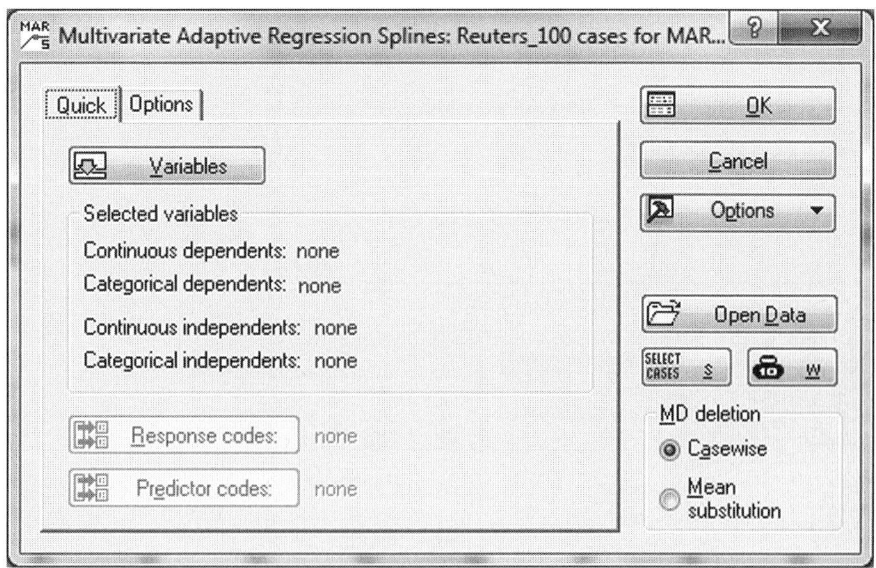

FIGURE X.27

MARSplines dialog window.

Next, the variables are selected as illustrated in Figure X.28.

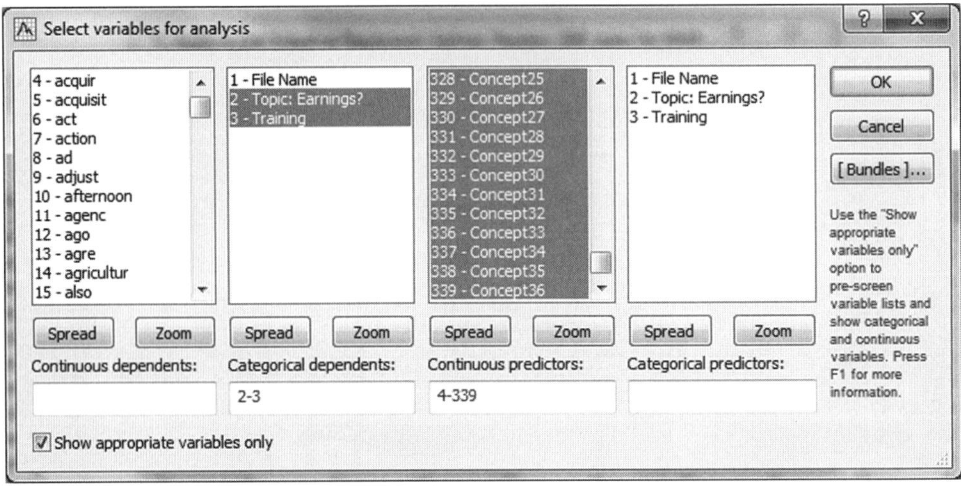

FIGURE X.28
Select variables window.

The results of this MARSplines analysis are seen in Figures X.28 and X.29.

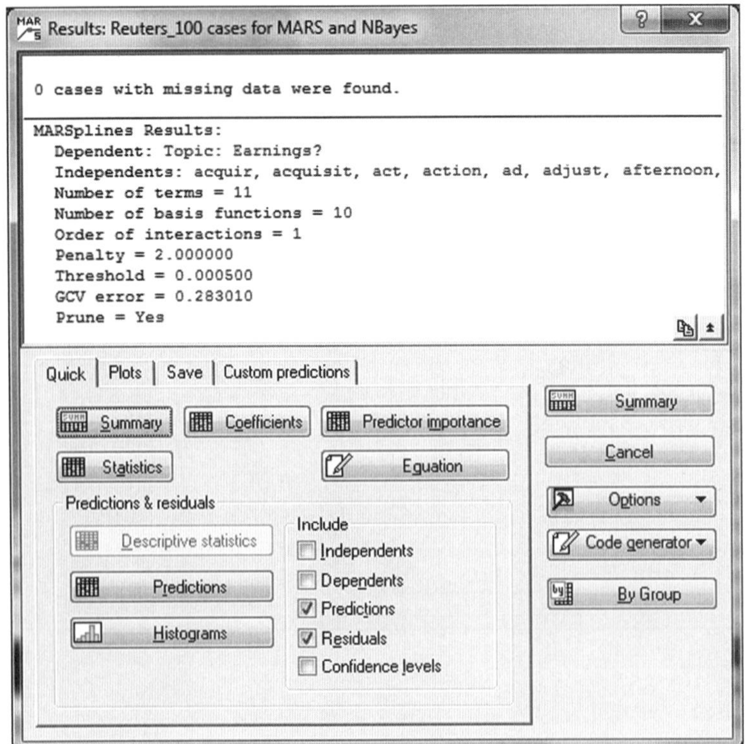

FIGURE X.29
Results window for MARSplines.

Class Predicted	Confusion matrix Topic: Earnings? (Reuters_100 cases for MARS and NBaye Predicted (rows) x Observed (columns)	
	No	Yes
No	67	7
Yes	3	23

FIGURE X.30
The resulting "Confusion matrix" shows that most of the documents were classified correctly as per the "earnings" target variable.

The Naïve Bayes Classifier will be computed next to compare with Trees and MARSplines.

The selection of NAIVE BAYES CLASSIFER from the Data Mining pull down menu is illustrated in the next series of figures, Figure X.31 through X.34.

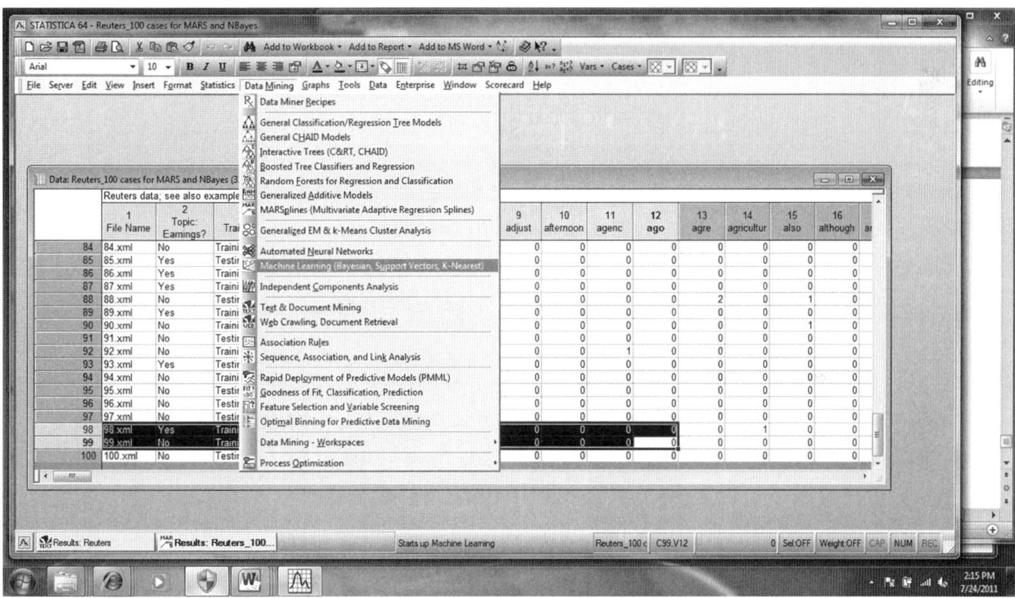

FIGURE X.31
Selecting the Naïve Bayes Classifier algorithm.

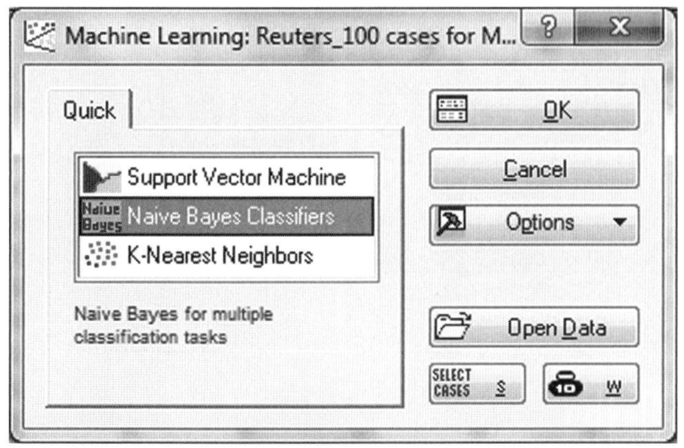

FIGURE X.32
Naïve Bayes Classifiers.

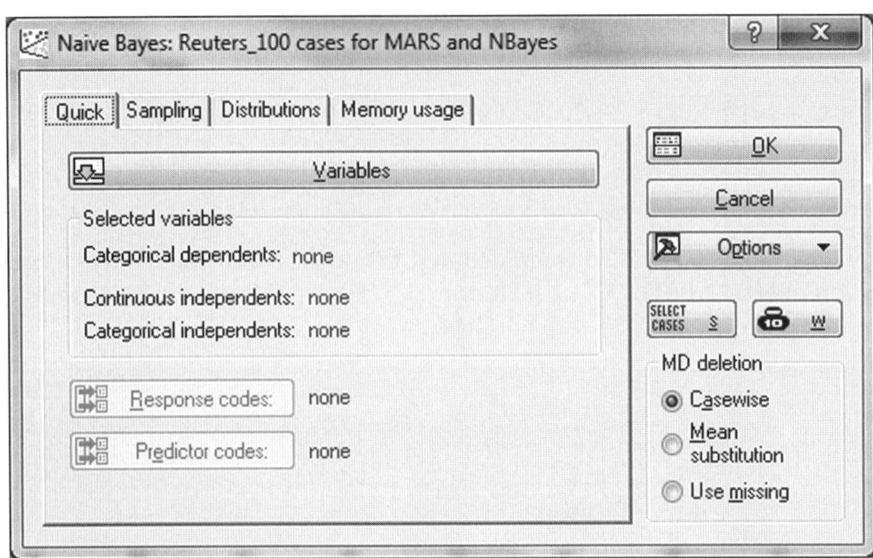

FIGURE X.33
Naïve Bayes dialog window.

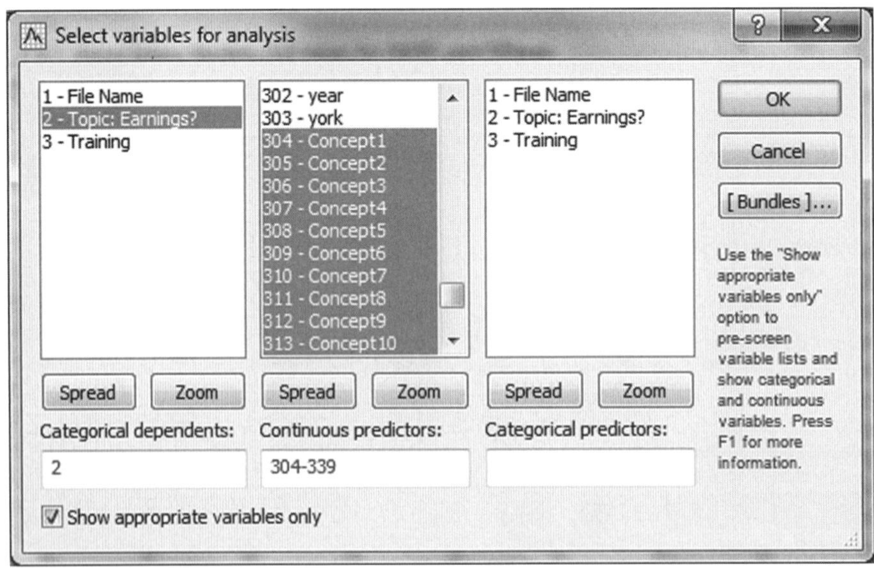

FIGURE X.34
Variables selected. We will use only the Concepts for this computation, since the concepts have extracted information from all of the 300 words.

Leaving all tabs of the Naïve Bayes dialog at their defaults, click on OK to run the computations. The results are shown in Figure X.35.

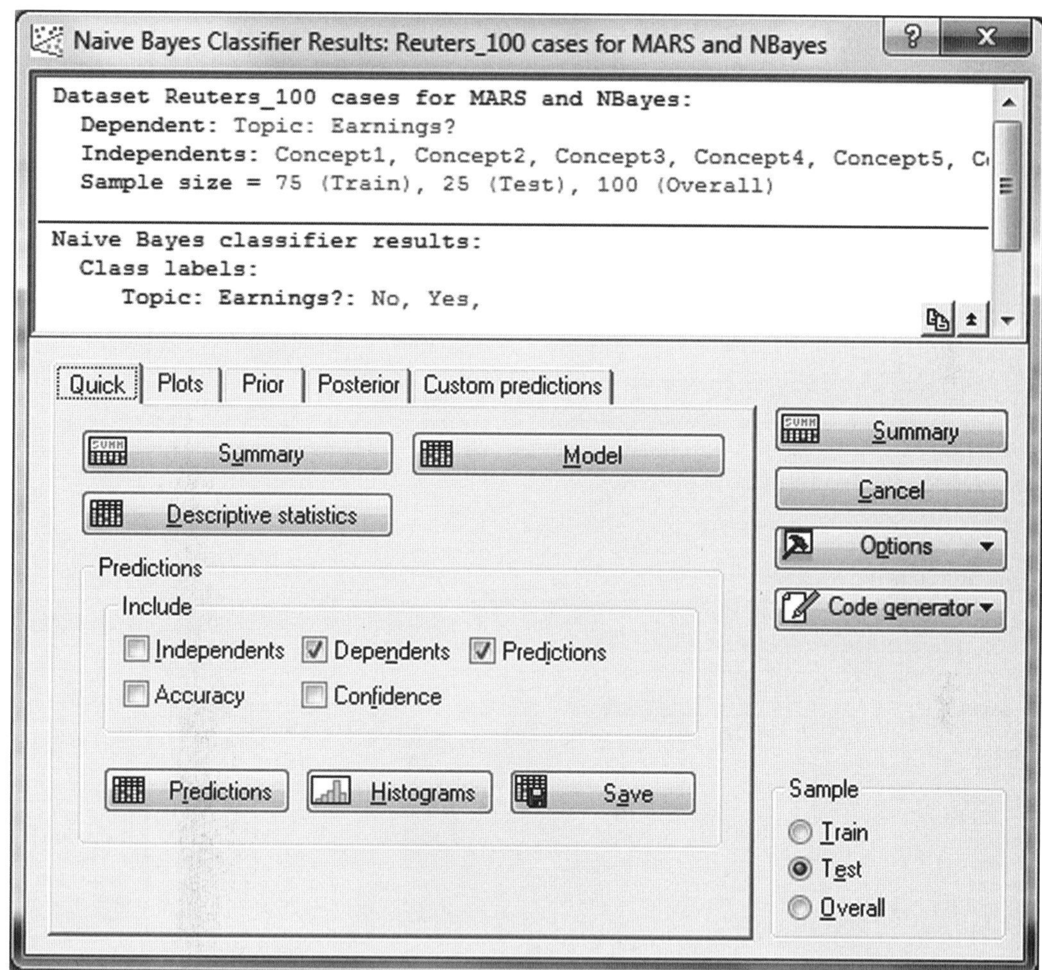

FIGURE X.35
Results of Naïve Bayes.

We will not go into all of the results here, but you can go to the DVD and get the "Naïve Bayes Workbook results.stw" file, open it in *STATISTICA*, and view some of the specific results obtained from selecting some of the "results buttons" seen in Figure X.35 above.

K-NEAREST NEIGHBORS MODELING

Let's try one more algorithm on this dataset: K-Nearest Neighbors as illustrated in Figures X.36 through X.38.

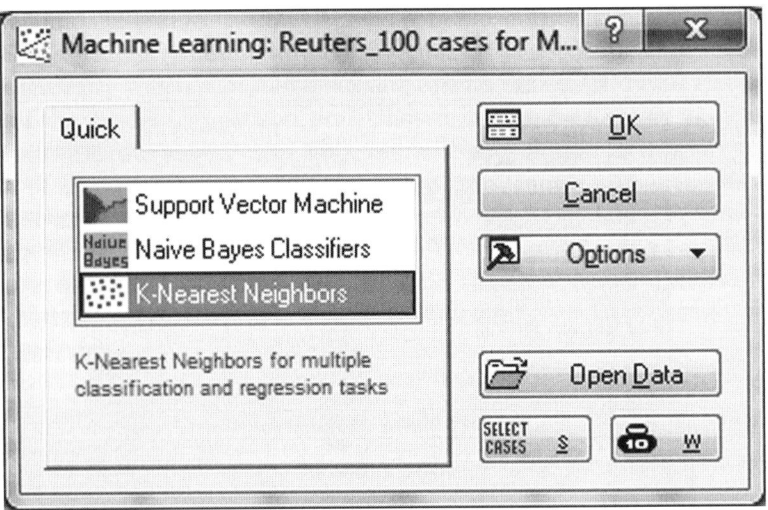

FIGURE X.36
Selecting the K-Nearest Neighbors algorithm in *STATISTICA*.

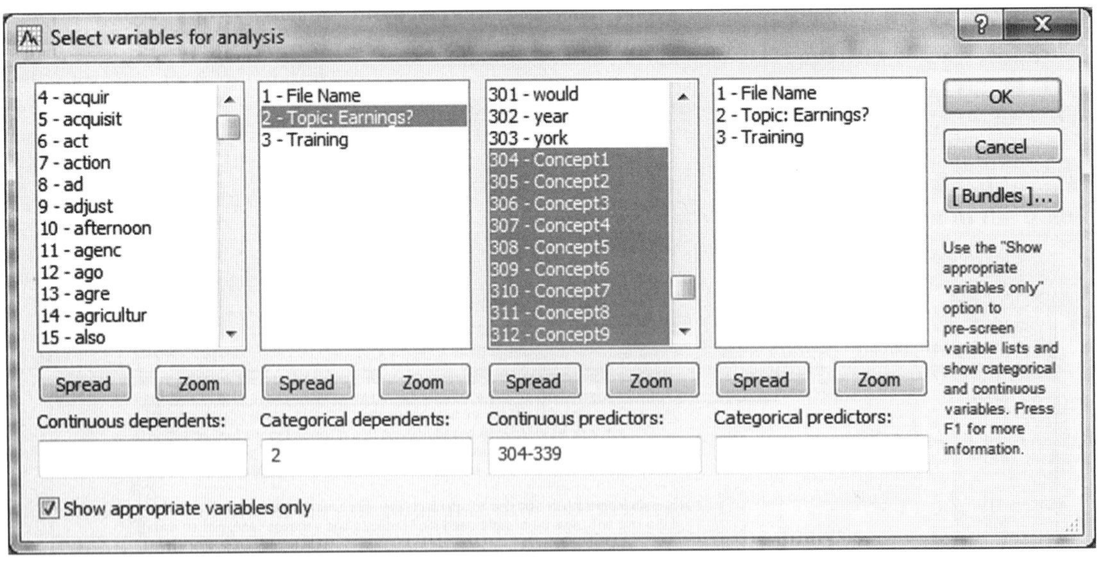

FIGURE X.37
Variable selection for the K-Nearest Neighbors computation.

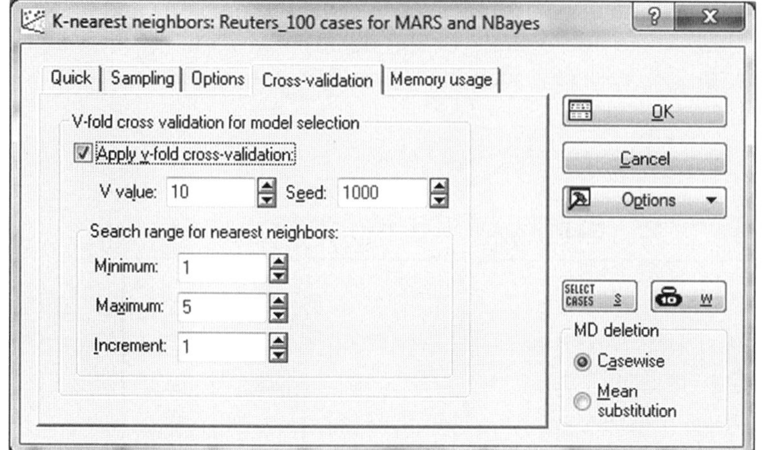

FIGURE X.38
Cross-validation selected on the appropriate tab.

Then click OK selected to compute the K-Nearest Neighbors Results (see Figure X.39).

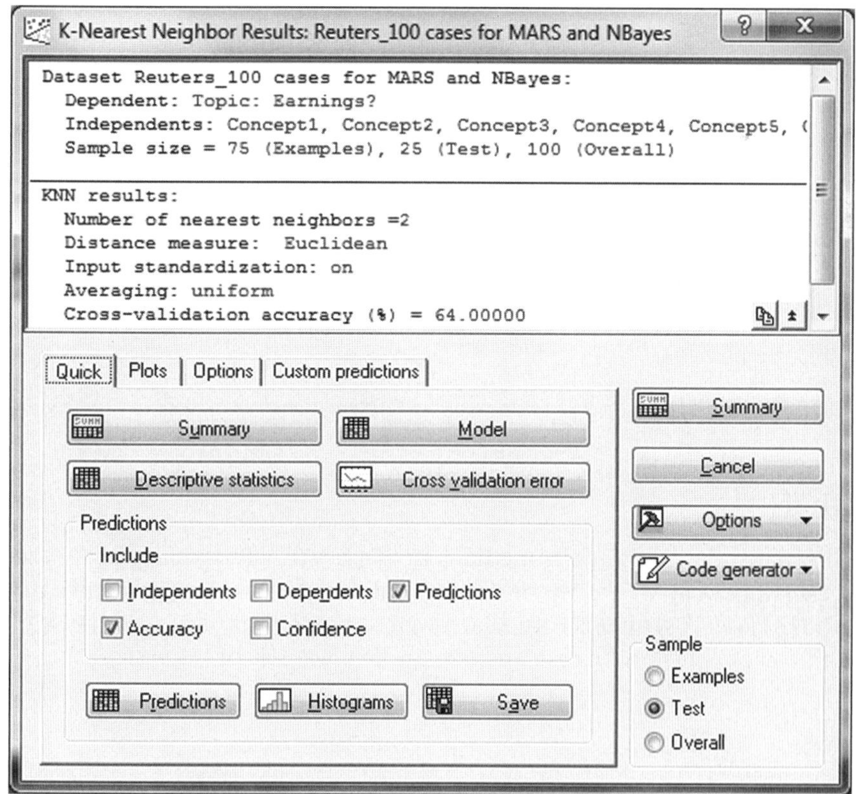

FIGURE X.39
K-Nearest Neighbors Results.

We won't go into this K-Nearest Neighbors Results in detail. You can go to the DVD and find the K-Nearest Neighbors Results Workbook to examine some of the details. Only one graph will be presented here, showing that this model correctly identified documents more often than not (see Figure X.40).

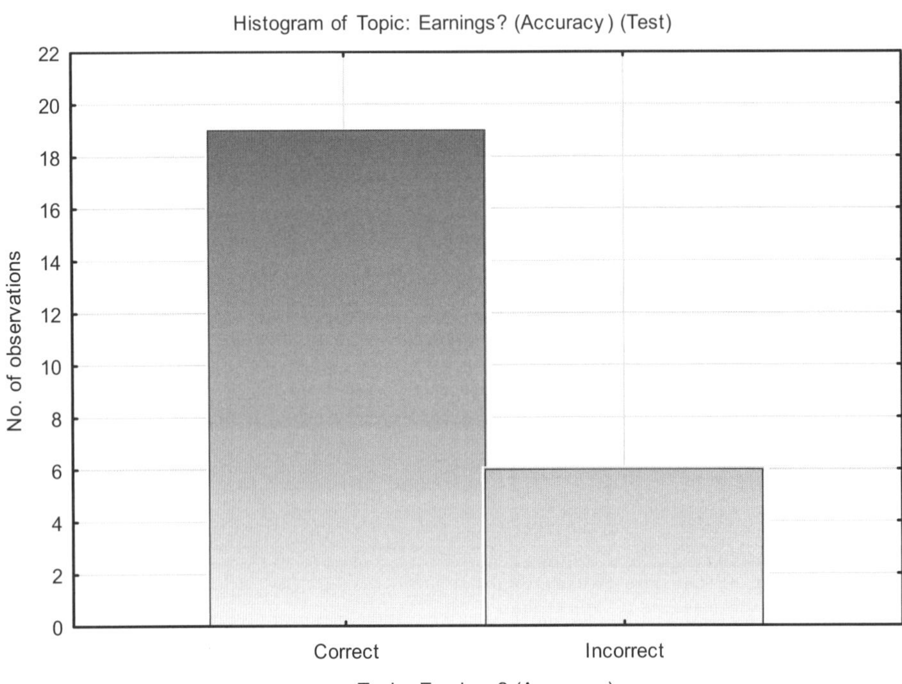

FIGURE X.40
Histogram of prediction of earnings from the Reuters documents, using K-Nearest Neighbors modeling.

CONCLUSION

This example illustrated how the various methods in *STATISTICA Text Mining and Document Retrieval*, along with *STATISTICA Data Miner*, can be used to build highly accurate predictive models for classifying text. The *STATISTICA* system is particularly well suited for this purpose because of the seamless integration of all components of the data and text mining facilities of the system.

Reference

Manning, C. D., and Schütze, H. (2002). *Foundations of statistical natural language processing* (5th edition). Cambridge, MA: MIT Press.

TUTORIAL Y

CASE STUDY: Predicting Exposure of Social Messages: The Bin Laden Live Tweeter

Tom Emerson
Senior Research Scientist, Topsy Labs

Rishab Ghosh
Cofounder and Vice President of Research, Topsy Labs

Eddie Smith
CRO, Topsy Labs

CONTENTS

INTRODUCTION

On April 30, 2011, at 7:08 PM, Sohaib Athar, an IT consultant residing in Abbottabad, Pakistan, tweeted, "Since taliban (probably) don't have helicopters", and since they're saying it was not 'ours,' so must be a complicated situation." Over the next 30 hours, Sohaib's tweets, under the Twitter handle @ReallyVirtual, were exposed over 82 million times to people around the globe.

Social media has shown in multiple instances the power a single person's communication can have when his or her communication is picked up and rebroadcast (retweeted in Twitter parlance) by many people. The ability to identify, track, and predict exposure of messages within Twitter has a number of benefits for commercial and other uses. Specifically, utility is extracted when exposure curves are computed based upon cumulative audience metrics for relevant keywords or phrases using census-based social data sets. Measuring an audience exposes the true reach of the target social communication and is especially valuable as events and happenings are unfolding within the social web.

It's easy to obtain counts of mentions within Twitter, but it's very hard to identify audience exposure within Twitter. Simple tweet counts do not represent the people that were actually exposed to a message, and there are no time-lapse metrics associated with counts, so you can't see how an audience and exposure is building over time. This tutorial uses the Bin Laden Live Tweeter example to discuss the methods, statistics, and data used to identify audience exposure curves within Twitter.

Practical Text Mining and Statistical Analysis for Non-structured Text Data Applications. DOI: 10.1016/B978-0-12-386979-1.00033-5

ANALYSIS

The data source for this analysis was a complete census of Twitter data. Queries were run for the account @ReallyVirtual, using Topsy Lab's proprietary fast-indexing tools, which output a complete census of tweets that were tweeted from this account from April 30, 2011, to May 5, 2011. Each tweet contained the tweet content, time stamp, and Twitter account name for each author.

Here's the sequence of tweets sent out by @ReallyVirtual the evening of April 30, 2011.

reallyvirtual: Helicopter hovering above Abbottabad at 1AM (is a rare event).
05/01/2011 2,706 retweet

reallyvirtual: Go away helicopter - before I take out my giant swatter :-/
05/01/2011 742 retweet

reallyvirtual: Since taliban (probably) don't have helicpoters, and since they're saying it was not "ours", so must be a complicated situation #abbottabad
05/01/2011 503 retweet

reallyvirtual: The abbottabad helicopter/UFO was shot down near the Bilal Town area, and there's report of a flash. People saying it could be a drone.
05/01/2011 366 retweet

Exposure analysis was performed on each tweet from @ReallyVirtual, starting with the initial tweet at 7:08 PM on April 30, 2011. As tweets were retweeted, they became exposed to each new user's follower list. Exposure was calculated by summing the number of followers within each new user and incrementing the total for each time period with fine-grained time accuracy—per second. For convenient graphing, this calculation was then aggregated for each hour after the tweet was originated (see Table Y.1 and Figure Y.1).

Hours After Tweet	Cumulative Exposure	Incremental Exposure
0	1,988,619	1,988,619
1	2,302,898	314,279
2	2,929,327	626,429
3	2,988,766	59,439
4	3,015,342	26,576
5	3,126,520	111,178
6	3,657,880	531,360
7	3,788,414	130,534
8	3,848,157	59,743
9	3,907,738	59,581
10	3,945,901	38,163
11	4,004,605	58,704
12	4,033,251	28,646
13	4,044,936	11,685
14	4,053,905	8,969
15	4,066,214	12,309
16	4,082,186	15,972
17	4,086,526	4,340
18	4,088,455	1,929
19	4,143,189	54,734
20	4,148,373	5,184
21	4,149,378	1,005

TABLE Y.1

Tweet exposure for the "Helicopter hovering above Abbottabad at 1 AM (is a rare event)" tweet.

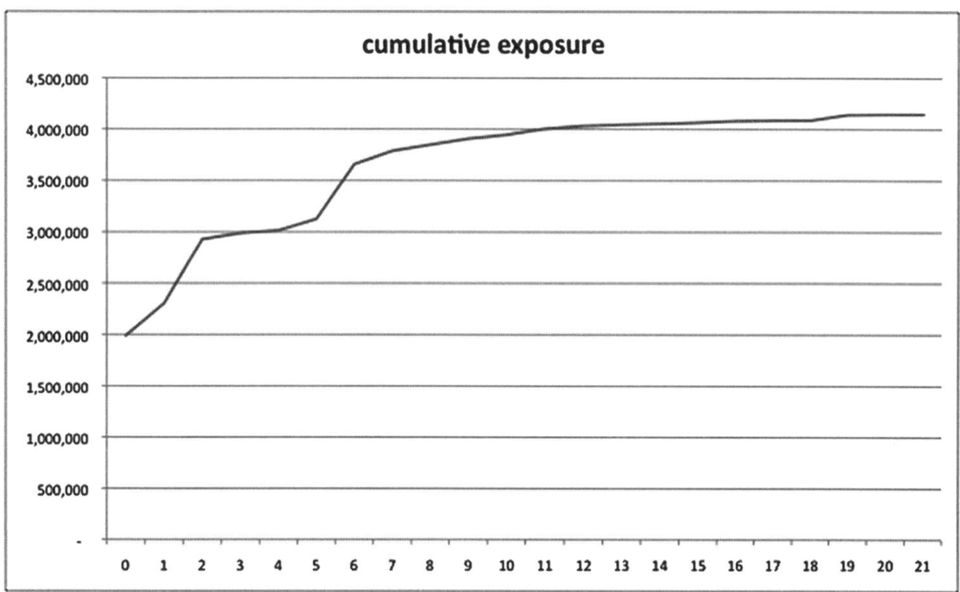

FIGURE Y.1

Graphical representation of the cumulative of the "Helicopter hovering above Abbottabad at 1 AM (is a rare event)" tweet.

This tweet was retweeted 5,102 times, for a total exposure of 4.89 million since May 1, 2011, at 21:42:49 (see Table Y.2 and Figure Y.2).

Hours After Tweet	Cume Exposure	Incremental Exposure
0	1,527,514	1,527,514
1	3,179,376	1,651,862
2	3,507,765	328,389
3	3,713,656	205,891
4	3,797,056	83,400
5	3,930,470	133,414
6	4,133,349	202,879
7	4,275,993	142,644
8	4,410,974	134,981
9	4,477,767	66,793
10	4,541,109	63,342
11	4,599,951	58,842
12	4,666,513	66,562
13	4,718,757	52,244
14	4,736,407	17,650
15	4,824,313	87,906
16	4,848,506	24,193
17	4,862,953	14,447
18	4,869,163	6,210
19	4,883,719	14,556
20	4,888,073	4,354
21	4,888,183	110

TABLE Y.2

Tweet exposure for the "Uh oh, now I'm the guy who live blogged the Osama raid without knowing it" tweet.

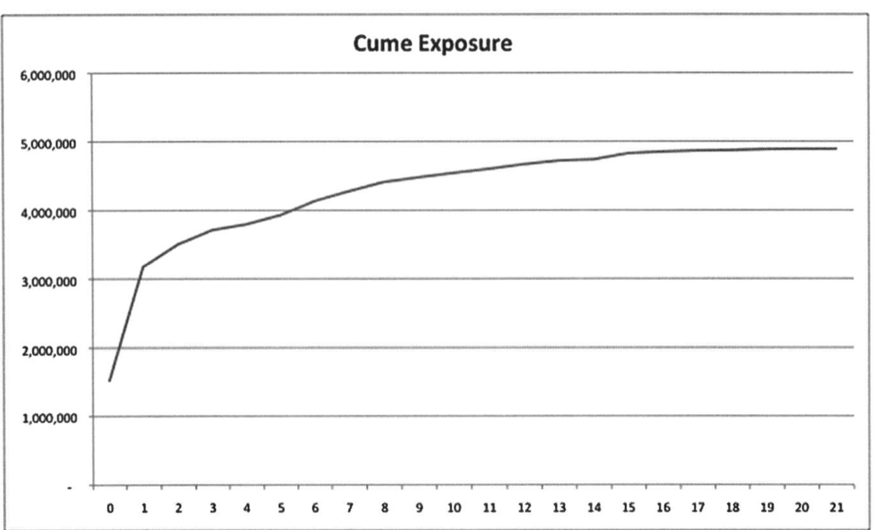

FIGURE Y.2

Cumulative exposure of the "Uh oh, now I'm the guy who live blogged the Osama raid without knowing it" tweet.

This tweet was retweeted 2,269 times, for a total exposure of 4.15 million since May 1, 2011, at 21:44:34 (see Table Y.3 and Figure Y.3).

Hours After Tweet	Cumulative Exposure	Incremental Exposure
0	414,024	414,024
1	663,369	249,345
2	663,369	-
3	745,790	82,421
4	745,790	-
5	745,790	-
6	748,017	2,227
7	1,079,117	331,100
8	1,648,207	569,090
9	23,481,688	21,833,481
10	29,388,813	5,907,125
11	35,055,528	5,666,715
12	41,604,527	6,548,999
13	42,671,480	1,066,953
14	46,709,163	4,037,683
15	50,215,634	3,506,471
16	52,860,567	2,644,933
17	54,602,690	1,742,123
18	55,657,690	1,055,000
19	58,947,851	3,290,161
20	62,187,413	3,239,562
21	63,576,909	1,389,496
22	67,298,030	3,721,121
23	74,612,099	7,314,069
24	78,799,104	4,187,005
25	81,690,143	2,891,039
26	81,919,153	229,010
27	82,457,834	538,681
28	82,607,095	149,261
29	82,676,524	69,429

TABLE Y.3

Tweet exposure for all mentions of @ReallyVirtual.

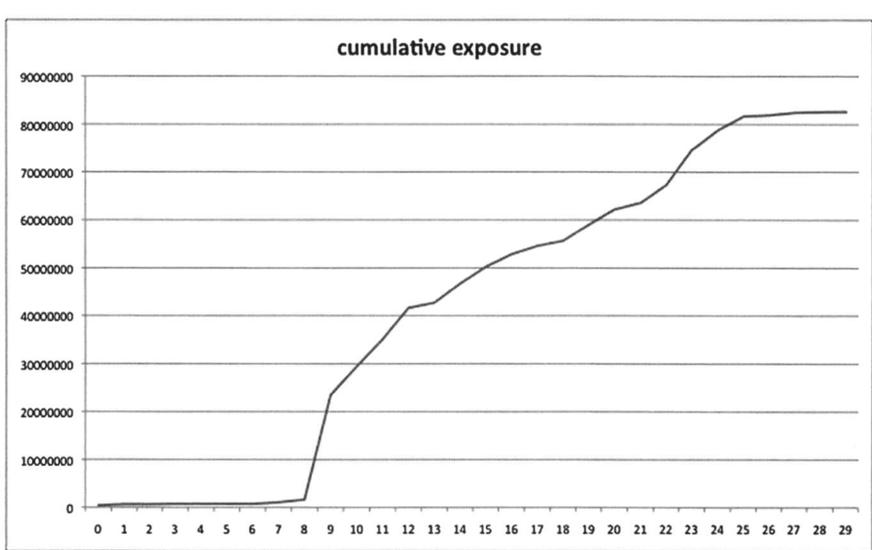

FIGURE Y.3

Cumulative tweet exposure for all mentions of @ReallyVirtual.

In the hour after his initial tweet at 7:08 PM on April 30 PST, Sohaib had very little exposure. But as Sohaib's tweets were retweeted and mentioned a large number of times (over 30,000), his exposure grew to a whopping 82.68 million unique tweets to people within 21 hours. As his tweets became more interesting to the Twittersphere, his exposure and influence grew dramatically, eventually capping out at a total exposure of 82.68 million. He went from 0 to 20 million in under 10 hours and over 82 million in just under 30 hours.

SUMMARY

Identifying and understanding when communication on the social web is "going viral" requires massive amounts of real-time data collection and statistical processing to ensure that only unique audiences are calculated. The extent to which a message is exposed to people can be measured by counting the unique number of followers exposed to a message as it's rebroadcasted (retweeted) within a social network. Discerning unique exposure over time requires that multiple calculations are done to compute cumulative exposure over time, and once this is accomplished, a time-lapse analysis of exposure is uncovered.

This type of time-lapse cumulative exposure analysis is valuable for identifying the extent to which messages are spreading across the social web and can be used to predict the "virality" of message exposure within the social web. While such analysis is pretty simple to compute given a period of time, Topsy's fast indexing and live ranking technology allows this to be done for thousands of keywords and phrases within seconds rather than hours.

TUTORIAL Z

The InFLUence Model: Web Crawling, Text Mining, and Predictive Analysis with 2010–2011 Influenza Guidelines—CDC, IDSA, WHO, and FMC

Benjamin R. Mayer, DO, Linda A. Miner, PhD, Gary D. Miner, PhD, Edward E. Rylander, MD, and Kristoffer Crawford, MD

In His Image Family Medical Residency Program, Tulsa, Oklahoma

CONTENTS

ABSTRACT

The challenge to differentiate between viral and bacterial infections has for decades led to missed diagnoses and unwarranted uses of medications. Influenza annually presents doctors with questions of using antibiotics instead of more symptomatic care. With the rise of antibiotic resistance and influenza morbidity and mortality, organizations such as the Centers for Disease Control (CDC), Infectious Disease Society of America (IDSA), and World Health Organization (WHO) annually research ways to aid clinicians to more accurately care for the patient populations.

Our goal was to compile current guidelines and recommendations from the leading scientific resources and objectively compare and evaluate the effectiveness of care at Family Medical Care (FMC). By doing so, we hoped to establish and implement these findings at FMC to improve the quality of care and reduce medical complications of mistreatment.

We formulated a chart review using all influenza-like illnesses (ILI) during the progression from beginning to peak flu season. Using *STATISTICA* software by StatSoft and these charts and current guidelines, we wanted to disprove the null hypothesis that FMC is effectively caring for and treating

Practical Text Mining and Statistical Analysis for Non-structured Text Data Applications. DOI: 10.1016/B978-0-12-386979-1.00034-7

patients based on the most current medical guidelines and recommendations. We hoped to find correlations and cause and effect in the style and demographic of FMC practice and create a new standard of care to improve its patient care.

We obtained the following results: Of the 481 charts reviewed, 195 were confirmed as ILI using combination of symptoms "fever plus 2" of cough, sore throat, myalgia, and/or headache. Of these, 71 were diagnosed with influenza using 127 Rapid Diagnosis of Influenza Test (RDITs). Incorporation of flu prevalence showed 62 percent of testing was done when not indicated, and 48 percent of tests were not done when indicated. Overall, however, testing scored 76 percent accurate based on all chart data (ILI and non-ILI) and 63 percent accurate based on ILI only charts. Treatment was indeterminable in 6 percent of cases because of no documented RDIT result. Ten percent of patients did not receive Tamiflu when they needed it, and 5 percent got it when it was not needed. Overall treatment was 85 percent correct when looking at all charts, but 70 percent correct looking only at ILI confirmed patients. P-value statistics for all guidelines other than testing and treatment equaled 0.000, but for testing was 0.143 and for treatment was 0.040. Data analysis using Support Vector Machine (SVM) and Feature Selection and Root Cause Analysis supported our use of criteria for diagnosing ILI. Their results suggested temperature, flu prevalence week, and chills might predict RDIT results with an accuracy of 78.151 percent. An importance plot was generated and showed temperature to be the most important variable in predicting RDIT. An ANOVA was performed and suggested that a temperature of between 99.8° and 100.6° F has a higher chance of being flu positive.

Overall, we concluded that FMC missed the mark on meeting the guidelines for flu as a whole, but did function at about 80 percent accuracy in flu testing and treatment. This is based on the selected influenza-like illness criteria and the prevalence of flu reported by the CDC during the early to peak flu season. P-value statistics demonstrated an average 0.0915 significance of data accuracy, while data mining was found to support the use of the "fever plus 2" criteria in combination with flu prevalence. The support vector machine showed 78.151 percent Positive Predictive Value (PPV) of RDIT using temperature, flu prevalence week, and chills. This helped confirm the PPVs demonstrated in the literature reviews. Of these, temperature was most important based on feature selection and root cause analysis, followed by ANOVA. Boosted Trees and Naïve Bayes Classification predictive models were also computed to compare the accuracies with SVM.

WEB CRAWLING AND TEXT MINING OF CDC DOCUMENTS ON FLU

In this section we will use the STATISTICA Web Crawler to find pertinent web information on flu, primarily from the CDC (Centers for Disease Control) web sites.

NOTE: Very little text dialog will be given in this section, as by this time, if you've done some of the tutorials prior to this, you should have a "good feel" of doing a text mining analysis. But all the screens of the computer analysis will be presented, so you can rapidly go from one to the next to re-create this analysis for yourself, using the data and other files located on the DVD that comes with this book; go to the folder on the DVD called "TUTORIAL data - etc" to get these datasets, plus some of the results workbooks, and even a "Data Miner Workspace" illustrating Feature Selection.

The first figure, Figure Z.1 below, shows the 'Text Mining' and the 'Web Crawling and Document Retrieval' selections on the STATISTICA Data Miner pull down menu.

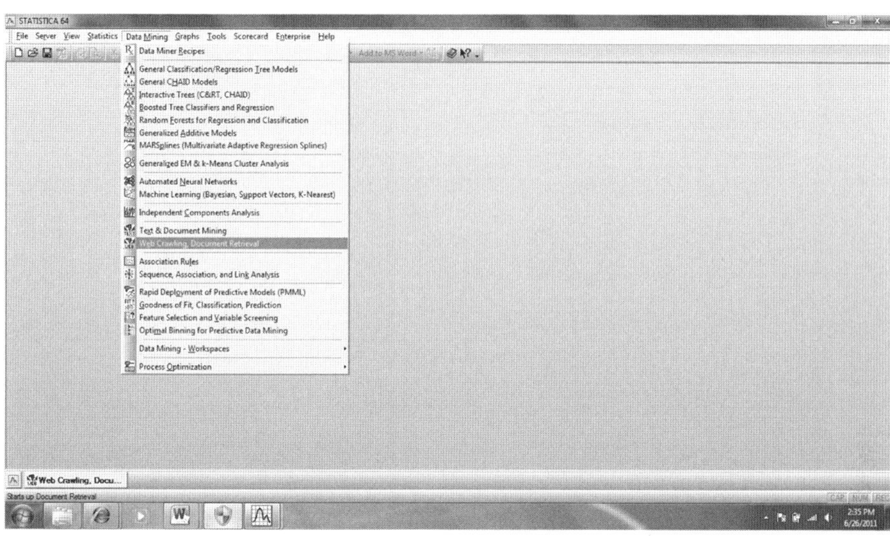

FIGURE Z.1
Getting the Web Crawling part of Text Mining from *STATISTICA* menus.

The following series of figures will illustrate the mechanics of Web Crawling to find documents of interest (see Figures Z.2 through Z.14).

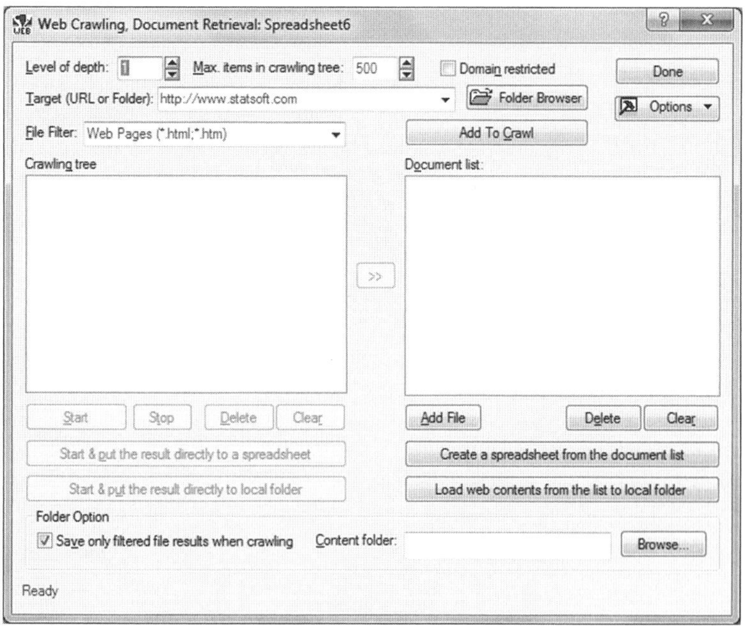

FIGURE Z.2
Web Crawling dialog window.

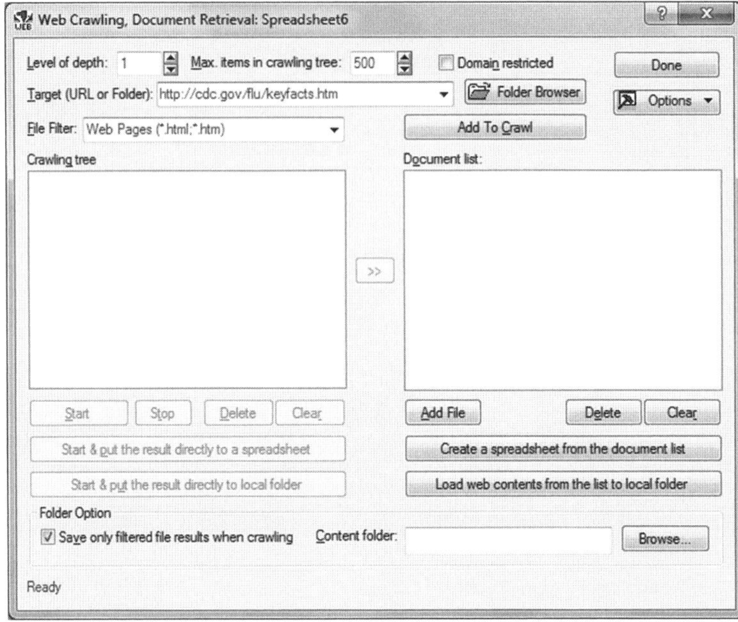

FIGURE Z.3

Adding the first primary web site: http://cdc.gov/flu/keyfacts.htm.

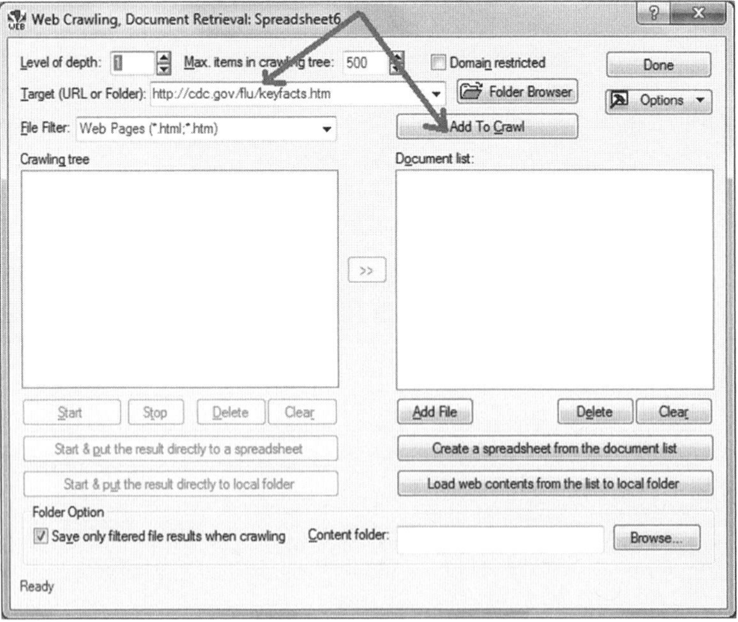

FIGURE Z.4

Paste in the URL web page address, then click the button "Add To Crawl."

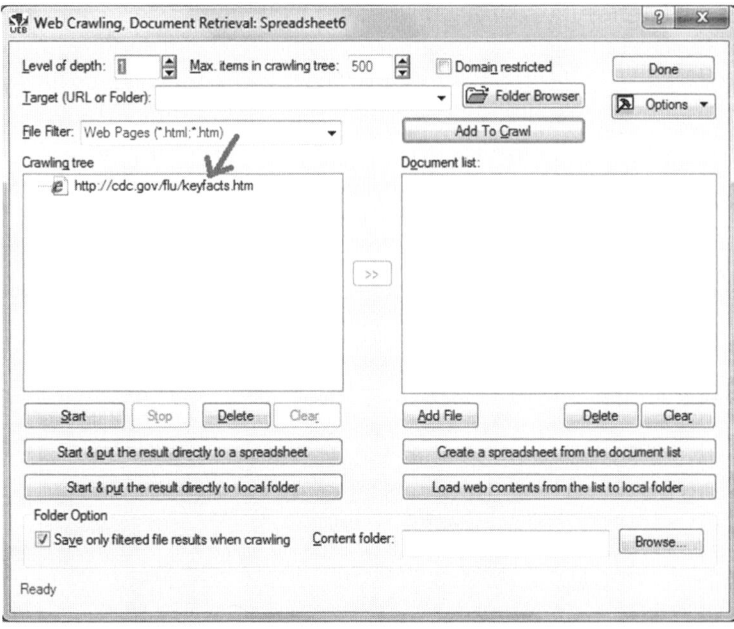

FIGURE Z.5
The pasted-in web page, after clicking "Add To Crawl" button, is placed in the left window.

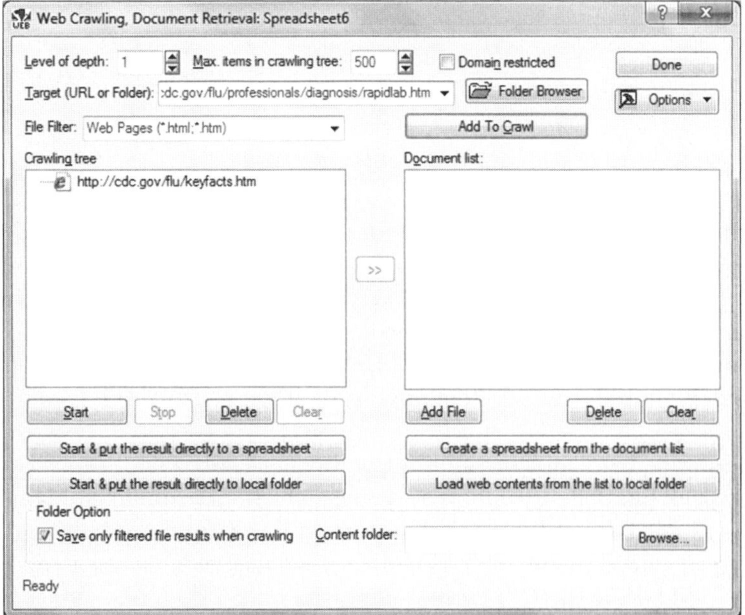

FIGURE Z.6
Second CDC web page for crawling: http://cdc.gov/flu/professionals/diagnosis/rapidlab.htm. Another primary Web site has been entered into the "Target" (URL or Folder) window.

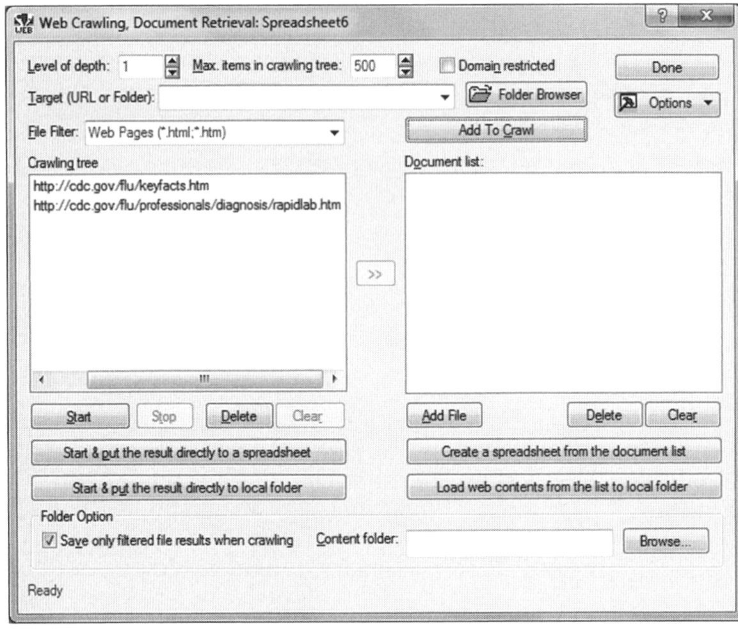

FIGURE Z.7

After selecting the “Add To Crawl” button, both CDC web pages show in left window.

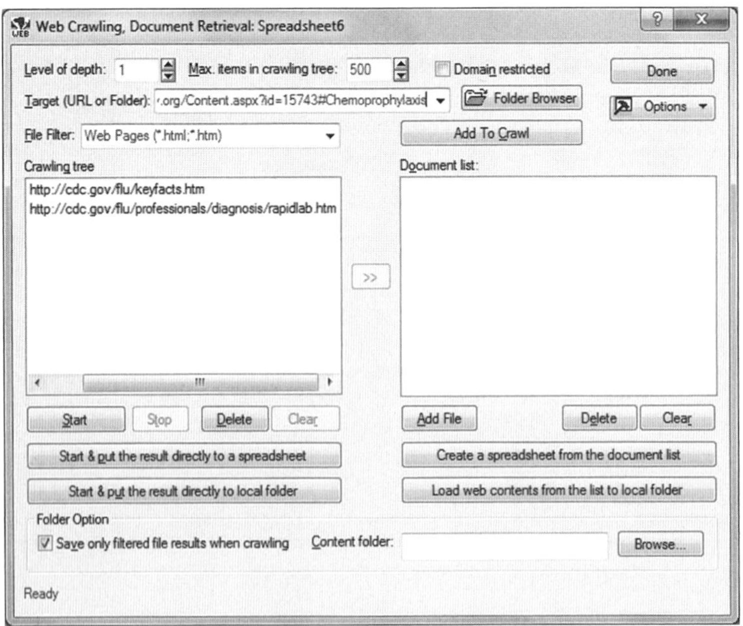

FIGURE Z.8

Adding third website, this one from the IDSA web: http://idsociety.org/Content.aspx?id=15743#Chemoprophylaxis.

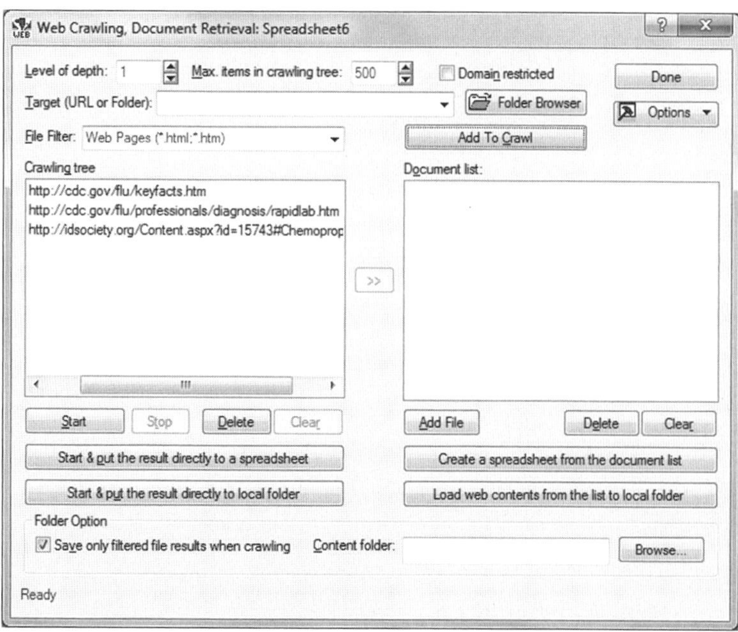

FIGURE Z.9
Clicking the "Add To Crawl" button adds this website to the left window.

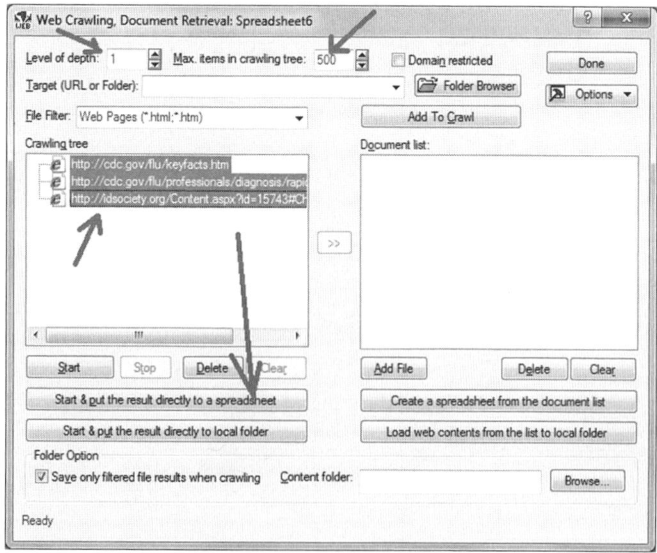

FIGURE Z.10
Select (highlight) all three primary websites; leave the depth to crawl at the default of 1 (this means that is will connect to links on each web page, but will *not* connect and add links within this first layer down of web pages); and leave maximum number at 500. Then select "Start & put the result directly to a spreadsheet" button.

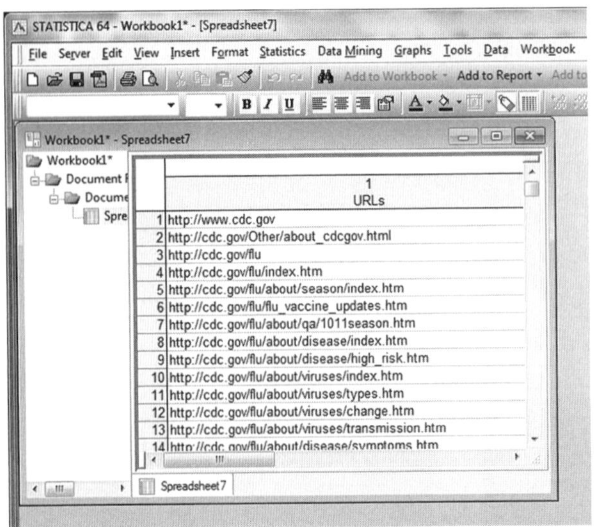

FIGURE Z.11
As the Web Crawling proceeds, a new data spreadsheet will pop up on the *STATISTICA* screen, showing you specific URLs that are being obtained from the "one layer deep."

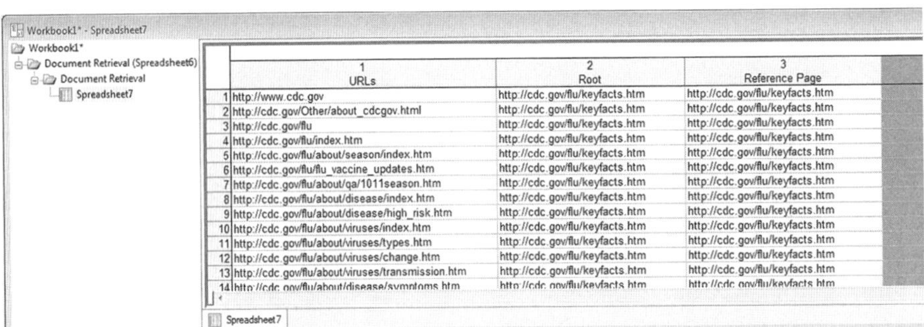

FIGURE Z.12
Pulling the right margin out to the right to see the entire spreadsheet, we see there is a "root" column and then the specific URLs are in Column 1.

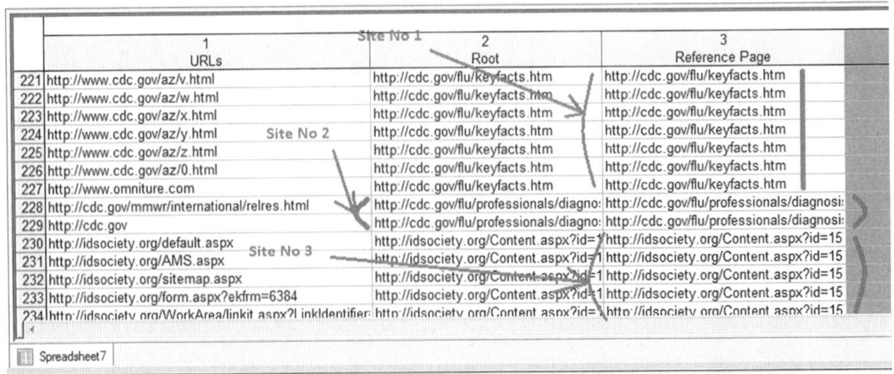

FIGURE Z.13
Scrolling down to about cases 220–230, we see the transition from the first URL reference link selected to the second URL reference link (which only returned two links), and finally to the third URL reference link.

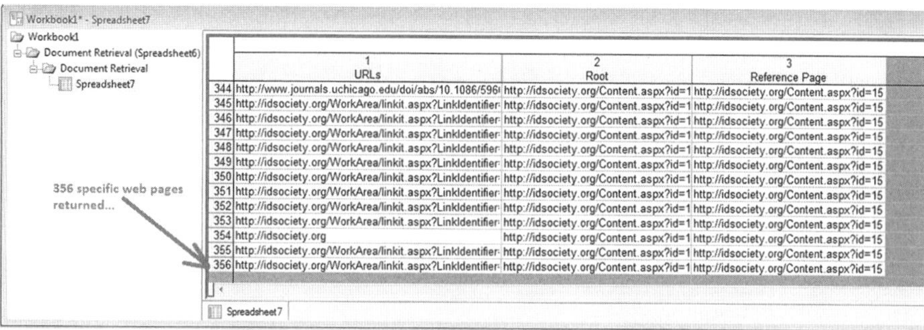

FIGURE Z.14

Then scrolling down to the bottom of this URL data spreadsheet, we see that only 356 unique web pages were returned. This is good, since by only selecting one link below the three primary pages, we have selected a workable number of pages from which to read and "Index" the words of the text.

Now we need to make this new spreadsheet the "Input Spreadsheet" so it can be used for further analyses. To do this, Select it as "Input Spreadsheet" from the DATA pull down menu (see Figures Z.15 and Z.16).

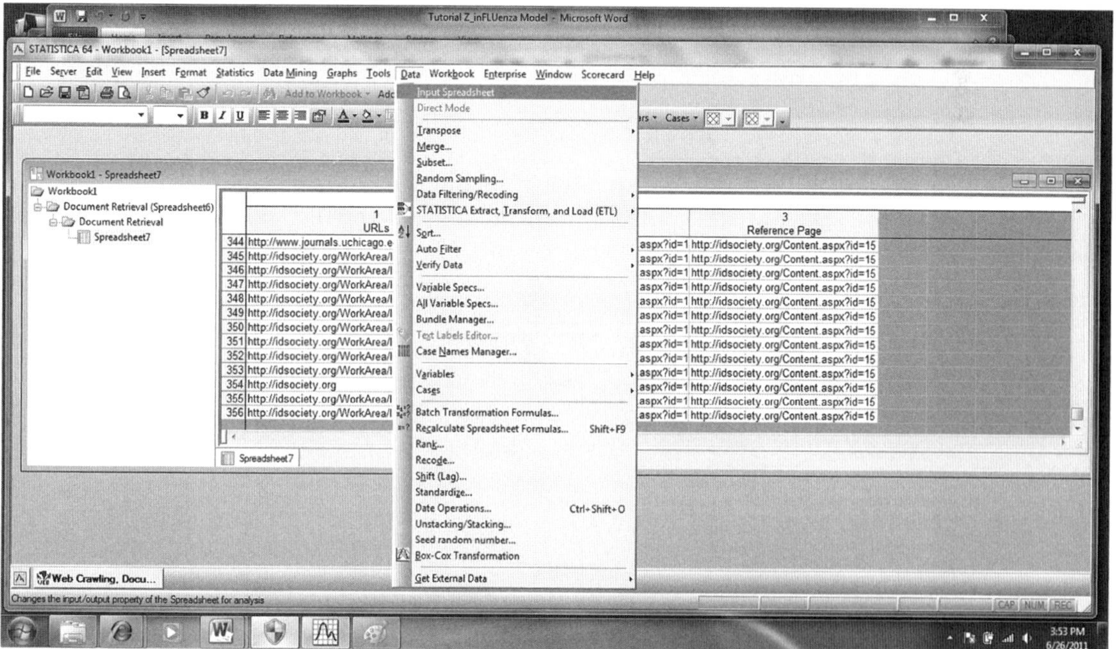

FIGURE Z.15

Data pull-down menu. Select "Input Spreadsheet" to make our new URL spreadsheet in the workbook our Input data sheet.

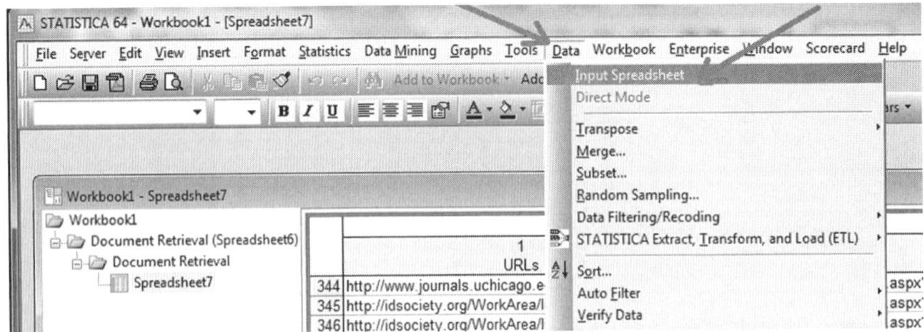

FIGURE Z.16
Selecting a spreadsheet as the Data Input Sheet, close-up view.

When this new spreadsheet is selected as "Input Spreadsheet", the icon representing this spreadsheet in the workbook will have a "red colored border on the computer screen" (see Figure Z.17).

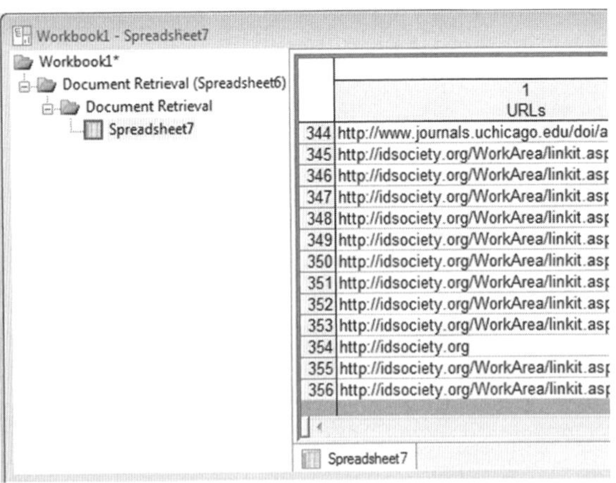

FIGURE Z.17
Gray outline will be red colored on the computer screen appears around the icon representing the spreadsheet in the Workbook Table of Contents when it is selected as "Input Spreadsheet."

Next we'll turn our attention to doing "text mining" on our newly created spreadsheet of web links. To do this, click on the DATA MINER pull down menu, and select "Text & Document Mining", as illustrated in Figure Z.18.

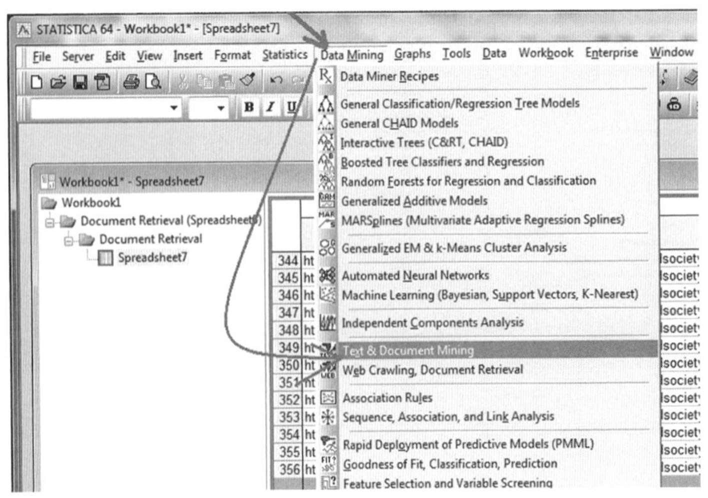

FIGURE Z.18
Selecting the Text Mining dialog from the Data Mining pull-down menu.

The next series of screens will take you through the "Text Mining" process (see Figures Z.19 through Z.57).

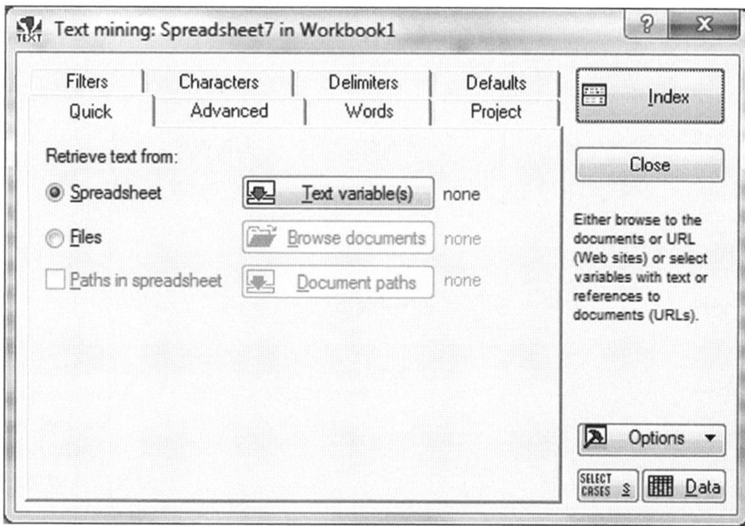

FIGURE Z.19

Text Mining primary dialog of *STATISTICA*. Select the spreadsheet and leave the defaults on all tabs. Then select Index to count to select words and count word frequencies in the web documents.

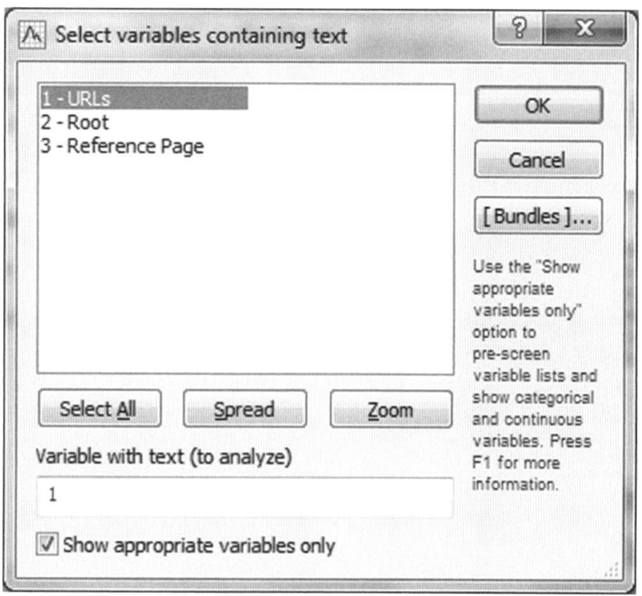

FIGURE Z.20

Select variables containing text dialog.

Click OK.

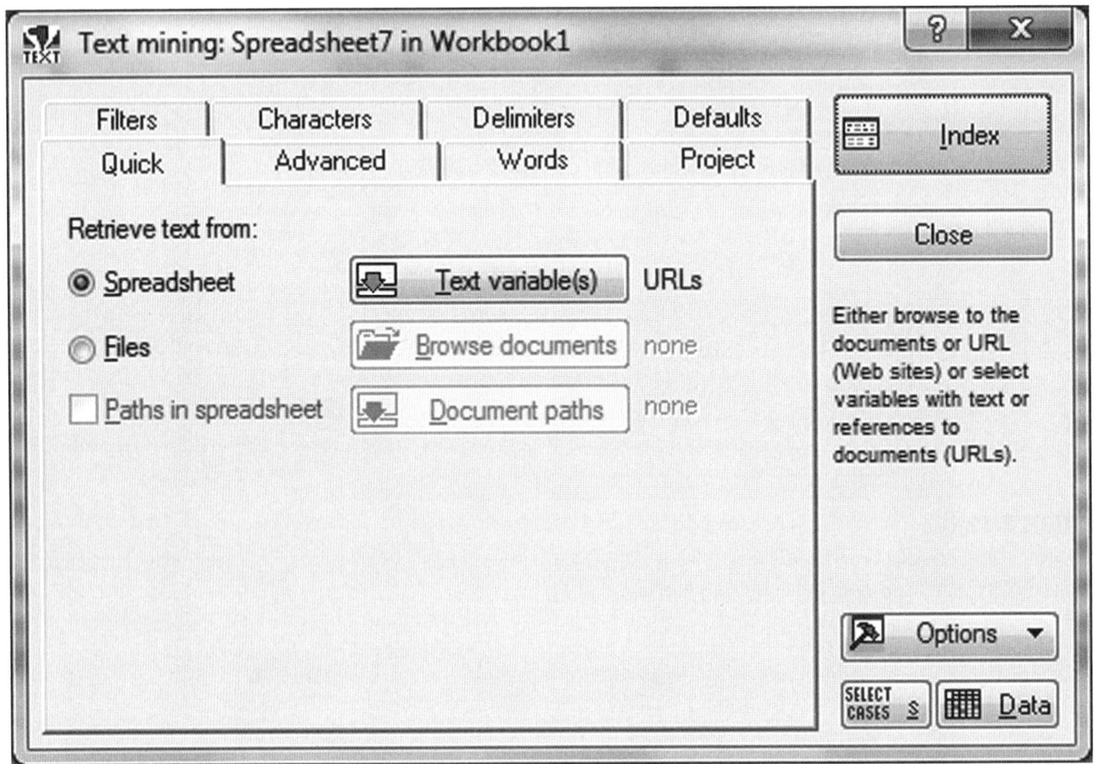

FIGURE Z.21
Text Mining window used to select text files in three different ways.

Click the Index button in the upper right-hand corner of the Text Mining window, and give it a few minutes to count and index all the words of the documents.

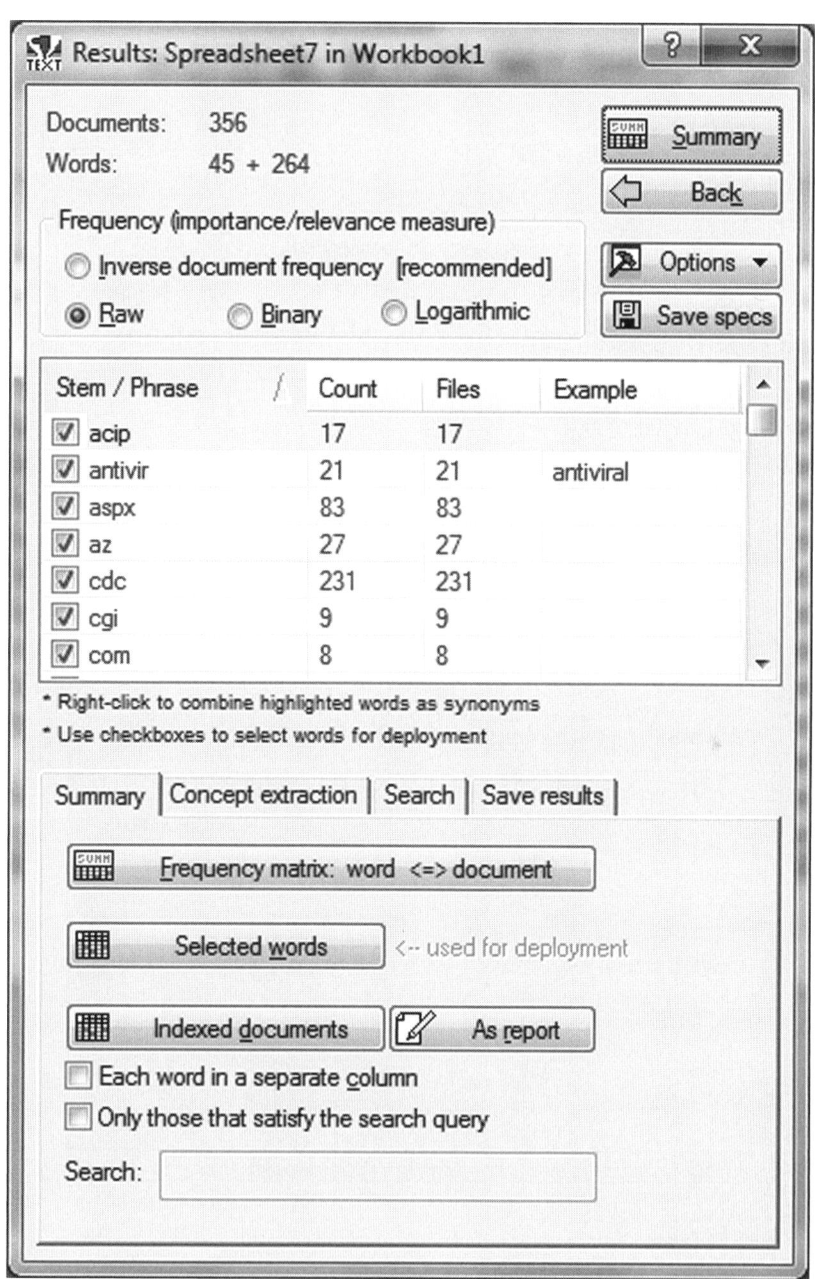

FIGURE Z.22
Results dialog following indexing.

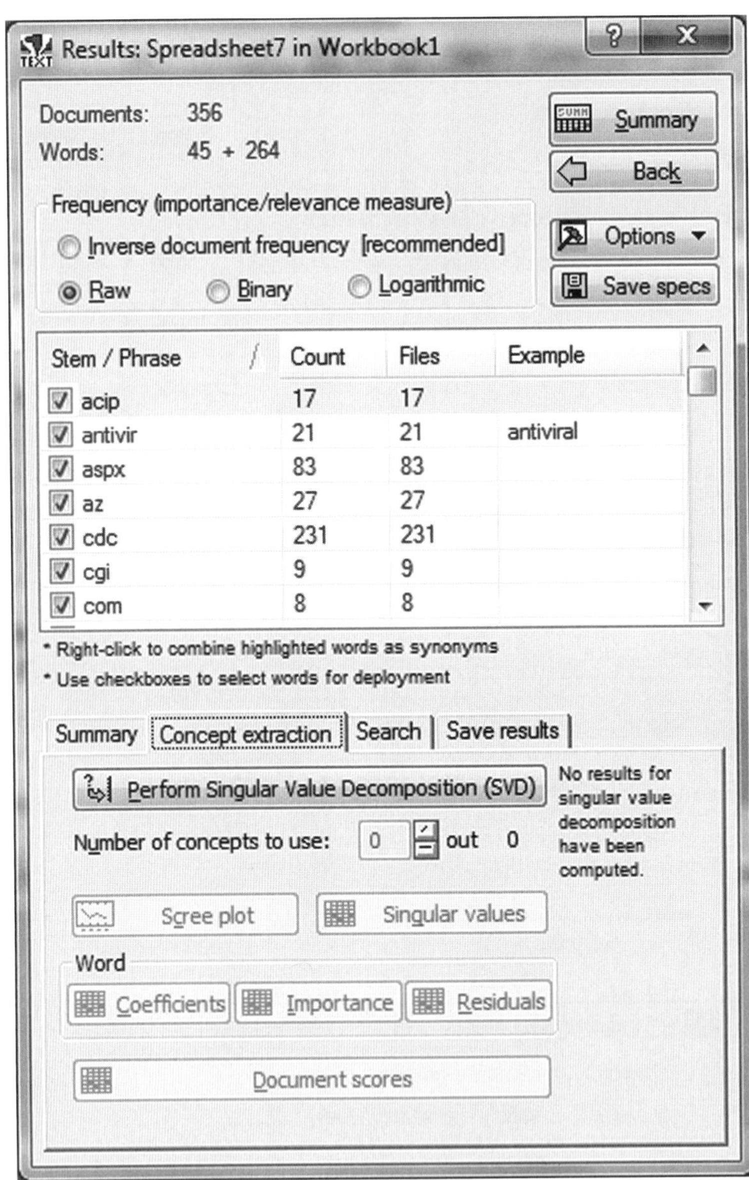

FIGURE Z.23
Concept Extraction tab of the Text Mining Results dialog.

Click on the "Inverse Document Frequency" radio button at the top of the Results dialog. Then click on the "Perform Singular Value Decomposition (SVD)" button in the lower part of the Results dialog/ Concepts tab. Note that the Results buttons of this SVD procedure are "grayed out" in the lower 1/3 of the Results dialog/Concept tab. These will become grayed-in after the SVD is computed.

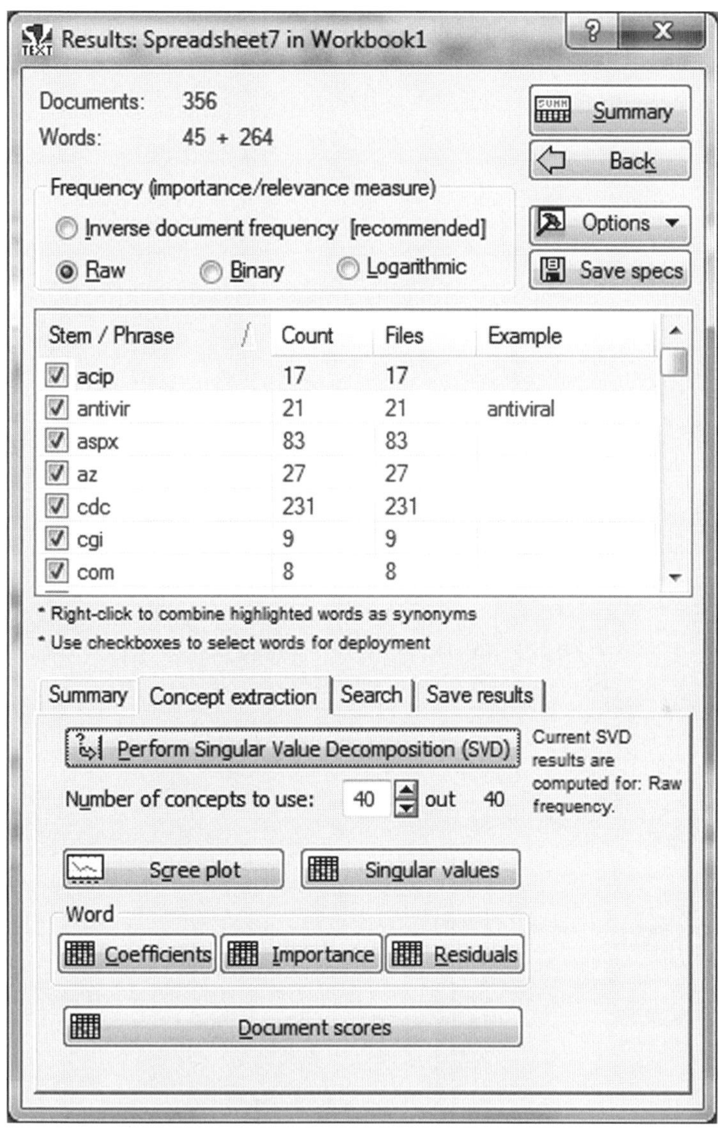

FIGURE Z.24
Results/Concepts dialog *after* the SVD is computed. Note that 40 concepts were extracted.

At this point click on the "Scree plot" button. This SVD is analogous to "Principal Components Analysis/Factor Analysis" from traditional statistics. Thus, we'd expect that maybe two or three concepts will extract "most of the variation" or "importance," or, as we call here, "concepts" from these indexed words, and the rest of the 40 concepts will trail off to the right in the Scree plot more or less horizontally. We'll click and see what is there.

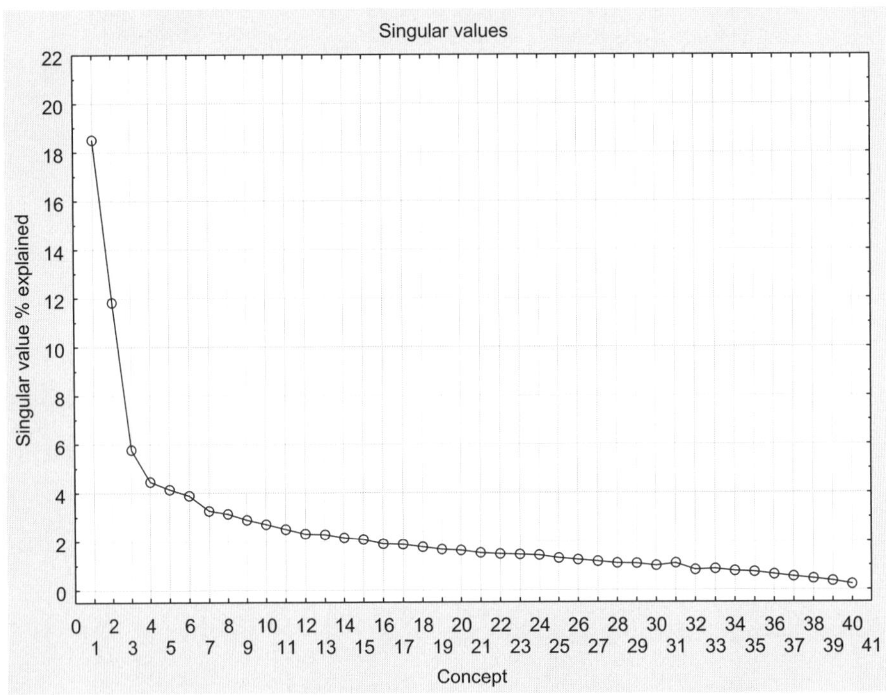

FIGURE Z.25
Scree plot from SVD.

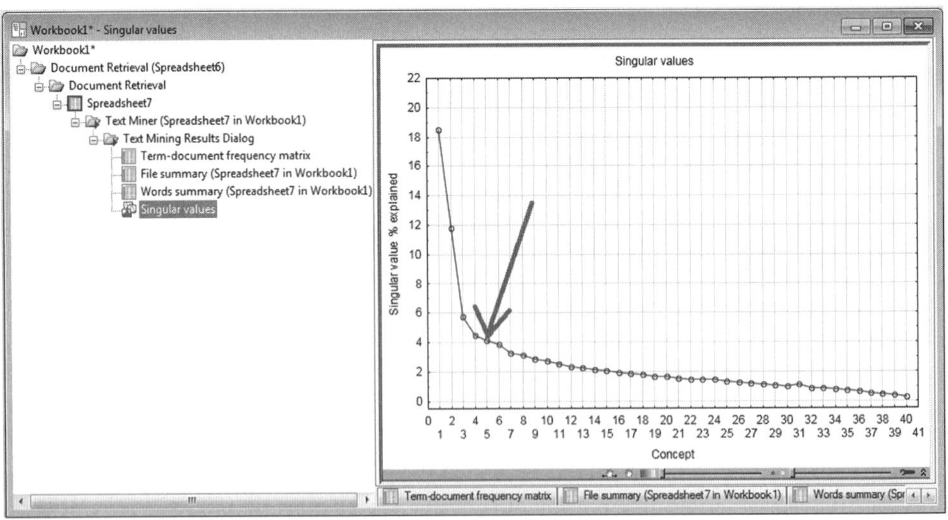

FIGURE Z.26
We can see that the "Inflection Point" on this Scree plot is at about Concept 4 or 5 (or at the most 6). Most likely plotting of the Concept Coefficients will show us that the most interesting words are pulled out by Concept 1 and Concept 2, so maybe adding in Concept 3 will add something, and at the most adding in Concept 4. It is doubtful that Concept 5 and 6 will add much to our understanding.

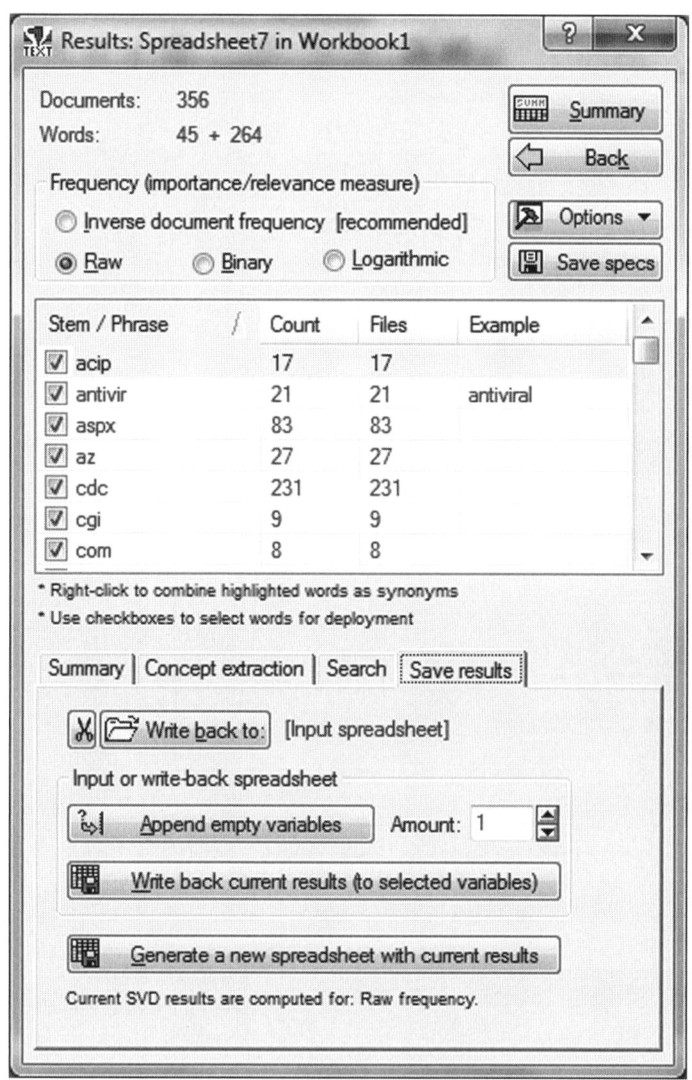

FIGURE Z.27
Save Results tab of Text Mining dialog.

We want to Save Back the results to the data spreadsheet. This means we need space—for example, new columns for the 45 words selected and also for the 40 concepts from SVD. Thus, we need 85 new columns. We also need to select the spreadsheet to which we will send this information. We'll use the spreadsheet we started with for this part. We could also add this to the "numeric FLU dataset" that will be used later in this tutorial. But for now we'll add to the URL spreadsheet. So we'll need to extract a Copy of the Workbook spreadsheet as an "external spreadsheet" and then Save it separately.

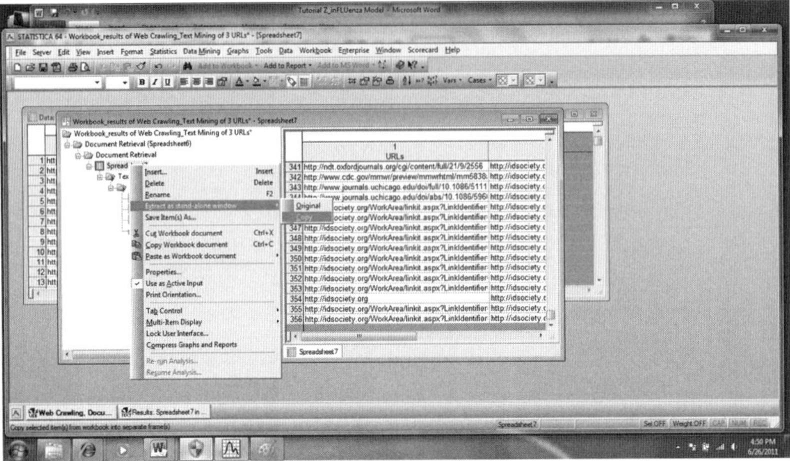

FIGURE Z.28
Results workbook.

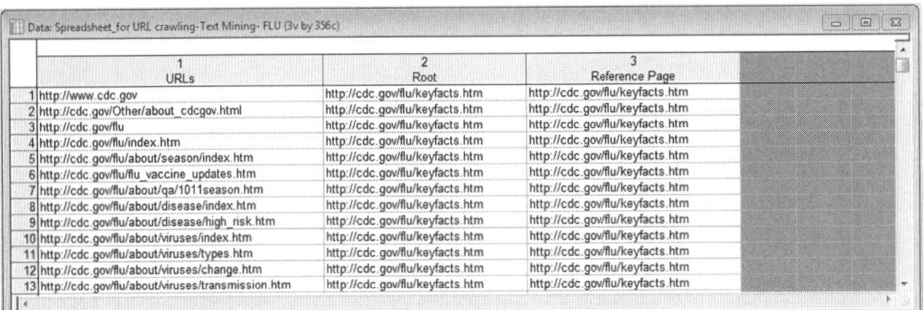
Data: Spreadsheet_for URL crawling-Text Mining- FLU (3v by 356c)

	1 URLs	2 Root	3 Reference Page
1	http://www.cdc.gov	http://cdc.gov/flu/keyfacts.htm	http://cdc.gov/flu/keyfacts.htm
2	http://cdc.gov/Other/about_cdcgov.html	http://cdc.gov/flu/keyfacts.htm	http://cdc.gov/flu/keyfacts.htm
3	http://cdc.gov/flu	http://cdc.gov/flu/keyfacts.htm	http://cdc.gov/flu/keyfacts.htm
4	http://cdc.gov/flu/index.htm	http://cdc.gov/flu/keyfacts.htm	http://cdc.gov/flu/keyfacts.htm
5	http://cdc.gov/flu/about/season/index.htm	http://cdc.gov/flu/keyfacts.htm	http://cdc.gov/flu/keyfacts.htm
6	http://cdc.gov/flu/flu_vaccine_updates.htm	http://cdc.gov/flu/keyfacts.htm	http://cdc.gov/flu/keyfacts.htm
7	http://cdc.gov/flu/about/qa/1011season.htm	http://cdc.gov/flu/keyfacts.htm	http://cdc.gov/flu/keyfacts.htm
8	http://cdc.gov/flu/about/disease/index.htm	http://cdc.gov/flu/keyfacts.htm	http://cdc.gov/flu/keyfacts.htm
9	http://cdc.gov/flu/about/disease/high_risk.htm	http://cdc.gov/flu/keyfacts.htm	http://cdc.gov/flu/keyfacts.htm
10	http://cdc.gov/flu/about/viruses/index.htm	http://cdc.gov/flu/keyfacts.htm	http://cdc.gov/flu/keyfacts.htm
11	http://cdc.gov/flu/about/viruses/types.htm	http://cdc.gov/flu/keyfacts.htm	http://cdc.gov/flu/keyfacts.htm
12	http://cdc.gov/flu/about/viruses/change.htm	http://cdc.gov/flu/keyfacts.htm	http://cdc.gov/flu/keyfacts.htm
13	http://cdc.gov/flu/about/viruses/transmission.htm	http://cdc.gov/flu/keyfacts.htm	http://cdc.gov/flu/keyfacts.htm

FIGURE Z.29
The spreadsheet showing the URL, the Root URL, and a Reference page.

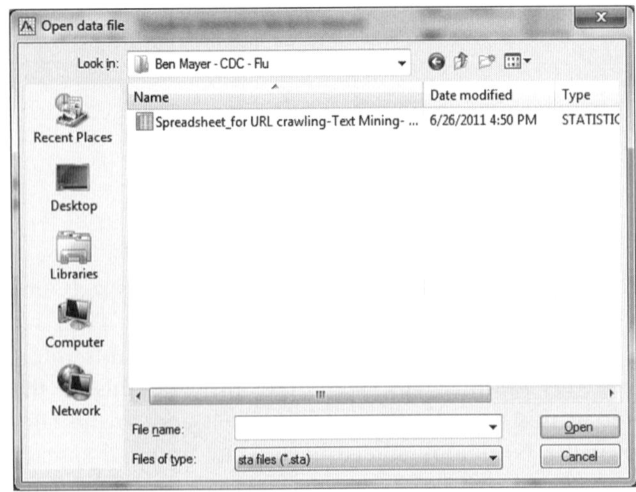

FIGURE Z.30
Select the spreadsheet after clicking on the "Write back to" button.

Then select the Index button on the Text Mining dialog so that the frequencies of words can be computed.

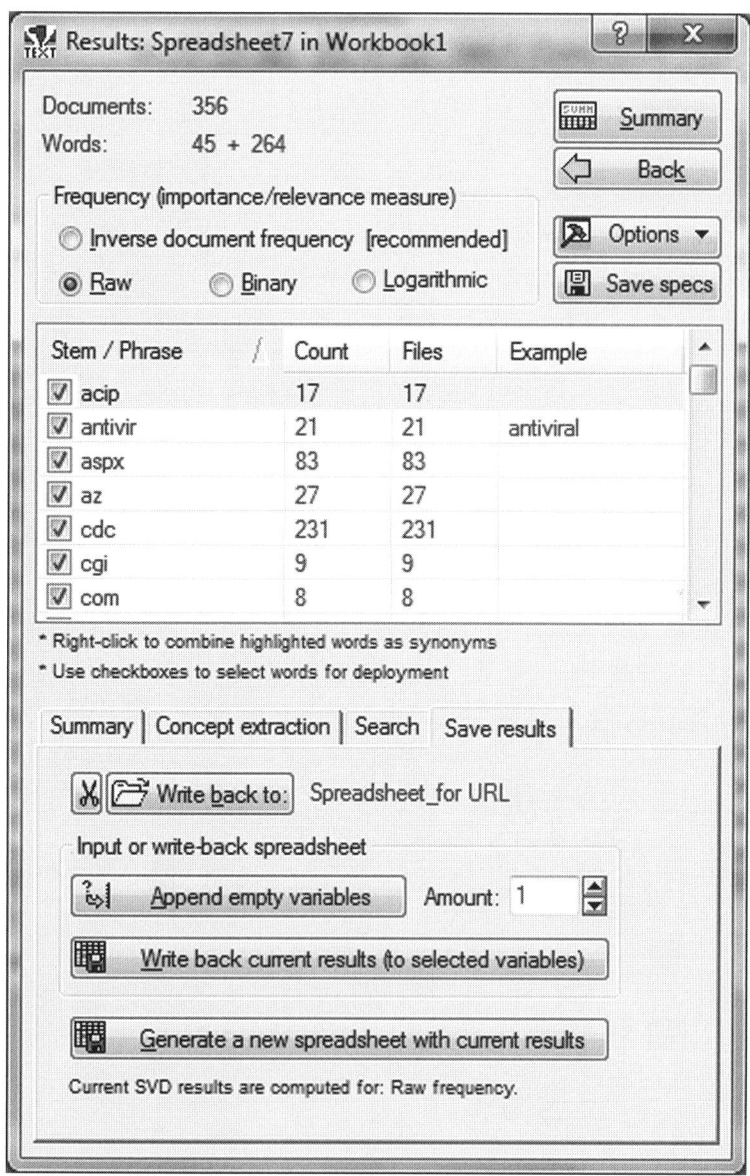

FIGURE Z.31
Results dialog of text miner after indexing the words.

Type in 85, since we need 85 new columns (45 for the words selected, and 40 for the 40 Concepts extracted).

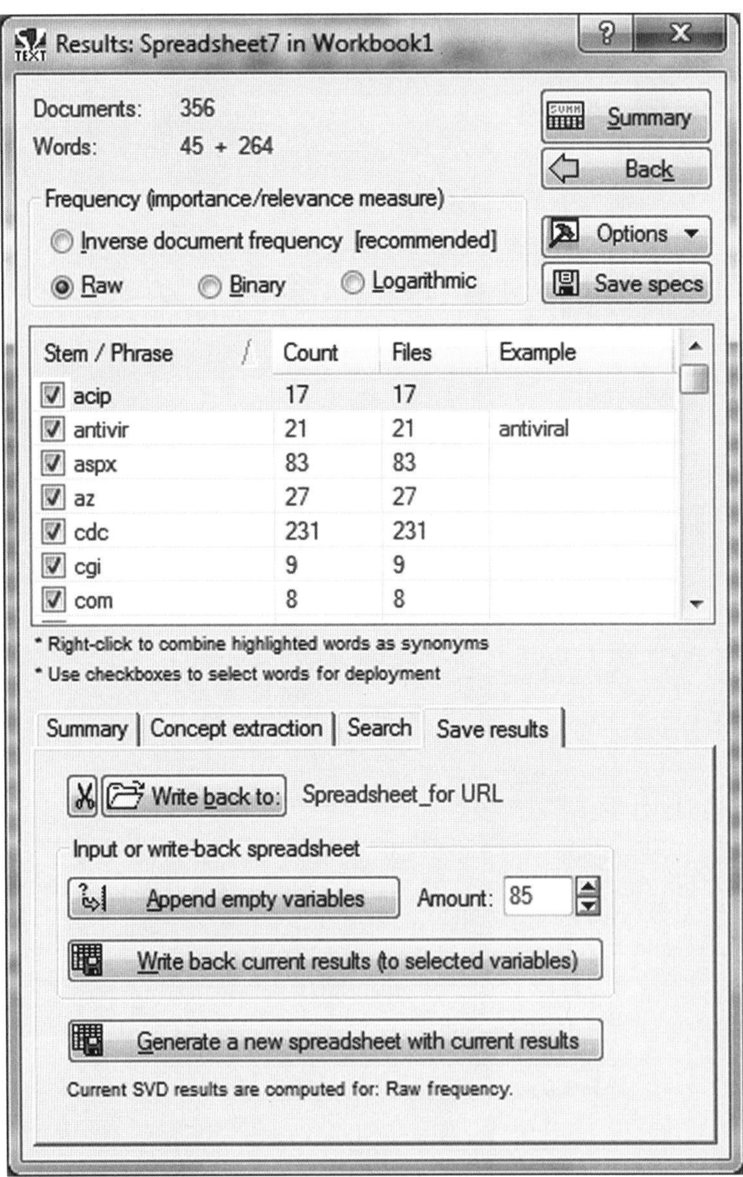

FIGURE Z.32
Amount set to 85 new columns for the 85 words selected.

Click the "Write back current results (to selected variables)" button.

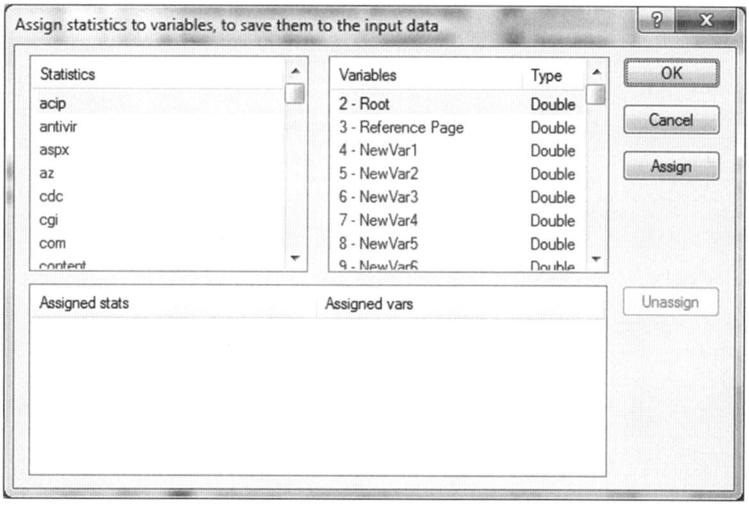

FIGURE Z.33
Assign statistics to variables dialog.

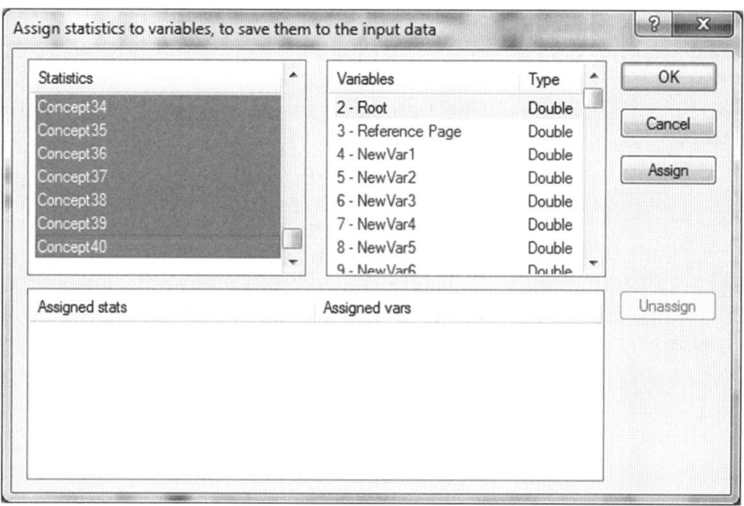

FIGURE Z.34
Select all of the concepts on the left panel.

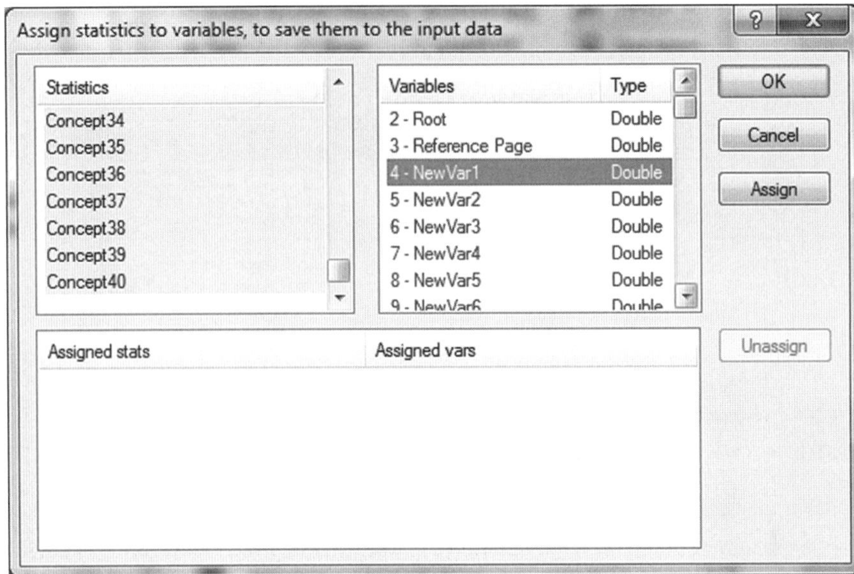

FIGURE Z.35

Then select the new variable numbers corresponding to the number of words that will be written to the master data file.

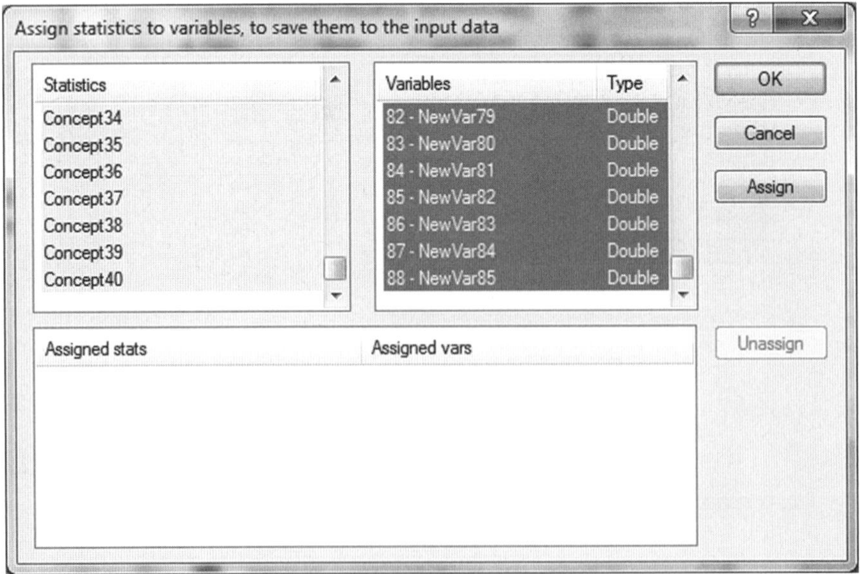

FIGURE Z.36

All of the new variable numbers selected; now click on "assign" button.

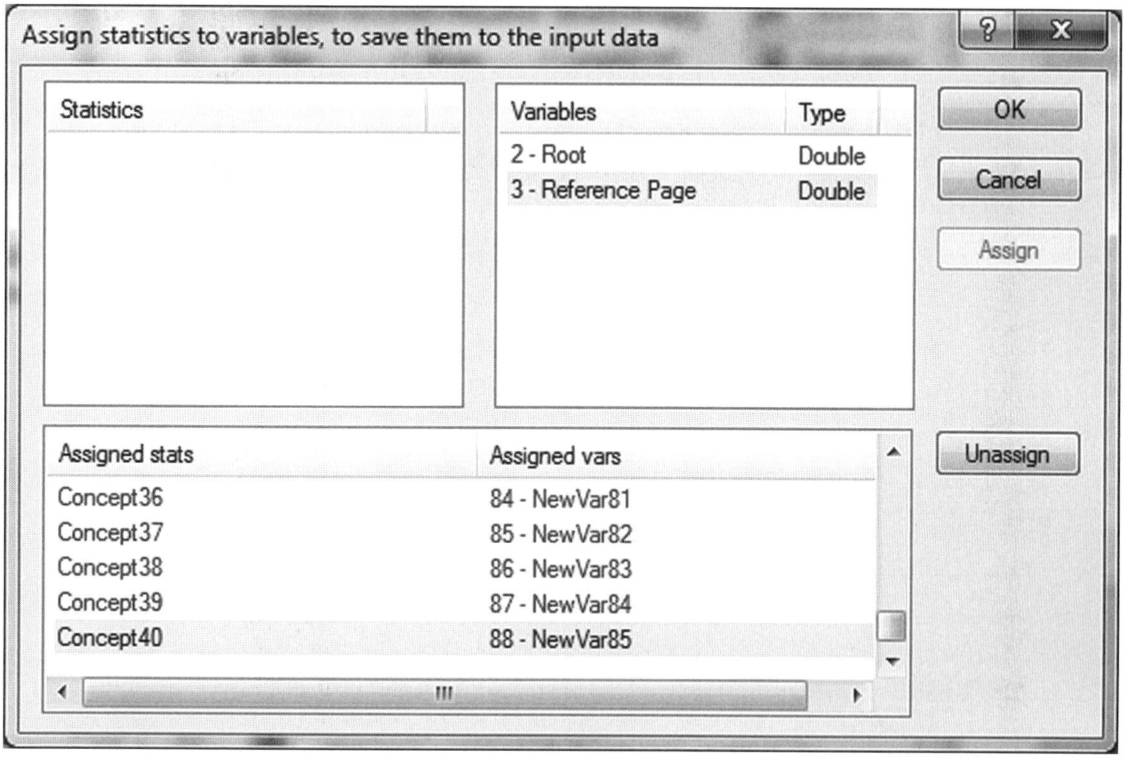

Assign statistics to variables, to save them to the input data

Statistics

Variables	Type
2 - Root	Double
3 - Reference Page	Double

OK

Cancel

Assign

Assigned stats	Assigned vars
Concept36	84 - NewVar81
Concept37	85 - NewVar82
Concept38	86 - NewVar83
Concept39	87 - NewVar84
Concept40	88 - NewVar85

Unassign

FIGURE Z.37
When Assign is selected, the stats and vars go to the bottom panel of the dialog.

Then click OK.

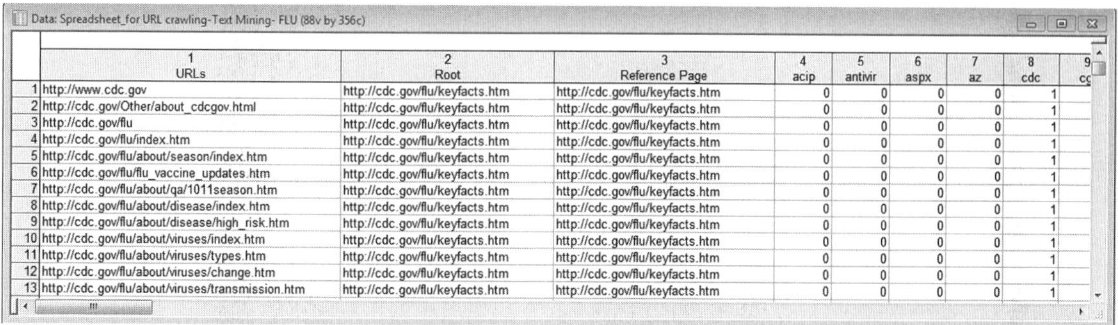

Data: Spreadsheet_for URL crawling-Text Mining- FLU (88v by 356c)

	1 URLs	2 Root	3 Reference Page	4 acip	5 antivir	6 aspx	7 az	8 cdc
1	http://www.cdc.gov	http://cdc.gov/flu/keyfacts.htm	http://cdc.gov/flu/keyfacts.htm	0	0	0	0	1
2	http://cdc.gov/Other/about_cdcgov.html	http://cdc.gov/flu/keyfacts.htm	http://cdc.gov/flu/keyfacts.htm	0	0	0	0	1
3	http://cdc.gov/flu	http://cdc.gov/flu/keyfacts.htm	http://cdc.gov/flu/keyfacts.htm	0	0	0	0	1
4	http://cdc.gov/flu/index.htm	http://cdc.gov/flu/keyfacts.htm	http://cdc.gov/flu/keyfacts.htm	0	0	0	0	1
5	http://cdc.gov/flu/about/season/index.htm	http://cdc.gov/flu/keyfacts.htm	http://cdc.gov/flu/keyfacts.htm	0	0	0	0	1
6	http://cdc.gov/flu/flu_vaccine_updates.htm	http://cdc.gov/flu/keyfacts.htm	http://cdc.gov/flu/keyfacts.htm	0	0	0	0	1
7	http://cdc.gov/flu/about/qa/1011season.htm	http://cdc.gov/flu/keyfacts.htm	http://cdc.gov/flu/keyfacts.htm	0	0	0	0	1
8	http://cdc.gov/flu/about/disease/index.htm	http://cdc.gov/flu/keyfacts.htm	http://cdc.gov/flu/keyfacts.htm	0	0	0	0	1
9	http://cdc.gov/flu/about/disease/high_risk.htm	http://cdc.gov/flu/keyfacts.htm	http://cdc.gov/flu/keyfacts.htm	0	0	0	0	1
10	http://cdc.gov/flu/about/viruses/index.htm	http://cdc.gov/flu/keyfacts.htm	http://cdc.gov/flu/keyfacts.htm	0	0	0	0	1
11	http://cdc.gov/flu/about/viruses/types.htm	http://cdc.gov/flu/keyfacts.htm	http://cdc.gov/flu/keyfacts.htm	0	0	0	0	1
12	http://cdc.gov/flu/about/viruses/change.htm	http://cdc.gov/flu/keyfacts.htm	http://cdc.gov/flu/keyfacts.htm	0	0	0	0	1
13	http://cdc.gov/flu/about/viruses/transmission.htm	http://cdc.gov/flu/keyfacts.htm	http://cdc.gov/flu/keyfacts.htm	0	0	0	0	1

FIGURE Z.38
Then looking at the spreadsheet, the words and concepts have been added.

Data: Spreadsheet_for URL crawling-Text Mining- FLU* (88v by 356c)

	83 Concept35	84 Concept36	85 Concept37	86 Concept38	87 Concept39	88 Concept40
1	0.0219051	0.0176624	-0.011101	-0.012225	-0.0007464	-0.0049618
2	0.0834918	-0.100362	0.0210214	0.0048667	-0.0016855	-0.0002517
3	-0.013794	0.0555956	-0.017813	-0.006692	-0.0097239	-0.0007577
4	0.0833243	-0.077399	0.0130262	-0.003315	-0.0056536	0.00136508
5	0.0833243	-0.077399	0.0130262	-0.003315	-0.0056536	0.00136508
6	0.1338441	-0.085521	0.0105553	0.0088773	0.02127015	0.00576722
7	-0.004127	-0.004549	0.0031452	0.0008772	-0.0115637	0.00006262
8	-0.049326	0.045924	-0.011676	-0.007859	0.00412378	-0.0000705
9	-0.044499	0.0273681	-0.003037	0.0012392	0.00007633	-0.0000401
10	-0.067851	0.0560096	-0.012116	-0.007519	0.00397095	-0.0000817
11	-0.063025	0.0374537	-0.003476	0.0015792	-0.0000765	-0.0000513
12	-0.063025	0.0374537	-0.003476	0.0015792	-0.0000765	-0.0000513
13	-0.063025	0.0374537	-0.003476	0.0015792	-0.0000765	-0.0000513
14	-0.044499	0.0273681	-0.003037	0.0012392	0.00007633	-0.0000401
15	-0.044499	0.0273681	-0.003037	0.0012392	0.00007633	-0.0000401
16	0.0881511	-0.095955	0.0216656	0.0057835	-0.0097011	0.00139552

FIGURE Z.39

Scrolling to the far right on the data spreadsheet, we see that Column 88 is the last, with Concept 40 put into this data column.

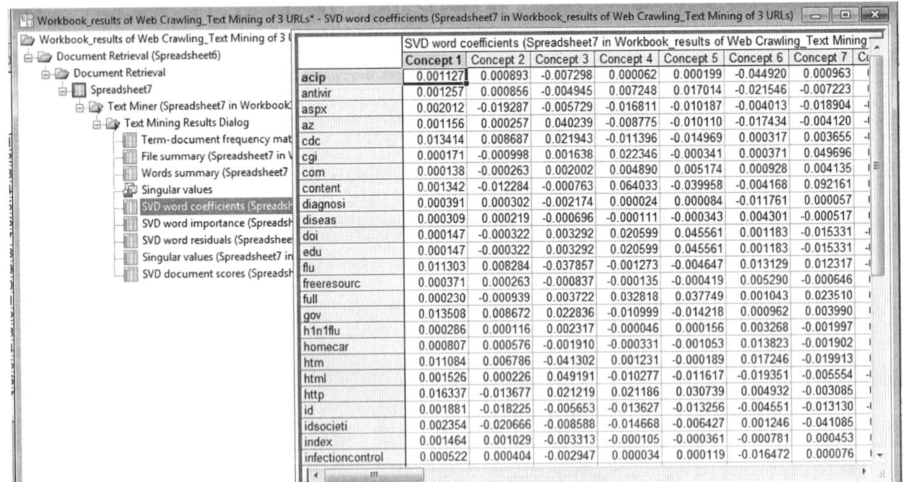

SVD word coefficients (Spreadsheet7 in Workbook_results of Web Crawling_Text Mining

	Concept 1	Concept 2	Concept 3	Concept 4	Concept 5	Concept 6	Concept 7
acip	0.001127	0.000893	-0.007298	0.000062	0.000199	-0.044920	0.000963
antivir	0.001257	0.000856	-0.004945	0.007248	0.017014	-0.021546	-0.007223
aspx	0.002012	-0.019287	-0.005729	-0.016811	-0.010187	-0.004013	-0.018904
az	0.001156	0.000257	0.040239	-0.008775	-0.010110	-0.017434	-0.004120
cdc	0.013414	0.008687	0.021943	-0.011396	-0.014969	0.000317	0.003655
cgi	0.000171	-0.000998	0.001638	0.022346	-0.000341	0.000371	0.049696
com	0.000138	-0.000263	0.002002	0.004890	0.005174	0.000928	0.004135
content	0.001342	-0.012284	-0.000763	0.064033	-0.039958	-0.004168	0.092161
diagnosi	0.000391	0.000302	-0.002174	0.000024	0.000084	-0.011761	0.000057
diseas	0.000309	0.000219	-0.000696	-0.000111	-0.000343	0.004301	-0.000517
doi	0.000147	-0.000322	0.003292	0.020599	0.045561	0.001183	-0.015331
edu	0.000147	-0.000322	0.003292	0.020599	0.045561	0.001183	-0.015331
flu	0.011303	0.008284	-0.037857	-0.001273	-0.004647	0.013129	0.012317
freeresourc	0.000371	0.000263	-0.000837	-0.000135	-0.000419	0.005290	-0.000646
full	0.000230	-0.000939	0.003722	0.032818	0.037749	0.001043	0.023510
gov	0.013508	0.008672	0.022836	-0.010999	-0.014218	0.000962	0.003990
h1n1flu	0.000286	0.000116	0.002317	-0.000046	0.000156	0.003268	-0.001997
homecar	0.000807	0.000576	-0.001910	-0.000331	-0.001053	0.013823	-0.001902
htm	0.011084	0.006786	-0.041302	0.001231	-0.000189	0.017246	-0.019913
html	0.001526	0.000226	0.049191	-0.010277	-0.011617	-0.019351	-0.005554
http	0.016337	-0.013677	0.021219	0.021186	0.030739	0.004932	-0.003085
id	0.001881	-0.018225	-0.005653	-0.013627	-0.013256	-0.004551	-0.013130
idsocieti	0.002354	-0.020666	-0.008588	-0.014668	-0.006427	0.001246	-0.041085
index	0.001464	0.001029	-0.003313	-0.000105	-0.000361	-0.000781	0.000453
infectioncontrol	0.000522	0.000404	-0.002947	0.000034	0.000119	-0.016472	0.000076

FIGURE Z.40

Concept coefficients spreadsheet.

We will make some scatterplots of Concept 1 versus Concept 2, and so on, to see if we can find "clusters of words" that seem to make sense. We need to either make this Coefficient spreadsheet the "Active Input" data set in the workbook or extract a copy of it to a separate stand-alone window, and then make it the "Active Input" spreadsheet to do the graphing.

By selecting it, and then going to the Data pull-down menu and releasing on "Active Input," the workbook copy of the Concept Coefficients becomes the "active spreadsheet," as denoted in the next figure where the icon in the TOC becomes outlined in red.

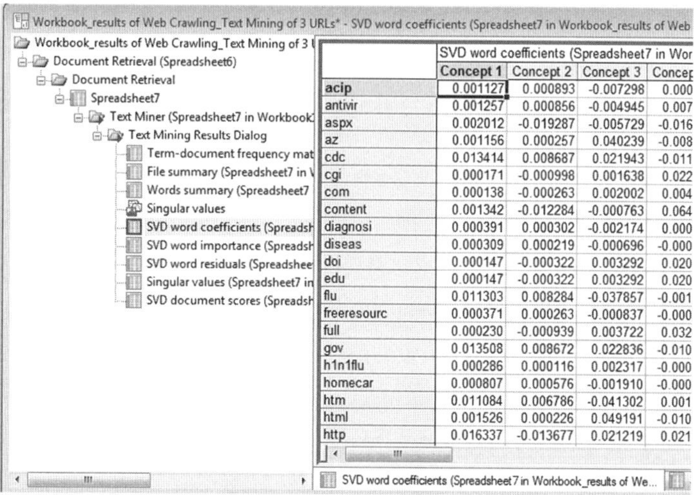

FIGURE Z.41

Concept Coefficient workbook spreadsheet made into "Active data input" spreadsheet.

Now we will make a scatterplot of Concept 1 versus Concept 2 by selecting the Graphs pull-down menu and releasing on Scatterplots in the 2D graphs part.

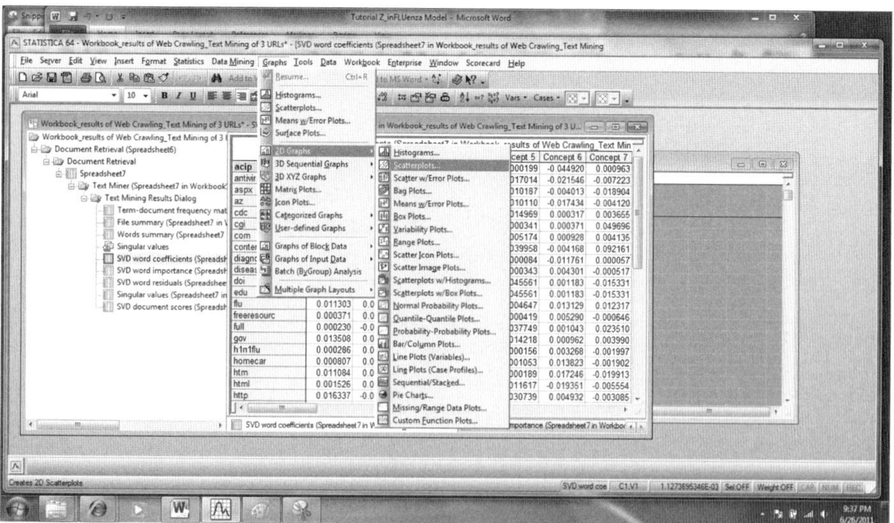

FIGURE Z.42

Selecting 2D Scatterplots from the Graphs pull-down menu.

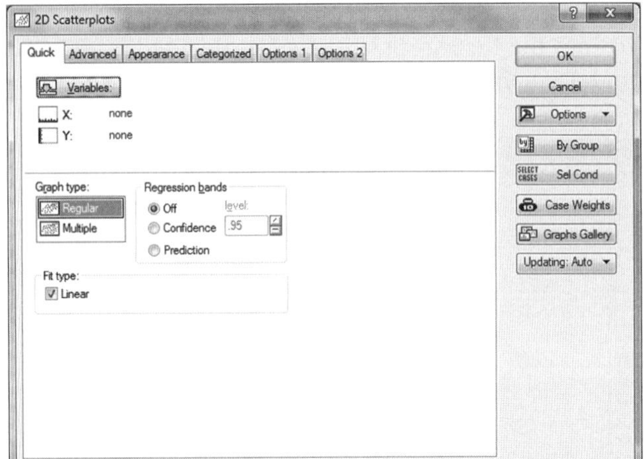

FIGURE Z.43
2D Scatterplots dialog.

Select Variables.

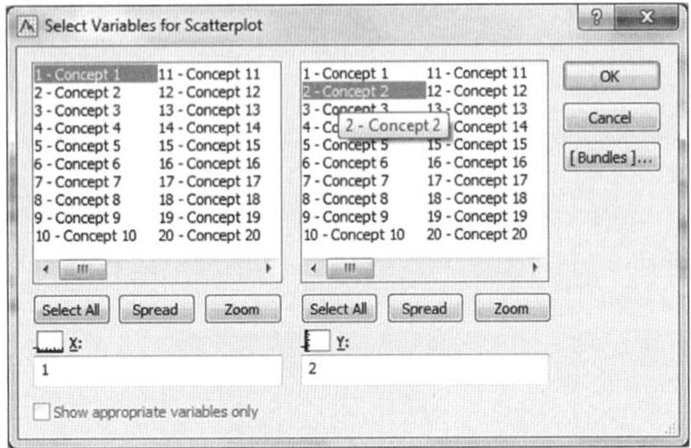

FIGURE Z.44
Select Variables dialog.

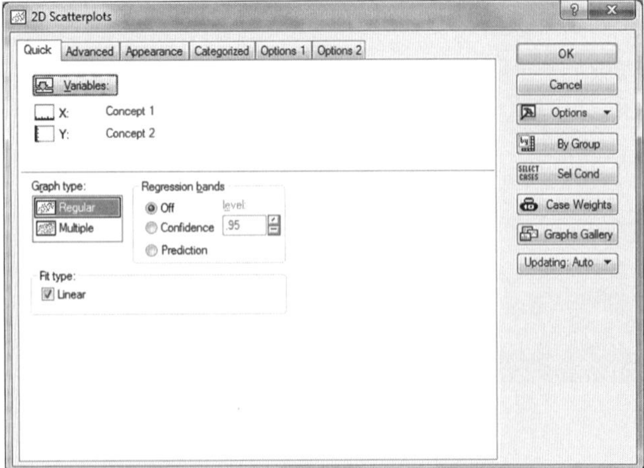

FIGURE Z.45
Concept 1 and Concept 2 are accepted for graphing, as shown selected in this figure.

Click OK to make the graph.

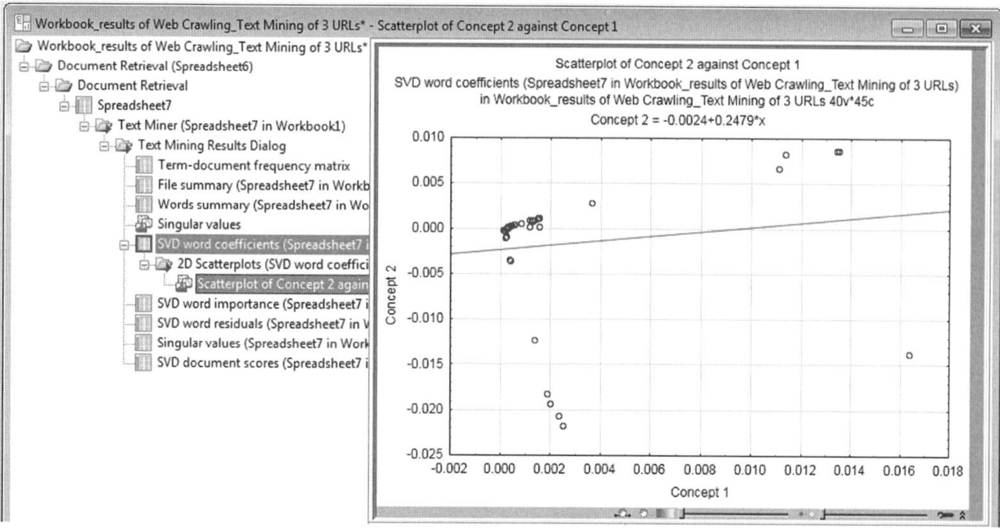

FIGURE Z.46

Scatterplot embedded in the workbook.

Now, using the Brushing tool with the "lasso" type, we will "lasso" or encircle "clusters" of points (words) on the graph to see if each cluster has similar words, and thus we can extract some "meaning" of these clusters.

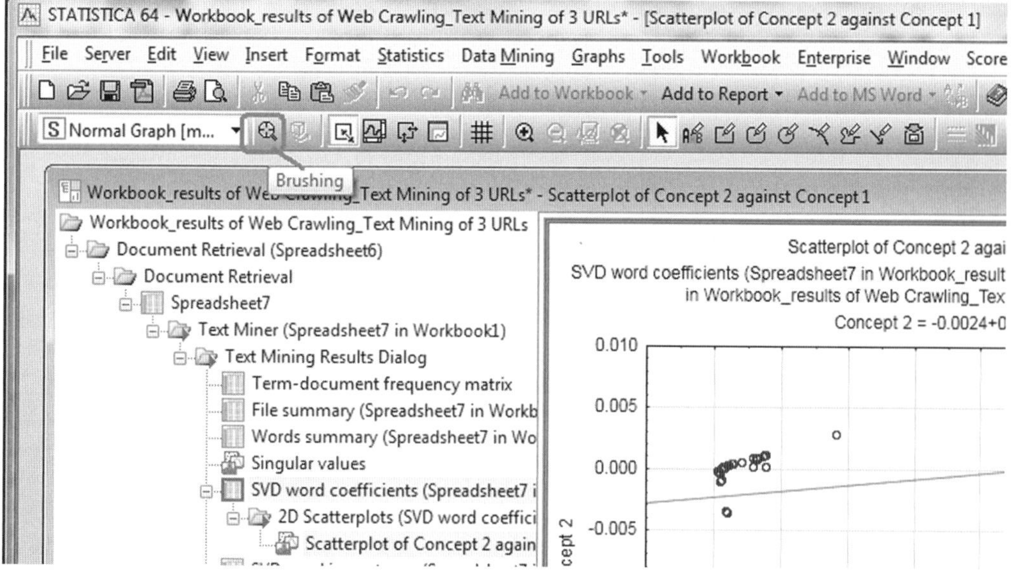

FIGURE Z.47

Brushing tool selection.

FIGURE Z.48
Brushing dialog.

Select "Label" and "Lasso."

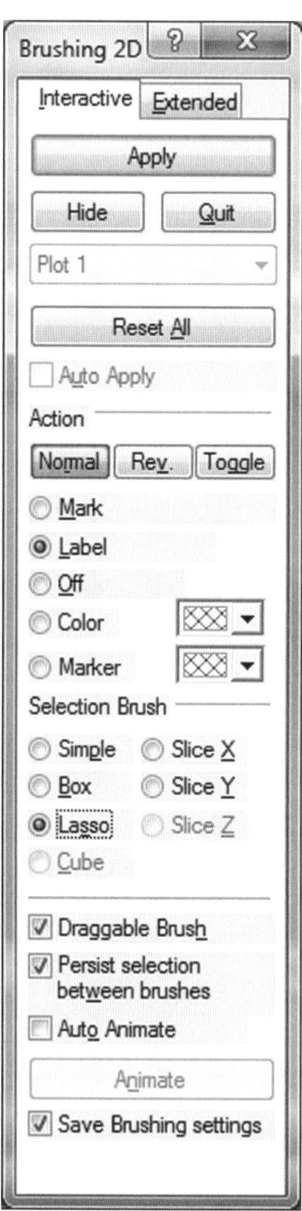

FIGURE Z.49
Brushing dialog with "Label" and "Lasso" selected.

Lasso a cluster, as shown in next figure.

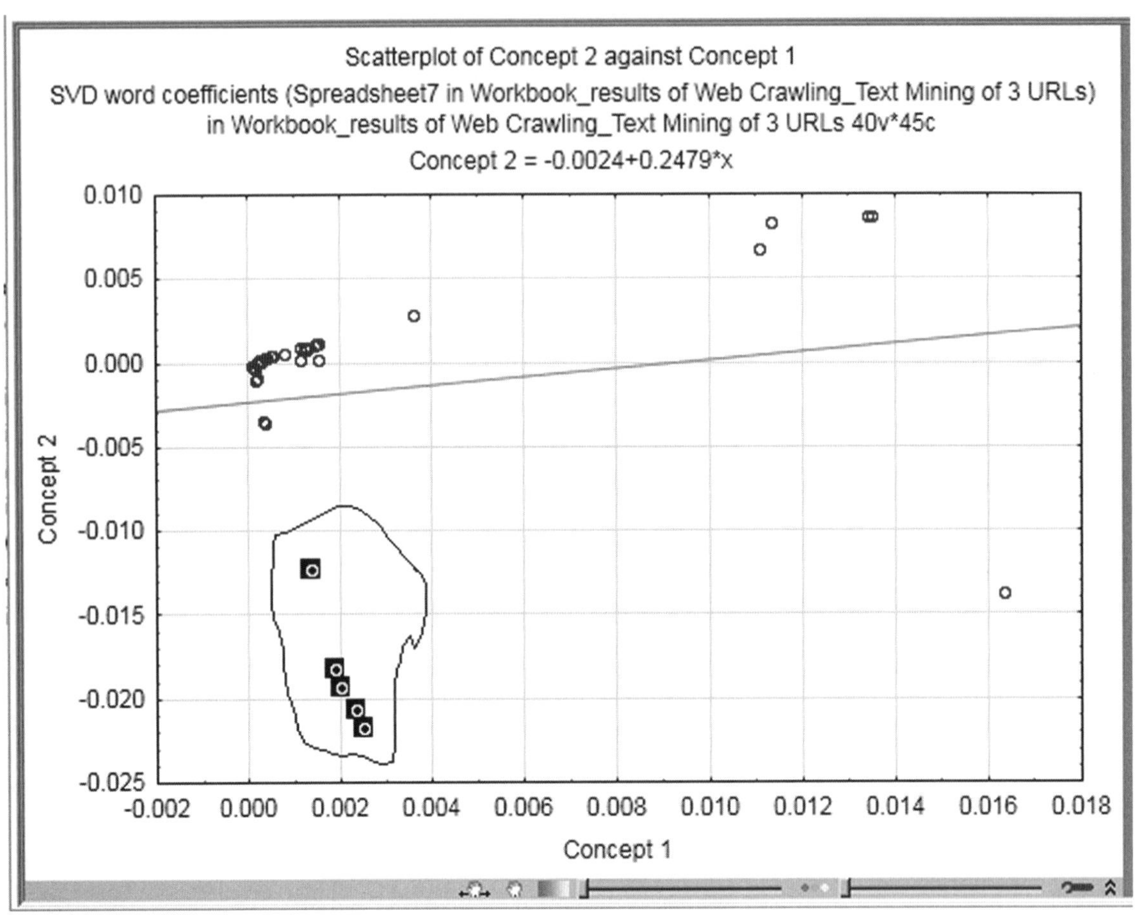

FIGURE Z.50

A cluster is surrounded by the "lasso" so that these points can be labeled as per word.

Now hit the Apply button at the top of the Brushing dialog.

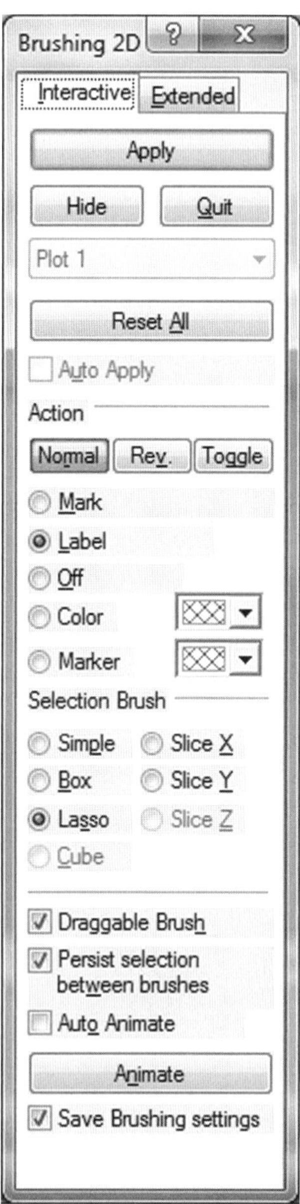

FIGURE Z.51
Apply button is at the bop of the Brushing dialog.

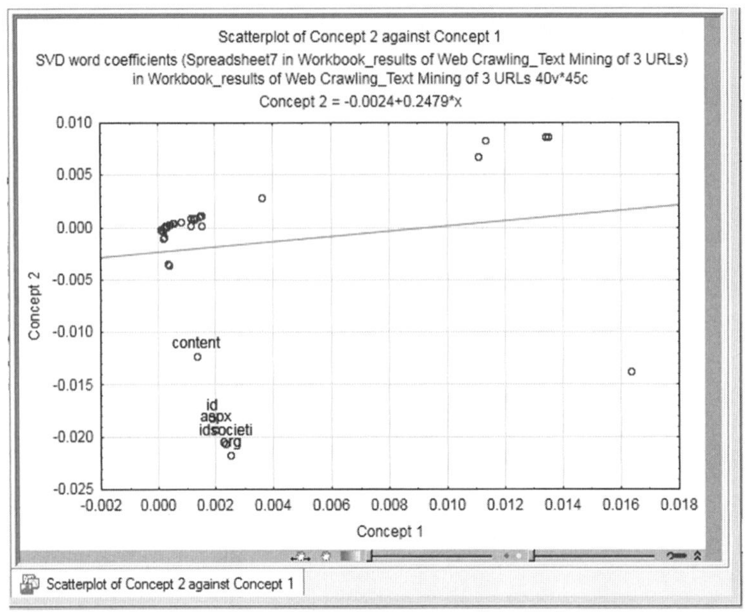

FIGURE Z.52

Scatterplot of Concept 2 versus Concept 1.

We see the points are now labeled with words, but some words overlap, so we need to catch each word with the mouse cursor and drag them apart so all of them can be read.

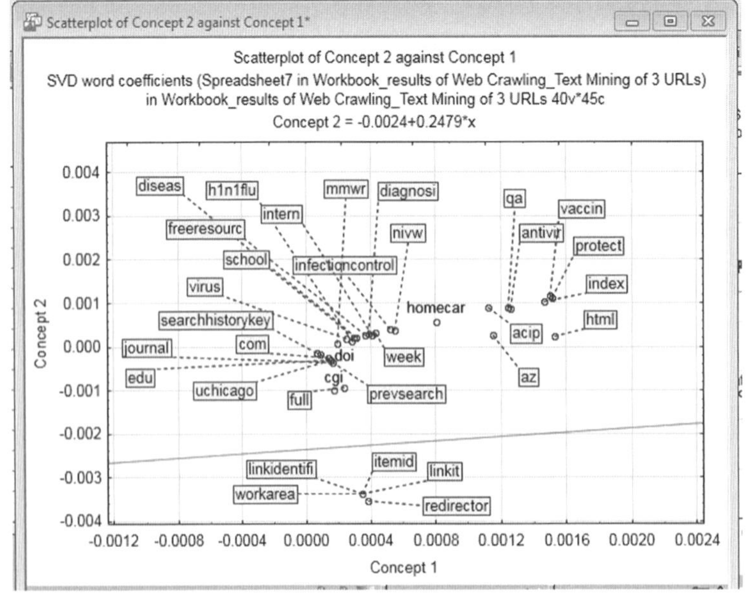

FIGURE Z.53

Words pulled apart in the 2D scatterplot to see if we can find "clusters" of similar words.

There appear to be about three clusters in "Concept 1 versus Concept 2." The cluster in the upper left is the largest and appears to have a lot of words related to *flu*, such as *H1N1-flu, disease, diagnosis, infection control, virus,* and *week* (or prevalance or flu season). This may mean that the CDC "week of flu season" is correlated with diagnosis, flu disease, and the need for infection control.

However, many words appear to be related to the HTML code—for example, the HTML coding that is used for web page construction (but that you do not normally see because it is "behind the scenes' of the web page). Thus, we really don't want these "HTML words" because they are not pertinent to our study. Thus, we can get rid of them by adding them to what is called the Exclusion List in text mining.

We can type these HTML words into the Exclusion List by just taking the "Exclusion List" provided as the default with the software and then adding them to the bottom of the list. Then this "modified list" can be saved under another name if we wish to keep it for future use.

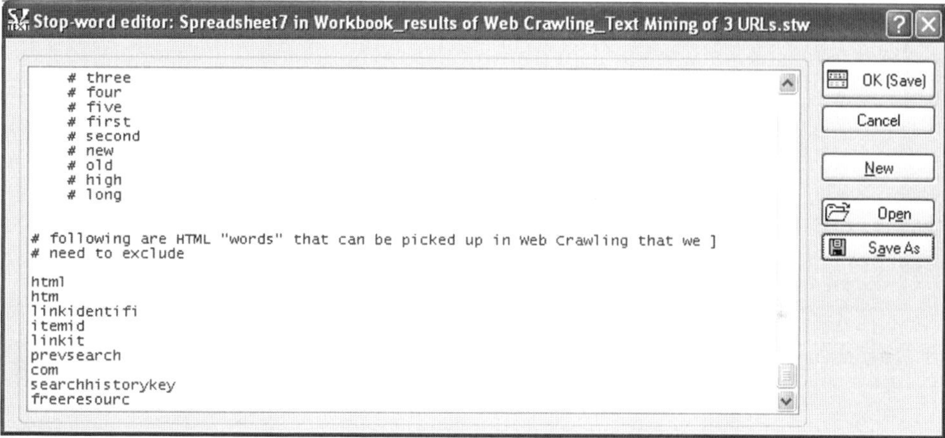

FIGURE Z.54
HTML words added to "Exclusion List."

This stopword list is saved as a special exclusion stoplist under the file name "BMayer Exclusion word."

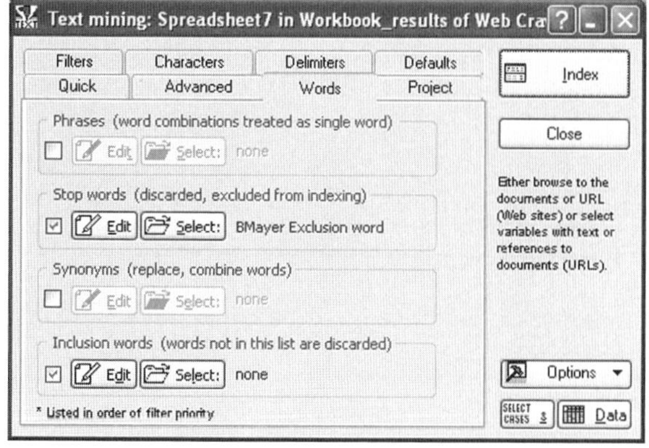

FIGURE Z.55
The BMayer Exclusion word list is selected.

We could also make an "Inclusion List." This would be a group of words we are particularly interested in and want to find out what their frequencies are in the documents, and maybe also generate concepts from them for 2_D graphing to find clusters of words.

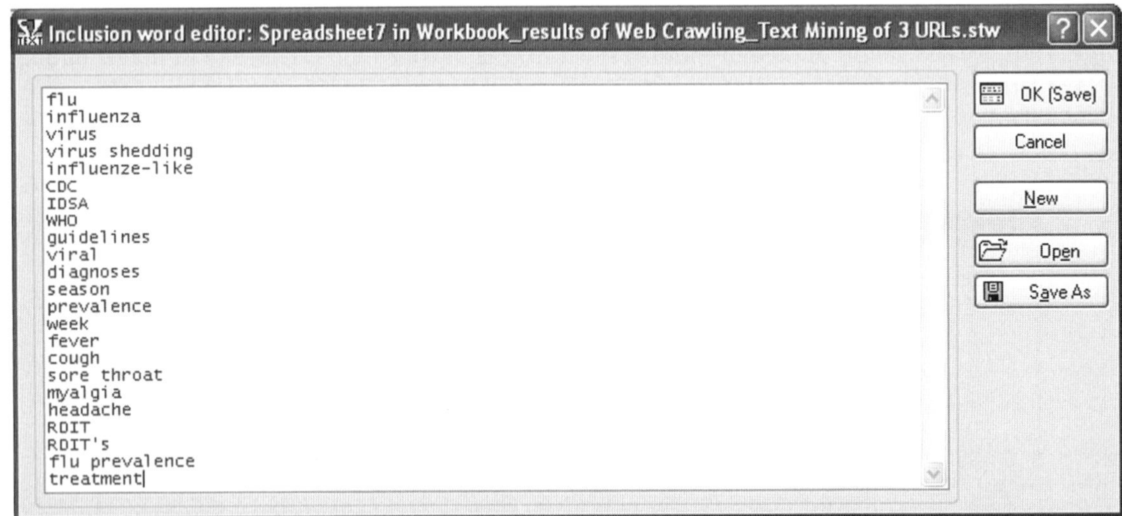

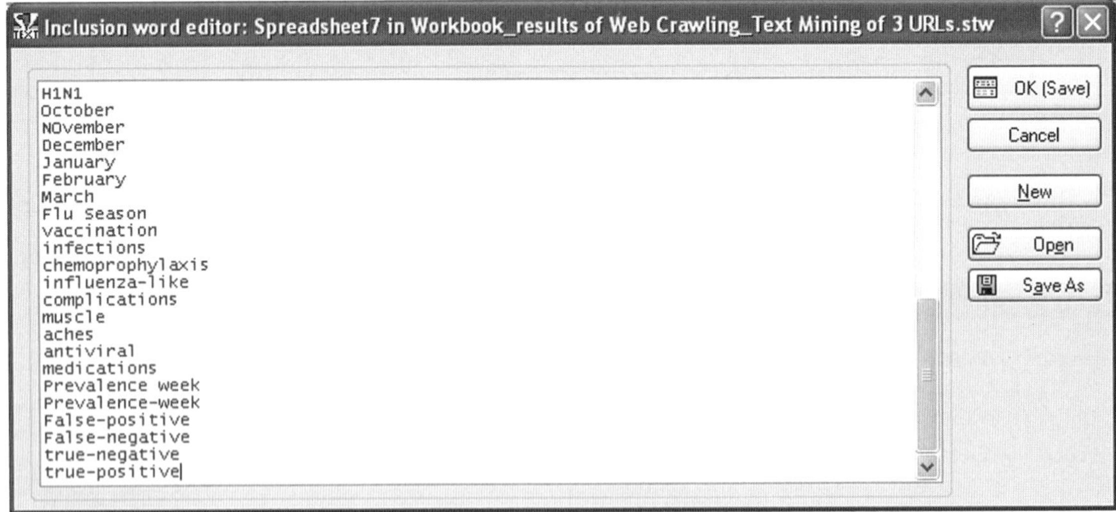

FIGURE Z.56
Inclusion List.

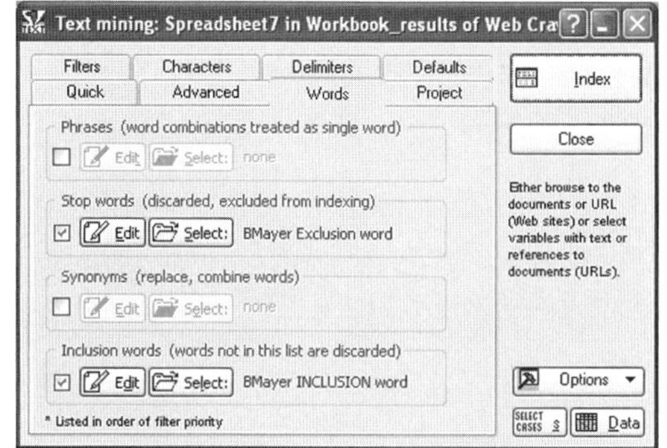

FIGURE Z.57

BMayer Inclusion word list is also now selected.

Then you would click on Index to create frequencies of words. We will not do this, however, in this tutorial.

Because web pages change with time, and by the time you are working through this tutorial, the CDC website and other websites may have changed the content on their web page, the results you'd get would be different from what we'd present in this tutorial. Thus, we made a "static group of documents" by copying the contents of 20 CDC-specific webpages into txt (Notepad) files. All of these files were put into one folder, as illustrated in Figure Z.58.

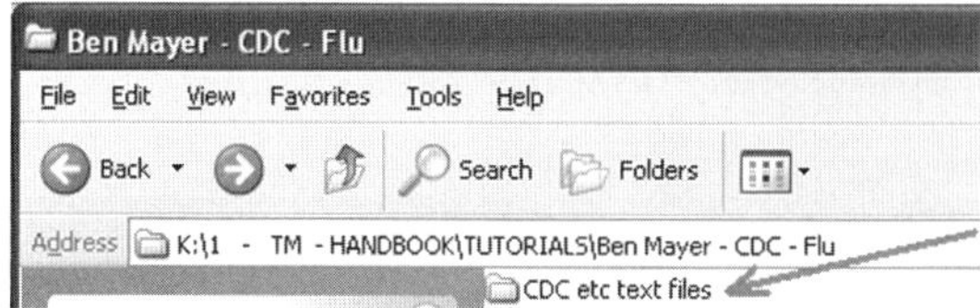

FIGURE Z.58

Folder containing the text of 20 specific CDC web pages.

You will need to get this folder from the DVD that comes with this book in order to recreate the rest of this tutorial (see Figure Z.58).

Recreating the CDC Text Files analysis

The next series of figures will walk you through text mining analyses on these CDC Text Files that were obtained via the Web Crawling illustrated in the previous part of this tutorial (see Figures Z.59 through Z.107).

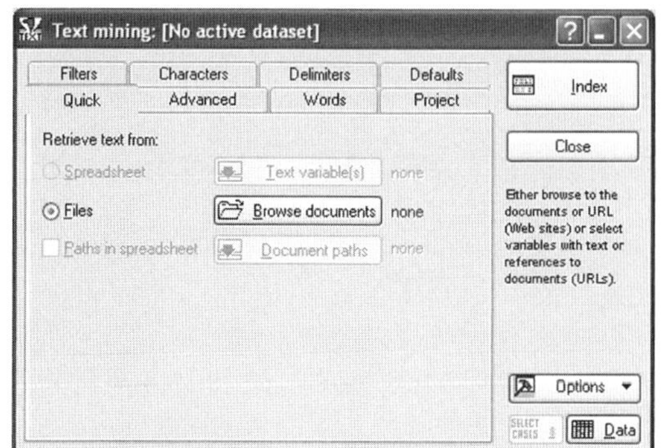

FIGURE Z.59

Text Mining dialog obtained from the Data Mining pull-down menu.

Click the "Browse documents" button as illustrated in Figure Z.59, to select the files.

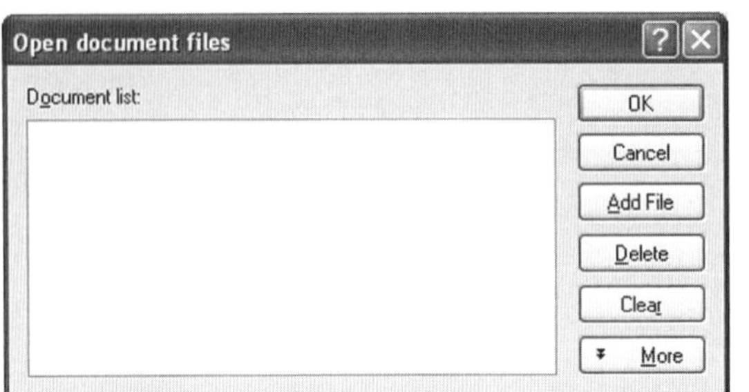

FIGURE Z.60
Open document files dialog.

Click Add File (see Figure Z.60) to get the files to list in the window. These txt files need to be in the same folder as where you are storing any information for this analysis.

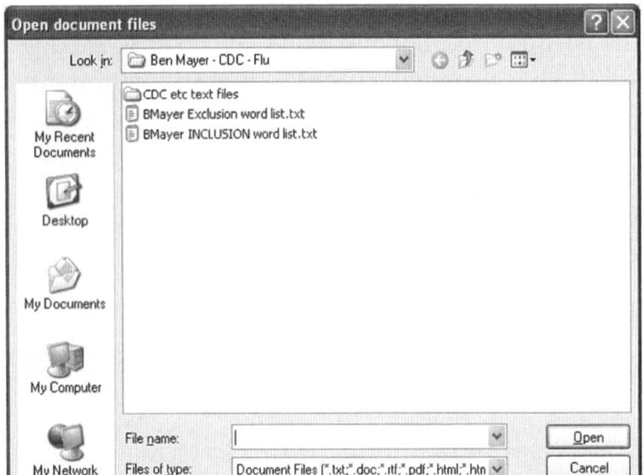

FIGURE Z.61
Window to select the file where the txt documents to be indexed are located.

Click on CDC etc text files (see Figure Z.61). Select all the files (see Figure Z.62).

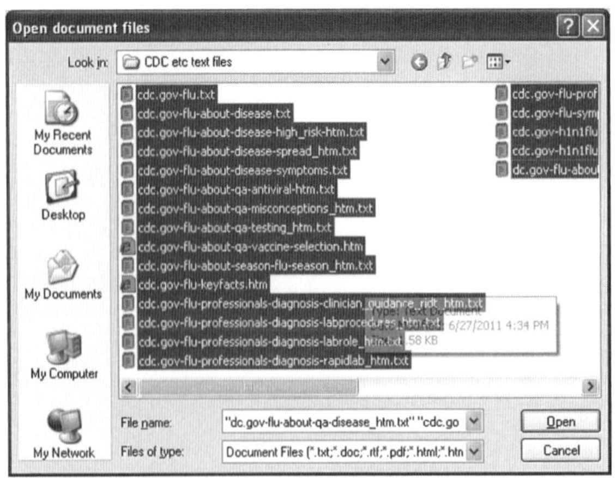

FIGURE Z.62
Files to be indexed selected (highlighted). Then click Open to put them into the text miner module.

Click Open. These files are then all put into this Text Mining files (see Figure Z.63) window.

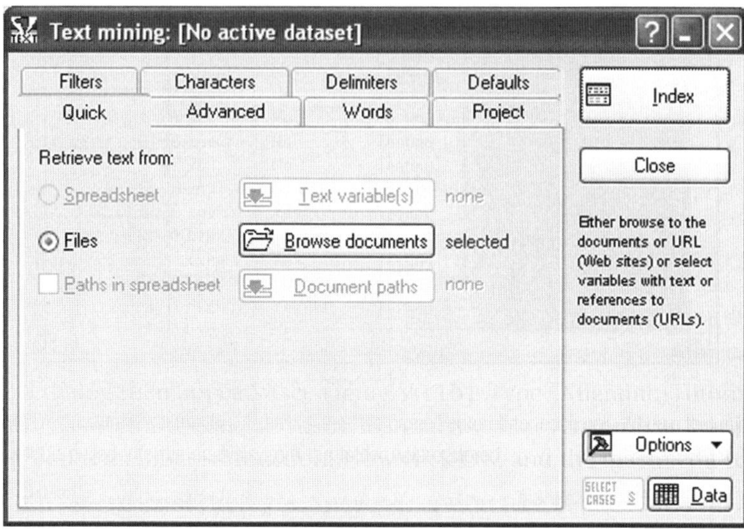

FIGURE Z.63
Text Mining main window, Quick Tab selected, showing that the documents have been "selected".

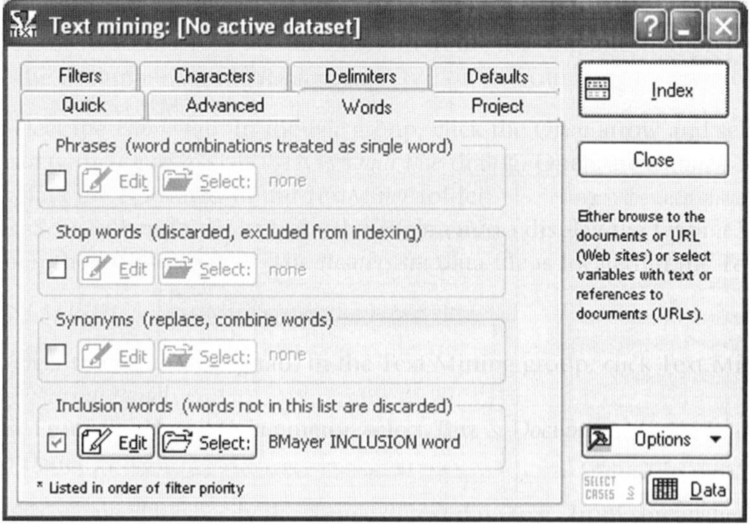

FIGURE Z.64
Using the BMayer Inclusion word list only.

Click the "Index Button" in the upper right corner of the Text Miner dialog to start the computation process.

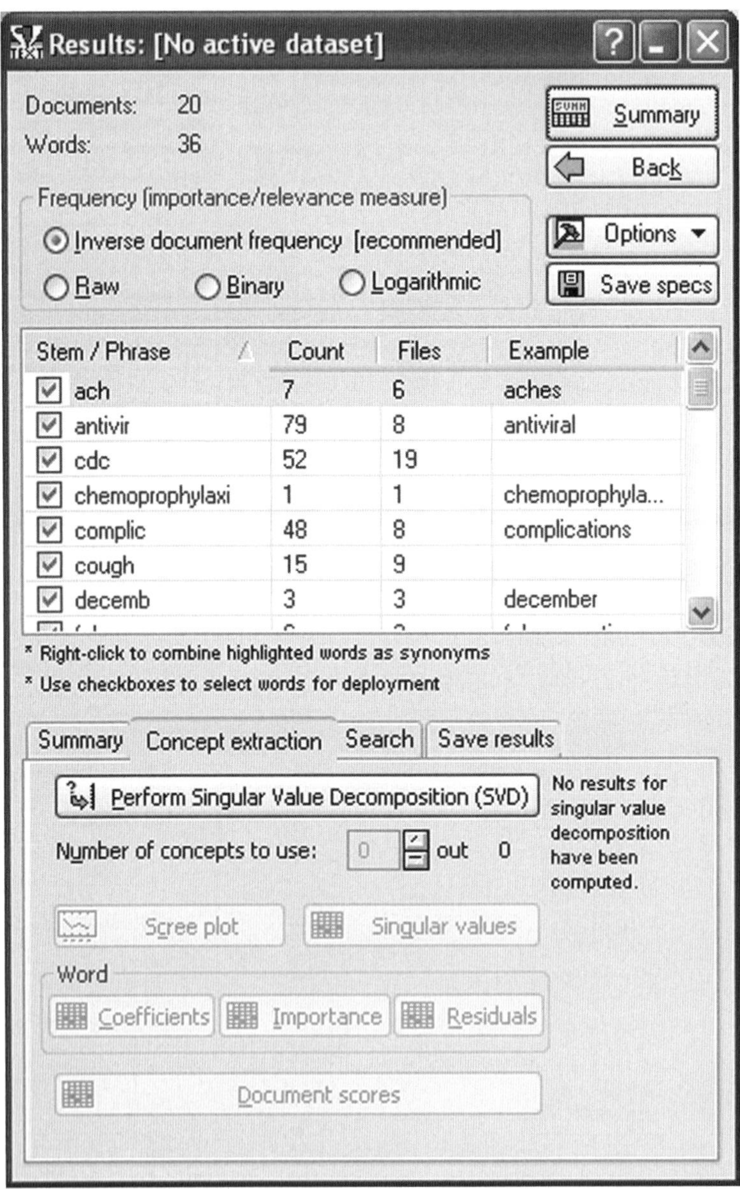

FIGURE Z.65
Results of the indexing of words.

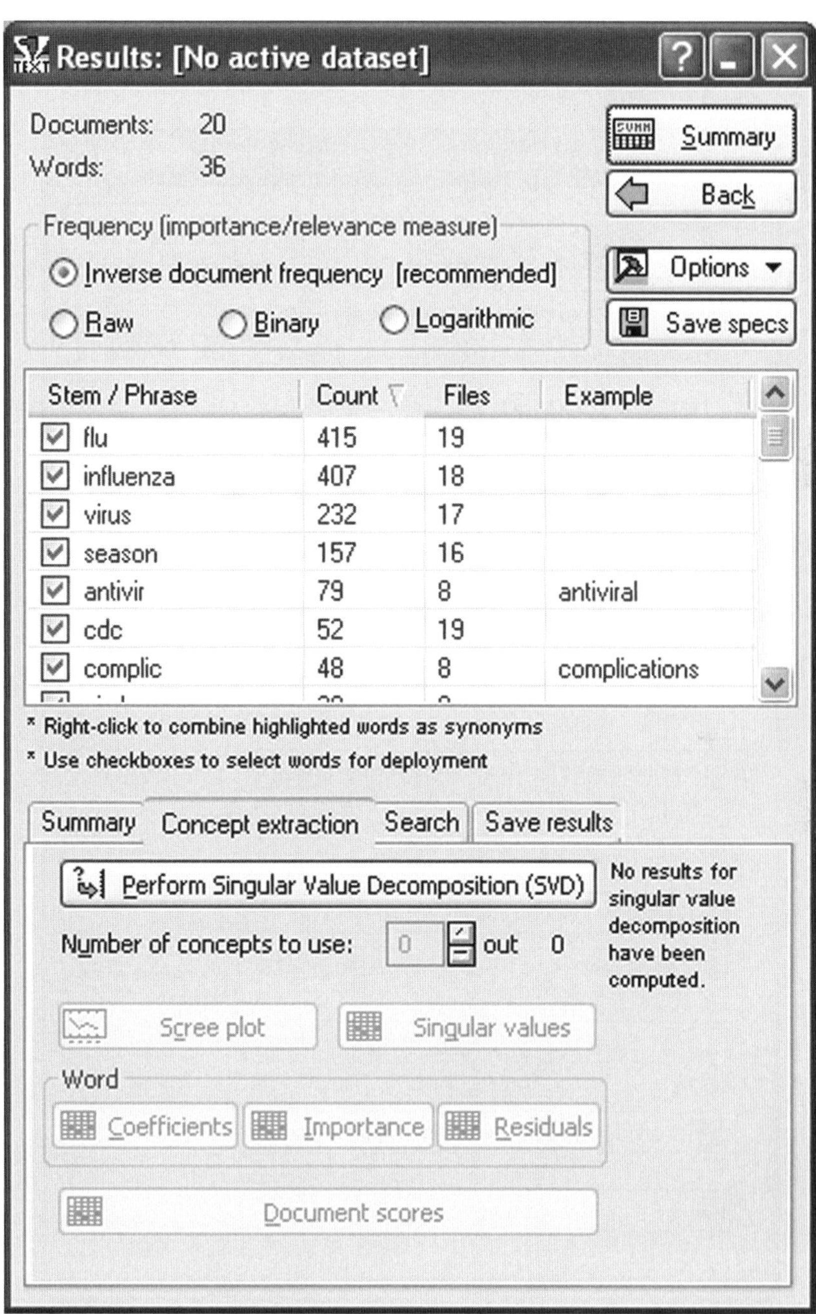

FIGURE Z.66
Sort the words by clicking at the top of the word list.

Now we can see the most frequent words at the top of the list, rather than the default alphabetical order (see Figure Z.67). Then perform SVD (see Figure Z.67).

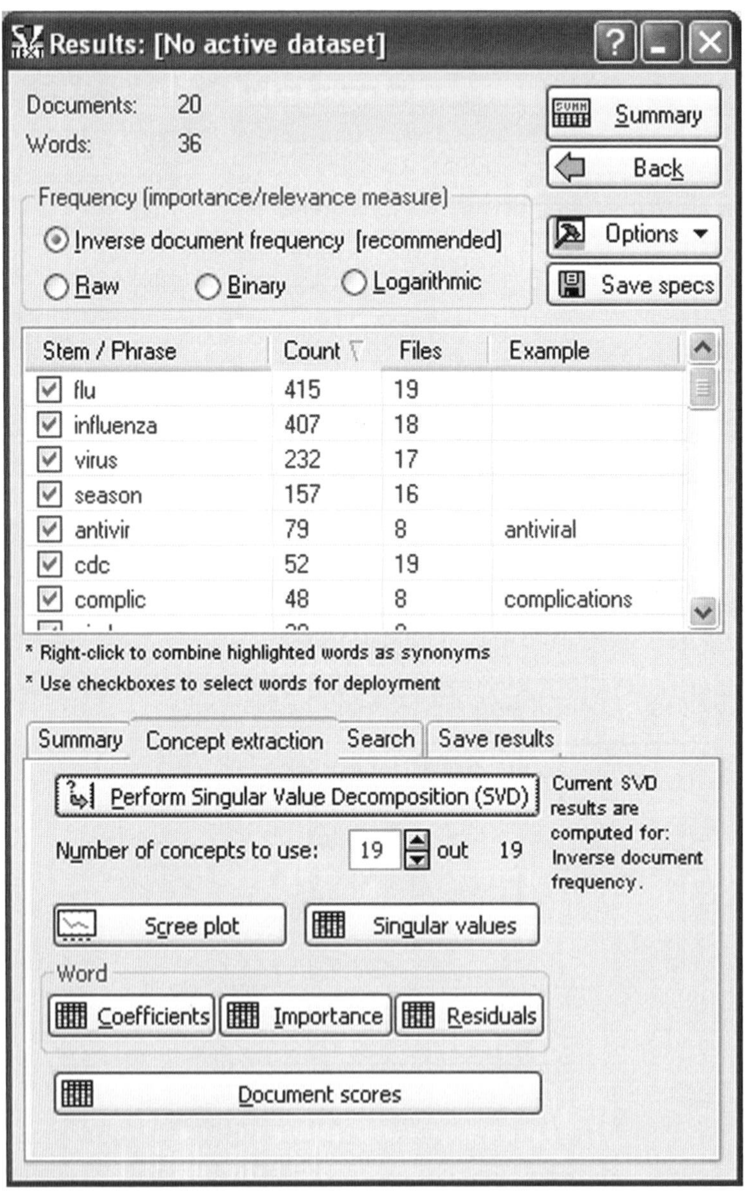

FIGURE Z.67

The (1) Inverse document frequency was selected, and then (2) the Concept Extraction tab was selected, and then (3) the SVD button was clicked to allow SVD to be performed. When this is completed, the bottom ¼ of the dialog is grayed-in, allowing you to click on various buttons to get the results put into a workbook.

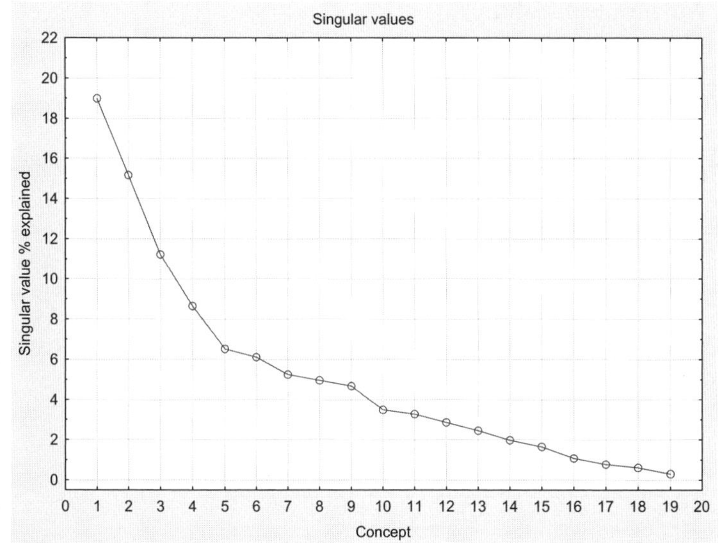

FIGURE Z.68
Click on the Scree plot button on the TM Results dialog to produce this graph.

We will select the top six concepts only, since an inflection point appears between the 5 and 7 area.

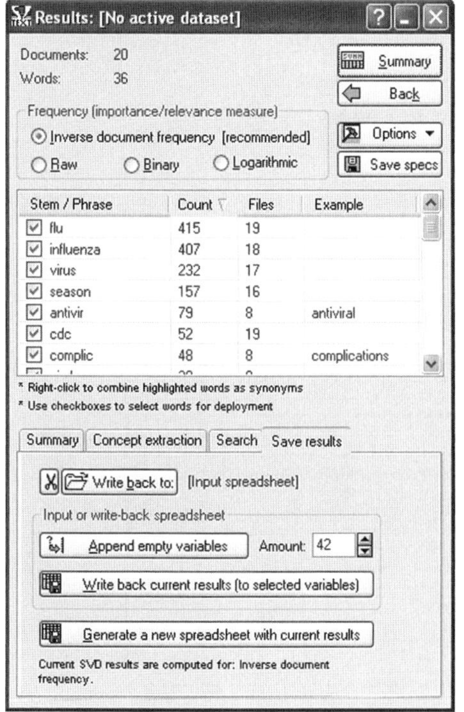

FIGURE Z.69
Amount set for 42: 36 words plus 6 concepts to be written to the master data file.

Click (see Figure Z.69 above) "Append empty variables."

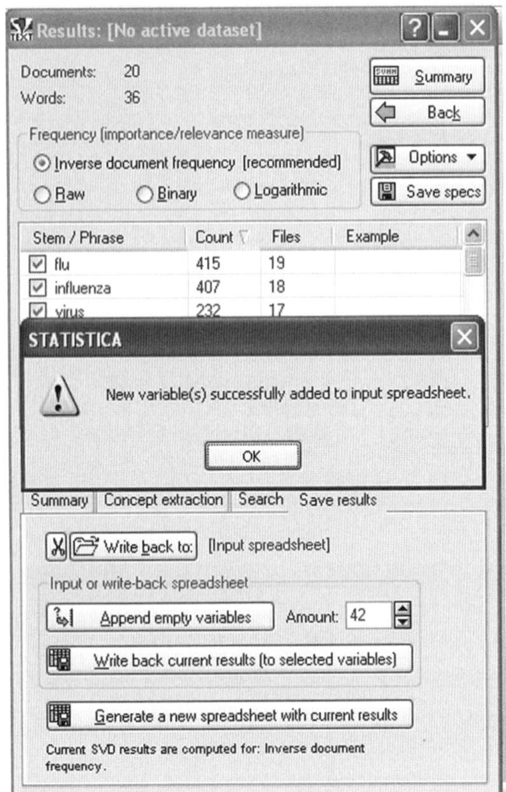

FIGURE Z.70
An intermediate window pops up to let the user know that the new variables were successfully added to the master data file.

Then go to the File pull-down menu and create a new blank spreadsheet of 20 cases and 42 variables as illustrated in the next Figure Z.71.

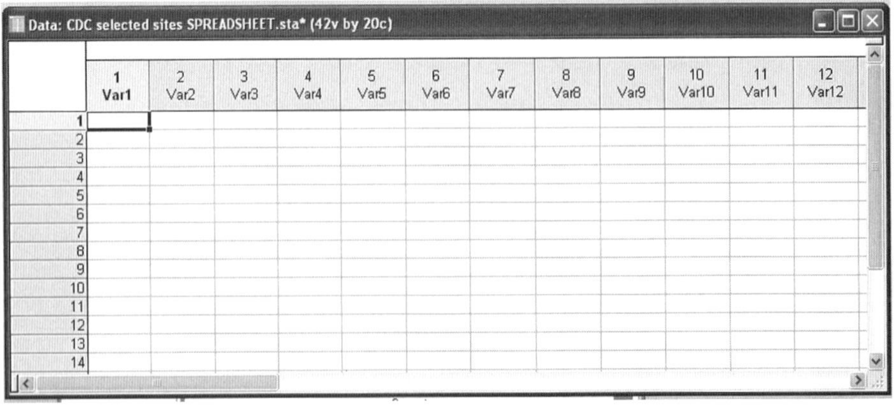

FIGURE Z.71
A new blank data sheet was made to hold the word frequencies and concepts.

Then click (see Figure Z.72) on "Write back to."

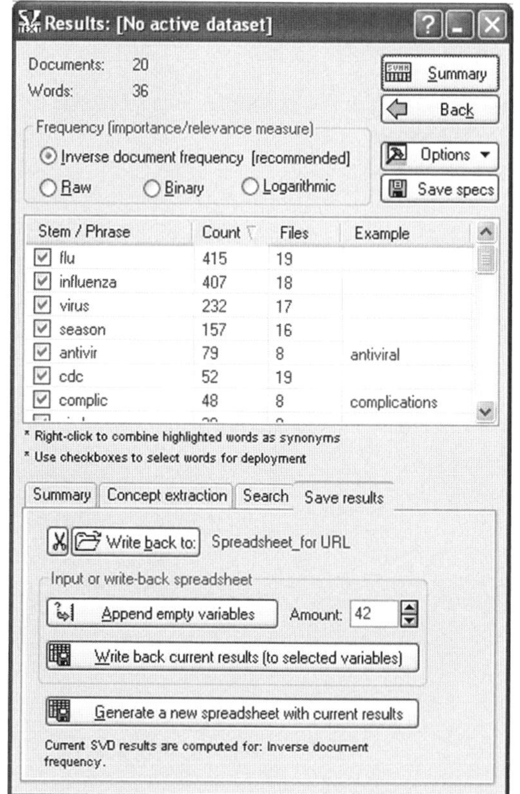

FIGURE Z.72
Click on "Append empty variables" to write the information back to the spreadsheet.

Then click on the spreadsheet to add these to the next figure.

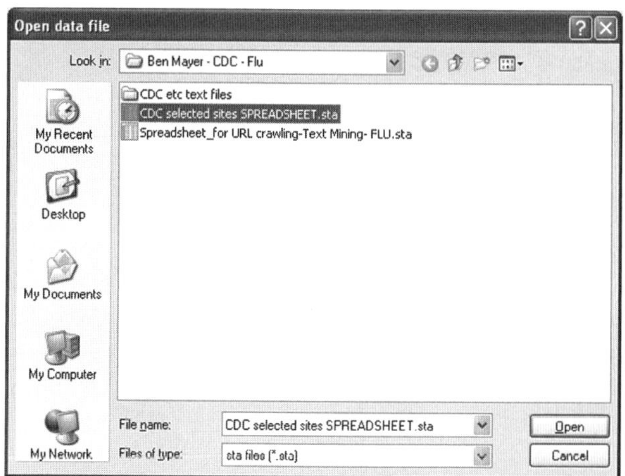

FIGURE Z.73
Data file name to which we want the word indexing results to be written.

Then click Open. Click on "Append empty variables." Note that these have been appended to the spreadsheet.

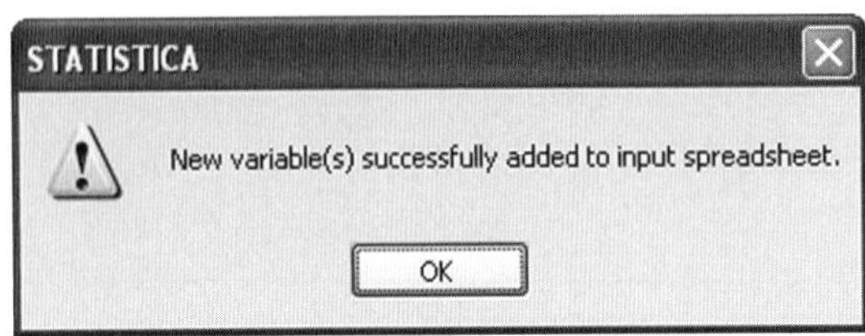

FIGURE Z.74
Intermediate dialog telling us that the variables were successfully added to the spreadsheet.

Then click on "Write back current results in the text mining results dialog."

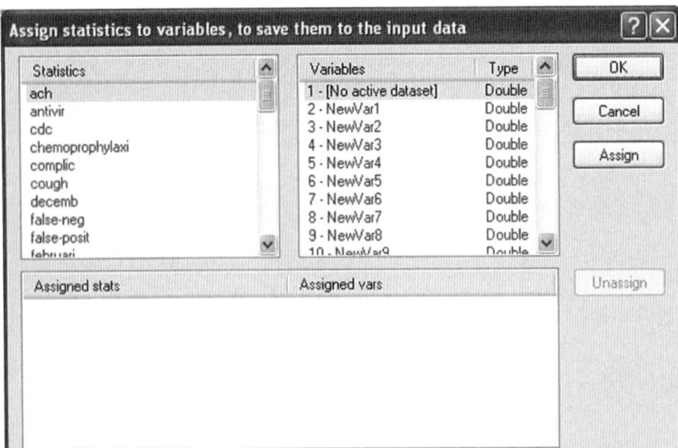

FIGURE Z.75
Assign statistics to variables.

Select words and concepts on the left side and then the New Vars to put these on the right side, and click Assign (see Figure Z.76).

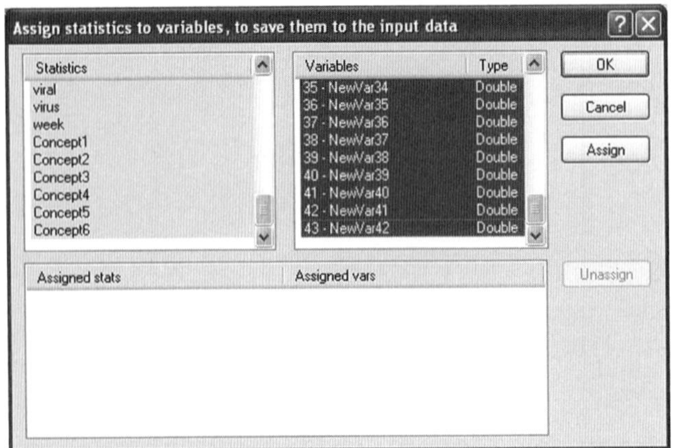

FIGURE Z.76
Assigning the new value to the variable columns.

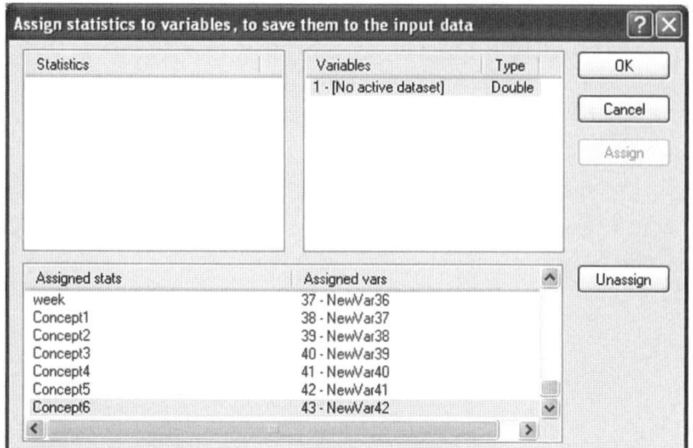

FIGURE Z.77
Stats and vars now assigned.

Then click OK.

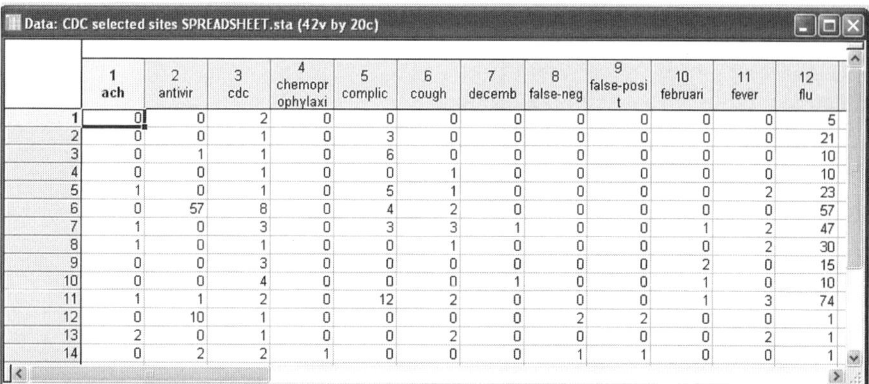

Data: CDC selected sites SPREADSHEET.sta (42v by 20c)

	1 ach	2 antivir	3 cdc	4 chemoprophylaxi	5 complic	6 cough	7 decemb	8 false-neg	9 false-posit	10 februari	11 fever	12 flu
1	0	0	2	0	0	0	0	0	0	0	0	5
2	0	0	1	0	3	0	0	0	0	0	0	21
3	0	1	1	0	6	0	0	0	0	0	0	10
4	0	0	1	0	0	1	0	0	0	0	0	10
5	1	0	1	0	5	1	0	0	0	0	2	23
6	0	57	8	0	4	2	0	0	0	0	0	57
7	1	0	3	0	3	3	1	0	0	1	2	47
8	1	0	1	0	0	1	0	0	0	0	2	30
9	0	0	3	0	0	0	0	0	0	2	0	15
10	0	0	4	0	0	0	1	0	0	1	0	10
11	1	1	2	0	12	2	0	0	0	1	3	74
12	0	10	1	0	0	0	0	2	2	0	0	1
13	2	0	1	0	0	2	0	0	0	0	2	1
14	0	2	2	1	0	0	0	1	1	0	0	1

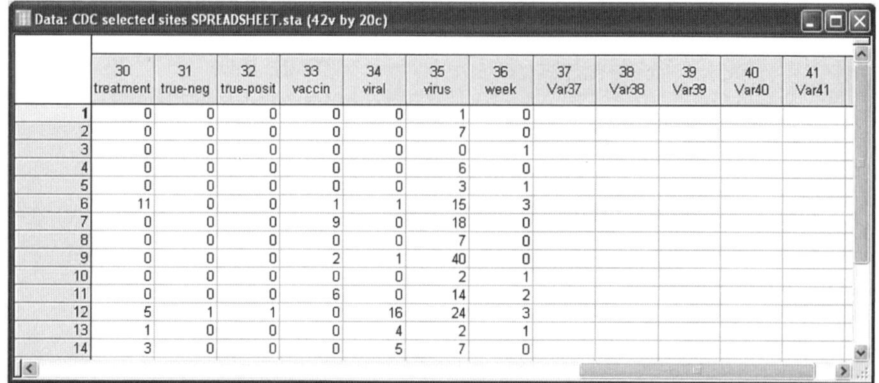

Data: CDC selected sites SPREADSHEET.sta (42v by 20c)

	30 treatment	31 true-neg	32 true-posit	33 vaccin	34 viral	35 virus	36 week	37 Var37	38 Var38	39 Var39	40 Var40	41 Var41
1	0	0	0	0	0	1	0					
2	0	0	0	0	0	7	0					
3	0	0	0	0	0	0	1					
4	0	0	0	0	0	6	0					
5	0	0	0	0	0	3	1					
6	11	0	0	1	1	15	3					
7	0	0	0	9	0	18	0					
8	0	0	0	0	0	7	0					
9	0	0	0	2	1	40	0					
10	0	0	0	0	0	2	1					
11	0	0	0	6	0	14	2					
12	5	1	1	0	16	24	3					
13	1	0	0	0	4	2	1					
14	3	0	0	0	5	7	0					

FIGURE Z.78
Word frequencies now inputted to the master data file, with room left for the concepts that we'll pull over later.

So we see that the word frequencies are inputted into this new spreadsheet. Now we need to form the Concepts and add them. Click the "Inverse document frequency" radio button (see Figure Z.79).

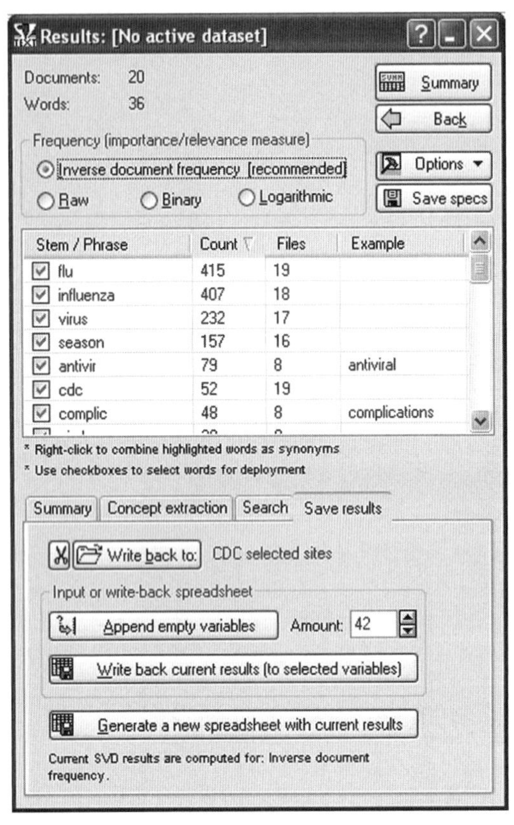

FIGURE Z.79

Inverse document frequency and SVD.

Then click on the "Concept extraction" tab. Since these have already been created, we can then click on Save and then "Write back current results."

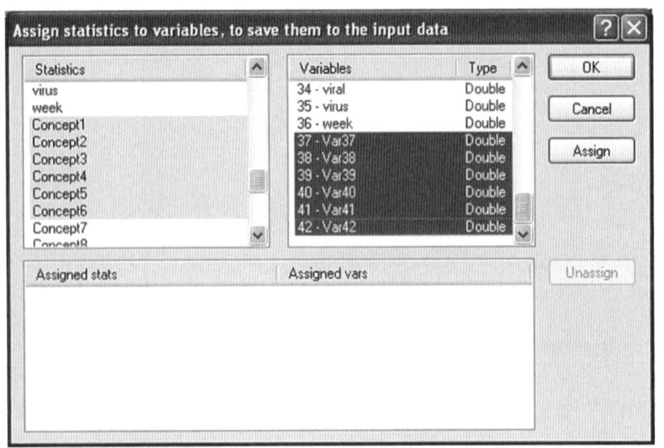

FIGURE Z.80

Selecting Concepts 1 through 6 and the remaining six variable columns.

Then click on Assign, and then OK.

Data: CDC selected sites SPREADSHEET.sta (42v by 20c)

	31 true-neg	32 true-posit	33 vaccin	34 viral	35 virus	36 week	37 Concept1	38 Concept2	39 Concept3	40 Concept4	41 Concept5	42 Concept6
1	0	0	0	0	1	0	0.004177	-0.00303	0.001732	-0.00452	-0.00022	-0.00402
2	0	0	0	0	7	0	0.055644	-0.0764	0.01967	-0.11708	0.019571	0.216769
3	0	0	0	0	0	1	0.074908	-0.09673	-0.00699	-0.02801	0.025542	0.391316
4	0	0	0	0	6	0	0.008877	-0.01311	0.00191	0.013111	-0.01309	0.000951
5	0	0	0	0	3	1	0.095005	-0.16675	0.095561	0.147728	-0.18926	0.349337
6	0	0	1	1	15	3	0.399651	-0.16164	-0.75397	0.218485	0.28288	0.071801
7	0	0	9	0	18	0	0.202217	-0.42763	0.28548	0.429813	0.032459	-0.36077
8	0	0	0	0	7	0	0.056902	-0.0979	0.060525	0.22019	-0.21611	-0.04026
9	0	0	2	1	40	0	0.097098	-0.13232	0.108734	-0.23338	0.132762	-0.47366
10	0	0	0	0	2	1	0.068995	-0.12124	0.115238	-0.29034	0.312544	-0.27735
11	0	0	6	0	14	2	0.228436	-0.3874	0.202178	-0.06775	0.010264	0.173276
12	1	1	0	16	24	3	0.560952	0.388528	-0.03707	0.006512	0.020414	-0.1285
13	0	0	0	4	2	1	0.144724	-0.1169	0.023649	0.163509	-0.60312	-0.02396
14	0	0	0	5	7	0	0.211234	0.084471	-0.13632	-0.07195	-0.43005	-0.20747

FIGURE Z.81

Concepts are now written back to master data set.

And the six concepts are now added to the spreadsheet.

Now go back to the Workbook and make the Concepts Coefficients the active input spreadsheet (see Figure Z.82).

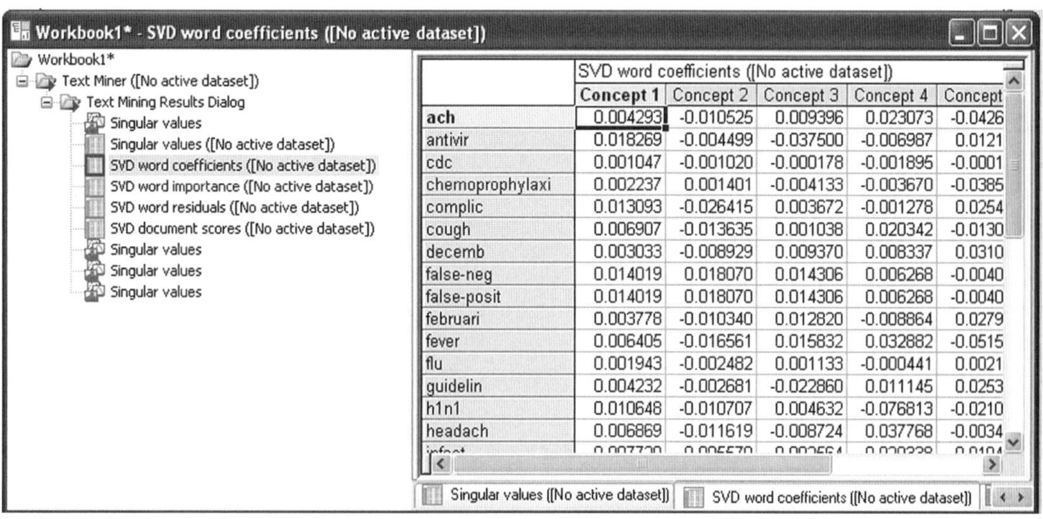

SVD word coefficients ([No active dataset])

	Concept 1	Concept 2	Concept 3	Concept 4	Concept
ach	0.004293	-0.010525	0.009396	0.023073	-0.0426
antivir	0.018269	-0.004499	-0.037500	-0.006987	0.0121
cdc	0.001047	-0.001020	-0.000178	-0.001895	-0.0001
chemoprophylaxi	0.002237	0.001401	-0.004133	-0.003670	-0.0385
complic	0.013093	-0.026415	0.003672	-0.001278	0.0254
cough	0.006907	-0.013635	0.001038	0.020342	-0.0130
decemb	0.003033	-0.008929	0.009370	0.008337	0.0310
false-neg	0.014019	0.018070	0.014306	0.006268	-0.0040
false-posit	0.014019	0.018070	0.014306	0.006268	-0.0040
februari	0.003778	-0.010340	0.012820	-0.008864	0.0279
fever	0.006405	-0.016561	0.015832	0.032882	-0.0515
flu	0.001943	-0.002482	0.001133	-0.000441	0.0021
guidelin	0.004232	-0.002681	-0.022860	0.011145	0.0253
h1n1	0.010648	-0.010707	0.004632	-0.076813	-0.0210
headach	0.006869	-0.011619	-0.008724	0.037768	-0.0034

FIGURE Z.82

Concept Coefficients. These will be used for 2D scatterplots to see if we can find "clusters" of similar words.

Then we will make scatterplots of Concept 1 versus Concept 2, and maybe some other comparisons, if it seems appropriate.

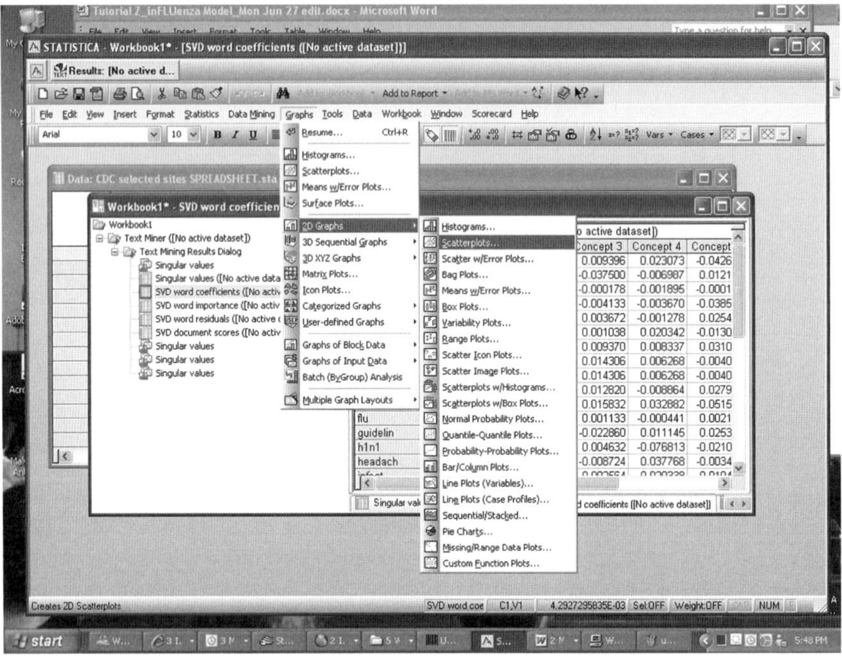

FIGURE Z.83
Selecting 2D Scatterplots.

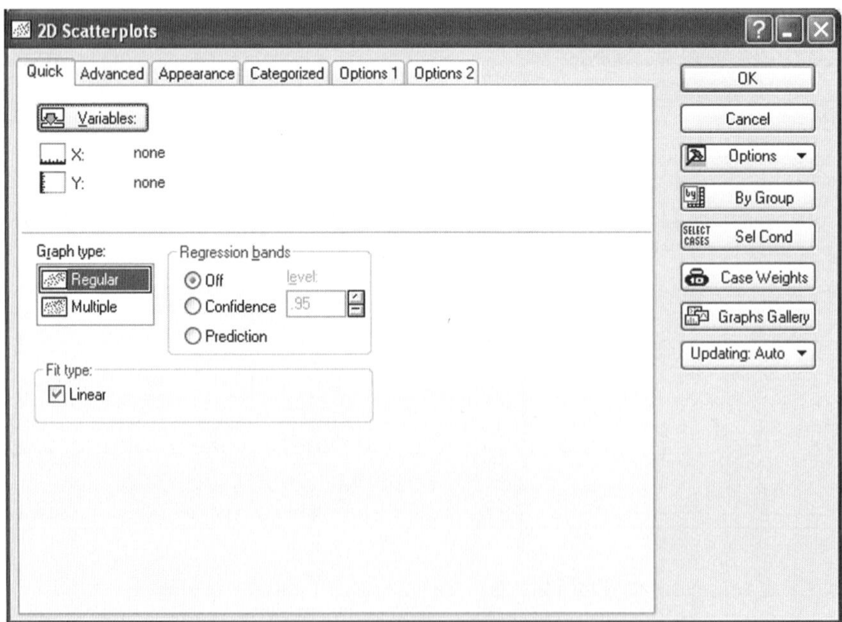

FIGURE Z.84
2D Scatterplot dialog.

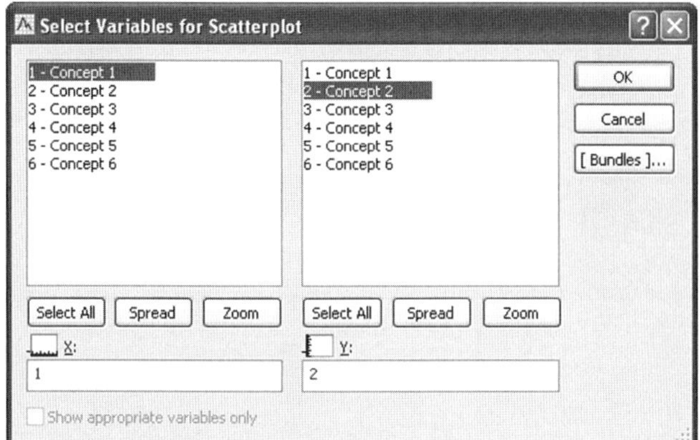

FIGURE Z.85
Selecting of Concept 2 versus Concept 1 to graph in a scatterplot.

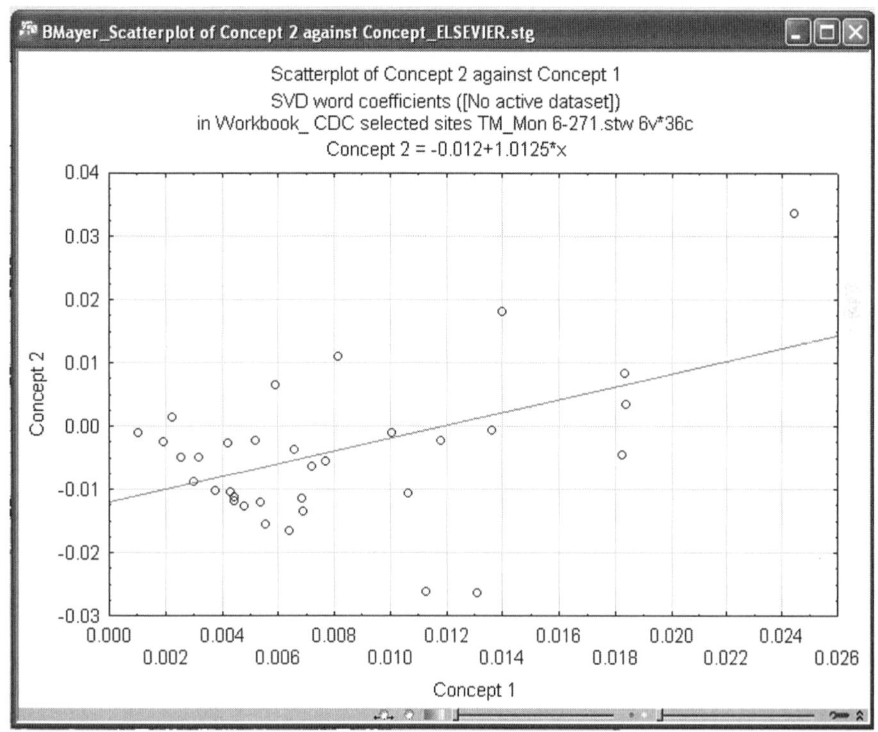

FIGURE Z.86
Scatterplot of Concept 2 versus Concept 1.

Now, using the Brushing tool, let's look at the "words" that are represented by each point in the graph.

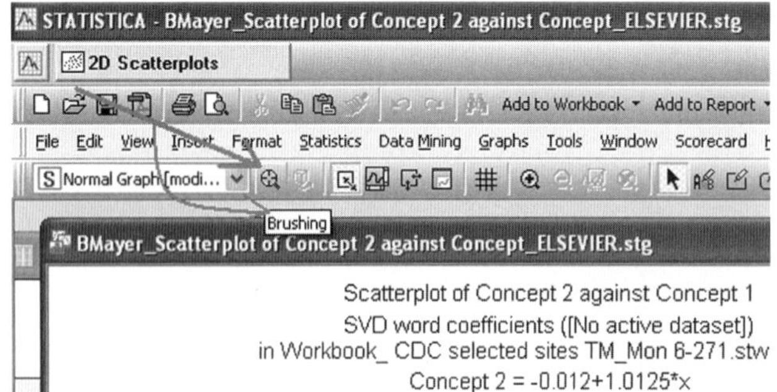

FIGURE Z.87

Brushing tool is selected from a toolbar on *STATISTICA*.

In the Brushing tool dialog, select "Label" and "Lasso."

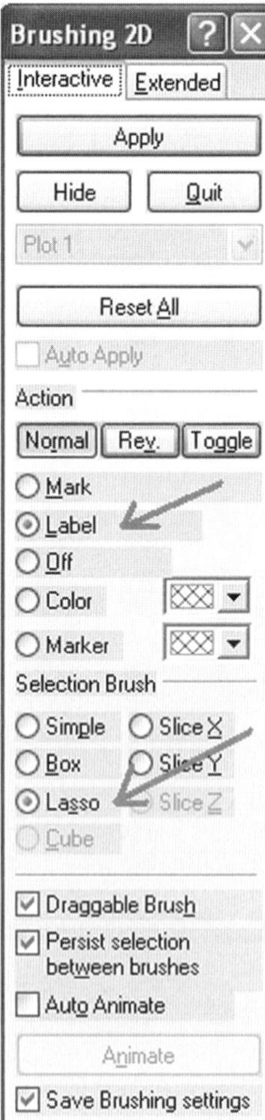

FIGURE Z.88

Brushing tool dialog.

We now can "draw – lasso" what we perceive as "different clusters" in the scatterplot. Here's what we'll call "Cluster No. 1."

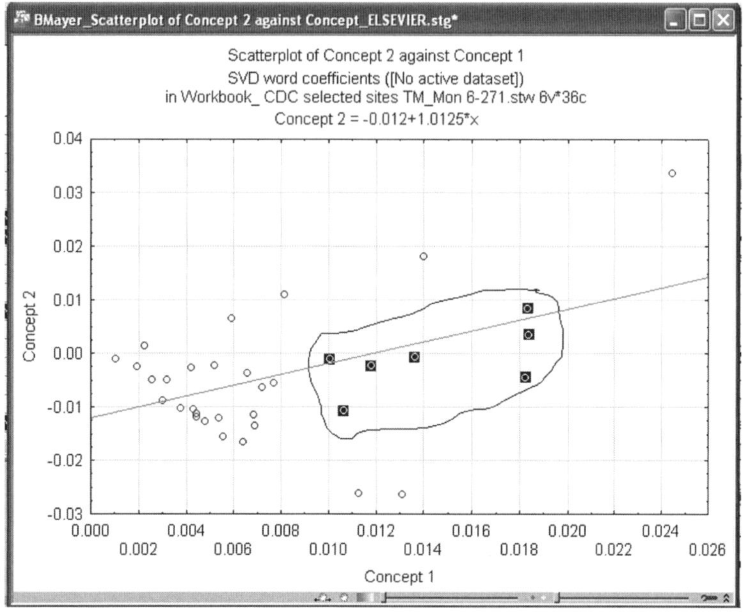

FIGURE Z.89

Lassoing a cluster of data points.

Now select the button "Apply" in the Brushing tool dialog to label these points in this cluster No. 1 with words.

FIGURE Z.90

After lassoing a cluster, click "Apply," and the words corresponding to the points will be placed on the graph.

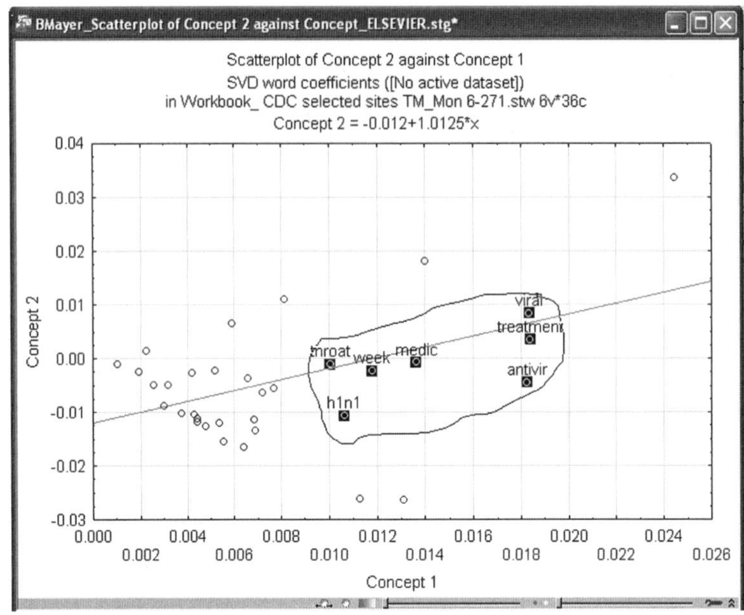

FIGURE Z.91
Scatterplot with words labeling data points.

Now we'll do Cluster No. 2.

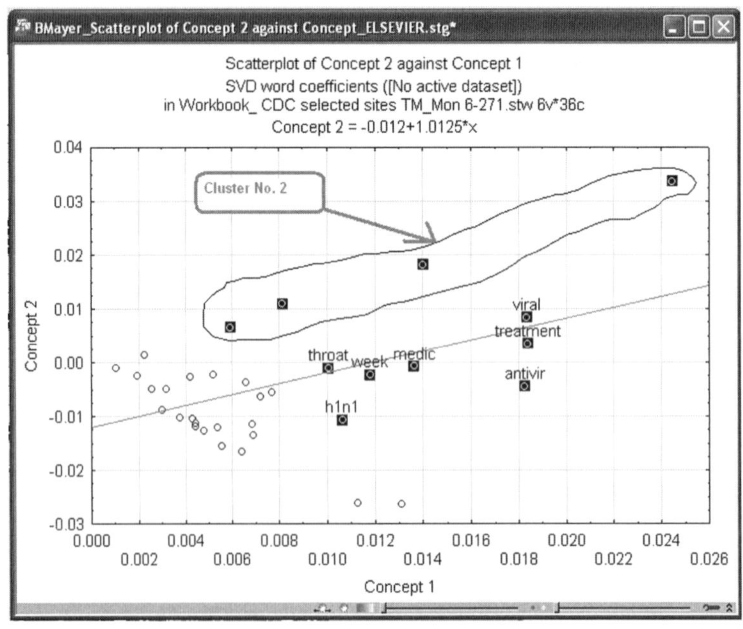

FIGURE Z.92
Cluster No. 2 selected.

And now Cluster No. 3.

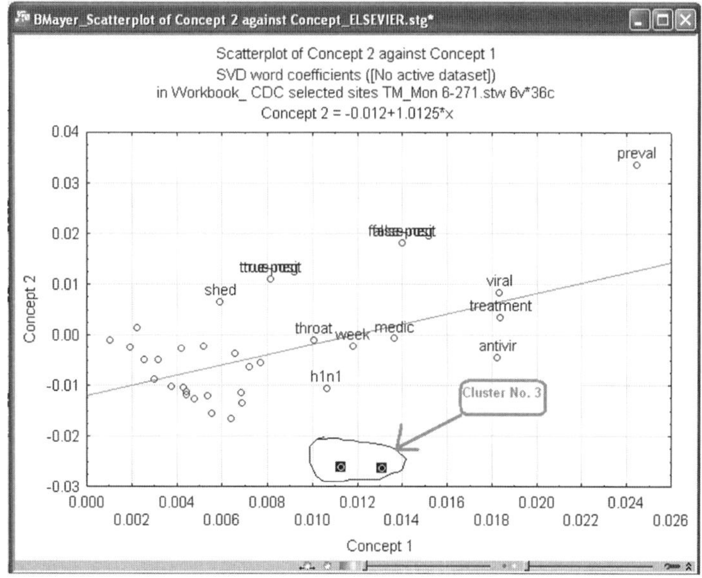

FIGURE Z.93
Cluster No. 3.

And Cluster No. 4.

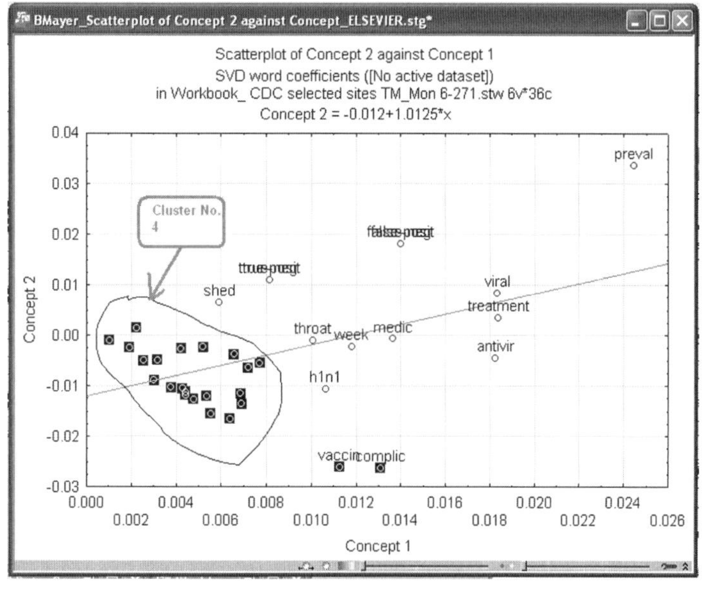

FIGURE Z.94
Cluster No. 4.

This is Cluster No. 4 with the "Apply" button on the Brushing dialog selected so that the points are labeled with words (see Figure Z.95). We note that there are many overlapping words here, so these will have to be "dragged apart" by clicking on the words, holding down the mouse, and dragging them away from each other.

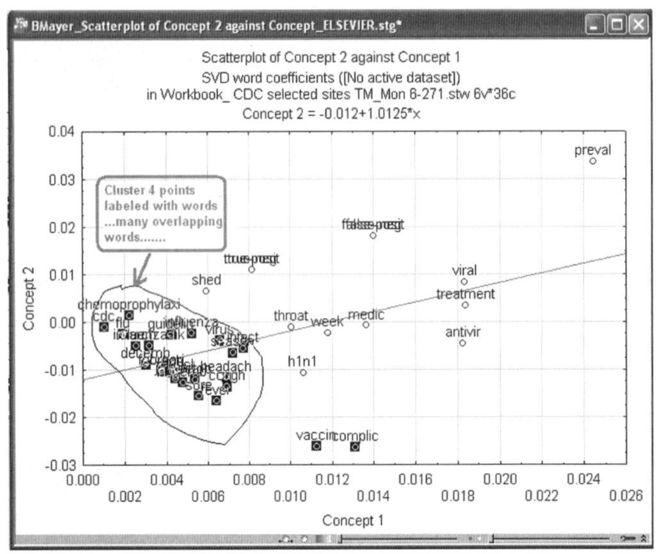

FIGURE Z.95

Cluster No. 4.

After the words were pulled apart, we decided to put them into five clusters, since "Prevalence" to the upper right seemed to be all by itself, so it was taken out of Cluster No. 1.

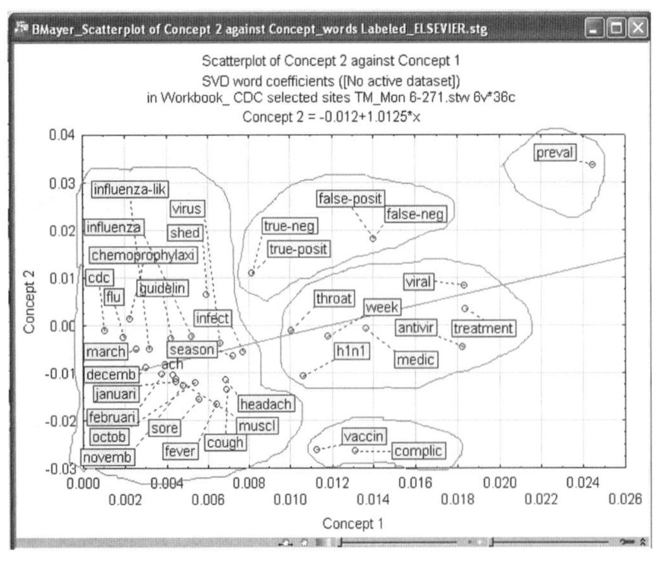

FIGURE Z.96

Five clusters of words selected.

The words do seem to form into somewhat meaningful clusters (although it did not pull out CDC "Prevalence" and "Accuracy of Diagnosis," which we'd hoped for). For example, during the very active weeks when Prevalence is high, doctors seem to diagnose "FLU" without doing all of the tests, including rapid flu tests, since they "suspect" flu because of the "prevalance figures" released by the CDC for their location. Thus they may give the Dx more often to someone who may have symptoms of "stomach flu" (which is not really viral influenza) or other symptoms of flu—something they would probably not do during the beginning of the flu season when prevalence is low. We can see *True-Positive/True-Negative* and *False-Positive/False-Negative* falling into the group located "top center." Some general terms like *throat, week, viral, treatment, H1N1, medications,* and so on fall into the "central group." *Vaccine* and *complications* fall into a lower central group by themselves, since they are not strongly associated with anything else. The big group at the lower left consists of all the months of the year that are generally the "flu season," plus associated symptoms of flu, like *sore muscles, fever, cough,* and *headache,* and another subset of words relating to the public health service processing of flu epidemics like *infection, flu, influenza, chemoprophylaxis, CDC, shed* (shedding of virus, when the person can infect others), and *infection.*

So in conclusion for this web analytics part of this tutorial, we see that the Web Mining has taken words from 20 CDC websites related to influenza and grouped them into meaningful clusters.

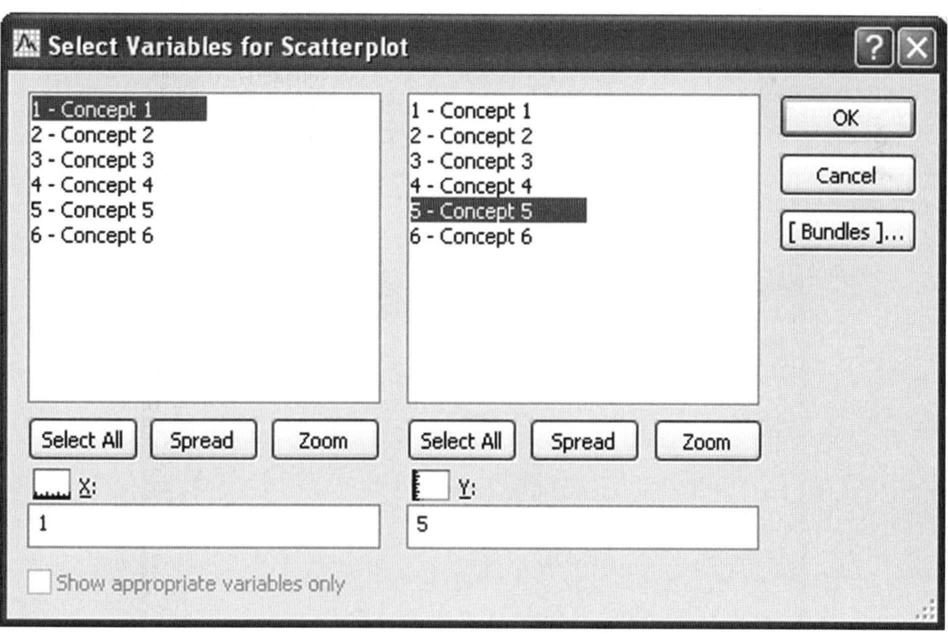

FIGURE Z.97

Graphing Concept 1 versus Concept 5.

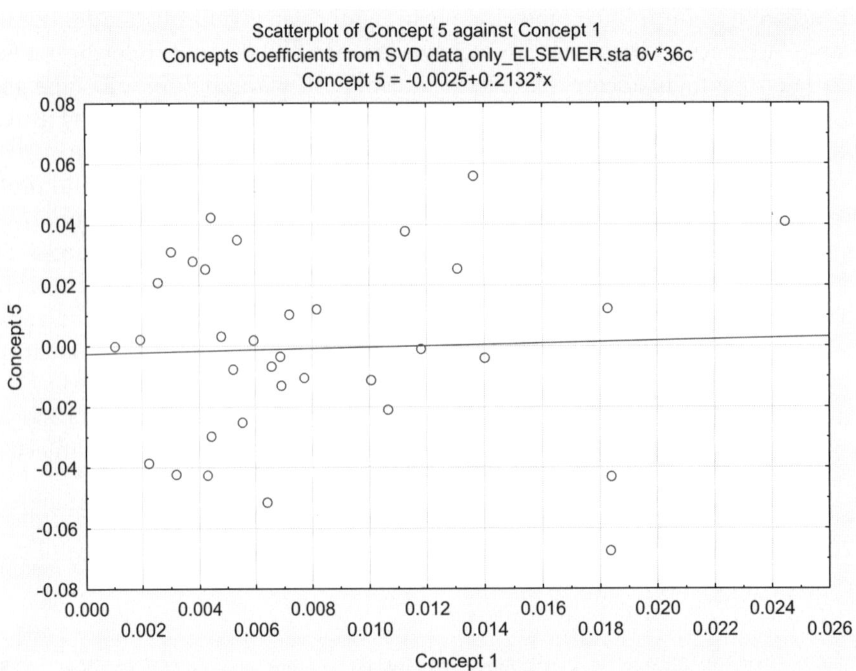

FIGURE Z.98

Scatterplot of Concept 5 versus Concept 1.

This does not look very interesting, so we won't pursue it further in this tutorial. Let's look at some other combinations of the concepts.

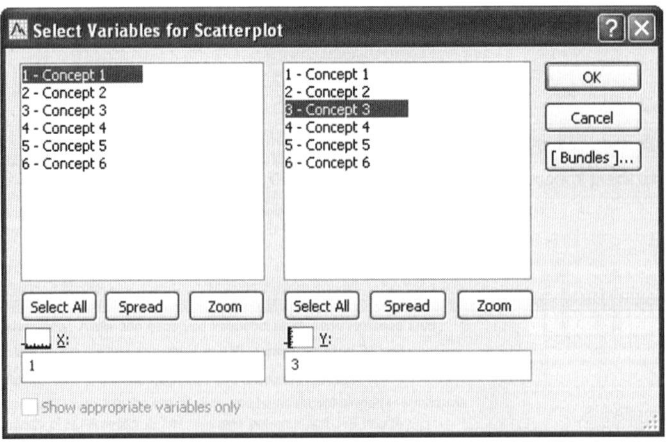

FIGURE Z.99

Concept 1 versus Concept 3.

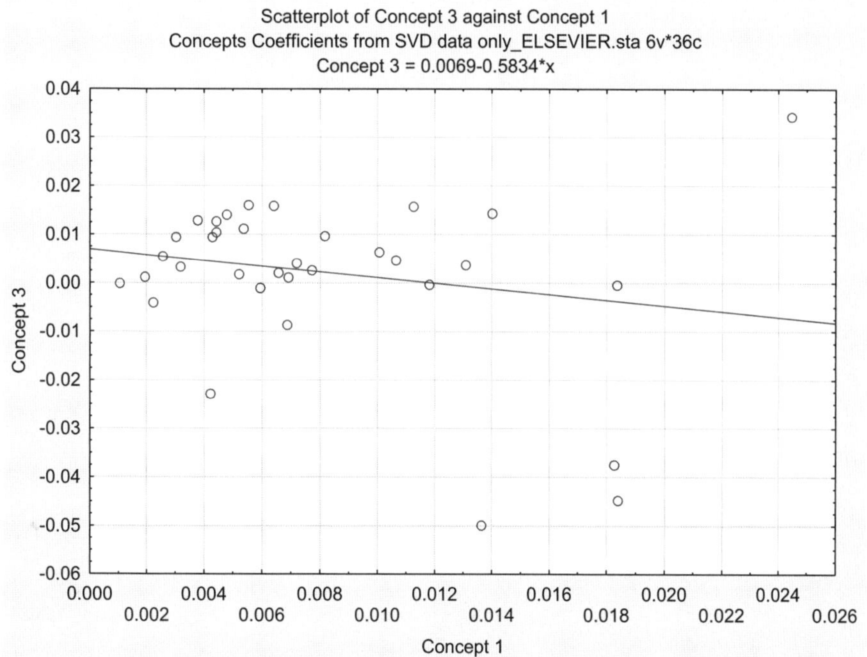

FIGURE Z.99
(*Continued*)

This might be interesting, but before labeling words on this one, let's look at some more combinations of the concepts to see if any can make a "more interesting" clustering pattern.

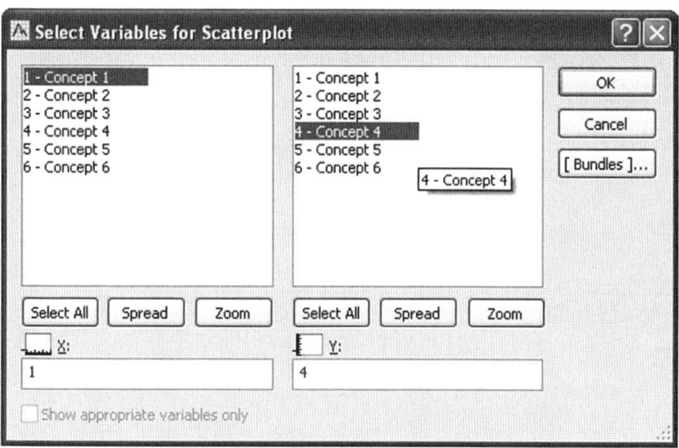

FIGURE Z.100
Concept 4 versus Concept 1.

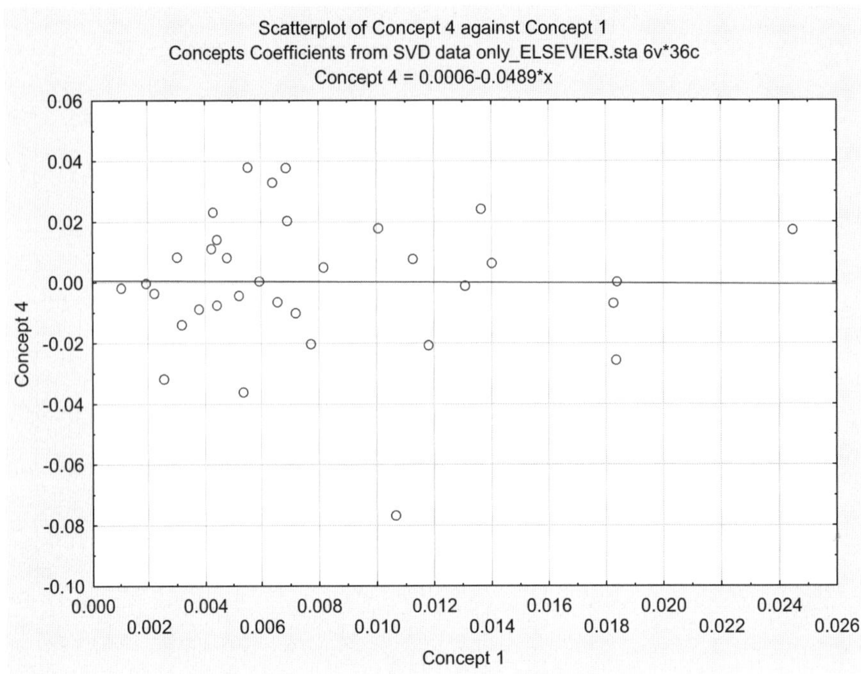

FIGURE Z.100
(*Continued*)

Okay, this is not too illuminating either, so let's look at another one.

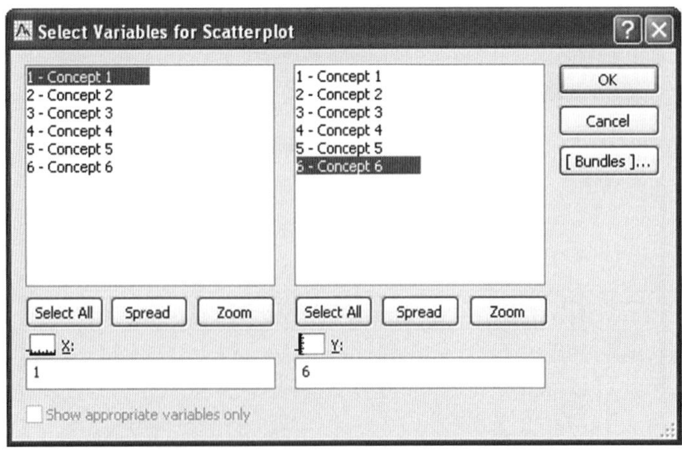

FIGURE Z.101
Concept 1 versus Concept 6.

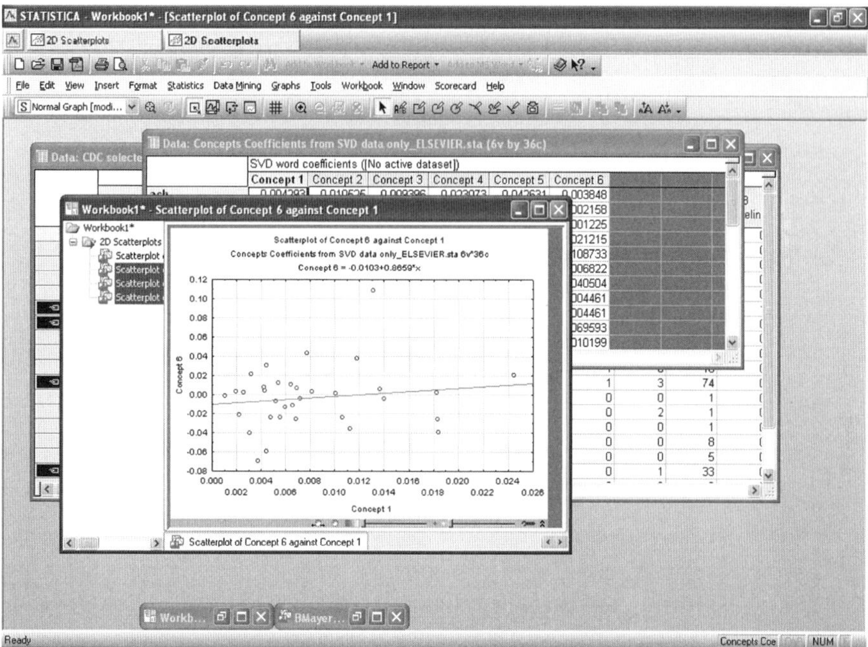

FIGURE Z.101
(*Continued*)

Now let's try Concept 2 versus some of the other higher-numbered concepts.

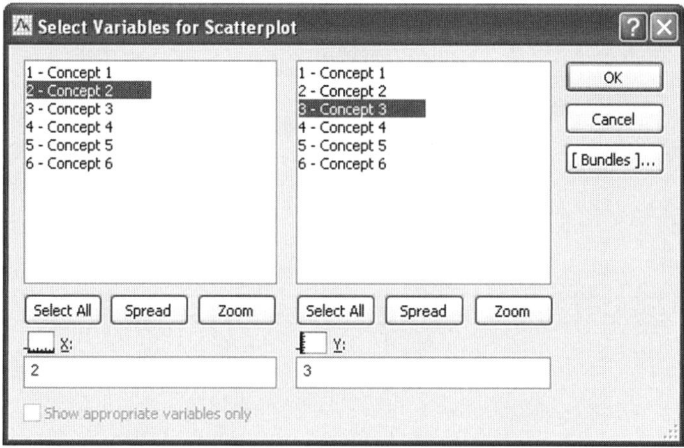

FIGURE Z.102
Concept 2 versus Concept 3.

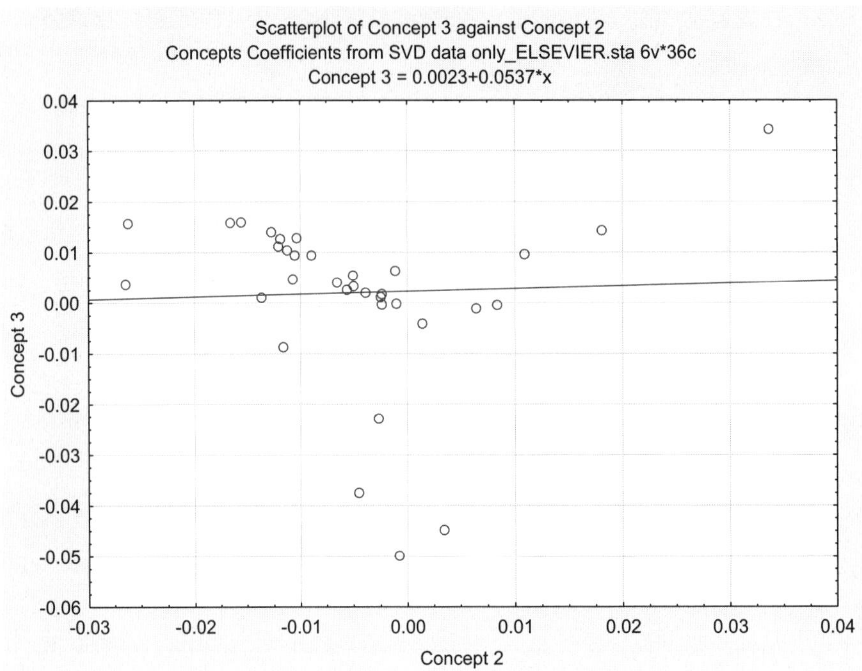

FIGURE Z.102
(*Continued*)

Again, this doesn't seem to show an interesting pattern.

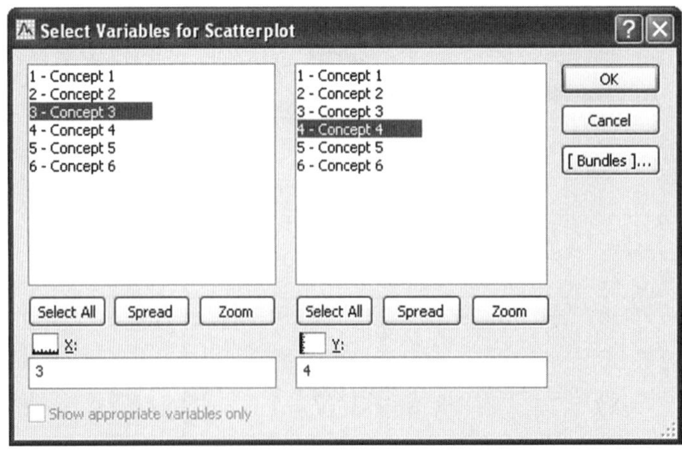

FIGURE Z.103
Trying Concept 3 versus Concept 4.

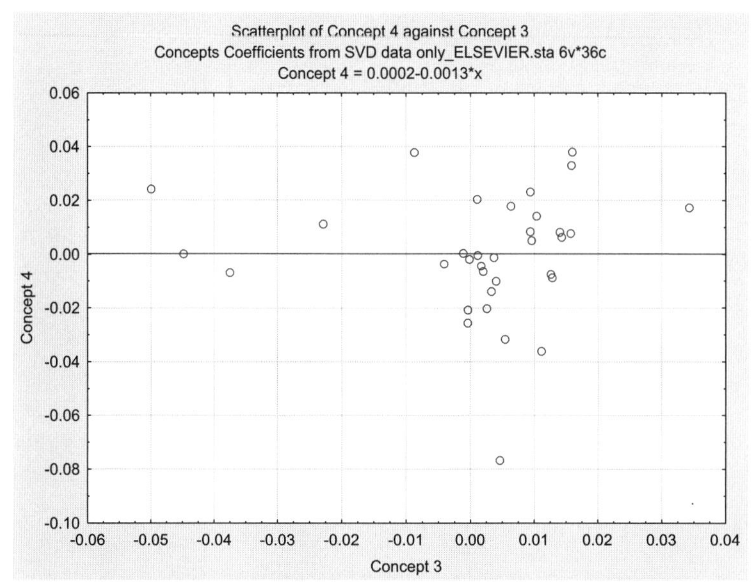

FIGURE Z.103
(*Continued*)

This gives a "good clustering" in the right center. So let's just take another look at the words to see if anything that we can attach meaning to appears.

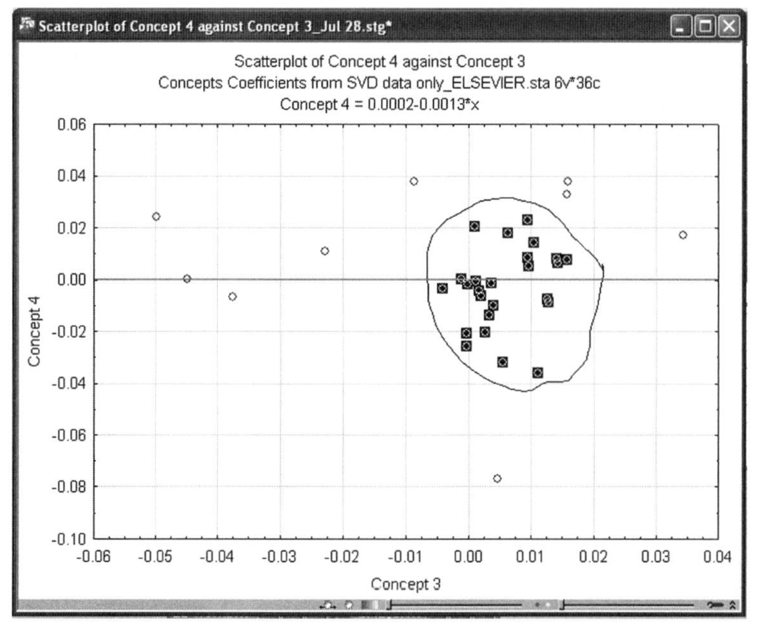

FIGURE Z.104
Concept 4 versus Concept 3.

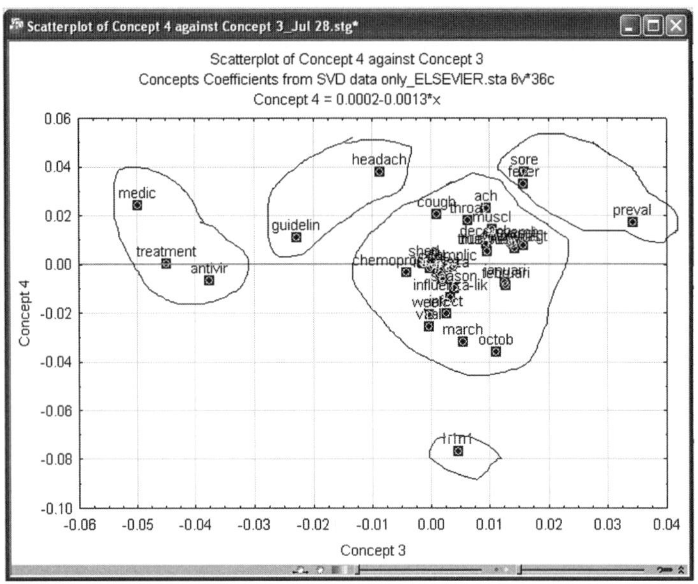

FIGURE Z.105

Concept 4 versus Concept 3 clusters.

We made four groups, although there are a couple of words in the "other groupings" that may belong with the heavy central grouping of words. Dragging the words apart, we see the following.

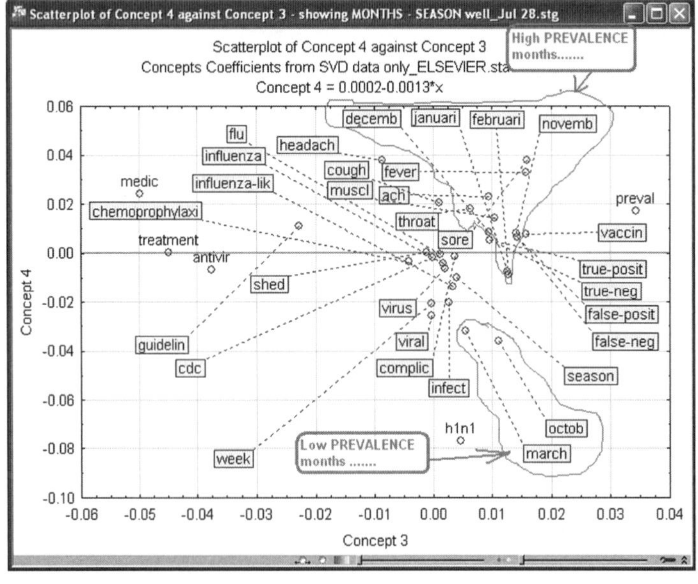

FIGURE Z.106

Concept 4 versus Concept 3 with word clusters pulled apart.

This is quite interesting. Note that October and March, the beginning and the end of the flu season, cluster together, but the high prevalence months, November, December, January, and February, cluster separately. We did not get this pattern in the Concept 1 versus Concept 2 plot, so looking at some of these additional plots can give us good information. If we look more closely, we can see that there are other clusters describing different aspects of the influenza story.

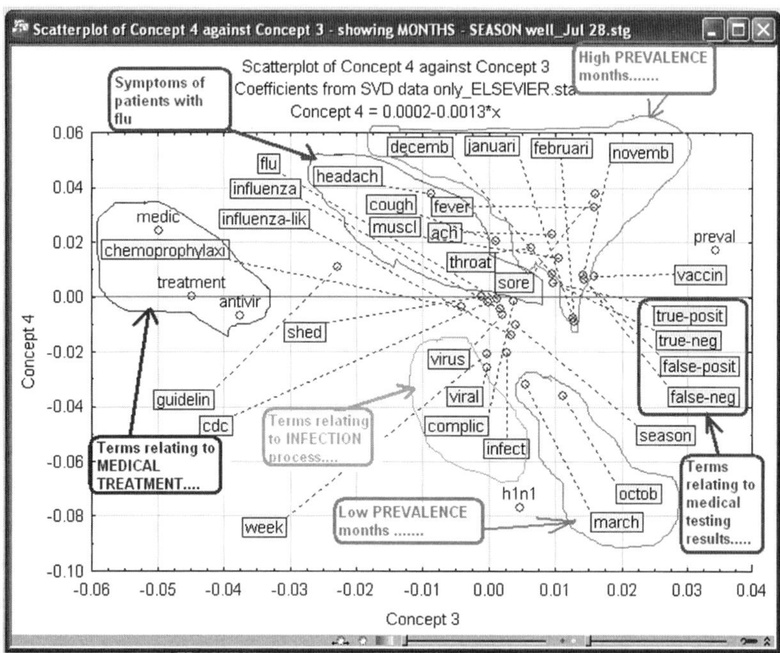

FIGURE Z.107

Meaningful clusters of words are found in the Concept 4 versus Concept 3 clustering.

In some ways, this Concept 3 versus Concept 4 scatterplot is more interesting and more descriptive in organizing the components of the influenza process than the Concept 1 versus Concept 2. However, neither pulled apart aspects that may give hints leading to a better understanding of the "Accuracy" of influenza diagnosis compared to different months or weeks, of the flu season. Thus, we will turn our attention to looking at numeric variable data collected at a medical clinic during the October to March influenza season in which we will apply "Predictive Analytic" data analysis to find which variables can predict an accurate diagnosis of flu.

FEATURE SELECTION

Now we will switch over from the CDC text mining to using a FLU data set to make some predictive models. We will not give the details of every step here, but if you need more information on steps of these processes, they can be found in the tutorials of the book *Handbook of Statistical Analysis and Data Mining Applications*, by Nisbet, Elder, and Miner.

Feature Selection is illustrated in the next series of figures (see Figures Z.108 through Z.110).

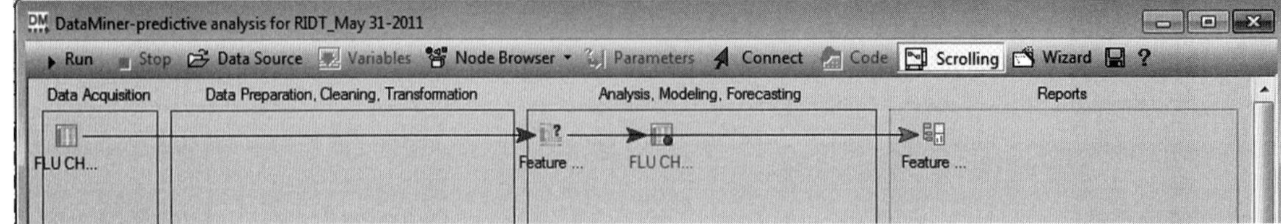

FIGURE Z.108
By using Feature Selection in the Data Miner Workspace to find which of the words and concepts are "most important," we may be able to reduce the number of predictor variables in a predictive model.

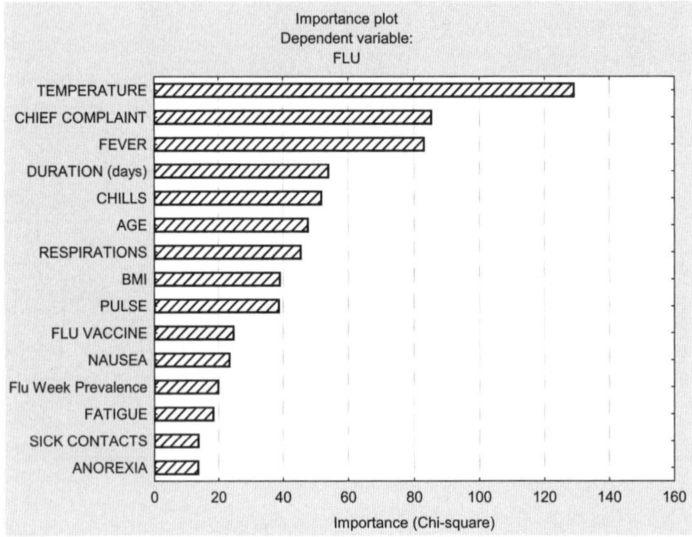

FIGURE Z.109
Importance plot from Feature Selection showing the relative importance of the numeric variables in the Flu data set.

	Best predictors for categorical dependent var: FLU (FLU CHART REVIEW+Fin	
	Chi-square	p-value
TEMPERATURE	129.1599	0.000000
CHIEF COMPLAINT	85.3439	0.000000
FEVER	83.0779	0.000000
DURATION (days)	53.9733	0.000003
CHILLS	51.7962	0.000000
AGE	47.6498	0.000767
RESPIRATIONS	45.4611	0.005136
BMI	39.1245	0.026499
PULSE	38.8374	0.002996
FLU VACCINE	24.7691	0.000377
NAUSEA	23.4711	0.000032
Flu Week Prevalence	19.9784	0.002794
FATIGUE	18.4240	0.000360
SICK CONTACTS	13.8254	0.003153
ANOREXIA	13.7196	0.003313

FIGURE Z.110
Best predictor list computed with the Feature Selection process.

MARSplines INTERACTIVE MODULE MODELING

The next series of figures will illustrate the use of MARSplines to make a predictive model (Figures Z.111 through Z.115).

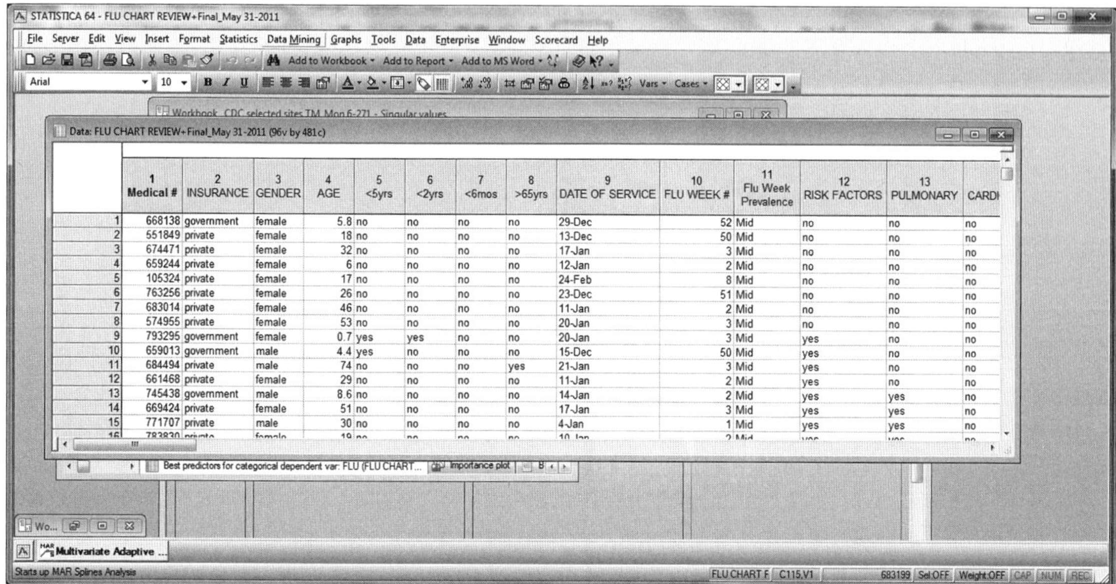

FIGURE Z.111

FLU data sheet that will be used for MARSplines and several other modeling algorithms.

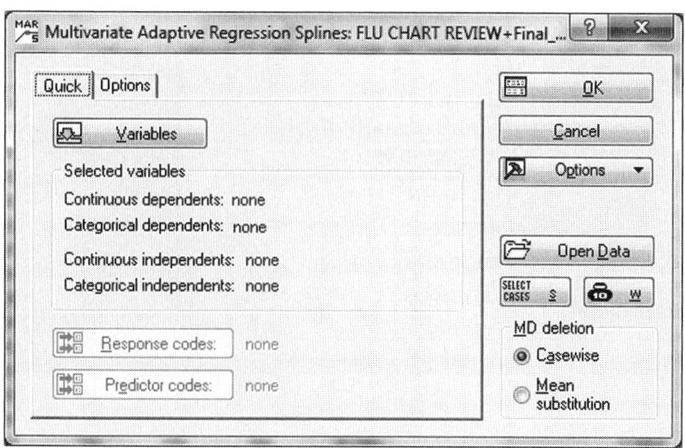

FIGURE Z.112

MARSplines dialog.

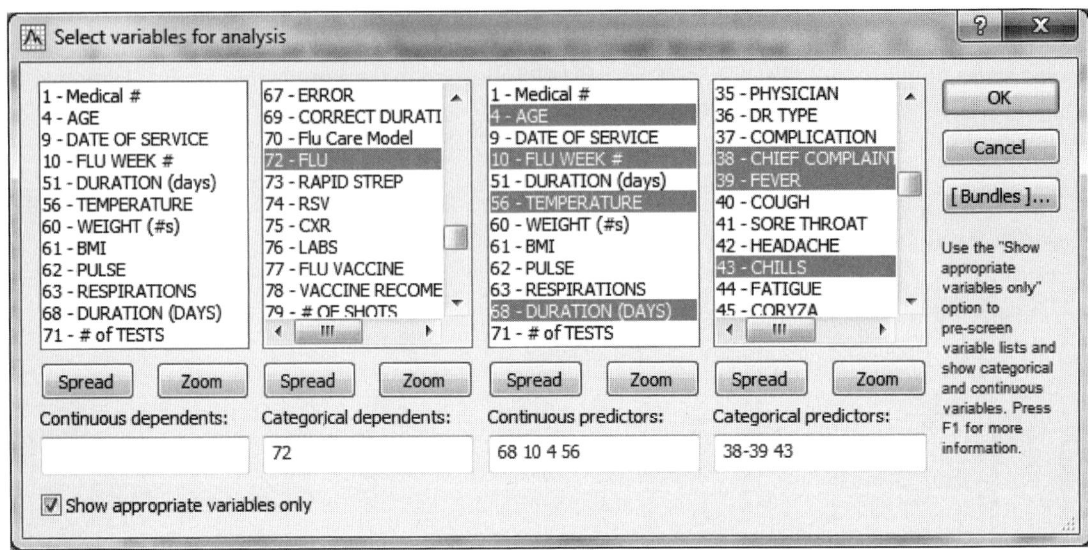

FIGURE Z.113
Variables selected.

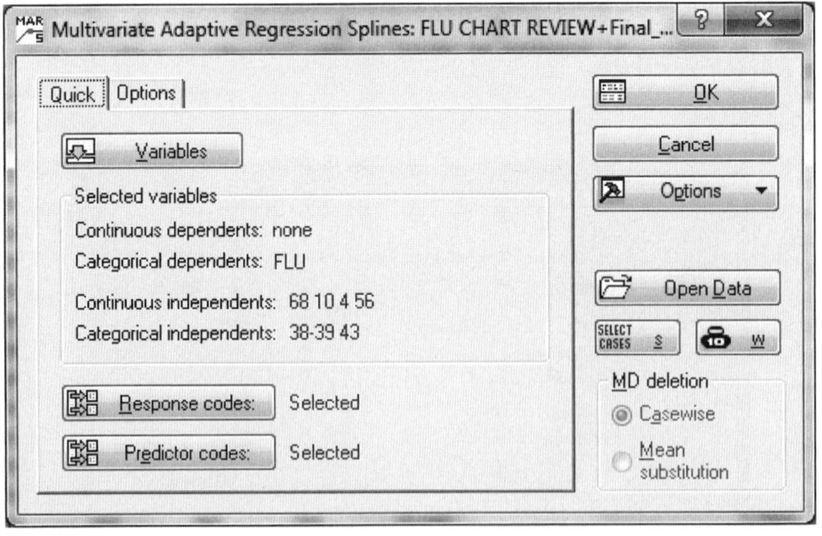

FIGURE Z.114
Variables selected are now showing in the dialog.

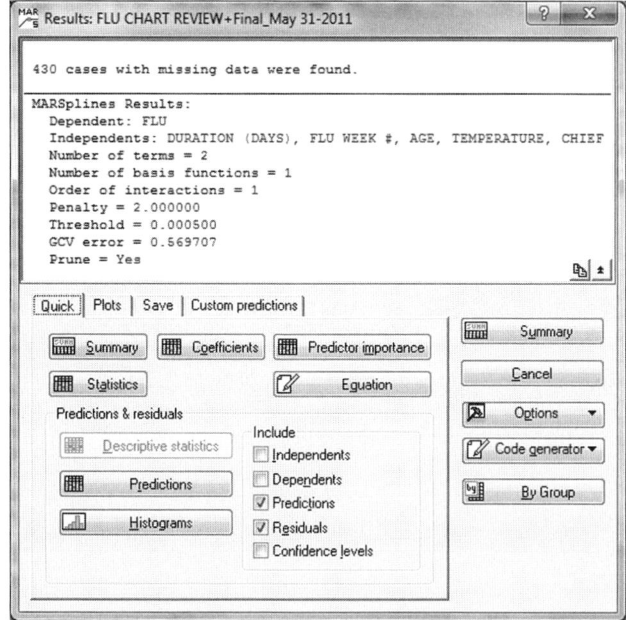

FIGURE Z.115
MARSplines results dialog.

See the Results Workbook for MARSplines. Since we don't see anything immediately interesting, we will stop MARSplines at this point and try another algorithm.

BOOSTED TREES

The next series of figures will walk you through using "Boosted Trees" with this dataset (see Figures Z.116 through Z.122).

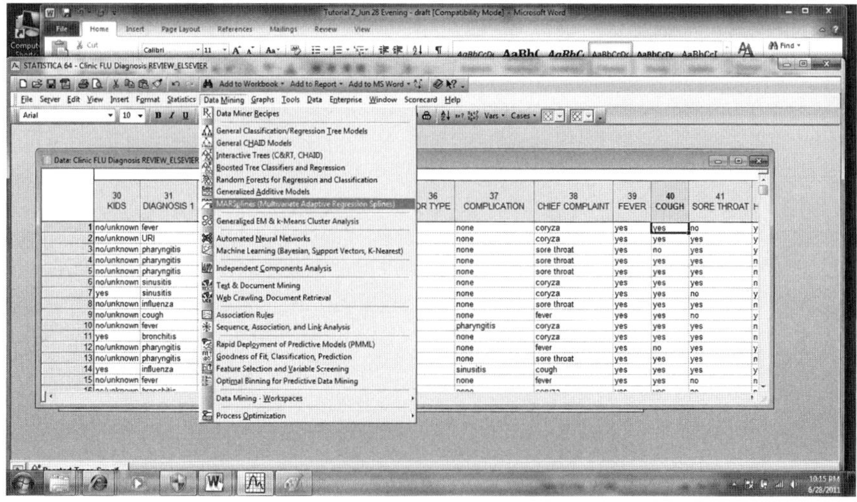

FIGURE Z.116
Selecting Boosted Trees algorithm.

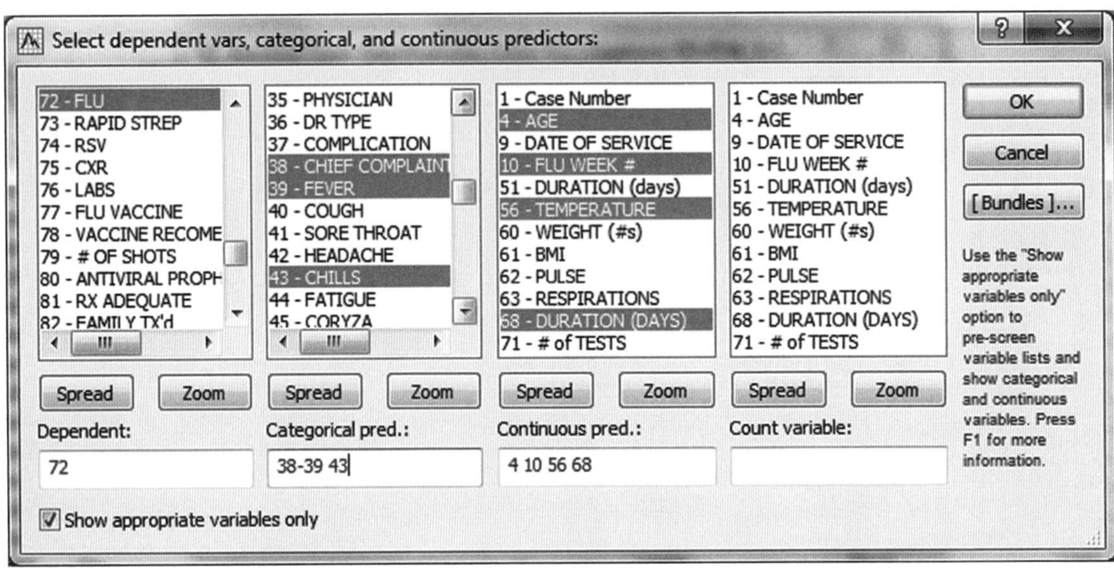

FIGURE Z.117
Variables selected.

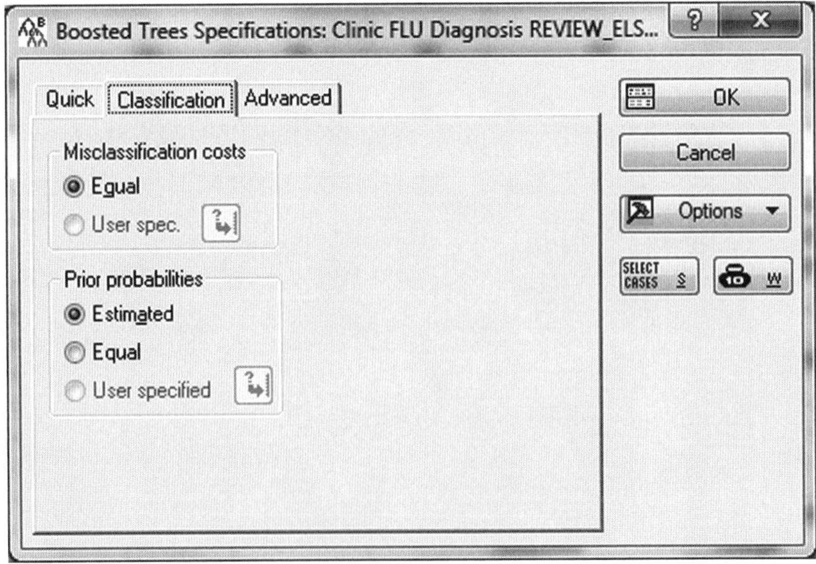

FIGURE Z.118
Boosted Trees Classification tab.

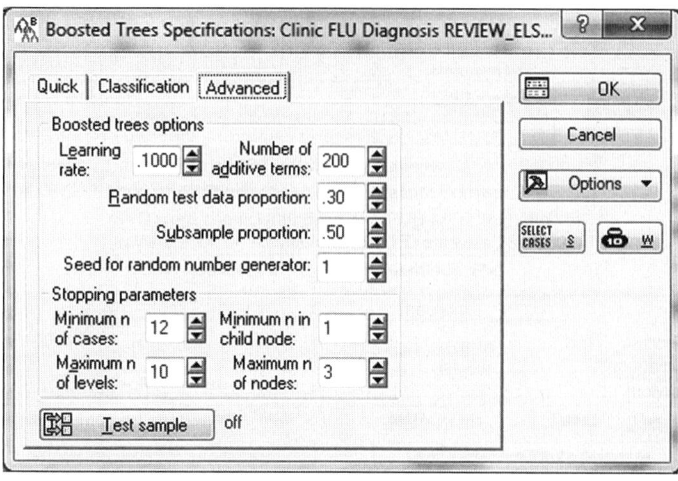

FIGURE Z.119
Boosted Trees advanced tab.

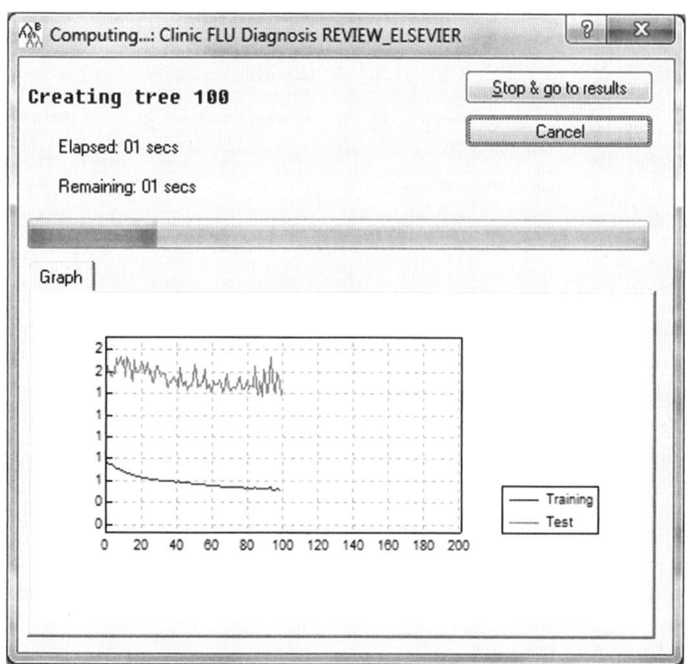

FIGURE Z.120
Boosted Trees computing.

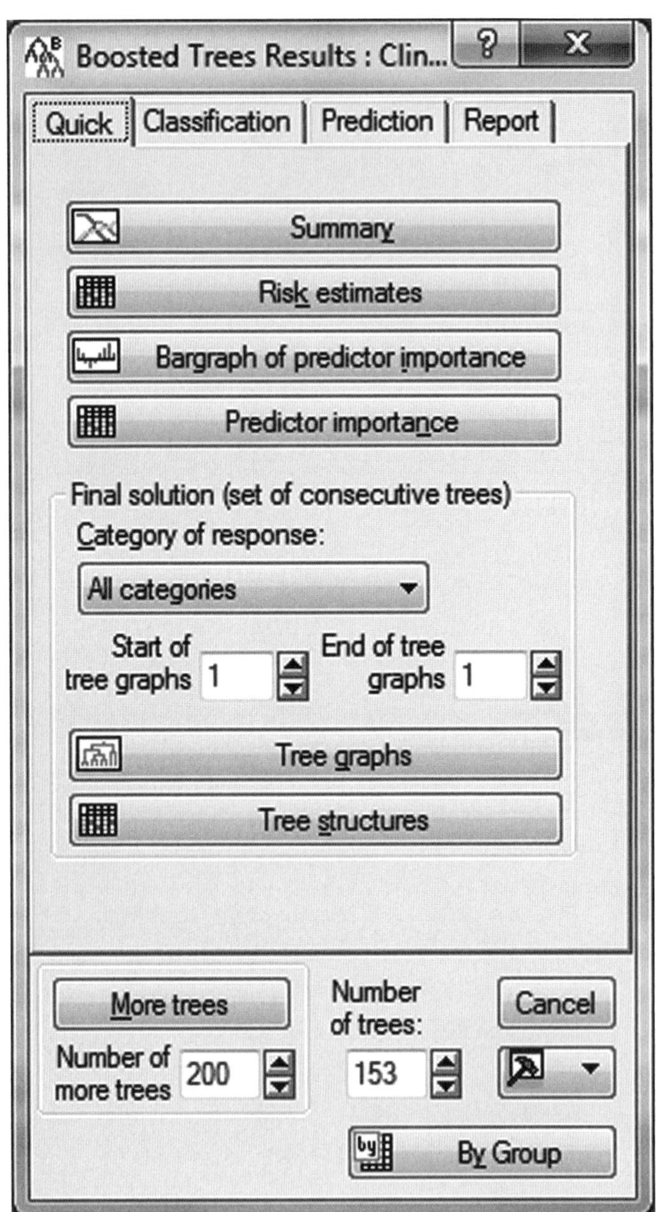

FIGURE Z.121
Boosted Trees Results dialog.

	Classification matrix (Clinic FLU Diagnosis REVIEW_ELSEVIER) Response: FLU Analysis sample;Number of trees: 153				
	Observed	Predicted positive	Predicted negative	Predicted unknown	Predicted not done
Number	positive	30	6	3	1
Column Percentage		52.63%	7.59%	5.26%	0.37%
Row Percentage		75.00%	15.00%	7.50%	2.50%
Total Percentage		6.48%	1.30%	0.65%	0.22%
Number	negative	7	27	5	9
Column Percentage		12.28%	34.18%	8.77%	3.33%
Row Percentage		14.58%	56.25%	10.42%	18.75%
Total Percentage		1.51%	5.83%	1.08%	1.94%
Number	unknown	1	1	18	3
Column Percentage		1.75%	1.27%	31.58%	1.11%
Row Percentage		4.35%	4.35%	78.26%	13.04%
Total Percentage		0.22%	0.22%	3.89%	0.65%
Number	not done	19	45	31	257
Column Percentage		33.33%	56.96%	54.39%	95.19%
Row Percentage		5.40%	12.78%	8.81%	73.01%
Total Percentage		4.10%	9.72%	6.70%	55.51%

FIGURE Z.122
Classification Matrix result from Boosted Trees computation.

Adding up the diagonal total percentages gives an accuracy overall of 71.71 percent, so this is not quite as good as the SVM.

NAÏVE BAYES MODELING

Next we'll look at a Naïve Bayes model (see Figures Z.123 through Z.127). Select Naïve Bayes, as shown in the next figure (see Figure Z.123).

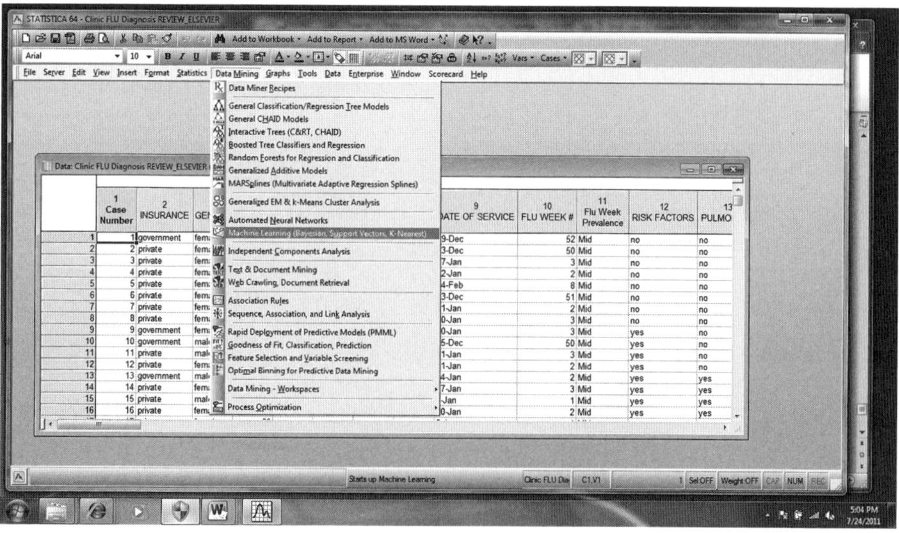

FIGURE Z.123
Selecting the Naïve Bayes algorithm from the Data Mining pull-down menu.

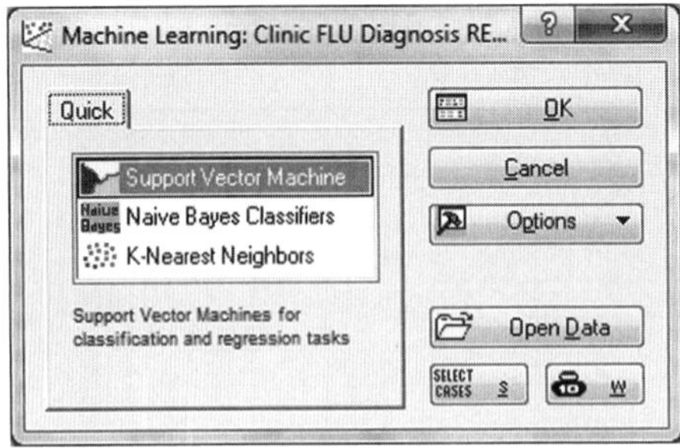

FIGURE Z.124
Naïve Bayes selected on this dialog.

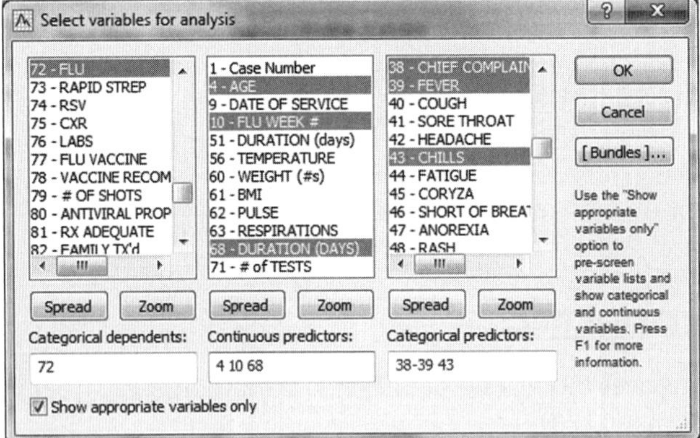

FIGURE Z.125
Variable selection.

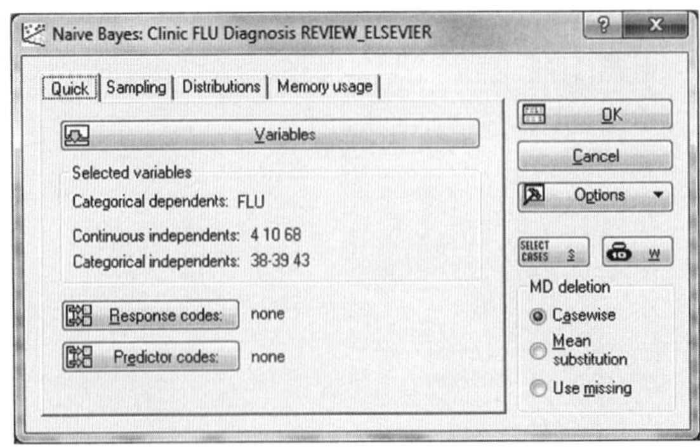

FIGURE Z.126
Naïve Bayes selections made.

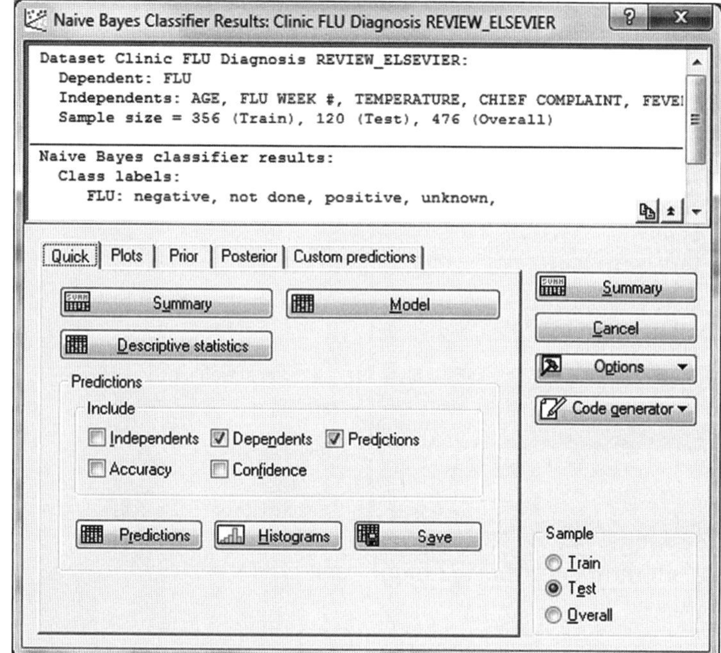

FIGURE Z.127
Naïve Bayes results dialog.

The results of this are not particularly interesting, but the Workbook of results is saved on the DVD bound with this book. Look for "Workbook for Naïve Bayes Results.stw."

K-NEAREST NEIGHBORS

We'll try one last algorithm: K-Nearest Neighbors, as illustrated in Figures Z.128 through Z.132.

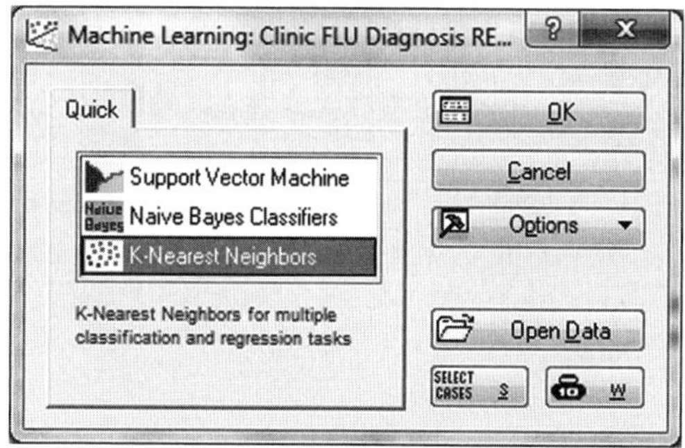

FIGURE Z.128
K-Nearest Neighbors selection dialog.

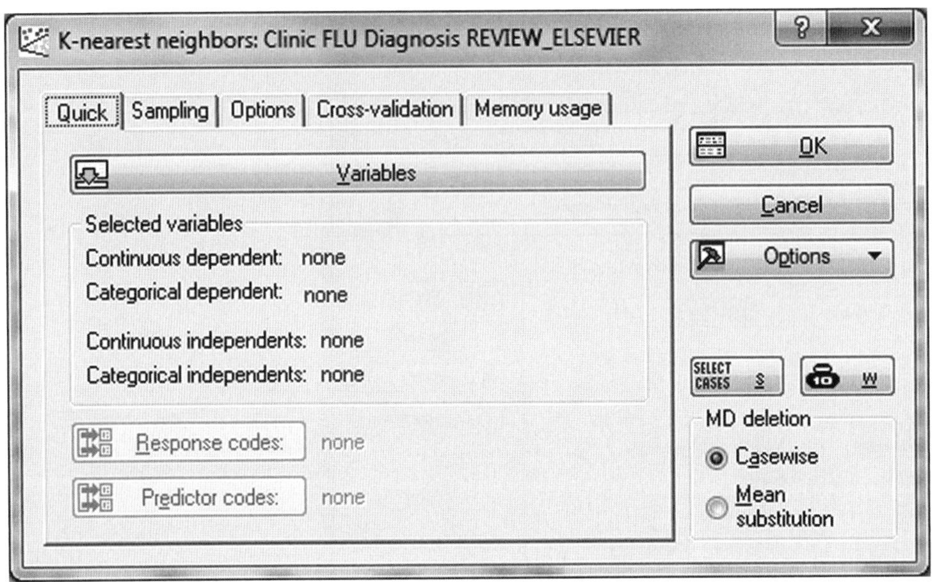

FIGURE Z.129
K-Nearest Neighbors dialog.

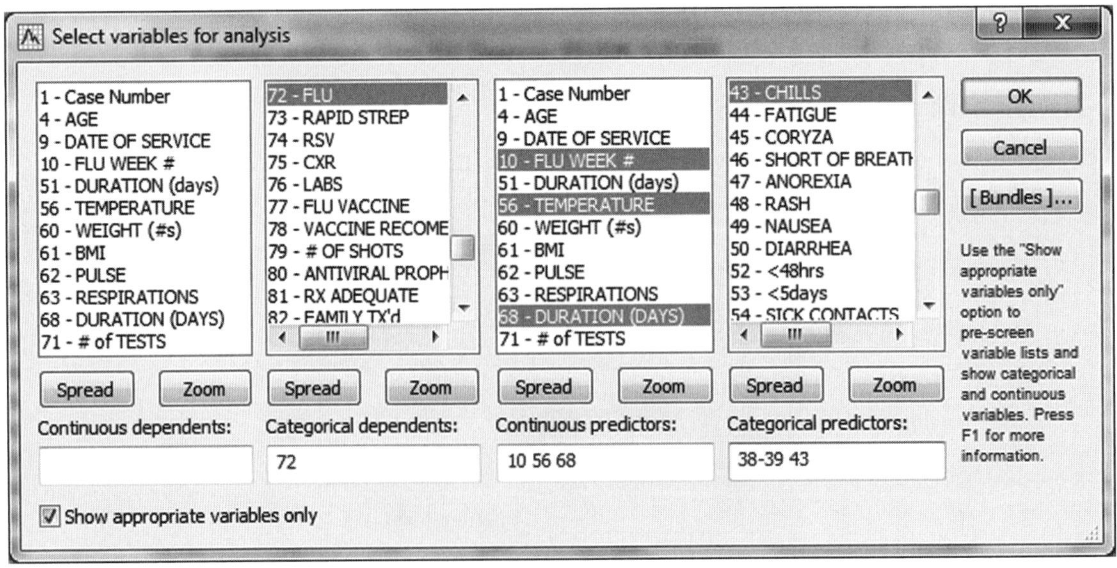

FIGURE Z.130
Variable selection.

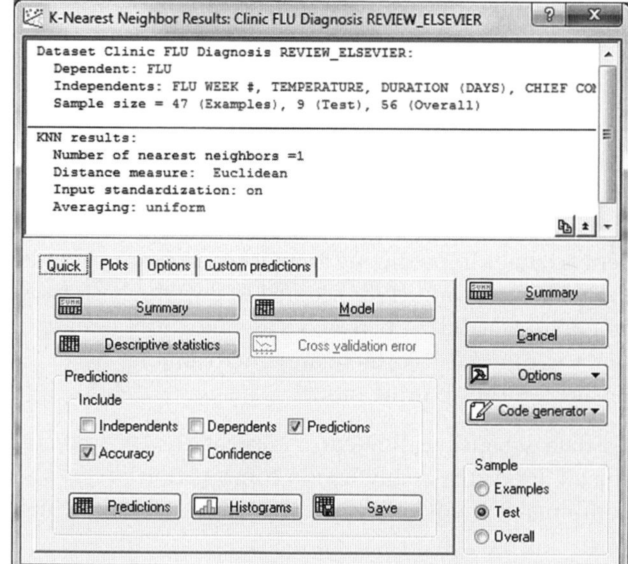

FIGURE Z.131
Results dialog for K-Nearest Neighbors.

This proved not to be a very good predictor of flu as per the specifications of this study, as observed by the following histogram.

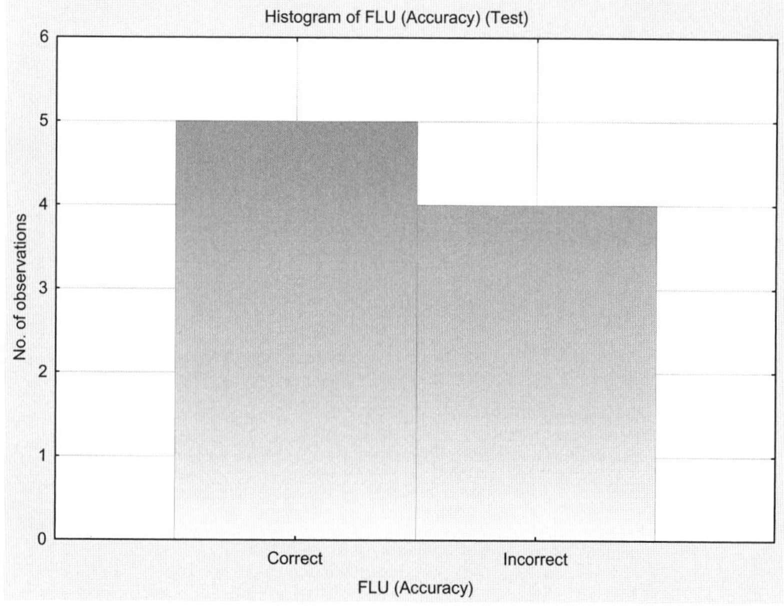

FIGURE Z.132
Histogram of predicting "real flu" in the diagnosis of a patient.

The Workbook of results is saved on the DVD bound with this book if you'd like to examine it further. In summary, three things came out of this study:

1. Interesting patterns/clusters of words from the CDC website.
2. A Predictive model using MARSplines that gave about a 71 percent accuracy in predicting a true case of flu from the predictor variables used in this study.
3. Using this same data set in a paper presented at the IHI Scientific Assembly, the authors found using Support Vector Machines that SVM modeling, using various combinations of Predictor Variables, that no combination got over about 78 percent accuracy in predicting accurate dx of flu in patients. Of the predictors used, temperature, flu prevalence week, and chills seemed to give the best, most consistent SVM model, with about 78 percent accuracy and 74 percent cross-validation accuracy. Thus, about 25 percent of the variation is not accounted for, so there are other variables or interactions of variables that are needed to make a more accurate model.

We suggest that you play around with this data set, using different combinations of predictor variables, to see what results you can obtain. After all, Predictive Analytics is a combination of "art and science," and trying these variables in different combinations will help you learn more about modeling.

Advanced Topics

CHAPTER 7

Text Classification and Categorization

CONTENTS

PREAMBLE

Text classification is the process of assigning text documents into two or more categories. The most common form is binary classification, or assigning one of two categories to all documents in the corpus. Text classification is often the first step in the selection of a set of documents to submit to further processing, or it can be the only step in text processing (e.g., spam filtering). The goal in text classification is not to extract information from text other than the category of the document. The basic approach to text classification is to derive a set of features to describe a document then apply an algorithm designed to process and use these features to select the appropriate category for a specific document.

INTRODUCTION

Text classification and categorization methods are used to automatically assign categories or keywords to text documents, the source of which depends on the problem and the domain. For example, one

Practical Text Mining and Statistical Analysis for Non-structured Text Data Applications. DOI: 10.1016/B978-0-12-386979-1.00035-9

popular practical application of text classification is email spam filtering. A spam filter examines an incoming email and determines whether to mark it as spam or let the message pass untouched into the inbox. This is a binary classification task, having two possible outcomes, or class labels: "spam" and "not spam." Another such task is alert notification from call transcripts or email tip lines. Text classification is not limited to binary class labels; algorithms can be used to assign multiple categorical labels or hierarchies of labels.

More formally, text classification is an analytical process that takes any text document as input and assigns it a classification from a predetermined set of class labels. Text classification algorithms typically employ a statistical model to assign labels, but rule-based approaches can also be used. These algorithms look at characteristics, or features, of a document, including its title, length, words used, file name, or URL. For spam filtering, example useful features include the presence of a dollar sign ($), or the number of exclamation points used (!!!!) in an email, since those are simple but strong indicators of spam. Multiple such features are then combined to make a decision regarding the label of the text.

Text features can be even more powerful predictors when used in combination with numeric features, as shown in Figure 7.1, which uses an example in the insurance industry. We can see that when just numeric variables are used in predictive modeling, we get a certain gain or "lift," as illustrated in a lift chart. When

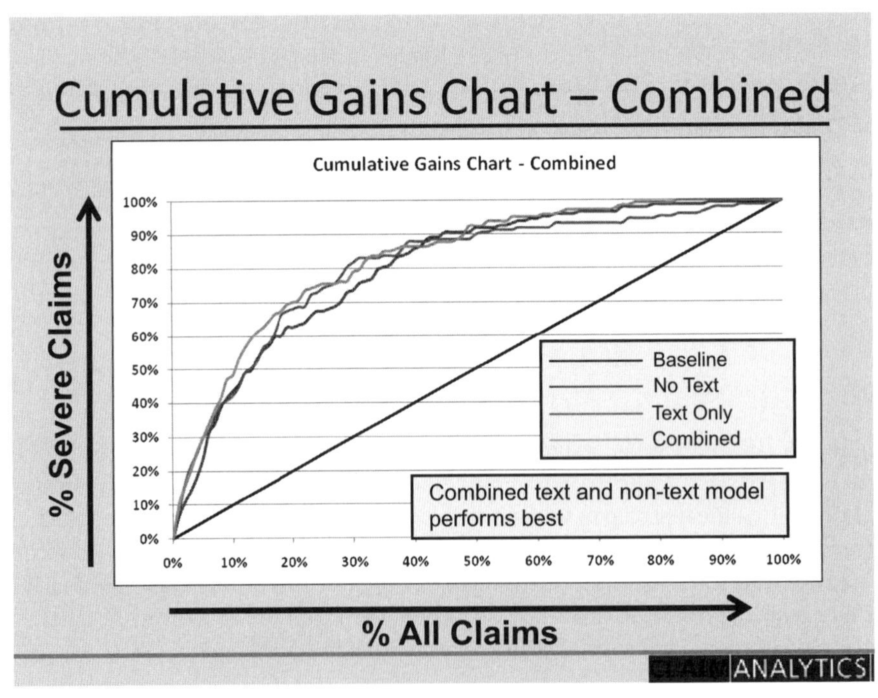

FIGURE 7.1

Expressing cumulative gains (showing effectiveness of ranking) is best when both text and numeric variables are used in predictive analytics modeling, as illustrated by the green top curve. *Source: From a case study prepared by Jonathan Polon of Claim Analytics (www.claimanalytics.com).*

just text variables are used, we also get a gain or a lift. But when both numeric and text variables are combined, we get the greatest improvement. Even such slight accuracy can result in serious cost savings for the company and its customers.

In this chapter, we describe the process of feature creation, examine the types of algorithms used for text classification, cover the evaluation of text classification, and conclude with a discussion of text classification in rich topic hierarchies.

DEFINING A CLASSIFICATION PROBLEM

Text classification is often used for automating or streamlining a business decision requiring processing or understanding text. Because manual text processing can be a massive task in terms of time and effort, it is often an ideal candidate for improving a business process. A text classification system can be used in place of human decision makers in order to make the overall process more efficient and to remove tedious work from human examiners by classifying the easy problems, allowing analysts to focus their effort on the most challenging tasks.

Given the growing wave of available textual data, having enough data to consider a text classification task is not usually a problem. However, having the right data to address your business problem can be more difficult. For a transition to text classification to be successful, there must be sufficient text data directly related to the core business problem, and the outcomes or labels used by the classification must be in line with the success criteria as well. When large amounts of data are available, it can be tempting to use the data whether it is tied to the business problem or not. Yet, it is best to start from a problem and then consider the suitability of the available data.

Throughout the remainder of this chapter, we will use an example from a pilot project engagement with the Social Security Administration (SSA) to apply text classification techniques to identify disability claims that are likely to be approved based mostly on the free text included in the claims form. This project had four primary characteristics that made it an ideal opportunity for applying text classification:

- **The problem was already massive and growing** – The SSA processes over 2.5 million claims annually for disability insurance under the SSDI and SSI programs. Each of these disability claims must be reviewed and then either approved for funding or denied.
- **There is room for improvement** – The time frame for an approval of disability in some cases exceeded over two years! In cases of severe illness, this approval often arrived long after it could have provided any help.
- **There is low-hanging fruit** – Through both law and leadership directive, the SSA had identified a number of diseases, such as amyotrophic lateral sclerosis (ALS or Lou Gehrig's disease), aggressive cancers, and end-stage renal failure, as those that should receive fast approvals. On the other end of the spectrum, there were many frivolous claims for disability that should be denied quickly.
- **The text tokens are highly informative** – The allegation field that holds text is very rich with significant and precise medical terms and conditions. This is much easier to work with than text "in the wild," such as blogs and discussions. Also, staffers entered the text from interviews, providing a baseline of useful filtering and keeping the text on topic.

The first challenge for transitioning to a text classification problem is finding the right text documents. To solve the business problem, the documents must be applicable to the task, and the labels must be tied to the business success. The SSA provided us with the free-text field containing the claims of each applicant, including a list of the diseases or symptoms the applicant had. These claims had been through an initial screening by a claims clerk, which reduced its messiness. Most importantly, the text was directly related to one key part of the business problem: making a decision about the level of health. The second part—deciding about the applicant's level of income (since one has to be both poor and sick to receive this type of taxpayer assistance)—could be handled well using the structured information in the application.

Another challenge for transitioning to a text classification solution is defining the class labels to be used and finding appropriate training data for the classifier. The labels should be directly tied to finding a successful business outcome. For a call center analyzing the transcripts of calls, the evaluation might be based on customer satisfaction. For a marketing department, the labels might be about whether a blog post represented a positive or negative reaction to your product. Historical data are an ideal source of labeled examples for training the classifier. The data are usually in the same format and have already been given a decision by the existing process. (This was the case with the SSA data.) In the absence of historical data, training data are usually obtained by tracking existing cases or labeling a number of documents manually. Once the training documents have been identified and their class labels assigned, the next step in the text classification process is creating features to input into the classification algorithms.

FEATURE CREATION

Text classification algorithms determine the final class label to assign based on the presence or absence of particular features of the text. For simple tasks, the classification may be as straightforward as a rule saying that if a certain keyword appears in the text, then you should assign a particular label. But few text classification problems are that easy!

Text Preprocessing

In the great majority of text mining tasks, individual words (often called *terms* or *tokens*) are the primary features used as input to the classification algorithms. Creating features from text requires applying standard text preprocessing actions, such as tokenization, stopping, and stemming. These put each of the words in a standardized form before classification. *Tokenization* is the process of dividing up raw text into discrete words called tokens. *Stopping* then removes extremely common words (or stopwords) that have limited predictive value for classification. *Stemming* normalizes word forms; for instance, *connects*, *connection*, and *connected* would all be reduced to the single feature *connect*. Finally, *case normalization* converts each token completely into either upper or lower case. Each of these steps is described in more detail in Chapter 3.

Spelling normalization is another key step that is often required. Misspellings were a large problem for processing SSA disability claims. Despite being entered by a trained clerk, there were large numbers of them. For instance, as shown in Figure 3.3 in Chapter 3, the phrase "learning disability" was spelled in over 50 different ways. The clerks used a drop-down menu to standardize data entry, but most found it overwhelmingly faster to just type it in. Such workarounds by those using a system day in and day out are very common and can frustrate data quality enforcement steps in a workflow. For instance, the authors once found an astonishingly large contingent of customers for a client that the data thought was over 90 years old (and precisely the same age). It turns out that many of the data entry staff were avoiding asking about customer birth dates—a required field—so they just entered all 1's.

Choosing Which Text to Use

Text documents that we wish to classify contain unstructured text, but they also may contain some structured information such as titles, abstracts, headers, or keyword metadata. Just as a reader will determine the content of documents and sections from the titles and headers, a text classifier can also use that information. Often, academic or technical literature can be classified based on the title and abstract alone. For example, the Association for Computing Machinery (ACM), the technical society for computer scientists, provides keywords with their articles that are designed to be very strong indicators for classification. These keywords are arranged in a hierarchy and could also be used as a set of hierarchical class labels. This scenario is discussed later in the chapter.

One of the primary strengths of text classification algorithms is their ability to use the entire text of documents to make a classification. While title and header text can provide informative features, the amount of text in those parts of the document is limited. Body text provides both strong and weak features and contains enough text to offset insufficient header text. Ideally, feature vectors for each document come from both sources. Then it is helpful to *prepend* a location indicator before each token to differentiate words appearing in both sources. For example, if the words *text mining* appeared in both the title and the body of a document, it is often useful to create two features for each token: *title:text, title:mining, text,* and *mining*. It is standard to leave tokens from the body text unannotated.

Other Feature Types

Nontext features describing the document can be useful for text classification. The length of the document is often valuable for determining its topic. For example, in news articles, different sections have different goals and word limits. Sports articles usually are either short recaps of earlier games or are longer feature articles, whereas articles updating current events provide more in-depth information. Other useful nonword features include the number of headers or sections, the length of the title, or the number of keywords.

Classification of web pages and other documents gathered from the Internet can employ a set of features unique to Internet sources. Every Internet document is uniquely addressed by its uniform resource locator (URL). URLs contain the protocol (e.g., http:// or ftp://), domain (e.g., www.textmininglab.com), and the top-level domain (e.g., .com, .edu, .gov), as well as additional information on where the file is stored on the web server. Each of these pieces of information can be easily converted into features, and the text of any links appearing in the document can be converted into useful features in a fashion similar to that for title words.

The essential stage of text mining is converting text into a numerical representation. Then powerful analytical techniques from data mining and statistics, and so forth, can make the connection between input features and class label (output). Much of the power and flexibility of text classification comes from the ability to creatively incorporate any available information into the classification process.

Vector Creation

The majority of text classification algorithms assume that each document has been converted to a vector of features. (That process is discussed in detail in Chapter 3.) There is then one feature vector for each document. If the goal is training a text classification model, then one must label each vector with the class of its document for a sufficient set of training cases and then build a model.

TEXT CLASSIFICATION ALGORITHMS

The reason for doing the hard working of converting text into a feature-based representation is to make it possible to apply the large number of classification algorithms designed for traditional structured data, which come out of the data mining and predictive analytics fields. They take as input numerical feature vectors and produce classifications. Any traditional classification algorithm can be used for text, but the extreme high dimensionality of text vectors clearly distinguishes between algorithms that are efficient and those that are not.

Two of the most popular algorithms for text classification are a Naive Bayes classifier and a logistic regression classifier (sometimes referred to as Maximum Entropy classifier or MaxEnt for short). These two algorithms are both efficient for high-dimensional data and have proven to be among the most accurate for text classification (see Manning et al., 2008, for more details about these algorithms).

To achieve the highest performance, classifiers should be *trained* for a specific task. Training is required to set the numerical parameters (weights and thresholds) of the model. These parameters reflect the contribution each feature makes to the final classification decision. Both Naive Bayes and logistic regression classifiers are supervised classifiers and require each case in the training data to be labeled with the correct class label for its document. These training cases are then used to estimate the parameters of the model. Methods for training Naive Bayes and logistic regression classifiers are provided in most major data mining and text mining packages.

Once a classifier has been trained, it can be used to assign categories to new, previously unseen text documents. The new text data must be run through the same feature creation process that was used for the training data. This will ensure that the set of features available during training are also available on the new documents. For text classification algorithms to be accurate out-of-sample, the training documents must be qualitatively similar to the new, unseen documents. Features that are highly correlated with the class label in the training data must also be highly correlated with the correct label in the new, unseen data. If not, the test classifier will not produce meaningful results.

Since the same feature calculation process must be done both on the training and evaluation data, it can be tempting to make the subtle mistake of doing it first before the data are divided into the two sets. (This is a feature creation version of the very common error of "peeking into the future," which is detailed in Chapter 20 of *Top Ten Data Mining Mistakes*, by Nisbet et al., 2009.) When mistakenly using *all* of the data to create features, the model will look better than it really is when it is tested on the evaluation data—a very dangerous situation, since this overestimate of accuracy is often not discovered until the model disappoints in the business setting after implementation. Instead, analysts must first split off the subset of evaluation data and then calculate training features (frequencies, word lists, principle components, etc.) based solely on the training data. If the modeling process "sees" the extremes and oddities of the evaluation data during training, and those data affect the features created and selected, then the model is less likely to be "astonished" by the evaluation data because those data are not fully new to the model. Unfortunately, the real world will be supplying data that are completely new in their details, and this will tax the unprepared model more severely than expected.

COMBINING EVIDENCE

One of the challenges of text classification is determining how much weight features should have in the final classification decision. This is particularly important with textual data, where there are a vast number of keywords and where many semantically similar words may appear in the documents. Should each word count equally or additively toward the decision, or should they be combined in some nonlinear manner? If semantically similar words—such as synonyms or alternate disease labels—are joined together into fewer keywords, then they will have more precise outcome information than if they had been left apart. If the semantic meaning of the words is tied to the class label, then dimensionality reduction techniques (see Chapters 10 and 11) can be used to extract features based on groups of words.

For the SSA, understanding both semantic similarity and disease severity was critical for success. Because they had to estimate the added dimension of severity for an illness allegation, word clustering algorithms for finding semantically similar words were insufficient. If all features are treated additively, when many features are "on" (present in a document), the likelihood that the document will be given a positive class label ("approval") increases, which is not exactly right. For the SSA, this presented a problem because many applications contain several minor health condition claims, each potentially contributing to the overall severity score. Textbook methods for combining evidence often don't work either, since they tend toward averaging severity. But minor complaints should not detract from more serious claims.

To address this problem, we carefully defined the desired properties that a function should have as it scores the severity of an application. Every part of an equation used for ranking cases has business implications, and the custom design of a function to best reflect the tradeoffs one is facing is very valuable to the final solution.[1] Our design review, confirmed by commonsense testing of several cases, led to the development of a joint probability method to combine weakly related complaints but consider strong complaints independently. Further, we first invented a Bayesian initialization strategy we call "nonzero initialization" to more robustly handle rare keywords.

The first step in the process was discarding concepts (terms or phrases) that appeared fewer than five times. This greatly enhanced training and evaluation speed, since typically about half of the terms found in mining a text document are unique—that is, they occur only once.

The second step is to calculate Bayesian priors to protect against paying too much attention to measured probabilities having only slight evidence. The technique was to start with a small initial count for each token when measuring how often a term is associated with the target (e.g., two rather than zero) having a density equal to the overall training probability of the target (e.g., 27 percent for one of the SSA approval rates). That is, with a nonzero initial weight of 2 and a prior of 27 percent, each term

[1] The business rewards and punishments should be reflected as closely as possible in the criterion being optimized. For instance, we know of a bank that incentivized its management to grow its credit business. Management did as rewarded and developed programs that brought in a flood of new customers. The problem was, the customers were of low quality, and they hurt the bank's profitability. When external analysts realized and reported this, the bank's shareholders lost over a billion dollars in the stock value in a very short time. Likewise, don't just use squared error as a fitting criterion or percent correct as an accuracy criterion without thinking of the solutions that it will drive the system toward.

would start out with a prior count of 2 * .27 = .54 positive, and 2 * (1 − .27) = 1.46 negative cases. (Note that the counts don't need to be integers.)

The first useful property of this method is to provide a very sensible answer (the prior) for tokens that have no evidence associated with them whatsoever. It is often the case that new tokens (words) appear in documents being evaluated after training. (In fact, if no new words appear, you may have looked ahead during training to see the evaluation data!) Implementations shouldn't blow up when calculating probabilities where unknown keywords may play a part, and nonzero initialization helps with this task.

On the other end of the spectrum, where there is massive evidence associated with a token, the small initial prior has very little effect. In between, where some evidence but not a lot exists, the technique has the property that a rare term's probability is discounted (or "shrunk" toward the prior) more than a term with more substantial evidence. For example, if a concept occurred five times with four positive cases, its raw measured probability would be 80 percent. But by adding the prior (nonzero initial count), we instead get (4 + 2 * .27)/(5 + 2) = 64.9 percent. This lower value is likely to be more realistic. It is known as "shrinkage" because the value to be used has been drawn away from 80 percent toward the prior (no information) value of 27 percent. Note that if a concept had 20 samples with the same measured density of 80 percent, its adjusted likelihood would be (20 * .8 + 2 * .27)/(20 + 2) = 75.2 percent, which is much closer to the measured value. With even more samples, the effect of the prior would fade almost entirely.

This simple but powerful idea appears often in science by different names. In mathematics, the adjustment is known as the "James-Stein Estimator"; in biostatistics, it is called "shrinkage"; in engineering, it is known as "ridge regression"; and, in neural networks, it is called "optimal weight decay." We have found it to be very useful even in structured data mining whenever the number of states of an input can be large. The value of the mass to be used initially for the prior weighted count is the one parameter to select. It can be determined through cross-validation experiments using subsets of the training data, and it is often much smaller than one might expect. For example, we found the value of 2 to work well for the SSA challenge.

After removing very rare tokens and initializing their counts through the Bayesian-type method described, we calculated their simple individual probability of being associated with the positive outcome. Then we developed a way to estimate the joint probability intelligently for combinations of terms. Most equations assume that each term is an independent measure of the same event. Thus, a low value could cancel out a high one. This is not true in this application, where a toothache (likely minor) should not distract from a cancer (major). Hence, our custom formula for joint probability became:

a. If (no data), then use *prior.*
b. Else if (max (probability) < 0.5), then use that *max.*

Else:

c. Ignore concepts with probability < 0.5.
d. Combine the remaining ones with a log-likelihood formula and use the resulting *joint probability*.

In case (b), the complaints are relatively minor (that is, have low probability of being associated with a positive value of the target), so the largest of them is used to represent the applicant. In case (c), the serious complaints add some evidence, logarithmically, to each other, so the joint probability asymptotically approaches 1. In no case does it go outside the bounds of (0, 1), and when no information is present (case (a)), it behaves as one would wish, using the overall prior probability.

The results of using this custom formula, which was designed to match the constraints of the specific SSA application, were very encouraging. This probability, which was extracted from the claims text field, was the single most useful variable, and it could be a viable model entirely on its own.

EVALUATING TEXT CLASSIFIERS

Evaluating the performance of text classifiers on new data can be performed with accuracy or precision and recall. Accuracy scores how often a document is classified correctly. This accuracy should be broken down by error type. A *false positive* error occurs when the algorithm incorrectly assigns a positive label to a document (for instance, labeling it as likely being "interesting" to an analyst doing a specialized search, when it turns out not to be). A *false negative* error occurs when the algorithm fails to assign it a positive label when it should have. For example, in spam filtering, a false positive occurs when the filter marks a message as spam when it is not, and a false negative occurs when a spam message successfully reaches the in-box. The costs of these two types of errors are rarely the same. For instance, in credit scoring it can take five to seven good customers (who pay their bills) to make up for one bad customer (who defaults on the loan). Thus, breaking out the two types of errors and penalizing them by their cumulative cost is much more superior to using overall accuracy, which implicitly assigns the equal costs to the two kinds of errors.

For the SSA, the costs of each type of error are affected by the time required to remedy the error. For a false negative error (a case that should have been approved but wasn't), the appeal process is lengthy, sometimes taking up to two years for the initial decision to be overturned, leading to undue burden on the applicant. However, false positive errors lead to large cost overruns on the part of the agency. The best decision point balances each of these different considerations. The error rate for the SSA as the decision threshold varies is shown in Figure 7.2.

Using *precision* and *recall* focuses the modeling process on the relative ranking of the documents for a given label. Instead of making a hard decision about the label for a given document, a text classifier can be used to compute a score (usually a probability) of a label being assigned to a document. The documents are then ranked by that score. *Precision* measures the proportion of documents that have been labeled correctly as you move down the ranked list. Or precision is reported as the actual number correct when measured at a particular document count (e.g., at 100 documents) or at a particular score threshold (e.g., precision at 0.5). *Recall* measures the proportion of documents with a given label that are found at the chosen threshold out of all the documents having that label. (Again, it too can be expressed as a count for a preagreed cutoff.) Precision and recall are the main alternative ways to measure false positive and false negative errors.

HIERARCHICAL TEXT CLASSIFICATION

In the previous sections of this chapter, we have focused on class labels that are "flat" and have no hierarchical structure. However, in many cases text can be naturally categorized into hierarchical topic trees, where branches in the tree indicate finer-grained categorical distinctions. Using a topic hierarchy for text classification can lead to improved performance by simplifying the decision at each step in the tree. To

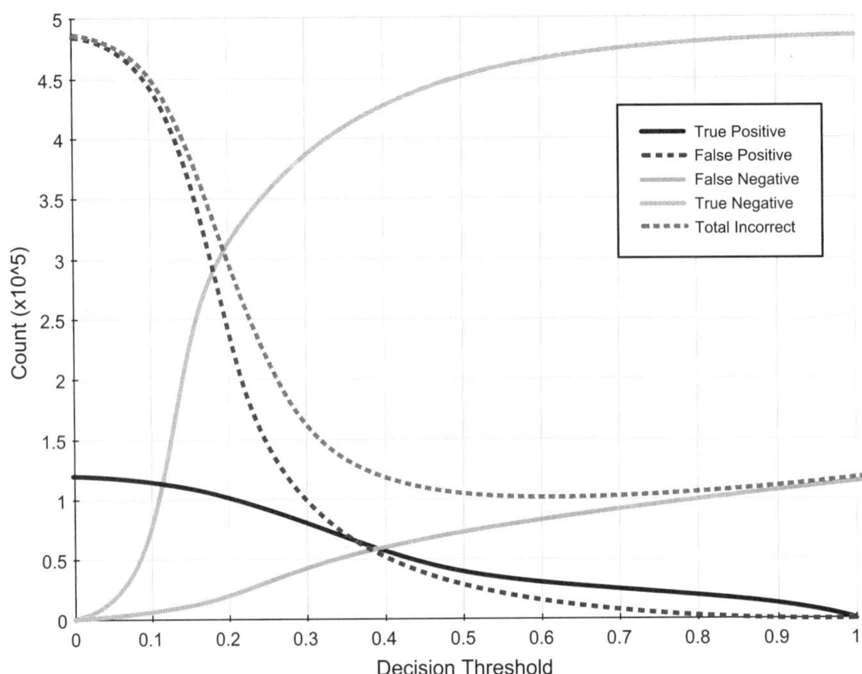

FIGURE 7.2
A visualization of the different error rates when the decision threshold is varied from 0 to 1 on the SSA disability approval problem.

illustrate how a hierarchy can help, imagine trying to create a text classifier to identify documents pertaining to large veterinarian offices. Rather than attempting to solve the problem in one step, it is easier to first separate out from the vast world of documents all those pertaining to veterinary services of any kind and then for that much smaller, more focused subset, build a second model to make the distinction between large and small veterinarian offices. If we used a "flat" classification (large vet services vs. everything else), the focus category would attract many uninteresting (to us) documents that deal with animals because those documents are much "closer" to the large vet category than they are to the set of "everything else." Using a topic hierarchy is a powerful way to manage the *negative space* for text classification—that space containing documents whose topics are not of interest. Breaking the negative space into two or more parts makes it easier for the text classifier to correctly identify documents that are near the goal class label.

There are many existing topic hierarchies to draw on when creating text classifiers. For classifying Internet pages, the Open Directory Project[2] created an extensive hierarchy of topics available on the web. This hierarchy is modeled after the original Yahoo! Search directory. The Open Directory contains nearly 5 million URLs and over 1 million topics. Many industrial societies such as the Association for Computing Machinists (ACM) and the Institute of Electrical and Electronics Engineers (IEEE) provide a rich topic hierarchy to classify documents within their topic areas. For nontechnical areas, both the Library of Congress and the Dewey Decimal System provide useable topic hierarchies. The challenge

[2] www.dmoz.org

with any hierarchy is finding suitable training data in an electronic format suitable for use as part of a text classification algorithm.

TEXT CLASSIFICATION APPLICATIONS

Applications of text classification and categorization algorithms are typically a form of a "needle in a haystack" type problem where the goal is to identify a small number of valuable documents ("needles") within a much larger collection of unimportant documents ("hay"). There are many examples of this type of problem throughout this book. In this chapter alone, we discussed email spam detection and claim prioritization for the SSA. In Chapter 14, we discuss applications in the insurance industry including the identification of fraud schemes from claim text. In Chapter 15, we discuss focused web crawling which combines web mining with text classification to identify relevant web pages on the Internet. Rather than revisit each of those applications here, we highlight two high-profile applications: eDiscovery and Customer Support.

eDiscovery

eDiscovery is a growing application area involving applying text classification algorithms in the legal domain. In many trials, litigants are required by the judge to exchange relevant documents during the initial stages of the trial, and counsel must sign that all pertinent documents have been identified. Traditionally, lawyers processed the available documents manually to find those of interest. A typical review involves about a million documents and costs about \$1 per document. With the massive amount of electronic data that is now available, including emails, chats, and other documents, manual processing is prohibitive, in both time and cost.

To streamline this costly process, eDiscovery uses predictive coding (classification) to identify documents relevant to each of the issues in the case. The goal is to identify a subset of documents that is sufficient to satisfy judicial scrutiny and also avoids costly attorney reviews. The primary challenge (both technically and legally) is demonstrating confidence that eDiscovery has found all of the relevant documents. At the time of writing, the proper use and bounds of eDiscovery is an active topic of discussion with the U.S. judiciary at both federal and state levels. eDiscovery techniques rely on many of the other techniques mentioned throughout the book including document pre-processing (Chapter 3), entity extraction (Chapter 9), and feature selection and dimensionality reduction (Chapters 13 and 14).

Customer Support

Many organizations record the text of any interaction with customers including transcriptions of calls to the call center, emails to customer support, and the text of live chat services. This text is often a valuable source of knowledge about the business process and customer sentiment. Three common text mining tasks using these data include: evaluating customer support personnel ("Does the customer leave satisfied?"), predicting customer transition ("churn") from client interaction, and automatically routing technical support tickets or issues to the proper staff.

Text classification algorithms can process the text interactions of customer support personnel. This text can be used to monitor customer attitudes, identify emerging sources of problems, or evaluate the performance of staff. In the latter case, the text of the call can be categorized into successful or unsuccessful categories based on whether the customer was presented with a satisfactory solution to their issue.

Predicting whether a customer will cancel their account ("churn") is a special case of analyzing customer support text. Dissatisfied customers are the largest source of churn; however, other reasons, such as inability to pay outstanding charges, are not satisfaction issues, but still result in a probable churn.

Finally, an increasingly popular area of text processing is routing issues (or "tickets") in a technical support system to the proper authority. For example, a large software company may have many different products or deploy on many different systems and support requests will be most efficiently handled if they are quickly routed to the appropriate support team.

SUMMARY

Text classification is a powerful tool for improving the efficiency and utility of processing text data. If the documents and the class labels are properly aligned with the business, there is potential for very large gains in efficiency. For the SSA, using text mining allowed 30 percent of the eventual approvals to be made without human intervention, leading to a dramatic speedup in overall processing time. This improvement was made possible by careful feature creation from the raw text data and by combining distinct features for improved performance.

POSTSCRIPT

As in traditional data mining, two of the most common application types are classification and prediction. Classification is the oldest type of text mining, but prediction is (arguably) the most interesting. We will describe prediction use cases and techniques in the next chapter. Prediction allows us to draw upon the rich capabilities of traditional data mining to calculate numerical prediction after the text documents have been preprocessed according to the methods presented in previous chapters. Thus, prediction use cases in text mining raise the question: Is text mining part of data mining, or is data mining part of text mining? One answer to this question is to suggest the term *text analytics* for the broadened form of text mining that uses advanced analytical algorithms developed originally to serve data mining purposes.

References

Christopher D. Manning, Prabhakar Raghavan, and Hinrich Schütze, *Introduction to Information Retrieval*, Cambridge University Press. 2008, New York.

Nisbet, Robert., John Elder, and Gary Miner. (2009). Handbook of Statistical Analysis and Data Mining Applications, Elsevier. Burlington, MA.

CHAPTER 8

Prediction in Text Mining: The Data Mining Algorithms of Predictive Analytics

CONTENTS

PREAMBLE

Predicting the future has always been at the center of man's fascination with the world around us. The Oracle of Delphi was consulted by the Ancient Greeks to learn if a war would be won. Nostradamus and (lately) the Mayan calendar claimed to predict the end of the world. During the last hundred years, this interest has developed in mathematics and statistics into two forms of forecasting the future state of something: classification and prediction. Classification is the oldest type of forecasting, in which the future state is a category, but when statisticians speak of prediction, they are referring to the future state of a real number. A real number is continuous throughout its range and can be expressed as a decimal value. While classification is the most common application in data and text mining, prediction is (arguably) the most interesting. Most parametric statistical routines require numbers to predict things. It is true that you can predict categorical things without numbers, but prediction in its purest form requires that all variables have real continuous numbers.

Practical Text Mining and Statistical Analysis for Non-structured Text Data Applications. DOI: 10.1016/B978-0-12-386979-1.00036-0

INTRODUCTION

Many of the chapters address in some detail how to convert text into numbers. Regardless of the specifics of the approach, in order to convert text into actionable information *without actually reading the text* requires two basic steps:

1. Converting each document into a sequence of numbers that summarizes the content of the text in a meaningful manner (meaningful with respect to the real-world decision or insight that is to be derived from the text)
2. Analyzing the numbers representing each document in a manner that derives meaningful actionable information and insights about the corpus of text or some other dimension or business key performance indicator for which the text is relevant

Chapters 4 and 5 describe many of the use cases for this process and the process steps in general, respectively. Once text data in a document corpus have been turned into numeric data, the information encoded into the numeric vectors for each document can be used in all data analysis, predictive modeling, or simple analytical reporting project.

This chapter provides a brief overview of the types of analytical approaches that are useful for leveraging text data and the "numericized" versions of the text data to extract useful, actionable information.

THE POWER OF SIMPLE DESCRIPTIVE STATISTICS, GRAPHICS, AND VISUAL TEXT MINING

As in any data analysis project, a first step toward extracting meaningful information from text that has been converted to numeric vectors is understanding the basic distributions and relationships between the respective numeric values. Chapter 4 includes a brief example that illustrates how simply counting the number of words, terms, or phrases in a document can quickly yield insights regarding the content of the text or corpus of text *without actually reading the text.*

To recap, that example dealt with an activity called "Bosseln" (or "Boßeln"). Instead of having to translate the entire text and then read it carefully, by simply tabulating and then translating the ten most frequent words and terms across the paragraphs of the Wikipedia article, the essence of what Bosseln is quickly became clear (it is a sport played mostly in the most northern part of Germany).

How Simple Counts Can Be Useful

In general, it is probably true that the majority of data analysis activities performed in support of solving some business or other problem involve simple tabulation and cross-tabulation of quantities. This is also true for text mining (or "text analytics"). Computing simple frequency tables or averages of terms can provide very useful information about the nature of the documents, the general sentiments or topics discussed in the documents, and the trends or relationships with other variables.

For example, a simple count of the number of times parts are mentioned among warranty narratives can be used to create sophisticated multivariate control charts (Hill et al., 2007; Sureka, De, and Varma, 2008). These charts can be used to support early-warning systems to detect increasing rates of part failure or when reports mention new combinations of parts. Simple text mining operations (involving simple indexing and tabulation operations) are particularly useful in this context because it is usually impossible to accommodate in structured report fields all parts and combinations of parts that *could* fail in a complex product (such as a computer or car). Using the simple indexing (of terms and words) and tabulation method described here, the unanticipated issues can be captured and tracked.

Similar simple analyses can be extremely useful for applications to track "buzz" on blogs and social networks. Regardless of "what it means" or "how it is discussed," the mere mention of a new product, technology, entertainer, book, and so on with significantly higher frequency than competing products indicates buzz—that is, an increased and perhaps increasing awareness among potential customers and clients of a specific product. Also, by also tracking the most frequently mentioned terms that co-occur with a product of interest, it is usually easy to determine if the buzz is generally associated with negative or positive sentiment or affect. For example, the website http://twittersentiment.appspot.com/ provides simple counts and percentages of positive versus negative sentiments for tracking a large number of concepts, politicians, and so on based on tweets on Twitter.

Graphical Summaries and Visual Data Mining

In addition to simple summary statistics, it is typically very useful to visualize the basic statistical summaries that are broken down by some meaningful dimensions or augmented in some way. For example, it is obviously useful to track the frequencies with which certain terms or groups of terms occur in a streaming corpus of text using graphical means—for example, to trend sentiments.

Bad Weather as the Cause of Plane Crashes

To illustrate how simple graphs based on word or term frequencies across the documents of a corpus can yield important insights, consider the graph shown in Figure 8.1. Figure 8.1 shows the average

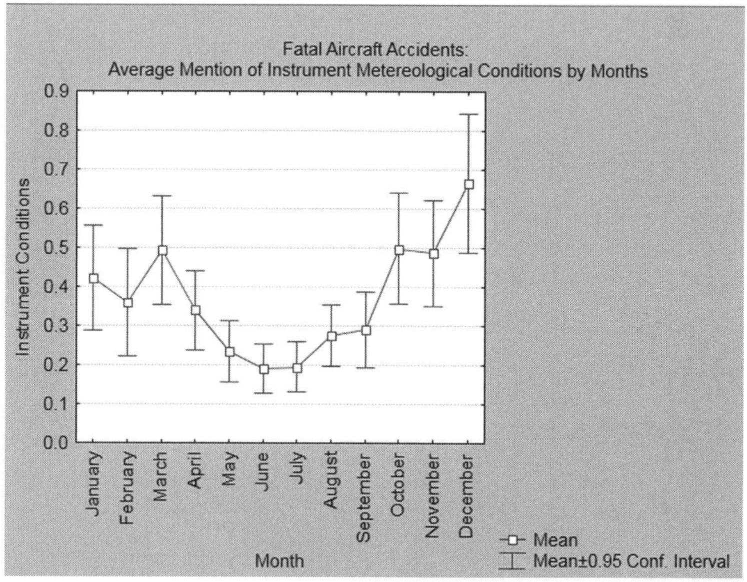

FIGURE 8.1

Mention of *instrument conditions* in fatal aircraft accident reports during the year.

relative frequencies with which the term *instrument conditions* (or synonyms of that term) was included in a selection of 3,235 aircraft accidents between 2001 and 2003. The term *instrument meteorological conditions* describes weather conditions that require a pilot to fly by using flight instruments alone because of clouds, fog, haze, or other weather characteristics that make it impossible to determine if the airplane is straight and upright by just looking out the window.

It is apparent that during the months of June and July, the percentage of accident reports (occurring in those months) that mention *instrument conditions* is much lower than during the months of November and December. This pattern suggests that "bad weather" plays less of a role in accidents that occur in the summer than in the winter. This conclusion can be supported simply by reviewing a simple means plot of the relative frequencies of specific terms across accident reports.

Warranty Claims and Cost of Repairs

To illustrate the "power" of simple graphical summaries based on text mining results, consider the two graphs in Figure 8.2 from a warranty claims application. The charts in Figure 8.2 were derived from 8,872 warranty claims for a type of tractor. The analyses leading to these results consisted of first classifying into categories of systems all the terms found in typical warranty claim write-ups. The claims can then be analyzed (in real time if required) to assign each warranty claim to one or more of these categories without requiring the attending technician to make the determination and classification himself or herself.

The next step in this application consisted simply of an analysis of the costs associated with each repair. For example, it is apparent that claims related to the *throttle systems* are on average more costly than claims relating to *instrumentation* (for example). In particular the Pareto Chart shown in Figure 8.2 is useful because it presents a very simple and clear picture of what components are most expensive to fix when they fail while under warranty. Again, simple graphical analyses and

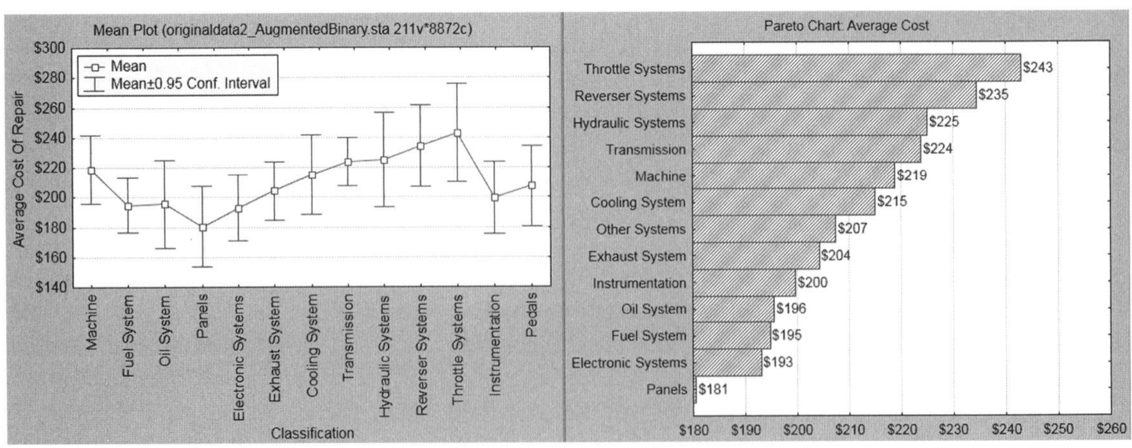

FIGURE 8.2

Average warranty repair cost by system.

presentation of results derived (at least partially) from the text can quickly identify important and actionable information.

VISUAL DATA MINING

The preceding examples are simple in the sense that once the basic frequency counts for key terms or phrases (often known a priori) are extracted from a corpus of text documents, it is easy to look for relationships of those terms to other important variables. Many graphics methods and tools are available to review data and to detect patterns in data through visual inspection, and all of them are applicable to data derived from text. As is the case with any data analysis project, an initial visual inspection of variables and their value ranges, interactions, and relationships is critical for data understanding. This is often particularly the case with data summarizing text—for example, the relative word or term frequencies across documents. Such graphical analyses may quickly identify clusters of documents or clusters of terms that vary across some other structured variable values, and a careful interactive graphical analysis or "mining" of the text data can often provide more relevant insights more quickly than applications of "formal" data mining algorithms.

PREDICTIVE MODELING (SUPERVISED LEARNING)

A major use for the application of text mining methods is to achieve better predictability of some important outcomes or key performance indicators. Numerous books have been published describing in detail the uses and approaches for predictive modeling in general (e.g., Nisbet et al., 2009; see also Hill and Lewicki, 2007) and the algorithms for building predictive models (e.g., Hastie et al., 2009). The purpose of this section is not to provide a comprehensive overview of those methods but rather to offer a broad survey of methods and their applicability in particular with respect to text mining.

In most general terms, the predictive modeling task can be summarized as finding a good representation of an outcome variable $y = f(x)$: The task is to predict an outcome y based on a vector of values for predictor variables $x_1, x_2, \ldots, x_n$. For example, y could be a discrete outcome such as credit default or fraudulent behavior, and the predictors x could consist of credit scores, coded properties of the credit application, or any other variable. The predictor variables may also include any number of variables derived by "numericizing" the text, using the methods described in Chapter 5.

Over the last ten years or so, the algorithms for predictive modeling and the hardware on which they are run have improved to the point where even very large prediction problems can be addressed, even though they have hundreds or thousands of predictors based on data sets with millions of observations. In many business domains, predictive modeling is now performed routinely to predict and then optimize various key performance indicators. In many of those applications, relatively little can be done to improve the predictive accuracy of existing models by "tweaking" (improving) the algorithms. However, available unstructured information that can be aligned with the data can be used to build more accurate models. Some opportunities include text mining analysis of insurance claim text narratives and adjuster notes that can be aligned with claims. Models improved by text mining results provide opportunities to enhance the effectiveness and efficiency of organizations or government institutions.

The Algorithms of Predictive Modeling

To reiterate, the purpose of this chapter is *not* to provide detailed overviews of all or even the most important predictive modeling algorithms. Many excellent books have been written on the subject, including those mentioned previously, and information can be obtained from the Internet, such as StatSoft's *Electronic Statistics Textbook* at www.statsoft.com/textbook and YouTube tutorials at www.youtube.com/user/StatSoft. Instead, the goal of this chapter is to provide a general overview and guidance regarding the different types of algorithms available for predictive modeling and their advantages and disadvantages when, for example, working with data derived (numericized) from text.

STATISTICAL MODELS VERSUS GENERAL PREDICTIVE MODELING

There are a number of important differences between traditional statistical analysis and modeling (e.g., multiple regression and logistic regression) compared to the general predictive modeling techniques that have been widely adopted across many domains over the past decade or so.

Statistical Data Analysis

Statistical analyses and modeling often focus on "hypothesis testing" and "parameter estimation." For example, in multiple regression, the parameters of a linear model are estimated that predict some outcome or response variable y as a linear function of the available predictor or x variables.

Often, only those parameters and predictors are retained for the analyses that are "statistically significant." This actually means that some x variables and associated parameters for the linear prediction equation are set to 0 (zero) because, based on the sample data and statistical/mathematical reasoning alone, there are insufficient evidence and confidence that the respective parameter is unequal to 0 in the population from which the analysis sample was drawn.

Predictive Modeling and General Approximators

In contrast to statistical modeling, the algorithms for general predictive modeling use a much more pragmatic approach: Is it possible to extract from the sample data repeatable patterns that allow for a more accurate prediction of y from the values of the predictors x? Ideally, a good predictive modeling algorithm would detect linear relationships, nonlinear relationships, interactive relationships, and so on; in other words, it could approximate any relationship between variables. In addition, such algorithms should detect *repeatable* patterns, and it would be important to demonstrate that accurate predictions could be obtained in a sample of cases that was *not* used for the predictive modeling itself (a "holdout" or "test" sample).

The most effective predictive modeling algorithms that often can derive accurate predictions from very complex data are general approximators of whatever repeatable relationships are found in the data. They provide also some mechanism to avoid *overlearning*—that is, the detection of relationships between variables in a specific analysis sample that do *not* replicate in another (holdout) sample of data. The reason that the relationships do not replicate to other data sets is that the model was trained to recognize very closely the relationships that are specific to the training data set and do not exist as strongly in other data sets.

Predictive Modeling and Text Mining

As mentioned before, the general approach to text mining involves converting the text documents in corpus to vectors of numbers, one for each document. In general, there is little advantage to do so "sparingly"—that is, to use some method that converts each text into a few numbers. Rather, it is usually advantageous to, for example, enumerate (count) as many words, word pairs, phrases, and so on as possible and, for example, to carry along for subsequent analyses the document coefficients for as many latent dimensions as possible (see Chapter 15). In most cases there are no a priori hypotheses or well-defined expectations about how specific components or aspects of text documents might contain useful information to address a specific problem. Therefore, the methods described in this chapter can handle large numbers of predictors to extract the important predictors and relevant information to solve specific analytical problems. Because of this ability, these methods are frequently used in text mining analyses that require predictive modeling operations with variables derived from the text. An important feature of these operations is that the outcome (that which is predicted) is known ahead of time; the model is built using these example outcomes as patterns. Methods like these are known as *supervised modeling algorithms*. Common supervised predictive text mining algorithms include the following:

- k-nearest neighbor and support vector machines (SVMs)
- Recursive partitioning decision trees
- Neural networks
- Ensembles

k-Nearest Neighbor Methods and Similar Methods

The k-nearest neighbor method is a good example of a "general approximator" that is entirely based on patterns in the data, without any specific "statistical model" that must be estimated. In fact, k-nearest neighbor methods do not rely on any "models" at all but instead simply use the existing data (or a sample of exemplars from the existing data) to "predict" new observations. Specifically, when predicting a new observation, the algorithm finds the most similar observations among the exemplars with respect to the available predictors and then makes the prediction that the new observation will have the same outcome (same predicted y, or predicted classification).

For example, consider the classification problem illustrated in Figure 8.3. In this case there are two predictors—*Predictor 1* and *Predictor 2*—and a number of exemplars or individual square (blue color in the computer screen) and round (red color in the computer screen) points that belong to category *A* or *B*, respectively. Now consider the problem of predicting new observations to belong either to category *A* or *B*. In Figure 8.3 the triangles (green color in the computer screen) show a few new observations.

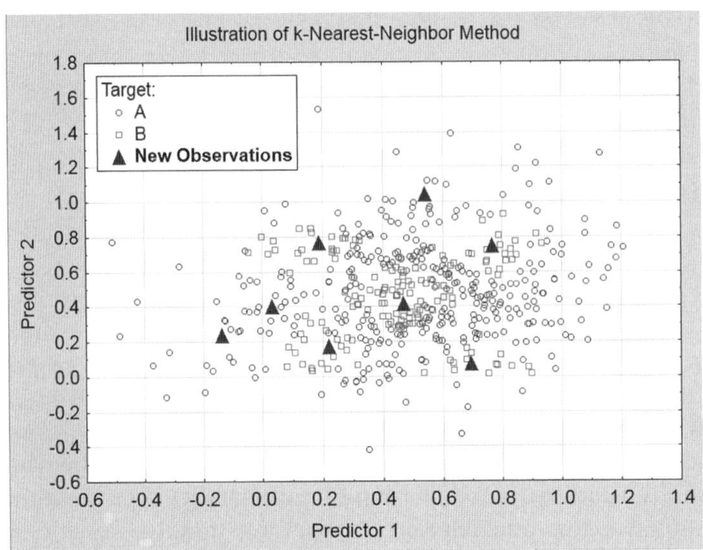

FIGURE 8.3
Classifying new observations using k-nearest neighbor method.

It is intuitively easy to classify the new points as either belonging to the cloud of square points or round points, based on the types of points in their respective "neighborhoods." Note that the scatterplot in Figure 8.3 actually depicts a fairly "difficult" prediction problem because it is very difficult, if not impossible, to build a linear model or any other simple model that could represent properly the obvious clustering of the points in the two-dimensional space.

There are a few additional considerations in the actual implementation of this algorithm—for example, how many exemplars to choose, how to determine "distance" between a new point and the "neighbors" (nearest points) among the exemplars, and how to compute the final prediction when more than one neighbor (i.e., k-nearest neighbors) is chosen to make the (e.g., average) prediction. This algorithm is also sometimes referred to as a "memory-based-learner" because it "memorizes" the exemplars and then makes predictions for new observations based on this memory.

Strengths of the k-Nearest Neighbor Algorithm

In practice, this algorithm is relatively simple and fast, and it scales well—in other words, it can be applied to score a large number of observations relatively efficiently. For that reason, it is often used to impute (or predict) missing data values by simply changing the missing values to the values observed among the k-nearest neighboring observations. Also, because of the nature of the algorithm, it can make accurate predictions even when there are highly nonlinear or interactive relationships between the predictor variables and the outcome of interest, as long as those relationships are reflected in the exemplars.

Weakness of the k-Nearest Neighbor Algorithm

This algorithm, however, has some obvious drawbacks:

- It is not clear how many exemplars to choose from the training data set.
- It is not clear how many nearest neighbors to choose to compute an average prediction.

- The algorithm will provide no insights into which predictors are the most important ones to achieve accurate predictions. This is perhaps the biggest disadvantage of the algorithm because of the high-dimensional data with large number of predictor variables often found in text mining operations.
- It is necessary to choose exemplars and important predictor variables ahead of time, following some *a priori* method.

Many of these disadvantages of k-nearest neighbor algorithms can be minimized by performing appropriate data preprocessing operations (i.e., singular value decomposition and PCA; see Chapter 11). Some of the small number of variables on the list output from these operations are linear combinations of original variables. There is a relatively large explanatory power of the fewer number of variables that result from this preprocessing, so the method can be very effective at determining similarity. For example, one might use the methods described in this book to "numericize" accident reports or insurance claims, have experts select typical "exemplar claims" of interest or belonging to a specific claim category (e.g., requiring specific expertise from the adjusters), and then use those exemplars and the numericized text to automatically assign new claims to categories of interest.

A number of refinements have been developed around the general ideas and approach of k-nearest neighbor methods. For example, one can try to "construct" the neighborhoods used for making predictions about new observations by dividing the training data into regions based on partitions or "lines" or optimal "hyperplane" defined by vectors of values (features) derived from the original inputs. For example, SVM (Support Vector Machines) algorithms will construct such optimal partitions. In some ways, SVMs are really more similar to neural nets, discussed following, because an SVM is not a simple memory-based learner as just described; rather, it requires a great deal of computations to build a prediction model. Nonetheless, the basic idea is the same in that the algorithm attempts to define the "neighborhoods" based on the predictor values around the points in the training sample so new observations can be classified or predicted consistent with the "neighborhoods" to which they are assigned.

Recursive Partitioning (Trees)

These algorithms have become very popular and have been applied successfully to virtually all domains where predictive modeling has been useful. The general method and algorithm are quite simple: Suppose the task is to classify cases into one of two categories *A* or *B*; also suppose that in the training data set, about 50 percent of the cases belong to A and 50 percent belong to B. In order to make a prediction from a training data set, the first step in the algorithm is to evaluate all predictors one at a time and then select the best one that allows to split the training data set into two groups, based on the values of the predictor. The result is to generate the two "purest" subgroups after the split with respect to the outcome of interest. This split results in two groups where one has as many *A* observations as possible and another with as many *B* observations as possible. This algorithm is applied recursively—that is, in the next round all predictors are again considered to split each of the subgroups resulting from the previous round of splits. This algorithm will create a (binary in this case) "tree." The algorithm is applicable to classification problems (predicting to which category an observation belongs), as well as to the prediction of continuous outcomes (regression problems). In the latter case, the notion of "purity of subgroups" is computed as a function of how different the means are in each subgroup relative to their respective standard deviations; in other words, the goal is to separate the value ranges of the outcome variable as much as possible in successive splits.

Example of a Simple Text Mining Application

Chapter 4 provided a relatively simple example of how to predict the asking price of a used Cessna aircraft based on the words and terms that are used in online advertisements. A simple regression tree result may look like that shown in Figure 8.4.

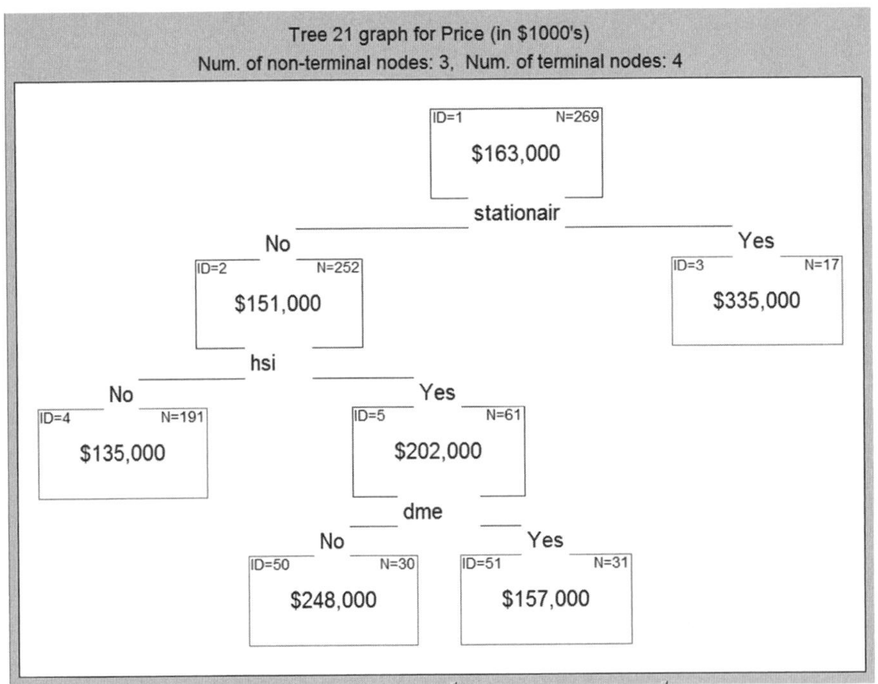

FIGURE 8.4
A simple regression tree model relating the presence of words in an airplane advertisement to the asking price.

Notice in Figure 8.4 that the average price across all Cessna airplanes that are advertised on the site is $163,000. However, if the word *Stationair* is used, then the average asking price for the subgroup of ($N = 17$) airplanes thus described is $335,000, while those that do *not* have that term in them only cost $151,000 on average. Furthermore, the $N = 252$ advertisements in the left node after the first split in Figure 8.4 can further be partitioned into $N = 61$ aircraft that have an HSI (horizontal situation indicator; average price $202,000) and those that do not ($135,000). (See Chapter 4.)

The advantages of recursive partitioning or tree algorithms become clear from this example. The analyses will result in relatively simple to interpret results of if-then "rules" for predicting the outcome of interest. For example, Figure 8.5 is a graph generated from the analysis of a credit problem.

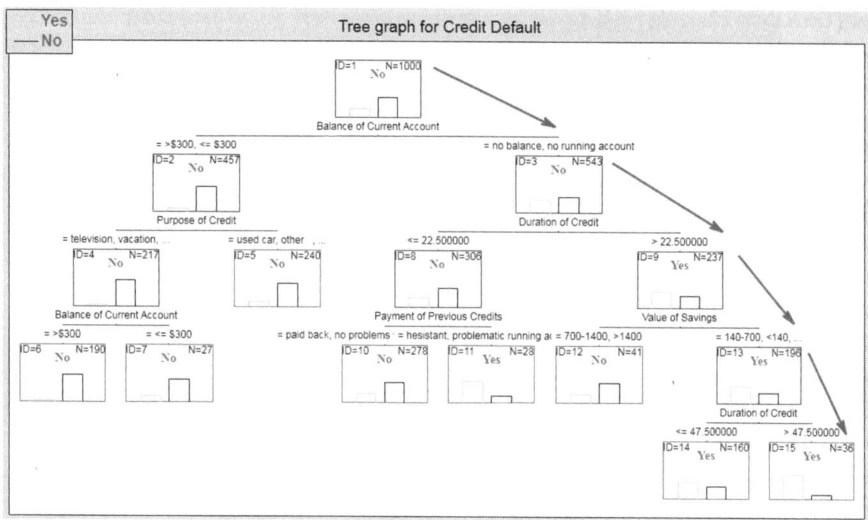

FIGURE 8.5
Tree model for predicting credit default.

Each box or "node" in the tree shown in Figure 8.5 contains two histo bars, indicating the relative frequencies of *Yes* and *No* (credit default or no credit default). Figure 8.5 also highlights a "path" or rule consisting of a combination of predictor values that will result in a likely *Yes* response (or a likely credit default). A default is very likely if a customer does not have an account or no balance in any account and the *duration of credit* is for longer than 22.5 weeks, and the *value of savings* is between \$140 and \$700 or less than \$140, and the *duration of credit* is greater than 47.5 weeks (see node $ID = 15$, with $N = 36$).

Implementations

The examples and figures shown so far only depicted so-called binary trees—that is, trees where each split results in exactly two subsamples. Several algorithms with names like C4.5; classification and regression trees, or CART; Chi-square Automatic Interaction Detection, or CHAID; and so on differ in the specific approach and calculations for how to recursively split the training data into increasingly homogeneous subgroups. These algorithms are, for example, described in Hastie and colleagues (2009). Some of these algorithms, like CHAID, will perform multiple splits in a single step, so the resulting tree is not binary (with only two subsamples resulting from each split of the previous node), but multiple subsamples (subnodes) can result at each split, making the resulting tree "wider" rather than "taller." However, each of these methods principally performs the same steps: Splitting a sample into ever more homogeneous subsamples based on the values in one or more predictors ("combination splits"), resulting in a "tree" of rules about how to predict the outcome of interest.

Strengths of Recursive Partitioning Algorithms

Recursive partitioning or tree algorithms are popular because they can easily approximate even very complex relationships between predictors and outcomes, yet yield simple to interpret results (rules). This is obviously very attractive in a number of applications where it is important not only to make an accurate prediction but also to justify it—for example, when making credit decisions where it is important to provide feedback to applicants whose credit applications are rejected. Also, the method is very efficient and can be applied to very large data sets. Some implementations also allow analysts to pick and choose by hand specific predictors and splits into the "model," which can be useful in order to build prediction models or rules that are more "acceptable" or perhaps even "legal" in applications that are subject to stringent regulatory oversight by government regulatory bodies (such as with financial institutions).

Weaknesses of Recursive Partitioning Algorithms

A main problem with recursive partitioning algorithms is that they can sometimes appear very "unstable." Consider the case at a particular point in the tree where two or more variables are exactly equal with respect to the quality (purity) of the subsamples resulting from a split based on their respective values. In that case, most algorithms will choose more or less randomly one predictor over the other, which may strongly alter the subsequent splits from the respective branch. This may not pose a problem if the main goal is to arrive at good predictive accuracy. But it is important to keep in mind that any tree resulting from this split is likely to represent a "locally optimal" solution, not a "globally optimal solution" for the prediction problem. This distinction means that the solution is the most accurate for this set of data but is much less accurate with other sets of data. Also, there may be other very different trees that might give a better predictive accuracy (a more general solution) with all data sets than the one resulting from the current analysis with the current analysis software. If that is a concern, then it is useful to do the following:

1. Understand the details of the respective algorithms one is using to see if there are any changes in their configuration that might create a more general solution.
2. Explore available diagnostics to identify other potentially capable predictors and splits.

When to Stop Splitting

Another very important consideration for recursive partitioning algorithms is to evaluate how detailed and "big" the final tree should be and, therefore, how complex the final rules should be to create the most accurate predictions for new data. In principle, one might be able to continue to split the data until every observation in the training data is correctly classified. However, in practically all cases this will yield rules that are "too specific" and will not generalize to new data—that is, giving yield predictions that are not as accurate as predictions made from a simpler tree. A number of methods can be used, depending on the specific implementation of the recursive partitioning algorithm, to address this issue. Most of these methods rely on some method of cross-validation, where the tree and rules of a specific complexity are applied to new data or subsampled data from the training sample to see how well each tree of a given complexity generalizes to other samples.

Tree Methods and Text Mining

Tree methods can be very useful in the context of text mining because (and as illustrated in Figure 8.4) when analyzing, for example, word or term frequencies as predictors, very interpretable models and the most "diagnostic" terms for the problem at hand can emerge. For example, imagine a study of physicians' notes and their value for predicting subsequent medical procedures. Using recursive partitioning methods applied to (relative) word or term frequencies extracted from the physicians' notes, it is possible to identify specific combinations of observations (terms reflected in the notes) that are associated with a higher probability of subsequent specific medical issues (for example, see Polon, 2011).

Neural Networks

The human brain is principally composed of a very large number (around 10 billion) of neurons that are massively interconnected with an average of several thousand connections per neuron, although this varies enormously. Each neuron is a specialized cell that can propagate a signal. From this biological "inspiration" has sprung a large number of algorithms that to some extent mimic this basic architecture. Inputs and outputs in a neural net are connected by activation functions through "hidden neurons." This is schematically shown in Figure 8.6.

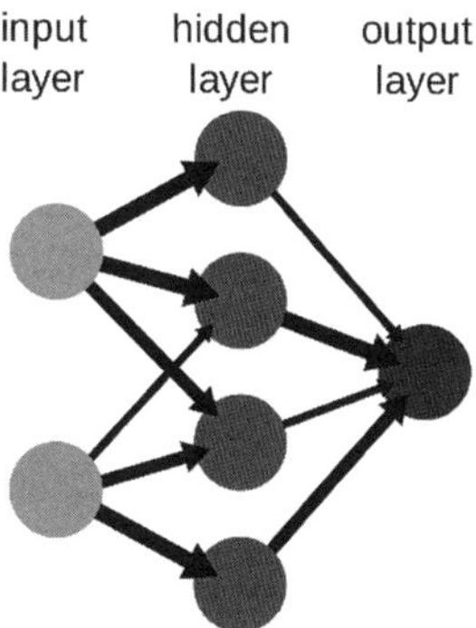

FIGURE 8.6
A simple neural network.

Specifically, at the *input layer* side of the network, the observed values of the predictor variables are recorded. These values are transformed via an activation function, such as a simple linear function, sigmoid function, and so on, to an "activation" value for the *hidden layer* neurons. These activation values are then transmitted to the *output layer* via another output function.

Another way to interpret this architecture is to consider neural nets as simply a set of nonlinear equations that connect the inputs (predictor variable values) to the output. The specific parameters or weights of the equations (the activation functions) are then estimated using general optimization algorithms that will minimize the prediction error for the response in the training data.

There are a large number neural network "architectures" that effectively represent various "arrangements" and/or constraints for nonlinear activation equations. Some of these architectures (arrangement of equations) have been developed to accomplish specific prediction tasks more effectively (e.g., in the time domain). However, in general, neural networks are best thought of as nonlinear equations fitted to observed data to achieve the most accurate predictions possible.

Strengths of Neural Network Algorithms

If allowed to grow to any required size (with as many hidden layers and hidden layer neurons as required), neural networks can represent any relationship between predictors and outcome variables, so they represent general approximators as described earlier in this chapter. Neural networks will produce (usually) smooth response functions (predicted values as a function of increasing or decreasing predictor values), which has made them very popular for various engineering applications where a smooth predicted response is required.

Weaknesses of Neural Network Algorithms

Neural networks of even moderate complexity (moderate numbers of nonlinear equation parameters that have to be estimated) can require significant computational resources before a satisfactory model can be achieved. Therefore, when a training data set has thousands of predictors and large numbers of observations, it is typically not practical to fit complex neural networks in order to build an accurate prediction model. Also, neural networks represent the proverbial "black box algorithm"—that is, it can be difficult to interpret the results of the analyses or understand the rules or logic of how the neural network model arrives at the prediction. There are methods and techniques to address that issue to provide insights into how predicted responses vary as a function of different input values. Compared to the simple results from analyses applying recursive partitioning algorithms, however, neural network results often tend to be "complex" and difficult to understand.

Neural Networks and Text Mining

Because text mining often results in large numbers of dimensions (e.g., word or term frequencies for large numbers of terms), neural networks may be less useful in predictive modeling tasks involving large numbers of numeric predictors derived from text. However, this can sometimes be remedied by applying appropriate data preprocessing methods, such as the singular value decomposition (SVD) methods described in Chapter 11. In that case, SVD was used to reduce the dimensionality of the predictor "space" by reducing large numbers of predictor variables that were the result of numericizing text to a few dimensions that summarized the information contained in the predictors.

Ensembles

In the predictive modeling disciplines, an "ensemble" is a group of algorithms that is used to solve a common problem (like a group of instruments playing the same musical composition). Each modeling algorithm has specific strengths and weaknesses, and each provides a different mathematical perspective on the relationships modeled, just like each instrument in a musical ensemble provides a different "voice" in the composition. Predictive modeling ensembles use several algorithms to contribute their perspectives on the prediction problem and then combine them together in some way. Usually, ensembles will provide more accurate models than individual algorithms, which are also more general in their ability to work well on different data sets. The most common method of combining algorithm predictions is to treat each one as a "vote," where the majority "wins." Seni and Elder (2010) provide details regarding the various methods and variation of methods (e.g., for combining predictions), but in general the approach has proven to yield the best results in many situations. For example, a team called "The Ensemble" recently achieved the best predictive accuracy in the open "Netflix million-dollar prize competition" and lost only because the respective prediction model was submitted 20 minutes after the winning entry that achieved the same accuracy; see Lohr, 2009).

Advantages and Disadvantages of Different Algorithms

In practice, the choice of which particular predictive modeling algorithm(s) to use is driven by general considerations that have less to do with the nature of the data—if they are derived from structured sources or unstructured text—but more with the applications and use cases themselves. Two of the main considerations when choosing a predictive modeling algorithm are if the results have to be interpretable (or if a "black box" is okay) and if the models will be used to make predictions based on interpolations or extrapolations of input data.

Interpretable Models versus "Black Box" Models

In general, recursive partitioning or tree algorithms as just described yield models that are very accessible and easily interpretable. That is perhaps one of the main reasons why tree models have become so popular in many domains. With respect to text mining applications, it is often useful to make predictions based on understandable "rules" that combine structured and unstructured text fields. For example, one might have a prediction model for expected health care expenses based on a patient's age, gender, and overall physical condition as stored in various structured data sources and based as well on physicians' text notes as to whether or not terms such as *heart disease* are present. If the physicians' notes include phrases like "perhaps early stage of heart disease," then subsequent health care costs are likely to increase.

Interpretability of models is often not just a preference but an actual requirement for a specific project or domain. For example, in insurance and financial applications (e.g., pricing of insurance policies; credit scores), there are regulatory constraints in place regarding what information can and cannot be used to, for example, approve or deny credit. Therefore, it is critical that whatever model is used to determine the price of an insurance policy is not based on information that is illegal to disclose. As we saw before, some models, such as neural nets, or ensembles of models usually do not lend themselves to easy "human inspection" and are therefore not used often for that reason.

Interpolation and Extrapolation

In some applications in predictive modeling, continuous "smooth" response functions are required—that is, where it is important that given continuously "sliding" values of some input, a continuously changing output function is observed. This is usually the case in engineering applications—for example, when building models to automatically manage some continuous damper or valve opening as a function of some inputs like a throttle setting. For example, imagine a sensor that meters fuel to the car's engine (determines the acceleration of the car) as a function of how much the driver presses the gas pedal. Obviously, it would be important that there be a smooth and continuous response for the power output from the engine in response to the smooth application of pedal force. A neural network model that would connect these two parameters would result in such a response function. On the other hand, suppose you had built a model based on a tree algorithm and training data that had only three distinct pedal inputs represented. In effect, such a model would only be able to put out three distinct values: low power, medium power, and high power. This of course would make for rather "jerky" driving.

In practice, and with respect to the application of text-derived data, there are likely to be few applications where a continuous response function is required. Instead, most predictive models are rather used to "select" new cases that will, for example, be more likely to default on credit or more likely to respond to a marketing campaign.

CLUSTERING (UNSUPERVISED LEARNING)

So far the discussion in this chapter has centered around predictive modeling, where based on some (historical) training data with known outcomes (e.g., credit default), predictive models are built to predict the outcome for new cases (expected probability of credit default). One way to frame this problem is as "supervised learning"—that is, the prediction models are built subject to the "constraint" or *supervision* imposed by the observed outcomes in the training data.

In contrast, often one wants to apply "unsupervised learning" strategies—that is, identify repeatable patterns in data, not necessarily to predict specific outcomes but simply to identify important segments, clusters, and stratifications in the data. For example, in fraud detection applications there may be a history of cases with known fraud, against which predictive (supervised learning) models can be built to predict the probability of fraud. But there are also likely many cases of fraud in the historical data that were never identified. In the latter case, it may be useful to apply unsupervised learning or clustering algorithms to identify the types or "buckets" of cases that are similar to one another. One can then inspect those buckets or clusters and their characteristics to detect unusual combinations of input variables or cases that cannot be easily assigned to any cluster and thus are somehow unusual compared to the other cases in the training data.

Unsupervised learning, or clustering, usually involves algorithms that assign observations to clusters (buckets) based on some shared characteristics or similarities across the input variables in the analyses. A number of methods and algorithms are used for that purpose.

k-Means Clustering and Expectation-Maximization Algorithms

The goal of these algorithms is to create a specific number of clusters that minimize the "distances" between the observations in the cluster and maximizes the distances between the observations in different clusters (or the distances between the cluster "centers"). For example, consider the hypothetical case of two input dimensions *x1* and *x2*, and two clusters of points, *Cluster A* and *Cluster B*. A scatterplot identifying the clusters is shown in Figure 8.7.

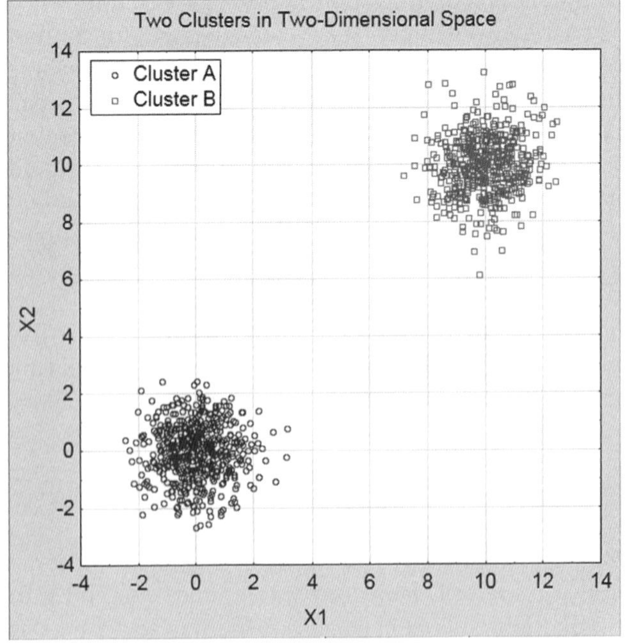

FIGURE 8.7
Two clusters of points in two-dimensional space.

Looking at the clouds of points in Figure 8.7, the clusters are easily identifiable. The k-means clustering algorithm would also (likely) correctly identify the points that belong to each cluster by performing these steps:

1. Choose two points at random; these are the temporary cluster centers for the two clusters.
2. Process all data and assign points to the closest cluster center, using, for example, a simple Euclidean distance criterion.
3. Recompute the new cluster centers as the means of all cases assigned to each of the clusters and go back to step 2.
4. Iterate steps 2 and 3 until no more points are reassigned to different clusters over consecutive iterations or some other criterion of "convergence" is reached.

These steps basically describe the logic of k-means clustering. Effectively, the algorithm simply continues to reassign points to clusters until in consecutive iterations no better cluster "solution" emerges where the points within each cluster are closer or more similar to each other than the points in different clusters.

Distances and Probabilities (EM Clustering)

The clustering method just described requires that a "distance" is computed between the points and the cluster centers. These distances can be computed as simple Euclidean distances between the respective cluster centers and the points or in any number of different ways. For example, suppose one assumes that the distribution of points within each cluster around its respective cluster center is multivariate normal; this is depicted in Figure 8.8.

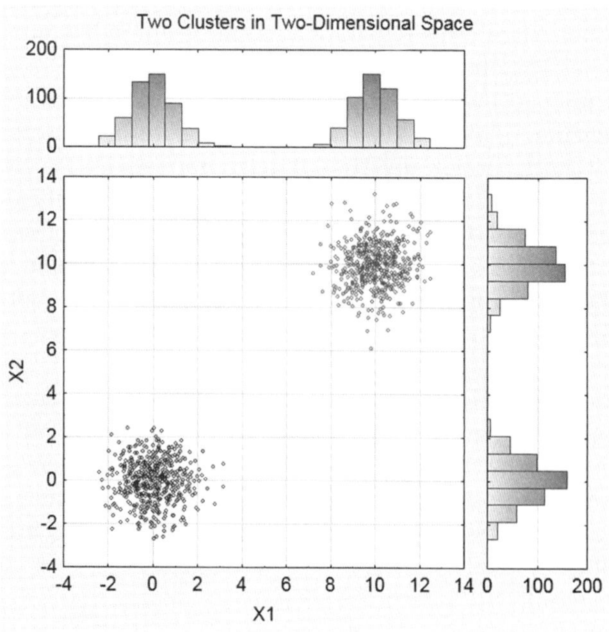

FIGURE 8.8
Distributions for two clusters in two-dimensional space.

In that case, the "distance" between the cluster centers and the points can be expressed as probabilities—that is, the probability or "expectation" that the respective points "belong to" (were sampled from) the respective distribution and the basic algorithm described earlier (the k-means algorithm) can be applied to maximize that probability (expectation).

An advantage of this, the EM algorithm, is that one does not need to assume that the normal distribution holds (fits well) the data within each cluster, but one may choose nonnormal distributions for continuous variables and/or also distributions for discrete variables. However, fundamentally, the algorithm for assigning points to clusters works the same.

Strengths of k-Means and Expectation-Maximization Algorithms

The k-means and EM algorithms for clustering are very efficient and generally scalable to large numbers of observations. In the context of text mining—for example, to cluster documents with similar combinations of phrases and terms—these algorithms can be used to identify clusters of similar insurance claims, product reviews, and so on. The distances of the individual observations from their respective closest cluster center (to which they are assigned) are diagnostic of how "atypical" the respective narrative is of the respective cluster of narratives. Therefore, this algorithm can identify claims narratives or accident reports, for example, that are "prototypical" of that cluster of narratives. On the other hand, if an observation is very far from any of the cluster centers, then the respective narrative would be very atypical or an outlier with respect to the clusters made up of the majority of the observations. In either case, those narratives might, for example, identify "novel" claims or narratives, new types of warranty claim reports, and so on.

Weaknesses of k-Means and Expectation-Maximization Algorithms

The main problem with these algorithms is that for a specific solution, the numbers of clusters must be known a priori. Put another way, the numbers of clusters parameter is not part of the algorithm itself, but it must be specified prior to invoking the algorithm to assign the observations to the respective numbers of clusters. Therefore, if the number of clusters sought is not known a priori, then other methods must be used to "guess" how many clusters are appropriate for the given data. In practice, this can be accomplished, for example, by using cross-validation methods, where a respective number of cluster "solutions" computed from one data sample is applied to a holdout sample. One can then plot, for example, the average Euclidian distances of points in the holdout sample from the assigned cluster centers against the numbers of clusters to identify the cluster number beyond which the quality of the cluster solution (the average distance from the cluster centers) no longer decreases as more cluster centers are added.

Similar to the situation described earlier in the context of predictive modeling methods, another problem of k-means clustering is that the cluster solutions (the specific assignment of points to clusters) can be affected significantly by the choice of initial cluster centers. In general, each specific solution is not necessarily the best solution possible for the given data (the solution may be "locally optimal" but not "globally optimal" over all possible solutions). As a practical matter, it is always a good idea to rerun analyses with a different choice for the initial cluster centers to ascertain that relatively similar (stable) solutions always result.

Hierarchical or Tree Clustering

Another popular clustering method is tree or hierarchical clustering. The goal of this method is to build a hierarchical tree that summarizes (reproduces) the distances between the items being clustered. A tree like that may look like the one in Figure 8.9.

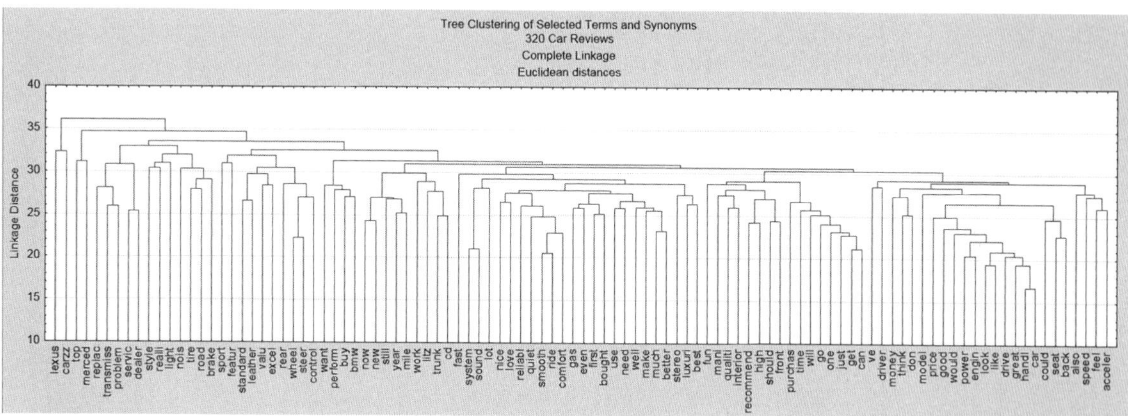

FIGURE 8.9

Hierarchical (tree) clustering of selected terms and synonyms used in car reviews.

Figure 8.9 shows a hierarchical tree clustering solution for selected terms and synonyms used in a corpus of 320 car reviews. This graph is constructed as follows. Suppose you started with a data matrix where each observation or case is a car review and each column is a relative term frequency for the terms used in the respective review (the frequencies with which they are used across documents). From that relative frequency, a distance matrix can be computed to express the degree of co-occurrences of terms across documents. For example, one could compute correlations to express the similarity of co-occurrences of terms and then rescale them so the resulting coefficients denote distance rather than similarity. For example, two terms, "performance" and "speed," that co-occur almost always together across the document corpus (the car reviews) would be very "close" in terms of distance, while two terms that almost never co-occur would be very distant.

The hierarchical tree algorithm will then process the distance matrix and combine the terms that are closest together or most similar. Next, the distance matrix is updated to reflect the distance between the combined terms and all other terms in the distance matrix. The process then repeats—that is, the algorithm will choose the next pair of items to join by finding the smallest distance between any two terms (or combined terms from a previous step). This process continues until all of the items are joined.

Figure 8.9 shows the Linkage Distance on the y axis—that is, the distance between individual terms (or combined terms) in the distance matrix at which the respective terms listed on the x axis are merged. For example, consider the subset of the hierarchical cluster tree shown in Figure 8.10.

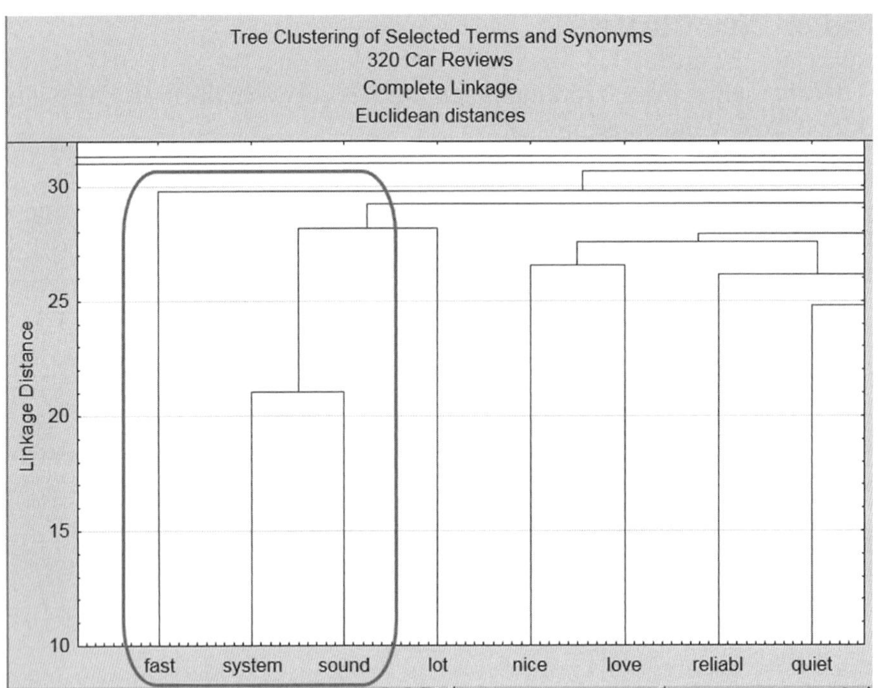

FIGURE 8.10
Small section of a hierarchical (tree) clustering solution.

Apparently, the terms *system* and *sound* co-occur frequently in the corpus of car reviews. In retrospect, most likely the term *sound system* should have been defined as a separate phrase. However, as illustrated in Figure 8.10, the analyses identified the correlation between the relative frequencies of those two terms and merged them at a relatively smaller linkage distance. In contrast, the term *fast* is not related to *sound system*.

Strengths of Hierarchical Tree Clustering Algorithms

The strength of the algorithm—especially in the context of text-derived data—is that it allows for the detailed inspection of a "rich" result to evaluate, for example, the details of the structure of how terms or phrases are used (co-occur) across the document corpus.

Weaknesses of Hierarchical Tree Clustering Algorithms

A weakness of the algorithm is that most full implementations begin with a distance matrix of items that are to be clustered. This means that it can be difficult to cluster, for example, hundreds of thousands of items because the distance matrix would become very large. Also, there are a number of options regarding how to compute "distance" and how to compute the distance between combined items to all other items in the matrix (linkage or amalgamation rule). Choices of those details will greatly affect the resulting cluster solution.

Kohonen Networks or Self-Organizing Feature Maps

Neural network methods can also be used for clustering. As described earlier in the context of predictive modeling techniques, neural networks fit systems of nonlinear equations to data to optimize the accuracy of prediction. In Kohonen networks, or self-organizing feature maps (SOFMs), the goal rather is to cluster observations into a "lattice" of "boxes" (clusters) to achieve maximum separation between the observations in different clusters. This "mapping" of observations to boxes in the lattice (clusters) is (typically) done using nonlinear activation functions.

In terms of interpretations, the results of Kohonen networks are very similar to those created via k-means clustering. Observations assigned to the same cluster tend to be more similar to one another than those in different clusters. However, Kohonen networks will typically use nonlinear activation functions to achieve the assignment of observations to clusters.

Strengths and Weaknesses

The Kohonen or self-organizing feature maps algorithms are generally more computationally "expensive" than, for example, the k-nearest neighbor methods. That means that for very large data sets, Kohonen networks may be impractical because of the computational effort involved to build the models. Because of the greater simplicity of k-nearest neighbor methods, the easier interpretation of results (identification of variables and variable values that define the respective cluster centers), and the better scalability to manage larger data problems, k-nearest neighbor clustering methods are probably the ones most commonly applied in many real-world applications.

Advantages and Disadvantages of Different Algorithms

From the foregoing discussion it should be clear that the k-nearest neighbor and EM algorithm clustering methods are the ones most commonly applied to cluster data sets, particularly when they have large numbers of observations and variables (dimensions). As a practical matter, hierarchical tree methods cannot be scaled up easily to cluster, for example, hundreds of thousands of observations, and the extra computational work required to fit Kohonen networks to such data sets usually is not justified by better "insights" with respect to the nature of the clusters that are revealed.

SINGULAR VALUE DECOMPOSITION, PRINCIPAL COMPONENTS ANALYSIS, AND DIMENSION REDUCTION

As described earlier in this chapter, text mining often yields for each document in the corpus vector numbers that reflect the word or term frequencies for different words and terms (or phrases) used in the documents of the corpus or some other characteristics that describe numerically the content of the documents. As a result, it is common for subsequent modeling or clustering analyses to find that the training data contain large numbers of variables with structured information, as well as large numbers of variables with values derived from text mining, yielding a data set with large numbers (perhaps thousands) of input variables available for subsequent modeling.

In general, when applying any of the methods described so far in this chapter, their performance (for prediction or clustering) will deteriorate when thousands or tens of thousands of predictors are available for modeling. For example, suppose that in a clustering task there are only a few variables of interest among hundreds available for modeling that actually are important for defining some segmentation of observations of (business) interest. Applying clustering methods to all variables will likely "hide" those important variables and result in inconclusive results.

Singular value decomposition and factor analysis methods are generally *dimension reduction techniques* that will take as input the large numbers of variables (available for modeling, e.g., the document by term or word frequency matrix) and derive much fewer new variables as linear combinations of the original variables. This smaller set of variables contains most of the relevant information from those original variables. More details on SVD are provided in Chapter 11. Principal components analysis (PCA) is a related and in fact very similar method that will accomplish the same goal.

Both SVD and PCA can be interpreted as strategies to rearrange the dimensions of the original input space (the variables) to a reduced space that summarizes most of the information contained in the original matrix. To illustrate, repeated here in Figure 8.11 is the simple example discussed in greater detail in Chapter 11. Suppose one conducted a study to measure the height and weight of 100 people and created a scatterplot of the results. The result might look like the one shown in Figure 8.11 (along with additional annotations).

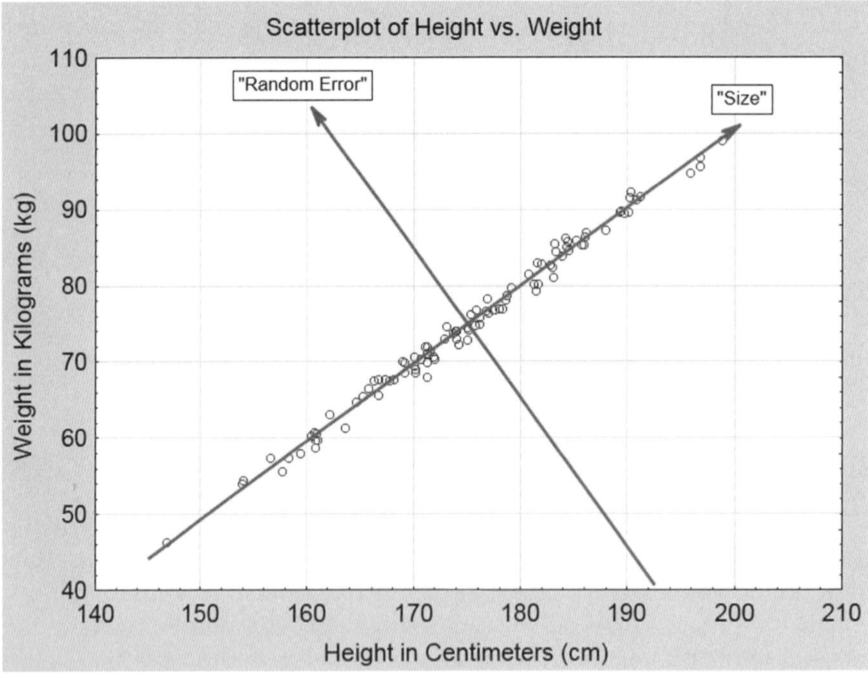

FIGURE 8.11
Rotation in two dimensions.

In short, it is evident that the variability across the individual observations in the study plotted in Figure 8.11 can be summarized with a simple new dimension, computed as the sum (or weighted sum) of the two dimensions *Height in Centimeters* and *Weight in Kilograms.* One might label that new dimension *Size*—that is, the "essence" of the information contained in the two original variables can be summarized via the single label and derived new variable *Size*. This new variable contains most of the relevant information to distinguish the two original variables, and it permits distinction between "small people" and "large people."

Singular Value Decomposition and Principal Components Analysis

SVD and PCA are very similar, and in fact they will yield identical results if these methods are applied to the covariance matrix derived from mean-adjusted document frequencies. In text mining using statistical natural language processing, this is, however, usually not done because it is easier to work with a sparse matrix of actual word frequencies across documents caused by the condition that some words will not appear at all in some or most documents. The alternative matrix to work with is the denser matrix that results by subtracting the means for each word or term frequency, which would lead to values in nearly every cell of the document by word or term data matrix.

Usually because it is easier to work with the sparse document by word/term frequency matrix (or matrix of adjusted frequencies), SVD is used to create a lower-dimensional space from the high-dimensional original data matrix. In terms of the nature and interpretation of these two methods, both are used to derive a small number of dimensions that summarizes the majority of the information contained in the original input matrix.

Partial Least Squares

Partial least squares (PLS) (also known as *projection to latent structure*) is a popular method for modeling industrial applications. It was developed in the 1960s as an economic technique, but soon its usefulness was recognized by many areas of science and applications.

In many ways, PLS can be regarded as a substitute for the method of multiple regression, especially when the number of predictor variables is large, as is often the case when modeling data derived from unstructured text. In such cases, there are seldom enough data to construct a reliable model that can be used for predicting the dependent ***Y-value*** (where Y is a matrix of cases by multiple outcome variables y_i) from the predictor variables X (a matrix of cases by multiple predictor variables x_j). Like PCA or SVD, PLS assumes that the prediction of Y-variables can be modeled with the aid of just a handful of components (also known as latent variables or components). This in fact is the same technique used by PCA or SVD for representing the X variables only, based on few components.

The idea of PLS is to construct a set of components that accounts for as much as possible of the variation in the data set while also modeling the Y variables well. The technique works by extracting a set of components that transforms the original data X to a set of x-scores T (as in PCA). Similarly, the Y data are used to define another set of components known as the y-scores U. The x-scores are then used to predict the y-scores, which in turn are used to predict the response variables Y. This multistage process is "hidden" in the sense that the outcome of this procedure is that for a set of predictor variables X, PLS predicts a set of relating responses Y. Thus, PLS works just as any other regression model. Figure 8.12 shows a schematic summarizing this approach.

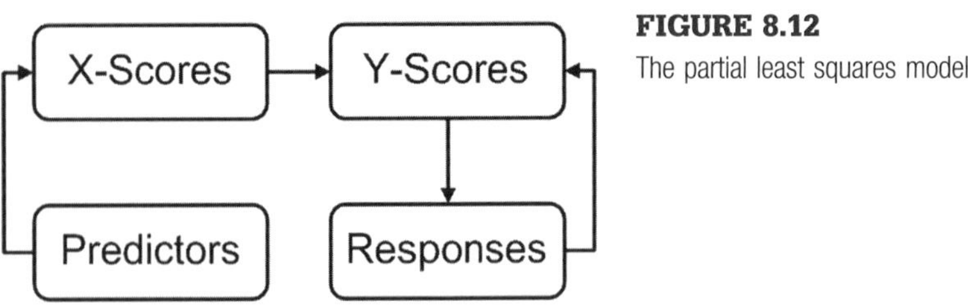

FIGURE 8.12
The partial least squares model.

PLS may prove to be a useful alternative to SVD when there is a specific set of outputs of *Y* variables of interest and the goal is to derive latent dimensions of "meaning," specifically with respect to those outputs of interest—that is, to derive dimensions that provide the greatest correlation (and predictive value) with respect to *Y*.

Feature Selection versus Feature Extraction

In general, PCA, SVD, and PLS can be considered *feature extraction* methods because they derive "features" (e.g., latent dimensions of meaning) from the higher-dimensional input matrix (e.g., the document by word or term frequency matrix containing many words, and thus columns). A general drawback of these methods is that they may "dilute" the diagnostic value for predictive modeling of just a few terms or words by extracting components defined by various words and terms that are not useful for improving a predictive model. For example, in fraud detection applications for insurance, there are typically just a few "red flag" terms that are related to the probability of fraud (see, for example, Francis, 2003).

In those cases it may be useful to directly use the columns of the original document by word or term frequency (or transformed frequencies) instead of components or latent dimensions derived from it. The task then is to apply effective *feature selection* methods that will identify the predictors (parameters, terms, words, phrases) that add to the predictive accuracy of the overall model. Many of the recursive partitioning or tree algorithms described earlier are often used for feature selection because they can efficiently handle large numbers of predictors and even detect interaction effects between predictors—for example, combinations of specific terms that, when both present, are diagnostic of some important outcome.

ASSOCIATION AND LINK ANALYSIS

The goal of the techniques often described as *association mining, link analysis*, or *sequence analysis* is to detect relationships or associations between specific values of categorical variables in large data sets (although the message can also be applied to continuous variables partitioned into discrete intervals). This is a common task in many data mining projects and often in text mining. For example, these methods may uncover frequently co-occurring terms or phrases in a document corpus and thus detect themes, names, places, and so on that are often mentioned in the same document.

How Association Rules Work: "Market Baskets"

Association rules mining algorithms were developed primarily to enable useful analyses of transactional databases and market baskets, illustrated by the following example. Suppose you are collecting data at the checkout cash registers at a large bookstore. Each customer transaction is logged in a database and consists of the titles of the books purchased by the respective customer, perhaps additional magazine titles and other gift items that were purchased, and so on. Hence, each record in the database will represent one customer (transaction) and may consist of a single book purchased by that customer or may consist of many (perhaps hundreds of) different items that were purchased and arranged in an arbitrary order, depending on the order in which the different items (books, magazines, and so on) came down the conveyor belt at the cash register. The purpose of the analysis is to find associations between the items that were purchased—that is, to derive association rules that identify the items and co-occurrences of different items that appear with the greatest (co)frequencies. For example, you want to learn which books are likely to be purchased by a customer who you know already purchased (or is about to purchase) a particular book. This type of information could then quickly be used to suggest to the customer that she might enjoy those additional titles. You may already be familiar with the results of these types of analyses if you are a customer of various online (web-based) retail businesses. Many times when making a purchase online, the vendor will suggest similar items (in addition to the ones you purchased) at the time of checkout based on rules such as "customers who buy book title A are also likely to purchase book title B," and so on.

Association Rules and Text Mining

In text mining, given a matrix of word and phrase counts extracted from a corpus of text (the document by word or phrase frequency matrix), it is often useful to derive association rules between the presence of specific terms or phrases. For example, in a warranty claims application, it may be useful to derive rules linking the mention of two parts (*fuel pump* and *fuel injector*). In other applications, the co-occurrences of specific words or terms may provide insights into the meaning of the text by identifying related themes. Two key statistics are often used to describe the strength of the relationship between two (or more) terms:

Support: First, there is the relative frequency across all documents (or transactions in nontext mining applications) with which the respective two words co-occur; this is referred to as *support*. Support is simply the relative frequency that an item (word, combination of two words, three words, etc.) occurs across the documents. Thus, the support value simply reflects how often a word or term or combination of words or terms co-occur.

Confidence: Second, there is the *confidence* statistic. Confidence is a conditional probability. Given a simple rule or association of the kind *if A, then B*, then given all cases (observations) with *A*, the confidence is the proportion of those cases that also have *B*. For example, in a corpus of text, among all documents that contain the word *Windows*, a certain proportion of documents may also contain the word *Microsoft*; that proportion is the confidence of the rule or association between the terms *Windows* and *Microsoft*.

Sequence Rules

Sometimes the specific *order* in which items or words co-occur is important. Note that *support* and *confidence* only reflect on the relative frequency with which two items or words co-occur across all observations (documents) or the conditional relative frequency, respectively. Algorithms exist to reflect the sequential order of items (terms) in these analyses, for example—returning to the market basket analysis example mentioned earlier—to distinguish between the case where the same customer first purchases a flashlight, followed by a purchase of batteries, versus a case when the customer first purchases batteries, followed by a purchase of a flashlight. Likewise, there are cases in (statistical natural language) processing where the specific order in which terms occur can be important. For example, in the analysis of warranty claims, it can be of critical importance to determine the specific order in which certain components are mentioned in consecutive failure reports by the same customer.

SUMMARY

Practically all text mining tasks and approaches in some way will numericize text—that is, turn the text documents in a corpus into vectors of numbers for each document. This process can add large numbers of new numeric variables to an analysis project that can be used for subsequent modeling to improve the accuracy of models or otherwise improve the usefulness of the results obtained from the analyses. This chapter provides a brief overview of the different types of analyses and most common algorithms used in data mining. Specifically, *supervised learning* algorithms are aimed at identifying repeatable patterns in historical data to predict some continuous or discrete (categorical) outcome. *Unsupervised learning* algorithms can identify repeatable clusters or dimensions in a data set based on the similarity of observations across variables. In addition, there are algorithms and methods to reduce the dimensionality of high-dimensional data sets with numerous variables, either by creating new derived variables that are "useful" for subsequent analyses (feature extraction) or by identifying the specific variables that are diagnostic or useful for further modeling (feature selection). Finally, algorithms for identifying rules and associations between items in a transactional data set (database) were discussed.

In general, the methods described apply learning *algorithms* to derive useful and valid information from data. Unlike statistical modeling, which involves parameter estimation and inferences from models based on statistical reasoning, in predictive modeling (for example) general approximators are used that can represent any type of relationship between predictors and outcomes and usually incorporate methods to establish the validity and repeatability of the patterns in holdout samples (to establish predictive validity).

POSTSCRIPT

Now we are prepared to dive a little deeper into several important areas of text mining. The first is entity extraction (Chapter 9), followed by feature selection and dimensionality reduction (Chapter 10), and singular value decomposition (Chapter 11).

With respect to text mining, many of the methods are particularly useful because of the large numbers of (additional) variables that often result from text mining.

References

Francis, L. (2003) FCAS, MAAA. Martian Chronicles: Is MARS better than neural networks?" March. Available at www.data-mines.com.

Hastie, T., Tibshirani, R., and Friedman, J. (2009). *The Elements of Statistical Learning: Data Mining, Inference, and Prediction*, 2nd ed. Springer Series in Statistics

Hill, T., and Lewicki, P. (2007). *STATISTICS Methods and Applications.* StatSoft, Tulsa, OK. (Also available as Electronic Version): StatSoft, Inc. (2011). *Electronic Statistics Textbook.* Tulsa, OK: StatSoft. http://www.statsoft.com/textbook/.

Hill, T., Lewicki, P., and Qazaz, C. (2007). Multivariate quality control. *Quality Magazine*, April, 38–45.

Lohr, S. (2009). A $1 Million Research Bargain for Netflix, and Maybe a Model for Others. New York Times, September 21, 2009.

Nisbet, R., Elder, J., and Miner, G. (2009). *Handbook of Statistical Analysis and Data Mining Applications.* Academic Press.

Polon, J. (2011). Text mining case study: Text mining for health insurance. Presentation delivered at the 2011 Society of Actuaries Health Meeting, Boston, MA.

Seni, G., and Elder, J. (2010). *Ensemble Methods in Data Mining: Improving Accuracy Through Combining Predictions.* Chicago, Morgan and Claypool Publishers.

Sureka, A., De, S., and Varma, K. (2008). Mining automotive warranty claims data for effective root cause analysis. *Lecture Notes in Computer Science*, Volume 4947/2008, 621–626.

CHAPTER 9

Entity Extraction

CONTENTS

PREAMBLE

We have introduced general classification and prediction operations in text mining and some of the common tools used to accomplish it. Perhaps, you have done one or more tutorials; if so, you have begun to learn by doing. We have kept you insulated from the advanced aspects of any text mining operation, because we wanted to build your text mining skills in layers. With the introduction and tutorials behind you, we can enrich your understanding of some of the advanced topics in text mining, the first of which is using entity extraction technologies for defining and extracting from the text documents names, places, organizations, and other text objects (entities) for further analysis.

INTRODUCTION

Entity extraction is the process of automatically identifying named entities from large collections of unstructured text. Named entities that are extracted in practice include the proper names of people, locations, and organizations. Several approaches for extracting named entities are discussed in this chapter, each of which may be composed of multiple phases of processing. The input to an entity extractor is a collection of text documents, typically contained in existing databases, that are gathered by hand and digitized or gathered automatically from the Internet. The entity extractor ingests such

Practical Text Mining and Statistical Analysis for Non-structured Text Data Applications. DOI: 10.1016/B978-0-12-386979-1.00037-2

a collection of documents and produces a list of named entities suitable for further analysis, including, for each entity, the document IDs and location of the named entities in each document.

Similar to other language tasks, entity extraction is a challenging problem because words can be associated with different meanings in different parts of the same document. For example, the word *general* can be used either as a title or an adjective. When used in a title, as in *General George Washington*, the trailing terms are almost always a person's name. However, when used as an adjective, as in a "general improvement in morale," it does not indicate a named entity.

For many tasks, it is necessary to categorize the named entities by type (e.g., person, location, organization). A category is a named class resulting from a classification operation. Words that are identified as named entities can fall into multiple categories. For example, the word *Washington* is almost always a part of a named entity, but it can also refer to a person or a location (city or state), and it could be used multiple ways in the same document.

In the following sections, we explore how entity extractors address these types of problems by using the context surrounding individual words to identify and label the named entities in a corpus of text.

TEXT FEATURES FOR ENTITY EXTRACTION

The goal of entity extraction is to decide whether or not a particular word or group of words (called the *target*) is part of a named entity. The decision to mark a given target as a named entity depends on the evidence provided by the *features* of that target and, depending on the method, the features of adjacent targets. Commonly used features include membership on a list of known terms, called a lexicon, "word shape" features, and grammatical features such as a part of speech. Each of these features is described in greater detail following. The process of entity extraction requires combining the information provided by the different features into a single prediction for each target.

Lexicons

Lexicon features are the first feature to consider when building an entity extractor. A *lexicon* is a list of words that are grouped by categories. Lexicons can be lists of known entities or may contain lists of "helper words" such as personal titles or corporate identifiers. Often, knowing only the word itself is enough to correctly identify a named entity. Many proper names (e.g., Microsoft) do not overlap with terms in the general English language and always indicate a named entity. If every named entity were unambiguous, then it would be possible to capture all of the named entities in a lexicon. However, this is not often the case.

Lexicons can be used as part of entity extraction in two ways. The first is to identify named entities directly. This is particularly useful for categories of words that correspond almost always to a named entity, such as country names or companies traded on the New York Stock Exchange. The second use of lexicons is to define a set of helper words for identifying named entities. Common helper word lexicons include honorifics (e.g., Mr., Mrs., Baron), street identifiers (e.g., Ave., Way), and company descriptors (e.g., Inc., LLP). An example of a lexicon of titles and honorifics is shown in Table 9.1, which lists those used by the British Airways Executive Club. (The full list contains over 200 possible honorifics and titles.)

Table 9.1 Honorifics and Titles Used by the British Airways Executive Club

Mr	Baron	Countess	Earl	Lieutenant	Prince	Sir
Mrs.	Baroness	Dame	Flying Officer	Lord	Princess	The Honorable
Ms.	Brigadier	Deacon	His Royal Highness	Marquis	Professor	The Right Honorable
Miss	Chief	Deaconess	Her Royal Highness	Marquess	Reverend	Viscount
Dr.	Commander	Duchess	Justice	Monsieur	Sheikh	Vicountess
Master	Count	Duke	Lady	President	Sheikha	Wing Commander

Helper words provide evidence that the target or a surrounding word should be considered as part of a named entity or be labeled with a particular category. However, in certain contexts, such evidence may be misleading. For example, using street identifiers to indicate location entities might be problematic. The phrase "Abbey Road" can refer to a location (the actual street in London), an organization (the famous musical studios of the same name), or another type of entity (the Beatles album recorded at the studios). The challenge of entity extraction is the need to determine when other evidence should overrule a category suggested by a lexicon.

In summary, lexicons can provide strong evidence that a particular term is a named entity, but they do not help in the analysis when a term is not on the list, is used in an uncommon or unfamiliar way, or is used in multiple ways. For most applications, lexicons alone are insufficient for identifying named entities from text, and other features are needed.

Word Shape Features

A critical piece of evidence used for entity extraction is the "word shape." This refers to the way the word appears to the entity extractor. In many languages, including English, capitalization is perhaps the most informative word shape. For example, a capital letter in the middle of a sentence typically indicates a proper noun. Acronyms are almost always listed in upper case. Organizations and product names may contain a mixture of uppercase and lowercase letters (e.g., ThinkPad). Other word shape characteristics include punctuation, such as hyphenation or being followed by a period; whether the word contains any digits; or whether the word contains any foreign characters. Occasionally, the input documents for entity extraction are stored in upper case. In this situation, capitalization must be ignored, putting more emphasis on other feature types.

Grammatical Features

When available, grammatical characteristics of words or phrases can be used to identify named entities. Because entity extraction is only one part of an information extraction process, the results of earlier elements of the extraction process can be helpful for entity extraction. Most information extraction processes include *part of speech tagging* and *shallow parsing* that provide word type and phrasing information, respectively. Shallow parsing techniques can be used to identify subcomponents of sentences, called phrases or chunks. Grammatical information that is valuable for entity extraction includes words that are tagged as nouns or are part of a noun phrase, which are much more likely to be part of a named entity than words that are tagged as verbs. The part of speech and phrase type of the surrounding words can also be helpful in determining whether a target belongs to a named entity. For example, proper nouns almost always indicate some form of named entity, and noun phrases containing proper nouns often indicate a company or place name.

STRATEGIES FOR ENTITY EXTRACTION

The key for successful entity extraction is determining how to combine these different kinds of information about the target and surrounding words into a consistent and accurate strategy for identifying named entities. Two strategies for entity extraction are *rule-based* and *statistical*.

- *Rule-based* approaches define a collection of conditional rules that are applied to the text to identify possible entities. The rule collections are usually built using a combination of automated approaches and hand tuning.
- *Statistical* approaches treat entity extraction as a sequence classification process. This approach uses a classification model to predict whether a word or group of words corresponds to a named entity.

Both approaches employ grammar and language features, along with the context of individual words, to identify named entities. This information can include the following:

1. The presence of the target in a lexicon
2. The word shape of the target including capitalization, hyphenation, and so on
3. Grammatical features such as part of speech
4. The context, including features of words before and after the target word or phrase

Rule-Based Approaches

Rule-based systems were the first automated systems developed for named entity extraction. They were built in the early to mid-1990s as large collections of text began to be collected and mined. In early evaluations of entity extraction technology, such as the Message Understanding Conferences (MUCs), highly tuned rule-based systems were consistently among the most accurate techniques used. These systems use a collection of stochastic grammar rules to combine lexicons, word shapes, and grammatical features to identify named entities. For example, a rule might indicate that a capitalized noun following "Mrs." is likely to be a named entity. These rules are typically weighted, and all matching rules are combined to determine the final label for a given entity (Krupka and Hausman, 1998). To achieve high performance, rules are tuned by hand for a particular corpus of text either by adjusting the weights of different rules or by changing the conditions inside the rule. The tuning process can require significant time by both linguistic and domain experts to identify and test rules for a new domain.

Statistical Approaches

Statistical approaches treat entity extraction as a sequence-labeling problem, with the goal of finding the most likely sequence of entity labels given an input sequence of terms. Statistical approaches have recently been shown to have become the more accurate approach to doing entity extraction (Ratinov and Roth, 2009). In these operations, the entity labels contain each of the categories of interest (e.g., Person, Location, or Organization) and one label indicating that the term is not a name. At the heart of these models is a set of probabilities capturing whether a particular feature of a word corresponds to a particular class label. The most popular statistical approach uses *supervised classification*. Supervised classification employs a set of training data containing the true labels of the target terms to estimate label probabilities. A model is built from these example relationships to be applied to new text to predict the label of words not used in training. Next, we describe two statistical approaches (Hidden Markov Models and Conditional Random Fields) that share a common approach but differ in the representation and procedure used to learn the probabilities and classify labels.

Hidden Markov Models

Hidden Markov Models (HMMs) were the first statistical model of sequences to be applied to entity extraction (Bikel et al., 1997). The assumption behind sequence models like HMMs is that text is originally generated in label-word pairs, but the labels have been lost (or "hidden"). Thus, entity extraction can be viewed as an attempt to recover the sequence of original labels associated with the text.

HMMs are a *generative* model—that is, they attempt to recreate the original generating process responsible for creating the label-word pairs. As a generative model, HMMs attempt to model the most likely sequence of labels given a sequence of terms by maximizing the joint probability of the terms and labels. A joint probability indicates the frequency of multiple events occurring together. These probabilities are estimated by counting the transitions between particular label states, including a null (or not-a-name) state and possibly an end-of-sentence state. Transition probabilities for each word are computed from observations in the training data. These learned transition probabilities are then used to estimate the most likely sequence of labels given new data. (This sequence estimation can be computed efficiently using the Viterbi algorithm [Viterbi, 1967].)

Since many of the possible word-label transitions are not observed during training, computing an empirical probability of observing a novel word sequence is difficult. Since the HMM cannot estimate this probability directly, some type of smoothing or "back-off" model is used to estimate the probabilities of unobserved transitions. A back-off model estimates the specific transition using other information, such as the probability of the word appearing in the corpus and the background probability of making a particular transition. (See Bikel et al., 1997, for information about models for back-off and smoothing.)

Conditional Random Fields

As a generative model, HMMs attempt to model all possible word and label sequences. However, in the average corpus, only a small subset of all possible sequences will ever be observed, meaning that HMMs waste effort on most unobserved sequences. Conditional Random Fields (CRFs) are a *discriminative* alternative to HMMs and are designed to model the conditional probability of a label sequence given a sequence of words directly, avoiding estimation of the full joint probability (Lafferty et al., 2001). Conditional probabilities indicate the probability of an event occurring given that other events have already occurred. The probability of the latter event is "conditioned" on the earlier events having already occurred.

CRFs are more expressive than HMMs, and they are consequently more powerful, allowing the model to be estimated from less training data. Modeling only the conditional probability and not the joint probability allows the CRF to focus its power on the observed sequences of words. Using a conditional model like a CRF also permits employing many overlapping features without a large efficiency penalty. CRFs improve upon previous conditional sequence models called Maximum Entropy Markov Models (MEMMs), also known as Conditional Markov Models (CMMs). Like an HMM, the parameters of a CRF model can be estimated from training data and then used to compute the most likely sequence of labels for the rest of the text.

CHOOSING AN ENTITY EXTRACTION APPROACH

Both rule-based and statistical approaches have been shown to perform at near-human levels if sufficient effort is invested into tuning or training the models. The effort required to reach this performance, however, is different in each approach. Rule-based approaches do not require any labeled training to start processing. Any analyst can begin using the tagging rules with only minimal training; yet, reaching the full performance can require extensive effort and linguistic capabilities to capture all of the special cases that routinely occur when tagging text. This tuning process is typically required whenever the rule engine is applied to a new corpus of text. In contrast, statistical approaches require significant effort up front to create tagged training data suitable for input into the statistical machinery. The tagging process does not require any advanced capabilities, but the number of documents required (typically several hundred) can be intimidating. Once the training data have been tagged, the statistical machinery is more flexible and can be quickly applied to new collections of text. Choosing an approach for entity extraction then depends on time constraints and the availability of labor and technical expertise.

Tagging Standards

Deploying an entity extractor typically requires either encoding input files in a particular format or creating output that will be input into other components of a system. The two standard tagging formats used for entity extraction are the Standard Generalized Markup Language and the Inside-Outside-Begin (IOB) tagging system.

Standard Generalized Markup Language

The most common format for tagged documents is the Standard Generalized Markup Language (SGML), in which documents contain inline XML-like tags around named entities. A simple SGML markup uses the entity type as the name of the tag, as in "<PERSON>General George Washington</PERSON>." A more sophisticated SGML markup has been developed for the MUC. These tags are of the form "<ENAMEX Type=PERSON>General George Washington</ENAMEX>." The standards vary on the question of whether or not to include honorifics and titles such as "General" as part of the tag. Depending on the task, the titles may or may not be counted as part of the entity in evaluation. Regardless of whether they are included in the evaluation, titles are always valuable as input features for detecting named entities.

The Inside-Outside-Begin Tagging System

An alternative tagging standard is the Inside-Outside-Begin (IOB) tagging system. It marks each term with an "I-<TYPE>" if the term is inside a tag of type <TYPE>, "O" if the term is outside of any tag, and "B-<TYPE>" if the term indicates the beginning of a tag. This tag format is often used in a file format where each term is on a single line followed by the tag.

EVALUATING ENTITY EXTRACTION

The efficacy of an entity extractor on a corpus is determined by scoring the output of the system against known labels for the same type of corpus. The following sections describe methods for scoring entity extraction and different ways to evaluate those scores.

Scoring Metrics

Entity extraction systems are scored using precision and recall. The *precision* of an extractor is the percentage of predicted named entities that are correct. *Recall* is the percentage of occurrences of a given named entity found by the system compared to total occurrences of the entity found in all of the data.

$$\text{Precision } (p) = \#\text{ correct}/\#\text{ found}$$
$$\text{Recall } (r) = \#\text{ correct}/\#\text{ in true data}$$

Usually, precision and recall are combined in a single measure called the *F-measure*. That is a weighted average of precision and recall given by the following equation:

$$f = (b+1)^{*}p^{*}r/((b^{*}p)+r)$$

where b represents the relative weight of recall over precision.

Most often, precision and recall are weighted equally, causing the preceding equation to collapse into the following equation:

$$f1 = (2^{*}p^{*}r)/(r+p)$$

Precision and recall can be computed for individual terms or complete entities. When computing by entity, special consideration must be taken to determine how to assign partial credit for variations in the start and end of an entity. The performance of the system may vary slightly depending on the method of computing the p and r metrics.

Determining a "Gold Standard"

Once scores have been determined, it is necessary to see how well they compare to baseline scores. Ideally, the true labeling is available. However, for many entity-extraction tasks, even experts may be unable to agree on the true labeling. In such cases it is best to use labeling supplied by *annotator agreement* to evaluate competing methods. In this agreement process, multiple analysts study the training and evaluation data and supply labels, and a committee of experts deal with the areas where individual analysts disagree.

Occasionally, this adjudication process can be quite difficult. One of the authors attended a workshop where one session focused on examining pilot transcripts and determining whether a spoken mention of, for example, "Baltimore" referred to the airport or the city. It turned out to be surprisingly difficult! To evaluate the automated systems, human annotators marked each transcript with their best decision. Then a committee of experts provided a second layer of review to adjudicate the differences among the human annotators. The best individual human was in agreement with the committee less than 80 percent of the time!

SUMMARY

Entity extraction is a mature subdiscipline in text mining. Both rule-based and statistical approaches are effective for creating high-performing entity extraction systems. These powerful systems have opened the door for analysts to perform entity extraction in more challenging environments, including many languages and extremely "messy" text.

POSTSCRIPT

After the named entities have been identified and extracted, we could have literally millions of terms to analyze. This large number of things to analyze poses two problems: The sheer number of items can take a very long time to analyze, and analysis accuracies will be not be optimal because the analysis technology cannot focus sufficiently on important items. In Chapter 9, we present some common feature selections and reductions of dimensionality technologies to sharpen the focus of the analysis on important keywords and permit it to happen in a reasonable time.

References

Bikel, D. M., Miller, S., Schwartz, R., and Weischedel, R. (1997). *Nymble: a high-performance learning name-finder.* In Proceedings of the Fifth Conference on Applied Natural Language Processing.

Krupka, G. R., and Hausman, K. IsoQuest Inc.: *Description of the NetOwl (TM) Extractor System as Used for MUC-7.* In Proceedings of MUC, Volume 7, 1998.

Lafferty, J., McCallum, A., and Pereira, F. *Conditional random fields: Probabilistic models for segmenting and labeling sequence data.* In Proceedings of 18th International Conference on Machine Learning, 2001.

Ratinov, L., and Roth, D. *Design challenges and misconceptions in named entity recognition.* In Proceedings of the Thirteenth Conference on Computational Natural Language Learning. Association for Computational Linguistics, 2009.

Viterbi, A. *Error bounds for convolutional codes and an asymptotically optimum decoding algorithm,* IEEE Transactions on Information Theory, Vol 13, No 2, pp. 260–269, 1967.

Feature Selection and Dimensionality Reduction

CONTENTS

PREAMBLE

Feature selection techniques remove features that have low information content, clarifying the data by identifying the more important features. Similarly, dimensionality reduction techniques extract new features from the data by combining two or more existing features. A simple example of dimensionality reduction strategy is to replace factors of "length" and "width" by multiplying them together to create the new factor, "area." Area has information elements related to length and width, in addition to elements related to their combination, and the combination factor (area) might be far more predictive than its components used separately.

INTRODUCTION

The most straightforward way to abstractly represent text data is to create a dimension for each unique word and then encode a phrase or even a full document as a single point in an extremely high-dimensional space. For instance, the sentence "The suspect ran swiftly down the street" becomes a point that is in the 1 location for the dimensions "the" through "street" and zero for every other dimension (word in the allowed vocabulary).[1] This abstraction is simple and powerful, it but has one huge

[1] In setting up the vector space representation, two design decisions must be made: whether to count common "stopwords" such as *the* and whether to count existence or frequency. In this example, stopwords are not preexcluded, and existence is noted. If frequency were used instead, the location along the *the* dimension of the point representing this phrase would be 2.

Practical Text Mining and Statistical Analysis for Non-structured Text Data Applications. DOI: 10.1016/B978-0-12-386979-1.00038-4

challenge: The resulting data dimensionality is far beyond what can be handled easily at the speeds required for many algorithms.

Feature selection and dimensionality reduction techniques are very useful for greatly reducing the size of the input for text mining algorithms. Usually, simple steps can speed up processing with only minimal reductions in accuracy. Done well, they can even improve accuracy by focusing the learning algorithms on the most important keywords and reducing the overfit that can occur when vast numbers of potential inputs can lead to apparent, but accidental, relationships between keywords and outcomes that do not generalize well to new data.

Many words occur either too frequently or infrequently to provide much value for text mining algorithms. Without selection or transformation, these features lead to extra processing, increased variance, and (typically) lower performance due to overfitting on the extra features. To limit the contribution of such "noise" features, one must identify the valuable features and remove the features with low value. Also, one can transform the feature space into a lower dimensionality space by combining features to summarize the original dimensions. One might start with, say, features A–Z and then identify that only seven $\{A, B, C, E, F, J, Z\}$ are needed, and then further reduce the dimensionality to three features α, β, χ, which are each a different equation using the subset of seven.

FEATURE SELECTION

Feature selection techniques are applicable in supervised text classification where a specific target variable (topic or favorable/unfavorable rating, etc.) can be assigned to each document. To prioritize features, one must have a way to measure their predictive strength. Feature selection techniques differ primarily in how they determine this.

Feature selection algorithms originated in the fields of statistics and traditional data mining. Text mining algorithms are data mining algorithms that have been applied to unstructured text data that have been translated into a structured, numerical representation. In data mining, two classes of feature selection algorithms have been considered: filters and wrappers. *Filters* use some criterion such as information gain or statistical correlation to filter out low-performing features. They can be applied separately from the predictive algorithm, performing the feature selection before the learning. In contrast, feature selection *wrappers* use a search procedure guided by a performance metric to find a high-performing subset of features. An example of a feature selection wrapper is a classification tree algorithm (e.g., as discussed in Chapters 7 and 8). Both filters and wrappers use similar measures to select features; their main difference is how they choose the threshold for discarding features.

As with traditional data mining, the top reason for introducing a feature selection stage to a text mining solution is to improve its computational efficiency. For this reason, we will focus primarily on filters for feature selection. The search process required for performing wrapper feature selection alone is typically too computationally intensive for high-dimensional textual data. If a wrapper algorithm is desired (for example, for learning an interpretable model), a feature filter can be applied as a pre-processing step before applying a wrapper model.

Removing features often leads to improvements in predictive accuracy of the overall model by reducing the overall variance of data. Feature selection prevents an algorithm from placing too much weight on "noise" features, called overfitting, when the feature accidentally has some value on the training data but does not generalize well to other data sets. Without noise features, the algorithm provides the most weight to features that carry over from data set to data set, which leads to greater accuracy.

FEATURE SELECTION APPROACHES

The following are the three primary categories of feature selection algorithms used in text mining, plus hybrid approaches that cross category lines:

1. Information theoretic approaches
2. Statistical approaches
3. Frequency-based approaches (term counts, document counts)

Information Theoretic Approaches

Information theory was developed by Claude Shannon in the 1940s and is concerned with how to most efficiently process signals, such as for data compression for storage or communication over a network (Shannon, 1948). Information gain (IG) is a widely used information theoretic approach for feature selection due to its success at removing noise features (Yang and Pederson, 1997). IG measures how much the uncertainty about the target variable, called *entropy*, is reduced when the feature is used. If the feature has high IG, it is useful for predicting the occurrence of the target. After all of the features have been ranked by the IG metric, either a fixed number of top features or features above a certain IG threshold are chosen for use. Yang and Pederson (1997) showed that selecting features by IG can reduce them by 98 percent, while still resulting in an increase in predictive performance. Their study showed that fewer than one thousand words (selected using IG) were sufficient to maintain the same level of predictive accuracy as all of the 50,000+ features. IG is effective for feature selection, yet its computational cost depends on both the number of documents and the number words in the vocabulary, and unfortunately it can be prohibitive for large collections.

Statistical Approaches

Statistical approaches for feature selection measure the statistical correlation between the features and the target labels. The most widely used measure is the χ^2 statistic, which measures the independence, or lack thereof, between the target label and a feature. Features that are completely independent of the target label provide no predictive value. As with the IG metric, features are ranked based on their χ^2 statistic. Though it is also possible to use the χ^2 statistic as part of a hypothesis test with a specific p-value to discard features that meet the strict standard of independence, this approach does not lead to reliable decisions about independence when the features are sparse (Manning et al., 2008). Instead, using the χ^2 statistic as a ranking metric performs similarly to the IG metric, though accuracy drops off faster when more features are removed (Yang and Pederson, 1997). The computational effort for computing the χ^2 statistic is proportional to the number of words in the vocabulary.

Frequency-Based Approaches

Frequency-based methods for feature selection are the simplest, and they also scale the best to large corpora due to their computational efficiency. Two frequency-based approaches are term frequency (TF) and document frequency (DF). The TF strategy consists of removing features that only occur a few times in the corpus. DF counts the number of documents in which a term appears and removes features that appear in only a few of the documents. Both strategies make the assumption that rare words are "noise" words and, therefore, not informative for prediction. In many cases this is true, but rare words can be highly informative in certain situations. Information theoretic and statistical approaches are better able to highlight the rare words that have predictive value.

Feature Selection Summary

Feature selection is an effective method for reducing the number of input dimensions, leading to improvements in computational speed of the analytical algorithms and often leading to improvements in performance of the learned model. Statistical and information theoretical approaches do a better job of selecting higher-value features, but they require more computation time compared with the frequency-based approaches.

DIMENSIONALITY REDUCTION

Dimensionality reduction techniques address the "curse of dimensionality" by extracting new features from the data, rather than removing low-information features. The new features are usually a weighted combination of existing features. Dimensionality reduction techniques for text mining are drawn from those for traditional structured data. Unlike the feature selection methods just described, dimensionality reduction techniques can be used with both supervised (classification) and unsupervised (clustering) analytical methods. As with feature selection, dimensionality reduction can decrease the size of the data without harming the overall performance of the analytical algorithm.

There are many different types of dimensionality reduction algorithms, and they can be categorized along three dimensions. The first is whether the feature extraction process uses linear or nonlinear combinations of existing features when creating the new features. Second, different dimensionality reduction techniques operate on either a covariance matrix between the features or on a term-document matrix (also known as a vector space; see Chapter 3 for more on vector spaces). Another type of matrix is the term-context matrix, which is useful for concept extraction. Third, reduction approaches are categorized as either second order or higher order methods. The order indicates how many variables are considered at once; "second order", for example, indicates that only pairs of variables are considered. Table 10.1 summarizes the main approaches to dimensionality reduction along these three dimensions; the algorithms are described in the following sections.

Table 10.1 Linear Dimensionality Reduction Techniques

Algorithm	Matrix	Order
Latent semantic indexing	Term-document	Second order
Principal components analysis	Coefficient	Second order
Factor analysis	Coefficient	Second order
Projection pursuit	Coefficient	Higher order

LINEAR DIMENSIONALITY REDUCTION APPROACHES

The rest of this chapter is devoted to linear dimensionality reduction techniques, since they are by far the most widely used in text mining today. We discuss the most popular approaches, including the following:

- Latent semantic indexing (LSI)
- Principal components analysis (PCA)
- Factor analysis (FA)
- Projection pursuit

Latent Semantic Indexing

LSI is a linear dimensionality reduction technique that operates on the term-document matrix. LSI uses a common matrix operation called the singular value decomposition to identify independent components of the data. LSI is one of the most popular dimensionality reduction algorithms for text because it often provides dimensions with semantic meaning; features in the same dimension are often topically related. Chapter 15 provides an in-depth discussion of the singular value decomposition (SVD) for LSI.

Principal Components Analysis

Like LSI, PCA relies on the SVD, but PCA operates on the covariance or correlation coefficient matrix between features, not on the term-document matrix. PCA was created by Karl Pearson (1901) for the purposes of transforming a data set with possibly many correlated variables into a simpler data set with fewer but uncorrelated variables. For many data sets, the largest principal components capture most of the relevant information in the data, allowing the other components to be discarded (Fodor, 2002). Because the components are based on feature correlation and not document co-occurrence (like LSI), the components extracted by PCA are not easily interpretable. For some applications, this makes PCA less desirable, despite its effectiveness at reducing the dimensionality of the data without harming performance.

Factor Analysis

FA comes from the field of psychology where it is used to identify variables that may have been caused by a shared, unobserved "factor" (Fodor, 2002). FA operates on the covariance matrix between features and uses matrix arithmetic to identify overlapping features. It is a second-order, linear method.

Projection Pursuit

Unlike the preceding methods, projection pursuit is a linear method that can incorporate more than second-order information, making it well suited for data that are not normally distributed (Fodor, 2002). Projection pursuit uses some measure of quality to determine which projections of the data to pursue. It can be thought of as a generalization of PCA, since running projection pursuit on the second-order variance matrix produces the same results as PCA.

POSTSCRIPT

Chapter 11 (which deals with singular value decomposition) could have been described in Chapter 10 as a feature selection method. But the complexity and the usefulness of it are so profound that we decided to dedicate a chapter to that technology alone.

References

Fodor, Imola K. *A Survey of Dimension Reduction Techniques*, Center for Applied Scientific Computing, Lawrence Livermore National Laboratory, 2002.

Manning, Christopher, Raghavan, Prabakhar, and Schutze, Heinrich. *Introduction to Information Retrieval*, Cambridge University Press, New York, USA, 2008.

Pearson, Karl. On lines and planes of closest fit to systems of points in space. *Philosophical Magazine Series*, 1901, vol. 2, no. 11, pp. 559−572.

Shannon, Claude. A Mathematical Theory of Communication. *Bell System Technical Journal*, 1948, v. 47, pp. 379−423.

Yang, Yiming, and Pedersen, Jan O. *A comparative study on feature selection in text categorization*. In Proceedings of the International Machine Learning Conference, 1997.

CHAPTER 11

Singular Value Decomposition in Text Mining

CONTENTS

PREAMBLE

Singular value decomposition is a form of dimensionality reduction, but it adds to the total information resident in the variables, rather than eliminates sources of information. It does this by

Practical Text Mining and Statistical Analysis for Non-structured Text Data Applications. DOI: 10.1016/B978-0-12-386979-1.00039-6

calculating linear combinations of existing variables. This process is analogous to calculating the variable "length" by variable "width" to derive variable "area." Area has information from both length and width, even though the original variables are eliminated. The result is to reduce the number of features submitted to the modeling algorithm, which will allow it to work more efficiently. You may be thinking, "Isn't this a form of data preparation?," and the answer is yes! Along with basic preprocessing steps (Chapter 3), parts of text classification (Chapter 7), entity extraction (Chapter 9), feature selection (Chapter 10), and clustering (Chapter 13), much of what you do in text mining (about 90 percent) is data preprocessing! Get used to it and treat it as a challenge for excellence, rather than view it as a necessary evil to "get over with" before the "fun" of model building can begin. Singular value decomposition may appear to be complicated, but you will find that a rather basic understanding of it can be quite rewarding in your model building activities. So spend some time digesting (rather than just reading) this chapter. You will find it worth the extra effort.

INTRODUCTION

Singular value decomposition (SVD) (see, for example, Manning and Schütze, 2002) can be used in data mining and predictive modeling to reduce the dimensionality of a data matrix. For example, when a data set includes 1,000 or more predictors, SVD allows analysts to construct a much smaller number of linear combinations of those predictors. The linear combinations or "weighted sums" of the values can then be used to build predictive models.

The SVD computations will create the linear combinations of variables subject to a number of constraints:

1. Each successively constructed linear combination of variables extracts from the data matrix the maximum amount of "information"; this will be discussed further following.
2. The linear combinations of variables are orthogonal to (or independent of) each other, so each linear combination contains "different information."

Therefore, SVD is a very useful *feature extraction* tool. It allows a very large number of variables to be reduced or summarized to a much smaller and more manageable number of linear combinations that extract the majority of information from the variables.

This method is particularly useful in the context of text mining and statistical natural language processing, where text documents are "numericized" so that each document is represented by a large number of numeric variables (e.g., with word counts). In fact, in the context of statistical natural language processing, the linear combinations themselves may sometimes clearly identify underlying or "latent" dimensions that can be interpreted as *latent semantic dimensions*—that is, as the important dimensions of "meaning" into which the corpus of documents can be mapped while maintaining the information necessary to differentiate between documents (so that each document can be assigned to a unique location in the reduced dimensional space).

This chapter elaborates on the SVD method, when it is and is not useful, and how the results can be interpreted and used in subsequent data analyses and modeling to improve domain insight, identify clusters of documents or words and phrases, or improve the accuracy of predictive models.

REDUNDANCY IN TEXT

The documents in most text corpora are highly redundant. For example, the consumer reviews published on popular websites for cars, electronics, books, and so on usually center around a few "themes" that are relevant for the respective types of consumer items. To illustrate this point, here are four recent (as of this writing) reviews of the Chevrolet Volt—an innovative hybrid gasoline/electric car)—published on Edmunds.com.

1. "My Volt is incredible. My lifetime mpg is 112 and rising. I've driven over 1,000 miles without any gas. I didn't buy any gas the first month I owned it. It has a sport mode that makes it fun and sporty to drive, with speed to jerk your head back. It's as powerful as a V-6. I'm loving it."
2. "I have only had my Volt #2499 for a week, and am I impressed. The car, the concept, everything about it is game changing. My other car is a CTS-V, and it has been parked ever since I got the Volt. The only problem is that my wife wants to trade in her SUV and get one, too. Saving gas $$ is not my objective; using the cutting-edge technology and proving it in everyday use is the fun part. What a tribute to the team that put this vehicle together and brought it to market. Let's hope the "halo effect" spills over and helps all of GM."
3. "The Volt is my first American car, ever. I currently also have a VW and SAAB. The ride and handling to me are very much like a European vehicle. The Volt is smooth and firm, with some heft to it, but is also comfortable with a compliant ride. It absorbs bumps very well and exceeded my expectations when I test drove it. The acceleration in "normal" mode is adequate. In "sport" mode, it is a blast to drive. The car is also eerily quiet in EV mode. The fit and finish are impressive. This is an extremely well-thought-out vehicle. The changeover to extended mode is absolutely seamless. Other than the display on the dash, there was no way to tell that the gas engine engaged."
4. "I purchased my Volt and took delivery on December 18, 2010. I've gone about 1,500 miles so far and burned about 20 gallons of gas, mostly due to a few long trips (200 miles each). Even on those trips, my mileage was in the 40- to 50-mile range. My overall mileage is about 75 mpg. But when I was not taking those trips, I could easily drive around town, to work, etc., on the 40 miles or so range, using all battery and electric drive. There is nothing better than that! The car drives solidly, crisp turning, and more than decent acceleration. The best acceleration is in LOW using SPORT mode. Seats are comfortable, and the fit and finish are quite good."

In total, these reviews contain 415 words. However, they can probably be summarized in one sentence:

> "The Chevrolet Volt delivers excellent gas mileage, is comfortable, delivers good performance, and is generally liked by reviewers."

This summary consists of a total of 18 words, and it probably communicates the core dimensions and evaluations contained in all the reviews. Put another way, much of the text in the individual reviews is redundant, describing just a few dimensions and characteristics of the car. SVD extracts the relevant dimensions from a corpus of text—that is, the common dimensions and concepts discussed across all of the documents (reviews in this case).

DIMENSIONS OF MEANING: LATENT SEMANTIC INDEXING

SVD is commonly used in statistical natural language processing to identify "latent dimensions of meaning" that organize the documents in the corpus and the words, terms, or phrases used in the documents. This is perhaps best illustrated with an example.

Example: Causes of Aircraft Accidents

The U.S. National Transportation Safety Board (NTSB) maintains a detailed database of aircraft accidents. When an aircraft accident occurs in the United States, at first a brief narrative is created and stored in the database describing the circumstances and facts observed during and immediately following the accident (such as the weather conditions, initial descriptions by witnesses of the accident sequence, etc.).

Suppose one analyzed the initial accident narratives and extracted all of the unique individual terms used in those narratives (typically, common and usually "nondiagnostic" words like *it, has, do,* and so on are excluded from the indexing of terms). As a result, a data matrix would be obtained where each row summarizes the word frequencies (the frequencies with which a respective term or phrase is used in the narrative), and the columns summarize the term frequencies (one column for each extracted term). Table 11.1 shows a small section of such a document by term data matrix.

Table 11.1 Documents by Words/Terms Frequency Matrix

Accident Report Number	Term-document frequency matrix										Term-document frequency matrix										
	1 abl	2 abnorm	3 aboard	4 abort	5 abrupt	6 acceler	7 accid	8 accomplish	9 accord	10 accumul	868 tri	869 trim	870 trip	871 tube	872 turbul	873 turf	874 turn	875 twice	876 two	877 type	878 unabl
1					1		1														
2																					
3																					
4							1		2								2				1
5				1																	
6																					
7																					
8		1		1					1			7						1			
9																					

Again, each row in this matrix represents the narrative associated with one aircraft accident or incident. Each column represents one word or term extracted from the corpus of aircraft accident reports. The entries in this matrix represent simple counts—that is, the frequencies with which the respective terms occur in the respective documents.

This data matrix that numericizes the accident reports can become rather large, so SVD may be useful here to reduce the dimensionality of these data. After applying SVD to these data, the patterns shown in Figure 11.1 emerge for the first two dimensions with respect to coefficients or weights that identify the degree to which the respective terms define the first and second components (dimensions).

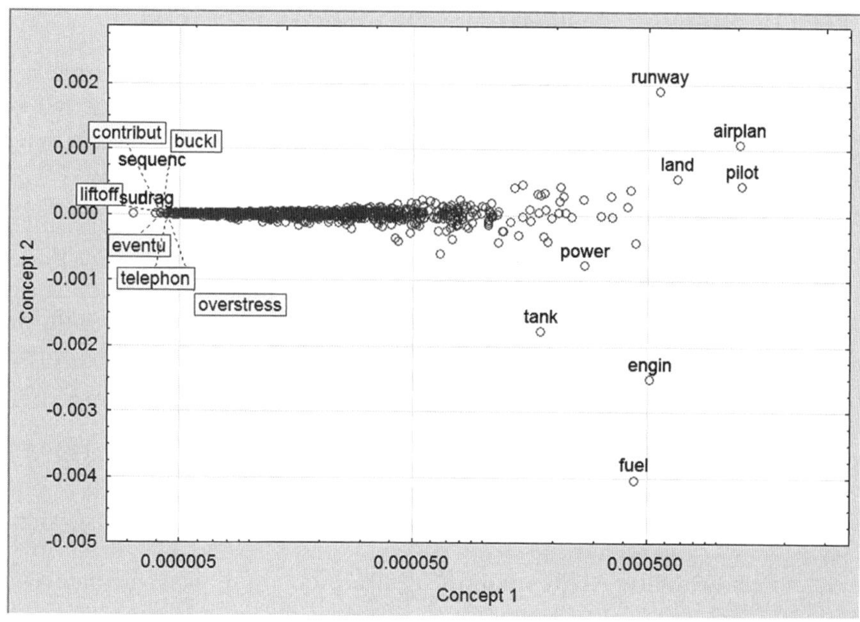

FIGURE 11.1
Scatterplot of selected words by two latent semantic dimensions.

Interpreting the Scatterplot

First, recall that this graph was derived from "real data," with practically no data preprocessing, so the results are usually not trivially obvious. Interesting patterns, however, still emerge.

The first dimension in Figure 11.1—Concept 1—uses log-scaling along the *x*-axis. To reiterate, in this example the simple word frequencies were analyzed. There are transformations of those frequencies that result in more evenly distributed word coefficients and more easily interpretable results. (These transformations were not used here; see, for example, Chapter 5.)

On the left side of Concept 1, we see terms like *liftoff* and *overstress*. On the right side are terms such as *pilot* and *engine*. So, subjectively, perhaps a good "label" for Concept 1 would be *general to specific* aspects and potential causes of the accident.

The second dimension—Concept 2—shows the terms *runway, airplane,* and *land* at the top, and *fuel, engine,* and *tank* at the bottom. So perhaps a good label for Concept 2 would be *accidents related to pilots landing on a runway* versus *accidents related to engine and fuel problems.*

To summarize, SVD was used to reduce the document by term frequency data matrix (where each document represented one accident narrative) to create a two-dimensional "space" where each term that was extracted receives a weight denoting its influence on the definition of the respective dimension. The actual interpretation of these dimensions is usually somewhat subjective. However, note how in this brief overview example, a complex topic (aircraft accident narratives) was analytically reduced or "projected" into two dimensions that appear to have reasonable interpretations.

Several issues must be considered when applying SVD to extract "dimensions of meaning" from a corpus of text. These will be discussed after a brief formal presentation of the SVD method.

THE MATH OF SINGULAR VALUE DECOMPOSITION

Technically, SVD will "rearrange" a matrix of *n* rows and *m* columns (e.g., of *n* documents by *m* words or terms, where the values in the matrix denote word or term frequencies, or relative frequencies such as inverse document frequencies; see Chapter 5). Specifically, the matrix is rearranged so that the following happens:

1. The values in each column are linear combinations of the values in the columns of the original matrix.
2. The values in consecutive columns (linear combinations) will provide the closest respective approximation of the original document by word or term frequency matrix; thus, the first dimension will provide the best approximation ("reproduction") of the document by word or term frequency matrix that is possible with one dimension, the second dimension will provide the best approximation possible with two dimensions, and so on. In that sense, the respective numbers of dimensions provide a reduction of the document by word or term frequency matrix to lower dimensional space, maximizing the information extracted from that matrix using the respective numbers of dimensions.
3. Consecutive columns are orthogonal to each other—that is, the consecutive linear combinations that extracted are unrelated to one another.

GRAPHICAL REPRESENTATIONS AND SIMPLE EXAMPLES

An intuitive way to illustrate the logic of this operation is to think of it as a rotation in *m* dimensional space. As an example, suppose you conducted a study to measure the height and weight of 100 people,

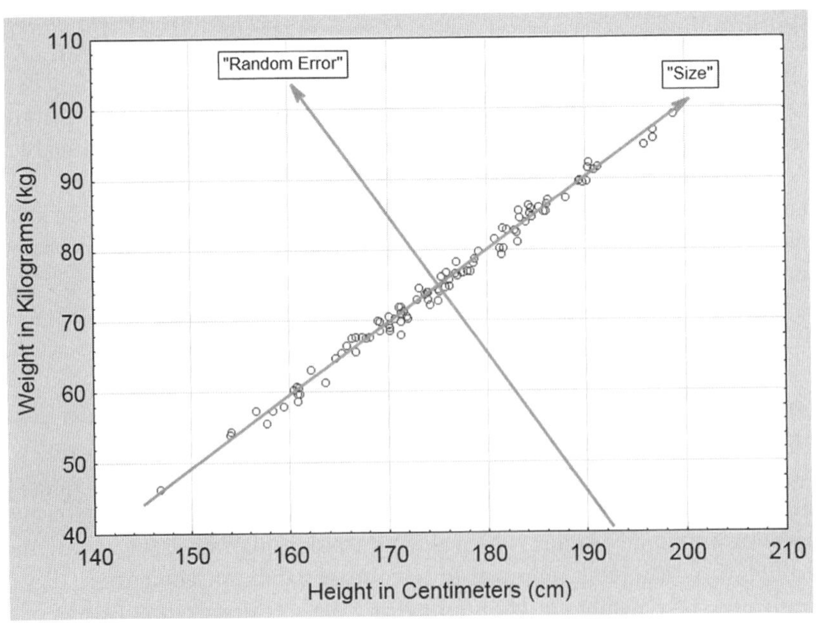

FIGURE 11.2
Rotation in two dimensions: height and weight.

and you created a scatterplot of the results. The scatterplot might look like the one shown in Figure 11.2 (along with additional annotations).

Figure 11.2 shows each individual in the study as a point against the horizontal *x*-axis "Height in Centimeters" and the vertical *y*-axis "Weight in Kilograms." As expected, people who are taller also tend to be heavier, and this is reflected in the strong positive linear relationship between height and weight.

Extracting Information from a Single Dimension

All of the points in Figure 11.2 can roughly be represented by a single line. This line is depicted as an arrow, labeled "Size" at the tip (upper right-hand corner of the graph). In fact, one may think of this arrow as representing a latent dimension that can be called *Size*, which approximates ("summarizes") the original data matrix consisting of the measurements of the individuals in the study on the two original variables. In that sense, the first dimension *Size* extracts most of the *information* from the data collected in the study.

The second arrow and dimension, perpendicular or orthogonal to the one labeled "Size," is annotated as *Random Error*. There is very little variability over the range of that dimension, and for practical and analytical purposes, it would be appropriate to just consider the remaining variability around the first *Size* dimension random "noise." As a practical matter, one could differentiate between practically all participants in the study by just using the *Size* dimension, while the *Random Noise* dimension does not contribute much useful information at all.

Rotation and Dimensionality Reduction

This simple example illustrates the basic mechanism and computations involved in SVD. Starting with a two-dimensional problem (height and weight), a single dimension can be defined that extracts most of the information from the original data matrix (the *Size* dimension). This new dimension can be expressed as a linear combination of the two variables, for example, as the simple sum of the two original dimensions.

As a result of these computations, the two-dimensional data have been reduced to a single dimension; this is the principle value of SVD for text mining, or for the reduction of any high-dimensional data to lower-dimensional space. SVD will identify consecutive linear combinations of the original variables, where after each consecutively extracted dimension there is minimal residual variability "left over" in the data matrix after reproducing the data values from the respective numbers of dimensions that are retained.

SINGULAR VALUE DECOMPOSITION IN EQUATION FORM

Given a matrix **A**(n,m), SVD will compute matrices **U**(n,m), **w**(m), and **V**(m,m) so that **A** = **U** * **W**(diagonal) * **V′** (where **V′** stands for the transposed **V** matrix, and **W**(diagonal) stands for a diagonal matrix(m,m), with the elements of vector **w** in the diagonal).

If matrix **A** is of reduced rank—that is, it has some columns or rows that are linear combinations of other columns or rows—then SVD will return as many nonzero singular values (in vector **w**) as there are independent column vectors that can be constructed from the columns and rows of **A**. Thus, the number of positive singular values in vector **W** is equal to the rank of the source matrix. Matrix **U** will contain a set of orthonormal vectors (so for each column vector $\mathbf{u}_i$ in

U, $\mathbf{u}_i * \mathbf{u}_i' = 1.0$, and for every pair of column vectors $\mathbf{u}_i$ and $\mathbf{u}_j$, $i \neq j$, $\mathbf{u}_i * \mathbf{u}_i' = 0.0$) for each singular value $w_i > 0.0$.

SINGULAR VALUE DECOMPOSITION AND PRINCIPAL COMPONENTS ANALYSIS EIGENVALUES

SVD is similar to principal components analysis (PCA). In fact, the two methods yield identical results if they are applied to the covariance matrix of the word or terms. Put another way, if the mean word or term frequencies were subtracted from the observed frequencies in each column of the document by word or term frequency matrix, both methods would produce the same results. However, SVD is usually applied to the relatively sparse document by word or term frequency (only a few documents have specific terms), while PCA is typically applied to the symmetric covariance matrix.

In terms of interpretation, PCA will maximize the variance of consecutively extracted dimensions, while SVD will minimize the residual sums of squares of the deviations of estimated values from the observed values in **A**, given the respective numbers of dimensions. However, the interpretations used in these two methods are very similar in that both extract underlying or "latent" dimensions that capture most information contained in the full data matrix.

SOME PRACTICAL CONSIDERATIONS

To summarize, SVD will transform a high-dimensional matrix ("space") of n rows and m columns into a usually much lower dimensional space of n rows and k columns, or k rows and m columns. In effect, the method creates a small(er) number of dimensions defined as linear combinations of row or column values, from which the original values in the data matrix can be approximated as closely as possible. Put another way, the method extracts latent dimensions from observed row and column vectors (e.g., the document by term matrix) that extract the most important information from the original matrix. Referring back to Figure 11.2 and the simple example discussed earlier, the method can identify a latent dimension *Size* from two observed dimensions *Height* and *Weight*. Referring back to Figure 11.1, which plotted the causes of aircraft accidents, the method extracted two (subjectively named) dimensions: *general versus specific causes of accidents* and *accidents related to pilots landing on a runway versus accidents related to engine and fuel problems*.

EXTRACTING DIMENSIONS

An obvious question at this point is, how many dimensions should one extract? Referring back to the Figure 11.1, Figure 11.3 is a graph of singular values plotted against the numbers of successively extracted dimensions. The relative sizes of the squared singular values for successively extracted dimensions reflect the proportion of the total variability in the documents by words or terms frequency matrix, "reproduced" or "accounted for" by the respective numbers of dimensions.

There appears to be a "knee" at the point of the second or third extracted dimension, where the squared singular values across subsequent dimensions become very shallow. A visual criterion similar to that sometimes used in principal components analysis, called the Scree test (Cattell, 1966), involves

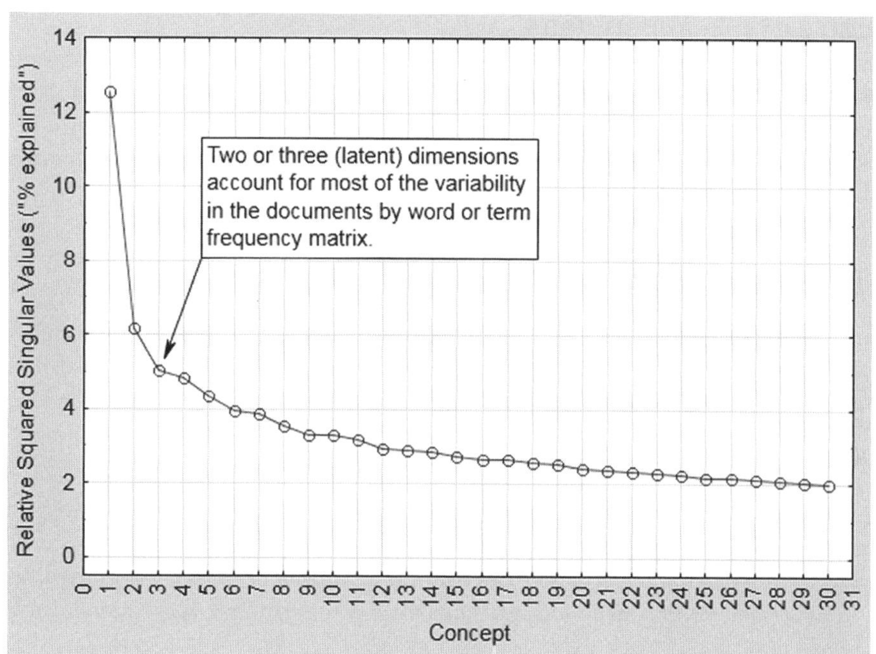

FIGURE 11.3
Plot of relative squared singular values by number of latent semantic dimensions.

finding the place where the smooth decrease of singular values appears to level off to the right of the plot. To the right of this point, presumably, you find only "factorial scree." *Scree* is the geological term for the debris that collects on the lower part of a rocky slope. Thus, most (or at least much) of the variability in the document by word or term frequency matrix is reproduced by the respective numbers of factors, because one may interpret the remaining dimensions to capture mostly much smaller noise variability.

In practice, when analyzing word or term matrices extracted from a corpus of text, one usually finds that no more than 5 to 20 dimensions or so extract most of the information from the document by word or term frequency matrix. This illustrates the point made earlier regarding redundancy in text in the context of owner reviews of the Chevrolet Volt automobile published on the Internet. The number of dimensions of meaning (and sentiments) expressed in such reviews is typically *much* smaller than the total numbers of words and terms used in the reviews. Figure 11.3 reflects this as a "knee" at a relatively small number of dimensions.

Retaining Many Dimensions

On the other hand, when the goal of the analyses is to extract dimensions for subsequent predictive modeling or clustering, it is often useful to work with larger numbers of dimensions than one would choose based on a plot of squared singular values. For example, predictive modeling techniques (see Chapter 11) can usually handle large numbers of predictors, and when the goal is to achieve the best predictive accuracy, it is advisable to capture information associated with singular values and dimensions that may only account for a small proportion of the total variability in the input data (the documents by word or terms frequency matrix).

Interpreting Latent Dimensions

SVD yields latent dimensions that reproduce the values and variability in the documents by word or term frequency matrix in the least numbers of dimensions. In order to interpret those derived latent dimensions—that is, to assign meaning to them—one can either review the scatterplot of terms/words against the latent dimensions to derive subjective labels or validate the dimensions by relating them to other available variables with known "meaning."

SUBJECTIVE METHODS: REVIEWING GRAPHS

As we saw in our aircraft accident example, a simple review of scatterplots of the weights (word coefficients) for the words and terms for the respective derived dimensions can yield useful insights about the nature and appropriate labels for the latent dimensions.

Labeling Dimensions

For example, Figure 11.4 shows a scatterplot of terms from car reviews from a website. While the interpretation of the first component in this case (and often) suggests some general versus specific topics discussed in the reviews, the second dimension ("Component 2") shows in the lower half of the graph terms like *problem, transmission,* and *warranty,* and in the top half, *sport, handle,* and *feel.*

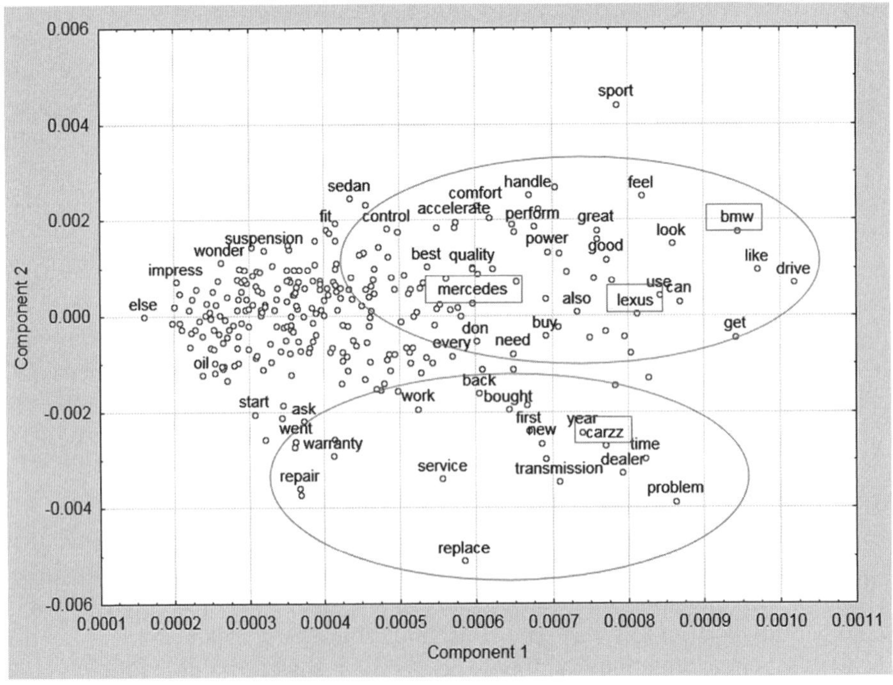

FIGURE 11.4
Scatterplot of words for the first two semantic dimensions: car reviews.

It might be justified, therefore, to name this dimension the "sportiness" (vs. "service problems") dimension. Interestingly, and perhaps in support of that interpretation, the car brand BMW scores relatively high on the second dimension, while Mercedes and Lexus are mostly neutral (near 0). A car labeled CarZZ, which was known for reliability issues at the time the respective car reviews were extracted, scores relatively low on "Component 2."

Identifying Clusters of Terms

Another pattern to look for is terms that "go together"—that is, clusters of terms or words that are located close together in the space defined by the derived latent dimensions of meaning. Looking at Figure 11.4, CarZZ is very close to *dealer, transmission*, and *problem*. This suggests that CarZZ maps into the two-dimensional space of "dimensions of meaning" in a manner that makes the car similar to those terms. Put another way, clusters of terms in relatively close proximity may suggest that the respective terms are associated with similar meaning, and, in this case, it was indeed the case that CarZZ had known problems with the reliability of the transmission.

ANALYTICAL METHODS: BUILDING MODELS FOR DIMENSIONS

Analytical methods can also be useful for the interpretation of the dimensions. The respective values plotted in Figure 11.4 are the weights that relate the terms to the respective dimensions. One can also compute the weights that relate the documents to the dimension; those scores are also called *document scores*. These document scores for each document and for each dimension can be used to relate (e.g., correlate) the dimensions to other structured information or variables of interest. For example, the car reviews may contain ratings of "sportiness" for each car review, and the values for the dimension that we think we should label *sportiness* should correlate with those ratings. Often, the product reviews by shoppers that are published on many websites provide explicit "recommend versus do not recommend" ratings as well, so dimensions extracted using SVD could also be validated against such ratings to derive "like versus dislike" dimensions (and the terms/phrases that define them).

USEFUL ANALYSES BASED ON SINGULAR VALUE DECOMPOSITION SCORES

Much of the discussion so far has centered around the word or term coefficients—that is, mapping the words and terms into the reduced space that results from the application of SVD to the documents by words/terms (relative) frequency matrix. However, as we just saw, the SVD document scores that map the documents into the same space are also of interest and useful for meaningful subsequent analyses. One can often include in subsequent (e.g., predictive) modeling the SVD document scores instead of the much higher dimensional original matrix with the original words or terms. Note that some of the methods that are useful for predictive modeling and clustering are also discussed in Chapter 11.

CLUSTER ANALYSIS

Cluster analysis can also be applied to the SVD document scores to identify homogeneous groups of documents with respect to the latent semantic dimensions extracted by SVD. This approach can be useful to identify clusters of similar warranty claims, accident reports, and so on. Also, clustering SVD scores may identify documents (observations) that cannot be assigned with high certainty to any cluster, so those documents represent "outliers" or "unusual observations" with respect to the latent semantic dimensions extracted via SVD. Such outliers could identify potential "unusual" claims or perhaps even fraud in insurance applications, new and unusual failure modes in warranty claims applications, and so on.

PREDICTIVE MODELING

In predictive modeling, the goal is to predict some important outcome or business key performance indicator, based on the available structured information (predictor variables) and unstructured information (text). SVD document scores can be used like any other continuous predictor variable to augment the predictive accuracy of models.

WHEN SVD IS NOT USEFUL

SVD extracts latent dimensions of meaning from the covariation of (relative) term frequencies across documents. For example, if half the documents have the terms *good, like,* and *convenient* in them, and the other half of the documents have the terms *bad, dislike,* and *inconvenient* in them, then a dimension may emerge with the first three terms on one end (of the dimension) and the other three terms on the other end of the dimension. Put another way, a "good-bad" dimension may emerge.

However, if hundreds of other terms were also included in the analyses, and you want to create a documents by term (relative) frequency matrix, you must look for one specific term or phrase that is actually of most interest and most diagnostic of some other important business key performance indicator (KPI). For example, in warranty claims or when text mining medical data (e.g., physician notes), what might be of greatest interest are not the "latent semantic dimensions" but the specific terms or phrases that are associated with high cost. In that case, deriving "dimensions of meaning" is not a relevant task, but it is important to find the specific words or combinations of words and phrases that identify new types of expensive warranty repairs or that predict likely medical expenses based on physicians' notes (e.g., see Polon, 2011).

To summarize, when the task is to identify the "few words that matter" with respect to some important business outcome, then SVD is *not* a useful tool. SVD will extract dimensions of meaning based on word/term frequencies without any consideration of how those dimensions relate to important business outcomes or insights. Other alternatives to SVD could be considered in that case, including partial least squares (PLS) methods, which can extract components and dimensions subject to the condition that the components correlate highly with some other outcome variables (see, for example, Vinzi et al., 2010).

SUMMARY

Given a matrix A(n,m), SVD will compute matrices **U**(n,m), vector **w**(m), and matrix **V**(m,m) so that **A** = **U** * **W**(diagonal) * **V′**, where **V′** stands for the transposed **V** matrix, and **W**(diagonal) stands for a diagonal matrix(m,m), with the elements of vector **w** in the diagonal. The positive values in vector **W** are the square root of the eigenvalues of **A′** * **A** or **A** * **A′**. Eigenvalues are related to the total amount of variability or "information" in the matrix **A′** * **A** or **A** * **A′**, and the relative size of each singular value is related to the information extracted by consecutive dimensions.

SVD will reduce the dimensionality of the documents by word/phrase (relative) frequency matrix to derive latent semantic dimensions that are useful for understanding the underlying dimensions of meaning that separate the documents and terms. The word or phrase coefficients for the latent semantic dimensions can be used to identify meaningful labels for the respective dimensions and thus to structure the dimensions of meaning along which the documents in the corpus can be organized. The SVD scores for the documents can be used for subsequent data analyses (predictive modeling, clustering) in order to improve the accuracy of prediction models, identify text "outliers," and so on.

SVD is not a useful technique if the purpose of the analytical project is to identify the specific phrases or terms that are important and related to key performance indicators (e.g., which phrases in physicians' notes are predictive of subsequent health care costs), rather than the dimensions of meaning that organize the documents in the document corpus.

POSTSCRIPT

The preliminaries are done. Now you are ready to study some particular use cases in depth. The next section of this book (Chapters 12–15) presents some in-depth discussions of several advanced use cases for text mining: web mining, clustering, insurance text mining, and focused web crawling.

References

Cattell, R. B. (1966). The Scree test for the number of factors. *Multivariate Behavioral Research, 1*, 245–276.

Manning, C. D., & Schütze, H. (2002). *Foundations of statistical natural language processing* (5th edition). Cambridge, MA: MIT Press.

Polon, J. (2011). Text mining case study: Text mining for health insurance. Presentation delivered at the 2011 Society of Actuaries Health Meeting, Boston, MA.

Vinzi, V. E. (Editor), Wynne W. C., Henseler, J., & Wang, H. (Eds.) (2010). *Handbook of partial least squares: Concepts, methods, and applications.* New York: Springer.

Web Analytics and Web Mining

CONTENTS

PREAMBLE

After a short introduction to the subject, the question is asked, "Why do it?" To many people, the subject of web analytics and web mining is a little esoteric, and the unspoken answer to this question is "I don't! I just avoid it." If that is not you, go on to Chapter 13. If this is you, reading this chapter might turn you on to web analytics. We will demystify the subject and provide you with a deeper understanding than you will gain solely from doing tutorials on the subject. After all, that is the goal of this book: to conduct you up the learning curve on text analytics and text mining applications as fast as possible.

WEB ANALYTICS

In this chapter, we give a very brief overview of web analytics and web mining so you can begin to get familiar with some of the terminology and concepts of these domains. If you want to explore web analytics and web mining more thoroughly, the references at the end of this chapter should be extremely helpful. We also provide a great blog site that will really get you immersed into web analytics.

After surveying several books on web analytics, the authors decided that with all the "chaos" out there in this rapidly developing but still "toddler" field, Avinash Kaushik has "cut through the chaff to find the wheat" better than anyone else. Thus, many of his ideas are summarized in this chapter in a format that will give you a good introduction to the field. After you understand the basic ideas presented here, you can use the references to fully catapult yourself into the big wide world of web analytics.

In his book *Web Analytics 2.0*, Avinash Kaushik made web analytics more understandable. His work at Google is the source of much of his perspective, and he is now called the "analytics evangelist for

Practical Text Mining and Statistical Analysis for Non-structured Text Data Applications. DOI: 10.1016/B978-0-12-386979-1.00040-2

Google." He has his own website (www.webanalytics10.com) and blog (Occam's Razor Blog at www.kaushik.net/avinash). He is also frequently asked to speak at key industrial conferences around the world. In 2009, he was awarded the Statistical Advocate of the Year award from the American Statistical Association.

THE VALUE OF WEB ANALYTICS

Why do we do web analytics? The following list helps to answer this question:

1. To *quantify* the economic value of a website
2. To create a "customer-centric" website
3. To enhance *profits* by using:
    a. Internal site searches
    b. Pay-per-click marketing
    c. Optimization of your search engines
4. To find the most important key performance indicators (KPIs)
5. And from the knowledge gained through numbers 1–4: to create an "actionable dashboard"

Figure 12.1 shows the processes of customer web searches, and Figure 12.2 shows the concept of the "web metrics life cycle." A good understanding of these concepts can help you to make decisions about eliminating or improving aspects of your website. Following are some of the main concepts and terminology in web analytics:

1. Web server log file analysis
2. Page tagging
3. Click analytics
4. Visitor geolocation
5. Customer life cycle analytics
6. Hybrid methods

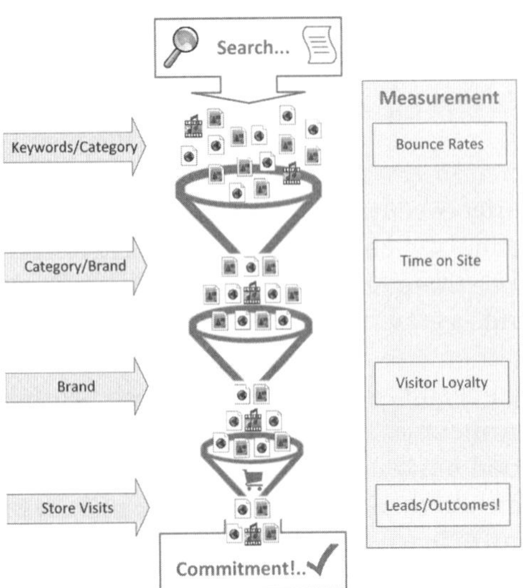

FIGURE 12.1
The process flow of customer web searches.

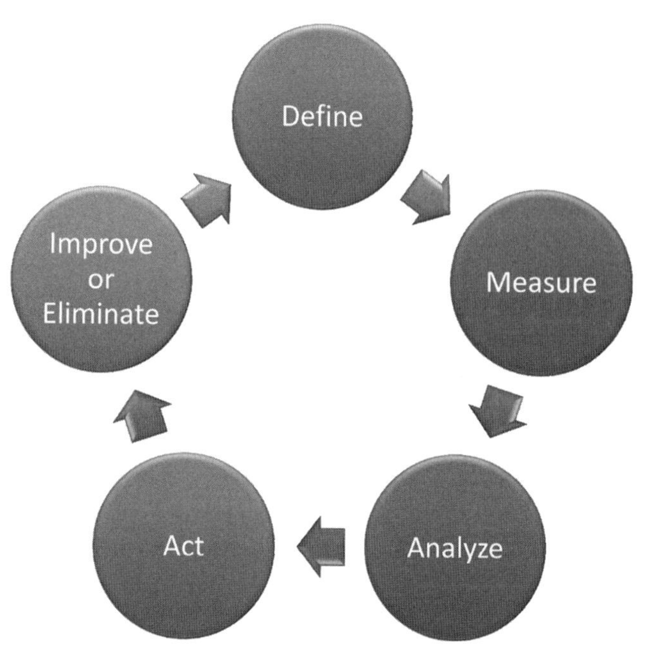

FIGURE 12.2
The web metrics life cycle.

Some of the main definitions in web analytics include the following, of which many of these concepts are illustrated in either Figure 12.2 above or Figure 12.3 below, and also in Figures 12.5 and 12.6:

- **Hit:** A request for a file from the web server that is found in log analysis. A single web page usually consists of multiple files, each of which is counted as a hit if the page is downloaded.
- **Visit/session:** A *visit* is defined as page requests from the same client (computer) with a time of no more than 30 minutes between each page request. A *session* is defined as a series of page requests from the same client with a time of no more than 30 minutes and no requests for pages from other domains intervening between page requests.
- **Visitor/unique visitor/unique user:** The client generating requests on the web server (log analysis) or viewing pages (page tagging) within a defined time period. A unique visitor counts once within the time scale. A visitor can make multiple visits. Identification is made to the visitor's computer, not the person, usually via cookies and/or computer IP addresses.
- **Impression:** An impression is each time an advertisement loads on a user's screen.
- **Bounce rate:** The percentage of visits where the visitor enters and exits at the same page without visiting any other pages on the site in between.
- **Visibility time:** The time a single page (or a blog, ad banner, etc.) is viewed.
- **Session duration:** The average amount of time visitors spend on the site each time they visit.
- **Page view duration/time on page:** The average amount of time visitors spend on each page.
- **Active time/engagement time:** The average amount of time visitors spend with content on a web page, based on mouse moves, clicks, hovers, and scrolls.
- **Page depth/page views per session:** Page depth is the average number of page views a visitor consumes before ending his or her session.

- **Frequency/session per unique:** Frequency measures how often visitors come to a website.
- **Click path:** The sequence of hyperlinks one or more website visitors follow on a given site.
- **Click:** A single instance of a user following a hyperlink from one page in a site to another.

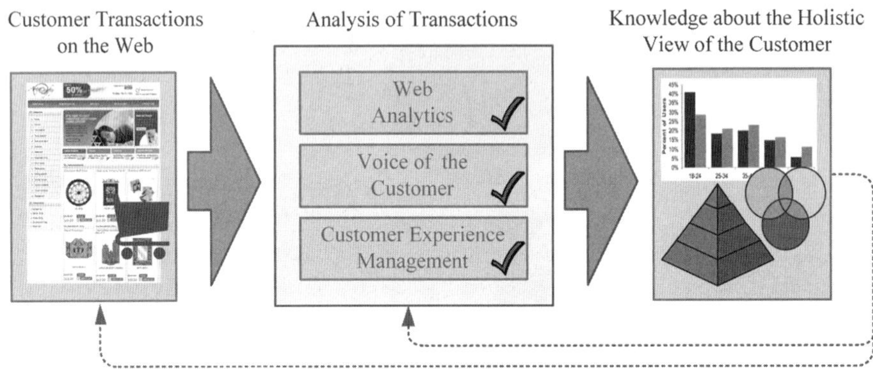

FIGURE 12.3
A process view to a website optimization ecosystem.

Another way of looking at this is via the circular model of Kaushik (2010). Figures 12.4, 12.5, and 12.6 illustrate this model.

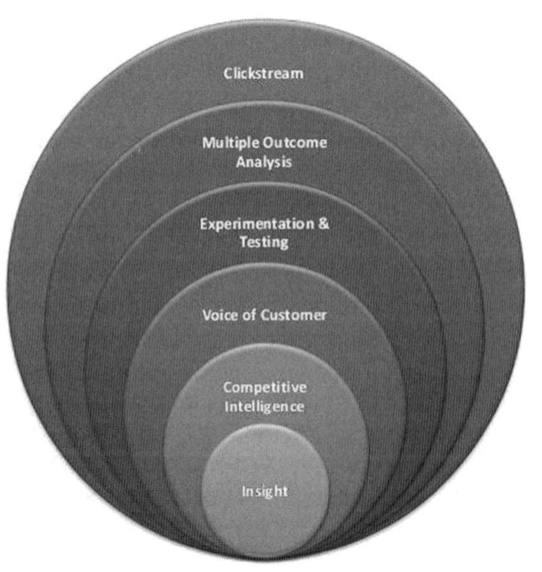

FIGURE 12.4
A circular model for understanding customer website interactions from initial "clicks" to understanding the voice of the customer and eventually insight about what is really important (insight).

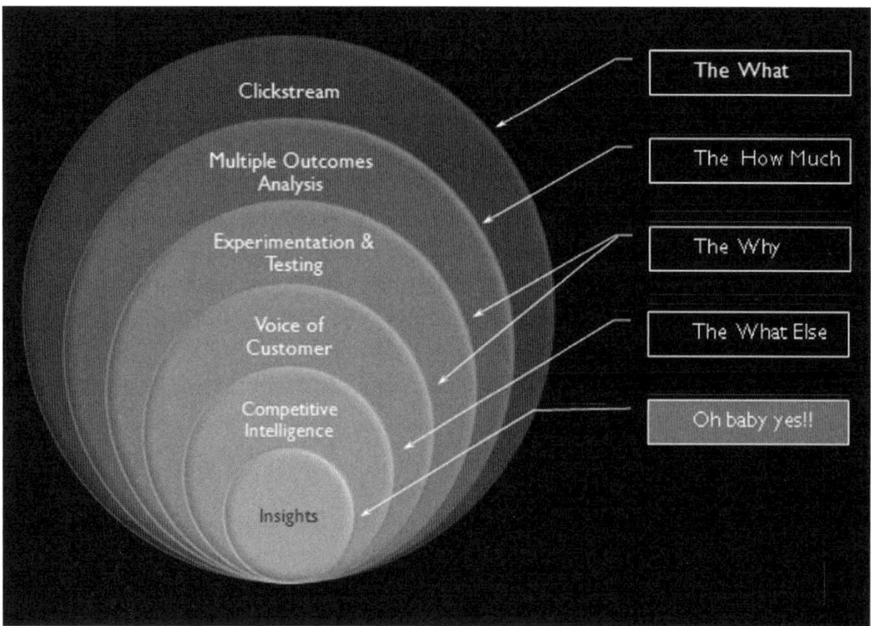

FIGURE 12.5
The what, why, how, and goal of the website attained.

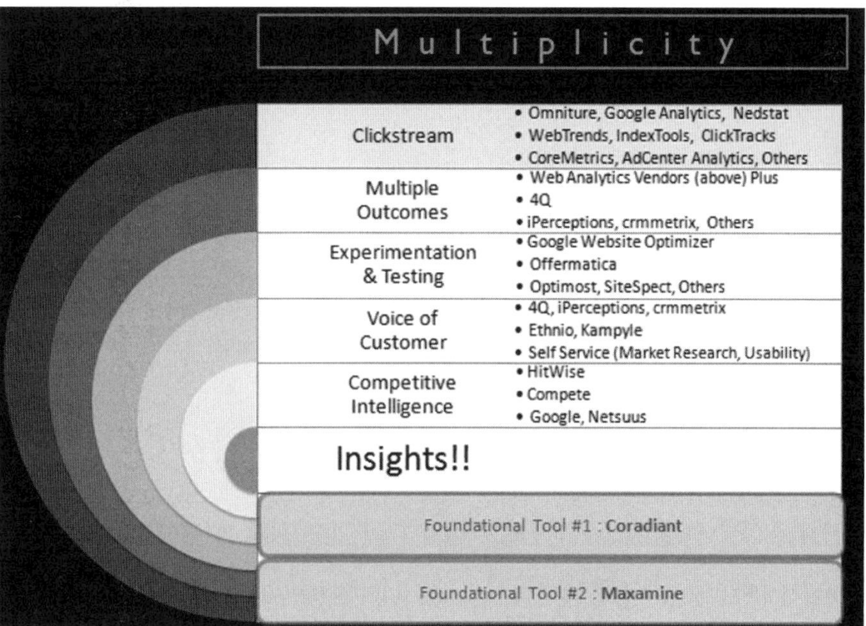

FIGURE 12.6
Tools that can jump the steps of the web analytics ladder to get to "insights" that can become actionable. *Source: Kaushik, 2010.*

Web Mining

The term *web mining* is reserved for text that comes from web pages. In one sense it is not much different from text mining in written documents, but a relatively unstructured mining and analytics field has developed around the web documents in particular, so web mining has developed in a rather confined niche.

Web mining is defined by many practitioners in the field as using traditional data mining algorithms and methods to discover patterns by using the web. Web mining can be divided into three different types: web usage mining, web content mining, and web structure mining. Figure 12.7 illustrates these three interrelated components of web mining.

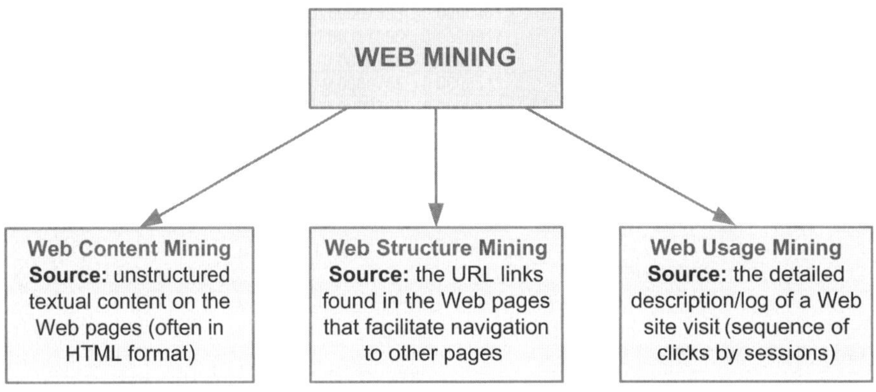

FIGURE 12.7
Three components of web mining.

Web usage mining is a process of gathering information from server logs to understand the history of users of the web.

Web content mining is the process of discovering what users are looking for on the Internet. Some users might be looking at only textual data, whereas others might be interested in multimedia data. This process tabulates information from text, image, audio, or video data on the web. Web content mining sometimes is also called web text mining because text content is used quite often.

Web structure mining uses graphing methods to illustrate connection structures of websites.

There are two types of web structure mining:

1. Discovering patterns from hyperlinks in the web page
2. Discovering treelike patterns of the web page structures that describe HTML or XML usage

THE FUTURE OF WEB ANALYTICS AND WEB MINING

The future for web analytics and web mining may be a long, hard road, but "long" may be misleading because it may not be so much in terms of time as in terms of effort as our current world changes and moves at record speeds that were unknown in previous centuries. However, certain challenges may slow down the attainment of the "optimized solutions" for focused and fully successful web analytics not yet obtained.

One of the challenges is the mind-set of business and industry. Companies today seem to have a difficult time spending dollars on expensive web analytic systems. This is due to several reasons:

1. Google, and Yahoo!, and other analytics are free.
2. Web analytics types often sit in a corner of the company; in other words, they are not integrated yet as important "cores" of a company.
3. There is a perceived lack of ROI for dollars spent on web analytics.

Thus, the real challenge is *quantifying* the economic impact that good web analytics can have on a company.

The absolutely humongous amounts of data available today (90 percent or more text—as we so eloquently discussed in the preface) is seen by many organizations as an almost insurmountable obstacle. Another problem for the future is called "dumb data," shown in Figure 12.8.

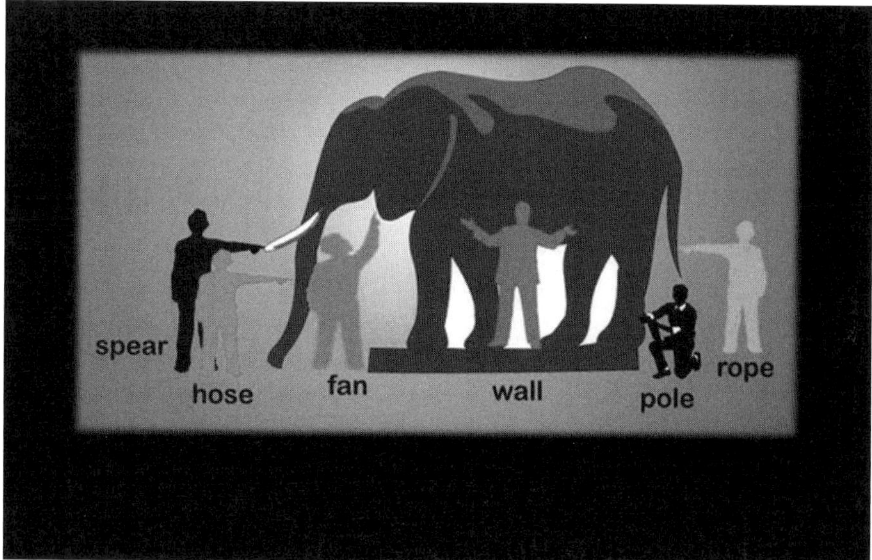

FIGURE 12.8
The elephant phenomenon. *Source: Kaushik, 2010.*

Figure 12.8 contains too much dumb data, and the output from many web analytic tools also has too much missing data. In other words, without seeing the whole picture, people misinterpret the parts of the elephant as something completely different. The blindfolded individuals in Figure 12.8 conclude that the elephant is a rope, a hose, a spear, and a wall. None of them has focused on the totality of the nature of what is before them: an elephant. Likewise, different web analytics tools show only part of the reality under study.

Web analytic tools push out a lot of data—in fact, so much that some in the field refer to it as "data puking." For example, if you get 50,000 keywords from a tool's search, which keywords should you concentrate on? That can be a big problem, right? Thus, new methods are needed to "categorize" these keywords and concentrate on a workable number. There is a great "redundancy" in the amounts of unstructured data, as you can see in many of the more traditional text mining tutorials in this book (see especially Tutorials C, E, I, and L for "redundancy" in text data).

Missing data is also a problem, and that translates into a challenge for the future. Often, data we really need are completely missing from today's tools. So web analytics still has a way to go compared to where the "written document" text analytics is at today. (You can work through some of these tutorials to see how text analytics analyzes a huge number of web pages of text, mostly redundant, and discover that in most cases only four, five, six, or seven words adequately describe the essence of what is being conveyed in the entire document.)

Figure 12.9 shows steps leading to a successful web analytics solution. Lower steps are for the most part free and do not take much effort or resources. Then, as we go up the ladder, increasing amounts of resources, people, money, and effort are required. Reaching the last step requires commitment from top executives down through the reporting structure to put web analytics in the center of their operations.

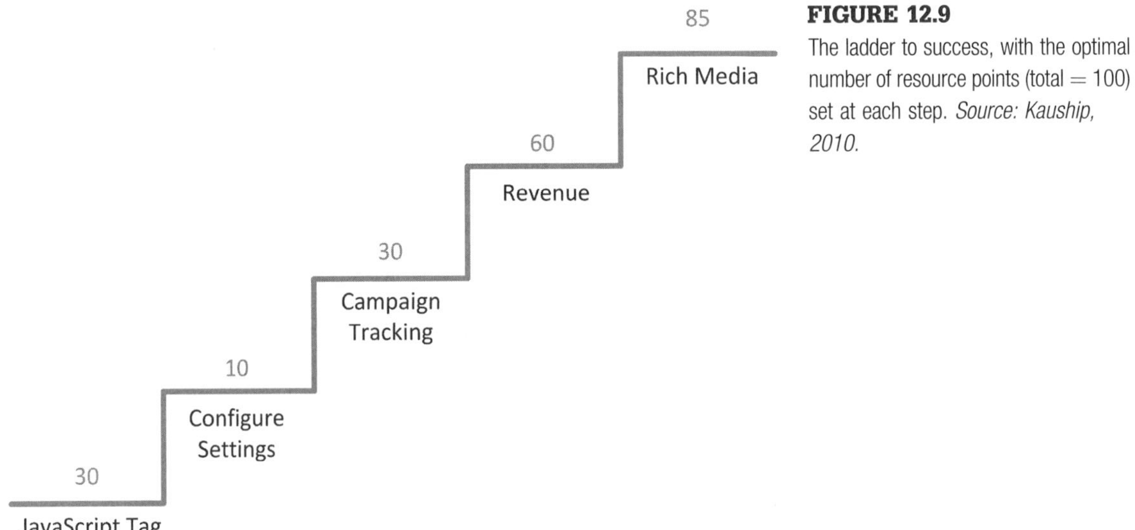

FIGURE 12.9
The ladder to success, with the optimal number of resource points (total = 100) set at each step. *Source: Kauship, 2010.*

Figure 12.9 illustrates, in a structured format, what some web analysts consider an "optimal path" to successful web analytics evolution in a company or organization. On a scale of 0–100, the bottom step, "JavaScript Tag," requires 30 effort–resource points but is mainly free. The second step, "Configure Settings," takes 10 effort–resource points. The third step, "Campaign Tracking," takes 30 effort–resource points. The fourth step, "Revenue," takes 60 effort–resource points. Finally, the top step, "Rich Media," takes 85 effort–resource points. This top step requires company commitment, money, people, and time. Let's take a closer look at what is required at each step of this ladder.

Ladder step 1: By adding a few lines of JavaScript code (tag) to every page, the analytics tools will automatically generate reports such as page views, bounce rates, and so forth.

Ladder step 2: Configuring of settings does not require programmers or the IT department. It does enable site searches and setting of goals.
Ladder step 3: Campaign tracking involves adding parameters and autotagging for AdWords campaigns in Google analytics.
Ladder step 4: The "revenue" step of the ladder involves enhancing or adding new JavaScript code. And at this point one needs to tell the script how to get data out of the page. This requires "systems" and "people," and it may require "legal review." So this step can take 30 to 40 percent or more of the "energy/money" needed to do good web analytics.
Ladder step 5: The "Rich Media" stage, where the real payoff is located, requires "systems," "people," and "processes." Web analytics tools require web page views, but this "Rich Media" stage does not generate multiple page views.
Potential Ladder step 6: There could be a 6th step to this Figure 12.9 ladder diagram, but it is not illustrated here. Additional work is needed at this stage to get multiple page views and the "focus" to hone in on the real meaning that should be available from mature web analytics. Reaching this utopia is still in the future.

In summary, it seems difficult to predict the future for web analytics in a world where things change so rapidly. But one thing seems sure: A new metric that focuses on what is important will be needed for this new world. *Smart people* who are *smart about data* will rule the world of the future. The future should be challenging, very exciting, and a lot of fun as the Internet, web, and social media play an increasing part in our lives.

POSTSCRIPT

What did you think about the ladder to success related to the number of resource "points" needed to accomplish each step in web analytics? For the most part, this ladder is analogous to the amounts of effort required to accomplish each step in other text analytics applications; much time but relatively few resource "points" are associated with early steps in the analytical process. This poses an interesting speculation for the future of text analytics of all kinds: Will many of the initial steps in analytical processes become progressively automated? We see this trend in place already in data mining; most commercial data mining tools have developed automated neural nets and boosting decision trees (for example), both of which automate much of the sampling and algorithm configuration steps that used to be performed manually. Will the same thing happen in text and web analytics? Probably it will.

References

Avinash Kaushik. *Web Analytics 2.0*. Wiley Publishing, Inc., by Sybex-R, An Imprint of Wiley, 2010. www.webanalytics20.com.

Bing Liu. *Web Data Mining: Exploring Hyperlinks, Contents, and Usage Data*. Springer-Verlag, Berlin & Heidelberg, 2007.

Dursun Delen, *Business Intelligence*, 2nd ed. Efraim and Turban (http://www.amazon.com/Business-Intelligence-2nd-Efraim-Turban/dp/013610066X/ref=sr_1_3?ie=UTF8&s=books&qid=1292358409&sr=8-3).

Occam's Razor: www.kashik.net/avinash.

CHAPTER 13

Clustering Words and Documents

CONTENTS

PREAMBLE

Clustering is arguably the oldest technology in text mining. Early uses of document clustering aided document retrieval systems by the military in World War II. Later, document search engines used document clustering as a preprocessing technique. Whether clustering is used to group documents in a corpus or words in a document, the technique is almost always used as a means to an end, rather than the end itself. Modern Internet search engines rely on document clustering techniques to aid in information retrieval. Therefore, we might consider clustering as a preprocessing technique in text mining. This concept was mentioned at the end of Chapter 11, in reference to singular value decomposition.

INTRODUCTION

Clustering, or cluster analysis, is the process of automatically identifying similar items to group them together into clusters. Clustering is an *unsupervised* learning method, which means no labeled training examples need to be supplied for the clustering to be successful. In other words, no "output" data are necessary, only "input" data. Unsupervised methods are less powerful than supervised methods, but they can be employed in a much wider range of problems. In text mining, clustering algorithms are used

Practical Text Mining and Statistical Analysis for Non-structured Text Data Applications. DOI: 10.1016/B978-0-12-386979-1.00041-4

to find similar documents or individual words. When documents are being clustered, it is typically called *document clustering* or *text clustering*. When focusing on words, it is called *concept extraction* or *topic modeling*, depending on the algorithm chosen for clustering.

Though clustering documents and clustering words are two separate tasks, they are closely related. After document clustering has been performed, a cluster is often identified by the most frequent words occurring within it. Likewise, word clusters, or concepts, can be used to categorize documents by indicating which documents mention certain concepts. In this chapter, we provide an overview of the algorithms used for clustering in general and then go deeper into strategies and algorithms for clustering both words and documents.

CLUSTERING ALGORITHMS

Clustering algorithms originated in the fields of statistics and data mining, where they are used on numerical data sets. In text mining, as with data mining, two components are needed for a clustering algorithm: a method for computing similarity between items (to determine when two items belong together in the same cluster), and an efficient method for comparing all of the items. The similarity metric depends on the type of data. Another key variable is the number of clusters to use. Most algorithms require the user to select this, but some can estimate the natural number of clusters, given the distance metric. All algorithms assign cases to clusters.

Distance Metrics

For text, the distance metric is often a variant of a vector distance method such as *cosine similarity*. This assumes that we are modeling a text document as a point in a high-dimensional vector space, as described in Chapter 3. For words, the vector captures the other words that appear in the context (local neighborhood) of the original word. For documents, feature extraction can be performed (see Chapter 10), and then numerical distance metrics such as Euclidean distance (root sum of squared distance) can be used.

Clustering Algorithms

The three main categories of clustering algorithms are hierarchical clustering, partitional clustering, and spectral clustering. *Hierarchical* clustering iteratively groups documents into cascading sets of clusters. This can be done in a top-down (divisive) or a bottom-up (agglomerative) manner, where items are either split or joined together based on their similarity measures. *Partitional* clustering algorithms are not incremental like hierarchical clustering, but instead produce all the clusters at once. k-means and its variants (k-medoids and k-medians) are the most popular type of partitional clustering algorithm. In the k-means algorithm, a random set of k points are selected as *centroids*. (The user chooses the number of clusters to create k.) The points are typically sampled from the data but can be invented as long as they are within the "space" of the data. The algorithm proceeds iteratively until conversion, alternating between two steps: assignment and recentering. In *assignment*, each of the data points is compared to each of the k centroids and assigned as a member of the closest cluster. In *recentering*, the location of the mean of each cluster is recalculated. The algorithm converges when no point changes cluster membership. *Spectral* clustering algorithms utilize matrix operations for dimensionality reduction and produce clusters based on those reduced dimensions. Spectral clustering is closely related to latent semantic indexing (LSI) and singular value decomposition (SVD), which is described in more detail in Chapter 11.

CLUSTERING DOCUMENTS

The goal of clustering documents is to group together documents with similar content into the same cluster. As with all text mining algorithms, document clustering requires converting the unstructured text in each document into a structured representation before applying the clustering algorithm. The most popular representation is the *vector space*, where each document is represented as a vector noting the words appearing in the document. Words can be represented in the vector as a binary value indicating whether the word occurred, an integer count indicating the number of times it occurred, or a weighted count value that discounts words that occur in many documents. More details about the vector space are provided in Chapter 3.

Given a vector space representation of a document database, there are two methods for clustering text documents: direct clustering and dimensionality reduction. Recall that every clustering algorithm requires both a distance metric and a comparison algorithm. The first method for clustering documents uses a vector distance metric such as cosine similarity or Jaccard's coefficient (Croft et al., 2009). The cosine similarity computes the angle between each document in an n-dimensional space (where n is the number of terms). This distance metric can be combined with a hierarchical or partitional clustering method.

The second method for clustering text first converts the text to a numerical representation using some form of dimensionality reduction or feature extraction and then uses a traditional numerical distance metric such as Euclidean distance with either hierarchical or partitional clustering.

CLUSTERING WORDS

Words can be clustered in two ways. The first is grouping together semantically similar words into a cluster called a *concept*. The second is grouping consecutive words together into a *collocation*.

Concept Extraction

J. R. Firth (1957) once said, "You shall know a word by the company it keeps." The goal of concept extraction is to identify words that share semantic meaning using the context in which the words appear. For example, a concept could be a collection of words defining household pets: dogs, cats, fish, guinea pigs, hamsters, birds, and so on. Each of these words is used in the same contexts. Concept extraction works by clustering on the similarity between the contexts of the words. A real-world example of an extracted concept is shown in Figure 13.1. In the figure, a term clustering algorithm was applied to a collection of PubMed abstracts, and concepts found to be close to the concept "cancer" are depicted in a spring-graph model in two dimensions.

Like document classification, concept extraction represents the text using a vector space. However, for concept extraction, the vector space is a term-context matrix rather than a term-document matrix. Instead of storing an entire document in a vector, a term-context vector stores the context for a single word across all vectors. The context is a window of size n around the given word, also commonly known as n-grams. N-grams can be counted across a single document database and enhanced by resources from the Internet.

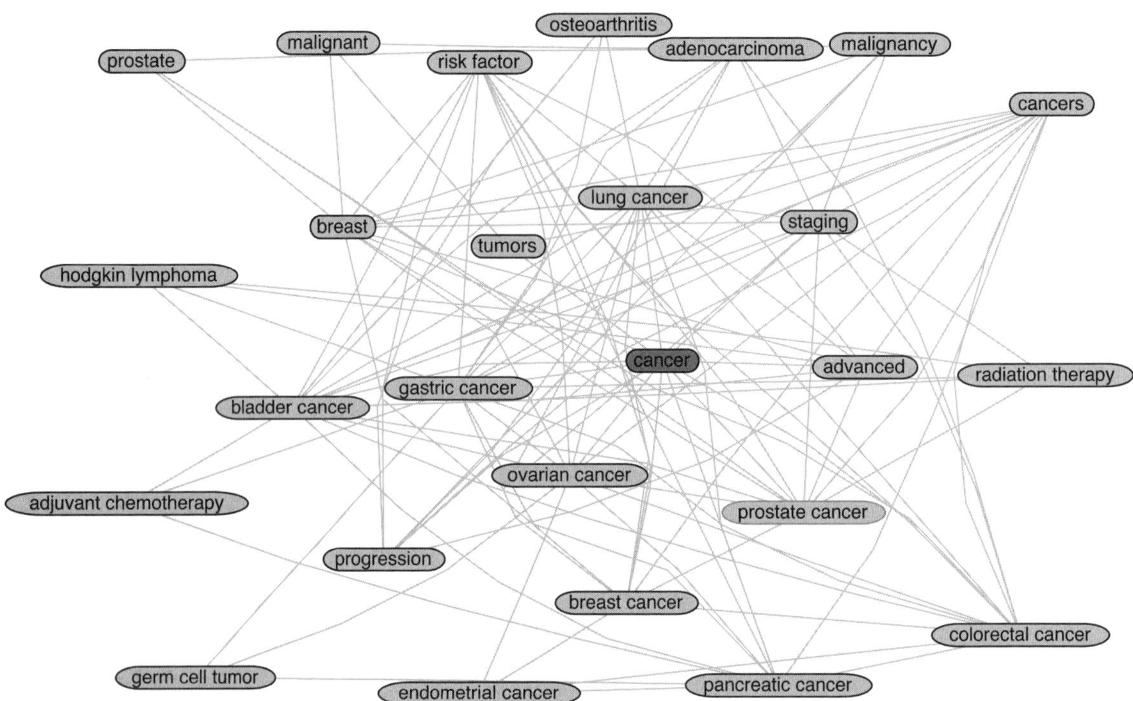

FIGURE 13.1
An example concept inferred from PubMed abstracts using a combination of concept extraction and collocation detection.

For example, Google has made available a large dataset capturing n-grams up to n = 5 for many languages, including English (both British and American dialects), Chinese, French, German, Russian, and Spanish.[1] These n-grams were collected during the scanning process for the creation of Google books.

Once the context has been gathered for each term, a clustering algorithm is applied to find words that appear in similar contexts. As with document clustering, any vector similarity metric can be used with the clustering algorithm.

Topic Modeling

Topic modeling algorithms are a closely related technology to concept extraction. Topic models differ from concept extraction in that they are more expressive and attempt to infer a statistical model of the generation process of the text (Blei and Lafferty, 2009). This model is then used to cluster words into topics. The clustering is considered to be a *soft* clustering—that is, words can probabilistically appear in multiple topics. In addition to clustering related words into topics based on the co-occurrence of words within documents, these *generative* models can also be used to learn correlations between topics. These clusters can then be used to represent documents in a lower-dimensional space. The topic distribution

[1] http://ngrams.googlelabs.com/

of words appearing in a document can also be aggregated into a single vector representing the contribution that each topic has in the document. This vector can then be used to compute similarity between documents and can be aggregated across tagged documents to represent the topical need of an individual.

Topic modeling is an active area of research that is rapidly gaining prominence in practical use. For example, it is used by Google to help improve search results. Despite this and other examples of topic models in practice, we have not emphasized these methods in this book because there are few commercial-grade implementations of topic modeling algorithms.

Collocation Detection

A collocation is a group of words that frequently occur together and represent a single idea. Some examples of two-word collocations include *middle management, nuclear power,* or *personal computer.* Like concept extraction and topic modeling, collocation detection attempts to uncover groups of words with coherent meaning from the text. Unlike concept extraction, a collocation consists of a group of words that appear consecutively. Often collocation detection is applied prior to concept extraction, so the concepts can include collocations as well as single words. Collocations can consist of proper nouns in addition to other phrase types, as shown in Figure 13.2.

The primary strategy for detecting collocations is to identify words that occur together more often than chance alone would suggest. This can be achieved by comparing the number of times two or more words appear together with the number of times those same words appear in other contexts. The challenge is determining the threshold that separates collocations from words that randomly appear together.

Three strategies are used commonly for detecting collocations:

1. Statistical hypothesis testing
2. Information theoretic analysis
3. Combination of a part of speech tagger and a simple frequency filter

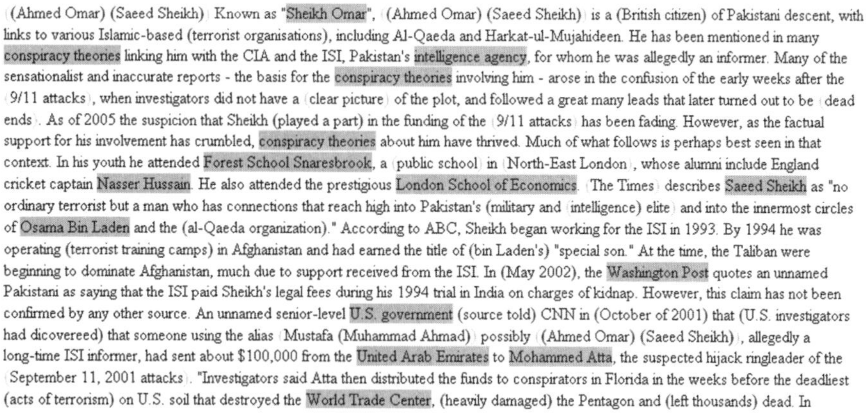

(Ahmed Omar) (Saeed Sheikh) Known as "Sheikh Omar", (Ahmed Omar) (Saeed Sheikh) is a (British citizen) of Pakistani descent, with links to various Islamic-based (terrorist organisations), including Al-Qaeda and Harkat-ul-Mujahideen. He has been mentioned in many conspiracy theories linking him with the CIA and the ISI, Pakistan's intelligence agency, for whom he was allegedly an informer. Many of the sensationalist and inaccurate reports - the basis for the conspiracy theories involving him - arose in the confusion of the early weeks after the 9/11 attacks, when investigators did not have a clear picture of the plot, and followed a great many leads that later turned out to be dead ends. As of 2005 the suspicion that Sheikh (played a part) in the funding of the 9/11 attacks has been fading. However, as the factual support for his involvement has crumbled, conspiracy theories about him have thrived. Much of what follows is perhaps best seen in that context. In his youth he attended Forest School Snaresbrook, a public school in North-East London, whose alumni include England cricket captain Nasser Hussain. He also attended the prestigious London School of Economics. The Times describes Saeed Sheikh as "no ordinary terrorist but a man who has connections that reach high into Pakistan's (military and intelligence) elite and into the innermost circles of Osama Bin Laden and the (al-Qaeda organization)." According to ABC, Sheikh began working for the ISI in 1993. By 1994 he was operating (terrorist training camps) in Afghanistan and had earned the title of (bin Laden's) "special son." At the time, the Taliban were beginning to dominate Afghanistan, much due to support received from the ISI. In (May 2002), the Washington Post quotes an unnamed Pakistani as saying that the ISI paid Sheikh's legal fees during his 1994 trial in India on charges of kidnap. However, this claim has not been confirmed by any other source. An unnamed senior-level U.S. government (source told) CNN in (October of 2001) that (U.S. investigators had dicovereed) that someone using the alias Mustafa (Muhammad Ahmad) possibly (Ahmed Omar) (Saeed Sheikh), allegedly a long-time ISI informer, had sent about $100,000 from the United Arab Emirates to Mohammed Atta, the suspected hijack ringleader of the September 11, 2001 attacks. "Investigators said Atta then distributed the funds to conspirators in Florida in the weeks before the deadliest (acts of terrorism) on U.S. soil that destroyed the World Trade Center, (heavily damaged) the Pentagon and (left thousands) dead. In

FIGURE 13.2
Example results of a collocation detection algorithm applied to text gathered from Operation Enduring Freedom.

Statistical Hypothesis Testing

This strategy uses one of three tests to detect collocations: the *t*-test, the chi-square test, and the likelihood ratio. For two-word collocations, these tests are based on a 2×2 contingency table where the cells of the table contain the counts of how many times each word appears both in isolation and together. If the pairing is statistically unlikely to have occurred by chance according to the test, then the word pair is marked as a collocation.

Information Theoretic Analysis

The second strategy for detecting collocations uses ideas from information theory. If the appearance of the first word provides a strong indication that the second word will follow, based on the principles of information theory, then those two words are marked as a collocation. Typically, the *mutual information* metric is the scoring function, and those word pairs that exceed a user-defined threshold are defined as collocations.

Combination of a Part of Speech Tagger and a Simple Frequency Filter

This approach for finding collocations proposes a few simple patterns for identifying collocations based on the part of speech of the words (Justeson and Katz, 1995). They noticed that most collocations consist of combinations of nouns and adjectives. Therefore, they propose using a part of speech tagger to identify certain phrases and then sorting based on frequency to identify collocations. The patterns used by Justeson and Katz (1995) are shown in Table 13.1 alongside examples of those patterns.

CLUSTER VISUALIZATION

Since clustering algorithms are unsupervised, it is often difficult to quantitatively evaluate the clusters. (That is, there is no right answer.) Instead, one can use some form of visualization to explore the clusters. The authors developed one such visualization tool, called *DocumentGraph*, to explore the relationship between documents via clustering. This tool is shown in Figure 13.3.

In place of a traditional clustering algorithm, DocumentGraph uses a spring-based physics model to identify clusters of documents. This approach lends itself well to visualization, since the spring-constant paradigm applies well to distances. Similar documents have a strong spring between them, pulling them closer together. Documents with weaker similarity have a weak

Table 13.1 Part of Speech Patterns for Collocation Detection

Pattern	Example
Noun, noun	news conference
Adjective, noun	third quarter
Adjective, adjective, noun	executive vice president
Adjective, noun, noun	tropical storm Josephine
Noun, adjective, noun	U.S. Supreme Court
Noun, noun, noun	World Trade Organization
Noun, preposition, noun	earnings per share

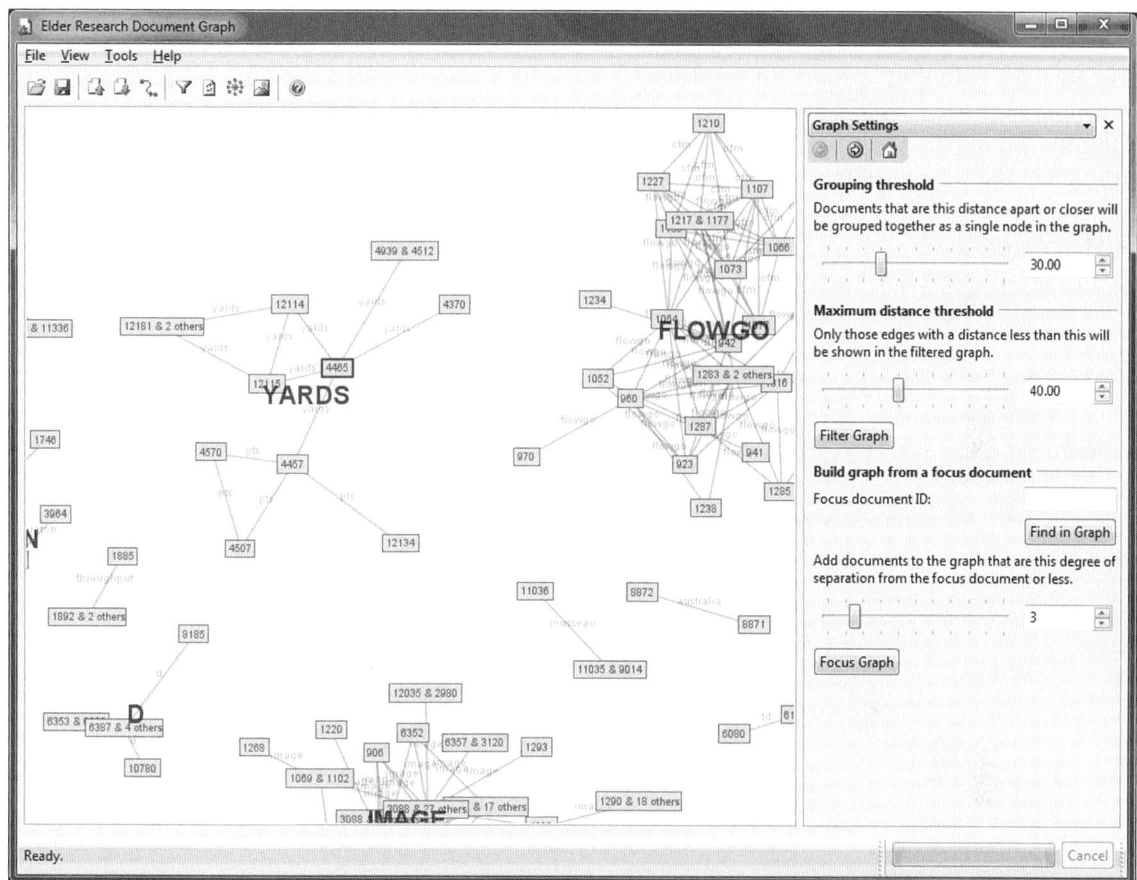

FIGURE 13.3
Example screen shot from the authors' DocumentGraph tool.

spring pulling them together. This keeps the document nearby on the screen but allows stronger strings to provide more pull. Weak connections with similarity below a given threshold are removed from the graph, though there may be some influence from those connections.

SUMMARY

Text clustering is used to group together words or documents based on similarity. Clustering of words can be used for the practice area of *concept expansion,* since words that appear in the same cluster often share topical or semantic meaning. Clustering of documents is its own practice area and is useful for exploring and understanding how documents are related.

POSTSCRIPT

We are done with the presentation of advanced concepts in text analytics. Now we can dive deeper into a couple of practical applications of text mining, beyond the depth presented in the tutorials. We will "plumb the waters" of text analytics used in the insurance industry (Chapter 14) and on the Internet using focused web crawling (Chapter 15).

References

Blei, D., and Lafferty, J. Topic Models. In A. Srivastava and M. Sahami, editors, Text Mining: Classification, Clustering, and Applications. Chapman & Hall/CRC Data Mining and Knowledge Discovery Series, 2009.

Croft, Bruce, Donald Metzler, and Trevor Strohman. Search Engines: Information Retrieval in Practice, Addison-Wesley, 2009. Boston, MA.

Firth, J. R. *Papers in Linguistics 1934–1951* (1957), Oxford University Press, London.

Justeson, John S., and Slava M. Katz. 1995. Technical terminology: some linguistic properties and an algorithm for identification in text. *Natural Language Engineering* 1:9–27.

Leveraging Text Mining in Property and Casualty Insurance

CONTENTS

PREAMBLE

The insurance industry (particularly the property and casualty part of it) is particularly conducive to the use of data mining for risk assessment, premium audit analyses, and insurance fraud. The basic tenet of insurance is that we can predict the future with a level of accuracy sufficient to decrease financial losses to a significantly greater extent than the cost of doing analysis. This is an ideal environment for data mining and text mining applications. This chapter conducts you into some of the otherwise little-known landscapes of insurance analytics for marketing and sales, underwriting and pricing, and various insurance operations aimed at increasing profitability.

INTRODUCTION

In the claims world of property and casualty (P&C) insurance, there is a well-known scheme for automobile insurance fraud known as the *swoop-and-squat*. It works like this. The claimants perpetrating such fraud usually pile up into a large, old car so as to fit as many people as possible. They target a vehicle—usually a newer, possibly luxury brand vehicle, since it is likely to have higher insurance coverage limits. They then create a situation where they pull in front of the target vehicle (*swoop*) and stop suddenly (*squat*), causing the target vehicle to rear-end them. This is best achieved near a highway exit because their story would be that they swerved and slowed to take the exit and the insured vehicle rear-ended them.

Practical Text Mining and Statistical Analysis for Non-structured Text Data Applications. DOI: 10.1016/B978-0-12-386979-1.00042-6

If they have planned and executed things right, the target vehicle will be held liable for the loss, and the fraudsters will be able to make claims against the insured's policy. Each of the many claimants in the vehicle will pretend to have soft-tissue injuries (neck and back sprain), but they will likely waive the EMTs, if they were to arrive, since they don't want to be taken to a hospital or a doctor lest their feigned injuries be exposed. With doctors, chiropractors, and attorneys as part of the scheme, the claimants will use many means available to them to inflate claim payouts. Schemes like this are estimated to cost the P&C insurance industry over $30 billion annually according to the Insurance Information Institute (2011).

Given that this scheme is so well known, it must be impossible for fraudsters to get away with such scams, right? Unfortunately, this is not the case. A number of factors contribute to keeping fraud a healthy, even growing, business. Junior adjusters new to claims handling may miss many of the telltale signs, while workload issues and frequent context switches between claims make things fall through the cracks even for seasoned adjusters. Thankfully, there are regulations and best practices in the insurance industry that require claims adjusters to electronically document progress on their claims handling. While these claim notes are often highly unstructured, grammatically unsound, and replete with typos and abbreviations, they are still incredibly valuable sources of insights not available elsewhere.

For instance, facts such as "the claimant vehicle braked suddenly in front of the insured vehicle, insured was forced to rear-end other vehicle as it braked suddenly, or claimant vehicle pulled in front of the insured vehicle and braked suddenly" might be thus described in the claim notes. Similarly, concepts denoting the loss occurred near a highway exit or that the claimants waived the ambulance might be documented by the adjuster in his or her notes. Using the art and science of text mining, such knowledge nuggets can be extracted and combined with other structured data such as multiple injured claimants with soft-tissue injuries to build powerful and systematic fraud detection capabilities.

The business of P&C insurance is replete with unstructured data, such as the claim notes, that afford numerous possibilities for text mining. Our attempt in this chapter is to offer a nonexhaustive list of areas that could benefit from such text mining insights.

PROPERTY AND CASUALTY INSURANCE AS A BUSINESS

The business of P&C insurance is very diverse and complex. The broad umbrella includes risk groups ranging from types of commercial insurance (for commercial/business ventures) through many types of personal insurance (home, auto, and other possessions). For a comprehensive discussion of the myriad types of insurance coverages and their nuances, the reader is referred to Vaughan and Vaughan (2008).

What makes the insurance business uniquely interesting is that there are no tangible widgets to sell; the insurance product is simply a promise to restitute an insured subject in the event of a *future* covered loss. Accurate predictions based on data and sophisticated analytics are therefore supremely critical in the insurance business, creating a high demand for actuarial and other quantitative skills in the P&C industry.

With increasing global pressures, particularly in *soft market* cycles (when insurance demand is low and carriers struggle to get good rates or increased premiums for their coverages), insurance carriers are particularly challenged to (1) identify and leverage high-impact opportunities; (2) strategically differentiate themselves from competition; (3) control and optimize costs; and (4) allocate resources and capital in increasingly smart ways.

With critical implications for the top and bottom lines, there is growing interest in harnessing every ounce of lift available from data and advanced analytics. It is no surprise then that text mining is increasingly being used to discover critical insights and drive systematic business actions in the P&C industry.

ANALYTICS OPPORTUNITIES IN THE INSURANCE LIFE CYCLE

In very broad terms, the business of P&C insurance can be viewed as three distinct but interfacing business activities that each offer numerous opportunities for analytics:

1. Marketing and sales: How to define target markets, channels, and customers; how to create appealing products; how to effectively reach and attract prospects to the carrier
2. Underwriting and pricing: How to understand, qualify, and assess risks; how to prevent or mitigate losses; how to accurately price risks
3. Operations: How to efficiently and effectively deliver claims, loss control, and other insurance services to policyholders; how to create superior customer experiences and minimize attrition of good customers

While there is no specific sanctity in having this three-bucket view, our experience suggests this as a logical demarcation of business emphasis—activities that bring the right business to the door (marketing analytics), activities that let the right risks through the door at the right price (underwriting and pricing analytics), and, finally, activities that service the right risks in the right way (operations analytics). Balakrishnan (2010) provides examples of these different types of analytics and their use in the P&C business (see Figure 14.1).

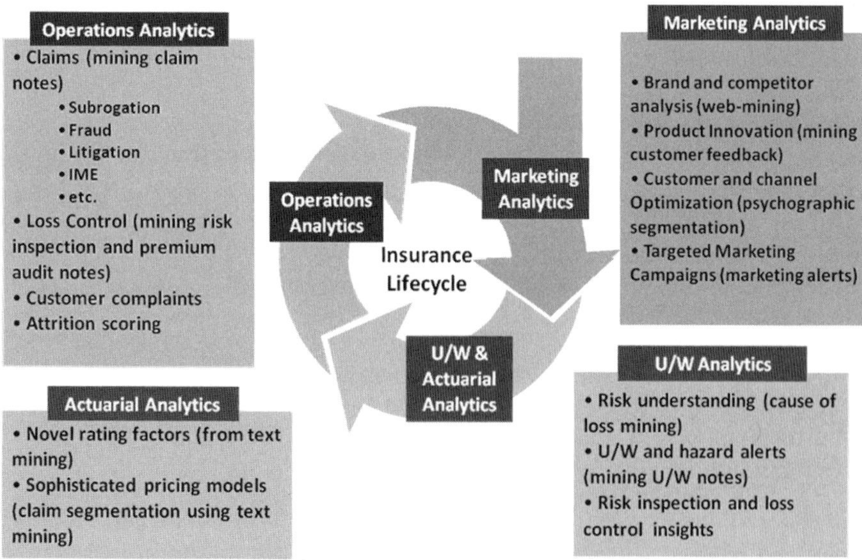

FIGURE 14.1

Some analytics opportunities in the P&C insurance life cycle. Text mining can impact and improve upon a number of these opportunities.

Figure 14.1 outlines some business opportunities that are likely to benefit from text analytics, from mining online blogs, press releases, and customer comments, to extracting actionable insights from underwriting and claims notes. Later in this chapter, we provide a number of examples of business opportunities with text mining. But before we get there, it will be helpful to describe some of the text sources available at a P&C company, along with the requirements they would impose on the suitability of a text mining tool.

Property and Casualty Textual Data and Associated Text Mining Requirements

Insurance carriers have access to a diverse array of textual data sources across various touchpoints in the insurance life cycle. Characteristics of these data sources, along with their intended business use, impose specific requirements on the necessary text mining competencies, as discussed following.

Marketing, Branding, and Customer Insights

Websites, media reports, and press releases, particularly from competitors, are largely textual assets that can be mined to increase the understanding of market needs, trends, and product directions. To serve these purposes effectively, a successful text mining tool must have a built-in web crawler, or at the very least have the ability to interface with one so it can be pointed to, automatically retrieve, and analyze results to produce insights of value to the company. This is a key component of the rich field of web mining, which is the subject of numerous books such as those by Russell (2011) and Liu (2010).

For carriers that engage focus groups or conduct customer touchpoint surveys, often there are unstructured text and comments that can be a valuable source of insight and possibly even innovation. Mining customer complaints effectively will help pinpoint areas of dissatisfaction, such as the lack of online bill payment options, which the carrier can act upon in a targeted manner. Similarly, customer focus group feedback, when mined, can generate new ideas for product or service innovation, such as perhaps the need to create smartphone *apps* for policy administration and claims reporting.

Underwriting and Risk Management

The underwriting process also results in a number of textual comments and notes, available in electronic form as *underwriter notes.* While the nature, content, and structure of these notes can vary by company (and possibly even by underwriter), they usually contain descriptive notes about the risk being underwritten. These can be mined for powerful business actions, as we will illustrate later.

Related to the underwriting process is the role of *risk inspection and audit* that may include:

- Building and home inspections to detect unusual risk factors and coverage limits
- Contents appraisal to assess high-value paintings, jewelry, and other valuables
- Business premises inspection to detect loss hazards and preventable claim situations
- Audits of business operations and accounts to determine payrolls, class codes, and revenues for premium estimation

The inspectors and auditors usually record detailed information on the subject risk, often including descriptive notes and reports. Text mining these notes can surface interesting insights, patterns, and trends that can effectively feed loss control and loss mitigation, as well as underwriting and pricing activities.

Claims and Customer Operations

Of all the text sources at a carrier's disposal, perhaps the most useful are the claims and customer service area notes, since they contain *direct* insights into the customer mind-set or specific loss activity. They can be mined to discover many kinds of interesting insights, from the true cause of the loss to who might be at fault to red flags for fraud and customer dissatisfaction. They can also serve as valuable sources of insight into the friction points in the customer experience and trigger points for new products and services.

Claim Notes

Claim notes are very detailed records of the claim adjustment process, from the first notice of loss to every action that an adjuster takes on the claim. Usually written in a conversational style, these electronic documents can run from a few pages to hundreds of pages for complex, long-tailed claims. While unquestionably replete with valuable insights, claim notes present unique challenges to the text mining process, including the following:

1. *Nonstandard abbreviations:* Because the notes are meant to be a quick and incremental documentation of the claim process, adjusters often use abbreviations. Unfortunately, these abbreviations are nonstandard, often across adjusters at the same carrier, and certainly across carriers. For instance, one adjuster might use the abbreviation CLM to refer to a *claim,* while another might use it to refer to a *claimant.* Yet another might use it to mean *called and left a message.* Any effective text mining capability will need to use smarter techniques, such as the *context,* to determine if that particular instance of CLM referred to a claim, a claimant, or something else.
2. *Sentence fragments and awkward grammar:* In an effort to be quick, some adjusters prefer to use sentence segments rather than fully formed, grammatically correct sentences. For instance, while one adjuster might record, "I called the insured and spoke to him about the loss facts and verified that he felt he was driving well under the posted speed limit of 35 and the claimant vehicle stopped suddenly in front for no apparent reason," another adjuster might document the same situation as "Called ins. ins spd < 35. clmt s/s. no reason." (Note: S/S is an abbreviation for *sudden stop.*) Most text mining systems are designed to work well with well-written text, meaning well-formed and grammatically structured documents. Claim notes will be a challenge for such systems.
3. *Noncorrectable errors and typos:* Since one of the key purposes of claim adjuster notes is to provide transparency into the claim handling process if a disputed claim goes into litigation, these electronic records are *once entered and done.* The ability to correct or modify is necessarily unavailable. This leads to numerous typos and other errors in the electronic notes. In extreme instances, we have also seen cases where details of a different, completely unrelated claim are erroneously entered into the notes of a claim. Most likely, the adjuster forgot to switch to the other claim before beginning the documentation process. While these extreme issues will pose a significant challenge to any text mining capability, at the very minimum an effective text mining system must be capable of detecting and working with common typos.
4. *Inconsistencies due to concepts evolving over time:* Finally, the feature that makes claim notes truly unique from a text mining perspective is that it is an incremental record of the claim process as it unfolds in real time across the duration of the process, from first notice of loss to the eventual closing of the file after payment or denial. This permits the overall claim document to have inherent inconsistencies, reflecting the evolution of the claim process.

For instance, an auto claim reported by the insured might have been a hit-and-run, and the adjuster might record it thus: "Insured struck by unknown at fault driver. Police informed but no record found." As the claim handing process proceeds over the next few days or weeks, it might so happen that the police may have tracked down the hit-and-run driver based on witness descriptions. When notified, the adjuster might have a new entry: "At-fault claimant driver identified by police." If a text mining system is used to mine the overall claim document for an *at-fault other party*, the system will likely struggle with this inconsistency in the document. Since each note entry from the adjuster is automatically date and time stamped, the text mining system will require the ability to decompose the document into time-stamped notes and place emphasis on more recent concepts.

Description of Loss

In instances where it may be difficult to get to the complete adjuster notes, a fair amount of business insight and value can be gleaned from the *description of loss*. Most claims systems have an unstructured field, usually called the description of loss, that allows an adjuster to *briefly* describe the type of loss. Generally populated by the adjuster early in the life of the claim, this field restricts the input to a maximum number of characters—say, 90. This allows the adjuster to describe certain aspects of the loss. "Freezer defrosted, causing water damage or clmt tripped on damaged carpet in store" might be typical examples describing a water loss in an insured home and a general liability loss in a retail store. While lacking in the richness of adjuster notes, these loss descriptions can nevertheless be usefully mined for actionable business insights.

Customer Contact Notes

While a claim experience is perhaps the biggest contact a carrier is likely to have with its insured, it is by no means the only one. In the context of billing and other policy inquiries, a carrier may have contact with their customers—the insured's as well as the agent's. For *direct writers* (companies that offer insurance directly to consumers, such as Progressive and GEICO) and *captive agent writers* (companies that have their own agents/agencies that exclusively represent them, such as State Farm and Allstate), these customer inquiries often translate into files and notes that can be mined for patterns and insights.

Sentiment analysis—the *tone* of a certain concept, positive or negative—would be a good feature for a text mining capability to have to make the best use of these kinds of text. For instance, identifying that a number of customers have *complained* about a recent rate increase might help the carrier make appropriate adjustments or tailor communications accordingly. Similarly, identifying *wants* and *needs* from customer feedback can be extremely helpful in identifying concepts for new products and coverages.

DRIVING BUSINESS VALUE USING TEXT MINING

In this section we present some examples of the use of text mining in various application areas within P&C insurance. This is by no means intended to be a comprehensive demonstration of all possible uses of text mining, but rather our attempt is to provide instances that readers can relate to, and perhaps draw inspiration from, in their own quest to drive novel business applications through the use of insights created via text mining.

Text Mining to Support Marketing and Sales

Customer, prospect, and competitor comments offer incredible sources of insight for carriers with the capability to tease them out. In addition to learning how customers and prospects perceive a company and the new products or services it may be considering, the right ability to harness insights from prolific information sources such as web media, blogs, and social networks can help companies brand themselves more effectively, market more efficiently, and safeguard against unfounded reputation attacks.

Mining Focus Group Transcripts and Survey Comments

A key capability in the market research tool kit is the ability to conduct user-group sessions to generate feedback around certain focused topics. Using an interview or group discussion type of a format, such *focus group research* aims to understand market reactions to a particular company or product, as well as identify behavioral and demographic factors that a company can use to better develop and target their products or service offerings.

Focus group sessions are often composed of open-ended questions such as the following:

1. When you think about insurance coverage, which companies come to mind?
2. As you look at your life, what kinds of risk concerns do you see that you would like to have peace of mind about?
3. What activities do you like to pursue outside of work or for leisure?
4. When you evaluate insurance companies or products, how critical is price versus the financial viability or reputation of the company?

As can be imagined, these sessions generate considerable verbal input that market researchers transcribe into textual reports and analyses with the goal of identifying *qualitative* insights for business consideration and action. Text mining can help augment these human insights, particularly in discovering new insights that may not have been the primary focus of prior research. For instance, a carrier may conduct market research for a new extended warranty coverage product. At a later point, the carrier may be toying with a new product idea—say, a travel insurance add-on to a homeowners policy—and may wish to know if such a product concept will have resonance in the market.

Text mining historic focus group reports and transcripts might help the company better understand what it has *already* heard from the market on this topic, even though the earlier focus groups were convened for a different topic. Text mining can also help reveal insights that might aid the carrier in better preparing for this new research. For example, text and data analysis on historic research notes might indicate that focus group participants most likely to mention *traveling* as a big part of their leisure activity were *married with older kids*. This insight cannot only help the carrier define its product concept to better fit the needs of this likely customer segment, but it can also recruit and qualify the right kinds of participants for in-depth focus groups on the travel insurance product concept. Text mining market research transcripts can also surface *new* ideas for product or service innovation.

Text Mining Web Data

The World Wide Web has evolved into an incredible tool and information source for businesses. The rise of social media such as blogs, chats, tweets, and social networks has led to a proliferation of *online opinions* from ratings and reviews, to recommendations and reputation attacks. While the ability to hone in on information of *relevance* in the sheer volume of data can provide significant business opportunities, it has considerable analytical challenges as well. A detailed discussion of these opportunities, capabilities, and challenges is beyond the scope of this book, and the interested reader is referred to Russell (2011) and Liu (2010).

A simple use case of web mining in insurance can work as follows. A carrier can initiate a web search with their company or product/service *name*. Depending on how well known the carrier is, this is likely to return thousands of page hits—certainly too numerous for effective human involvement and assessment. This is where text mining, and in particular, sentiment analysis, can help. A text mining system could be set up to classify these pages into different groups of *content*. For instance, groupings could include pages where:

1. The carrier is mentioned along with one or more of their competitors.
   - For example, "Unlike Carrier X, *Carrier Y* and *Carrier Z* have filed rating plans that are …"
2. Specific products/services are referenced along with an *opinion* (sentiment analysis of good or bad).
   - For example, "The MileageAware product from Carrier X is an *excellent* one for good drivers …"
   - Or "Carrier X's MileageAware product is a *poor* substitute for …"
3. Certain *entities* (people's names, organization/area, office, etc.) are referenced.
   - For example, "The *adjusters* in the *Atlanta office* were terrible: condescending, unhelpful, and had little sensitivity to the trauma the loss had caused my family …"
4. Specific *geographies or locations* are referenced.
   - For example, "In California, Carrier X offers fewer coverages and lower limits for the same price as Carrier Y …"
5. Reputations are under attack.
   - For example, "Carrier X is a horrible company. Don't ever get insurance from them …"

With classifications like these, the human task of sorting through and analyzing the relevant information becomes exceptionally focused and efficient. For instance, pages related to items 1 and 2 could be reviewed by staff in the product management or competitive intelligence functions to more capably identify opportunities, distinctions, and concerns with products in the competitive landscape.

Pages with content as in items 3 and 4 should ideally be routed to the business area most likely to benefit from the insight and most capable of taking action. For instance, a page with reference to an *adjuster and his/her insensitivity* should be sent to the claims area for solution and action, while a page with reference to *coverage inadequacy and pricing* should be routed to the product management area. Text mining profiles to further distinguish and classify page contents into business function, type of product, type of concern, immediacy of required action, and so on can significantly aid and optimize this routing.

Similarly, pages containing reputation attacks—a growing concern—can be quickly detected and routed to the marketing, product, and/or perhaps the legal department for assessment and intervention.

Text Mining to Support Underwriting and Pricing

Underwriters, particularly in commercial lines or *high-touch* personal lines (e.g., affluent/high-net-worth writers), maintain detailed electronic notes on risks they have underwritten. These notes often contain intricate details on the subject risks not captured in a structured fashion elsewhere in the policy quoting or policy administration systems.

For instance, an underwriter might document the fact that the policyholder is considering adding a swimming pool in her backyard. From the perspective of reducing liability and preventing losses, it would be a great benefit to have an alert during the renewal cycle that allows the underwriter and/or agent to ensure the policyholder has taken all the necessary steps to minimize her liability exposure, such as installing a fence around the pool area. This might even allow the carrier to proactively offer discounts to the policyholder for the right measures they may have taken. Without text mining, this is not easily or systematically possible.

Similarly, most homeowners policies offer discounts for real-time burglary monitoring. Text mining profiles can be set up to automatically identify policies where homeowners have received the discount while promising to activate such monitoring later. Targeted follow-ups are then possible with customers or their agents to determine if monitoring was indeed activated.

On the flip side, imagine the customer satisfaction/retention benefits accrued from setting up alerts to identify customers that may have indicated at some prior point that they were thinking of removing the pool in their backyard. Reaching out to these customers proactively to learn if they have indeed removed the pool and therefore might deserve a lower rate is sure to generate considerable customer goodwill.

In another example, the underwriter or agent notes might indicate that the policyholder is getting married soon. Setting up simple text mining filters can trigger a targeted customer service or agent call to ensure that their coverage remains adequate after their wedding. This is particularly critical if they get items such as jewelry, china, and silverware as wedding gifts, since these must often be insured separately as *scheduled valuables*.

Where available, underwriters also document the preexisting conditions of the subject risk. For instance, a home may have some suspect areas on the roof, or a car may have preexisting damage. Most often, these types of remarks are only available in a textual form in the policy systems. If a loss were to be reported later, these preexisting conditions should be factored in to reduce the extent of the loss, or perhaps even deny the entire claim. Unfortunately, this is something many carriers struggle with. Using text mining constructs to extract and flag policies with such preexisting conditions can lead to significant contributions to business results at the time of the claim.

Mining for Risk and Loss Control Insights

In P&C insurance, inspection of risks is a key part of the business, particularly in commercial lines where no two risks are likely to be the same. Before insuring business premises, operations, or commercial property, an insurer might inspect the subject risk to better assess and evaluate the specific risk exposures critical to pricing. For instance, an inspection of a manufacturing facility might show that used cleaning rags are not properly collected and disposed of. Since such rags lying around pose an increased fire hazard, this is critical information that the insurance carrier will use in its rating. They could factor this into the underwriting of the risk and place the manufacturing facility in a higher experience rating tier or might insure it conditional on the manufacturing facility implementing better processes for handling used rags. Of course, it could also choose to not insure this risk based on these findings.

Loss control is generally a function associated with risk inspections, with the goal of minimizing or preventing common losses through better practices and processes. In the preceding example, the loss control area might provide recommendations to the manufacturing facility to guide them on best practices for used rag disposal.

Risk inspectors visit client premises and document their findings. While certain attributes are structured (industry, employees, size of premises, location, etc.), much of the inspector's findings are in a textual report form. Generally, the inspectors document their findings and then summarize or highlight their key takeaways. The rest of their findings remain as unstructured data.

This is where text mining can help. For instance, building and mining the concept of *worker fatigue* might reveal insights such as the increasing frequency of observed fatigue by industry or region. Overlaying with business practices for minimizing worker fatigue might yield clues into practices that work and those that might require changes.

Effective text mining can also point to new and emerging trends that will require business action. For instance, text mining inspection reports for hotels with pools might yield nuggets, such as broken glass pieces in the pool area, that increase the hazard for cuts and scrapes. Such insights can allow a carrier to devise novel ways to deal with the risk exposures. For instance, insurers may stipulate that no glass containers are allowed in the pool area and require the hotel to ensure that guests transfer their beverages to plastic cups installed near doorways leading to the pool area.

Mining for Loss Understanding

The business of insurance is all about risk assessment and management. To be successful, a carrier must identify, understand, and measure the true drivers of loss. Most claim systems have a *cause of loss* field that allows the adjuster to assign a predetermined set of loss types to a given claim. For instance, typical loss types for a home loss might include such things as fire, water, wind, hail, and theft. In commercial general liability claims, loss types would include slip and fall, trip and fall, and cuts and abrasions, among others.

While useful in generally understanding the losses, these loss types do not specifically indicate *what caused the loss*. For example, in a homeowner claim, *water* simply indicates the peril that caused the corresponding loss. It doesn't explain *why* water caused the loss or *where* it came from. Was it a burst pipe, a faulty washer, an overflowing bathtub, or rainwater through an open window that resulted in the water causing damage? This level of loss understanding has immense implications for the business because it will allow them to prevent losses from occurring, enhance their screening of risks in the underwriting stage, or refine their pricing mechanism to appropriately account for the losses.

Figure 14.2 shows some examples of loss descriptions and the result of text mining to identify the true causes of loss. The examples, all related to water losses in homes, illustrate the challenges inherent

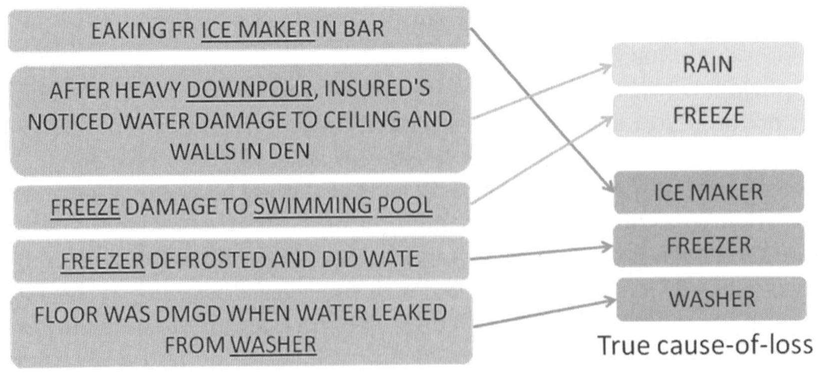

FIGURE 14.2
Text mining the *description of loss* field can yield real insights into the true causes of water damage. The descriptions illustrate typos and abbreviations common in the claims world. Phrases with an underline represent domain-specific concepts that can be used to classify losses.

in claims text, including typos, abbreviations, and truncated text. It must be noted that while the cause of loss end results look appealing, the process of getting to these results is not straightforward. Given sentence fragments, typos, and other issues, most NLP-based text mining tools fail when faced with text of this type. Text mining tools that allow interactive extraction of concepts or phrases and also provide some *semantic generalization* to expand the concepts to include other similar ones, work best in this situation. There is also a strong need to have custom dictionaries for common abbreviations and typos.

Once the text mining is complete and each historical claim can be tagged with its true cause of loss, a carrier can benefit from these deep insights in a number of ways, from simple analysis of key loss drivers and trends to more sophisticated characterizations by geography and type of customer segment. For instance, frozen and bursting pipes might drive water losses in Idaho, while water losses in California might be the result of malfunctioning washers and dishwashers. These insights can drive new underwriting qualification questions such as *age of pipes* for homes in Idaho and *age and type of washer* for the risks in California.

Novel applications of these insights can also drive unique business actions. For instance, if certain key customer segments (e.g., young, middle-class families) are more prone to water losses due to washer malfunction/leaks, the carrier may choose to undertake a campaign to install plastic water-overflow trays *under* washers for their target customer segment. While the cost of this program might be considerable in the short term, it would likely pay for itself by eliminating a key source of water-related losses. It might also result in improved customer relationship benefits because the carrier could position this as a *peace of mind* activity targeted at their best customers.

Mining to Support Rating Models

Figure 14.3 shows an example of separating homeowners' water losses into those caused by weather-related events such as rain and ice and non-weather-related events such as those caused by faulty appliances, burst pipes, overflowing bathtubs/sinks, and so forth. Being able to decompose losses in this way has significant implications for a carrier's pricing plan.

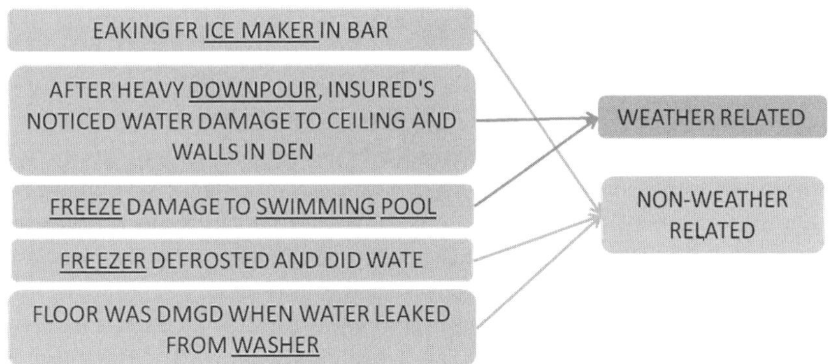

FIGURE 14.3
Text mining to separate water losses into weather-related versus non-weather-related events. This distinction is very helpful in building accurate predictive models for weather- and non-weather-related water losses.

Imagine a carrier building sophisticated models to predict, and thus price, its homeowners product. The attributes of a home risk that will correlate strongly with weather-related water losses are likely to be attributes such as the weather patterns at the risk location, age and type of roof, and so on. This is likely to be different from attributes that will correlate to non-weather-related losses, such as number of bathrooms, number of floors, age and quality of pipes, size of the family, and so on. By being able to separate the water losses into these two very distinct sources of loss, the carrier can build effective and accurate models for predicting losses and pricing their homeowners policies. Similar opportunities to build loss prediction models segmented by core *types of claims* exist in other lines of business as well.

Text Mining to Support Claims Processes

Claims are a key business area for any carrier, with significant impact on bottom-line results. Much rides on the expertise and follow-through of the adjusters, who not only have to negotiate a fair and equitable settlement but also make a number of other assessments such as subrogation potential, suspicion level, and the need for an independent medical exam. They have to make decisions regarding the need for a nurse/case manager for significant injuries and whether a professional engineer should oversee reconstruction of property losses. Missed opportunities and bad decisions can significantly impact business results.

The claim adjustment life cycle is highly nonlinear. Adjusters don't have the luxury of working a claim from start to finish in a defined and timely manner. Much depends on when insureds or claimants return their calls and when and what kinds of required documentation they submit. It is typical for an adjuster to work on tens of open claims at any given time. Furthermore, while working on a particular claim, an adjuster might be interrupted by a call from an involved party on another claim and will have to *context switch*. The constant stop-and-start nature of work, combined with the volume of claims an adjuster handles, results in situations where key facts and decisions on the claim might be missed. For example, another party may have been responsible for a loss suffered by the insured. The adjuster may have investigated and arrived at this conclusion and may even record the intention of referring the claims to the subrogation unit for recovery proceedings. However, the adjuster might forget to do so due to the frequent interruptions. Text mining can be used to uncover insights from adjuster notes and aid the systematic detection and referral of claims to such specialists.

For instance, Figure 14.4 contrasts claims notes from two different adjusters. While one is well written and basic English, the other is rife with abbreviations. Both, however, contain rich information not found in structured fields in the claims system such as *low-impact* accident, *questionable injuries,* and *exaggerated treatment*—concepts that will be useful in building models for systematic fraud detection.

Furthermore, using the art of text mining adjuster notes, one can build profiles for key concepts such as *excessive treatment* by reconciling related concepts such as *overtreatment, exaggerated treatment, unnecessary treatment,* and domain-specific references such as *buildup.*

Once built, these textual profiles can be applied to each claim. If a concept is discoverable in the adjuster notes, a structured value of "yes" (or "no") can be created in the database. Since these are structured attributes derived from the text mining of the claim notes, we refer to them as *structurized data*. These structurized data elements can be used just like structured data in building decision models for fraud, subrogation, and other claims and underwriting analytics.

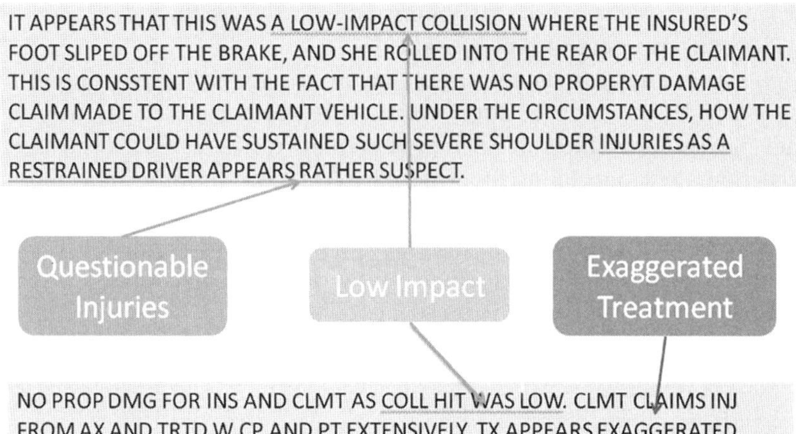

FIGURE 14.4

Snippets from claims adjuster notes show the rich concepts that can be extracted via text mining. Typos and abbreviations offer significant challenges to effective text mining.

Figure 14.5 shows examples of unique concepts related to fraud detection, such as sudden braking near a highway exit and that the injured claimants waived ambulance/EMR. Combining these concepts with other structured data yields predictive models with high coverage and accuracy in identifying suspicious claims.

Text mining can also help identify subrogation concepts such as *other party at fault, other party identified,* and *insured not at fault.* In this case, building a comprehensive dictionary of other parties (driver, repairman, roofer, etc.) can help the system determine if someone else was at fault. For instance, *furnace repairman incorrectly installed, other driver struck our insured, insured struck by adverse party,* and *caused by roofer* would all trigger the concept of *other party at fault.* Once such concepts have been created, decision models can be developed for the identification and systematic referral of cases with subrogation opportunity.

In addition to the big opportunities with fraud and subrogation detection, adjuster notes can be text mined for concepts that can support a wide variety of other business actions. For instance, finding concepts related to *questionable injuries* or *excessive or unusual treatment,* coupled with some thresholds on the medical bills, can trigger a recommendation to the adjuster to consider an *independent medical*

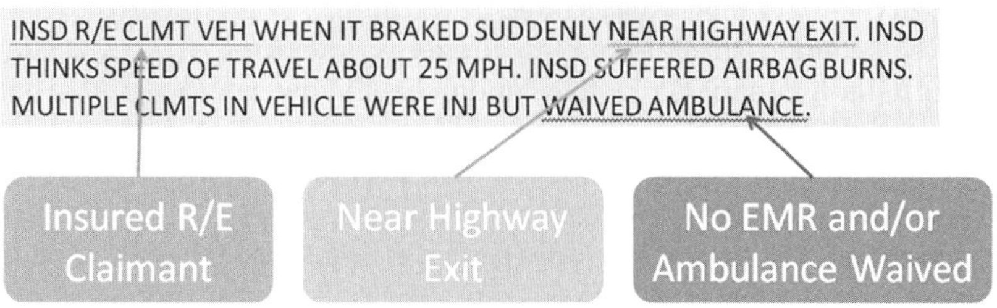

FIGURE 14.5

Unique concepts for fraud detection.

examination (IME). An IME is where a doctor contracted with the carrier independently examines and validates the injuries and treatment of the claimant.

Adjusters often document in their notes if an involved party is angry or upset with the claims handling process. Since an unhappy or angry claimant might engage an attorney, which generally results in a higher payout for a like-kind of injury, a smart carrier can set up text mining filters to automatically trigger an alert to a claims supervisor. Directed and prompt intervention by the supervisor, or other specially trained claims staff, can help manage claimant upsets and keep them from engaging attorneys. When the upset party happens to be an insured, again, early detection and systematic engagement can help turn the experience around in a positive way so the situation does not degrade into a *bad faith* lawsuit and/or attrition of a good customer.

Finally, text mining filters can also be used to effectively eliminate missed opportunities due to process workflows. For instance, a number of insurance companies still use legacy claims handling systems. While an adjuster might identify that a subrogation opportunity exists on a claim and might have every intention of sending it to the recovery unit, most legacy systems either require the adjuster to navigate to a different section of the legacy application to trigger the routing or, worse, use an offline database to enter the claim information for referral to the recovery unit. With such discontinuous processes, it is little surprise that many valid referral opportunities get missed.

A simple solution to minimize such missed opportunities could be to develop and train the adjusters on the use of a predetermined set of phrases. For instance, *Route to Subro* or *Route to SIU* might be agreed-upon phrases that adjusters might insert into their claim notes if they would like the claim sent to the recovery unit or the SIU, respectively. Once they document this, the adjusters can continue on with their claim handling. An offline, nightly batch process can look for these specific phrases and automatically route the claims to the recovery or SIU unit. Even simple implementations like these can drastically reduce missed opportunities while also increasing adjuster productivity by helping them to focus on loss investigation and assessment rather than coding claims for additional attention.

Text Mining for Customer Management and Innovation

Customer contact areas at insurance companies typically deal with issues outside of claims. These issues can range from billing concerns to policy- and coverage-related inquiries. Insurance companies that are direct-to-customer, in particular, collect significant amounts of textual information from their contacts with their customers. These comments, usually buried in their CRM systems, can be extracted and systematically mined to discover emerging patterns and novel insights.

For instance, a customer might call the service center to inquire about the liability coverage on their homeowners policy in the event they were to build a swimming pool in their backyard. Assuming the service representative documents this information in the customer contact file, it would be immensely useful for the carrier to set up a simple text extraction query to generate an alert for the underwriter to inquire about the swimming pool during the next policy renewal cycle.

Figure 14.6 shows some other examples of comments captured by the customer service representatives. Using simple text mining and classification capabilities, carriers can leverage this information in a number of ways. For instance, the first two notes pertain to *billing*; the third comment relates to *policy, billing, and claims*; and the last one is specific to *claims*. Setting up text profiles to classify customer comments into the *type* of issue allows the carrier to efficiently route the customer concern to the correct customer support area for resolution. For instance, billing concerns should be routed to agents with the

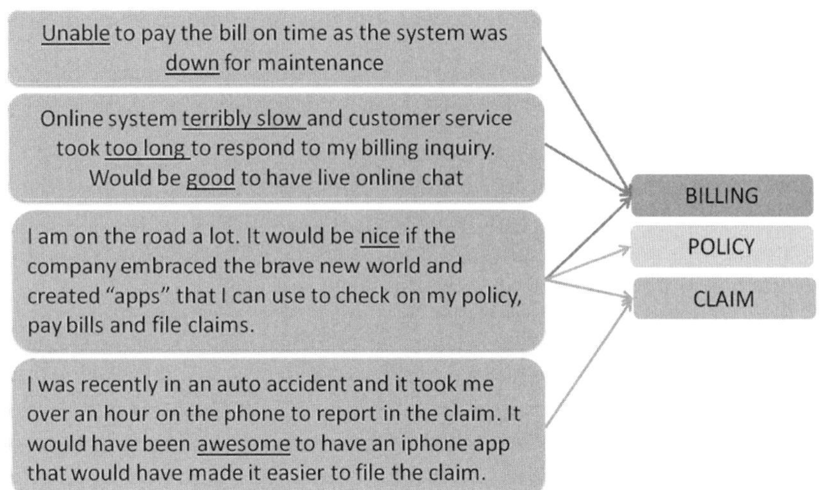

FIGURE 14.6

Examples of typical customer comments from a CRM-type system, along with their classification into the *type of issue*. Attributes that point to the *tone* of the customer issue are highlighted with an underline.

most knowledge of billing system issues and progress, while policy- and coverage-related issues should be routed to representatives with that domain knowledge.

Particularly with the advent of the Internet, it is commonly the case that customers don't speak to a customer representative directly. Instead, they are often asked to email their concerns, which can then be text mined, classified, and routed to the right customer representative for follow-up and resolution.

Customer comments can also be systematically mined to discover persistent issues and new ideas using *sentiment analysis*—the ability to discover opinions and tones in text. For instance, it would be fairly straightforward to assess that the first two customer comments in Figure 14.6 are customer complaints regarding the billing system (giveaways of this sentiment are *dissatisfaction* words such as *unable, down, terrible, slow,* etc.), while the latter two comments are positive remarks or suggestions for improvement (with *encouragement* words such as *good, nice, awesome,* etc.).

Armed with the type of issue and the sentiment, a carrier can take appropriate action. While the persistent billing complaints can be used to make the business case for prioritizing resources and budget investments to improve the billing system, customer suggestions can be effectively mined to create new services and products. For instance, the ability to chat online with a live customer representative rather than use the phone could be an actionable business direction. Similarly, suggestions to create iPhone-type apps for policy administration and claims reporting might be useful service delivery ideas for the company to pursue.

SUMMARY

As this chapter illustrates, text mining is a very relevant and useful capability that can be easily leveraged to create value across multiple functional areas of an insurance business. It can bring to light critical insights from textual notes of any kind, including claims, underwriter, loss control, auditor, and customer representative notes, in addition to numerous possibilities offered by the ever-innovating social web.

It is important to understand that the examples mentioned here merely scratch the surface. Novel business applications of insights are continually emerging, as are the technologies to make more effective use of unstructured data. As an example, emerging advances in *speech recognition* can open up entirely new possibilities for text mining in the insurance world. Effectively converting speech—conversations with insureds, claimants, and other parties—into text holds the possibility of adding to the richness of textual assets that smart-thinking carriers can use for text mining.

POSTSCRIPT

The discussion in this chapter of how text mining can bring to light critical insights from textual notes and the emerging techniques of speech recognition in the insurance world points to an overall trend in text analytics: Our computer-based analysis techniques are approaching more closely the ability to gather and analyze information gathered in very human ways. These ways are analogous to human senses and intuition. The next step is to combine some sort of automated response system to be triggered by text information gathered in these ways. If *that* can be done, then we will have taken a large step in the direction of developing a communications processing machine that truly *thinks*.

References

Karthik Balakrishnan.* Analytical Opportunities in General Insurance. *Journal of the Insurance Regulatory and Development Authority*, India. Vol VIII, No 9. September 2010.

Insurance Information Institute. *Fraud Report*. 2011. (see: http://www.iii.org/)

Bing Liu. *Web Data Mining: Exploring Hyperlinks, Contents, and Usage Data*. Springer-Verlag: Berlin, Heidelberg, New York. 2010.

Matthew A. Russell. *Mining the Social Web: Analyzing Data from Facebook, Twitter, LinkedIn, and Other Social Media Sites*. O'Reilly: Bejing, Cambridge, Famham, Kola, Sebastopol, Tokyo. 2011.

Emmett J. Vaughan & Therese M. Vaughan. *Fundamentals of Risk and Insurance*, John Wiley and Sons: Hoboken, NJ, USA. 2008.

* Karthik Balakrishnan, Ph.D., is Vice President of Analytics at ISO Innovative Analytics (IIA). A unit of Verisk Analytics (VRSK), IIA is focused on delivering advanced predictive analytics solutions to the property/casualty insurance, health care, and mortgage verticals. He can be reached at kbalakrishnan@iso.com.

CHAPTER 15

Focused Web Crawling

CONTENTS

PREAMBLE

Focused web crawling stands one step above the other techniques discussed in this book. It is an integrated technology that combines two base technologies: classification and web analytics. This combination of basic capabilities enables more complex decision making and provides information that is more specific to its intended use. Figure 15.1 shows a complete analysis system that "feeds" automatically on Internet data and outputs information that can be used to make predictions. This general architecture follows accepted educational theory proposed by Bloom (1956), which describes and promotes higher-level questioning to produce higher-level thinking. All components of the system are coordinated and focused on a single outcome. As you read this chapter, start thinking about how the text analytical systems of the future might improve on this architecture in search of the goal of higher-level decision making.

Practical Text Mining and Statistical Analysis for Non-structured Text Data Applications. DOI: 10.1016/B978-0-12-386979-1.00043-8

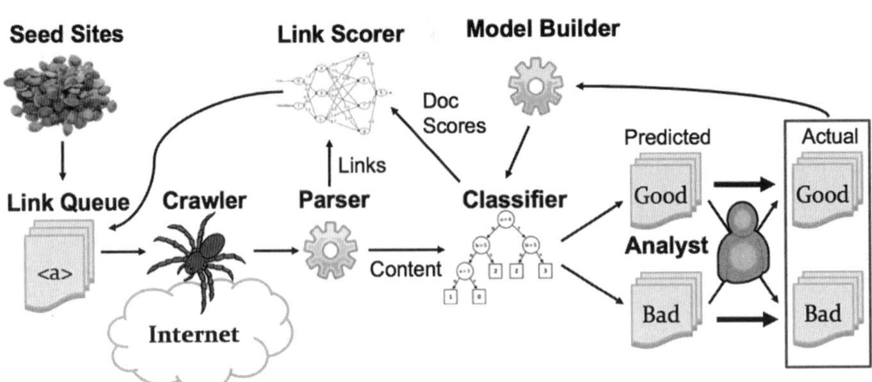

FIGURE 15.1
Example architecture for a focused crawler.

INTRODUCTION

A focused web crawler is a type of web mining system that combines document classification techniques (see Chapters 7 and 8) with a conventional web crawler to efficiently find new, on-topic content from the Internet. To be able to search the entire online universe about anything, leading search engines, such as Google, Yahoo!, and Bing, rely on many warehouses of computers to crawl and index vast swaths of the Internet. However, these search engines are tuned for general queries and often do not perform well for specific needs. A focused crawler can queue up specifically relevant documents by using document classification to only crawl and index pages that match a desired "focus topic." Using document classification with a focus topic allows the crawler to follow hyperlinks from on-topic pages to other on-topic pages. A good focused crawler can allow a user to search out new, relevant content using only a single machine.

A focused crawler consists of two primary parts with a number of optional components. First, a web crawler is used to find and download pages from the Internet. This component includes a link queue for tracking pages to be downloaded and a document parser for extracting the main text, or *payload*, of a web page, as well as any hyperlinks to other pages for processing. The second part is a document classifier to determine whether or not a page matches the desired focus topic. Users define the focus topic by providing a set of initial training documents; the crawler will then seek out more documents like those supplied.

Using a document classifier differentiates the focused crawler from other methods of collecting information from the web, including web scraping and web harvesting. These other methods require human intervention to point the crawler toward the right content, usually by selecting the sites to be scraped. Once started, a web scraper uses only simple rules to guide the crawling process. These rules usually limit the crawl based on distance from the starting page, or they limit the crawler to a certain domain. Using a document classifier as part of the crawling allows more complex decisions to be made about which pages to continue crawling.

To aid in this goal, the document classifier can optionally include a topic hierarchy and allow for user feedback. A *topic hierarchy* defines a more detailed hierarchy or taxonomy of topics to help guide the crawler toward documents about the focus topic. User feedback is a powerful way to quickly point the crawler toward the correct content by approving or rejecting early decisions made by the document classifier. Figure 15.1 illustrates the interaction of these components. The components and their uses are described in detail in this chapter.

THE FOCUSED CRAWLING PROCESS

Figure 15.1 shows all the parts of a focused crawling system. The process starts with a collection of seed sites that provide links to the pages that will initiate the focused crawling process. Typically, a combination of training documents for the document classifier and general sites from the Internet are used as seed sites to provide very targeted pages as well as opportunities to find new pages. The links extracted from these pages form the link queue, also called the *frontier*. The link queue is a list of the pages remaining to be explored by the crawler. Each link in the link queue is given a weight indicating its likelihood of being related to the focus topic.

Once the link queue has been populated, the crawler can begin fetching the pages that the links have pointed to, starting with those with the largest weight. After the page has been fetched, it is parsed to extract payload text and outlinks. The payload is passed on to the topic classifier to be assigned a score indicating how on-topic the page is. This score is then attached to outlinks extracted from the page, and those links are added to the link queue. The goal is to put links found on highly relevant pages at the front of the queue so those links are explored first. This is the primary difference between a focused web crawler and a regular web crawler, which makes no distinction between outlinks.

This process can continue as long as there are links in the link queue to explore. In practice, however, two types of limits are used both individually and in combination to make focused crawling more efficient. First, a weight threshold is used to remove links with low weights from the link queue. Though the removed links will be the last to be explored due to their low weights, the link queue can become quite large; removing the low-weight links makes searching and sorting the link queue more efficient. The second form of limitation is a distance limit, called the *hop count*. A *hop* is each time the crawler follows a link from one page to the next. The hop count measures how many hops there are between two pages. The assumption is that pages too far away from a seed page are not likely to continue to produce valuable content.

THE OPPORTUNITIES AND CHALLENGES OF MINING THE WEB

Web pages differ significantly in both style and form from text documents used for text mining. Instead of plain text, most web pages use Hyper Text Markup Language (HTML) to display their content. HTML easily allows a page to be a combination of textual information, called the page payload; formatting information such as tables and headers; multimedia elements such as pictures or video (including advertisements); and hyperlinks to other HTML pages. This combination of features provided by HTML provides both opportunities and challenges for text mining algorithms.

Hyperlinks between pages and the defined HTML structure are the two largest positives of mining web text. Hyperlinks (or links) connect related pages together. A single HTML page has both *inlinks* and *outlinks* associated with it. Inlinks are links on other pages that point in to the current page. Outlinks are links on the current page that point out to other pages on the Internet. Both inlinks and outlinks are used by the focused crawler to find on-topic content.

A focused crawler starts at a set of seed pages and explores the outlinks on those pages, repeating the process until enough on-topic documents have been found or only off-topic documents are left to explore. In addition, these linked pages can be used to improve document classification performance (Chakrabarti et al., 1998). In the process of crawling the web, the focused crawler can also examine the *anchor text* for a page. The anchor text is the text associated with an inlink to a page, and it often provides clues to its content.

The structure of HTML documents can also provide rich clues to a text mining algorithm. Often titles and headers contain the most important words for describing a section of text. Since HTML clearly marks the headers and titles using <header> and </header> tags, this information can easily be used automatically. In addition, HTML defines tags for tables and ordered and unordered lists. These structural tags provide hints for finding meaningful words on the page.

In addition to their main content, most web pages include large amounts of unimportant information, including advertising, legal "small print," and "boilerplate" navigation elements. Each of these elements adds text to the page that is not directly related to its content, but it is often difficult to avoid when mining web documents. Advertising is the easiest to remove for text mining purposes because most advertising is in graphic or video form. The focused crawler and other text mining algorithms already ignore these formats, since they don't consist of text. Legal qualifiers and "boilerplate" navigation elements, however, do contain text, and they can be more difficult to remove. Fortunately, these elements are typically situated in predictable locations on the page and can usually be differentiated from payload text.

TOPIC HIERARCHIES FOR FOCUSED CRAWLING

Using a conventional web crawler on the Internet will lead to finding pages from many different topics, making it hard to identify on-topic pages. Rather than using a binary classifier to separate on-topic from off-topic documents, a topic hierarchy can be used to sort the documents into finer-grained, hierarchical topics, which can improve document classification accuracy. One such topic hierarchy for the Internet is freely available as part of the Open Directory Project (www.dmoz.org), which also supplies example pages in each of the topics.

The main motivation for incorporating a topic hierarchy into a focused crawler is to help the system better differentiate between related topics. For example, the authors once had a client who was interested in monitoring livestock diseases. When using a binary classifier comparing Internet pages pertaining to livestock diseases with the rest of the Internet, we found that the classifier made many false positive errors, such as marking pages about animals, veterinary practices, and other closely related topics as being "on-topic" for livestock diseases. Because these pages were closer in content to the focus topic than to a typical page on the Internet, they were being lumped together into the livestock diseases topic.

To solve this problem, a topic hierarchy was used to first direct all of the related animal content into a single topic and then differentiate the content further into subtopics, including one for the focus topic. It is much easier for an automatic text classifier to make the fine-grained distinction between personal pets and livestock diseases when it can focus on those two topics and not worry about the other unrelated topics. Different features become informative when looking at smaller collections of pages.

TRAINING THE DOCUMENT CLASSIFIER

The first step for performing a crawl with a focused crawler is defining a focus topic. The crawler's ability to discover on-topic documents depends on the document classifier "understanding" the focus topic. The document classifier must therefore be trained with handpicked documents that exemplify the focus topic and with negative examples that represent other topics. These documents, called *training documents*, are the building blocks with which the classifier defines the boundaries it uses to classify newly crawled web pages and other documents, as illustrated in Figure 15.2.

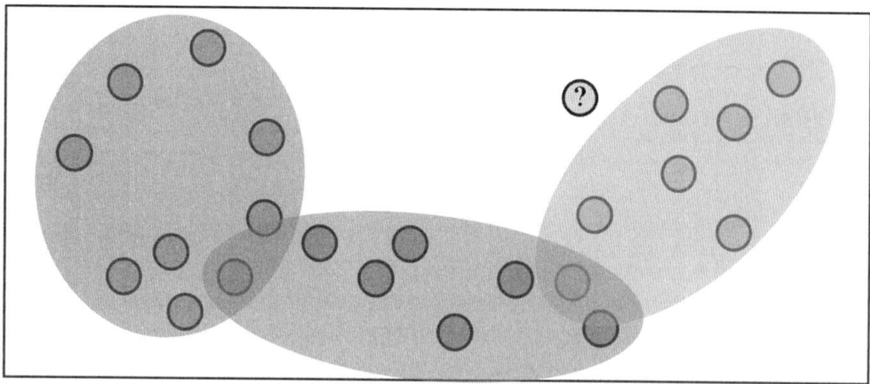

FIGURE 15.2
An illustration of how training documents (represented by circles) define the boundaries in the "topic space" so that the classifier can accurately classify documents into topics. In this illustration, the crawled document (with the question mark) lies closest to the cluster on the right and will be classified as such.

The accuracy of the document classifier depends on the representation quality of the training documents, the organization of the topic taxonomy, and the quality of the classification modeling algorithm, such as decision trees or logistic regression. Often, documents, particularly web pages, have text that is not part of the information-rich payload—for example, copyright notices, navigation menus, or banner advertisements or sidebars. If included, this distracting text inhibits the ability of the classifier to define clear boundaries between topics, since this text is often common among completely unrelated topics. When handpicking training documents, users should strive to identify high-quality documents with a high proportion of payload text.

The position of the focus topic in the taxonomy also affects the performance of the crawler. In Figure 15.2, there is good separation between the different topic clusters, but in real projects there is often significant overlap in the content of training documents from different topics. A well-organized topic taxonomy can help a classifier disambiguate confusing documents by breaking up a document's classification into multiple levels of detail.

For an example of disambiguation, the word *jaguar* could refer to multiple topics: a species of large cat, a manufacturer of luxury cars, a professional football team, a German World War II antitank vehicle, or a Cray supercomputer in Oak Ridge, Tennessee. If the Jaguar supercomputer were the focus topic for a crawl, the classifier would need training documents for each use of the word *jaguar* to correctly distinguish truly on-topic documents from similar but off-topic documents (Figure 15.3).

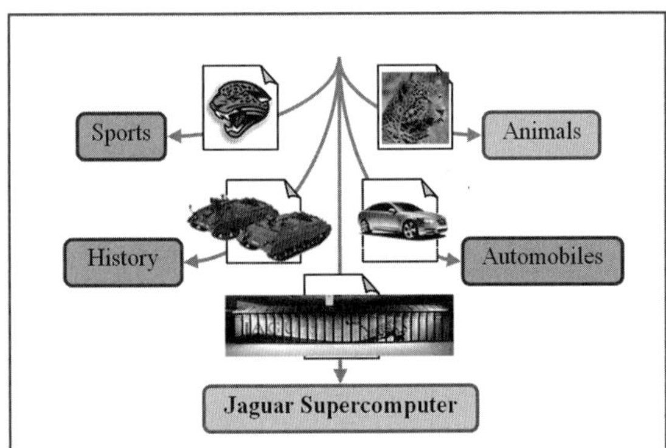

FIGURE 15.3
Crawled pages with different uses of the word *jaguar* are classified correctly because training data were provided for each use of the word.

However, in this example, documents discussing the IBM supercomputer Roadrunner would likely be incorrectly classified as being on-topic. This would be because the Roadrunner documents are more related to the Jaguar supercomputer topic than any of the other available topics. The solution to this problem is to generalize the focus topic to include all documents relating to supercomputing and then train a second-level, or *child,* classifier to distinguish between the Jaguar and non-Jaguar (e.g., Roadrunner) *sibling* topics after they have been classified as belonging to the more general *parent* supercomputing topic. Because the classifier picks the "closest" topic for a given web page, the more choices the classifier has for alternative "buckets" to put pages in, the fewer similar but off-topic pages that will leak into the focus topic's bucket.

There is no good rule as to the number of training documents you must provide. The more training data you provide, the more the classifier will be able to derive more accurate characteristics of on-topic documents. It is important, however, not to sacrifice quality (purity of the examples) to achieve greater quantity. Also, the crawler uses websites that link to and link out from the training web pages as the starting point for the crawl, so providing a higher quantity of training documents creates a broader *link frontier* that increases the chances of finding relevant web pages.

CAPTURING USER FEEDBACK

Because focused crawling is an ongoing process, there is an opportunity to dynamically update the model as it runs over time. If the focused crawler is working correctly, it will be fetching additional documents that are on-topic according to the classification model. These additional documents are either truly on-topic or are incorrectly marked as on-topic. User feedback to the crawling process allows an analyst to improve the classifier model by evaluating the returned documents and adding them to the training data by marking them as either on-topic or off-topic. The new training data from this feedback are then incorporated into the document classifier through updating the model, improving how links are weighted (ranked).

User feedback is an extremely valuable feature. It allows the system to take maximum advantage of the work the user just undertook to understand and evaluate the document read. Existing systems are close to being able to reorder the document queue in real time after any "thumbs up" or "thumbs down" is given a document! The system is essentially "learning" from subtle clues that the user provides about what types of documents are valuable for a given search. This occurs even if the user is not explicitly aware of his or her criteria or how to define it in a query or search language. Furthermore, it is straightforward to combine labeled data (rated documents) from multiple cooperating analysts to accelerate the learning process, leading to greatly enhanced efficiency and effectiveness when searching the vast web for rare but valuable on-topic documents.

SUMMARY

Focused crawling is a powerful web mining tool for collecting specific information automatically from the Internet. By combining a document classifier, topic hierarchy, and web crawling machinery, a focused crawler can improve on the performance of a general search engine by focusing on a single topic and crawling that topic in depth, rather than skimming across the surface of many irrelevant topics. In addition, by ignoring irrelevant topics, a focused crawler does not require large numbers of servers for crawling.

Because the focused crawler relies on a document classifier, a user can provide feedback to the models, adding additional training documents based on the early pages returned by the crawler. This can help the crawler quickly focus on the right information, saving the analyst a great deal of time and effort.

POSTSCRIPT

We put this chapter at the end of the book because of its integrative nature and because it points to the future. Text analytics will become increasingly integrated with other data manipulation and modeling tools that are represented currently by separate tools. Eventually, all data preprocessing, modeling, and application-specific uses of text data will be combined into integrated application-specific text processing tools. Get ready for the future!

References

Bloom, B. S. (Ed.). 1956. *Taxonomy of Educational Objectives.* David McKay, Inc. New York, NY. p. 207.

Chakrabarti, S., Dom, B., and Indyk, P., *Enhanced hypertext categorization using hyperlinks*, ACM SIGMOD Record, 1998, Vol 27, No 2, pp. 307–318.

CHAPTER 16

The Future of Text and Web Analytics

CONTENTS

We thank coauthor Gary Miner's mentor, Dr. Irving Gottesman, for reviewing this chapter and providing important edits. Dr. Gottesman is also a guest-author on Tutorials I and U.

TEXT ANALYTICS AND TEXT MINING

Text analytics and text mining are at the stage that data mining was in the late 1990s. At that time, several tools were available for data mining. At conferences such as the KDD-Data Mining Conference, there would be 100 or more exhibitors, each offering their "tool" or their "algorithm" for doing some aspect of data mining. By 2003, this large number of exhibitors at KDD had been reduced to only about 15. This reduction was caused by winnowing in the marketplace, which also caused the further development of a smaller number of tool kits with increased capabilities and usefulness.

Practical Text Mining and Statistical Analysis for Non-structured Text Data Applications. DOI: 10.1016/B978-0-12-386979-1.00044-X

Table 16.1 Commercial Text Mining/Text Analytics Software

- *ActivePoint* offers natural language processing and smart online catalogues, based contextual search, and ActivePoint's TX5(TM) Discovery Engine.
- *Aiaioo Labs* offers distributed corpus annotation tools and services for use in machine learning, relation and event extraction, transliteration, part of speech tagging, and classification.
- *Alceste* is software for the automatic analysis of textual data (open questions, literature, articles, etc.).
- *Arrowsmith software* supports discovery from complementary literatures.
- *Attensity* offers a complete suite of text analytic applications, including the ability to extract "who," "what," "where," "when," and "why" facts and then drill down to understand people, places, and events and how they are related.
- *Basis Technology* provides natural language processing technology for the analysis of unstructured multilingual text.
- *Clarabridge* is text mining software that provides end-to-end solutions for customer experience professionals wishing to transform customer feedback for marketing, service, and product improvements.
- *ClearForest* is tools for analysis and visualization of your document collection.
- *Compare Suite* compares texts by keywords and highlights common and unique keywords.
- *Connexor Machinese* discovers the grammatical and semantic information of natural language.
- *Copernic Summarizer* can read and summarize document and web page text contents in many languages from various applications.
- *Crossminder* is natural language processing and text analytics (including cross-lingual text mining).
- *DiscoverText is* a powerful and easy-to-use set of text analytic solutions for eDiscovery and research.
- *dtSearch* is for indexing, searching, and retrieving freeform text files.
- *Eaagle text mining software* enables you to rapidly analyze large volumes of unstructured text, create reports and easily communicate your findings.
- *Enkata* provides a range of enterprise-level solutions for text analysis.
- *Entrieva,* which consists of patented technology indexes, categorizes and organizes unstructured text from virtually any source.
- *Expert System* uses a proprietary COGITO platform for the semantic comprehension of the language to do knowledge management of unstructured information.
- *Files Search Assistant* provides quick and efficient searches in text documents.
- *IBM Intelligent Miner Data Mining Suite,* which is now fully integrated into the IBM InfoSphere Warehouse software, includes data and text mining tools (based on UIMA).
- *IBM-SPSS Modeler Premium* enables you to extract key concepts, sentiments, and relationships from call center notes, blogs, emails, and other unstructured data, and then convert it into structured formats for predictive modeling.
- *Intellexer* is a natural language searching technology for developing knowledge management tools, document comparison software and document summarization software, custom built search engines, and other intelligent software.
- *Insightful InFact* is an enterprise search and analysis solution for mining text, images, and numerical data.
- *ISYS Search Software* is an enterprise search software supplier specializing in embedded search, text extraction, federated access solutions, and text analytics.
- *IxReveal* offers uReveal "plug-in" advanced analytical platform and uReka! desktop "search and analyze" consumer products, based on patented text analytical methods.
- *Kwalitan 5 for Windows* uses codes for text fragments to facilitate textual search, display overviews, build hierarchical trees, and more.
- *KXEN Text Coder (KTC)* is a text analytics solution for automatically preparing and transforming unstructured text attributes into a structured representation for use in KXEN Analytic Framework.
- *Lexalytics* provides enterprise and hosted text analytics software to transform unstructured text into structured data.
- *Leximancer* makes automatic concept maps of text data collections.
- *Lextek Onix Toolkit* is for adding high-performance full-text indexing search and retrieval to applications.

Table 16.1 Commercial Text Mining/Text Analytics Software *(Continued)*

- *Lextek Profiling Engine* is for automatically classifying, routing, and filtering electronic text according to user-defined profiles.
- *Linguamatics* offers natural language processing (NLP), search engine approach, intuitive reporting, and domain knowledge plug-in.
- *Megaputer Text Analyst* offers semantic analysis of freeform texts, summarization, clustering, navigation, and natural language retrieval with search dynamic refocusing.
- *Monarch* is a data access and analysis tool that lets you transform any report into a live database.
- *NewsFeed Researcher* presents a live multidocument summarization tool, with automatically generated RSS news feeds.
- *Nstein* provides enterprise search and information access technologies. On your public website, Nstein will guide your customers to the most relevant information more quickly than other solutions.
- *Odin Text* provides actionable do-it-yourself text analytics, with a focus on market research.
- *Power Text Solutions* offers extensive capabilities for "free text" analysis, offering commercial products and custom applications.
- *Readability Studio* offers tools for determining text readability levels.
- *Recommind MindServer* uses PLSA (probabilistic latent semantic analysis) for accurate retrieval and categorization of texts.
- *SAS Text Miner* provides a rich suite of text processing and analysis tools.
- *Semantex from Janya Inc.* is an enterprise-class information extraction system, detecting entities, attributes, relationships, and events.
- *SPSS LexiQuest* is for accessing, managing, and retrieving textual information; it is integrated with the SPSS Clementine data mining suite.
- *STATISTICA Text Miner* enables efficient extraction of word counts, key concepts and relationships, and fast predictive modeling with text converted to structured formats.
- *SWAPit, Fraunhofer-FIT's text and data analysis tool*, an updated version of DocMINER, offers visual text mining and retrieval capabilities, including search, term statistics, and summary; it also visualizes semantic relationships among text documents.
- *TEMIS Luxid*® is an information discovery solution serving the information intelligence needs of business corporations.
- *TeSSI*® is software components that perform semantic indexing, semantic searching, coding, and information extraction on biomedical literature.
- *Textlifter* streamlines the process of sorting large amounts of unstructured text with the Public Comment Analysis Toolkit (PCAT), DiscoverText and Sifter, off-the-shelf, and enterprise-class business process applications.
- *Text Analysis Info* offers software and links for text analysis and more.
- *Textalyser* is an online text analysis tool that provides detailed text statistics.
- *TextPipe Pro* provides a text conversion, extraction, and manipulation workbench.
- *TextQuest* is text analysis software.
- *Readware Information Processor*, for intranets and the Internet, classifies documents by content, provides literal and conceptual searches, and has a ConceptBase with English, French, or German lexicons.
- *Quenza* automatically extracts entities and cross-references from free text documents and builds a database for subsequent analysis.
- *VantagePoint* provides a variety of interactive graphical views and analysis tools, with powerful capabilities to discover knowledge from text databases.
- *VisualText*™ by TextAI is a comprehensive GUI development environment for quickly building accurate text analyzers.
- *Xanalys Indexer* is an information extraction and data mining library aimed at extracting entities, and particularly the relationships between them, from plain text.
- *Wordstat* is an analysis module for textual information such as responses to open-ended questions, interviews, and so on.

Table 16.2 Free and Open Source Text Mining/Text Analytics Software

- *GATE* is a leading open source tool kit for text mining, with a free open source framework (or SDK) and graphical development environment.
- *INTEXT* is the MS-DOS version of TextQuest, which has been in public domain since January 2, 2003.
- *LingPipe* is a suite of Java libraries for the linguistic analysis of human language.
- *Open Calais* is an open source tool kit for including semantic functionality within your blog, content management system, website, or application.
- *RapidMiner Text Mining* is an efficient "workspace" interface similar to Clementine, but with almost any algorithm anyone would want for the predictive analytic phase; it contains a large library of text analytics procedures.
- *S-EM (Spy-EM)* is a text classification system that learns from positive and unlabeled examples.
- *The Semantic Indexing Project* offers open source tools, including Semantic Engine, which is a standalone indexer/search application.
- *uReka!* is aimed at helping consumers, enterprise searchers, students, and researchers rapidly find information on topics from the search results and websites that they are inundated with and to understand them with automatic idea extraction and the ability to customize search sources (the free part of the ixReveal company—http://www.ixreveal.com/company/aboutUs.htm).

Today, a plethora of unclassified tools are available for text analytics and web analytics. In fact, the list is currently so long that it is almost impossible to fully document. Some of the tools are commercially available, and others are open source freeware tools. Over the years, the KDnuggets™ News has provided lists of these tools in their e-newsletter. These listings, plus a few additions and a statement by the vendor website, are provided in Tables 16.1 and 16.2 above.

Based on what happened with data mining tools, it is expected that the lists in Tables 16.1 and 16.2 will shorten significantly over the next few years to form a much smaller group of tools with proven accuracy, reliability, and usefulness through the peer review process. This "winnowing" process is already in place, evidenced by the fact that the latest copyright date on websites for some tools as early as 2006.

Some of the free and open source tools are illustrated in some of the tutorials in this book. GATE and WEKA were used very effectively in Tutorial N, "Deception Detection," and RapidMiner Text Miner was used in Tutorials H and O. GATE has unique features, and RapidMiner's Text Miner is fast and efficient to use with its fun workspace and plethora of algorithms for the predictive part of text mining. However, when we subjected these tools to "heavy loads"—for example, upscaled large data sets—they "balked" and stopped running. Additionally, we found that RapidMiner Text Miner would not create all of the SVD graphs that were selected; it seemed to be a "hit or miss" whether "Concept 'X'" versus "Concept 'Y'" would draw a scatterplot. These same data sets worked to completion with the commercial tools used in our tutorials. This means that the free and open source tools will need to be scaled up to work on the humongous data sets of commercial and governmental interests if these open source tools want to remain viable in the workplace.

Some of the commercial text mining applications rely on proprietary algorithms for presumably extracting "concepts" from text and may even claim to be able to automatically summarize large numbers of text documents, retaining the core and meaning of the text. But this type of technology of extracting "meaning" from documents is very much still in its infancy; the goal to make truly meaningful automated summaries of large numbers of documents may remain elusive in the near future.

THE PROS AND CONS OF COMMERCIAL SOFTWARE VERSUS OPEN SOURCE SOFTWARE

The authors have used some of the open source free data/text mining tools (Weka, RapidMiner, YALE, Gate, etc.) and offer the following assessments:

Pros

- No cost; you can even use them for commercial purposes. Some of the commercial data/text mining tools that we have at Oklahoma State University can only be used for academic purposes (we need to get a different license if want to use them for a funded project, which makes sense).
- They tend to have a large number of algorithms—larger than any commercial tool.
- For the open source ones, you can develop your own algorithm and integrate it into their environment, compare it against the others, and if it is good, it can become a part of the next release.
- They tend to have free community-based support (blogs, discussion forums, video tutorials).
- They also have intuitive interfaces (in the case of RapidMiner, it has a very logical and pleasant interface).

Cons

- These tools are not computationally efficient. It takes a long time to run an NN with relatively large number of cases/variables. Users with supercomputer access will not feel this criticism.
- They do not pay enough attention to memory management; for relatively large data sets/complex calculations, they run out of memory (they don't tell you that until after it works on it for a few hours).
- The algorithms are not reliable (since it is open source, they can come from anywhere).
- The tools may have hidden code (spy) in them (although we have not had this experience ourselves).
- Often, you don't get a professionally designed help documents.
- They do not have professional support people whom you can call for help and speak to.

THE FUTURE OF TEXT MINING

Some of the commercial text mining applications claim to be able to extract meaning from text with proprietary algorithms and automatically summarize large numbers of text documents, retaining the core and meaning of the text. However, extraction of meaning from text is very much still in its infancy; the goal to make truly meaningful automated summaries of large numbers of documents has not yet been achieved, and it possibly never will.

The great practical and commercial value of traditional text mining (word counts, exclusion lists, inclusion lists, synonyms, etc.) combined with the concept of "redundancy of text" has been substantiated by some of the predictive text analytics tutorials presented in this book. Gains in accuracy rates of predictive models were obtained by adding text to insurance industry data sets. Even the identical (monozygotic) twin tutorial presented in Tutorial I found a remarkable 92 percent accuracy rate in predicting a diagnosis of schizophrenia based on just an inclusion list of a small number of words. Many more techniques like these will be developed in the future.

Sentiment Analysis

It will be exciting to see how this field unfolds in the decade ahead. Some sentiment analysis is being performed now, particularly on data from social networks like Twitter. We expect that this type of analysis will be applied in the future to extract sentiments from online computer reviews for marketing purposes.

Text Mining of Mixed Messages

In the Introduction to this book, the three phases of development in text mining were outlined. A fourth possible phase is proposed that may focus on meaning and text mining tools to read "between the lines" of text to distinguish:

- Idioms and colloquialisms
- Sarcasm, puns, and other ambiguous verbal forms
- Innuendo and euphemisms

These meanings are part of a broader discipline of the study of mixed messages in sociology and political science. Similar studies of the dominance of indirect communications and mixed messages are recounted in American politics by Luck (1999). Rockwell (2006) analyzed the social complications arising from the ambiguous ways that people use language, particularly sarcasm and other mixed messages like innuendos, which can complicate spoken communications.

A recent paper by González-Ibáñez and colleagues (2011) on distinguishing sarcasm in Twitter data is one of the rare attempts to apply text mining technology to finding meaning in direct textual communications. We believe this area will probably be developed further in the future.

Web Analytics and Web Mining

Web analytics is probably at an even earlier stage of infancy than text analytics. The advance of the Internet and web pages extending into the commercial world has brought the need for web mining and web analytics. Today almost every company and retail business has its own web page *"shopping cart,"* where items can be purchased and delivered to one's home, in addition to their traditional *"in store"* inventory and checkout counters. But these businesses do not *"see"* their customers come into their web shopping carts, so new methods of identifying their customers on the web, understanding which customers are *"good"* or return/repeat customers, and other classification and segmentation of their *"web hits"* needed to be developed. This is what web analytics and web mining are all about.

Tables 16.3, 16.4, and 16.5 are the combined list of web mining and web analytics software, both commercial and free open source, compiled from the lists presented in KDnuggets™ News from time to time. This list is not as extensive as the previous text analytics/text mining tool list, but it is still impressive.

Table 16.3 Web Mining Software: Commercial

- *11Ants Model Builder* will mine your web usage data in Excel. It is a powerful but simple-to-use data mining tool.
- *AlterWind Log Analyzer Professional* is a website statistics package for professional webmasters, with standard log analyzer features and unique features for pSEO (search engine optimization) and website promotion.
- *Amadea Web Mining* includes multiple transformations, reports, and parametric and modular marketing indicators for an effective CRM.
- *ANGOSS KnowledgeWebMiner* combines ANGOSS KnowledgeSTUDIO with proprietary algorithms for clickstream analysis, Acxiom Data Network, and interfaces to web log reporting tools.
- *Affinium NetInsight* is for Unica enterprise-class web analytics.
- *Azure Web Log analyzer* gives you key information about your site.
- *ClickTracks* displays visitor patterns directly on the pages of your website.
- *ConversionTrack from Antssoft* is for web log analysis and reports on visitor conversion ratios.
- *Download Analyzer* can track visitors, hits, downloads, referring sites, and search phrases, and it can provide traffic analysis data for web promotion and search engine optimization.
- *Megaputer WebAnalyst* integrates the data and text mining capabilities of Megaputer's analytical software directly into your website.
- *Nihuo Web Log Analyzer* provides a comprehensive analysis of the who, what, when, where, and how of customers who visit your website.
- *prudsys ECOMMINER* is combined clickstream and database analysis for e-commerce.
- *SAS Webhound* analyzes website traffic to tell you who visited your site, how long they stayed, and what they looked at.
- *SPSS Web Mining for Clementine* enables you to extract web events, including online campaign results, and to use this online behavior in Clementine's predictive modeling environment.
- *STATISTICA Web Crawler* can be used to crawl websites to any depth of links listed in the sites and then impute the data directly into STATISTICA Text Miner for analysis.
- *Surf Pattern Visual Analyzer* provides web navigation visual link analysis tools that show web page viewing patterns of a website.
- *The Data Miner* is a tool for automating web data extraction and manipulation.
- *WebLog Expert 2.0 for Windows* is an easy-to-use and feature-packed web log analyzer.
- *WebTrends* is a suite for data mining web traffic information.
- *XAffinity*™ is for identifying affinities or patterns in transaction and click stream data.
- *123LogAnalyzer* is simple to use and offers high-speed processing, low disk space requirements, filtering, and built-in IP mapping.

Table 16.4 Web Mining Software: Free and Open source

- *AlterWind Log Analyzer Lite* quickly generates all traditional reports, supporting 430+ search engines from 120 different countries.
- *Analog* (from Dr. Stephen Turner) is a free and fast program to analyze web server log files (Win, Unix, more).
- *htminer* supports the analysis of web logs (including unique visitors, sessions, transactions) and organizes the data in a PostgreSQL data warehouse.
- *jwanalytics* is a Java utility for the storage of information in a dimensional model and is useful for storing web analytics data for Java web applications. Web real-time data mining functionality is being built.
- *Visitator* is for clustering and visual presentation of visitor groups based on access patterns.
- *WUM: Web Utilization Miner* is an integrated, Java-based web mining environment for log file preparation, basic reporting, and discovery of sequential patterns and visualization.

Preceding from http://www.kdnuggets.com/software/web-mining.html.

Table 16.5 Web Content Mining Software

Commercial
• *Bixolabs* is an elastic web mining platform built with Bixo, Cascading & Hadoop for Amazon's cloud (EC2). • *Metafy Anthracite Web Mining Software* is for visually constructing spiders and scrapers without scripts (requires MacOS X 10.4 or newer). • *Mozenda* is for more Zenful data and web content mining. • *Screen Scraper* allows users to scrape structured and unstructured data from websites and format it (free download). • *WebQL* is for creating turnkey web extraction applications, such as price collectors, patent information aggregators, and so on. • *XML Miner* is a system and class library for mining data and text expressed in XML, extracting knowledge and reusing that knowledge in products and applications in the form of fuzzy logic expert system rules.
Free and open source
• *Bixo* is an open source web mining tool kit that runs as a series of cascading pipes on top of Hadoop. • *DEiXTo* is a powerful tool for creating "extraction rules" (wrappers) that describe what pieces of data to scrape from a web page; consists of GUI and a stand-alone extraction rule executor. • *GNU Wget* is a command line tool for retrieving files using HTTP, HTTPS, and FTP.

Preceding from http://www.kdnuggets.com/software/web-content-mining.html.

THE FUTURE OF WEB ANALYTICS

In regard to its stage in the usual life cycle, web analytics is in the same situation as text analytics, although at a much earlier phase of the *growth curve*. Web analytics will need to focus on a number of innovations to reach its goals.

Integration across Multiple Social Networks

The online community has never seen an application become so popular so fast as Twitter. This social networking web application has spread almost as fast as a viral video. Twitter and other social networking sites (e.g., Facebook and LinkedIn) are in the very early stages of their evolution. Primarily, they have been used for sharing small bits of conversations (Twitter limits their "tweets" to 140 characters). These "microblogs" are becoming one of the dominant forms of communication in some circles.

Bollen and colleagues (2011) used the Profile of Mood States (POMS) methodology to assign general moods to tweets on a given day and related those moods to a stock market index. The tracking of shifting moods across regional or national segments may become very popular for predicting other social responses, such as box office successes.

Kaushik (2010) provides some insight into the future of web mining applications in the form of free tracking services across multiple social networks (Twitter, Facebook, and LinkedIn). Rather than require you to do the web mining, Klout (http://klout.com) does it for you and displays results in the form of a dashboard (still in beta, October 2011). Klout calculates your degree of interaction with people across multiple social networks (user selectable) and calculates your "influence" on people in your networks. Figure 16.1 shows the dashboard.

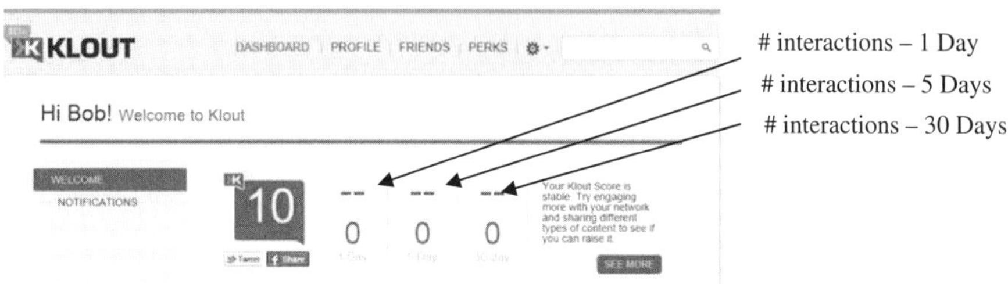

FIGURE 16.1
The Klout dashboard, showing the calculated Klout score and the number of network interactions over the last day, 5 days, and 30 days.

Klout also provides a number of dimensional metrics, described by Kaushik (2010):

- *Engagement:* A measure of how diverse your interactions are.
- *Reach:* The degree to which your interactions (e.g., tweets) are enough of interest to build an audience.
- *Velocity:* Measures how likely you are to receive replies (e.g., retweets).
- *Demand:* Determines the number of people you followed to build your count of followers.
- *Network strength:* Measures how influential the people who @ message you are and how influential the people who retweet you are.
- *Activity:* Tell you if you are tweeting too little or too much for your audience and if your tweets are effective in generating new followers, retweets, and @ replies.

MULTISESSION PATHING

Currently, web analytics tools keep track of paths through web pages for one session only. But most people visit a website multiple times before they make the tracked action (e.g., buy). The future web analytics tools may have the capability to track multiple sessions for a given visitor and relate all of the sessions to the converting action.

Sullivan (2009) said the following:

> Struggling to analyze data and prove campaign performance from websites, Facebook, Twitter, iPhone, and BlackBerry applications, U.S. companies in aggregate will more than double investments in web analytics during the next five years....

INTEGRATION OF WEB ANALYTICS WITH STANDARD BI TOOLS

Data mining capabilities are becoming integrated with standard business and office tools. Several data mining packages are available as MS-Excel add-ons (e.g., XLMiner). Several data warehousing systems vendors have integrated data mining directly into their software offerings (Teradata and Oracle). A similar integration of web analytics software is likely to follow the same path of development.

Recently, Legutko (2008) asked the thorny question "To what extent will web analytics continue to exist as a separate discipline, and to what degree will it be absorbed into broader disciplines around BI, business strategy, and general marketing?" Time will tell when and how far down this path of development that web analytics will go.

ATTRIBUTION ACROSS MULTIPLE SESSIONS

Currently, web analytics tools can't track a customer across multiple sessions, each of which may have a different web page as the point of entry. If your web analytics platform tells you that you sold five widgets in a transaction, and the customer came to your website via Bing as the last click before the sale, then Bing gets the credit for the sale of the five widgets in a pay. In Google Analytics, the last click that led to the sale gets full credit, and this is known as last-click attribution. If you don't believe this is true, you are doomed to lower quality decisions regarding your marketing. It appears clear that many web analytics tools in the future will offer mechanisms to track customers across a series of websites (Bennett, 2010). Additional information on the future of web analytics is presented in Chapter 12.

THE FUTURE: WHAT DOES IT HOLD?

It is very difficult to predict the future, in specifics, for fields that are in their infancy to teenage years and have such volatility. We can't tell what exact tools and processes will be predominantly in use ten years from now. But I think we can say with some certainty that during the next few years, the tools that are workable, among the current very large number of tools, some commercial, some open source, will undoubtedly decrease in number. Those that survive will be the ones that have stood the test of time and have proven themselves to be valid, reliable, and efficient, and thus are practical and useable. And with almost 100 percent certainty we can predict that web analytics will *not* go away. Web analytics and web mining will be part of our global society for the foreseeable future.

NEW AREAS THAT MAY USE TEXT ANALYTICS IN THE FUTURE

In the last 20 years, following the creation of the web, dramatic changes have occurred in how people communicate with each other. The primary mode of human communication was by talking with one another, initially in person and later via the telephone. Today people are having conversations by writing in a growing variety of social media. And this is not limited to the Western world but is occurring in Third World countries (even in and helping create the "Arab Spring"), basically around the Globe. Emails are exchanged each day in the billions. Even Facebook has a billion or more users today.

During this time we have also seen the growth of speech recognition technology, which is allowing computer systems to automatically transcribe spoken conversations. We even talk to "smartphones" that transcribe our words into text messages. We can shout at our car's navigation system, asking how to go to a particular destination; it responds with a map, but it also speaks to us, telling us which turn to take at every fork in the road.

The net result is that an increasing portion of human conversations are available in *text* form, either because they were originally written or because they were originally spoken and then automatically transcribed. These text conversations can be processed by adapting the natural language processing

(NLP) and text analytics techniques discussed in this book for conversational data. In fact some of the tutorials and case studies in this book did just that (see the Twitter tutorials AA, BB, P, and Y). The upper right quadrant of Figure 16.2 shows the research space that Carenini and colleagues (2011) believe to be the most promising for future work on summarizing conversations.

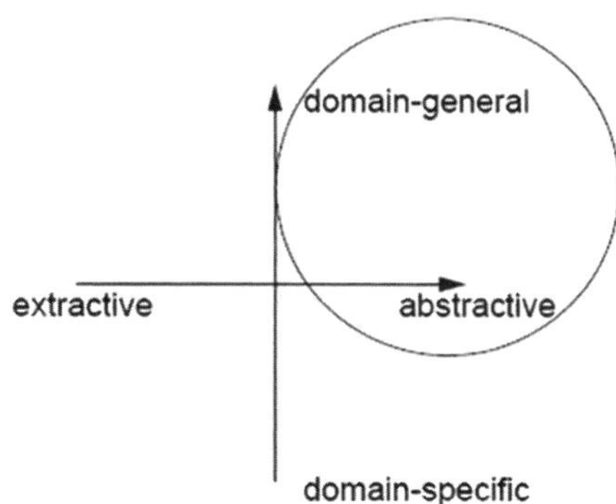

FIGURE 16.2
One way of looking at the research space for summarization. *Source: Carenini et al., 2011.*

IBM WATSON

Watson is a system that brings together four things: text data from millions of documents stored in its own hardware system (Ferrucci, 2010); a computer hardware system comprised of clusters of servers with massive parallel processing capability that can process possibly upward of 500 gigabytes of information per second equivalent to a million books per second (Renee, 2011; UMBC, 2011); the use of more than 100 different (apparently already available) software processes to sequentially analyze text, generate hypotheses, and score and rank hypotheses using distributed computing (Jackson, 2011; Novell, 2011; Takahashi, 2011); and the hardware ability to store all of the text content on RAM for the game show (Rennie, 2011). All of this was put together in a combination of advanced hardware and the linking together of available software, all architecturally designed to rapidly answer questions posed in natural human language. Historically it was developed over a period of years and finally pulled together as "Watson" in 2010 by an IBM research team as part of the DeepQA project. It was named after IBM's first president, Thomas J. Watson.

About three years ago, IBM Research was looking for a major research challenge to rival the scientific and popular interest of Deep Blue, the computer chess-playing champion, which would also have clear relevance to IBM business interests. The goal was to advance computer science by exploring new ways for computer technology to affect science, business, and society. Accordingly, IBM Research undertook a challenge to build a computer system that could compete at the human champion level in real time on the American TV quiz show *Jeopardy*. The extent of the challenge includes fielding a real-time Robot Human-like Machine contestant on the show, not merely a laboratory exercise.

In 2011, as a test of its abilities, Watson competed on *Jeopardy*, which was the first ever human versus machine match-up for the show. In a two-game, combined-point match (broadcast in three

Jeopardy episodes, February 14–16), Watson beat Brad Rutter, the biggest all-time money winner on *Jeopardy*, and Ken Jennings, the record holder for the longest championship streak (75 days). In these episodes, Watson consistently outperformed its human opponents on the game's signaling device, but had trouble responding to a few categories, notably those having short clues containing only a few words. Watson had access to 200 million pages of structured and unstructured content consuming four terabytes of disk storage. During the game, Watson was not connected to the Internet.

Meeting the *Jeopardy* challenge required advancing and incorporating a variety of QA technologies, including parsing, question classification, question decomposition, automatic source acquisition and evaluation, entity and relation detection, logical form generation, and knowledge representation and reasoning. Winning at *Jeopardy* required accurately computing confidence (i.e., probabilities) in your answers. The questions and content are ambiguous and "noisy," and none of the individual algorithms is perfect. Therefore, each component must produce a certain confidence in its output, and individual component confidences must be combined to compute the overall confidence in the accuracy of the final answer. The final confidence is used to determine whether the computer system should risk choosing to answer at all. In *Jeopardy* parlance, this confidence is used to determine whether the computer will "ring in" or "buzz in" for a question. The confidence must be computed during the time the question is read and before the opportunity to buzz in. This is roughly between one and six seconds, with an average around three seconds.

The system behind Watson, which is called DeepQA, is a *massively parallel* probabilistic evidence-based computational architecture. For the *Jeopardy* challenge, Watson used more than 100 different techniques for analyzing natural language, identifying sources, finding and generating hypotheses, finding and scoring evidence, and merging and ranking hypotheses. What is far more important than any particular technique utilized was how they combine them in DeepQA such that overlapping approaches can bring their strengths to bear and contribute to improvements in accuracy, confidence, or speed.

DeepQA is an architecture with an accompanying methodology, which it is not specific to the *Jeopardy* challenge. The overarching principles in DeepQA are massive parallelism, many experts, pervasive confidence estimation, and integration of shallow and deep knowledge.

- *Massive parallelism:* Exploit massive parallelism in the consideration of multiple interpretations and hypotheses.
- *Many experts:* Facilitate the integration, application, and contextual evaluation of a wide range of loosely coupled probabilistic question and content analytics.
- *Pervasive confidence estimation:* No component commits to an answer; all components produce features and associated confidences, scoring different question and content interpretations. An underlying confidence-processing substrate learns how to stack and combine the scores.
- *Integrate shallow and deep knowledge:* Balance the use of strict semantics and shallow semantics, leveraging many loosely formed ontologies.

Figure 16.3 illustrates the DeepQA architecture at a very high level. More details about the various architectural components and their specific roles can be found in Ferrucci et al., 2010. The "Query Decomposition" box (left side of Figure 16.3) and the "Synthesis" (right side of Figure 16.3) are apparently where the unique or new features of Watson reside. It is the subarchitecture provided by these two parts of DeepQA that gives Watson the advantage seen in *Jeopardy* demonstrations. At this point of what has been publicly revealed about the mechanics of Watson, it appears that the "novelty" and "most critical part" of Watson lie in the "Query Decomposition" and "Synthesis" parts of its architecture. Thus, it appears that we can say that the main innovation in Watson is the ability to execute many (100 or more) text analysis algorithms to get a correct answer, but it does not appear that this is done in the "usual ensemble method" but probably as a "score" of how many come to the same answer. Thus, at this point we have no evidence of a new algorithm being part of the DeepQA system (Thompson, 2010; Ferrucci, 2011).

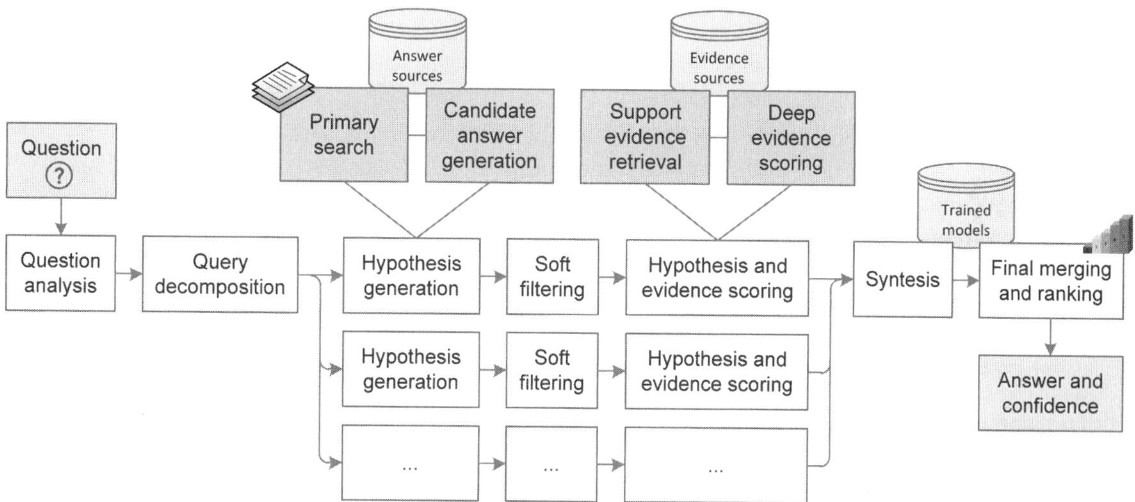

FIGURE 16.3
A high-level depiction of DeepQA architecture.

What are possible commercial uses for Watson? It has been suggested that Watson DeepQA technology could be very useful in the field of medicine as a clinical decision support system for the diagnosis and treatment of patients. Others have suggested it could be used for legal research. The announcement in early September 2011 (Mathews, 2011) that Wellpoint (one of the largest health insurance companies in the United States) is licensing the system validates that it is not being considered as "just a toy" but could have significant impact on commercial and government concerns. We would expect that Watson will not be used as it currently exists for Wellpoint, but instead be custom developed to provide right answers to the unique questions in the health industry. We predict that additional partnerships between IBM-Watson and other industries will soon develop.

In summary, for IBM-WATSON: The *Jeopardy* challenge helped IBM address requirements that led to the design of the DeepQA architecture and the implementation of Watson. After three years of intense research and development by a core team of about 20 researchers, Watson is performing at human expert levels in terms of precision, confidence, and speed on *Jeopardy*.

IBM claims to have developed many computational and linguistic algorithms to address different kinds of requirements in QA. Even though the internals of these algorithms are not known, IBM plans to publish many of them in detail in the future. During the development of Watson, the impact of any one algorithm on end-to-end performance changed over time as other techniques were added and had overlapping effects, resulting in a final Watson with exceptional speed in finding answers to questions.

SUMMARY

No matter where we are going in the future in text analytics, statisticians have big roles to play in data analysis and predictive analytics. As Vincent Granville (2011) said so eloquently in a recent article:

> Soon, a new type of statistician will be critical to optimize "big data" business applications. They might be called *data mining statisticians*, *statistical engineers*, *business analytics statisticians*, data or *modeling scientists*, but essentially they will have a strong background in many technologies and processes.

One of the coauthors replied to Granville's statement:

> Yes, this is why I refer to myself now as a "data scientist," not just a data miner. I do data quality, fuzzy data merging, and data integration projects, as well as data mining projects. My modeling activities have come almost full circle back to similar activities I performed as a research professor in forest growth modeling at the University of California–Santa Barbara over 17 years ago. I have been preaching this "doctrine" to data mining tool companies for the last 15 years; they needed to become more "data cosmopolitan." It all seems to be coming together now as we view this in late 2011, while completing the editing of this book.

References

Bennett, T. (2010). Web Analytics and Attribution. *MoreVisibility Blog*, August 12, 2010. http://www.morevisibility.com/analyticsblog/web-analytics-and-attribution.html.

Bollen, J., Mao, Huina, and Zeng, Xiao-Jun. (2011). Twitter mood predicts the stock market. *Journal of Computational Science*, 2(1):1–8.

Carenini, G., Murray, G., and Ng, R. (2011). Methods for Mining and Summarizing Text Conversations; Chapter 5, page 103+ Morgan & Claypool Publishers: San Rafael, CA USA.

González-Ibáñez, R., Muresan., S., and Wacholder, N. (2011). Identifying sarcasm in Twitter: A closer look. Proceedings of the 49th Annual Meeting of the Association of Computational Linguistics.

Granville, V. Statisticians Have Large Role to Play in Web Analytics. 2011. *AMSTAT NEWS*, September, pp. 31–32.

Kaushik, Avinash. 2010. Web Analytics 2.0. SYBEX: An Imprint of Wiley: Indianapolis, Indiana. www.webanalytics20.com.

KDnuggets™ News; newsletters of the past two years listing various Text Analytics and Web Analytics tools. http://www.kdnuggets.com/

Legutko, P. (2008). The future of web analytics consulting. Web Analytics Applied Blog, February 22, 2008.

Luck, E. (1999). Mixed Messages: American Politics and International Organizations. Washington D.C.: Brookings Inst. Press.

Rockwell, P. (2006). Sarcasm and other mixed messages: The ambiguous ways people use language. Lewiston, New York: Edwin Mellen Press.

Sullivan, L. (2009). Web Analytics Marketing Estimates rise. Online Media Daily, May 28, 2009.

IBM-Watson References

DeepQA (2011). DeepQA Project: FAQ, IBM Corporation. http://www.research.ibm.com/deepqa/faq.shtml, retrieved on September 17, 2011.

Ferrucci, D. 2011. Will Watson Win on Jeopardy? Nova ScienceNOW (Public Broadcasting Service). http://www.pbs.org/wgbh/nova/tech/will-watson-win-jeopardy.html.

Ferrucci, D., Brown, E., Chu-Carroll, J., Fan, J., Gondek, D., Kalyanpur, A. A., Lally, A., Murdock, J. W., Nyberg, E., Prager, J., Schlaefer, N., and Welty, C. (2010), "Building Watson: An Overview of the DeepQA Project." *AI Magazine* 31 (3). http://www.aaai.org/Magazine/Watson/watson.php.

Jackson, J. 2011. IBM Watson Vanquishes Human Jeopardy Foes, PC World. http://www.pcworld.com/businesscenter/article/219893/ibm_watson_vanquishes_human_jeopardy_foes.html.

Mathews, A. W., 2011. Wellpoint's New Hire. What Is Watson? The Wall Street Journal. http://online.wsj.com/article/SB10001424053111903532804576564600781798420.html.

Novell, 2010. Watson Supercomputer to Compete on "Jeopardy!"—Powered by SUSE Linux Enterprise Server on IBM POWER7; The Wall Street Journal. http://online.wsj.com/article/PR-CO-20110202-906855.html.

Renee, J. 2011. How IBM's Watson Computer Excels at Jeopardy. PLoS blogs. http://blogs.plos.org/retort/2011/02/14/how-ibm%E2%80%99s-watson-computer-will-excel-at-jeopardy/.

Takahashi, D. 2011. IBM researcher explains what Watson gets right and wrong. VentureBeat. http://venturebeat.com/2011/02/17/ibm-researcher-explains-what-watson-gets-right-and-wrong/.

UMBC. 2011. Is Watson the smartest machine on earth? Computer Science & Electrical Engineering Department, UMBC, February 10, 2011. http://www.cs.umbc.edu?2011/02is-watson-the-smartest-machine-on-earth/.

CHAPTER 17

Summary

CONTENTS

WHY ARE YOU READING THIS CHAPTER?

It is likely that you are reading this chapter for one of two reasons:

1. You read through all of the previous chapters, and this chapter is last. If this is your reason, you are following the pedagogical (teaching) method that is implicit in the structure of the book:
   a. Tell them what you plan to tell them (in the Introduction).
   b. Tell them all about it (in the chapters and tutorials).
   c. Tell them what you told them (in the Summary chapter).

This is one way to learn a new subject, but it is not the only way.

2. The other possibility for why you are reading this summary chapter is that you want a quick overview of the book beyond what is in the Table of Contents and the Introduction, but not as deep as that presented in the other chapters.

This is the most likely scenario. You may have chosen this path, because you want to devote the bulk of your time presently to learning about how to do text mining. This approach is called

Practical Text Mining and Statistical Analysis for Non-structured Text Data Applications. DOI: 10.1016/B978-0-12-386979-1.00045-1

meta-learning. You may choose this meta-learning path because you want to scope out the book to see if it is worth your while to dig deeper.

If this is indeed your choice, and you decide to read further in the book, we suggest strongly that you approach the task by reading through the chapters in Part I and do at least one tutorial, read (at least selectively) Part III, and then come back to the Summary chapter to help put all your learning experience into a proper perspective, as described following.

OUR PERSPECTIVE FOR APPLYING TEXT MINING TECHNOLOGY

Text mining is a means of analyzing the communications of other people. This analysis draws data and methods from many other fields, and to that extent, text mining is an *eclectic* discipline. The challenge of doing text mining in the context of so many disciplines comes not in the performance of specific methodological tasks but in the *blending* of information and methods from many different disciplines.

Thus, the perspective we seek to inculcate in the minds of the readers is that text mining employs processes of eclectic blending of diverse textual communications for the purpose of discovering commonalities among them and inferring insights that were not obvious at first glance. The application of text mining results within this perspective will permit you to make the most of those results. To serve that purpose, it is helpful to read this summary before tackling the material in the rest of the book.

PART I: BACKGROUND AND THEORY

The background for the development of this perspective is presented as a short history of text analytics and text mining. Chapter 1 discussed the three phases of development in text analytics: information retrieval, information extraction, and information discovery. Related to the time-course of this development path are nine enabling technologies that support text mining, described and associated with specific steps in the data mining process (this process is described in greater detail in Chapter 5). These enabling technologies are:

- Impact of domain knowledge
- Machine learning in TM
- Bag-of-words versus high-dimensional spaces
- Kernel methods
- Neural nets
- Feature selection
- Automation
- Semantic mapping
- Sentiment analysis

Chapter 1 ends with the question "Is information discovery the last phase in text mining development?" The answer is "Probably not." A fourth possible phase is proposed that helps text mining tools to read "between the lines" of text to distinguish:

- Idioms and colloquialisms
- Sarcasm, puns, and other ambiguous verbal forms
- Innuendo and euphemisms

WHAT IS TEXT MINING?

Chapter 2 poses this question and presents seven practices of text mining to answer that question.

1. **Information retrieval (IR):** Storage and retrieval of text documents, including search engines and keyword searches.
2. **Document clustering:** Grouping and categorizing terms, snippets, paragraphs, or documents using data mining clustering methods.
3. **Document classification:** Grouping and categorizing snippets, paragraphs, or document using data mining classification methods.
4. **Web mining**: Data and text mining on the Internet, with a specific focus on the scale and interconnectedness of the web.
5. **Information extraction (IE)**: Identification and extraction of relevant facts and relationships from unstructured text; the process of making structured data from unstructured and semistructured text.
6. **Natural language processing (NLP):** Low-level language processing and understanding tasks (e.g., part of speech tagging); often used synonymously with computational linguistics.
7. **Concept extraction:** Grouping of words and phrases into semantically similar groups.

The challenge before the text miner is to understand which practice area of text mining fits the problem. Five questions are posed, the answers to which can "diagnose" the appropriate practice area in text mining. Then you are ready to review the theoretic foundations and required preprocessing steps to create the analysis file related to that foundation in Chapter 3 (e.g., the generalized vector-space model).

Basic preprocessing steps in text mining include:

1. Choice of scope
2. Tokenization
3. Stop word removal
4. Stemming

Option preprocessing steps include:

1. Sentence boundary detection
2. Case normalization

The last thing that is done to fit the vector-space model is to create appropriate vectors of variables to submit to the text mining engine. Before performing text mining analysis, it is important to make sure that you understand the tools available in various applications to enable text mining. These tools include:

1. Common techniques for extraction of meaning from text
2. Techniques for doing trend analysis
3. Techniques of text summarization
4. Techniques for creating dictionaries and lexicons
5. Techniques for process automation
6. Techniques for determining similarity and relevance
7. Techniques for model evaluation.

When your "quiver" is full of appropriate techniques, you are ready to consider how you will perform the text mining project. Chapter 5 discussed one of the common process flow methodologies (CRISP-DM) to set the stage for suggesting a proposed linear text mining process flow plan. The complexity of the TM process flow requires that we decompose the linear flow into phases before showing interactions and feedbacks in the process flow.

1. Phase 1: Determine the *purpose* of the study.
2. Phase 2: Explore the *availability* and the nature of the data.
3. Phase 3: Prepare the data.

This phase is decomposed further into a deeper discussion of activities and tasks in this phase:

4. Phase 4: *Develop and assess* the models.
5. Phase 5: *Evaluate* the results.
6. Phase 6: *Deploy* the results.

WHAT TOOLS CAN I USE?

Chapter 6 presents general introductions to the graphical user interface (GUI) of three common data mining tools with text mining capabilities:

1. IBM-SPSS Modeler
2. SAS-Enterprise Miner
3. *STATISTICA* Data Miner

The purpose of the tool descriptions is to help you understand how to use the GUIs to perform the text mining tutorials. All of the features and functions are not presented, but rather discussion is focused on the use of the text mining capabilities.

PART II: THE TEXT MINING LABORATORY—28 TUTORIALS

The tutorials are placed in the middle of the book for a reason. You are conducted through the background (history), general theory, common applications, and common tools of text mining in Part I. Now you are ready to enter the laboratory and put this general knowledge to work by going through at least several of the tutorials nearest to your area of interest. The tutorials are not confined to the use of the tools describe in Chapter 6 but include five other tools that you can use:

- *STATISTICA* Data Miner......................... 16 tutorials
- IBM-SPSS Modeler............................... 2 tutorials
- SAS-Enterprise Miner .... 2 tutorials
- RapidMiner... 2 tutorials
- Topsy .... 2 tutorials
- Weka .... 1 tutorial
- R (with *STATISTICA* Data Miner).................. 1 tutorial
- Salford Systems .... 1 tutorial

PART III: ADVANCED TOPICS

This section contains introductory discussions of the four general modeling types in data mining and text mining.

- Classification in text mining: An unsupervised assortment of all documents into a number of classes (bins) according to their similarities to one another.
- Categorization in text mining: A supervised assortment of all documents into a number of named classes, according to their similarities to a group of previous named documents.
- Clustering in text mining: A form of unsupervised classification
- Prediction: Assignment of a decimal value to a target variable, based on the predictive value of a number of variables.

The second group of advanced topics covers four enabling technologies, including:

- *Named entity extraction:* The process of finding and collecting specific words or phrases that are associated with labels (names).
- *Feature selection and reduction of dimensionality:* Processes of choosing the best predictors for modeling.
- *Singular value decomposition:* An approach for reduction of the dimensionality of the data set.
- *Focused web crawling:* A process of directed searching on the Internet.

The last group of advanced topics includes a chapter on web mining and an exquisite chapter on applications of text mining in a specific industry, property, and casualty insurance, by Karthik Balakrishnan, who is an expert in the field.

OUTLINES OF CHAPTER 7–15

For Part III, short outlines are provided to summarize the many subjects and issues present in these chapters. The reader can refer to these outlines while reading each chapter. This approach to reading this book is one of the advantages of starting with the Summary chapter.

Chapter 7: Text Classification

1. More formally, text classification is an analytical process that takes any text document as input and assigns a label (or classification) from a predetermined set of class labels.
   a. For example, one familiar application of text classification in practice is email spam filtering. A spam filter examines an incoming email and uses:
2. Feature creation
3. Text preprocessing
4. Choosing which text to use
5. Algorithms
   a. Naïve Bayes
   b. Logistic regressions (MaxEnt)

6. Evaluation of classifiers
   a. Overall (or global) accuracy
   b. Precision and recall
7. Hierarchical text classification

Chapter 8: Prediction in Text Mining

1. The power of simple descriptive statistics, graphics, and visual text mining
   a. How simple counts can be useful
   b. Graphical summaries: visual data mining
2. Predictive modeling (supervised learning): Using text to predict important outcomes
   a. The algorithms of predictive modeling
      i. Statistical models versus general predictive models
      ii. K-nearest neighbor and similar methods
      iii. Recursive partitioning (trees)
      iv. Neural networks
      v. Ensembles
   b. Advantages and disadvantages of various algorithms
3. Clustering (unsupervised learning): Identifying clusters of similar documents and outliers
   a. Clustering algorithms
      i. K-means clustering algorithm; Expectation mazimization (EM) algorithm
      ii. Hierarchical or tree clustering
      iii. Kohonen networks (self-organizing feature maps)
   b. Advantages and disadvantages of various algorithms
4. Singular value decomposition and similar techniques
   a. Singular value decomposition
   b. Principal components analysis
   c. Partial least-squares
   d. Feature selection and feature extraction
5. Association and link analysis
6. Terms that go together

Chapter 9: Named Entity Extraction

1. What is named entity extraction?
2. Text features for NEE
   a. Lexicons
   b. Word shape features
   c. Grammatical features
3. Strategies for entity extraction
4. Rule-based approaches (change "strategies" to "approaches")
5. Statistical approaches
   a. HMMs using joint probability
   b. Conditional random fields using conditional probability

6. Tagging standards
   a. Standard generalized markup language (SGML)
   b. IOB tagging system
7. Evaluating entity extraction
   a. Scoring metrics
   b. Determining the "gold standard" metric

Chapter 10: Feature Selection and Reduction of Dimensionality

1. Reduction of dimensionality
   a. Reasons for doing so
      i. Focusing learning algorithms on the most important keywords
      ii. Reducing the problem of overfitting
      iii. Simplify and speed up processing
2. Feature selection
3. Feature selection approaches
   a. Information-theoretic methods
   b. Statistical methods
   c. Frequency-based methods
4. Dimensionality reduction methods
   a. Latent semantic indexing (LSI)
   b. Principal components analysis (PCA)
   c. Factor analysis (FA)
   d. Projection pursuit

Chapter 11: Singular Value Decomposition in Text Mining

1. Redundancy in Text
2. Dimensions of Meaning: Latent Semantic Indexing
3. Example: Causes of Aircraft Accidents
4. Interpretation
5. The Math of Singular Value Decomposition
6. Graphical Representation and Simple Example
7. Singular Value Decomposition in Equation Form
8. Some practical considerations
   a. How Many Dimensions to Extract?
   b. Interpretation of Latent Dimensions
      i. Subjective Methods, Reviewing Graphs
      ii. Analytical Methods, Building Models for Dimensions
   c. Useful Analyses Based on SVD Scores
      i. Cluster Analysis
      ii. Predictive Modeling
   d. When is SVD not a good Choice/Method

Summary
References

Chapter 12

1. Web Analytics
   a. The Value of Web Analytics (Why do it?)
   b. Main topics in Web Analytics
   c. Main definitions in Web Analytics
2. Web Mining: 3 Types of Web Mining
   a. Web usage mining
   b. Web content mining
   c. Web structure mining
3. The Future of Web Mining and Web Analytics
   a. Need to quantify the benefit of web analytics to a company
   b. Need to separate a relatively small amount of truly useful data from a large amount of rather useless data.
   c. Development of ways to handle missing unstructured data, similar to those used with structured data.
   d. Need to follow an ordered development process for web mining and web analytics, in which successive steps are increasingly expensive in terms of time and resources.

Chapter 13: Clustering

1. Clustering is the process of grouping similar items together into clusters.
2. Clustering requires two things:
   a. A method of computing similarity between two items (whether topics or words)
   b. An algorithm for searching through all of the items to find similar objects
3. Clustering documents
4. Clustering words
   a. Collocations
   b. Synonyms
5. Creating the term-document matrix
6. Cluster visualization

Chapter 14: Leveraging Text Mining in Property and Casualty Insurance

(Note: An expanded outline is presented here to present additional information about the application of text mining methods to the property and casualty insurance segment of the Insurance industry.)

1. Introduction: The business of P&C insurance is replete with unstructured data, such as claim notes, that afford numerous possibilities for text mining. Our attempt in this chapter is to offer a nonexhaustive list of areas that could benefit from such text mining insights.
2. P&C insurance as a business
   a. The business of P&C insurance is very diverse and complex.

   b. What makes the insurance business uniquely interesting is that there are no tangible widgets to sell; the insurance product is simply a promise to restitute an insured subject in the event of a *future* covered loss.
   c. With critical implications for the top and bottom lines, there is growing interest in harnessing every ounce of lift available from data and advanced analytics. It is no surprise then that text mining is increasingly being used to discover critical insights and drive systematic business actions in the P&C industry.
3. Analytics opportunities in the insurance life cycle
   a. Marketing and sales: how to define target markets, channels, and customers; how to create appealing products; how to effectively reach and attract prospects to the carrier
      i. Customer, prospect, and competitor comments offer incredible sources of insight for carriers with the capability to tease them out.
   b. Underwriting and pricing: how to understand, qualify, and assess risks; how to prevent or mitigate losses; how to accurately price risks
      i. The inspectors and auditors usually record detailed information on the subject risk, often including descriptive notes and reports. Text mining these notes can reveal interesting insights, patterns, and trends that can effectively feed loss control, loss mitigation, as well as underwriting and pricing activities.
      ii. Underwriting risk management
   c. Operations: how to efficiently and effectively deliver claims, loss control, and other insurance services to policyholders; how to create superior customer experiences and minimize attrition of good customers
      i. Claims notes
      ii. Description of loss notes
      iii. Customer contact notes
4. Driving business value using text mining
   a. As this chapter illustrates, text mining is a very relevant and useful capability that can be easily leveraged to create value across multiple functional areas of an insurance business.
   b. It can bring to light critical insights from textual notes of any kind, including claims, underwriter, loss control, and auditor and customer representative notes, in addition to numerous possibilities offered by the ever-innovating social web.
   c. It should be kept in mind that the examples mentioned here merely scratch the surface; novel business applications of insights are continually emerging, as are the technologies to make more effective use of unstructured data.

Chapter 15: Focused Web Crawling

(Note: An expanded outline is used to summarize this chapter also.)

1. A focused web crawler combines document classification techniques (see Chapters 7 and 8) with a conventional web crawler to efficiently find new, on-topic content from the Internet.
2. Search engines are tuned for general queries and often do not perform well for specific needs. A focused crawler can queue up specifically relevant documents by using document classification to

only crawl and index pages that match a desired "focus topic." Using document classification with a focus topic allows the crawler to follow hyperlinks from on-topic pages to other on-topic pages.
3. Web pages differ significantly in both style and form from text documents used for text mining. Instead of plain text, most web pages use HyperText Markup Language (HTML) to display their content.

Overall Ideas

1. The huge body of unstructured text in the world is by far the most promising "sandbox" in which to pursue the primary goal of artificial intelligence: the intelligent machine. Developments in speech recognition—"Hal" in Arthur C. Clarke's book *2001: A Space Odyssey*, and "Skynet" of the *Terminator* series—are but reflections of one of the strongest drives of technological humans: to reproduce themselves in technical terms.
2. Computer scientists, artificial intelligence practitioners, and even physicians are chasing the goal of the intelligent machine in order to understand how we work and how to fix problems that lead to sickness and death. The development of text mining technology in computer science is analogous to the discovery of DNA, how it works, and the mapping of the human genome. Gene mapping is aimed at understanding the bewildering complexity of the functioning of the human body for the (primary) purpose of building the "workshop manual" that will permit us to fix its physical problems and prolong its life. The discipline and practice of text mining are geared toward understanding the most complex analytical processing system in the universe through the analysis of written language. This book is one step in the direction of building the text mining "workshop manual" that will enable us to "fix" many problems of miscommunication and serve as the basis for extending justice and equity in all mediums of human communication involving written text.

Chapter 16. The Future

1. The Future of Text Analytics and Text Mining
   a. Commercial Text Analytics and Text Mining Software
   b. Free and Open-Source Text Analytics and Text Mining Software.
   c. Pros and Cons of Commercial vs Open-Source Software
   d. The Future: What Lays Ahead for Text Mining?
2. The Future of Web Analytics and Web Mining
   a. The Present background.
      i. Commercial General Web Mining Software
      ii. Open-Source Web Mining Software
      iii. Commercial Content Mining Software
      iv. Open Source Content Mining Software
   b. Development Pathways in the Future
      i. Shakeout of the existing software packages
      ii. Integration Across Multiple Social Networks
      iii. Multi-session Pathing
      iv. Integration of Web Analytics with BI Tools
      v. Attribution Across Multiple Sessions
3. The Future Overall - What Does it Hold?

Glossary

Algorithm A proscribed set of steps or operations. A recipe or procedure.

Artificial intelligence A field of computer science focused on the development of intelligent-acting agents. Often guided by the theory of how humans solve problems. Has a reputation for overpromising. Wryly definable as all computational problems not yet solved.

Assessment Determining how well a model estimates outputs from input data that were *not* used during training, by scoring it on completely new cases (test or evaluation data) having the input variables needed by the model.

Association rules A data mining process used to discover interesting pairs in transactions, lists of items, unique phrases, and so forth. The *a-priori* algorithm (Witten and Frank, 2000) is a popular algorithm for deriving association rules.

Bagging Bootstrapped aggregating—combining the results from more than one model as a final model. Bagging can improve accuracy, especially when data sets are small. Done by creating multiple new data sets through repeated random copying of cases in the data set (sampling with replacement), fitting a model to each, and then averaging or voting on the outputs of the separate models together.

Bag-of-words See *vector space*.

Basis functions Building blocks of an algorithm, such as powers of a polynomial in regression, or unit splines in MARS (Multivariate Adaptive Regression Splines).

Bayesian statistics Inference based on Bayes's law. The estimated probability of a factor depends on its prior probability (supplied from outside of the data) and its measured value on known data. With no data, the estimate is the prior value. With a lot of data, the measured value takes over (the Frequentist approach). In middle regions, the measured and prior values compromise, with the influence of the prior fading as data accumulate.

Binary variable A variable that contains two discrete values (e.g., 0/1 or yes/no).

Blurring Adding random noise to input patterns during training to prevent overlearning.

Bootstrapping A resampling process, where a procedure is performed several times with a different random sampling of the data to get a more accurate expected value (mean) of the result, as well as a confidence (standard deviation of the resulting distribution). See also jacknife, *V-fold cross-validation*, and leave-1-out methods.

BPMD Business Process Modeling Notation.

Branch In a decision tree, the arc from a parent node to a child node.

CART (Classification and Regression Trees) An influential decision tree algorithm developed by Breiman and colleagues (1984). Splits are binary, inputs and outputs can be either categorical or real, cross-validation is used to decide where to stop splitting, and missing cases are handled by surrogate split variables (backup questions to ask when no value exists for the primary split question).

Case normalization The process of converting text into either all lowercase or all uppercase.

Categorical variable One measured on an unordered or nominal scale, such as color: {red, green, blue}.

Categorization of text material Assigning documents to one or more categories based on content. See *text classification* and *document classification*.

CHAID (Chi-square Automatic Interaction Detection) A decision tree algorithm developed by Kass (1980), where all variables are categorical, splits can be multivariate, and the significance level of a chi-square test is used to decide when to stop tree growth.

Champion model The currently best predictive model developed. Used in production while a "challenger" model is constructed.

Clustering Divides data (typically by cases, but can be by variables) into groups such that members of a group are as close to one another as possible, but the groups are as far apart from one another as possible. Usually, one has to specify the number of clusters to find.

Collocation A group of words that work together to represent an entity or idea, such as "United States of America." Typically found by identifying words that are near one another, as opposed to far apart, in statistically significant frequencies.

Computational linguistics Field of automated analysis of languages.

Concept linking Connecting related documents by identifying their shared concepts, thereby helping users find information that is hard to find using traditional search methods.

Concept mining Mapping text snippets to "concepts," where multiple expressions can represent the same idea. One example tool is Princeton's WordNet.

Confusion matrix A table of counts of predicted versus actual classes for a model at a given decision threshold. If there are two classes—for example, fraud/nonfraud—the diagonal cells count the correctly classified cases, and the off-diagonal cells count false positives and false negatives. With more classes, the off-diagonal error types grow as the square of the number of classes. Each error type should be analyzed for its cost, and the modeling should be guided to minimize cost. A confusion matrix represents a point along an *ROC* or *lift chart* curve.

Co-occurrence Occurrence of similar or same words/concepts in the same unit of text—typically a sentence, but sometimes as large as an abstract—suggesting a relationship between them.

Coreference Grouping similar mentions of an entity in ground truth units. Entity resolution.

Corpus (plural *corpora*): A large and structured set of texts (now usually stored and processed electronically) prepared for the purpose of conducting knowledge discovery.

Cosine similarity A metric measuring the similarity of two points in high-dimensional space by the cosine of the angle between them (where the third point is the origin). For instance, if the dimensions are words and a point represents a document, two points with similarity of 1.0 would have a zero angle between them—that is, the same collection of words.

CRISP (Cross-Industry Standard Process for data mining) A nonproprietary process model for data mining, proposed in the mid-1990s by a European consortium of companies. See also alternative models for data mining DMAIC: the Six Sigma methodology, involving the steps Define→Measure→Analyze→Improve→Control, and the SAS-developed model SEMMA.

Cross-validation See *V-fold cross-validation*.

C-SVM classification A support vector machine (SVM) method for solving multiclassification prediction problems.

Customer relationship management (CRM) Processes a company uses to manage its contact with its customers. CRM software is used to maintain records of customer addresses, quotes, sales, and estimated future needs so customers can be easily and effectively supported.

Data mining Field of analytics with structured data. The model inference process minimally has four stages: data preparation, involving data cleaning, transformation and selection; initial exploration of the data; model building or pattern identification; and deployment, putting new data through the model to obtain their predicted outcomes.

Databases Mechanism for the storage and retrieval of data.

Date tagging Recognizing and extracting dates from unstructured text.

Document classification Automated classification and organization of text documents using data mining classification methods on unstructured text. See *categorization of text material*.

Document clustering Automated process of grouping documents into topics based on textual similarity. *Document matching*.

Document matching See *document clustering*.

Document similarity Techniques for identifying similar documents in a corpus; they include shared word count, *cosine similarity*, and *locality sensitive hashing (LSH)*.

Document standardization Automatic standardization of the formatting of documents.

Document summarization Automatic creation of text summarizing multiple documents.

eDiscovery The process of identifying useful documents or passages and quantifying available electronic information in response to a legal or judicial query.

Ensemble A collection of different models fit toward the same target that are combined to form a model likely more accurate than its components. For example, see *bagging*. Extremely powerful tool; the best techniques up to 2010 are detailed by Seni and Elder (2010).

Entity resolution See *co-reference*.

Entity extraction Extracting proper names, places, and organizations from unstructured text. *Named entity* recognition.

Epoch In neural networks, a single pass of the training data set through the model during training, where each case slightly affects the internal weights. Many thousands of epochs may be required for the parameters of the neural network to converge.

F1-measure A measure of document retrieval accuracy; the harmonic mean of *precision* and *recall.*

FACT A decision tree algorithm developed by Loh and Vanichestakul (1988).

Feature creation Low-level process of translating unstructured text into higher-order forms more suitable for use with data mining techniques such as classification and clustering. Vector generation. See also *feature extraction*.

Feature extraction Techniques to combine predictors (features) to extract their common information or lower the dimensionality of the information. Methods include factor analysis, *principal components analysis (PCA)*, correspondence analysis, multidimensional scaling (MDS), partial least squares (PLS), and *singular value decomposition (SVD)*. See also *feature creation*.

Feature selection Winnowing down the candidate inputs (data features) to obtain a smaller set more suitable for accurate modeling. Useless features may appear to help during training but will hurt on evaluation (new data). For methods like neural networks, where the model structure is fixed, useless features can even hurt training accuracy. And even for input-selecting methods like stepwise regression, they can distract a model, increase its effective complexity, and lead to worse out-of-sample performance. Also called variable screening.

Focused web crawling A combination of document classification and web crawling for targeted information acquisition on the web.

Gains chart A lift chart where the *y*-axis is the absolute or proportional return. A 45-degree angle line is the expected (random) return; the actual gains, due to sorting the cases so the model-predicted best is leftmost, should be above that line.

Gene mention Keeping tabs on the mention of genes in the text. An example tool for identifying a range of biomedical categories, such as diseases, drugs, chemicals, and methods of treatment, is the National Library of Medicine's MetaMap.

Generalization The degree to which a model, trained on one set of data, works on new data similar to training but that it has never seen.

Generalized regression neural network (GRNN) A nonlinear model in network form using kernel-based features and regression (Speckt, 1991; Patterson, 1996; Bishop, 1995).

Genetic algorithm An optimization method that attempts to find optimal binary strings by analogy with the theory of natural selection (Goldberg, 1989). An initially random population of strings is modified through mutation (random changes), breeding (crossover swaps with other strings), and selection (scoring against the training data). A population of "fit" entities (strings that score well) is retained, and the process is repeated until convergence.

Geotagging Geo-locating documents and place mentions from unstructured text.

Gini measure A purity-of-node measure used by decision trees in classification problems.

Grammar The structure of a language reflecting its normal association of word forms.

Grammatical parsing Full automatic sentence diagramming.

Hazard rate The probability per unit time that a case that has survived to the beginning of an interval will fail during that interval.

Hidden Markov model (HMM) A process to determine hidden parameters from observable parameters where the system is being modeled as a Markov process.

Hidden nodes In neural networks, the nodes that are between input and output nodes. Their outputs to the next layer of nodes are nonlinear (logistic) transformations of their inputs.

Hold-out data That data removed from the initial whole data set and set aside so it can be used as test data to benchmark the fit and accuracy of the predictive model produced from the training data set. Also called *validation*, evaluation, test, or out-of-sample data.

Homonyms See *polysemes*.

Hybrid methods for text analysis Systems that combine list-based methods (such as taxonomies) with statistical-based methods (such as collocation detection from word pair frequencies. For example, a hybrid n-gram/lexical analysis tokenization algorithm, or *HybGFS*, a hybrid method for genome-fingerprint scanning.

IE Information extraction.

Imputation An algorithmic way to fill in missing data, such as by using the variable mean, a fixed value, a special class, or an inferred value from other variables with known values. Relatively difficult to do well, but it enables one to employ valuable cases that would otherwise be thrown out for being incomplete.

Incremental algorithms A suboptimal but fast learning algorithm that only requires one or a few passes of the data to set its parameters. Reviving in popularity in the current age of vast streaming data sets.

Indexing Creating an inverted index to support rapid keyword querying.

Information extraction (IE) Identifying and extracting relevant facts and relationships (structured data) from unstructured or semistructured text.

Information retrieval (IR) Storage and retrieval of text documents.

Information visualization Visual exploration, management, and discovery using unstructured text documents.

Input variable A variable that is known at the time a decision needs to be made and is therefore eligible to be used to predict the value of a target variable.

Interval variable A continuous or real-valued variable containing values over a range; for example, heights of 5.5 feet, 6 feet, 6.1 feet, and so on.

Inverse document frequency Transformation of raw word frequency counts used to measure the frequencies with which specific terms or words are used in a collection of documents, as well as the extent to which particular words are used only in specific documents in the collection (see Manning and Schütze (2002) or the online help in *STATISTICA Data Miner* on the DVD enclosed with this book).

Inverted index A data structure used to support a search where each item is a list of documents in which a particular term appears.

Jogging weights Adding a small random amount to the weights in, say, a neural network to help avoid becoming trapped in local optima in performance space.

KDD Originally, *knowledge discovery in databases*; currently, *knowledge discovery and data mining*.

KDM Knowledge discovery metamodel.

Kernel A function receiving two vectors and returning a scalar. For example, a kernel membership function represents the affinity of a point and an archetype (a meaningful point in space) by calculating the distance of the two points and transforming that by a Gaussian-shaped decay. The effect is to make a greater distinction, for a given absolute distance, when the points are close rather than when they are far apart.

Keyword search Searching a corpus of documents for a specific keyword query.

K-means clustering An algorithm used to divide *N* points into *K* clusters so points in a cluster are more similar to one another than to points in different clusters. The algorithm repeatedly iterates through two steps: assignment, where each point is put in the cluster to which it is closest, and adjustment, where the mean of each cluster is recalculated. The initial centers are typically randomly selected points, and *K* must be selected by the user. The algorithm converges when the assignments don't change. This heuristic algorithm is fast, yet suboptimal; however, finding the optimal solution is NP-hard.

K-nearest neighbor An algorithm assigning the answer for a new case to that of the average (or vote, or weighted average) of the *K* cases nearest to it. The most important analytic choice is the selection of the dimensions (input variables to be used).

Latent semantic indexing (LSI) A method for deriving underlying dimensions of "meaning" from the documents-by-words (terms) data matrix extracted from a collection of documents. Typically, *singular value decomposition (SVD)* is applied to create a mapping of the words and documents into a common space, computed from the word frequencies or transformed word frequencies (e.g., inverse document frequencies, IDF).

Leaf In a decision tree diagram, any node that is not further divided. A terminal node.

Lemmatization The process of identifying the root concept behind a word (closely related to *stemming* from IR) so different inflected forms, for instance, can be analyzed as a single item. Can be inflectional or root stemming.

Lexicon A list or vocabulary for a specific concept.

Library and information sciences Field around the organization, storage, and retrieval of information. (Many information retrieval techniques originated in the field of library sciences.)

Lift The expected return of an action divided by the default (random) return. For example, a model prioritizing customers who are likely to respond to an offer would have a lift of 2.5 at the 10% depth if that group provided 25% of all sales.

Lift chart A graph revealing the quality of a model's ability to sort cases, such as customers who do or do not purchase products. The *x*-axis contains the cases, sorted from left to right, by most to least interesting. The *y*-axis has a measure of quality, such as *gain*, lift, or profit. If lift is plotted, it will typically begin very high at the left-hand side and drop to 1.0 at the far right.

Likelihood The probability of an event based on current observations.

Link analysis In a web-based document search, using the links on the web to understand and improve searching or advertising on the web.

Locality sensitive hashing (LSH) A method for mapping items from a high-dimensional space into a reduced-dimensional space while preserving, as much as possible, their original interpoint distances. LSH is similar to *principal components analysis (PCA)* and *singular value decomposition (SVD)*, but with a computer science vocabulary and perspective.

Lookahead For time series analysis, the number of time steps ahead of the last input variable value that the output variable values should be predicted.

Logistic regression Regression analysis in which the target (response) variable is binary. The weighted sum of the inputs goes through a sigmoidal (S-shaped) logistic transformation so the output is bounded by 0 and 1. Used as the nodes of neural networks, where the weights are found iteratively, but used alone, the weights are found instantly, as in linear regression.

Machine learning Data mining done by computer scientists.

Market basket analysis Association rules for product purchases.

Markov process A mathematical model for the randomized evolution of a memoryless system—that is, a process that depends only on present, not any past, events. HMMs were first applied to speech recognition problems in the mid-1970s and to biological processes, particularly DNA, in the 1980s.

Metadata Data about the data; a description or definition of the rows, columns, and/or links in a data set.

Morphology A branch of the field of linguistics and a part of *natural language processing (NLP)* that studies the internal structure of words (patterns of word formation within a language or across languages).

Multilayer perceptron (MLP) The most frequently used form of neural network, with at least two layers of logistic nodes (weighted sums of inputs put through a sigmoidal "activation function"). The outputs of an early layer are the inputs into the next layer.

Naive Bayes classifier A simple classifier where the inputs are (naively) assumed to be independent. Thus, only the mean and standard deviation of each is estimated from data, and not the whole covariance matrix. Under Bayes's rule, the probability of a class given the data is proportional to the probability of the class times the probability of the data given the class. With independence assumed, these probabilities are easily estimated. This classifier is a good baseline method. It is robust, and it works relatively well in high dimensions with sparse data; yet, if more data are available, other methods are usually more accurate.

Named entity recognition See *entity extraction.*

Natural language A language that has evolved naturally, such as English or Chinese, contrasted with artificially constructed languages such as Esperanto or Latino sine flexione, or to technical grammars, such as computer programming or mathematical languages.

Natural language processing (NLP) A theoretically motivated collection of computational techniques for analyzing natural texts for the purpose of achieving human-like language processing for a wide range of applications. Low-level language processing tasks (e.g., *part of speech tagging*); often used synonymously with computational linguistics.

Neural network An interconnected feed-forward network of nonlinear nodes, as described for a multilayer perceptron. Inspired by the interconnectivity and hypothesized activation rules of the human brain. (However, the analogy is weak; the human brain contains billions of neurons, each better modeled as a supercomputer than a simple equation.)

Noise Extraneous text that is not relevant to the task at hand, which must be filtered out from important text.

Ontology A structured description of the concepts and relationships that can exist for a given domain.

Outlink extraction Building of a list of URLs that point away from a given web page.

Overfit Overlearning the training data. Occurs when the model is more complex than the data can sustain. This can happen when the candidate input variables or parameters to be fit are too numerous for the unique cases constraining the model. *Cross-validation* is a good preventative.

PageRank An algorithm for determining the value of a web page based on the structure of hyperlinks on the Internet.

Parsing The act of identifying the grammatical structure of text with respect to a particular grammar.

Part of speech detection Detection of a grammatical part of speech (noun, verb, adverb, etc.).

Part of speech tagging The process of marking up the words in a text as corresponding to a particular part of speech (such as nouns, verbs, adjectives, and adverbs) based on a word's definition and the context in which it is used.

Payload/content detection Extraction of the content of a web page, removing the advertising and navigational boilerplate text.

PCA Principal components analysis.

Phrase chunking See *shallow parsing.*

Polysemes Also called *homonyms*, are syntactically identical words (spelled the same) with different meanings (e.g., "bow" can mean to bend forward, the front of a ship, a weapon that shoots arrows, or a kind of tied ribbon). (See *synonyms* for contrast.)

Precision An evaluation metric for text mining algorithms that counts the number of correct answers out of all the selected items. (See also *recall.*)

Predicted value The output of a model; the estimated value for the target variable given the input variable values for that case.

Principal components analysis (PCA) A linear dimensionality reduction technique, which identifies orthogonal directions of maximum variance in the original data and projects the data into a lower-dimensionality space formed of new dimensions that are the linear combinations of the original dimensions that best retain the variance of the original data.

Probabilistic neural networks (PNN) A neural network using kernel-based approximation to form an estimate of the probability density functions of classes. One of the Bayesian networks (Speckt, 1990; Patterson, 1996; Bishop, 1995).

Profit chart A *lift chart* where the y-axis is net profit. It will typically rise at the beginning, when the returns of a model-driven campaign outweigh its costs, then fall below zero as one goes deeper into the list (heads right on the x-axis).

Profit matrix A table of expected revenues and costs for each level of a target variable. See also *confusion matrix.*

PubMed/MEDLINE A database of publications in medicine and biomedicine/bioscience that can be accessed and searched online. Particularly useful for genomics-related publications.

Querying Using query text to find relevant documents in a corpus (not limited to a keyword search).

QUEST A decision tree program by Loh and Shih (1997).

R A popular freeware programming language for statistics and graphics. Created by Ross Ihaka and Robert Gentleman at the University of Auckland and kept in development by the R Development Core Team. Designed based on the S language of Bell Labs. See www.r-project.org/.

Radial basis functions A model of a form similar to a neural network where the first layer (set of transforming functions) consists of kernels (radially symmetric multilinear functions) whose outputs are combined into a logistic regression node (Broomhead and Lowe, 1988; Moody and Darkin, 1989).

Ranking The process of ranking a set of documents based on a query.

RDF Resource description framework.

Recall An evaluation metric for text mining algorithms = the number of desired items that were found divided by all the desired items that could have been found.

Regularization Modifying the accuracy metric while training a model to prevent overfitting by building in a penalty factor for model complexity. Usually penalizes large weights.

Relationship extraction The process of labeling named relationships between entities in text. See also *link analysis.*

ROC (receiver operating characteristic) curve: From signal detection theory, illustrates a binary classifier's discrimination power by plotting how the classification threshold changes the trade-off between the rates of opposing types of errors. The curve can be a graph of the false alarm rate versus the false dismissal rate, or the *sensitivity*, or true positive rate, versus the false positive rate (1 − specificity or 1 − true negative rate). The relative area under the ROC curve (in the latter form) is often used as a value of model quality, but Hand (2009) proved this to be an unsound measure. Better to discern the relative costs of the two types of errors and choose a classifier with the best cost-weighted performance.

Root node The first node in a decision tree; contains the first splitting condition.

Rule-based or knowledge-based system One of two main approaches to text mining, the other being statistical- or machine-learning-based systems. Rule-based systems use some sort of prior knowledge, often in the form of lists. This might take the form of general knowledge about how language is structured, specific knowledge about how a dicipline's relevant facts are stated in the literature, knowledge about the sets of things that a displine discusses and the kinds of relationships that they can have with one another, the variant forms by which they might be mentioned in the literature, or any combination of these.

Scoring Computing the values of a model for new cases.

Semantics The meaning of text.

Semantic specificity See *inverse document frequency.*

Semistructured data Unstructured text data appearing within a structured format such as a relational database or XML file.

SEMMA A standard process model for mining, proposed by SAS, involving the steps Sample→Explore→Modify→Model→Assess. (See alternate model *CRISP.*)

Sentence boundary detection Automatically sectioning unstructured text into smaller sections corresponding to sentences. Also called *sentence segmentation.*

Sentiment analysis Determining whether text contains positive, negative, or mixed opinions about the subject or product mentioned in the text. A challenging text mining task.

SGML Standard generalized markup language.

Shallow parsing Identifying, for example, noun phrases or verb phrases. *Phrase chunking.*

Singular value decomposition (SVD) Used in data mining and predictive modeling to reduce the dimensionality of a data matrix. For example, SVD will reduce the dimensionality of the documents-by-word/phrase-(relative) frequency matrix to derive latent semantic dimensions that are useful for understanding the underlying dimensions of meaning that separate the documents and terms. Computationally, given a matrix A(n,m), SVD will compute matrices **U**(n,m), vector **w**(m), and matrix **V**(m,m), so that A = **U*****W**(diagonal)***V**′ (where **V**′ stands for the transposed **V** matrix, and **W**(diagonal) stands for a diagonal matrix(m,m), with the elements of vector **w** in the diagonal). The positive values in vector **W** are the square root of the eigenvalues of A′ * A or A * A′. Eigenvalues are related to the total amount of variability or "information" in the matrix A′ * A or A * A′, and the relative size of each singular value is related to the information extracted by consecutive dimensions.

Spam filtering Identifying spam email and forum posts automatically.

Spelling correction Identifying and correcting misspelled words based on lemmatization, part of speech analysis, and word disambiguation.

Statistics Field of modeling and reasoning with uncertainty.

Statistical- or machine-learning-based systems One of the two main approaches to text mining, the other being rule-based or knowledge-based systems. Statistical systems do not require preloaded lists of topics or terms but store their knowledge in the form of equations relating measurable traits of an entity to its target label. This knowledge is gained from a model built on known cases whose labels have been provided by the results or by an expert's judgment.

Statistical significance True if a measured relation has a sufficiently small (e.g., 5%) probability of having occurred by chance. The threshold probability is called the *p*-level.

Stemming A text preprocessing procedure applied to a document corpus in order to reduce different grammatical forms (i.e., declinations and conjugations) of a word to its common roots (e.g., map all instances of *walking, walked, walk* to the common root *walk*), so that each occurrence of the grammatical form is counted as an instance of the root.

Supervised learning An approach to mining where the model can be trained from data with known answers contained in a target variable.

Support vector machine (SVM) A classification method based on the maximum margin hyperplane.

SVD Singular value decomposition.

Synonyms Syntactically different words (i.e., spelled differently) with identical or at least similar meanings (e.g., movie, film, and motion picture). (See *polysemes* for contrast.)

Syntax The structure of text.

Target variable The output or dependent variable that the model is trying to estimate or classify.

Taxonomy A hierarchical structure of concepts.

Term-by-document matrix A common representation schema of the frequency-based relationship between the terms and documents in tabular format where terms are listed in rows, documents are listed in columns, and the frequency between the terms and documents is listed in cells as integer values. Also called occurrence matrix.

Term clustering Grouping terms that frequently co-occur in text or are used in similar conditions. See *collocations.*

Text analytics A term that broadly defines many analytic techniques used for intelligent analyses of textual data—for example, extracting information and discovering knowledge, summarizing documents, grouping documents, or facilitating smart searches. Synonymous with text mining for most, but for some, a larger set of related activities around the technical core inference steps of text mining.

Text classification Grouping and categorizing snippets, paragraphs, or documents using data mining classification methods.

Text clustering Grouping and categorizing terms, snippets, paragraphs, or documents using data mining clustering methods.

Text mining The semiautomated process of extracting patterns (useful information and knowledge) from large amounts of unstructured data sources. See also *text analytics.*

Tokenization The process of dividing an unstructured text document into single word tokens.

Transformation Applying a function to a variable in order to adjust its range, skewness, granularity, and so on to make it more likely for the modeling algorithm to find useful. For example, it is a good idea to transform real-valued variables (like Income) to be roughly normally distributed and to pare down excess categories for a categorical variable (like State of Residence) by forming sensible geographic or demographic groupings beforehand.

UIMA (Unstructured Information Management Architecture) An open-source set of frameworks, tools, and annotators for facilitating the analysis of text (Ferrucci and Lally, 2004, Mack et al., 2004).

Universal grammar A theory proposing that all natural languages have certain underlying rules that constrain their structure. Some focus on language acquisition, in particular, during child development.

Unstructured data Raw textual data with no inherent structure—for example, a Word document rather than a tabular Excel document.

Unsupervised learning Techniques such as clustering or principal components that do not require a target variable.

Validation data Data held out from training to test the resulting model to see if it generalizes.

Variable A feature or column in a data set.

Vector generation See *feature creation.*

Vector space A representation of text where documents are a point in a high-dimensional feature space, where each word appearing in the document is a dimension of that space.

V-fold cross-validation Repeated samples are drawn randomly from the data for analysis (number of replicates $= V$), and the algorithm is then applied to compute predicted values, classifications, and so on. Summary indices of the accuracy of the prediction are computed over the V replications; thus, this technique allows the analyst to evaluate the overall accuracy of the prediction algorithm through a distribution of out-of-sample results, rather than a single one. This provides a more accurate estimate of the mean error as well as a confidence (standard deviation) for the result. Note that the model is different in detail for each fold of the cross-validation (CV). Thus, what is tested is the modeling procedure, not a single model—that is, what was held constant for each experiment. Analysts usually fit a final single model with all the training data and assign the accuracy of the CV to it. (Though ensemble methods like bagging are an even better way to use the multiple models; see Seni and Elder, 2010.)

Web mining Data and text mining on the Internet, with a specific focus on the scale and interconnectedness of the web.

Web structure mining Using the hyperlink structure of the web for organizing and finding information on the web.

Word cloud A specific visualization technique for text where the frequent words are displayed prominently.

Word disambiguation Identifying the current usage from multiple senses of a particular word.

Word shape The pattern of capitalization, hyphenation, and other punctuation in a particular word.

References

Bishop, C. (1995). *Neural Networks for Pattern Recognition.* Oxford: University Press.

Breiman, L., Friedman, J. H., Olshen, R. A., & Stone, C. J. (1984). *Classification and regression trees.* Monterey, CA: Wadsworth & Brooks/Cole Advanced Books & Software.

Broomhead, D.S. and Lowe, D. (1988). Multivariable functional interpolation and adaptive networks. *Complex Systems 2,* 321–355.

Cunningham, Hamish (2002) GATE, a general architecture for text engineering. *Computers and the Humanities* 36(2): 223–254.

Ferrucci, David; and Adam Lally (2004) UIMA: An architectural approach to unstructured information processing in the corporate research environment. *Natural Language Engineering* 10(3/4): 327–348.

Goldberg, D. E. (1989). *Genetic Algorithms.* Reading, MA: Addison Wesley.

Hand, D. J. (2009). Measuring classifier performance: a coherent alternative to the area under the ROC curve, *Machine Learning* 77: 103–123.

Haykin, S. (1994). *Neural Networks: A Comprehensive Foundation.* New York: Macmillan Publishing.

Kass, G. V. (1980). An Exploratory Technique for Investigating Large Quantities of Categorical Data. *Applied Statistics,* Vol. 29, No. 2 (1980), pp. 119–127.

Loh, W.-Y, & Shih, Y.-S. (1997). Split selection methods for classification trees. *Statistica Sinica, 7,* 815–840.

Loh, W.-Y., & Vanichestakul, N. (1988). Tree-structured classification via generalized discriminant analysis (with discussion). *Journal of the American Statistical Association, 83,* 715–728.

Mack, R.; S. Mukherjea; A. Soffer; N. Uramoto; E. Brown; A. Coden; J. Cooper; A. Inokuchi; B. Iyer; Y. Mass; et al. (2004) Text analytics for life science using the unstructured information management architecture. *IBM Systems Journal* 43(3):490–515.

Manning, C. D., & Schütze, H. (2002). *Foundations of statistical natural language processing* (5th printing). Cambridge, MA: MIT Press.

Moody, J. and Darkin, C.J. (1989). Fast learning in networks of locally tuned processing units. *Neural Computation* 1(2), 281–294.

Patterson, D. (1996). *Artificial Neural Networks.* Singapore: Prentice Hall.

Seni, G. and J. Elder (2010). *Ensemble Methods in Data Mining: Improving Accuracy Through Combining Predictions.* Morgan & Claypool.

Speckt, D.F. (1990). Probabilistic Neural Networks. *Neural Networks* 3(1), 109–118.

Speckt, D.F. (1991). A Generalized Regression Neural Network. *IEEE Transactions on Neural Networks* 2(6), 568–576.

Witten, I. H., & Frank, E. (2000). *Data Mining: Practical Machine Learning Tools and Techniques.* New York: Morgan Kaufmann.

Index

Note: Page numbers followed by *f* indicate figures and *t* indicate tables.

A

D

I

T

X

Y

Z

How to Use the Data Sets and the Text Mining Software on the DVD or on Links for *Practical Text Mining*

I. DATA SETS FOR THE TUTORIALS IN *PRACTICAL TEXT MINING*

Data sets for each tutorial in *Practical Text Mining* are located in the folder called "Datasets" found on the DVD that is bound with this book. Open this folder, and inside will be a subfolder for each tutorial. Inside each tutorial folder will be the data set needed, plus in some cases results workbooks and in other cases complete DMRecipe projects (for *STATISTICA*- based tutorials only) that can be used as a guide but are also the results of doing the steps in that particular tutorial. *Note:* A few of the "tutorials" are case studies, which do not have data sets; they have been included to use for discussions about the processes for those particular subjects.

II. SAS TEXT MINER SOFTWARE

To gain experience using SAS® Text Miner™ for the desktop using tutorials that take you through all the steps of a data mining project, visit:

http://support.sas.com/PracticalTM.

The tutorials include problem definition and data selection, and continue through data exploration, data partitioning, modeling, and model comparison. The tutorials are suitable for data analysts, qualitative experts, and others who want an introduction to using SAS® Text Miner™ for the desktop.

III. SALFORD SYSTEMS SOFTWARE, INCLUDING A NEW TEXT MINER MODULE MADE FOR THIS BOOK (*30-DAY FREE TRIAL AVAILABLE*)

To gain experience with Salford Systems software, load the DVD bound with this book into your computer's DVD reader and find the "Salford Systems" folder. Open the folder, and you will see five documents.

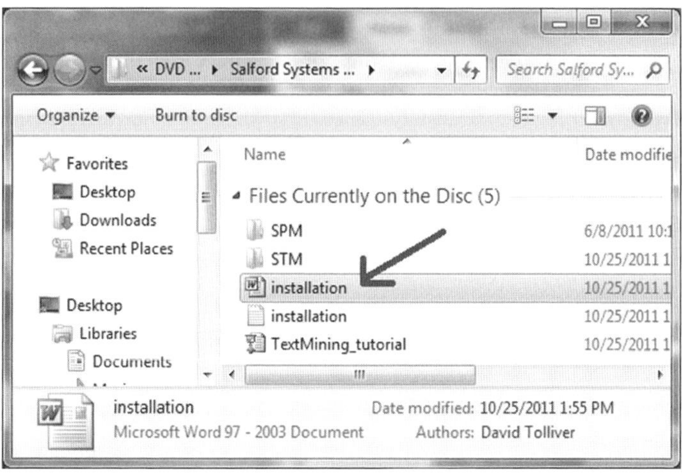

The "TextMining_tutorial" is a PowerPoint with over 100 slides that is the tutorial for this software (this PPt is also found in the "Tutorial—Data Sets" folder on this book's DVD). The Word "installation" document has complete instructions for installing the Salford Systems software, but part of this is also presented here. The Salford Systems software must be installed and licensed. No-cost license codes for a 30-day period are available on request to visitors of this tutorial.

Installing the Salford Predictive Modeler (SPM)

1. Locate the folder **"Salford Systems Tools/SPM"** located on the supplied DVD.
2. Double click on the **"Install_a_Transform_SPM.exe"** file located in the **"SPM"** folder to install the specific version of SPM used in this tutorial.
3. Follow the simple installation steps on your screen.
4. When you launch the Salford Systems Predictive Modeler (SPM), you will be greeted with a License dialog containing information needed to secure a license via email.

Licensing the Salford Predictive Modeler (SPM)

1. Execute the STM executable (**SPM.EXE**). You will be presented with a License Information window in which to put pertinent information and email to Salford Systems (see the "Instructions" Word document on the DVD to see this window and get full information).
2. Copy and send the necessary **System ID** information to Salford Systems' unlock@salford-systems.com to secure your license by entering the "Unlock Code," which will be emailed to you.

Note: The software will operate for three days without any licensing, but you can secure a 30-day license on request.

Installing the Salford Text Miner (STM)

In addition to the Salford Predictive Modeler (SPM), you will also work with the Salford Text Miner (STM) software:

1. Locate the folder **"Salford Systems Tools/STM,"** located on the supplied DVD.
2. Unzip the contents of the archive STM.zip into a directory of your choice. We recommend the **"stmtutor/STM/"** folder to follow along with the tutorial material.
3. The executable **STM.EXE** resides in the **"stmtutor/STM/bin"** folder and executes as a command line application (DOS window).

Expect to see several folders and a large number of files located under the **"stmtutor/STM"** folder. It is important to leave these files in the location to which you have installed them.
Note: Please do not move or alter any of the installed files other than those explicitly listed as user-modifiable!

4. **STM.EXE** requires its own no-cost license. After launching the application in a DOS (Command) window, follow the directions for what you must send to Salford Systems to secure the license. (See the following licensing information.)

Licensing the Salford Text Miner (STM)

1. Execute the STM executable (**STM.EXE**). You will see something like the following:

```
Unable to find, open, or read the following license file: license.txt

    Press ENTER to continue.

No valid license found. Please contact Salford Systems to
confirm/arrange licensing, noting the following parameters:

    HOSTID=7CE07262    MODULE=STM    VERSION=1.5
```

2. Provide the information on the final line before the ENTER prompt (HOSTID, MODULE, VERSION) to unlock@salford-systems.com. Once the license is verified, a license string will be emailed to you.
3. The license string will be emailed in the form of a text file (license.txt) that must be placed in the STM application's "**/bin**" directory.

IV. STATISTICA TEXT MINER SOFTWARE (30-DAY FREE TRIAL ON THE DVD THAT ACCOMPANIES THIS BOOK)

To gain experience using *STATISTICA* Text Miner and Data Miner and Optimization for the desktop using the tutorials in this book, put the DVD that comes with this book in your computer's DVD reader, and open the "*STATISTICA*" folder, which should look like this:

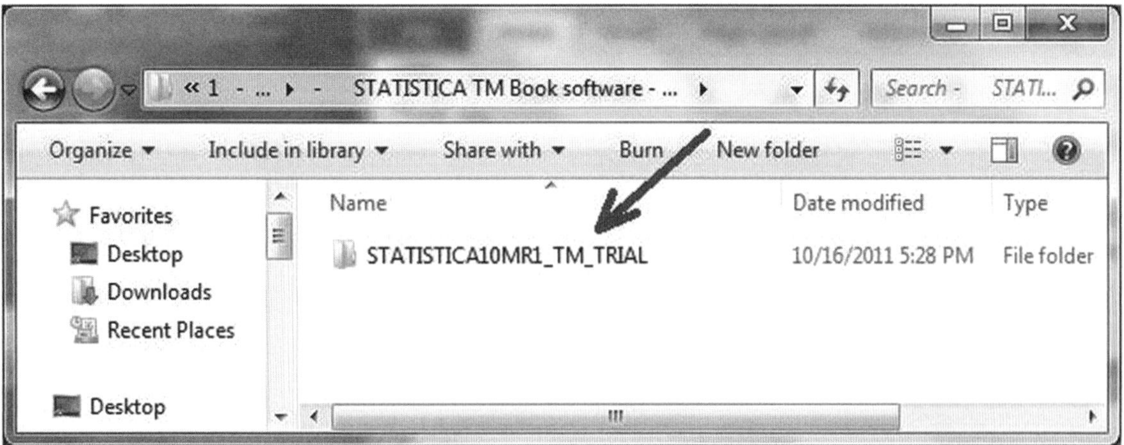

Then open this "*STATISTICA*" folder, which will display the following subfolders or documents:

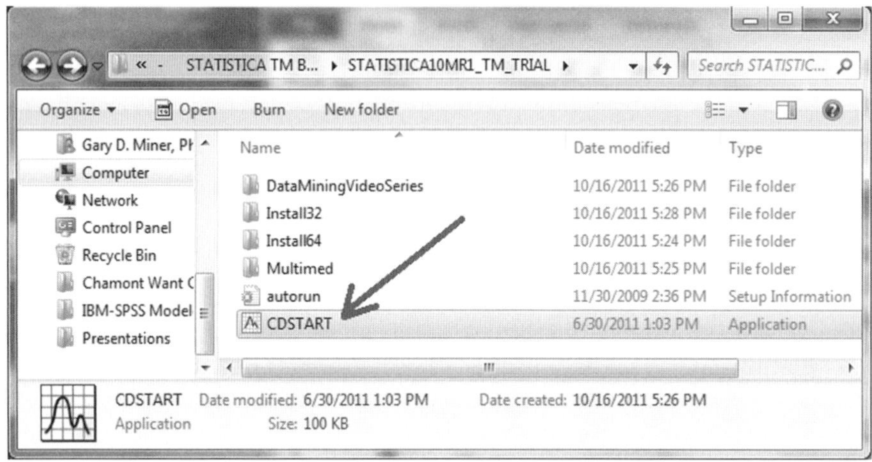

If you wish, you can click on the "Install32" or the "Install64" folder if you know that your computer is a 32-bit or 64-bit computer, and then find the "setup" executable application icon, click on this, and install the program. However, if you click on the "CDSTART" selection, this will automatically determine if your computer is a 32-bit or 64-bit system, and it will install the correct *STATISTICA* Text Miner application for your computer. At this point, a bluish gray *STATISTICA* install window should appear on your computer screen:

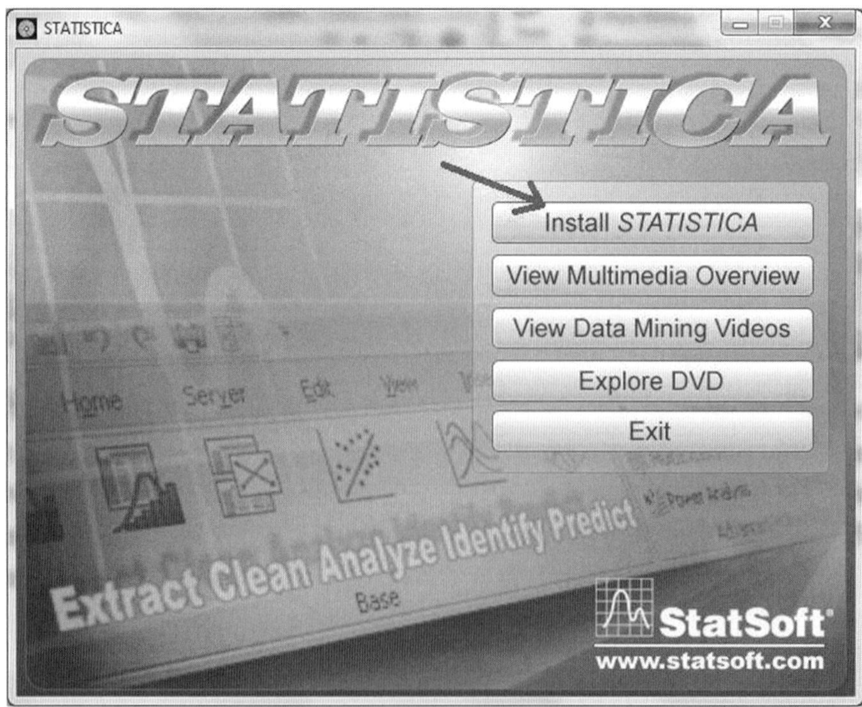

If you wish to explore some of the short topics in the Data Miner Video series, click on that button on this bluish gray install dialog or the "View Multimedia" button, but if you wish to install the *STATISTICA* software, click on the "Install *STATISTICA*" button (the button with the arrow pointing to it above), and then follow the prompts and questions on the screen to complete the installation.

Important note: Your computer *must be connected to the Internet* when you do this install to allow the software to run. Your installation will send a message to the StatSoft—*STATISTICA* licensing server to get a license to allow this free trial to run for 30 days.

After the initial install process checks out your computer to discover parameters and if it is 32-bit or 64-bit, the following dialog screen should appear:

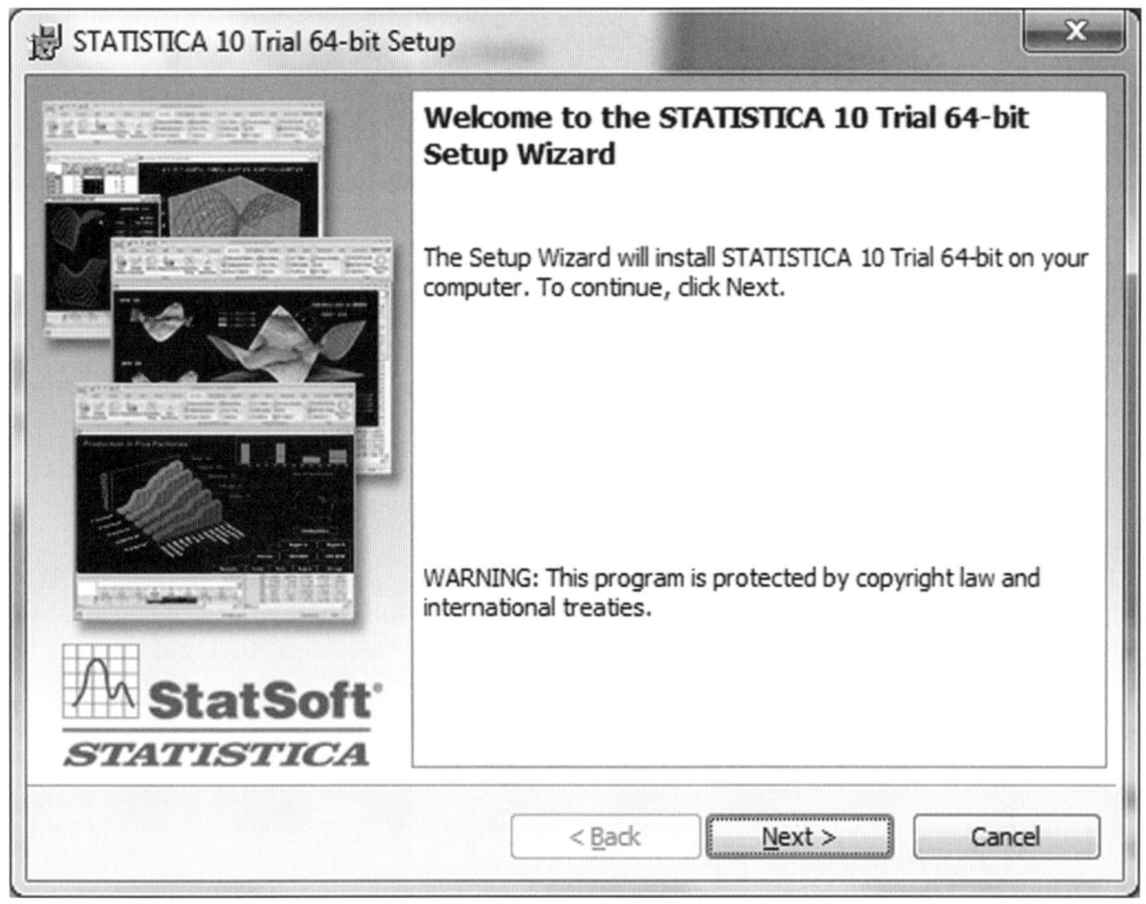

Click the "Next" button, and then continue the installation, watching the process and responding to any queries that may appear on some of the install screens. One screen will ask you for some information, and then it will send a message to the licensing server. Immediately, a response screen will appear that looks like this:

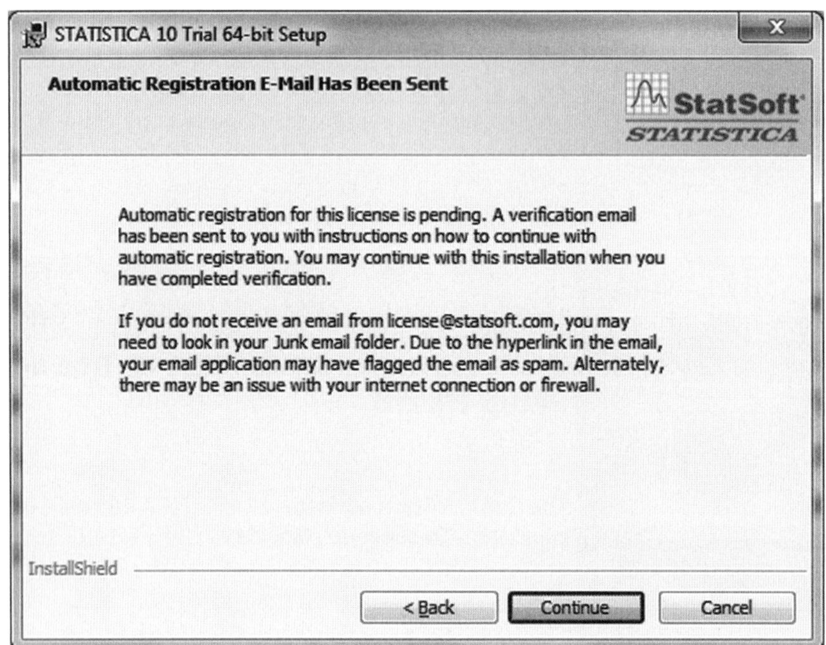

The email message will look like this:

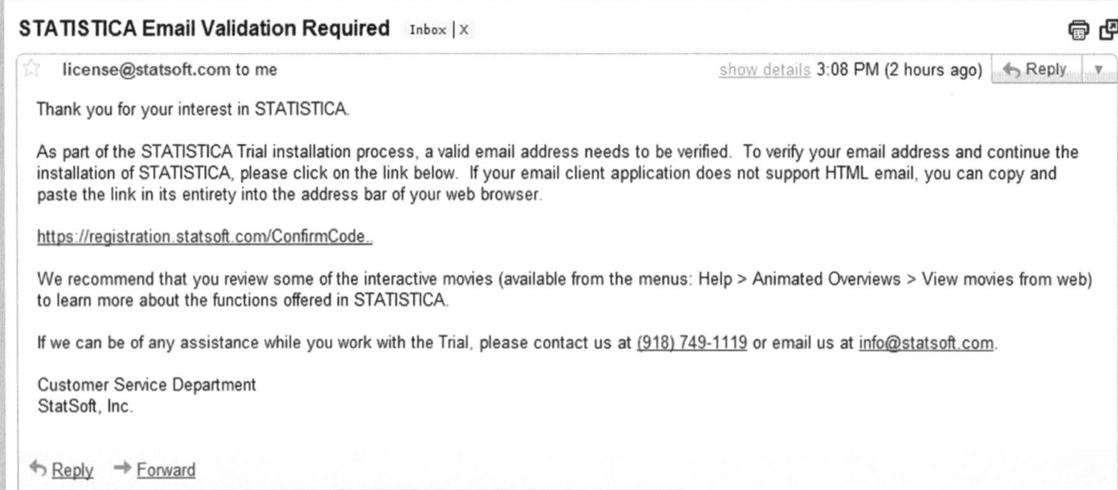

STATISTICA Email Validation Required Inbox | X

license@statsoft.com to me show details 3:08 PM (2 hours ago) Reply

Thank you for your interest in STATISTICA.

As part of the STATISTICA Trial installation process, a valid email address needs to be verified. To verify your email address and continue the installation of STATISTICA, please click on the link below. If your email client application does not support HTML email, you can copy and paste the link in its entirety into the address bar of your web browser.

https://registration.statsoft.com/ConfirmCode..

We recommend that you review some of the interactive movies (available from the menus: Help > Animated Overviews > View movies from web) to learn more about the functions offered in STATISTICA.

If we can be of any assistance while you work with the Trial, please contact us at (918) 749-1119 or email us at info@statsoft.com.

Customer Service Department
StatSoft, Inc.

Reply Forward

Click on the link, and the following message will appear:

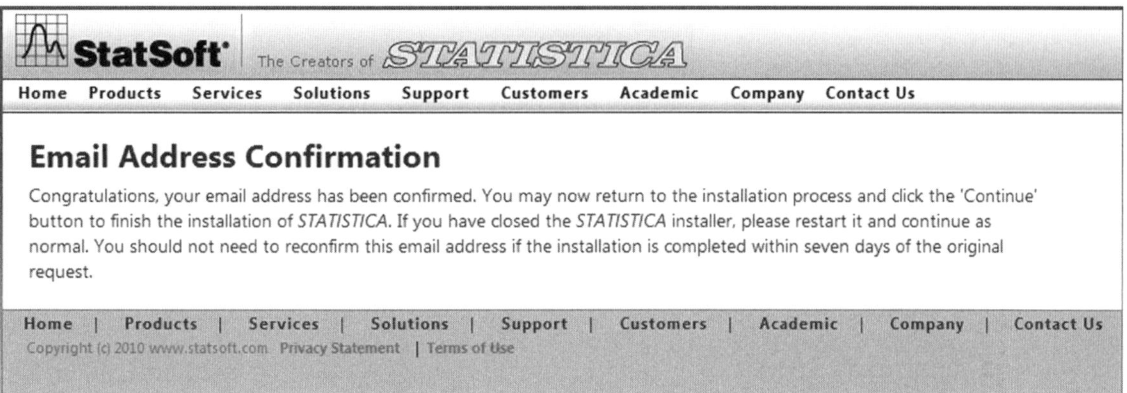

Then continue on with the install by clicking the "Continue" button on the install dialog. This will bring up a "success" screen:

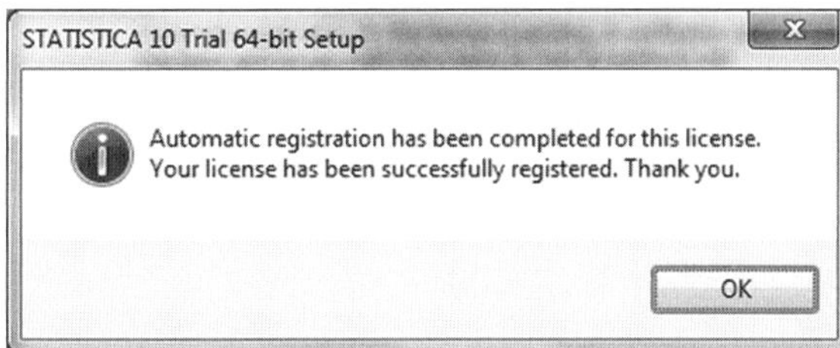

Click OK, and continue through the installation of *STATISTICA* Text Miner by following the dialogs until completed.